# Geology of North Africa

"No limit may be set to art, neither is there any craftsman that is fully master of his craft."

(Instructions of Ptahhotep, XXIV Century B.C.)

# Geology of North Africa

E. Edward Tawadros

 **CRC Press**
Taylor & Francis Group
Boca Raton London New York

CRC Press is an imprint of the
Taylor & Francis Group, an **informa** business
A BALKEMA BOOK

CRC Press
Taylor & Francis Group
6000 Broken Sound Parkway NW, Suite 300
Boca Raton, FL 33487-2742

First issued in paperback 2018

CRC Press/Balkema is an imprint of the Taylor & Francis Group, an informa business

© 2012 Taylor & Francis Group, LLC

Typeset by Vikatan Publishing Solutions (P) Ltd, Chennai, India

No claim to original U.S. Government works

ISBN-13: 978-0-415-87420-5 (hbk)
ISBN-13: 978-1-138-07726-3 (pbk)

Published by: CRC Press/Balkema
P.O. Box 447, 2300 AK Leiden, The Netherlands
e-mail: Pub.NL@taylorandfrancis.com
www.crcpress.com – www.taylorandfrancis.co.uk – www.balkema.nl

Library of Congress Cataloging-in-Publication Data

Tawadros, E. (Ezzat)
  Geology of North Africa / E. Edward Tawadros. -- Expanded and updated ed.
    p. cm.
  Rev. ed. of: Geology of Egypt and Libya / E. Tawadros. 2001.
  Includes bibliographical references and index.
  ISBN 978-0-415-87420-5 (hardback) 1. Geology--Egypt. 2. Geology--Libya.
  3. Geology--Africa, North. I. Tawadros, E. (Ezzat) Geology of Egypt and Libya.
  II. Title.

  QE328.T39 2012
  556.1--dc23

                                                    2011038735

**Visit the Taylor & Francis Web site at**
**http://www.taylorandfrancis.com**

**and the CRC Press Web site at**
**http://www.crcpress.com**

# Contents

# Preface

This book is a revised and expanded new edition of the Geology of Egypt and Libya (Tawadros, 2001). It follows the same structure as the first edition. New chapters covering the geology of Algeria, Tunisia and Morocco have been added to cover the geology of all of North Africa. The petroleum systems of North Africa are discussed in a new chapter. In addition, a new chapter dedicated to the history of geological ideas and oil exploration in North Africa has been included. The book reviews the tectonic elements, the geology of the Pan-African Shield, and Phanerozoic geological evolution of the five North African countries, and gives a description of most of the lithostratigraphic units in the region. It includes a bibliographic list of more than 2500 references, which will guide the interested reader to further information. The main objective still remains the same; to give a quick reference to the geology of North Africa that can be useful to both professionals and students. The first edition contains more than 20 tables that show stratigraphic nomenclature used by various authors for different areas. These tables are not reproduced in the present edition. The present work is an attempt to compile a large portion of the published data to summarize the geology of North Africa.

The thickness of the rock-stratigraphic units is expressed in both Imperial and SI units. However, distances are expressed in the SI units only.

This compilation does not contain proprietary data. It is based entirely on published data. However, in writing the book, the author was guided by the many years of experience gained while working in and on North Africa.

# Acknowledgements

I thank Sebastian Lüning who supplied me generously with copies of his publications, videotapes, and pictures from his fieldtrips. I am indebted to the people who provided me with photographs of outcrops in many parts of North Africa. Prof. Jochen Kuss (University of Bremen) kindly provided the picture for the book cover. Prof. Baher El Kaliouby (Ain Shams University) contributed pictures from Sinai and read the chapter of the Pan-African basement. Steven Aitken (Calgary) provided the picture of the Hamar Laghdad buildups of Morocco. I am also obliged to Kent Wallace and John Harper for the thorough reviews of the first edition. I am grateful to the late Prof. Manuel S. Pinto (INHIGEO) who invited me to write two chapters on the history of geological ideas and oil exploration in Egypt and Libya for a planned publication, which form the basis of the history chapter in this edition. I am also grateful to Dr. A.A. Ismail (Ain Shams University) for providing me copies of his publications that otherwise would not have been available to me. Andras Zboray drew my attention to the photograph of the Staff of the Desert Survey Department, and Peter Clayton kindly gave me the permission to reproduce it here. Extensive discussions with the staff of Nexen in Calgary helped me to further my education about the geology of North Africa. Two positive book reviews of "Geology of Egypt and Libya" by Richard Moody and A.J. Reedman pointed out some of the short comings of the first edition, which I tried to rectify in the current edition. Initial correspondence with A. Michard and D. Frizon de Lamotte put my unrealistic ambitions into check and helped me to restructure the chapter on the geology of Morocco.

My thanks are also due to CRC Press, particularly Germaine Seijger and Lukas Goosen for their dedication and support during the process of producing this volume.

# About the author

Ed. Tawadros, *Ph.D., P. Geol.*, is an international geological consultant based in Calgary, Alberta, Canada. He received a B.A.Sc. degree in geology from Ain Shams University, Cairo, in 1968, followed by a post-graduate Diploma in applied geology from Cairo University, a M.Sc. degree in geology from Université de Montréal, Canada, and a Ph.D. in geology from the University of St. Andrews, Scotland. He also holds a BA Major in Spanish from the University of Calgary, Alberta. Since graduation he has worked for a number of consulting and major oil companies, including the General Petroleum Company, Cairo, Egypt, Géophysique France-Québec, Montreal, Robertson Research Company, Canadian Superior Oil, Mobil Oil Company, and Nexen, Calgary, Canada, and Sirte Oil Company, Libya, and more recently in Argentina. He has authored and co-authored a number of papers dealing with the geology of parts of Canada and Libya. He also co-authored papers on the comparative geology of North Africa and the Caucasus with colleagues from Russia and Azerbaijan. He is a member of the Canadian Society of Petroleum Geologists (CSPG) and the Association of Professional Engineers, Geologists and Geophysicists of Alberta (APEGGA).

# PART I

## History of geology and hydrocarbon exploration in North Africa

# Chapter 1

# Introduction

The history of geological and petroleum exploration in parts of North Africa has been treated by Gentil (1902), Demoulin (1931), Desio (1967, 1968), Meckelein (1975), Said (1990b), EGPC (1992, 1994, 1996), Klitzsch (1994), Legrand (2002), Bettahar (2007), and Missenard et al. (2008), in addition to brief reviews of the history of specific geological points in the numerous publications on the geology of North Africa. The history of geological research in Egypt and Libya is treated in detail by (Tawadros, in preparation). The five North African countries treated here share many aspects in their geological make up, yet the pace and timing of the development of the geological ideas vary widely.

## 1.1 EGYPT

In the 1700s, Egypt was visited by many famous travellers, such as Thomas Shaw (1692–1751), Fredric Hasselquist (1722–1752), Richard Pococke (1704–1765), and James Bruce (1730–1794).

The end of the 18th and the beginning of the 19th Centuries were the most significant times for the development of scientific research in Egypt. The results of the French Expedition led by Napoleon Bonaparte in 1798–1801 were included in the monumental publications of the "*Description de l'Egypte*" and the "Memoirs relative to Egypt" (1799, 1800). Napoleon also established the *Institut d'Egypte*, which was probably the first truly scholastic institute in North Africa since the Ancient Library of Alexandria. The Expedition also produced the first two reliable maps of Egypt; one at the scale of 1:100,000 in 47 sheets; the other at the scale of 1:1,000,000 (published in 1828). One of the earliest descriptions of *Nummulites* in Egypt was that by Monge (1746–1818).

In 1820, Hemprich and Ehrenberg from the Zoological Museum in Berlin collected land animals and plants from the Gulf of Suez, Sinai, Nubia, and Lebanon. In 1824–1825, Klunzinger and Ehrenberg collected marine animals from the Red Sea. Ehrenberg wrote on coral reefs of the Red Sea in 1832 and 1834. Cailliaud (1826) described Eocene fossils in Egypt. Eduard Rüppell from the Senkenberg Museum, Frankfurt, began his work in Egypt in 1826. In 1837, Russegger introduced the term "Cataract Sandstone" in Upper Egypt, which he subsequently changed in 1842 into the "Nubian Sandstone" to initiate one of the most enduring controversies in the history of geology in North Africa. In 1841, D'Hericourt made a few geological observations on the Red Sea in Egypt and the Gulf of Aden. Other geological observations were made by Orlebar (1845). Bellardi (1851 and 1854) described and catalogued the Egyptian *Nummulites* collection at the Mineralogical Museum of Turin. Lieutenant Newbold (1848a) wrote a brief note on the geology of Egypt. O. Fraas (1867) described the Eocene outcrops between Assiut and the Pyramids.

In 1868, the first shafts and wells for oil were dug at Gebel Zeit in the Southern Gulf of Suez. During that time, H. Bauerman (1869) discovered the Carboniferous in Sinai. Schweinfurth discovered the Carboniferous sediments at Wadi Araba (Schweinfurth, 1885) (described by Walther in 1890), studied the Mokattam Hills (Schweinfurth, 1883), fossil vertebrates in Fayium (Schweinfurth, 1886), and the Cretaceous at Abu Roash (Schweinfurth, 1889), He compiled his observations in Egypt in a book (Schweinfurth, 1921). E.T. Hamy (1869) wrote on the Quaternary of Egypt. Owen (1875) described a Sirenian mammal (*Eotherium aegyptacum*) from the Eocene Nummulitic Limestone of the Mokattam Hills near Cairo. Duncan (1869) described echinoderms, molluscs, and other fossils from the Cretaceous rocks of Sinai.

A number of Expeditions were sent by the Khedive of Egypt and the Royal Geographical Society to explore for the sources of the Nile, which led to the discovery of the sources of the Nile in 1862 by John Speke (1827–1864) and James Grant (1827–1892). An expedition to Sinai and Palestine in 1871 was sponsored by the Palestine Exploration Fund and led by officers of the Royal Engineers (Major C.W. Wilson, 1872–1873). Another expedition was sent to the Eastern Desert in 1874–1876; the result of which was a geological and mineralogical reconnaissance of the region carried out by Mitchell (War Office, 1876) between Rudeis and Qoseir on the Red Sea coast and Qena on the Nile. A report on the geology and petroleum occurrences of Ras Gemsah and Gebel Zeit was published (Mitchell, 1887). Many samples were collected during that expedition and the final report included a geological map and a profile. The southern part of the Eastern Desert between Aswan and Qena in the Nile Valley and Berenice and Qoseir on the Red Sea coast was visited by the Scientific Expedition dispatched by the Khedive of Egypt in 1891 under the leadership of E.A. Floyer (1893) who made a few geological observations in the area. Felix (1884, 1904) carried out detailed studies of Egyptian corals. Antonio Figari Bey (1804–1870) summarized his observations in his book (Figari, 1864) and collected many Tertiary and Cretaceous fossils samples that were described several decades later by Greco (1915, 1916, 1917, 1918) and Stefanini (1918).

The last quarter of the 19th Century saw another turning point in the development of geological research in Egypt with the accomplishments of the Rohlfs Expedition in 1874–1876. The Expedition was the first comprehensive and multidisciplinary work on the Western Desert and included, in addition to Rohlfs, many other renowned scientists, such as Ascherson (botanist), Jordan (geodesist and astronomist), Zittel (geologist and paleontologist), and G. Remelé (photographer). The results of the Expedition were produced in three volumes of the "*Expedition zur Erforschung der libyschen Wüste*" (1875–1883). The results were also published by Zittel (1883a, b) in Palaeontographica, volume 30 "*Beiträge zur Geologie und Paläontologie der libyschen Wüste und der angrenzeden Gebiete von Aegypten (unter Mitwirkung mehrerer Fachgenossen herausgegeben von Karl A. Zittel*), along with contributions from Jordan, Ascherson, Schenk (1883), Mayer-Eymar (1883, 1886, 1898), Schwager (1883), de la Harpe, Quaas (1902), Wanner (1902), de Loriol (1864, 1881, 1883),

Pratz (1883), Fuchs, Dacqué, and Oppenheim (1903, 1906). This work also produced the first map of northern Egypt at a scale of 1:300,000.

In 1882, the British occupied Egypt, but apparently the occupation did not slow down geological exploration activities. In 1883, Zittel subdivided the Cretaceous and Tertiary rocks in Egypt into units that persist until today. Meyer Eymer conducted many studies on the paleontology of Egypt from 1883 until 1903. Dames (1883, 1894) examined Tertiary vertebrates in Fayium which were housed in the Berlin Museum of Natural History. Dawson, J.W. (1884) discussed the Geology of the Nile Valley in a brief note, and in 1888 he published the book "Modern Science in Bible Lands". Von Nötling (1885) studied Tertiary crustacean. In 1893, Suess introduced the name Tethys, which was followed by many discussions, debates, and controversies that continue until today (e.g. Ball, 1910, 1911). In 1894, Schellwein identified Upper Carboniferous brachiopods in exposures along the Gulf of Suez. Seward (1907, 1935) identified fossil plants from Egypt. Osborne & Granger (1907) made an expedition to Fayium and collected vertebrate samples for the American Museum of Natural History.

In 1886, the discovery of the Gemsa Field in the Southern Gulf of Suez in Miocene dolomitic reefal carbonate (later Nullipore Limestone) reservoirs signalled the beginning of oil exploration in Egypt. Ardagh (1886) made the first attempts to explain the origin of the oil seeps in Gebel Zeit.

Egypt entered the 20th Century with a giant step for geology. In 1896, the Geological Survey of Egypt was founded by Colonel Henry George Lyons (1864–1944). Hume, Beadnell, and Barron joined the Survey. Since that time the Geological Survey played a crucial role in the development and advancement of Egyptian geology and the development of the mineral wealth of Egypt. Barron discovered phosphate in Gebel Qurn near Qift in Upper Egypt 1896 and in Wadi Hammama in 1897. During that time, Gregory (1896) introduced the "African Rift Valley" concept and the debates the ensued about the origin of the Gulf of Suez (Hume, 1901, Ball, 1910, 1911, 1920, Gregory, 1920a, b, 1921). The celebrated Edward Hull (1896) wrote the "Geology of Nile Valley". E. Fraas (1900) constructed a cross-section across the Eastern Desert from the Nile to the Red Sea, and reported a new whale (Zeuglodon) species from the Middle Eocene of the Mokattam Hills in Cairo (Fraas, 1904).

In 1901, Hume wrote "Rift valleys and geology of Eastern Sinai" for the International Geological Congress in response to Gregory's arguments. Barron & Hume (1900, 1902) studied the geology of the Eastern Desert. Ball (1900) published "Topography and Geology of Kharga Oasis". Blanckenhorn joined the Survey in 1900 and published on the geology of Egypt and the origin of the Nile, where he introduced the term "Urnil" to define an ancestor of the Nile. Blanckenhorn (1900) described fossils collected by Schweinfurth from Tubruk and Porto Bardia in Libya and correlated them with Miocene rocks in Egypt. Douvillé (1900, 1901a, b) studied the foraminifera collected by Fourtau (1902) and reported the discovery of Orbitolina at Gebel Genefe. In 1902, Beadnell described Cretaceous rocks in Abu Roash west of Cairo and divided them into lithostratigraphic units that are still in use today, albeit under different formational names. Ball (1902) published "Gebel Garra and the Oasis of Kurkur" in Southwestern Egypt. C.W. Andrews accompanied Beadnell on two expeditions (1902 & 1903) to the Fayium where he identified new species of vertebrate fossils. Ball & Beadnell (1903) published the treatise on the "Baharia Oasis". Their findings were only surpassed by those of Ernst Strömer von Reichenbach almost thirty years later. Strömer made the most significant discoveries of vertebrate fossils in Egypt (from 1910 until 1936). His collection was sent to the Berlin Museum in 1922, but was destroyed when the Museum was bombed by the allies in 1944. Strömer was accompanied by Markgraf in 1910. Strömer introduced the *Baharijhe Stufe* (Baharia Formation) (1914) and confirmed its marine origin. The fossil plants collected by Strömer were described by Hirmer (1925). At the same time, Harding King (1916, 1918) made an expedition to the Dakhla Oasis in the Western Desert.

In 1905, Beadnell published the results of his studies on the Eocene-Cretaceous systems in the Esna-Aswan areas in Upper Egypt, and introduced the Esna Shale for a succession that straddles the Cretaceous-Tertiary boundary, which triggered an enduring controversy (reviewed by Tawadros, 2001). In 1907 Ball examined the first Cataract at Aswan following the pioneering work by Hawkshaw (1867). In 1906, Barron mapped the western portion of Sinai, and Hume & Beadnell mapped Southeastern and Central Sinai. In 1908, Weill published "La Presqui'ile du Sinai" which included descriptions of sedimentary and igneous rocks.

Cairo (Fouad I) University, the first university in Egypt, was established in 1908.

In 1909, Hume became head of the Geological Survey (Fig. 1.1), and systematic geological studies and mapping of Egypt began. In 1910, Schweinfurth produced independently the first modern map of Egypt.

*Figure 1.1* Photograph of members of the Geological Survey of Egypt in 1925. Front row: Patrick Clayton, George Murray, John Ball, W.F. Hume, H.L. Beadnell, O.H. Little, Hassan Sadek, George Walpole (Courtesy of Peter Clayton).

In 1911, the Nakhla Meteorite fell in Egypt; Hume collected samples and Ball (1912) described them in "The Meteorite of El Nakhla el Baharia". Recently, samples of this meteorite were subjected to various analytical tests and were subject of numerous debates, including the origin of life (Jull et al., 2000, Meyer, 2004).

Edward Hull joined Kitchner's Expedition (1892–1895) and carried out geological mapping in Sinai. Hull (1886) regarded Lower (Northern) Egypt and Sinai as traversed by a series of north-south-oriented folds. He postulated that the Gulf of Suez was formed due to faulting along folds (the Syrian Arc folds).

In 1911, Production started from the Gemsa Field (depleted in the 1930s after producing 1.4 mmbbls of oil). Shell entered Egypt, with the Anglo Egyptian Oil Fields Company "AEO" (50% with BP). In 1913, the Hurghada Field was discovered by Shell in Carboniferous "Nubian" sandstones by drilling a surface anticline, which constituted the first major oil discovery in Egypt. The field was put on production the following year and produced nearly 48 million barrels by the time it was depleted in 1996. In 1918, the Abu Durba Field (depleted in the early 1940s) onshore the Gulf of Suez was discovered in Nubian Sandstone near oil seeps by the Egyptian Government. No trap or seal were detected in that field (EGPC, 1992). Reserves were small and the Government handed over operations to the Egyptian Oil Syndicate in 1923 after drilling 10 wells.

The Petroleum Development Committee was formed in 1913 following the discovery of the Hurghada Field and Hume and Ball became members. Barthoux & Douvillé (1913) discovered Jurassic sediments at Maghara in Northern Sinai (Douvillé, 1916). Ball (1913) wrote "Topography and Geology of the Phosphate District" and "A Brief Note on the Phosphate Deposits of Egypt". Eck (1914) described the cephalopods collected by Schweinfurth from the Cretaceous chalks.

In the period 1914–1918, WWI interrupted the geological exploration activities in most of North Africa. However, new information was added during that time thanks to the efforts of the soldiers/scientists who participated in the war. In 1916, the British Light Car Patrol was formed, of which Claude H. Williams and John Ball were members. The use of motor cars for reconnaissance in the Libyan Desert contributed to a better understanding of the Egyptian deserts and the distribution of sand dunes. Hassanein Bey reached the Kufra Oasis accompanied by Rosita Forbes in 1920 (Forbes, 1922); one of the earliest female travellers in Egypt.

Moon, Madgwick and Sadek joined the Survey in 1917, and in 1920, Hassan Sadek became head of the Petroleum Research Board. In 1922, Egypt gained independence from Britain, and Fouad I became king of Egypt. In 1923, Schweinfurth compiled his works in Egypt in "Auf unbetreten Wegen in Aegypten". Hassanein Bey (1924) crossed the desert from Sallum to Kufra and Sudan. The samples collected from Kufra and Oweinat during that expedition were described by Hume (1924) and Moon (1924). The Anglo-Egyptian Oil Company (AEO) introduced the Eöstöv torsion balance into Egypt for the first time, and Gemsa and Hurghada fields were surveyed. Moon & Sadek (1923) carried out a study on the Gulf of Suez and introduced the name Nullipore Rock for a Middle Miocene *Lithothamnion* Limestone unit. Hume (1924) wrote "The Egyptian Wilderness". In 1925, Hume published his monumental work, the "Geology of Egypt" in three volumes, which was completed in 1945. Barthoux (1922) gave a detailed account of the Precambrian basement in Sinai.

In 1927, Beadnell published an account of his studies in Central Sinai in "Wilderness of Sinai", which was preceded in 1909 by a review in the Geographical Journal.

In the meantime, H. Sadek and P.A. Clayton studied the Northern Galala Plateau and Gebel Ataqa on the western side of the Gulf of Suez and George Walpole (1927) discovered the Qattara Depression in the Western Desert. Debates on the origin of the desert depressions and their potential for power generation started (Serry Bey, 1929, Anderson, 1947, Pfannenstiel, M., 1953, Gindy & El-Askary, 1969, Albritton et al., 1990, 1991, Gindy, 1991). In 1927, Ball published his article "Problems of the Libyan Desert" in The Geographical Journal. Bagnold traversed the Western Desert from Cairo to Siwa in motor cars. Beadnell surveyed the Western Desert south of Kharga Oasis and drilled the Bir Missaha well. In 1928, the Geological Survey of Egypt made a new geological map of Egypt, which was included in the "Atlas of Egypt" presented to International Geographic Congress in Cambridge. Gebel Oweinat (Uweinat) was shown on that map for the first time.

Bagnold led another expedition in 1938, during which Tomkeief & Peel (1942) described the igneous rocks at Gebel Oweinat. In 1978, a multidisciplinary group, led by Farouk El-Baz (El-Baz et al., 1980), Vance Haynes and William McHugh retraced Bagnold's 1938 expedition to Gilf Kebir and Uweinat (Bagnold et al., 1939).

In the 1930s, a group of scientists-soldiers, that included Beadnell, Ball, Bagnold, and Clayton, as well as members of the royal family, including Prince Kamal el-Din Hussein (1875–1932) (1928), the son of Sultan Hussein Kamel and grandson of Khedive Ismail, and Hassanein Bey (another member of the royal family), ventured into the heart of the Western Desert. They were joined by Sir Clayton-East-Clayton (1908–1932)[1] and his wife Lady Dorothy (nicknamed Peter) Clayton (1908–1933), as well as the controversial Ladislau Almàsy. Rosita Forbes, who accompanied Hassanein Bey (1924), disguised herself as a Bedouin woman in order to achieve her mission. Their enthusiasm was also fired by the desire to locate the legendary Oasis of "Zerzura", the first mention of which was in the Arabic fiction book, "Book of Hidden Pearls".[2] Its existence and location made the subject of long discussions in the Journal of the Royal Geographical Society, which involved Almàsy (1936), Harding King (1928), Wingate (1934),[3] and Ball (1928). In addition, Almàsy was occupied with the search for Cambyses army, believed to have vanished in a sandstone storm. In 1929, Almàsy drove from Wadi Halfa to Selima Oasis and along Darb el Arbain in the Eastern Desert. The same year (1929), Bagnold explored the Sand Sea in the Western Desert. Bagnold made another expedition in 1930 to continue his study of the desert sands (Bagnold & Harding King, 1932).

In 1932, Almàsy, Sir Robert Clayton, P.A. Clayton, and Penderel explored the western side of the Gilf Kebir in the southwest corner of Egypt. The same year,

---

1  Sir Clayton died of polio, and Lady Clayton died a year later in a plane crash after her return to England.

2  According to Ahmed Bey Zaki (1907) (cited by Ball, 1927b), this book has been the cause of more damage to the Ancient Egyptian monuments since the Arab invasion of Egypt than all the ravages of war and the and of time.

3  Wingate (1903–1944) was a Lieutenant with the British army in the Sudan when he left to Egypt in January of 1933 in search of Zerzura. His fame came after he left Egypt and was transferred to Israel (before its foundation), then to Ethiopia, where he led the force against the Italians. He then worked in Burma, organizing the Chindits, the commandos who operated behind the Japanese lines. Wingate died in an air crash in Burma and was buried in the United States.

Bagnold led another expedition to Gebel Uweinat, and P.A. Clayton achieved the first east-west crossing of the Great Sand Sea (Clayton, 1933). In his trip, P.A. Clayton discovered the enigmatic Libya Desert Glass, which was the subject of a number of studies (Clayton & Spencer, 1934, Cohen, 1959, Jux, 1983, Bigazzi & de Michele, 1996, Barrat et al., 1997, Osterlund et al., 1997, Müehle, 1998, Frischat et al., 2001, Koeberl et al., 2003). Interestingly, the desert glass was used by prehistoric inhabitants and the Ancient Egyptians (Roe et al., 1982, George, 2000). In 1933, Wingate crossed the Sand Sea from Abu Mungar to Dakhla on camelbacks. An Italian mission led by Marchesi mapped Gebel Uweinat.

In addition to these desert exploration missions, other key scientific studies were published in the early 1930s. MacFadyen studied the coral reefs of the Red Sea (1930) and Miocene foraminifera from the Gulf of Suez and Sinai (1931). Vadasz (1933) studied iron ores in Egypt (re-examined by Nassim, 1950, and assigned an Upper Cretaceous age by Attia, 1955). Cuvillier (1930) published his treatise "*Nummulitique Egyptien*", in which he treated the geology of Egypt in precise details. Gardner & Caton-Thompson (1926) Caton-Thompson & Gardner (1932), the first female geologists to work in Egypt, published the results of their combined geological-archeological studies in the early 1930s; thus establishing the geoarcheology discipline in Egypt. Caton-Thompson (1952) published her work on the Kharga Oasis.

In 1935, King Farouk succeeded his father Fouad, and the Alexandria (Farouk I) University; the second in Egypt, was established in 1938. Cayeux (1935) described the Egyptian phosphates and compared them with those of Palestine and Jordan. Bagnold published his results in "Libya Sands".

In 1937, the South Mediterranean Oil Co. (Catex or SOMED), Anglo Egyptian Oil Co. (Shell or AEO), and Standard Oil of Egypt (Standard Oil of New Jersey) acquired several concessions in the Western Desert west of the Nile Delta.

The first refraction seismic survey in Egypt was carried out at Abu Shaar in the Gulf of Suez by the Egyptian Government in 1929. Seismic surveys in Egypt were made by Socony (later Mobil) in Northern Sinai and by SOMED in the Western Desert. The Ras Gharib Field was discovered in the Miocene Nullipore Carbonates with original oil in place (OOIP) of 290 mmbbls oil.

In 1938–1945, WWII was another period when geological exploration was interrupted. On the other hand, thanks again to a group of scientists who participated in the war continued their scientific investigations. Bagnold formed the Long Range Desert Group (Bagnold, 1945, Clerk et al., 1945) and in the meantime examined the desert sands.

Scientific interest in sand dunes probably started in Egypt by V. Cornish (1897) who discussed extensively their shapes and formation modes in "On the formation of sand dunes", followed by Johannnes Walther (1900) in "*Das Gesetz der Wustenbildung*". H.J.L. Beadnell (1910) and John Ball (1927a, b), officers of the Geological Survey, wrote on the sand dunes of the Libyan Desert. Harding King (1916), a member of the British Light Car Patrol suggested that the dunes "have a curious power of collecting sand" and explained the origin of the seif and barchan sand dunes. R.A. Bagnold (1941) published his classical book "The Physics of Blown Sand and Desert Dunes", based on his observations in the deserts of Egypt. Other studies of sand dunes were made by Ladislas Kádár (1934) in Egypt and di Caporiacco (1934) in Libya.

In 1944, exploration for hydrocarbons in Egypt entered a new phase with the Dabaa-1, the first well to be drilled in the Western Desert by Anglo-Egyptian Oil Company (AEO), followed by the Khatatba-1, the second well in the Western Desert, spudded by SOMED, and the Abu Roash-1 (1946) and Abu Roash-2 wells (1947).

In 1945, the Sudr Field was discovered by Mobil in Miocene reservoirs in two onshore fault blocks east of the Gulf of Suez. In 1947, the Asal Field in the onshore of the Gulf of Suez was discovered in Miocene sands and fractured Eocene limestones by Mobil. Stainforth (1949) studied the foraminifera of the Oligocene, Miocene and Pliocene in the Gulf of Suez. In 1949, the Feiran Field, onshore the Gulf of Suez, was discovered in Miocene and Turonian-Cenomanian sands by Standard Oil. The Ras Matarma Field was discovered in the Rudeis Formation and fractured Eocene carbonates. In 1946, Awad gave the first description of Triassic rocks at Arif el Naga, and Eicher (1946) described the conodonts in the same succession. Spath (1946) described Middle Triassic cephalopods in Sinai. In 1951, the first mention of the Brown Limestone as a major source rock for hydrocarbons in Gulf of Suez was cited by Tromp (1951).

In 1952, a revolution ousted King Farouk and a republic was proclaimed in Egypt and in 1954, Nasser became president of Egypt. In 1953, a new mining law was issued to encourage new companies to conduct exploration. Sahara Petroleum Company was awarded a large concession covering the Western Desert north of Lat. 28° and large scale exploration was carried out which included areal photography, surface mapping, and refraction seismic and gravimetric surveys. Nakkady (1952) published key papers on the micropaleontology of Egypt. Leroy (1953) studied the stratigraphy and paleontology of the Maqfi section in the Farafra Oasis. The Geological Survey started an aggressive program of exploration. However, exploration was hampered by land mines left from WWII.

In 1955, the first offshore seismic survey was conducted by COPE in the Gulf of Suez. The Belayim Land Field was discovered by IEOC (Eni) in Miocene sandstones, limestones, and conglomerates. The field proved to contain 11 reservoir intervals (with > 1Bbbls of oil reserves), including reservoirs within the South Gharib Formation evaporites (Rohrback, 1983, El Ayouti, 1961, 1990). Sahara Oil Company started its drilling operations in the Western Desert (1955–1958), which led to the drilling of the Mersa Matruh-1, Burg El Arab-1, and Mamura-1 wells (in 1958); all proved dry, after which Sahara relinquished its concessions.

In 1956 war erupted between Egypt and Israel, which led the Egyptian president at the time, Gamal Abdel Nasser, to nationalize the Suez Canal and the majority of foreign assets. This war was followed by the defeat in Yemen in the 1960s and another defeat in the 1967 (Six Days War), during which Sinai was lost and occupied by Israel until 1985. In order to achieve his objectives, which included the building of the Aswan High Dam, Abdel Nasser switched his alliance to the Communist Block. As part of the Russian-Egyptian cooperation, Russian scientists and Engineers were brought into Egypt to replace the American and Western Europeans. However, the Russians' contributions, as far as geology is concerned, was minimal. The only visible contributions are the Tectonic Map of Egypt by Sigaev (1959) and in addition to a few palynological studies. In July 1972, Sadat expelled the Russian experts along with the Russian military from Egypt.

Following the 1956 War, the Arab countries imposed an oil embargo on the West. In 1957, the Egyptian General Petroleum Corporation (EGPC) was established to control all aspects of oil activities, including exploration, production, transportation, and refining. It was followed by the establishment of the General Petroleum Company (GPC). The Abu Zeneima SE Field in the offshore of the Gulf of Suez was discovered in Miocene rocks by Petrobel and the Abu Rudeis-Sidri Field in the basal Miocene Nukhul conglomerates (abandoned in 1998). In 1958, the Bakr Field which straddles the onshore/offshore of the Gulf of Suez was discovered in Eocene and Miocene sediments by GPC. The Kareem Field was discovered by the Egyptian General Petroleum Corporation (EGPC) in the Gharandal Group sandstones. The Rahmi Field was discovered in 1960 in Lower Eocene sediments in the onshore of the Gulf of Suez. Production of the Field started in 1975, but was abandoned in 1988 after it produced only 3 mmbbls of oil. The Belayim Marine, the first offshore field in Egypt, was discovered by Petrobel in Miocene and Upper Cretaceous (on stream since 1962); with ultimate reserves of about 5 billion barrels of oil similar to the July and Ramadan oil fields. The Morgan Field was discovered by Pan America (later Amoco, Gupco) in the offshore of the GOS in 1965 based on aeromagnetic and reflection seismic surveys (Brown, 1980), and production started in 1967; reserves were estimated at 2.5 billion barrels of oil in place, and yearly production reached about 14 million barrels in the year 2000.

Ain Shams (Ibrahim Pasha) University, the third in Egypt, was established in 1957. Nakkady (1957) discussed the biostratigraphic characteristics of the Cretaceous-Danian succession in Egypt, recognized the Maastrichtian and Danian, and correlated the succession in nine sections in Egypt.

In 1959, Sigaev compiled the first tectonic map of Egypt at the scale 1:2,000,000. Synelnikov & Kellerov (1959) carried out the earliest palynological work on Carboniferous coal samples at Um Bogma, Egypt, and assigned them a Visean age. Kummel (1960) described Middle Triassic nautiloids from Sinai. Commercial coal deposits were discovered in Middle Jurassic beds (Safa Formation, equivalent of the Khatatba Formation in the Western Desert) at Gebel Maghara in Northern Sinai by the Geological Survey of Egypt (current production 58,000 tons/year). In 1965, Al Far et al. (1965) studied the coal and stratigraphy of Jurassic deposits in Sinai, and as a consequence, Al Far (1966) established Jurassic stratigraphic nomenclature in Sinai, and divided the succession into six formations: Mashabbah, Rajabiya, and Shusha (Early Jurassic), Bir Maghara and Safa (Middle Jurassic), and Masajid (Late Jurassic).

The threat of the drowning of many of the Ancient Egyptian monuments in Upper Egypt by the Aswan High Dam led many international organizations to spend huge sums of money and efforts to salvage some of those monuments, including the cutting and re-assembling of the colossal Abu Simbel Temple. In an effort to salvage prehistoric data, Butzer (1960) published a paper under the title "Archeology and geology in Ancient Egypt" in collaboration with the renowned archeologist Kaiser, thus leading the way to modern geoarcheology in Egypt. Butzer & Hansen (1968) published the book "Desert and River in Libya". The Combined Prehistoric Expedition (1968–1976), led by Wendorf, started its systematic work in Egypt. In 1972, The Combined Prehistoric Expedition made its first excavation.

Modern Egyptian geology profited greatly from the contributions of Rushdi Said (Fig. 1.2), a Harvard graduate geologist, geoarcheologist, professor, and politician, who started his carrier as early as the late forties, which culminated with the heading of the Geological Survey. He authored and co-authored numerous papers and books since 1957.

*Figure 1.2* Rushdi Said (reproduced with permission of Prof. Said).

In 1961, Said examined the tectonic framework of Egypt and its influence on the distribution of foraminifera. In 1962, Said published his "The Geology of Egypt", the first detailed book on the subject since Hume's work in the 1930s. Said (1990a) edited the second edition of "The Geology of Egypt", a collection of papers by various authors.

In 1963, aeromagnetic surveying was introduced into Egypt and the entire Western Desert was covered over the next 10 years. BP also started exploring the Gulf of Suez in that year (1963). In 1966, hydrocarbon exploration in Egypt took a major leap with the discovery of the Alamein Field, the first field in the Western Desert by Phillips Petroleum in fractured Aptian dolomites. Later, oil, gas, and condensate were found in the Alamein Field in the Upper Cenomanian Baharia sands, Aptian Kharita sands, Abu Roash sands, and the Jurassic Safa sands.

Heybroek (1965) published his classical study on the evaporites in the Gulf of Suez.

In 1967, the Suez (Six-Day) War with Israel led to the loss of Sinai and the Eastern Gulf of Suez, and the closing of the Suez Canal. Another oil embargo was imposed by the Arab oil producing countries. The July Field was discovered in the Miocene of the offshore of the Gulf of Suez by Gupco, with estimated reserves of 730 mmbbls OOIP. Production of the July Field started in 1973. The discovery of the Abu Madi Gas Field in 1967, the first discovery in the Nile Delta, by IEOC in the Abu Madi Formation sands established a new hydrocarbon province in Egypt. Subsequent discoveries led to the delineation of the Messinian Abu Madi Channel in the Central Nile Delta. In 1969, the Abu Qir Gas Field was discovered northeast of Alexandria in the Abu Madi and Sidi Salem formations. During the occupation of Sinai, Israeli geologists carried out systematic geological studies. An extensive drilling program onshore and offshore Sinai led to the discovery of a number of small oil and gas fields, such as the offshore Sadot and Raad fields in 1975 in Cenomanian sands, Wakar Field (later named North Port Said) in Middle Miocene turbidite sands in 1983, the Kersh Field in 1983, and the Abu Zakn Field in 1986 in Tortonian-Serravallian sands (EGPC, 1994). IEOC discovered the Tineh Field in 1981 in Oligocene sands and the Bougaz Field in 2006. The Mango Field was discovered by Total in 1986 in lower Cretaceous sands.

In the Western Desert, Amoco discovered gas at Abu Gharadig in 1969 in the Cretaceous Kharita and Baharia Formations, and oil was discovered in other Cretaceous intervals (Abu Roash Formation) by GUPCO in the early 1970s. Philips discovered the Umbaraka Field in seven Cretaceous sand intervals (Debeis et al., 1988).

In 1968, the Organization of Arab Petroleum Exporting Countries (OAPEC) was established. The same year (1968), Rushdi Said became head of the Egyptian Geological Survey. Rushdi Said (1971) wrote "Explanatory Notes to Accompany the Geological Map of Egypt", scales 1:1,000,000 and 1:2,000,000, which included definitions of many lithostratigraphic units.

The Amal Field in offshore Gulf of Suez was discovered in 1968 in Miocene sandstones, which was put on production in 1989 by Amapetco (now operated by Pico Petroleum)(EGPC, 1996). In 1970, the offshore Badri Field was discovered in the Gulf of Suez in Miocene sands by the Gulf of Suez Oil Company (GUPCO). In 1971, the Shoab Ali Field was discovered by Tor Sinai Oil (EGPC, Amoco & IEOC) in Miocene reservoirs (mainly in the Rudeis Group).

In the Western Desert, oil was discovered in the Khalda area in 1971 by WEPCO in Cretaceous sandstones, but was considered uneconomical and abandoned (EGPC, 1992). A small discovery was made in 1971 in the Yadma Field (abandoned in 1971) by WEPCO in the Aptian dolomites based on geology. Amoco discovered the Razzak Field in Cretaceous sands and dolomites in three structural culminations. Additional discoveries were made in the Upper Cenomanian Baharia Formation in the Wadi, Meleiha (on production at 17,000 bopd in April 2006), Meleiha NE, and Meleiha SE fields.

In October 1973, the Yum Kippur War erupted. Oil embargo was imposed briefly by the Arab countries (ended the following year), but the following years and after the signing of the peace treaty with Israel, hydrocarbon exploration in Egypt entered a new phase. Egypt became member of OAPEC in 1973. The same year (1973), the July Field was discovered in Miocene sands by GUPCO, with estimated reserves of 625 mmbbls. Amoco made a small heavy oil discovery in the Ahram Field (shut- in 1974) in the Albian Kharita Formation in the Western Desert. In 1974, another giant, the Ramadan Field was discovered in Cretaceous sands by GUPCO in the Gulf of Suez (reserves of 668 mmbbls). Shell returned to Egypt in 1974 through a partnership with Deminex and BP. In 1976, the Zeit East Field was discovered by GUPCO in Miocene and Nubian sandstones. The Zeit East Field had cumulative production of 88.6 mmbbls in 2002. The horizontal well East Zeit A-19 reached a record length of horizontal section of 2700 ft (Blanchard et al., 2002). Oil was discovered in 1976 in the Qantara Formation (Qantara-1) in the Eastern Nile Delta. In 1977, the October Field (Abdine et al., 1992), the third largest oil field in Egypt after Belayim and Morgan fields, was discovered by GUPCO.

In 1978, Petrobel Petroleum Company was founded as a joint venture between Eni and EGPC for the exploitation of the oilfields in the Gulf of Suez. The Bakr W G1 and Bakr W H fields were discovered onshore the Gulf of Suez in Eocene reservoirs by the General Petroleum Company (GPC). The Ras Fanar Field was discovered with the KK84-1 well in the Gulf of Suez in Middle Miocene coralline buildups of the Nullipore Carbonates by Deminex GmbH (SUCO). The trap in the Ras Fanar Field is a tilted horst-block. The field started production in 1984 and produced 96 mmbbls of oil by 2005. The Ras Budran Field was discovered by Deminex (SUCO) in Paleozoic and Lower Cretaceous Nubian "A" sandstones in the Gulf of Suez, with estimated reserves of 150 million barrels of oil. In the Western Desert, the East Razzak Field was discovered in the Upper Cenomanian Abu Roash "G" and Baharia formations. In 1979, the Bakr W K Field was discovered in Miocene sands by the GPC onshore the Gulf of Suez. Exploration by Shell on its two concessions; Badr El Din and Sitra, in the Western Desert acquired in 1979, led to the discovery of the two fields in the early eighties.

In 1980, 3D seismic surveys were run for the first time in Egypt in the Mediterranean Sea and the Gulf of Suez. The El Khaligue Field onshore Gulf of Suez was discovered in Miocene sands by the GPC (EGPC, 1996). The Zeit Bay Field was discovered in Cretaceous by SUCO in the Gulf of Suez.

In 1981, the Amer North Field offshore the Gulf of Suez was discovered in the Belayim Formation by the GPC and the Esh El Mellaha Field was discovered onshore in Miocene sands by the Canadian Superior Group in a stratigraphic trap. Scimitar discovered the Issaran heavy oil field onshore the Gulf of Suez in the same year. The Zeit Bay Field was discovered by Deminex and partners Shell and BP. The appraisal well QQ89-1 encountered gas in basement rocks, while the appraisal well QQ89-2 encountered 253 m (830 ft) column of oil in basal Miocene Nukhul Formation (heavy oil), Middle Miocene sands, Nubian sandstones, and fractured and weathered Precambrian granites (EGPC, 1996). The Zeit Bay Field started production in 1984. The Tineh-1 oil discovery in Oligocene sands and the Temsah Field by Mobil Oil in Miocene sediments offshore Sinai by IOEC added to the prospectivity of that area. Agiba Petroleum was formed following the discovery of Meleiha Field in the Western Desert as an operating company for EGPC, IEOC, IFC, and Lukagip. Agiba also operates the Ashrafi offshore field in the Gulf of Suez.

In 1982, the offshore Hilal Field (on the B-Trend) (Helmy, 1990) was discovered in Senonian by Torsina Oil (a joint venture between BP and Amoco). The Badr el Din (BED-2) oil and gas field was discovered by Shell in Cretaceous sands and the Aghar Field by Agiba in the Western Desert. Originally, the dry well Aghar-1 was drilled by Wepco in 1970, but re-examination of well logs by Agiba led to the discovery of Aghar Field in Upper Cenomanian Baharia Formation, with estimated reserves of more than 27 mmbbls of recoverable oil (EGPC, 1992).

Egypt's daily production in 1980 was estimated at 584,148 million barrels of oil/day, 210,972 million cubic feet of gas/day (mmcfgd), and 6,952 barrels of condensate/day. In 1983, Egypt's daily production increased for oil, gas and condensate, with estimated 652,921 bopd, 257,990 mcfgd, and 7,501 bcd.[4]

_____

4 AAPG Bull., 1983, v. 67/10, p. 1796–1797.

Minor discoveries made in 1983 included the North Shadwan (GH452 1A) Field by Amoco in Miocene sands and GS327 by GUPCO in the Gulf of Suez.

In 1984, the NE Abu Gharadig (NEAG) Field was discovered in the Abu Roash (Turonian) and Kharita (Albian) sands in an anticline bounded by faults to the north. The recoverable gas reserves were estimated at 44 bcf and recoverable oil at 1.4 mmbbls. In 1985, the Salam Field was the first discovery in Jurassic sediments of the Western Desert. The discovery well flowed >6000 bopd from the Cretaceous reservoirs and >5000 bopd from the Jurassic. Conoco Inc. bought into the Khalda Concession and the Khalda Petroleum Company was formed as a joint venture between Conoco, Texas International, and EGPC. These discoveries were followed by the Yasser-1 which flowed 8941 bopd and 13.6 mmcfgd from two separate Cretaceous intervals. The Amoun-2 well tested 4160 bopd and 12.5 mmcfgd from Jurassic sands (EGPC, 1992). The Hayat Field and Kenz Field encountered hydrocarbons in the Upper Baharia. Other discoveries were made by Agiba included Meleiha SE, Aman, Emry, and Lotus fields in the Western Desert. In 1986, the Ashrafi Field was discovered offshore the Gulf of Suez by IEOC (with partner Marathon who sold interest to Eni in 1999) in fractured granites, Nubian sandstones and Miocene sandstones. The Field has been on production since 1992, with a rate of 8100 bopd in April 2006).[5] Other Jurassic and Cretaceous discoveries by Agiba in the Western Desert followed, which included the Falak (Jurassic & Neocomian sands), Dorra (Jurassic), Zahra-1X (Aptian and Upper Cenomanian sands), Karnak, and Bardy (Baharia sands) fields.

In 1987, the Gazwarina Field was discovered in the Gulf of Suez by Marathon and partners in Middle Miocene sands, and the Badri Field was discovered by GUPCO in Miocene reservoirs.

In 1988, oil prices crashed to about $10 a barrel in July, down from $30 the year before, which had a global effect on oil exploration activities and professionals in the field. In the meantime, Shell signed the first gas clause (amended in 1993) in Egypt with EGPC for Bed-3 in the Western Desert (EGPC, 1992). The purpose of the "gas clause" was to tie gas prices to crude prices and encourage companies to invest in gas. The Abu Rudeis Marine 3 Field was discovered in 1988 in the Gulf of Suez in Turonian sandstones by Petrobel and the Esh El Mellaha East Marine Field offshore the Gulf of Suez in Miocene sands by Magapetco. In 1989, the Gemsa SE Field was discovered by Shell in the Miocene dolomitic reefal "Nullipore limestones" (Belayim Formation) in the Gulf of Suez. The Ashrafi Island Petroleum Company (Asheptco) was founded in 1989 as a joint venture between EGPC, Agip, Agiba, and partners.

In the 1990s, hydrocarbon exploration continued along with the appearance of new technologies. In 1990, the first horizontal wells in Egypt were drilled by the Khalda Oil Company (Schlumberger, 1995, El-Refaie, 1993). In 1991, the Badri E Field in offshore the Gulf of Suez was discovered in Miocene sands by GUPCO. The offshore Warda (Zaafarana) Oil Field was discovered in Miocene reservoirs by British Gas in 60 m (197 ft) of water. The discovery well Hb78-2 tested oil in five separate sandstone intervals with a cumulative rate of 7000 bopd. Consequently, the Zaafarana Oil Company was formed between EGPC and BG.

In 1992, gas discoveries continued to be made in the Western Desert. The Kanayes Field was discovered in Jurassic sands by IEOC and the Obaiyed oil and gas Field by

---

5 Production figures for Ashrafi, Raml, El Faras, Zarif, W. Razzak, and Meleiha are from www.Agiba.com.

Shell in the Middle Jurassic Lower Safa sands; the discovery well Obaiyed 2-2 tested 13.7 mmcfgd and 1470 bcd (production started in 1999). The latter led to the formation of the Obaiyed Company. In 1993, the Ras Qattara Petroleum Company was created between EGPC, Agip, Agiba, and partners. In 1997, Repsol discovered the Shams Field in Jurassic sandstones and Apache made a discovery in the North Alamein Field in 1998 in Alamein Dolomite and Dahab sands in the Western Desert.

The year 1994 signaled the beginning of a "Gas Boom" (Hataba, 2005) in Egypt when the proven gas reserves of about 19.7 TCF in that year rose gradually to 66.8 TCF in 2005. In the meantime, oil reserves were estimated at 3.3 Billion barrels (Oil & Gas J., 2005). The same year (1994), Apache entered Egypt with the farm-in of a 25% non-operated interest in the Western Desert's Qarun Concession. In 1995, the Ras El Ush Field was discovered by Marathon in lower Senonian and Nubian sandstones. The Denise Field, the first field drilled with Pliocene objective in the Mediterranean offshore was discovered by IEOC (Eni in Egypt, 2006). Apache also drilled the wells Shaqiq 1X (dry) and Marakia 1X (oil) in Cretaceous AEB.

In 1996, Ashrafi SW Field offshore the Gulf of Suez was discovered in Middle Miocene sediments by Ashpetco. The Ashrafi SW 1X encountered 82 m (269 ft) of net pay in the Cretaceous Nubian Sandstone and 22.3 m (73 ft) of net pay in fractured granites (Oil & Gas J., Sept. 1996). The El Faras Field was discovered the same year. Raml Petroleum Company (Rampetco) founded between EGPC, Agip, Agiba, and partners. The Raml Field was put on production at 2100 bopd in 1997.[6] In 1998, the Ashrafi S. Field offshore the Gulf of Suez was discovered in Middle Miocene reservoirs by Lukoil and the July North Field by Fanar Oil. In the meantime, Tanganyika acquired the West Gharib Block in GOS, and the Hana Field was discovered in 1999 on the West Gharib Block by Dara (Dublin/EGPC) in the Miocene Karim Formation. In 1999, the Ashrafi W Field offshore the Gulf of Suez was discovered in Middle Miocene sediments by IEOC.

Large concessions were granted in the offshore of the Nile Delta. The discovery of Ha'py Field by BP Amoco/Agip in the Ras el Barr Concession in Pliocene slope channel offshore the Eastern Nile Delta (operated by GUPCO) was made based on a high-amplitude (bright spot) anomaly. The structure in the Ha'py Field is related to the NW-SE-trending growth faults of the Bardawil Line (Abdel Asal et al., 2000, Samuel et al., 2003). The Ha'py Field was put on production in 2000 at a rate of 280 mmcfgd. In 1997, BG discovered the Rosetta gas and condensate field in Pliocene channel sands, about 60 km north of Alexandria its Rosetta Concession in the Mediterranean offshore in 61 m (200 ft) of water. The First well to be drilled in deep water offshore Nile Delta was the East Delta Deep Marine-1 (EDDM-1) in water depth 750 m (2460 ft).

In 1997, Egypt was consuming all the gas it produced. The Abu Marwa Field in the Gulf of Suez was discovered in 1997 in Lower Miocene reservoirs. Yet a new hydrocarbon province was proven in Egypt in 1997 when Apache (50% with Seagull) discovered the Beni Suef Field in Upper Egypt, with 42° API oil, in seismically identified structures, on the East Beni Suef Concession (Dolson et al., 2001, Tawfik et al., 2005). The Beni Suef-1X flowed 5,200 bopd from a 12 m (40 ft) interval at 133 m (7000 ft) depth from the Albian Kharita Formation. The Beni Suef Field is located near

6 (www.bg-group, 2011).

the northern limit of the Nile Basin as defined by Tawadros (2001). In 1998, another field was discovered near the southern edge of the Nile Basin at Kom Ombo. The Komombo-1 (re-named al-Baraka by the Petroleum Minister in 2007) was drilled by Repsol on the Ganope Block 2 and recovered 39°API light oil on test in Early Cretaceous sandstones. The Komombo Basin is a rift basin. The oil is believed to have originated from Jurassic type II–III lacustrine source (Dolson et al., 2001). In 1998, Shell signed the Northeast Mediterranean Deepwater Concession (NEMED) Agreement.

Apart from the successes achieved in hydrocarbon exploration, other major contributions to our geological and exploration knowledge were made in the field of geoarcheology during the later part of the 20th Century. The Dakhla geoarcheological multidisciplinary project (DOP) started in 1978, under the leadership of Maxine R. Kleindienst, a geoarcheologist and Pleistocene prehistorian, which passed to her successor Antony J. Mills in 2005, together with archeologist Mary A. McDonald and geoarcheologist C.S. Churcher. The team recognized a number of paleolakes and tufa terraces deposits in the Dakhla Oasis. Pleistocene terraced gravels and tufas were also mapped in the Kharga Oasis by the team members of the Kharga Oasis Project (KOP), Jennifer Smith and R. Giegengack of the University of Pennsylvania.

The 21st Century opened with one of the most horrific terrorist attacks in recent history; the attack on the twin towers in New York on September 11, 2001. The attack strained the relation between the West and East. In Tunisia and Egypt, the unexpected events of January 25, 2011 and the subsequent outcome brought new hopes and fears to the countries as well as the whole Middle East. The civil war in Libya, which is still ranging at the time of writing, has and will have a negative effect. It is not known yet how these events will affect geological and hydrocarbon exploration in the area and around the world.

Extensive efforts made in the late 1970s and early 1980s were consecrated to the solution of the Nubian Sandstone controversy. Klitzsch et al. (1979) and Ward & McDonald (1979) attempted to subdivide the Nubian Sandstone in the Western and Eastern Deserts, respectively, and represented a major step toward solving the Nubian Sandstone controversy. In 1985, a German multidisciplinary team made regional studies of the Nubian Sandstone in southern of Egypt. The results of the German Special Research Project, supported by the German Research Foundation until 1987 were published in volumes 50 and 75 of the *Berliner Geowissenschaftliche Abhandlungen*.

In 2001, Tawadros published "Geology of Egypt and Libya", with the main objective to review the geology of the two countries in the context of global and local tectonics and the stratigraphic nomenclature, and to tackle some of the chronic stratigraphic problems, such as the Nubian Sandstone, the Esna Shale, and Paleozoic and Lower Cretaceous quartzites and sandstones. The latter was fully treated in Tawadros et al. (2001).

Hydrocarbon exploration and production in Egypt entered a new era in the beginning of the century, especially in the gas sector. In 1999, The Scarab-1 and Saffron-1 wells discovered gas in two separate accumulations in the West Delta Deep Marine Concession (WDDM) in water depths of 250–850 m (820–2789 ft) (Samuel et al., 2003). The well Scarab-1 tested more than 30 mmcfgd; the Saffron-1 tested 90 mmcfgd (reserves in the two fields are >4tcf of gas). The Scarab-Saffron Field is the largest gas field development in Egypt, on production in 2004. The Simian

Field was discovered in 1999 in Pliocene sediments in the Mediterranean by British Gas. The Simian/Sienna fields began production in 2005 with a capacity of 565 mmcfgd.

The 21st Century opened with the announcement of new discoveries in 2000 in Egypt (AAPG Explorer, Jan. 2001). The Akik-1 well tested 54 mmcfgd and 543 bopd. The GS 302-2 tested 4000 bopd. The Karama-1 tested 1450 bopd and 500 mcfgd. The Lagia-6 tested 170 bopd and 300 mcfgd. The Neama-1 tested 159 bopd. The Sapphire-1 tested 35 mmcfgd and 1100 bcpd. The Tawoos-1 tested 833 bopd. The Ha'py Field came on line with 280 mmcfgd. The Karama Field was discovered by Apache in the Eastern Abu Gharadig Basin in 2001. The real "Gas Boom" started in Egypt in 2000 following the discovery of the Scarab-Saffron and Sapphire gas fields and the beginning of production from the Ha'py Field in the offshore of the Mediterranean Sea, which led the government to restructure its oil and gas industry the following year. Gas production which equalled consumption in 2001, 2002 and 2003 at 2.1 bcfgd, each; surpassed it since 2004 (graphs in Global Research; Egypt, Nov. 2006).

The Egyptian Government restructured the oil industry with the formation of three separate entities. The Egyptian Natural Gas Holding Company (EGAS) was established in 2001 as a new state-owned body, separate from the EGPC, to manage the natural gas sector. The Egyptian Petrochemical Holding Company (EChem) was established in 2002 to manage the petrochemical industry. The Ganobe El Wadi Holding Co. (previously known as the South Valley Development Company "SVDC") was established in 2003 to manage exploration and production activities in Upper Egypt south of Lat. 28°, while the EGPC controls activities north of that latitude.

In 2002, Apache discovered El Max-1X, with a 152 m (500 ft) gas column, in 945 m (3100 ft) of water, the Abu Sir-1X with 95 m (311 ft) gas column, the Al Bahig-1X in the Upper Paleocene Kafr el Sheikh sands with 75 m (247 ft) gas column in water depth of 1070 m (3510 ft). The Ozoris-1X discovery tested 2504 bopd of 38.5° API oil from the Lower Cretaceous AEB Sand. The Khepri-9 well on the South Umbarka lease tested 29.5 mmcfgd and 220 barrels of condensate per day from the AEB. The Farag Field onshore the Gulf of Suez was discovered in Eocene reservoirs by Tanganyika. In 2002, Egypt and EGAS offered 34 blocks in a bid round (4 more were added in 2003). The Scarab/Saffron Fields began production in 2003 with a capacity 700 mmcfgd. The El Wastani Field of Centurion Petroleum in the El Manzala Concession in the Eastern Delta started production at 12 mmcfgd (Centurion Oil). Eni, in partnership with BP, announced the discovery of the Tennin-1 in the East Delta Deep Block in water depth of 300 m (984 ft), which tested 24.7 mmcfgd, with estimated gas reserves of 0.5–1.0 bcf (billion cubic feet).

In 2003, Apache discovered the Qasr-IX, identified by a 3D seismic survey, and tested 51.5 mmcfgd and 2,688 bcd with a net pay of 185 m (606 ft) in six Jurassic Lower Safa and Ras Qattara reservoirs (according to Apache public announcements). The Jurassic continued potential was attested by the JG discovery by Shell, and the SD-9 by HBSI. Apache's El King-1X tested gas in Miocene and Pliocene reservoirs, and constituted the first Miocene Abu Madi deepwater 720 m (2361 ft) discovery offshore the Nile Delta. The well tested 31 mmcfgd and 757 bcd. BP announced the discovery of a large new field in the Gulf of Suez, the Saqqara Field near the Morgan Field, with estimated peak production of 40–50,000 bopd. The Egyptian government began several rounds of new concession awards in 2003, and by 2005 it had signed 35 new oil and gas exploration agreements.[7]

In 2004, 16 oil and gas discoveries were made in Egypt and six in Libya (AAPG Explorer, Jan. 2005). BP discovered Taurt 1, which tested 22 mmcfgd in one of four intervals, and was expected to go on stream in 2008. In the Ras El Barr Concession in the Eastern Nile Delta, the Raven Field (BP and RWE) was discovered NW of Rosetta which tested 37.44 mmcfgd and 741 bcd in 650 m (2133 ft) water with estimated reserves of TCF of gas. RWE Dea Egypt and partner BP discovered gas in the Polaris 1X Field in the North Alexandria Block in the West Med Deep Water (WMDW) Concession. The Polaris 1X was the first Miocene gas and condensate discovery in the Concession, in water depth of 1162 m (3813 ft). The exploratory well tested a rate of 26.7 mmcfgd from a 42 m (138 ft) with about 19 m (62 ft) of net pay in a Middle Pliocene slope channel/levee play at a depth of 2178 m (7146 ft). In 2008, BP signed a deal with EGPC and EGAS to supply natural gas to the Damietta LNG Plant and to purchase LNG from EGAS. BG also signed a deal with EGPC to supply natural gas from the Scarab/Saffron fields in WMDW to the Damietta LNG Plant.

The Bakr E1 Field offshore the Gulf of Suez was discovered in Miocene sands by GPC in 2004 and tested 2000 bopd. Shell drilled four ultra-deep wells in the NEMED Concession, Kg45-1, kj49-1 & La52-1 and made two gas discoveries. Melron (later Melrose) discovered the Aga-1 in N. Egypt which tested 9.3 mmcfgd and 19 bcd (first test of condensate in El Mansoura Concession) and the Batra South 14 in the Delta which tested 9.7 mmcfgd and 150 bcd in Pliocene sands. El Amir (NW Gemsa) was discovered in the Gulf of Suez by Vega Oil & Gas in the Ras Gharib Formation. The Fadl Field in West Gharib Block was discovered by Tanganyika. The West Ashrafi Petroleum Company was created between EGPC, Agip, Agiba, and partners. The Russian Lukoil discovered oil in the NE Geisum Offshore Block. Petrobel made three additional discoveries in the Gulf of Suez in BL 113-A-24, BM 85, and BM W1. Five exploration contracts were awarded to a newly-formed, state-owned upstream company, Tharwa Oil, four of which were in the Western Desert and one offshore the Mediterranean. In 2004, TransGlobe Petroleum and Arsenal Energy acquired 50% from Quadra Resources of Calgary share and operatorship of the Nuqra Block, which covers part of Komombo Basin in Upper Egypt, east of the Nile (see above). In the meantime, RWE Dea acquired operatorship of Al Amriya Concession in shallow waters in the West Nile Delta following its bid during the EGAS 2004 Bid round. RWE Dea AG, via its wholly owned subsidiary RWE Dea Nile GmbH, signed a new concession agreement with the government of Egypt and the Egyptian Natural Gas Holding Company (EGAS) for the Disouq Area. RWE Dea's Ras Fanar B10 well tested 3000 bopd. Shell's Sheiba 18-3 well, the first commercial discovery in the eastern part of NE Abu Gharadig Concession, tested 1600 bopd and 0.9 mmcfgd. Selpetrol's discovery, Ferdaus 1 tested 3250 bopd. Khalda's Mihos 1 well tested 41.8 mmcfgd and 1419 bcd.

Other discoveries were made in 2004; oil in West Geisum offshore block by the Russian Lukoil, the Hoshia-1 and Naiem fields by Tanganyika in Rudeis Formation, and the El Diyur Field in the Western Desert with 1000 bopd. The El Tamad Field was discovered by Melrose[8] in July 2005 in the Miocene Sidi Salem Formation in the El

---

7 Economic Trend Report, April 2006, Embassy of the United States, Cairo, Egypt, p. 51.
8 Melrose acquired Melron in 2006.

Mansoura Concession in the Nile Delta. The discovery was followed by the Tamad-2 well which encountered a 7.3 m (24 ft) oil column. A 2 m (7 ft) interval at the top of the oil column tested at 1,041 bpd of 45.6° API oil and 780 mcfgd. Apache made a discovery with the Syrah 1X well, northwest of the Qasr Field in 2005. The well tested 46.5 mmcfgd from Cretaceous and Jurassic horizons. The Qasr and Syrah fields represented the largest gas discoveries in the onshore of Egypt. Another discovery by Apache was the Tanzanite-1X tested 5,296 bopd and 7.7 mmcfgd in multiple pay zones in the Lower Cretaceous Alam El Bueib (AEB) sands. The Tanzanite-2 tested 2846 bopd in the Alamein Dolomite.

By the end of 2004, the remaining reserves of Egypt were estimated (Hataba, 2005) at 23 MMBOE, divided as follows: Oil: GOS 43%, Western Desert 26%, Mediterranean and Delta 13%, Eastern Desert 11%, and Sinai 7%. Gas: Mediterranean 77%, Western Desert 12%, GOS 8%, and Delta 3%. Egypt's production was about 594,000 bopd and 2.6 bcfgd, exceeding gas consumption at 2.5 bcfgd, for the first time. At the end of 2005, the estimated proven reserves were 3.71 billion barrels of oil and 66.84 bcf of gas.[9]

The First liquefied natural gas (LNG) export began in January 2005 from the Damietta and Idku facilities. The Arab Gas Line to Jordan consists of three sections. The first section was completed in July 2003 and runs from Al Arish in Egypt to Aqaba in Jordan. The second section is from Aqaba to Rehab in Jordan. An Egyptian consortium will operate the pipeline for a period of 30 years with optional extension period of 10 years after this period the ownership of the project will be transferred to Jordan. The third line extends north of the Jordanian-Syrian border to the Turkish-Syrian boarders and from the west to Banias and Tripoli in Lebanon. The Arab Gas Pipeline project was designed supply the region with Egyptian gas. However, as for the year 2007, it was facing a problem because of differences between Jordan and Egypt over prices (Alarab Alyawm, April 10, 2007).

The West Khilala Field was discovered by Melrose with the West Khilala No.1 well in October 2005 in Miocene turbidite sands,[10] with proven reserves of 218 bcf, and a production rate of 63.2 mmcfgd in 2007. The West Dikirnis Field was discovered in December 2005 with the West Dikirnis No.1 well, which encountered gas and condensate in the Miocene Qawasim Formation. The West Dikirnis No. 2 well, drilled in May 2006 tested at over 5,000 bopd. Other discoveries by Melrose include the Rummay Field in the Nile Delta and the Ferdous Field and wildcat discovery Rayan 1X in the Western Desert (OGJ, Sept. 1, 2006).

The West Manzala Concession was awarded to Centurion in 2004, but 50% was farmed out to Shell in 2006, and oil was discovered in 2006 in the Luzi-1 well with a flow rate of 9 mmcfgd and 123 bcd from Abu Madi Formation. Two wells were drilled and tested gas from Sidi Salem Formation by previous holders of the concessions: Abu Naga-1 tested gas at a rate of 5.7 mmcfgd and 360 bcd; Matariya-1 encountered high pressured gas (drilled in 1976). Dana Gas PJSC of Sharja, UAE acquires Centurion Energy International in November 2007 (Centurion Energy News Press release, 2007).

---

9 World Oil, Sept. 2006, p. 57.
10 Source: Melrose Resources, 3 April, 2007. Preliminary results for the year ended 31 Dec. 2006.

In 2006, Ganope offered International 2006 Bid Round 8 blocks in the Red Sea and Western Desert, including the new Mesaha Block in the extreme southwest, which was awarded to a consortium led by Melrose Resources, and the Dakhla and Diyur blocks in the Western Desert. The West Esh El-Mellaha Block was awarded to the British Aminex Petroleum, and the W. Komombo Block was awarded to the Australian Pan Pacific. In July 2006, Groundstar Resources acquired 60% working interest on the West Komombo Block (Block-3, west of Centurion's block) from Pan Pacific Petroleum Egypt Pty. In the Western Desert, Sipetrol made a discovery in the East Ras Qattara Block with the Shahd-1 well in 2006. INA partnership with RWE Dea made an oil discovery with the Sidi Rahman-IX on the East Yidma Concession in the Cretaceous Kharita and Baharia sands.

In 2007, Sipetrol made a second oil discovery in the East Ras Qattara block in Western Desert with the Ghard-1 well, which encountered 40.5° API oil in the lower Baharia Formation at a rate of 2026 bopd and 2.6 mmcfgd. The Imhotep Field was discovered by Apache in 2004 in the Jurassic Upper Safa Formation. Hathor Deep 1X on the offset of Khalda Concession tested gas in Lower Cretaceous Alam El Bueib (AEB) at a rate of 12 mmcfgd and Oil from AEB 3D at 1237 bopd. Apache made the Alexandrite 1X discovery which tested 20 mmcfgd and 4045 bopd. The Jade-1X drilled by Apache in 2007 encountered 217 ft of AEB pay and 20 m (66 ft) in the Jurassic Upper Safa Formation. In September 2007, Centurion (Sharja's Dana) discovered a new oil field in Upper Egypt. The discovery well tested 37° API oil from a Cretaceous reservoir. The West Dikirnis Field came on production in November.

In 2008, Hess and partners RWE Dea and Kufpec discover the Dekhila-1X well in the West Mediterranean (Block 1) Deepwater Concession in 1183 m (3883ft) of water. Gas de France with partner Dana discovered gas in the WEB-1X in the W. Burullus Concession, offshore the Nile Delta, which tested 27 mmcfgd from Pliocene turbidites in 62 ft of water (World Oil, July 2008). Sipetrol discovered oil in the Shahd SE-1 in the East Ras Qattara Concession in 2008, which tested 39°API oil at a rate of 2500 bopd (World Oil, June 2008). Melrose discovered 13 mmcfgd and 182 bcpd from 38 ft of pay in E. Abu Khadr-1 well and the Damas-1 which tested 14.3 mmcfgd and 105 bcpd in Sidi Salem, both in the El-Mansoura Block (World Oil, June 2008). RWE Dea discovered gas and condensate with the N. Sidi Ghazi-1X well in the Desouq Concession onshore Nile Delta, which tested 37 mmcfgd from Upper Messinian sands (World Oil, June 2008). Vegas tested 3388 bopd of 41° API oil and 4.25 mmcfgd in the Al Amir-SE-1 on the NW Gemsa Concession onshore the Gulf of Suez (World Oil, Nov. 2008). BP made other discoveries in the Northern Shedwan Block in the Gulf of Suez at a rate of 10,000 bopd and the Satis-1 in the North El Burg Block offshore Nile Delta, in 61 m (200 ft) water in Oligocene reservoir (World Oil, March 2008). Apache tested 41.6 mmcfgd and 1313 bcd from a 178 ft (54 m) pay zone in the Jurassic Lower Safa and probable 14 m (45 ft) oil pay in Jurassic AEB-Unit 6 and 9 m (30 ft) oil pay in the Lower Cretaceous Alam El Bueib in the Shushan Concession (World Oil, March 2008).

In 2009, Improved Petroleum Grp (IPR) (in partnership with Sojitz Corp) found oil and gas in the Zain-1X in the Yidma-Alamein Block, which tested 5414 bopd and 16 mmcfgd from two Jurassic intervals (World Oil, Jan. 2009). Vegas Oil & Gas discovered 5785 bopd and 7.8 mmcfgd with the Amir South-2X in the NW Gemsa Concession onshore the Gulf of Suez (World Oil, April 2009) and tested 2809 bopd and 3 mmcfgd from the Kareem Shagar Sand in the Geyad-1X on the N West Gemsa Concession and

1174 bopd and 1.3 mmcfgd in Lower Kareem Rahmi sands (World Oil, July 2009). Dana Gas discovered gas with the Azhar-1 well at a rate of 15.1 mmcfgd and 444 bcpd from a 30 m (98 ft) interval in the Upper Sidi Salem Formation; the W. Manzala-2 well in the Nile Delta (World Oil, April 2009), and gas and condensate with the Salma-1 well, from an 25 m (82 ft) pay in the Abu Madi Sandstone and 4 m (13 ft) pay in the Kafr El Sheikh Sandstone on the W Qantara Concession (World Oil, Feb. 2009), and tested 11.4 mmcfgd and 381 bcpd with Tulip-1 from Abu Madi sands (World Oil, July 2009). BP discovered gas with the Ji 50-2 Ruby-3 well on the WMDW Concession in Pliocene sands in 3018 ft of water. The Ruby-1X and Ruby-2 wells were drilled by RWE/BP in 2002 and 2003, respectively. Apache tested 8279 bopd and 0.4 mmcfgd from Cretaceous Alam El Bueib and the Jurassic Safa formations at Phiops on the S. Umbarka Concession (World Oil, July 2009). Apache mad a number of discoveries in the Western Desert in 2009, including three Jurassic discoveries which tested a total of 80 mmcfgd and 5909 bopd, the Sultan-3X which tested 5021 bopd and 11 mmcfgd from the Alam El Bueib (AEB-6) and the Safa sands on the Khalda Concession, and the Adam-1X and Maggie-1X on the Matruh Block (World Oil, Feb. 2009).

In 2010, Eni[11] discovered oil in the Arcadia Field in Alam El Bueib Formation on the Meleiha Concession in the Arcadia IX well, with additional gas potential in the Khatatba Formation. The Arcadia Filed was put on production on July 28, 2010 (World Oil, August 2010).

## I.2   LIBYA

One of the earliest voyages to Libya was made in 1797–1798 by Fredrick Hornemann who travelled by Caravan from Cairo via Gialo and Murzuk to the Niger River (Hornemann, 1802) and gave the first description of Jabal Haruj. In 1817, the Italian physician-naturalist Paolo Della Cella traveled from Tripoli to Cyrene where he recognized fossiliferous limestones with *Cardium, Pecten* and *Nummulites*.

In 1850, the British Government sent an expedition led by James Richardson, accompanied by Overweg and Barth to discover caravan routes in the Southern Sahara and to conclude treaties with the chieftains. Overweg identified different rocks in Tripolitania, described the Gharyan volcanics, recognized the Cretaceous rocks in Jabal Nafusah, and collected the first Carboniferous fossils in West Libya (Overweg, 1851). The latter were described by Beyrich (1952). Unfortunately, Richardson and Overweg died prematurely. James Richardson died of fever in 1851 and Overweg continued the expedition, but died in 1862 of malaria on Lake Chad.

In 1865, Rohlfs travelled across Libya during his first expedition to North Africa and described Jabal Assawda in Central Libya. In 1867, G.B. Stacey described several outcrops and discussed the subsidence of the coast near Benghazi (Stacey, 1867). Gustav Horneigal was the first European to travel from Murzuq to Tibesti in 1869 (Von Bary, 1877). In 1876–1877, Von Bary travelled over North Africa to Ghat and Tuareg (account published in German in 1890 and in French in 1898) (Von Bary, 1898).

---

11 Eni is Egypt's leading producer with about 230, 000 boepd, including 37,000 boepd from the Western Desert (WO, August 2010).

In 1880–1895, Rolland summarized the geological history of North Africa and provided the first geological maps of western Fezzan (Rolland, 1880, 1889). Rolland also identified Devonian and Cretaceous rocks at Jabal Akakus and Hamada al Hamra area. In 1893, Elisée Reclus (1893) wrote an exhaustive account on the geography, culture, and people of North Africa, with the exclusion of Egypt.

In Libya, Krumbeck (1906) studied the geology and paleontology of Tripolitania and identified Carboniferous fossils from the Southern Ghadames Basin. In 1911, J.W. Gregory wrote the most significant paper on the geology of Cyrenaica and summarized the results of the 1908 expedition (Gregory, 1911a, 1916). Gregory introduced a number of formational names, subdivided the Eocene succession into Apollonia, Derna, and Slonta limestones, and included the Aquitainian in Oligocene. R.B. Newton, Chapman (1911), and Gregory described the fossils collected during that expedition (Gregory, 1911b). The molluscs were studied by Bullen-Newton (1911).

In 1911, the Italians invaded Libya and the Turks surrendered in 1912. The same year (1912), the first geological map of Libya was made. In 1913, extensive geological studies of Fezzan were started by Italian geologists, notably among them are Crema, Parona (1914, 1928), Zaccagna (1919a, b), Sassi, Coggo, Stefanini, Franchi, Sanfilippo, Migliorini (1914, 1920), and Checchia-Rispoli (1913) (Crema et al., 1913). Vinassa de Regny (1912) produced a geological map of Libya at a scale of 1:6,000,000. The WWI (1914–1918) hampered geological exploration in Libya and the only apparent achievement was that of Parona (1914) who wrote a summary on the geology of Libya. In 1920, Italian Touring Club organized an excursion to Cyrenaica led by Marinelli, made geological observations, morphological studies and collected fossils, and interpreted two escarpments that border the Cyrenaica Jabal (Jabal Akhdar) as raised beaches (Marinelli, 1920, 1921, 1923). Fossils collected were later studied by Raineri, Zuffardi-Comerci (1929, 1934, 1940), Cipolla, and Checchia-Rispoli. Stefanini (1918, 1921, 1923) presented a sketch map of Cyrenaica, where he grouped all the Eocene rocks in one unit and included the Aquitainian in the Miocene instead of Oligocene. Crema (1922) studied Maastrichtian sediments of Cyrenaica and in 1926; he reported traces of oil in a well, which led to the drilling of a well in Tripolitania by Agip (the Libyan subsidiary of ENI), although it was dry.

One of the most prominent geologists of that period was Ardito Desio (1897–2001). Desio wrote extensively during his career in Libya (Ettalhi et al., 1978 show a list of more than 122 of his publications). Desio (1926–1927) led an expedition to Jaghbub Oasis. He explored the then unexplored Marmarica Plateau. Pfaltz (1930) suggested that the Cyrenaica terraces were fault controlled. Silvestri (1934) examined the fossils collected from nummulitic rocks in Derna by Desio and indicated the presence of Upper Eocene in Cyrenaica for the first time. Chiarugi (1929) studied the silicified woods found south of Wadi Faregh and Maaten Risan and correlated them with those of Jaghbub.

It was in 1930 that a more active and systematic exploration of Libya was carried out. Dalloni led a scientific mission organized by the French Government in 1930–1931 to study the entire Tibesti Massif, and produced a map at a scale of 1:500,000. Dalloni divided the Precambrian complex into Archean and Algonkian Series. Dalloni introduced the name Olochi Sandstone for a conglomeratic sandstone unit separating the Precambrian complex from the overlying "Nubian Sandstone" (Dalloni, 1934, 1945).

The Italians occupied Kufra in 1931and the Turkish control of Libya ended in 1932. During that time, Desio made an expedition in 1931 on camel backs from Jaghbub to Kufra, Arkenu, and Gebel Uweinat (Desio, 1931, 1934). Desio published three volumes[12] between 1934 and 1939, which contained the results of an expedition organized by the Royal Academy of Italy to the Kufra Basin in 1931. These reports appeared in 1934 (v. 3), 1935 (v. 1), and 1939 (v. 2). The first contained the geological data and a geological map at the scale of 1:250,000; the second volume dealt with the paleontological findings, and the third summarized the geomorphological results. During the expedition to Kufra, the crystalline massifs of Arkenu and Uwaynat (Oweinat) were described and mapped in more detail. Petrographic description of igneous and metamorphic rocks was carried out by Gallitelli (1934), detrital rocks were described by De Angelis (1934), and fossil plants were identified and illustrated by Negri (1934) and Chiarugi (1934).

The first bone fragments of large mammals were collected by Desio during his 1931 trip and were studied by Airaghi (1934). Chiarugi (1934) described silicified wood from the Kufra. D'Erasmo (1934) described Tertiary vertebrates from Sirtica during the Kufra Mission.

Desio (1931) predicted the potential of the Sirte Basin as a hydrocarbon province (today the Sirte Basin is known to contain about 80% of Libya's oil reserves and accounts for 90% of its oil production). In 1932, Desio explored the area south of Cyrenaica and collected many Oligocene and Miocene samples (Desio, 1935). The 45th Congress of the Italian Geological Society and the first to be held in one of Italy's colonies convened in Tripoli under the leadership of Paolo Emilio Vinassa de Regny (1871–1957) who wrote a brief guide for the geological excursion in Libya and constructed a new map of Libya in 1912 at a scale of 1:6,000,000. This congress was attended by some of the most prominent Italian geologists at the time, including Vinassa de Regny and Federico Sacco (1864–1948) in which Vinassa de Regny(1932) wrote a field guide. In 1933, Serra (1933–1937) described Maastrichtian mollusks from Tripolitania. Machete led a geological expedition to Cyrenaica (1934–1938). Floridian (1933, 1934) dated nummulitic limestone samples collected by Desio near Brace, Cyrenaica, as Upper Eocene; thus confirming the presence of Upper Eocene sediments in the area. Trellis (1933) was the first European to visit the north-eastern spur of the Tibesti Massif in 1930, and noticed that the area was composed of quartzite sandstones of various colors. Desio (1933) produced another map of Libya at a scale of 1:4,000,000, the first since that of Vinassa de Regny in 1912.

In 1934, Chiesa who worked on the geology of West Libya and Sirtica in Central Libya from (1934–1954) provided a geological sketch map at the scale of 1:5000,000. The 48th Annual Meeting of the Italian Geological Society was held again in 1935 in Cyrenaica, and a guidebook was compiled by Breccia et al. (1935).

Machete (1934, 1935a, b) reported the discovery of additional Cretaceous outcrops and a possible Maastrichtian limestone near the coast between Apollonia and Derna, and Senonian and Maastrichtian beds near Jardas al Abid and Majahir. He also reported the presence of vast Oligocene rocks in the area southeast of Al Makili. Machete interpreted the Cyrenaica Plateau as a double plunging anticline toward the

12   Reviewed by Sandford (1939).

east and west. Monterin (1935) described the distribution of the Nubian Sandstone in north-eastern Tibesti and divided it into two units separated by a horizon of limonitic sandstone.

During the WWI (1938–1945), geological activities resumed in Libya with the publication of the "Annals of the Libyan Museum of Natural History" in 1939. The first volume contained a paper by Desio (1939b) and a map of Libya at the scale of 1:3,000,000, as well as a report by Lapparini on the Oligocene foraminifera of the Derna area. Desio confirmed the observations of Gregory (1911a) and Pfalz (1938) on Cyrenaica and suggested that the Cyrenaica terraces represented step faults. Tavani (1938,1939, 1946) described pelecypods from the Miocene of Cyrenaica and identified new species. Coggi, L. (1940) described Triassic fossils from Jifarah. In 1940, Desio (1942) made another journey to north-eastern Tibesti sponsored by the Italian Royal Geographic Society. During that mission, he discovered Eocene rocks in the Black Jabal in the middle of Serir Tibesti. Interestingly, Desio used cars and an aircraft during the last expedition instead of camels. Desio, like Dalloni, divided the Precambrian complex into Archean and Algonkian series. In 1943, Desio published a detailed account of the mineral deposits in Libya. The following year, in 1944, led a scientific mission to Fezzan in West Libya (1944–1945) (Dalloni, 1945).

In the early 1950s major political changes took place in Libya, Egypt and neighboring countries which led to the restructuring of most of the geological and petroleum exploration organizations and laws. Libya gained independence from Italy on Christmas Eve of 1951 and a federal constitutional monarchy was formed under King Idrisi.

Active exploration for oil started in Libya in 1954 after the discovery of oil in neighboring Algeria and a Minerals Law for allocating permits to foreign oil companies was introduced. Initially, drilling was not permitted. Esso, Shell, BP and CFP were among nine international companies that obtained permits during that time. The first Libyan Petroleum Law was introduced in June 1955. According to that law, the government was to receive 12.5% royalty on the realized price of oil produced, and profits were to be divided 50-50 (Clarke, 1963). Concession 1 was awarded to Esso in West Libya and Concession 2 was awarded to Nelson Bunker Hunt in Northern Cyrenaica. Exploration for oil in Libya in 1953 led to the discovery of fossil (non-renewable) ground fresh water, about 4000 years old, in the Nubian Sandstone in Libya. The first Libyan university, the University of Libya (now El-Fateh) was established in 1955. In January 1956, 45 concessions were awarded. The first wildcat well was drilled under the new Petroleum Law, A1-18; by Libya-American Oil Company to test a surface mapped anticlinal structure in northern Cyrenaica. All subsequent wells drilled onshore Cyrenaica proved dry. Esso drilled two wells, the second well discovered gas in Concession 1 in West Libya. In 1957, the Libyan Government granted Concession 65 in the Sirte Basin to Nelson Bunker Hunt, before the discovery of oil in the Basin. The same year, BP acquired 51% in the concession and became operator.

Mobil drilled A1-57 (dry); the first well in the Sirte Basin on the Dahra Platform in 1958. The same year, the Oasis A1-32 well discovered the first commercial oil in Paleocene carbonates in the Bahi Field, followed by the B1-32 well in the Dahra Field. Mobil drilled the A1-11 discovery well in the Hofra Field, also in Paleocene carbonates. Gulf drilled three holes in the Murzuq Basin, of which A1-68 tested oil. CFP discovered oil in Devonian sandstones at Tahara. Shell made the first discovery in the northern Ghadames Basin at Tlaksin in the Late Silurian Akakus Sandstone. The Atshan Field (on trend with the Adjeleh structure in Eastern Algeria) was discovered in Lower Devonian Tadrart sands in the western Ghadames Basin by Esso, with a potential of 500 bopd. In 1958, William Berry of the University of Houston described Silurian graptolites and Lower Silurian spores from the Atshan-1 well, which was probably the first palynological study in Libya. The German company Wintershall entered Libya in that year.

In 1959, Esso discovered the giant Zelten (Nasser) Field by the C1-6 well, on the Zelten Platform in Late Paleocene reefal carbonates (Fraser, 1967, Bebout & Pendexter, 1975), with reserves of 2 billion barrels of oil. The field has been on production since 1961 and produced 2426 mmbbls of oil and 1.5 TCF of associated gas by the end of 2005 (SOC website). The Zelten reef was the subject of the classical paper by Bebout & Pendexter (1975). Also in 1959 Oasis drilled the Waha Field discovery well A1-59 in the Upper Cretaceous Waha carbonates, the Defa Field discovery well B1-59 in Early Paleocene carbonates, and the F1-32 Dahra East Field in Paleocene shelf-edge carbonates. Esso discovered the Mabruk Field in Concession 17 in heavily faulted and structurally complicated Paleocene reservoirs. Eni signed its first exploration and production agreement for Concession 82 in NE Libya in 1959, where the Ar Raml Field was discovered in Nubian Sand in 1965. Mobil discovered the Amal Field with multiple reservoirs in fractured Precambrian basement rocks, Cambro-Ordovician quartzites, Upper Cretaceous, Paleocene, Middle Eocene, and Oligocene carbonates and sandstones.

In 1960, OPEC was established. Libya joined the organization in 1962, but Egypt was not a member since it was not an oil exporting country. An active exploration phase started in Libya which led to a string of significant discoveries. By June 1960, 70% of Libya had been awarded in licence agreements. Amosea discovered the Al Kotlah Field in Concession 47 in the Sirte Basin in Upper Cretaceous reservoirs. Esso discovered the Assumood Gas Field by the H1-6 well, which produced 75 mmcfgd and 103 bcf of gas by the end of 2005 (SOC website). Esso also discovered the Raguba Field in Upper Cretaceous carbonates in Concession 20, on production since 1963 with a total of 787 mmbbls of oil and 859 bcf of associated gas by the end of 2005 (SOC website). Shell discovered the Antalat Field near Benghazi in Eocene carbonates.

The first lexicon of stratigraphic nomenclature in Libya was compiled by Burollet (1960), followed by another by the geologists of oil companies (Compagnies Pétrolieres, 1964). In 1966, a field conference on South-Central Libya and Northern Chad was held (Williams, 1966). In 1968, a field conference on Jabal Nafusah and another on Cyrenaica were held. The General Petroleum Corporation (Lipetco) (the predecessor of NOC) was established by a royal decree to participate in all phases of petroleum operations in Libya. In 1968, Libya became member of OAPEC, and Joint production sharing agreement begun. A field conference (Kanes, 1969) was held with an objective on the geology, archeology and prehistory of Southwestern Fezzan. Pesce (1969) compiled a volume of Gemini Space Photographs. In addition, a number of key papers on oil fields in Libya were published in the period 1967–1975. In 1962, a British Royal Military Academy expedition was led by Williams and Hall to explore Jebel Arkenu in Southeast Libya. Beuf et al. (1971) published the classical book on glacial deposits in Algeria and West Libya. BP discovered the Messla Field in Nubian Sandstones. Libya became the second country in the world, after Algeria, to export LNG from Marsa El Brega (capacity 125 bcf per year). A symposium was held at the University of Libya. The proceedings of the symposium were published in 1971 under the title "Geology of Libya", edited by C. Gray. Hea (1971) attempted to differentiate between Nubian and Gargaf quartzites based on petrography, which inspired the paper by Tawadros et al. (2001). The Gargaf Group was introduced by Burollet (1960) for Cambro-Ordovician succession in West Libya to include the Hassaouna, Haouaz, Melez Chograne, and Memouniat formations. The Gargaf Group was extended by Barr & Weegar (1972) into the Sirte Basin and was designated as Cambro-Ordovician. Recent studies proved the presence of Devonian, Carboniferous, Permian, Triassic, Jurassic, and Cretaceous sands (mostly based on palynology) (Tawadros et al., 2001).

The Gargaf Group was also extended to the subsurface of the Jifarah Plains and the northern Ghadames Basin in Northwest Libya by Deunff & Massa (1975) who divided it into Sidi Toui, Sanrhar, Kasbah-Liguine, Bir Ben Tartar, and Djeffara formations. Massa (1988) made a detailed study on these areas. Palynology of Cambro-Ordovician was carried out by Vecoli et al. (2000, 2009). Bryozoans were studied in detail by Butler & Massa (1996) and Butler et al. (2007). One of the key geological publications on the geology of Libya, a lexicon of stratigraphic terminology in the Sirte Basin, was compiled by Barr & Weegar (1972) under the auspices of the Petroleum Exploration Society of Libya. The lexicon included the definition of 40 new formations used by oil companies and reviewed many of the existing names. This lexicon established the stratigraphic framework of the Sirte Basin and is widely accepted among workers in Libya. A second edition of Conant & Goudarzi's 1964 Geological Map of Libya was published (1972). In 1973, A French Study Group examined the Ash Shati iron ores in West Libya.

The discovery of anhydrite in the Mediterranean Sea during the Deep Sea Drilling Program (DSDP) led to the recognition of the Messinian Salinity Crisis (Hsü, 1972, Ryan et al., 1973, Cita, 1982), a subject that is still heavily debated until today (Clauzon et al., 1995, Rouchy et al., 2006, Roveri & Manzi, 2006). Barr & Walker (1973) authored a pioneering paper on the buried Messinian Sahabi Channel[13] in Northeast Libya. In 1982, An international symposium entitled "The Sahabi Research Project, 1982" on the geology, flora and fauna of the Pliocene Sahabi Formation in the Sahabi area was held at the Naturhistorisches Museum in Main, Germany in 1981. The results appeared in Special Issue No. 4 of the University of Garyounis, Benghazi.

In 1961, BP discovered the giant Sarir 'C' Field in Concession 65 in the Eastern Sirte Basin in Lower Cretaceous "Nubian" sands based on reflection and refraction seismic. The discovery well C1-65 tested 3900 bopd. In the meantime, Amoseas discovered the Beda Field in Concession 47 and the Italian company CORI discovered oil in the deep well R1-82 in Lower Cretaceous sandstones. Oasis discovered the Giant Jalu Field in Eocene and Oligocene reservoirs, as well as the Samah, Masrab, Harash, Zaqqut, and Kalifa fields. Consequently, Oasis established a facility at Ras el Sidra and Mobil constructed another facility at Ras Lanuf to export oil from the Bahi and Dahra fields. A pipeline with a capacity of 300,000 b/d was constructed the following

---

13 See also Abu Madi Channel, Bahr bela ma'a, ancient desert channels, SIRX-Radar channels in Egypt. In 1979, Satellite and radar (SIR-C) images were used to interpret buried channels.

year to connect the Bahi and Dahra Fields to Ras el Sidra. Esso discovered the Meghil Field, which was put on production in 1969 and produced 392 bcf by the end of 2005 (SOC website). In 1961, Esso constructed a facility to deliver liquefied natural gas (LNG) and a 36-inch pipeline from Zaltan to Al Burayqah (Brega). The first Libyan oil was shipped in September 1961. Amendments to the 1955 Petroleum Law were introduced following these discoveries and passed in 1965.

In 1962, the National Oil Company of Libya (NOLCOL) was formed, of which 51% of the total shares were to be held by Libyan nationals (Clarke, 1963). Esso made a series of significant discoveries in 1962. The Jebel Field was found in Concession 6 in the Maastrichtian Waha carbonates, which was put on stream in 1965 with a total production of 201 mmbbls of oil and 352 bcf of associated gas by end 2005 (SOC website), the Sahel Gas Field in Concession 6 in the Upper Cretaceous carbonates, with estimated 1.54 TCF OGIP, and produced 103 bcf of gas by the end 2005 (Sirte Website), and the Hateiba Field in Upper Cretaceous carbonates, which was put on production in 1978. Oasis discovered the Belhidan Field with the well V1-59 with estimated OOIP >315 million barrels in Cambro-Ordovician "Gargaf" sandstones. The main reservoir in the Belhidan Field was recently defined as the Belhidan Formation and assigned a Devonian-Carboniferous age by Tawadros et al. (2001). Oasis also discovered the Kalanshiyu Field in the same year. Philips and Pan American (later Amoco) made a small discovery at Umm al Furud in Paleocene carbonates. Mobil had a number of discoveries in 1962, such as the Abu Maras in Paleocene carbonates and the Al Facha in Lower Eocene carbonates in Concession 11, the Ora (Tibesti) Field in Gargaf quartzites in 1962 and in Upper Cretaceous carbonates in 1964 on the Bayda Platform, and the Rakb Field in Upper Cretaceous carbonates in 1962, with additional Upper Eocene and Cambro-Ordovician quartzite reservoirs were confirmed in 1970.

In 1963, Philips and Pan American made a discovery in the E-92 in Paleocene carbonates. Oasis found oil in well YY1-59 in Middle Eocene and Oligocene reservoirs. Gulf, CFP and Oasis made other discoveries in the Ghadames Basin.

In 1964, Esso discovered the giant Attahaddy Field (developed by Sirte Oil Company and became fully on production in 2005 at 270 mmcfgd and 36,000 bcd) in fractured Cambrian quartzites in a horst fault block within the Wadyat Trough in the Sirte Basin. Tawadros et al. (2001) showed that there are two Upper Cambrian fractured quartzite reservoir intervals in the Attahaddy Field, separated by a vari-colored, argillaceous, arkosic sandstone interval, commonly known as 'Red Shale'. The lower interval contains volcanic intervals that have been interpreted as shale on logs. A symposium on the Hon Graben and the Western Sirte Basin was organized in 1964. Conant & Goudarzi (1964) constructed a geological map of Libya at scale 1:2,000,000, which formed the bases of all the subsequent maps. Philips and Pan American made a discovery at Al Khuf (1964, Upper Cretaceous). Mobil made other discoveries at Ad Dib (Lower Eocene, 1964) and Farrud (Paleocene, 1964) on the Dahra-Hufra Platform. Amoseas discovered the Al Kotlah Field in Upper Cretaceous reservoirs and the Haram Field in Upper Cretaceous and Cambro-Ordovician sediments. Esso discovered the Ralah Field in Upper Cretaceous in 1964.

In 1965, the Libyan government issued new concessions. Occidental, Phillips, Aquitaine, Wintershall, and Agip, Shell, and Union Reinische were the major winners. Esso discovered the Lahib Field in Waha carbonates, which was put on production in 1967 and produced 20 mmbbls of oil and 292 bcf of associated gas by the end of 2005 (SOC website). Amoseas discovered the Nafurah Field in the Eastern Sirte Basin with the G1-51. Eni discovered the Ar Raml Field in Concession 82 in tilted fault blocks in Nubian Sand. Esso found the Lahib Field in Upper Cretaceous Waha Limestone in 1965.

In 1966, Occidental discovered the Augila Field with the D1-102 well, which tested 14,860 bopd from Precambrian, Cambrian, Middle Eocene, Paleocene, and Upper Cretaceous reservoirs. Wintershall discovered the Hamid Field in Upper Cretaceous Lidam Dolomites and Lower Cretaceous "Nubian" sandstones in Concession 97. BP discovered the Sarir L Field in Nubian Sandstone. Amoseas discovered the giant Augila-Nafurah Field in Gargaf fractured quartzites, Middle Eocene carbonates, Upper Cretaceous Bahi sandstones and Paleocene carbonates. Aquitaine made discoveries at the Mansur, Majid, D-1044, and East Masrab fields. Pan American discovered the As Sahabi Field in Concession 95 and Amoseas found the Dur Field. Mobil also made discoveries at Al Farigh, Chadar, and Marada. Esso discovered the Sorra Field in Upper Cretaceous carbonates in Concession 6, which was put on stream in 1971 with a total of 204 bcf gas by the end of 2005 (SOC website). Libyan Atlantic tested gas in B1-88 in offshore Libya.

In 1967, Occidental discovered the Intisar (Idris Field) A-D Fields in Upper Paleocene pinnacle reefs in Conc. 103 based on seismic reflection data. The Intisar A Field flowed at a rate of 43,000 bopd and the Intisar D at a rate of 74,867 bopd (the highest rate ever reported in Libya).

Desio (1968) suggested that the silicified woods west of Jaghbub originated from alluvial and gravel sediments transported at the end of the Miocene and during Pliocene by northward flowing streams from the Libyan Desert to Jaghbub. Jardiné & Yapadjian (1968) and Jardiné et al. (1974) conducted palynological studies in Eastern Algeria which was applied widely in Libya. Thusu & Owens (1985) and El-Arnauti et al. (1988) edited two volumes on the palynology of NE Libya. Massa et al. (1974) and Massa et al. (1980) on the Carboniferous of West Libya, and Massa & Moreau-Benoit (1976) on the palynology of Devonian of West Libya.

In September 1969, a revolution was led by Qaddafi, and immediately production cuts were imposed on foreign companies.

In 1969, Wintershall discovered the Jakhira Field in Lower Cretaceous sandstones in Concession 96 in the Sirte Basin, the Hamid Field in the Upper Cretaceous Lidam Formation in Concession 97, the Nakhla Field, and the Tuama Field in Lower Cretaceous sandstones. The same year, concessions were awarded to the NOC's wholly owned subsidiary AGOCO.

In 1970, Libya became one of the top 10 oil-producing countries, with a record production of an average of 3.3 million bopd (Nelson, 1979). Agip discovered the giant Abu Attiffel Field in Concession 100 (awarded in 1966) in Nubian Sandstone in the Eastern Sirte Basin, with estimated 1.1 Bbbls of oil recoverable. The Abu Attiffel Field has been on production since October 1972. The giant Wadi Field (on production since 1986) was discovered by Esso in Concession 104 (later Concession NC149) in fractured "Nubian" quartzites in the Hagfa Trough, with OOIP of more than 1.0 Bbbls of oil. The Wadi Field produced 91 mmbbls oil and 52 bcf of associated gas at end of 2005 (SOC website). The stratigraphy and nature of the Nubian Sandstone in the Wadi Field has been debated since its discovery. Bonnefous (1972), postulated correctly, albeit based on erroneous data, that the upper part of the quartzite succession in the Wadi Field, previously assigned to the Cambro-Ordovician, to be of Cretaceous age. The Nubian section in the Wadi Field was thought to contain a shale interval which El-Hawat et al. (1996) correlated with the middle variegated shale in the Eastern Sirte Basin. Tawadros et al. (2001) proved that this interval in the Wadi Field to be a volcanic succession. Aquitaine discovered Al Mheirigah Field in Paleocene reefs similar to the Intisar Reefs. Amoseas discovered the Alwan and Warid fields on the Dahra and Al Bayda Platforms, respectively, in the Sirte Basin.

In 1970, Libya nationalized the oil industry, and the National Oil Corporation (NOC), a state-owned company, was created to replace the older Libyan General Petroleum Corporation (Lipetco) and to control Libya's oil and gas production. In 1971, BP's share in the Sarir Field was nationalized. In 1973, the assets of Hunt, Amoseas, Shell, and Bunker Hunt's (BP partner) share in the Sarir Field were nationalized. In the meantime, the Libyan Government acquired 51% in concessions of Esso, Mobil, Oasis, and Occidental. Libya responded to the October 1973 war between Egypt and Israel by cutting down production by 5% and an oil embargo against the United States and the Netherlands. Disputes over oil prices that followed led to a decline in exploration activities.

In 1974, the EPSA I was introduced to replace the Joint Venture Agreement. In the new formula the Government would share 85%. The first block was awarded to Occidental. Esso and Agip (Eni) were awarded offshore concessions, including NC41. Braspetro was awarded a concession in the Murzuq Basin. In the meantime, Amoco surrendered its holdings. A new concession numbering system began with the designation NC (New Concession).

In 1976, Agip drilled the well B1-NC41 in offshore Libya in New Concession NC-41which tested 4457 bopd. It was the first commercial discovery in the offshore in Lower Eocene Nummulitic buildups. The discovery was named Al Bouri Field and came on production in 1988. Also in 1976, BP discovered the giant Messla Field north of the Sarir Field stratigraphic traps in the Nubian Sandstone. In the meantime, Gulf, CFP and Oasis made discoveries in the Ghadames Basin.

Mapping of Libya at a scale of 1:250,000 started by IRC in 1974 (completed 66 sheets in 1985). The second edition of the Geological Map of Libya (1977) was updated by Maghrabi & Cheshitev of the Industrial Research Center. Fournié (1978) published the Lexicon of offshore Tunisia, which was used equally in the offshore of Libya and formed the bases for the Libyan Lexicon of Hammuda et al. (1985). In 1980, Banerjee compiled a Lexicon of stratigraphic nomenclature of Libya.

The Geological Map of Libya (IRC) project which started in 1974 completed (66 sheets at a scale of 1:250,000). Hammuda et al. (1985) compiled a lexicon of stratigraphic nomenclature for the offshore of Libya.

In 1974, Britain severed its ties with Libya over the shooting of a policewoman in London. Deterioration of Libyan-USA relation increased when the USA imposed a ban in 1978 on sales of aircrafts and electronic equipments to Libya. In 1979, Occidental sold its interest to the government. Waha and Zueitina companies were established to run the operations of Oasis and Occidental, respectively. In 1986, USA cut its relations with Libya and sanctions were imposed by the US government, and American companies withdrew from Libya. Benghazi and Tripoli were bombed by US planes.

Rampetrol discovered a third oil pool in NC115 in peri-glacial sands in the Murzuq Basin in West Libya. In 1988, the EPSA III was offered under modified terms to allow exploration cost to be recovered from output. Petrofina of Belgium, Lasmo from UK, International Petroleum Company from Canada, INA from Croatia, OMV from Austria, Shell, and Braspetro were awarded concessions. The British Lasmo signed an exploration agreement for Concession NC174 in the Murzuq Basin in 1988 and its efforts were rewarded by the discovery of the Elephant Field 1996. The Bouri Field came on production the same year.

The EPSA II was offered in 1980. Rompetrol and BOCO signed two acreages in the Murzuq Basin; NC115 and NC101. In 1981, the USA asked all US citizens to leave Libya. Esso closed down its operations, which were transferred to the NOC in 1982, and the Sirte Oil Company was established. Mobil left Libya and handed over its operation of Amal Field to its partner the German company Veba. In the meantime, the Bulgarian State company discovered oil in NC100 (awarded under ESPA II) near the Tunisian border. The Romanian company Rompetrol (in partnership with OMV and Total) encountered oil in NC115 in peri-glacial sands in the Murzuq Basin. A total of six fields designated A to H were discovered and later grouped in the Sharara Field.[14] Production started in December 1996. The field is currently operated by Spain's Repsol and partners Total and OMV, and produced approximately 200,000 bopd in 2007.

Libya's daily production in 1983 was about 1,158,000 bopd and approximately 1.1 mmcfgd (the actual figures are not released by the Libyan government and vary considerably among the various organizations).

Harding (1984) divided the Sirte Basin succession into pre-rift, syn-rift, and post-rift sequences; a scheme that proved useful in the study of the basin. The proceedings of the symposium of 1987 were published in 1990 (Geology of Libya, volumes 4–7). Proceedings of the 1993 symposium on geology of the Sirte Basin were published in 1996 in three volumes. Tawadros (2001) published "Geology of Egypt and Libya" which reviewed the stratigraphy, tectonics, and geological history of the two countries. Hallett (2002) published Petroleum Geology of Libya, with an extensive review of the petroleum systems of Libya. A symposium on the Murzuq Basin was held in 2000 (Sola & Worsely, 2000) and another symposium was held in 2003 on the geology of Northwest Libya published in three volumes. Another geological symposium was in 2005 on NE Libya.

In 1984, NOC/Agip drilled the A1-NC120 well offshore Cyrenaica which tested 5263 bopd of 36° API Oil in Lower Cretaceous and Turonian carbonates. In 1985, Waha/Agip drilled a dry hole A1-128 offshore Cyrenaica.

Work on the Great Man-made River Project (GMR) began in 1984, at an estimated cost of $25 billion, to deliver water from the Kufra Basin to the Mediterranean coast towns. Excavations of Phase I of the project started in 1991. Phase II of the project started in 1996 with the first water delivery to Tripoli. However, the origin and exploitation of the water of the Nubian aquifer system in Northeast Africa has been disputed among hydrogeologists (Puri & Aureli, 2005). The extracted

14 Sharara Field straddles Concessions NC115 and NC186.

groundwater formed mostly during several humid phases and has been depleting for several thousand years (Pallas, 1980, Heinl & Brinkmann, 1989).

In 1991, Sirte Oil discovered the Al Wafa Field in Middle Devonian sands, on the Libyan-Algerian border, near the Algerian Alrar Gas Field (discovered by Sonatrach in 1980 with estimated reserves of 4.7 TCF of gas). Production started in 2004 at 23,000 bopd, 450 mmcfgd, and 15,000 bcd. An agreement was signed between NOC and Sonatrach in September 2006 to study the link between the Al Wafa and Alrar fields. Wintershall's As Sarah Field in Concession 96 in the Eastern Sirte Basin west of the Amal Field, was discovered in Triassic sandstones and came on production in 1991.

New confrontations between Libya and the West were renewed at the end of the 1980s. The bombing of Pan Am Flight 103 over Lockerbie, Scotland, by Libyan terrorists in 1988 strained the US-Libyan relations further. In 1992, the UN imposed sanctions on Libya for refusal to extradite two Libyan nationals accused of bombing of the Pan Am flight. Sanctions, which lasted for 10 years, prohibited all flights into and out of Libya, froze Libyan assets overseas, and banned weapons sales to Libya.

In 1994, the Elephant Field was discovered in Concession 174 in the Murzuq Basin in Late Ordovician Memouniat peri-glacial sands. Production started in 2004 with an initial flow of 10,000 bopd, and ENI as operator. The field has an estimated OOIP of 1.2 Bbbls of oil and 680 mmbbls recoverable oil (OGJ, March, 6, 1995). ENI entered the NC174 joint venture by signing a farmout agreement with Lasmo. Following the takeover of Lasmo in 2000, ENI became operator of the concession. Production started from the Sharara Field in the Murzuq Basin in 1996. In 1997, Agip (Eni) made a new discovery with the F1-NC174 well in the Elephant Field which tested 7,500 bopd of 38° API oil. Agip made other discoveries with the A1-NC175 in the Ghadames Basin at 16,260 bopd, and the B1-NC125 in the Hammeimat Trough in the Sirte Basin, which tested at 5.0 mmcfgd and 3,903 bopd. OMV acquired Concessions NC186 and NC197.

Libya entered the 21st Century with new political and economic objectives. Libya handed over the two suspects in the Lockerbie bombing for trial in 1999. The UN suspended the sanctions in 2002 and the UK re-established relations with Libya. In December 2003, the Libyan government announced its decision to eliminate materials used in weapons of mass destruction. In 2004, USA lifted the sanctions on Libya, and the UN lifted the sanctions in 2003. In May 2006, the US restored relations with Libya.

Elf Equitaine made a new discovery in 2002 in the offshore Block NC137 with the B3-NC137 well, which tested 3,600 bopd of 32° API oil; the first well by Elf Aquitaine since its return to Libya in 1996.

In the year 2000, the Ministry of Energy was abolished and the NOC became in charge of the oil sector, but was reinstated in 2005. Repsol discovered the A-Field in NC186 in the Ordovician Memouniat sands, which was put on production in 2003 at a rate of 25,000 bopd. An agreement was signed between EGPC and NOC for the construction of a pipeline between Egypt and Libya (Arab Company for Oil and Gas Pipeline). Repsol YPF discovered the D-Field in NC-186, which started production at a rate of 23,000 bopd in 2004, and made a new discovery in H1-NC186 in Ordovician Memouniat in 2005, which tested 1350 bopd. In 2003, Woodside was awarded the onshore blocks 35, 36, 52 & 53. Total announced the start of production of Al Jurf Field on the offshore Block C137.

In January 2005, Round 1 of EPSA IV offered 15 blocks in different parts in Libya. Another 17 blocks were offered in Round 2. ExxonMobil Libya began exploration in Block 44 offshore Cyrenaica awarded in Round 2. Occidental resumed its operations in the Libyan contract areas abandoned in 1980s. The Japanese Nippon Oil Exploration Ltd (NOEX) was awarded Blocks 1 & 2 in the offshore Area 2, near the Libyan/Tunisian border in Round 2 and blocks 3 & 4 in Area 40 offshore Cyrenaica, northeast of A1-NC120. BG Group PLC acquires onshore Blocks 123 (1 & 2) in the Sirte Basin, and Area 171 (Blocks 1–4) in the Kufra Basin (OGJ, Dec. 9, 2005). Norway's Hydro was awarded Block 146-1 in the Murzuq Basin. Fourteen areas onshore and offshore, with 41 blocks, were offered by NOC in Round 3 of ESPA IV (OGJ, Sept. 4, 2004).

Shell signed an agreement in May 2005 with NOC to upgrade the Marsa El Brega LNG plant operated by Sirte Oil Co. Eni won four exploration permits in the Murzuk and Kufra basins. Repsol YPF (with partner Hydro) made a total of seven discoveries including I-1, J-1, and K-1 on NC-186 in the Murzuq Basin. The I-1 and J-1 discoveries gave preliminary production rates of 2060 and 4650 bopd, respectively. The K1-NC186 tested 2300 bopd in January 2007. The field extends over the NC 186 and NC 115 licences, and may contain a total of 1.26 billion barrels of oil, with 474 million barrels of recoverable oil.

Oil exports in 2006 generated $28.3 billion for the government and were forecast to be $31 billion.[15] Estimated proven reserves (2006) were 34 billion barrels of oil and 51,500 bcf of gas (World Oil, Sept. 2006, p. 57). Average annual production in 2005 was 1.63 mmbopd (World Oil, Dec. 1, 2006).

In 2005, the $5.5 billion West Libya Gas Project (WLGP) led by Eni was inaugurated. The WLGP was designed to export gas from the Al Wafa Field in West Libya and the offshore Bahr Essalam Field (in Block NC41) to Italy, in addition to the subsea 32 inch pipeline, Greenstream (75% owned by Eni), which exports gas and condensate from North Africa to Europe. Libya planned to increase oil production to 3 million barrels/d by 2010–2911 (OGJ, Jan. 23, 2006) and embarked on an extensive exploration and production program.

Total of France struck first oil in Block NC191 in the Murzuq Basin (OGJ, July 2006 & Total's website), awarded in 2001. The well tested 675 bopd. Woodside's A1-NC210 well on NC210 in West Libya encountered oil in several pay zones and tested 5.5 mmcfgd from two zones. Well B1-NC210 encountered several hydrocarbon-bearing zones in the primary Ordovician Formation objective. The deepest zone tested 11 mmcfgd through a 56/64-inch choke.

RWE discovered oil in Concession 193 in the Sirte Basin which was awarded in May 2003 under EPSA III. The A1-NC193 well tested oil in the Paleocene Dahra Formation at a flow rate of 410 bopd (OGJ, Oct. 31, 2006). RWE Dea announced a second oil discovery on Concession NC193 awarded in 2003 in the Sirte Basin with the well B1-NC193. The well tested 933 bopd from two oil intervals in the Upper Satal and Dahra formations. A second discovery was made in the Eocene Gir in C1-NC193 which tested 393 bopd. In 2010, the RWE Dea well E1-NC193 tested 704 bopd in the Dahra Formation, and the F1-NC193 tested 439 bopd in Upper Dahra Formation (World Oil, Oct. 2008). The G1-NC193 tested 426 bopd of 34° API from

---

15  Nickle, J., 2006. African Focus, Libya. Petroleum Africa, September 2006, p. 31–40.

Upper Satal Formation (World Oil, Nov. 2008). In 2007, RWE Dea third onshore discovery, the A1-NC195, tested 1981 bopd in the Dahra and Beda (World Oil, Nov. 2007). The B1-NC195 tested 1044 bcpd and 15.4 mmcfgd from the two formations (World Oil, June 2008).

ConocoPhilips and partners Marathon and Amerada Hess re-entered the Waha Concession, which was producing 350,000 bopd at the time (OGJ, Dec. 29, 2005 & OGJ, Jan. 11, 2006). Russia's Gazprom acquired 49% in concessions N-97 and N-98 after swabbing assets with Wintershall. Repsol's new discovery in NC115 in Murzuq Basin tested 2300 bopd.

Repsol YPF and partner Woodside Petroleum reported that the C1-NC210 in the Murzuq Basin tested 5.7 mmcfgd with an Absolute Open Flow at 10.7 mmcfgd from the Devonian Awaynat Wanin (Ouinat Ouneine) Formation. Another production test of the Mrar M7 reservoir tested 5.8 mmcfgd. The F1-NC210 well tested 280 bopd and 10 mmcfgd from the Ordovician Memouniat and Devonian sands (World Oil, March 2009).

In 2009, Repsol and partner OMV discovered oil in the A1-NC202 offshore the Sirte Basin. The well was drilled to a total depth of 15815 ft and tested 1264 bopd and 0.6 mmcfgd from the Eocene Dernah Formation (World Oil, May 2009).

In the Ghadames Basin, the A1-47/02 well in Area 47 awarded to Verenex in January 2005 tested 5,172 bopd of 46.5° API oil and associated gas at 6.7 mmcfgd in the Lower Acacus Formation. A second discovery was made by B1-47, and another discovery by D1-47/02 which tested 7742 bopd of oil from two intervals in the Lower Acacus Formation (World Oil, March 2008). The F1-47 well tested 7215 bopd. In 2010, Verenex Energy tested 4.2 mmcfgd and 900 bopd from the Ordovician Memouniat Formation, and 4.5 mmcfgd and 150 bopd from the Lower Acacus Formation. The G1-47/02a well flowed 1739 bopd from the Lower Acacus (World Oil, Nov. 2008).

Amerada Hess was awarded the offshore Block 54 north of Ras Lanuf in Round 1 of ESPA IV. In 2008, the company discovered oil in several intervals in the well A1-54/1(Arous Al-Bahar), in 2807 ft of water (Hess Corp News Release, Dec. 17, 2008). Hess estimated the reserves at 1–5 TCF of gas in 2010.

However, all exploration efforts came to a halt in the beginning of 2011 with the eruption of the civil war.


## 1.3   ALGERIA

The history of geological ideas in Algeria during various periods has been summarized by Larnaude (1933), Legrand (2002), Bettahar (2007), and Taquet (2007).

The first crossing of the Sahara by a European traveler took place in 1825 and 1826 by the Englishman (Scottish) Major Alexander Gordon Laing (1793–1826) of the Royal African Colonial Corps, on a mission opening up commerce and endeavouring to abolish the slave trade in that region. Leaving Tripoli, he arrived in Timbuktu (in Mali) on 18 August 1826, but he was murdered on the way back. The following year, the Frenchman René Caillé (1799–1838) crossed the Sahara from south to north via Timbuktu (Taquet, 2007). Denham, Oudney and Clapperton reached Murzuq in 1822, and then left Tripoli and travelled to Lake Chad from 1822–1825. The German Heinrich Barth, accompanied by the geologist Overweg, joined the James Richardson Expedition in 1850 and travelled to Tripoli. Overweg was the first to make geological observations and to note the presence of granites and sandstones in the Central Sahara.

Algeria was occupied by the French in 1830. The earliest geological studies were carried out by military officers, such as Rozet who established the *Commission Scientifique* in 1839 which was headed by E. Renou. In 1852, the Service des Mines was formed of a group of mine engineers and university professors. The latter included Auguste Pomel and Conrad Kilian who carried out reconnaissance geological studies in North Africa, including the Sirte Basin (Pomel, 1878). Pomel also played a role in the organization of the *Service de la Carte Géologique de l'Algérie*. He also headed the University of Algiers; followed by Ficheur in that post in 1891.

The first geological congress was held in Algiers in 1881. A geological map of Algeria at the scale of 1:800,000 was prepared for that congress (Bettahar, 2007). However, Péron established the stratigraphy of Algeria in 1883 by classifying rocks in a chronological order. His paper also included a geological map.

Rolland (1890) examined all the by then available geologic and geographic data from the Atlantic Ocean to the Red Sea and compiled what is probably the first geological map of Northwest Africa (Wendt et al., 2009). He recognized that the plateau of the Tassilis is composed of Devonian sandstones into which he included the older sedimentary cover of the Precambrian granites and metamorphic rocks of the Hoggar Massif (Tuareg Shield) in Southern Algeria.

The Foureau-Lamy expedition of 1898–1900 recognized the presence of a crystalline basement under sandstone formations (Foureau, 1898, cited by Legrand, 2002). His collection was studied by Gentil (1909). From about 1900 until the First World War, the geological explorations followed the axes of the French military progression (Ouzegane et al., 2003).

Three exploration missions to Morocco were organized by Brives (1905) from 1902 to 1904. After the First World War, the pace of geological studies in Hoggar quickened. Bütler's memoir (1922, 1924) is the first really precise description of Hoggar geology (Ouzegane et al., 2003).

Flamand (1911) made the first detailed study of the geology of the south of Algeria under the protection of 140 men of the Camel Corps (Taquet, 2007). Conrad Kilian (1898–1950), who was abandoned by his military companions, completed his exploration of the Tassilis des Ajjers and of the Central Hoggar. Among his achievements; he examined the main features of the geology of the Hoggar Massif, discovered the graptolitic shales (1928), and divided the Tassilis into two units; the Internal Tassilis which he regarded as being 'pre-Gothlandian' (pre-Silurian) comprised Lower Ordovician sandstones and the External Tassilis which consisted of upper Devonian sandstones. Kilian also indicated that the basement was Precambrian in age and divided the Precambrian sequence into two series; the Suggarian and Pharusian, separated by conglomerates. The results were submitted to the Acadamy of Sciences in a short note entitled "Aperçu general de la structure des Tassilis des Ajers" and again to the International Geological Congress in Brussels in 1925. Kilian also explored the borders of the Italian Fezzan and Algeria and discovered the Monts Doumergue. Kilian (1931) also recognized the *Continental intercalaire* with reptile and fish fossils between the marine Upper Carboniferous and the marine Middle Cretaceous, and the *Continental terminal* of Upper Cretaceous and Cenozoic age; a concept that is still in use today (Lefranc & Guiraud, 1990).

D. Dussert (1924) published on the Algerian phosphates. In the period of 1924–1929, work on the Atlas of Algeria and Tunisia, was published by Augustin Bernard & R. de Flotte de Roquevaire in 8 volumes; the first section deals with geology (published 1925) (Lespès, 1931). Jacques Levainville (1925) showed that the two horizons were of Lower Eocene" *Phosphate algerien*". J. Savornin (1931) published "*La géologie algerienne et nord-africaine depuis 1830*".

Théodore Monod (1902–2000) was another pioneer Saharan geologist and pale-ontologist who discovered an intermediate series between the Precambrian and the Paleozoic of the NW Hoggar (Monod & Bourcart, 1931, 1932), which he named Pur-ple Series, the first known example of Pan-African molasse (Taquet, 2007). In 1934, he also discovered Silurian graptolitic rocks in the Tassilian borders of Ahnet.

Albert Félix de Lapparent (1905–1975) studied the stratigraphy of the Mesozoic Saharan basins, starting in 1946. De Lapparent gave a detailed account of the stratig-raphy of the Gourara, of Touat and Tidikelt. During his trips, he studied the dinosaurs of the Sahara (de Lapparent, 1960) and gave the first interpretation of the geology of the Edjeleh region. He discovered the bones of a giant crocodile in an underground irrigation canal, which was later named *Sarcosuchus imperator*. His discoveries were the starting point of the study of vertebrate fauna of the Sahara.

The first oil discovery in Algeria occurred at the end of last century in the Chelif Basin (Ain Zeft). However, due to the great geological complexity of Northern Alge-ria, it remained mostly unexplored (Askri et al., 1995). In 1948, oil was discovered in the Oued Gueterini Field in Eocene reservoirs in Tellian nappes,[16] and gas was discov-ered at Berga-1 in 1954. Subsequently, the Edjeleh Oil Field was discovered in Lower Devonian sands; the Hassi Messaoud Oil Field in the Cambrian sandstones, and the Hassi R'Mel Gas Field in the Triassic in 1956.

The 19th International Geological Congress of Algiers (1952) marks the starting point of the geological exploration of the Algerian Sahara (Legrand, 2002), followed by the publication of the Lexique International (1956).

Following the discovery of oil in 1954, an enormous amount of stratigraphic and sedimentologic data were accumulated by oil companies, but unfortunately remained inaccessible. Some results were published by Legrand (1967). Askri et al. (1995), Zeroug et al. (2007) provided valuable overviews especially of the hydrocarbon systems. In the years 1974–1975 a series of geological maps of 26 sheets at the scale of 1:200,000 was published by Sonatrach-Beicip, unfortunately they were not accompanied by individual explanations. Outcrop studies of the Lower Paleozoic were made by Dubois et al. (1968), Beuf et al. (1971), and Eschard et al. (2005), while the Middle Devonian to Carbonifer-ous rocks were examined by Remack-Petitot (1960) and, Chaumeau et al. (1961).

The Alrar Field was discovered in 1961 with the well Alrar East-1 in Middle Devonian sandstones (F3). Oil was discovered three years later with the wells Nord Alrar 103 and Nord Alrar 106 (Askri et al., 1995). The Zarzaitine Field was discov-ered in 1957 and production started in 1960. The gas-cap of Tin Fouye Tabankort

---

16 Messaaoudi, M., undated. Northern Algeria, A general overview of hydrocarbon prospectivity. Africa Session, Forum 21 Poster. WPC-cd250.

Field was discovered in 1961 in the well TFE-, but oil was found only in 1965 in the well TFEZ (Askri et al., 1995). The Rhourde El Baguel Field was discovered in 1961. It was put into production in August 1963 with well RB-1 and produces from a large accumulation set within Cambrian quartzite sandstones.

Recent oil and gas discoveries in Algeria include the Anadarko Petroleum oil discovery of oil in the Berkine Basin Block 403c/e, where the well tested 870 bopd (WO Nov. 2007). In 2008–2009, Repsol discovered gas in the KLS-1 in the Reggane Basin, which tested 22.2 mmcfgd; the OTLH-2 in the Ahnet Basin, which tested 8.8 mmcfgd on the M'Sari Akabli Block, and the A1-2 in the Berkine Basin with 5.58 mmcfgd on the Gassi Chergui Block (WO March 2009), and discovered gas in the TGFO-1 with 12.8 mmcfgd in the Emsian in the M'Sari Block (WO May 2009). Petrovietnam (PVEP) discovered oil in the BRS-6 on Block 433a in the Oued Maya Basin (WO April 2009). In the Illizi Basin, Sonatrach's Issaouane NO-1 well tested 6.22 mmcfgd from the Ordovician in the Amenas Block 240b and the Tin Dadda Sud-1 which tested 6.33 mmcfgd from Silurian and Devonian sands. In the Oued Maya Basin, the Nechou Nord-3 tested 2189 bopd and some gas, and the well Madjbeb-1 which tested 332 bopd and some gas from Triassic sands. StatoilHydro tested 7.76 mmcfgd in Devonian-Carboniferous reservoirs on the Hassi Mouina Block in Timimoun Basin (WO Sept. 2008), and in the TNK-2 which tested gas in the Lower Carboniferous (WO Oct. 2008). Sonatrach discovered gas in the Tirehoumine-2 on the Ahnet Block in the In Salah region which tested 1.75 mmcfgd from the Gedinnian (WO July 2008). Sonatrach Araret-2 (OTS-2) tested 8.3 mmcfgd in the Ahnet Basin. Sonatrach also tested 5.9 mmcfgd from a 66 ft (20 m) pay of Carboniferous reservoirs in the Tinerkouk-1 (TNK-1) (WO Jan. 2008). PetroVietnam International (PIDC) made an oil discovery on Blocks 416b and 433a with the well BRS9 with a potential flow rate of 5000 bopd (WO May 2008).

## 1.4  TUNISIA

In Tunisia, the first exploration expedition was the voyage of Peyssonel & Desfontaines in 1725 across Algeria and Tunisia (Burollet, 1995). Peyssonel (1738, cited by Burollet, 1995) described quarries and economic minerals, as well as the historic variations in the shorelines of Tunisia. At the same time, M.D. Shaw (1738, 1743) published an illustrated description of many fossils and minerals collected during his voyage to the Berbers region. He pointed out the Cenomanian coral *Aspidiscus cristatus*, which was later described by Lamarck.

In 1830, S.E. Hebenstreit published the results of his voyage to Alger, Tunis and Tripoli, sponsored by King Frédéric Auguste of Polonia. In 1877, Pomel carried out the first systematic geological work in Tunisia. Overweg (see Demoulin, 1931, for an account of the trip) during his travel from Philippeville to Murzuk via Tunis and Tripoli mentioned for the first time the Strombus beds of the Tunisian Sahel. Edmond Fuchs explored Tunisia from 1873–1877. He mentioned the Middle Triassic of Zaghouan and the associated faults. G. Stache also crossed Tunisia about that time and confirmed the presence of the Turonian near Gabès. The Italian G. Perpetua published the *Compendio della Geografia della Tunisia* (1880, 1883) and *Geografia della Tunisia* (1883), where he gave a detailed description of the shorelines, the relief, the islands, the Chottes, and the Sebkhas. He considered the Bou Kornine as a dormant volcano. M. Blanckenhorn (1888, cited by Burollet, 1995) presented a study on the Atlas Chain, especially the Algerian part, in *"Die geognostischen Verhältnisse von Afrika. I. Teil: Der Atlas, das nord-afrikanische Faltengebirge"*.

The Scientific Exploration Mission of Tunisia *"Mission Scientifique de Tunisie"* (1885–1887) was led by the botanist Ernest Cosson (d. 1889). During that mission, Georges Rolland explored Central Tunisia, Phillippe Thomas (d. 1910) travelled in Southern Tunisia, and Georges Le Mesle (d. 1895) studied Northern Tunisia in 1889–1890. They provided a good description of the Jurassic of Zaghouan and the Eocene of Maktar and Kairouan. Philippe Thomas discovered the phosphate deposits of Jebel Tselja. The report of Valéry Mayet in 1887, although botanical and zoological in nature, contains geological information on the gypsum deposits of Oued Leben, south of Jebel Meheri, the Cretaceous of Bou Hedma, the Cenomanian *Exogyras* to the west of Orbata, and the Miocene fossils west of Chemsi. Ph. Thomas published the paleontological works made by Victor A. Gauthier (Echinoids), Arnould Locard (Molluscs), Auguste Péron (Brachiopods, Bryozoans, and Pentacrines), and H.E. Sauvage (fishs). Francis Aubert, a mining engineer, who commenced his geological research in 1884, compiled the works made by members of the Mission and constructed a map at the scale of 1:800,000 in 1892. Georges Rolland (1891) published the *"Aperçu sur l'histoire du Sahara depuis les temps primaires jusqu'à l'époque actuelle"*, which included a map.[17] In 1896, Capitaine E. de Larminat studied the geography of Southern Tunisia and published the results in the *"Etude sur les formes du terrain dans le sud de la Tunisie"*.

The phosphates were treated by M. G. Bleicher (1890), David Levat (1894–1895), and E. Vassel (1887, 1899), as well as E. Trodos (1898) on the phosphates of Kalaa Djerda, Edmond Nivoit (1897) on the Metlaoui and on Moulares (1903). A. Pomel defined *Dyrosaurus thevestensis* that was erroneously identified as *Crocodilus phosphaticus* by Ph. Thomas. F. Priem (1903) studied the vertebrates and the fishs in particular of the phosphates. G. Di Stefano (1903) identified new reptiles in the phosphates. Priem (1909) described an Upper Senonian fish from the Abiod Formation west of Metlaoui. V.A. Gauthier (1892) described the Cretaceous echinoids collected by M.F. Aubert and the Jurassic echinoids collected by G. Le Mesle (1896).

From 1896, the geology of Tunisia was studied extensively by Léon Pervinquière (1873–1913) (Burollet, 1995). His work was published in 1903 and included a map of Central Tunisia at the scale of 1:200,000. He continued with a number of paleontological studies in Tunisia that included the Mesozoic cephalopods in 1907 and the Cretaceous gastropods and pelecypods in 1912, as well as Quaternary studies of shorelines of Tunisia. In 1909, Pervinquière led a scientific mission in Algeria and in 1911 accompanied another mission to determine the boundary between the French Tunisia and the Italian Tripolitania. During the latter, he described the geography and geology of Ghadames (1911) and Tripolitania (1912).

Ch. Monchicourt studied the Maktar Massif in Central Tunisia. In 1902, Ph. Thomas compiled the geological description of Tunisia with the collaboration of A. Gaudry, A. Péron, and Paul Bursaux. The results were published between 1907 and 1910. The work was continued by Pervinquière and later by Emile Haug.

17 Bull. Soc. Géol. France, v. 3, no. 19, p. 237–246.

In 1912–1913, L.F. Spath described the ammonites of Jurassic Zaghouan. His observations were later confirmed by Biely & Rakus (1969–1970). In 1896, the archeologist R. Cagnat described ancient quarries and mines, in particular the lead and zinc at Jebel Ressas, and the marble at Chemtou. M. Gerest (1889) attributed an Upper Cretaceous age to Jebel Tebaga. The presence of the Triassic at Jebel Rehach was confirmed for the first time by A. Joly (1906–07) who assigned a Permian age to the lower sands. The Tunisian Atlas was studied at Bou Kornine by Janko in 1890, the area around El Kef by P. Mares in 1884, the discovery in 1893–95 of Liassic fossils by A. Baitzer in the Zaghouan Limestones which were determined by Ch. Mayer-Eymar, and the Gafsa area by E. Koken in 1909. C. de Stefani described Jurassic fossils at Jebel Aziz in 1907. Several alpine geologists also visited Northern Tunisia; among them were Léonce Joleaud (1880–1938) who identified the Lower Cretaceous at Jebel Bou Kornine (1901) who also wrote on the tectonics of Northwestern Tunisia from 1913 until the 1930s, and Pierre Termier who described many ore occurrences. Ch. Monchicourt (1913) published his thesis on the geography of the *"Haut Tell Tunisien"* using the geology of Pervinquière. Solignac (1927, cited by Boughdiri et al., 2005) carried out work in Northern Tunisia, and published his work in 1929, in which he included a map at a scale of 1:200,000. In 1934, F. Bonniard published his report on the geography of the Northern Tell based on the geology by Solignac (1929).

In the 1950s, Mattieu (1950) carried out work in Tunisia using local stratigraphic names. C. Castany (1951–1955) studied Central Tunisia. 1956 signaled the beginning of oil exploration in Tunisia. Burollet (1956) was the first to establish a systematic stratigraphic nomenclature in Tunisia. Castany (1962) included a part on Tunisia in the International Stratigraphic Nomenclature and mentioned the stratigraphic terms in use at that time.

In 1964, Glaçon & Rovier (1964) indicated the presence of limestone micro-breccia at Kroumirie, and Jauzein & Rovier (1965) defined the Adissa and Ed Diss formations at Kroumirie. Busson (1967a, b) studied Southern Tunisia. Rakus & Biely (1971) defined the *"Formation calcaire de l'Ouest"* in the Lower Liassic of the Tunisian Dorsal. Bajanik et al. (1972) revived the names of the Miocene Beglia and Saouf formations and introduced the Mahmoud Formation, Cap Bon Group, Sehib Formation, and the Bou Sefra Facies. In a jubilee volume dedicated to Solignac in 1973, many new formations were introduced by various authors, such as the Tithonian Ressas Formation, Djebel Siouf Limestone, Cherahil Formation, Haffouz Facies, Korbous Limestone (Comte & Dufaure, 1973), and Oued Hamman Formation (Hooyberghs, 1973). Biely et al. (1974) revised the Neogene stratigraphy in Northern Tunisia and introduced the names Bejaoua Group, Medjerda Group, and Oued Mellegue Group with its two formations, Tessa and Oued Djouana. Turki (1975) introduced Gridja Formation for the Aptian at Djebel Bargou, and Khessibi (1975) defined the continental Kebar Formation for the Middle Cretaceous at Djebel Kebar. Dominique Massa carried out many studies on the Paleozoic of Libya and Southern Tunisia since the 1960s.

In May 1964, Tunisia's first oil field, El Borma, was discovered in Triassic reservoirs near the Algerian border. Other fields in Tunisia include the 7 November, Ashtart, Bouri, El Biban, Ezzaouia, Sidi El Kilani, El Menzah, Cercina, Miskar, and Tazarka. The *Entreprise Tunisienne d'Activités Petrolières* (ETAP) was established in March of 1972 to manage exploration and production activities in the country.

The Isis Field in the offshore of Tunisia was discovered in 1974, but remained undeveloped until December 2001.

Bishop (1975) published his classical paper on the geology of Tunisia and adjacent parts of Algeria and Libya, followed by a second paper (Bishop, 1988) on the hydrocarbon geology of East-Central Tunisia. Robinson & Wiman (1976) reviewed the stratigraphic subdivisions of the Miocene rocks in the sub-Dorsale in Tunisia and placed the top of Saouf Formation in the late Tortonian based on planktonic foraminifera. The Petroleum and Exploration Society of Libya publishes "Guidebook to the Geology and history of Tunisia for its 9th Annual Field Conference (1976), where Burollet (1976) reviews the Tertiary geology of Tunisia. Fournié (1978) compiled the "Lithostratigraphic Nomenclature of the Upper Cretaceous and Tertiary Series of Tunisia" sponsored by Serept and Elf-Aquitaine where he revised the stratigraphic nomenclature of Tunisia and introduced 10 new formations. Ferjani, Burollet & Mejri (1990) review the petroleum geology of Tunisia under the auspices of the *Entreprise Tunisienne d'Activités Petrolières*.

In 2007, Eni's Karma-1 well on Adam Concession tested more than 4000 bopd. Nakhil-1 well on Bordj el Karma Permit produces 1200 bopd (Eni's Website, Jan. 2007). PA Resources AB began producing 20,000 bopd from Didon Field offshore Tunisia (World Oil, May 2007). Pioneer Natural Res made a discovery in Shaheen-1 on Jenein Block, with daily gross production of 8000 boe from multiple zones (World Oil, July 2007). In 2008–2008, PA Resources made an oil discovery with Didon North-1 on the offshore Zarat Block, which tested 46 ft of pay in the El Guaria Formation (World Oil, Jan. 2009). BG discovered a 200 ft gas column in the RM-1 well on the Hassi Ba Hmaou Permit; the well tested 8.7 mmcfgd at depths 2917-2943' (World Oil, Aug. 2008).

Like Egypt, Tunisia was shaken by unexpected events in January 2011, which led to a change in government.

## 1.5  MOROCCO

One of the earliest scientific reports on Morocco was that of Hodgkin in 1864 who wrote an account on Sir Moses Monteflore's Mission to Morocco. In 1868, Rohlfs visited Morocco before he embarked on his famous voyages in North Africa. In 1870, Mourlon published in the Bulletin De l'Academie Royale de Belgique a description of fossils collected by Desquin and stored at the Museum of Brussels.

In 1871, a botanical expedition into Morocco was led by Joseph D. Hooker (1817–1911) and John Ball (1818–1889), accompanied by the tile-maker, botanist and geologist George Maw (1832–1912). The results of the expedition were published in 1878 in "Journal of a Tour in Marocco and the Great Atlas". The book included an appendix by G. Maw entitled "Notes on the Geology of the Plains of Marocco (*sic.*) and the Great Atlas" and included the first geological cross-section of the High Atlas.

Lieut. Washington wrote some notes on the geological features of the district between Tangier and Morocco and published in the first volume of the Journal of the Royal Geographical Society an article entitled "Geographical Notice of the Empire of Marocco" (Hooker et al., 1878). A geological memoir by Henri Coquand

(1811–1881) was published on the environs of Tangier and the northern part of Morocco (Coquand, 1847). A paper was read by G. Maw before the Geological Society of London in 1872 on the northern promontory of Morocco, facing the strait of Gibraltar. South of Tangier, M. Desquin (Hooker et al., 1878, p.448) collected the Upper Cretaceous fossils *Inoceramus*, *Ostrea nicaisei*, *O. syphanx*, *Globiconcha ponderosa*, *Trigonia castas*, and undet. *Echiondermata*. James Smith (J. Geol. Soc., v. 2, p. 41, also in Hooker et al., 1878, p. 448) mentioned *Terebratula fimbriata* and *T. concinna* from the "Lower Oolite" in the Gibraltar Limestone, which Coquand compared with the Jurassic Tetuan Limestone near Tetuan (Tetouan) and divided it into four units of marls, dolomites, calcareous sandstones with odor of petroleum, and lithographic limestones with siliceous concretions. Maw suggested that the Tetuan Limestone is separated from the Cretaceous Series to the northwest by a major N-S fault that runs in the middle of the Tangier promontory. Maw also described a 200 m (656 ft) cliff near Cape Cantin which consists of alternating beds of grey and reddish fine-grained sandstones and beds of ferruginous carbonates. Maw collected specimens of *Exogyra conica*, *Ostrea leymerii*, and *O. Boussingaulti* near Staffi, which were dated as Neocomian. Maw also identified Miocene and Pliocene beds near Staffi and Eocene beds of clays and red ferruginous marls with oyster beds containing *Teredina personata* in an escarpment at Sidi Ammar, overlain by a bed with *Balanus sulcatus*, *Pecten beudanti*, *Arca*, *Buccinum prismaticum*, and *Conus*, which were assigned to the Miocene. In addition, Maw identified metamorphic rocks north of the City of Marrakech, and micaschists intruded by porphyritic dikes at Djebel Tezah.

In 1883, Charles-Eugène de Foucauld wrote "*Reconnaissance au Maroc*" (1883) where he described Jebel Sirous. In 1889, Joseph Thomson published "Geological Map of Southwestern Morocco 1:500,000". The Expedition of Emin Pacha left in April 1890 for Bagamoyo, the results of which were reported in "*Resultat de la dernière expedition de Emin Pasha*" by B. Auerbach (1894). The results of another expedition were published in 1891 "*Expedition du Comte Telek (1887–1888)*". In 1892, P. Schnell wrote an account titled "*Marrokisches Atlas*". In 1897, M.R. de Flotte de Roquevaire published topographic maps at the scale of 1:1,000,000 (1897 & 1904). In 1899, Joseph Thomson wrote the paper "The Geology of Southern Morocco and the Atlas Mountains".

In 1900, the first collection of Carboniferous material from the Zousafana Valley was made by officers of Colonel Bertrand's detachments. The paleontological results were published by Ficheur (1900).[18] In 1901, the Carboniferous was recognized in outcrops at Tidikelt, NW of Hoggar by Commandant Cauvet, at Tademait by Lieutenant Besset, at Touat by Commandant Deleuze, and at Ahnet by Lieutenant Mussel. In 1904, R. De Flotte (1904) made a map of Morocco at a scale of 1:1000,000. Louis Gentil began his expeditions in 1904, and in 1905, he published the "*Mission de Segonzac: Dans le Bled es Siba*" in the Western Atlas. Paul Lemoine introduced the "*Carte géologique schmatique 1:2,000,000*". Gentil (1906) gave an account of the geological contributions in Morocco and discussed the results of A. Brives' 1905 work in the Atlas Mountains. In 1907, Gentil wrote a note entitled "*Esquisse géologique du Haut Atlas occidental*" to accompany a map at the scale of 1:750,000, where

18 Weynat (1985, p. 300).

he discussed again the stratigraphic and geographic subdivisions of the High Atlas proposed by Brives. We see also the first mention of Carboniferous at Djerada in Morocco by Gentil (1908). In 1912, Gentil published *Le Maroc Physique* (1912), where he presented the first geological map of Morocco. In 1911, G.B.M. Flamand carried out surveys in the area NW of Hoggar.

In 1912, Morocco was occupied and became a French protectorate.

The Carboniferous was recognized at In Tedreft, south of Hoggar by Chudeau (1913), and in Central Morocco by Gentil (1914). Carboniferous plants collected by Commandant Carrier were studied by Fritel (1925), and the Carboniferous was identified at Iullemedden by Boureau (1953). In 1918, the first geological map of the Western High Atlas made by Louis Gentil was published by E. Suess in *La Face de la Terre* and the French edition *Das Antlitz der Erde*. In 1924, exploitation of phosphate started at Khourbiga (Charles & Charles, 1924). In 1927, Bourcart discovered *Archaeocyathus* in the Moroccan Anti-Atlas. In 1936, H. Termier (1936) published *"Etudes géologiques sur le Maroc central et le Moyen Atlas septentrional"* in four volumes in *Notes Mémoires de la Service Géologique Maroc*. In 1950, E. Roch (1950) published *Histoire stratigraphique du Maroc* to accompany the geological maps at scale of 1:500,000 published by the Geological Survey of Morocco.

In March 1956, Morocco gained its independence from France to become the Kingdom of Morocco, with Mohammed V as the king of the new monarchy.

In 1960, a book entitled *Livre à la mémoire du Professeur Paul Fallot* was published in 1960–62, and offers a summary of the geology of the Atlas and Rif Domains (Missenard et al., 2008). A. Michard (1976) summarized the geological knowledge of Morocco in his *"Eléments de Géologie Marocaine"*. In 1980, Gabriel Suter (1980) constructed new geological and structural maps of the Rif at a scale of 1:500,000.

In the 1990s and the beginning of the 21st Century, a tremendous amount of research papers and post-graduate thesis have been produced. Worthy of mentioning are two major compilations: Jacobshagen (1988) edited a book on the Atlas System of Morocco, and most recently, Michard et al. (2008) compiled in one volume a variety of papers written by many experts in the geology of Morocco.

From 1900 to 1928, petroleum exploration was carried out in the vicinity of oil seeps, which led to the discovery of the Ain Hamra oil pool in the Rharb Basin in 1923. The *Bureau de Recherches et d'Exploitations Minières* (BRPM) was created in 1928 and the *Société Chérifienne de Pétrole* (SCP) was established in 1929 and carried out most of petroleum exploration activities. Hydrocarbon exploration continued in 1937 in the Rifian and Prerifian zones of Morocco by the *Societé Chérifiane des Pétroles* with a drilling campaign of 160 shallow wells; only 12 wells were >1000 m (3281 ft), to solve tectonic problems (Célérier, 1937). Four occurrences of hydrocarbons were located as a result of that campaign; two at Djebel Tselfat, one at Djebel Bou Draa, and one at Djebel Outita. All occurrences were considered uneconomical, but the potential of the Lower Jurassic (Domerian) limestones was shown. Exploration efforts since that time led to the discovery of a number of oil and gas fields. The most important fields are situated in the Essaouira and Gharb (Rharb) Basins in Western Morocco. Morocco also has oil shale deposits (Kolonic et al., 2002) at Tangier, Timahdit and Tarfaya in the Atlas Mountains which were studied in the 1930s. Seismic reflection techniques were first introduced in the Rharb and Prerif basins in 1935 and in the

Essaouira, Souss and Guercif basins in 1955 (Onhym, 2000, www.onhym.com). Oil and gas discoveries were made in the Prerif Ridges and in the Sidi Fili Trend. A new Hydrocarbon Law was passed in 1958. Wells drilled between 1958 and 1981 led to further discoveries in the Essaouira and Rharb Basins. In 1981, the *Office National de Recherches et d'Exploitations Pétrolières* (ONAREP) was created by the Moroccan government. A total of 85 wells have been drilled in the period 1981–2003, and led to the discovery of Meskala Field in the Essaouira Basin, in addition to several biogenic gas accumulations in the Rharb Basin. The Hydrocarbon Law was amended in 1992, and in 2003 ONAREP and BRPM were merged into a newly formed entity, ONHYM.

In 2008–2009, Circle Oil made a gas discovery with CGD-0, which tested 3.9 mmcfgd from the Lower Guebbas Formation (World Oil, Jan. 2009). Repsol-YPF discovered gas in the offshore Anchois-1 well in the Tanger-Larache area; about 40 km offshore Morocco from two gas pays totaling 427 ft, with estimated recoverable reserves of 100 bcfg (World Oil, May 2009). In January, 2011, Circle Oil announced that the ADD-1 exploration well on the Sebou Permit, Rharb Basin, has tested 3.57 mmcfgd on a 24/64' choke from the Hoot and Guebbas formations.

# PART 2

# Tectonic framework of North Africa

# Chapter 2

# Northeast Africa's Basins

The tectonic setting of North Africa is an integral part of global tectonics. It is related to the formation and break-up of the super-continents Rodinia, Gondwana and Pangaea, the rotation of the different continents relative to each others and the opening and closing of major present-day and paleo-oceans, such as the Iapatus, Atlantic, Pacific, and Tethys. These events led to rifting, shear movements, uplifts, volcanic activities and thermal subsidence throughout the geologic history of the area. The tectonic history of the area becomes less clear as we move down the geologic column due to the lack of data on one hand, and the overprinting of the older tectonics by younger ones.

North Africa can be divided into six major domains; each is subdivided into various tectonic elements (Fig. 2.1): 1) The Precambrian basement, which includes the African-Nubian Shield and the West African Craton. 2) The Sahara Platform that includes cratonic basins, tectonic highs, arches, and platforms. 3) Rifted marginal basins. 4) Rift basins, 5) The Atlas/Alpine folded belts. 6) Passive continental margins.

The cratonic basins include the Nile Basin in Egypt, and the Kufrah, Murzuq, and Ghadames basins in Libya. Algeria encloses the Ghadames/Illizi/Berkine, Oued Mya, Mouydir, Ahnet, Timimoun, Bechar, and Tindouf basins. Tunisia includes the northern part of the Ghadames Basin, the Pelagian Shelf, and Western Tunisia (Atlas Basin). Morocco contains numerous basins, although much smaller than those of Egypt, Libya, and Algeria; such as the Anti-Atlas, Prerif, Boudenib, Tafilalt, Maider, Ouarzazate, Zag, Tindouf, Draa, Missour, High Plateaux, Saïs, Tadla, Haouz, Bahira, Tagalft, Guercif, Beni Znassen-Beni Bou Yahi, Souss, Argana, and Taoudenni basins; the latter is located mostly in Mauritania. The rifted marginal basins include the Northern Egypt Basin, the offshore Nile Delta Basin, the Cyrenaica Basin, and the Jifarah Basin. Rifted basins include the Red Sea and Gulf of Suez in Egypt and the Sirte Basin in Libya, and the Mediterranean Basin. The passive margins include the Mediterranean and Atlantic margins and their offshore basins, such as the Agadir, Rharb (Gharb), Doukkala, Essaouira, Tarfaya, Safi, and the Atlantic Deepwater Frontier. The Alpine/Atlas folded belts are located mostly in Morocco and northern Algeria and Tunisia.

## 2.1  MEDITERRANEAN BASIN

The Mediterranean Sea is an east-west elongated trough extending for 3500 km (2175 miles) separating the continents of Africa, Europe, and Asia (Fig. 2.1).

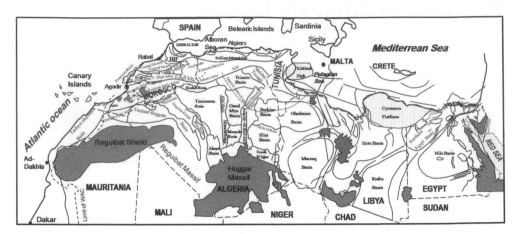

*Figure 2.1* Tectonic map of North Africa (compiled from Tawadros, 2001, Cochran & Peterson, 2001, Piqué, 2003, Michard et al., 2008). International boundaries are approximate (See color plate section).

The Mediterranean is bordered on the north by the Alpine Chain and to the south by the African Platform and the Atlas Chain. It is connected to the Atlantic Ocean via the Strait of Gibraltar and with the Black Sea in the northeast via the Straits of Dardanelles and Bosporus. The man-made Suez Canal connects it to the Red Sea.

The Mediterranean Sea is divided into several subbasins separated by ridges, escarpments, and deep channels. The Peninsula of Italy, Sicily, Malta, Pantelleria, and Cap Bon Peninsula in Tunisia, mark the limit between the Eastern Mediterranean and the Western Mediterranean (Alboran) basins. The southeastern part of the Eastern Mediterranean is known as the Levant Sea. The Ionian Abyssal Plain, the Pelagian Sea, the Messina Cone, the Hellenic Trough, and the Mediterranean Ridge are often referred to as the Central Mediterranean.

The continental shelf, delimited by the 200 m (656 ft) contour, is narrow offshore Egypt and Eastern Libya and wide off West Libya and Tunisia. The eastern margin of the Mediterranean is steep and marked by recent faulting and submarine canyons (Derin & Garfunkel, 1988) and by the Oligocene Dead Sea fault system and the Syrian Arc fold system of Northern Sinai. The latter extends offshore the continental slope, resulting in deep, broad NE-SW-trending anticlinal features (Chanliau & Bruneton, 1988, Yousef et al., 2010). In Tunisia, the Pelagian Platform extends offshore to about 120 km. The Algerian margin is about 5 to 10 km in average and is 20–30 km wide off Morocco.

The Pelagian Sea is occupied by the mostly submerged Pelagian Platform (Burollet, 1979), also known as the Sicilio-Tunisian Platform (Biju-Duval et al., 1974). The Pelagian Platform is limited to the north by the Sicily Channel and the Pelagian Islands and to the south by the Tripolitania Trough (Sabratah Basin). The Jarrafa Graben, which is located to the north of the Tripolitania Basin, is an asymmetric, NW-SE-trending trough, parallel to the other Pelagian channels. Its northern flank is a monocline formed by Miocene sediments dipping to the south which are overlain by a thick Pliocene-Quaternary sedimentary succession; its southern flank is made up of faulted Miocene and thin post-Miocene rocks (Bellaiche & Blanpied, 1979). The Malta Escarpment, an abrupt N-S-oriented submarine cliff with elevations of more than 1000 m (3281 ft), marks the limit between the Ionian Sea and the Pelagian Sea. The water depth in the Pelagian Sea varies from 0–400 m (0–1312 ft), although the NW-SE oriented grabens, such as the Pantellaria and Linosa channels which occur within the Sicily Channel may reach depths of more than 1300 m (4265 ft) and 1600 m (5250 ft), respectively (Winnock, 1979, 1981). These channels have been interpreted as rift basins delimited by dip-slip bounding faults, representing a short rifting episode during the Maghrebian folding and thrusting episode (Argnani & Torelli, 2001). The continental shelf area is delimited by the 200 m (656 ft) contour and it includes the Gulf of Hammamat, the Gulf of Gabes, and the Adventure, Malta, and Medina banks. The Pelagian Platform extends inland northwards into the Ragusa region in Southern Sicily and westwards into Tunisia and Northwest Libya. The Pelagian Platform represents the extension of the African Platform and it is probably underlain by a continental crust (Biju-Duval et al., 1974, Burollet, 1979, Bishop & Debono, 1996). It formed part of a subsiding passive margin since about the Late Early Triassic (Liassic) (Jongsma et al., 1985) and it was dominated by shallow-water Mesozoic and Cenozoic carbonate sediments. The northern part of the Pelagian Platform was affected by NE-SW-oriented horizontal Alpine movements and the southern part by vertical displacements associated with N-S, E-W and NW-SE-oriented old basement faults (Bellaiche & Blanpied, 1979). The Azizia-Tebaga Fault in NW Libya and Tunisia separates the African Shield from the Pelagian Block (Conant & Goudarzi, 1964). This fault was active as a large right-lateral shear during the Atlas compression phase, and probably represents a basement feature (Burollet, 1991).

## 2.2 ALBORAN SEA

The Alboran Sea (Fig. 2.2) is the westernmost portion of the Mediterranean Sea, lying between Spain on the north and Morocco and Algeria on the south. The Strait of Gibraltar, which lies at the west end of the Alboran Sea, connects the Mediterranean with the Atlantic Ocean. Its average depth is 445 m (1461 ft) and maximum depth is 1500 m (4920 ft). The Alboran Sea is limited on the west by the Strait of Gibraltar and on the East by a line passing through Capo de Gata in Spain. Mud diapirism and mud volcanism occur in the West Alboran Basin major sedimentary depocenters (Sautkin et al., 2003). The sedimentology and sequence stratigraphy of the West Alboran Basin have been the subject of various studies (Comas et al., 1992, 1999, Watts et al., 1993, Soto et al., 1996, Chalouan et al., 1997, Perez-Belzuz et al., 1997).

The Alboran Sea (Fig. 2.2) falls at the boundary of two lithospheric plates (Europe and Africa) and marks the limit between an oceanic domain on the west and a continental domain to the east (Hatzfeld & Frogneux, 1980).

A variety of models have been proposed to explain the synchronous extension and compression that occurred in the Alboran region and its surroundings during the Early Miocene despite the continued convergence of Africa and Iberia (Hatzfeld & Frogneux, 1980, Calvert et al., 2000). The Alboran Sea appears to be a complex of pull-apart

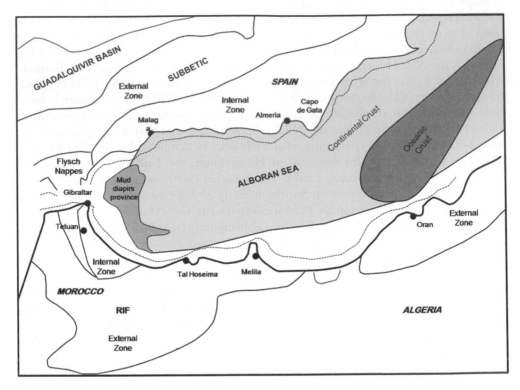

*Figure 2.2* Alboran Sea (after Sautkin et al., 2003, Crusts are after Calvert et al., 2000).

basins that forms a zigzag relay between the Gulf of Cadiz and Algeria (Ammar et al., 2007). The present compressional deformation of the Northwestern African Mediterranean margin is dominated by strike-slip movements on the Algerian margin (Mauffret, 2007). North of Boumerdes-Zemmouri, the oceanic crust is deformed by south-dipping blind thrusts. Off Eastern Algeria and Tunisia, the deformation is more intense but limited to the north by the continental slope. After the attachment of the Kabylies to the Africa Plate around 18 Ma, the crust has been thinned and the Algerian Basin opened during the middle-late Miocene. From the late Miocene to the Present the margin has been thickened through transpression and uplifted (Mauffret, 2007).

The nature of the crust in the Mediterranean Basin is still debated. The crust is believed to be variously continental (Hirsch, 1984, Hirsch et al., 1995, Ricou et al., 1984), oceanic surrounded by a thinned granitic continental crust (Biju-Duval et al., 1974, Schuster, 1977, Le Pichon, 1988), or transitional (oceanized) between a typically continental (sialic) and a typically oceanic (simatic) crust (Menard, 1967, Smith & Woodcock, 1982). Bijou-Duval et al. (1974) suggested an oceanic crust underneath all of the Mediterranean. Smith & Woodcock (1982), based on seismic data, favored an oceanic or a transitional (i.e. oceanized) crust for the deeper parts of the Levant Basin. However, they speculated that the Levant Sea is either a remnant of the Neo-Tethys Ocean or a subsiding portion of a continental crust. Hirsch (1984) and Ricou et al. (1984) interpreted the Levant Sea as part of the Gondwana platform. Biju-Duval et al. (1974), Burollet (1979), and Bishop & Debono (1996) believe that the Pelagian block is underlain by a continental crust. Therefore, generalizations about the nature of the crust in the Mediterranean can not be made. It appears, however, that the crust is oceanic (or oceanized) in the center of the basin and continental around the periphery and in the Pelagian Sea. A Neogene oceanic crust occur beneath the majority of the Western Mediterranean, although most of the Alboran Sea is underlain by both Neogene oceanic and continental crusts (Calvert et al., 2000, Fig. 1.1).

The evolution of the Mediterranean Basin and the problems associated with the different interpretations have been reviewed by Biju-Duval et al. (1974), Smith & Woodcock (1982), and Robertson & Dixon (1984). There are disagreements on the times, number, and models of rifting in the Mediterranean. At least nine rifting episodes which took place during the period from the Late Permian to Maastrichtian have been suggested in the literature. During that time, the African and European plates were moving relative to each other, due to lateral shear stresses combined with the rotation of Africa. These processes were attributed to the opening of the Atlantic Ocean (Biju-Duval et al., 1974), which resulted in an equal number of rifting episodes in Atlantic and the Atlas Domain in NW Africa. The Southern, Central, and Northern Atlantic opened at different times. These movements and rifting phases interfered with one another to give a very complex history (Sengör et al., 1984). The history of the Northern Alpine tectonics (Mesozoic-Cenozoic) is even more complex.

Virtually nothing is known about the pre-Permian history of the Mediterranean Sea. However, the Mediterranean was a part of Gondwana during that time, and probably the Paleozoic epeiric seas which covered most of Gondwana also covered the Mediterranean. The Taconian, Caledonian, and Hercynian events must have had their effects on the area.

Gondwana was dominated during the Middle Carboniferous Hercynian (Variscan) Orogeny by a compressional regime caused by the collision between Africa and Eurasia on one hand and the subduction of the oceanic plate along the Paleo-Pacific on the other (Visser & Praekelt, 1996). This phase resulted in the formation of a number of uplifts on the northern edge of Gondwana, such as the Gargaf Arch, the Tibesti Arch (Goudarzi, 1980), the Sirte Arch, the Gharyan Dome (Burollet et al., 1971, Koehler, 1982) in Libya, strike-slip faulting and uplifting in the Meseta, Anti-Atlas, and Atlas Domains in Morocco (Michard et al., 1982, Saber et al., 2007), and the Helez Geanticline in Northern Sinai (Gvirtzman & Weissbrod, 1984, Weissbrod & Horowitz, 1989). Compressional regimes resulted also in the formation of mega-shear systems, such as the Falkland-East Africa-Tethys Shear (Visser & Praekelt, 1996). Renewed movements occurred along the Trans-Africa Lineament (Nagy et al., 1976) and the Pelusium Shear (Neev, 1977). Thick Carboniferous-Permian shallow-marine detrital and minor carbonate sediments were deposited in the Jifarah Basin in NW Libya north of the Azizia Fault (Koehler, 1982), the Permian Basin in Tunisia (Burollet et al., 1971, Bishop, 1975), Cyrenaica (El-Arnauti & Shelmani, 1988), and the Northern Western Desert (Keeley, 1989, 1994).

Late Permian-Middle Triassic dextral (clockwise) motion of Africa relative to Eurasia created a wedge-shaped westward-narrowing arm of the paleo-Tethys (Robertson & Dixon, 1984). This phase of rifting led to the collapse of paleohighs, subsidence, and widespread basin formation in North Africa, especially around the margins of the Tethys. The Pelagian block, Jifarah Plains, Cyrenaica Basin, Northern Egypt Basin and Northern Sinai probably subsided at that time and received thick sediments. The Pelagian Shelf was dominated by carbonates (Van Houten & Brown, 1977). Late Permian shallow-water carbonates were deposited in the southeastern part of the Eastern Mediterranean (Levant Sea) and in northern Sinai (Derin & Garfunkel, 1988, Druckman, 1974a & b, Bartov et al., 1980). Permian clastic deposition dominated in the southwestern Atlas Chain, Tebaga in Tunisia (Kilani-Mazrzoui et al., 1990, Toomey, 1991), the Tripolitania Basin, Jifarah Plains (Del Ben & Finetti, 1991), Northern Cyrenaica, Central Sinai, and the Gulf of Suez region. Carbonates with local reef development (Toomey, 1991, Rigo 1995, Rigby et al. 1979, Boote et al., 1998) and evaporites were deposited in the Gulf of Gabes and Tunisia, where halite forms domes and diapirs (Gill, 1965, Burollet, 1979, 1991, Bishop, 1988), and Algeria (Galeazzi et al., 2010, Bourquin et al., 2010). Middle Triassic sandstones were deposited in the Eastern Sirte Basin (Shelmani et al., 1992, Thusu, 1996, Thusu & Mansouri, 1998) contemporaneous with the collapse of the Kalanshiyo Arch. Polymictic conglomerates were formed along the coastal plains of the Levant Sea as a result of foundering of the Helez Geanticline (Druckman, 1984, Gvirtzman & Weissbrod, 1984, Weissbrod & Horowitz, 1989). In the northern Ghadames Basin, polymictic conglomerates were associated with the collapse of the Gharyan Dome and movements along the Azizia Fault (Mokaddem, 1995).

The fate of the Paleo-Tethys Ocean[1] and the exact timing of the opening of the Neo-Tethys are not known. It has been suggested that the Paleo-Tethys was destroyed before the opening of the Neo-Tethys in the Late Triassic. Its demise was the result of the convergence of the African and European plates toward each other and the northward subduction of the oceanic plate (Dewey et al., 1973, Smith & Woodcock, 1982). Sengör et al. (1984) and Robertson & Dixon (1984) argued that the disappearance of the Paleo-Tethys was coeval with the propagation of the Neo-Tethys. According to their model, the Neo-Tethys started to form in the Middle Triassic, but the closure of the Paleo-Tethys was not complete until the Late Triassic. The present Mediterranean basins are believed, in either case, to be relics of the Mesozoic and Cenozoic Neo-Tethys basins.

The northern margin of the Tethys Ocean, now located in Turkey, is demarcated by a series of ophiolitic belts. Arguments exist about their origin, time of emplacement, and whether they belong to one oceanic basin or to a number of basins (Spray et al., 1984, Knipper et al., 1986, Robertson, 1994). These ophiolites have a complex history and most of them probably resulted from multi-step development (Knipper et al., 1986). Ophiolites in the northern Mediterranean probably represent ancient oceanic crust of the paleo-Tethys and neo-Tethys oceans formed during Triassic-Cretaceous rifting phases (Biju-Duval et al., 1974). All these ophiolitic massifs were thrust over the continental margin at the end of Cretaceous (Whitechurch et al., 1984). Spray et al. (1984) and Knipper et al. (1986) suggested that the ophiolites were generated at a spreading centre in the Middle Jurassic, and were thrust onto the Arabo-African continental margin during the Upper Jurassic-Early Cretaceous. Robertson (1994) suggested that the Eastern Mediterranean ophiolites were formed in

---

1 Ruban (2007) argued that since the "Tethys" was made up of several oceans, there was no unique ocean which might have been called "Tethys Ocean".

a supra-subduction zone setting during the Maastrichtian, because of the continuing narrowing of the Neo-Tethys.

A renewed rifting phase took place during the Late Triassic-Early Jurassic. Opening of the Central Atlantic Ocean by Middle Triassic time led to a left-lateral (sinistral) motion of Eurasia relative to Africa (Robertson & Dixon, 1984, Zülke et al., 2004). Rifting led to the formation of the Essaouira, Tarfaya, Agadir, Haha, Doukkala, Souss, and Aaiun basins in Morocco (Medina, 1995, Le Roy et al., 1998, Le Roy & Piqué, 2001, Hafid et al., 2008, Zühlke et al., 2004, Davison & Daily, 2010). A shallow-water carbonate platform was formed in the Pelagian Sea (Bosellini & Hsü, 1973). This platform was dominated by dolomites and oolitic limestones (Bishop, 1988, Bishop & Debono, 1996). The platform was separated from the North African continental facies belt by the Tripolitania and Jarrafa basins. These two basins subsided thermally (Benniran et al, 1988) and received thick carbonate, marl, and shale deposits. Volcanics were encountered in boreholes in the offshore of Libya, such as in the A1-NC35a well (Winnock, 1981, Jongsma et al., 1985, Bishop & Debono, 1996) and point to extensional tectonics. Extensional features consist mainly of tilted basement blocks, covered by Jurassic and Cretaceous sediments (Burollet, 1991). Salt swells and diapirs are common in the Guercif Basin in Morocco (Bernini et al., 1999), the Atlantic margin of Morocco (Lancelot & Winterer, 1980, Hinz et al., 1982, Holik & Rabinowitz, 1992), the Tripolitania Basin offshore Libya (Bishop & Debono, 1996) and Tunisia (Bishop, 1988) due to sediment loading. Diapirs pierce the Mesozoic and Cenozoic successions, often forming long subsurface salt walls (Burollet, 1991). These salts are the extension of the evaporite deposits of inland NW Libya and Tunisia. The development of horsts and grabens, the formation of sea channels, crustal thinning, and magmatism dominated in the Ionian Sea and the Sirte Rise (Del Ben & Finetti, 1991). Another shallow-water carbonate shelf composed of high-energy sediments and fringing reefs was formed in the Levant Sea; they grade inland into low-energy carbonate muds and dolomites (Derin & Garfunkel, 1988) and then into continental deposits.

The Middle-Upper Jurassic time appears to have been a quiet period in North Africa. Pelagic nodular limestones, fossiliferous limestones and dolomites with salt diapirs are common in these sediments on the eastern platform of Tunisia and Algeria (Bishop & Debono, 1996). Shallow-marine carbonates and clastics were deposited in the northern Western Desert, and Northern and Central Sinai.

During the Late Jurassic-Early Cretaceous time, activity of intra-plate hotspots caused extensive magmatism, uplift, and erosion in the offshore of the Mediterranean Sea, the Sirte Basin, and NW Africa (Westphal et al., 1979, Van Houten, 1983, Garfunkel & Derin, 1984, Derin & Garfunkel, 1988, Garfunkel, 1991, Wilson & Guiraud, 1998, Guiraud et al., 2005). The South Atlantic opened during the Early Cretaceous (Neocomian-early Aptian), accompanied by anti-clockwise rotation of Africa (Maurin & Guiraud, 1990, Guiraud & Maurin, 1991, 1992, 1993). These events led to the formation of thick continental clastics of the so-called Nubian Sandstone. Basalts and breccias are commonly associated with these deposits in the Pelagian Platform (Jongsma et al., 1985, Bishop & Debono, 1996), the Sirte Basin (Massa & Delort, 1984, Tawadros et al., 2001), and the Northern Egypt Basin (Keeley, 1994). The Atlas Mountains in Morocco and Algeria were dominated by the deposition of continental redbeds with dinosaur bones and tracks (Mekahli et al, 2004, Charrière et al., 2005).

This phase was followed by a period of faulting and magmatic activities associated with extension and rifting (Del Ben & Finetti, 1991, Mattoussi Kort et al., 2009) during the Late Cretaceous. Eastern Tunisia onshore and offshore display crustal thinning induced from the Tethyan rifting, which affected the subsequent evolution of the North African passive margin during the Late Cretaceous and the creation of the fold-thrust belt and associated foreland deformations. The thinned crust led to the rise of basalt magma and hydrothermal fluids (Mattoussi Kort et al., 2009). A combination of subsidence and sea-level rise led to a new phase of carbonate deposition (Le Pichon, 1988). In Malta (Pedley et al., 1976) and in the offshore of Libya (for example in the Tama-1 well) (Bishop & Debono, 1996), these carbonates are made up of shallow-water high-energy limestones and dolomites. The shallow-water carbonates around the flanks of the Tripolitania Basin change facies into deep-water limestones and shales in the centre of the basin (Bishop, 1988, Bishop & Debono, 1996, Benniran et al., 1988). Early Aptian to Maastrichtian shallow-water carbonates containing zones of rudistid reefs were deposited in SE Tunisia and extended offshore onto the Pelagian Platform, such as in the A1-NC35a well (Jongsma et al., 1985). These rudistid build-ups form hydrocarbon reservoirs in the Isis Field (34°37′N, 12°41′E) (Bishop & Debono, 1996). Late Cretaceous carbonates on the Cyrenaica Platform, the Northern Western Desert, and Northern Sinai, change facies seaward into deep-water shales Bauer et al., 2001, 2002, 2003, 2004).

Communication between the Mediterranean and the South Atlantic took place intermittently during Late Cretaceous time (Reyment, 1966, 1971, 1981, Reyment & Reyment, 1980, Collignon & Lefranc, 1974, Kogbe, 1980). The Cenomanian-Turonian interval reflects poorly oxygenated condition in the Tethys (Oceanic Anoxic Event "OAE") (Schlanger & Jenkyns, 1976, Kuhnt et al., 2004, Kolonic et al., 2005, Keller et al., 2009, Sepúlveda et al., 2009), which resulted in the deposition of dark-colored marly limestones proven to be a source rock for hydrocarbons (Bishop, 1988, Kolonic et al., 2002, Lüning et al., 2004a, b). This phase was terminated by Senonian and later deformation related to plate collision and the Alpine Orogeny (Derin & Garfunkel, 1988, Le Pichon, 1988). Basin inversion during the Santonian compression phase resulted in the formation of a number of anticlinal structures, such as the Atlas Chain (Beauchamp et al., 1996, 1997), the Al Jabal Al Akhdar in Northern Cyrenaica and Esh el Mellaha in the Gulf of Suez Region (Bosworth et al., 1999). These anticlinal structures with faulted northern flanks (Conant & Goudarzi, 1967, Röhlich, 1974, 1980) continue into the Mediterranean offshore area, where a deep trough with more than 5000 m (16,405 ft) of sediments is present (Biju-Duval et al., 1974).

The uppermost Cretaceous-Early Tertiary compressional tectonics were associated with the middle Alpine (Laramide) Orogeny. This compressional regime led to dextral movements in the Mediterranean and the northward subduction of the Tethys oceanic crust, and eventually to the closing of the Mediterranean (Biju-Duval et al., 1974). The Cyprus Arc and the Mediterranean Ridge mark the convergence between the Eurasian and African plates. They divide the Eastern Mediterranean into a thrust-belt with Cenozoic sedimentary basins to the north, and a passive and subsiding African continental margin with Mesozoic sediments to the south (Le Pichon, 1988). Important tectonic movements took place in the Sirte Rise and the Pelagian Platform during that time (Schuster, 1977). These stresses caused uplifting accompanied by northward tilting of North Africa. The sea transgressed southward and

Maastrichtian chalks covered most of North Africa. Maastrichtian sediments were deposited unconformably on older rocks in platform areas in the Sirte Rise, the Sirte Basin, and Egypt, which remained high areas during most of the Late Cretaceous (Del Ben & Finetti, 1991). In the Ragusa Platform in southern Sicily, Maastrichtian limestones overlie a volcanic formation dated at around 71 Ma (van den Berg & Zijderveld, 1982). Maastrichtian volcanics are also present in the B1-NC35a well offshore Libya (Jongsma et al., 1985).

Uplift during the Uppermost Maastrichtian-Early Tertiary resulted in the development of a regional unconformity at the Cretaceous-Tertiary boundary (Keller, 1988, Keller et al., 1995, Adatte et al., 2002, Bensalem, 2002, Gardin, 2002, Karoui-Yaakob et al., 2002). In the offshore well B1-NC35a, Early Maastrichtian shallow-water algal limestones with *Orbitoides* are overlain by pelagic Danian sediments with karst fill (Jongsma et al., 1985). Karstification on top of similar Maastrichtian carbonates was also reported from the subsurface of the Sirte Basin (Jones, 1996), and attests to an unconformity at the K/T boundary. In Northern Sinai, Maastrichtian chalks are separated from the Paleocene shales by paleosoil horizons (Hirsch, 1984).

The Tertiary and Quaternary tectonics in the Pelagian Sea acted along WNW-ESE and ENE-WSW axes (Blanpied & Bellaiche, 1983). Subduction (and/or thrusting) of the Tethys oceanic crust in the Northern Mediterranean continued during the Tertiary. In the meantime, reactivation of ancient faults and subsidence in the southern Mediterranean and North Africa took place due to extensional stresses. Subsidence led to the deposition of Tertiary carbonates on the platform areas and thick shale sections in the troughs (Fournié, 1978, Bishop, 1988, Bishop & Debono, 1996, Schuster, 1977). Nummulitic banks developed in the Pelagian Platform during the Early Eocene, especially in Egypt, Libya and Tunisia, and form reservoirs in the Ashtart Field in the Gulf of Gabes, at the Sidi El Itayem Field in the Sfax area, Eastern Tunisia, and the Bouri Field offshore Libya (Bishop, 1988, Ben Ferjani et al., 1990, Bernasconi et al., 1991, Loucks et al., 1998a & b). Deep-marine sediments persisted in the Tripolitania Basin (Jongsma et al., 1985, Benniran et al., 1988).

The Late Oligocene was also dominated by the deposition of carbonates on the Pelagian Platform. Coralline carbonates crop out in Malta (Pedley et al., 1976) and locally in NW Libya (Hladil et al., 1991). However, over most of Libya and Egypt Oligocene deposits are dominated by continental and fluvio-marine clastics. Tectonic movements accompanied by a global relative drop in sea level (Vail et al., 1977, Haq et al., 1987) led to a hiatus during the Oligocene. This hiatus was detected in the A1-NC35a well offshore Libya (Jongsma et al., 1985) and Egypt (Dolson et al., 2005, Kuss & Boukhary, 2008, Boukhary et al., 2008). The Dead Sea faults which delimit the southeastern Mediterranean (Druckman, 1974) were formed. In the Atlas System, the widely recognized angular unconformity between the Oligocene and older sediments coincides with the Atlas Orogeny. This unconformity pre-dates the emplacement of the Tell units. This unconformity is observed in the Tell below the M'sila Basin (Bracène & Frizon de Lamotte, 2002), but is absent in the Tell nappes where the passage from Upper Eocene to Oligocene is transitional (Benaouali-Mebarek et al., 2006). Late Tertiary clastics were shed from the Tellian Atlas in Northern Tunisia (Benniran et al., 1988, Bédir et al., 1991).

The Pelagian Shelf was a subsiding stable area during most of the Miocene with less than 200 m (656 ft) of Miocene sediments (Winnock, 1981). Miocene shallow-water carbonates crop out on the surface in the islands of Malta and Gozo (Pedley et al., 1976), the Ragusa region in Southern Sicily, as well as the Island of Lampedusa (west of Malta) (Burollet, 1979). Early Miocene carbonates consist of deep-water *Globigerina* limestones, which contain shallow-water fauna and phosphatic material. This mixing of deep and shallow-water faunas was a result of slumping (Pedley et al., 1976), probably caused by movements on the Malta Escarpment. Movements along the Malta Escarpment also caused gravity sliding and the displacement of allochthonous masses during the Miocene (Schuster, 1977). This sliding led to the formation of submarine topographic features, such as the Mediterranean Ridge and the Messina Cone in the northern Ionian Sea. These gravity flow deposits are also known on land in Italy and Sicily (Biju-Duval et al., 1974), where they overlie a thick Mesozoic-Cenozoic sedimentary succession. The Cenozoic sediments extend southward toward the Gulf of Sirte, but no allochthonous masses are present.

Locally on the southern edge of the Jarrafa Trough reefs were developed during the Late Miocene (Blanpied & Bellaiche, 1983). The formation of the deep-sea channels in the Pelagian Sea is related to extensional tectonics during the Miocene and post-Miocene times (Burollet, 1979, 1991). Rifting was due to basement extension associated with shear faults, which border these channels (Burollet, 1991).

During the Messinian, the isolation of the Mediterranean and the fall of sea level led to desiccation of the Mediterranean Sea (Ryan et al., 1973, Ryan, 1978, Hsü, 1972a & b, Hsü, 1977, Hsü et al., 1973). This drying-up period resulted in the deposition of evaporites, which include salts of more than 2,000 m (6562 ft) in thickness in some areas (Le Pichon, 1988). Discovery of these anhydrites during the DSDP (Ryan et al., 1973) triggered controversies around what is known as "the Messinian Salinity Crisis". Many attempts have been made to explain the origin of these enigmatic evaporites encountered in such a deep basin like the Mediterranean Sea (e.g. Ryan et al., 1973, Cita, 1982, Sonnenfeld, 1975, Fabricius, 1984, Roveri & Manzi, 2006, Roveri et al., 2006). The origin of evaporites in the Gulf of Suez and the Red Sea Basin is also controversial, but the connection between the processes which led to the formation of these evaporites in the two areas is rarely made. A comparative study of the two evaporitic basins may prove to be helpful in solving some of the problems.

The Messinian sediments encountered in all the wells in the Pelagian Sea contain more than 10 m (33 ft) of anhydrite (Jongsma et al., 1985), and they are overlain unconformably by open-marine Pliocene sediments. On seismic sections, the Pliocene-Quaternary sediments have low velocities and the Miocene evaporites have high velocities (Winnock & Bea, 1979). For this reason, the top of the Messinian represents a good seismic marker, known as the "M" reflector (Ryan et al., 1973). Over large areas on the Pelagian Platform, the "M" reflector is absent and the Pliocene-Quaternary sediments are reduced in thickness (Jongsma et al., 1985). Erosional channels are observed in the "M" reflector on the southern flank of the Pelagian Platform. Erosional effects of a river system at the same level in the Tripolitania and Jarrafa troughs are also believed to have taken place at the end of the Messinian. The "M" reflector in many places in the Levant Sea marks the top of the evaporites, but where the evaporites are absent; it denotes an erosional unconformity (Ben Avraham & Mart, 1981).

Major post-Miocene tectonics are reflected in the topography of the "M" reflector in the Central Mediterranean. An extensional phase at the end of the Miocene was followed by a compressive phase at the end of the Pliocene, followed by another extensional phase in the Quaternary (Bellaiche & Blanpied, 1979). E-W dextral wrenching took place in the Ionian Sea (Jongsma et al., 1985). Dense block faulting affected the Pelagian Platform (Burollet, 1979, 1991, Winnock, 1981). This faulting was accompanied by basic volcanism, which resulted in the formation of the Pantellaria and Linosa islands. More than 1000 m (3281 ft) of Pliocene-Quaternary sediments accumulated in the Tripolitania Basin. These sediments are probably the result of continental or shallow-marine deposition (Winnock & Bea, 1979).

The Western Mediterranean (Alboran Sea) was a land area in early Tertiary times and was subject to strong volcanic and seismic activities. The area was dominated by convergence and compression during the Alpine/Atlas Orogeny, which lead to the formation of the Atlas Mountains and Rif domains. The Western Mediterranean basins, such as the Alboran, Balearic and Tyrrhenian basins were formed during the Neogene. Their origin was attributed to vertical foundering (Laubscher & Bernoulli, 1977) associated with rifting (Hsü, 1972). The Maghrebian Cenozoic Orogenic Domain consists of two different systems (Benaouali-Mebarek et al., 2006): The Tell-Rif or Maghrebides and the Atlas, extending from Morocco to Algeria and Tunisia as a result of the closure of an oceanic domain (Fig. 2.1).

Communication between the Mediterranean and Red Seas took place during the Lower and Middle Miocene (Hughes et al., 1992), and ceased in the Pliocene as a result of the uplifting of Sinai.

The ancestral Nile Delta has been active since the Oligocene and migrated eastward during the Miocene, but deltaic sediments of the present-day delta were deposited only from the latest Miocene time onward (Said, 1981, Harms & Wray, 1990). The Middle Oligocene and earliest Miocene were times of low sea level and erosion in the Nile Delta region. This phase led to the development of an extensive erosional unconformity (6 Ma hiatus) in the south of the delta. Marine transgression took place southward only in Burdigalian. The platform margin trends E-W and is located in the Central Nile Delta (Harms & Wray, 1990). Another hiatus occurred during the early Middle Miocene (Serravallian-Tortonian), followed by subsidence during the late Tortonian and early Messinian, especially in the Eastern delta.

The present-day Nile Delta extends for a long distance offshore Egypt in the form of an asymmetric cone, known as the Nile Cone (Biju-Duval et al., 1974) (Fig. 2.1). Recent sediments are relatively thin in the east (Damietta Fan) and pinch out in the north on the Erathostheus Plateau. A thick Pliocene-Pleistocene deltaic wedge was deposited when sea level was restored to near-present level (Harms & Wray, 1988). These sediments thicken to more than 2000 m (6560 ft) to the west in the Rosetta Fan and gradually thin northward on the Mediterranean Ridge (Biju-Duval et al., 1974). Large folds, developed between the Rosetta Fan and the Mediterranean Ridge, are probably due to Messinian salts. On the Mediterranean Ridge, Messinian evaporites are overlain by Pliocene-Quaternary pelagic marls, clays, silts, and sandstones interbedded with volcanic ash (Ryan et al., 1973, Biju-Duval et al., 1974). Major gas discoveries have been made in the Nile Delta (Abu El -Ella, 1990, Abdel Aal et al., 2000, Samuel et al., 2003, Sharp & Samuel, 2004), the offshore deep marine slope channels of the Nile Delta, such as the Abu Qir, Sequia, King, Max, Scarab, Saffron, Sapphire, Serpent, and Simian in the Western Delta (Sharaf, 2003, Samuel et al., 2003, Sharp & Samuel, 2004, Whaley, 2008) and the Ha'py, Denise, Temsah, Karous, and Darfeel fields offshore the Eastern Nile Delta (Heppard & Albertin, 1998, Barsoum et al., 1998, Abdel Aal et al., 2000, Sharaf, 2003).

The Pliocene-Quaternary succession, between 1000–2000 m (3281–6562 ft) in thickness in the SE Mediterranean is predominantly detrital and is composed of marls, shales, and sandstones. Its major source is the sediments of the Nile (Mart & Ben Gai, 1982). This succession forms a sedimentary prism along the continental margin of the SE Mediterranean. It overlies a thick Miocene succession between 1000–1500 m (3281–4922 ft) in thickness (Biju-Duval et al., 1974). Diapirs are common (Ben Avraham & Mart, 1981) as a result of loading of the evaporitic sequence.

## 2.3 NILE BASIN

The Nile Basin (Fig. 2.3) corresponds to the "Stable Shelf" of Said (1962a) and partly to the Nile Basin of Kostandi (1963) and the Assiut Basin of Klitzsch et al. (1979), Hendriks et al. (1990), and Hermina (1990). It covers the southeastern part of Egypt and Southern Sinai. It is bordered in the east and south by the Precambrian basement complex. In the west and north, it is bounded by the Northern Egypt Basin. The Nile Valley runs in a N-S direction in the middle of the basin. The basin-fill is dominated by Cambrian to Eocene continental and shallow marine sediments up to 4000 m (13,124 ft) in thickness, and is composed of sandstones, shales, marls, and limestones. Tectonic elements consist mainly of folds and mild flexures bounded by faults. The boundary between the Nile Basin and the Northern Egypt Basin is marked by NE-SW-trending domal structures, such as the Abu Roash, Bahariya, and Kharga uplifts, and multi-directional faulting. This boundary corresponds to the Trans-Africa Lineament (Nagy et al., 1976) and the Pelusium Shear (Neev, 1977). This major NE-SW lineament is associated with the conjugate NW-SE Tibesti Lineament (Guiraud et al., 2000), which extends from Southwestern Algeria to Kenya with a possible Proterozoic origin. The Nile Basin is divided into a number of sub-basins separated by arches. A number of small basins have been also identified in the present work. These basins are designated informally as the Esna Basin with thick Cretaceous sediments, the Mut Basin with a thick Danian shale section, and the Farafra Basin with a thick Landenian shale section. However, the limits and extent of these elements are not clear. Oil and gas discoveries were made in the Beni Suef and Kom Ombo basins (Dolson et al., 2002), located at the northern and southern edges of the basin as delineated by Tawadros (2001).

## 2.4 NORTHERN EGYPT BASIN

This is a marginal basin, which represents a transitional zone between the North African and the Mediterranean Basin. It corresponds to the "Unstable Shelf" of Said (1962a), which covers most of the northwestern part of Egypt, the Nile Delta, and Northern Sinai (Fig. 2.3). It includes major basins and uplifts, such as the Ghazalat, Tenehue, Qattara, Abu Gharadig, Alamein, Rissu, Shushan, and Fayium (Gindi) basins. Northern Sinai consists of several basins and highs trending NE-SW. It represents the second largest hydrocarbon province in Egypt. Active exploration since the late 1960s lead to the discovery of numerous oil, gas, and condensate fields in the area.

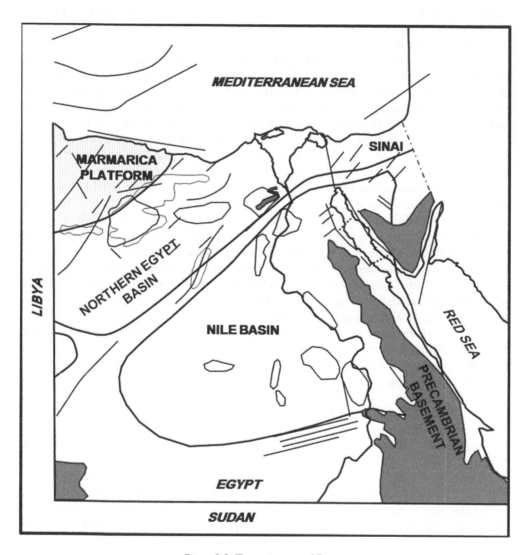

*Figure 2.3* Tectonic map of Egypt.

The onshore of central Sinai, includes the South Yelleg, North Fallig, East Maghara, North Maghara, and NE Sinai basins. The offshore Sinai includes the Walker-Port Fouad, Sadot, Tineh, and Mango basins. Oil and gas were discovered in the Tineh, Port Fouad, Abu Zakin, Walker, South Rafah, and Mango fields (Alsharhan & Salah, 1996). The Northern Egypt Basin is floored by Pan-African basement rocks. The sedimentary section is approximately 183 m (600 ft) in the south and increases to more than 7620 m (25,000 ft) in the coastal areas in the north. The succession is made up of sandstones, shales, and carbonates ranging in age from Cambrian to Tertiary. The northern margin of the Paleozoic Ghazalat Basin is formed by the Umbaraka ridge striking E-NE. The eastern and western boundaries are determined by N-S syn-depositional faults associated with basic intrusives. It is bordered on the west by a regional basement ridge and on the south by the NE-striking Siwa High.

The Northern Egypt Basin and the Sirte Basin in Libya form a triangular area bordered on the southeast by the Tibesti-Kalanshiyo-Bahariya Arch, on the southwest by the Ben Ghanimah-Brak and Gharyan uplifts, and on the northern side by the Jifarah-Cyrenaica-Matruh strike-slip faults (Fig. 2.1). The Egyptian part of the triangle is dominated by NE-SW trending folds and E-W and NE-trending faults, while in Libya it is dominated by NW-SE faults. The Tibesti-Kalanshiyo-Bahariya Arch coincides with the western boundary of the "Trans-Africa Lineament" of Nagy et al. (1976) and its northern extension, the Pelusium Shear of Neev (1977). These lineaments represent a Pan-African zone of crustal weakness within the African-Nubian Shield. This lineament probably acted as a path for the migration of a hotspot (the Darfur hotspot) between the Late Jurassic-Aptian in the Levant and the Late Oligocene-Early Miocene in northwestern Sudan (Garfunkel, 1991). This lineament includes the Bahariya Fault, which may extend southwestwards into the Farafra Oasis, and an inferred fault east of the El Guss Abu Said section. According to Said (1962a) folding in the Bahariya Oasis took place in post-Cenomanian time. Both the southeastern and southwestern borders of the triangle are characterized by Tertiary volcanic activities. The Levant is underlain by a NNE-SSW-trending extensional margin of Late Triassic-Early Jurassic age. The Eastern Mediterranean Basin initially developed as a left-lateral ocean-continent transform boundary separating oceanic crust of the Southern Tethys from the mildly-extended continental crust of Northern Egypt (Longacre et al., 2007).

The Northern Egypt Basin was active throughout the Phanerozoic and connected to the sea in the north. Deposition ended in the Ghazalat Basin during the Strunian (Keeley, 1989) and started at that time in the Tenehue Basin and continued until the Early Permian. The Mesozoic passive margin in the Western Desert region is developed above a broad NW-trending basement arch of Late Carboniferous age (Aadland & Schamel, 1988).

The Abu Gharadig and Fayium (Gindi) basins are separated structurally by the Kattaniya Ridge (Abdallah & Moustafa, 1988). Both basins are bounded on the north by another major ridge associated with Cretaceous and Tertiary faulting. A major platform with a thin sedimentary cover forms the southern margin (Awad, 1984). The Abu Gharadig Basin is an E-W trending structurally-controlled basin developed as a result of deep crustal extensional tectonics that affected northern Egypt during the Mesozoic (Bayoumi & Lotfy, 1989). The stratigraphic sequence includes Early Paleozoic sediments overlain by Mesozoic and Tertiary deposits with more than 2000 m (6562 ft) of fluvio-marine, shallow marine, and deep marine Upper Cretaceous sediments. The basin was initiated in response to the left-lateral motion of Africa relative to Europe during the Late Jurassic, which reached its acme in the Upper Cretaceous, and led to the opening of the Tethys. This phase resulted in the formation of the two E-W listric faults that bound the basin on the northern and southern sides. The development of the basin halted as a result of the dextral shear movement of Africa that led to the closing of the Tethys and basin inversion during the Santonian (Keeley, 1994, Guiraud, 1998, Guiraud & Bosworth, 1997, Bosworth et al., 1999, Guiraud et al., 2005). The basin was uplifted and tilted northward as a result of the NW-SE compressive stresses during the Uppermost Cretaceous-Early Tertiary Laramide Orogeny. Oil has been discovered recently by Apache Oil in the North Qarun Field in the Gindi Basin from the Lower Cretaceous Kharita and Cenomanian Baharia sandstones. The oil pools in the field are associated with two sets of faults, trending NW-SE and NE-SW (Oil & Gas Journal, 1996a). The four Mesozoic subbasins, Shushan, Alamein, Abu Gharadig, and Gindi, contain up to 300 mmbbls of proven reserves.

## 2.5   GULF OF SUEZ & RED SEA

The Red Sea region (Fig. 2.4) was a stable craton from Early Paleozoic to Late Eocene (Plaziat et al., 1990). Compressive strike-slip faulting was initiated during the Oligocene.

The Gulf of Suez is an elongated NW-SE-trending trough, extending over 320 km in length, and bounded by the Pan-African basement on both sides. The Gulf of Suez is divided into three provinces (Moustafa, 1976): A Northern Province, a Central Province, and a Southern Province. These zones are separated by transform faults or "hinge zones" (accommodation zones of Bosworth & McClay, 2001 and Younes & McClay, 2002) and are characterized by a change in the general structural dip. They can be studied in outcrops in the Gebel Gharamul area (Coffield & Schamel, 1989). The Northern and Southern provinces dip to the southwest, whereas the Central province dips to the northeast. The Northern and Central Provinces are separated by the Zaafarana Hinge Zone. The Morgan Hinge/Accommodation Zone (Younes & McClay, 2002) separates the Central Province from the Southern Province. The three provinces are termed the Ataqa, Gharib, and Zeit Zones, respectively. A third zone in the Northern Red Sea, the Duwi Accommodation Zone, has been described by Younes & McClay, 2002). The surface expression of the accommodation zones, as in the Gebel Gharamul region, is dominated by a basement promontory, where a south-west-dipping half-graben and a northeast half-graben intersect (Coffield & Schamel, 1989). The draping over of the pre-rift and synrift successions over these fault blocks results in an inhomogeneous stratigraphic succession within the accommodation zones. The structure of the Gulf of Suez area is governed by normal faults and tilted blocks whose crests constitute the main target for hydrocarbon exploration. The fault pattern consists of two major sets of trends (Colleta et al., 1988); longitudinal faults parallel to the rift axis, created in an extensional regime during Neogene time, and transverse faults with N-S to NE-SW dominant trend, which are inherited passive discontinuities in the Precambrian basement. The Precambrian basement displays a system of regional deep-seated, left-lateral, NW-oriented faults and shear zones (Younes & McClay, 2002). One of these is the Najd Shear Zone, which extends for a distance of 1200 km in a NW-SE direction from Saudi Arabia to the Central Eastern Desert of Egypt (Sultan et al., 1988). The Zaafarana Accommodation Zone separates the SW-dipping Darag half-graben in the north from the NE-dipping October graben to the south. When the rift faults transect the pre-existing Precambrian shear zone, the polarity (or dip direction of fault blocks) changes (Younes & McClay, 2002). The Suez vertical rift movements were explained by two hypotheses: a) the advection of a hot mantle away from the center of the rift combined with regional stretching of the lithosphere (Chenet et al., 1987). This advection was responsible for the uplift of the rift shoulders. At present, the Lower Miocene coarse clastics near the border faults of the rift are 300 m (984 ft) above sea level. b) Large doming with erosion before

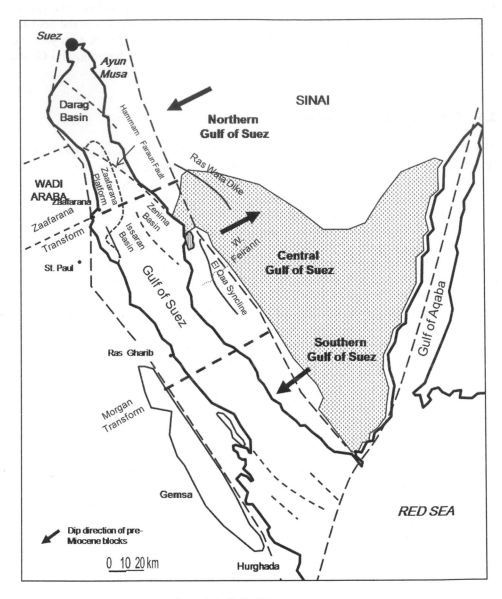

*Figure 2.4* Gulf of Suez structure.

the rift. However, the presence of the Eocene deposits in the center of the Gulf rules out this hypothesis (Chenet et al., 1987).

The stratigraphic section of the southern part of the Gulf of Suez (Fig. 2.5) consists of pre-rift, proto-rift, syn-rift, and post-rift sequences bounded by unconformities (Garfunkel & Bartov, 1977, Orszag-Sperber et al., 1990, Schütz, 1994). The pre-rift sequence unconformably overlies the Precambrian basement. It is approximately

| RIFTING PHASE | ERA | PERIOD | FORMATION |
|---|---|---|---|
| Post-Rifting | Tertiary | Post-Miocene | Zaafarana Wardan |
| Syn-Rifting | Tertiary | Miocene | Zeit |
| Syn-Rifting | Tertiary | Miocene | South Gharib |
| Syn-Rifting | Tertiary | Miocene | Belayim |
| Syn-Rifting | Tertiary | Miocene | Kareem |
| Syn-Rifting | Tertiary | Miocene | Rudeis |
| Syn-Rifting | Tertiary | Miocene | Nukhul |
| Syn-Rifting | Tertiary | Oligocene | Red Beds |
| Pre-Rifting | Tertiary | Eocene | Thebes |
| Pre-Rifting | Tertiary | Paleocene | Esna |
| Pre-Rifting | Mesozoic | Upper Cretaceous | Sudr |
| Pre-Rifting | Mesozoic | Upper Cretaceous | Duwi |
| Pre-Rifting | Mesozoic | Upper Cretaceous | Matulla |
| Pre-Rifting | Mesozoic | Upper Cretaceous | Wata |
| Pre-Rifting | Mesozoic | Upper Cretaceous | Abu Qada |
| Pre-Rifting | Mesozoic | Upper Cretaceous | Raha |
| Pre-Rifting | Mesozoic | Lower Cretaceous | Malha |
| Pre-Rifting | Mesozoic | Jurassic | |
| Pre-Rifting | Mesozoic | Triassic | Qeseib |
| Pre-Rifting | Paleozoic | Permian | Rod el Hamal |
| Pre-Rifting | Paleozoic | Upper Carboniferous | Abu Durba |
| Pre-Rifting | Paleozoic | Lower Carboniferous | Um Bogma |
| Pre-Rifting | Paleozoic | Ordovician | Naqus |
| Pre-Rifting | Paleozoic | Cambrian | Araba |
| Pre-Rifting | Pre-Camb. | | |

*Figure 2.5* Stratigraphic column of the Gulf of Suez.

914 m (3000 ft) in thickness. The sequence consists of Early Cretaceous non-marine, well sorted, arkosic and kaolinitic sandstones, 152–488 m (1500–1600 ft) thick, Turonian-Late Senonian marine deposits, approximately 229 m (750 ft) thick made up of limestones, shales, and sandstones, and Maastrichtian chalks. These rocks

are overlain unconformably by Paleocene calcareous marine shales and Ypresian cherty limestones. The facies constitute a platform opening northwards to the Tethys (Orszag-Sperber et al., 1990). The pre-rift sequence is separated from the Proto-rift sediments by an angular unconformity. The proto-rift sediments are of Late Oligocene-Early Miocene age, and they are localized within the earliest grabens. The syn-rift sequence is variable in thickness and reaches up to 3657 m (12,000 ft) in low areas. It is composed of Lower-Middle Miocene open-marine clastics and Middle to Upper Miocene evaporites. The sedimentary sequence is cyclic, reflecting episodic uplift, erosion, and gradual in-filling of the basin. Fine-grained clastics are dominant in the deeper troughs, shallow marine reefal carbonates on the crests and submarine highs, and sands in the marginal seas (Hughes et al., 1992). The Lower Miocene Gharandal Group, which includes the Abu Zneima, Nukhul, and Rudeis formations represent the syn-rift sequence (Winn et al., 2001, Malpas et al., 2005). It overlies the early syn-rift Oligocene redbeds and conglomerates of the Tayiba Formation exposed at Wadi Tayiba on the eastern side of the Gulf of Suez (Jackson, 2008). The Gharandal Group starts with redbeds and basic volcanics of the Abu Zneima Formation, followed upward by conglomerates, sandstones, and carbonates of the Nukhul Formation and the *Globigerina* marls, sandstones, and conglomerates of the Rudeis Formation (Jackson et al., 2005, Malpas et al., 2005). The post-rift sequence is composed of post-Miocene sands and limestones, separated from the syn-rift sequence by an unconformity (Orszag-Sperber et al., 1998). During the Pliocene, the rift opened to the Indian Ocean and connection with the Mediterranean ceased.

## 2.5.1   Evolution of the Red Sea and the Gulf of Suez

The evolution of the Red Sea, the Gulf of Suez, and the Gulf of Aqaba rifts has been discussed by scores of workers and is still in much debate. Some of the main points of controversy are: 1) Whether structure is the product of strike-slip faulting (Makris & Rihm, 1991, Girdler, 1991a & b, Rihm et al., 1991, Makris et al., 1991) or normal faulting (Heybroek, 1965). 2) The amount of offset along the rifts (Makris & Rihm, 1991, Girdler, 1991a & b, Rihm et al., 1991, Makris et al., 1991, Mart, 1991). 3) Whether displacement was continuous (Bartov et al., 1980) or occurred in more than one discontinuous phase (Quenell, 1958, Makris & Rihm, 1991, Girdler, 1991a & b). 4) The nature of the crust underlying the rifts and whether it is oceanic (Makris & Rihm, 1991, Girdler, 1991, Rihm et al., 1991, Makris et al., 1991) or continental with limited oceanic crust (Morgan, 1983). 5) Inherited from these problems are the tectonic models of the evolution of the Red Sea. For example, one model suggests a pull-apart basin with the formation of a new oceanic crust (Makris & Rihm, 1991, Girdler, 1991, Rihm et al., 1991, Makris et al., 1991). Another model proposes continental stretching accompanied by intrusion of volcanics (Coutelle et al., 1991).

The tectonic history of these rifts can be summarized as follows:

1   The Red Sea region in Egypt was a stable craton from Early Paleozoic to Late Eocene (Plaziat et al., 1990), probably forming the eastern extension of the Nile Basin. The Sinai was a part of Arabia and the Red Sea Rift was already a zone of structural weakness or a suture zone during the Pan-African Orogeny approximately 600 Ma (Harris & Gass, 1981, Makris & Rihm, 1991).

2    Rifting and magmatism started in the southern Red Sea around 30 Ma ago (lower Oligocene) and propagated in the central Red Sea around 25 Ma ago (upper Oligocene) (Perry & Schamel, 1990, Bohannon & Eittreim, 1991, Makris & Rihm, 1991, Girdler, 1991a & b). As rifting began, very thick Miocene salt and evaporite sections accumulated in the trough.

3    The geometry of the initial rift line was controlled by the combination of extensional stress fields due to the anticlockwise rotation of Arabia with the pre-existing structures. Lineament analysis of the bathymetry in the northern Red Sea axial valley and reflection seismic show good correlation with the bordering eastern and western continental plates (Coutelle et al., 1991).

4    In the Red Sea the directions of the master faults and plate motion are oblique to one another which led to the development of pull-apart basins (Makris & Rihm, 1991). Pull-apart basins form where the direction of motion of continental plates differs from those of the master fault (Mann et al., 1983). The amount and rate of creation of oceanic sea floor decrease from south to north consistent with the anticlockwise rotation of Arabia (Girdler, 1991). The two sides of the Red Sea display two different tectonic processes. The formation of pull-apart basins and oceanization took place on the western flank. Stretching occurred on the eastern flank. Therefore, the eastern and western flanks are asymmetrical (Makris & Rihm, 1991). The thickness of the eastern cratonic continental crust thins eastwards from more than 35 km to less than 20 km over a distance of 50 km. The western continental crust on the other hand changes abruptly eastward into an oceanic crust less than 10 km thick (Rihm et al., 1991, Makris et al., 1991).

There is a very strong support for the presence of an oceanic floor in the Red Sea. In the northern Red Sea, marine magnetic anomalies suggest the occurrence of sea floor spreading (Girdler, 1991a & b). The Bouguer gravity anomalies (Girdler, 1991a & b, Makris et al., 1991) are zero to +20 mGal near the coast and increase to a maximum of 60 to 100 mGal over the deep water suggesting an oceanic crust. The seismic refraction data (Makris et al., 1991) indicate that the continental crust thins rapidly toward the coast and high velocities are present beneath the main axial trough (Girdler, 1991a & b). Heat flow systems show values 10 times higher than average in the center of the rift and 2 times higher on the flanks (Makris et al., 1991).

5    Intensified rifting and magmatism started in the northern Red Sea/Gulf of Suez around 20 Ma (Burdigalian), followed by extensional shoulder uplift and basin subsidence (block faulting) in the Gulf of Suez (Garfunkel & Bartov, 1977). A series of NW-SE-trending diabase dikes occur in southeastern Sinai (Bartov et al., 1980, Eyal et al., 1991, Baldridge et al., 1991, Stern et al., 1988). The dikes are tens to hundreds of kilometers long and commonly occur as swarms. The dikes are intersected and laterally offset by a series of N-S trending faults. The age of these dikes is approximately 20 Ma (Baldridge et al., 1991). Magmatism was followed by a phase of subsidence and extension in the Gulf of Suez between 19 and 16 Ma (Burdigalian) (Chenet et al., 1987, Moretti & Chenet, 1987, Moretti & Colleta, 1987, 1988). Large doubly-plunging folds formed in pre-rift strata during the Late Oligocene-Miocene extension of the Red Sea rift, such as the asymmetric syncline of the Duwi and Hamadat areas near Safaga (Khalil & McClay, 2002).

6   Approximately 14 Ma ago (Serravallian) the strike-slip movements switched from the Gulf of Suez to the direction of the Gulf of Aqaba-Dead Sea Rift (Ten Brink & Ben Avraham, 1989). Quennell (1958) suggested that the two series of faults on both sides of the Dead Sea Rift were offset 107 km sinisterly by the rift boundary fault, and argued that horizontal displacement took place in two steps. During the first (latest Oligocene/early Miocene) the movement was 62 km, and the second (Pliocene-Pleistocene) was 45 km. However, the 107 km of sinistral offset along the Gulf of Aqaba-Dead Sea Rift was challenged by Mart (1991) who argued that when the depositional facies and isopach contours are interpolated across the Rift, they appear to be continuous and indicate that lateral displacement was only 10 km. Bartov et al. (1980) and Eyal et al. (1981) suggested that left-lateral displacement of 105–110 km between Sinai and Arabia occurred in one continuous tectonic phase starting in the Middle Miocene.

7   At the end of the Miocene, the Strait of Bab al Mandab opened and waters from the Indian Ocean flooded the Rift ending the evaporite basin (Girdler, 1991a & b).

8   The last phase of evolution of the Red Sea is the seafloor spreading which commenced 5 Ma ago (Pliocene) in parts of the central and southern Red Sea and is still propagating northward. The exposed part of the rift was extensively intruded by gabbroic to dioritic plutons and dyke swarms 5-2 Ma ago (Bohannon & Eittreim, 1991). Pliocene-Pleistocene oozes were deposited on top of the Miocene evaporites.

## 2.6   CYRENAICA/MARMARICA PLATFORM

The Mediterranean margin of Libya consists of three physiographic provinces, the Pelagian Shelf, the Sirte Embayment, and Offshore Cyrenaica. Cenozoic strata along much of the Libyan margin have a progradational character punctuated by surfaces of erosion and margin failure, seismically coincident with the top of Messinian unconformity (Fiduk, 2009).

The Cyrenaica platform in Northeast Libya (Fig. 2.6) is separated from the Sirte Basin to the west by an arcuate fault (Antelat Fault) of Tertiary age (Burollet et al., 1971, Goudarzi, 1980) and from the Mediterranean Basin by the Jabal Akhdar uplift. The boundary between the Jabal Akhdar and the Cyrenaica platform is marked by a deep-seated fault which rises up to 882 m (3894 ft) above sea level (Röhlich, 1974). It extends eastward into Egypt under the name of Marmarica Plateau. The geology of the Jabal Akhdar and Cyrenaica was studied by Gregory (1911), Klitzsch (1970), Röhlich (1974, 1980), Klen (1974), Zert (1974), and El-Hawat & Shelmani (1993). The subsurface geology has been reviewed by El-Arnauti & Shelmani (1988), Thusu et al. (1988), Grignani et al. (1991), and Hallett (2002).

The Cyrenaica platform was the site of a marginal basin (the Al Jabal Al Akhdar Trough) with accumulations of thick sediments ranging from Paleozoic to Tertiary and probably formed a part of the Northern Egypt Basin. Up to 3000 m (9843 ft) of Mesozoic marine sediments have accumulated in the Jabal Akhdar Trough (Klitzsch, 1970). These rocks are mostly limestones, with some sandstones and shales. They include turbidites, debris flows and slumps. Deep-marine shales accumulated

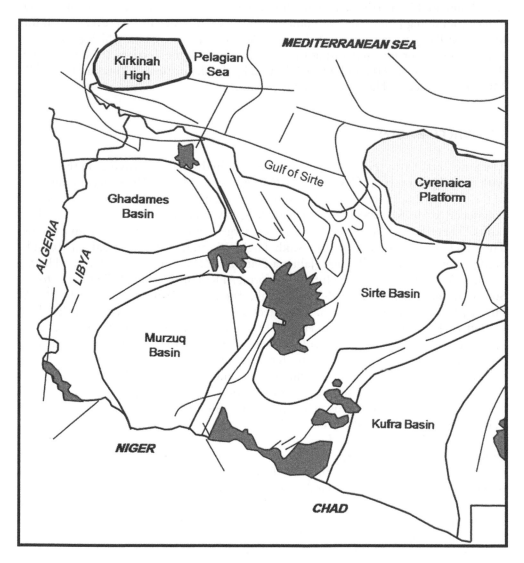

*Figure 2.6* Tectonic map of Libya.

mostly in the basin to the north. The exposed section in the Jabal Akhdar ranges in age from Cenomanian to Eocene. Röhlich (1974, 1980) divided this geologic section into two depositional cycles: The first cycle includes Cenomanian to Coniacian sedimentation which was ended by an orogenic phase during the Santonian, leading to basin inversion. Stresses produced high-angle faults striking NE-SW, E-W and NW-SE, and an ENE-WSW-striking arch. The latter emerged as an island during the Santonian and it was deeply eroded. North of the arch, deposition continued uninterrupted. The second cycle includes Campanian to Landenian sediments. Deposition was dominated by neritic sediments during Late Senonian time and by chalky limestones containing planktonic foraminifera during the Paleocene. This cycle ended following an Early Eocene uplift of the area and subsequent erosion which produced a regional unconformity separating the Cretaceous from the Tertiary (Röhlich, 1980, Barr & Berggren, 1980), and it was followed by the deposition of Ypresian to Middle Miocene sediments. The latter form the plateau.

## 2.7   SIRTE BASIN

The Sirte (Sirt) Basin (Fig. 2.6) is the sixth largest hydrocarbon province in the world, with oil and gas reserves of 33–45 BBOE (Macgregor & Moody, 1998). It contains most of the hydrocarbon reserves in Libya, and approximately 29% of Africa's total reserves (Chatelier & Slevin, 1988). It consists of NW, EW and NE-SW arms with attendant faults. These faults divide the basin into a large number of platforms and basins. They include the Amal Platform-Rakb High, Maragh Trough, Ajdabiya Trough, Suluq Basin, Hameimat Trough, Jahama Platform-Zelten Platform, Marada (Hagfa) Trough, AzZahra (Dahra)-Al Hufrah Platform, Al Kotlah Trough, Al Bayda Platform, Southern Shelf, Dur Al Abd-Zellah Trough, Tumayyim Trough, Waddan Uplift, Al Hulayq High, and Hun Graben. These tectonic elements are described in detail by Hallett & El-Ghoul (1996), Anketell (1996), and Abadi et al (2008). Sediments in the Sirte Basin range in thickness from 1500–7000 m (4921–22,966 ft) (Goudarzi, 1980), and range from Cambrian to Tertiary. The intersection of faults of different orientations created a complex system of horsts and grabens, which controlled hydrocarbon accumulations and source rock distribution. In the Sirte Basin, oil is dominant, gas occurs only in three fields. The oil has 32°–44° API, and a sulphur content of 0.15–0.66% (Parsons et al., 1980).

The Sirte Basin is bounded on the west side by the Hun (Hon) Graben, a NW-SE elongated graben, which appears as a huge furrow cutting the present-day relief (Cepek, 1979), and separates the Sirte Basin from the Hamada al Hamra Platform (Ghadames Basin). The Hun Graben is one of the youngest tectonic elements and the only outcropping graben in the Sirte Basin. The graben is floored with basement schists and it contains about 1500 m (4922 ft) of sediments ranging in age from Triassic to Oligocene (Anketell, 1996, Hallett & El-Ghoul, 1996). The fault zone on the west boundary of the graben consists of several sub-parallel faults accompanied by brecciated zones. The nature and origin of the Hun Graben is controversial. Klitzsch (1967, 1970) assumed that the block of the Hun Graben sunk 500 m to 800 m (1641–2625 ft) against the Hamada al Hamara block. Cepek (1979), on the other hand, estimated a throw of 100 to 120 m (328–394 ft). Klitzsch (1967, 1970) suggested that the tectonic development of the Hun Graben started prior to the Late Cretaceous. He also postulated that the western boundary of the graben is Paleocene in age. However, Cepek (1979) inferred that the graben is probably Oligocene in age.

The Zellah Trough contains over 3600 m (11,811 ft) of sediments (Hallett & El-Ghoul, 1996). It is bounded to the west by the Waddan Uplift and to the east by the AzZahrah-Al Hufrah Platform. The Zellah Trough is divided into several sub-basins, such as the Facha Graben, Ar Ramlah Syncline, Ayn An Naqah subbasin and Maamir Graben (Schröter, 1996, Johnson & Nicaud, 1996). Northwards it extends into the Dur al Abd Trough and southwards it connects with the Tumayyim Basin. The southern part of the Zellah Trough is floored with basement volcanics and the northern part with quartzites. The Tumayyim Basin contains over 3000 m (9843 ft) of sediments overlying basement metamorphic rocks. The Zellah Trough did not develop before 60 MM years ago (Gumati & Nairn, 1991).

The AzZahrah-Al Hufrah Platform is separated from the Al Bayda Platform by the NNE-trending Kotlah Graben. The eastern edge of the AzZahrah-Al Hufrah Platform is marked by the Manzila Ridge. It remained as a land area until the onset of the Cenomanian transgression where sandstones were deposited (Barr & Weegar, 1972). Isolated parts of the platform remained as land areas during the Santonian-Coniacian and Maastrichtian marine transgressions. Carbonates and shales are the dominant lithologies on the platform. The Kotlah Graben contains a thick Upper Cretaceous (Cenomanian-Maastrichtian) section of clastics, evaporites, and shales. The Al Bayda Platform is connected to the Southern Shelf by a NW-SE-trending horst (Sinha & Mriheel, 1996).

The Hagfa Trough (also known as Maradah Trough or Kalash Trough) is a deep, narrow graben, located between the AzZahrah-Al Hufrah and the Jahamah platforms in the north, and between the Al Bayda and Zelten platforms in the south. It is bounded in the offshore by the As Sidrah High. Displacement on the faults in the centre of the basin reaches 1800 m (5906 ft) (Hallett & El-Ghoul, 1996). The Hagfa Trough contains a sedimentary sequence of more than 6000 m (19,685 ft). The succession is composed of Cambro-Ordovician quartzites, Upper Jurassic-Early Cretaceous sandstones, Cenomanian sandstones and shallow-water carbonates, Turonian evaporites, with more than 295 m (968 ft) of halite in the southern part of the trough, Coniacian-Santonian shales, limestones, and dolomites, deep-water Campanian-Maastrichtian organic-rich shales, and deep-water Maastrichtian-Paleocene argillaceous carbonates. The trough was active between 60 and 40 MM years ago (Gumati & Nairn, 1991).

The Ajdabiya Trough (also known as the Maragh Trough) is the largest and deepest of the Sirte Basin troughs. It contains over 6000 m (19,685 ft) of sediments, including 2000 m (6562 ft) of Miocene rocks. Upper Jurassic-Early Cretaceous sandstones were encountered in places (Anketell, 1996). The trough was relatively inactive during the Paleocene, but was reactivated again during the Eocene (Hallett & El-Ghoul, 1996). The Ajdabiya Trough is bounded to the west by the Assumood Ridge, and further south by the Zelten Platform. To the east, it is flanked by the Cyrenaica Ridge and the Amal High. Southwards it terminates against the Messlah-Sarir High. Northwards it widens into the offshore and onto the Sirte Rise (Del Ben & Finetti, 1991). The Wadayat Trough is a small asymmetric half-graben extending westwards from the Ajdabiya Trough to the south of the Assumood Ridge, from which it is separated by a major fault. The trough contains over 4500 m (14,764 ft) of sediments.

The Hameimat Trough in the eastern part of the Sirte Basin is located between the Amal and Messlah highs. Eastwards it connects with the Jaghbub-Siwa Basin through the Abu Attifel Subbasin (El-Arnauti & Shelmani, 1985) and westwards into the Ajdabiya Trough. The trough is floored by Precambrian basement rocks, which are overlain by more than 2500 m (8200 ft) of Upper Jurassic-Lower Cretaceous to Oligocene rocks, and possibly includes Permo-Triassic rocks (Thusu, 1996, Gras, 1996). The trough contains several of the largest oil fields in Libya, such as the Jalu, Sarir, and Abu Attiffel fields, as well as the An Nafurah and Awjilah fields. The Abu Attiffel Basin is an ENE-WSW depression bounded by several pre-Mesozoic basement highs, such as the Al Jaghbub, Kalanshiyu and Majid uplifts. A system of syn-sedimentary listric normal faults has affected the sedimentation at different times (Rossi et al., 1991). The Sarir Trough (equivalent to the Qattarah Graben, El-Arnauti & Shelmani, 1985) is a relatively shallow trough located between the Messlah and Sarir highs. It contains about 3800 m (12,467 ft) of sediments. The trough is floored with Pre-cambrian basement rocks, and by a variable thickness of Nubian Sandstones. The southern margin of the granitic and metamorphic Messlah High is dissected by an *en échelon* fault pattern caused by a right-lateral sense of shear contemporaneous with basement controlled extensional faults (Gras, 1996). The Hameimat Trough may bear genetic relationships with the northern Western Desert basins, such as the Abu Gharadig Basin.

## 2.7.1   Evolution of the Sirte Basin

Because of the absence of deep seismic refraction data, one can only hypothesize about the origin of the Sirte Basin. Since Harding (1984) divided the sedimentary fill of the Sirte Basin into pre-graben, syn-graben, and sag basin successions, several attempts have been made to refine these subdivisions. There is no general agreement on what constitutes the pre-rift and syn-rift sequences. Disagreements are probably due to the multiple rifting stages that led to the formation of the Basin. For example, the Nubian sandstones (Late Jurassic-Early Cretaceous) were considered as pre-rift by Harding (1984), Baird et al. (1996), and van der Meer & Cloetingh (1996), and syn-rift by Gras (1996), Schröter (1996), Tawadros et al. (2001), among many others. It is possible that the Sirte Basin area went through multiple stages of rifting since the Middle Triassic and that different parts of the basin acted differently at different times. Only in the Upper Paleocene-Eocene time the whole basin acted as one unit, and showed the highest rate of thermally-driven subsidence (downwarping) (Gumati & Kanes, 1985, Gumati & Nairn, 1991, Schröter, 1996, Bender et al., 1996, Baird et al., 1996). This phase of subsidence and burial was the major process of hydrocarbon generation in the Sirte Basin (Bender et al., 1996).

The Sirte Basin, with its three arms, has been viewed by many workers as a failed triple-junction system. This model has been opposed by Baird et al. (1996) because of their belief that the southeast sector of the basin, i.e. the Hameimat-Sarir Basin, is older than the two other sectors. The three arms, directed NW-SE, SW-NE, and E-W, correspond to three pre-existing arches, the Sirte Arch, the Tibesti Arch and the Kalanshiyo Arch, respectively. The meeting point of these three arms is probably the Baydah Platform, the centre of the Al Haruj al Aswad and the Jabal asSawda volcanics. The NW-SE arm is on trend with the South Sudan rift basins (Schull, 1988).

In spite of recently published extensive works (including three volumes in 1996), the origin of the Sirte Basin is still not very well documented and very little is known about the history of the basin in pre-Cenomanian times. The Precambrian basement structure in the Sirte Basin was probably the main controlling factor on the subsequent tectonic history of the region. NW-SE-oriented pull-apart basins probably formed in the region during the Pan-African Orogeny (Schandelmeier, 1988), followed by extension and the formation of horsts and grabens during the Cambrian (Klitzsch,

1971, Schandelmeier, 1988). Subsidence was accompanied by the deposition of shal-low-marine Late Cambrian sediments and the emplacement of acidic volcanic rocks (Tawadros et al., 2001). This extension phase continued into the Early Carboniferous and it was succeeded by thermal uplift and erosion during the Late Carboniferous and Permian. Thermal uplift was caused by the Hercynian Orogeny, on one hand, and the subduction of the Proto-Pacific oceanic plate underneath the Gondwana continent (Visser & Praekelt, 1996), on the other. These events must have affected the Sirte Basin area but were camouflaged by the later events. There is now ample evidence that Devonian, Carboniferous and Triassic sediments occur in several parts of the Sirte Basin (Thusu, 1996, Wennekers et al., 1996, Tawadros et al., 2001).

Central Libya remained apparently a positive area (the Sirte Arch) until the Early Cretaceous when the arch collapsed. The grabens were filled with predominantly con-tinental sandstones of the "Nubian" type (Tawadros et al., 2001), and remained sub-aerially exposed until the Cenomanian.

The evolution of the Sirte Basin is marked by various rifting phases from the Early Cretaceous to early Eocene (Abadi et al., 2008). The rifting phases accompanied changes in the plate motions of Africa relative to Europe, and the rifting in other Afri-can basins. Subsidence initially occurred in narrow troughs, which gradually widened to incorporate platforms until finally the whole basin subsided at the end of each rift cycle (thermal subsidence).

The origin of the basin can be explained in terms of mantle upwelling following melting due to hotspot activities. Strike-slip faulting and extension of the region took place above a fixed mantle "hotspot" over which Africa rotated in Late Jurassic-Early Cretaceous times (Van Houten, 1983). Asthenospheric upwelling yielded volu-minous basalt magmas (Tatsumi & Kimura, 1991), especially during the Jurassic (Guiraud, 1998). Rhyolites and basalts in the Sirte Basin gave K-Ar dates of 161 Ma and 151 Ma (Bathonian and Tithonian), respectively (Hallett & El Ghoul, 1996). However, these ages should be considered as minimum due to argon (Ar) evapora-tion. Jurassic volcanics are also common in the subsurface of the northern Western Desert of Egypt (Bayoumi & Lotfy, 1989). This upwelling resulted in thermal thin-ning of the lithosphere due probably to subcrustal erosion. The increased volume associated with subcrustal erosion and the heat generation from magma resulted in the doming and expansion of the crust. This process could have produced a crustal uplift of approximately 1–2 km (Arthyushkov et al., 1991) and the formation of the Sirte Arch. This uplift was accompanied by the cracking of the rigid crust in response to the rotation of Africa relative to the European block. Strike-slip movement along the Jifarah-Cyrenaica shear zone associated with the opening of the Mediterranean Sea resulted in the collapse of the Sirte Arch and the formation of the Sirte Basin. The motion was sinistral during the Early Cretaceous (Burollet, 1993, Anketell, 1996). This motion was translated eastwards into the Sirte region. The collapse of the Sirte Arch probably took place in different steps. The initial collapse started in the Triassic in the Eastern Sirte Basin and led to the initiation of the Sarir and Hameimat troughs, followed by the collapse of the main arch during the Lower Cretaceous (Tawadros et al., 2001). Only the northeastern end of the Tibesti arm collapsed at that time. Contemporaneous subsidence of half-graben and pull-apart basins in the Benue area, northern Cameroon, and southern Chad took place (Maurin & Guiraud, 1990, Guiraud, 1990, Guiraud & Maurin, 1991, 1992, 1993). The strike-slip motion of

Africa was followed by an N-E motion until Campanian times (Anketell, 1996). The evidence for this motion is supported by the change of magmatic activities from older to younger along the track of the Darfur hotspot (Garfunkel, 1991) in a southwest direction. A new phase of extension and subsidence, with faulting and graben formation, occurred from Cenomanian to Campanian time (Gumati & Kanes, 1985, Gumati & Nairn, 1991, Schröter, 1996, Baird et al., 1996) when the sea encroached on it (Barr & Weegar, 1972). The formation of horsts and grabens was induced by the first collision between Eurasia and Africa, which led to NW-SE compressional stress and associated SW-NE distension (Burollet, 1993). A quiescent period followed before convergent movement in a northerly direction was resumed in the early Eocene (Dewey et al., 1973). Thermally-driven subsidence dominated (Gumati & Kanes, 1985, Gumati & Nairn, 1991, Schröter, 1996, Baird et al., 1996), and the former structures were sealed by platform sediments (Burollet, 1993). In the Late Eocene, the motion of Africa relative to Europe changed from compressional to right lateral, then into a NW-SE compressional regime in the Early Miocene (van der Meer et al., 1993). The Alpine-related tectonic pulses in the late Eocene resulted in northward tilting of the Sirte Basin (Gruenwald, 2001). A regional unconformity separates the Eocene carbonates from the overlying Oligocene clastics.

## 2.8 MURZUQ BASIN

The Murzuq Basin in Southwest Libya (Fig. 2.6) is a Paleozoic cratonic basin. It is bordered on the north by the Gargaf Arch, on the east by the Tibesti Massif, and on the west by the Akakus-Tadrart Range where Early Paleozoic sediments are exposed. The Precambrian basement of the Hoggar Massif which crops out further to the southwest is separated from the Illizi Basin in Algeria by the Tihemboka Arch.

The basin is floored by Archean and Proterozoic basement rocks of the East Saharan Craton (Vail, 1991, Bauman et al., 1992). The basement is composed of Precambrian high-grade metamorphic rocks associated with plutonic rocks as well as low-grade metamorphic to unmetamorphosed rocks of the Infracambrian Mourizidie Formation (Belaid et al., 2009). Its eastern rim coincides with the boundary between a Precambrian oceanic basin to the west and a platform area to the east (Nagy et al., 1976, Ghuma & Roger, 1980).

During the Caledonian, Hercynian and younger tectonics, the Murzuq Basin was divided into various uplifts and troughs with different orientations (Fürst & Klitzsch, 1963). The basin-fill consists of marine sediments ranging in age from Cambrian to Carboniferous, followed by continental Mesozoic sediments.

The Murzuq Basin contains approximately 600 MMBOE of recoverable reserves (Boote et al., 1998). Reservoirs are provided by Cambrian, Devonian, and Carboniferous sediments, and hydrocarbons originated from Silurian and Devonian source rocks (Lüning et al., 2000c, d). However, Silurian Hot Shales are missing in few wells, such as the A1-NC190 and A1-NC186 (Belaid et al., 2009).

Pull-apart basins, oriented NW-SE probably formed during the later stages of the Pan-African Orogeny in the Early Cambrian (Schandelmeier, 1988, Klitzsch, 1971). From the Middle Cambrian, vertical movements along the old basement faults took place due to the extensional stress regime which dominated North Africa during that time. During the Middle to Late Cambrian times, the area subsided and it was covered by shallow epicontinental seas, where shallow-marine and locally fluvial clastic sediments were deposited (Korab, 1984, Radulovic, 1984).

The area was uplifted in the Lower Ordovician, followed by the formation of hills and valleys, which were later filled with peri-glacial deposits during the Late Ordovician glaciation episode. The Caledonian tectonics acted along a NW-SE axis (Klitzsch, 1963, Fürst & Klitzsch, 1963, De Lestang, 1965). The Brak-Bin Ghanimah High divided the basin into two subbasins after the Ordovician (Fürst & Klitzsch, 1963), the main Murzuq Basin and the Dur al Qussah Basin. The Tihemboka Arch is Middle Devonian in age. Movement on the Arch started during the Givetian-Frasnian (Legrand, 1967, Burollet & Manderscheid, 1967, Burollet et al., 1971) and was submerged in the middle Famennian. In eastern Algeria, the Lower Devonian sediments are reduced in thickness near the Tihemboka Arch and they are absent on top of the Arch (Legrand, 1967). Five main unconformities have been recognized in the Murzuq Basin (Belaid et al., 2009):

1   Alpine and Austrian events: these events occurred during the Cretaceous (Austrian) and during the Tertiary (Alpine) after the deposition of the Mesak Formation.
2   Base Jurassic unconformity: this event is represented by Jurassic sediments unconformable overlying Lower Carboniferous (Marar Formation) in the eastern and north-eastern parts of the Murzuq Basin.
3   Hercynian tectonics: this event greatly affected Carboniferous sediment thickness. Upper Carboniferous (Tiguentourine Formation) is missing as result of erosion.
4   Caledonian event: the Caledonian period is of major importance in the Murzuq Basin and during the uplift of the Al Gargaf Arch. The Silurian section comprises the Tanezzuft and Acacus formations in the eastern part of the basin.
5   Taconian unconformity: this event at the base of the Silurian exerts important controls on the distribution of the Hot Shale source rock. In some parts of the Murzuq Basin, the Ordovician Memouniat sandstones were partially eroded.

The Hoggar Massif to the southwest and the Tibesti range to the east were covered with ice (Beuf et al., 1971), probably during the early parts of the Ordovician. The melting of the ice cap during the Late Ordovician resulted in the deposition of glacio-marine deposits (Radulovic, 1985, Protic, 1985). Epeirogenic uplift and erosion followed.

The Silurian transgression which took place probably during a brief period of tectonic stability led to the deposition of graptolitic shales, which provide source for hydrocarbons in West Libya and Eastern Algeria (Lüning et al., 2000c, d, 2004a, b, Sikander et al., 2003). Epeirogenic uplifting of the area caused a regional regression during the upper part of the Silurian, and the whole area was transformed into dry land by the end of the Silurian. Continental, deltaic and shallow-marine conditions dominated during the Early Devonian. By the Upper Devonian time, the Gargaf Arch was developed and separated the Murzuq Basin from the Ghadames Basin (Seidl & Röhlich, 1984). The two basins were connected through a passage along the western side of the plunging arch. Along the northern shore of the Murzuq Basin, shallow marine conditions resulted in the formation of oolitic ironstones of the Wadi Ash Shati (Goudarzi, 1970, Seidl & Röhlich, 1984). Marine conditions persisted to the end of the Carboniferous until the Hercynian (Variscan) movements converted the area to a dry land, which remained exposed until the Triassic.

The orientation of late Hercynian tectonism was perpendicular to that of the early Caledonian, thereby producing a network of grabens and horst (Klitzsch, 1963, 1971, De Lestang, 1968). In the SW Murzuq Basin (Ghat area) fractures are dominated by a NW-SE (Caledonian) trend in the pre-Silurian sediments and by NE-SW (Hercynian) trend in the Silurian-Devonian rocks (Radulovic, 1985). This event was caused by the collision between Gondwana and Eurasia.

The Tikiumit and the 10°30' strike-slip faults (Protic, 1985) (Fig. 2.6), both are oriented approximately N-S. They were formed during Late Carboniferous (Moscovian) Hercynian movements. These compressional structures resulted in the formation of secondary folds, such as the Bilqaghamr Dome and the NW-SE-trending Tikatanhar Monocline, to the west and east of the Tikiumit Fault zone, respectively.

The Dur al Qussah Basin ceased to exist during the late Hercynian, and the eastern edge of the Murzuq Basin was uplifted to become a mountain chain (Klitzsch, 1963).

The opening of the Tethys Ocean during the Mesozoic and the global relative drop in sea-level (Haq et al., 1987), accompanied by a northward tilting of the African continent, and led to the deposition of extensive siliciclastic deposits in the Murzuq Basin. Continental Mesozoic sediments were deposited by braided streams during the Upper Jurassic-Lower Cretaceous. The Upper Cretaceous marine transgression did not reach the Murzuq Basin. In Tertiary-Quaternary times, escarpments were formed in the eastern side of the basin and the depressions were dominated by continental and brackish-water sediments. This phase was followed by arid conditions which have continued until present time (Korab, 1984).

## 2.9   GHADAMES BASIN

The Ghadames Basin (Al Hamada Al Hamra Basin) (Fig. 2.6) is a cratonic Paleozoic basin located in the northwestern part of Libya and extends into Tunisia and Algeria. It is bordered by the Sirte Basin to the east, the Edjeleh Anticline along the Libyan-Algerian border to the west, the Gargaf (Al Qarqaf) Arch to the south, and the Jifarah (Nafusah) High to the north. The latter is an east-west trending prominent tectonic feature that marks the limit between the Sahara craton and the Jifarah Basin and the central Tunisian trough. The NW-trending Hun (Hon) Graben marks the transition between the Ghadames Basin and the Sirte Basin. The Ghadames Basin is separated from the Murzuq Basin to the south by the east-west trending Gargaf Arch and its SW subsurface continuation, the Atshan or Awbary Saddle (Burollet et al., 1971). The Gargaf Arch was probably uplifted during the Caledonian (or even Taconian) Orogeny and remained a positive feature throughout its geologic history. Only Cambrian sediments, with rare Oligocene deposits are preserved on the crest (Jurak, 1978). The extension of the Brak-Ben Ghanimah High separates the Ghadames Basin from the Zamzam subbasin to the east. The Ghadames Basin continues westwards into Algeria where it opens into the Illizi (Fort-Polignac) Basin. Mesozoic and Paleogene sediments crop out around the periphery of the basin. Tertiary volcanics crop out along the eastern edge of the basin. Sediments vary in thickness between 500–5000 m

(1641–16,405 ft) and range in age from Cambrian to Tertiary. Paleozoic sediments are mostly sandstones and shales in the south, and they include continental, shallow marine and glacio-marine sediments. They grade into deltaic and shallow marine sediments in the north. Dolomites and evaporites are common in the upper part of the section. Mesozoic sediments are predominantly continental sandstones in the south. They thicken northward and become marine sandstones, carbonates and evaporites. In the north, they are overlain unconformably by Tertiary sediments.

The present structure of the basin is the combined effect of the Taconian, Caledonian, Hercynian and Alpine orogenies. Its history follows closely that of the Murzuq Basin.

The Ghadames Basin in Libya and Algeria contains some 160 oil pools with approximately 9.5 BBO in place and recoverable reserves of about 3.5 BBO (Echikh, 1998). Production is predominantly from Cambro-Ordovician, Devonian, and Triassic sediments (Echikh, 1998, Sikander et al., 2003).

## 2.10  KUFRA BASIN

The Kufrah Basin in southeastern Libya (Fig. 2.6) is a NE-SW elongated depression in SE Libya, bordered on the west by Jabal Nuqay (Eghi), on the north by the Jabal az-Zalma uplift, and on the east by the Jabal Arknu-Jabal Oweinat-Bahariya uplift. The latter separates the Kufrah Basin from the Northern Egypt Basin. The sediments are up to 1700 m (5578 ft) in thickness, and reach a maximum of 4500 m (14,765 ft) in the centre of the basin, and range in age from Cambrian to Cretaceous (Lüning et al., 2003a, b, c).

The geology of the Kufrah Basin has been discussed by Bellini & Massa (1980), Bellini et al. (1991), Grignani et al. (1991), Lüning et al. (2000a, b), and Ghanoush & Abubaker (2007). This basin is the least explored of all the Libyan basins. No hydrocarbon discoveries have been made in that basin, although Silurian Hot Shales and potential Ordovician reservoirs have been identified in the Basin (Lüning et al., 2003c, Le Heron & Howard, 2010).

The Kufrah Basin is on trend with the Benue Basin in Nigeria, both of which falls along the Trans-Africa Lineament, associated with lineament swarms (Bellini et al., 1991), and appear to share a common origin. Small pull-apart basins probably formed within the basin during the Pan-African Orogeny (Schandelmeier, 1988). The Pan-African phase was followed by extension and the formation of horsts and grabens during the Cambrian (Klitzsch, 1971, Schandelmeier, 1988). Extension continued into the Early Carboniferous until it was halted by thermal uplift and erosion during the Late Carboniferous and Permian Hercynian Orogeny. Structural elements within the basin are NE-SW oriented faults (Hercynian trend) and strike-slip faults, NNE-SSW synthetic faults, and NW-SE antithetic faults (Bellini et al., 1991). NW-trending fractures and faults appear to have originated from Caledonian tectonics. The Jabal Arknu uplift probably took place in late or post-Carboniferous time and may be related to the Hercynian Orogeny (Goudarzi, 1980). Tectonically, the Kufrah Basin may represent an ancient tectonic plate (Fig. 2.1). It is underlain by Archean and Proterozoic basement rocks of the northern part of the Nile craton (Vail, 1991) and extends southward into Chad (De Lestang, 1965). Subsidence within the Kufrah

Basin during the Lower Cretaceous was related to sinistral strike-slip motion along the Trans-African Lineament. These movements probably resulted in the formation of pull-apart basins similar to and on trend with the Benue Trough in Nigeria (Guiraud, 1998, Wilson & Guiraud, 1998, Guiraud et al., 2005), accompanied by northward tilting of Africa.

## 2.11   JIFARAH BASIN

The Jifarah Plains (Fig. 2.6) is a low-lying area in NW Libya. It rises gradually from sea level along the Mediterranean coast to 200 m (656 ft) at the foot of the Jabal Nafusah (or Jifarah) escarpment, which extends from Jabal Tebaga in Tunisia to Jabal Gharyan in Libya. It is underlain by the Jifarah Basin, which extends westward into the Permian Basin in Tunisia, where a considerable thickness of sediments is known in Tebaga and in the subsurface (Burollet et al., 1971). It is bordered on the north by the Mediterranean and on the south by a pre-Miocene fault and the Jabal Nefusah, the Azizia Fault (Conant & Goudarzi, 1967, Koehler, 1982). The continental margin is abrupt with no transitional zone between the African Platform and the Mediterranean Basin. Basin-fill is between 3000–5000 m (9843–16,405 ft) increasing in thickness toward the Mediterranean (Goudarzi, 1980). The Jifarah Basin was a part of the marginal trough from the Paleozoic to Early Cretaceous. In Late Carboniferous and Permian sedimentation accompanied subsidence in the Jifarah Plains (Burollet et al., 1971). It became a stable platform from Late Cretaceous to Miocene. Downwarping in Miocene time resulted in the deposition of Miocene sediments.

# Chapter 3

# Tertiary volcanics

Tertiary volcanics are common in North Africa (Fig. 3.1). They occur in the Gebel Oweinat-Arknu complex, Jabal Hassawnah, Jabal asSawda, and Jabal Al Haruj al Aswad in Libya, Algeria, Tunisia, and Morocco. They include the Canary Islands in the Central Atlantic offshore Northwest Africa (Figs. 3.1 and 3.2a, b). Volcanic rocks are reported in Egypt also from Kom Ombo, Gulf of Suez, Bahariya oasis, west-central Sinai, Wadi Araba, Gebel Qatrani north of Fayium, and Gebel Khesheb west of Cairo.

Periodical reactivation of Precambrian fracture systems throughout the Phanerozoic supplied conduit for different types of plutonic and volcanic rock assemblages (Meneisy, 1990). During the Late Eocene (41 Ma) to Recent, magmatic activity became widespread on a regional scale and extensive basaltic volcanic occurrences formed in North Africa (Wilson & Guiraud, 1998). These occurrences reflect a change in the plate tectonic regime induced by the alpine collision. They are associated with the opening of the Red Sea and the Gulf of Suez, and with domal uplifts of the basement.

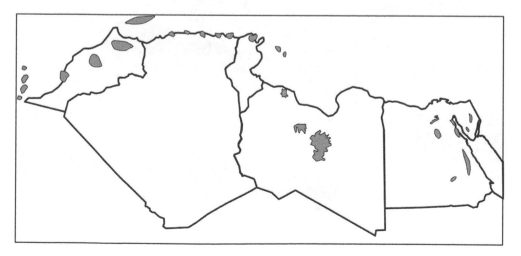

*Figure 3.1* Cenozoic igneous volcanics (after Tawadros, 2001 and Lustrino & Wilson, 2007, Figure 1).

*Figure 3.2a*  The El Teide Volcano, Tenerife (courtesy of Rebecca Pitre) (See color plate section).

*Figure 3.2b*  Lava flows, El Teide, Tenerife (courtesy of Rebecca Pitre) (See color plate section).

Tertiary basaltic rocks are widely distributed north of Latitude 28°N in Egypt. Basaltic rocks also cover a large area beneath the Nile Delta and adjacent parts of the Western Desert. In Sinai, several minor Tertiary basaltic outcrops occur. Tertiary basaltic rocks occur mainly in the form of sheets, dikes, and sills. The basalts of the Abu Zaabal, Abu Rawash, and Gebel Qatrani are doloritic basalts. Volcanics of the Gebel Qatrani overlie the sandstones of the Lower Oligocene Gabal Qatrani Formation and they are overlain by the Lower Miocene Khashab sandstones. They are composed predominantly of amygdaloidal, altered, olivine basalts, which represent tholeiitic to transitional-continental within plate basalts (El Bayoumi et al., 1998).

Basaltic dikes from the southern Western Desert gave ages around 40 Ma (Meneisy & Kreuzer, 1974a). Volcanic activities related to the opening of the Red Sea took place in a series of successive pulses ranging in age from the Late Oligocene to the Middle Miocene. The Ragabat el Naam dike trends E-W parallel to a large fault belonging to the central Sinai-Negev shear zone. It gave an isochron age around 25 Ma (Upper Oligocene) (Steinitz et al., 1978).

Five linearly arranged ring complexes straddle the Libyan-Egyptian-Sudanese borders. These are Jabal Kissu in Sudan, Jabal Oweinat in Libya and Egypt, Jabal Arknu, Jabal Bahri and Jabal Babein in Libya. The complexes have intruded the Precambrian metamorphic basement and the overlying Cambro-Ordovician sediments. In addition to these, there are alkaline and basic plugs, dykes, and cone sheets. The latter cut the sediments in the Kufrah Basin including the Cretaceous. Therefore, they are probably Late Cretaceous or Tertiary (Flinn et al., 1991).

Recent age determinations, e.g. K-Ar and Rb-Sr indicate Tertiary age for the Jabal Arknu and Jabal Oweinat complexes. They are of Eocene age and are made up of rocks with markedly varied textures. The coarse grained rocks vary from feldspar-augite to feldspar-bearing alkali syenite[1], alkali syenite, alkali quartz syenite to alkali granite (Flinn et al., 1991). The fine-grained rocks include various saturated and undersaturated trachytes and phonolites (Atherton et al., 1991). Emplacement of the Jabal Arknu complex probably started with an explosive volcanic phase, followed by the emplacement of ring dykes of phonolite and microsyenite, and cone sheets (Flinn et al., 1991).

Tertiary (Oligocene-Miocene) alkaline lava cropping out in the Jabal Hassawnah, west-central Libya, includes basanites, phonolites, olivine basalts, and gabbros and syenites intrusions. Sr/Sr ratios of 0.7033–0.7039 and Nd/Nd ratios of 0.51285–0.51294 indicate derivation from a depleted upper mantle source with no significant crustal contamination (Oun et al., 1998a). These volcanics are probably related to the Tertiary reactivation of two intersecting major elements in west Libya, an early Paleozoic NNW-SSE trend and the Hercynian ENE-WSW (Jurak, 1978).

In Jabal asSawda, Tertiary volcanics overlie rocks ranging in age from Cambro-Ordovician to Oligocene, suggesting that these volcanics are Oligocene in age or younger. The most common types of volcanics are basalt lava flows composed of alkali olivine basalts, with variations from olivine basalts to nepheline basanites (Woller & Fediuk, 1980).

---

1 The name syenite comes from the ancient name of the City of Aswan in Upper Egypt. Delesse (1851) explained the origin of the rose color of syenite.

The Al Haruj al Aswad volcanics range in composition from olivine basalt to hawaiite, which belong to the alkali olivine basalt association (Busrewil & Wadsworth, 1980). K-Ar radiometric age data indicate that volcanic activity in the Al Haruj al Aswad area continued over the period 6.0–4.0 Ma ago. Most of the flows presently exposed are younger than 2.2 Ma (Upper Pliocene or younger) (Ade-Hall et al., 1974a).

During the Cenozoic, the Circum-Mediterranean Anorogenic Cenozoic Igneous Province (Lustrino & Wilson, 2007) was characterized by igneous activity with an extremely wide range of chemical compositions. Both orogenic subduction-related and anorogenic intra-plate igneous rocks were emplaced over a wide region, contemporaneous with the Alpine collisional tectonics associated with the convergence of Africa and Europe (El Azzouzi et al., 1999, Duggen et al., 2003, 2005, Lustrino & Wilson, 2007). The anorogenic igneous rocks are moderately to strongly $SiO_2$ undersaturated and show incompatible trace element ratios typical of magmas emplaced in intra-plate settings. The most typical rock-types are basanites and nephelinites, followed by hawaiites, alkali basalts, tephri-phonolites and phonolites. Sr-Nd isotopic ratios suggest derivation of the magmas by mixing of partial melts from both lithospheric and EAR-like mantle sources with $^{87}Sr/^{86}Sr$ ratios of 0.7029–0.7049 and $^{143}Nd/^{144}Nd$ of 0.51255–0.51303 (Lustrino & Wilson, 2007).

Cenozoic magmatism in Southern Morocco is dominated by alkaline to hyperalkaline rocks and includes basalts, phonolites, and trachytes, and minor plutonic rocks of nepheline syenites with carbonatite occurrences (Teixell et al., 2005). It occurred synchronically with the Atlas compression, a fact that has intrigued previous authors. Available Sr-Nd isotopic data for Neogene to Quaternary lavas suggest a depleted asthenospheric mantle source for the magmas, which later interacted with lithospheric mantle. This magmatism defines a separate district from that of the westernmost Mediterranean (Betics, Alboran, Rif), which shows a shift from calc-alkaline to alkaline character in Pliocene times (Duggen et al., 2003, and references therein). On the other hand, the Atlas magmatism is roughly coeval and has the same petrological and isotopic signature as that of the Canary Islands (Teixell et al., 2005).

The Cenozoic volcanism of the Atlas System is exclusively of intraplate alkaline type, composed of alkali basalts, basanites, nephelinites, and associated intermediate and evolved lavas, whereas in the Rif's volcanism evolved through time from calc-alkaline to shoshonitic and alkaline (Frizon de Lamotte et al., 2008). The Atlas volcanism is located within a SW-NE trending strip, underlain by thinned lithosphere (Frizon de Lamotte et al., 2004, 2008, Teixell et al., 2005, Zeyen et al., 2005, Missenard et al., 2006), for which (Frizon de Lamotte et al., 2008) proposed the name "Morocco Hot Line". This trend extends towards the Mediterranean Coast near Oujda and the Oran areas in Algeria (Frizon de Lamotte et al., 2008).

The NE-trending thinned lithosphere line coincides with the distribution of Miocene to Recent alkaline volcanism of the Morocco Hot Line. This lithospheric structure is oblique on the structural grain of the crust as it crosscuts not only the SAF, but also the Jurassic Atlantic margin up to the Canary Islands to the south, and the Oligocene-Miocene Alboran Basin up to eastern Spain and possibly the French Central Massif to the north (Chalouan et al., 2008, Michard et al., 2008).

The lithospheric thickness varies from about 90 km under the Variscan Iberian Massif in agreement with earlier modeling results, to 160–190 km underneath the Gulf of Cadiz and Gharb Basin, decreasing strongly to 70 km underneath the Middle Atlas and thickening gradually toward the Sahara Platform to 180 km (Zeyen et al., 2005, and reference therein).

Mantle xenoliths are locally included in the Neogene-Quaternary alkali-basalt/ hawaiite lavas of the Gharyan volcanic field (NW Libya). They mostly consist of spinel lherzolites. REE patterns of clinopyroxenes are characterized by flat HREE distribution (7.5–15.3) and variable LREE enrichment (Beccaluva et al., 2008).

The mantle xenoliths included in Quaternary alkaline volcanics from the Manzaz-district (Central Hoggar) are spinel lherzolites (Baccaluva et al., 2007). Major and trace element analyses on bulk rocks and constituent mineral phases show that the primary compositions are widely overprinted by metasomatic processes. Therefore, the Hoggar volcanism, as well as other volcanic occurrences in the Saharan belt, are likely to be related to passive asthenospheric mantle uprising and decompression melting linked to tensional stresses in the lithosphere during Cenozoic reactivation and rifting of the Pan-African basement. This can be considered a far-field foreland reaction of the Africa-Europe collisional system since the Eocene (Baccaluva et al., 2007).

In the Jabal Nafusah-Gharyan area, basalts occur on top of Cenomanian-Turonian sediments in the upper part of the Jabal Nafusah escarpment. Phonolite plugs, domes, ring dykes and pyroclastic rocks crop out also on the top of the Jabal and within Upper Triassic and Lower Jurassic limestones at the base of the escarpment (Ade-Hall et al., 1975). The Gharyan volcanics represent three modes of occurrences (Almond et al., 1974): 1) an early phase (Ypresian, about 50–55 Ma) of extensive plateau lava (basaltic hawaiite) eruptions. 2) A phase of emplacement of small phonolite domes (Bartonian, around 40 Ma). 3) A third phase of activity (Miocene-Pliocene, between 12-1 Ma), which resulted in small composite central volcanoes and lavas. The latter are composed of subalkaline mildly basaltic andesites (Busrewil & Wadsworth, 1996). The pyroclastic rocks are black, massive and poorly sorted deposits made up of fragmented products of explosive eruptions and matrix-supported blocks. The southeastern corner of Sicily is the Ragusa-Iblei carbonate platform, which has been affected by Alpine faulting and which appears to be a prolongation of the Saharan platform. At Capo Passero, on the southeastern tip of the island, a volcanic formation (approximately 71 Ma) is overlain by Maastrichtian limestone. Paleomagnetic studies show that the paleomagnetic direction of these volcanics is similar to those from Africa and suggest that Sicily is a part of the African plate (van den Berg & Zijderveld, 1982).

# Chapter 4

# Northwest Africa's Basins

## 4.1 ALGERIAN BASINS

The continental shelf offshore Algeria consists of Tertiary and Quaternary sediments, more than 3500 m (11,480 ft) overlying a metamorphic basement (Askri et al., 1995). Onshore, the main tectonic elements of Algeria are the Atlas Mountains in the north and the Sahara Platform over the rest of Algeria (Fig. 4.1).

In Northern Algeria are the Tellian Atlas, the Hodna Basin, the High Plateau, the Saharan Atlas, and the Chott Melrhir basins (Askri et al., 1995). The Tellian Atlas is a nappe domain, with intermountain basins, such as the Chelif Basin. The sedimentary succession extends from Jurassic to Miocene. Oil has been found at Ain Zeft, Tliouanet, and Oued Guettirini. The Hodna Basin is a foredeep basin with a sedimentary infill dominated by Eocene and Oligocene continental sediments, overlain by Miocene marine sediments. The High Plateaus are the foreland of the Alpine Atlas Mountains, with intermountain basins, such as the Telagh and Tiaret basins. The Saharan Atlas developed over the site of an elongated extensional trough between the High Plateaus and the Saharan Platform. The sedimentary succession in the Saharan Atlas is up to 9000 m (29,530 ft) in thickness (Askri et al., 1995). Tertiary compressive tectonics led to the formation of the mountain range. The Chott Melrhir basins in the SE Constantinois area developed during the Tertiary. The Cretaceous sedimentary sequence is about 5000 m (16,405 ft) in thickness and form hydrocarbon reservoirs at Djebel Onk, Rass Toumb, and Guerguet El Kihal North.

In Northern Algeria, the Kabylides comprise the Dorsale Kabylie (Dorsale Calcaire) and the upper Kabylie Massif. The Dorsale Kabylie complex consists of a pile of thrust sheets involving Lower Jurassic platform carbonates overlying Upper Paleozoic and Triassic detrital sediments and a thin cover of Jurassic-Cretaceous pelagic and Cenozoic siliciclastic sediments (Benaouali-Mebarek et al., 2006). The Dorsale Calcaire is considered the former northern margin of the Tethys. The underlying Kabylie Massif consists of Paleozoic greenschist facies phyllites overlying amphibolite facies rocks which constitute the Kabylian basement. The Kabylides are covered by the so-called "Oligo-Miocene Kabylie molasses" deposited during rifting which preceded the opening of the Algerian Basin. The Oligo-Miocene Kabylie molasses are overlain by Flysch nappes formed as a result of back-thrust over the Kabylies Domain, followed by Langhian-Early Serravallian marine sediments (Benaouali-Mebarek et al., 2006).

The Saharan Platform consists of numerous Paleozoic cratonic basins separated by tectonic highs and arches (Figs. 2.1, 3.1, 4.2). The basins include the Tindouf and

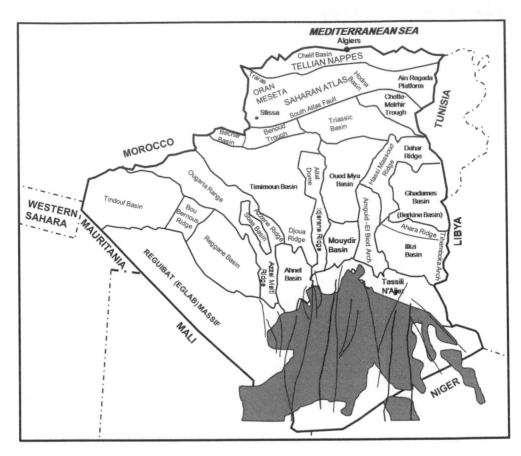

*Figure 4.1* Tectonic map of Algeria (after Askri et al., 1995).

Reggane Basins on the northern and northeastern side of the Reguibat Shield, which extend into Morocco (Fig. 4.2). The sedimentary succession is more than 8000 m (26,248 ft) in the Tindouf Basin and 6500 m (21,327 ft) in the Reggane Basin (Askri et al., 1995). The Bechar Basin is limited to the north by the High Atlas and to the south and to the west by the Ougarta Range. It contains more than 8000 m (26,248 ft) of sediments. The Ahnet-Timimoun Basin is bordered to the north by the Oued Namous shoal, to the west by the Ougarta Range, to the south by the Touareg Shield, and to the east by the Idjerane-Mzab Ridge. The sedimentary succession is up to 4000 m (13,124 ft).

The Mouydir and Aguemour-Oued Mya basins (Fig. 4.1) are bordered to the west by the Idjerane-Mzab Ridge and to the east by the Amguid-El Biod High. Paleozoic rocks crop out in the Mouydir area to the south. The Aguemour-Oued Mya Basin contains a thick Paleozoic to Cenozoic sedimentary succession up to 5000 m (16,405 ft) in thickness. The basin contains the giant Cambrian Hassi Messaoud Field (Balducchi & Pommier, 1970) and the Triassic Hassi R'mel Field.

The Illizi-Ghadames Basin in Algeria (Fig. 4.1) is the extension of the Ghadames Basin of Libya and Tunisia. It is bordered to the west by the Amguid-El Biod High and

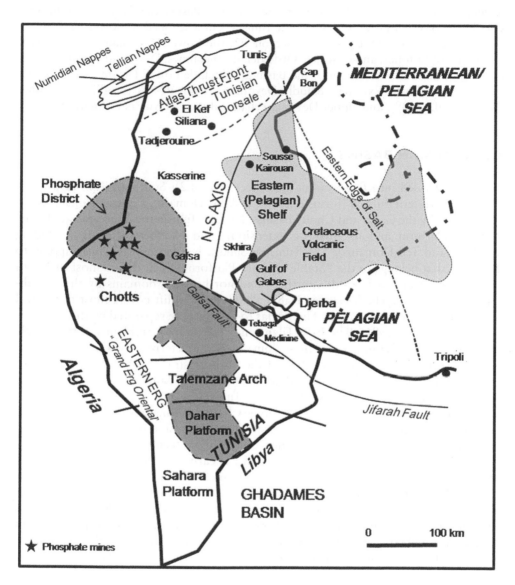

*Figure 4.2* Tectonic map of Tunisia (Cretaceous volcanics after Mattoussi et al., 2009).

to the east by the Tihemboka uplift at the Algerian-Libyan border. The sedimentary succession in the Ghadames Basin exceeds 6000 m (19,686 ft) in thickness. The Illizi Basin was folded at during the Hercynian Orogeny at the end of the Paleozoic and was subsequently overin unconformably by Triassic to Cretaceous sediments which exend over large tracts of the Sahara (Tissot et al., 1973). The Illizi is bounded to the south by the Precambrian rock of the Hoggar Massif, to the west by the El Abiod-Amguid High where only Cambro-Ordovician sediments remain, and to the east by the Tihemboka

Ridge, and to the north by the Ahara Ridge. The Illizi Basin extends eastwards into the Ghadames Basin in Libya and Southern Tunisia. Oil is trapped in Lower Carboniferous reservoirs B and D, Devonian reservoir F2, the Lower Devonian-Upper Triassic F4 and F6 sandstones, and Camrian-Ordovician sandstones. Devonian sediments crop out along the southern margin of the Basin and include Lower Devonian sandstones of the Upper Tasilis Externes, Middle Devonian carbonates of the Illizi Formation (Wendt et al., 2009), and Upper Devonian shales of the Tin Meras Formation.

## 4.2 TUNISIAN BASINS

Tunisia is divided into five tectonic provinces (Fig. 4.2); Southern Tunisia (Sahara Platform), Western Tunisia (forms part of the Atlas Domain), Central Tunisia (occupied mostly by the Gafsa and Chotts basins), Eastern Tunisia (occupied mostly by the Pelagian Platform and its offshore extension), and Northern Tunisia (including the Atlas and Tell Domains). The Tunisian Sahara is a large monocline dipping to the south that belongs to the "Stable" Sahara Platform which covers most of North Africa. It is separated from the northern and northeastern domains by the Jifarah-Gafsa fault system. The Mesozoic Dahar Platform or Uplift covers most of Central Tunisia. Eastern Tunisia and the High Atlas to the west are separated by the N-S Axis (NOSA) (Burollet, 1991). The E-W Tilamzane (Talemzane) Arch runs in the middle of the Saharan Domain. Each of these domains is characterized by a distinctive geological history and stratigraphy. Most of the onshore stratigraphic nomenclature of Tunisia and Libya extend to the offshore area. The Ghadames Basin in West Libya and Eastern Algeria extends to southernmost Tunisia.

Northern Tunisia forms part of the Alpine Chain (the Maghrebides) of the Western Mediterranean and is occupied by the easternmost extension of the Atlas Mountain Chain, including the Tell and the Dorsale Domains. Four tectonic phases can be recognized in the Tunisian Atlas (Snoke et al., 1988): 1) early Mesozoic rifting of the North African continental margin. 2) Diapiric emplacement of Late Triassic-Liassic evaporites into the overlying strata beginning in the early Cretaceous. 3) Folding of the cover strata in response to regional compression culminating in the Middle Miocene Atlas Orogeny. 4) The development of transvers grabens during latest Miocene to Recent orogenic normal faulting. The principal driving mechanism of regional crustal shortening was the reactivation of early Mesozoic normal faults as reverse faults. The shortening recorded in the Tunisian Atlas has been attributed to structural thickening of basement and concurrent deformation of the sedimentary cover (Snoke et al., 1988).

Western Tunisia is part of the Eastern Atlas Domain and is characterised by NE-SW trending folds, strike-slip faults and diapirism (Burollet, 1990, Bouaziz et al., 2002). It consists of several basins with thick successions of Mesozoic and Cenozoic sedimentary rocks. Two major tectonic cycles have been recognized in the area; the first tectonic cycle was related to the NE-SW to N-S extensional tectonics that led to the opening of the Tethys in the Late Permian to Early Cretaceous and the formation of rifts, tilted blocks, horsts and grabens. The second tectonic cycle was characterised by NW-SE compression in the Cenozoic and caused the inversion of normal faults, reactivation of uplifts and intrusion of diapirs of Triassic evaporites (Ben Ferjani et al., 1990, Grasso, 1999, Bouaziz et al., 2002).

## 4.3  MOROCCAN BASINS

The main tectonic domains in Morocco include the Atlas Mountains, Anti-Atlas, the Rif, the Mesetas, and the Mediterranean and Atlantic margins (Fig. 4.3). The Anti-Atlas Domain is discussed in the chapter on the Pan-African Shield. There are more than 30 sedimentary basins in Morocco. They range from cratonic basins of the Sahara Platform; many of which straddle the Algerian-Moroccan or Mauritanian-Moroccan borders; inter-mountain basins in the Rif, Meseta, and Atlas domains, and rifted-margin basin, particularly on the Atlantic side. The basins include the Bechar (mostly in Algeria), Reggane Basin, Taoudenni Basin (mostly in Mauritania), Agadir, Zag Basin, Tadla Basin, Haouz Basin, Bahira Basin, Tagalft Basin, Prerif Basin, Doukkala Basin, Essaouira Basin, Tarfaya Basin, Rharb (Gharb) Basin, Boudenib Basin, Tafilalt Basin, Maider Basin, Ouarzazate Basin, Tindouf Basin, Draa Basin, Missour Basin, High Plateaux, Saïs Basin, Guercif Basin, Beni Znassen-Beni Bou Yahi Basin, Souss Basin, Argana Basin, Mediterranean Offshore Basins, Atlantic Offshore Basins (extension of Rharb, Safi, Agadir-Essaouira, Tarfaya, Aaiun (offshore), Atlantic Deepwater Frontier.

### 4.3.1  Evolution of the Atlas Mountains

The Atlas structure and evolution have been the subject of many studies since the initial observations by Thomson (1899) and Gentil (1906, 1907, 1912), among them are those by Choubert & Faure-Muret (1962), Mattauer et al. (1972, 1977), Beauchamps et al. (1997), Frizon de Lamotte et al., (2000, 2005, 2008), Piqué et al., (1998), Teixell et al. (2003, 2007), Arboleya et al. (2004), Laville et al. (2004), Benaouali-Mebarek et al. (2006), Missenard et al. (2007), Barbero et al. (2007), and Bejjaji et al. (2010).

The history of the Atlas Mountains encompasses two major episodes starting with rift basins during the Mesozoic, evolving into compressional and inverted belts from Cenozoic to present times. The Atlas System is developed on the site of the Late Permian-Late Triassic rift domain (Laville et al., 2004, Barbero et al., 2007, Frizon de Lamotte et al., 2008). The West Moroccan Arch (WMA) was submerged and became a submarine high during Late Triassic-Liassic. Extensional tectonics continued during the Liassic. Variscan faults were reactivated during extension and transtension associated with the opening of the Central Atlantic and the left-lateral movements between European and African plates. A second rifting episode took place in late Liassic-Dogger time (Barbero et al., 2007). Compressional tectonic processes were not progressive from Late Cretaceous to present, but they occurred in steps.

Mattauer et al. (1972, 1977) and Laville & Piqué (1992) proposed that in the late Jurassic there was a major phase of compressive folding and erosion. However, Frizon de Lamotte et al. (2008) did not find any evidence of this compression. Frizon de Lamotte et al. (2008) propose two steps: during Middle-Late Eocene and Pliocene-Quaternary. They attributed these steps to spatial and temporal pattern of the Mediterranean subduction and coupling between the African and European plates. The Atlas Mountains have been considered intracontinental chains because they are located far from collisional zones and because of the lack of fold nappe structures, regional metamorphism, granitoid intrusions and other features typical of interplate orogens (Beauchamps et al., 1997, Barbero et al., 2007, Frizon de Lamotte et al., 2008).

The amount of tectonic shortening and relief associated with the Alpine compression are controversial. Beauchamp et al. (1996) estimated an Alpine shortening of 36 km in the Central High Atlas from a restored section west of Beni Mellal. Teixell et al. (2003) estimated 26 km at Errachidia to the east and 30 km at Imilchi and 13 km at Demnanti in the west. Tesón & Teixell (2006) suggested a shortening of about 7 to 8 km for the Sub-Atlas.

Shortening in the whole Atlas System began in Late Cretaceous due to convergence between Europe and Africa. The Atlas relief is not related to crustal shortening alone, but thermal factors are involved in the Atlas uplift, related to thinned lithosphere shown by the Moroccan Hot Line. The timing of the Cenozoic inversion in the Atlas also remains a matter of debate. The Atlas experienced moderate crustal shortening and exhumation inspite of their high elevation (Barbero et al., 2007). Many authors argue that the height of the Atlas Mountains, their shallow roots, and the moderate crustal thickening cannot be attributed to compression alone, and therefore, they envisage the involvement of the mantle or a thermal component (Wigger et al., 1992, Ayarza et al., 2005, Teixell et al., 2003, 2005, Zeyen et al., 2005). The timing, causes, and amount of shortening in Atlas Mountains are a matter of debate. Mattauer et al. (1972, 1977), Laville & Piqué (1992), and Laville et al. (2004) argued that total shortening includes a pre-Cretaceous component which resulted in folding, cleavage and erosion of Jurassic igneous and sedimentary rocks, and a late Cretaceous to Miocene component which led to uplift of the High Atlas and basin development to the south of the Atlas Chain. On the other hand, Gomez et al. (2000, 2002) consider the shortening of the Moroccan Atlas the result of Cenozoic plate convergence, with no significant pre-Cretaceous shortening. Barbero et al. (2007) agree with the latter and rule out any significant uplift, exhumation, or erosion during the late Jurassic and argue that the main uplift of the Central High Mountains occurred in the Miocene around 20 Ma (Burdigalian). Several authors have proposed different phases for the Cenozoic deformation and uplift from the Late Eocene to the Quaternary (Fraissinet et al., 1988, Görler et al., 1988, El Harfi et al., 1996, Frizon de Lamotte et al., 2000, 2008, Morel et al., 2000). Based on AFT and (U-Th)/He dating on apatite crystals from Paleozoic to Jurassic igneous complexes and sedimentary rocks of the Atlas Mountains, Barbero et al. (2007) also preclude the role of thermal components and argue that the data from Paleozoic basement massifs indicate a long residence at low temperatures in the Mesozoic rift troughs and minor exhumation. Their models suggest a slow cooling trend from ca.80 Ma (date of Jurassic intrusions) to ca.50 Ma ago which they attributed to post-rift thermal relaxation. This cooling is followed by a stability period of ca.30 Ma and then by final cooling due to continued exhumation.

Frizon de Lamotte et al. (2008) postulate that the Paleozoic basement in the Central High Atlas is divided into blocks separated by Mesozoic rifting normal faults that were partially inverted. In contrast, Teixell et al. (2003, 2007) and Arboley et al. (2004) believe that all the pre-existing faults were completely inverted. Syn-tectonic low grade metamorphic conditions in the Central Atlas are postulated by Brechbühler et al. (1988) based on illite crystallinity.

A tectonic domain between the Western and Central High Atlas has been called the Marrakesh High Atlas (MHA) (Fig. 4.3). It is also known as the Paleozoic High Atlas

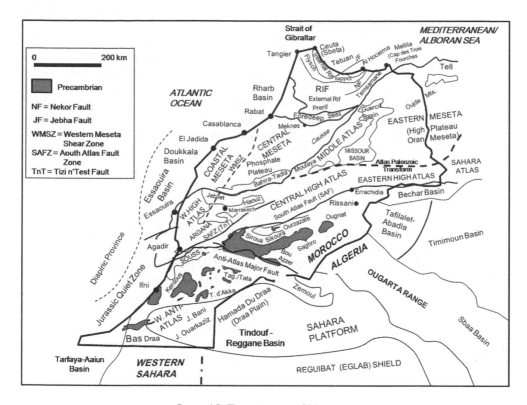

*Figure 4.3* Tectonic map of Morocco.

or the *Massif Ancien* (Proust et al., 1977, Missenard et al., 2007). It is characterized by the abundance of Paleozoic and Precambrian rocks in the Sirwa Inlier and was part of the West Moroccan Arch (WMA) during the Triassic and Jurassic (Choubert & Faure-Muret, 1962, Jacobshagen et al., 1988, Michard et al., 2008). During the Jurassic the Sea transgressed simultaneously from the Tethys Ocean in the northeast and the Atlantic Ocean to the west. However, the WMA acted as a barrier between the two domains during that time (Michard et al., 2008).

In the South Atlas Front (SAF) zone, major thrust faults carried allochthonous units onto the foreland basins, such as the Souss Basin in the west and the Ouarzazate Basin in the central part of the SAF (Frizon de Lamotte et al. 2008). In other areas, the South Atlas is either in direct contact with the Sahara Platform, for example the Boudenib-Errachidia Basin, or separated from it by uplifts, such as the Siroua (Sirwa) Plateau.

During the Mesozoic, the High Atlas experienced extension and rifting. The first rifting period occurred during the Triassic and led to the accumulation of thick red beds "Couches Rouges" and tholiitic basalts Couches Rouges (Monbaron et al., 1990, Fiechtner et al., 1992, Laville et al., 2004). Gentil (1912) postulated that the thick Pemo-Triassic conglomerates and Red Beds at Ait Khzame in the Southern High Atlas were eroded from the Atlas Chain and then depositd by rivers or flash floods accompanied by trachytes, andesites, and basalts. A post-rift Liassic platform was drowned (Rabat, 1992, Lachkar et al, 2009) and disrupted during the late Liassic-Dogger by a renewed rifting episode (Warme, 1988). Subsidence of the High and Middle Atlas area led to the formation of basins filled with up to 5000 m (16,400 ft) of marls, calciturbidites, and reefal limestones from Toarcian to Bajocian in the central High Atlas. The Red beds marked a regional regression and became widespread again during the Bathonian. Alkaline magmatic rocks crop out along the axis of the Central High Atlas. Three magmatic formations can be distinguished; epizonal plutonic complexes composed of syenites-monzodiorites, olivine dolerite dikes intercalated within mid-Jurassic deposits, and plateau basalts produced by fissure volcanism. K-Ar ages of the Mesozoic intrusions of the High Atlas range from 119 to 173 Ma (Bajocian-Albian).

The Late Cretaceous is represented by another basal red bed formation followed by a platform limestone of Cenomanian-Turonian age. The two units rest on either earlier Jurassic deposits or basement rocks to the north and south of the High Atlas. Cretaceous rocks are interpreted as post-rift deposits (Teixell et al., 2003). The uppermost Cretaceous (Maastrichtian) consists of fine-grained red beds coeval with the Alpine compression (Monbaron et al., 1990, Amrhar, 1995). Maastrichtian-Middle Eocene marine limestones overlie the continental red beds. Upper Eocene-Pliocene Atlas compression led to the deposition of continental sediments and volcanic rocks. Cenozoic alkaline to hyperalkaline magmatic rocks are composed of basalts and phonolites and minor subvolcanic rocks, such as syenites and carbonatites, dated at 40–45 Ma (Lutetian-Bartonian) and 15–0.5 Ma (Langhian-Pleistocene) (Tisserant et al., 1976, Harmand & Cantagrel, 1984, Berrahma and Hernandez, 1985, El Azzouzi et al., 1999). This magmatism has been attributed to a mantle upwelling which accompanied the Cenozoic Atlas compression (Teixell et al., 2005).

Relative motion of peri-Atlantic plates changed in the Late Cretaceous, leading to inversion of faults and basins as early as Albian-Cenomanian and block motions in the Atlas and Meseta domains (Frizon de Lamotte et al., 2008). During the Santonian inversion which prevailed from Sinai to the Atlas Mountains, the depocenters of the Middle Atlas area were connected to the Atlantic Ocean through the Phosphate Plateau. Tectonic shortening led to folding, invesion of faults, formation of breccias along faults, and a regional unconformity at the base of the Eocene (Frizon de Lamotte et al., 2008).

The Atlas Mountains contain a variety of structural styles that resulted from reactivation of pre-existing faults as well as the generation of new faults (Beauchamps et al., 1997). Thin-skinned structural styles are characterized by bedding which is parallel to hanging wall ramps and flats of thrust faults. Deformation in the Atlas has also occurred by thick-skinned deformation with shortening of the basement through reverse faults (Michard, 1976, Frizon de Lamotte et al., 2004, 2008, Michard et al., 2005, 2008a-d, Crespo-Blanc & Frizon de Lamotte, 2006), with the reactivation of pre-existing syn-rift faults and previously deformed syn-rift rocks. On seismic reflection data these features include non-parallel bedding and stratigraphic thickening of syn-rift rocks in the hanging wall of thrust faults. The base of the syn-rift sedimentary sequence is marked by strong reflections in the Triassic generated by interbedded volcanics, evaporites, and clastic rocks. An example of this phase is the Ait Attab syncline which contains Lower-Middle Jurassic syn-rift rocks and Upper Cretaceous rocks deposited in an extensional half-graben (Beauchamps et al., 1997).

## 4.4   WESTERN HIGH ATLAS

The Western High Atlas is a mountainous region composed of Neoproterozoic and Paleozoic formations folded during the Variscan Orogeny and lifted up by the Atlas (Alpine) Orogeny tectonic events. It extends from the Atlantic Coast to the Argana Valley (at the junction of the Western High Atlas and the Middle Atlas) (Fig. 4.3), for a distance of about 70 km. The Western High Atlas (also known as the Paleozoic High Atlas) is more related to the Atlantic margin of northwest Africa than to the rest of the High Atlas (Schlüter, 2008) and forms part of the Atlantic Domain of Frizon de Lamotte et al. (2008). In the Western High Atlas, Mesozoic sediments are preserved in a number of basins, such as the Argana Basin, Oued Zat Basin, Souss Basin (Ida Ou Ziki and Ida Ou Zal Subbasins), and Haouz Basin.

The Mesozoic High-Atlas Basin opened north of the Anti-Atlas during the Late Triassic rifting phase (Soulaimani & Burkhard, 2008), accompanied by doleritic intrusions. Cretaceous to Neogene sediments in the Eastern Anti-Atlas show many unconformities and tilts caused by the inversion of High-Atlas during the Eocene-Oligocene (Syrian-Arch/Alpine phase) and Pliocene-Pleistocene times (Frizon de Lamotte et al., 2000).

The main rifting in the Western High Atlas took place at the end of Triassic, which led to the opening of the Central Atlantic and the Western Tethys Ocean. The structural development of the Western High Atlas was dominated by strike-slip faulting since the end of the Hercynian Orogeny until Triassic times, as a result of several tectonic phases (Saber et al., 2007): 1)A major compressive Hercynian cycle (Namurian-Westphalian) characterized by the formation of tectonic ductile structures and micro-structures led to N20°-N30° folds and asymmetric shear zones trending N60°-N70° (Imin-Tannout and Tizi n'Test faults) and N90°-N110° (Tizi Maâchou Fault). 2) A NW-SE to NNW-SSE extension phase started during the Stephanian, which led to the formation of a number of basins in the Western High Atlas, such as the Argana Basin, Souss Basin (Ida Ou Ziki and Ida Ou Zal Subbasins), Oued Zat Basin, and Haouz Basin. Another extensional phase in a NW-SE direction led to the formation of normal faults trending N20° to N40° also affected the Argana, Tizi n'Test and Oued Zat basins. These basins were filled with red conglomerates and conglomeratic sandstones at the base, and silty sandstones and siltstones towards the top.

Volcanic rocks associated with rifting contemporaneous with the Central Atlantic Magmatic Province "CAMP" (Frizon de Lamotte et al., 2008) floor the Lower Liassic evaporitic basin, which is a sag basin related to thermal relaxation.

## 4.5   ARGANA BASIN

The Argana Basin is located in the Western High Atlas of southern Morocco about 30 km northeast of the port city of Agadir (Fig. 4.3) and extends 85 km from the village of Ameskoude in the south to Imi n'Tanoute in the north (Brown, 1980). The tilted-blocks pattern within Triassic redbeds in the Argana Basin appears to be related to a NW-SE extension during the early rifting of the Central Atlantic. The amount of extension measured on these structures reaches 15% (Medina, 1988).

Volcanic rocks may reach 350 m (1148 ft) in thickness in the Argana Basin (Aït Chayeb et al., 1998, Brown, 1980, Olsen et al., 2000, Le Roy et al., 1998). The Triassic syn-rift sediments crop out in the Argana Corridor *"Couloir d'Argana"*. The succession in the southwesternmost part of the High Atlas extends from the Late Permian to the Early Jurassic (Saber et al., 2007).

## 4.6  SOUSS BASIN

The Souss Basin is a Mesozoic-Cenozoic rift basin located between the southwestern end of the Western High Atlas Mountains and the Southwestern Anti-Atlas Chain (Fig. 4.3). The triangular shape of the Basin is due to pre-existing structures (Frizon de Lamotte et al., 2008). The Kléa Fault (Triassic) is a normal fault that crosses the Souss Basin and connects to the Tizi n'Test Fault. The difference between the Souss Basin and the Ouarzazate Basin on trend to the north is the lack of Triassic beds in the latter. The Souss Basin consists of two subbasins; the Ida Ou Zal and Ida Ou Ziki sub-basins. They represent remnants of a former larger basin. The Souss Basin and the Ouarzazate Basin are separated by the Siroua Plateau and the Miocene Siroua Volcano.

## 4.7  HAOUZ BASIN

The Haouz Basin is an intermountain basin (Frizon de Lamotte et al., 2008) at the front of the Jebilet and the Western High Atlas. The basin shows thin Mesozoic and Cenozoic units and Recent sediments resulting from the erosion of the Atlas Chain (Jaffal et al., 2010). Neogene and Quaternary alluvial sediments filled up the Paleozoic or Mesozoic paleotopography. The stratigraphic sequence is relatively complete. The Hercynian basement crops out in the Jebilets and the Guemassa massifs, north and south of the Haouz Basin, respectively. In the Jebilet Massif, the basement is made up of metapelites intercalated with sandstones and limestones and assigned to Middle and Upper Visean (Essaïfi et al., 2003). Several Carboniferous intrusions of gabbros and granites cut through this series (Essaïfi et al., 2003, Lagarde et al., 1990). Southwards, the Guemassa Massif is composed mainly of flysh sequences of interbedded sandstones, shales, and limestones. In the Guemassa Massif to the West, the Hercynian structures are oriented NE-SW to NS and in the N'fis Domain in the East, the structure is mainly in a NNW-SSE direction. The Oligocene Atlasic compression led to uplift of the Paleozoic massifs and resulted in faulting in an ENE-WSW direction.

### 4.7.1  Tectonic history of the Meseta Domain

The location of the Meseta Domain relative to Gondwana during the Paleozoic is controversial. Stampfli & Borel (2002) suggest that the Moroccan Meseta drifted away from Gondwana at ca. 490 Ma (Lower Ordovician), while others postulate that the Meseta Domain remained close to the West African Craton (WAC) during the Paleozoic (Piqué & Michard, 1989, Dostal et al., 2005, Hoepffner et al., 2005, 2006, Soulaimani & Burkhard, 2008, Baidder et al., 2008). The Meseta Domain is also interpreted as the distal parts of the African passive margin, which remained close to the Anti-Atlas throughout Paleozoic times (Soulaimani & Burkhard, 2008, Pouclet et al., 2008), but was subsequently thrust onto the Anti-Atlas Domain during the Late Carboniferous-Early Permian time.

Neoproterozoic rocks crops out only in a few places, such as in the cores of anticlines at El Jadida, Rehamna, Jebilet, and in the eastern part of the Central Massif in the Central Meseta. Paleozoic massifs are well developed in the Western Meseta, but they are smaller in the Eastern Meseta, which straddles the Algerian-Moroccan border (Soulaimani & Burkhard, 2008).

The northern border of the Meseta is characterized by Cambrian rocks and granites of the Sehoul Block (see Hoepffner et al., 2006, Fig. 1.2), which were thrust southward over Ordovician, Silurian and Devonian sediments and volcanics of the Bou Regreg Corridor (Tahiri et al., 2010). The age of the Tiflet granitoids in the Sehoul Block is unclear; they were assigned a Late Devonian (367 Ma) based on U-Pb data, but recently, Tahiri et al. (2010) assigned the Rabat Granites a Devonian age of 430 Ma and the Tiflet Granites a Late Proterozoic age of 605–609 Ma. The two granites belong to Andean-types generated in subduction zones. Cambrian rocks which crop out in the Central and Northwestern Meseta (Pouclet et al., 2008) overlie Late Precambrian volcano-plutonic rocks. Basaltic rocks occur at the base of the Cambrian formations in the Sidi-Said Maachou, Bou Acila and Oued Rhebar areas (Gigout 1951, Morin 1962, Cornée et al. 1984). They are composed of continental tholeiites and alkali basalts dated Early to Middle Cambrian (El Attari et al., 1997, Ouali et al., 2000, 2003, El Hadi et al. 2006). They were emplaced during active rifting (Pouclet et al., 2008).

Calc-alkaline volcanics associated with Middle Cambrian sediments in the Western Meseta occur in the Oued Rhebar Complex and originated in an orogenic setting. High La/Nb ratios of about 5.2 suggest a lithospheric mantle origin, while La/Ta ratios, higher than 26 and a negative Nb anomaly indicates contamination by the continental crust inherited from the partial melting of a metasomatized mantle (El Hadi et al., 2006).

The Meseta and Anti-Atlas domains underwent a separate tectonic history from the Ordovician onward (Piqué, 1989, 2001, Hoepffner et al., 2006, Soulaimani & Burkhard, 2008). Hence, the two domains differ in the amount of tectonic shortening, degree of metamorphism, and the presence of Variscan granites. Orogenic shortening is more significant in the Meseta Domain than in the Anti-Atlas Domain. Regional metamorphism reaches medium grade (upper greenschist to amphibolites facies) in places in the Meseta, but is very low to absent in the Anti-Atlas. Variscan granitoids are present in the Meseta, but are absent in the Anti-Atlas. These differences between the Meseta and Anti-Atlas domains are attributed to the role of the South Atlas Fault (SAF) (Mattauer et al., 1972, 1977, Ouanaimi & Petit, 1992). The Paleozoic expression of this fault has been named the "Atlas Paleozoic Transform Fault Zone" (APTZ) or the "Atlas Paleozoic Fault Zone" (APFZ) by Piqué & Michard (1989) and Piqué (2003).

Within the Meseta Domain, the Eastern Meseta differs from the Western Meseta by their early metamorphic evolution and the degree of tectonics. In the Western Meseta, the deformation is heterogeneous and concentrated within shear zones (Lagarde et al., 1990, Essaïfi et al., 2001), while it is homogeneous and higher in the Eastern Meseta (Michard, 1976, Piqué & Michard, 1989, Hoepffner et al., 2005, Soulaimani & Burkhard, 2008).

The Meseta Domain subsided to >3 km depth during the Late Triassic-Middle Jurassic, followed by two episodes of subsidence and exhumation during the Jurassic-Early Cretaceous and the Late Cretaceous-Neogene (Ghorbal et al., 2008), related to the opening of the Atlantic and the Alpine Orogeny, respectively. It became stable after the Paleozoic Hercynian Orogeny (Schlüter 2008). Paleozoic sediments were then covered by thin tabular successions of Mesozoic and Cenozoic sediments.

The Azores-Tunisia region is located at the western plate boundary between Eurasia and Africa, which was in proximity to the pole of rotation of the African plate that led to crustal extensions and normal faulting at the Azores archipelago, transcurrent motion with strike slip faulting at the center of the Azores-Gibraltar Fault, and plate convergence with reverse faulting (Buffron et al., 2004). El-Harfi et al. (2007), based on seismic refraction data and gravity modeling, postulated the existence of a thrust fault that may penetrate the lower crust and offset the Moho. They attributed the Neogene and Quaternary volcanism along the High Atlas Lineament to northward continental subduction of the WAC plate under the Moroccan microplate. They also interpreted the South Atlas Fault Zone (SAFZ) as the northwestern boundary of the African Plate instead of the Azores-Gibraltar Fault.

## 4.8 THE ATLANTIC PASSIVE MARGIN

A major magmatic event has been linked to the rifting stage at the Triassic-Jurassic boundary, which led to the opening of the Central Atlantic and Western Tethys Oceans. This event has been labeled the Central Atlantic Magmatic Province "CAMP", and affected the entire Morocco. Basaltic flows emplaced within the Triassic-Liassic basins reach 350 m (1148 ft) in thickness. They occurred in short pulses mainly at the Triassic-Jurassic boundary at around 200 ± 1 Ma, based on $^{40}Ar/^{39}Ar$ (Knight et al., 2004). CAMP lavas in Morocco are found at the top of the Triassic sequence filling large Triassic-Jurassic synrift basins. They are divided into four basic units (the lower, intermediate, upper, and recurrent units) by Knight et al. (2004). Magnetostratigraphic results suggest that the lower unit and part of the intermediate unit were erupted before a brief period of reversed magnetic polarity (*ibid.*). These magmatic activities are believed to be responsible for climatic and biotic crisis around the Triassic-Jurassic boundary (Marzoli et al., 2004). In the Southern High Atlas, basalts are dated at about 196 Ma (Verati et al. (2004). In the southeastern Middle Atlas, they are dated Lower Liassic (Hettangian-Sinemurian)[1] (Ouarche et al., 2000). The basalts are characterized by weak to moderate environments of incompatible elements and negative niobium anomalies, which indicate contamination by continental crust. Some basalts show hydrothermal alterations, as indicated by amethyst druses.

The break-up of Pangaea continued with the opening of the Central Atlantic and the Tethys Oceans during the Late Triassic-Early Jurassic (Michard et al., 2008, Schettino & Turco, 2009). The opening of the Central Atlantic was marked by basaltic magmatism of the Central Atlantic Magmatic Province (CAMP)[2]. This event probably caused the climatic-biological crisis at the Triassic-Jurassic boundary. Magmatism on

---

1 This would be quivalent to 199.6–189.6 Ma according to IUGS (2009).
2 For CAMP in Algeria, see Chabou et al. (2010).

the African side is recorded by dikes, sills and lava flows, mostly dated at $200 \pm 1$ Ma (Michard et al., 2008). By the end of the rifting stage, Morocco was located opposite Nova Scotia (Sahabi et al., 2004). During the Jurassic, the Central Atlantic and the Tethys were connected through a transform fault system north of Morocco. The opening of the North Atlantic during the Cretaceous resulted in the eastward displacement and anticlockwise rotation of Iberia (Rosenbaum et al., 2002). The opening of the South Atlantic resulted in the closure of the Tethyan basins and the building of the mountain belts in Morocco. The inversion of the Atlas Rift and the subsequent formation of the Atlas Mountain Belt occurred during the Oligocene-early Miocene time interval. In the central Atlantic, this event was associated with higher spreading rates of the ridge segments north of the Atlantic Fault Zone. An estimated 170 km of dextral offset of Morocco relative to northwest Africa in the central Atlantic has been envisaged (Schettino & Turco, 2009).

Morocco is bounded by the active Mediterranean margin to the north and the passive Atlantic margin to the west. The two margins are different due to their structural or tectonic settings (Michard et al., 2008). The Atlantic passive margin of Gondwana is probably buried beneath the Mesozoic onshore and offshore basins, such as the Aaiun.

## 4.9   MOROCCAN MARGINAL BASINS

## 4.10   AGADIR, TARFAYA, AND ESSAOUIRA BASINS

The Moroccan Continental Margin is divided into three major structural-tectonic provinces defined by geophysical lineaments (Holik & Rabinowitz, 1992), from west (most offshore) to east (landward) (Fig. 4.3): 1) The Mesozoic Magmatic Lineation which correspond to sea-floor spreading magnetic anomalies (ouside the limit of the figure). 2) The Jurassic Quiet Zone (Lancelot & Winterer, 1980) with subdued magnetic anomalies. 3) The Diapiric Province characterized by salt diapirs and rotated continental basement fault blocks (Hinz et al., 1982). Seismic stratigraphic studies show a thick Mesozoic succession (Holik & Rabinowitz, 1992, Fig. 2.1). The Triassic high-velocity sequence overlies the irregular surface of the oceanic basement. The Upper Jurassic is characterized by deep-water carbonates represented by parallel continuous reflectors. The Cretaceous-Tertiary boundary is marked by a prominent erosional surface on top of a chaotic seismic unit, which may be caused by reefs, gravity slides, or volcanics (Lancelot & Winterer, 1980, Hinz et al., 1982, Holik & Rabinowitz, 1992).

The Atlantic Province or Domain includes the Westernmost High Atlas, Essaouira Basin, Tarfaya, Agadir Basin, Haha Basin, Doukkala Basin, Souss Basin, and Aaiun Basin, including their offshore and onshore parts. The Tarfaya Basin, the Essaouira Basin, the Aaiun Basin, and Agadir Basin were formed during the break-up of Pangaea and the opening of the Central Atlantic Ocean and the Atlas rift basins in Late Triassic-Early Jurassic times (Medina, 1995, Le Roy et al., 1998, Le Roy & Piqué, 2001, Hafid et al., 2008, Zühlke et al., 2004, Davison & Daily, 2010). This extension phase affected both the crystalline basement and the folded Paleozoic cover of the Anti-Atlas and is manifested by the intrusion of NE-SW-trending dikes and associated sills of gabbro and dolerite (Sebai et al. 1991).

## 4.11   ESSAOUIRA-HAHA BASIN

The Essaouira Basin together with the Haha Basin form a triangular area bordered by the Jebilet Massif to the north and the Western High Atlas to the south (Fig. 4.3). They extend westward into the offshore of the Atlantic Ocean. The western border of the Essaouira Basin is the break of slope of the continental shelf approximately at the 200 m (656 ft) isobaths (Le Roy et al., 1998). The onshore part of the Essaouira Basin is a wide syncline between the Jebilet and the Western High Atlas, with post-Creta-cous uplift of the Jebilet Massif and large evaporites-cored compressional structures. The Essaouira Basin and Haha basins lay within an area of the continental margin situated between Casablanca and Cape Sim (Le Roy et al., 1998). The basement is made up of Paleozoic successions affected by the Hercynian Orogeny. The offshore part of the basins contains numerous salt domes and diapirs. The timing of salt move-ments (halokinosis) in the Essaouira Basin began in the Early Jurassic and lasted until the lower Cretaceous (Frizon de Lamotte et al., 2008).

The development of the Essaouira Basin is linked to the structural evolution of the Central Atlantic Ocean (Le Roy et al., 1998) (Fig. 4.3). Two phases can be recognized (Laville et al., 2004): (1) a rifting phase dominated by normal growth faulting during the Late Triassic that led to the formation of numerous half-grabens in the Atlas rift, and (2) a postrift stage characterized by an aggradational sedimentary sequence which extends over the onshore part of the Basin. The Upper Cretaceous sequence is transitional into a progradational system. Subsidence curves show differences between the onshore and offshore areas, with low subsidence rates in the onshore (Le Roy et al., 1998).

## 4.12   GUERCIF BASIN

The Guercif Basin (Figs. 4.3, 16.6) is a Neogene basin located in northern Morocco between the eastern termination of the Rifian front and northern termination of the Middle Atlas System. It is separated from the Foredeep by the late Neogene Tezzeka spur of the Middle Atlas (Sani et al., 2000). This area is known as the South Rifian Corridor (named by Hodell et al., 1989), a seaway which connected the Mediterranean Sea and the Atlantic Ocean prior to the opening of the Strait of Gibraltar, or the Miocene foredeep of the Rif (Gelati et al., 2000). The basin was the result of the interaction between the Late Cenozoic deformation of the Middle Atlas and the thin-skinned thrust tectonics of the Rif Mountains (Gomez et al., 2000). The Guercif Basin has the charac-teristics of both a foredeep and a basin which developed in a strike-slip setting (Bernini et al., 1999). The Guercif Basin was developed on top of a late Cretaceous inverted Jurassic failed rift (Gomez et al., 2000, Sani et al., 2000), and experienced considerable subsidence during the late Neogene and Quaternary. The boundary between the Guercif Basin and the Middle Atlas is marked by E-W to ENE-WSW normal faults.

The history of stratigraphic nomenclature in the Guercif Basin has been summa-rized by Bernini et al. (1999). The pre-Neogene basement consists of a Paleozoic sec-tion overlain unconformably by a Permo-Triassic continental sequence of red marls and basalts up to 500 m (1641 ft) in thickness. Red marls and salt diapirs occur along faults in the basin. The Jurassic succession is made up essentially of dolomites, lime-stones and marls. No younger Mesozoic sediments have been found.

| | | ESSAOUIRA | Atlantic domain |
|---|---|---|---|
| Cenozoic | Compression | | Atlas uplift |
| | ～～～～～～～ | Open Marine conditions | Conversion between N. Africa & Iberia |
| Upper Cretaceous | Westward progradation | | |
| Cenomanian | | | Magnetics in North Atlantic |
| Aptian-Albian | Subsidence | | Oceanic accretion N. & S. Atlantic |
| | Block faulting ～～～～～～ | | |
| Barremian | | | |
| Hauterivian | Uplift in western part _ _ _ _ _ _ _ | Thermal relaxation | Rifting in N. & S. Atlantic |
| Valang-Berriasian | Subsidence ～～～～～～～ | | Distension between Iberia & Grand Banks |
| Upper Jurassic | Low subsidence | | |
| Oxford-Callovian | | ——— | Oceanic accretion in Central Atlantic |
| Middle Jurassic | Differential subsidence | Platform | |
| Lower Jurassic | | | |
| | ～～～～～～～～ | ——— | |
| Upper Triassic | WNW-ESE Extension ～～～～～～～ | Rifting | Rifting in Central Atlantic |
| | Block faulting | | |

*Figure 4.4* Evolution of Essaouira Basin and the Atlantic margin (after Le Roy et al., 1998).

The Neogene succession is up to 1500 m (4922 ft) in thickness (Bernini et al., 1999). The depocenter migrated through time. The succession (Bernini et al., 1999) includes the continental Draa Sidi Saada Formation at the base, overlain successively by the transitional and shallow-marine Ras el Kasr Formation, the thick open-marine Melloulou Formation, and the continental Kef ed Deba and Bou Irhardainene Formations.

The Guercif Basin development records a period of extension in late Miocene followed by an episode of contraction at the end of the Miocene. Most of the late Miocene deposition took place in a narrow graben (the Guercif Graben of Gomez et al., 2000). Two periods of volcanic activities are reported in the Guercif Basin: Shoshonite volcanism (8–5 Ma) (Tortonian-Messinian) and alkali basaltic volcanism (5–2 Ma) (Pliocene).

Two stages of late Cenozoic basin development were recognized by Gomez et al (2000) in the Guercif Basin: 1) During Tortonian, extension in the Guercif Basin, as well as in the tabular Middle Atlas and the Moroccan Meseta, resulted from sinistral shear zone and WSW-verging thrusting of the Rif. 2) During the Messinian, the change in the Betic-Rif plate motion resulted in kinematic changes in the Rif and Guercif Basin. Thrusting in the Rif became southward, and the Guercif Basin contracted accompanied by the closer of the marine seaway (South Rifian Corridor or Fore-deep of the Rif).

Only four exploratory wells have been drilled in the Guercif Basin; some reaching the Paleozoic section (Gomez et al., 2000). The GRF-1 well penetrated about 2 km of Neogene and Quaternary section, while the TAF-1X encountered less than 100 m of Cenozoic strata. According to Gomez et al. (2000), although the four exploratory wells did not encounter any hydrocarbons, the basin still has potential because of the presence of source rocks in the Middle Carboniferous Namurian shales in the Missour Basin (11% TOC), Lower Liassic and Toarcian, and Pliensbachian calcareous mudstones (4% TOC) successions (Beauchamps et al., 1996). Potential reservoirs include Triassic clastics, Lower Jurassic reef complexes, and lower Tortonian clastics. Seals include Triassic evaporites, Middle Jurassic mudstones, and Tortonian mudstones.

The above formations are disconcordantly overlain by sandstones and carbonates of the Messinian Kef ed Deba Formation (equivalent of the Laguno-lacustre of Kandek el Ourish of Benzaquen, 1965), several hundred meters in thickness (Sani et al., 2000); the Irhardaiene Formation (Pliocene) (equivalent to El Mongar Formation of Benzaquen, 1965 and the *Grès et Conglomerats Continentaux* of Colletta, 1977); continental conglomerates, calcarenites, and lacustrine calculutites of the Grès et Conglomerats Continentaux (Colletta, 1977) more than 400 m (1312 ft) in thickness, and finally by Plio-Quaternary alluvial conglomerates, sandstones, and lagoonal deposits.

All the major Neogene unconformities at the base of the Tortonian, the base of Messinian (Kef ed Deba) and the base of the Pliocene (Bou Irhardaiene) sequences observed on the surface and in the subsurface can be recognized in seismic sections (Sani et al., 2000).

## 4.13 TAFILALT & MAIDER SUB-BASINS

The eastern portion of the Anti-Atlas includes the sub-basins of Tafilalt, and Tizimi (Fig. 4.3). The Tlafilalt Sub-basin is one of the unexplored sub-basins in the Eastern Anti-Atlas of Morocco (Toto et al., 2008). The basin has a multi-phase deformation history. No wells have been drilled in the basin. Extension took place at the end of the Pan-African Orogeny in the Early Cambrian, followed by thermal subsidence till the end of the Silurian. However, major extension and subsidence occurred in the Middle Devonian. The Carboniferous compressional phase produced thin-skinned folds and thrust faults.

The Tafilalt-Maider region has been the subject of numerous studies. The Cretaceous deposits were studied by Lavocat (1948, 1951, and 1954). The Paleozoic sediments were examined by Destombes (1963, 1985a, 1985b, 1985c, 1987), Hollard (1967), Bultynck & Hollard (1980), Alberti (1981, 1982), Destombes et al. (1985), Bultynck (1989), Belka & Wendt (1992), Becker & House (1994), Belka et al. (1999), Wendt et al. (2001), Aitken et al. (2002), Klug (2005a, 2005b, 2005c), Wendt & Kaumann (2006), and Toto et al., (2008). Tectonic evolution has been discussed by Jacobshagen et al. (1988a, b), Piqué (1994), Cartig et al. (2004), and Helg et al. (2004). The Tafilalt Sub-basin is an oval shaped graben-like depression, bordered by two uplifted zones: one trending E-W-trending that separates this sub-basin from the Tizimi Sub-basin to the north, and NW-SE trending separates the Maider and Tafilalt sub-basin (Toto et al., 2008).

The succession overlying the Precambrian basement ranges from Cambrian to Upper Visean. The Cambrian comprises interbedded sandstones and shales. Silurian deposits have been described by Schoeffler (1950), Destombes (1985a, 1985b, 1985c, 1987), and Destombes et al. (1985). The Devonian and Carboniferous successions have been examined by Wendt et al. (1984, 2001), Wendt (1985, 1988), Belarbi (2000), Wendt & Belka (1991), and Aitken et al. (2002). The Devonian consists of limestones and shales with crinoids and brachiopods. Mud mounds are common in the Devonian and Carboniferous.

The most prominent structures in the Tafilalt area are E-W to NW-SE-trending faults (Toto et al., 2008). The faults are predominantly left-lateral strike-slip. South of Rissani, Ordovician rocks overlie Devonian rocks, indicating the presence of reverse faults. The Deformation has been attributed to either thin-skinned deformation (Miller 2003) or strike-slip movements (Mattauer et al., 1972, 1977, Jeannette, 1981). The late Paleozoic sedimentary cover is detached over a Siluro-Ordovician décollement with a northward displacement (Toto et al., 2008).

# PART 3

## Pan-African Shield & West African Craton

# Chapter 5

# Precambrian

## 5.1 STRATIGRAPHY

The exposed basement rocks of the Pan-African Shield cover approximately 10% of the area of Egypt and only 1% of the area of Libya, as well as Algeria and Morocco (Fig. 5.1). They outcrop in southern Sinai, the Red Sea mountain range of the Eastern Desert, and the Gebel Oweinat (Jabal Awaynat) area in the Western Desert which straddles the borders of Egypt, Libya and Sudan. They also occur in the Tibesti area in southern Libya and northern Chad, in numerous small isolated outcrops in the Jabal Al Qarqaf area in west-central Libya, in the Hoggar Massif (Touareg Shield) in the extreme southwest corner of Libya, Algeria, and the Anti-Atlas of Morocco. Radiometric dating indicates that the basement rocks are Archean to Early Cambrian in age. The depth to the basement in Egypt increases northward to more than 4.3 km (14,000 ft) (Rabeh & Ernst, 2009).

Magnetic, gravity, and seismic data show NW, NE, and E-W structural trends in the southern part of Sinai Peninsula, Gulf of Suez and western part of Gulf of Suez (Rabeh, 2003). The NW and E-W trends may be related to the Gulf of Suez and Red Sea stresses.

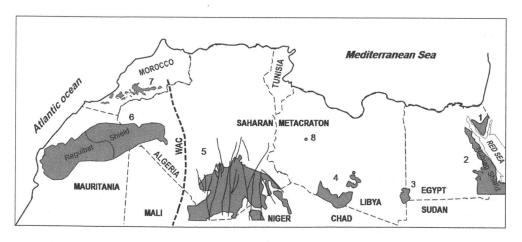

*Figure 5.1* Distribution of Precambrian basement outcrops in North Africa. 1. Sinai. 2. Red Sea Hill (Easten Desert). 3. Tibesti Massif. 4. Gebel Oweinat. 5. Hoggar Massif. 6. Reguibat Shield. 7. Anti-Atlas Domain. WAC = West African Craton.

The post-Pan-African evolution of the African-Nubian Shield in the Eastern Desert of Egypt was dominated by a late stage of vertical motion and the exhumation of a large area of the Shield (Bojar et al., 2002). Apatite, zircon and sphene fission track analyses from the Neoproterozoic basement indicate two major episodes of exhumation. An intraplate thermotectonic episode during the Late Devonian-Early Carboniferous, during which the intraplate stresses led to deformation, uplift and erosion, were induced by the collision of Gondwana with Laurussia which started in Late Devonian times. Apatite fission track data indicate that the second cooling phase started in Oligocene and was related to extension, uplift, and erosion along the margin of the Red Sea. Two stages of rift formation took place. 1) Cretaceous strike-slip tectonics resulted in formation of small pull-apart basins parallel to the trend of Pan-African structural elements. 2) NE-SW extension occurred during the Oligocene to Miocene period.

The basement rocks were grouped on the Geological Map of Egypt (1981, in Said, 1990a and references therein) into 13 units (Fig. 5.2). This grouping is followed here with some modifications: 1) Gneisses and migmatites, 2) Geosynclinal metasediments, 3) Geosynclinal Shadli, 4) serpentinites, 5) Metagabbros-diories, 6) Syntectonic- late tectonic granites, granodiorites and diorites (older granitoids), 7) Dokhan volcanics, 8) Hammamat Group, 9) younger gabbros, 10) younger granitoids, 11) Trachyte plugs (Cambrian-Cretaceous), and 12), swarm dikes, and 13) ring dike complexes (mostly Mesozoic). Said (1971), El Gaby et al. (1990) and Hassan & Hashad (1990) give detailed descriptions of these units. Bentor & Eyal (1987) give an extensive description of the Precambrian basement in southern Sinai.

The Geological Map of Libya (1985) shows five main assemblages of basement rocks: 1) The Archean assemblage includes migmatitic and granitic gneisses with local intercalations of amphibolites and diopside-hornblende gneisses, and it is restricted only to the Gebel Oweinat area (Fig. 5.1). 2) The Lower Proterozoic comprises medium- to coarse-grained, strongly foliated quartz-feldspathic gneisses with bands of pyroxene gneisses, amphibolites, marbles and ferruginous quartzites in the Gebel Oweinat. In the Tibesti Massif, amphibolites, basic metavolcanics and serpentinites with bands of mica-schists, phyllites, marbles, graphite schists, quartzites, and conglomerates are dominant. 3) The Upper Proterozoic encompasses massive metagreywackes and feldspathic sandstones, with limestones, siltstones, mudstones, shale interbeds, and conglomerate bands. This group of rocks is common in the Tibesti area in southern Libya, in a very restricted area in the northwestern Gebel Oweinat, and in the Hoggar Massif in the southwest corner of Libya. 4) The older granites/granodiorites unit is limited to the eastern Tibesti area. 5) Undifferentiated Precambrian granites and granodiorites are common in the southern Tibesti area and the southwest corner of Libya. The age and stratigraphic position of the last two groups are not clear from the Geological Map of Libya (1985), but they are believed to have formed between 600–500 Ma. Wacrenier (1958a, b) described and discussed the origin of the Precambrian basement rocks of the Tibesti Massif in northern Chad, which extend into Libya. Rogers et al. (1978), Ghuma (1975), Ghuma & Rogers (1980), and El Makhrouf (1988) discuss the geology of the Tibesti Massif in Libya.

There is still much confusion and debate about the definition of the different units which constitute the basement in North Africa, their mutual stratigraphic relations, their absolute ages, and the nature of their contacts (Figs. 5.3, 5.8). What

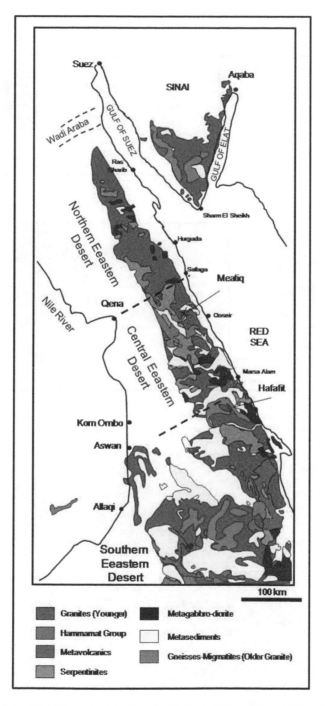

*Figure 5.2* Precambrian Shield, Egypt (mainly after the Geological Map of Egypt, 1971 and Liégeois & Stern, 2009) (See color plate section).

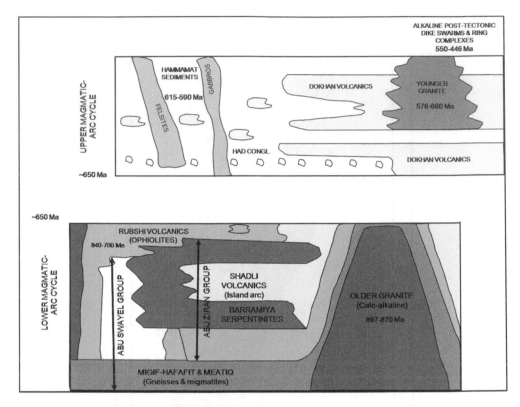

*Figure 5.3* Schematic cross-section showing the stratigraphic relations between the Precambrian units in the Eastern Desert, Egypt (based on many sources).

*Figure 5.4* Granite outcrop, Gebel Oweinat, SE Libya (courtesy of Sebastian Lüning) (See color plate section).

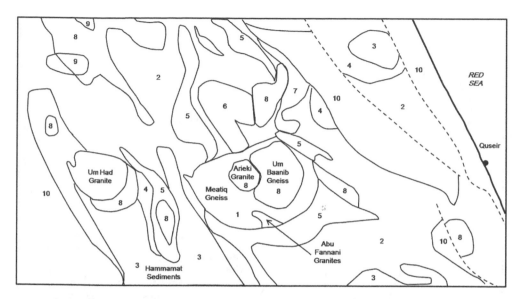

Figure 5.5 Meatiq map in Central Eastern Desert (after Geol. Surv. Egypt, 1971, Fowler & Osman, 2001, Andresen et al., 2009). 1. Gneisses, schists & migmatites. 2. Arc Metavolcanics. 3. Hammamat Group. 4. Dokhan Volcanics. 5. Ophiolitic melange. 6. Ophiolitic gabbros. 7. Post-Hammamat Felsites. 8. Younger Granites. 9. Post-tectonic Gabbros. 10. Post-tectonic cover.

Figure 5.6 Ancient Roman mine, Wadi Sikait, Eastern Desert (Courtesy of Baher El Kaliouby).

have been perceived by some workers as stratigraphic contacts are considered by others as structural or tectonic. Many of the units are believed to be allochthonous and the processes which led to their formation are still debated. Examples of these differences in points of views can be seen in the Kid Complex in east central Sinai and the Meatiq Complex in the central Eastern Desert. Many of the isotopic age

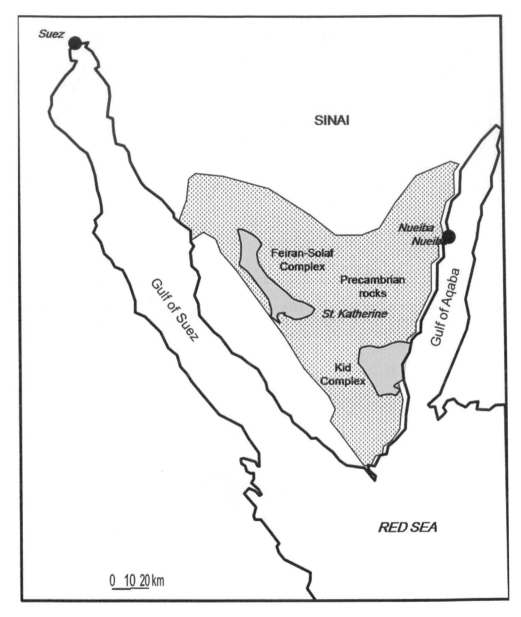

*Figure 5.7* Location of the Kid and Feiran Solaf complexes, Sinai (after Eliwa et al., 2008).

determinations probably reflect the ages of metamorphism, migmatization, or granitization of the rocks and may represent minimum ages for these rocks. Other age determinations, especially of metamorphic rocks of sedimentary origin, may indicate the age of the provenance of the sediments, giving, therefore, an age that is too old for these rocks. Also the age of allochthonous units may indicate the time of their

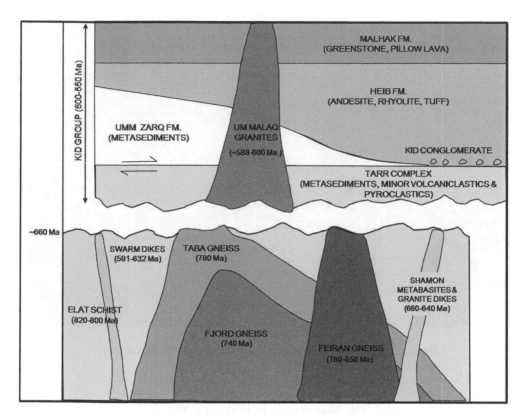

*Figure 5.8* Schematic cross-section showing the stratigraphic relations between the Precambrian units in Sinai, Egypt (based on many sources).

formation and not necessarily the time of their emplacement. Age determinations are also affected by the resetting of radiometric ages caused by the loss of Argon during thermo-tectonic events.

In the following discussion, the subdivision of the Precambrian (Plumb, 1991) into Archean (>2500 Ma), Paleoproterozoic (2500–1600 Ma), Mesoproterozoic (1600–1000 Ma), and Neoproterozoic (1000 Ma-base of Cambrian) has been adopted.

## 5.1.1 Gneisses and migmatites

This is a group of rocks ranging in age from 2900 Ma to 650 Ma and includes the gneisses and migmatites of the Migif-Hafafit Dome southwest of Marsa Alam, the Meatiq dome west of Qoseir, the Feiran-Solaf gneisses in Sinai, and migmatites of the Gebel Oweinat Complex, the Tibesti Massif, and the Hoggar Shield.

Archean rocks are known from outcrops in the Gebel Oweinat area (Figs. 5.1, 5.4), spanning the borders of Egypt, Libya and Sudan, the Reguibat and Eglab

*Figure 5.9* Recumbent fold, Taba, Sinai (Courtesy of Baher El Kaliouby).

Shields in Algeria and NW Mauritania. In that area, the basement is made up of Archean and Proterozoic folded gneisses, granites, migmatites, granulites, and metasediments, intruded by Mesozoic and Tertiary volcanic ring complexes and Eocene granites, granodiorites, and syenites. These rocks are strongly folded with a predominant fold axis trend in a NE-SW direction (Richter & Schandelmeier, 1990) in Egypt and Libya. Two metamorphic units were recognized on the Libyan side of the complex (Klerkx, 1980): 1) The Karkur Murr Series[1] is composed of gneisses with dips of 40° to 60° to the north, overlain unconformably by horizontal quartzites and intruded by alkaline ring intrusions. 2) The Ayn Daw Series consists of migmatitic granitic gneisses and it overlies the gneisses of the Karkur Murr Series. The contact between the two series is a NE mylonitic zone. Richter (1986) and Richter & Schandelmeier (1990), on the other hand, distinguished three basement formations in southwestern Egypt and northwestern Sudan, from base to top: 1) A high-grade granulite "Granoblastite Formation", equivalent to the Karkur Murr Series of Klerkx (1980), consists of grey-green to brownish-grey, fine to medium-grained, well layered gneissic granulites and granoblastites. 2) A remobilized "Anatexite Formation", equivalent to the Ayn Daw Series of Klerkx (1980), is characterized by the predominance of migmatitic gneisses with predominantly normal-granitic composition and minor syenitic and monzonitic varieties.

1  Note that the name Karkur Murr is used for a Precambrian series as well as a Carboniferous formation.

3) A bedded "Metasedimentary Formation" is made up of ensialic clastic rocks of psammitic and subordinate pelitic character. It is exposed mainly in northwestern Sudan, with no equivalent on the Libyan or Egyptian sides. In the northwestern part of the Gebel Oweinat Massif, beds of green, well-foliated, fine-grained, and frequently cataclastic paragneisses prevail. Orthogneisses of originally magmatic material occur in all three formations. These rocks are intruded by Precambrian grey-green calc-alkaline granitoids and red alkaline granites, and Paleozoic and Cenozoic porphyritic calc-alkaline granitoids and alkaline syenitic to alkali-granitic ring complexes. Radiometric age compilations by Richter & Schandelmeier (1990) give an Archean age of 3200–2556 Ma for the granulitic gneisses, a Paleoproterozoic age of 2399–1839 Ma for the migmatites, 1878 Ma for the granodiorite gneisses, 1784 Ma for the migmatitic biotite gneiss, and Mesoproterozoic age of 1446 Ma for the granodiorites. In addition, Late Neoproterozoic migmatites were dated at 673 Ma and granodiorites at 581 Ma east of the Gebel Oweinat Massif. Charnokites gave modal Nd ages of 3000 and 3200 Ma (Harris et al., 1984).

In the southern Western Desert of Egypt, the Bir Safsaf Complex, the Gebel el Asr Complex, and the Gebel Umm Shghir Complex (Fig. 5.2) form a large E-W trending system of basement uplifts. Uplift of these basement highs probably occurred during the Paleozoic, followed by subsidence during the Mesozoic and subsequent uplift during late Cenozoic times (Huth & Franz, 1988). The three basement inliers all belong to the same type of basement. In the Bir Safsaf Pluton, two different major dike rocks (calc-alkaline, tholeiitic) are typical of continental margins and extensional intracontinental regimes, respectively. The dikes of Bir Safsaf are basalts to rhyolites and have calc-alkaline and tholeiitic compositions. The dikes were intruded contemporaneously and shortly after the intrusion of granitoids, with tholeiitic dikes representing the final magmatic activity (Pudlo & Franz, 1994).

In the Tibesti Massif in southern Libya, there are two Precambrian units separated by an unconformity (Wacrenier, 1958, Klitzsch, 1966). These two units were called the *Tibestien inférieur* and *Tibestien supérieur* by Wacrenier (1958). Both units are metamorphosed, highly deformed, and intruded by granitic batholiths. They are believed to represent the lower Proterozoic (Paleoproterozoic) and upper Proterozoic (Neoproterozoic), respectively. The Lower Tibestian in Libya outcrops only in the east of the Tibesti Massif and is composed of amphibolites, basic meta-volcanics and serpentinites with bands of mica schists, phyllites, marbles, graphite schists, quartzites, and conglomerates, intruded by calc-alkaline porphyritic granites, granodiorites, and quartz diorites, and dikes of microgranites and pegmatites (Wacrenier, 1958, The Geological Map of Libya, 1985). The series is folded into NNE-SSW-oriented tight folds, and it is migmatized in part and metamorphosed into the low grade greenschist facies near the Ben Ghanimah Batholith (El Makhrouf, 1988). The Lower Tibestian metavolcanics of Jabal Eghei gave a Rb/Sr age of 939 ± 52 Ma and an initial $^{87}Sr/^{86}Sr$ ratio of 0.70304 ± 0.00023. The $^{87}Sr/^{86}Sr$ ratios suggest an upper mantle origin (El Makhrouf, 1988). Metasediments yielded an age of 873 ± 12 Ma and an initial $^{87}Sr/^{86}Sr$ ratio of 0.70736 ± 0.00023. The Upper Tibestian is made up of schists, sandstones, arkoses, rhyolites and conglomerates. A whole rock Rb-Sr analysis from a single rhyolite sample yielded an age of 788 ± 78 Ma (Schürmann, 1974). It is intruded by alkaline granites (Wacrenier, 1958) and calc-alkaline granites and granodiorites of the Ben Ghanimah Batholith dated at 550 Ma in the western Tibesti

Massif (Rogers et al., 1978, Ghuma, 1975, Pegram et al., 1976, Ghuma & Rogers, 1980, Fullagar, 1980), and the Eghei Magmatic Series (El Makhrouf, 1988) in the eastern Tibesti Massif. The Upper Tibestian series is metamorphosed into the upper greenschist to almandine-amphibolite facies (El Makhrouf, 1988). The Lower and Upper Tibestian series are separated by an aeromagnetic lineament. The presence of the aeromagnetic lineament and the change in grade of metamorphism suggests the presence of a suture zone between a continental margin of the Lower Tibestian (Jabal Eghei area) and a mature island arc of the Upper Tibestian (Ben Ghanimah Batholith area) (El Makhrouf, 1988).

In Egypt, the gneisses and migmatites were called the "Fundamental gneiss" by Hume (1932, 1934, 1935, and 1937). They are exposed in two major areas in the central Eastern Desert, the Wadi Hafafit and Meatiq anticlines (Fig. 5.5). These metamorphic core complexes occur parallel to the strike of the Eastern Desert Orogeny. They exhibit polyphase metamorphism and deformation (Habib et al., 1985, Fritz et al., 1996). NW-trending orogeny-parallel, sinistral strike-slip faults define the western and eastern margins of the domes. North and south dipping low-angle normal faults developed along the northern and southern margins of the domes and form extensional bridges between them (Fritz et al., 1996). The gneisses and migmatites of the Migif-Hafafit anticline, west of Marsa Alam, constitute a 4200 m (13,780 ft) sequence bounded on the north and east by a group of faulted and partly dismembered low-grade metasediments, metavolcanics, and plutonic rocks, and on the south by older granitoid batholiths (Greiling et al., 1984). Hedge et al. (1983) obtained an age of 680 Ma for the granitic gneisses in the core of the Hafafit dome. Akaad & Noweir (1969) described the Meatiq Group, which they believed to represent the oldest rocks in the Meatiq dome area, as composed mainly of 2500 m (8203 ft) of granulites (used in a textural sense), and minor pelitic and semipelitic schists, displaying sedimentary structures, such as graded bedding, current ripples, and ripple marks. According to Sturchio et al. (1983) however, the group is composed of granite gneisses, intruded by granitic plutons and overlain, with a shear contact, by quartzo-feldspathic mylonites and quartz phyllonites of the Abu Fannani Formation. Amphibolites occur in the granitoid gneisses, within the sedimentary cover of the gneiss, and within the shear zones bordering the Meatiq dome (Neumayr et al., 1996). Ages of these gneisses in the Eastern Desert average 670 Ma (Stern & Hedge, 1985). More recent Rb/Sr whole-rock dating yielded ages of $629 \pm 14$ Ma for the Ghadir Gneisses, $594 \pm 9$ for Wadi Sikait, $646 \pm 23$ Ma, and $608 \pm 18$ Ma for the Watir Gneisses (Basta & Stern, 1998).

At Wadi Sikait, emerald and other green beryls occur within the contact zone between schists and intrusive quartz and pegmatite veins and have been exploited by the Romans (Harrell, 2004) (Fig. 5.6).

Southern Sinai is occupied by large late to post-orogenic granites that intruded the metamorphic sequences and led to the formation of several isolated metamorphic complexes. They include the Taba Metamorphicc Complex in the east and the Feiran-Solaf Metamorphic Complex in the west (Fig. 5.7). The Precambrian basement rocks of Sinai (Fig. 5.7) consist of schists, orthogneisses, and plutonic rocks ranging in composition from gabbros to granites (Halpern & Tristan, 1981, Stern & Manton, 1987, Bentor & Eyal, 1987, Kröner et al., 1990).

The Feiran migmatites-gneisses are associated with granodiorites and delimited to the north and south by granites. Neither the age nor the origin of the gneisses is known. The gneisses were dated at approximately 782 Ma by Stern & Manton (1987), while Bentor (1985) reported a Rb/Sr age of 650 Ma, however, this age may represent the age of metamorphism and migmatization. Schürmann (1966) and Shimron (1980, 1983), on the other hand, consider the Feiran gneiss to have formed prior to 1100 Ma. Hassan & Hashad (1990) believe that the Feiran migmatites-gneisses were part of a sedimentary succession that was metamorphosed and migmatized as a result of batholitic granite intrusion during the Pan-African Orogeny (1100–500 Ma).

## 5.1.2  Metasediments

There are two groups of metasediments, an older group which includes the Wadi Mubarak (Group) geosynclinal sediments (Said, 1971), metasediments of the Abu Fannani Formation and the Abu Ziran Group (Akaad & Noweir, 1969), parts of the Wadi Abu Swayel Group and the Wadi El Mueilih Formation (El-Shazly, 1977) in the Eastern Desert, and the Elat Schist in Sinai, and a younger group which includes the metasediments of the Kid Group (Navon & Reymer, 1984) in southeastern Sinai. Volcanogenic clastic metasediments are the most common type of the younger metasediments with affinities to andesitic and dacitic volcanic rocks (Stern, 1979, in Stern & Manton, 1987, Bentor & Eyal, 1987). The age equivalent of the Kid Group in the Eastern Desert is the unmetamorphosed sediments of the Hammamat Group (Akaad & Noweir, 1969, Said, 1971).

The Wadi Mubarak Group was described by Said (1971) as a succession of more than 10,000 m (32,810 ft) at Wadi Mubarak, Eastern Desert, and composed of a succession of metamorphosed wackes, mudstones, slates, chlorite and biotite schists, and phyllites. The name is obsolete (B. El Kaliouby, personal communication), although it has been used by Shalaby et al. (2005). It appears to be equivalent to the Geosynclinal Metasediments on the Geological Map of Egypt (1981) and discussed by Hassan & Hashad (1990, p. 217–221). The sediments underlie the Shadli Metavolcanics and overlie the Migif-Hafafit Gneisses and Migmatites (Shalaby et al., 2005). The Wadi Mubarak belt in Egypt strikes west-east and crosscuts the NW trend of the Najd Fault System in the Central Eastern Desert of Egypt; therefore, the belt probably postdates the Najd Fault System. These intrusions include older gabbros, older granites, younger gabbros, and younger granites. Several major thrusts in the belt are correlated with the main deformation event in the Eastern Desert, known as D2. The Wadi Mubarak belt was subsequently affected during D3 by NW-SE sinistral conjugate strike-slip shear zones related to the formation of the Najd Fault System. The El Umra Granite has been dated by single zircon at 654 and 690 Ma (Shalaby et al., 2005). The Wadi Hafafit Dome is one of these domes and occupies the southern part of the Central Eastern Desert of Egypt. The Dome is cored with five separated gneissic domes ranging in composition from orthogneisses to paragneisses (Shalaby, 2010). They are overthrust by a low-grade, volcano-sedimentary association which constitutes the Pan-African cover nappes. The emplacement of the gneissic core probably occurred during accretion of the

Pan-African nappes (Shalaby, 2010). The strike-slip shear zones of the Najd Fault System postdate the emplacement of the dome.

### 5.1.3   Old metasediments

Akaad & Noweir (1969) divided the section overlying the Meatiq Group granite gneiss in the Meatiq Dome area into the Abu Fannani Formation and Abu Ziran Group. The Abu Ziran Group consists of metasediments, metavolcanics and serpentinites. It includes the metasediments of the Wadi Mubarak Group, the Barramiya serpentinites, the Shadli metavolcanics, and the Rubshi volcanics, and is partly equivalent to the Atalla Series of Schürmann (1966). Shackleton et al. (1980), Ries et al. (1983), and Sturchio et al. (1983) believe that the metasediments, the metagabbro-diorite complex of the Rubshi Group, and the Barramiya serpentinites constitute an ophiolitic mélange and not a stratigraphic sequence. Therefore, it is recommended that the name Abu Ziran Group be discarded. The Abu Fannani Formation is composed of schists, quartzo-feldspathic mylonites, and quartz phyllonites and appears to form a tectonic mélange zone along a thrust separating the Meatiq Group and the Rubshi volcanics (Shackleton et al., 1983, Ries et al., 1983, Sturchio et al., 1983). Muscovites within the shear zone bordering the Meatiq dome gave an $^{40}Ar/^{39}Ar$ age of 595–588 Ma (Fritz et al., 1996).

El-Shazly (1977) divided the sequence underlying the Barramiya Group serpentinites and Shadli Group volcanics in the southernmost part of the Eastern Desert into a pre-older granite Abu Swayel Group and a post-older granite Wadi El Muweilih Formation. The Abu Swayel Group is composed of a shallowing-upward sequence ranging from deep-marine to fluvial sediments and includes Said's (1971) Wadi Mubarak metasediments, Hume's (1934, 1935, 1937) Fundamental Gneiss, and the migmatites of the Wadi Hafafit and Meatiq Group. The Wadi El Muweilih Formation is dominant in the central part of the Eastern Desert and consists of more than 2700 m (8859 ft) of polymictic conglomerates with igneous rock fragments. El-Shazly (1977) further subdivided the Abu Swayel Group, in an ascending order, into the Wadi Haimur Formation composed of a 5000 m (16,405 ft) thick sequence of deep-marine calcareous pelitic sediments, the Wadi Mereikha Formation, a 1000–3000 m (3281 ft–9843 ft) thick section of pelitic sediments, and the Wadi Nagib Formation, 1000 m (3281 ft) in thickness and consists of shallow-marine psammitic sediments. The age of metamorphism of the metasediments is assumed to be ±1195 Ma (El-Shazly, 1977).

In southern Sinai, the Elat Schist consists primarily of quartz, oligoclase-andesine, biotite and some muscovite, metamorphosed to the amphibolite and locally to the granulite facies. It represents the oldest lithostratigraphic unit in the area, ranging in age between 820 Ma and 800 Ma (Halpern & Tristan, 1981, Stern & Manton, 1987, Kröner et al., 1987, Kröner et al., 1990, Eyal et al., 1991) and is probably equivalent to the Wadi Mubarak Group and Abu Swayel Group metasediments in the Eastern Desert. These rocks are believed to comprise vestiges of an ancient continental margin (Halpern & Tristan, 1981, Stern & Manton, 1987, Shimron, 1983).

### 5.1.4   Old metavolcanics

Two units of metavolcanics can be recognized, old metavolcanics (OMV) and young metavolcanics (YMV). Ali et al. (2009) pointed out that the subdivision of metavolcanics in the Central Eastern Desert into Older and Younger Metavolcanics is flawed since they have geochronological expressions of the 750 Ma crust-forming events, and the ophiolitic basalts previously considered as Old Metavolcanics may be slightly younger than the mafic-intermediate Younger Metavolcanics. Accordingly, they proposed that a new stratigraphic subdivision of Eastern Desert metavolcanics is needed, but until such time, the traditional simple subdivision will be used.

The old metavolcanics are widely distributed in the Eastern Desert of Egypt and probably have no equivalent in Sinai or in Libya. Hume (1935) applied the name Shadli geosynclinal metavolcanics after the type locality at Bir Shadli, 70 km southwest of Marsa Alam, Eastern Desert. They are approximately 10,000 m (32,810 ft) thick and composed of metavolcanics, and submarine volcanic effusions of basic and acidic composition, including rhyolites, andesites, basalts, and volcaniclastics (Hassan & Hashad, 1990). They overlie the Barramiya serpentinites and are overlain either by the Barramiya serpentinites or the Dokhan volcanics (according to Said's 1971 description). Searle et al. (1976) subdivided the Shadli metavolcanics into two units, a lower Wadi Um Samuiki Volcanics unit made up of basic and intermediate lavas with rhyolites and acid metavolcanics, and an upper Hammamid Group composed of a cyclic sequence of island-arc type metavolcanics. Radiometric ages for these volcanics range from 1070 Ma to 842 Ma (Hassan & Hashad, 1990). The Wadi Um Samuiki volcanics represent a highly dismembered ophiolitic unit and includes metagabbros, metadolerites, massive and pillowed spilitic metabasites, with thin beds of metasediments and chert (Takla et al., 1998). The ophiolites and island-arc units are invaded by younger Dokhan-type volcanics. The island-arc unit is intruded by post-tectonic granites in the northern part of the Um Samuiki area, west of Aswan, and by younger gabbros in the southern part.

In the Central Eastern Desert, two volcanic complexes at Wadi El Dabbah (amygdaloidal meta-andesite and metadiabase) and Wadi Kareim (metagabbro and metabasalt) have been studied recently by Ali et al. (2009). The Wadi El Dabbah lavas have affinities to the volcanic arc, whereas the Wadi Kareim rocks are similar to the Mid-Ocean Ridge Volcanics (MORB). However, the Wadi Kareim and El Dabbah metavolcanics contain abundant pre-Neoproterozoic zircons which might be derived from underlying sediments or older continental crust. Ali et al. (2009) proposed that a major crust-forming event involving ophiolite generation and eruption of juvenile melts took place from 736 Ma to 750 Ma.

## 5.1.5  Serpentinites

Serpentinites are widely distributed in the ophiolite assemblage of the basement complex of the Eastern Desert (Ritmann, 1958, Garson & Shalaby, 1976, El Sharkawy & El Bayoumy, 1979, Takla et al., 1980, Amstutz et al., 198, Habib, 1988, El Gaby et al., 1984, Khalil, 1994, Akaad & Abu Ella, 2002, Azer & Stern, 2007, Farahat et al., 2010).

Serpentinites occur in a number of outcrops in the Eastern Desert where they form the bulk of the Barramiya Group and parts of the Abu Ziran Group (Akaad & Noweir, 1969, Said, 1971). They also include the harzburgite-serpentinite unit in Sinai (Shimron, 1984). Ritmann (1958) suggested that the Barramiya serpentinites were related to ophiolites. The Barramiya Group serpentinites occur as steeply inclined sheets or as minor dike-like bodies or lenses. They underlie the metagabbros and diorites complex of the Rubshi Group or the Shadli metavolcanics and overlie either the Shadli metavolcanics or the Wadi Mubarak metasediments. The peripheral parts are altered into talc-carbonate rocks as a result of $CO_2$ metasomatism (Said, 1971). The serpentinites were believed to be ultrabasic intrusive rocks by the earlier workers, but they are probably allochthonous masses dismembered from mafic-ultramafic ophiolite complexes (Shackleton et al., 1980, Ries et al., 1983, Sturchio et al., 1983, Shimron, 1984, Church, 1986, Hassan & Hashad, 1990, Takla et al., 1998, Abd El-Karim et al., 2008).

## 5.1.6   Gabbros

There are essentially two types of gabbros which occur in the basement complex in Egypt, ophiolitic metagabbros and gabbros-diorites complexes. The metagabbros are believed to be synorogenic plutonites and older than the serpentinites and metavolcanics, but older than the older granites. The original gabbros and dolerites were subjected to a low grade regional metamorphism between the greenschist and the amphibolite facies.

The metagabbros and diorites occur in several outcrops in the Eastern Desert and Sinai, the largest of which is near Marsa Alam on the Red Sea coast. They are represented in the central Eastern Desert by the Rubshi Group, the type locality of which is located northwest of Wadi Dabr, West of Marsa Alam. They are composed of a heterogeneous assemblage of rock types of mainly metamorphosed basic rocks, including gabbros, norites, dolerites, and basalts. They commonly contain rafts of metasediments and form dike-like bodies and minor intrusive masses cutting the metasediments (Hassan & Hashad, 1990). The gabbroic rocks of Wadi Dabr include pyroxene-hornblende gabbro, hornblende gabbro, quartz-hornblende gabbro, metagabbro and amphibolite (Abu El-Ela, 1997). The composition of the pyroxenes ranges from diopside to augite. The plagioclases are $An_{34-75}$ for the gabbros and $An_{11-18}$ for the metagabbros. The gabbroic rocks are relatively enriched in large ion elements (K, Rb, Sr, and Ba) and depleted in Nb, Zr, Ti and Y which suggest subduction-related magma. They are enriched in light Rare Earth Elements (REE) [(La/Yb) N = 2.67–3.91] and depleted in heavy REE [(Tb/Yb) N = 1.42–1.47], which suggest the parent magma was relatively primitive mantle source. They also include the Dahab gabbro-diorite complex in Sinai (Shimron, 1984), and the Sherira Gabbro/Diorite Formation (Furnes et al., 1985). Recently the older gabbros were interpreted as a part of an ophiolitic sequence (Shackleton et al., 1980, Ries et al., 1983, Sturchio et al., 1983, Shimron, 1984, Church, 1986, Hassan & Hashad, 1990). Small masses of gabbros and diorites occur in the extreme SW corner of Libya (Radulovic, 1984, Grubic, 1984).

A very few radiometric age determinations were reported from the metagabbros and diorites. U-Pb ages for ophiolitic gabbros, diorites, and plagiogranites in western Arabia range from 840 Ma to 700 Ma (Pallister et al., 1987).

## 5.1.7   Granitoids

Two types of granitic rock associations occur in the Pan-African Shield, the first is pre-Hammamat (older) and the other is post-Hammamat (younger). The older granitoids are of calc-alkaline composition formed before ca. 670 Ma and younger granitoids of predominantly alkaline character, but also include calc-alkaline types, formed after 670 Ma. However, absolute radiometric ages of the two types of granites often overlap (cf. Ragab et al., 1989, 1991, Ragab & El-Gharabawi, 1989).

Hume (1935) considered the Egyptian granites to be of Ghattarian (Late Precambrian) age and younger than the Dokhan volcanics and Hammamat sediments. Schürmann (1953) added the Shaitian Granite to Hume's classification and considered it as the oldest plutonic cycle preceding the older granites and not younger than the Dokhan. El Ramly & Akaad (1960) later divided the Egyptian granites into two groups, the "older granite" of grey color and tonalitic to granodioritic composition and the "younger granite" of pink and red colors and of granitic and alaskitic composition. The two granites are separated by the Dokhan volcanics and Hammamat sediments. The older granites have ages between 987–670 Ma (Ries et al., 1983, Stern et al., 1984, Stern & Manton, 1987). Diorites in the Ras Gharib area gave an absolute age of 881 ± 50 Ma (Abdel Rahman & Doig, 1987). Granodiorites in the central Eastern Desert yielded an age of 670 Ma (Stern et al., 1984). Granodiorites and diorites in the Ras Gharib area were dated at 881 ± 50 Ma (Abdel Rahman & Doig, 1987). The Elat Schist is intruded by calc-alkaline plutonic rocks ranging from gabbroic to granitic orthogneisses, such as the Feiran, Fjord, and Taba gneisses (Fig. 5.9). Granodiorites at Wadi Feiran, Sinai, gave an age of 780 Ma (Stern & Manton, 1987), which is the same age as the Taba gneiss (Kröner et al., 1987, Kröner et al., 1990, Eyal et al., 1991).

The older group is also described as the arc granitoids (Syn-Orogenic Granitoids) that include the older grey granites and Shaitian granites, predominantly grey color and range from quartz diorite, trondhjemite, tonalite, granodiorite to quartz monzonite. The Shaitian granite, best developed in the area of Wadi Shait in the south Eastern Desert is a cataclased granodioritic rocks composed principally of oligoclase, potash feldspar, quartz and biotite with minor chlorite after biotite, epidote and opaque. The Shait granite may represent an early phase of the arc granitoids (B. El Kaliouby, personal communication). They were emplaced during the main phase of orogeny and were classified as subduction related plutonic rocks of the ensimatic island arc stage. Chemically, they are calc- alkaline. The age of the arc Granitoids ranges between a maximum of 850 Ma and a minimum of 614 Ma.

## 5.1.8 Dokhan volcanics

The Dokhan volcanics represent an assemblage of predominantly calc-alkaline volcanic rocks ranging in composition from intermediate andesites to acidic rhyolites, rhyodacites, alkaline trachytes, and welded tuffs (Ries et al., 1983). They are characterized by the dominance of porphyritic types and a purple color. In the Eastern Desert, they overlie the older granites and are overlain by and interfinger with the Hammamat Group sediments (Gass, 1982, Furnes et al., 1985). They include the purple Imperial Porphyry *profido rosso antico* in Wadi Hafafit in the Eastern Desert and Wadi Feiran in Sinai and have been exploited since ancient times. Radiometric age determinations from the Dokhan volcanics range from 639 Ma to 581 Ma (Stern & Hedge, 1985) and 639–602 Ma (Ries et al., 1983). Rb-Sr whole-rock dating gave an age of 620 ± 16 Ma for the Dokhan extrusive rocks in the Ras Gharib area

(Abdel Rahman & Doig, 1987). In the subsurface of Cyrenaica and the Sirte Basin, andesites, rhyolites and chloritic schists in the greenschist facies are common. Single sample K-Ar and Rb-Sr dating gave ages of 673–460 Ma (Schürmann, 1974). The Dokhan volcanics are probably equivalent to the Heib volcanics in southern Sinai.

Basaltic to rhyolitic lavas from the Dokhan Volcanics in Wadi Um Sidra and Wadi Um Asmer areas, NW of Hurghada, are made up predominantly of intermediate, medium- to high-K calc-alkaline lavas with subordinate adakitic lavas and minor alkali basalt (Eliwa et al., 2006). Moderate Mg and elevated Cr and Ni in adakites suggest that they originated from slab melts that later interacted with mantle peridotites. Alkali basalt has little Nb-Ta negative anomaly suggesting an OIB mantle source (Eliwa et al., 2006). The Dokhan Volcanics appear to have formed in the transition stage between subduction and post-collision tectonics. The Origin and tectonic setting of the late Neoproterozoic DokhanVolcanics (ca. 610–560 Ma by whole rock Rb-Sr and ca. 600–590 Ma by SHRIMP U-Pb zircon) in the Egyptian Eastern Desert is debated. They range from medium- to high-K basalts to rhyolites and comprise medium- to high-K calc-alkaline lavas, subordinate adakitic lavas, and minor alkali basalts. Adakitic lavas contain high Cr and Ni and low Nb, Rb, and Zr contents compared to coexisting calc-alkaline lavas. These alkali basalt, calc-alkaline, and adakitic lavas all show enrichment in large ion lithophile elements (LILEs) and show negative Nb-Ta anomalies with high Th/Zr, which suggest involvement of subducted slab melts in their petrogenesis (Eliwa et al., 2006).

The Fatira area, located on the Qena-Safaga road, consists of volcanic rocks that comprise island arc Metavolcanics and the Dokhan Volcanics. Three volcanic sequences have been recognized (Khalaf, 2010), ranging from continental alluvial fan and submarine shelf deposits. The sequences are the Fatira El Beida tholeiitic basalts and turbiditic volcaniclastics, which were deposited in an extensional back-arc basin; Fatira El Zarqa, which preserved the calc-alkaline volcanism and sedimentation in an alluvial fan basin; and the youngest domal alkali felsic volcanic and dyke-sill swarms of the Gabal Fatira sequence that formed in an anorogenic setting. Stacking of lithofacies associations is aggradational and is consistent with limited lateral migration of facies tracts in fault-confined basins.

The Fatira El Beida volcanic sequence (Khalaf, 2010) is of limited distribution. The lithofacies of the sequence include four groups of volcanics, including sheet lavas, pillow lavas, pillow breccias, and stratified volcanogenic sediments. The Lower volcanic suite comprises black, green and grey varieties of pyroxene andesites, hornblende andesites, aphyric and amygdaloidal andesites. The volcaniclastic deposits are course- to fine-grained, evenly bedded and poorly sorted. Occasionally, volcaniclastic breccias up to one meter (3.3 ft) in thickness occur within the sequence. They are composed of disorganized subangular to angular, dominantly cogenetic lithic clasts up to 30 cm in size. The volcaniclastics are composed of unsorted lapilli and tuff-sized crystal and lithic clasts with a fine-grained matrix. The Upper volcanic suite is composed mainly of brown, pink and red dacites and rhyolites. The pumice-rich felsic tuffs are made up of fine-grained, massive, with rare lithic fragments. Framework grains in these tuffs are dominantly broken feldspar clasts, volcanic lithic fragments and vitriclasts with minor volcanic quartz and opaque.

The Fatira El Zarqa sequence (Khalaf, 2010) includes subaerial calc-alkaline intermediate to felsic volcanics and an unconformably overlying siliciclastic succession comprising clast-supported conglomerates, massive sandstone sheet floods and mudstones, and a lateritic argillite paleosol top formed in an alluvial-fan system. The youngest rock of Gabal Fatira sequence comprises anorogenic trachydacites and rhyolites. Fatira El Zarqa sequence forms high mountainous ridges with uneven topography and includes voluminous lava flows and volcaniclastics unconformably overlain by alluvial fan volcanigenic sediments.

The Gabal Fatira sequence (Khalaf, 2010) is the youngest sequence in the Fatira area. It includes both dacitic/rhyolitic domes and bimodal sill/dike rock swarms. These rocks can be subdivided texturally into dacite porphyry, aphyric and granophyric rhyolites.

Younger sediments include the Hammamat Group in the Eastern Desert and the Kid Group in Sinai. Although the Kid sediments are metamorphosed, the Hammamat sediments are regionally unmetamorphosed (Said, 1971, Akkad, 1972, cited by Hassan & Hashad, 1990, p. 229). The Hammamat Group has been described as a foreland molasse (Fritz & Messner, 1999, Osman et al., 2001, Shalaby et al., 2006, Abd El-Rahman et al., 2010).

The name Hammamat Series was first applied by Hume (1934) to a 1500 m (4922 ft) thick sedimentary succession that constitutes a part of the Precambrian basement in the Eastern Desert and Sinai. The Hammamat Group (Said, 1971, p. 65) is made up of a succession of unmetamorphosed clastic beds consisting essentially of arenites, wackes, and conglomerates. Akaad & Noweir (1969, 1980) estimated a thickness of 4,000 m (13,124 ft) for the group and divided it into two formations. The upper Shihimiya Formation comprises, from top to bottom, the Umm Hassa Greywacke Member, the Umm Had Conglomerate Member (*breccia verde antico*), and the Rasafa Siltstone Member. The lower Igla Formation is composed of arenites, conglomerates, wackes, and red-purple to brick-red hematitic siltstones. It overlies unconformably the Dokhan volcanics. El Shazly (1977) applied the name Hammamat Supergroup and divided it into the Wadi Kareem Group at the base and Wadi El Mahdaf Formation at the top. The Wadi Kareem Group was further divided, in an ascending order, into the Wadi Melgi Formation, a 983 m (3225 ft) conglomerate section, the Wadi Tarfawy Formation, 2186 m (7172 ft) in thickness and composed of wackes, mudstones, and siltstones, and the Wadi Abu Ghadir Formation, made up of 2700 m (8859 ft) of conglomerate. The Wadi El Mahdaf Formation is 2700 m (8859 ft) in thickness and contains pebbles of the younger granites.

The Um Hassa Greywacke Member, the uppermost unit of the Hammamat Group in the Wadi Hammamat area, overlies conformably the polymictic conglomerates of the Umm Had Member. The greywackes are poorly sorted and composed mainly of quartz, lithic fragments, and plagioclase grains (Abd El-Rahman et al., 2010). The lithic fragments are predominantly intermediate to felsic volcanic rocks. The provenance of the Um Hassa sandstones was probably the continental arc volcanic rocks and oceanic island arc ophiolites. The Umm Hassa greywackes are believed to have been deposited in a retro-arc foreland basin behind the Dokhan continental arc developed over a west-dipping subduction zone (Abd El-Rahman et al., 2010).

The Hammamat Group sediments overlie and intercalate with the Dokhan volcanics (Gass, 1982, Furnes et al., 1985). The conglomerates contain pebbles of Dokhan-type volcanics (Ries et al., 1983), suggesting that the Hammamat sediments and the Dokhan volcanics are probably contemporaneous. The age of the Hammamat Group sediments is between 616 Ma and 590 Ma, based on Rb-Sr whole rock analysis of the underlying Dokhan volcanics and of granites which intrude the sediments (Ries et al., 1983). These age determinations are comparable with the ages of the Dokhan volcanics from 639 Ma to 581 Ma (Stern & Hedge, 1985). A sample from near the base of the sedimentary Hammamat Group at Gebel Umm Tawat, North Eastern Desert, contains detrital zircons with sensitive high-resolution ion microprobe (SHRIMP) U-Pb ages as young as $585 \pm 13$ Ma (Wilde & Youssef, 2002).

The Kid Group in southern Sinai (Figs. 5.7 & 5.8) is a complex metavolcanic and metasedimentary terrane with various related subvolcanic intrusions, and reaches a thickness of more than 12,000 m (39,372 ft) (Fig. 5.8). The relation between the different lithologies within the Kid Group and equivalents in the Eastern Desert is intricate and not yet clear. For this reason, both the volcanic and metasedimentary units of the group will be described here. Navon & Reymer (1984) originally divided the rocks in southeastern Sinai, from base to top, into the Umm Zariq Formation, Tarr Formation, and Heib Formation. Subsequently, Shimron (1987) viewed the group as composed of four major tectono-stratigraphic units, from south to north and in an ascending order: The Tarr Complex, Heib Formation, Umm Zariq Formation, and Malhak Formation. Some of these units may be allochthonous (Shimron, 1983, 1984, 1987). Metapelitic rocks of the Kid Complex gave a Rb-Sr date of about 600 Ma (Halpern & Tristan, 1981). The whole sequence is folded into NNE-striking folds and NW-trending cross-folds (Fig. 5.9). Two deformation phases have been recognized. The first phase, D1 (pre-620 Ma), produced regional foliation and upright folds, and was attributed to subduction in an island-arc setting (Shimron, 1980, 1983, Reymer, 1983, 1984, Blansband et al., 1997). The second phase, D2 (post-620 Ma), is expressed by expressed by the widespread development of thick, sub-horizontal mylonitic zones. Shimron (1980, 1983) interpreted the D2 phase as due to thrusting, while Reymer (1983) related the D2 structures to diapiric emplacement of granites. Blansband et al. (1997) interpreted the mylonitic zones as due to low-angle normal shear in a northwest direction, which were related to core-complex development during an extensional event, and postulated that doming induced reversal of transport direction to the southeast.

The name Tarr Formation was proposed by Reymer & Yogev (1983) for a unit composed of tuffs, flows, greywackes, pyroclastics, calc-silicates, and Mg-rich marble, which they divided into two members. The lower Umm Barram Member is more than 200 m (656 ft) thick and is composed of light brown to dark grey fine to medium-grained pyroclastics and some fine grained greywackes, and an upper El-Arkana Member, resting with a conformable contact on the Umm Barram Member. The El-Arkana Member shows great variations in thickness, reaching more than 800 m (2625 ft), and it is composed of cross-bedded greywackes, with intercalations of slates and marbles in the lower part and pyroclastics and intermediate acid flows in the upper part. Due to the intricate and complex nature of the Tarr Formation, Shimron (1983) proposed the name Tarr Complex for this unit. The Tarr Complex forms the southernmost unit of the Kid Group and it is composed of highly deformed, tectonically amalgamated blocks of metasedimentary rocks, and minor amounts of volcanic rocks of oceanic crust and mantle affinities (Shimron, 1984). The Umm Zariq Formation in the centre of the area is composed of cross-bedded and laminated wackes with abundant pelitic layers, and isolated marble lenses. It has a minimum thickness of 1500 m (4922 ft) (Reymer & Yogev, 1983). Detrital components in the wackes are quartz, andesine, and volcanic rock fragments. It is in fault contact with the Tarr Complex. The Heib Formation is more than 1000 m (3281 ft) thick and overlies unconformably both the Umm Zariq Formation and Tarr Complex. It forms a sequence of amygdaloidal andesites and rhyolitic tuffs and rhyodacitic flows with sedimentary and pyroclastic intercalations, conglomerates, and minor clastics. The contact between the Heib Formation and the Tarr Complex is a major mega-shear zone marked by mylonites, mélanges, albitites, and some ultramafic rocks of oceanic crust and upper mantle affinities (Bentor & Eyal, 1987). The Heib Formation is probably equivalent to the Dokhan volcanics in the Eastern Desert, but the exact relation between the two volcanic suites is not yet clear. The Kid Conglomerate (Shimron, 1987), up to 1000 m (3281 ft) in thickness, separates the Tarr Complex and the Heib Formation. Most of the pebbles in the conglomerates are of volcanic origin, with some granitic and sedimentary rock fragments (Reymer & Yogev, 1983, Bentor & Eyal, 1987). The northern part of the Kid metamorphic complex comprises a thick sequence of volcano-sedimentary association of the Malhaq Formation. The Malhaq Formation forms a greenstone belt to the north composed of intercalated marine sediments, pillow lavas, pillow breccias, and hyaloclastics of subaqueous origin (Furnes et al., 1985, Shimron, 1987). It is composed predominantly of rhyodacitic to andesitic, less commonly of subalkaline basaltic metavolcanics and meta-tuffs interbedded and intercalated with metapelites and metagreywackes. Their bulk rock chemistry indicates that the magmatic rocks have calc-alkaline affinity, and derived from island arc- mid-oceanic ridge transitional regimes, presumably in a back-arc setting (Abu El-Enen, 2008).

## 5.1.9 Younger gabbros

The younger gabbros form small intrusive discordant masses of fresh unmeta-morphosed gabbros and norites which were collectively grouped into one unit on the geological map of Egypt (Hassan & Hashad, 1990). They include norites, gabbro-norites, olivine norites, and olivine gabbros. These gabbros belong to a younger group of basic and ultrabasic intrusions, such as the Umm Bassilla Gabbros in the Red Sea range, which are believed to be post-tectonic intrusions of post-Hammamat age (Said, 1971). However, Ghoneim (1989) concluded that the younger gabbros were intruded in an arc-related sub-alkaline tectonic environment based on the presence of Ca-rich plagioclase ($An_{55-80}$), Mg-rich olivine ($Fo_{76.61-77.06}$), and Ca-rich pyroxenes. Younger gabbros also occur in small, isolated blocks in the Ben Ghanimah Batholith in the Tibesti Massif in west-central Libya. Most of these intrusions are now hornblende gabbros to amphibolites composed of brown and green amphiboles and calcic plagioclase ($An_{30-52}$) (Ghuma & Rogers, 1980). The gabbros have been dated at 556 Ma (Ghuma & Rogers, 1980) and 545 Ma (Fullagar, 1980).

## 5.1.10 Younger granites

The younger granites (El Ramly & Akaad, 1960) are widely distributed in the Egyptian basement and constitute a series of predominantly pink and red alkaline granites

(Fullagar, 1980, Greenberg, 1981), dominated by syenogranites and monzogranites with a lesser amount of granodiorites and tonalites. A large number of radiometric age determinations have been reported in the literature. A compilation of radiometric dates of the younger granites by Hassan & Hashad (1990) gives ages between 620 Ma and 530 Ma. Greenberg (1981) reported an age ranging between 603 Ma and 575 Ma. Granodiorites and granites in Sinai have been dated at 570–600 Ma (Halpern & Tristan, 1981, Stern & Hedge, 1985). The sediments of the Kid Group are intruded by the Umm Malaq Granite at about 590 Ma, resulting in a contact metamorphism up to the pyroxene hornfels facies (Halpern & Tristan, 1981). Younger granites which intrude the Hammamat sediments in the Central Eastern Desert were dated at 580 Ma (Engel et al., 1980, Sturchio et al., 1983) and at 615–570 Ma (Ries et al., 1983). The Aswan granite in the southern Eastern Desert and the Nakhil granite in the central Eastern Desert yielded ages of 595 Ma and 578 Ma, respectively (Sultan et al., 1990). Rb-Sr whole-rock isochron age determinations on basement igneous rock suites from the Ras Gharib area gave an age of 620 Ma (Abdel Rahman & Doig, 1987), and granodiorites of the Northern Eastern Desert were dated at 620–515 Ma (Stern et al., 1984).

Younger granitoids were also reported from the subsurface of eastern and central Libya. In Cyrenaica and the Sirte Basin area, many wells intersected the basement complex. Granodiorites and granitic intrusions are common. Single sample K-Ar and Rb-Sr dating gave ages of 672–530 Ma for these rocks (Schürmann, 1974). A few wells in the southern part of the Sire Basin have also reached granodiorites, migmatites, and gneisses. These rocks may represent the northern extension of the Tibesti Massif. Williams (1972) reported granitic rocks from the Augila oil field in the southeastern Sirte Basin composed of calc-alkaline granophyric and fine-grained granites with K-feldspars, plagioclase and biotite. The granites yielded K/Ar ages of 568 Ma and 400 Ma.

The granitic rocks from the Ben Ghanimah Batholith in the northwestern part of the Tibesti Massif have been dated at 600 ma to 500 Ma (Fullagar, 1980, Pegram et al., 1976, Nagy et al., 1976, Rogers et al., 1978, Ghuma & Rogers, 1980, Almond, 1991). They are mostly calc-alkaline in composition. The granitic rocks from the Ben Ghanimah Batholith in the northwestern part of the Tibesti Massif are typical calc-alkaline granitic suites (Rogers et al., 1978, Ghuma & Rogers, 1980). Hornblende-biotite granites, biotite granites, and minor intrusions of two-mica granites and monzonite granites intrude the Lower Tibestian Series at Jabal Eghei (El Makhrouf, 1988). They were dated at $554 \pm 6$ to $528 \pm 7$ Ma. Their initial $^{87}Sr/^{86}Sr$ ratios range from 0.70436 to 0.70791. Granites in the subsurface of the Murzuq Basin yielded ages of 554 Ma to 551 Ma (Bauman et al., 1992).

Precambrian rocks of the northeastern tip of the Hoggar Massif make up the Tassili Tafasasset range in the extreme southwest corner of Libya. These rocks consist of five assemblages (Radulovic, 1984, Grubic, 1984): 1) Migmatitic granites with zones of leucocratic granites, granodiorites, granodiorites, and diorite, intruded by aplite and pegmatite veins. 2) Mixed granites (with both intrusive and partly migmatitic characteristics), porphyritic, calc-alkaline biotite granite. 3) Intrusions of coarse-grained, porphyritic, calc-alkaline biotite granites (Taourirt Granite), which cut the older granites. The intrusions are dissected by veins of quartz, microgranites, and microdiorites. 4) Small masses of diorites and gabbros oriented in a NW-SE direction and cut by mixed granites. 5) Volcanic rocks, which include rhyolites and dacites, composed of perthitized orthoclase and quartz. The ages of these rocks are uncertain, but they are believed to have formed between 600 Ma-500 Ma (Radulovic, 1984, Grubic, 1984). The Touareg Shield is separated from the >2700 Ma West African Craton by a major Pan-African suture line (Dostal et al., 1996).

Precambrian basement rocks crop out also in a large number of small, isolated exposures in the Jabal Hassawnah (Al Qarqaf) area (Jurak, 1978). They consist mainly of medium to coarse-grained, porphyritic, biotite or two-mica granites composed of approximately 50% K-feldspars, 10% albite-oligoclase, and 40% quartz, with small inclusions of biotite schists, and they are intruded by aplite granites and pegmatites. K/Ar age determinations on biotite granites yielded an age of 554 Ma. Rb/Sr isochron ages are 519 Ma and the initial $^{87}Sr/^{86}Sr$ ratio is 0.7116 (Oun et al., 1998a, b).

## 5.1.11  Trachyte plugs (post-Hammamat felsites)

The Trachyte plugs have been previously known as the post-Hammamat felsites (Hume, 1937, Akaad & Noweir, 1969, Said, 1971, Hassan & Hashad, 1990, p. 236). Although the latter is no longer used (Dr. B. El-Kaliouby, personal communication), it still appears occasionally in literature, such as Greiling et al. (1994) in Wadi Queih area northwest of Qoseir, Abd El-Naby et al. (2008) at Hafafit Dome, and Abdeen & Greiling (2005) at the Meatiq Dome. They include effusive felsites, felsite porphyry and quartz porphyry sheets, plugs, and breccias which occur in small outcrops throughout the Precambrian basement. The felsites are composed predominantly of rhyolites and rhyodacites (Stern et al., 1988). The largest of these bodies is the Atalla felsite intrusion forming the summit of the Gebel Atalla in the central Eastern Desert. They form an elongate belt in the form of a sheet-like body injected along a thrust plane trending NNW-SSE. They intrude the ophiolitic mélange, island arc volcanics, and serpentinites (Hassan, 1998). The felsites are intruded by post-orogenic granites of the Um Had and Um Effein (Hassan, 1998), and rhyolitic flows and tuffs of the Dokhan volcanics. The Rb/Sr isochron age is 588 ± 12 Ma and the initial $^{87}Sr/^{86}Sr$ ratio is 0.7043 ± 0.0005 and they suggest derivation from the basal crust or upper mantle (Hassan, 1998). The felsites are probably directly related to or co-magmatic with the younger granites and Dokhan volcanics (Greenberg, 1981, Stern et al., 1988).

## 5.1.12  Swarm dikes

The Pan-African Shield is intensely intruded by post-tectonic dikes, commonly referred to as dike swarms (Fig. 5.10) and ring complexes. They are particularly very common in the Niger and Nigeria (Schlütter, 2008). In Egypt, this group includes the "Ghattarian plutonites and dikes", muscovite trondjhemites, dikes and peralkaline rocks, granodiorites, adamellites, and leucogranites in the Ras Gharib area (Abdel Rahman & Doig, 1987) and Phanerozoic alkaline rocks in the Eastern Desert (El-Gaby et al., 1990, Ries et al., 1983). The dikes are narrow, steeply dipping bodies, a few meters thick and several kilometers in length. They strike predominantly NE-SW to E-W; N-S-trending dikes are less common (Stern et al., 1988). They are peralkaline to alkaline igneous rocks, bimodal in composition (Stern et al.,

*Figure 5.10* Dike swarms, Sinai (Courtesy of Baher El Kaliouby) (See color plate section).

1988) and vary from mafic, medium to high-K andesite and alkali basalts to felsic dikes composed mainly of rhyolites. The dikes are mainly Late Precambrian-early Phanerozoic (600–500 Ma).

### 5.1.13   Ring dike complexes

Alkaline intrusions and granites of the Wadi Dib ring complexes in the Eastern Desert gave an average age of 554 Ma (Serencsits et al., 1981, Frisch, 1982). They are probably contemporaneous with the emplacement of the younger granites and associated with the Najd transcurrent fault system developed around 580–530 Ma (Stern, 1985, Fleck et al., 1976). Intrusion of the dikes was controlled by the pre-existing structure and individual igneous bodies were emplaced in an extensional tectonic regime during a phase of fracturing and uplift at the end of the Pan-African orogeny (Azer, 2006). Ring dike complexes with a central volcanic pile of alkaline rocks (Shimron, 1983) and N30°E-trending dike swarms dated at 591 Ma (Stern & Manton, 1987) occur in Sinai. Similar rocks, which consist of dacitic to rhyolitic flows and tuffs, and rhyolitic, microgranitic, granophyric, pegmatitic and aplitic dikes occur at Jabal Eghei, on the eastern side of the Tibesti Massif, were dated at 530–500 Ma (El Makhrouf, 1988).

The West African Craton (WAC) extends over millions of kilometers in the Sahara Desert (Fig. 5.11). The Archean and Proterozoic crystalline basement crops out in the Reguibat Shield of NW Africa and the Leon Shield in SW Africa and is covered by a thick pile of undeformed Neoproterozoic-Cenozoic sediments in the Tindouf, Reggane and Taoudenni basins. Early Neoproterozoic quartzites and stromatolitic limestones, 1000–700 Ma in age, are exposed on the southern border of the Reguibat Shield and in the Zemmour region on the northwestern side.

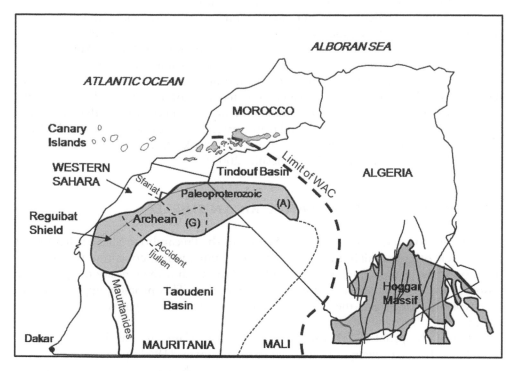

*Figure 5.11* Schematic geological map of West Africa and the West African Craton (WAC), after Bronner et al. (1980), Schofield & Gillespie (2007) and Schofield et al. (2006). (A) Amsaga Group. (G) Gallaman Group.

The Reguibat Shield is composed of two groups; the Amsaga and Gallaman groups, which represent two contrasting domains: a western Archean terrane (3960–2200 Ma) and an eastern Eburian terrane (2200–2000 Ma), respectively (Schofield et al., 2006, Schofield & Gillespie, 2007, Gasquet et al., 2008, Key et al., 2008, Hoepffner et al., 2005).

The Archean Amsaga Group (3.04–2.83 Ga.) is dominated by gneisses and granitic rocks with scattered lenses of metagabbros and serpentinites, laminated and ferruginous quartzites, felsic gneisses, pyroxene-rich gneisses, cherts, and impure marbles. Granitic intrusions yielded a Mesoarchean age of 3.04–2.83 Ga (Michard et al., 2008).

The Eburian (Paleoproterozoic) Gallaman Group is made up of granitic and metasedimentary rocks, such as mylonites and paragneisses with calcsilicate nodules, amphibolites, cherts, bimodal metavolcanics, such as rhyolites and basalts, ferruginous quartzites, and marbles. Banded iron formations are exploited in the Zouerat area in Mauritania. These iron deposits have been interpreted as chemical deposits precipitated in an aerobic environment in the presence of high concentrations of iron and cyanobacteria. Lemoine et al. (1990) recognize two separate units in the Paleoproterozoic Gallaman Group: A lower unit (equivalent to the Eburian I) of low to medium metamorphic grade, composed of basic bimodal magmatic rocks and some trondhjemitic to tonalitic anatectic gneisses and mica-schists, dated at 2183 Ma, and an upper unit (equivalent to the Eburian II), starting with polymictic conglomerates, with a low metamorphic grade and bimodal volcanics. They suggest that Eburian I is an independent orogenic cycle that took place around 2100–1800 Ma and proposed the name "Burkinian Cycle" for it. For the sedimentary and magmatic formations of the upper unit, they applied the stratigraphic name "Birimian". The contact between the Archean and Paleoproterozoic terranes occurs in the Sfariat region (Fig. 5.11). The Paleoproterozoic continental margin succession in that area was intruded by synorogenic granitoids and transported southwestward onto the Archean foreland during an oblique sinistral collision that took place around 2.12–2.06 Ga (Schofield et al., 2006, Schofield & Gillespie, 2007).

## 5.2    EVOLUTION OF THE PAN-AFRICAN SHIELD

Any attempt to interpret the geological history of the Precambrian Shield is hampered by the lack of precise definition of the different units which constitute it, a lack of understanding of their mutual stratigraphic relations, and the scarcity of systematic absolute age determinations and geochemical studies. This problem applies equally to Egypt and Libya and the Anti-Atlas basement and the Reguibat Shield in NW Africa.

One of the problems regarding these rocks is related to the definition of the term Pan-African orogeny and its time-span. Kennedy (1964) proposed the name "Pan-African" to denote a thermo-tectonic event at the close of the Proterozoic at about 500 Ma which caused the remobilization of Archean and Proterozoic rocks, their deformation, metamorphism, and migmatization, anatexis, and wide-scale intrusion of granites. Kennedy (1964) did not, however, specify a lower limit to this orogeny. The term was subsequently used by different workers to denote different time intervals. The majority of workers, for example El-Gaby et al. (1990), Vail (1991), and many others, restrict the term to the time span between 650 Ma and 550, and still others extend its upper limit to about 450 Ma. Gass (1977) suggested expanding the Pan-African event to about 1100–1000 Ma. Kröner (1980) accepted this concept and applied the term to the time interval starting from 1100 Ma to about 500 Ma. This expanded definition has been adopted in this work.

Early interpretations of the evolution of the Precambrian basement were based on the geosynclinal theory. The basement complex rocks in Egypt were thought to be represented by two orogenic cycles (Said, 1962a, 1971, El-Shazly, 1977): a lower flysch cycle and an upper molasse cycle, each cycle began presumably with deposition of geosynclinal sediments accompanied by intrusions of basic and ultrabasic rocks, followed by folding, metamorphism, and granitization. The mafic and ultramafic rocks of Egypt were interpreted by earlier workers as geosynclinal submarine flows (Rittman, 1958), or intrusive bodies emplaced into geosynclinal metasediments and metavolcanics (Akaad & Noweir, 1969).

During the last two decades, however, the evolution of the shield has been interpreted in the light of plate-tectonic models. Greenwood et al., (1976) and Gass (1977, 1981) visualized the Arabian-Nubian Shield as composed of a number of intra-oceanic island arcs that have been swept together during subduction of oceanic basin margins, along with their sedimentary aprons and slices of the oceanic lithosphere, to form new continental crusts. Gass (1977) defined five of these island arcs within the Nubian Shield in the Eastern Desert of Egypt and Sudan. The gabbros, diorites, tonalites, and serpentinites were subsequently described as oceanic substrata (Engel et al., 1980), or ophiolites (Shackleton et al., 1980, Stern et al., 2004) and the calc-alkaline volcanics were interpreted as island-arc assemblages (Searle et al., 1976, Gass, 1977, 1981). These earlier papers stimulated the reinterpretation of the Pan-African Shield in terms of modern plate-tectonic concepts. The interpretation of the geological history of the shield is therefore facilitated by grouping the different rock units into tectono-stratigraphic assemblages, each representing a specific depositional and tectonic setting within a certain time interval.

The basement rock units which form the African-Nubian Shield in Egypt and Libya can be grouped into five major tectonic cycles: 1) a lower Gebel Oweinat cycle (Archean, >2500 Ma), 2) an upper Gebel Oweinat cycle (Paleoproterozoic, 2500–1000 Ma), 3) a lower magmatic-arc cycle (Neoproterozoic, 1000–650 Ma), 4) an upper magmatic-arc cycle (Late Neoproterozoic-early Paleozoic, 650–550 Ma), and 5) a post-tectonic cycle (early Paleozoic, <550 Ma). The events of cycles 1 and 2 led to the establishment of the Nile Craton. The last three cycles constitute the Pan-African Orogeny.

The late evolutionary stages of the Pan-African Shield (650–545 Ma) document the growth and maturation of the continental crust from orogenic to cratonic. During this period magmatism changed from calc-alkaline to alkaline and the tectonic setting changed from collision to post-collision, to within-plate extension and finally to a stable platform setting (Be'eri-Shlevin et al., 2009). The nature of transition from calc-alkaline to alkaline magmatism is ambiguous. Microprobe (SHRIMP) U-Pb dating of zircons from the Dokhan Volcanics of the Eastern Desert suggests a significant inherited component, thus questioning the validity of previous Rb-Sr age determinations of these rocks (Wilde & Youssef, 2000). Geochronology based on high spatial resolution secondary ionization mass spectrometry (SIMS) dating of zircons from 24 plutons in the Sinai Peninsula (Egypt) and southern Israel (Be'eri-Shlevin et al., 2009) shows the overlap between a major pulse of felsic calc-alkaline magmatism and the onset of alkaline magmatism during the period ca.608–590 Ma. This suggests that a large proportion of calc-alkaline magmas were also intruded in a syn-extensional environment. The two suites probably represent different magma-source regions, however, calc-alkaline and alkaline magma chambers existed coevally at different levels of lithosphere and some interaction between them may have occurred.

## 5.2.11. Lower Gebel Oweinat cycle

The oldest rocks in Egypt and Libya are late Archean and represented only at the Gebel Oweinat and includes the granulitic gneisses of the Karkur Murr Series, dated approximately at 2700 Ma (Klerkx, 1980) and its equivalent gneisses, granites, migmatites, granulites, and metasediments of the Granoblastite Formation (Richter & Schandelmeier, 1990). This cycle coincides with a period of accelerated crustal growth which took place between 3000 and 2000 Ma (Meissner, 1986, p. 376). Although the information is very limited and scanty, it can be speculated that this cycle represents an earlier volcanic-arc, collision, and cratonization stage. It probably involved the deposition of sedimentary rocks during the Archean (>2500 Ma) in an oceanic basin, followed by subduction, granitization, migmatization, and metamorphism. Whether arc-type subduction did actually take place during the late Archean or these terranes were formed as a result of ensialic orogenies is still not clear. Meissner (1986) and Kröner (1991) asserted that subduction-related tectonics could have taken place during that period. The granulite facies of metamorphism may indicate the site of a continent-continent collision (Meissner, 1986, Morgan, 1990) and this process could be subduction related. Granulites are formed as sedimentary sequences are transported down to a great depth in collision zones and transformed into metasediments and then into sialic granulites (Meissner, 1986). The gneisses of the Karkur Murr Series and the migmatitic granitic gneisses of the Ayn Daw Series are folded in a NE-SW direction (Richter & Schandelmeier, 1990) and the contact between the two series is a NE mylonitic zone (Klerkx, 1980). Intracrustal shear zones consisting of mylonite belts are reported from different continents and have been attributed to large-scale transform faults or rotation of crustal blocks (Kröner, 1991). In the absence of ophiolites and other subduction-related phenomena in these rocks, intraplate tectonics related to hotspot activities remain a possibility.

## 5.2.2 II.   Upper Gebel Oweinat cycle

This episode took place during the Paleoproterozoic (2500–1000 Ma) and corresponds to one of two major world-wide crust forming (cratonization) episodes during the Proterozoic (Meissner, 1986, p. 375, Kröner, 1991). It is represented in the Gebel Oweinat area and the Tibesti Massif. It includes the Ayn Daw migmatitic Series (Klerkx, 1980) and its equivalent the Anatexite Formation migmatite gneiss, dated approximately at 1784 Ma in the Gebel Oweinat area (Richter & Schandelmeier, 1990), and the Lower Tibestian schists, quartzites, very coarse-grained feldspathic amphibolites, gneisses, pyroxinites, and marbles in the Tibesti Massif (Wacrenier, 1958, El Makhrouf, 1988).

The Gebel Oweinat and Tibesti terranes were subjected to NW-SE-oriented stresses which resulted in a major folding phase. The predominant fold axis trend is in a NE-SW direction (Richter & Schandelmeier, 1990) in the Gebel Oweinat area to NNE-SSW in the Tibesti Massif (Wacrenier, 1958). This folding stage was accompanied by the intrusion of calc-alkaline granitoids and migmatization. These late Archean-early Proterozoic rocks probably represent the eastern margin of the pre-1000 Ma Nile Craton (Rocci, 1965, Gass, 1982, Vail, 1976, 1985, 1991) onto which the Pan-African terrane accreted in the late Proterozoic time. Black & Liégeois (1993) believe that the Nile Craton, which they called the "Central Saharan Ghost Craton", was decratonized (oceanized?) during the Pan-African event.

Probably at the end of this period, during between approximately 1800 Ma and 1000 Ma, i.e. during the late Paleoproterozoic and all of the Mesoproterozoic, extension of the craton led to rifting and the formation of a graben within which a new oceanic lithosphere began to form. Stoeser & Camp (1985) proposed a rifting episode of the African craton between 1200 Ma and 950 Ma, and the Touareg Shield was presumably dominated by extension up to ca. 800 Ma (Dostal et al., 1996).

### 5.2.2.1    Rodinia Supercontinent (1300–1000 Ma)

Rodinia is defined as a supercontinent that contained all the continental fragments around Laurentia. Recent studies show that Rodinia was assembled during the Mesoproterozoic[2] through worldwide orogenic events between 1300 Ma and 1000 Ma and remained stable between 1000 and 750 Ma (Dalziel, 1997, Weil et al., 1998, Meert & Lieberman, 2007). According to the model adopted by Li et al. (2008), Rodinia was assembled by the accretion or collision of continental blocks around the margin of Laurentia and lasted about 150 million years after complete assembly. At about 1100 Ma many continental blocks were still separated from Laurentia by oceans. For example, Cordani et al. (2003) postulated the presence of an ocean which separated the South American and African cratons from the Laurentia-Amazonia-West Africa margin. This ocean was closed between 940 and 630 Ma (or between 1300–750 Ma according to other estimates). Widespread continental rifting occurred between ca. 825 Ma and 740 Ma, with episodic plume events at ca. 825 Ma, ca. 780 Ma and ca. 750 Ma, which resulted in widespread bimodal magmatism. The break-up of Rodinia occurred diachronically. The first major break-up event occurred along the western margin of Laurentia as early as 750 Ma. Rifting between the Amazonia Craton and the southeastern margin of Laurentia was completed at ca. 600 Ma. In the meantime, the western Gondwana continents continued to join together until the end of the Pan-African Orogeny ca. 530 Ma, which led to the final assembly of Gondwana (Meert & Lieberman, 2007). Dalziel (1997) postulated that the accretion process led to the formation of another supercontinent, Panntia, which included Gondwana, Laurentia, Baltica and Siberia, but this hypothesis did not meet much acceptance.

### 5.2.3 III.    Lower magmatic-arc cycle

Following a long period of extension and rifting dominated by deposition of shelf and basin sediments, probably between 1000 Ma and 650 Ma, the Pan-African Atlantic-type passive margin was converted into a convergent Andean-type margin with the onset of the Pan-African Orogeny. This event coincides with the second major worldwide crust-forming episode from 950 Ma to 650 Ma (Kröner et al., 1990, Kröner, 1991) (Figs.). Volcanic-arc activities and emplacement of calc-alkaline plutonic batholiths accompanied subduction of the oceanic crust underneath the Archean and early Proterozoic foreland. This cycle includes platform and basinal sediments, ophiolites and calc-alkaline volcanics, and associated granitic rocks which were deformed or emplaced during the Neoproterozoic between 1000 Ma and 650 Ma. It corresponds to the flysch cycle of earlier workers.

### 5.2.3.1    Platform and ocean basin assemblages

During the initial Pan-African stage, the Proterozoic Pan-African Shield in Egypt and Libya was a site of small ocean basins bordering the Nile and West African

---

2 Mesoproterozoic record (1600–1000 Ma) is absent over most North Africa. This period is equivalent to the Upper Gebel Oweinat Cycle (1800–1000 Ma) (Tawadros, 2001).

cratons characterized by island-arc orogenic activities (Gass, 1977, Greenberg, 1981). A shelf sequence probably passed basinward into a thick wedge of continental slope and rise deposits. Shales and fine quartzo-feldspathic sediments, the protoliths (parent rocks) of the Migif-Hafafit, Meatiq, Abu Swayel, and Abu Ziran groups in the Eastern Desert and the Elat Schist in southern Sinai, the Upper Tibestian schists, sandstones, arkoses, and conglomerates in the Tibesti Massif, and the psammitic and pelitic rocks of the Metasedimentary Formation in northwestern Sudan were deposited in these basins between 1000 Ma and 650 Ma, or even earlier.

The age of metamorphism of the Abu Swayel Group sediments is believed to have taken place ca. 1195 Ma (El-Shazly, 1977). However, an age of $768 \pm 61$ Ma was obtained from the Abu Swayel rhyodacite by Stern & Hedge (1985). The younger ages of the Migif-Hafafit granitic gneisses and migmatites of 680 Ma (Hedge et al., 1983) and the Meatiq Group gneisses may indicate the time of their granitization and migmatization.

The source of the Elat and Feiran-Solaf metasediments was probably the old Nile Craton (Bentor & Eyal, 1987) that bordered the African-Nubian Ocean, of which the Gebel Oweinat complex formed a part. The Meatiq Group is composed predominantly of continental quartzo-feldspathic sediments deposited on a continental shelf (Ries et al., 1983). Rare cobble beds in the metasediments in the Eastern Desert contain granitic and quartzite clasts (2300–1100 Ma) which were probably derived also from the Nile Craton (Engel et al., 1980).

The evolution of the Meatiq Gneiss Dome has been deduced by Andresen et al. (2009) by dating pre-, syn-, and post-tectonic igneous rocks in and around the Dome. The Um Ba'anib Orthogneiss, comprising the deepest exposed structural levels of the dome, has a crystallization age of $630.8 \pm 2$ Ma. The overlying mylonites are interpreted to be a thrust sheet/complex (Abu Fannani Thrust Sheet) of highly mylonitized metasediments, migmatitic amphibolites, and orthogneisses with large and small tectonic lenses of less deformed intrusives. Two syn-tectonic diorite lenses in this complex have crystallization ages of $609.0 \pm 1.0$ and $605.8 \pm 0.9$ Ma, respectively. The syn-tectonic Abu Ziran diorite, cutting across the tectonic contact between mylonite gneisses of the Abu Fannani Thrust Sheet and a structurally overlying thrust sheet, has a magmatic emplacement age of $606.4 \pm 1.0$ Ma. Zircons from a gabbro (Fawakhir ophiolite) within the thrust sheet yielded a crystallization age of $736.5 \pm 1.2$ Ma. The post-tectonic Fawakhir monzodiorite intrudes the ophiolitic rocks and has an emplacement age of $597.8 \pm 2.9$ Ma. Two other post-tectonic granites, the Arieki Granite that intrudes the Um Ba'anib Orthogneiss, and the Um Had granite that cuts the Hammamat sediments (Figs. 5.4, 5.6), have emplacement ages of $590 \pm 3.1$ and $596.3 \pm 1.7$ Ma, respectively (Andresen et al., 2009). The formation of the Meatiq Gneiss Dome is probably a young structural feature of 631 Ma contemporaneous with the folding of the Hammamat sediments around 605–600 Ma during oblique collision of East and West Gondwana and the formation of the Greater Gondwana (Liégeois & Stern, 2010).

The Touareg Shield was dominated by a period of thermal metamorphism which took place around 793 Ma, corresponding to a time of regional extension and widespread magmatism in West Africa, associated with the opening of a pre-Pan-African ocean to the west (Dostal et al., 1996).

### 5.2.3.2 Ophiolite assemblages

There are many ophiolite belts in North Africa (Fig. 5.1), which may represent suture lines. Examples of possible suture zones exist in Central Sinai, the Fawakhir Ophiolites, the Allaqi-Heiani and Hodeini trend in SE Egypt, the Tibesti Massif in Libya, the Hoggar/Touareg Massif and the West African Craton (WAC) in Algeria, and the Bou Azzer in the Anti-Atlas Domain in Morocco. However, there are controversies regarding the locations of these sutures, structures of the collision zones, and the polarity of subduction zones (El-Kazzaz, & Taylor., 2001, Abdeen et al., 2008).

Ophiolites are widespread in the southern and central Eastern Desert of Egypt, but are much reduced in the northern Eastern Desert and Southern Sinai, and they are virtually absent in Libya. They include the gabbros, diorites, granodiorites, and tonalites of the Rubshi Group and the Barramiya Serpentinites in the Eastern Desert (Shackleton et al., 1980), the Wadi Ghadir Ophiolites (Abd El-Rahman et al., 2009, Farahat et al., 2010, Basta et al., 2011), Umm Shagir (Abdel-Karim et al., 2008), Wadi Hodein (Abdeen et al., 2008), and the Dahab gabbro-diorite complex and the harzburgite-serpentinite units in Sinai (Beyth et al., 1978, Shimron, 1984, Abu El-Enen, 2008). Ophiolites were also reported along NW-SE-oriented shear zones from the Arabian-Nubian Shield on the eastern side of the Red Sea and dated at 840–700 Ma, based on their stratigraphic position (Pallister et al., 1987), and from NE-SW-oriented shear zones in northeastern Sudan, emplaced at their present level probably at about 700 Ma (Almond et al., 1989).

The origin of the ophiolites in the Pan-African Shield is also a subject of debate. Shackleton et al. (1980) and Ries et al. (1983) interpreted the ultramafic rocks of the Rubshi Group and the Barramiya Serpentinites in the Central Eastern Desert as allochthonous fragments of ophiolites. They were probably formed in a continental back-arc basin (Farahat et al., 2004, 2010). Beyth et al. (1978) and Shimron (1984) also believe that the Dahab gabbro-diorite complex and the harzburgite-serpentinite units in Sinai are ophiolites formed in a back-arc basin. On the other hand, it has been argued that not all ultra-mafic assemblages are ophiolites (Bentor, 1985, Stern et al., 1984, Church, 1988). Church (1976) observed that the ophiolites in the Eastern Desert occupy certain stratigraphic positions, i.e. autochthonous, and regarded them as eroded anticlinal structures. Church (1986), however, assumed that the ophiolites were emplaced during obduction of the oceanic crust and that their present distribution was controlled by deformational structures, although some of them may be intrusions (Church, 1988). Shackleton (1992) postulated that the ophiolites and ophiolitic mélanges in the Central Eastern Desert were thrust sheets of back-arc basin origin that were obducted northwestwards, and were later thrust contemporaneously with the north-south Hamisama transpressional shear zone (Kröner et al., 1987) east of the Gerf-Heinai ophiolites (Shackleton, 1992, Abdeen et al., 2008).

Another point of argument is whether collisions took place between two oceanic blocks, two continental blocks, or between continental and oceanic blocks. Ultramafic ophiolitic bands are generally interpreted as representing suture zones, either between intracratonic plates or continental and oceanic island-arc terranes (Gass, 1981, Vail, 1983, 1985, 1991, Stoeser & Camp, 1983, Kröner, 1985). There is no firm evidence to support one hypothesis or the other (see discussions under granitoids). In western Arabia, the ophiolitic gabbro, diorite, and plagiogranite assemblages occur within six major fault zones believed to mark sutures between crustal blocks that were accreted between about 633 Ma and 715 Ma to form the Arabian Shield (Pallister et al., 1987). Vail (1983, 1985) indicated that the ophiolite lineaments in the Arabian-Nubian Shield may represent ancient subduction zones. Harris et al. (1991) suggested that sharp isotopic variations coincide with structural discontinuities between terrane boundaries. The ophiolites in Sinai are believed to represent a suture line which extends in northwestern Arabia and the Eastern Desert (Beyth, 1981, Shimron, 1984). The present writer believes that this suture line probably extends southwestward in the subsurface of the Western Desert to the east of the Gebel Oweinat Complex (Fig. 2.1).

The serpentinites, metagabbros, and diorites in the Eastern Desert and southern Sinai represent ophiolitic oceanic plates probably generated during rifting and sea-floor spreading at the end of the Upper Gebel Oweinat cycle between 1800 Ma and 1000 Ma. The metagabbros probably originated from subalkaline low K-tholeiitic magmas in an ocean floor setting (Takla et al., 1998). The ophiolites were later obducted over the old continental margin (Kröner, 1985). With continued compression, thrusting and shearing, faults developed and portions of the ophiolites were detached from the main ophiolite bodies and displaced downslope for long distances to form mélanges in front of the obducted masses. Ries et al. (1983) observed that the contact between the Meatiq gneisses and migmatites and the ophiolitic mélanges is tectonic rather than stratigraphic, and concluded that the ophiolites were dismembered and moved by gravity sliding to be included in an extensive mélange. Shackleton et al. (1980) estimated that ophiolite masses in the Eastern Desert have moved by gravity sliding and later thrusting for distances of more than 200 km. Habib et al. (1985) recognized five thrust sheets in the Meatiq dome area. In that area, the mylonitic zone of the Abu Fannani Formation occupies a ductile shear zone between the gneisses and the overlying ophiolites (Habib et al., 1985, Fritz et al., 1996). Mylonitization, phyllonitization, and tectonic mixing of rock types occur within this zone (Sturchio et al., 1983). On the other hand, Church (1988) proposed that the displacement of the ophiolites also took place along strike-slip faults. Fritz et al. (1996) and Fowler & Osman (1998) postulated that the Meatiq Group gneisses were folded and sliced due to NW-oriented low-angle thrust systems. These thrust systems emplaced ophiolitic mélanges over the paragneisses such as Gebel Meatiq, El-Sibai, and El-Shalul. Doming of the shear zone was a result of a later ENE-WSW tectonic shortening event, which produced NW-striking thrust faults. This event was responsible for the thrusting of the serpentinites, Dokhan Volcanics over the Hammamat metasediments. These thrust systems represent the extension of the Najd Fault System wrenching (Stern, 1985, Sultan et al., 1988, Fritz et al., 1996, de Wall et al., 2001, Abdeen & Greiling, 2005, Abdeen et al., 2008). There is not enough information in the literature to indicate the formation of nappes similar to those described by from the Touareg Shield by Dostal et al. (1996).

### 5.2.3.3 Volcanic-arc assemblages

The partial melting of the oceanic crust under high pressure and shear stresses probably provided a source for calc-alkaline magmas to form island arcs. The Rubshi volcanic sequence consists of weakly metamorphosed, calc-alkaline intermediate to acid volcanics composed of andesites, dacites, and volcaniclastics typical of modern subduction related magmas. Bimodal volcanic associations of basalts and rhyolites and rhyodacites are subordinate. These volcanics were probably emplaced between 1070 Ma and 842 Ma (Hassan & Hashad, 1990). Erosion of the island-arcs supplied detritus to the basin which led to the deposition of the Elat Schist, composed predominantly of quartz and oligoclase-andesine plagioclase (Halpern & Tristan, 1981, Kröner et al., 1990, Eyal et al., 1991) and similar sediments in the Eastern Desert. The volcanogenic wackes are often banded, with graded bedding, and frequently intercalated with metamorphosed iron-ore. The island-arc association was later obducted or thrust over the ophiolite sequences, accompanied by the formation of mélanges of the Abu Fannani Formation (Ries et al., 1983) as discussed above. The opening of a pre-Pan-African ocean to the west of the Touareg Shield was followed by the development of an intraoceanic arc between 730–630 Ma and the formation of obduction/subduction-related calc-alkaline volcanic complexes (Dostal et al., 1996). Vein-type gold mineralization that constituted the main target for gold since ancient times is associated with this orogenic stage (Botros, 2002). It is hosted in stratiform to stratabound Algoma-type banded iron formations associated with tuffaceous sedimentary rocks (Sims & James, 1984, Botros, 2002, 2003, El Nagdy, 2004). In the meantime, economic talc deposits, such as those of the Um Saki area, are associated with intensively altered volcanic rocks (Botros, 2003).

### 5.2.3.4 Older granitoids, gneisses & migmatization (987–670 Ma)

The metasediments, ophiolitic and mélange sequences, and coeval calc-alkaline metavolcanic island-arc sequences in the Eastern Desert and Sinai were intruded by syntectonic calc-alkaline granitoid bodies (mostly granodiorites and diorites) of the (grey) older granites. The older granites yielded ages between 987–670 Ma (Ries et al., 1983, Hedge et al., 1983, Stern et al., 1984, Stern & Manton, 1987, Abdel Rahman & Doig, 1987, Kröner et al., 1990, Eyal et al., 1991).

The age and origin of the older granitoids and the associated migmatitic rocks are controversial. The gneisses and migmatites are believed to comprise vestiges of an older sialic continental margin (Halpern & Tristan, 1981, Stern & Manton, 1987, Shimron, 1983, Dixon, 1981, Sturchio et al., 1983, Habib et al., 1985) consider the Meatiq Group gneisses to be example of an older sialic sequence in the Egyptian basement. Shackleton et al. (1980) and Sturchio et al. (1983) believe that the complex of granitoid-cored domes in southeastern Egypt apparently evolved in a dominantly compressional tectonic environment during the late Proterozoic (630–580 Ma ago). Similar features which occur in the Wadi Hafafit structures were dated at 680 Ma by Hedge et al. (1983). These ages are more compatible with the ages of the younger granites. Greiling et al. (1984) postulated that the structures and metamorphic evolution of these domes involved at least 11 phases of deformation and two metamorphic events, the earlier of which led to sillimanite-grade conditions and migmatization. The amphibolites associated with the Meatiq basement rocks (Neumayr et al., 1996) consist mainly of plagioclase, hornblende, quartz, and pyroxene. Whole rock chemistry indicates tholeiitic basalt to basaltic andesite composition of most of the amphibolites. Immobile trace elements such Zr, Y, and Ti suggest that one group of amphibolites is derived from within-plate basalts and another belongs to the mid-oceanic ridge basalts (MORB) tectonic setting.

The Meatiq and Hafafit Neoproterozoic gneisses give Rb-Sr and U-Pb zircon ages of 750–600 Ma. The initial $^{87}Sr/^{86}Sr$ of $0.70252 \pm 0.00056$ and eNd value of $6.4 \pm 1.0$ suggest derivation from depleted asthenospheric mantle sources during Neoproterozoic time (Liégeois & Stern, 2010). There is no support for the hypothesis that ancient crust lies beneath the Eastern Desert (Liégeois & Stern, 2010). The Sr and Nd isotopic data for Meatiq and Hafafit gneisses indicate that these gneisses are juvenile Neoproterozoic crustal additions and that the important metamorphic event recorded in the Eastern Desert gneissic domes is related to the main Neoproterozoic Pan-African Orogeny at ca. 600 Ma corresponding to the formation of Greater Gondwana (Liégeois & Stern, 2010).

The older granites were probably formed during subduction of oceanic plates underneath the continental crust. The heat flow generated by the rise of the basaltic and calc-alkaline magmas during the formation of the volcanic arcs led to the development of expanding domes. Calc-alkaline gabbroic and granodioritic bodies of the older granites occupied the cores of these domes. High temperatures and deformation associated with these granitic bodies caused the migmatization of the basinal and platform sedimentary rocks deposited during the passive margin stage. Such a sequence of events was visualized by Dewey and Bird (1970) to show the conversion of an Atlantic-type continental margin into a Cordilleran-type mountain belt. According to the present interpretation, migmatization was contemporaneous with the emplacement of the older granites. However, the migmatites should be expected to change facies into metasediments away from the granitoid intrusions and to be of an older age.

## 5.2.4 IV.  Upper magmatic-arc cycle (650–550 Ma)

The late Pan-African tectonic stage (late Neoproterozoic) involves the formation of a second cycle of possible magmatic-arc-related activities, the eruption of calc-alkaline volcanics, deposition of more than 4,000 m (13,124 ft) of molasse-type sediments, and the intrusion of younger granitoids (both alkaline and calc-alkaline), rhyolites, and younger gabbros. This cycle is characterized by local occurrences of ophiolites in Sinai and their virtual absence at this level in the Eastern Desert and Libya. This stage corresponds to the Pan-African Orogeny of most authors and the molasse cycle of earlier workers.

### 5.2.4.1  Upper magmatic-arc sequence of the Northern Eastern Desert

The upper magmatic-arc cycle (Fig. 5.3) is best represented in the northern Eastern Desert and includes four principal rock associations: the Hammamat Formation sediments, the Dokhan volcanics, the younger granitoids, and younger gabbros.

### 5.2.4.2  Hammamat sediments (616–590 Ma)

Following the metamorphism and migmatization episode at the end of the lower magmatic-arc cycle, domal structures such as the Meatiq dome and the Migif-Hafafit underwent uplift and erosion (Sturchio et al., 1983, Liégeois & Stern, 2010). More than 4,000 m (13,124 ft) of the Hammamat arenites, wackes, and conglomerates (Akaad & Noweir, 1969) were deposited between 616–590 Ma (Ries et al., 1983) in fluviatile to shallow marine environments (El-Shazly, 1977) and probably also in extensive alluvial fans. The Hammamat sediments interfinger with the Dokhan volcanics (Gass, 1982, Furnes et al., 1985) and contain pebbles derived from Dokhan-type volcanics (Ries et al., 1983) which reflects the contemporaneity of the two units. Greenberg (1981) believes that the Hammamat sediments and the associated Dokhan volcanics and Younger Granites were deposited in isolated fault-bounded basins. Stern et al. (1984) also argued that the northern Eastern Desert of Egypt was dominated between 670–550 Ma by an extensional tectonic setting analogous to that of the late Paleozoic Oslo Rift of Norway. In support of this argument is the virtual absence of ophiolites, mélanges, and banded iron in the northern Eastern Desert.

The Hammamat sediments in the Eastern Desert of Egypt and Sinai and their equivalents in the Tibesti Massif in south-central Libya include thick sequences of polymictic conglomerates. The pebbles in these conglomerates consist predominantly either of granitic or volcanic pebbles which may reflect two different episodes of deposition associated with two different tectonic domains. The Umm Had conglomerate (*breccia verde antico*) of the Hammamat sediments east of Wadi Qena in the Eastern Desert is composed mainly of granite pebbles, as well as minor quartzite and rhyolite pebbles and preserved only in topographic lows (Said, 1962a, Akaad & Noweir, 1969). These conglomerates were derived probably from the erosion of the older granites of the lower magmatic cycle during the early phase of the upper magmatic-arc cycle.

### 5.2.4.3 Younger granites (615–550 Ma)

The Hammamat sediments, as well as the earlier units, were contemporaneously intruded by younger pink granite plutons, rhyolites, and younger gabbros.

Two main types of granitoid rocks can be recognized: Calc-alkaline and alkaline, believed to be syn-tectonic (orogenic), and post-tectonic (anorogenic), respectively. Data regarding the stratigraphic position, petrographic and geochemical characteristics, and absolute ages of these rocks are still scant and conflicting.

Most of the isotopic age dating range from 615 Ma to 550 Ma for the younger granitoids (Greenberg, 1981, Engel et al., 1980, Halpern & Tristan, 1981, Sturchio et al., 1983, Ries et al., 1983, Abdel Rahman & Doig, 1987, Shimron, 1987, Sultan et al., 1990, Richter & Schandelmeier, 1990, Fritz et al., 1996).

Two conflicting trace element results were reported which lead to two different interpretations of the tectonic setting in general and of the origin of the younger granitoids in particular[3]. The low $^{87}Sr/^{86}Sr$ (0.7016–0.7061) and Nb concentrations preclude a crustal origin for the granites (Halpern & Tristan, 1981, Shackleton et al., 1983, Fullagar, 1980, Harris et al., 1984, 1991, Stern & Hedge, 1985). Stern et al. (1984), based on low $^{87}Sr/^{86}Sr$ ratios and the presence of non-radiogenic lead, suggested that anatexis of metasomatically enriched mantle and/or first-cycle geosyn-

---

3 This is consistent with Liégeois & Stern's (2010) observations.

clinal sediments and igneous rocks were responsible for magma production. On the other hand, Mittlefehldt & Reymer (1986) believe that it is unlikely that the granites were derived from the mantle, but that they are probably remobilized lower crustal material. Evidence for the involvement of older crustal rocks includes: 1) U/Pb zircon ages up to 1.6 Ga (Sultan et al., 1990). This age is comparable to the age of quartzite pebbles found in the metasediments (Engel et al., 1980) of the lower magmatic-arc cycle. 2) A relatively high $^{207}Pb/^{204}Pb$ (Abdel Rahman & Doig, 1987, Sultan et al., 1990). 3) High $^{87}Sr/^{86}Sr$ ratios (0.7110–7136) (Abdel Rahman & Doig, 1987, Sultan et al., 1990). Yonan (1998) suggested that the younger granites of Urf Abu Hamam, southern Eastern Desert, were derived from a mantle source with minor additions of the crust during lithospheric extension. Contradicting these results are a low $^{87}Sr/^{86}Sr$ for the Aswan granite (0.7029) and a high initial $\varepsilon_{Nb}$ (+5.7)[4] for the Nakhil granite (Sultan et al., 1990) which have been interpreted as indicating a depleted mantle source (Harris et al., 1984).

Ensuing from the variable results, two tectonic models for the origin of younger granites have been proposed, Andean-type subduction related magma and within-plate, hot-spot-related magma. Greenberg (1981) envisaged that the younger granites were formed at the end of a cratonization process between 603 Ma and 575 Ma and the beginning of an anorogenic or tensional environment as a result of reactivation of structural weaknesses where suturing had occurred. Stern et al. (1984, 1988) also proposed a within-plate, tensional environment in rifted basins. On the other hand, Ghuma & Rogers (1980), Ragab et al. (1989a, b), Ragab (1991), and Abdel Monem et al. (1998a, b) suggested an active continental-margin, Andean-type magmatic-arc setting for the origin of these granites.

It is possible that both tectonic styles existed and were contemporaneous, i.e. extensional accompanied by subduction. There is now ample evidence that both magmatic-arc and extensional tectonics operate simultaneously. For example, during the Mesozoic-Cenozoic accretionary processes in the Antarctic Peninsula (Garret & Storey, 1987) and in the Basin and Range Province of the North American Cordillera (Coney, 1987, Hamilton, 1987). Unfortunately, petrogenetic processes operating at both extensional and convergent settings can generate similar sequences of rocks (Dostal et al., 1996) and the effect of one process may be overshadowed by the other.

A mantle source would suggest formation in an oceanic island-arc, while a crustal or mixed origin would suggest a continental volcanic-arc underlain by a continental crust. A mixed source could result also from the melting of the oceanic crust and its sedimentary cover. Primary magmas are unique in different arcs because of the variety of components and processes involved in their generation, but they should show the common characteristic of high concentrations of Mg, Ni and Cr, reflecting their mantle origin (Smith et al., 1997). According to these authors, magmas with primitive characteristics are rare and occur only in extensional tectonic settings where the magma is allowed to ascend rapidly to the surface. On the other hand, if the magma is trapped beneath the crust, fractionation of the magma and assimilation of the country rock will result in the enrichment of Sr, Rb, B, K, U, and the light

---

4 Deviation from the value expected in a chondritic reservoir (CHUR) at time T.

rare-earth elements (DeBari, 1997) and the magma will contain a few traces of its mantle origin.

Almond et al. (1989) noticed that a change from calc-alkaline to alkaline magmatism took place in the African-Nubian shield at around 600 Ma, followed by a change to less siliceous magmatism at about 500 Ma. The change from calc-alkaline to alkaline or to bimodal (basalt-rhyolite) volcanism indicates a change from subduction to extensional regimes (Garret & Storey, 1989, Hamilton, 1987).

### 5.2.4.4   Dokhan volcanics

The Dokhan volcanics range in composition from intermediate andesites to acidic rhyolites and alkaline trachytes. They were dated at 639–602 Ma (Ries et al., 1983) and 620 Ma (Abdel Rahman & Doig, 1987). The presence of pebbles derived from the Dokhan-type volcanics within the Hammamat Group sediments (Ries et al., 1983) and the interfingering of these volcanics with the Hammamat sediments, as well as the presence of ignimbrites indicate subaerial volcanic activities (Gass, 1982, Furnes et al., 1985) contemporaneous with the sedimentation of the Hammamat clastics.

The Dokhan volcanics are believed to be co-magmatic with the younger granites and share a common origin (Ries et al., 1983). Gass (1982) suggested that these volcanics erupted along an active continental margin during subduction, while Stern et al. (1984, 1988) suggested that the crust was extensively extending and undergoing rifting similar to the Oslo Graben.

### 5.2.4.5   Younger gabbros

The younger gabbros, such as the Umm Bassilla Gabbros (Said, 1971) in the Red Sea range form small intrusive discordant masses of fresh unmetamorphosed norites, gabbro-norites, olivine norites, and olivine gabbros believed to be post-tectonic intrusions of post-Hammamat age (Said, 1971). On the other hand, the presence of Ca-rich plagioclase ($An_{55-80}$), Mg-rich olivine ($Fo_{76.61-77.06}$), and Ca-rich pyroxenes led Ghoneim (1989) to believe that the younger gabbros of the central Eastern Desert were formed in an island-arc ocean floor tectonic setting. Coexisting Ca-rich plagioclase ($An_{95.6-91.6}$), Mg-rich olivine ($Fo_{91-80}$), and pyroxene in gabbroic cumulates of the Mersin ophiolites, Turkey, were considered to indicate formation from a basaltic magma in an arc setting (Parlak et al., 1996). However, because of the marked differences between the two gabbros, it can be assumed that the younger gabbros of the Eastern Desert were derived from andesitic magmas.

### 5.2.4.6   Upper magmatic-arc sequence in Southern Sinai

A volcano-sedimentary sequence similar to that of the northern Eastern Desert, in addition to ophiolites, occurs in southeastern Sinai (Fig. 5.8). The volcano-sedimentary sequence of the Kid Group comprises four major allochthonous tectono-stratigraphic units, the Malhak Formation, the Umm Zariq Formation, the Heib Formation, and the Tarr Complex with a combined thickness of 12,000 m (39,372 ft). These units probably represent magmatic-arc assemblages (Shimron, 1980, 1983, 1984, 1987, Furnes et al., 1985). The Malhak Formation forms a greenstone belt composed of pillow lavas, pillow breccias and hyaloclastics of subaqueous origin and intercalated marine sediments (Shimron, 1983, Furnes et al., 1985).

Blasband et al. (1997) and Blasband (1998) recognized two deformation phases in the late Precambrian history of the Wadi Kid area. The first phase (pre-620 Ma) produced steep regional foliation and upright folds in a compressional regime related to subduction in an island-arc setting. This phase was associated with calc-alkaline magmatism. The second phase (post-620 Ma) led to widespread development of mylonitic zones along low-angle shear zones related to extension. These rocks were intruded by A-type alkaline granites and NW-NE-trending dikes. The dike swarms, dated at 590–580 Ma, indicate extension in NW-SE direction (Blasband et al., 1997).

The work of Blasband (1998) and Blasband et al. (1997) in Sinai, and Fowler & Osman (1998) in the Eastern Desert of Egypt, suggest that the Lower Magmatic-Arc cycle was dominated by compression, and the Upper Magmatic-Arc cycle was dominated by extension.

Following northward subduction of the oceanic lithosphere beneath the early Pan-African (late Proterozoic) continental margin (the Elat Schist), an Andean-type magmatic-arc assemblage of the Heib Formation volcanics composed of amygdaloidal andesites and rhyolitic and rhyodacitic flows with sedimentary and pyroclastic intercalations was developed (Shimron, 1983, 1984). Uplift and rapid erosion of this volcanic-arc assemblage led to the deposition of the Tarr Complex mélange-olistostrome system (Shimron, 1983). Gravitational slumping carried down highly deformed, tectonically amalgamated blocks of metasedimentary rocks composed of pebbly mudstones, megabreccias, olistostromes, and minor volcanic rocks of oceanic crust affinities into submarine fans. The sequence passes from turbidites to shallow marine and fluvial sediments in Southern Sinai. This sequence is similar to that described from the lower magmatic-arc cycle in the central Eastern Desert (Ries et al., 1983, Shackleton et al., 1980, Sturchio et al., 1983) and may represent deposition in fore-arc basins. The Kid Conglomerate (Shimron, 1987) (probably equivalent to the Umm Had Conglomerate) of up to 1000 m (3281 ft) in thickness and composed of pebbles of volcanic origin with some granitic and sedimentary fragments (Reymer & Yogev, 1983, Shimron, 1983) separates the Tarr Complex and Heib Formation. The origin of this conglomerate is not clear, but it may have been deposited in alluvial fans in extensional rifts and grabens following the erosion of the volcanic arc assemblage.

Continued uplift, faulting and erosion of the volcanic-arc complex also led to the deposition of more than 1500 m (4922 ft) of the Umm Zariq Formation (probably equivalent to the Hammamat Group). The sediments consist of cross-bedded and laminated wackes composed of quartz, andesine plagioclase and volcanic rock fragments, which suggest a provenance from the island-arc volcanics (Reymer & Yogev, 1983).

The Umm Zariq Formation was thrust along E-W and NE-SW-trending shear zones marked by mylonites (Shimron, 1983, 1984, 1987), deformed, and metamorphosed in the greenschist to amphibolite facies between 650–600 Ma (Halpern & Tristan, 1981). The Feiran gneiss (equivalent to the alkaline younger granites), dikes, and associated granodiorites in west-central Sinai were dated at 632 Ma by Stern & Manton (1987) who considered them to be post-tectonic. They were believed to be the oldest rocks in Sinai (older than 950 Ma) by Shimron (1983). These rocks were subsequently intruded by late- to post-orogenic younger granites and volcanics of the Feirani and Catherina cycles around 600–550 Ma (Halpern & Tristan, 1981, Shimron, 1987, Eyal et al., 1991) such as the Umm Malaq granite and resulted in contact metamorphism of the country rocks up to the pyroxene hornfels facies.

A parallel dispute about the origin of the magmatic activities also took place among geologists working in Sinai. Reymer & Yogev (1983) and Navon & Reymer (1984) disputed the presence of a thrust fault between the Tarr Complex and Umm Zariq Formation, and rejected the island-arc model proposed by Shimron (1983). They suggested that a shallow continental-margin or intracratonic rifted basin with active volcanism existed in the area during the late Pan-African times where the Kid Complex was deposited and was later uplifted and gently folded. A similar setting has been suggested by Greenberg (1981) and Stern et al. (1984) for the upper magmatic-arc assemblages in the Eastern Desert of Egypt.

### 5.2.4.7  Upper magmatic-arc sequence of Libya and South-Central Algeria

The Upper Tibestian Series in SE Libya (Wacrenier, 1958) was intruded by calc-alkaline granites and granodiorites of the Ben Ghanimah Batholith dated at 550 Ma (Rogers et al., 1976, Ghuma, 1975, Pegram et al., 1976, Ghuma & Rogers, 1980, Fullagar, 1980) in the western Tibesti Massif, and by the alkaline granites of the Eghei Magmatic Series (El Makhrouf, 1988) in the eastern Tibesti Massif. The Upper Tibestian series is metamorphosed into the upper greenschist to almandine-amphibolite facies (El Makhrouf, 1988). The Upper Tibestian Series is separated from the Lower Tibestian by an aeromagnetic lineament. The presence of the aeromagnetic lineament and the change in grade of metamorphism suggests the presence of a suture zone between a continental margin of the Lower Tibestian (Jabal Eghei area) to the east and a mature island arc of the Upper Tibestian (Ben Ghanimah Batholith area) to the west (El Makhrouf, 1988).

The granitic rocks from the Ben Ghanimah Batholith in the western part of the Tibesti Massif have been dated at 600 Ma to 500 Ma (Fullagar, 1980, Pegram et al., 1976, Nagy et al., 1976, Rogers et al., 1978, Ghuma & Rogers, 1980, Almond, 1991). The granitic rocks from the Ben Ghanimah Batholith in the northwestern part of the Tibesti Massif are typical calc-alkaline granitic suites (Rogers et al., 1978, Ghuma & Rogers, 1980) emplaced between two tectonic plates. They show a change in chemical composition from relatively mafic granodiorites on the eastern side to relatively felsic granitic rocks on the western side. This change has been attributed to intrusion along a continental margin (Ghuma & Rogers, 1980). The Ben Ghanimah granitic rocks probably represent a westward subduction-related island-arc assemblage which was intruded between shallow-water sediments in the west and deep-water sediments in the east (Rogers et al., 1978, Ghuma & Rogers, 1980, Fullagar, 1980). Alkaline hornblende-biotite granites, biotite granites, and minor intrusions of two-mica granites and monzonite granites intrude the Lower Tibestian Series at Jabal Eghei in the eastern Tibesti Massif (El Makhrouf, 1988). The Eghei alkaline magmatic rocks were emplaced between $554 \pm 6$ to $528 \pm 7$ Ma during a period of stress relaxation (extensional) in the eastern part of the Tibesti Massif, contemporaneous with the subduction of the oceanic crust in the Ben Ghanimah area (El Makhrouf, 1988), dated at

556 Ma (Ghuma & Rogers, 1980) and 545 Ma (Fullagar, 1980). The initial $^{87}Sr/^{86}Sr$ ratios range from 0.70436 to 0.70791 for the Jabal Eghei granites (El Makhrouf, 1988), and average 0.706 for the Ben Ghanimah granites (Pegram et al., 1976). These $^{87}Sr/^{86}Sr$ data suggest that the magma was derived from mantle or oceanic crustal material, with some degree of contamination with older sialic material (Pegram et al., 1976, Fullagar, 1980, Ghuma & Rogers, 1980).

The Ben Ghanimah Batholith also includes gabbros, which occur in small isolated blocks. Most of these intrusions were later converted into hornblende gabbros to amphibolites composed of brown and green amphiboles and calcic plagioclase ($An_{30-52}$) (Ghuma & Rogers, 1980).

Conglomerates at the base of the Upper Tibestian contain fragments of granite and perthitic microcline which have no equivalent in the area (Wacrenier, 1958). The provenance of these clasts is unknown, but it is probably buried underneath the Upper Tibestian complex. They were intruded by dolerites and rhyolites followed by metamorphism and the intrusion of calc-alkaline granites. The conglomerates were probably deposited in extensional basins (Wacrenier, 1958) during a period of regional extension which affected most of North Africa.

The Upper Magmatic-Arc Cycle is represented in the Al Qarqaf Arch (Jabal Hassawnah) area predominantly by porphyritic granites, which yielded a K/Ar age of 554 Ma (Jurak, 1978), with small inclusions of biotite schists, and intrusions of aplite granites and pegmatites. They gave a Rb/Sr date of 519 Ma and a $^{87}Sr/^{86}Sr$ ratio of 0.7116 (Oun et al., 1998a, b). The Nd isotope modal ages gave 1476–1619 Ma crustal residence ages which suggest that these granites were probably derived from mixed source juvenile Pan-African mantle material and older meta-sediments, and emplaced in a within-plate tectonic setting.

Equivalent rocks occur in the subsurface of the Murzuq Basin as granites dated at 554 Ma to 551 Ma (Bauman et al., 1992). The initial $^{87}Sr/^{86}Sr$ of 0.706 ratio of the Egyptian Younger Granites ranges between 0.7016 and 0.7025 (Fullagar, 1980). This low ratio suggests that the granitic magma was not contaminated by older sialic rocks. The initial Sr isotope ratios indicate thickening of the crust in Libya, as compared to Egypt, with greater assimilation of crustal materials.

The Hoggar Massif (Touareg/Tuareg Shield) has been divided into four structural domains, separated by N-S lineaments (Black et al., 1979) (Fig. 5.12), from west to east, the suture zone, the Pharusian belt, the Central Hoggar, and the Eastern Hoggar. The suture belt is characterized by island arc volcanics, marginal basin, clastics, ultrabasic and basic rocks, and internal nappes emplaced onto the craton margin (Lesquer et al., 1989, Dostal et al., 1996). The Pharusian belt includes a green belt facies (*Série verte*) of volcanic and detrital rocks, and a relict granulitic facies (2000 Ma) that has not been affected by the Pan-African events (Lesquer et al., 1989). The Pharusian terrane that is described in detail, is floored in the east by the Iskel basement, a Mesoproterozoic arc-type terrane cratonized around 840 Ma and in the southeast by Late Paleoproterozoic rock sequences (1.85–1.75 Ga) similar to those from northwestern Hoggar (Caby, 2003). Unconformable Late Neoproterozoic volcanosedimentary formations that mainly encompass volcanic greywackes were deposited in troughs adjacent to subduction-related andesitic volcanic ridges during the c.690–650 Ma. The Central Hoggar is composed of granulites approximately 2075 Ma in age intruded by younger

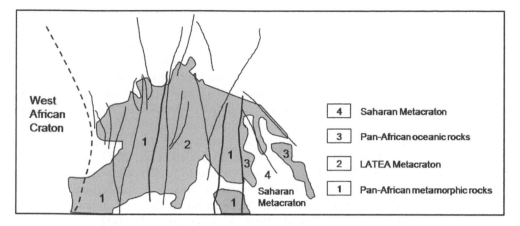

West
African
Craton

| 4 | Saharan Metacraton |
| 3 | Pan-African oceanic rocks |
| 2 | LATEA Metacraton |
| 1 | Pan-African metamorphic rocks |

Saharan
Metacraton

*Figure 5.12* Tuareg (Hoggar) Shield in Central-Southern Algeria (after Askri et al., 1990, Black et al., 1990, Azzouni-Sekkal et al., 2003, Abdallah et al., 2007).

Pan-African granitoids (Bertrand et al., 1986). The Eastern Hoggar belt was stabilized and metamorphosed around 725 Ma (Lesquer et al., 1989). Collision between the Touareg Shield and the West African Craton took place around 600 Ma (Dostal et al., 1996). This collision also produced a N-S intraplate deformational belt, emplacement of nappes, lateral displacement along megashear zones, and intrusion of granites. The northeastern tip of the Eastern Hoggar domain crops out in the extreme southwestern corner of Libya. The late Pan-African episode is represented by syn-tectonic migmatitic granites with zones of leucocratic granites, granodiorites, and diorites intruded by aplite and pegmatite veins, followed by the emplacement of porphyritic, calc-alkaline biotite granites (Taourirt Granite) (Radulovic, 1984, Grubic, 1984) into the older granites around 560–540 Ma (Lesquer et al., 1989, Liégeois & Black, 1989, Liégeois et al., 1996). The Taourirt granites were considered to be post-tectonic alkaline, and anorogenic by Radulovic (1984), and Azzouni-Sekkal & Bonin (1998), but they could be also syntectonic because of their calc-alkaline character. These granites were dissected themselves by veins of quartz, microgranite, and microdiorite. These assemblages probably represent terranes that were accreted to the West-African Craton during the late Pan-African events, between 650–500 Ma (Radulovic, 1984, Grubic, 1984). Rhyolites, dacites, and small masses of diorites and gabbros were intruded in a NW-SE direction (Radulovic, 1984) at the end of this cycle.

The Touareg Shield (Fig. 5.12), located between the Archean-Paleoproterozoic Saharan Metacraton[5] and the West African Craton (WAC), is composed of 23 recognized

5 The term "Saharan Metacraton" was introduced by Abdelsalam et al. (2002) to refer to "a craton that has been remobilized during an orogenic event but that is still recognizable dominantly through its rheological, geochronological and isotopic characteristics', to replace other terms, such as the "Nile Craton" (Rocci, 1965), "Saharan-Congo Craton" (Kröner, 1977), "Eastern Sahara Craton" (Bertrand & Caby, 1978), and "Central Sahara Craton" (Black and Liègeois, 1993).

terranes that welded together during the Neoproterozoic (850–520 Ma) Pan-African Orogeny (Azzouni-Sekkal et al., 2003). Final convergence occurred mainly during the 620–580 Ma period with the emplacement of high-K calc-alkaline batholiths, and continued into the Cambrian until 520 Ma with the emplacement of alkali-calcic and alkaline high-level complexes. The last plutons emplaced in Central Hoggar at 539–523 Ma are known as the Taourirt Province. Three geographical groups are identified (Azzouni-Sekkal et al., 2003): Silet-, Laouni-, and Tamanrasset-Taourirt. The Taourirts are high-level subcircular calc-alkaline, occasionally alkaline, complexes. They are aligned along mega-shear zones. Nd isotopes indicate a major interaction with the upper crust during the emplacement of highly differentiated melts. All the Taourirt plutons are strongly contaminated by the lower crust (Azzouni-Sekkal et al., 2003). The Touareg Shield formed during the Pan-African Orogeny (850–520 Ma) by the welding of ophiolite-bearing juvenile terranes between two major Archean-Paleoproterozoic cratons, the West African Craton and the Saharan Metacraton to the west and to the east, respectively (Boullier, 1991, Black et al., 1994, Abdelsalam et al., 2002). The Touareg Shield was subjected to post-collisional and post-orogenic phases around 560–540 Ma (Liégeois & Black, 1987, Liégeois et al., 1996). By definition, a metacraton is a craton partially remobilized during an orogeny, keeping large parts of its cratonic characters but not all (Abdelsalam et al., 2002, Liégeois et al., 2003). Nd isotopes demonstrate the presence of the old LATEA basement below the whole province, implying that the early island arc terrane is thrust onto LATEA. Caby & Andreopoulos-Renaud (1987) concluded that the Eastern block was cratonized around 730 Ma.

## 5.2.5V. Post-Tectonic Cycle

Extensional tectonics existed during the Late Proterozoic and continued during the early Paleozoic. Pull-apart basins and horst and graben structures probably developed due to compressional and extensional processes affiliated with the late stages of subduction. Associated with extensional tectonics are ring dike complexes and swarm dikes in Egypt and Sinai, the intrusion of granites, gabbros, granodiorites, pegmatites, and aplites in the Tibesti Massif, and the deposition of feldspathic and quartzose sandstones in the Eastern Desert, Sinai and the Tibesti Massif.

### 5.2.5.1 Ring dike complexes and Swarm dikes

The basement in the African-Nubian Shield is intensely intruded by post-tectonic dikes, commonly referred to as dike swarms and ring complexes (Said, 1971, Stern et al., 1988) (Figs. 5.10, 5.14). The dikes vary in composition from acidic to basic and alkaline. The dikes are narrow, steeply dipping bodies, a few meters in thickness and several kilometers in length. They are mainly Late Precambrian-early Phanerozoic (600–500 Ma) peralkaline to alkaline igneous rocks. This group includes muscovite trondjhemite, dikes and peralkaline rocks, granodiorites, adamellites, and leucogranites in the Ras Gharib area (Abdel Rahman & Doig, 1987), and Phanerozoic alkaline rocks in the Eastern Desert (El-Gaby et al., 1990, Ries et al., 1983). Alkaline intrusions and granites of the Wadi Dib ring complexes in the Eastern Desert gave an average age of 554 Ma (Serencsits et al., 1979, Frisch, 1982, Frisch & Abdel-Rahman,

*Figure 5.13* Pan-African granites, Mount Sinai, City of St. Katherine, Sinai (Courtesy of Baher El Kaliouby) (See color plate section).

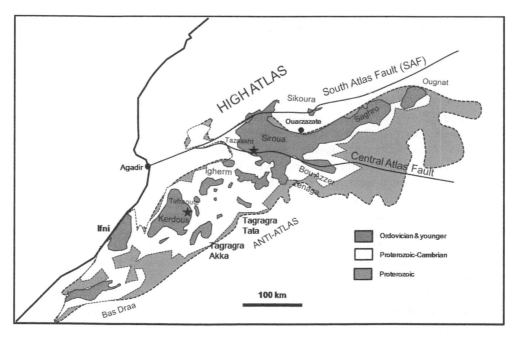

*Figure 5.14* Geological map of the Anti-Atlas (after Piqué, 2003, Touil et al., 2008) (See color plate section).

1999). Ring dike complexes with a central volcanic pile of alkaline rocks (Shimron, 1983), and N30°E-trending extensional dike swarms dated at 591 Ma (Stern & Manton, 1987), occur in Sinai. The Katharina (Saint Catherine/Katherine) Ring Complex of Gebel Musa (Fig. 5.13) is made up of syenogranites grading upward into alkali-feldspar granites, for which a Rb-Sr dating yielded an age of 593 ± 16 Ma (Katzir et al., 2007). The dikes represent evidence of NW-SE directed crustal extension accompanied by bimodal dike emplacement (Stern et al., 1988, Iacumin et al., 1998), and are probably contemporaneous with the emplacement of the Younger Granites and associated with the Najd transcurrent fault system developed around 580–530 Ma (Fleck et al., 1976, de Gruyter & Vogel, 1981). It has been suggested (de Gruyter & Vogel, 1981) that the origin of these complexes is due to alkaline melts formed in the asthenosphere by shear heating caused by plate motion. The alkaline complexes show a change in the degree of alkalinity and silica saturation with age. The change from silica-oversaturated, low alkaline older rocks to silica-saturated, silica-rich younger rocks was probably the result of the cooling of the lithosphere following the Pan-African thermal event (de Gruyter & Vogel, 1981). The Ben Ghanimah Batholith (Nagy et al., 1976, Ghuma & Rogers, 1980, Almond, 1991) of the Tibesti Massif in Libya is intruded by younger rocks which include granites (446–539 Ma), gabbros and granodiorites (545 Ma), pegmatites (522), and aplites (545 Ma) (Fullagar, 1980). In the Hoggar Massif rhyolites, dacites, and the small masses of diorites and gabbros were intruded in a NW-SE direction (Radulovic, 1984). The orientation of these masses coincides with NW-SE-oriented faults in the area (Grubic, 1984) and may represent a post-tectonic extensional regime (Bradley, 1986).

### 5.2.5.2   Feldspathic and quartzose sandstones

Peneplanation and erosion at the end of the Pan-African Orogeny led to the deposition of thick sections of sandstones. Thick coarse-grained quartzose and arkosic sandstones of the Olochi Sandstone, with interbeds of marble and schist (Dalloni, 1934) of the Tibesti Massif and more than 285 m (935 ft) of wine-red, fine grained, conglomeratic and arkosic sandstones and siltstones of the Mourizidie Formation (Jacqué, 1962) in the Jabal Al Qarqaf area were deposited on folded and partially granitized basement rocks. Bellini & Massa (1980) postulated that the Mourizidie sediments may represent the Precambrian glacial cycle known in Algeria and Mauritania, but no evidence has been presented. Equivalent sediments of medium- to coarse-grained feldspathic sandstones with interbeds of siltstone, chert, marble, calcareous sandstone and banded ironstone with ferruginous and manganiferous concretions were deposited in the Jabal Arknu area (Mamgain, 1980). In the subsurface of Cyrenaica, fining-upward marine sequences, up to 1059 m (3475 ft) in thickness, composed of reddish-brown, very fine to medium-grained quartz arenites with abundant feldspars and micas (Early to Late Proterozoic) (Baudet, 1988, El-Arnauti & Shelmani (1988), with acritarchs, are probably equivalent to the Mourizidie and Olochi formations and may also extend into the subsurface of the Sirte Basin, but no work has been done so far to confirm this possibility.

The Anti-Atlas basement is a complex assemblage of Paleoproterozoic-Neoproterozoic and Lower Paleozoic crystalline, metamorphic and sedimentary rocks (Burkhard et al., 2006, Soulaimani & Burkhard, 2008). The Anti-Atlas marks the northern boundary of the Eburnean (ca. 2 Ga) West African Craton (WAC) (Ennih & Liégeois, 2001). The oldest Precambrian rocks exposed in the Anti-Atlas belt represent fragments of the Paleoproterozoic (Eburnean) WAC (Ait Malek et al., 1998, Thomas et al., 2002, Walsh et al., 2002, Gasquet et al., 2008). In contrast to the WAC (stable since about 2 Ga), the Anti-Atlas inliers display two superimposed pre-Variscan orogenic events: 1) A Neoproterozoic compressive event, related to the Pan-African Orogeny (Leblanc & Moussine-Pouchkine, 1994, Saquaque et al., 1989), and 2) A late Neoproterozoic-Early Paleozoic extensional event (Piqué et al., 1999, Soulaimani et al., 2003, El Aloui et al., 2010).

A considerable volume of research on the Anti-Atlas Orogeny and its role in the evolution of the belt have been proposed (for example, Saquaque et al., 1992, Fekkak et al., 2000, Ennih & Liégeois, 2001, Hefferan et al., 2002, Thomas et al., 2002, Thomas et al., 2004, Walsh et al., 2002, Burkhard et al., 2006, Soulaimani & Burkhard, 2008, and Gasquet et al., 2008). Most of these studies have concentrated on the sedimentological-volcanological, structural, geochemical and geochronological aspects. With the exception of the work by Thomas et al. (2002, 2004), few stratigraphic compilations have been published. However, their work has not been widely accepted and yet there is no accepted lithostratigraphic framework for the Anti-Atlas Domain.

One of the earliest attempts to classify the Precambrian basement rocks of the Anti-Atlas Domain was made by Choubert (1953) and later elaborated upon by Choubert (1963) (Table 5.1). According to Choubert's classification, the Precambrian of Morocco is subdivided chronostratigraphically into three "Stages", designated as PI (Archean or Paleoproterozoic), PII (Lower and Middle Neoproterozoic), and PIII (Upper Neoproterozoic) (Choubert, 1963). Each of these stages was further subdivided into "Series", such as PIII1, PIII2, PIII3, etc. This system, though debated, still the unifying point in all discussions of the Precambrian successions in the Anti-Atlas. One of the setbacks of this nomenclature is that it is based on the grade of metamorphism; the higher the metamorphic grade of the rocks, the older they were assumed to be. This simplistic picture has been challenged by Thomas et al. (2004), who consider it misleading and often erroneous. Nonetheless, this nomenclature still provides a basic framework for the construction of a valid stratigraphic nomenclature, as well as regional correlations of the Precambrian basement rocks over most of North Africa.

*Table 5.1* Comparison of Megasequences in Egypt and Libya (Tawadros, 2001), Precambrian subdivisions of the Anti-Atlas (Choubert, 1963), and major lithostratigraphic units (Thomas et al., 2004). Note the period 2000–1000 Ma (Mesoproterozoic) is not represented.

| AGE | Egypt & Libya (Tawadros, 2001) megasequences | Anti-Atlas systems (Choubert, 1963) | Anti-Atlas (Thomas et al., 2004) |
|---|---|---|---|
| 650–540 Ma | Upper Megacycle (Late Neoproterozoic) | PIII | Ouarzazate Supergroup |
| 1000–650 Ma | Lower Megacycle (Early Neoproterozoic) | PII | Anti-Atlas Supergroup |
| 2200–2000 | Upper Gebel Oweinat Cycle (Paleoproterozoic) | PI | Basement Complex |

They are easily correlated with the Megasequences of Tawadros (2001) in Egypt and Libya (Table 5.1).

As is the case of the West African Craton and over most of North Africa, the Mesoproterozoic (2000–1000 Ma) rocks are absent in the Anti-Atlas and represent a gap between 2000 Ma and 1000 Ma. Paleoproterozoic basement rocks are, therefore, unconformably overlain directly by Neoproterozoic metasedimentary and magmatic units, which were subsequently affected by the Pan-African Orogeny (Clauer, 1974, Hassenforder, 1987).

A more recent stratigraphic scheme for the Anti-Atlas Domain has been worked out by Thomas et al. (2002) and revised by Thomas et al. (2004). The first is chronostratigraphic; the second is lithostratigraphic and includes all the major rock units identified in the literature. All the lithostratigraphic units of the Anti-Atlas Orogeny are well represented in the Sirwa (Siroua), Kerdous, and Sarhro (Saghro) inliers, as well as the Zenaga Inlier (Thomas et al., 2004).

Two groups of Precambrian inliers can be recognized on either sides of the Anti-Atlas Major Fault (AAMF) (El Aloui et al., 2010, Gasquet et al., 2008, Ennih & Liégeois, 2001 (Fig. 5.14):

The first group located southwest of the AAMF Fault includes the Bas Draa, Ifni, Kerdous (with spectacular outcrops of the Tafraoute Granite) (Fig. 5.15), Ighern, Tagragra of Akka and Tata, Iguerda and Zenaga. In that region, the Precambrian basement consists of Paleoproterozoic basement (PI) dated at ca. 2 Ga and correlates with the Eburnean basement of the Reguibat Shield (Fig. 5.11). It is overlain by Early Neoproterozoic shallow-water metasediments (PII) (1000–880 Ma) and Late Proterozoic Pan-African granites and rhyolites dated at 630–560 Ma (PII-PIII). Equivalent Proterozoic granitoids in the Western High Atlas are dated at 625 ± 5 Ma (Eddif et al., 2007). A detailed study of the Tafraoute Granite in the Kerdous Inlier by Charlot (1976) suggests the existence of a Cambrian granitization in the Moroccan Anti-Atlas, dated at 530 ± 15 Ma. The southwestern part of the Anti-Atlas corresponds to the autochthonous northern margin of the WAC.

The second group of inliers is located northeast of the AAMF (Fig. 5.11) and includes J. Siroua, Ouzaellarh, J. Saghro, and Ougnat. In that group, the Neoproterozoic is made up of oceanic island-arc lithologies, including ophiolites which occur on both sides of the AAMF. The ophiolites are mainly 760–700 Ma in age. Calc-alkaline mafic rocks in the Igherm, Ifni, and Kerdous inliers (Figs. 5.15, 5.16) composed of doleritic dikes and gabbros bodies are characterized by enriched LILE, Th, Ce, P, Sm, a high La/Nb ratio and a low HFSE content, and negative Nb, Zr, and Ti values. Subsequently, during the Cryogenian extension (850–630 Ma), the enriched mantle contributed the calc-alkaline magma (El Aloui et al., 2010). The Eburnean (PI) basement does not crop out, but its presence underneath the Neoproterozoic domain has been postulated by Ennih & Liégeois (2001) and Gasquet et al. (2008).

The Kerdous Massif (Soulaimani et al., 2004), south of the AAMF (Fig. 5.15), includes the Paleoproterozoic (PI) Kerdous Schists and granites, the Quartzite Series (PII), the Neoproterozoic Tafraoute Granite, the volcaniclastics of the Anezi Series (PII-PIII) and the Tanalt Series (PIII), and the Ida Ougnif and Jbel Kerkar basalts. The complex is capped by sandstones and limestones of the Adoudounian Formation (basal Cambrian).

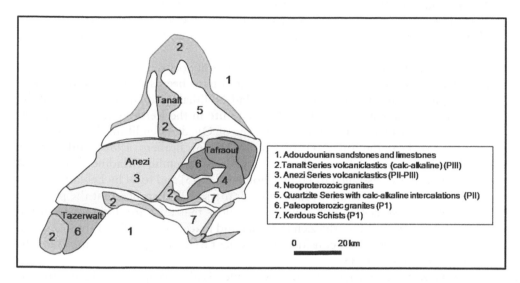

*Figure 5.15* Geological map of the Kerdous Inlier (after Soulaimani et al., 2004) (See color plate section).

The Siroua Massif, north of the AAMF, is made up of Paleoproterozoic and Cryogenian (850–635 Ma) basement complexes (PI and PII and a cover of Ediacaran and late Neoproterozoic (PIII) rocks, followed by Mesozoic-Cenozoic sequences (Touil et al., 2008). The Eburnean Orogeny (PI) (ca. 2000 Ma according to Thomas et al., 2004) is only locally present as lenses of gneisses and strongly foliated amphibolites found to the SE of Jbel Siroua (N'Kob District). However, the Pan-African units (ca. 750–520 Ma) crop out widely throughout the Siroua District. The Siroua Neoproterozoic plutons (PII-PIII) evolved from a subduction-type (calc-alkaline) in the northern part of the inlier to a within-plate sub-alkaline in the southern part (Touil et al., 2008). The first magmatic rocks in the north represent two tectonic episodes; the first was dominated by gabbros and basalts associated with the opening of the Khzama oceanic basin; a second collision episode was associated with low potassium calc-alkaline magmatic assemblages of gabbro, diorites, tonalities, and trondhjemites of the Nebdas Pluton, Askaoun, and Ifouachguel plutons. These two cycles ended with crustal uplift and extension marked by sub-alkaline magmatism of the Ida ou Illoun, Imdghar (the Ouafalla Granite of Thomas et al., 2002, 2004), and Affela N'ouassif granites in the south.

Three main assemblages are distinguished in the Siroua Inlier (Touil et al., 2008):

I The Paleoproterozoic basement rocks (PI) (2200–2000 Ma)

These rocks are the products of the Eburnean-Birimian Orogeny recognized in many parts of the West African Craton (Thomas et al., 2002, Thomas et al., 2004).

The Paleoproterozoic (PI) rocks in Northwest Africa form the northern margin of the West African Craton (WAC) and crop out mainly in the Reguibat Shield in

Mauritania and Algeria. The granitoids, gneisses and metamorphic rocks of the West African Craton are attributed to the Eburnean Orogeny at around 2000 Ma (Thomas et al., 2002, Thomas et al., 2004, Walsh et al., 2002, Gasquet et al., 2008).

In the Anti-Atlas Domain, the Paleoproterozoic rocks are grouped in the Basement Complexes by Thomas et al. (2004) and have been subdivided into a series of Complexes. They include the Zenaga, Had-N-Tahala and Draa Groups, and the Ifzwani and Toudma Suites. The Tahala Granite in the Kerdous Inlier, the augen gneisses of Jbel Ouiharem, and the gneisses of Oued Assemlil are believed to be the oldest rocks in the Anti-Atlas by Charlot (1976), Brabers (1988), and Schlüter (2008). The gneisses of Tamazight, Imghlay, Oued Assemlil, Tiffadine, Ourika and E1 Hara have been assigned to P0 by Choubert (1963) and to PI by Leblanc (1981) because of their high metamorphic grade. Brabers (1988) argued that the presence of ultramafics associated with the gneisses show that they are different from the Eburnean basement (PI) of the Zenaga Inlier and placed them in PII (Table 5.2). The Oued Assemlil gneisses are considered equivalent to the Amguemerzi and Taznakht granites by Brabers (1988).

In the Kerdous Inlier, the Kerdous Complex is also composed of Paleoproterozoic rocks comparable to those of the Zenaga Complex, though of lower metamorphic grade. The oldest rocks are phyllites and psammites of the Had-n-Tahala Group and are of probable turbiditic origin (Thomas et al., 2004). Calc-alkaline granites which intrude these rocks have been dated by the U-Pb zircon method as follows; the Amaghouz Granite (2042 ± 1 Ma), the Anammr Granite (2058 ± 11 Ma) and the Tahala Granite (2060 ± 12 Ma) (dated at 1920 Ma by Charlot, 1976). These granites were also included in the Tazenakht Suite. Other granites gave older dates, such as the Agrsiyf Granite (2263 ± 38 Ma) and the Zawyat Granite (2187 ± 33 Ma). All these granites have been included in the Tazeroualt Suite. The Paleoproterozoic basement inliers of Bas Draa, Tagragra d'Akka, Ighrem and Tagragra de Tata also contain older low-grade (mainly greenschist facies) metamorphic rocks intruded by porphyritic calc-alkaline granites. The granites gave U-Pb dates between 1987 Ma and 2046 Ma, for example, the Sidi Said Granite in the Bas Draa (1987 ± 20 Ma), Ait Makhlouf Granite in the Ighrem inlier (2050 ± 6 Ma), Targant and Oudad Granites in the Tagragra de Tata Inlier gave 2046 ± 7 Ma and 2041 ± 6 Ma, respectively (Aït Malek et al., 1998, Walsh et al., 2002). These granites are included in the Kerdous Suite by Thomas et al. (2004).

The Zenaga Group (Neoproterozoic according to Thomas et al., 2004, Soulaimani & Burkhard, 2008) is a unit of schists, gneisses, metadolerites and metamorphic rocks. Equivalent rocks are the granites at Azguemerzi (Thomas et al., 2004) and Tazenacht in the Central Anti-Atlas and farther west in the Lower Draa Inlier. Northwest of the Zenaga, the Jbel Amsemsa tuffs are unconformably overlain by volcanics of the Ouarzazate Series (Upper PIII) (Brabers, 1988). The schists are intruded by calc-alkaline granites which yield U-Pb zircon ages between 2032 Ma and 2037 Ma (Thomas et al., 2002): The Assourg Tonalite (2037 ± 7 Ma), the Azguemerzi Granodiorite (2032 ± 5 Ma), and the Tazenakht Granite (2037 ± 9 Ma). The latter is equivalent to the Tamazzarra Granite of Thomas et al. (2002). Thomas et al. (2004) proposed to group all these rocks into the Tazenakht Suite.

Mesoproterozoic rocks (2000–1000 Ma) between PI and PII are not known in the Anti-Atlas belt.

II Early Neoproterozoic sequences (PII) (800–650 Ma)

All the Neoproterozoic rocks (1000–650 Ma) are included in the Anti-Atlas Supergroup by Thomas et al. (2004), which includes five groups (Taghdout, Jbel Lkst, Sarhro, Iriri, and Bou Azzer Groups) composed of volcano-sedimentary units and various intrusive igneous rocks (e.g. Ifzwane and Toudma Suites). These units are related to the earliest passive margin oceanic, island-arc, and flysch sedimentation phase of the Pan-African Orogeny between 800 and 650 Ma. The earliest Pan-African deformation (ca. 650 Ma) resulted from the closure of the ocean basin (Khzama Basin), SW-directed thrusting and accretion of the island-arc remnants.

The oldest rocks of the Anti-Atlas Supergroup comprise three discrete sequences related to the rifting and break-up of the northern margin of the West African Craton (represented by a passive margin sequence), the formation of oceanic crust (ophiolitic remnants) and island-arcs (thrust sheets of calc-alkaline meta-volcanic and plutonic rocks) (Leblanc & Moussine-Pouchkine, 1994, Thomas et al., 2002). These three broadly coeval sequences were grouped originally by Thomas et al. (2002) into the "Bleida Group". However, Thomas et al. (2004) proposed that the allochthonous ophiolitic and island-arc units should not form part of the same lithostratigraphic group.

The Bleida Group (Leblanc & Moussine-Pouchkine, 1994, Thomas et al., 2002) (early Neoproterozoic) is represented mainly in the eastern part of the Siroua area, where plutonic formations (e.g. the Tourtit Orthogneiss of Samson et al., 2004). It comprises the Khzama and N'Kob Ophiolitic Complex and the volcaniclastic formations of the Imghlay Series (El Boukhari, 1991). A tonalite protolith from the ophiolitic sequence has been dated by Sm-Nd at 743 Ma (Cryogenian) and a plagiogranite from the Tasriwine Ophiolite has been dated at 762 Ma (Samson et al., 2004). The Imghlay series is affected by the major Pan-African event (660 Ma) (Thomas et al. 2004), which was responsible for the obduction of ophiolitic complexes on the West African Craton (Villeneuve & Cornée, 1994). The Khzama Ophiolite has been dated at 663 ± 13 Ma (De Beer et al., 2000). In the western part of the Siroua Massif, the Bleida Group is represented by quartzites, serpentinites and mafic to ultramafic amphibolites enclosed within the plutonic formations (Regragui, 1997).

The passive-margin shallow-water sequence along the northern edge of the Zenaga Complex has been termed the Taghdout Group (Taghdout Subgroup of Thomas et al., 2002), which formed part of the Bleida Group. However, Thomas et al. (2004) rejected the term Bleida Group and upgraded the Taghdout Subgroup to group status, which they subdivided into three formations composed of basalts, dolomite-shales, and quartzites, respectively. The Taghdout Group is well exposed in the type Area near Taghdout village, where it has a thickness of 1500 m (4922 ft).

Swarms of doleritic dikes, sills, and irregular sheet-like bodies intrude into the Zenaga Complex and the lower part of the Taghdout Group. These continental tholeiites are termed the Ifzwane Suite by Thomas et al. (2004). The Ifzwane

Table 5.2 Stratigraphic classification of Precambrian units in the Anti-Atlas (based in part on Brabers, 1988). (See also Piqué et al., 1999, Burkhard et al., 2006, Soulaimani and Burkhard, 2008, Gasquet et al., 2008).

| Choubert (1953, 1963) Anti-Atlas Domain | Leblanc (1981) (Bou Azzer Block) | Brabers (1988) (Taznakht Block "SE extension of Sirwa") | | Thomas et al. (2004) (Anti-Atlas Domain) | Soulaimani et al. (2004) (Kerdous Inlier) |
|---|---|---|---|---|---|
| PIII | | | | | |
| Infracambrian-Cambrian: Adoudounian Series | Adoudounian | Adoudounian | | Adoudounian | J.Kerkar & Ougnidif, |
| PIII: Ouarzazate Series | PIII: Ouarzazate Series | PIII | Upper PIII: Ouarzazate Series | Ourzazate Super Group = Volcano-sedimentary Group | Tanalt Series |
| PII-PIII: Sirwa/Saghro Series | PII supérieur: Tidiline Formation | | Lower PIII: Sirwa-Saghro Series | | |
| | Metavolcanics | PII-PIII | Jbal Mimount Quartzites | | Anezi |
| | Detrital sediments | | | | |
| PII | PII inferieur | PIIc: Jbal Amsemsa ash flows tuffas | | | |
| | PII2: Volcano-sedimentary sequence of Ousdrat | PIIb: Tachackcht (Tachoukacht) Formation | | | |
| PII: Taghdout Formation (including Jbel Mimount Quartzites) | Ophiolites sequence of Bou-Azzer | Calc-alkaline volcanic | | Anti-Atlas Super Group = | |

| PI | | | |
|---|---|---|---|
| | PIII: Tachdamt Formation | Ophiolitic rocks | Granites & quartzites |
| | | PIIa: Ophiolitic mélanges: Tamazight & Imglay gneisses  Continental shelf formation: Taghdout dolomites, Tizi Wadou andesites, shales and meta-arkoses | Sarhro (Saghro) Group + Taghdout Bou Azzer Group + Iriri Group |
| | PI: Oued Asemlil gneisses | PI: Azguemerzi and Taznakht granites | |
| | | Metaschists | |
| PI | PI: Bou-Azzer & Tachoukacht Amphibolites  Tamazight and Imghlay Gneisses Azguemerzi and Taznakht Gneisses Timjicht Micaschists | | Basement Complexes = Zenaga Group + Had-N-Tahala Group + Draa Group  Kerdous Granites & Schists |

Suite and the Agouniy Formation of the Taghdout Group are believed to be associated with the early rifting of the West African Craton (Thomas et al., 2004).

The Tizi n'Taghtrine Group (Bouougri & Saquaque, 2004) includes 12 formations, and represents the Neoproterozoic volcano-sedimentary cover (ca.0.8 Ga) deposited on the northern passive margin of the West African Craton in the Central Anti-Atlas inliers. It is a ca. 2 km thick succession. The group comprises two sedimentary sequences (lower and upper) separated by a middle volcanic unit. The Tizi n-Taghatine Group is subdivided into 12 Formations, 10 of which are defined in the lower sedimentary package while the middle volcanic unit and the upper sedimentary package constitute the two uppermost Formations. The lithostratigraphic subdivision of the Tizi n-Taghatine Group constitutes a stratigraphic record of major paleogeographic settings and large cyclic changes in depositional system due to relative sea-level variations.

In the Kerdous inlier, shoreline mature quartzitic sandstones equivalent to the *Serie des Calcaires et Quartzites* of Choubert (1963) were termed the Jbel Lkst Group by BGS (2001c, cited by Thomas et al., 2004) and is probably equivalent of the Taghdout Group. This group is intruded by dolerites and gabbros which are grouped in the Toudma Suite; probable equivalent to the Ifzwane Suite. In the Bas Draa, Tagragra d'Akka and the Tagragra de Tata inliers, comparable swarms cut across the Paleoproterozoic basement.

Ophiolites coeval with the Taghdout and Jbel Lkst Groups occur in the Bou Azzer Inlier; the type area of the Bou Azzer Group (Thomas et al., 2004, Bousquet et al., 2008), which embraces all the ophiolitic fragments of the Anti-Atlas. The Bou Azzer Group consists of harzburgites, gabbros, sheet dikes, and pillow basalts. The northern part of the Bou-Azzer Inlier consists of a meta-sedimentary sequence with subordinate volcanic units with calc-alkaline and island arc tholeiitic composition (Naidoo et al. 1991). The ophiolites underwent greenschist metamorphic conditions (Bousquet et al., 2008). Orthogneisses of the Bou Azzer Inlier were formed around 750–700 Ma and are believed to be allochthonous as the ophiolite (D'Lemos et al., 2006). Associated plagiogranites near Bou Azzer has been dated at about 760 Ma (Admou et al., 2002). Other ophiolites crop out in the Sirwa region at Khazama and Nqob (Thomas et al., 2002).

The Tachoukacht Schists and tonalitic orthogneisses of the Iriri Migmatite form an island arc suite composed of medium-grade biotite-rich and andesitic schists. Thomas et al. (2004) proposed the term Iriri Group to include the Tachoukacht Schists and the Iriri Migmatite, as well as the Ourika Complex (named after the Ourika Valley in the High Atlas). The Iriri Migmatite has been dated at $743 \pm 14$ Ma (Thomas et al., 2002, Admou et al., 2002).

The distal equivalent of the Taghdout Group is the Sarhro (Saghro) Group (Late Neoproterozoic), which may attain a thickness of more than 8000 m (26,248 ft) (Thomas et al., 2002). The Sarhro Group (Saghro) comprises a volcaniclastic sequence (the Siroua Series) overlying unconformably the Bleida Group and is intruded by granites and diorites emplaced between 680 and 600 Ma (De Beer et al., 2000, Thomas et al., 2002, 2004). The Sarhro Group is subdivided into six formations. The lowermost four formations are composed of deep-water flysch sediments and volcanic and volcaniclastic rocks derived from the Iriri island arc, while the upper two formations are characterized by coarse, immature clastic sediments (conglomerates and arkoses) (Thomas et al., 2002, 2004). The stratigraphic position of the Saghro Group is controversial. The Group is assigned to the PII by Brabers (1988), to PII-PIII by Burkhard et al. (2006), and to PII3 by Soulaimani et al. (2001, 2004). It was considered early Precambrian (PII) by Saquaque et al. (1992) based on structural data and Rb/Sr dating of quartzitic diorite intrusions which gave dates between 754–722 Ma. These data led Saquaque et al. (1992) to redefine PII and to restrict PIII to the rocks overlying the basement conglomerates. The volcaniclastics may reach 2000 m (6562 ft) in thickness and are composed of basalts, andesites, sandstones, silts and conglomerates. They were accumulated in a back-arc basin with horst and graben structures (El Boukhari, 1991, Regragui, 1997, Touil et al., 2008). The lowermost flysch formations contain glacial diamictite horizons (Thomas et al., 2002), assumed to be equivalent to the widespread Neoproterozoic diamictites found throughout Gondwana (700 Ma).

III Late Neoproterozoic (650–540 Ma)

### Pan-African Orogeny in the Anti-Atlas (PIII)

The Pan-African Orogeny is represented in the Anti-Atlas by the closure of the ocean basin and subsequent accretion of the island arcs onto the northern rifted edge of the West African Craton. The collision probably occurred between 660 Ma (metamorphic zircons, Thomas et al., 2002) and 680 Ma (Leblanc & Lancelot, 1980), followed by a hiatus between 660 Ma and 625 Ma (Thomas et al., 2004).

The Tidiline Series (Upper PII or PIII) (Leblanc, 1973, Thomas et al., 2004) in the Bou-Azzer Inlier overlies the ophiolites and has a similar paleogeographic setting similar to that of the Saghro Group (Saquaque et al. 1989, Leblanc & Moussine-Pouchkine, 1994, Villeneuve & Cornée, 1994), and it has been dated at 615 ± 12 Ma (Clauer et al., 1982) and 579 ± 1.2 Ma (Inglis et al., 2004). Thomas et al. (2004) proposed to include the Tidiline Series, the Sidi Flah, Kelaat Mgouna, Boumalne and Imiter Groups of Fekkak et al. (2000) and the Habab Group of BGS (2001e) in the Sarhro Group. The volcaniclastic series is intruded by numerous granitic plutons which make up the Assarag Suite (Thomas et al., 2004), such as the Askaoun and Ida ou Illoun granites. In the Sirwa Inlier, two high-K calc-alkaline batholiths of gabbro-diorite-granodiorite granites were emplaced into the deformed Sarhro Group rocks. They are dated at 515 Ma in the Ida Ou Illoun batholiths and at 580 Ma in the Askaoun Batholith (Thomas et al., 2002).

According to Saquaque et al. (1992), the main schistosity in the Saghro Inlier is oriented NE-SW and cut by a conjugate set of strike-slip system attributed to a NW-SE compression. On the other hand, faults and dikes are attributed to late Pan-African deformation caused by NE-SW compression phase.

The final Pan-African stage (Ediacaran-earliest Cambrian, 600–540 Ma) was dominated by extension and the formation of pull-apart basins where thick volcano-sedimentary sequences have been deposited (Azizi Samir et al., 1990, Thomas et al., 2002, Soulaimani et al., 2004). In the Anti-Atlas Domain, these sequences include the Ouarzazate Supergroup (PIII) and the Adoudounian Formation, and associated tholeiitic and alkaline volcanic, such as Jbel Boho Syenite (Ducrot & Lancelot, 1977, Soulaimani et al., 2004, Álvaro et al., 2006). The Ouarzazate Supergroup is separated from the crystalline basement by a major unconformity, and shows rapid thickness variations across fault-bounded basement blocks (Piqué et al., 1999).

Ouarzazate Supergroup (PIII)

All the volcano-sedimentary rocks that were deposited after the Anti-Atlas Supergroup (Middle Neoproterozoic) have been grouped into the Ouarzazate Supergroup by Thomas et al. (2004). In the Sirwa region, about 2000 m (6562 ft) of equivalent sediments have been designated as the Bou Salda Group (Bou Salda Formation of Thomas et al., 2002). Two associated rhyolites, the Tadmant and Tamriwine Rhyolites, gave identical U-Pb SHRIMP ages of 605 Ma (Thomas et al., 2002).

The Ouarzazate Group (Late Neoproterozoic or PIII) was raised to supergroup status by Thomas et al. (2004). It consists of volcanic and volcaniclastic rocks and conglomerates (PIII Conglomerate) unconformably overlying the Sarhro Group (PII). These are intruded by pinkish granites and microgranites (Amassine, Tifnout and Immorghane granites) emplaced between 595 and 520 Ma (De Beer et al. 2000; Thomas et al., 2002, 2004).

In the Tiwit/Ikniwn area of the Jebel Sarhro Massif, several granites, granodiorites, and igneous charnockites (Errami, 2001, Errami et al., 2002) intrude the Sarhro Group. They are called the Bardouz Suite, while their equivalents in the Kerdous Inlier are termed the Guellaba Suite (615–580 Ma). Zircon dates of the granites range from 615 to 580 Ma (Aït Malek et al., 1998, De Wall et al., 2001, Levresse et al., 2001, Thomas et al., 2002, Inglis et al., 2004). The PIII Conglomerate has not been dated paleontologically, but it was assigned to the terminal Proterozoic based on the age of the associated volcanics, dated between 586 Ma and 563 Ma (Piqué et al., 1999). The upper part of the Ouarzazate Supergroup is associated with multiple high-level alkaline granite intrusions and sub-volcanic caldera complexes. In the Sirwa Inlier, these complexes are grouped in the Toufghrane Suite (dated between 575–560 Ma (Thomas et al., 2002). Equivalent granites, dated between 545–579 Ma, are termed the Guellaba Suite in the Kerdous and Bas Draa inliers and the Tanghourt Suite in the Iknwin/Tiwit area of Jbel Sarhro.

Similar sequences have been recorded in the Jbel Sarhro (Saghro) Massif near Imiter, where huge-boulder conglomerates are developed. The sequence is called the Mgouna Group by Thomas et al. (2004). In the Kerdous Inlier, the Enezi Group ('Serie d'Anezi"of Choubert et al., 1973) occupies a similar lithostratigraphic position. This volcaniclastic succession is subdivided into a basal Tafrawt (Tafraoute) Group (dominated by dacitic to rhyolitic ignimbrite) and an upper Anezi Group (dominated by sandstones and conglomerates derived from the Tafrawt Group). They are associated with alkaline sub-volcanic granite complexes, basic and acid dike swarms, various rhyolitic plugs, domes and necks and gabbroic bodies. The plutonic rocks of the Assarag Suite and the sediments of the Mgouna, Tafrawt, Anezi and Bou Salda Groups are overlain unconformably by the upper part of the Ouarzazate Supergroup.

The Quartzite Series (PII) (late Neoproterozoic-Infracambrian) north of Kerdous is a detrital unit with siltstones, pelitic sandstones, conglomerates, thick quartzitic layers (which yielded Neoproterozoic ages), stromatolitic limestones, and sills and batholiths of basic intrusions of dolerites and tholeiitic gabbros. In the Bou Azzer inlier, the Quartzite Series is replaced by an ophiolitic complex of the Bou Azzer Group of Thomas et al. (2004). Similar conditions occur at Jbel Siroua. At Jbel Saghro, they are replaced by sandy-pelitic terrigenous formations.

Epigenetic deposits occur in the inlier and are probably the result of hydrothermal activity linked to the granites of the Assarag Suite. The most characteristic mineralizations are the Ag ±/Cu-U deposits of Zgounder (Petruk, 1975, Thomas et al., 2002). In the Assarag Valley, the Imourkhsane Granite contains disseminated molybdenite in quartz veinlets and small pegmatites. Kutina (1986) emphasize the role of deep-seated faults in the concentration of molybdenum ore, and postulate that the intersection of N-S faults with the South Atlas Fault (SAF) may present exploration targets for metals. The silver ore at Imiter, situated on the northern border of Jbel Saghro Inlier, is a world class deposit with reserves of 8.5 million tons at 700 g/t. It consists essentially of an amalgam of Silver (Ag) and mercury (Hg), Ag sulphate salts, arsenopyrite, and base-metals-bearing sulphur (Tuduri et al., 2006). The ore occurs in epithermal veins within the Precambrian sandstones and shales and the overlying volcanic along regional faults oriented N70°E to N90°E. Two major tectonic episodes have led to the formation of the ore: 1) The first episode was associated with extensional tectonics oriented NNW-SSE to N-S, which led to the formation of quartz veins, 2) the second episode was dominated by left-lateral strike-slip faulting which led to the opening of previous fractures and the deposition of gangue dolomite. Recent studies (Tuduri et al., 2006) show that the silver mineralization was associated with Late Neoproterozoic felsic volcanic; subsequent alteration processes contributed to Ag enrichment.

The Neoproterozoic-Lower Cambrian (equivalent to the Adoudounian Formation) succession in the Kerdous area has been assigned a Lower Cambrian age by Geyer & Landing (1995). The sedimentary succession at the Ait Abdallah and the Kerdous areas (Figs. 5.14, 5.16) has been subdivided by Benssaou & Hamaoumi (2001) into several informal formations:

The Basal Conglomerate Formation in the Ait Abdallah area consists of two fluvial depositional facies; the first includes lenticular conglomeratic channel bars; the second contains alluvial plain fine- to medium-grained sandstones. In the northeastern Kerdous area, this formation is subdivided into two members. The lower member is made up of conglomeratic fan-delta deposits, alternating sandy micrites, sandstones, and laminated argillite-rich lacustrine evaporites. The upper member comprises alternating microconglomerates and sandstones, interpreted as transitional between fan-delta and marine deposits (Benssaou & Hamaoumi, 2001).

The Basal Limestone Formation is made up of sandstones with mud drapes, flaser bedding, and laminated muddy beds with lenses of siltstone and laminated sandstones, interpreted as tidal-flat deposits. In the Ighir Ifrane area, it is dominated by shallow water stromatolitic carbonates, changing northward into sandstones and conglomerates. In the Ait Baha area, the formation consists of inversely-graded conglomerates alternating with carbonates, sandstones, and flat-pebble conglomerates, interpreted as reworked storm deposits (Benssaou & Hamaoumi, 2001).

The Basal Siltite Formation is represented by two facies; the first consists of graded and hummocky, cross-stratified sandstone beds, and microconglomeratic bars intercalated within graded beds. The facies form a coarsening-upward and thickening-upward succession (Benssaou & Hamaoumi, 2001).

The Stromatolitic Dolomite Formation consists of two members; the lower member contains domal and planar stromatolitic carbonates, locally interbedded with silicified and brecciated evaporitic layers. The upper member comprises alternating planar and domal stromatolitic carbonates and reworked siliciclastic sediments (Benssaou & Hamaoumi, 2001).The Thrombolitic Limestone and Dolomite Formation contain alternating stromatolitic beds; the former is composed of vertically stacked hemispheroidal cyanobacterial laminites and thrombolites; the latter shows evidence of episodic emergence (Benssaou & Hamaoumi, 2001).

The Basal Conglomerate Formation comprises a fluvial system in the Ait Abdallah area. Along the northeastern border of the Kerdous Inlier, the formation is made up of fan-delta sediments (Benssaou & Hamaoumi, 2001).

The Reef Limestone Formation consists of limestone reef buildups of Archeocyaths and bioherms forming shallowing-upward, meter-scale cycles (Benssaou & Hamaoumi, 2001).

During the extension phase, the Anti-Atlas Domain experienced rifting, tilting of basement blocks and the creation of half grabens. Extension is also documented by the Volcaniclastic Series of the Latest Proterozoic-Earliest Cambrian 'Ouarzazate Supergroup' (Thomas et al., 2004), as well as the Early Cambrian marine 'Taroudant Group' (Buggisch & Siegert, 1988, Algouti et al., 2001, Benssaou & Hamoumi, 2003) (equivalent to the Adoudounian Formation and the Tata Group). More than 10 km of shallow-marine sediments were deposited in an intracratonic setting on top of these synrift successions during the remainder of the Paleozoic (Burkhard et al. 2006). The Precambrian and early Paleozoic structures had a strong influence on the structures of the Late Carboniferous Variscan compressive episode (Michard, 1976, Jeannette & Piqué, 1981, Soulaimani et al., 1997), which resulted from the reactivation of basement faults (Leblanc, 1972, 1975, Donzeau, 1974, Jeannette & Piqué, 1981, Hassenforder, 1987, Soulaimani, 1998, Belfoul et al., 2001).

The tectonic evolution of the Anti-Atlas Basin and fold belt in relation to global plate tectonics is summarized by Burkhard et al. (2006) as follows:

1   During the Pan-African Orogeny, a series of terranes are accreted to the West African Craton (WAC) on its northwestern side. The suture (Bou Azzer Ophiolites) and the terranes to the northeast remained in place. The Western Anti-Atlas, however, remained attached to Gondwana at the margin of the Saharan Metacraton (Abdelsalam et al., 2002).

2   In Late Proterozoic-Early Cambrian, the Anti-Atlas area was under extension with the formation of grabens and half-grabens [cf. Tawadros, 2001 and the rest of

North Africa]. These grabens were filled with coarse clastics (PIII or Ouagarta Series) mostly of igneous origin accompanied by calc-alkaline to tholeiite-alkaline volcanics (Soulaimani et al., 2004). However, Burkhard et al. (2006) postulated that the Anti-Atlas was the location of a major broad thermal dome with limited grabens and half-grabens, followed by thermal subsidence throughout the Paleozoic.

3  From the Middle Cambrian through the Middle Carboniferous, the Western Anti-Atlas Basin was characterized by a strong subsidence, leading to the accumulation of more than 10 km of mostly fine-grained clastic sediments, shed into an epicontinental sea from the African Craton. There is little evidence in the stratigraphic record for tectonic events along the active northwestern margin of Gondwana (Stampfli & Borel, 2002). The drifting of Avalon, Armorica and Hunic terranes from this margin (during the Cambrian and Ordovician?) connected the Anti-Atlas Sea with the Paleotethys Ocean. From Silurian times onward, the Anti-Atlas Basin probably represented the passive margin of the Paleotethys Ocean, but very little evidence of this passive margin is preserved anywhere (with the possible exception of terranes west of Guelmin in the Western Atlas (Belfoul et al., 2002) and near Tineghir (Michard et al., 1982).

4  The Late Carboniferous-Permian (Hercynian-Variscan) compression led to inversion and folding (Beauchamps et al., 1997, Soulaimani & Burkhard, 2008). The basement is uplifted and folded into huge antiformal culminations (*boutonnières*) which punctuate the Anti-Atlas fold belt. The structural relief of the basement culminations is in excess of 10 km, and minimum estimates of total shortening are between 15 and 25 km. Similar time-equivalent belts occur further east in the African Craton, Ougarta, Ahnet, and others in Egypt and Libya.

## 5.3 SUMMARY

In summary, the Pan-African basement rocks can be grouped into five major tectono-stratigraphic assemblages: 1) Lower Gebel Oweinat assemblage (>2500 Ma). 2) Upper Gebel Oweinat assemblage (2500–1800 Ma). 3) Lower magmatic-arc assemblage (1000–650 Ma). 4) Upper magmatic-arc assemblage (650–550 Ma). 5) Post-tectonic assemblage (<550). The cycles 2 and 3 were probably separated by a period of extension and rifting. The temporal and spatial boundaries of these assemblages, however, are imprecise and more systematic work on the stratigraphy, geochemistry, and absolute age determination is needed before a clearer picture can emerge. Nevertheless, the evolution of the Pan-African Shield can be depicted tentatively as follows:

1  The Archean lower Gebel Oweinat assemblage (>2500 Ma) includes granulitic gneisses in the Gebel Oweinat area and represents an ancient craton in Northeast Africa, upon which the younger terranes have been accreted. Equivalent rocks include the Gallaman Group (2183–1800 Ma) in the Reguibat Massif and the Zenaga and Kerdous groups in the Anti-Atlas.

2  The upper Gebel Oweinat assemblage (2500–1800 Ma) includes migmatitic rocks in the Gebel Oweinat area and its equivalent of the Lower Tibestian in the Tibesti Massif. In the Anti-Atlas Domain, it includes the so-called "Basement Complex" (PI).

The lower and upper Gebel Oweinat assemblages form the Nile Craton. They, along with their counterparts in the West African Craton, belong to two major global crust-forming periods during the Archean and Early Proterozoic.

The period from ca. 1800 Ma to 1000 Ma was dominated by the formation of supercontinent Rodinia. In many parts of North Africa, rocks from that period are absent. Other parts were subjected to extension, rifting, and the formation of oceanic crusts represented by ophiolitic sequences composed of serpentinites, gabbros, and diorites. This event probably led to the formation of horst and graben associated with the emplacement of rhyolites and bimodal volcanics in the cratonic areas.

3   The lower magmatic-arc assemblage (1000–650 Ma) comprises the Anti-Atlas Supergroup (PII) and its component groups, such as the Saghro, Taghdout, Bou Azzer, and Iriri groups. It includes:

a   Deposition of shallow shelf and basin sequences in shallow to deep-marine environments and represented by the protoliths of the migmatites and metasediments in the Eastern Desert and the Upper Tibestian in the Tibesti Massif. Deposition of these sediments was probably partly contemporaneous with the formation of the oceanic serpentinites and ophiolites.

b   Subduction of the oceanic crusts beneath the ancient cratons led to the formation of a series of island arcs represented by calc-alkaline volcanic suites. Erosion of these island-arcs led to the deposition of volcanogenic sediments in different areas.

c   The island-arc volcanic activities were associated with the emplacement of a series of older calc-alkaline granites and gabbros into the older units, the formation of domal structures exemplified by the Migif-Hafafit and Meatiq domes in the Eastern Desert and the Feiran and Taba Gneiss in Sinai, and the migmatization of the country rocks. They are also represented in Northwest Africa by the Tachakacht Formation in the Anti-Atlas.

d   An ophiolitic-mélange sequence of serpentinites, metavolcanics and metasediments which represent an ancient oceanic crust that was subducted, then obducted and thrust over younger volcanics, granitoids, and migmatitic rocks of the late Proterozoic craton. In the Anti-Atlas, ophiolites occur in the Bou Azzer Group.

4   The upper magmatic-arc assemblage (650–550 Ma) is related to the main Pan-African Orogeny. In Northwest Africa, the time interval is represented by the PIII Sequence or the Ouarzazate Supergroup. It was dominated by compression and continued subduction of the oceanic crusts underneath the various cratons. However, it was represented by different tectonic styles in different parts of North Africa:

a   Domination of compressive orogenic tectonics in some areas, similar to the lower magmatic-arc cycle, such as in Sinai, with the formation of volcanic arcs, ophiolites and mélanges assemblages, and the emplacement of younger calc-alkaline granites and gabbros, such as in the northern Eastern Desert.

b   Extensional anorogenic tectonics which were probably contemporaneous with the above tectonic event and led to the deposition of volcanogenic sediments, such as the Hammamat sediments, in grabens and the intrusion of younger alkaline granites, rhyolites, and dacites. This phase is also recognizable in the Anti-Atlas Domain of Morocco.

5   The post-tectonic assemblage (<550) represents deposition during a period of extension which started during the previous cycle and is characterized by the deposition of quartzose and quartzo-feldspathic sandstones and conglomerates with granitic rock fragments. Extensional tectonics continued through the Early Paleozoic with the emplacement of swarm dikes, ring complexes, and the inception of the Paleozoic extensional and sag basins, such as the Kharga, Kufrah, Murzuq, Ghadames, Illizi, Tindouf, and Taoudini basins. The Hoggar Massif was buried during Cambrian times beneath 3–7 km of Paleozoic sediments that have been subsequently eroded (Lesquer et al., 1989). In the Anti-Atlas, the period was dominated by the Adoudounian and Ouarzazate groups, which straddle the Precambrian-Cambrian boundary.

6   The Hercynian-Variscan compression during the Late Carboniferous-Permian led to inversion and folding of many basins in North Africa. The Pan-African basement was uplifted and folded which led to the exhumation of the basement in Southern Sinai and the Red Sea Hills, the Tibesti Complex, the Hoggar Massif, the Touareg Massif, and the formation of the *boutonnières* in the Anti-Atlas belt.

# PART 4

# Phanerozoic geology of North Africa

# Chapter 6

# Phanerozoic geology of Egypt

## 6.1 INTRODUCTION

Egypt is located in the NE corner of Africa and bordered on the north by the Mediterranean Sea, on the east by the Red Sea, on the west by Libya, and on the south by Sudan. The Gulf of Suez separates Sinai from the mainland of Egypt.

Egyptian geology practically started on the hands of German geologists and took a giant step with the advent of Rohlf's expedition. The results were published in 1883 and contained the first geological map of Egypt. Among the geologists who accompanied that expedition and have notable contributions were Zittel and Schweinfurth.

The Egyptian Geological Survey was founded in 1896 by H.G. Lyons. He contributed much knowledge to the geology of the River Nile (Lyons, 1894, 1906, 1908, 1909). In his report (Lyons, 1894) of the stratigraphy of the Libyan Desert of Egypt, he presented a geological map of Egypt at the scale of 1:3000,000. A more detailed geological map of Egypt was published by the Survey in 1928. Among the most prominent geologists of that period were Beadnell, Hume, Blanckenhorn, Barthoux, and Cuvillier. Hume published his "Geology of Egypt" between 1925 and 1937. Following Hume's work, most of the information remained dispersed until Said (1962a) compiled a large portion of the work published before 1962 and summarized the ideas in his classic book "The Geology of Egypt", which has been considered the source book on that subject until now. Readers are urged to consult this book for a summary of the pre-1960's ideas. Said (1990a) also edited a book with the same title which contains a collection of papers by various authors.

Two major problems existed with the Egyptian geology and persisted for over a century. The first of these two problems is famous worldwide and deals with the Nubian Sandstone enigma. Russegger introduced the name in 1837 to describe some sandstones exposed in Nubia, but soon thereafter, the term was applied to any problematic sandstones throughout the stratigraphic column in areas extending over the stretch of North Africa, Sudan, Jordan, and Saudi Arabia. The second problem deals with the Upper Cretaceous-Lower Tertiary boundary which revolves around the definition of the so-called "Esna Shale". Pomel (1877) realized that the Nubian Sandstone included sediments of various ages. This name was used by Beadnell, Baron, and Hume in the International Geological Conference in Paris in 1900 and by Blanckenhorn (1902). However, it was not until 1905 that Beadnell introduced the name formally. It is of archeological and historical interest that the mostly hard and massive Nubian Sandstone of Jebel Silsila, north of Aswan, was heavily quarried by the Ancient Egyptians for the construction of temples (Beadnell, 1905a, Klemm & Klemm, 2001).

Ironically, Egypt had to wait for another German "expedition" to tackle the Nubian puzzle. The results of the German Special Research Project, supported by the German Research Foundation until 1987 was published in volumes 50 and 75 (three parts) of the *Berliner Geowissenschaftliche Abhandlungen*.

Exploration for hydrocarbons in the Gulf of Suez, the Western Desert, the Nile Delta and its offshore area has added much to our knowledge, although much of the information still remain confidential. Exploration in Egypt started in 1868 with the drilling of the first well near an oil seep at Gebel Zeit on the Gulf of Suez. Today, oil and gas have been discovered in the Gulf of Suez, the Western Desert, the Nile Delta and the offshore area, and in Upper Egypt, with more than 30 giant fields (Dolson et al., 2001). For a review of the history of geological and hydrocarbon exploration in Egypt, see Part 1, and Tawadros (in preparation).

Mineral resources in Egypt include oil and gas, coal, and phosphate, in addition to other mineral deposits (Hussein & El Sharkawi, 1990). Estimated proven recoverable reserves in Egypt (2005) are approximately 3.71 billion barrels of oil and 66.84 bcf of gas (World Oil, Sept. 2006). Daily production is about 579,000 bopd and gas production is 3.6 bcfd (OGJ, 2005).

Coal was discovered in Sinai at Gebel Maghara in Jurassic sediments in the 1950's, and development of mining operations was carried out from 1964 until 1967 when it was interrupted by the Gulf of Suez war. Reserves available to modern mining techniques from the main coal bed are about 23 million short tons (eia, 2009). Present production is at a rate of 125,000 tons per year. Members of the Geological Survey of Israel and several universities conducted intensive geologic mapping in the Sinai between 1967 and 1984.

Geoarcheology in Egypt probably started with Gertrude Caton-Thompson's archeological work in the Fayium and Kharga Oasis in the 1920s and 1930s. Caton-Thompson (1888–1985) worked with a number of the most prominent archeologists in Egypt, such as Flinder Petrie (the Father of Archeology). Recent progress in the Quaternary geology of Egypt was made during the investigation of the prehistoric settlements along the Nile Valley that began in 1961, by K.W. Butzer (1960a, b, 1976), C.L. Hansen (Butzer & Hansen, 1968), and W. Kaiser. These investigations were carried out during an international campaign to salvage the archeological material, which was to be inundated by the reservoir of the Aswan High Dam. The most important of these expeditions was the Combined Prehistoric Expedition sponsored by the Southern Methodist University, the Polish Academy of Science, and the Geological Survey of Egypt. The results of this work were discussed extensively by Wendorf & Schild (1970, 1976, 1980) and Said (1981). A project to define the late Quaternary geological evolution of the lower Nile Delta of Northern Egypt, taking into account both natural and anthropogenic factors, was initiated in 1985 by the National Museum of Natural History, Smithsonian Institution. Stanley et al. (1996) presented lithologic logs of the 87 cores, which were cut during that project, and were later used in numerous publications. Other important geoarcheological works in Egypt include the Dakhla and Kharga oases Projects by M.R. Kleindienst since 1978, together with M.A. McDonald, C.S. Churcher, J. Smith and R. Giegengack.

Other than the work of Tawadros (2001) and this edition, there is no formal lexicon of stratigraphic nomenclature for Egypt. Egyptian nomenclature contains numerous units that are ill-defined. Many of the formational boundaries are still not defined and different formations are given the same name.

## 6.2 PALEOZOIC

Cambrian and Carboniferous outcrops have been known for a long time from Western Sinai, Wadi Araba on the eastern side of the Gulf of Suez, and Gebel Oweinat in the extreme southwest of Egypt. Recent work in the southern Western Desert, such as the Abu Ras-Wadi Malik area and northern Wadi Qena in the Eastern Desert, indicated the presence of Lower Paleozoic deposits ranging in age from Cambrian to Carboniferous. Exploration drilling for hydrocarbons showed that Paleozoic rocks are also widespread in the subsurface of Northern Egypt and the Gulf of Suez area.

## 6.3 CAMBRIAN AND ORDOVICIAN

Cambro-Ordovician rocks crop out in very limited areas in Egypt. Hassan (1967) divided the pre-Carboniferous section in the Abu Durba area, on the western side of the Gulf of Suez, into a lower Araba Formation and an upper Naqus Formation.

The Cambrian Araba Formation (Hassan, 1967) also crops out in Western Sinai in the Um Bogma-Wadi Feiran area and northern Wadi Qena (Somr el Qaa) in the Eastern Desert of Egypt (Bandel et al., 1987, Klitzsch, 1990a, Klitzsch et al., 1990). The succession overlies the Precambrian basement directly, and is approximately 80 m (262 ft) in thickness. At Wadi Feiran, the Araba Formation is up to 190 m (623 ft) in thickness and composed of sandstones and conglomerates at the base, followed upwards by interbedded cross-bedded sandstones and sandstones with abundant *Skolithos* (Klitzsch, 1990a). Seilacher (1990) identified the Lower Cambrian trilobite tracks *Cruziana aegyptica, C.* cf. *nabataeica, C. salamis,* and other trace fossils from this unit. Allam (1989) reported volcanic rock fragments from the sandstones of the Araba Formation in the Wadi Feiran-El Tor area. At Taba, the Araba Formation is about 70 m (230 ft) thick and is dominated by braided channel deposits, grading upward into shoreface sandstones (El-Araby & Abdel-Motelib, 1999). Petrographic studies indicate that the Araba Formation sands in Eastern Sinai were derived from the exposed basement shield area (Akarish & El-Gohary, 2008).

Omara (1972) described a limestone bed, approximately 4 m (13 ft) thick, at the base of the Araba Formation which contains Lower Cambrian stromatolites and small archaeocyathids. Brenckle (in Keeley, 1994) believes that these are actually Cretaceous-Tertiary red algae. However, it is clear that the Araba Formation is overlain by the Naqus Formation of possible Ordovician age, or other sediments of definite Carboniferous age. At Somr el Qaa, in Wadi Qena, the Araba Formation overlies the Precambrian basement. It is 25 m (82 ft) in thickness and is composed of arkosic conglomerates, fine to very fine to coarse-grained sandstones, siltstones, and shales, with various species of *Cruziana* (Klitzsch et al., 1990a).

The Araba Formation, at Wadi Araba and at Um Bogma in Sinai, as well as at Taba on the Gulf of Aqaba (El-Araby & Abdel-Motelib, 1999), is overlain by 150 m–200 m

(492–656 ft) of fluvial sandstones of the Naqus Formation (Weissbrod, 1969). The Naqus Formation was assigned a Cambrian age by Weissbrod (1969) and an Ordovician age by Issawi & Jux (1982). Allam (1989) described the Formation in the Wadi Feiran-El Tor area as 280 m (919 ft) in thickness and composed of pebbly, medium- to coarse-grained, well sorted quartzarenites, with festoon cross-bedding, and truncated and overturned cross-beds. The Formation conformably overlies the Araba Formation, and is unconformably overlain by the Early Carboniferous Abu Durba Formation.

At Um Bogma, Sinai, Weissbrod (1969) divided the Paleozoic section below the Um Bogma Formation into a lower Yam Suf Group and an upper Netafim Formation. The Yam Suf Group is 62 m (203 ft) to 137 m (449 ft) in thickness and consists of a lower member of arkosic sandstones overlying the Precambrian basement and an upper member of variegated sandstones. The age of the group is assumed to be Paleozoic, and the lower part is believed to be of Cambrian age. Weissbrod (1969) subdivided the Yam Suf Group into five formations, in an ascending order: Paleozoic Nubian Sandstone, Hakhlil Formation, Nimra Formation, Mikhrot Formation, and Shehoret Formation. The type sections of all these units fall outside the borders of Egypt, but reference sections were designated at Um Bogma for each of these formations. All these formations can be considered as members of the Araba Formation. The Paleozoic Nubian Sandstone was described as reddish brown with some white, bedded, fine- to coarse-grained, occasionally micaceous, arkoses and arkosic sandstones. A thin layer (20 cm) of conglomerate, consisting of subangular to rounded magmatic pebbles, occurs at the base of the unit. It unconformably overlies the Precambrian rocks. The Formation varies in thickness in SW Sinai between 10–45 m (33–148 ft). The Hakhlil Formation is composed of an alternation of arkosic sandstones and micaceous shales. The sandstones are thinly bedded, red and brown, very fine-grained with some coarse sand bands and lenses of clays. The shales are red-brown and green. Groove casts, load casts, burrows, and trilobite tracks are common. Impregnations of manganese minerals occur in layers. The unit varies in thickness from 6–14 m (20–46 ft). The Nimra Formation consists of a series of sandy dolomites and dolomitic sandstones, alternating with arkosic sandstones and micaceous shales. The dolomites are purple-red or white-grey, with varying amounts of quartz and feldspars. The sandstones are greenish-white, well-bedded, and very fine-grained with some bands of granules. The Formation is 6 m (20 ft) in thickness at Um Bogma. The Mikhrot Formation is made up of a thin series of violet and green micaceous shales, approximately 2.6 m (9 ft) in thickness. The Hakhlil, Nimra, and Mikhrot formations have not been recognized in SW Sinai. The Shehoret Formation is made up of a series of white and brown, thinly bedded, occasionally cross-bedded, arkosic sandstones. Micaceous shales and manganese and iron impregnations occur in the lower part and layers of quartz pebbles occur in the upper part. The thickness of the Formation is 57.7 m (189 ft) at Um Bogma, but varies from 30–70 m in SW Sinai. It is absent in Gebel Nukhul. The overlying Netafim Formation is about 30 m (98 ft) in thickness and composed of massive, fine- to coarse-grained sandstones with some shale, ferruginous siltstone, and quartz pebble layers. Calcitic and manganiferous veins traverse the Formation in various directions. Copper mineralization occurs in the uppermost part at Gebel Adeidiya, Wadi Natash, and Gebel Maghara. It unconformably overlies the Yam Suf Formation and it is unconformably overlain by the Early Carboniferous Um Bogma Formation. In NE Sinai, the Yam Suf Formation is absent and the Netafim Formation directly overlies the Precambrian basement. Weissbrod (1969) included this Formation in the Cambro-Ordovician, as did Mart & Sass (1972), but Said (1971) considered it as Lower Carboniferous. Weissbrod & Perath (1990) included it in the Cambrian. The Yam Suf Group and the Netafim Formations are equivalent to the Lower Sandstone Series of Attia (1954), the Lower Sandstone of the Um Bogma Series of Kostandi (1959), and the Araba and Naqus formations of Hassan (1967), respectively. The Netafim Formation is probably Ordovician in age.

At Gebel Oweinat (Jabal Awaynat) in the Karkur Talh area, the Karkur Talh Formation is composed of shallow marine sandstones interbedded with conglomeratic fluviatile sandstones, with *Cruziana rouaulti*, which is taken as evidence for Ordovician age (Klitzsch, 1990a). This Formation is probably equivalent to the Naqus Formation.

El Dakkak (1988) recognized Cambrian sequences in the Gibb Afia-1, Ghazalt-1, Bahariya-1, and Betty-1 wells, based on lithological criteria and the presence of the trilobite *Ptychaspis* sp. in the Gibb Afia-1 well.

Keeley (1989) proposed the name Siwa Group to include the Shifa, Kohla, and Basur formations in the subsurface of the Ghazalat Basin in the northwestern Western Desert. The Siwa Group is unconformably overlain by the Faghur Group or Mesozoic strata east of 26°E. The age of the group is from Late Cambrian to Late Silurian.

The Shifa Formation in its type section in the Siwa-1 well (29°07'18"N, 23°25'50"E), at a drill depth of 3058–3373 m), is 315 m (1034 ft) thick and composed of sandstones, conglomerates, shales, fine glauconitic sandstones, claystones and skeletal carbonates, and coarse arkosic sandstones. The thickness of the unit ranges from 300–1500 m (984–4921 ft) in the Ghazalat Basin. Keeley (1989) suggested glacial erosion at the top. The Shifa Formation is overlain by the Kohla and Basur formations of Silurian age. It was assigned a Middle Cambrian-Middle Ordovician (Llanvirnian) age by Keeley (1989). Gueinn & Rasul (1986) recognized Middle Cambrian-Ordovician palynomorphs from this Formation. They reported the Middle Cambrian acritarchs *Eliasum asturicum* and *Skiagia* sp. in the Bahrein-1, Gibb Afia-2 and Kohla-2 wells which they designated as Zone WD1, and the Middle-Late Cambrian (Zone WD2) acritarchs *Timofeevia phosphoritica*, *T. lancarae*, and *Cristallinium cambriense*. They also recorded the Early Tremadocian (Zone WD3, subzone A) acritarchs *Saharidia fragile*, *S. downiei*, *Vulcanisphaera* cf. *simplex* in the NWD-302–1 well, the Late Tremadocian (Zone WD3, subzone B) acritarchs *Acanthodiacrodium tremadocum* and *A. ignoratum* in the East Faghur-1 well, and the Llanvirnian (Zone WD4) acritarchs *Frankea sartbernadense*, *Veryhachium subglobosum*, *Baltisphaeridium klabavense*, and *Stelliferidium striatulum* in the Gibb Afia-2 and East Faghur-1, which they compared with the top of Zone E of Jardiné et al. (1974) in Eastern Algeria. This Formation is equivalent to the Gargaf Group (*s.l.*) in Libya.

## 6.4 SILURIAN

Silurian strata occur in southwestern Egypt and the northwestern Western Desert. In the Abu Ras Plateau area, north of the Sudanese border, approximately 400 m (1312 ft) of sandstones occur near the Um Ras Passage (Klitzsch, 1990a). These sandstones are interbedded with rare silt to fine sand layers, locally containing abundant *Cruziana accacensis*, *Harlania harlani* and *Skolithos* sp. Klitzsch & Lejal-Nicol (1984) applied the name Um Ras Formation to these sandstones. The unit was assigned a Lower Silurian (Llandovery) age and correlated with the Acacus Sandstone Formation in Libya by Klitzsch (1990a) and Wycisk (1990). However, if this Llandovery age is accepted, this Formation is probably a nearshore equivalent of the Tannezouft Formation. In the Abu Ras Plateau, the Formation overlies Precambrian metamorphic rocks; however, near Gebel Oweinat at Karkur Talh, it overlies Ordovician sandstones of the Karkur Talh Formation. The Um Ras Formation is made up of shallow marine coarsening-upward sequence in the lower part, followed by a predominantly fluvial sequence in the upper part (Wycisk, 1990). The sequence starts with a basal conglomerate and medium to coarse grained sandstones, followed upward by another coarsening-upward sequence of interbedded, laminated siltstones and fine-grained sandstones with wavy lamination, ripple cross-lamination and trace fossils, such as *Cruziana accacensis*, *Harlania harlanai* and abundant *Skolithos*, followed by cross-bedded, *Skolithos*-bearing, medium- to coarse-grained, partly conglomeratic sandstones. The upper part of the succession is represented by fluvial, medium- to coarse-grained sandstones with small and large-scale tabular cross-bedding.

In the northwestern Western Desert, the Silurian is represented by the Kohla and Basur formations of the Siwa Group (Keeley, 1989).

The Kohla Formation in the type well Zeitoun-1 (29°14′43″N, 25°43′32″E), at a drill depth of 2616–3242 m, is represented by 626 m (2054 ft) of siltstones, with minor sandstones and shales. Its thickness varies between 80–626 m (262–2054 ft) in the Ghazalat Basin. It is conformably overlain by the Basur Formation. Keeley (1989) assigned it a Silurian age (Llandovery to early Ludlovian), based on unspecified acritarchs. The Silurian microflora in the Foram-1 well, near the Egyptian-Libyan border, is characterized by a very restricted assemblage which includes Leiospheres and a few other palynomorphs (Schrank, 1984, 1987), such as *Leiofusa* cf. *tumida*, *Micrhystridium* sp., and spore-like microfossils, cf. *Tetrahedraletes* of possible Early Llandovery age. The Kohla Formation is probably equivalent to the Tannezouft Formation in Libya (Keeley, 1994).

The type section of the Basur Formation is in the El-Basur-1 well (29°54′23″N, 25°50′08″E), at a drill depth of 2549–3165 m, where it is 616 m (2021 ft) in thickness and consists of interbedded siltstones and conglomerates. In the Ghazalat Basin, the unit varies in thickness between 400–700 m (1312–22–97 ft). It is overlain with a sharp contact by the Devonian Zeitoun Formation. However, palynomorphic associations continue across the boundary between the two units. Gueinn & Rasul (1986) recognized two Silurian palynological zones in this Formation. Zone WD5 (?Llandovery-Ludluvian) characterized by the acritarchs *Neoveryhachium carminae*, *Gorgonisphaeridium saharicum*, and *Baltisphaeridium gueltanese*, in addition to the chitinozoa *Ancryochitina fragilis* and *Sphaerochitina sphaerocephala*, and Zone WD6 (Ludluvian-Early Gedinnian) dominated by the acritarchs *Cymbosphaeridium pilaris* and *C. carniosum*. This Formation is probably equivalent to the Acacus Formation in Libya (Keeley, 1994).

El Shazly (1977) used the name Bahrein Formation for the Silurian sediments in northern Egypt, a name which is now applied to an Early Jurassic unit in the subsurface of the Western Desert (cf. Hantar, 1990). It is equivalent to the Kohla and Basur formations of Keeley (1989).

## 6.5 DEVONIAN

Devonian sediments are present in the subsurface of the Ghazalat Basin in the Western Desert in the north and crop out in the Gilf Kebir-Abu Ras area in the south. They are probably present in the subsurface of the Hurghada area along the Red Sea coast. Keeley (1989) proposed the name Faghur Group to include the Zeitoun, Desouqy, Dhiffah, and Safi formations in the subsurface of the northwestern Western Desert in Egypt. The group thins or becomes eroded eastwards. The age of this group ranges from Devonian to Permian.

The Devonian is represented by the Zeitoun Formation. The thickness of the Zeitoun Formation in the type well Zeitoun-1, at a drill depth of 1670–1958 m, is 288 m (945 ft), but a thickness of 759 m (2490 ft) has been encountered in the Faghur FRX-1 well. The Formation is composed mostly of sandstones and conglomerates forming a coarsening-upward sequence, with common limestone beds near the base. The Formation unconformably overlies the Silurian Basur Formation and is unconformably overlain by the Carboniferous Desouqy Formation. It was assigned a Gedinnian-Strunian age by Keeley (1989) based on acritarchs. Strata of Late Emsian to Early Givetian age were also identified in the subsurface in the Foram-1 well, at a drill depth of 2547–2550.2 m, using palynomorphs (Schrank, 1984, 1987), such as the acritarchs *Polydryxium fragosulum, Veryhachium downiei, Duvernaysphaera tessella*, and *D.* cf. *tenuicigulata*, in addition to large spores of the *Grandispora* type.

El-Ghazaly & Ali (1985) examined spores assemblages from the Devonian-Carboniferous section in the well Kohla-2. They reported *Punctatisporites planus, Retusotriletes* cf. *triangulates, Apiculiretusispora angulata, Acanthotriletes parvus, A.* cf. *echinatus, Verrucosisporites nitidus, Convolutispora vermiformis, Emphanisporites rotatus, E.* cf. *radiatus, Densosporites* sp., *Vallatisporites* sp., and *Hymenozonotriletes* sp., Gueinn & Rasul (1986) recognized four palynological zones within the Devonian (Gedinnian-Strunian) of the northern Western Desert. Zone WD7 (Gedinnian-Givetian) is characterized by the acritarchs *Polyedryxium fragosulum, P. carnatum, Tunisphaeridium tentaculiferum*, and *Diexallophasis remota*. They compared the top of this zone with Zone 6 of Massa & Moreau-Benoit (1976) in western Libya. Zone WD8 (Givetian-Frasnian) is illustrated by the spores *Grandispora inculta* and *G. libyensis*. They correlated the top of this zone with Zone 8 of Massa & Moreau-Benoit (1976) in western Libya. Zone WD9 (Frasnian-Famennian) is characterized by the spores *Elektoriskos* cf. *deconinckii*, and *Unellium winslowae*, and the acritarchs *Gorgonisphaeridium ohioense*, and *Veryhachium pannuceum*, recorded from the West Faghur-1X, Faghur (FRX)-1 and NWD-302–1 wells. They correlated the top of this zone with zone L6 of Jardiné et al., 1974). Zone WD10 (Late Famennian-Strunian) includes the spores *Spelaeotriletes lepidophytus* and *Horologinella* cf. *quadrispinosa*, encountered in the Faghur (FRX)-1 well. They equated this zone with Zone XI of Massa et al. (1980) in western Libya.

It should be noted that El Shazly (1977) used the name Zeitoun Formation to denote the Cambo-Ordovician succession in the subsurface of the northern Western Desert, a name which has been used by Keeley (1980) to describe the Devonian succession. This Formation is equivalent to the Shifa Formation of Keeley (1989).

In the Gilf Kebir-Abu Ras area, Devonian sediments are believed to be present above the dated Silurian (Klitzsch, 1990a). These sediments are exclusively fluvial and were compared with the fluvial part of the Tadrart Formation of southern Libya by Klitzsch (1990a) and Wycisk (1990). They are composed of pebbly, medium to coarse-grained sandstones with tabular cross-bedding and minor horizontal and trough cross-bedding (Wycisk, 1990).

Jongsmans & van der Heide (1953) suspected the presence of Late Devonian sediments in the subsurface of the Hurghada area, along the Red Sea coast, based on the presence of plant fragments of *Pseudobornia* sp., *Platyphyllum williamsoni* and *P. brownianum*.

## 6.6   CARBONIFEROUS

Carboniferous rocks are known from Sinai, Wadi Abu Ras-Wadi Abd el Malik area of SW Egypt, and also from the Eastern Desert (northern Wadi Qena). They also occur in the subsurface of the Tenehue Basin in NW Egypt. Carboniferous sediments are exposed on the eastern and western sides of the Gulf of Suez at Um Bogma and Gebel Nukhul in Sinai, and Wadi Araba and Wadi Qiseib in the Eastern Desert. They were also reported from the subsurface of the Hurghada area on the Red Sea coast and the Ayun Musa-1 well in Sinai. Carboniferous sandstones, usually designated as "Nubia Sandstone", are oil-bearing in many fields, such as Ramadan, West Abu Rudeis, and July in the Gulf of Suez (Brown, 1980, Abdine et al., 1992, 1994, Schütz, 1994, Alsharhan & Salah, 1997a, b).

Jongsmans & van der Heide (1953) reported Early Carboniferous flora from a number of wells in the Ras Gharib area, along the Red Sea coast, and from the Ayun Musa-I well in Sinai. The flora is dominated by species of *Lepidodendropsis*, such as *Lepidodendropsis fenestrata* and *L. hirmeri*, as well as a few specimens of *Cyclostigma aegyptiaca* and the fern *Sphenopteris whitei*. In addition, they reported Early Carboniferous brachiopods, lamellibranchs, and ostracods, as well as a pygidium of a trilobite and a large number of fish scales. The faunas include *Lingula parallela*, *Orbiculoidea cineta*, *O. newberryi ovata*, *Schellwienella* sp., *Rhadinichthys canobiensis*, and *R. laevis*, all of Early Carboniferous age, as well as the Carboniferous *Lingula mytilloides*, *Posideoniella laevis*, *Leperditia okeni*, and *Elonichthys aitkeni*.

At Um Bogma, the early Carboniferous section overlies Early Cambrian sandstones and siltstones of the Araba Formation or the Ordovician Naqus Formation. At Gebel Nukhul, Carboniferous sediments overlie directly the Precambrian basement. Attia (1955) divided the Paleozoic section exposed at Um Bogma into a Lower Sandstone Series, a Limestone Series, and an Upper Sandstone Series.

The name Um Bogma Formation was originally introduced by Kostandi (1959) as the Um Bogma Series to include the carbonate unit and the underlying sandstone unit. The Um Bogma Formation was divided into upper and lower sandy dolomite members and a middle member of interbedded calcareous shales, sandy dolomites and sandy dolomitic limestones by Brenckle & Marchant (1987). Both the lower sandstone member and the middle member contain manganese and iron ore deposits. Mamet & Omara (1969) concluded that the oldest possible age for the Um Bogma Formation (or their Dolomitic Limestone Formation) in the Um Bogma area is Late

Middle Visean (V2b, Zone 13), based on the presence of *Omphalotis* sp., *Endothyra bowmani*, and *Draffania quasibiloba*. Brenckle & Marchant (1987) assigned the Um Bogma Formation in the Nukhul area, Sinai, a Lower Carboniferous, but not older than the early Middle Visean, based on the foraminifera *Koninckopora* sp., *Uralodiscus adindanii, Planoarchaediscus aegyptiacus, Viseidiscus umbogmaensis, V. monstratus,* and *Endothyra bowmani,* among others. They also pointed out that the age of *Draffania* extends below the late Middle Visean. Kora (1995) assigned the Formation a Middle-early Late Visean age based on a coral/brachiopod assemblage.

Kostandi (1959) introduced the name Ataqa Formation (Series) for an Upper Carboniferous unit overlying Um Bogma Formation in the subsurface of the Eastern Desert. The type section of the Ataqa Formation is in the Ataqa No. 1 well (29°54′N, 32°24′E), where it has a thickness of 227 m (745 ft) and is composed predominantly of shale, with sand and limestone intercalations. It is overlain by Permo-Triassic or other Mesozoic sandstones. The Formation is also recognized in outcrops at Wadi Araba and Um Bogma. At the latter location, it has a thickness of 190 m (623 ft) (Weissbrod, 1969). It contains silicified trunks of *Araucoxylon.* In the southeastern scarp of the Northern Galala, at Rod el-Hamal, the Formation is 62 m (302 ft) thick (Said, 1962) and overlain by Cenomanian marls and limestones (Cuvillier, 1937) of the Raha Formation. In Gebel Nukhul it has a thickness of 150 m (492 ft) and contains *Lepidodendron* and *Sigillaria.* Weissbrod (1969) divided the section equivalent to the Ataqa Formtion into a lower sandy clayey member and an upper sandy quartzitic member. The precise age of the Ataqa Formation is not known. The Formation is intruded by black, finely crystalline basalts showing columnar jointing. The basalts gave a K/Ar date of 178±12 Ma (Late Triassic-Early Jurassic) (Weissbrod, 1969). The Formation overlies the Visean Um Bogma Formation and it is overlain in places by the Permo-Triassic Qiseib Formation. Its age is probably Namurian.

Schürmann et al. (1963) reported red shales and sandstones with *Cordites megaflora* and associated microflora of latest Carboniferous-earliest Permian age red shales and sandstones with *Cordites megaflora* and associated microflora of latest Carboniferous-earliest Permian age at the eastern end of Wadi Qiseib, north of Abu Darag. At Wadi El Dakhel, 40 km south of Wadi Araba, Issawi & Jux (1982) recorded *Calamites* trunks, *Equisetites* stems and shark remains from cross-bedded sandstones, which they attributed to the Permo-Carboniferous. These sediments are probably equivalent to Abu Darag Formation or the Wadi Qiseib Formation.

The Um Bogma Formation is followed by sandstones, siltstones, and shales of the Carboniferous Abu Thora Formation (Weissbrod & Perath, 1990). The succession is separated from the Early Cretaceous sediments by basalt sheets of Jurassic age. Omara & Schultz (1965) previously called this unit "the continental coal-bearing Upper Sandstone Formation" in the Abu Thora area and assigned it a Visean age based on microspores recovered from the coal seams. The name Abu Thora Formation is a homonym of the Ataqa Formation. Kora (1995) assigned the Formation a Serpukhovian-Bashkirian (approximately equivalent to Namurian-Westphalian) age based on a brachiopod/trace fossil assemblage.

The Carboniferous succession exposed at Wadi Araba, on the western side of the Gulf of Suez, consists of a lower unit composed of shales and sandstones, followed upward by crinoidal, dolomitic limestones and sandstones, which are overlain in turn by marls and limestones. It has a thickness of approximately 770 m (2526 ft).

Abdallah & El Adindani (1965) divided this section, in an ascending order, into Rod el Hamal Formation, Abu Darag Formation, and Ahmeir Formation. Each formation occurs as an isolated outcrop. The exact superposition of these formations is unknown.

The Rod el Hamal Formation is 350 m (1148 ft) thick and composed of variegated shales, sandstones, and crinoidal-bryozoan limestones. Omara & Kenway (1966) suggested a lower Stephanian age for the Rod el Hamal Formation, based on the occurrence of Ammodiscidae, Linuoidea, and Tertulaniidae. Said (1971) reported a thickness of 720 m (2362 ft) in the type locality at the junction of Wadi Araba and Wadi Rod el Hamal, on the western side of the Gulf of Suez, and suggested a Lower Carboniferous age for the Formation.

The Abu Darag Formation is 170 m (558 ft) thick and made up of shales, sandstones, and sandy limestones near the top. Omara & Vangerow (1965) reported the foraminifera *Ammodiscus roessleri*, *A. ovalis*, *Glomospirella unangularis*, *G. umbilicata*, *Agathamminoides gracilis*, *A. milioloides*, and *Hyperammina clavacoides recta*, along with the conodonts *Streptognathus opletus* and *S. excelsus*, and assigned the Abu Darag Formation a Visean age. Further analysis by Omara et al. (1966) suggested a Upper Carboniferous (Westphalian) age for the Formation based on the presence of the arenaceous foraminifera *Hippocrepina conica* (Lower Carboniferous), *Hyperammina clavacoidea*, *Glomospira articulosa*, *G. compressa*, *Ammovertella prodigalis*, *Spiroplectammina clavata*, *S. extrayensis*, *Textularia bucheri*, *Bigenerina ciscoensis*, *B. virgilensis*, *B. anglifera*, *Cribrobigenerian jeffersonensis*, *Trochammina arenosa*, and *Climacammina schwartzbachi*.

The Ahmeir Formation is 250 m (820 ft) thick and consists of shales, sandstones, and oolitic limestones, with algal fragments, arenaceous foraminifera, conodonts, reef-building corals, bryozoans, crinoids, calcareous algae, and brachiopods. The type locality of the Ahmeir Formation is at Wadi Ahmeir, south of Ain Sokhna, Eastern Desert (Said, 1971). The Formation contains *Neospirifer* sp., *Rhipidomella* sp., and *Lophophollidium* sp. Said (1971) equated the Rod el Hamal and Ahmeir formations with the Ataqa Formation.

Said & Eissa (1968) reported the following microfauna from the Rod el Hamal, Abu Darag, and Ahmeir formations, Ammobaculites pensilis, Ammodiscus arabicus, Biseriammina conica, Bigenerina abudaragensis, Climacammina biserialisissima, Earlandinella irregularis, Lugtonia rectangula, Nodosinella aegypticae, Palaeotextularia awadi, and Toypammina aheimerensis, and suggested that the three formations are probably equivalent. However, Druckman (1974a), and Weissbrod & Horowitz (1989) arranged the three formations in chronological order. Druckman (1974b) correlated the three formations in the Wadi Araba and Abu Hamth areas with the Upper Carboniferous Sa'ad Formation in the Ramon-1 well (29°58′N, 33°38′E). Allam (1989) described the Ahmeir Formation in the Wadi Feiran-El Tor area, west-central Sinai, as being 90 m (295 ft) in thickness and composed of shales at the base, changing upwards into shales, limestones, and dolomites, then into massive sandstones, greenish-grey shales, and hard, sandy, highly fossiliferous limestones at the very top. He correlated it with the Ataqa Formation.

Hassan (1967) divided the Carboniferous succession at Abu Durba (28°28′N, 33°19′E), SW Sinai, into a lower Abu Durba Formation and an upper Huswa Sandstone. The type locality of the Abu Durba Formation is at Gebel Abu Durba, on the eastern side of the Gulf of Suez. The Formation consists of cross-bedded sandstones with *Lepidodendron* and *Calamites* tree trunks in the lowers parts and alternating shales and sandstones in the upper parts. A coquina bed with bryozoans and spiriferids is present at the top. The lower part was considered by Issawi & Jux (1982) to be a fluvio-marine equivalent of the Um Bogma Formation. The Abu Durba Formation and Huswa Sandstone of Hassan (1967) are probably equivalent to the Um Bogma and Ataqa formations, respectively. Allam (1989) estimated a thickness of 110 m (361 ft) for the Abu Durba Formation in the Wadi Feiran-El Tor area and reported the brachiopods *Spirifer striatus* and *Productus semireticulatus* from the Formation. According to Klitzsch (1990a), the Abu Durba Formation in the Wadi Feiran, as well as in the type area, is underlain by sediments of early Carboniferous age. These are composed of 50–80 m (164–262 ft) in thick fluvial and shallow-marine sandstones, which he correlated with the Ataqa Formation or the Thora Formation. Kora (1995) assigned the Formation an Early Moscovian age based on brachiopods and bryozoans. If this age dating is correct, then the Abu Durba Formation is equivalent in part to the Ataqa Formation of Kostandi (1959). The Huswa Sandstone is 148 m (486 ft). Said equated it with the Ataqa Formation.

Soliman & El Fetouh (1970) raised the Ataqa Formation to group status (Ataqa Group) to include three formations, from base to top: The Hashash Formation, composed of cross-laminated sandstones, and up to 6 m (20 ft) in thickness, the Magharet el-Malah Formation, made up of 33 m (108 ft) of carbonaceous shale, and the Abu Zarab Formation, composed of 100 m (328 ft) of friable sand. The Hashash and Magharet el-Malah are equivalent to the Rod el-Hamal Formation and the Abu Zarab Formation is equivalent to the Ahmeir Formation (Said, 1971). This nomenclature has not been accepted by subsequent workers. It is recommended that these names be deleted from the Egyptian stratigraphic nomenclature.

El Shazly (1977) used the name Siwa and Kohla formations for two Carboniferous units in the subsurface of the northern Western Desert. These two units are equivalent to the Desouqy, Dhiffah, and Safi formations of Keeley (1989).

Kora & Jux (1986) divided the Paleozoic rocks, which are about 370 m (1214 ft) in thickness, in the Um Bogma area into six rock units from base to top, the Sarabit El Khadem, Abu Hamata, Nasib, Adedia, Um Bogma and Abu Thora formations. The lower four units are composed predominantly of clastics and assigned to the Cambro-Ordovician (i.e. equivalent to the Yam Suf Group of Weissbrod, 1969, or the Araba and Naqus formations of Hassan, 1967). They assigned the Um Bogma and Abu Thora formations an Early Carboniferous age. According to these authors, the lower part of the Um Bogma Formation is made up of oolitic dolomites with marine fossils, such as tabulates (corals or sponges) and trepostamates (bryozoans). The middle part is composed of marly dolomites with abundant small corals, brachiopods, some bryozoans, and a few molluscs. The top part consists of poorly fossiliferous dolomites, with pelecypods, tabulate corals, crinoidal columns, and oolites.

Weissbrod (1969) grouped the Netafim, Um Bogma and Ataqa formations into the Negev Group. Alsharhan & Salah (1994, 1995, 1996), and Salah & Alsharhan (1996), following Soliman & El Fetouh (1970), raised the Ataqa Formation to group status, the Ataqa Group, and divided it into the Abu Durba and Rod el Hamal formations. This nomenclature is not recommended, because in other areas the group will include the Ataqa Formation.

Carboniferous exposures on the eastern and western sides of the Gulf of Suez are very similar. Kostandi (1959) and Said (1962) considered the sections on both sides of the Gulf to be equivalent. Omara & Kenawy (1966) and Mamet & Omara (1969) on the other hand considered them to belong to two separate lithostratigraphic units. They assigned the Wadi Araba exposures a Westphalian to Lower Stephanian (Upper Carboniferous) age, and the Um Bogma exposure a Tournaisian-Visean (Lower Carboniferous) age. Omara & Vangerow (1965) also assigned a Westphalian age to the Carboniferous section at Abu Darag, north of Abu Durba. Carboniferous exposures on the two sides of the Gulf of Suez probably represent a single basin, separated due to movements along strike-slip faults (Youssef, 1968).

Siliciclastic Carboniferous rocks at Wadi Qena, Eastern Desert, were called the Um Bogma Formation by Bandel et al. (1987) and Somr el Qaa Formation by Klitzsch et al. (1990). In the Wadi Qena area, Bandel et al. (1987) described the Um Bogma Formation (Somr el Qaa Formation) as consisting of basal beds composed of coarsely angular material reworked from the granitic basement which it overlies, followed by reddish, fine-grained, laminated sandstones, then by sandstones intercalated with thin conglomeratic beds. Iron-stained *Skolithos* burrows (pipe-rocks) and *Cruziana*-like trails are abundant in the upper part of the section. The Formation is overlain by Early Cretaceous sandstones.

In the subsurface of the Ghazalat-Tenehue Basin in the northwestern Western Desert, the Carboniferous succession is divided, from bottom to top, into the Desouqy Formation, Dhiffah Formation, and Safi Formation (Keeley, 1989).

The type section of the Desouqy Formation is in the Desouky-1 well (28°22′23″N, 25°56′36″E), at a drill depth of 1933–2295 m, where the unit has a thickness of 362 m (1188 ft). It was assigned a Tournaisian-Early Visean age by Keeley (1989). It rests unconformably on the Devonian Zeitoun Formation. The Formation is made up of high-energy fluvial sandstones, shoreface-delta front sandstones and siltstones, and pro-delta shales. It varies in thickness from 100–300 m (328–984 ft). Gueinn & Rasul (1986) reported the spores *Acyrospora* sp., *Dictyotriletes fimbriatus*, and *Lophozontriletes cristifer* from Zone WD11 in the Faghur (FRX)-1 well, which they assigned a Tournaisian age. Said & Andrawis (1961) described Lower Carboniferous (Visean) from the Faghur-1 and Mamura-1 wells. This Formation is approximately equivalent to the Abu Durba Formation in the Gulf of Suez region, and Unit C-I of El-Arnauti & Shelmani (1988) in the subsurface of Cyrenaica.

The Desouky Formation is conformably overlain by the Dhiffah Formation (Early Visean-Late Namurian). The type section of the Dhiffah Formation is in the West Faghur-1X well (30°54′18″N, 25°11′22″E), at a drill depth of 2641–2245 m, where it has a thickness of 396 m (1299 ft). Elsewhere, the Formation ranges in thickness from 300–45 m (984–148 ft), and is composed of calcareous and carbonaceous shales, sandstones, and oolitic and bioclastic limestones. This Formation is probably equivalent to Units C-II and C-III of El-Arnauti & Shelmani (1988) in NE Libya. Gueinn & Rasul (1986) recorded the spores *Archaeoperisaccus* sp. and *Densosporites variomarginatus* from Zone WD12 (Visean), which they compared with Zone XIV of Massa et al. (1980) in West Libya, the spores *Spelaeotrilites owensi*, *Verrucosisporites* sp., *Diatomozonotriletes fragilis*, and *Radiizonates* sp. from Zone WD13 (Late Visean-Namurian), which they compared partly with Zone XV of Massa et al. (1980), and the spores *Spelaeotrilites arenaceus*, *S. pretiosus*, and *Lycospora pusilla* from Zone WD14 (Namurian-Westphalian), recorded from the West Faghur-1X, Faghur (FRX)-1, Kohla-2, and Siwa-1 wells. They compared this assemblage with that of Zone XVII of Massa et al. (1980). Visean palynomorphs were recognized in the Foram-1 well and include *Punctatisporites*, *Radiizonates*, *Vallatisporites*, and *Spelaeotrilites* (Schrank, 1984, 1987). Similar assemblages have been reported by El-Ghazali & Ali (1985) from the Kohla-2 well.

The type section of the Safi Formation (Keeley, 1989) is in the Siwa-1 well (29°07'18"N, 23°25'50"E), at a drill depth of 976–812 m, where the unit is 155 m (509 ft). It rests conformably on the Dhiffah Formation and is overlain by Cenomanian or younger rocks. The Safi Formation is made up of sandstones and siltstones, which change facies into shales and oolitic limestones northwards. Gueinn & Rasul (1986) reported the pollen *Vittania* sp. and the monosaccate *Cannanoropollis janakii*, in addition to *Distriatites* cf. *dettamannae* from their Zone WD15 in the West Faghur-1X well, to which they assigned a ?Late Carboniferous-Early Permian age. The Safi Formation was given a Late Namurian-Early Permian age by Keeley (1989). The Formation is present only in the westernmost part of Egypt due to erosion. A Permian succession, 230 ft (70 m) in thickness, was recognized in the Faghur-1 well by El Dakkak (1988) based on the occurrence of *Waadenoencha montepelierensis* and *Anisopyge* cf. *perassulata*. It is probably equivalent to the Ataqa Formation in the Gulf of Suez region, and Units C-IV and C-V of El-Arnauti & Shelmani (1988) in NE Libya.

Vittimberga & Cardello (1963) and Burollet (1963b) used the term Kurkur Murr Formation (also spelled Karkur Murr) for quartzitic, conglomeratic sandstones and siltstones overlying the Precambrian basement in the central part of Jabal Awaynat in SW Egypt. The Formation has a thickness of approximately 80 m (262 ft). The basal 30 m (98 ft) is made up of siltstones and white and pink sandstones, which contain *Archaeosigillaria* aff. *vauxemi*, suggesting an Upper Devonian-Lower Carboniferous age. The Formation is intruded by rhyolitic veins (Menchikoff, 1927a, b, Burollet, 1963b) and overlain by thick syenites. The Formation is probably partly equivalent to the Dalma Formation in SE Libya, and partly to the Wadi Malik Formation which also contains *Archaeosigillaria* aff. *vauxemi* in SW Egypt and was assigned a Tournaisian-Visean age by Klitzsch (1990a).

Burollet (1963b) subdivided the Carboniferous section overlying the Kurkur Murr Formation in the Gebel Oweinat (Awaynat) into the Ouadi Oaddan Argillites (Wadi Waddan), "cima" (top) Sandstones, and Plateau silts. The Ouadi Ouaddan Argillites are exposed at the center of the Gebel Oweinat, and are composed of steeply-dipping, alternating siltstones, shales, and sandstones overlain by quartzites. The Formation is approximately 280 m (919 ft) in thickness. Burollet (1963b) applied the informal name "cima" Sandstones to a very thick Carboniferous series of quartzitic sandstones, overlying the Ouadi Ouaddan Argillites. The thickness of this unit is 370 m (1214 ft) at Karkur Murr. The informal name Plateau Silts was given to a Carboniferous unit overlying the "cima" sandstones. The unit is approximately 60 m (197 ft) thick, and composed of black silts interbedded with mudstones and coarse grained beds. The silts are intercalated with trachytes and overlain by 500 m (1641 ft) of purple to green volcanic breccias of andesites and rhyolites. These three units combined are probably equivalent to the Dalma Formation in southeastern Libya, and the Wadi Malik and North Wadi Malik formations in southwestern Egypt.

Klitzsch (1990a) subdivided the Carboniferous succession in the Gebel Oweinat and Abu Ras areas into the Wadi Malik Formation (Lower Carboniferous) and North Wadi Malik Formation (Upper Carboniferous). The type area for both formations is in Wadi Abdel Malik in the Abu Ras Plateau. These two formations are probably equivalent to the Wadi Waddan (Ouadi Ouddan) Argillites, "cima" Sandstones, and Plateau Silts formations of Burollet (1963b).

The Wadi Malik Formation consists of planar cross-bedded, medium to coarse-grained sandstones at the base, followed by fine- to medium-grained sandstones, siltstones, and shales, then by a fining-upward sequence of ripple cross-laminated fine-grained sandstones and shaly siltstones at the top. The sequence is capped by Late Visean, heavily bioturbated sandstones, with brachiopods and trace fossils, such as *Bifungites* (Klitzsch, 1990a, Wycisk, 1990).

A rich fauna was reported from the Wadi Malik Formation type area, including Archaeosigillaria minuta, Lepidodendropsis cf. sinaica, Lepidosigillaria intermedia, Prelepidodendron lepidodendroides, P. rhomboidales, Rhacopteris ovata, and Triphyllopteris gothani, suggesting a Tournaisian to Visean age (Wycisk, 1990). At Gebel Oweinat, a similar assemblage was reported from the Formation, in addition to Cyclostigma ungeri, L. cf. Lepidodendropsis vandergrachti, L. aff. rhomoiformis, and Precyclostigma sp. (Lejal-Nicol, 1990, Klitzsch, 1990a). Similar fauna were also reported from the Somr el Qaa Formation in the Wadi Qena-Wadi Dakhla area, and from a continental sandstone interval in the Ataqa Formation in the Um Bogma and Wadi Mukattab areas of Sinai.

The North Wadi Malik Formation rests unconformably on the Wadi Malik Formation. This Formation is characterized by diamictitic sediments interpreted as glaciofluvial deposits (Wycisk, 1990, Klitzsch, 1990a). The Formation consists of three interfingering different facies: 1) tillites in the north composed of matrix supported fine sandstones with blocks several meters in size. These tillites are topped by erosional channels. 2) glacio-fluvial deposits made up of braided channel sediments. These sediments are medium to coarse grained, pebbly sandstone, with tabular- and trough cross-bedding, horizontal bedding, pebble lag deposits, and scour-and-fill structures. 3) Distal deposits made up of coarsening-upward sequences of glacial lake sediments at the base grading into delta-fan succession upward (Wycisk, 1990). The sediments grade into varves in the south in Sudan (Klitzsch, 1983, 1990a).

# Chapter 7

# Mesozoic

## 7.1 TRIASSIC-LOWER CRETACEOUS

## 7.2 TRIASSIC

Continental and marine Triassic sediments occur in Wadi Araba along the western Red Sea coast, northern and central Sinai, the Nile Delta region (Klitzsch, 1990b), and the subsurface of the Ghazalat Basin in the northern Western Desert (Keeley et al., 1990, Keeley & Massoud, 1998). They have not been identified in Southern Egypt, although rocks of probable Permo-Triassic age, such as the Lakia Formation, were identified in NW Sudan, near the Egyptian border (Klitzsch, 1990b, Wyscik, 1990).

Continental Permian sediments are difficult to differentiate from the overlying Triassic deposits, and they are often included in them. In the northern Western Desert, they form part of the Safi Formation. In central and northern Sinai, they are incorporated in the Wadi Qiseib Formation and its equivalent the Budra Formation. Only the marine section in northern Sinai can be separated from Carboniferous or Triassic units.

In central and southern Sinai and the Gulf of Suez region, the Triassic sequence is represented by two equivalent units: The Wadi Qiseib Formation and the Budra Formation. The name Wadi Qiseib Formation was introduced by Abdallah et al. (1963) for a 50 m (164 ft) unit in Wadi Malha, in the North Galala escarpment, on the western side of the Gulf of Suez. This unit is composed of fossiliferous red shales and sandstones, overlain by thin orange to yellow marls and limestones. It overlies the Abu Durba Formation in the Wadi Feiran-El Tor area on the eastern side of the Gulf of Suez. The sandstones form a fining-upward sequence, with trough cross-bedding, ripple cross-stratification, and horizontal bedding (Allam, 1989). The age of the Formation is not certain. It was considered to be Permo-Triassic in age by Abdallah et al. (1963) and Gvirtzman & Weissbrod (1984), and Triassic by Said (1971). Klitzsch (1990b) identified the Qiseib Formation at Wadi Araba as an 8 m (26 ft) thick unit which overlies Late Carboniferous ferruginous sandstones, shales, salts, and gypsum, and assigned it a Permian age based on the following flora: *Asterotheca* aff. *leeukuilensis*, *Callipteris conferta*, *Dorycordaites* sp., cf. *Lobatannularia*, cf. *Sphenophyllum*, and *Thinnfeldia* aff. *decuurens*. The Wadi Qiseib Formation was considered as Permo?-Triassic by Abdallah et al. (1963) and Lower Permian by Klitzsch (1990b). It is probably of Permian age in the lower part (equivalent to the Yamin Formation) and of Triassic age in the upper part.

Triassic rocks at Um Bogma, central Sinai, are represented by the continental sediments of the Budra Formation. The type section of the Budra Formation is at Wadi Budra, Sinai, where it has a thickness of 327 m (1073 ft) (Weissbrod, 1969). The Formation is composed of brown, purple, grey, coarse- to fine-grainned sandstones, variegated shales and sandstones, with ripple marks, mud cracks, cross-bedding, and silicified tree trunks several meters in length (Druckman, 1974b). Its upper boundary is marked in the type section by a ferruginous quartzitic layer (Weissbrod, 1969). Weissbord & Perath (1990) reported a 300 m (984 ft) thick section of quartzarenites, with shales and siltstones in the upper part. In the subsurface of northern Sinai, the Budra (Qiseib) Formation reaches a thickness of 376 m (1234 ft) in the Abu Hamth-1 well (29°57′N, 33°38′E) in central Sinai. The upper 36 m (118 ft) are made up of limestones, which contain Middle Triassic marine fossils (Druckman, 1974b). At Abu Hamth, the Budra Formation overlies the Upper Permian Yamin Formation. In the Um Bogma area, the Yamin is absent and the Formation overlies directly the Carboniferous Ataqa Formation (Abu Thora Formation) (Druckman, 1974b). In the Ataqa-1 well (29°54′N, 32°24′E), Triassic sediments are 73 m (240 ft) in thickness, and in the Nukhul-1 well, they are 36 m (164 ft). The Qiseib Formation and the Budra Formation are equivalent and their names are used interchangeably in the literature.

Marine Triassic rocks occupy the core of the Arif el-Naga anticline in northern Sinai. They crop out also at Um Bogma and occur in the subsurface of central and northern Sinai.

Early Triassic marine sediments are represented by the Zafir Formation and the Ra'af Formation, both of which are known only from the subsurface in the Halal-1 well (Druckman, 1974b). The Zafir Formation (Scythian) is 196 m (643 ft) in thickness and is composed of alternating dark grey shales, fossiliferous limestones and sandstones, with molluscs, echinoids, agglutinated forams, conodonts, and ostracods. The Ra'af Formation (Upper Scythian-Lower Anisian) is 50 m (164 ft) and consists mainly of fossiliferous limestones, biomicrites and biosparites, with echinoids, brachiopods. The limestones, become oolitic to the northwest and dolomitic to the south.

The exposure at the core of the Arif el-Naga anticline (30°21′N, 34°28′E) was first described by Awad (1946). The section is 200 m (656 ft) thick and consists of cross-bedded sandstones, with plant remains, and interbeds of variegated shales and limestones, grey to brown fossiliferous limestones and brown to greenish grey shales. The limestones contain pelecypods, gastropods, crinoid stems, and echinoid spines. The fossils reported by Eicher (1946) are *Pseudoplacunopsis fissistriata*, *Pogonocertites primitivus*, *Adontophora munsteri*, *Myophoria laevigata*, *M. blakei*, *Ostrea montis caprilis*, *Avicula* sp., *Germanonautilus biodorsatus*, *Ceratites* sp., *Terebratula julica*, and *Paraceratites binodosus*, in addition to conodonts.

The Gevanim Formation is exposed over a small area in the core of the Arif el Naga anticline (Karcz & Zak, 1968, Bartov et al., 1980). The exposed 68 m (223 ft) of the upper part of the Formation consist of dark-colored, fine- to medium-grained, cross-bedded sandstones, with intercalation of shaly, marly and silty beds, and a few limonitic limestone beds. The shales and sandstones contain plant imprints and fragments of petrified wood. Bartov et al. (1980) reported the ammonite *Benckeia levantina*, and the bivalves *Trigonodus tenuidentatus* and *Neoschizodus orbicularis* from the carbonate beds and assigned the Formation a Middle to early Late Anisian age.

The Saharonim Formation consists of fossiliferous limestones, marls, shales, and a few beds of dolomite and gypsum (Karcz & Zak, 1968, Druckman, 1974b, Bartov et al., 1980), with ammonites, nautiloids, pelecypods, brachiopods, echinoids, miliolids, ostracods, and conodonts. The *Ceratites* beds occur in the lower part of the Formation. The Formation is 116 m (381 ft) at Gebel Arif El Naga and 275 m (902 ft) in the Halal-1 well, and represents the main Triassic marine transgression (Jenkins, 1990). It was assigned an Upper Anisian-Carnian age by Druckman (1974b). An olivine basalt dyke intrudes the lowest limestone member of the Formation (Bartov et al., 1980).

The uppermost Triassic sequence is represented by the Mohilla Formation which is 50 m (164 ft) in the Halal-1 well (Druckman, 1974b), but it is absent at Gebel Arif El Naga. The Formation is made up of dolomicrite, dolomitic shales, dolomitic limestones, algal stromatolites and anhydrites (Jenkins, 1990). Druckman (1974b) reported *Spiriferina lipoldi* and *Myophoria inequicostata* and assigned it a Carnian-Norian age. These two species were reported from the Azizia Formation in NW Libya.

Said (1971) gives the name Arif El Naga Formation to the section at Arif el-Naga and divides it into four rock units, 1 to 4. The basal unit-1 (subunit A) is 50 m (164 ft) and made up of cross-bedded sandstones with fossil plants. Unit-2 (subunits B & C) is a sandstone-shale section, with poorly preserved fossils. Subunits A and B are equivalent to the Gevanim Formation and subunit C is equivalent to the Saharonim Formation. Unit-3 (subunit D), also known as the *Ceratites* beds, is a 45 m (148 ft) thick section composed of limestones and marls, with Middle Triassic ammonites. The top unit-4 consists of hard compact, non-fossiliferous lithographic limestones.

In the subsurface of the Matruh and Gebel Rissu basins, in the northern Western Desert, the Early Triassic is represented by carbonates, evaporites and sands of the Behir Formation (Early to Late Triassic). Keeley et al. (1990) introduced the name Fadda Formation for a subsurface unit composed of a heterogeneous mixture of limestones and dolomites, with thin-interbedded sandstones, shales, and anhydrites. They designated the interval between the drill depths of 4745 m and 4802 m in the Fadda-1 well (30°51′N, 28°03′E) as the type section. They assigned it a Late Triassic (Norian-Rhaetian) age, and equated it with the Eghei Group in the Western Desert and the Gevanim and Saharonim formations (both are equivalent to the Arif el Naga Formation in Sinai. According to them, the unit is overlain in the Gulf of Suez and Sinai region by the Rajabiah Formation. This unit is most likely equivalent to the Mohilla Formation in the subsurface of Sinai, and the Bu Sceba Formation in NW Libya. Keeley (1994) used the name Ras Qattara Formation for equivalent sediments of the Qiseib Formation, composed of cross-bedded sandstones with plant remains, variegated shales, and fossiliferous limestones with pelecypods, gastropods and crinoids. It grades into open-marine sediments in northern Sinai. However, Keeley et al. (1990) previously threw doubt over the definition of this Formation and recommended the name to be dropped. The Ras Qattara Formation contains gas in the Shams-1X well, in the Khalda Block (World Oil, 1995). The top of the Eghei Group in the Western Desert is marked by a red horizon known as the Yakout Red Shale.

## 7.3 JURASSIC

Continental and marine Jurassic sediments crop out in Northern Sinai and on the western side of the Gulf of Suez. They also occur in the subsurface of the

Northern Western Desert and the Nile Basin (Tawadros, 2001, Dolson et al., 2001). Predominantly continental Jurassic sediments crop out in southern Egypt and Central Sinai. Al Far et al. (1965) and Al Far (1966) provided the first systematic studies of the Jurassic sediments in Sinai. The earliest attempt to classify the Jurassic sediments in the northern Western Desert was carried out by Norton (1967). Detailed studies of Jurassic sediments in northern Egypt were carried out by Bartov et al. (1980), Keeley et al. (1990), Keeley & Wallis (1991), and Bagge & Keeley (1994). Systematic work on the continental Jurassic sediments in southern Egypt was undertaken by Klitzsch et al. (1979), and extensive palynological analysis was performed by Schrank (1984, 1987, 1992).

### 7.3.1   Jurassic of Sinai

Jurassic sediments are exposed in northern Sinai at Gebel Maghara (30°42'N, 33°21'E) and Gebel Minsherah (30°18'N, 33°42'E). These outcrops represent the thickest Jurassic exposure in Egypt, with thicknesses up to 2000 m (6562 ft) (Said & Barakat, 1958, Said, 1962). The succession is unconformably overlain by Lower Cretaceous sediments. The uppermost Jurassic seems to be missing due to erosion at the end of Jurassic time (Cimmerian unconformity of Keeley et al., 1990).

The outcrop at Gebel Maghara has been extensively studied and many attempts have been made to subdivide it. Al Far (1966) described a 1507 m (4944 ft) section at Gebel Maghara, from base to top the succession is as follows (see also Said, 1971): The Mashabba, Rajabiah, Bir Maghara, Safa, and Masajid formations. The type section of the Mashabba Formation (Lower Liassic) is at Wadi Sadd El Mashabba, Gebel Maghara, where it is made up of 100 m (328 ft) of alternating quartzitic sandstones and fossiliferous marly limestone beds. The Rajabiah Formation (Lower Liassic) is 293 m (961 ft) in thickness in its type locality at Wadi Sadd El Mashabba and at Wadi Rajabiah, Gebel Maghara, and consists of greyish, stylolitic coralline algal limestones, becoming sandy and marly upwards. Keeley et al. (1990) restricted the Mashabba Formation to the lower sandstone unit of the original Mashabba Formation as defined by Al Far (1966) and included it as a member in the Rajabiah Formation. They assigned it a latest Pliensbachian-earliest Toarcian age. The sandstones of the "Mashabba Member" grade westward into evaporites of the Kattaniya Member.

The Shusha Formation (Pliensbachian) is 272 m (892 ft) thick and composed of sandstones and clays with coaly material and abundant plant remains. The Bir Maghara Formation (Late Liassic-Late Bajocian) is made up of 444 m (1457 ft) thick massive marine limestones and clays. The Bir Maghara Formation is divided into three members, from base to top: the Mahl Member is composed of 94 m (308 ft) of hard coralline massive limestones, the Mowerib Member (Middle Bajocian) is a 135 m (443 ft) section of shales, clays and marls with *Normannites brakenridgei*, and the Bir Member (Upper Bajocian) is composed of 216 m (709 ft) of marls and shales with *Ermoceras* sp., *Thamboceras* sp., and *Oppelia subradiate*, and a hard sandstone bed at the top with *Thambites*.

The Safa Formation (Bathonian) is 215 m (705 ft) thick and consists of cross-bedded sandstones and shales, wih economically important coal beds at Gebel Maghara (Ghandour & Maejima, 2007). Jurassic rocks of probable equivalence to the Safa Formation are exposed in faulted outcrops on the western side of the Gulf of Suez at El Galala El Bahariya (Northern Galala Plateau). The section is composed of 170 m (558 ft) of Bathonian limestones, marls, and sandstones (Arkell, 1956). At Khashm el Galala, sandstones rich in fossil plants overlie marine Bathonian argillaceous sandstones and limestones with *Nucula variabilis, Trigonia pullus, Thamastrea crateriformis,* and *Rhynchonella asymmetrica* (Nakkady, 1955). This section rests unconformably on a 50 m (164 ft) succession of non-fossiliferous sandstones of presumed Triassic age, and is overlain by 150 m (492 ft) of Nubian sandstones. The coal deposits of the Safa Formation contain at least 11 coal seams ranging in thickness from 130 cm (52 in.) to 2 cm (0.8 in.) (Baioumy, 2009). The geochemistry of the coal has been treated by Mostafa & Younes (2001) and Baioumy (2009).

The Masajid Formation (Bathonian-Kimmeridgian) is 576 m (1890 ft) thick and composed of shallow marine limestones. The Formation is divided into a lower Kehailia Member and an upper Arousiah Member. At Gebel Maghara, the Kehailia Member (Bathonian-Callovian) is composed of fossiliferous shales, white limestones, cherty, sandy and silty shales, oolitic marly limestones, and glauconitic limestones. The following brachiopod assemblage was reported by Hegab (1989) from calcareous shales in that member: *Russirhynchia fischeri fischeri Rouillier, Rhactorhynchia pinguis ukrainica Makridin* and *Eudesia,* and rare *Eligmus* and *Terebratulida.* These shales contain hard fossiliferous limestone beds rich in *Ostrea,* with rare *Eligmus, Africogryphea, Eudesia,* and *Rhynchonellida,* in addition to corals. Rosenfeld et al. (1987) reported the ostracod species *Glyptogatocythere magharaensis* and *Ektyphocythere zoharensis* from the Kahailiah Member (Sherif and Zohar formations). The Arousiah Member (Oxfordian-Kimmeridgian) is 443 m (1453 ft) thick and made up of stylolitic, coralline and/or algal limestones with a few marl and clay interbeds. An 80 m (262 ft) section of Bajocian shales (equivalent to the Bir Maghara Formation), sandstones and limestones is exposed at Gebel Minsherah in northern Sinai (Farag & Shata, 1954, mentioned by Said, 1962, p.233). In the subsurface at Ayun Musa (29°52′N, 32°39′E) Jurassic clastics contain coal seams and are overlain directly by Miocene sediments. The section attains a thickness of up to 855 m (2805 ft) (Said, 1962). The Katib el Makhazin-1 well (30°30′45″N, 32°51′20″E), SW Sinai, encountered approximately 890 m (2920 ft) thick Jurassic section (equivalent to the Masajid Formation). The base of the sequence was not penetrated. The section is composed of flinty limestones with Upper Jurassic fauna, believed to be characteristic of the Bathonian, in addition to *Involutina aegyptiaca, Haplophragmoides misrensis, Nonion sinaiensis, Turrilina andreaei* var. *punctata,* and *Theelia sinaiensis.* This fauna suggests Middle-Upper Jurassic (Bathonian-Kimmeridgian) (Abdou & Marzouk, 1969). The Jurassic sediments in this well are overlain by Lower Cretaceous, ferruginous claystones, iron bands, and hard limestones.

Said & Barakat (1958) described 128 species of foraminifera and twelve species of holothuroid sclerites from the Jurassic rocks outcropping at Gebel Maghara and they came to the conclusion that the Jurassic rocks in Egypt represent the platform deposits of the Tethys sea deposited away from the deep-sea area. In northern Sinai, in the El Mazar-1 well (31°05′N, 33°9′E), the Masajid Formation changes facies into shales (Keeley et al., 1990).

Bartov et al. (1980) divided the Jurassic succession in the Gebel Arif el Naga area, in an ascending order, into Mish'hor, Ardon, and Inmar formations. The Mish'hor Formation (Liassic) is about 2 m (6.5 ft) in thickness and unconformably overlies the

Saharonim Formation and is in turn gradationally overlain by the Ardon Formation. The Ardon Formation (Liassic) is 19 m (63 ft) thick and composed of ferruginous silty shales, alternating with limestones, dolomites, marls, and variegated sandstones. The Ardon Formation is probably equivalent to the Mashabba, Rajabiah, and Shusha formations. Southward, all these formations grade into the predominantly continental sandstones of the Amir Formation.

The Inmar Formation (Middle Liassic-Lower Bajocian) consists of reddish, variegated, cross-bedded sandstones, with a few layers of brown and dark grey shales, plant remains, and ferruginous black crusts. It is overlain with angular unconformity by the Lower Cretaceous Arod Conglomerate. Rosenfeld et al. (1987) reported the ostracod species *Ektyphocythere bucki* from the Toarcian Inmar Formation (lower part of Bir Maghara Formation) and *Glyptogatocythere magharaensis* from the Bajocian Daya Formation (upper part of Bir Maghara Formation). The Inmar Formation is overlain by the Kidod Shale (considered a member of the Masajid Formation by Keeley et al., 1990) of early Oxfordian age. In northern Sinai, it overlies Callovian limestones of the Kehailia Member and is overlain by "Bir Sheiba" Oxfordian limestones of the Masajid Formation. The base of the Kidod Shale (top Callovian) is marked by hardgrounds encrusted by ammonites (Keeley et al., 1990). Rosenfeld et al. (1987) reported the ostracod *Exophthalmocythere kidodensis* from the Kidod Shale.

Jurassic sediments occur also in the subsurface of the northern Eastern Desert, on the western side of the Gulf of Suez. Aboul Ela et al. (1998) established 9 Jurassic dinoflagellate zones (Toarcian-Late Kimmeridgian). These zones, in an ascending order, are *Parvocysta ampulla* (Toarcian-Alaetian), *Pareodinia ceratophora* (Bajocian), *Dichadogonyaulax sellwoodi* (Bathonian-Early Callovian), *Ctenidodinium continuum-Ctenidodinium ornatum* (Middle-Late Callovian), *Wanaea digitata* (Latest Callovian-Early Oxfordian), *Gounyaulacysta jurassica-Epiplosphaera reticulospinosa* (Middle Oxfordian), *Epiplosphaera bireticulata-Acanthaulax granuligera* (Late Oxfrodian), *Aphmorula dodekovae* (Early Kimmerdgian), and *Gochteodinia mutabilis* (Late Kimmeridgian). Ibrahim et al. (1998) recognized 3 miospore biozones in the Jurassic section in deep wells in the northern part of the Eastern Desert. These miospore zones are *Classopollis/Circulina-Deltaoidospora* sp. (Toarcian-Alenian), *Verrucosisporites* sp.-*Coverrucosisporites* sp.-*Trilobosporites* sp. (Early Bajocian-Callovian), and *Cicatricosisporties* sp.-*Contignisporites* sp. (Oxfordian-Kimmeridgian).

## 7.3.2 Jurassic of the Northern Western Desert (Northern Egypt Basin)

In the subsurface of the Western Desert, Jurassic sediments are up to 1444 m (4738 ft) (in the type Khatatba well) (Said, 1962). The Jurassic succession is represented by the Bahrein, Wadi Natrun, Khatatba, Masajid, and Sidi Barrani formations.

The Bahrein Formation was introduced by WEPCO (cited by Hantar, 1990) for the Jurassic continental sediments to replace the Eghei Group of Norton (1967) who assigned it a Permo-Jurassic age (the Eghei Group is believed to be Triassic by Keeley et al., 1990). The type section of the Bahrein Formation is in the interval between 3888 m and 4437 m in the Khatatba well. The unit is composed of red fine- to coarse-grained quartzose sandstones with interbeds of conglomerates, siltstones, and shales. A few anhydrite beds are present in the Yakout-1 well. The Formation unconformably overlies different units of Paleozoic age or the Precambrian basement and is overlain by marine sediments of the Middle Jurassic Khatatba Formation, the Cretaceous Betty Formation, or the Alam El Bueb Member of the Burg el Arab Formation. Toward the east and north, the unit changes facies into the marine facies of the Wadi Natrun and Khatatba formations (Hantar, 1990). Westward, it pinches out against the Paleozoic high or grades laterally into the Alam El Bueb Member. The age of the Formation ranges from Callovian in the type section to late Jurassic, but its age may span the entire Jurassic time from Bathonian to Kimmeridgian. However, Keeley et al. (1990) believe that the Hettangian-Bajocian and Late Kimmeridgian-Tithonian sediments are missing in Egypt. The Bahrein Formation in the Zeitun-1 well yielded several miospores and dinoflagellate cysts assigned to the Bajocian-Bathonian including *Crassitudisporites problematicus, Staplinisporites caminus, Neoraistrikia truncata, Matonisporites crassiangulatus, Baculatisporites comaumensis*, and *Leptolepidites bossus* (El-Beialy et al., 2002).

The Wadi Natrun Formation[1] was proposed by Norton (1967) for a middle Jurassic marine carbonate and shale section in the subsurface of the northern Western Desert. Evaporites are present locally. The type section is in the Wadi Natrun-1 well (30°23′N, 30°18′E), int a drill depth interval of 3594 m to 4056 m. The age of the Formation is assumed to be early to middle Jurassic based on unspecified palynological assemblages (Hantar, 1990). The unit unconformably overlies the Bahrein Formation and is overlain by the Khatatba Formation. The unit is of limited geographic distribution and occurs mainly in the area west of the Nile Delta and along the coastal area of the northern Western Desert. It apparently increases in thickness in the offshore area (Hantar, 1990, Fig. 14.5).

The Khatatba Formation was proposed by Norton (1967) for a unit composed of grey shales, fine- to medium-grained sandstones, and a few shallow-marine limestone beds. The type section is in the Khatatba-1 well (30°13′N, 30°50′E), in the interval 355 m to 1536 m. The unit rests conformably on the Wadi Natrun Formation or the Bahrein Formation, and is conformably overlain by the Masajid Formation. Southward, it grades into the Bahrein Formation where it is unconformably overlain by the lower Cretaceous Burg El Arab Formation (Hantar, 1990). The Khatatba Formation was assigned a Bathonian-early Callovian age by Keeley (1994) and Keeley et al. (1990). The latter divided it into Lower Safa, Kabrit, Upper Safa, and Zahra members. The Safa Member (Bathonian) has a thickness of 150 (492 ft) to 900 m (2953 ft) in Gebel Rissu Basin in the Western Desert. The basal part is marked by red, silty, lateritic clays and sandstone red beds, interbedded with thin coal beds. In the northern regions of Sinai and the Western Desert the Safa Member is divided into lower and upper parts by the Kabrit Member. The Kabrit Member was introduced by Keeley et al. (1990) for a unit dominated by limestones, with subordinate shales, sandstones, and coal beds. The type section is in the Kabrit-1 well (30°12′N, 32°29′E), at a drill depth interval of 2451 m to 2577 m. It ranges in thickness from 0–191 m (0–627 ft) in the Gebel Rissu Basin. The name Kattaniya Member was introduced by Keeley et al. (1990) for the equivalent of sandstones of the Mashabba Member (*sensu stricto*) in the Gebel Rissu Basin. They designated the drill interval between 3799 m and 3838 m (12,465–12,592 ft) in the Kattaniya-1 well

1 Said (1990) described another Wadi Natrun Formation of Pliocene age.

(29°49′N, 30°10′E) as the type section. The unit consists of anhydrites interbedded with thin limestone and shale beds. It has an average thickness of approximately 30 m (98 ft). Keeley et al. (1990) assumed a latest Pliensbachian-earliest Toarcian age for the Kattaniya Member. The Zahra Member was proposed by Keeley et al. (1990) for a transitional unit between the Masajid and Safa formations. The type section of the member is in the Zahra-1 well (30°38′44″N, 27°00′28″E), at a drill depth interval of 3360–3541.5 m. The unit is equivalent to the upper Safa Formation and the Kehelia Member of the Masajid Formation of Al Far (1966). The Zahra Member is composed of thinly bedded shales and limestones. Sandstone beds are common in the south. The Khatatba Formation is a major reservoir and source rock in the Northern Egypt Basin and the Nile Basin (El Sisi et al., 2002, Dolson et al., 2001). The source rock potential of the Khataba Formation is discussed in more detail in chapter on the petroleum systems.

The Masajid Formation was proposed by Al Far (1966) for Jurassic rocks exposed in Wadi Masajid, north of Gebel Maghara, Sinai. In the subsurface of the Northern Western Desert, it is composed of marine cherty limestones and shales. In the Wadi El Natrun-1 well, where the unit attains a thickness of 450 m (1476 ft) (Hantar, 1990), Upper Jurassic sediments are predominantly clastics and include an abundance of the reef-forming foraminiferal species *Kurnubia palastinensis, Valvulinella jurassica, Pseudocyclammina virguliana*, and *P. personata*. This assemblage is known from many wells in the Western Desert (Said, 1962). Keeley et al. (1990) subdivided the Masajid Formation, in an ascending order, into a Lower Limestone Member, Kidod Shale Member, Bir Sheiba (spelled Beer Shiva in Bartov et al., 1980) Limestone Member, Abu Hammad Member of interbedded sandstones and shales, with subordinate limestones, and an Upper Limestone Member. The Abu Hammad Member of the Masajid Formation was proposed for a unit dominated by thinly interbedded sandstones and shales, with subordinate limestones. The type section of the unit is in the Abu Hammad-1 well (30°34′07″N, 31°50′04″E), at a drill depth of 1990 m–2138 m. Its lower and upper contacts are gradational with the Upper Limestone and Bir Sheiba or Darduma limestone members, respectively. The type section of the Kidod Shale is in the Kidod-2 well in eastern Palestine. It was extended into the subsurface of the Western Desert by Keeley et al. (1990), as far as 29°30′E. It has a thickness between 70 m (230 ft) and 200 m (656 ft). Its lower boundary with the Lower Limestone Member is marked by a hardground with pyritized ammonites. In the northern Western Desert, the Lower Limestone Member, Kidod Shale, and Bir Sheiba Limestones grade westward into undifferentiated limestones of the Darduma Limestone Member (early to middle Callovian-late Oxfordian). It is best developed in the Darduma-1A well (31(14′50″N, 26(5719‴E) in the Matruh Basin. This unit is probably equivalent to the Sidi Barrani Formation.

The type section of the Sidi Barrani Formation is in the Sidi Barrani-1 well at a drill depth interval of 1899 m to 4301 m, where it attains its maximum thickness of 2402 m (7881 ft). It is limited in distribution to the northwestern part of the Western Desert (Hantar, 1990). The Formation is composed mainly of dolomites with a few interbeds of sandstones, shales, and anhydrites. It conformably overlies the Khatatba Formation and is overlain by the Alam El Bueb Formation. It is assumed to be middle Jurassic (Callovian-Kimmeridjian)-Lower Cretaceous in age based on its stratigraphic position. It is probably equivalent to the Masajid Formation.

Norton (1967) grouped the Rajabiah, Shusha, Bir Maghara, Khatatba, and Masajid formations in the Maghara Group. Keeley et al. (1990) proposed to change the name to the Gebel El Maghara Group, and selected the interval from drill depths of 1503 m to 3536 m in the Gebel Rissu-1 well (29°57′N, 30°24′E) in the Western Desert as the proto-type section.

## 7.4  UPPER JURASSIC-LOWER CRETACEOUS

### 7.4.1  Upper Jurassic-Lower Cretaceous of the Southern Western Desert

Strata of Upper Jurassic to Lower Cretaceous age cover large areas in SW Egypt. They form a thick succession of continental sandstones and shales, which cannot be readily differentiated and were previously included in the Nubian Sandstone. Barthoux & Frittel (1925) reported Cretaceous plants from the Nubian Sandstone, and later, Attia & Murray (1951) discovered Lower Cretaceous ammonites in the Nubian Sandstone in the Eastern Desert. The work of Klitzsch et al. (1979) enabled the recognition of a number of units within this predominantly continental sequence. Marine Upper Jurassic strata were also identified in water wells near the Kharga Oasis based on foraminifera and palynomorphs by Soliman (1975, 1977). Marine Aptian sandstones were identified in outcrops based on brachiopods and palynomorphs (Barthel & Boettcher, 1978, Schrank, 1984, 1987, 1992, Ibrahim et al., 1995, Schrank & Mahmoud, 1998a, b). In the southern Western Desert, the succession is represented by the Gilf Kebir, Six Hills, and Abu Ballas formations. These formations are exposed on the northeast side of Gebel Oweinat, near the Sudanese-Libyan border.

The typical area of the Gilf Kebir Formation is near the Akaba passage (23°02′N, 25°51′E) on the Gilf Kebir plateau (Klitzsch et al., 1979). Its thickness varies from a few meters to approximately 300 m (984 ft). The upper part of the Gilf Kebir Formation lies directly on the Precambrian surface (Wycisk, 1990). It is composed of medium to coarse-grained, well-sorted sandstones, with bimodal tabular and trough cross-stratification overlying a thin basal conglomerate. These sediments grade upwards into fine- to coarse-grained, cross-bedded sandstones, with wavy, flaser, and lenticular lamination, and intensively burrowed intervals with spongebiomorphs and vertical burrows. The Gilf Kebir Formation in NE Gebel Oweinat contains abundant flora (mainly ferns) of Jurassic to Lower Cretaceous age near the base. This flora includes *Cladophlebis* aff. *oblonga*, *C.* aff. *patagonia*, *Pagiophyllum* sp., *Phlebopteris* aff. *muensteri*, *Podozmites* sp., *Weichselia reticulata*, and *Xylopteris* sp.

The Six Hills Formation was named by Klitzsch (1978, in Barthel & Boettcher, 1978). It is equivalent to the Basal Clastic unit of Klitzsch et al. (1979) and grades laterally into the Gilf Kebir Formation. The Formation is up to 500 m (1641 ft) thick in its type area at Six Hills (24°21′N, 29°15′E), about 100 km south of Mut, but thins to the east and south to 100 m (328 ft). The unit unconformably overlies Precambrian basement and is disconformably overlain by the Abu Ballas Formation. It is composed of cross-stratified, medium to coarse-grained sandstones forming fining-upward cycles. Each cycle starts with an erosional base, a lower unit of tabular cross-bedded sandstones, followed by kaolinitic sandstones, and topped by root horizons.

Ferns and other plant remains are locally abundant. A thin zone at the top of the unit is marked by straight vertical tubes or U-shaped burrows similar to *Skolithos* and *Rhizocorallium*, respectively. Locally, the basal strata of this Formation consist of fanglomerates up to 5 m (16 ft) thick and rest unconformably on the Precambrian basement. The Formation was assigned an Upper Jurassic-Late Cretaceous age by Klitzsch et al. (1979), and Klitzsch & Nicol-Lejal (1984). According to Mahmoud & Soliman (1994), the upper part of the Six Hills Formation west of the Abu Tartur Plateau is Aptian in age and contains exclusively land-derived palynomorphs, such as spores, gymnospores, angiosperms, and a few fungal spores. Schrank & Mahmoud (1998a, b) assigned the Six Hill Formation in the Dakhla area an Early Neocomian to Late Barremian-Early Aptian (?) age. They identified a rich, marine dinoflagellate fauna, in addition to the land-derived miospores in the upper Six Hills and lower Abu Ballas formations.

The type area of the Abu Ballas Formation (Barthel & Boettcher, 1978) is near Abu Ballas (24°18′N, 27°33′E). The Formation is equivalent to the *Lingula* Shale unit of Klitzsch et al. (1979). Its thickness varies from a few meters to 100 m (328 ft) and wedges out to the east and west. It unconformably overlies the Six Hills Formation and is disconformably overlain by the Sabaya Formation[2] (Lower Cenomanian, Klitzsch, 1990b), or the Desert Rose Beds of Klitzsch et al. (1979). The Abu Ballas Formation is composed of coarsening-upward sequences starting with purple and green shales 20 m–30 m (66–98 ft) thick, followed upwards by siltstones and fine-grained, tabular cross-bedded sandstones, with wave ripples and burrows, and topped by thick paleosols (Klitzsch, 1990). The unit contains abundant brachiopods, such as *Lingula* sp., gastropods, pelecypods, plant remains, sea urchins, insects and *Archaeoniscus* (Barthel & Boettcher, 1978). Remains of fruits are typical of this Formation at the type area. The Abu Ballas Formation in the type area was initially assigned an undifferentiated Upper Jurassic-Early Cretaceous age, but more detailed paleontological and palynological studies proved an Aptian age (Böttcher, 1982, Schrank, 1983, 1987, 1991, 1992). Mahmoud & Soliman (1994) assigned the Abu Ballas Formation an Early Aptian age, based on palynomorphs. The Abu Ballas and Six Hills formations are replaced westward by the Gilf Kebir Formation.

## 7.4.2 Upper Jurassic-Lower Cretaceous of the Northern Western Desert

Lower Cretaceous sediments are extensively developed in the subsurface of the Western Desert. A thick Lower Cretaceous section of more than 3100 m (10,171 ft) was encountered in the Matruh-1 well (Metwalli & Abd el-Hady, 1975). Upper Jurassic-Early Cretaceous sediments have been identified in the Foram-1 well (Schrank, 1984, 1987, 1992) near the Egyptian-Libyan border and the Betty-1 and Ghazalat-1 wells (Abdelmalik et al., 1981). Locally, however, the whole

---

2 Note that the name Sibaiya and Sabaya are used for two different formations which can be confused with each other. The first designates phosphatic and phosphate-bearing sediments of Maastrichtian age overlying the Qoseir Variegated Shales and underlying the Dakhla (or Sharawna) Shales, while the latter denotes an Albian-Cenomanian unit overlying the Abu Ballas Formation and is overlain by the Maghrabi Formation.

Mesozoic section is missing, such as in the Rabat-1 well and Tertiary sediments rest unconformably on Paleozoic rocks.

The Early Cretaceous sediments are represented by the Burg El Arab Formation. The type section of the Burg El Arab Formation (Norton, 1967) is in the Burg El Arab-1 well (30°55′N, 29°31′E), at a drill depth of 2305 m to 4054 m, where the Formation is 1749 m (5738 ft) thick. It unconformably overlies the Jurassic Masajid Formation, Wadi Natrun Formation, Bahrein Formation, Paleozoic sediments, or the Precambrian basement (Hantar, 1990), and is overlain by the Cenomanian Baharia Formation. The thickness of the Formation varies between 500 m (1641 ft) and 2000 m (6562 ft), and increases northward to approximately 3057 m (10,030 ft) in the Matruh-1 well. In the Abu Gharadig and Fayium basins, the Formation is up to 1981 m (6500 ft) in thickness, consisting of continental to shallow marine deposits (Awad, 1984). Rare brackish-water fossils were encountered in the Burg El Arab-1 well. In the Abu Roash-1 well (29°59′N, 31°04′E) the unit is made up entirely of sandstones and it is unconformably overlain by Cenomanian shales and limestones of the Baharia Formation (Said, 1962). The Burg El Arab Formation includes five members (Norton, 1967), from base to top, the Alam El Bueb and its lateral equivalent the Matruh Shale, Alamein Dolomite, Dahab Shale, and Kharita members. All these members were later raised to formation status. The Alam El Bueb Formation is a sandstone unit with frequent shale and occasional limestone beds, both of which increase in abundance to the northwest. The type section of the member is in the Alam El Bueb-1 well, at a drill depth interval of 3927 m and 4297 m. The unit ranges in age from Barremian to Aptian (Hantar, 1990). Schrank (1984, 1987) and Abdel Kireem et al. (1996) identified a number of terrestrial microflora in the Neocomian cores in the Foram-1 well, such *Gleicheniidites*, *Spheripollenides* and *Corollina*. The thickness of the member varies from 200 m (656 ft) to a maximum of 1820 m (5971 ft) in the Alamein-1 well. The Alam el Bueb shales (Betty Formation) probably provided the source of the gas in the Abu Gharadig Field (Khaled, 1999). The Alam El Bueb Member changes facies in the Matruh area into the Matruh Shale (Neocomian-Aptian) composed of dark brown to dark grey, pyritic shales, with carbonaceous and lignitic layers in the upper part. The type section of the Matruh Shale is in the Matruh-1 well at a drill depth interval of 2585 m to 4572 m, where the unit is 1987 m (6519 ft) in thickness. The Matruh shales constitute a good source rock for hydrocarbons in the area (Amine, 1961, El Ayouty, 1990). The Alamein Dolomite Formation is made up of light brown, hard crystalline dolomite with vuggy porosity. The type section is in the Alamein-1 well, at a drill depth interval of 2489 m to 2573 m, where the unit is 84 m (276 ft) in thickness. It grades into the Matruh Shale northwards. The Alamein Dolomite overlies the Alam El Bueb Member and it is overlain by the Dahab Member. The Alamein Dolomite forms hydrocarbon reservoirs at the Alamein, North Alamein, Razzak, and Yidma fields. The dolomite is medium to coarsely crystalline, with intercrystalline and vuggy porosity. The rock includes diagenetic products, such as saddle dolomite, quartz, pyrite, glauconite, anhydrite, and dedolomitization (Metwalli & Abd el-Hady, 1975). In other fields, such as the Mubarka and Abu Gharadig, the Alamein Dolomite is unproductive, and production is from Aptian clastics. The Dahab Formation is composed of grey to greenish grey shales, with thin interbeds of siltstones and sandstones. The type section is in drill interval 3180 m to 3354 m in the Dahab-1 well (30°51′N, 28°42E), where the unit has a thickness of 174 m (571 ft). The Kharita Formation (Aptian-Albian) is made up of fine to coarse-grained sandstones with subordinate shale and carbonate beds. The type section is in the Kharita-1 well, at a drill depth interval of 2501 m to 2890 m, where the unit is 389 m (1276 ft) in thickness. El-Beialy (1994) suggested an Albian age for the Kharita Member in the Badr El Dein-1 (Bed 1–1) well (28°31′N, 29°51′E) (drill depth interval 3498 m to 3626 m) based on occurrence of the palynomorphs *Cretacacisporites scabratus* and *Trilobosporites laevigatus*. The Kharita Member forms the reservoir in the Qarun Field in the Gindi Basin, SW of Cairo (OGJ, 1996). Previously, Metwalli & Abd el-Hady (1975) treated the Dahab and Kharita members as one unit, the Abu Subeiha Formation, which they dated as Aptian, based on the presence of *Orbitolina discoidea*.

Aboul Ela et al. (1998) reported three Early Cretaceous marine dinoflagellate cysts zones, ranging in age from Barremian-Early Albian, from deep wells in the northern part of the Eastern Desert. These zones are *Pseudoceratium anaphrissum-Moderongia simplex* (Barremian), *Pseudoceratium securigerum* (Aptian), and *Subtilisphaera senegalensis-Dinopterygium cladoides* (Early Albian). They suggested that the Jurassic-Cretaceous boundary in these wells is unconformable. Ibrahim et al. (1998) recognized three miospore biozones in the Early Cretaceous section. The miospores are dominated by pteridophytic spores, gymnosperm pollen, and angiosperm pollen. The miospore zones are *Stellatopollis* sp.? *Schrankipollis* sp. (Barremian), *Afropollis operculatus-Brenneripollis* sp.-*Tricolpites* (Aptian), and *Crybelosporites pannuceus*-Afropollis *jardimis-Tricolporopollenites* (Earl Albian). These zones span the Alam El Bueib, Alamein, and Kharita formations.

## 7.4.3   Upper Jurassic-Lower Cretaceous of Sinai

In Central Sinai and the Gulf of Suez area, Early Cretaceous rocks (partly equivalent to the Nubian Sandstone) (Fig. 7.1) are represented by continental sediments of the Malha Formation. The Malha Formation was introduced by Abdallah et al. (1963) for a Lower Cretaceous section exposed at Wadi Malha (29°20′N, 32°30′E) in the SE corner of the North Galala Plateau. In the type area, the Malha Formation is 120 m (394 ft) in thickness. It is composed of white, light violet, fine-grained sandstones with a number of kaolinitized clay beds. The Formation overlies the Permo-Triassic Qiseib Formation and is overlain by the Cenomanian Raha Formation. The Malha Formation in the Tih scarp forms a section of 500 m (1641 ft) composed of white, brown, red and purple, medium-grained sandstones, intercalated with thin carbonaceous shales (Said, 1962). It overlies the Carboniferous Ataqa Formation and is overlain by Cenomanian marls of the Raha Formation. The age of the Malha Formation is considered Lower Cretaceous by Abdallah et al. (1963), Said (1971), Gvirtzman & Weissbrod (1984), Kora et al. (1994), and Alsharhan & Salah (1995, 1997).

In the Gebel El Minshera area, Northern Sinai, the Malha Formation and the overlying Galala Formation span the Albian-Cenomanian interval. The Malha Formation represents a clastic facies deposited in a continental fluvial environment. The spore and pollen association and palynofacies suggest a depositional setting close to the vegetational source (El-Beialy et al., 2010).

Two units can be recognized in the Malha Formation in Central Sinai, which are equivalent to the Amir and Hatira formations (Weissbrod, 1969, Bartov et al., 1980). The reference section of the Amir Formation is in Gebel Raqaba, Sinai (Weissbrod, 1969), where it has a thickness of 52.7 m (173 ft). The Formation is well exposed at Gebel el Tih and extends from Wadi Hamr in the west to Gebel Dhalal in the east. It pinches out

*Figure 7.1* Honey-comb structure (Arabesque), Nubian Sandstone, Sinai (courtesy of Baher El Kaliouby) (See color plate section).

toward the south. It unconformably overlies the Permo-Triassic Qiseib Formation or the Budra Formation and it is unconformably overlain by the Lower Cretaceous Hatira Formation. It was assigned a Jurassic-Lower Cretaceous age by Weissbrod (1969) based on its stratigraphic position. It is composed of white, cross-bedded, kaolinitic, fine-grained sandstones, alternating with thin silty beds (Weissbrod & Perath, 1990). The burrows *Diplocraterion* are widely distributed, ripple structures occur only in the lower part (Weissbrod, 1969). The reference section of the Hatira Formation is in Gebel Raqaba, Sinai, where it attains a thickness of 210 m (689 ft) (Weissbrod, 1969). It is exposed in SW Sinai in the cliffs of Gebel el Tih, Gebel Sarbut el Gamal, and Gebel Musaba Salama, and at Wadi Ba'ba, Wadi Budra, and Wadi Sidri. The Hatira Formation has been divided into two members. The Avrona Member is made up of massive and thickly bedded, coarse- to medium-grained, kaolinitic sandstones, with large-scale cross-bedded sandstones at the base, followed upward by variegated sandstones (Weissbrod & Perath, 1990). Fossil trees are common in the Avrona Member. The upper member is unnamed and it is composed of variegated, grey, violet, and red fine- to medium-grained sandstones (Weissbrod, 1969). It has a thickness between 60–80 m (197–262 ft). It is slightly impregnated with manganese and iron minerals. At Gebel el Tih it consists of a red silty-clayey bed, approximately 10–15 m (33–49 ft). This member is overlain by Cenomanian sediments of the Raha Formation. The Hatira may contain a conglomeratic unit at the base (Arod Conglomerate), such as in the Arif en Naga area. The conglomerate varies in thickness between 0–15 m (0–49 ft), with quartzite pebbles up to 30 cm in size (Bartov et al., 1980).

Marine Lower Cretaceous sediments crop out in Northern Sinai on the flanks of the Maghara and Gebel Halal structures. The section at Risan Aneiza (30°55′N, 33°46′E), a hill north of Gebel Maghara, is made up of more than 250 m (820 ft) of non-fossiliferous variegated sandstones, alternating with fossiliferous marly beds (Said, 1962). Said (1971) applied the name Risan Aneiza Formation to this succession. The type locality of the Formation is at Lagama-Minsherah (30°18′N, 33°42′E), northern Sinai, where it is more than 110 m (361 ft) in thickness. The Formation is divided into several units, from base to top, 1) limestones and marls, 2) variegated sandstones, 3) oolitic limestones and marls, 4) sandstones, and 5) sandstones with intercalations of oolitic limestones carrying marine fossils. It was assigned an Aptian-Albian age by Said (1971). Fossils reported from this Formation include *Exogyra boussingauli, Trigonia orientalis, Nerinella algarbiensis, Knemiceras gracilis, K. spathi, K. rittmanni, Salenia humei, Pseudodiadema deserti*, and rudistids. The foraminifera from this exposure (Said & Barakat, 1957) are shallow-water and include the Aptian fossils *Pseudodiadema julii, Zeilleria tamarindus, Terebratula sella, Trochodiadema libanotica, Typociaris proxima, Trigonia* cf. *quadrata, Cyprina lagamensis, Holectypus macropygus, Pecten syriacus, Orbitolina lenticularis, Flabellammina aegyptica, Siroplectammina arabica, Lingulina sadeki*, and *Discorbis beadnelli*. This section is followed upward by a 110 m (361 ft) thick section of Albian oolitic and sandy limestones, with some shale and fossiliferous beds, followed by a 400 m (1312 ft) section of the Cenomanian Raha Formation or Halal Formation white limestones.

A 520 m (1706 ft) section of Lower Cretaceous rocks exposed at the core of Gebel Halal (30°38′N, 34°00′E) (Said, 1962), the type section of the Halal Formation. It is composed of unfossiliferous sandstones, and fossiliferous interbeds of shale and siltstone, with corals, pelecypods, gastropods, and ammonites. These sediments are equivalent to the Risan Aneiza Formation of Said (1971). Lower Cretaceous rocks are also exposed at the core of Gebel Shabrawet anticline (30°17′N, 32°17′E), between the cities of Ismailia and Suez. These rocks are composed of variegated shales and marls of Albian age, containing *Hemiaster cubicus* and *Knemiceras syriacum*, and are overlain by the Cenomanian Raha Formation. The lithostratigraphy and biostratigraphy of the Aptian-Albian succession at Risan Aneiza have been studied by Ismail & El Saadany (1995), Bachmann et al. (2003), and Abu Zied (2008). The molluscs of the Risan Aneiza Formation were studied by Hamama (2010). The Malha, the Risan Aneiza and the Halal formations are rich in cephalopods. Eight ammonite zones ranging in age from Barremian to late Albian have been designated by Abu Zied (2008), from base to top: *Subpulchellia oehlerti* Zone (late Barremian), *Barremites difficilis* Zone (late Barremian), *Deshayesites deshayesi* Zone (early Aptian), *Aconeceras nisus* Zone (early middle Aptian), *Epicheloniceras tschernyschewi* Zone (late middle Aptian), *Acanthohoplites nolani* Zone (late Aptian), *Knemiceras gracile* Zone (early Albian), and *Mortoniceras inflatum* Zone (late Albian).

## 7.5   UPPER CRETACEOUS

### 7.5.1   Upper Cretaceous of the Northern Western Desert

Upper Cretaceous rocks are exposed in eroded anticlines, such as the Abu Roash complex southwest of Cairo and the Bahariya High in the central Western Desert.

The Upper Cretaceous succession in the Abu Roash area, approximately 10 km SW of Cairo, was divided by Beadnell (1902) into the following units, from top to bottom:

1 Gebel Khashab red beds (Oligocene) (Top)
2 Basalt flows
3 Limestones and sandy limestones (Eocene)
4 Chalk (Maastrichtian-Campanian)
5 *Plicatula* Series (Santonian)
6 Flint Series (Santonian)
7 *Acteonella* Series (Santonian)
8 Limestone Series (Turonian)
9 *Rudistae* Series (Turonian)
10 Sandstone Series (Cenomanian) (Base not exposed)

Dacqué (1903) examined the rudistids and later Jux (1954) examined the geology of the Abu Roash area and Said & Kenawy (1958) examined the foraminifera of the Turonian rocks in the area. More recently, Mansour (2004) examined the rudistids of the Abu Roash area. In spite of the importance of this area, from the geological and archeological point of views, there is no accurate map of the area, except for the one presented by Said (1962) and Said & Martin (1964) (Fig. 7.2).

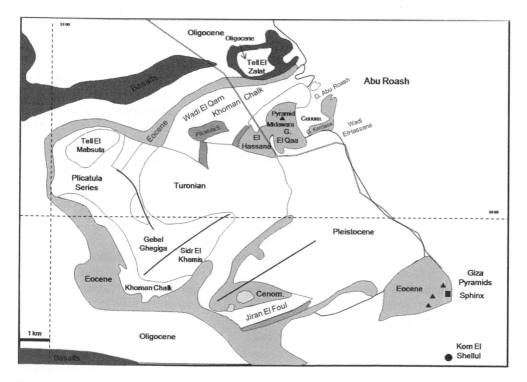

*Figure 7.2* Map of the Abu Roash area (based on Said, 1962, Said & Martin, 1964, and De Casatro & Sirna, 1996) (See color plate section).

Beadnell's (1902) subdivisions were widely accepted and used until Norton (1967) divided the Upper Cretaceous sequence in the northern Western Desert into the Abu Roash Formation at the base and the Khoman Formation at the top. The former includes rocks of Cenomanian to Santonian age and includes units 5–10 of Beadnell (1902), the latter includes those of Campanian to Maastrichtian and is equivalent to unit 4 of Beadnell (1902). The discussion of the Upper Cretaceous-Tertiary sequence in Egypt in the present work led itself to two age subdivisions, Cenomanian-Santonian and Campanian-Landenian.

### 7.5.2   Cenomanian-Santonian

The Abu Roash Formation was named after its type locality in the Abu Roash area (Norton, 1967). As originally defined, it ranges in age from Cenomanian to Lower Santonian, and includes the *Plicatula* Series, Flint Series, *Acteonella* Series, Limestone Series, *Rudistae* Series, and the Sandstone Series of Beadnell (1902) (subdivisions 4–10).

Aadland & Hassan (1972) divided the Abu Roash Formation in the subsurface of the Abu Gharadig Basin, in the northern Western Desert, into seven members designated by the letters A to G. These units correspond approximately to Beadnell's subdivisions and form both reservoir and source rocks in the Western Desert (Abdine et al., 1993, Wever, 2000, Zobaa et al., 2008). Chatelier & Slevin (1988) introduced formal names for these units, but the letter designation continues to be favored and used.

Unit G (Cenomanian) is composed of reddish brown bioturbated sandstones, sandy argillaceous limestones, gypsiferous green shales, with the pelecypods *Durania* sp. and *Hipurites* sp., and glauconitic sandstones. It is equivalent to the Sandstone Series and the Rudistae Series of Beadnell (1902), and the Abyad Member of Chatelier and Slevin (1988). Its base is not exposed in the Abu Roash area.

Unit F (Cenomanian) is 41.5 m (136 ft) thick and composed of thick-bedded, white limestones, highly argillaceous, with foraminifera in the lower part, and cherty limestones, with oyster beds, pellets, and intraclasts, and glauconitic lower limestones in the upper part. The unit contains *Cyphosoma abatei* and *Ostrea flabellata*. It is equivalent to the Limestone Series of Beadnell (1902). Chatelier & Slevin (1988) applied the name Mansour Member to this unit, and assigned it a Turonian age.

Unit E (Turonian) is approximately 28.5 m (95 ft) in thickness and consists of reddish brown sandstones, yellowish green argillaceous limestones, yellowish brown shales, and white limestones, becoming sandy and gypsiferous at the base. This unit is equivalent to the upper Limestone Series of Beadnell (1902). It was called Miswag Member by Chatelier & Slevin (1988).

Unit D (Turonian) is 80 m (262 ft) in thickness. It forms high cliffs and ridges at Abu Roash. It is composed of white, dolomitic, cherty limestones, with thin sandstone and shale beds. The limestones contain *Acteonella salamonis*, and *Nerinea requienieana*. This unit is equivalent to the *Acteonella* Series of Beadnell (1902). It was called the Meleiha Member by Chatelier & Slevin (1988).

Unit C (Turonian) is 60 m (197 ft) thick and made up of shales, white, sandy and highly fossiliferous, bryozoan limestones, and sandstones. The limestones become cherty in the middle part, and locally form *Durania* biostromes. Red siltstones occur within this unit at El-Hassana, SW the village of Abu Roash. This unit is equivalent to the *Acteonella* Series of Beadnell (1902). It was called the Abu Sennan Member by

Chatelier & Slevin (1988). The rudist buildups (biostromes) at the El Hassana Dome crop out in the form of an isolated mound and consist of rudist-coral communities in which rudists are dominated by radiolitids (De Castro & Sirna, 1996), such as *Durania arnaudi*. De Castro & Sirna (1996) also reported *Ceriopora multiformis* and a new species of *Acicularia? aegyptiaca* n.sp., and assigned the interval to the Coniacian. El-Sabbagh & El-Hedeny (2005), El-Hedeny & El-Sabbagh (2005), and El-Hedeny (2006) reported seven radiolitid species belonging to the genera *Durania, Lapeirousella,* and *Sauvagesia*. The rudist shells are present in the form of isolated specimens or small clusters embedded in marly limestone beds. The size of the shells may reach up to 20 cm, and are associated with coral heads of hemispherical growth forms that may reach 30 cm in size (Mansour, 2004). The rudist-coral dominated beds occur and are exposed on the top of the *Actaeonella* Series, flanking the crest of the El Hassana Dome. They are underlain by limestones crowded with gastropod shells of *Actaeonella (Trochactaeon) salomonis* (Mansour, 2004). For a review of the rudistids reported from Egypt, see Zakhera (2010). The diagenesis of the rudistid clasts has been discussed by Mansour (2004).

Unit B (Upper Turonian) is 35 m (115 ft) thick and consists of white, chalky limestones, with flint bands in the upper part. The lower part is highly argillaceous, sandy, with pellets and intraclasts of gypsiferous limestone. This unit is equivalent to the Flint Series of Beadnell (1902) and the Rammak Member of Chatelier & Slevin (1988).

Unit A (Coniacian-Santonian) is 50 m (164 ft) in thickness and is composed of white, partly dolomitized, and chalky limestones, with bryozoans, pellets and intraclasts. The limestones contain shale intercalations and become highly argillaceous, glauconitic, and sandy in the lower part. It is equivalent to the *Plicatula* Series of Beadnell (1902). In general, the Santonian cannot be differentiated from the Coniacian. Possible Coniacian beds are represented a few kilometers north of the pyramids of Giza by marly limestones with *Echinobrissus waltheri, Orthopsis milliaris, Leicocidaris crameri*, the ammonite *Tissotia* and other molluscs (Cuvillier, 1930). Unit A was called the Ghorab Member by Chatelier & Slevin (1988).

In the subsurface of the northern Western Desert, the Turonian (equivalent to units B-E) is 352 m (1155 ft) thick and consists of very fine calcareous glauconitic sandstones in the upper part and oolitic/glauconitic limestones in the lower part. The limestones are pyritic and dolomitic in places, and interbedded with shales, sandstones, and anhydrites. This interval was assigned a Turonian age by Metwalli & Abdel-Hady (1975) based on the presence of *Discorbis turonicus* (the range of this genus is believed to exend from Turonian to Santonian).

Dacqué (1912) published a review of the fossil turtles of the Tertiary of Egypt in which he attempted an outline of the phylogenetic relationships with the living forms.

Strömer (1914) studied the geology of the Bahariya Oasis and introduced the name *"Baharjhestufe"* or the Baharia Formation for a succession of Cenomanian crossbedded sandstones and variegated shales with ferruginous sandstone concretions, and contains a mixture of fresh and marine fossils which indicates fluvio-marine conditions (Cuvillier, 1930, Said, 1962), such as echinoids, the molluscs *Exogyra columba* and *Ostrea flabellata*, and a number of vertebrate fossils, such as fishes, plesiosaurs, crocodiles, turtles, and the snake *Symoliophis*. Smith et al. (2001) reported the discovery of *Paralitatan stromeri*, the first tetrapod described from the Baharia Formation since 1930s, in addition to fish, turtles, plesiosaurs, squamates, crocodyliforms, and four dinosaurs including *Spinosaurus, Carcharodontosaurus,* and *Bahariasaurus* originally identified by Strömer (1936).

The Baharia Formation ranges between 45–303 m (148–994 ft) in thickness (Strömer, 1914, Said, 1962). As originally defined, these rocks are probably equivalent to those exposed at the base of the Abu Roash massive, and unit G (Abyad) in the subsurface of the northern Western Desert. In the Khalda Field in the northern Western Desert the upper part of the Baharia Formation is made up of marine shales and limestones (Conway et al., 1988a, b). The shales of the Baharia Formation (G Member of the Abu Roash Formation *sensu lato*) is considered to be the source of the oil trapped in the Wata reservoirs (C-D members of the Abu Roash Formation *sensu lato*) in the Abu Gharadig Field (Khaled, 1999).

Foraminifera and ostracods were described from the Baharia Formation and basal Abu Roash "G" Member by Ismail & Soliman (1997). The foraminifera include *Thomasinella punica, Nazzazatinella adhami, Nazzazata convexa conica,* and *N. convexa convexa.* They assigned Baharia and Abu Roash "G" a Cenomanian age, but suggested that the top of the latter exends into the Turonian. Recovered dinoflagellates from the same interval in the Razzak-7 well, southwest of Horus-1, include *Coronifera oceanica, Cyclonephelium vannophorum, Florentinia cooksoniae, Palaeoperidinium cretaceum,* and *Subtilisphaera perlucida,* and these also confirm Albian-Cenomanian age. However, the presence of the spore species *Cicatricosisporites orbiculatus* and the dinoflagellate *Cyclonephelium vannophorum* indicates an age not younger than middle Cenomanian for the Baharia Formation (Zobaa et al., 2008).

Columnar Oligocene basalt sheets cover the Baharia Formation at Gebel Mendisha (El Sisi et al., 2002) (Fig. 7.3).

El-Akkad & Issawi (1963) recognized a unit overlying the Baharia Formation for which they introduced the name El Heiz Formation. The Formation was described as a 21.4 m (70 ft) section composed of two dolomitic units, separated by a clastic carbonate section. They assigned it an Upper Cenomanian age. The type section of the Formation is at the El Tibnia promontory, along the western scarp of the Bahariya Oasis. The thickness of the El-Heiz Formation varies from 0.5–30 m (1.6–98 ft) (Strömer, 1914, Said, 1962). This unit is equivalent to unit F (Mansour) of Beadnell (1902) in the Abu Roash area. The ammonites of the Baharia Formation in the Bahariya Oasis outcrops were studied by Dominik (1985) who treated the El Heiz Formation as a member of the Baharia Formation. He divided the Baharia Formation into three members, which include from base to top, the Gebel Ghorabi Member, Gebel Dist Member, and El Heiz Member. The Gebel Ghorabi Member[3] is made up of fluviatile cross-bedded, coarse-grained sandstones, the Gebel Dist Member consists of estuarine, fine grained, well bedded, ferruginous clastics, with vertebrate fossils and oysters, and the El Heiz Member is composed of lagoonal dolomites, sandy dolomites and clastic limestones.

Tawadros (2001) suggests that units B-E (Rammak, Abu Sennan, Meleiha, and Miswag members) of the Abu Roash Formation be treated as a separate unit of Turonian age, equivalent to the Wata Formation. Units F and G (Mansour and Abyad

---

3 The name Ghorabi Iron Ore Member was used by Akkad & Issawi (1963), Said & Issawi (1964) and Said (1971) for a Middle Eocene (Lower Lutetian) unit in North Gebel Ghorabi, Bahariya Oasis.

*Figure 7.3* Bahariya basalts, Western Desert (Courtesy of Baher El Kaliouby).

members) which are equivalent to the El Heiz and Baharia formations, respectively, should be excluded from the Abu Roash Formation, and restrict the latter to rocks of Coniacian-Santonian age (unit A, or Ghorab Member). The Abu Roash Formation can be recognized in the Northern Egypt Basin at Abu Roash and Bahariya Oasis and is probably present in the subsurface of the Farafra Oasis. It is also present in the subsurface of the Nile Delta, the Gulf of Suez area, and western Sinai (Matulla Formation). It is absent in the stable shelf (the Nile Basin) probably due to either non-deposition or erosion.

In the Abu Gharadig and Fayium basins, the Cenomanian is represented by fluvio-marine to open marine sandstones in the lower part and shales with some carbonate interbeds in the upper part (Awad, 1984). In the Alamein-IX well, the Cenomanian section is 650 m (2133 ft) in thickness. Metwalli & Abdel-Hady, (1975) divided the section into two units, equivalent to the El-Heiz and Baharia formations, respectively. The upper unit (El-Heiz Formation) is composed of carbonate rocks and sandstones with minor shales and characterized by *Thomasinella punica*. The lower unit (Baharia Formation) is made up of clastics and contains *Thomasinella fragmentaria*. Gohrbandt (1966a) reported these two *Thomasinella* species from the Ain Tobi Formation in Libya and indicated that they are the same. Therefore, these two species cannot be used to differentiate between Lower and Upper Cenomanian. Sharabi (1998) recorded the Early Cenomanian *Rotalipora reicheli* and *Favusella washitensis*, *Charentia cuvillieri*, and *Mayncina orbignyi* from the Baharia Formation in the Abu Gharadig and Gindi/Fayium basins, and the Late Cenomanian *Rotalipora cushmani* and *Whiteinella archaeocretacea* from the El Heiz Formation.

Production in the Khalda and Rezzak fields in the Western Desert is from the Cenomanian Baharia Formation where it is composed of a sequence of tidal flat deposits at the base, grading into mixed flats, mud flats, and shallow marine sands (Conway et al., 1988a, b). Production in the Qarun Field, SW of Cairo is from the Baharia and Abu Roash formations (*s.l.*) (OGJ, 1996).

Said (1962) introduced the Hefhuf Formation for a succession of hard crystalline limestones in the lower part and variegated shales and sandtones with phosphatic beds in the upper part. It overlies the Baharia Formation, or the Wadi Hennis Formation of Dominik (1985). It is overlain in turn by the "Chalk" (Khoman Formation of Norton, 1967). The unit was assigned a Turonian age by Said (1962) and a Turonian-Santonian age by Said (1971). Dominik & Schaal (1984) assigned it an Upper Campanian age based on ammonites. The upper part contains *Gigantychthys pharao* and is correlatable with the Duwi Formation, while the lower part is correlatable with the Abu Roash Formation (*s.l.*). The Hefhuf Formation, therefore, includes the Duwi Formation, the Abu Roash Formation, and older rocks in that area. In the Bahariya and Farafra oases, the Hefhuf Formation is considered the equivalent of the Duwi Formation (Hermina, 1990). It is recommended that the name Hefhuf Formation be discarded. Kamal El-Din (2003) describe the petrified woods *Celastrinoxylon celastroides* and *Ficoxylon cretaceum* from the Late Cretaceous Hefhuf Formation in the Farafra Oasis.

### 7.5.2.1 Upper Cretaceous (Cenomanian-Santonian) of Sinai and the Gulf of Suez region

The stratigraphy of the Cretaceous outcrops in Sinai have been studied by Ghorab (1961), Al Far (1964), Lewy (1975), Bartov & Steiniz (1964), Kuss & Schlagintweit (1988), Cherif et al. (1989), Shahin & Kora (1991), Ziko et al. (1993), Ismail (1993, 2000), Akarish (1998), Bachmann & Kuss (1998), Issawi et al. (1999), Lüning et al. (1998a–d), Samuel et al. (2009), Wanas (2008), Abu Zied (2008), Bishta & Aita (2009), and Hamama (2010). The Upper Cretaceous succession in Sinai and the Gulf of Suez region includes, in ascending order, the Raha, Abu Qada, Wata, and Matulla formations.

The name Raha Formation was introduced by Ghorab (1961) for a 120 m (394 ft) thick, Early to Late Cenomanian section in the Scarp of the Raha Plateau, in west-central Sinai. Said (1971) reported a thickness of 230 ft (70 m) for the Formation. In the type locality along the scarp of the Raha plateau, the unit is composed of limestones, marls, shales, and sandstones, and contains *Thomasinella aegyptia*. Ghorab (1961) divided the Formation into two members in the fault blocks of the Esh Mellaha range (27°35′N, 33°30′E) in the Eastern Desert. The lower Abu Had Member, with its type section at Gebel Abu Had (27°41′N, 33°04′E), ranges in thickness between 100 ft (30 m) and 140 ft (43 m) and is composed of a sequence of shales and marls, with a few streaks of sandstones and sandy shales, followed by a bed of partly siliceous or dolomitic limestones and marly limestones with *Thomasinella aegyptia*. The upper Mellaha Sand Member, with its type section at Bir Mellaha, varies in thickness between 40 ft (12 m) and 70 ft (21 m), but is only few feet thick in the Esh Mellaha range. The sands are devoid of fossils and the unit was assigned a Cenomanian age based on its stratigraphic position. The Raha Formation overlies sandstones of the Malha Formation and underlies Turonian (Abu Qada Formation) sediments. The Raha Formation is equivalent to the Hazera Formation of Late Albian-Cenomanian age in the Arif Al Naga outcrops (Bartov & Steinitz, 1977, Bartov et al., 1980). The exposed part of the Raha Formation in the Gebel Somar area, north-central Sinai, yielded several species of planktonic and benthic foraminifera, suggesting a late Cenomanian age. This age is supported by the presence of the oysters *Ceratostreon flabellatum* and *Ilymatogyra Africana* and the echinoid *Hemiaster cubicus* (Samuel et al., 2009). In Central Sinai, the Raha Formation was encountered in the subsurface in the Darag, Nekhl and Abu Hamth wells with thicknesses between 310 m (1017 ft) and 326 m (1070 ft) (Kerdany & Cherif, 1990).

El-Shinnawi & Sultan (1972b) examined the foraminifera of the Upper Cretaceous section in five offshore wells, the Ghareb-1, Alef-1, J-1, Morgan-1 and H-1, and two onshore wells, Bakr-6 and Ras Ghareb-7. They divided the Raha Formation into the Abu Had Member[4] which they assinged a lower Upper Cenomanian age (*Thomasinella punica* zone), and the Mellaha Member[5] designated an upper Upper Cenomanian, based on its stratigraphic position, since it is barren of fossils. In the meantime, Twefik & Ebeid (1972) examined the foraminifera of Upper Cretaceous surface and subsurface samples on both sides of the Gulf of Suez and along the Red Sea coast. They divided the Raha Formation into the Abu Had Member (Cenomanian), where they recognized the *Thomasinella fragmentari* and *T. punica* zones, and the Mellaha Member (upper Upper Cenomanian), based on its stratigraphic position (barren of fossils). Cherif et al. (1989) divided the Raha Formation into three units, in an ascending order: 1) The Abu Had Member (early late Cenomanian) is composed of siliciclastics and carbonates. 2) The Mukattab Member is a sequence of hard carbonates. 3) The Ekma Member is a predominantly clastic sequence. The last two members are equivalent to the Mellaha Member of other authors.

In Northern Sinai, the Cenomanian is represented by the Halal Formation. This unit was introduced by Said (1971) for a 550 m (1805 ft) section at Gebel Halal (30°38′N, 34°00′E) composed of resistant dolomitic and recrystallized limestones with minor marl intercalations. The lower part contains *Manelliceras* fauna and the upper part *Neolobites fourtaui*. The unit overlies sandstones of the Early Cretaceous Malha Formation and is overlain by marine marl beds of the Abu Qada Formation (Wanas, 2008). This unit is equivalent to the Raha Formation of Ghorab (1961), although the Halal Formation is more dolomitic, and the Galala Formation according to El-Beialy et al. (2010). Cenomanian rocks equivalent to the Raha Formation at Gebel Shabrawet and Gebel Maghara in Sinai include *Ostrea flabellata*, *O. africana*, *O. olisiponensis*, *Exogyra columba*, *O. mermeti*, and *Venus reynesi*, *Neolobites fourtaui*, *Hemiaster cubicus*, *H. pseudofournelli*, and *Tissotia tissoti*, among others (Cuvillier, 1930), and they are overlain by a unit of poorly fossiliferous limestones of Turonian age. More recently, Gendi (1998) examined the macrofossils in the Raha and Halal formations and assigned them a Cenomanian age based on the presence of *Hemiaster cubicus* near the base, *Hymatogyra africana* and *Ceratostreon flabellatum* in the middle part, and *Exogyra olisiponensis* near the top. In the Gebel El Minshera area, Northern Sinai, the Upper Albian-Lower Cenomanian Galala Formation signals

---

4 The name Abu Had Member was also used by Said (1990, p.457, Table 24.2) for an Eocene unit. Bandel et al. (1987) also used the name Abu Had Formation for a Campanian unit, equivalent to the Duwi and Qoseir formations.

5 James et al. (1988) used the name Mellaha Member for a unit of a Miocene age.

the Cenomanian marine transgression, with the establishment of proximal carbonate ramp sedimentary facies (El-Beialy et al., 2010). The spore-pollen assemblage near the base of the Galala is similar to that of the Malha Formation, but with the addition of *Classopollis* pollen, which suggests coastal vegetation.

The name Abu Qada Formation was introduced by Ghorab (1961) to describe a unit which overlies the Mellaha Sand Member of the Raha Formation composed of brown to very dark grey to black marls, crowded with *Heterohelix globulosa* fauna, and contains *Ostrea africana*. Its thickness ranges from 15 m (50 ft) to 24 m (80 ft). The type section of the Abu Qada Formation is at Wadi Gharandal (29°16′N, 32°58′E), east-central Sinai, where the unit is 16 m (53 ft). It was assigned a Cenomanian age by Ghorab (1961) and Said (1971). It was designated an Early Turonian age by El-Shinnawi & Sultan (1972b) and Twefik & Ebeid (1972) in offshore wells in the Gulf of Suez area based on fauna of the *Heterohelix globulosa* zone (Caron, 1985, considers the range of this fossil as Campanian-lowermost Maastrichtian), and by Jenkins (1990). It was designated a Late Cenomanian-Early Turonian age by Kora et al. (1994). According to Kora et al. (1994), the lower part of the Abu Qada Formation contains oolite shoals and reefal facies with orbitolinids and miliolids, and the upper part is dominated by planktonic foraminifera, such as heterohelecids, praeglobotruncanids and hedbergellids, associated with rare calcareous agglutinated benthonic foraminifera, abundant oysters, ammonites, and ostracods. At Gebel Arif El Naga, the Formation is 68 m (223 ft) to 141 m (463 ft) thick and composed of thick gypsiferous shales with thick limestone beds. It unconformably overlies the Lower Cretaceous Malha Formation and it is overlain by the Wata Formation. The Abu Qada Formation is equivalent to the Ora Shale (Bartov et al., 1980), which contains ammonites, such as *Mammites nodosoides, Neoptychites cephalotus*, and *Choffaticeras luciae trisellatum*. It was assigned to the Lower Turonian-basal Upper Turonian. Wanas (2008) considered the Abu Qada to be Turonian in age. The lower part of the Abu Qada Formation in the Gebel Somar area, north-central Sinai, yielded several species of agglutinated benthonic foraminifera that suggest a late Cenomanian age, while the upper part of the Formation contains *Whiteinella archaeocretacea*, which marks the beginning of the Turonian (Samuel et al., 2009). Similar-aged planktonic foraminifera are recorded from the underlying Raha Formation (Samuel et al., 2009). It appears then that the age of the Abu Qada Formation is uncertain, but it may straddle the Cenomanian-Turonian boundary.

Ghorab (1961) proposed the name Wata Formation for a unit which consists of brown to light yellow, hard, siliceous, and partly dolomitic limestones with *Discorbis turonicus* and *Spiroplectammina arabica* in the type section at Wadi Wata, in the Raha plateau, west-central Sinai. He reported a thickness of about 150 ft (46 m) for this unit. It contains the fossils *Durania gaensis, Acteonella salamonis*, and *Nerinea requieniana*, and was assigned a Turonian age by Ghorab (1961), Said (1971), Kora et al. (1994), and Ismail (2001). A similar assemblage has been reported from units C and D of the Abu Roash Formation (*s.l.*) in the Abu Roash area. The Wata Formation overlies the Abu Qada Formation and is overlain by the Matulla Formation. At Gebel Arif El Naga, the unit is made up of a succession of massive, well-bedded limestones and dolomites with minor amounts of marls, shales and chert, ranging in thickness from 132 m (433 ft) to 170 m (558 ft). El-Shinnawi and Sultan (1972b) assigned the Wata Formation an Upper Turonian age based on a fauna of the *Discorbis turonicus* zone in the Red Sea offshore wells. Twefik & Ebeid (1972) also assigned the Wata Formation an Upper Turonian age, where they recognized the foraminiferal *Ammo-marginulina blanckeni* and *Gaudrina matullaensis* zones in surface and subsurface samples on both sides of the Gulf of Suez and along the Red Sea coast. The Wata Formation is equivalent to the Gerofit Formation in the Arif Al Naga outcrops (Lewy, 1975, Bartov et al., 1980).

In the St. Anthony section at Wadi Araba on the western side of the Gulf of Suez, the Raha, Abu Qada, and Wata formations (Cenomanian-Turonian) are represented by a 120 m (394 ft) section composed of coarsening-upward cycles (Kuss, 1986). This succession can be divided into a basal limestone-dominated and an upper siliciclastic-dominated units. The basal unit (probably equivalent to the Abu Had Member) is composed of oncolitic, oobiomicritic limestones with the codiacian algae *Bouenia* sp. and *B. pygmaea*, miliolids, ostracods, and a few echinoderm fragments and other bioclasts. The upper unit (probably equivalent to the Mellaha Member) is made up predominantly of bioturbated, glauconitic, silty sandstones, with a few horizons of *Turitella*-bearing beds, glauconitic sandy marls with bivalves, cross-bedded ferruginous sandstones, glauconitic shaly siltstones with planktonic foraminifera, such as *Praeglobotruncana* and *Galvelinella*, and nodular limestone beds composed of cortoids, brachiopods, miliolids, ostracods, and ophiurid fragments, set in a micritic groundmass, with intense *Ophiomorpha* bioturbation and soil horizons. The Turonian is probably absent in this section. The Cenomanian-? Turonian section is unconformably overlain by the Campanian St. Paul Formation. The Matulla Formation (Coniacian-Santonian) is missing in the St. Anthony section.

Ghorab (1961) designated the name Matulla Formation (Coniacian-Santonian) for a succession of sandstones, marls and shales which forms the highest Upper Cretaceous unit in the Ras Gharib area on the western side of the Gulf of Suez. Said (1971) reported a thickness of 558 ft (170 m) for the Matulla Formation in the type section at Wadi Matulla (29°03′N, 33°09′E), west-central Sinai. It has a thickness of 200 ft (61 m) in the Ras Gharib-93 well. The Formation contains a fauna of the *Discorbis turonicus* zone and was assigned a Santonian age by Ghorab (1961). When a complete section of the Formation is present, its age extends from Coniacian to Santonian. The Coniacian-Santonian contact can be recognized in many places in Sinai, either paleonologically or by the presence of an erosional surface (El-Azabi & El-Araby, 2007). The Matulla Formation overlies the Wata Formation and is overlain by the Sudr Formation. The upper part of the Matulla Formation is eroded in the Ras Gharib area (Ghorab, 1961). The whole Coniacian-Santonian section (Matulla Formation) is missing in the St. Anthony section in the Southern Galala Plateau (Kuss, 1986). At Esh el Mellaha, the Matullah Formation is tightly folded about north-trending axes (Bosworth et al., 1999) and is unconformably overlain by Campanian sediments.

El-Shinnawi & Sultan (1972b) assigned a Coniacian-Santonian age to the Matulla Formation in the offshore of the Gulf of Suez based on fauna of the *Discorbis turonicus* zone. Twefik & Ebeid (1972) also assigned a Coniacian-Santonian age and recognized two zones, the *Discorbis simplex* and *Hedbergella hansbolli* zones.

The rich assemblage of ostracods recovered from the Matulla Formation in the Abu Zeneima area, which includes *Bythocypris windhami, ovocytheridea caudate, O. reniformis, Brachycyrhere ledaforma, Pterygocythere raabi, Metacytheropteron berbericum, Cythereis rawashensis kenaanensis, C. r. silicea, Anticythrereis gaensis,* and *Loxoconcha striata,* have been given a Coniacian-Santonian age by Ismail (1993).

The Upper Turonian-Coniacian is missing in the Gebel Musabaa Salama area, Western Sinai (Kassab & Ismael, 1994), and only the Santonian part is present, as in the case of the Matulla Formation type section. The Santonian part is composed of cross-bedded sandstones with gypsum bands, and it contains the gryphoid *Pycnodonte* in its lower part. At Gebel Magmar (29°28'N, 33°30'E), the Matulla Formation is 52 m (171 ft) thick, and is believed to be Turonian-Santonian in age by Lewy (1975). At Gebel Minsherah the Matulla Formation is 148 m (486 ft) in thickness and is made up of a lower marly sequence and an upper chalky limestone sequence with glauconitic interbeds. El-Dawy (994) divided the Coniacian-Santonian Matulla Formation exposed at Wadi El Seig (29°29'N, 33°08'E), west-central Sinai into three planktonic foraminiferal zones: *Dicarinella asymetrica Zone* (Santonian), *D. concavata cyrenaica* Zone, and *D. concavata concavata* Zone (Coniacian), and two benthonic biozones: *Discorbis simplex* Zone (Santonian) and *D. bakrensis* Zone (Coniacian). The occurrence of *Dicarinella asymetrica* marks the Coniacian/Santonian boundary (Ayyad et al., 1996, Kassab & Obaidalla, 2001, Obaidalla & Kassab, 2002). The boundary is also marked by a regional relative sea-level fall (El-Azabi & El-Araby, 2007).

The Matulla Formation is equivalent to the Zihor Formation and Menuha Formation in the Arif Al Naga outcrops (Lewy, 1975, Bartov et al., 1980), and the Abu Roash Formation (*s.s.*) near Cairo. The Zihor Formation (Late Turonian-early Late Coniacian) consists mainly of chalky bioclastic limestones and marls, with ammonites such as *Roemeroceras parnesi, R. tunisiense* and *Barroisiceras onilahvense* in the upper part. The uppermost part contains *Pycnodonte vesicularis* and fish teeth. The Menuha Formation (topmost Coniacian-Santonian) is made up of a massive chalk sequence, up to 180 m (591 ft) in thickness. It was divided by Bartov et al. (1980) into a lower chalk and an upper chalk separated by a marlstone bed containing phosphatic grains and corroded oysters, fish teeth and *Spinaptychus spinosus,* which indicate an unconformity. The marlstone bed and the lower chalk unit are probably Campanian in age.

The Matulla, Raha and Abu Qada sandstones produce oil from the Belayim Marine, October and Ras Budran fields, in the Gulf of Suez. The Wata and Matulla reefal carbonates are the main reservoir in the Sadot Gas Field and the Raad discovery in NW Sinai (Alsharhan & Salah, 1996).

### 7.5.2.2   Cenomanian-Santonian of the central Nile Valley and the Eastern Desert

At Wadi Qena, Eastern Desert, the Cenomanian to Maastrichtian stratigraphic column is made up of the following formations, from base to top: Wadi Qena Formation, Galala Formation, Umm Omeiyad Formation, Qoseir Formation, Duwi Formation, and Sharawna Shale. This is one of the most difficult and complicated successions of the Phanerozoic sediments, due to intense faulting of the area. The section has been examined by a large number of workers and the stratigraphy is still controversial and unclear.

Cenomanian sediments continue into the Eastern Desert in the Wadi Qena, in the Nile Valley, and at the foot of El Galala el Bahariya escarpment along the Red Sea coast, and contain the characteristic fossils *Ostrea flabellata* and *O. africana*. In the central parts of Wadi Qena, the Cenomanian is represented by the Wadi Qena and Galala formations. The Wadi Qena Formation (Klitzsch & Wycisk, 1987) is made up of massive fluviatile and deltaic sandstones, associated with mottled paleosols and root horizons (Hendriks et al., 1990). Its thickness varies between 10 m (33 ft) and 64 m (210 ft). It is overlain by marine strata of the Upper Cenomanian Galala Formation (Klitzsch, 1990b). Along the El Sheikh Fadl-Ras Gharib road, in the northeastern Desert, the Wadi Qena Formation overlies basement rocks directly (Kassab, 1994). Klitzsch (1990b) reported the ammonites *Angulithes mermei*, *Metengonoceras dumpli* and *Neolobites brancai*. Klitzsch et al. (1990) and Klitzsch (1990b) assigned it a Lower Cenomanian age and correlated it with the Sabaya Formation (Lower Cenomanian). Bandel et al. (1987) applied the name Dakhal Formation to these sediments in Wadi Qena. The Dakhal Formation is equivalent in part to the Wadi Qena Formation and the Baharia Formation.

The Galala Formation[6] (Klitzsch & Wycisk, 1987) overlies the Wadi Qena Formation and it is overlain with an erosional contact by the Umm Omeiyad Formation in the central parts of the Eastern Desert, such as at Wadi Qena. Its thickness varies from 10 m (33 ft) to 64 m (210 ft) (Hendriks et al., 1990). The Formation is rich in ammonites. The Lowermost ammonite horizon contains *Neolobites vibryeanus* and *Neolobites* sp., and the nautiloid *Angulithes mermei*, overlain by sandy limestones with *N. vibrayeanus* and *Pseudocalycoceras* cf. *haugi (Pervinquiere)*, together with *A. mermei*. The upper part of the Galala Formation is made up of yellow, massive marls and marly limestones, containing hermatypic corals at the base, with *Metengonoceras* cf. *acutum*. The Galala Formation, along the El Sheikh Fadl-Ras Gharib road, in the northern Eastern Desert, consists of fossiliferous limestones, marly limestones, and dolomitic limestones, containing ammonites, nautiloids, oysters, gastropods, echinoids, and ichnofossils, with intercalations of poorly fossiliferous to unfossiliferous sandy shales and shales. Kassab (1994) recognized five ammonite zones, *Neolobites vibrayeanus* (I), *Meoiceras geslinianum* (II), *Vascoceras cauvini* (III), *Pseudaspindoceras flexuosum* (IV), and *Choffaticeras segne* (V). He assigned the Galala Formation an Upper Cenomanian-Lower Turonian age. The Galala Formation is equivalent to the Atrash Formation and the overlying Tarfa Formation of Bandel et al. (1987) in the Wadi Qena area and the El Heiz Formation.

The Umm Omeiyad Formation (Klitzsch & Wycisk, 1987) is composed of fluvial sandstones in the lower part, followed by shallow marine glauconitic sandstones, with *Coilopoceras requienianum*, and then by marly limestones in the upper part. Its thickness ranges from 21.5 m (71 ft) to 44 m (144 ft). It was assigned a Turonian age by Hendriks et al. (1990) and a Late Turonian age by Klitzsch (1990b). The Umm Omeiyad Formation is overlain by the Qoseir Formation. Kassab (1994) recognized two ammonite zones in the Umm Omeiyad Formation along the El Sheikh Fadl-Ras Gharib road, in the northern Eastern Desert, *Coilopoceras requinianum* (VI) and *Metatissotia fourneli* (VII). He assigned the Formation an Upper Turonian-Middle Coniacian age, based on these ammonites. Bandel et al. (1987) assumed that the Umm Omeiyad Formation is missing in the Wadi Qena area.

---

6 Not to be confused with the Southern Galala Formation (Paleocene-Ypresian) at Wadi Araba, Gulf of Suez (Kuss et al., 2000, Boukhary et al., 2009).

### 7.5.2.3　Cenomanian-Santonian of Southern Egypt

Upper Cretaceous (Cenomanian-Santonian) sediments in southern Egypt are composed mainly of sandstones which were included in the "Nubian Sandstone" by earlier workers. Recent work however has shown that these sandstones can be divided into several distinct mappable units, hence eliminating much of the confusion in the Egyptian geology which resulted from the use of the term Nubian Sandstone. Klitzsch et al. (1979) suggested the use of the name "Nubia Group", to include Jurassic to Late Cretaceous rocks overlying the Precambrian basement or Paleozoic beds and underlying Lower Maastrichtian or Tertiary sediments in Southern Egypt. This succession is more than 2000 m (6562 ft) in thickness. They rejected Issawi's (1973) choice of the Barget el Shab as a type section for the Nubian, because, according to them, the rate of sedimentation was slow in the upper Nile region during Mesozoic time and periods of erosion or non-deposition were common, therefore, the section is incomplete.

In the area between Qena and Aswan the Upper Cretaceous succession unconformably overlies Precambrian basement. The sequence is made up of the Abu Agag, Timsah, and Umm Barmil formations.

The Abu Agag Formation (El-Naggar, 1970) overlies basement in the area between Qena and Abu Simbel, and is overlain by the Timsah Formation. Basal conglomerates fill the irregular topography of the basement and grade upward into coarse-grained, cross-bedded sandstones with tetrapod trackways (Demathieu & Wycisk, 1990). The uppermost part of the Abu Agag Formation consists of paleosols and channel sandstones. The Formation was assigned a Turonian age by El-Naggar (1970). There is no paleontological proof for the age of the Abu Agag Formation.

The Timsah Formation (El-Naggar, 1970) consists of marine oolitic iron-bearing shales. In the Aswan-Abu Simbel area, the Formation varies in thickness between 5–50 m (16–164 ft). The Formation was divided by Hendriks et al. (1990) into a "lower marine facies" of deltaic, beach, distributary channels, lagoons, levees and pond sediments, and an "upper emergence facies" of intensely mottled paleosols. South of Aswan, the Formation is replaced by backshore and shoreface sediments. A Coniacian-Santonian age was established for the Timsah Formation based on palynological studies (Sultan, 1985), and the presence of *Inoceramus balli*. The Formation is overlain by the Umm Barmil Formation.

The Umm Barmil Formation (El-Naggar, 1970) attains a thickness of 15–50 m (49–164 ft), and is made up of alluvial-plain deposits which grade northwards into deltaic sand bars and tidal channel deposits of the "Hawashya Formation" (Hendriks et al., 1990) or the Kiseiba Formation (Campanian-Maastrichtian). The Formation was assigned a Santonian-early Campanian age by El-Naggar (1970).

The Upper Cretaceous (Cenomanian-Santonian) succession on both sides of the Nile Valley is similar and consists of the Sabaya Formation, Maghrabi Formation, and Taref Formation. In the southern Western Desert, the sequence overlies the Abu Ballas and Six Hills formations or the Gilf Kebir Formation of Jurassic-Lower Cretaceous age in the Gebel Kamil area. In the Eastern Desert, the Upper Cretaceous succession crops out on the western side of the Red Sea hills, rests unconformably on Precambrian basement (Van Houten et al., 1984, Ward & McDonald, 1979), and in fault blocks on the eastern side of the mountains in the Safaga-Qoseir area (Van Houten et al., 1984).

The Sabaya Formation (Barthel & Boettcher, 1978) is equivalent to the Desert Rose unit of Klitzsch et al. (1979). The typical area of the Sabaya Formation is at Sabaya, near km. 100 of the Dakhla-Kharga road (25°14′N, 29°43′E) in the Western Desert (Klitzsch et al., 1979)[7]. Its thickness varies from 30 m (98 ft) to more than 300 m (984 ft) and wedges out both eastwards and westwards. The Sabaya Formation makes up the higher parts of the Gilf Kebir Plateau and Gebel Kamil. It rests on the Gilf Kebir or the Abu Ballas formations with erosional contact (Wycisk, 1990, Klitzsch et al., 1979), and it is conformably overlain by the Maghrabi Formation (Barthel & Hermann-Degen, 1981). The Sabaya Formation contains relatively rare plant remains like *Frenelopsis* aff. *parceramosa*, *F.* aff. *ramossissima*, *Paradoxopteris stromeri*, and *Phlebopteris* sp. It is composed of medium- to coarse-grained, tabular and trough cross-bedded, kaolinitic sandstones (Wycisk, 1990, Klitzsch, 1990b, Klitzsch et al., 1979). Mottled horizons are common and may represent paleosols (Hendriks et al., 1990).

In the Eastern Desert, the Sabaya Formation is equivalent to the "trough cross-bedded conglomeratic sandstone unit" of Ward & McDonald (1979) and "Facies 1" of Van Houten et al. (1984). It unconformably overlies Precambrian basement. Its thickness is up to 100 m (328 ft). The Formation is made up of brown to red brown, trough cross-bedded intervals which grade upward into ripple-laminated fine-grained sandstones and siltstones. Paleosol horizons are common. The only fossils present are silicified and hematized wood fragments. In the Aswan area, the unit interfingers with basaltic lava flows (Van Houten et al., 1984, Ward & McDonald, 1979). This unit is absent in the Safaga area (Ward & McDonald, 1979). The age of the unit is not precise. It was assigned an Albian to Cenomanian age by Van Houten et al. (1984), Lower Cenomanian by Klitzsch (1990b), Lower Cretaceous by Klitzsch et al. (1979), and Albian by Wycisk (1990). An Albian-Cenomanian age seems reasonable, based on corrrelations with SE Libya.

The Maghrabi Formation (Klitzsch & Wycisk, 1987) is equivalent to the "Plant Bed unit" of Klitzsch et al. (1979). For a brief time it was called the Kharga Formation by Klitzsch (1978, in Barthel & Boettcher, 1978). The typical area of the Formation is between Gebel Taref near Kharga and the southern edge of Abu Tartur in the Western Desert (25°21′N, 29°37′E). Its thickness varies from several meters to 200 m (656 ft). It wedges out eastwards and westwards. The Maghrabi Formation is composed of fine-grained sandstones and shales, with horizontal and ripple cross-laminations, and plant remains. Continental sediments rich in paleosols occur south of Gebel Kamil and cover parts of the Gilf Kebir Plateau (Hendriks, 1986). A rich flora, including *Credneria* sp., cf. *Celastrophyllum* sp., *Ficophyllum* sp., *Laurophyllum* sp., cf. *Liridendropsis* sp., *Magnoliaephyllum* sp., and cf. *Sassafras* sp. was reported by Hendriks (1986). It was assigned a Late Cenomanian-Early Turonian age by Hendriks (1986) and Klitzsch (1990). In the Abu Tartur area, the Maghrabi Formation contains brachiopods (*Lingula* sp.), lamellibranchs, and remains of fish (Klitzsch, 1990b, Klitzsch et al., 1979). The presence of tricolpate

---

7 Hermina (1990) mentions the type locality at Qulu El Sabaya hills on the Kharga-Dakhla road (25°21′N, 29°43′E), where the unit measures 170 m (558 ft) in thickness.

and tricolporate miospores in the Dakhla Oasis area and the absence of triporates led Mahmoud & Soliman (1994), and Schrank & Mahmoud (1998a, b) to assign the Maghrabi Formation an Albian-Early Cenomanian age. However, this age dating would make the Formation time equivalent to the Sabaya Formation. Triporates are present in the Kharga area, and the age of the Magharbi Formation may range from late Cenomanian to Turonian (Schrank & Mahmoud, 1998a, b). This age appears to be more reasonable, since the the Maghrabi Formation overlies the early Sabaya Formation.

The Maghrabi Formation is equivalent to the "grey and red shale and thin sandstone unit" of Ward & McDonald (1979) and "Facies 2" of Van Houten et al. (1984) in the Eastern Desert. Its thickness is approximately 50 m (164 ft). It is made up of several cycles of grey, red, and purple shales and sandstones, which grade upward into ripple-laminated siltstones, then into trough cross-bedded, fine sandstones. Thin layers of red-brown chamositic-hematitic oolitic ironstones associated with phosphatic and conglomeratic lag deposits are common (Van Houten et al., 1984, Ward & McDonald, 1979). Freshwater species of the bivalve *Unio*, and the marine bivalves *Inoceramus balli, Cyprina humei*, and *Isocardia aegyptiaca* (Abbass, 1961) were recovered from this Formation. Van Houten et al. (1984) assigned a Cenomanian age to this unit.

The Taref Formation was established by Awad & Ghobrial (1966). The typical area is on the southwestern slope of Gebel Taref near Kharga Oasis, in the Western Desert. It disconformably overlies the Maghrabi Formation and is conformably overlain by the Qoseir Formation (variegated shales). Issawi (1972) considered the Taref Formation and the Qoseir Variegated Shale as two members of the "Nubia Formation" in the Kurkur-Dungul area. The Taref Formation is a few meters to 130 m (417 ft) thick and is composed of medium to coarse-grained sandstones with small to large scale cross-laminations, occasional paleosols and root horizons (Wycisk, 1990, Hendriks et al., 1990).

The Taref Formation corresponds to "Facies 3" of Van Houten et al. (1984) and the "tabular cross-bedded sandstone unit" of Ward & McDonald (1979) in the Eastern Desert. These authors also pointed out the correlation between their units and the Taref Formation. It is composed of large scale planar cross-bedded, medium- to coarse-grained brown sandstones, ranging from quartzarenites to subarkoses, rippled and burrowed at the top. Petrified wood and plant remains are common in this unit. The base of the unit is scoured and contains quartz pebbles, shale rip-up clasts, teeth, bone fragments, and oolitic ironstones. The Formation has a thickness of 120 m (394 ft) (Van Houten et al., 1984). The Taref Formation was assigned a Turonian age by Barthel & Boettcher (1978) and Klitzsch et al. (1979), a Late Turonian-Coniacian age by Van Houten et al. (1984), and a Santonian-Campanian age by Ward & McDonald (1979). However, the latter included the Qoseir Variegated Shale Formation in this unit. The Taref Formation is equivalent to the Tarfa Formation (Turonian) of Bandel et al. (1987).

Late Turonian to Middle Campanian strata are eroded in the area immediately west of the Nile. The Taref and Maghrabi formations are missing (Hendriks, 1986).

### 7.5.3   Campanian

Campanian sediments in Egypt are represented by two main formations, the Qoseir Variegated Shales and the Duwi Phosphate Formation. In the subsurface of the northern Western Desert, Campanian sediments are included in the basal part of the Khoman Formation. In Sinai and the Gulf of Suez region these two formations become predominantly chalks and no acceptable stratigraphic nomenclature has been established in those areas. Some of the proposed formation names are Gebel Thelmet Formation, St. Anthony Formation, St. Paul Formation, Menuha Formation, Mishah Formation, and the lower part of the Sudr Formation.

The name Qoseir Formation was given by Youssef (1957) to the variegated shales in the Qoseir area, near the Red Sea coast. In that area, the Formation is composed of blue-grey, yellow-brown, ripple-laminated shales, siltstones, fine-grained sandstones, and oyster-rich lenses. Hermina (1990) divided the Qoseir Formation in the Dakhla area into two units, a lower brick red unit, 30 m (98 ft) in thickness, composed of red and green shales and white, very fine sandstones, and rare current-rippled sandstones, and an upper unit made up of alternating ferruginous-glauconitic sandstones and brown and grey sandy clays, 20–35 m (66–115 ft) thick. The two units are separated by a distinctive green siltstone bed. At Abu Tartur, the upper part consists of massive red mudstones with scattered carbonate nodules. The lower part consists of bedded, locally coarse-grained, very fine sandstones, with plant fragments and large burrows (Klitzsch et al., 1979, Hendriks et al., 1987). Its thickness in Gebel Kamil area ranges from 50 m (164 ft) to 100 m (328 ft). It conformably overlies the Taref Formation (Turonian) and it is conformably overlain by the Duwi Phosphate Formation (Campanian). The age of the Qoseir Formation is probably Campanian. In the Bahariya Oasis, the Formation may be represented by a thin variegated shale interval on top of the Abu Roash Formation.

In the Dakhla-Farafra area, the Qoseir Formation (Campanian) is known under different names. It was called the Umm Barmil Formation by El-Naggar (1970) and the Mut Formation[8] by Barthel and Herrmann-Degen (1981). In that area, the Qoseir Formation has a thickness of 70–80 m (230–262 ft) and consists of claystones and sandstones at the base, grading upward into coarsening-upward sequences composed of varicolored, mottled, silty and sandy claystones. The claystones and sandstones at the base contain faunal and floral remains (Hendriks et al., 1990). In the Farafra area, a 35 m (115 ft) thick clay and sandstone sequence forms the core of an anticlinal feature in the northeastern part of the depression between Wadi Hennis and Maqfi. This sequence was named the Wadi Hennis Formation by Dominik (1985) and is considered to be the stratigraphic equivalent of the Qoseir Formation of Upper Campanian age. It had been assigned a Cenomanian age by previous authors.

The name Hawashya Formation was applied to rocks equivalent to the Qoseir Formation by Kallenbach & Hendriks (1986), but Klitzsch & Wycisk (1987), Klitzsch (1990b), and Hendriks et al. (1990) extended its age from Coniacian to Early Campanian. This unit consists of alternating regressive sandstones and transgressive sandstones and limestones. Massive oyster beds with *Metatissotia founeli*, *M*. cf. *ewaldi*, *M*. sp., and *Subtissotia africana*, and minor limestone beds with pelecypods and gastropods occur in places. The Hawashya Formation is equivalent to the Qoseir Formation of Youssef (1957). The latter has priority and will be used throughout the present work.

---

8 The Mut Formation is equivalent to the Umm Barmil Formation, and both are equivalent to the Qoseir Formation which has priority. The latter will be used throughout the present work.

In St. John's (Zabargad) Island (23°37′N, 36°12′E), located in the Red Sea, a 150–200 m (492–656 ft) succession consisting of alternating strata of silicified limestones and quartzitic sandstones with intercalations of black shales and phosphatic beds, crops out above Precambrian metamorphic rocks (Bonatti et al., 1983). The limestones are about 2–3 m (7–10 ft) thick and they are composed of silicified micrites with scattered pyrite crystals. The sandstones are 2–3 m (7–10 ft) thick and consist mainly of massive, medium- to coarse-grained clasts of quartz. The black shales are finely laminated; 20 cm–3 m (0.7–10 ft) thick, and they contain quartz, muscovite, chlorite, and feldspar grains. Lenses of phosphates occur with the succession. El Shazly & Saleeb (1977) applied the name Zabargad Formation to this succession. The age of this Formation is unknown. It overlies Precambrian metamorphic rocks and underlies Miocene evaporites. The phosphorites include fossil fish, such as the teleostean (jawed fishes) *Biryciformes* group of Cretaceous or Paleocene age. Bonatti et al. (1983) assigned an upper Cretaceous and/or Paleocene age to the succession and correlated it with the succession cropping out in the Qoseir area. The latter includes Qoseir, Duwi, and Sudr formations, which range in age from Campanian to Maastrichtian.

At Arif el Naga, central Sinai, the Campanian is represented by the Menuha and Mishash formations (Bartov et al., 1980). The two formations are equivalent to the Duwi and Qoseir formations, respectively, and they were included in the "Chalk" by Said (1962). They show rapid facies variations from a chert and chalk sequence to bioclastic phosphatic and glauconitic, partly silicified limestones with oysters, such as *Pyconodonte vesicularis*, *Lopha villei*, *Lopha*. cf. *morgani* and fish teeth, intercalated with chalk and gypsum. Close to the southern flank of the uplift, the two formations are made up of conglomerates and bioclastic limestones up to a few meters in thickness, with Cenomanian and Turonian limestone and dolomite pebbles.

The name Duwi Formation was introduced by Youssef (1957) for a carbonate and phosphate unit in the Qoseir area composed of limestones, marls, and three horizons of economically exploited phosphate rocks (Fig. 7.4). Oyster limestone beds rich in *Ostrea villei* biostromes occur in the upper part of the Formation. The phosphatic beds contain vertebrate remains, coprolites, and fish teeth. The marls contain pelecypods, gastropods, and cephalopods, e.g. *Libycoceras ismaeli*. In the Safaga area to the north, the shales interbedded with the phosphate horizons contain *Pecten farafrensis* and *Exogyra overwegi*, as well as *Libycoceras ismaeli* (Said, 1962). The Duwi Formation overlies the Qoseir Formation and it is overlain by the Sharawna Shale. The Duwi Formation is of Late Campanian-Early Maastrichtian age (Hendriks et al., 1987, Schrank, 1984, Hendriks, 1985, 1986). It was assigned a Middle Campanian to Lower Maastrichtian age by Luger & Gröschke (1989) based on ammonites. This Formation and its equivalents represent the main phosphorite-bearing sequence throughout Egypt (Glenn, 1990, Glenn & Arthur, 1990).

In the Red Sea coast area, the Duwi Formation consists of three phosphate members (Said, 1971) from top to bottom: The Atshan "A" Member, the Duwi "B" Member, and the Abu Shgeila "C" Member. The Atshan and Duwi members are separated by an oyster bed, and the Duwi and Abu Shgeila are separated by a shale-marl unit. The late Campanian-early Maastrichtian age of the phospahes in Egypt has been confirmed by Hendricks & Luger (1987), Luger & Gröschke (1989), and Ganz et al. (1990). The ammonites from the lower part of the phasphates of the Duwi Formation,

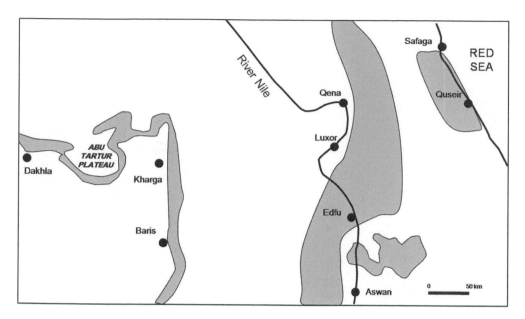

*Figure 7.4* Distribution of phosphate deposits in Egypt (after Germann et al., 1987, Awadalla, 2009).

the Hamadat "C" bed or Abu Shegeila bed in the Quseir-Safaga district are of late Campanian age, whereas the middle part, the Duwi "B" bed, is late late Campanian. The uppermost part of the Atshan "A" bed yielded *Rosita* cf. *patelliformis, Achae-oglobigerina blowi, Rugoglobigerina rugosa,* and heterohelicids (Ganz et al., 1990), which assigns it a late Campanian-early Maastrichtian age.

In the Dakhla, Kharga, and Abu Tartur areas in the Western Desert, the Duwi Formation shows wide variations in thickness. It overlies the Qoseir Formation and underlies the Sharawna Shale. It consists of phosphate beds interbedded in a sequence of alternating claystones, sandstones, glauconitic sandstones, siltstones, and conglomerates. The oyster bed is absent and the three phosphate members are separated by shale beds. The Duwi Formation in these areas was divided into two phosphate-bearing horizons by Hermina (1990), a lower A-Horizon and an upper B-Horizon, and a calcareous cap rock of chalks and marls of restricted occurrence. The latter is equivalent to Qur El Malik Member of the Dakhla Formation of Barthel & Hermann-Degen (1981). In the Kharga area, the A-Horizon is either missing or is less than 1 m (3 ft) thick. In the Abu Tartur area, the A-Horizon, probably equivalent to the Abu Shgeila "C" Member, is developed into an economically important phosphate bed 4.5–5 m (15–16 ft) thick. The late Campanian/early Maastrichtian Phosphate (Duwi) Formation in the Western Desert of Egypt represents a heterogenous marine facies that involve marine phosphorites, black shales and glauconitic sandstones. This group of facies represents the initiative stages of marine flooding over the study area (Rifai & Shaaban, 2007). Glenn (1990) postulated that both phosphorites and glauconitic greensands are the result of current winnowing and reworking of authigenic grains from previously deposited sediments.

*Figure 7.5* St. Anthony and South Galala formations, St. Paul Monestary, Wadi Araba.

The Abu Tartur phosphorites (Fig. 7.4) are considered the highest grade phosphates in Egypt (Awadalla, 2009). The REE in the Maghrabi-Liffiya Phosphates of Abu Tartur area suggest deposition on a broad shallow shelf during the Upper Cretaceous Tethyan transgression. The REE distribution and the presence of pyrite as diagenetically produced framboids within the phosphatic bone fragments and within the teeth and coprolites grains suggest a relatively suboxic to anoxic environment. The Abu Tartur phosphorites (Fig. 7.5) show a $P_2O_5$ content ranging from 19.8 to 29.8% with an average 24.6%, while showing a low U content ranging from 15 ppm to 34 ppm with an average 22 ppm (Awadalla, 2009).

In the Wadi Qena, Bandel et al. (1987) grouped the Qoseir and Duwi formations as members in what they called the "Abu Had Formation". The Duwi Formation consists of two phosphatic sequences rich in ammonites separated by unfossiliferous shales. The Duwi Formation in the southern Wadi Qena is characterized by sand-filled channels which are cut into bituminous shales, marls, and silts of the Qoseir Formation. These channels are believed to be the result of synsedimenaty strike-slip-fault movements (Bandel et al., 1987). The upper part of the Duwi Formation is up of hard phospates, with sandstone, shale, and siltstone intercalations, and bonebeds and limestone pebble layers with ammonites. In the central and northern Wadi Qena, the Qoseir and Duwi formations cannot be differentiated. These sediments were called the Rakhiyat Formation[9] by Kallenbach & Hendriks (1986). The thickness of the Duwi Formation in the Wadi

---

9  The Rakhiyat Formation is equivalent to the Duwi Formation which has priority. The latter will be used throughout the present work.

Qena is 1.5–70 m (5–230 ft). Hammama & Kassab (1990) reported thicknesses of 75 m (246 ft) at Wadi Hamama and 105 m (345 ft) at Gebel Abu Had. Ammonites include *Canadoceras cottreaui, Manabolites piveteaui,* and *Baculites ovatus* in the lower part and *Nostoceras* sp., *Solenoceras humei, Baculites subanceps,* and *Libycoceras ismaeli* in the upper part (Luger & Gröschke, 1989). The Late Campanian Qoseir Formation is separated from the Maastrichtian Duwi Formation by an erosional unconformity with conglomeratic and phosphatic marls containing the pelecypod *Grypghea vesicularis,* vertebrate remains, and planktonic foraminifera of the *Globotruncana falsostuarti* zone. Hammama & Kassab (1990) divided the Duwi Formation in the Wadi Qena into three units. The lower phosphatic unit contains *Libycoceras ismaeli.* The middle unit is composed of shales and cross-bedded siltstones and iron bands. No ammonites were recorded from this unit, but it contains the forams *Trochammina wickendeni, T. undulosa, T. borealis, Haplophragmoides galbra, H. rota,* and *Lituola taybrensis.* The upper phosphatic unit is composed of marly phosphatic beds, sandstones, marls, and limestones. The top 20 cm contain the ammonites *Bostrycoceras polyplocum, Solenoceras humei, Baculites scotti,* and *Exiteloceras unciforme,* in addition to other species belonging to the genera *Bostrycoceras, Nostoceras, Colenoceras,* and *Baculites.*

Phosphatic lenses near the base of the Sharawna Shale (Dakhla Shale according to Issawi, 1972) in the Kurkur-Dungul area may represent the southern extension of the Duwi Formation.

In the subsurface of the Northern Egypt Basin (northern Western Desert), Campanian sediments are 157 m (515 ft) in thickness and consist of highly fossiliferous limestones characterized by fauna of the *Globotruncana fornicata* Zone such as *G. lapperenti, G. tricarinata,* and the *G. elevata* Zone (Metwalli & Abd el-Hady, 1975). This interval is equivalent to the lower part of the Khoman Formation and it may include rocks equivalent to the Qoseir and Duwi formations, although the whole section is made up of carbonates. In westcentral Sinai and the Gulf of Suez region the Campanian is also represented by the lower part of the Sudr Formation.

The Campanian sediments in the subsurface of the Northern Egypt Basin do not show the source rock quality of either the Brown Limestone in the Gulf of Suez or the Sirte Shale in the Sirte Basin in Libya.

The succession across the Upper Cretaceous-Tertiary boundary includes in an ascending order, the Sharawna Shale, and its lateral equivalent the Khoman and Sudr formations, Dakhla Shale, Tarawan Chalk, Esna Shale, Ypresian Shales, and the Thebes-Farafra Formation.

The classification and correlation of the sedimentary units which straddle the Cretaceous-Tertiary boundary has been in controversy for a long time. In many places in Central and Upper Egypt this boundary occurs within thick shale sections with no discernible lithological breaks, although it is characterized by the appearance of new taxa (Morsi & Scheibner, 2009). Similar shales that occupy apparently the same stratigraphic position were found to be of different ages. The idea that these sediments were deposited on a stable shelf (Said, 1961, 1962) underplayed the effect of tectonics on the distribution of these units. In fact, Cretaceous and Tertiary shale sections of more than 190 m (623 ft) are believed to be missing in places, either due to erosion or non-deposition as a result of tectonic movements. In the meantime, successions that were believed to represent continuous sedimentation across the Late Cretaceous-Tertiary boundary have been shown to contain unconformities along this boundary.

The Maastrichtian is represented over most of Egypt by two lithofacies: chalks of the Khoman and Sudr formations, and shales of the Sharawna Shale Formation. Dolomites and sandstones occur locally.

### 7.5.3.1   Maastrichtian chalks

The term "Chalk" was applied by Said (1962) and many other authors to any Cretaceous or Paleocene chalks or chalky limestones, which again resulted in confusion in regional correlation. As defined, this unit was assigned different ages in different places (Said & Kenawy, 1956, Nakkady, 1957, Said & Kerdany, 1961, Said, 1961, 1962).

The name Farafra Chalk was given by Youssef & Abdel Aziz (1971) to the soft white chalks that make up the floor of the depression in the Farafra Oasis. However, this name may be confused with the Eocene Farafra Limestone of Said (1960), therefore, it should be abandoned. In the Farafra Oasis, the base of this section is not exposed. The exposed section north of the Ain El Wadi (Wadi Hennis) depression, Farafra Oasis, is more than 160 m (525 ft) in thickness. The base of this exposure is Lower Maastrichtian. The upper part of the chalk (exposed in the Maqfi section and north of Gunna) is Danian (Youssef & Abdel Aziz, 1971) and probably belongs to the Dakhla Shale. This unit was assigned a Maastrichtian age by Leroy (1953) in the Farafra Oasis. Foraminifera recovered from this unit by Youssef & Abdel Aziz (1971) are *Bolivinoides draco draco*, *Globotruncana fornicata cesarensis*, *G. fornicata fornicata*, *G. fornicata ackermani*, *G. gansseri*, *G. gagnebini*, *G. aegyptiaca aegyptiaca*, *G. leupoldi*, *Heterohelix navaroensis*, *H. reussi*, *H. striata*, *Pseudotextularia* cf. *elegans*, *Rugoglobigerina rugosa*, and *Pseudoguembelina exaltata*. This assemblage belongs to the *G. aegyptiaca* and *G. gansseri* zones of Middle to Upper Maastrichtian age. The Uppermost Maastrichtian *Abathomphalus mayaroensis* Zone is missing. The Farafra Limestone is equivalent to the Khoman Formation in the Abu Roash, Bahariya Oasis, and the subsurface of the Western Desert and to the Sudr Formation in Wadi Qena and Sinai. The name Farafra Chalk is redundant.

The Khoman Formation was introduced by Norton (1967). The designated type locality is located at Ain Khoman (28°07′N, 28°45′E), southwest of Bahariya Oasis. In its type locality, the Formation consists of more than 50 m (164 ft) of highly fossiliferous, fractured chalky limestones. The contact between the Khoman and Abu Roash formations is erosional. It is equivalent to the Campanian Chalk of Beadnell (1902) in the Abu Roash area.

The Khoman Formation in the subsurface of the Western Desert has an average thickness of 244 m (5800 ft) (Hosney 2005) and composed of fine-grained limestones and shales characterized by the presence of *Globotruncana stuarti* (Metwalli & Abd el-Hady, 1975). The Formation is unconformably overlain by Paleocene sediments. The Khoman Formation was subdivided by Aadland & Hassan (1972) into an upper Khoman "A" and a lower Khoman "B" members. The "A" Member is made up of chalky limestones, and the "B" Member consists of shales and calcareous siltstones. The age of the Khoman Formation is Campanian-Maastrichtian. Planktonic foraminifera from the Khoman Formation, in the area between the Farafra and Bahariya oases belong to the middle-late Maastrichtian *Globotruncana aegyptiaca* and *Gansserina gansseri* zones (Samir, 1994). The Khoman Formation in the Badr El Dein-1 (Ded 1-1) well overlies the Abu Roash Formation and is composed of chalky limestones and shales. The lower part was assigned a Campanian-Maastrichtian age by El-Beialy (1994), based on the palynomorphs *Dinogymnium acuminatum*, *D. euclaensis*, *D.* cf. *westralium*, and *Isabelidinium cooksoniae*. Schrank (1987) reported a similar assemblage from the Maastrichtian of the Kharga Basin. This assemblage is considered most likely to be of Maastrichtian age (Rasul, person. Comm. 1998). The upper part is devoid of palynomorphs. On the other hand, Obaidalla & Kassab (1998) assigned the Khoman Formation a Maastrichtian-Danian age based on the presence of the Maastrichtian fauna *Globtruncana aegyptiaca*, *Gansserina gansseri*, and *Abathomphalus mayaroensis*, and the Danian foraminifera *Guembelitria cretacea* and *Globoconusa conusa*. Reworked foraminifera are abundant in the earliest Paleocene. The Danian part of this Formation is equivalent to the Danian chalks of the Dakhla Shale, overlying Maastrichtian chalks in the Farafra Oasis (Youssef & Abdel Aziz, 1971).

In other parts of Central and Southern Egypt, the Khoman Formation changes facies laterally into shales of the Sharawna Shale of El-Naggar (1966) and Tawadros (2001). The Khoman Chalk interfingers with the Sharawna Shale to the northwest of Qena. The diversified *Globotruncana* assemblage in the Khoman Chalk unit indicates deposition in an open-sea environment (Said & Kerdany, 1961). In contrast, the Sharawna Shale Maastrichtian fauna (El Naggar, 1966) contain fewer planktonics and seem to have been deposited in a near-shore muddy facies (Said & Kerdany, 1961).

The name Sudr Formation was introduced by Ghorab (1961) after the type locality at Wadi Sudr (29°38′N, 32°44′E), west-central Sinai. This unit which overlies the Matulla Formation and underlies the Esna Shale, and is composed mainly of chalks, changing partly into marls or argillaceous to crystalline limestones. Ghorab (1961) assigned the Sudr Formation a Campanian-Maastrichtian age and divided it into a lower Markha Member and an upper Zeneima Member. However, these names have been used for two Miocene units in the Gulf of Suez and may cause confusion and should not be used. They are equivalent to the St. Paul and the Thelmet/St. Anthony formations, respectively. They are mentioned here for historical interest and ther paleontological information. The Markha Member (Campanian), named after its type locality at Markha (28°54′N, 33°12′E), east central Sinai, is composed of brown, hard, poorly bedded, cherty, partly crystalline to chalky or marly limestones, it was assigned to the "*Neobulimina canadinsis* zone". The Markha member is equivalent to the St. Paul Formation and the Khoman "B" Member. The Abu Zeneima Member (Maastrichtian), named after Abu Zeneima (29°03′N, 33°06′E), is made up of white, bedded, chalky and marly limestones, it was assigned to the "*Bolivinoides draco* zone". El-Shinnawi & Sultan (1972b) assigned the Markha Chalk Member to the "*Globotruncana stuarti*" and "*G. fornicata*" zones (Campanian). Twefik & Ebeid (1972) placed the Markha Chalk Member in the "*Lacostina maquawilensis*" and "*Globotruncana cesarensis*" zones (lower Campanian). El-Shinnawi & Sultan (1972b) designated the Abu Zeneima Chalk Member to the *G. gansseri* zone (Maastrichtian). The latest Maastrichtian *Globotruncana (Abathomphalus) mayaroensis* zone is missing, and the Maastrichtian sediments are unconformably overlain by Paleocene sediments. Twefik & Ebeid (1972) designated the Abu Zeneima Chalk Member to the "*Globotruncana rosetta*", "*G. fornicata fornicata*", and *G. gansseri* zones (upper Campanian-Maastrichtian). Cherif & Ismail (1991) detected a sedimentary gap within the Campanian and Maastrichtian succession of the Abu Zeneima Member at Esh el Mallaha-Gharamul area.

In the Southern Galala Plateau at Wadi Araba, Campanian-Maastrichtian rock units overlying the Matulla Formation are represented by the Sudr Chalk and its partly equivalents, the St. Paul Formation and the Gebel Thelmet Formation (St. Anthony Formation) (Fig. 7.5).

The Upper Campanian-Late Maastrichtian section near the Monastery of St. Anthony, Wadi Araba, was named Gebel Thelmet Formation by Abdallah & Eissa (1971) and assigned a Campanian age. Kuss (1986) divided the section into Upper Campanian-Early Maastrichtian and Late Maastrichtian units based on the presence of *Orbitoides* and globotruncanids. The lower unit is 120 m (394 ft) thick and consists of cyclic alternations of marly-sandy limestones and hard massive, partly dolomitic carbonates and sandy limestones, with *Orbitoides* bearing limestones, marly wackestones, and beds with baculites, ammonites, echinoids and oysters. The upper unit is 75 m (246 ft) thick and is composed of sandstones intercalated with marly-silty layers or clastic-dominated, occasionally dolomitic limestones with planktonic foraminifera of the *Globotruncana* group overlain by Paleocene limestones. The lower part of the Sudr Chalk section was called the St. Paul Formation (latest Campanian) by Kuss (1986), which is composed of 30 m (98 ft) of white, massive, finely crystalline, and dolomitic limestones. The base of the Formation is marked by a fine-grained, bioturbated, sandy horizon, and phosphate layers with black phosphate grains, phosphatized fragments, and fish teeth. A conspicuous layer with *Pycnodonte vesicularis* occurs about 10 m (328 ft) above the base. The St. Paul Formation unconformably overlies the Turonian Abu Qada and Wata formations in the Gulf of Suez region. In the Wadi Qena, the St. Paul Formation overlies the Duwi Formation or the Duwi Member of the Abu Had Formation of Bandel et al. (1987) and is overlain by the "Esna Shale" (Maastrichtian-Lower Eocene, according to these authors).

The upper unit of the Sudr (Chalk) Formation was called the St. Anthony Chalk (Upper Campanian-Maastrichtian) by Abdallah & Eissa (1971), after the Monastery of St. Anthony[10], Wadi Araba. The St Anthony Formation (Upper Campanian-Upper Maastrichtian) near the Monastery of St Anthony disconformably overlies chalks of the Sudr Formation and is overlain in turn by carbonates of the Southern Galala Formation (Paleocene-Lower Eocene). The upper part of the St Anthony Formation contains *Exogyra overwegi*. The base of the St. Anthony Formation is of Late Campanian age (Kuss, 1986, Kulbrok, 1996). The upper sandy-dolomitic units have been assigned a Maastrichtian age (Scheibner et al., 2001a, b). Abdel-Kireem & Abdou (1979) described Upper Maastrichtian planktonic foraminifera of the *Abathomphalus mayaroensis* Zone at Gebel Thelmet. The St. Anthony Formation is equivalent to the Thelmet Formation, but because the section at the St Anthony Monastery is more accessible and thicker than the type section of the Thelmet Formation of Abdallah & Eissa (1971), Scheibner et al. (2001a, b) favored the use of the name St. Anthony Formation. The Gebel Thelmet/St. Anthony Formation at Gebel Thelmet, Southern Galala, contains abundanant *Orbitoides media* and *Omphalocyclus macroporus*. The presence of *Orbitoides media* and *Omphalocyclus macroporus* assign the upper part of the Gebel Thelmet Formation to the Late Campanian (Ismail et al., 2007). In the Umm Khayshar section (on the norhetrn edge of the Southern Galala Plateau), this

10 St. Anthony is considered to be the father of monasticism; he was a contemporary and a friend of the elder St. Paul.

interval is followed upward by a horizon with Late Campanian-Early Maastrichtian foraminifera, such as *Haplophragmoides calculus, H. excavata, H. globulosa, H. hausa, Ammobaculites subcretaceus, A. texanus, Cyclammina cancellata, Heterohelix glabrans, Globigerinelloides prairiehillensis, Hastigerinoides subdigitata, Rugoglobigerina macrocephala, Bulimina kickapooensis, B. reussi, Discorbis beadnilli, Anomalina umbonifera* and *Gyroidinoides goodkoffi* (Ismail et al., 2007). The Gebel Thelmet/St. Anthony Formation in the St. Anthony Monastery area is overlain unconformably by the Tertiary Southern Galala Formation (Danian is missing, as well as the Maastrichtian to the west) (Kuss et al., 2000a, b, Ismail et al., 2007), whereas at St. Paul and Bir Dakhl to the south, its equivalent the Sudr Formation is overlain by the Danian shales of the Dakhla Formation (Kuss et al., 2000a, b).

Although the Khoman and Sudr Formations are equivalent, and they may be in fact the same formation, the present writer suggests that both names should be retained until correlations become better established. The use of the name Sudr Formation should be restricted to the Maastrichtian chalks in Sinai and the western side of the Gulf of Suez.

In the Arif el Naga area, northern Sinai, the Maastrichtian is illustrated by the Ghareb Formation (Bartov et al., 1980) composed mainly of chalks and some argillaceous chalks. A basal quartzitic pebble conglomerate is present at the base of the Formation. At Gebel Umm Mafruth (30°50′N, 33°35′E), on the northern flank of Arif el Naga, a rich fauna occurs in the Formation and includes ferruginous internal molds of solitary corals, bivalves, gastropods, echinoids, and crinoids. The Ghareb Formation is a homonym of the Sudr Formation and should not be used in that area. The Sudr Formation in the Gebel Somar in north-central Sinai has been described recently by Samuel et al. (2009) who assigned an early to late Maastrichtian age, belonging to the *Globotruncanella havanensis* Zone (early earlyMaastrichtian), *Globotruncana aegyptiaca* Zone (early Maastrichtian) and *Gansserina gansseri* Zone (late Maastrichtian).

### 7.5.3.2 Maastrichtian-Paleocene shales

Maastrichtian shales are common in Egypt, especially in the central part of the country. However, they were often misidentified and included with Paleocene shales as one unit, or treated as members of lithostratigraphic units that span the Late Cretaceous-Early Tertiary boundary. Maastrichtian shales are represented mainly by the Sharawna Shale Formation.

Paleocene shales are found in the Farafra, Dakhla, Kharga, Kurkur, and Dungul oases (Fig. 7.6) in the Western Desert, in the Nile Valley near Esna and north of Thebes, in the Eastern Desert north of Qena, and in Sinai. Approximately 100 m (328 ft) of Paleocene shales were encountered in the Natrun-1 well (30°23′N, 30°18′E). They are generally absent or very thin north of a line passing approximately south of Maghara, Suez, Cairo, and Bahariya Oasis. This line represents the boundary between the Nile Basin and the Northern Egypt Basin, or between the stable and unstable shelf areas. Where Paleocene deposits are absent, Cretaceous sediments are unconformably overlain by younger sediments.

The Paleocene is generally represented by two shale units, the Esna Shale (Landenian) and the Dakhla Shale (Danian), separated by a thin chalk unit, the Tarawan Chalk (Landenian). They are overlain by Lower Eocene limestones or Early Eocene

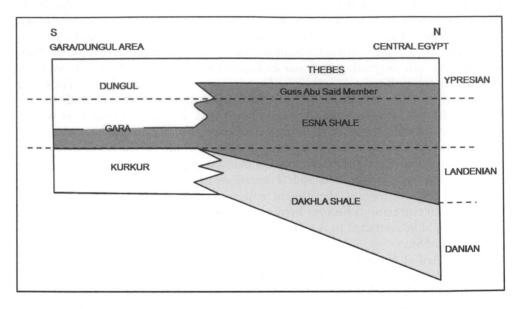

*Figure 7.6.* Possible correlation between the Paleocene-Ypresian succession in Central Egypt and the Gara/Dungul area (See color plate section).

shales (Guss Abu Said Member of Tawadros, 2001). The Dakhla Shale unconformably overlies the Maastrichtian Sharawna Shale or the Khoman and Sudr limestones. When shales overlie shales, the differentiation of the succession into mappable units may prove to be cumbersome without a detailed paleontological examination. The similarity of these shales, even with the help of fossils, has been the source of a long and laborious controversy.

Delanoüe (1868) gave a description of a geological section in the Thebes area and drew attention to a slope-forming, 31 m (102 ft) thick, highly fossiliferous unit of foliated marls, with abundant concretions and brown ironstone fossils, overlain by Eocene white fossiliferous chalky limestones. D'Archiac (1868) identified over 40 forms of small molluscs, echinoids, crinoids, and solitary corals from Delanoüe's collection and assigned the shales an Eocene age. Zittel (1883) subdivided the slope-forming shales underlying the chalk[11] and overlying (Qoseir) variegated shales in the Dakhla Oasis, into two units: an upper unit, the *"Aschgraue Blätterthone"*, and a lower unit rich in *Exogyra overwegi*, the *"Overwegischichten"*. The latter included phosphatic beds (of the Duwi Formation) at the base. Zittel (1883) also described the section at El-Guss Abu Said (27°40′N, 27°08′E), in the Farafra Oasis, and assigned the whole section to the Lower Eocene, based on the presence of *Nummulites deserti*, *N. fraasi*, and *Operculina libyca*. He also considered the upper unit to be equivalent to the foliated shales *"Blättermergel"* of the Thebes section. At the International Geological Congress in Paris in 1900, Beadnell, Barron and Hume, and Ball applied the name Esna Shale to these foliated shales. Barron & Hume (1900) divided the "Eocene" rocks in

---

11 This is Eocene limestone.

Wadi Qena into two units, an upper unit of limestones, 225 m (738 ft) thick, which they called the Serai Limestone (Ypresian), and a lower unit of shales, marls, and marly limestones, 122 m (400 ft) thick, which they considered to be the Esna Shale. Fourtau (1900) and Blanckenhorn (1900, 1902) suggested that these foliated marls represent a pelagic facies of the Lower Eocene, which further to the south, develop into a littoral shale facies. The latter consists of 5 m (16 ft) of yellow marly shales with *Bothriolampas*, echinoids, oysters, and ochrous yellow molluscs. This unit was termed the *Kurkur Stufe* by Blanckenhorn (1900), following Mayer-Eymar (1886) and Fourtau (1899). It occurs in the Kurkur Oasis and Gebel Garra west of Aswan, as well as the Dungul Lake area. Blanckenhorn (1902) suggested that the fossils of the foliated marls, above the white chalk[12] with *Anachytes ovata* (*Echinochorys fakhreyi*) and *Schizorahbdus libycus*, and below the "*Libysche Stufe*" of Zittel with *Operculina libyca*, Alveolinids and Nummulites, were not Eocene but Danian in age.

Beadnell (1905a) described the section in Gebel Aweina (25°15′N, 32°49′E), near Esna, in detail and formally established the name Esna Shale. He divided the section, in an ascending order, into a lower unit of oyster limestones with bone and phosphate beds, a unit of dark shales, a chalk bed, a unit of green and grey shales, and an upper unit of hard limestones with chert concretions. He considered the Esna Shale as "transitional beds" between the Lower Eocene limestone with flint concretions and the underlying "Danian" chalk bed. The dark shale unit between the chalk bed[13] and the phosphatic beds were assigned a Campanian age. Beadnell (1905a) considered the 100–150 m (328–492 ft) thick shale below the Alveolinid Chalk or the Plateau Limestone of Zittel's "*Libysche Stufe*" as equivalent to his "Eocene" Esna Shale. Hume (1911, 1912) expanded the term Esna Shale to include the lower shale unit below the chalk bed (Paleocene). He used the terms lower and upper Esna Shale for the two shale units, below and above the chalk, respectively. Cuvillier (1930) considered the Esna Shale, including both shale units, below and above the "Danian" chalk, to be of Danian age. However, he suggested that the term Esna Shale should be restricted only to the upper shale unit. He also pointed out the difficulty in differentiating the units based solely on lithological similarities.

Nakkady (1950, 1957), following Hume (1911, 1912) and Cuvillier (1930), applied the name Esna Shale to the whole succession between the Cretaceous chalk and Eocene formations. He examined the foraminifera from five different sections in Egypt and concluded that the "Esna Shale" fauna is transitional; with Cretaceous affinities below and Eocene affinities above, and that the strata were conformable at all five localities. Youssef (1954, 1957) similarly applied the name Esna Shale Formation to the whole succession of shales between the phosphate-bearing formation and the Eocene limestones. Youssef (1954) assigned the lower shale unit and the overlying chalk bed in Gebel Aweina a Maastrichtian-Danian age, and the upper part a Paleocene age. Leroy (1953) studied the foraminifera of the Maqfi section in the Farafra Oasis. He divided the "Esna Shale" section into three units, a lower unit (unit IV), containing *Globorotalia velascoensis* fauna, a middle unit (unit III) containing *Nummulites deserti*, *N. fraasi*, and *Operculina libyca*, and an

---

12 This a Paleocene chalk.
13 Apparently this is a Maastrichtian chalk. Note that there was confusion between the Paleocene and Maastrichtian chalks.

upper unit (unit II), with *Eponides lotus* fauna, which inculde *Operculina libyca*, *Globigerina bulloides*, *N. deserti*, and *N. fraasi*. Said & Kenawy (1956) examined the foraminifera from two Upper Cretaceous-Lower Tertiary sections in the northern part of the Sinai Peninsula. They gave the "Esna Shale" a Maastrichtian to Lower Eocene age. Said (1960) gave a description of the foraminifera of the Esna Shale in Luxor and assigned the unit a Landenian age, based on the presence of a sharp-keeled *Globorotalia* assemblage, including *Globorotalia velascoensis*, *G. simulatilis*, *G. pseudotopilensis*, *G. interposita*, *G. pentacamerata*, *Globigerina triloculinoides*, *G. inaequispira*, and *G. eocaena*. Said & Kerdany (1961) studied material collected from the Maqfi section, Farafra Oasis. They divided the Esna Shale (between the Maastrichtian Chalk and the Ypresian Farafra Limestone) into two foraminiferal zones, a lower *Globorotalia velascoensis-simulatilis* zone, and an upper *Globorotalia colligera-esnaensis-pentacamerata* zone. The uppermost part of the latter zone contains the *Eponides lotus* fauna of Leroy (1953). They assigned the Esna Shale a Landenian age. They pointed out the presence of an unconformity between the Maastrichtian Chalk and the Esna Shale. A thin limestone bed in the lower part of the Esna Shale, with *Operculina libyca*, *Nummulites deserti*, *N. fraasi*, *N. solitarius*, *Alveolina ovulum*, and *Globorotalia velascoensis*, among others, was called the Maqfi Limestone Member by Said & Kerdany (1961). The upper 10 m (33 ft) of the shale section contains limestone beds, with numerous Alveolinid and Nummulites species, *Alveolina decipiens* is particularly abundant. This shale section grades upward into the Farafra Limestone and was assigned an Ypresian age.

Said (1961) initially introduced the name Dakhla Shale to replace the *Aschgraue Blätterthone* of Zittel (1883) or the lower Esna Shale of Hume (1911), and selected the type locality at Dakhla, to the north of Mut (25°29′N, 28°59′E). The type section of the Dakhla Formation is at Gebel Gifata, about 15 km north of Mut in the Dakhla Oasis (Tantawy et al., 2001). According to Said (1962, p.133), the Dakhla Shale is divisible into two parts, a lower calcareous marl (the so-called *Pecten* marls) which may contain biostromal developments of *Exogyra overwegi*, and an upper shale unit. These two parts correspond to Zittel's (1883) *overwegschichten* and *Aschgraue Blätterthone*, respectively. Said (1961) and Said & Sabry (1964) assigned a Maastrichtian age to the lower part of the Dakhla shale, and a Danian age to the upper part. The Esna Shale was given a Landenian age in the type locality, but was believed to be of different ages (Danian to Landenian) in other places. Awad & Ghobrial (1965) subdivided the Dakhla Shale in the Kharga Oasis into three units; the lower Mawhoob Shale Member, middle Baris Oyster Member, and upper Kharga Shale Member. The Mawhoob Member contains foraminifera of the *Globotruncana gansseri* Zone, suggesting a Late Maastrichtian age for this unit (Hendriks et al., 1984). The Baris Oyster Member contains *Exogyra overwegi* and *Libycoceras ismaeli*, and was assigned a Middle Maastrichtian age. The Kharga Shale Member is Maastrichtian in the lower part and Early-Middle Paleocene in the upper part. This unit was correlated with the Kurkur Formation by Hermina (1990).

Said (1961, 1962) in a study of the Gebel Aweina oucrop, restricted the use of the term Esna Shale to the uppermost shale beds below the Thebes Formation, and designated this section as the type section of the Esna Shale.

The Danian is represented by the upper part of the Dakhla Formation (Said, 1990c, Speijer, 2003), or Dakhla Formation *s.s.* of Tawadros (2001). A thin prominent "event bed" occur within the marly Dakhla Formation close to the boundary of the P3a-P3b planktonic subzones (Sprong et al., 2009, Bornemann et al., 2009). The lower contact of the event deposit is sharp and possibly unconformable (Speijer 2003, Sprong et al., 2009). At Gebel Qreiya, Danian-Ypresian sediments have a thickness of about 100 m (328 ft) (Said 1990) and consist of laminated dark brown-pinkish marls rich in coprolites and planktonic foraminifera, clay pebbles, and fish remains. The marls contain up to 3.5% total organic carbon (TOC) (Bornemann et al., 2009). In the Gebel Aweina section (Speijer, 2003, Sprong et al., 2009), the upper Danian event beds occur about 25 m (82 ft) above the Cretaceous-Paleogene (K/P) boundary (Speijer, 2003) within the marls of the Dakhla Formation. The event deposit is represented by a single light grey sandy marl bed of 10 cm thickness with fish remains and coprolites (Bornemann et al., 2009).

The contact between the Duwi and Dakhla *s.l.* (Sharawna & Dakhla *s.s.*) formations in the Western Desert marks the Campanian-Maastrichtian boundary (zone CF8a/b boundary), dated at about 71 Ma (Tantawy et al., 2001). The age of the Dakhla Formation is estimated to span from 71 Ma at the base to about 63 Ma at the top (Zones CF8a-Plc). Tantawy et al. (2001) placed the Cretaceous/Tertiary (K/T) boundary at Gebel Gifata, the type locality of the Dakhla Formation, within the upper Kharga Shale Member, which is marked by a hiatus that spans from the lower Paleocene (base Plc) to base *M. prinsii* Zone (CF2) in the upper Maastrichtian. According to their interpretation, the Bir Abu Minqar Horizon, (Plc(l) Zone, directly overlies the K/T boundary hiatus.

A nearly complete sucession of Paleocene nannofossil biozones has been recorded from the Esna Formation at G. El Mishiti (east-central Sinai) by Faris & Abu Shama (2007), which include from base to top; *Markalius inversus* (NP1), *Cruciplacolithus tenuis* (NP2), *Chiasmolithus danicus* (NP3), *Ellipsolithus macellus* (NP4), *Fasciculithus tympaniformis* (NP5), *Heliolithus kleinpellii* (NP6), *Discoaster mohleri* (NP7/8) and *Discoaster multiradiatus* (NP9) zones. Faris & Abu Shama (2007) placed the Danian/Selandian[14] boundary at the base of Zone NP5 at the first appearance of *Fasciculithus* taxa, and the Selandian/Thanetian boundary at the base of the *Discoaster mohleri* Zone (NP7/8). They located the Paleocene/Eocene boundary at the boundary of the NP9a/NP9b Subzones which is marked by the first appearances of *Discoaster araneus*, *Rhomboaster calcitrapa*, *R. bitrifida* and *R. cuspis*. Faris & Abu Shama (2007) traced the Paleocene/Eocene boundary in the upper part of the Esna Formation within a thin yellowish brown argillaceous marl bed. However, for mapping purposes, the very thin layer of conglomeratic chalk layer between Zone NP9 and NP10 or the very thin brownish yellow limestone bed (Faris & Abu Shama, 2007) could probably be used as the P/E boundary, which also marks the base of the Guss Abu Said Member of the Thebes Formation (Tawadros, 2001).

In the present work, the use of the name Esna Shale is restricted to those Upper Paleocene (Landenian) shales which conformably overlie the Tarawan Chalk. They are unconformably overlain by the Thebes Formation or its equivalent the Farafra Limestone of Ypresian age. The Esna Shale (*sensu* Said, 1962) is 190 m (623 ft) at El Guss Abu Said, or 150 m (492 ft) (restricted), and 40 m (131 ft) in the North Gunna section. It is 75 m (246 ft) thick in its type section at Oweina. Youssef & Abdel-Aziz

---

14 Upper Danian.

(1971) assigned it a Late Paleocene to early Eocene age. The thickness of the Esna Shale ranges from 120–160 m (394–525 ft) (Said & Kenawy, 1961). These shales grade upwards through the "Guss Abu Said Member" into the Thebes or Farafra Limestones. Said (1960) reported the following assemblage of *Globorotalia* from the Esna Shale in its type locality: *G. velascoensis*, *G. simulatilis*, *G. pseudotopilensis*, *G. interposita*, *G. pentacamerata*, *in addition to Globigerina triloculinoides*, *G. inaequispira*, and *G. eocaena*, as well as *G. esnaensis*, *G. aequa*, and *G. pusilla pusilla*, suggesting a Landenian age. In the Farafra oasis, parts of the upper Danian and lower Landenian are missing, e.g. *Globorotalia (Planorotalites) pusilla pusilla* Zone and part of *G. (Morozovella) uncinata uncinata* (Youssef & Abdel Aziz, 1971), and the Esna Shale rests unconformably on the Dakhla Shale.

El-Naggar (1966) examined various outcrops in the Esna-Idfu region, and suggested raising the Esna Shale to group status. He divided the Esna Group into two formations, a lower Sharawna Shale (Maastrichtian) and an upper Oweina Shale (Paleocene), separated by a conglomerate. The conglomerate contains reworked Maastrichtian and Danian fossils which he considered to represent the boundary between the Cretaceous and Tertiary. He also excluded the Lower Eocene calcareous shale member from the Esna Shale and assigned it to the Thebes Formation as a new member. Although this terminology did not receive wide acceptance among his contemporary and later workers, it appears to have a lot of merit.

The Sharawna Shale Formaion was named after the type section at Wadi El Sharawna (25°15′N, 32°45′E). In the type section the Formation is 117 m (384 ft) in thickness. It overlies the Sibayia Phosphate Formation (Duwi Formation) and is overlian by the Oweina Shale Formation (Dakhla Shale of the present study). El-Naggar (1966) divided the Sharawna Shale Formation into 3 units or members: the Lower Sharawna Shale Member is 57 m (187 ft) in thickness in the type section at Wadi El Sharawna, and is composed of ash-grey, iron-stained, locally ferruginous, paper-like shale, with thin intercalations of marls and pyrite nodules. It is characterized by the presence of *Pecten mayereymari* and *Pycnoconta vescicularis*, in addition to abundant dwarfed limonitic megafossils. Planctonic foraminiferas are abundant and include a rich assemblage of the "*Globotruncan fornicata* Zone" of Lower Maastrichtian age, such as *G. stuarti*, *G. havenensis*, *G. aegyptiaca aegyptiaca*, and *G. tricarinata tricarinata*, among others. The Middle Sharawna Shale Member is 12 m (39) in thickness in the type section at Wadi El Sharawna, and consists of pale yellowish to pale grey, locally ferruginous, marly clay, clayey marl and marl, with *Pecten mayereymeri* and *Pycondonta vesicularis*, as well as dwarfed limonitic megafossils. Planktonic foraminiferas include species of the *Globtruncanta gansseri* Zone, which include *Globotruncana gansseri gansseri*, *G. contus contusa*, *G. contusa patelliformis*, *G. conica*, and *G. esnehensis*, among many others. The Upper Sharawna Shale Member is 48 m (157 ft) in the type section at Gebel Oweina, and composed predominantly of iron-stained, locally ferruginous, dark grey to greyish black shales, with a few prints of *Pecten mayereymari* and dwarfed megafossils. It was assigned a middle to late Maastrichtian age by El-Naggar (1966) and El-Naggar & Haynes (1967) based on the planktonic foraminifera *Globotruncana gansseri* and *G. esnehensis*. The top of the member is marked by a lithological break, which is characterized by a conglomerate with reworked Maastrichtian ammonites. The Sharawna Shale also contains Maastrichtian nannofossils in the Kharga area (Faris, 1984). In Sinai, the Maastrichtian is represented by the chalks of the Sudr Formation in the Gebel Nekhl area (29°50′N, 33°34′E), and shales (Sharawana Shale) in the Gebel Giddi (30°09′N, 33°15′E), both of which are characterized by the presence of species of *Globotruncana* and *Gümbelina*. Said and Kenawy (1956) included the Maastrichtian shale section in the Esna Shale.

The Dakhla Shale is equivalent to the Lower Oweina Member of El-Naggar (1966). It has its maximum thickness in the Maqfi section in the Farafra Oasis.

Fauna from the Dakhla escarpment near the Dakhla Shale type section were collected and studied by Youssef (in Youssef & Abdel Aziz, 1971). The lower part of the sequence is lower to Middle Maastrichtian and is separated from the Upper Danian by a disconformity. The Dakhla Shale in its type area, therefore, includes a part of the Sharawna Shale. The Dakhla Shale in the Farafra area is represented by a 12 m (39 ft) sequence of shales and marly shales overlying the Chalk and underlying the Tarawan Chalk. The age of these Dakhla shales is Danian. The lower part grades into chalks in places and differentiation between these Danian chalks and the underlying Maastrichtian chalks becomes almost impossible without micropaleontological examination.

It is in the present writer's opinion that the Mawhoob and Baris members, as well as the Maastrichtian part of the Kharga Member of the Dakhla Shale (as defined by Awad & Ghobrial, 1965), are equivalent to the Sharawna Shale of El-Naggar (1966). The Paleocene part of the Kharga Member is equivalent to the Dakhla Shale of this study.

Barthel & Herrman-Degen (1981) included in the "Dakhla Formation in the Farafra and Dakhla oases, sediments ranging in age from Middle Campanian to Ypresian. They divided it, from base to top, into three members, El Hindaw Member, Qur el Malik Member and Dakhla Shale Member. The El Hindaw Member, which they assinged a Middle to early Late Campanian age, is equivalent to the Duwi Formation. The Qur el Malik Member consists of whitish, soft limestones, chalks and siltstones, approximately 14 (46 ft) in thickness. It was assigned an Upper Campanian age. The Dakhla Shale Member contains ammonites of Late Campanian age at the base and *Globorotalia angulata* at the top. The lower part of this unit is equivalent to the Mawhoob Member of Awad & Ghobrial (1965). The top of the Dakhla Shale member is marked by the Bir Abu Minqar Horizon, a horizon about 1.5 m (5 ft) thick which marks the Cretaceous-Tertiary boundary (Barthel & Herrmann-Degen, 1981, Hermina, 1990). This horizon is composed of greyish to yellowish sandy marls, with glauconite and ferruginous ooids, phosphatized fossils and large gastropods. Westward, Barthel & Herrman-Degen (1981) divided the Dakhla Formation into two informal units, the Ammonite Hills and Peak Hill members. The Ammonite Hills Member was named after the Ammonite Hill scarp, and it consists of a series of fossiliferous and heavily bioturbated, calcareous siltstones and sandstones, intercalated with foliated clays, and contain phosphatic nodules and phosphatized fossils. The Peak Hill Member was named after an interdunal channel marked by a peak, about 60 km southwest of El Guss Abu Said, in the Farafra Oasis. It is composed of oolitic limestones, clays and calcareous siltstones, with small patch reefs. They assigned this member to the Late Danian-Ilerdian (Ypresian). Correlations between these two members and the Dakhla and Tarawan formations are ambiguous. Barthel & Herrmann-Degen (1981) combined the Tarawan Formation and the Farafra Chalks of Youssef and Abdel-Aziz

(1971) under the name Tarawan Formation, and included Campanian rocks in the Dakhla Shale. Along the Ain Dalla scarp, west of Qasr El Farafra (27°03′N, 27°58′E) and northeast of El Guss Abu Said (27°40′N, 27°08′E), the Esna Shale changes facies into white to yellow, well-bedded limestones and marl, 47 m (154 ft) thick, with abundant echinoids, molluscs and nummulites, such as *N. Deserti, N. fraasi, N. solitarius*, as well as *Operculina* sp. Planktonic foraminifera include *Globorotalia subbotinae, G. marginodentata, Acarinina wilcoxensis* and *Pseudohastigerina wilcoxensis*. The top of the succession is composed of calcarenitic limestones with hermatypic corals. This part of the succession was referred to as the Ain Dalla Formation by Barthel & Herrmann-Degen (1981) who assigned it a Middle Ilerdian age. They selected the type setion in a scarp-spur 10 km north of Ain Dalla.

At Wadi Qena, Hendriks et al. (1990) grouped the Dakhla Shale, Tarawan Chalk, and Esna Shale in one unit, the Sharib Formation. The Formation overlies the Rakhiyat (Duwi) Formation and it is overlain by the Thebes Formation. Bandel et al. (1987) included these sediments in the Esna Shale.

The name Tarawan Chalk was introduced by Awad & Ghobrial (1965) for a persistent chalk unit that separates the Dakhla Shale from the Esna Shale in the Kharga Oasis. It is equivalent to Zittel's (1883) *Anachytes ovata* chalk and was included in the "Chalk" of Said (1962). The unit is composed dominantly of white chalks, becoming marly in places. The unit has rare macrofossils, but locally it contains abundant *Schizorhabdus libycus* (*Echinocharys fakhryi*), *Ventriculites poculum, Terebratula libyca* and *Ostrea hypoptera* (El-Naggar, 1966, Cuvillier, 1930, Nakkady, 1957). It conformably overlies the Dakhla Shale and conformably underlies the Esna Shale. It was assigned an Upper Paleocene age by most of the authors. It contains *Globorotalia pseudomenardii* and *G. velascoensis* (Youssef & Abdel Aziz, 1971, also Said & his students). The Tarawan Chalk in the Farafra oasis is 5.5 m (18 ft) in the North Gunna section. In the southern part of the Farafra depression, as well as at the El Guss Abu Said section, the Tarawan Chalk contains *Schizorhabdus libycus* and *Ventriculites poculum*.

The Paleocene sediments overlying the Late Maastrichtian sandstones and dolomites in the St. Anthony section consist of a mixture of two facies of limestones. The first is composed of peloidal grainstones, with bryozoan fragments, gastropods, miliolids, *Nummulites* sp., *Discocyclina* sp., and *Miscellanea* sp., together with *Archaeoclithothamnium* sp., the second include deeper-shelf deposits with echinoids, brachiopods, and planktonic foraminifera (Kuss, 1986). El-Shinnawi & Sultan (1972a) examined the foraminifera of the Paleogene section in five offshore wells, the Ghareb-1, Alef-1, J-1, Morgan-8 and H-1, and the onshore well Bakr-6. They assigned the Dakhla Formation a Danian age, the Tarawan Formation a lower Landenian age (*Globorotalia pseudomenardii* zone), the Esna Shale (including the Guss Abu Said Member) an upper Landenian age (*G. velascoensis* zone) and a lower Ypresian (*G. subbotinae-G. formosa* zone), and the Thebes Formation an upper Ypresian-Lutetian (*G. bullbrooki* zone).

The Danian-Early Eocene succession at the Saint Paul Monastery section, Southern Galala, includes in an ascending order by the Dakhla, the Southern Galala, and the Thebes formations (Galal & Kamel, 2007).

The Paleocene section near the St. Paul Monastery unconformably overlies Cretaceous rocks, which belong the latest Maastrichtian *Abathomphalus mayaroensis*.

Strougo et al. (1992) showed that this hiatus spans the basal Paleocene *Globigerina fringa* Zone and the lower part of *Globigerina eugubina* Zone. This interval is overlain by approximately 100 m (328 ft) of marls and limestones with fossils of the *Morozovella pseudobulloides, M. trinidadensis, M. uncinata, M. angulata, P. pusilla,* and *M. velascoensis* zones. The Late Paleocene *Planorotalites pseudomenardii* Zone (P4) interval consists of a lower marly unit with wavy bedding and an upper unit of boulder-size carbonate clasts floating in a marly matrix, associated with lenses of cross-laminated sandstones, interpreted as mass flow deposits. The P4 Zone, as well as the P5-P6a (*Morozovella velascoensis-M. edgari*) Zone, is marked by important changes in the depositional setting in a large number of sections from the western Desert, the Gulf of Suez/Red Sea area, and the Nile Valley. These changes are mainly controlled by syndepositional faulting. These changes have been termed the *velascoensis* Event by Strougo (1986).

At Wadi Araba, the St. Anthony Formation is overlain by the Southern Galala Formation (Upper Paleocene-Lower Eocene) which was introduced by Abdallah et al. (1971) and redefined by Kuss & Leppig (1989). The reference section is situated in a north-south-trending wadi, 6 km west of the Monastery of St Anthony (Kuss & Leppig, 1989, Scheibner et al., 2001a, b). It is composed of massive sandy fossiliferous limestones, with alveolinas, Nummulites, and coralline algae, and conglomeratic sandstone intercalations, changing into limestones with flint southward. It has a thickness of about 250 m (820 ft). It has been assigned an Upper Paleocene-Lower Eocene age by Scheibner et al. (2001a, b) based on the presence of foraminifera of the *Glomalveolina levis* Zone (Upper Thanetian) at the base and the *Alveolina dainelli* Zone (Cuisian) at the top. In Bir Dakhl and St Paul, the Southern Galala Formation overlies the Dakhla Formation. The Southern Galala Formation interfingers with the Esna Formation. In St Anthony, the Southern Galala Formation overlies the St Anthony Formation (Upper Campanian-Upper Maastrichtian), whereas to the north it unconformably overlies Turonian strata (Kuss & Leppig, 1989, Scheibner et al., 2001a, b). In the Northern Galala area, the Southern Galala Formation overlies the Paleocene Dakhla Formation and is overlain by the Thebes Formation. In the area west of the St. Anthony Monestary, the Early-Middle Paleocene deposits are only recorded in the Umm Damaranah section and occupy the lower part of the Southern Galala Formation (Paleocene-Lower Eocene) (Ismail et al., 2007). The thickness of this horizon is 127.5 m (418 ft). It overlies the *Orbitoides media* and *Omphalocyclus macropora* assemblage zone of the Late Campanian (of the Thelmet/St. Anthony Formation) and underlies the *Morozovella angulata* zone of the Middle Paleocene. This horizon is characterized by a large number of white globular larger foraminifera *Glomalveolina dachelensis,* in addition to *Miscellanea rhomboidea* and *Fallotella kochanskae persica.* In the Umm Damaranah section, this zone underlies the *Morozovella angulata* zone of Middle Paleocene age. The Middle Paleocene is recorded in the Umm Damaranah section, which occupy the middle part of the Southern Galala Formation (Ismail et al., 2007). The thickness of this zone is 6 m (20 ft). This zone contains *Morozovella angulata* and *Morozovella conicotruncata.* The Upper Paleocene in the Umm Damaranah section comprises the upper part of the Southern Galala Formation and is 14 m (46 ft) thick, with *Planorotalites pseudomenardii, Morozovella angulata, M. trinidadensis, M. uncinata, Acarinina primitiva, Nodosarella mappa, Reussoolina apiculata, Globigerina triloculinoides, Cibicidoides padella* and *Cibicidoides* cf. *succedens* (Ismail et al., 2007). The Ypresian uppermost part of the Southern Galala Formation in the Southern Galala Plateau contains *Nummulites* cf. *subramondi, Alveolina pasticellata, Fabularia zitteli* and the algae *Ethelia alba*. Paleocene-Early Eocene ostracod fauna retrieved along a platform-basin transect in the Southern Galala Plateau area, show that the P/E transition is characterized by the appearance of new taxa rather than extinctions (Morsi & Scheibner, 2009). From Early to Late Paleocene, the ostracod assemblages were dominated by middle-outer neritic taxa. In the latest Paleocene and Early Eocene, they changed from deeper marine environments in the south to shallower marine environments in the north.

In the Northern Galala, Boukhary et al. (2009) identified the larger foraminifera species *Nummulites* aff. *nemkovi, Nummulites partschi, Nummulites bassiounii, Nummulites* cf. *campesinus, Assilina* aff. *major, Decrouezina aegyptiaca*, and *Operculina* sp. from the Early Eocene of the Gebel Umm Russeies succession. The succession unconformably overlies Upper Cretaceous rock; the Paleocene is completely missing. However, at Wadi Naot in the Northern Galala, the Paleocene section is well represented. In the Gebel Ataqa area to the north, the Paleocene is missing again. In the Southern Galala Plateau, Boukhary et al. (1998) identified *Nummulites praeatacicus, N. saharaensis* and *Bassiounina sanctipauli* the Lower Eocene section. Early Ypresian conditions of deposition varied greatly between the Northern and Southern Galala areas from a reefal shallow environment in the south, where *Nummulites* flourished to an open platform in the north where Nummulites are replaced by ?*Operculina*. The Gebel Umm Russeies area was probably uplifted following the deposition of the Upper Cretaceous sediments of the Adabiya Formation (El Akkad & Abdallah, 1971).

Strougo & Haggag (1983) reported Paleocene sediments at Jiran el Ful (29°58′N, 31°04′E) in the Abu Roash area, west of Cairo. The section unconformably overlies the Maastrichtian Khoman chalks and is overlain by Middle Eocene sediments. They recognized three calcirudite bands in the section, the lower one lies at the Maastrichtian-Paleocene boundary (probably equivalent to the Abu Minqar Horizon), while the middle and upper bands are intra-Paleocene conglomerates. The total thickness of the Paleocene section is 3.5–6 m (11–20 ft). The lower part of the Paleocene section, 2.5–4.5 m (8–15 ft) in thickness, is made up of white chalks of Danian age, with foraminifera of the *Globorotalia trinidadensis* and *G. uncinata* zones, unconformably overlain by a discontinuous band of limestone conglomerate (0–45 cm), with molds of nautiloids, gastropods, solitary corals, echinoid spines, and planktonic foraminifera of the *G. velascoensis* zone. The upper unit is a limestone conglomerate, 0.5–0.8 (2–26 ft) in thickness, passing laterally into fine, marly chalks, with fauna of the *G. velascoensis* zone. This succession is, therefore, equivalent to the entire Dakhla Shale, Tarawan Chalk and Esna Shale succession, or the Sharib Formation of Hendriks et al. (1987) and Hendriks et al. (1990) in the Wadi Qena.

### 7.5.3.3  Upper Cretaceous-Tertiary succession in extreme Southern Egypt

West of Aswan, in the Dungul, Kiseiba and Garra areas, the Maastrichtian to Eocene sequence is made up of the Kiseiba, Kurkur, Garra, and Dungul formations. The succession in that area represents another problem due to the change of the section into nearshore facies.

The succession was divided by Cuvillier (1930) into the following units, from base to top: 1) the Nubian Sandstone, 2) yellow shales with *Gitolampas abundans*, 3) foliated marls with *Ostrea overwegi*, and 4) white limestones.

The Kiseiba Formation is Campanian-Maastrichtian in age (Hendriks et al., 1987). It has been divided, in the Bir Dungul-Bir Kiseiba area, in southern Egypt, into the Shagir Member and the Shab Member. The Shagir Member is equivalent to the Qoseir Formation, and the Shab Member is equivalent to the Sharawna Shale.

The Shab Member of the Kiseiba Formation (Sharawna Shale) is unconformably overlain by reef-like, thick-bedded limestones, sandy in part, with minor shale intercaltations of the Kurkur Formation (Issawi, 1968) of Early-Late Paleocene age (Hendriks et al., 1987). The name *Kurkur Stufe* was originally introduced by Fourtau (1899) for this unit and later established by Blanckenhorn (1900) and assigned a Paleocene age. Blanckenhorn (1902) and Said (1962) estimated a thickness of 5 m (16 ft) for the Kurkur Formation, while Hermina & Issawi (1971) reported a thickness between 11–50 m (36–164 ft), and Hendriks et al. (1987) estimated a thickness between 6.5–140 m (21–459 ft). Blanckenhorn (1902) considered the Kurkur Formation to be the littoral equivalent of the pelagic "*Blätterrmergel*" or the "Esna Shale" (*s.l.*). The Formation in south Kharga oasis contains *Globorotalia uncinata*, *G. pseudobulloides* and *G. trinidadensis* (Hermina, 1990). Defined as such, this Formation is probably equivalent to the Dakhla Formation and Tarawan Chalk.

The Kurkur Formation grades upward into the Garra Formation (Issawi, 1968) (Fig. 7.6), 4–74 m (13–242 ft) thick, composed of well-bedded and massive limestones (Hendriks, 1985). It contains *Morozovella uncinata*, *M. angulata*, *M. pseudomenardii* and *M. velascoensis*. It was assigned a Late Paleocene-Early Eocene age by Issawi (1972), Hendriks et al. (1984), and Hermina (1990). This Formation is equivalent to the Esna Shale and the Lower Eocene shales (Guss Abu Said Member, Tawadros, 2001). Said (1990c) believes that the Garra Formation is equivalent to the lower members of the Thebes Group.

### 7.5.3.4   The nature of the Upper Cretaceous-Early Tertiary contact (unconformities)

The Upper Cretaceous-Early Tertiary contact is marked by a regional unconformity, which has been manifested in different parts of Egypt and North Africa, either by the presence of conglomerates, or by the absence of certain Maastrichtian and Danian foraminiferal zones. It has also been suggested that the magnitude of this unconformity varies with the position of the succession relative to structural high (Said & Kerdany, 1956, Shukri, 1954, Kostandi, 1963, Salem, 1976). This unconformity probably reflects the combined effect of tectonics and sea level fluctuations.

Zittel (1883) and Cuvillier (1930) considered the Farafra Oasis as a place of continuous sedimentation from the Cretaceous to the Eocene. However, subsequent micropaleontological studies on the Cretaceous-Tertiary succession revealed the presence of a number of unconformities, some of these are difficult to detect in the field. In the Farafra oasis, the uppermost Maastrichtian *Abthamophalus mayaroensis* Zone, as well as the *Globigerina-Globorotalia* assemblage of the Danian is missing (Said, 1960, Said & Kerdany, 1961, Youssef & Abdel Aziz, 1971). Youssef (in Youssef & Abdel Aziz, 1971) reported small (2–4 mm) brown and polished pebbles and foraminifera at the unconformity surface. A paleontological break between the Maastrichtian and Danian in the Qoseir-Safaga area along the Red Sea coast is marked by the absence of the Upper Maastrichtian *Abathomphalus mayaroensis* zone and the Danian *Globigerina pseudobulloides* Zone (Issawi, 1972). Girgis (1989) studied the nannofossils from two Maastrichtian-Tertiary outcrops around St-Paul Monastery, Gulf of Suez region. He subdivided the Maastrichtian section into four nannofossil zonal units: *Quadrum trifidum, Arkhangelskiella cymbiformis, Lithraphidites quadraus*, and *Nephrolithus frequens*. The presence of the latest Maastrichtian *Nephrolithus frequens* Zone and the earliest Danian *B. sparsus* Zone, together with the absence of any sign of break in sedimentation, suggested to him that the sequence across the Maastrichtian-Danian boundary is more or less continuous. However, Strougo et al. (1992) showed that the basal Paleocene, equivalent to the *Globigerina fringa* Zone and the lower part of *Globigerina eugubina* Zone, is absent in this section. The *Arkhangelskiellla* Group displays a decrease in size in the latest Maastrichtian, associated with an increase in the abundance of the calcareous dinoflagellate *Thoracosphaera operculata*. Girgis (1989) interpreted this feature as a reflection of stressful environments for nannoplanktons prior to the Cretaceous-Tertiary boundary. In the offshore of the Gulf of Suez, the Cretaceous-Paleocene boundary is also marked by an unconformity (El-Shinnawi & Sultan, 1972a). The earliest Tertiary assemblage of *Globigerina eobulloides* zone is missing in all wells. The magnitude of this unconformity increases southward in the Gulf of Suez. In the Oweinat area and along the Nile Valley, the unconformity is also marked by a thin conglomerate (El-Naggar, 1966, Issawi, 1972). In the Garra-Kurkur-Dungul area, the Danian is absent and the unconformity is marked by a thin conglomerate bed (Issawi, 1972). In the Ain Amur and Ghanima sections in the Kharga Oasis area, the uppermost Maastrichtian and the basal Danian are not represented (Faris, 1984). This gap is manifested by the absence of the nannofossil zones *Micula prinsii* (uppermost Maastrichtian) and *Markalius inversus* and *Cruciplacolithus tenuis* zones (Danian). In the Ghanima section, this boundary is also marked by the presence of phosphatic beds. In the Farafra and Dakhla oases, the Cretaceous-Tertiary boundary is marked by 1.5 m (5 ft) of sandy marls, with glauconite and ferruginous ooids, phosphatized fossils and large gastropods of the Bir Abu Minqar Horizon (Barthel & Herrmann-Degen, 1981, Hermina, 1990). At Abu Roash, the Paleocene section unconformably overlies the Maastrichtian Khoman chalks (Strougo & Haggag, 1983).

# Chapter 8

# Tertiary

## 8.1 EOCENE

The three stages of the Eocene Epoch, Lower, Middle, and Upper, are well represented in Egypt. They are exposed along the stretch of the Nile Valley, the Western Desert, the Eastern Desert, and Sinai. They also occur in the subsurface of the northern Western Desert (Northern Egypt Basin), the Nile Delta, and Northern Sinai.

Two main sedimentary facies are present in the Eocene rocks in Egypt, a deep-marine facies, dominated by marls and shales, and a shallow platform facies, represented by limestones rich in Nummulites and other large benthonic foraminifera (Said, 1962a, Boukhary & Abdelmalik, 1983). Because of the uncertainties of age dating by means of large foraminifera and the predominance of long-ranging micro- and macrofossils, a large number of formational names have been proposed. Authors often do not agree on the Nummulites of a given formation, even at the same locality (Strougo et al., 1990). The recent trend in using ostracods to date these rocks (cf. Boukhary et al., 1982a, b, Bassiouni et al., 1984, Bassiouni & Luger, 1990) probably added more to the confusion than to the solution. Confusion arises because the exact stratigraphical ranges of many Eocene ostracod species are not well defined (Keen et al., 1994, El-Khoudary & Heldmach, 1981).

Zittel (1883) considered all the Eocene sediments in Egypt to be of Ypresian age and subdivided them into a lower unit which he termed the *Libysche Stufe*, and an upper unit, the *Mokattam Stufe*. The *Libysche Stufe* was further subdivided into upper *Libysche Stufe* and Lower *Libysche Stufe* characterized by the presence of *Alveolina frumentiformis* and *A. oblonga*, respectively. The *Mokattam Stufe* was divided in the type secction at Gebel Mokattam (Fig. 8.1), east of Cairo, into Lower *Mokattam Stufe* and Upper *Mokattam Stufe*. The former includes the *Nummulites gizehensis*-bearing beds, and the latter encompasses a series of brownish limestones and shales with sand beds. Blanckenhorn (1902) mapped the *Mokattamstufe* from Helwan to Maghagha and pointed out that the best development of the Lower *Mokattamstufe* is at Wadi el Sheikh (28°41′N, 31°02′E), opposite Maghagha. This section was described also by Cuvillier (1930).

Hume (1911) subdivided the *Libysche Stufe* in the Western Desert into three zones: the *Operculina libyca* and *Ostrea multicostata* Zones (Lower Libyan) and the *Callianassa* Zone (Upper Libyan). He also subdivided the *Mokattam Stufe* into *Nummulites gizehensis* beds (Lower Mokattam), *Exogyra* beds (Middle Mokattam), and *Carolia* beds (Upper Mokattam).

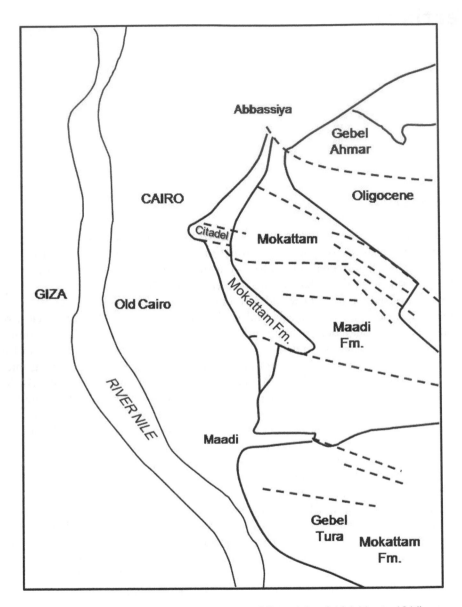

*Figure 8.1* Map of the Mokattam area, east of Cairo (after Said & Martin, 1964).

Boussac (1913), based mainly on macrofossils, believed that the Ypresian did not exist in Egypt and considered Zittel's Libyan to be Lutetian. Douvillé (1920a, b, 1924) examined Eocene foraminifera from various parts of Egypt.

Cuvillier (1930) assigned the Libyan formations between the "Upper Cretaceous" and the beds with *Orbitolites complanatus*, which mark the base of Lutetian, an Upper Londinian (Ypresian) age. He subsequently subdivided the Middle Eocene (Lutetian) into Lower Lutetian, characterized by the appearance of *Orbitolites complanatus*, accompanied by *Assilina frumentiformis*, *A. praespira*, and *Alveolina* cf. *subpyrenaica*, and Upper Lutetian (Lower Mokattam), characterized by the appearance of *Nummulites gizehensis*, *N. uroniensis*, *N. Lucasi*, and *Operculina libyca*.

Zittel's *Libysche Stufe* was raised by Said (1960) to a group status, the Libya Group, to include, from base to top, the Esna Shale, Farafra Limestone, Thebes Formation (Lower Libyan), and Minia Formation (Upper Libyan). He also assigned the *Mokattam Stufe* to a group status, the Mokattam Group, to include the Mokattam Formation (Lower Mokattam) and the Maadi Formation (Upper Mokattam). It is obvious that the use of the same name for a group, as well as for a formation, can and has caused a plethora of confusion. That practice is still carried out today.

Numerous recent paleontological studies were carried out to classify and date the Eocene rocks in Egypt. The ostracods of the Eocene rocks were studied by Bassiouni (1969a, b, 1971), Boukhary et al. (1982a, b), Bassiouni et al. (1984), Bassiouni & Luger (1990), Cronin & Khalifa (1987), and Elewa et al. (1995). The Nummulites were studied by Blondeau et al. (1984, 1987), Boukhary et al. (1982a, b), and Strougo et al. (1990). Boukhary & Abdelmalik (1983) and Strougo & Boukhary (1987), and Boukhary et al. (2009) attempted to revise the stratigraphy of the Eocene rocks in Egypt. The authors of the new Geological Map of Egypt (1987) also reviewed the stratigraphy of the Eocene rocks and recognized several rock units, which overlap and interfinger with one another. Said (1990d) and Hermina (1990) reevaluated the Eocene stratigraphy based on the new data. The dynamics of Nummulites accumulations were discussed by Aigner (1982, 1983 & 1984), who was the first to realize the importance of the sedimentological factors in their distribution. Facies mapping of Eocene rocks in the surface and subsurface was carried out by Kostandi (1963) and Salem (1976).

Pioneering work on the geology of the Fayium Province was carried out by Dames (1883), Schweinfurth (1886), Andrews (1901), Beadnell (1901, 1905b), and Osborn & Granger (1907) (Morgan & Lucas, 2002). More recent studies were done by Fleagle et al. (1986), Bown & Kraus (1988), Simons & Rasmussen (1990), Gingerich (1992, 1993), De Blieux et al. (2006), Gunnell et al. (2008), and Smith et al. (2008), among many others.

Bassiouni et al. (1984) described Eocene fauna from the Fayium Province and, based on ostracods, redated several formations previously regarded as Late Eocene and assigned them to the late Middle Eocene, and others of Middle Eocene to the Lower Eocene. These revisions not only resulted in several ostracod species which have been regarded as typical Upper Eocene to move into the Middle or Lower Eocene (Keen et al., 1994), but also in making correlations between Egyptian and Libyan units more difficult.

## 8.1.1    Lower Eocene (Ypresian)

Said (1960) introduced two Ypresian formations: the Thebes Formation and the Farafra Formation. The former represents an open sea facies, the latter a shallow shelf facies. On the new Geological Map of Egypt (1987), the Thebes Formation was raised to a Group Status (Thebes Group) and includes the Serai, Farafra, Rufruf, Dungul and Drunka formations which overlap or interfinger with one another. In reality, all these formations are merely facies variations of the Thebes Formation of Said (1960). Hermina (1990) and Said (1990d) summarized the characteristics of these units.

The name Thebes Formation was introduced by Said (1960) to replace the terms "*Operculina* limestone", "Lower Libyan", and "Limestone with flint" which were in common use at that time. The type locality of this Formation is at Gebel Gurnah, behind the famous temple of Deir el-Bahari (Hatshepsut)[1] (Figs. 8.2 & 8.3) in Thebes, on the western side of the Nile, facing the city of Luxor[2]. The unit consists of a 290 m (951 ft) thick limestone section with many bands of flint concretions. The Formation is Ypresian in age at the type locality and probably represents the open sea facies of this age in Egypt (as opposed to the reefal facies of the Farafra Formation). The unit contains the following fossils: *Lucina thebaica, Heterospatangus lefebvrei, Nummulites praecursor, N. subramondi, Ostrea multicostata,* and *Gryphaea pharaonus.* Planktonic foraminifera include *Globorotalia planoconica, G. conicotruncata, G. thebaica, G. imitata, Hastigerina aspera, H. micra,* and *Globigerina triloculinoides.* The unit overlies conformably the Esna Shale. Said (1990d) considered it as equivalent to the Serai Formation and assigned it an Ypresian age. However, *Globorotalia conicotruncata and Globigerina triloculinoides* are indicative of a Paleocene age (P3-P4) according to Toumarkine & Luterbacher (1985) and the Thebes Formation, as originally defined, may include rocks of Paleocene age or a part of the Esna Shale.

The name Farafra Formation was originally introduced by Said (1960) as the Farafra Limestone to designate buff-colored beds of hard, medium to coarse-crystalline, alveolinid limestones that overlie the Esna Shale. The type locality of the Farafra Formation is at El Guss Abu Said (27°40'N, 27°08'E) in the Farafra Oasis. In the type locality it has a thickness of 34 m (112 ft). It contains numerous specimens of alveolinids, especially *Alveolina schwageri*, together with *Nummulites* species. The Farafra Limestone is equivalent in age and stratigraphic position to the Thebes Formation, but differs in its faunal content and lithology from the Thebes Formation. This facies is spotty in its geographic distribution and has been interpreted by Said (1960) as due to the development of "table reefs" on the submerged highs of the Egyptian Ypresian sea. Approximately 25 km to the east, the limestones overlying the Esna Shale in the North Gunna section are bluish grey, devoid of alveolinids, and contain a few Operculinas and Nummulites. These limestones are similar to the Thebes Formation and represent a deeper water facies of the alveolinid-nummulitic limestones of the Farafra Formation. Youssef & Abdel Aziz (1971) referred to both facies collectively by an informal formational name, the Thebes-Farafra Limestone. The Formation was dated Ypresian by all workers. The upper part, however, is typically composed of allochthonous limestones, which contain rare, badly preserved planktonic foraminifera of Middle Eocene affinity.

In the central and southern Western Desert, Lower Eocene (Ypresian) shales with thin limestone and marly beds overlie the Esna Shale and underlie limestones of the Thebes Formation. This unit contains *N. deserti, N. fraasi, N. burdigblensis, Operculina libyca, Globorotalia wilcoxensis, Discocyclina nudimargo* in the Farafra Oasis (Youssef & Abdel-Aziz, 1971), as well as *Morozovella (Globorotalia) subbotinae, Tribrachiatus contortus, Discoaster binodosus,* and *N. praecursor* (top part of NP9 and NP10) in the Ghanima section, Kharga Oassis (Faris, 1984, Faris et al., 1998).

---

1  Also made famous by the tragic act by fundamentalists in Nov. 1997.
2  The temple is completely incised into the Esna shale of Gebel Gurnah and is overlain by highly fractured Thebes Limestone which threatens its stability (Abdallah & Helal, 1990).

This unit is referred to in the present work as the Guss Abu Said Member of the Farafra/Thebes Formation. It is also present at the base of the Dungul Formation in the Garra/Dungul area to the south. It has been called the Farafra Shale by Kostandi (1963). This unit is absent in many places due to facies changes, non-deposition, or erosion.

Barron & Hume (1902) introduced the name Serai Limestone for a 265 m (870 ft) unit composed of jointed and cavernous limestones with flint, named after the Gebel Serai in the Wadi Qena area. The lower 15 m (949 ft) is composed of pink limestones. The type section of this unit is at the southern end of the Gebel Abu Had (26°36'N, 32°57'E) in Wadi Qena, where it overlies the Esna Shale. The limestones are devoid of fossils, but become nummulitic near the top. The name Serai Limestone fell into disuse as it was replaced by the more distinguished name, Thebes Formation.

Bandel et al. (1987) reinstated the name Serai Formation according to the rule of priority. However, although it may have been an oversight by Said (1960) when he introduced the new name of Thebes Formation, the name Thebes Formation gained a very quick and wide acceptance among geologists. Thus, it would be imprudent to drop the name or tamper with it. No continuous section of the Eocene succession is exposed at Wadi Qena. The succession starts with 35 m (115 ft) of massive chalks with flints, overlain by a 25 m (82 ft) section of marly chalks with flint intercalations, then by 10 m (33 ft) of chalky marls with tube-like (*Skolithos?*) burrows. The upper part of the Eocene succession (Bandel et al., 1987) consists of 42 m (138 ft) of escarpment-forming chalks with flint nodules and beds, followed upward by soft marly chalks, 30 m (98 ft) thick, with a synsedimentary slump fault, then by soft marly chalks, 18 m (59 ft) in thickness. All the Serai beds lack macrofossils. At Wadi Tarfa, the succession is made up of nummulitic limestones, capped by an 80 m (262 ft) section of flint-rich chalks and marly chalks. Bandel et al. (1987) assigned these nummulitic limestones an Upper Paleocene-Lower Eocene age, based on alveolinids, and considered them equivalent to the Minia Formation.

Bishay (1961) subdivided the Eocene succession overlying the Esna Shale in the Eastern Desert, between Samalut and Assiut into the Assiuti Chalk Formation, Manfalut Nummulitic Limestone Formation, Minia Formation, and Samalut Formation. He introduced the name Assiuti Chalk Formation for a unit composed of well-bedded chalk with flint, devoid of large foraminifera, and approximately 150 m (492 ft) in thickness. The type section is in Wadi el Assiuti (approximately 27°24'N, 32°29'E). It overlies conformably the Esna Shale and it is overlain by the Manfalut Formation. He assigned the Formation a Lower Eocene age. The name Manfalut Nummulitic Limestone Formation was also introduced by Bishay (1961) for a limestone unit which overlies the Assiuti Formation and underlies the Minia Formation. The unit is made up of well-bedded limestones with occasional flint nodules. There is no complete exposed section of the unit. The exposed part of the Formation in the type section at Gebel Himu (27°20'00"N, 31°20'30"E), approximately 40 km west of Manfalut is 120 m (394 ft) thick, but its compound thickness is approximately 350 m (1148 ft). It was assigned a Lower Eocene age based on the large foraminifera *Discocyclina nudimargo*, *D. archiaci*, and *Nummulites atacicus*. The upper boundary of the Formation is not clear and may extend into the Lutetian. This Formation is equivalent to the Serai Limestone of Barron & Hume (1902). The Assiuti and Manfalut formations correspond to the Thebes Formation of Said (1960).

The name Rufruf Formation was proposed by the authors of the new Geological Map of Egypt (1987). The unit crops out along a belt 10 to 25 km wide at Naqb El Rufruf in the Kharga scarp, and consists of marls grading upward into marly limestones and thick-bedded chert limestones. The Formation is up to 145 m (476 ft) in thickness. Abundant *Nummulites* sp. is present in the limestones, and *Globorotalia esnaensis* and *G. gracilis* in the shales, indicating a Ypresian age. *Morozovella (Globorotalia) subbotinae, Tribrachiatus contortus, Discoaster binodosus,* and *N. praecursor* (top part of NP9 and NP10) have been reported from the Ghanima section by Faris (1984) and Faris et al. (1998), confirming the Ypresian age. This Formation is very similar to the Thebes Formation of Said (1960) and the name appears to be redundant. To the north of the Kharga Oasis, the upper parts of the Rufruf and Serai formations pass laterllay into the Drunka Formation.

El Naggar (1970) introduced the Drunka Formation to replace the Minia Formation of Said (1960), since the type section of the Minia Formation is incomplete (although Bishay, 1961, showed a composite section of the entire Minia Formation in the type area). The Drunka Formation overlies the Luxor Formation (Ypresian) of El Naggar (1970) in the type locality at Drunka, west of Assiut. Hermina (1990) reports a thickness of 200 m (656 ft) at the type locality. Said (1990d) divides the Formation into two units and estimates thicknesses of 360 m (1181 ft) for the lower unit and 175 m (574 ft) for the upper unit. The Formation has a thickness of 275 m (902 ft) in the subsurface in the Assiut-Kharga well. The Drunka Formation is composed of dense, thick-bedded, locally "reefal" or lagoonal limestones, with characteristic chert concretions and chert bands (McBride et al., 1999). Said (1990d) equated the two units of the Drunka Formation to the Assiuti and Manfalut formations of Bishay (1961), both of which are Ypresian in age. Calcareous algae are common, dominated by dasycladacean green algae, Halimedacean and udoteacean green algae, and gymnocodiacean red algae, and include *Acicularia valeti, A.* n.sp., *Sandalia pavisici, Clypeina accidentalis, C.* cf. *rotella, Terquemella bellovacensis, Ovulites Arabica, O. pyriformis, O. margalitula, O. elongata, Halemida tuna, H. opuntia, H. fragilis,* and *Niloporella subglobosa* n. gen. n.sp. (Dragastan & Soliman, 2002).

In the extreme southern part of Egypt, such as in the Garra-Dungul area west of Aswan, the Paleocene Garra Formation grades upward into the Dungul Formation. The name Dungul Formation was introduced by Issawi (1968) for a unit composed of shales and limestone interbeds with *Nummulites deserti, N. irregularis,* species of *Operculina, Ostrea multicostata, Sondylus alexandrae, Conoclypeos delanouei,* and *Schizaster* sp. in the lower part, and limestones with *Lucina thebaica* and flint bands in the upper part. It is 127 m (417 ft) in the type locality at Wadi Dungul (23°26′N, 31°37′E). It was assigned a Lower Eocene age by Issawi (1968) and a late Paleocene-Early Eocene age by Hendriks et al. (1987). At Barqat El Shab, the Formation is represented by 5 m (16 ft) of limestones with *Lockhartia* sp., overlying a Nubian Sandstone section (Luger, 1985). Said (1990d) included the Garra Formation in the upper members the Thebes Group.

Lower Eocene sediments in the subsurface of the northern Western Desert, the Nile Delta, and northern Sinai are variable in thickness and they are missing on the crests of major subsurface structural highs. They are composed dominantly of carbonates and include cherty, dolomitized, and argillaceous limestones. Their thicknesses vary between 122 m (400 ft) in the Abu Sennan-1, 244 m (4082 ft) in the Diyur-1, and

930 m (3051 ft) in the Gindi-1 wells (Salem, 1976). Formations have not been established for the Eocene in the subsurface, and surface formations were not extended to the subsurface, although both Egyptian and Libyan terminologies have been used in some cases (cf. Sestini, 1984).

## 8.1.2  Middle Eocene

The Middle Eocene (Lutetian) does not extend southward beyond latitude 27°N (Cuvillier, 1930, Said, 1962a). The Lower Lutetian includes the Minia, Mokattam, Qazzun, Samalut, Qarara, Darat, Maghagha, and Sheikh el Fadl formations. The Upper Lutetian includes the Gebel Hof, Observatory, Giushi, El Fashn, Khaboba, Tanka, and Schaibon formations.

Said (1960) introduced the name Minia Formation to replace the *Alveolinen Kalk* or *Oberlibysche Stufe* of Zittel (1883). It designates beds of white Alveolinid limestone, 30 m (98 ft) thick. The type section of this unit is at Zawiet Abu Saada (28°04′N, 30°49′E) south of Minia (Said, 1962). It underlies the nummulitic limestones of the Mokattam Formation and overlies the Thebes Formation. The Minia Formation has a spotty distribution. It is best developed between Assiut and Minia in the Nile Valley. The unit was given an early Middle Eocene (Lower Lutetian) age by Said (1960, 1971), based on *Alveolina frumentiformis* and *Orbitolites complanatus*, but recently Boukhary et al. (1982) and Strougo et al. (1990) suggested a late early Eocene (Ypresian) age. Boukhary et al. (1982) reported *Nummulites pomeli*, *N. rollandi*, *N. partschi* and *N. irregularis*, as well as *N. praecursor ornatus*, *Orbitolites* sp., and abundant corals and shell debris from the basal part of the Minia Formation, in the area between Minia and Samalut. They assigned this interval an Ypresian age.

Said (1990d) considered the Minia Formation and Naqb Formation to be equivalent and of Lutetian age, although he showed them to span the Nummulites zones N(G), *N. rollandi*, *N. pomeli*, *N. partschi*, and *N. praecursor ornatus* of Strougo & Boukhary (1987).

In the scarp on the eastern side of Minia, the Minia Formation consists of (Cuvillier, 1930) limestones with chert concretions. The unit is very fossiliferous, with *Nummulites lucasi*, and the echinoids *Conoclypeus delanouei*, *Leiocidaris miniehensis*, *Echinopsis libyca*, *Heterspatangus lefebverei*, and *Pleocarpilius* sp., followed upward by limestones with molluscs and gastropods, limestones with *N. atacicus* and echinoids, and then by limestones with *Alveolina frumentiformis* and *Orbitolites complanatus*. The classic Minia Formation section at Beni Hassan (ancient Speos Artemidos), to the north of Assiut, facing Abu Qurqas, was described by Cuvillier (1930). At the base, a 100 m (328 ft) of white limestones with small Nummulites, such as *N. atacicus* (*Obsoletes* according to Strougo et al., 1990), and *N. obesa*, *Orbitolites complanatus*, *Alveolina frumentiformis*, *Operculina canalifera*, as well as the small echinoids *Sismondia logotheti*, and the crustecea *Callianassa*. This unit is overlain by the Mokattam Formation.

Strougo et al. (1990) described four sections between Minia and Assiut, exposed at Gebel Drunka, Wadi Gabrawi, Tell Amarna, and Beni Hassan. They described the Upper Libyan succession (equivalent to the Minia Formation) as neritic limestones, devoid of significant planktonic microfossils, but with abundant large benthonic foraminifera (mainly *Nummulites*). They assigned the basal beds of the 200 m

(656 ft) section with *Nummulites* gr. *praecursor* an early Ypresian age, and the strata with *N. migiurtinus* and *N. Zitteli* a Lutetian age. They provisionally considered the basal Mokattamian (traditionally assigned to the Upper Lutetian, or the base of the Mokattam Formation) with *N. mariettti* as a transitional level between the Ypresian and the Lutetian. Nevertheless, their conclusions are based on meager evidence, and they failed to identify the *Alveolina* and *Orbitoliites* to species level. Bishay (1961) recongized two *Alveolina* zones in these sections: a lower *Alveolina frumentiformis* and an upper *A. ellipsoidales*.

According to Strougo et al. (1990), the stratigraphic range of *O. complanatus* extends from the early Ypresian to late Lutetian and *A. frumentiformis*, which they reported from the top unit at Beni Hassan, straddles the Ypresian-Lutetian boundary. Accordingly, the Minia Formation and its equivalents were assigned to the Ypresian. However, the age of these two species is mainly Lutetian. According to Jones & Racey (1994), *Orbitolites complanatus* ranges from late Early to early Late Eocene. In western Libya, the Lutetian age of *Orbitolites complanatus* and *A. frumentiformis*, reported from the Bir Zidan Member of the Wadi Tamet Formation, is constrained by the presence of *Dictyoconoides cooki*, which ranges from late early Eocene (P9) to top Middle Eocene (P14) (topmost Ypresian-top Lutetian in Oman, according to Jones & Racey, 1994). The latter was reported, along with *Alveolina frumentiformis* and *N. gizehensis*, from the base of the Lutetian Seeb Limestone in Oman by White (1994), and from the Lutetian part of the Dammam Formation of Kuwait, Oman, and Saudi Arabia by Jones & Racey (1994). The co-existence of *O. complanatus* and *N.* gr. *Praecursor* in the Gebel Drunka section is puzzling, although reworking of the latter should not be ruled out. The evidence for a Ypresian age for the entire Minia Formation is not convincing. The age of the Formation appears to be Lutetian, although the age of its lower part may extend into the Ypresian.

Said (1962a) applied the name Mokattam Formation to the Lower *Mokattam-stufe* of Zittel (Figs. 8.2, 8.4). The Formation overlies the Minia Formation and it is overlain by the Maadi Formation. The Formation crops out over large areas in the Western Desert, the Nile Valley, the Gulf of Suez, the Eastern Desert, and Sinai. The base of the Formation contains *Nummulites gizehensis*, and the top contains *Operculina pyramidum*. This unit was assigned an Upper Lutetian age by Cuvillier (1930) and an upper Middle Eocene-lower Upper Eocene by Said (1962). It has a thickness of 133 m (436 ft) at the type section near the Citadel, Cairo. The section described by Cuvillier (1930) at Gebel Mokattam starts with sandstones and conglomerates, approximately 3 m (10 ft) in thickness, with ferruginous nodules, followed upward by detrital limestones with *N. gizehensis, N. globulus, N. guettardi, Echinolampas africanus, Orthechinus mokattamensis, O. humei, O. schweinfurthi, Coptosoma aegyptiacum*, in addition to molluscs, a marly limestone bed with *Lithothamnium*, limestones with Nummulites, gastropods, and bivalves, followed by limestones with *N. atacicus, N. lucasi, N. uroniensis, N. curvispira*, limestones with corals, crustacea, *Operculina pyramidum*, orbitoides, *Lucina pharaonis, vasum frequens*, and numerous species of *Pecten* and *Cardita, Schizaster mokattamensis, S. foveatus, S. africanus, S. humei, S. libycus, S. deserti, Brissopsis excentrica, B. lamberti, B. lorioli, Echinolampas africanus*, and *Nautilus* sp. with diameters up to 50 cm (20 in), the succession ends with limestones formed entirely of molluscs, with *Nummulites atacicus*. This succession is overlain by Upper Eocene limestones rich in bryozoans.

The Mokattam Formation at the type locality consists of three members (Said, 1962, 1971, Said & Martin, 1964), in ascending order:

1   Gizehensis Member composed of limestones crowded with *N. gizehensis* and numerous gastropods, 53 m (174 ft) in thickness. This unit is equivalent to the Gebel Hof Member in Helwan (Said, 1971).
2   Cairo Buildingstone horizon, white massive limestones with gypsum, and contains *Nummulites biarritzensis*, shark teeths, *Oxyrhina* sp., *Carchardon* sp., *Ceriphium giganteum, Natica cornuammonis, Lobocarcinus* sp., and *Nautilus* casts. It has a thickness of 33 m (108 ft). A bed crowded with small *N. biarrizensis* occurs at the top. This unit is equivalent to the Observatory Member in Helwan (Said, 1971). Bitner & Boukhary (2009) recorded the brachiopod species *Terebratulina tenuistriata* in the Bartonian nummulitic limestone of the Mokattam Formation. That was the first that brachiopods were reported in Eocene sediments in North Africa.
3   Giushi Member or Giushi Formation, a white, occasionally compact, limestone unit (the *Operculina pyramidum* Zone), with *N. contortus-striatus, Dictyoconus aegyptensis, Echinolampas fraasi, E. africana, Euspatangus formosus, Porocidaris schmiedelii*, and abundant pelecypods and bryozoans. The unit grades upward into clayey gypsiferous beds (celestine is quarried in these beds). Its thickness is 60 m (197 ft). It was assigned a Bartonian age. This unit is overlain by the Maadi Formation. The type section is at Gebel Giushi, east of Cairo.

*Figure 8.2*  Thebes Formation Type section and the contact between the Thebes Formation and the Esna Shale, Der El-Bahari Temple, built by Queen Hatshepsut (ruled c.1490-1468 BC) on the west bank of the Nile, facing Luxor. The temple is incised into the Esna Shale. For mechanical properties of the rocks and their effect on the temple, see Abdallah & Hilal (1990).

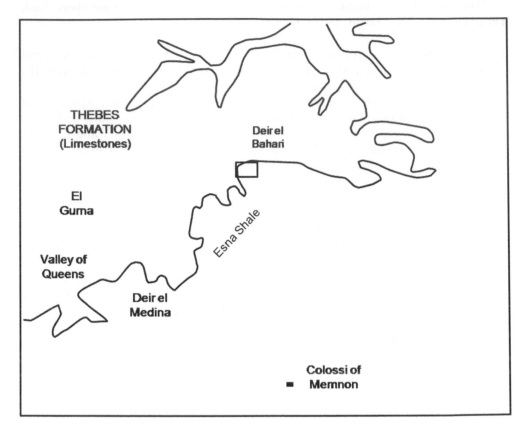

*Figure 8.3* Geological map of the Deir El-Bahari area, Thebes (after Said & Martin, 1964).

The Great Sphinx of Giza (Fig. 8.4) was carved in Middle Eocene rocks (Said, 1962). Its is carved out of the *Operculina pyramidum* Horizon of the upper Mokattam Formation and its body is made up of Nummulitic limestones of the lower Mokattam Formation (Said & Martin, 1964). Although many scholars believe that the Sphinx depicts the Pharaoh Khephren who ruled Egypt c.2558-2532 BC, the age of its construction is still being debated.

The Mokattam Formation exposed at the scarp on the eastern side of Minia is made up of limestone beds with gastropods and bivalves, devoid of large foraminifera, and limestones with *N. gizehensis, Alveolina* cf. *subpyrenaica, Echinolampas africanus, Schizaster fourtaui, Cassidulus* sp., and *Nautilus* sp. (Cuvillier, 1930). The unit is approximately 160 m (525 ft) in thickness. At Beni Hassan, north of Assiut, the Mokattam Formation is composed of white limestones, with *N. gizehensis, N. curvispira,* and *Echinolampas africanus, E. perrier, Schizaster mokattamensis,* and the molluscs *Vulsella* sp., *Carolia placunoides, Rosetellaria* sp., *Mesalia hofama, Turitella pharaonica,* followed by white limestones with *N. gizehensis* and *N. atacicus,* and yellow limestones with abundant *N. gizehensis* in addition to *Echi-*

*Figure 8.4* The Great Sphinx of Giza and the Pyramid of Khefren.

*nolampas africanus, E. perrier, Schizaster mokattamensis,* and *Ostrea pharaonum,* and *Corbula* sp. (Cuvillier, 1930). Bishay (1961) described the lower 50 m (164 ft) of this unit which he called the Samalut Formation. He estimated a thickness of 160 m (525 ft) at the type locality, opposite Samalut, near Minia. He reported common *N. gizehensis, N. curvispira, N. guttardi, N. subdiscorbinus,* and *N. subatacicus,* and dated it Lower Lutetian. The Samalut Formation is partly equivalent to the Sheikh Fadl Formation.

The section north of Samalut was divided by Bishay (1966), from base to top, into Maghagha Formation, Qarara Formation, and El Fashn Formation.

The Maghagha Formation was proposed by Bishay (1966) for a unit exposed in the Gebel El Sheikh Fadl and Gebel Mereir which overlies the Thebes Formation and consists of white, chalky limestones and marls, with a maximum thickness of 81 m (266 ft) (Elewa et al., 1995). It has been dated middle Eocene by Boukhary et al. (1982). It is considered partly coeval with the Sheikh Fadl Formation and the Samalut Formation, although the upper beds may be younger (Said, 1990e). Similarly, the Maghagha Formation was considered equivalent to the Qarara Formation by Elewa et al. (1995), although the latter may include younger beds.

The Qarara Formation was proposed by Bishay (1966) for a carbonate unit which overlies the Samalut Formation. The type section is in the Qarara butte opposite Maghagha, where the Formation attains 170 m (558 ft). The Formation is made up of 20 m (66 ft) basal shales, grading upward into siltstones with occasional carbonaceous beds, with brackish-water molluscs (Strougo & Azab, 1982). The upper part is 150 m (492 ft) thick and consists of massive banks of *Nummulites gizehensis.*

Elewa et al. (1995) described the Formation from the El Sheikh Fadl area, northeast of Minia, as composed of yellowish-white, very hard calcareous sandstones, clayey limestones, and greenish-white to grey shales rich in macrofossils and *Nummulites*.

The overlying El Fashn Formation (Bishay, 1966) is composed of white limestones with flint band intercalations, rich in *Nummulites* and bryozoans. It has a maximum thickness of 20 m (66 ft) (Elewa et al., 1995). It was assigned a Late Lutetian age by Bishay (1966) based on *Nummulites*.

Cronin & Khalifa (1979) reported *Truncorotaloides rohri, T. topilennsis, Globigerina linaperta, G. parva, G. yeguaensis,* and *Globorotalia centralis* from the Qarara and El Fashn formations from the Gebel Mereir, east of Beni Mazar, and assigned them a late Lutetian age.

The Naqb Formation was introduced by Said & Issawi (1963) for a unit in the type section at Naqb Ghorabi, Bahariya Oasis. It is made up of a lower unit of non-fossiliferous dolomitic limestones, followed by an upper unit of fossiliferous limestone beds with minor shale intercalations, with pelecypods, gastropods, *Operculina* sp., *Alveolina* sp., *Nummulites* sp., and coralline algae. The limestones are mostly intensely bioturbated dolomitized lime wackestones (Said, 1990e). Its thickness is 13 m (43 ft) in the type section, but reaches up to 68 m (223 ft) in boreholes (Said, 1971). The Naqb Formation is overlain by the Qazzun Formation. It was assigned a Lower Lutetian age and is considered an equivalent of the Minia Formation by Said (1971).

The iron ore deposits of the Bahariya Oasis are layered lenticular bodies within the Naqb Formation (Nakhla, 1961, Said & Issawi, 1963, Said, 1990e), and are, therefore, Lutetian in age. Mesaed & Surour (1998) discussed the origin of these iron ores, but considered them to be Bartonian in age. At Gebel Ghorabi in the northern part of the Bahayria Oasis, Eocene sediments host pisolitic and pseudo-pisolitic iron ore beds, which cap the Gebel. These ores are exploited in the El Gedida Mine. The ore is in the form of stratabound and stratiform ironstone bands. The stratabound are lateritic muddy and sandy ironstones, the stratiforms are glauconitic ironstone pockets and concretions, developed as a result of subaerial exposure and laterization. Basta & Amer (1969) postulated a hydrothermal origin for these ores.

In the Farafra Oasis, the Naqb Formation forms the top of the scarp and the plateau surface to the north of the Farafra depression. It overlies the Esna Shale except north of Ain Maqfi where it overlies directly the Tarawan Chalk (Hermina, 1990). Youssef & Abdel Aziz (1971) called this unit the Thebes-Farafra Formation and assigned it a Ypresian age.

In the Bahariya Oasis, the Lutetian is represented by two formations, the Qazun Formation and the Hamra Formation.

The Qazzun Formation was introduced by Said & Issawi (1964) for a carbonate unit in the north Bahariya plateau, which overlies the Naqb Formation and unconformably underlies the Hamra Formation. The type section is at Ghard Qazzun, where the Formation is made up of thick white nummulitic limestones with *Nummulites cailliaudi, N. atacicus* and *N. variolarius*. It was assigned an Upper Lutetian age by Said & Issawi (1964) and Said (1971). Said (1990d) considers it as equivalent to the lower part of the Samalut Formation.

The Hamra Formation was proposed by Said & Issawi (1963) for a 75 m (246 ft) thick unit in the type section at Gar El-Hamra (28°30'N, 29°40'E), northern Bahariya plateau, and consists of limestones with marl intercalations, with *Nummulites champollioni, N. gizehensis, Ostrea multicostata,* and *Turitella angulata.* It overlies the Qazzun Formation and unconformably underlies the Radwan Formation (Oligocene) (El Akkad & Issawi, 1963). It was assigned an Upper Lutetian-Bartonian age, although there is no proof for the Bartonian age.

In the Fayium district, the Muweilih Formation and Midawara Formation were introduced by Beadnell (1905b). The Muweilih Formation is 30 m (98 ft), base unexposed, and composed of limestones rich in *N. zitteli* and *N. delaharpei.* The Midawara Formation is 15 m (57 ft) in thickness, and consists of a lower shale unit, occasionally lignitic, a middle carbonate bed rich in *N. gizehensis,* and an upper unit rich in macrofossils. The two formations are probably equivalent to the Samalut and Qarara formations, respectively (Said, 1990e).

Boukhary & Abdelmalik (1983) introduced the name Sannor Formation for a section at the Wadi Sannor-Wadi El Mawathil area (28°31′N, 31°37′E) on the eastern side of the Nile Valley, composed of 484 m (1588 ft) of white to yellowish white limestones intercalated with very hard, dark grey dolomitic limestones, and white crystallized limestone beds, with *N. striatus, N. beaumonti, Gypsina carteri,* and *Dictyoconus aegyptiensis,* among others. They assigned it a middle Middle Eocene age.

Middle and Upper Eocene sediments crop out in Sinai at Gebel Krar (29°17′N, 33°07′E), on the eastern side of the Gulf of Suez. Lutetian marly limestones contain the following fossils (Cuvillier, 1930): *Bolivina puncatata, Globigerina bulloides, G. conglobata, G. cretacea. Discorbina rugosa, D. globularis, Rotalia calcaliformis, Nummulites subdiscorbina, N. curvispira, N. gizehensis, Orthophragmina dispansa, O. ephippium,* and *O. papyracea.* These fossils fossils are also characteristic of the Middle Eocene around Cairo.

At Gebel Tanka (29°05′N, 33°04′E) and Wadi Gharandal (29°15′N, 32°58′E), on the eastern side of the Gulf of Suez, Hume et al. (1920) and Moon & Sadek (1922) divided the Upper Eocene section, from base to top, into the Cardita series, Green beds, gypsiferous marls, Tanka beds, and Red beds. The *Cardita* Series is 160 m (525 ft) at Gebel Tanka and 15–30 m (49–98 ft) thick at Wadi Gharandal, and it is made up of limestones and chalky marls interbedded with flint beds in the upper part. It contains *Cardita viquesneli, C. depressa, C. mutabilis, C. subcomplanata, Lucina blanckenhorni, L. mutabilis, L. pharaonica* and *Leda phacoides.* Cuvillier (1930) suggested a Bartonian age for this unit. The Green Beds are made up of green marls, 15–30 m (49–98 ft) thick at Wadi Gharandal, rich in fish teeth of *Carchardon auriculatus, Odontaspis* sp., *Oxyrhina Desori* reported from Wadi Gharandal (Moon & Sadek, 1922), as well as *Galeocendo latidens, Hemipristis curvatus,* and *Lamna* sp. reported from Gebel Tanka (Hume et al., 1920). The unit also contains *Nummulites beaumonti* and a large number of mollusc species Cuvillier (1930). The green color of the marls is caused by glauconite (Said, 1962a). The Gypseous Marls are non-fossiliferous at Gebel Tanka and consist of 80 m (262 ft) of interbedded shales and argillaceous limestones, fissile shales, and gypsiferous marls. At Wadi Gharandal, it contains *Ostrea aschersoni, Axinus schweinfurthi, Lucina polythele,* and *L. blanckenhorni,* among other species. The Tanka Beds are composed of non-fossiliferous marls and bedded limestones at Wadi Gharandal. They form a part of the Gypseous Marls at Gebel Tanka. The Red Beds consist of sandstones, conglomerates, marls and shales. It is often overlain by Miocene sediments and overlies the Khaboba Formation.

Its thickness varies from a few centimeters to 20 m (66 ft) at Gebel Tanka and up to 30 m (98 ft) at Wadi Gharandal. The marls and shales contain *Nummulites beaumonti, N. contortus, N. striatus*, and many mollusc species, such as *Vulsella crispata, Pecten cailliaudi*, and *Ostrea stanleyi*. Cuvillier (1930) suggested a Bartonian age for this unit.

In the Gebel Nukhul area (29°03′N, 33°16′E), on the east coast of the Gulf of Suez, Viotti & El-Demerdash (1969) proposed the names Darat Formation for the Green Beds and the Khaboba Formation for the Gypseous Marls, and raised the Tanka Beds (Upper Eocene) to a formation status. They assigned a Middle Eocene age to the Darat and Khaboba formations. The Darat Formation contains the pelagic foraminifera *Globorotalia bullbrooki* and *G. bolivariana*, the Khaboba Formation contains *Truncorotaloides pseudodubius* and *T. topilensis* (upper Lower-upper Middle Eocene according to Toumarkine & Luterbacher, 1985). The Tanka Formation contains *Globigerapsis semiinvolutus* and *Globorotalia cerroazulensis* "zones" assemblages. These "zones" were also reported from the Eocene of the subsurface of the Nile Delta by Viotti (in Viotti & Mansour, 1969). It was dated Middle-Upper Eocene by Viotti & El-Demerdash (1969). At Burnubul, east of Wasta (29°20′N, 31°12′E), approximately 95 km south of Cairo; the equivalent of the Darat Formation is composed of marls with thin beds of gypsum. The marls contain teeth of *Ginglymostoma blanckenhorni, Oxyrhina desori, Odontaspis venticalis, Lamna macrota, Galeocerado latidens, Apriondon frequens*, and *Amblypristis cheops*, among others, as well as abundant mammal vertebrae (Blanckenhorn, 1902, Cuvillier, 1930). This succession was dated Bartonian by Cuvillier (1930).

El-Akkad & Abdallah (1971) divided the Eocene section in the Gebel Ataqa area (29°59′N, 32°22′E), near the city of Suez, into a lower Suez Formation composed of 224 m (735 ft) of dolomitic limestones and marls, and an upper El Ramiya Formation, 78 m (256 ft) thick and and made up of clastics and marls. The two formations are Middle Eocene in age and were correlated with the lower part of the Khaboba Formation by Said (1990d).

Said (1990d), following the authors of the new Geological Map of Egypt (1987), divided the Mokattam Group (Middle Eocene) into a lower Mokattam Subgroup comprising the *Nummulites gizehensis*-bearing formations, such as the Mokattam, Qazzun, Samalut, Qarara, and Darat formations, and an upper Observatory Subgroup (Middle Eocene) which includes the Gebel Hof, Observatory, Giushi, El Fashn, Khaboba and Tanka formations. It is however, a bad practice to use the same name for a group as well as for a formation (see North American Stratigraphic Code, 1983).

The type section of the Observatory Subgroup is at the Observatory Plateau at Helwan (29°51′N, 31°20′E). Previously Farag & Ismail (1959) subdivided the Mokattam Formation at Helwan into a lower Gebel Hof Member and an upper Observatory Member. These two members were raised to formation status by the authors of the new Geological Map of Egypt (1987) and Said (1990d). The Gebel Hof Formation is 121 m (397 ft) thick (base not exposed) at the type locality at Gebel Hof, east of Helwan. It is made up of non-fossiliferous limestones, becoming heavily burrowed and rich in *Nummulites gizehensis* upward. The Observatory Formation consists of limestones and chalky limestones. Farag & Ismail (1959) estimated a thickness of 80 m (262 ft) for the unit at the type locality at the Observatory Plateau, Helwan, while Said (1990d) reported a thickness of 136 m (446 ft) for the Formation. It is overlain by the Qurn Formation (upper Middle Eocene).

The Qurn Formation (Farag & Ismail, 1959) is well developed in the Helwan area. It reaches 97 m (318 ft) in thickness in the type locality at Qurn, Helwan, and it is made up of marly and chalky limestones alternating with shales, sandy marls and shell banks rich in *Ostrea reili, Nummulites beaumonti, N. striatus, N.* aff. *pulchellus,* and *Operculina pyramidum.* It overlies the Observatory Formation and it is overlain by the Wadi Garawi Formation. The Wadi Garawi Formation (Farag & Ismail, 1959) is made up of a 25 m (82 ft) section of poorly fossiliferous sandy shales with a hard, highly fossiliferous middle bed with *Plicatula polymorpha* and *Nicaisolopha clotbeyi.* It overlies the Observatory Formation and is overlain by the Wadi Hof Formation. The type locality is at Wadi Garawi, Helwan. The Qurn and Wadi Garawi formations were dated Upper Eocene by Farag & Ismail (1959) and Middle Eocene by Strougo & Boukhary (1987). They were correlated with the Tanka Formation in the Gulf of Suez by Said (1990d).

In the Fayium district, north of Birket Qarun, the Eocene succession is composed of the Wadi Rayan Formation and Gehannam Formation (Middle Eocene), and the Birket Qarun and Qasr el Sagha formations (Upper Eocene). The stratigraphy and history of Eocene nomenclature in the Fayium and Mokattam area have been thoroughly reviewed by Ginegrich (1992). Marine mammals including cetaceans and sirenians (Gingerich, 1992) and bats (Gunnell et al., 2008) are common in these two areas.

Gunnell et al. (2008) reported four new genera and six new species of microchiropteran bats from the late Eocene (37–34 Ma) of the Fayum Depression, including new taxa and the first and only African record of a fossil rhinopomatid and the earliest African records of megadermatids, emballonurids, and vespertilionids, as well as a new genus and two new species of the Afro-Arabian bat family Philisidae. Bats and primates may have entered Africa during the same dispersal event early in the Paleogene. Upper Eocene species include *Witwatia schlosseri, Witwatia eremicus,* and *Qarunycteris moerisae* (Umm Rigl Member of the Birket Qarun Formation), *Dhofarella sigei, Saharaderma pseudovampyrus,* and *Khonsunycteris aegypticus* (Jebel Qatrani Formation).

Beadnell (1905b) applied the name Wadi Rayan series (Upper Lutetian) to an Eocene succession composed of glauconitic limestones with *N. gizehensis, N. curvispira, N. atacicus,* and abundant corals, followed upward by marls and shales and sandy shales, argillaceous sands with *Ostrea* sp. and *Carolia placunoides,* and then by white limestones with *N. gizehensis, Lucina mokattamensis,* and abundant gastropods. Said (1962a) denoted the unit as the Wadi Rayan Formation. To the west of the Birket Qarun lake, at Gar Gehannam, the Wadi Rayan Formation is 70 m (230 ft) in thickness and overlain by the Ravine beds. The Ravine beds of Beadnell (1905b) are 50 m (164 ft) and composed of shales and marls with marine vertebrates in the lower part and sandy limestones and sandstones with *Zeuglodon* (later renamed *Basilosaurus*) and *Dorudon atrox* (Dolson et al., 2002), in addition to shark and fish remains in the upper part. The Ravine beds were called the Gehannam Formation by Said (1962a). Strougo & Haggag (1984) recognized two biostratigraphic units in the Gehannam Formation; the *Globorotalia lehneri* Zone and the *Hantkenia alabamensis/Truncorotalides rohri* Zone. They suggested that the Gehannam Formation belongs entirely to the Middle Eocene and correlated it with the Giushi Member of the Mokattam Formation of the Nile Valley. However, Said (1971) postulated that the Giushi Member is Upper Eocene in age, based on the presence of *N. contortus-striatus* and *Opeculina pyramidum* (renamed *Nummulites* aff. *pulchellus* by Strougo, 1986), Boukhary & Abdelmalik (1983) assigned it a late Middle Eocene age, and Said (1990d) gave it a Middle Eocene age.

In the subsurface of the northern Western Desert, the Middle Eocene is more terrigenous than the Lower Eocene. In the Abu Sennan-1, Abu Gharadig-1 & 2, and WD58-VH-1 wells, the Middle Eocene is composed of thick sand, silt and shale beds (Salem, 1976).

### 8.1.3    Upper Eocene (Bartonian)

The Upper Eocene does not extend south of the Fayium Province (Said, 1962a, p.24). The Upper Eocene sequence maintains the same lithological characteristics in the Red Sea area, the Gulf of Suez area, and Sinai, except that marls increase in abundance in these areas. Sandstones are common only in the Fayium area.

The Upper Eocene is marked by the appearance of *Nummulites contortus* and *N. striatus*. Cuvillier (1930) suggested that in the absence of diagnostic fossils, the horizons rich in bryozoans could be taken as the base of the Upper Eocene.

In Gebel Gehannam (29°19′N, 30°09′E), the Gehannam Formation is overlain by a 50 m (164 ft) thick unit of sandstones, with shale intercalations at the base, and fossiliferous limestones at the top, with *N. beaumonti*, *N. contortus*, *N. striatus*, *N. atacicus*, *Operculina* cf. *discoidea*, *Ostrea multicostata*, *O. plicata*, *O. longirostris*, *O. reili*, *O. fraasi*, and *Carolia placunoides*. This unit was called the Birket Qurun series by Beadnell (1905b) and was assigned a Lower Bartonian age by Cuvillier (1930) and Said (1962a). Said (1962a) changed the name to Birket Qarun Formation.

Blanckenhorn (1900) designated Upper Eocene (Upper Mokattam) fresh and brackish water sediments in the Faiyum area the name "*Carolia Stufe*" which he assigned a Bartonian age. Beadnell (1905b) also gave a Bartonian age to the fluvio-marine deposits with terrestrial vertebrate and marine invertebrate fossils in the Gebel Qatrani area (29°40′N, 30°36′E) in the Fayium province which he called the Qasr el Sagha series. The Qasr el Sagha Formation (Said, 1962) starts at the base with limestones, followed upward by argillaceous and gypsiferous limestones, shales, siliceous limestones rich in molluscs and nummulites of the Birket Qarun Formation, and ends with sediments with *Carolia placunoides*, *Ostrea Clotbeyi*, *O. reili*, and *O. multicostata* at the top. The succession is 180 m (591 ft) thick in the type section, in the northern Fayium district, and overlain by the Oligocene Qatrani Formation. In the Qattara Depression to the northwest, the Qasr el Sagha Formation is 36 m (118 ft) in thickness and consists of gypsiferous shales, interbedded with calcareous sandstones and sandy limestones, and contains marine invertebrate and land vertebrate fossils (Albritton et al., 1990). The Temple Member of the Qasr El Sagha Formation contains spheroidal calcite-cemented concretions ranging from 70 cm (28 in.) to 9 m (30 ft) in diameter, although most are 1–4 m in diameter. Miocene marine carbonate shells are preserved inside the concretions (Abdel-Wahab & McBride, 2001).

Strougo & Haggag (1984) and Strougo & Boukhary (1987) correlated the Wadi Rayan Formation, Gehannam Formation, and Birket Qarun Formation, with the three members of the Mokattam Formation, *Gizehensis* Limestone Member, Building Stone

Member, and Giushi Member in the Nile Valley, respectively. They also correlated the Qasr el Sagha Formation with the Maadi Formation.

Said (1962a) introduced the name Maadi Formation to replace the Upper *Mokattam Stufe* of Zittel (1883). The type locality of the Formation is near Maadi, south of Cairo. The unit is composed of brown sandy limestones with *Nummulites contortusstriatus* in the lower part and shales and sandy shales with *Carolia placunoides, Ostrea clotbeyi, O. fraasi*, and *Plicatula polymorpha* banks in the upper part. The top of the Formation is composed of a limestone bed rich in echinoids (Ain Musa bed), such as *Echinolampas crameri* and *Anisaster gibberulus*. The Maadi Formation varies in thickness from 64 m (131 ft) in Maadi to 180 m (591 ft) in Fayium. It does not extend beyond the latitude of Fayium. Said (1990d) raised the unit to a group status, the Maadi Group, to include the Maadi Formation in the Gebel Mokattam, the Qurn, Wadi Garawi and Wadi Hof formations of Farag & Ismail (1959) in the Helwan and Giza areas, Gehannam and Qasr El Sagha formations of Beadnell (1905b) in the Fayium district, and the Tanka Formation in the Gulf of Suez area. Said (1962a) originally assigned the Maadi Formation an upper Upper Eocene age, and Farag & Ismail (1959) gave an Upper Eocene age to their Qurn and Wadi Garawi formations. However, recent studies have shown that the two latter formations are Middle Eocene in age, and the Wadi Hof Formation is of an Upper Eocene age (Strougo & Boukhary, 1987).

At Gebel Mokattam, east of Cairo, the Maadi Formation is made up of (Cuvillier, 1930): 1) Bryozoan limestones rich in molluscs and gastropods, shales, and brown siliceous limestones, with *N. atacicus, N. contortus*, and *N. striatus*, coral fragments, and annelides. 2) Foliated gypsiferous shales. 3) Brown limestones with *N. contortus, N. striatus*, fish remains, shells, and coprolites. 4) Coquinoidal limestones with *Carolia placunoides*. 5) Siliceous limestones with abundant molluscs and gastropods. 6) Siliceous limestones with echinoids, e.g. *Echinolampas crameri*, as well as fragments of *Callianassa* and abundant corals. 7) Calcareous sandstones with abundant *Callianassa*. The total thickness of this sequence is approximately 90 m (295 ft).

The Maadi Formation section at Gebel Kibli el Ahram, west of Cairo, was described in detail by Blanckenhorn (1902) and Cuvillier (1930). The section consists of, from base to top: 1) Fissile marls with *Textularia, Rotalina, Globigerina, Operculina*, ostracods, and teeth of *Carcharias frequens, Propristis schweinfurthi, Saurocephalus fayumensis*, and *Lamna* sp. 2) Fossiliferous marly limestones and siliceous limestone beds with *Echinolampas globulus* and numerous mollusc species. 3) Gypsiferous marls and brown limestones with *Carolia placunoides*, oyster banks with *Ostrea clotbeyi, O. elegans, Plicatula polymorpha, Turitella polytoeniata*, and teeth of *Propristis scweinfurthi*. 4) Marly limestones rich in molluscs. 5) Brown sandy limestones rich in the echinoids *Echinolampas humei, Thagastea luciani* and *Brisspospis* sp., fragments of *Callianassa*, and *Ostrea clotbeyi* and other molluscs.

Omara et al. (1977) introduced the name El Mereir Formation for a 90 m (295 ft) Eocene section at Gebel Mereir, 80 km east of Beni Mazar, composed of limestones, arenaceous marls, marly limestones, and shales. Bishay (1966) previously included this section in the Qarara Formation. Cronin & Khalifa (1979) reported a foraminiferal assemblage from that section, which includes *Globigerina ciperoensis angustiumbilicata, G. parva, G. rohri, G. ampliapertura, Globorotalia opima nana, G. cocoaensis*, and *Hantkenina alabamensis*, in addition to *Nummulites striatus, N. lucasi*, and *N. discorbinus*, and designated it a Late Eocene age.

Youssef et al. (1984) divided the Maadi Formation in the Sakkara area (29°51′N, 31°14′E), south of Cairo, into two members, the Sakkara Member at the base and the Giran El Ful Member at the top. The Sakkara Member consists of a basal bed, approximately 4 m (13 ft) in thickness, exposed only at the Abu Sir plateau, and it is composed of greyish-green, marly shales, with some gypsum veins, and an upper bed of alternating light yellow, hard limestones and yellow marls, approximately 22 m (72 ft), and few fossils. The Giran El Ful Member is approximately 17.4 m (57 ft) thick and composed of sandy to marly limestones and shales, highly fossiliferous with miliolids, ostracods, and shallow water benthic foraminifera, and banks of *Carolia planucoides* near the top. It is overlain by the Early Pliocene Kom El Shelul Formation. The name Saqqara Limestone was given by Hume (1911) in the Saqqara area to a 25 m (82 ft) thick unit composed of brown, siliceous, dolomitized, non-fossiliferous limestones overlying the Mokattam Formation and marked by a hard *Carolia*-rich bed at the top. Said (1962a) included this unit in the Minia Formation. It is equivalent to the Giran El Ful Member of Youssef et al. (1984).

The information about the Eocene in the offshore of the Gulf of Suez is scanty. Ansary et al. (1961) reported Upper Eocene from the Bakr and Karim fields in the central Gulf of Suez. They also form potential reservoirs in the Sudr, Asl and Matarma oil fields on the eastern side of the northern Gulf of Suez, where they consist of fractured marine carbonates (Alsharhan & Salah, 1995). The Eocene sediments are formed mainly of dolomitic limestones in the lower part and grade into cherty, shaly limestones in the upper part. Two faunal assemblages have been recognized. The lower assemblage is probably Lower Eocene in age and includes species of *Anomalina* and *Globorotalia*, as well as large foraminifera, such as *Operculina libyca*, *Nummulites deserti*, *N. atacicus*, and *Alveolina oblonga*. The upper assemblage is probably Upper Eocene in age and characterized by *Bolivina moodysensis*, *Robulus trompi*, *Globigerina triloculinoides*, *Globorotalia crassata*, *Gumbelina barnardi*, *Bulimina jacksonensis*, *Plectofrondicularia californica*. The absence of the Middle Eocene was attributed to uplift and erosion of fault blocks (Ansary et al., 1961).

## 8.2 OLIGOCENE

Oligocene rocks in Egypt include predominantly siliciclastic sediments of sandstones, shales, and gravels, and minor marls and limestones. They include the Qatrani Formation in the Fayium Province, Radwan Formation in the Bahariya Oasis, the Gebel Ahmar sands and gravels near Cairo, the Nakheil Formation along the western Red Sea coast, the Tayiba Red Beds in Sinai, and the Dabaa Formation in the subsurface of the northern Western Desert.

Oligocene rocks are exposed north of the Fayium depression, southwest of Cairo, and they comprise the Qatrani Formation. Beadnell (1905b) first applied the name "fluvio-marine series" or "Qatrani beds" to a unit of sandstones, shales and argillaceous marls overlying the Qasr el Sagha "series" at Gebel Qatrani on the northern side of Birket Qarun. These rocks consist of cross-bedded sandstones and gravels with interbeds of shales and limestones, rich in vertebrate fauna and fossil wood. The Qatrani Formation is rich in vertebrate fauna and fossil wood, and contains the earliest fossil record of African anthropoid primates (monkeys and apes), such as

*Catopithecus, Proteopithecus, Oligopithecus, Apidium, Simonsius, Propliopithecus,* and *Aegyptopithecus* (Fleagle et al., 1986, Gingerich, 1993). It also yielded a nearly complete skull of *Parapithecus grangeri* (Simons, 2001), from a quarry discovered in 1962, where thousands of vertebrate fossils, including primate maxillae and mandibles of the late Eocene elephant-shrews *Herodotius pattersoni* and an upper dentition of the early Oligocene taxon Metoldobotes were found (Simons et al., 1991). The Formation is 340 m (1116 ft) (Simons & Rasmussen, 1990), but thins to 50 m (164 ft) at Widan el Faras. The Qatrani Formation is overlain by the Widan el Faras Basalts dated at 31 ± 1 Ma (Fleagle et al., 1986). This age provides a minimum age for the Qatrani Formation of late Early Oligocene. The Oligcene age of the Gebel Qatrani Formation has been confirmed by Ginegrich (1993). The Qatrani Formation is also exposed in the southwestern part of the Qattara Depression, but extends into the subsurface on the eastern side. The exposed section is 41.6 m (136 ft) and made up of variegated sandstone and shale with numerous marl beds with fossil wood and verebrate remains (Albritton et al., 1990). It overlies conformably the Qasr el Sagha Formation and is unconformably overlain by the Lower Miocene Moghra or Mamura formations.

In the Bahariya oasis, the Oligocene Radwan Formation (El-Akkad & Issawi, 1963) is composed of non-fossiliferous, ferrugenous quartzites and coarse grained sandstones. The type locality of the Formation is at Gebel Radwan, Bahariya oasis, where it has a thickness of 45 m (148 ft). It overlies unconformably the Cenomanian Baharia Formation or Upper Eocene sediments. The Formation is spotty in distribution and its thickness varies from 1 m (3 ft) to 45 m (148 ft).

East of Cairo, Oligocene quartzitic and reddish sandstones were exposed at Gebel Ahmar (Red Mountain)[3]. These sands extend to Lake Timsah (Cuvillier, 1930) and are associated with petrified wood and chert conglomerate. They contain miniature volcanoes (Rennebaum Volcano) (de la Roche, 1936, Tosson, 1953) and associated sandstone pipes (Fig. 8.5). The Rennebaum Volcano lay approximately on longitude 31″ 17′ and latitude 30″ 02′, on the northern slope of Gebel Mokattam, about three kilometers east of the Citadel of Cairo. It was marked on the map of Cairo, sheet N″ 81/630, as an extinct gyser Tosson (1953). It owed the denomination "Volcano" to its conical form, the black debris cover, and the presence of a 9 m (30 ft) deep and 50 m (164 ft) wide crater-like depression at its top. It had a height of about 20 m (66 ft) and a diameter of 110 m (361 ft) at the base.

In the Qoseir-Safaga district, the Thebes Formation is unconformably overlain by the Nakheil Formation (El-Akkad & Dardir, 1966) assigned to the Oligocene based on its stratigraphic position. The type area of the Formation is Wadi Nakheil, Qoseir, where it has a thickness of up to 120 m (394 ft) and composed of coarse breccia beds and lacustrine variegated calcareous shales.

In the Tiran Island, at the southern end of the Gulf of Aqaba, Oligocene sediments are composed of 60–100 m (197–328 ft) of polymictic conglomerate composed of pink granite, rhyolite, anhydrite, and dolomitic sandstone pebbles with arkosic sandstone lenses (Goldberg & Beyth, 1991). They overlie Precambrian volcanics and underlie Miocene sediments.

---

3 Presently, the Green Mountain. The outcrops disappeared under the recent development of Nasr City.

*Figure 8.5* Hydrothermal sandstone pipes, former Gebel Ahmar. Leaning on the pipe on the right is the author; sitting is the late Fawzi Hamza (photo taken by Baher El Kalioubi).

Oligocene sediments in Sinai are represented by the Tayiba Red Beds (Lower Oligocene) (Said, 1971, 1990d) and the Wadi Arish Formation (Upper Oligocene). The Tayiba Red Beds are composed of variegated sandstones and shales, and are injected and capped by basalts. They are also known from boreholes in the extreme north and the offshore of Sinai (Jenkins, 1990). They are approximately 91 m (299 ft) in the type section at Wadi Tayiba on the western margin of the Hammam Faraun Fault block on the eastern side of the Gulf of Suez. Their thickness increases from 152 m (499 ft) in northern Sinai to more than 1800 m (5906 ft) in the offshore. Based on sedimentological and stratigraphic studies of the Wadi Tayiba type-section, Jackson (2008) concluded that the Wadi Yayiba Formation was deposited in a major paleovalley system at the base of an early syn-rift succession. Only moderate amplitude variations in relative sea-level occurred during the Eocene to Early Oligocene within the Suez Rift. A major relative sea-level fall occurred during the middle Oligocene and a regionally developed erosional surface associated with this event marks the contact between late pre-rift and early syn-rift strata. Jackson (2008) postulated that the Tayiba Formation (Lower Oligocene) is a prerift unit and the non-marine Abu Zenima Formation is an early syn-rift unit (cf. Nukhul Formation).

The Wadi Arish Formation crops out along the easternmost flank of the Risan Aneiza Mountains (Kuss & Boukhary, 2008). The formation is composed of a carbonate-dominated succession. It has a thickness of 77 m (253 ft). It disconformably overlies Jurassic to lower Cretaceous carbonates. The Wadi Arish Formation consists of three members: 1) The Lower Wadi Arish Member consists of 42 m (138 ft) of a gypsiferous sandstone unit overlain by two limestone units, with foraminifera and corallinaceans. 2) The Middle Wadi Arish Member is represented by a conspicuous marl unit about 10 m (33 ft), with frequent larger foraminifera. 3) The Upper Wadi Arish Member is 25 m (82 ft) thick and made up of two limestone units, with larger foraminifera and algal rhodoliths. Foraminifera recorded from the Wadi Arish Formatiom include *Nephrolepidina sinaica n.sp., Miogypsinoides complanatus, Eulepidina elephantine,* and *Globotextularia* sp. (Kuss & Boukhary, 2008), as well as *Risananeiza pustulosa* (Boukhary et al., 2008). The Wadi Arish Formation was assigned an upper Oligocene (Chattian) age by Kuss & Boukhary (2008).

In the subsurface of the northern Western Desert, the Oligocene sediments are composed of siltstones, shales, and minor sandstones. The name Dabaa Formation was proposed by Norton (1967) for a subsurface unit of Eocene-Oligocene marine, light grey to greenish grey shales with thin beds of limestone. The type section of the Formation is in the Dabaa-1 well at a drill depth of 579 m to 1021 m. The thickness of the Formation varies from 200 m (656 ft) to 400 m (1312 ft). Hantar (1990) reported a maximum thickness of 828 m (2717 ft) from the WD-7-I well. Salem (1976) reported thicknesses for other Oligocene sections as 77 m (253 ft) in the Diyur-1, 154 m (505 ft) in the Abu Sennan-1 well, and 250 m (820 ft) in the WD-8 well. In the offshore of Sinai, the Tineh-1 well encountered significant oil indications in Upper Oligocene sand reservoirs, interbedded with shales (Youssef et al., 2010).

Miocene rocks crop out over much of the northern Western Desert, along the Cairo-Suez road, and along the western and eastern sides of the Gulf of Suez, as well as in a small area in north central Sinai. Miocene sediments also occur in the subsurface of the Gulf of Suez, the Nile Delta and its offshore area, and the northern Western Desert. The Miocene sediments rest unconformably on older rocks, ranging in age from Precambrian to Oligocene.

## 8.2.1 Miocene of the Northern Western Desert of Egypt

In the northern Western Desert, the Miocene forms a thin blanket deposit and overlaps unconformably older sediments. In the east, the Lower Miocene sediments rest over basalt flows or Oligocene rocks. In the west they rest unconformably on Middle and Upper Eocene sediments or Turonian dolomites (Said, 1962). The Miocene rocks are uncovered, except along the coast where they are unconformably overlain by Pliocene rocks. The Miocene succession includes four main units, the Moghra Formation, Mamura Formation, Gebel Khashab Formation, and Marmarica Formation.

The Moghra Formation is a thick unit of clastics and minor carbonate interbeds, with plant and vertebrate fragments, which underlies the Marmarica Formation. The type locality is at the cliff north of the Moghra Oasis, on the eastern side of the Qattara Depression. The section was described in detail by Said (1962a, p. 204). It becomes shaly westward and contains limestone beds. The thickness of the unit is 203 m (666 ft) in the type locality, but it generally varies between 53–215 m (174–705 ft) and it thins westward. The Moghra Formation was divided by Said (1962) into two main facies, a fluvio-marine of clastics connected to a river system that drained the North African continent, and a marine facies of reefal, open-bay, and open-marine sediments. It was assigned a Lower Miocene age by Said (1962). In the Qattara Depression, the Moghra Formation rests unconformably on Oligocene and Eocene sediments. It contains large accumulations of petrified logs up to 10 m (33 ft) in length with a general E-NE orientation. Sequence stratigraphy of the Moghra Formation in Wadi Qattara was treated by El-Barkooky et al. (2009) who interpreted it as estuarine and marginal marine deposits.

The fist systematic work on the Moghra fauna was made by Fourtau (1918, 1920). More recently, extensive research was carried out by Albretton et al. (1990), Miller (1996, 1999), Miller & Simons (1996), Morlo et al. (2007), and Pickford et al. (2010). Large mammals and rodents were reported by Albritton et al. (1990) from Wadi Moghra. Rassmussen et al. (1989) described the Hayenodontid creodont *Megistotherium osteothlastes* from the Moghra Formation. This species was reported from Gebel Zelten by Savage (1973). Miller (1996) identified *Haena fourtau* and *M. osteothlastes*. Morlo et al. (2007) reported new occurrences of the carnivorous mammals from the Moghra Formation, including *Buhakia moghraensis*, gen. et sp. nov., and *B.* cf. *Teratodon*, *Cynelos*, sp. nov., *Herpestides aegypticus*, sp. nov., *Ketketictis solida*, gen. et sp. nov., and *Moghradictis nedjema*, gen. et sp. nov., in addition to the two large creodonts *Hyainailouros fourtaui* and *Megistotherium osteothlastes*.

In the subsurface of the Western Desert, the Moghra Formation is made up of sandstones, shales, and thin limestones. The limestones disappear eastwards (Salem, 1976). It has a thickness of 615 m (2018 ft) in the WD-5 well and 307 m (1007 ft) in well Wadi el Natrun-1. Marzouk (1970) treated the western marine facies of the Moghra Formation as a separate formation and applied the name Mamura Formation to it. Consequently, the Moghra Formation is restricted to the fluviatile and fluvio-marine clastic facies of Said (1962). The boundary between the two formations is gradational and arbitrary. The type section of the Mamura Formation is in the Mamura-1 well at a drill depth of 114–401 m (374–1316 ft). A maximum thickness of the Formation of 964 m (3163 ft) was reported from the Dahab-1 well (Hantar, 1990). The Moghra Formation grades southwards into the Gebel Khashab Red Beds (Beadnell, 1905b, Said, 1961) composed of coarse-grained, non-fossiliferous sandstones. The type locality of this unit is at the Gebel Khashab, west of Cairo (Said, 1961). It is 67 m (220 ft) thick and composed of vividly colored sands and gravels, with silicified tree trunks and *Scutella* remains in the basal beds. It overlies basalt lava flows and it is overlain by Pliocene-Pleistocene gravels.

The Marmarica Formation in the type locality at the northern escarpment of the Siwa Oasis is made up of an upper, white limestone member and a lower member of richly fossiliferous, grey limestones with some shale intercalations. The Formation has a thickness of 78 m (256 ft) (Said, 1962a, b). The Formation contains a rich neritic and reefal assemblage of invertebrate fossils and foraminifera. In the subsurface of the Western Desert, the Marmarica Formation is made up also of limestones (Salem, 1976). It was assigned a Middle Miocene age by Said (1962a, b, 1990d).

Gindy & El Askary (1969) subdivided the Miocene succession in the Siwa Depression into three lithostratigraphic units. The lowermost unit was named the Oasis Member. It forms the floor of the depression and the flanks of escarpments and consists of alternations of carbonate beds and greenish, bluish, or blackish shales and marls. The carbonates are made up of cross-bedded coquinas and other organoclastic limestones. Cross-bedding shows paleocurrent directions predominantly to the north and northwest, and less commonly in the opposite direction. Fossils in this unit are mainly species of the echinoids *Scutella*, *Echinolampas* and minor *Clypeaster*, as well as benthonic foraminifera, bryozoan and algal fragments, and plant remains. Abraded fragments of colonial corals occur locally, such as at Gebel Migahhiz and Gebel Drar Ensbar. The middle unit was called the Siwa Escarpment Member. It is exposed mainly in the middle and upper slopes of the escarpments. It is made up of thick, snow-white, chalky beds, more than 15 m (49 ft) thick, alternating with minor shale, marl and fossiliferous limestone beds. Cross-lamination is present only in the lower part of the unit, but becomes more common north of Salum. The uppermost unit was designated as the El Diffa Plateau Member. The member is composed of white to cream, thin chalky limestone beds, rich in bivalves and gastropods and a few colonial coral fragments, alternating with poorly fossiliferous chalky marls and chalks. It forms the cap rock of the escarpments and mesas within the depression. They equated the Oasis Member with the Moghara Formation of Said (1962a, 1990d), however, the former is devoid of sandstones. It is probably equivalent to the Mamura Formation of Marzouk (1970). The upper two members were compared with the Marmarica Formation. The three members can be traced to the cliffs on the Mediterranean coast, north of the town of Salum near the Libyan border.

In the Qattara Depression the Marmarica Formation rests conformably on the Moghra or Mamura formations, except at the eastern tip of the depression where the contact becomes unconformable (Albritton et al., 1990). Its thickness varies from 78–95 m (256–312 ft) in the Qattara Depression, and increases to the north near Salum to 158 m (518 ft) and to 167 m (548 ft) in the subsurface in the Faghur-1 well (Said, 1962).

The Marmarica Formation extends westwards into Libya. It is equivalent to the Regima Formation in NE Libya and the upper part of the Marada Formation in the eastern Sirte Basin.

## 8.2.2   Miocene in the subsurface of the Nile Delta and its offshore area

Active exploration for reservoirs and and source rocks has been active since the discovery of the Abu Madi Field in 1969.

In the northern part of the Nile Delta, the Miocene is represented by more than 3100 m (10,171 ft) of terrigenous sediments, evaporites, and dolomites (Salem, 1976). They occur in the North Delta Embayment of Said (1981, 1990d); a basin with thick Neogene sections the northeastern Delta. In the offshore, the thickness of Miocene beds is highly influenced by rotational block faulting in the eastern and west-central parts of the Delta (Harms & Wray, 1990). The Abu Madi and Sidi Salem formations are the main targets of hydrocarbon exploration (Abu El-Ella, 1990). In recent years, Miocene fluvial, deltaic and turbidite deposits have proven to form prolific gas, condensate, and oil reservoirs in the onshore and offshore areas of the Nile Delta, such as the Akhen and Temsah fields (Bertello et al., 1996, Dolson et al., 2001), the Tennin-1 discovered by Eni and BP in 2001, and the Polaris 1X discovered by Apache in 2003. The Akhen and Temsah fields are located on the Ras El Bar Concession in about 100 m (328 ft) of water. The fields are situated on the Temsah-Akhen structure, which is more than 40 km (25 miles) in length (Marten et al., 2004). The primary reservoirs in these fields are Serravallian (Miocene) in age. The reservoirs are overpressured.

Lower and Middle Miocene sediments are mainly non-marine in the southern part of the Delta and change into shelf and slope deposits in the north, and in Late Miocene time, progradation of paralic and and shelf facies occurred in the northern part of the Delta (Harms & Wray, 1990). During the Messinnian Salinity Crisis, incession led to the formation of channels and valleys, such as the hydrocarbon-bearing Abu Madi Channel.

The Middle Miocene is represented by up to 931 m (3055 ft) of shales, sandy shales, and dolomites of the Sidi Salem Formation. The type section of the Sidi Salem Formation (Rizzini et al., 1978) is the Sidi Salem well #1 well, at a drill depth of 3592 m–4038 m. The Formation is composed of clays with a few intervals of dolomitic marls and rare interbedding of sandstones and siltstones. Clays are grey-green and consist mainly of kaolinite with montmorillonite and illite. It overlies the Moghra Formation, Gebel Khashab Formation, or older sediments and it is overlain by sands and conglomerates of the Qawasim Formation, the Rosetta anhydrite, or Pliocene clays of the Kafr el Sheikh Formation in the offshore area of the Nile Delta. Its age is Serravallian-Tortonian and contains fauna of the *Globorotalia gr. fohsi* and *G. menardii* zones. The unit is absent in the El Hagar-1 well in the northeastern Delta (Cherif et al., 1994). In 2003, Melrose Resources plc made three gas discoveries in its El Mansoura concession in wells Mansouriya-1, South Batra-1, and South Mansoura-2. The first oil discovery in the area was the El Tamad Field discovered by Melrose in the same concession.

The type section of the Qawasim Formation (Rizzini et al., 1978) is in the Qawasim #1 well at a drill depth 2800 m–3733 m where it is made up of sands, sandstones, and conglomerates interbedded with clay layers. Its thickness is approximately 700 m (2297 ft). Benthonic foraminifera include species of *Cibicides, Bolivina,* and *Epinoides.* Its age is Messinian. It is overlain by the Rosetta Formation anhydrites, Pliocene clays with *Sphaeroidinellopsis* of the Kafr el Sheikh Formation, or sands of the Abu Madi Formation. A sharp angular unconformity characterized by emersion and slumps marks the boundary between the Qawasim Formation and the Lower Pliocene Abu Madi Formation.

The type section of the Rosetta Formation (Rizzini et al., 1978) is in the Rosetta well #2 well, at a drill depth of 2678 m–2718 m. It is only a few tens of meters thick and composed of anhydrite interbedded with thin layers of clays. The unit is barren of fossils, but its age was interpreted as Messinian based on its stratigraphic position and paleogeographic considerations. It is developed only in the northern Delta and the offshore area.

The defunct Nile branches have played an important role in the development of the landscape of the present Nile Delta, its continental shelf, and distribution of sediments. The Messinian Eo-Nile canyon could be delineated in the eastern part of the Nile Delta by using seismic reflection data, while the historical Late Quaternary Neo-Nile channels could be delineated in the eastern Nile Delta using geoelectric resistivity surveys (El Mahmoudi & Gabr, 2009). The buried channels in eastern Nile Delta are tributaries of the Pelusiac, Seppenitic, and Tanitic arms. There is a positive correlation between Late Quaternary channels and the Messinian canyons in the eastern Nile Delta, which indicates that the Messinian canyons controlled the formation of the historical branches (El Mahmoudi & Gabr, 2009).

## 8.2.3  Miocene of the Gulf of Suez and Red Sea region

The Miocene rocks crop out on both sides of the Gulf of Suez, and in several islands in the Red Sea, such as the St. Jones (Zabargad) and Tiran islands. They show great lateral lithological and thickness variations. Numerous workers have described and classified the Miocene sediments in the Gulf of Suez area, such as Moon & Sadek (1923), Stainforth (1949), Souaya (1963), the EGPC (1964, 1974), Said & El-Heiny (1967), El-Heiny & Martini (1981), and Philip et al. (1997).

Shallow marine Miocene sandstones, gravels, and limestones are exposed along the Cairo-Suez road, and unconformably overlie Oligocene basalts or gravels and sands (Said, 1990e, Ismail & Abdelghany, 1999, Abdelghany, 2002, Kroh & Nebelsick, 2003)[4]. Their thickness varies between 30–140 m (98–459 ft). The stratigraphy of that area is not very clear and the subdivisions of the different formations are confusing because of the simultaneous use of the same lithostratigraphic names for formations and members (cf. Ismail & Abdel Ghany, 1999). In general, four formations have been recognized within the Miocene sediments between Cairo and Suez: Sadat and Gharra formations (Lower Miocene), and the Hommath and Genefe formations (Middle Miocene). The latter two formations are overlain unconformably by the Pliocene Hagul Formation. The Sadat Forrmation was introduced by Abdallah & Abdelhady (1968) in the Sadat area, southwest of Suez. The Sadat Formation has a thickness of 98 m (322 ft) and has been divided into two members by Ismail & Abdel Ghany (1999); a lower Taratir Member composed of coralline algal limestones rich in the larger foraminifera *Planostegina* sp., echinoderms, and molluscs, with gypsiferous shale layers, and an upper Quarry Member of about 50 m (164 ft) of coralline algal limestones intercalated with course-grained sandstones. In addition to *Planostegina* sp., the Quarry Member contains abund planktonic foraminifera, such as *Globigerina ciperoensis ottangiensis*, *Globigerinella obesa*, and *Globigerinoides altiaperturus*. The Gharra Formation is made up of a lower unit made up of sandy limestones, shales, and marls, equivalent to the Gharra Formation, a middle unit composed of carbonates, and an upper unit of sandstones with fossil wood. The first two units are separated by a thin ledge-forming *Scutella*-bearing dolomite (Said, 1971). The lower part was correlated with the Lower Miocene Moghra Formation of the Western Desert by Said (1962a, b) and the upper part with the Middle Miocene Marmarica Limestone. Souaya (1963) studied the foraminifera separated from the Miocene section at Gebel Gharra, northwest of Suez. The section is made up of limestones, marly limestones, arenaceous limestones, and salty marls and shales. The marls and shales are common in the lower part of the section. The foraminifera include *Miogypsina intermedia*,

4  See also Abdelghany & Piller (1999) and Mandic et al. (2001).

*Operculina complanata*, and *Heterostegina heterostegina*, in the lower part of the section, and they are probably Burdigalian in age, *Miogypsina cushmani, Operculina carpentari,* and *Heterostegina frizzzelli* of probable Helvetian (Serravallian) age, and *Neoalveolina melo, Amphistegina lessonii, heterostegina costata,* and *Discorbis bybeei* in the upper part of the section and they are of probable Vandobonain (Serravallian-Tortonian) age, as well as various species of miliolidae. The lower part of this section was called the Gharra Formation by Said (1971) (see Mandic & Piller, 2001 for paleoenvironmental interpretations based on echinoids). The unit overlies the Abu Zaabal basalts. The upper part includes the Genefe and Hommath formations (Said, 1990e).

The Lower Miocene echinoid fauna from Gebel Gharra have been examined by Kroh & Nebelsick (2003) and include seven assemblages: 1) a *Parascutella* assemblage with sand dollars accumulated by proximal storm deposits and winnowing; 2) a Cidaroid-Echinacea assemblage deposited in a slightly deeper, moderate-energy environment; 3) a Spatangoid assemblage with a diverse fauna of burrowing echinoids; 4) a transported assemblage of allochthonous echinoids; 5) a mixed assemblage of shallower, low- to moderate energy environment; 6) a *Clypeaster martini* assemblage deposited in a shallow, higher-energy environment, and 7) a poorly diverse *Phyllacanthus* assemblage of shallow water carbonates. The Hommath Formation (Middle Miocene) was introduced by Abdalla & Abdelhady (1968) at Wadi Hommath in the Ataqa area. It consists of sandstones, siltstones, shales, and fossiliferous limestones in the lower part, and highly fossiliferos limestones, marls with brozoans, and siltstones in the upper part. The Genefe Formation was considered as a member of the Gharra Formation by Ismail & Abdel Ghany (1999) who described it as a 16.7 m (55 ft) section of coral and coralline algal limestones and sandy bioturbated limestones with shale intercalations with foraminifera, ostracods, and small molluscs.

The surface Miocene section at Wadi Gharandal was described by Moon & Sadek (1923) who subdivided it into 1) a Burdigalian unit which included basal Miocene clays overlying flints and basalts, and 2) a Vindobanian (approximately equivalent to the Serravallian-Tortonian) unit made up of grits, marls, and gypsum. The latter was subdivided, in an ascending order, into: Miocene Grits, Miocene Marls, Gypsum 1, Lower intergypseous Marl with Gypsum 2, Gysum 3, Upper intergypseous Marl, Gypsum 4, Nullipore Rock, and Gypsum 5. The Nullipore Rock is composed of *Lithothamnion* limestones. It is 130 m (427 ft) thick in Gebel Hammam Faraun. The *Lithothamnion* Limestone unit was considered to be upper Middle Miocene in age by Said (1962) and of a Vindobonian age (*Neolaveolina melo* zone) by Souaya (1963) in the Gebel Zeit, on the eastern side of the Gulf of Suez. Stainforth (1949) divided the Miocene succession into a Lower *Globigerina* Marl (Aquitainian and pre-Aquitainian), an Upper *Globigerina* Marl, and an Evaporite Series (Burdigalian-Helvetian).

The Miocene sequence was subdivided by the EGPC Stratigraphic Committee (1964, 1974) into the Gharandal Group, consisting of the Nukhul, Rudeis, and Kareem formations, and the Ras Malaab Group, consisting of the Belayim, South Gharib, and Zeit formations. At the Wadi Gharandal type section, the Gharandal Group is composed mainly of sandstones, shales and limestones. It unconformably overlies Eocene sediments and conformably underlies the predominantly evaporitic Ras Malaab Group. The Gharandal Group was assigned an Aquitainian age by Said & El-Heiny (1967), Lower-Middle Miocene by Andrawis & Abdelmalik (1981),

Burdgalian-Langhian by El-Heiny & Martini (1981), and Lower Miocene-basal Middle Miocene by Hughes et al. (1992). The Ras Malaab Group in the type at Wadi Gharandal is composed mainly of evaporites, with minor clastics. The Ras Malaab Group in the Abu Rudeis well #2 well is composed of alternating evaporites (evaporites 1 to 5), 11–190 m (35–625 ft) in thickness, and marls (interevaporite Marls), 21–91 m (70 ft–300 ft) in thickness (Said & El-Heiny, 1967). In the Wadi Gharandal surface section, the Evaporite Group is 683 m (2241 ft). The evaporites are mainly halite in outcrop, but in the subsurface they are mainly anhydrite (Said & El-Heiny, 1967). It was designated a Burdigalian-Helvetian age by Said & El-Heiny (1967), Serravallian by El-Heiny & Martini (1981), and Langhian-Tortonian by Hughes et al. (1992). In the Wadi Gharandal surface section, the Ras Malaab Group is 683 m (2241 ft) in thickness (Said & El-Heiny, 1967). In the South Bakr Field, the Ras Malaab Group is 515 m (1690 ft) in thickness (Elzarka & Wally, 1987). The Gharandal Group is 55 m (180 ft) in the South Bakr Field (Elzarka & Wally, 1987), and up to 600 m (2000 ft) in the Abu Rudeis #2 well (Said & El-Heiny, 1967).

The Nukhul Formation was introduced by Waite & Pooley (1953, cited by Hughes et al., 1992) for the basal marine Miocene beds in the Sinai and Gulf of Suez regions. The Formation is of Early Miocene age and consists of carbonates, sands, and anhydrites. The distribution of facies was controlled by the paleorelief of the basin. Anhydrite is common in the central and southern parts of the region. Reefs and platform carbonates were deposited on submerged paleohighs. The Formation is poorly fossiliferous, which include *Nonion boueanum*, *Eponides subhaidingeri*, *Quinqueloculina* sp., and *Ammonia beccarii*, as well as *Globigerinella obesa*, *Globigerinoides triloba*, and *Globigerina atypica*. Philip et al. (1997) recorded the benthonic foraminifera *Bolivina dilalatata*, *Cibicides ellisi*, *Chilostomella ovoidea*, *Elphidium* sp., *Nonion boueanum*, and *Siphonina reticulata* from the Formation at Wadi El-Tayiba, Gabal Sabut El-Gamal, Wadi Nukhul, Wadi Baba, Wadi Alaqa, and Wadi Sidri in west central Sinai. It also contains reworked pre-Miocene fossils. Hughes et al. (1992) noted the presence of abundant ostracods and the calcareous algal cyst *Chara charoides* in the lower part of the Nukhul Formation. Ghorab (1961) proposed the name Ras Gharib Flint Conglomerate Formation for the basal beds of the Miocene succession of variable thickness in the Ras Gharib area composed mainly of pre-Miocene reworked material with subordinate amount of Miocene shales and occasional thin streaks of anhydrite and dirty chalky Miocene marls. The latter may reach a thickness of 550 ft (168 m), such as in the Ras Gharib-126 well. Boulders of hard blue siliceous limestone, igneous and metamorphic rocks are present in places. This unit is probably equivalent to the Nukhul Formation.

The Nukhul Formation represents a shallow marine, early syn-rift sequence (Winn et al., 2001, Malpas et al., 2005). In the Wadi Baba area, on the eastern side of the Suez of Suez, it is highly bioturbated and shows changes in trace fossils within the depositional system (Malpas et al., 2005). The succession is made up of interfingering, calcareous mudstones and calcarenites forming coarsening-up units of up to 30 m (98 ft) thick, bounded by marine flooding surfaces (Malpas et al., 2005). The succession grades upwards from basal offshore mudstones with concretions and *Planolites-Chondrites*, into a transitional coarsening-up succession of alternating calcarenites and mudstones with *Thalassinoides*-mottled sediments, then into a lower to upper shoreface sequence with *Ophiomorpha irregulaire*, *Ophiomorpha nodosa-*

*Thalassinoides, Thalassinoides-Taenidium,* and *O. nodosa* (upper shoreface) ichno-fabrics. *Gastrochaenolites* is common in hardgrounds (Malpas et al., 2005).

The type locality of the Rudeis Formation is in the Abu Rudeis #2 well (28°51′04″N, 33°10′37″). It is of Early Miocene age. The Formation was subdivided by the EGPC Strat. Com. (1974), from base to top, into the Mheiherrat, Hawara, Asl, and Mreir members. The lithology of the Formation is variable and composed of carbonates, marls, shales, and sandstones. The Formation is rich in planktonic foraminifera (Hughes et al., 1992), such as *Globigerina ciperoensis atypica, G. praebulloides, Globigerinella obesa, Globigerinoides quadrilobatus,* and *G. trilobata.* Benthonic forms include *Ammonia beccarii.* El-Heiny & Martini (1981) recognized three foraminiferal biozones, in an ascending order: *Globigerinoides trilobus trilobus* Zones, *Glonigerinoides sicanus* Zone, and *Praeeorbulina glomerosa* Zone, and assigned it a Burdigalian-early Langhian age. The oil and gas discovery in the Asl sandstones by Amoco in 1995 produced 38° API oil from 121 ft (37 m) of pay, at a rate of 5500 bopd and 5.9 mmcfgd (World Oil, August, 1995).

El-Haddad et al. (1984), Coniglio et al. (1988), and James et al. (1988) carried out detailed study of the Middle Miocene platform carbonates exposed along the Red Sea coast at Gebel Abu Shaar (James et al., 1988) and West Gemsa area in the Esh el Mellaha range. The age and stratigraphic position of these rocks are uncertain, but they may be partly correlatable with the Rudeis Formation (James et al., 1988). The carbonates are entirely dolomitized. They consist of the Kharasa, Mellaha[5], and Bali'h members. The platform carbonates of the Kharasa Member occur on horst blocks of Precambrian volcanics. They are composed of platform margin reefs and grade westward into bioclastic carbonates. The southern end of Esh el Mellaha range is characterized by alluvial conglomerates and sands. The Mellaha Member contains reefs, coral biostromes and associated sediments up to 50 m (164 ft) in thickness (James et al., 1988, Perrin, 2000). Stromatolites and karsting occur at the top of the Mellaha Member. The Bali'h Member is 20–50 m (66–164 ft) in thickness and consists of shallowing-upward sequences of stromatolitic carbonates and evaporites.

The Kareem Formation is of Early to Middle Miocene age and composed of interbedded sandstones, shales and carbonates, and minor anhydrites in the lower part of the Formation. It is equivalent to the Upper *Globigerina* Marl of Stainforth (1949), the Evaporite I, Interevaporite Marls, Evaporite II, and B Interevaporite Marls of Said and El-Heiny (1967). The Formation was divided by the EGPC Stratigraphic Committee into a lower Markha Member (also known as Rahmi Member) and an upper Shagar Member. The type section of the Markha Member is the Zeneima-1 well (24°03′41″N, 33°04′59″E), and consists of thin anhydrite beds intercalated with shales and marls. The member may be locally absent due to facies change or faulting. The evaporites of the Markha Member are interpreted as salina deposits (deep-marine evaporites) (Hughes et al., 1992), based on deep outer neritic environments suggested by planktonic, such as species of *Globigerina* and *Globigerinoides,* and the benthonic foraminifera *Uvigerina semiornata, Nonion boueanum, Bulimina ovata* and *Puullenia bulloides.* The type section of the Shagar Member is in the Gharib North-2 well (28°21′30″N, 33°06′16″E), West Sinai, and consists of fossiliferous shales and marls,

---

5 Note that the name Mellaha Member is also used for a member of the Cenomanian Raha Formation (El-Sinnawi & Sultan, 1972, Twefik & Ebeid, 1972, Kora et al., 1994).

with occasional limestone and fine- to coarse-grained sandstone beds. The unit is widely distributed, but may be absent locally due to faulting. El-Heiny & Martini (1981) placed the Kareem Formation in the *Orbulina suturalis-Globorotalia fohsi peripheroronda* Zone and assigned it a late Langhian age. Hughes et al. (1992) placed the Lower-Middle Miocene boundary at the first downhole occurrence of *Praeorbulina glomerosa curva*, or at the last downhole occurrence of common *Orbulina suturalis* and *O. bilobata* which takes place within the Shagar Member of the Kareem Formation. The sandstones of the Kareem Formation form one of the most prolific reservoirs in the Gulf of Suez. They produce oil and gas in almost 30 fields, with average net pay of 195 m (640 ft), porosity up to 33%, and permeabilities of up to 33% (Salah & Alsharhan, 1997).

In several marginal areas in the Gulf of Suez, the Kareem Formation is composed mainly of sandstones, and other areas, such as the northern part of the Gulf, the Formation is represented by carbonates. In areas, where it is difficult to differentiate between the Kareem and Rudeis formations, the sequence is called the Ayun Musa Formation[6].

The Belayim Formation is of Middle Miocene age, and represents the beginning of the main Miocene evaporite cycle. The type section is in the Belayim 112–12 well, where it has a thickness of 302 m (991 ft). The EGPC Strat. Com. (1974) subdivided the Belayim Formation into four members, from base to top, Baba, Sidri, Feiran, and Hammam Faraun. The Baba Member and Feiran Member are composed mainly of evaporites, with minor clastics. The type section of the Baba Member is in the Baba-1 well (28°58′26″N, 33°08′59″E). It unconformably overlies the Kareem Formation and underlies the Sidri Member. The type section of the Sidri Member is in the Sidri-3 well (28°50′54″N, 33°10′56″E). The Sidri Member is composed mainly of shales with minor carbonates and sands. It unconformably overlies the Baba Member and it is overlain by the Feiran Member. The type section of the Feiran Member is in the Feiran-2 well (29°43′12″N, 33°13′04″E), where it has a thickness of 143 m (469 ft). The Hammam Faraun Member is equivalent to the Nullipore Rock of Moon & Sadek (1923). The type section is located north of Wadi Gharandal, on the eastern side of the Gulf of Suez, where it has an estimated thickness of 43 m (141 ft). The Hammam Faraun Member is either predominantly clastics or carbonates. Foraminifera include the planktonic forms *Globigerinoides quadrilobata*, and *Globigerina obesa*, and the benthonic forms *Borelis melo melo*, *Borelis melo curdica*, and *Ammonia umbinata*. The Middle Miocene planktonic foraminifera index species *Praeorbulina glomerosa circularis* occurs in the Hammam Faraun Member (Hughes et al., 1992). The Hammam Faraun Member unconformably overlies the Sarbut El-Gamal Formation (Philip et al., 1997) in west central Sinai. In marginal areas of the Gulf of Suez, the Belayim Formation is composed predominantly of sandstones, while reefal and oolitic carbonates developed on submerged paleohighs. Where the Formation is made up entirely of carbonates, it has been called the Gemsa Formation. Said & El-Heiny (1967) assigned the Belayim Formation to the Lower Burdigalian, while El-Heiny & Martini (1981) placed it in the *Globorotalia siakensis* Zone and assigned it a Serravallian age. The large foraminifera *Borelis melo curdica* and *Borelis melo melo* are

---

6 The name Ain Musa Bed has been used for a Middle Miocene unit within the Maadi Formation by Said (1962a).

present in the Belayim Formation (Hughes et al., 1992). In the South Bakr Field, the Belayim Formation is 34 m (112 ft) in the No.19 well to 112 m (367 ft) in the No.8 well (Elzarka & Wally, 1987), it is absent in the west of the field. Anhydrite is common, but gypsum is rare.

The name South Gharib Formation was introduced by the EGPC Stratigraphic Committee (1974) for a thick evaporitic sequence, with thin intercalations of shales and sands in the type well South Gharib-2 (28°16′42″N, 33°08′54″E). The Formation is Middle-Upper Miocene in age. Said & El-Heiny (1967) assigned the South Gharib Formation (Evaporite IV) to the Helvetian. It was designated a late Serravallian-late Tortonian age by Burchette (1986). In the type section, the Formation consists of 71 m (233 ft) of massive halite and anhydrite, with minor shales and very fine- to fine-grained sandstone beds. The evaporites are composed of anhydrite and shale in the marginal areas, and halite in the depocentres. The Formation is missing in some areas due to erosion (Hughes et al., 1992). In the South Bakr Field, the South Gharib Formation unconformably overlies the Gharandal Group (Elzarka & Wally, 1987). It is 453 m (1486 ft) thick in the South Gharib #22 well, but it thins eastwards. The Formation consists mostly of evaporites (gypsum and anhydrite), with occasional shale and limestone beds.

The name Zeit Formation was introduced by the EGPC Stratigraphic Committee (1974) and it is equivalent to the Upper evaporite of Said and El-Heiny (1967). Said & El-Heiny (1967) assigned it a Helvetian age. It was designated a Tortonian-Messinian age by Burchette (1986). The type section is in the Gebel Zeit-1 well (27°55′00″N, 33°24′38″E). It forms a 941 m (3087 ft) thick section of evaporites and shales, with minor sandstones, representing alternating open and restricted shallow marine facies (Hughes et al., 1992). Sandstones predominate in marginal areas. The Formation is missing in extreme northern wells due to erosion. In the South Bakr Field, the Zeit Formation is composed of anhydrites and shales, with thin fine- to medium-grained sandstones and argillaceous, glauconitic limestones (Elzarka & Wally, 1987). Sandstones are restricted to the western part of the area and limestones to the eastern part. The average thickness of the Formation is 150 m (492 ft), with a maximum thickness of 250 m (820 ft) in the Gebel Zeit-47 well.

The Miocene sediments form a long strip along the Red Sea coast. They are essentially littoral in character (Said, 1990e). Miocene sediments are represented by the Ranga, Um Mahara, Abu Dabbab, Um Gheig, Samh, and Gabir formations. El-Akkad & Dardir (1966) originally proposed the name Gebel El Rusas Formation (equivalent to the Basal Limegrit unit of Said, 1962) for the Miocene section at Gebel El Rusas, on the Red Sea coast, and divided it into two members, a lower member composed of sandstones, conglomerates, and clay interbeds, with occasional limestone intercalations, and an upper member of limestone with intercalations of marls and clays. The two members were later called the Ranga and Um Mahara formations, respectively, by Samuel & Saleeb-Roufaiel (1977). The Ranga Formation overlies sediments ranging in age from Precambrian in the south to Cretaceous in the north. The lowermost beds are 8–16 m (26–52 ft) thick and composed of polymictic conglomerates, derived mainly from the basement, with boulders up to 1 m (3 ft) in diameter embedded in a red matrix, followed upward by unfossiliferous fine- to medium-grained sandstones and shales. The Formation ranges in thickness from 186 m (610 ft) at Ranga to 103 m (338 ft) in the Um Ghusun area. The Um Mahara Formation rests unconformably on the Ranga Formation. The contact is marked by conglomerates. In the Um Ghusun area, the Formation is made up of a lower sandy member and an upper gypsiferous fossiliferous limestone member. The rocks contain coral reefs and they are partly dolomitized, with Mn, Pb and Zn mineralization. The Formation is 181 m (594 ft) thick in the Um Ghusun area, but thins northwards to 60 m (197 ft) at Um Gheig. It was assigned a Middle Miocene (Langhian) age based on molluscs and corals. The Abu Dabbab Formation (equivalent to the Evaporite series of Said, 1962a & b, and the Gypsum Formation of El-Akaad & Dardir, 1966) was introduced by Samuel & Saleeb-Roufaiel (1977) in the Abu Ghusun area for a unit composed of white gypsum, with common lenticular masses of dolomitic and microcrystalline limestones. The Um Gheig Formation was proposed by Samuel & Saleeb-Roufaiel (1977) for a dolomite unit, 8–10 m (26–33 ft) in thickness, which overlies the Abu Dabbab Formation. The Samh Formation was introduced by El-Akkad & Dardir (1966) for a unit at Wadi Samh composed of marls, sandstones, shales, intraformational conglomerates and occasional limestone beds. The unit is 32 m (105 ft) at the type locality. It overlies the Abu Dabbab Formation and underlies the Pliocene Ras Shagra Formation. It contains casts of the freshwater species *Melania* (Said, 1990e).

The Miocene sequence exposed in west central Sinai, at Wadi El-Tayiba, Gabal Sarbut El-Gamal, Wadi Nukhul, Wadi Baba, Gabal Abu Alaqa, and Wadi Sidri, has been been divided by Philip et al. (1997) in ascending order into the Early Miocene Nukhul and Rudeis formations and the Middle Miocene Sarbut El-Gamal Formation and the Hammam Faraun Member of the Belayim Formation. The Sarbut El-Gamal Formation is composed of an undifferentiated coarse clastic sequence of polymictic conglomerates composed of cherts, limestones, and reefal-algal limestone clasts, embedded in a sandy calcareous matrix, interbedded with limestone beds. It is highly fossiliferous with large benthonic foraminifera, such as *Borelis melo*, *Heterostegina* sp., and *Operculina* sp. It unconformably overlies the Rudeis Formation and unconformably underlies the Hammam Faraun Member of the Belayim Formation. The Sarbut El-Gamal Formation is equivalent to the upper part of the Rudies Formation (Mreir Member), Kareem Formation, and the Baba, Sidri, and Feiran members of the Belayim Formation.

The Miocene section exposed in the Tiran Island at the southern end of the Gulf of Aqaba is approximately 1400 m (4593 ft) thick and it overlies Oligocene or Eocene polymicitic conglomerates (Goldberg & Beyth, 1991). These sediments overlie a peneplained surface of Precambrian volcanics. The Miocene succession consists of a section of 600–700 m (1969–2297 ft) fine-grained sandstones with igneous rock pebbles and a few sandy dolomitic limestone beds (Aquitanian-Burdigalian), overlain by dolomites, shales, silts, and laminated and cross-bedded sandstones, 0–50 m (0–164 ft) in thickness (Burdigalian-Langhian), followed upward by a 0–200 m (0–656 ft) thick section of white massive anhydrite and gypsum with lenses of dolomite up to 5 m (16 ft) in thickness (Serravallian-Tortonian), a 0–200 m (0–656 ft) thick section of shale, partly dolomitic with gypsum and halite veins, intraformational conglomerate, fine-grained sandstones and conglomerates with igneous rock pebbles (Tortonian), unconformably overlain by 118–226 m (387–742 ft) dolomitic organogenic limestones grading into limy dolomite, and a few beds of shales and conglomerates with igneous rock pebbles. Fossils in the organogenic limestones include echinoderms, pelecypods, gastropods, and foraminifera, such as miliolids, *Elphidium* sp., *Textularia* sp.,

*Amphistegina* sp., *Neoalveolina* ex.gr. *pygamea*, and *Neorotalia* cf. *viennoti*. The Miocene section is overlain by a sequence of Quaternary to Recent marine terraces of detrital limestones similar to the Recent fringing reefs.

In the St. John's (Zabargad) Island, middle-upper Miocene evaporites are exposed at several places. They consist of nodular recrystallized gypsum, with lenses of selenite and levels of reddish Fe hydroxides (Bonatti et al., 1983). They form a syncline with an E-W axis and dips about 40°. El Shazly & Saleeb (1977) called these evaporites the Gypsum Valley Formation. They have a thickness of 35–40 m (115–131 ft) and they reach their maximum thickness in the north of the island.

In the continental margin of the southeast Mediterranean off Sinai, the Miocene sediments are represented by the Maviq'im Formation of late Tortonian-Messinian age (Ben Avraham & Mart, 1982, Mart & Ben Gai, 1982) and form an evaporitic sequence up to 1400 m (4593 ft) in thickness. Two facies can be recognized in the Formation (Mart & Ben Gai, 1982), a shelf facies composed mainly of evaporites, and a basinal facies made up of gypsum, anhydrite and halite, with interbeds of shales, sands and marls. The base and the top of the Formation are marked by seismic reflectors, the "N" and "M" reflectors of Ryan et al. (1973), respectively. The "M" reflector marks the top of the evaporites, or where the evaporites are absent, an unconformity. The Formation is unconformably overlain by the Yafo Formation of Pliocene-Pleistocene age.

# Chapter 9

# Tertiary-Quaternary

## 9.1  PLIOCENE-PLEISTOCENE

Pliocene-Pleistocene sediments crop out along the Nile Valley as far south as Aswan, along the Red Sea coast, in the Cairo-Suez district, and the northern Western Desert. Pliocene-Pleistocene sediments also occur in the subsurface of the Nile Delta region, offshore Sinai and in the Nile Cone in the Mediterranian Sea. Pliocene and Pleistocene deposits of Wadi Qena were reviewed extensively by Sandford (1929).

### 9.1.1  Pliocene-Pleistocene sediments of the Nile Valley

The Nile has passed through five main episodes since the valley was cut down in late Miocene time along a tectonic valley, which was invaded by the Mediterranean Sea. Each of these episodes was characterized by a major river system. These five rivers were termed by Said (1981) the eo-Nile, paleo-Nile, proto-Nile, pre-Nile, and neo-Nile. The eo-Nile was excavated during the Messinian time (Qawasim Formation, Rosetta evaporites). The paleo-Nile was generated during the Upper Pliocene (Gar el Muluk [Qaret El-Muluk], Helwan, Madamud, and Kafr el Sheikh Formations) (see El-Shahat et al., 1997). The proto-Nile was a highly competent river, which carried cobble and gravel-sized sediments made up mainly of quartz and quartzites (Armant, Issawia, and Idfu formations). The last three rivers were active successively during the Pleistocene. Their deposits are separated from one another by unconformities and long periods of recession. The sediments of the pre-Nile are composed mainly of massive cross-bedded sands (Qena Formation). The neo-Nile deposits are indistinguishable from those of the present-day river.

Pliocene deposits of the upper parts of the Nile canyon are composed of sands and clays, with thin lenses of fine polymictic sands and sandy loams rich in authigenic minerals, such as glauconite, pyrite and siderite (Said, 1990e). Toward the periphery of the Nile canyon and in the Nile Valley, the Pliocene can be differentiated into a lower marine sequence (Kom el Shelul Formation) of early Pliocene age and an upper fluviatile sequence (Gar el Muluk and Helwan formations) of late Pliocene age. The marine deposits are represented by the Kom el Shelul Formation (Said, 1971). The type section of the Formation is at Kom El Shelul, in the Abu Roash area (31) near the Pyramids of Giza. This section is 25 m (82 ft) in thickness and made up of coquinoidal limestones, marls, and fossiliferous sandstones in the lower part and non-fossiliferous sandstones with flint pebbles in the upper part. The fauna in the lower marine part of the Kom el Shelul Formation are *Pecten benedictus, Clypeaster aegyptiacus, Chlamys scabrella,* and *Strombus coronaturs,* in addition to a rich foraminiferal and ostracod fauna (Said, 1981). Bassiouni (1965) studied the ostracods of the Kom el Shelul Formation. The Formation unconformably overlies the Late Eocene Maadi Formation and unconformably underlies the Idfu gravels.

The Gar El Muluk Formation was introduced by Said (1981). The Gar El Muluk Formation is best exposed at the Gar El Muluk Hill in the Wadi Natrun Depression, where it consists of 35 m (115 ft) of shales, calcareous sandstones and thin limestone beds. It was assigned a Lower-Upper Pliocene age. The type section was described and illustrated by Blanckenhorn (1902b) and Said (1981). The succession includes, in a ascending order: 1) Black carbonaceous shales with plant fragments, green sandy micaceous shales with *Ostrea cucullata* and fish bones. 2) Cross-bedded sandstones and shales with remains of *Hipparion* and *Hippopotomus* sp. 3) A 30 cm bed of calcareous sandstones with numerous molds of ostracods. 4) A 10 m (33 ft) interval composed of green sands and shales, where a skull of a crocodile was found. 5) A 10 cm limestone bed with the ostracod *Cytherida mulukensis* and fish bones. 6) Dark shales and green gypsiferous sands with chert pebbles and gypsum breccia. Said (1981) postulated that this fluvio-marine sequence formed the earliest deltas of the paleo-Nile.

The Helwan Formation (Pliocene) was proposed by Said (1971) for a 120 m (394 ft) thick section of alluvial sands, sandstones, shales and a few marl interbeds, with *Melanopsis* sp. This Formation was called the *Melanopsis Stufe* by Blanckenhorn (1901). The type section is at Wadi Garawi, south of Helwan, south of Cairo. It underlies alluvial terraces of fluviatile origin and unconformably overlies the Upper Eocene Maadi Formation. Said (1981) assigned it a late Pliocene age and correlated it with the paleo-Nile fluvial sediments.

The Madamud Formation (Said, 1981) consists of chocolate brown and rhythmically banded fine sand and silt laminae. The type section is in Wadi Madamud, east bank of the Nile, Qena Province. The Formation butts against the sides of many of the wadis draining into the Nile from the Eastern Desert. It overlies older bedrock and underlies the early Pleistocene Armant Formation. The Formation is unfossiliferous, but it was considered to be coeval with the Helwan Formation, and other sediments of the late Pliocene paleo-Nile.

Pliocene deposits at Gebel el-Rus, Fayium district, were named the Seila Formation by Hamblin (1987), after the town of Seila near the eastern edge of the Fayium Depression. The third dynasty Seila Pyramid lies on top of the type section of the Seila Formation. The Formation is more than 72 m (236 ft) in thickness and rests unconformably on Eocene sediments. It consists of three lithological units, the basal unit "breccia member" overlies Eocene sediments and it is made up of channeled and sheet-like debris flow deposits composed of subangular blocks of middle and upper Eocene calcareous sandstones set in a muddy matrix. The middle unit "the draped sand and mud member" consists of unfossiliferous, pale yellowish orange and greyish orange fine-grained, well-sorted sands and muds, which drape the breccia. The upper unit is "the sand and gravel member" composed of poorly lithified, lenticular, gravelly sandstones with oysters, snails and silicified wood fragments 10 cm to 50 cm (4–20 in.) in length, and common silicified burrows less than 10 cm (4 in.) in length.

## 9.1.2    Pliocene-Pleistocene in the subsurface and the offshore of the Nile Delta

In the subsurface of the Nile Delta, the sediments overlying the Miocene beds with *Globorotalia fohsi peripheroronda*, were considered as Pliocene (Viotti & Mansour, 1969, Mansour et al., 1969) and cover three "zones", *Globigerina nilotica, G. nepenthes* and *Sphaeroidinellopsis seminulina grimsdalei,* indicating an open-sea planktonic facies. The Pliocene-Quaternary sediments form a shallowing-up cycle composed of the Lower Pliocene sands of the Abu Madi Formation, open-marine shales of the Kafr el Sheikh Formation, and the fluvio-deltaic sediments of the El Wastani and Mit Ghamr formations. The sequence is overlain by the Holocene Bilqas Formation and present-day coastal and lagoonal sediments. Pliocene deltaic and turbidite sandstones are the main targets of exploration in the offshore area. The Ha'py and Timsah in the offshore of the eastern side of the Delta and the King, El Bahig, Al Max, Abu Qir, Abu Sir, Scarab-Saffron, and Simian fields on the western side are examples of Pliocene fields. Hydrocarbons occur in structural and combination traps in deltaic and turbidite sediments (Dolson et al., 2001, Abdel Aal et al., 2001, Samuel et al., 2003, Sharp & Samuel, 2004).

The type section of the Abu Madi Formation (Rizzini et al., 1978) is in the Abu Madi #1 well (31°26'17"N, 31°21'14"E) at a drill depth of 3007–3329 m. Said (1981) mentioned the interval 2007–3129 m. No reason was given for this discrepancy. The unit is composed of thick, rippled and bioturbated beds of sand, rarely conglomeratic, and interbedded with clay layers. Its thickness is approximately 300 m (984 ft). The Formation is overlain by the Kafr el Sheikh Formation. Foraminifera belong to the *Sphaeroidinellopsis* sp. Zone typical of the Mediterranean fauna. Said (1981, 1990d) assigned the Formation an Early Pliocene age. Ince et al. (1988) considered the Abu Madi Formation to be Upper Miocene in age. The Formation is coeval with the Kom El Shelul Formation. The Abu Madi Formation is the main reservoir in the onshore Abu Madi Gas Field, west of Port Said in the Delta region and the offshore Abu Qir Gas Field, northeast of Alexandria. It forms more than 200 ft (61 m) of gross pay (Abdine, 1981), as well as the other fields along the Baltim Trend (Dolson et al., 2001).

The type section of the Kafr el Sheikh Formation (Rizzini et al., 1978) is in the Kafr el Sheikh #1 well (31°10'23"N, 31°04'55"E) at a drill depth of 1277–2735 m. It is composed of soft clays with some interbeds of sands, overlain by the El Wastani Formation sands. Its age is Lower Pliocene (*Globorotalia puncticulata* and *G. margaritae* zones) to Middle Pliocene (*G. aemiliana* and *G. crassaformis* zones) according to Rizzini et al. (1978). Said (1981), on the other hand, mentioned the interval 975–2735 m as the type section. The Kafr el Sheikh Formation provides secondary reservoirs in the Abu Madi Field in the Delta region and the offshore Abu Qir Field. In the Burg el Arab #1 well (30°55'N, 29°31'E), a 46 m (150 ft) section of Pliocene clays and shales contains abundant planktonic foraminifera (Omara & Ouda, 1972), such as *Globorotalia puncticulata, Orbulina bilobata, O. universa, Globigerinoides conglobata, G. cyclostomata, G. obliqua, G. trilobata trilobata, G. trilobata immatura, G. elongata, G. ruber, G. sicamus,* and *G. sacculifera.* These sediments rest unconformably on Middle Miocene limestones, and they are overlain by Pleistocene oolitic limestones. These clays are probably equivalent to the Kafr el Sheikh Formation.

Rizzini et al. (1978) introduced the name El Wastani Formation for thick sand beds interbedded with thin clay levels of probable Upper Pliocene age in the El Wastani #1 well (31°24″08″N, 31°35″46″E) at a drill depth of 1009–1132 m. It has a thickness of approximately 300 m (984 ft). Said (1981) considered the interval 886–1009 m as the type section, which gives the unit a thickness of 123 m (404 ft). Said (1981) correlated it with the proto-Nile early Pleistocene Idfu gravels.

The type section of the Mit Ghamr Formation (Rizzini et al., 1978) is in the Mit Ghamr #1 well (30°41′44″N, 31°16′26″E) at a drill depth of 20–483 m. It is composed of thick layers of sand with pebbles of quartzite, chert, and dolomite in the lower part and sporadic shells and interbeds of coquina and peat levels. Its thickness is up to 700 m (2297 ft). It was assigned an Upper Pliocene-Quaternary age. The Formation also crops out as isolated yellow mounds in the green agricultural fields, where they were described as "turtle-backs" by Sandford & Arkell (1939). The turtle-backs rise 10–12 m above the surface. They are present only to the southeast of a major fault that crosses the Nile at Mit Ghamr and runs in a southwest direction (Issawi, in Wendorf & Schild, 1976). Said (1981) correlated the Formation with the pre-Nile Qena Formation.

### 9.1.3   Pliocene-Pleistocene of the Western Desert (Northern Egypt Basin)

In the northern Western Desert, the Pliocene is developed only in a small area. Along the coast, the Pliocene is represented by the Hagif Formation. The Formation was named by the authors of the new Geological Map of Egypt (1987) after the Hagif area to the northwest of Wadi Natrun. It consists of a pink dolomitic limestone sequence with a few gypsum beds, which overlies the Marmarica Formation and underlies oolitic limestones. Fossils are rare, but a few benthonic foraminifera and abundant stromatolitic calcareous algae are present. The Formation is 80 m (262 ft) in the type locality, but thins to 25 m (82 ft) along the coast. This Formation was originally named the El Dabaa Formation (Thiele et al., 1970, Said, 1971), a name that is now applied to a late Eocene-Oligocene unit (Dabaa Formation) in the subsurface of the northern Western Desert. At Salum, the Formation is made up of 78 m (256 ft) of sandy limestones (Mansour et al., 1969).

In the Qattara Depression, the Kalakh Formation (Pliocene-Pleistocene) rests unconformably on the Middle Miocene Marmarica Formation (Albritton et al., 1990). Its average thickness is 14 m (46 ft) and it is made up of a basal conglomerate, 1–2 m (2–7 ft) thick, and composed of cobbles and pebbles of the Marmarica Limestone with scattered quartz pebbles and some oyster banks. It is equivalent of the Alexandria Formation, which forms a calcareous ridge along the coast and is considered as Pleistocene by Hantar (1990). Ibrahim & Mansour (2002) reported the planktonic foraminifera *Globigerinoides trilobus trilobus*, *G. trilobus immaturus*, *G. obliquus extremus*, *G. obliquus*, *Globorotalia inflata*, and *Orbulina universa* from the basal part of the Alexandria Formation in shallow boreholes drilled in the El Dabaa area, and confirmed its Pliocene-Pleistocene age.

In the northern part of the Farafra Oasis in the Ain el Wadi Depression, there is a series of cross-bedded fine-grained sandstones, siltstones, silty shales and shales capped by a thin allochthonous, fossiliferous limestone bed. This sequence was called the Ain Wadi Formation by Youssef & Abdel Aziz (1971). The type section lies to the north of the Ain El Wadi shallow water well. The exposed part of the succession in this locality is about 30 m (98 ft). It contains a mixed Danian-Landenian planktonic foraminiferal assemblage, which suggests that these beds are reworked Paleocene rocks. They fill ancient depressions and were formed as a result of dissolution (karsting) of the Farafra Chalk, probably during the Pleistocene.

## 9.1.4 Pliocene-Pleistocene of the Gulf of Suez and Red Sea region

In the Cairo-Suez district, the Hagul Formation is made up of 20 m (65 ft) thick, nonfossiliferous fluviatile deposits of a late Pliocene age, with a large number of rolled fragments of silicified tree trunks (Said, 1990e). In the type section at Wadi Hagul, the top 2 m (7 ft) of the Formation is composed of a sandy limestone bed with flint bands. The overlying Hamzi Formation (Said, 1971) is a 35 m (115 ft) thick sequence of porcellaneous limestones with numerous beds of flint. The base is marked by a conglomerate consisting of pink limestone fragments. Fossils consist mostly of the freshwater snail *Pirenella conica*, suggesting deposition in freshwater lagoons. The type section of the Hamzi Formation is at Gebel Hamzi, 35 km west of the Barage.

In the Gulf of Suez region, the Pliocene is made up of limestones with *Metis papyracea* and pseudo-oolitic quartz sandstones. The latter grade into conglomerates with *Pecten vasseli, P. fischeri*, and oyster beds with *Ostrea gryphoides*. This fauna is of Indo-Pacific affinity. Pliocene deposits in the north of the Gulf of Suez are represented by thin continental deposits.

Along the Red Sea coast, in the Safaga-Qoseir area, the Pliocene sediments include the Gabir Formation and Ras Shagra Formation (El-Akkad & Dardir, 1966). The Gabir Formation was introduced for a 44 m (144 ft) unit composed mainly of sandstones with marls and a few intraformational conglomerates. The unit is unconformably overlain by the Ras Shagra Formation, and it unconformably overlies the Samh Formation. The Ras Shagra Formation was proposed for a 102 m (334 ft) thick unit at Wadi Shagra, composed of coarse-grained sandstones and grits underlain by reefal limestones and sandstones. It was divided into two members. The lower member is 80 m (262 ft) thick and composed of sandstones, marls, reefal limestones and calcareous grits. The upper member is 22 m (72 ft) thick and consists of sandstones and gravels with marly intercalations and reefal limestones near the top. Fossils recorded include, among others, the Indo-Pacific forms *Ostrea cucullata* and the echinoids *Clypeaster scutiformis* and *Laganum depressum* (Said, 1962). The lower member of the Ras Shagra Formation was included in the underlying Gabir Formation by Said (1990e), making its thickness to 124 m (407 ft). He also restricted the Ras Shagra Formation to the upper member, which reduces its thickness to 22 m (72 ft) at the type locality.

The Pliocene succession in the Hurgada Field (Kostandi, 1963) is made up of marly beds at the base, which unconformably overlie Miocene evaporites, followed upward by variegated sands and marls, calcareous sandstones with casts of the echinoid *Laganum, Pecten* beds, calcareous sandstones, and cross-bedded grits. This succession is overlain by a Pleistocene-Recent section composed of reefs, gravels, raised beaches, old Pleistocene gravels, and young alluvials. Coral reefs were well developed during the Pleistocene of the Red Sea region and are now well-exposed along the coastal plain in Egypt, adjacent to the modern fringing reefs (Tucker, 2003). The Marsa Alam reefs were deposited during inter-glacial sea level highstands. The Pleistocene carbonates contain shoreface to foreshore and fluvial well-sorted and cross-bedded lithic sands and pebble conglomerates (Tucker, 2003).

In the Tiran Island, the Pliocene-Recent sediments form five main raised marine terraces, which extend from sea level up to 500 a.s.l. (Goldberg & Beyth, 1991). These terraces are built of detrital limestones similar to that of the Recent fringing reefs. The formation of these terraces was attributed to tectonic uplift. The oldest terrace, between 60–502 m a.s.l., is made up of blocks of older sediments composed of massive, organogenic-coralline limestones with bivalves, gastropods, and echinoderms. The following terrace is composed of a number of minor sandy conglomeratic river terraces, with pebbles of magmatic rocks, gypsum, calcite and barite, and sandstones. The third terrace consists of marly fossiliferous reefs, tabular, with magmatic pebbles, 5–20 m a.s.l. The following terraces include fossiliferous reefs, and a few minor terraces between 4–60 m a.s.l. The marine terrace at 40 m a.s.l. gave a date of 130,000 yr BP. The youngest terrace consists of alluvium deposits of sands, pebbles, boulders, telus, and loess.

Pleistocene-Holocene limestones crop out on the northern and eastern sides of the St. John's (Zabargad) Island. The Pleistocene Old Reef Limestones (Bonatti et al., 1983) consist of 20–40 m (66–131 ft) thick fossiliferous biocalcarenites and reef limestones which are recrystallized and tectonized to various degrees. They occur up to 80 m above sea level. They change from back reef and/or lagoonal, calcareous sands near the base, into fossiliferous reef flat and reef edge limestones near the top. They directly overlie peridotites of the igneous basement. The fossils which occur in these limestones belong to forms which are still living. They include the scleractinid corals *Fungia* sp., *Goniastrea* sp., *Acropora* sp., and *Porties* sp., the gastropods *Gasmaria* cf. *ponderosa*, *Strombus gibberulus*, *Torchus* sp., *Cypraea* sp., and *Conus* sp., the bivalves *Lithophaga* sp., Pectinids, *Codokia tigerina*, *Pteria* cf. *penguin*, *Barbatia* cf. *lacrata*, and Lucinids, and the echinoid cf. *Herocentrotus mammillatus*. The limestones have been subjected to post-tectonics, which resulted in tilting, faulting and folding, and N-S microfracturing. The Old Reef Limestones are overlain by conglomerates and breccias (Old conglomerate and breccias of Bonatti et al., 1983), up to 4 m (13 ft) in thickness, and contain large clasts of the older formations. These conglomerates are overlain by the Young Reef Limestones, which occur around the edges of the island, and consist of typical coastal reefs, with lagoonal, reef-flat, and reef-slope deposits. They are believed to have been deposited during the last interglacial (Bonatti et al., 1983).

### 9.1.5 Pliocene-Pleistocene of Sinai

In northern Sinai, the Pliocene-Quaternary deposits are represented by continental to littoral sediments which are approximately 3000 m (9843 ft) in thickness (Jenkins, 1990). In northern Sinai, Pliocene sediments occur in a small area near Awlad Ali (30°52′N, 34°04′E). The section is composed of 15 m (49 ft) of marls known as the Saqia Marls (Said, 1962a). It has a thickness of 42 m (138 ft) in the El Khabra well (30°55′N, 34°15′E). The unit overlies Upper Eocene rocks and it is overlain by reddish conglomerate beds. Said (1971) assigned the unit a Pliocene age and correlated it with the Hamzi and Helwan formations. In the continental margin of the southeast Mediterranean off Sinai, the Pliocene-Pleistocene sediments are represented by the Yafo Formation (Ben Avraham & Mart, 1982, Mart & Ben Gai, 1982). The Formation is composed mainly of clastic sediments derived predominantly from the Nile River and some wadis in northern Sinai. It forms a continental prism along the continental margin of the southeast Mediterranean and is composed of marls, shales and sandstones. Its thickness ranges from 1000 m (3281 ft) to 2000 m (6562 ft). Diapirs are common due to salt flow caused by loading of the underlying evaporitic sequence. The Formation unconformably overlies the Tortonian-Messinian Maviq'im Formation.

# Chapter 10

# Quaternary (Pleistocene-Holocene)

The Quaternary was a period dominated by the development of the River Nile and the Nile Delta, the appearnce of modern man, and the establishment of the Ancient Egyptian civilization. These developments are summarized in Table 10.1. The Quaternary sediments include the older Nile deposits in the Nile Valley. From oldest to youngest, they are the Armant, Issawia, Idfu, Qena, Abbassia, Dandara, Korosko-Makhadma-Ikhtiariya, Masmas-Ballan, Deir Fakhuri, Sahaba-Darau, Dishna-Ineiba, Arkin, and El-Kab formations.

The Armant Formation (Said, 1981) is made up of alternating beds of locally derived gravels and fine-grained clastic rocks. The gravels are cemented by tufaceous material and the pebbles are subangular and poorly sorted. The fine-grained clastic beds are calcareous, sandy, shaly, or phosphatic depending on the provenance of the sediment particles. The type section of the Formation is in Wadi Bairiya, opposite Armant, Luxor district. No thickness has been given. The Formation butts against the sides of the wadis draining into the Nile or forms inverted-wadi ridges. It overlies older formations, and it is overlain by the Qena Formation. The Formation is unfossiliferous, but it was assigned an early Pleistocene age based on its stratigraphic position.

In the area between Esna and Manfalut, the Issawia Formation (Said, 1971) is made up of 16 m (52.5 ft) of chocolate brown clays, followed by a 7 m (23 ft) thick tufaceous hard limestone bed, and a 22 m (72 ft) of red limestone breccia. The breccia contains a hard cemented bed in the higher levels known as the brocatelli Limestone. The type locality of the Formation is at the Issawia quarry, Akhmim, in Upper Egypt. The Formation overlies older strata of mostly Eocene age, and it is overlain by Pleistocene alluvial sediments. Said (1971) assumed a Pliocene (?) age for the Formation. Said (1981) designated it an Early Pleistocene age.

The type locality of the Idfu Formation (Said, 1971) is at Wadi el-Hassayia, Darb El-Gallaba, west of Idfu. It is composed of more than 15 m (49 ft) of fluviatile gravels and sands embedded in a red-brown matrix. The coarse sands and gravels consist mostly of rounded flint, and they are covered with red-brown soil. Its stratigraphic position is unknown, but it is assumed to underlie the Dibeira-Ballana Formation and to overlie the Qena Formation. Said (1981) an early Pleistocene age for the Formation.

The Qena Formation (Said, 1971) is composed of alternating, cross-bedded, occasionally consolidated, coarse-grained sands and thin beds of grits, and gravels, with casts of shells of *Unio abyssinicus* and *Aspatharia cailliaudi*. Its thickness is

Table 10.1 Paleolithic, Neolithic and Predynastic settlements in the Nile Valley and Western Desert of Egypt[1].

| | Absolute Dating[2] | Fayium[3] | Archeological Settlement (Nabta)[4] | Archeological Settlement (Dakhla) | Archeological Settlement (Nile Valley) Lower Egypt | Archeological Settlement (Nile Valley) Upper Egypt |
|---|---|---|---|---|---|---|
| Predynastic | 4500–3100 BC (6500–5100 BP) (or 4800–3050 BC or 5500–3100 BC in other estimates) (Calcolithic, 4500–3300 BC) | | | | Maadi (3400–3200 BC) El Omari (3700–3400 BC) Merimda (4300–300 BC) Fayium A (4400–3900 BC) (Bard, 1994) Lake Moeris reflooded[5] | Naqada III (Gerzean B = Semainean) (3250–3050 BC) = Dynasty '0'[6] Naqada II (Gerzean A) (3700–3250 BC) Naqada I (Amratian) (4200–3700 BC) Badarian (4400–4000 BC)[7] |
| Late Neolithic | 7500–6200 BP (5500–4200 BC) | Lake Moeris dried up (Moerian Group) | Hyperaridity (6700–6500 BP[8]) Late Neolithic settlements (Time of Megaliths[11], quartzite slabs, "calendar circle" @ 6270 BC, buried cows (domesticated cattle)[12] (7400 BP wet period according to Schild & Wendorf, 2004) | Sheikh Muftah (6000–2200 BC)[9] | | Earliest domesticated cattle at Merimda (6400–5400 BP)[10] and in Upper Egypt at Khattana (5300 BP) |

| Period | | | | | |
|---|---|---|---|---|---|
| **Middle Neolithic** 8300–7600 BP | Youngest Moeris (ca.4910 BC) (6910 BP)<br><br>Proto-Moeris (ca.5190 BC) (7190 BP)<br><br>(Fayium Group)<br><br>Pre-Moeris (ca.6150 BC) (Qarunian = Fayium B)<br><br>Aggradations of Lake Moeris Paleo-Moeris (ca.7000 BC) (Fayium A) (Earliest farmers, cultivated wheat, barley & flax; also used goat & sheep[18]) | Arid (7600–7500 BP) [or 7300–7100 BP[13]]<br><br>Middle Neolithic settlements (8700–7300 BP)<br><br>Gap (period 8700–7800 BP not accounted for)<br><br>El Kortein Phase (Jerar Assemblage type) (8800–8700 BP)<br><br>El Nabta Phase (9100–8900 BP) (Deep water wells) | Bashendi-B (6500–5200 BP)[14] Equivalent unit is present at Farafra[15]<br><br>Beshandi-A (7600–6800 BP)[16]<br><br>Gap (8500–7600 BP) [Dry 7900–7700 BP]<br><br>Sheep & goats introduced into Nabta from SW Asia[19] | Qarunian (Fayium B) (8000–7000 BP) | Elkabian (ca.8000 BP) [8400–7980 BP][17] Three occupation levels at El Kab, radiocarbon dated (Vermeer-sch, 1970) |
| **Early Neolithic** 10,800–8900 BP[20] | Arid (9200–9100)<br>El Ghorab Phase (9600–9200 BP)<br>Arid (9802–9600 BP)<br>El Adam Phase (10,800–9800 BP) (Pastoralists) | Masara A, B & C (Dakhla) (9200–8500 BP)[21] | | | Arkinian (ca.10,580 BP) |

*(Continued)*

Table 10.1 (Continued).

| Absolute Dating[2] | Fayium[3] | Archeological Settlement (Nabta)[4] | Archeological Settlement (Dakhla) | Archeological Settlement (Nile Valley) Lower Egypt | Archeological Settlement (Nile Valley) Upper Egypt |
|---|---|---|---|---|---|
| Upper Paleolithic (Late Pleistocene)<br>30,000–10,800 BP | | "Abbassia Pluvial" Desert conditions in the Sahara | | | Makhadma 1[22]<br>Makhadma 2<br>Makhadma 3<br>Makhadma 4 (13,500–12,500 BP)<br>Gebel Sahaba (?)<br>Makhadma 5?<br>Makhadma 6<br>Esnan (13,000–12,500 BP)<br>Silsilian<br>Afian (ca.14,000 BP)<br>Sebilian (16,000–11,000 BP)<br>Edfu (17,800–17,000 BP)<br>Fakhurian (18,000–17,600 BP)<br>Kubbaniyan (19,000–17,000 BP)<br>Early Kubbaniyan 21,000–19,000 BP |

| | | | | |
|---|---|---|---|---|
| Middle Paleolithic | 90,000–30,000 BP | "Acheulian" from Abu Simbel | Bir Tarfawi Bir Sahara (Wendorf et al., 1993) | Shuwikhat Nazlet Khater 4 (see also McPharsen et al., 2003) (Abydos) Pre-Nile started (Said, 1993) |
| Lower Paleolithic (Pleistocene) | 250,000–90,000 BP | | | Khormusan Makhadma 6 Nazlet Khater 1–3 |

1 See Close (1980, 1984). Dates are widely varied, even by the same authors and are generally unreliable.

2 Dates are according to Wendorf & Schild (1998), but vary widely among various authors.

3 Wenke (1989), Midant-Reynes (2000).

4 Wendorf & Schild (1994).

5 Caton-Thomson & Gardner (1934).

6 El-Amra cemetery at Abydos (Abjdu), west of Qena, burial place of King Scorpion, Mena, and other first dynasy kings. The Amratian (Naqada I) derives its name from that location.

7 Bard (1994).

8 Malville et al. (1998). Two wet periods at 6600-5400 BP (Megaliths) and 700 BP (cattles) according to Schild & Wendorf, 2004). This would cover most of Predynastic period.

9 McDonald (1998). Units equivalent to Bashendi-B are present at Farafra (Alessio et al., 1992).

10 Wendorf & Schild (1994).

11 Discovered by Malville et al. (1998). See Brophy & Rosen (2005) for alignment of Megaliths.

12 Schild & Wendorf (2004).

13 Malville et al. (1998).

14 McDonald (1998).

15 Alessio et al. (1992).

16 McDonald (1998).

17 Vermeersch (2001).

18 Midant-Reynes (2000)

19 Haynes (2001)

20 Wendorf & Schild (1998).

21 McDonald (1998).

22 Partly contemporaneous with the Kebaran and Natufian cultures in SW Asia.

over 20 m (66 ft). It unconformably overlies the Dandara Formation (according to Wendorf et al., 1970, Said, 1971, and Wendorf & Schild, 1976), or the Idfu Formation (according to Said, 1981), and it is overlain by gravels embedded in a red soil matrix of the Abbassia Formation. At the type locality along the Abu Mana'a Bahari quarry face in the Nile Valley, the Formation is 20 m (66 ft). The base is unexposed. At Wadi Nafukh east of Qena, the Formation overlies the late Pliocene paleo-Nile sediments of the Madamud Formation, and unconformably underlies the Abbassia gravels (Said, 1981). It includes some late Acheulian implements in the upper levels.

The type locality of the Dandara Formation (Said, 1971) is 3 km south of the Temple of Hathor, at Dandara, opposite Dishna, in the Nile Valley. The Formation is approximately 15 m (49 ft) in thickness. It unconformably overlies the Abbassia or the Qena Formation and it is overlain by the Ikhtiariya dunes (Said, 1981). The Dandara Formation is overlain by the Qena Formation according to Said (1971) and Wendorf & Schild (1976). It is composed of a grey, loose and fine sandy silt bed at the base, followed by brown silts with thin carbonate interbeds and occasional lenses of gravel. The top of the Formation often has a distinct red soil (Wendorf & Schild, 1976). It was dated more than 39,000–40,000 years B.P. (Wendorf & Schild, 1976, Said, 1981), and it includes late Acheulian (Said, 1971) or Levalois (Said, 1981) tools. It was assigned an upper Pleistocene age. The Dandara Formation was equated by Said (1968) with the Koroso Formation of Butzer & Hansen (1968).

The Abbassia Formation (Said, 1981) is composed of massive, loosely consolidated gravels of polygenetic origin. The pebbles are rounded to subrounded. They were derived from the uncovered basement of the Eastern Desert. The type section of the Formation is in a quarry of gravel pit at Rus Station, halfway along the Fayium-Wasta railway line. The Formation unconformably overlies the Qena sands. The upper boundary has not been observed. The sediments include late Acheulian hand axes, and it was assigned a middle Pleistocene age.

The Korosko Formation (Late Pleistocene) was introduced by Butzer & Hansen (1968). The type section is in Wadi Ayed, New Korosko village, Kom Ombo. The Formation is made up of a lower unit of subangular pebbles of local derivation, embedded in a matrix of coarse sand, and an upper bed of light grey to light brownish grey, poorly-sorted, medium-grained sand to sandy marl (Said, 1981). The Formation overlies bedrock or older Nile deposits, and it is unconformably overlain by the Ikhtiariya dunes or the Masmas-Ballana silts. The Korosko Formation occurs at Kom Ombo and in Egyptian Nubia. It does not extend north of the Kom Ombo area (Wendorf & Schild, 1976). Adequate chronological controls are not available for the Korosko Formation. The aggradation represented by the Korosko Formation was dated by radiocarbon around 25,250 BC (Wendorf & Schild, 1976). The Formation is equivalent to a thick unit of slopewash along both sides of the river. This slopewash is composed mainly of gravels, pebbles, and cobbles derived from the gravel mantle covering the Qena hills and contain at the base numerous artifacts of Middle Paleolithic aspect. Near the top it becomes a true colluvium, with extensive settlement of the early Late Paleolithic, dated at 20,000 BC (Wendorf & Schild, 1976). These sediments may represent an episode contemporaneous with the downcutting that separated the Korosko and Masmas formations. Said (1981) showed that this unit underlies the older neo-Nile deposits and that it was formed as a result of sheet floods, caused by torrential rains at the mouth of the wadis which drained the Eastern Desert.

The type section of the Makhadma Formation (Late Pleistocene) (Said, 1981) is in Makhadma, east of the Nile, opposite Dandara. In that area, sheet wash deposits, 20–30 m (66–98 ft) above the Valley plain, are composed of gravels, pebbles, and boulders, and carry middle Paleolithic tools (Wendorf & Schild, 1976). They rest unconformably over silts of the Dandara Formation or sands of the Qena Formation. It is overlain by sediments of the neo-Nile carrying fresh late Paleolithic artifacts (Wendorf & Schild, 1976, Said, 1981). The Formation is probably contemporaneous with the Korosko Formation. It was deposited probably during the wet Mousterian-Aterian interval (Said, 1981).

The Ikhtiariya Formation (Said, 1981) is made up of well-sorted, massive, dune sands, with a thickness of 4 m to 6 m (13–20 ft). The type section is at Site 34 of the Combined Prehistoric Expedition in Dibeira west, Sudanese Nubia. It contains middle Paleolithic implements and a few mammal bones. The Formation overlies eroded bedrock, the Korosko Formation with which it also interfingers, or the Dandara silts. It is conformably overlain by the Masmas-Ballana Formation or the Dibeira-Jer aggradational silts and fluviatile sands. The Formation is assumed to represent the eolian deposits contemporaneous with Mousterian-Aterian pluvial dated at 80,000–40,000 BP. A soil profile characterized by the presence of gypsum and salt pans of later genesis is associated with the Formation. This soil profile extends along the western bank of the Nile, from Assiut to Cairo, and it is particularly well developed at Darb Gerza.

The Dibeira-Jer Formation, as originally defined by de Heinzelin (1968), represents the first episode of silt deposition by the modern Nile. It forms terraces about 36 m (118 ft) above the modern flood plain in Nubia and about 8 m above the modern flood plain in Upper Egypt. The type locality of the Formation is in Sudanese Nubia at Site 1017 in the Khor Musa, just south of Wadi Halfa (Wendorf & Schild, 1976). It has been divided into several aggradation episodes separated by periods of regression. The former is represented by aggradational silts and the latter by interfingering dunes. It has an estimated thickness of more than 8 m (26 ft). Radiocarbon ages are between 16,070 BC and 15,000 BC (Wendorf et al., 1970). However, its age is assumed to extend between 25,000 and 15,000 years B.P. Wendorf and Schild (1976) recognized two aggradational units, a lower unit carrying Middle Paleolithic artifacts, and an upper unit carrying Late Paleolithic (Levallois) technology. The latter underlies aggradational silts of the Sahaba Formation. Said (1981) correlated the lower unit with the Dandara Formation and the upper silts with the Masmas-Ballana Formation.

The Masmas-Ballana Formation (Said, 1971, 1981) (late Pleistocene) is composed of fine-grained, well-sorted dune sands which interfinger with silts, topped by a podzol soil. The silts are typical flood-plain deposits. It overlies the Dibeira-Jer Formation, the Koroso Formation, or the Ikhtiariya Formation. It is overlain by the El Fakhuri Formation. It has a thickness of more than 5 m (16.4 ft) at its type locality at Ballana, Nubia. Its age is assumed to be between 17,000 and 15,000 years BC, based on radiocarbon dating. The top of the dune deposits contains rich late Paleolithic implements. This interval was a major episode of Nile aggradation accompanied by dune migration. Butzer & Hansen (1968) introduced the name Masmas Formation for silts and channel beds in the Kom Ombo area in Upper Egypt. The thickness of the Formation was estimated to be more than 43 m (141 ft). The sediments contain a rich molluscan fauna, such as *Planorbis ehrenbergi, Bulinus truncatus, Vallata nilotica, Unio willococksi, Corbicula fluminalis,* and *Cleopatra bulimoides.* De Heinzelin (1968) applied the name Ballana Formation to dune sands which interfinger the upper part of the Dibeira-Jer silts in the Egyptian Nubia. The type locality of the Ballana Formation is just across the Sudanese-Egyptian frontier. Charcoal radiocarbon dating gave an age of 16,650 BC ± 550 years (Wendorf & Schild, 1976). The dune deposits of the Ballana Formation of de Heinzelin (1968) and the interfingering Nile silts of the Masmas Formation of Butzer & Hansen (1968) are now regarded as different lithological facies of the same chronological episode. This episode was termed Ballana-Masmas aggradattion by Wendorf & Schild (1976).

The Deir El-Fakhuri Formation was originally proposed by Wendorf et al. (1970) and Said et al. (1970) for diatomites and pond sediments interrupted by silt units, which overlie the Ballana Formation, and underlie the Sahaba Formation. The Formation has an estimated thickness of more than 6 m (19.7 ft) in the type locality, at Deir Fakhuri, Esna. They also occur at Tushka in Egyptian Nubia. The age of this Formation is estimated between 15,000 and 12,000 B.C., based on the ages of the underlying Dibeira-Jer Formation and the overlying Sahaba Formation (Wendorf et al., 1970). The Deir El-Fakhuri deposits formed in ponds that developed on the Ballana dunes. The Formation overlies a noncalcic soil horizon, which was developed over the stabilized dunes. The origin of the noncalcic soil is not known. Wendorf & Schild (1976) believe that they formed due to oxidation caused by subsequent ponds that developed in the area. The pollen and spores, as well as diatomite flora separated from the Deir El-Fakhuri Formation suggest an arid grassland environment (Wendorf & Schild, 1976). Said (1981) argued that the Deir El-Fakhuri Formation represents a recessional interval which followed the Masmas-Ballana aggradation. Wendorf & Schild (1976), on the other hand, do not believe that the Formation represents a period of Nile recession, but an early phase of rising of the Nile of the subsequent Sahaba-Darau aggradation.

The Sahaba-Darau Formation (Said, 1981) consists mainly of fluviatile silts. It has a thickness of more than 6 m (19.7 ft) at its type locality at Sahaba, Nubia. The Formation overlies the recessional pond deposits of the Deir El-Fakhuri Formation. Its age is assumed to be between 12,000 and 9,700 BP (Wendorf & Schild, 1976). Said (1971), however, estimated an age between 12,060–13,700 BP. Carbonaceous silts from a marker bed caused by brush fire yielded a date of 10,550 BC ± 230 years (Wendorf et al., 1970). The Sahaba-Darau Formation is equivalent to the Gebel Silsila Formation of Butzer & Hansen (1968) in the Kom Ombo area. The Sahaba Formation as described by de Heinzelin (1968) was divided into two aggradational units separated by an episode of downcutting (Deir El-Fakhuri Formation). Wendorf & Schild (1976) considered the early part of the Sahaba Formation to be contemporaneous with the Masmas Formation of Kom Ombo. They correlated the upper part of the Formation with the Darau Member of the Gebel Silsila Formation of Butzer & Hansen (1968). Accordingly, they proposed to name this episode of aggradation the Sahaba-Darau aggradation. The Sahaba Formation yielded typical Sebilian assemblages characterized by abundant Levallois artifacts (Wendorf et al., 1970).

The type locality of the Dishna-Ineiba Formation (Holocene) (Said, 1981) is near Dishna, Qena Province, in the Nile Valley. Said et al. (1970) introduced the name Dishna Formation (designated as member by Said, 1981) to a succession of playa deposits of unknown thickness with Nile silt interbeds, with gravels and pebble sheets embedded in fine sand at the base. The Dishna Formation is coeval with the Malki

Member of Butzer & Hansen (1968), and slope-wash debris of the Birbet Formation of de Heizenlin (1968). The Formation overlies the Sahaba Formation and it is overlain by the Arkin Formation. *Unio* shells assciated with these sediments yielded a date of 9610 BC ± 180 years (Wendorf et al., 1970), but its age was estimated between 10,000 and 9200 BP (Said, 1971) and 12,000–11,000 BP (Said, 1981). The Ineiba Formation was introduced by Butzer & Hansen (1968) for a widespread wadi accumulation with a lower conglomeratic bed composed of small pebbles of polygenetic origin, and brown clays in the upper part. They divided the Formation into a lower Malki Member and an upper Sinqari Member. The type section of the Ineiba Formation (designated as a member of the Dishna-Ineiba Formation by Said, 1981) is in the new Ineiba village, Kom Ombo plains. It is composed of playa deposits of brown clays and interfingering gravel beds of local derivation. Its age was estimated at 9000–7000 BP (Said, 1981). The Dishna-Ineiba Formation represents the deposits, which formed during the recession following the Sahaba-Darau aggradation. The playa deposits accumulated behind the natural levees and in abandoned channels of the Sahaba-Darau aggradation (Said, 1981).

The Arkin Formation (Holocene) (Said, 1971) is made up of silts and fine-grained micaceous sands, 6 m (19.7 ft) in thickness. The name was originally introduced by de Heinzelin (1968) as the Arkin aggradation. Its type locality is at Arkin, Wadi Halfa, Sudanese Nubia. The Formation overlies the Dishna Formation and underlies post-Arkin sediments. Its age is assumed to be 9200 to 7200 BP, based on radiocarbon dates. The Arkin Formation is coeval with the Arminna Member of the Gebel Silsila Formation of Butzer & Hansen (1968) in the Kom Ombo area.

The El-Kab Formation was introduced by Vermeersch (1970, cited by Wendorf & Schild, 1976, p.310) for a series of Nile sediments, now under cultivation, on the east bank of the Nile River, from El-Kilh (about 15 km north of Idfu) at the Old Kingdom fortress of El-Kab. Radiocarbon dates on charcoals yielded ages between 6400–5980 BC.

In the Delta region the Holocene is represented by the Bilqas Formation. The type section of the Bilqas Formation (Rizzini et al., 1978) is in the Bilqas-1 well (31°10'12"N, 31°30'22"E) from 0–25 m drill depth. It is made up of alternating fine and medium-grained sands, interbedded with clays rich in pelecypod, gastropod and ostracod fragments, plant material, and peat levels. It occurs in the subsurface of the northern Delta region, where it overlies the Mit Ghamr Formation. This unit crops out in the whole area of the Delta and represents Holocene coastal lagoon, brackish swamp, and beach deposits.

Quaternary deposits in northern Egypt are represented by elevated offshore bars, lagoonal beds, evaporites and marls (Butzer, 1960, Said, 1983). The Mediterranean coastal plain, such as the Gulf of Arabs to the west of Alexandria, is characterized by the presence of a number of elongated ridges, such as the Kurkar ridges, or the Alexandria Formation, which run parallel to the coast, separated by longitudinal depressions. They appear to represent successive fossil bars that were formed in the receding Mediterranean during the Pleistocene. These ridges are 10 m (33 ft), 25 m (82 ft), 35 m (115 ft), 60 m (197 ft), 80 m (262 ft), 85 m (279 ft), 90 m (295 ft), and 110 m (361 ft) high. The depressions between the ridges contain lagoonal deposits, such as evaporites and marls. Three calcareous oolitic ridges with two intervening lagoonal depressions have been recognized in the coastal plain in the Salum area

(Selim, 1974). They range in age from Late Monasterian to Tyrrhenian. The ooids in the oldest (Tyrrhenian) ridges have developed micritic envelopes and are probably recycled detrital grains. The younger deposits (Late Monasterian and main Monasterian) have well-developed oolitic texture and were probably deposited in shallow, agitated marine waters.

The Mediterranean Coast of Egypt between Alexandria and Salum (at the Egyptian-Libyan border) was dominated by carbonate sedimentation during the Quaternary. The carbonate form parallel limestone ridges dominated by carbonate dunes. Oolites dominate the eolianites of the first and second ridges, whereas bioclastics with abundant coralline algae, benthonic foraminifera, molluscs, echinoderms and intraclasts prevail in the eolianites of the third and fourth ridges (El-Shahat, 1995). Diagenetic alterations and cementation are concentrated below exposure surfaces (pedogenic calcrete horizons).

In the Qattara Depression, Quaternary sediments unconformably overlie the Miocene Moghra Formation, and consist mainly of wind-blown sands and sabkha deposits, composed of fine sands, silts, and clays, with crusts of salt and gypsiferous silts. A 3 m (10 ft) thick bed of halite, cut by numerous gypsum veins caps the Moghra Formation on the western side of the depression. Albritton et al. (1990) interpreted these Quaternary sediments as erosional remnants of sabkha deposits. The evaporites form two terraces at 10 m (33 ft) and 20 m (66 ft) above the modern sabkhas. They consist of about 2 m (6 ft) of laminated gypsum, rich in algal filaments at the base, reddish muddy sands in the middle, and nodular gypsum at the top (Hemdan & Aref, 1998).

Said et al. (1970) grouped the Holocene lacustrine deposits of the Fayium Depression into the Fayum Formation. The Formation includes four succeeding and disconformable members. The older three are exposed in isolated hillocks and wind-scoured depressions to the east of the Qasr el Sagha ridge. The youngest is well exposed in the dissected terrace along the southern flank of this ridge. These members record four lakes: The paleo-Moeris Lake, pre-Moeris (Terminal Paleolithic) Lake, proto-Moeris (Terminal Paleolithic) Lake, and the Youngest (Neolithic-Old Kingdom) Moeris Lake. The deposits of the paleo-Moeris Lake are composed of two diatomites separated by a layer of sand rich in aquatic snails. The upper surface of this unit shows polygonal mud desiccation cracks. This unit did not yield any archaeological material. The deposits of the pre-Moeris Lake occur along the northeastern edge of the Fayium depression. The thickness of these deposits is 3.5 m (11 ft) (base unexposed). The deposits consist in the lower part of alternating thin silts and black peat-like sediments with the snail *Pila ovata* and thicker, clean, fine- to medium-grained sand deposited in nearshore and swampy environments, followed by laminated and partly cross-laminated, clean, fine to medium-grained sand in the upper part. The latter are believed to be forest beds which were seasonally reworked by the lake water. Charcoal $C^{14}$ dating yielded ages between 5190 B.C. ± 120 years and 6150 B.C. ± 130 years. The deposits of the proto-Moeris Lake are made up of lacustrine deposits. A shell yielded an age of 5190 B.C. ± 120 years. They grade updip into channel sediments composed of interbedded thin black peat-like deposits, sandy silts, and sands, approximately 7.5 m (25 ft) in thickness. The youngest (Neolithic-Old Kingdom) Moeris Lake sediments are composed of layers of fine- to medium-grained lacustrine sands, deposited in nearshore environment. They contain Neolithic artifacts. Charcoal dating yielded an age of 3860 B.C. ± 115 years. The type section of this unit is in a gully directly south of the Qasr el Sagha temple.

Three levels of raised beaches skirt the Red Sea coast. The 1 m (3.3 ft) beach is dated 2500 to 6500 BP, while the 7 m (23 ft) and 11 m (36 ft) beaches are dated 81,000 to 141,000 years BP. The Pleistocene section of the Red Sea coastal strip seems to have been deposited in a continuously subsiding basin (Said, 1990f). The alternating coral reefs, sands and conglomerates were deposited in littoral to beach environment during alternating arid and pluvial episodes.

# Chapter 11

# Phanerozoic geology of Libya

## 11.1  INTRODUCTION

Libya is located in North Africa between longitudes 9°50′E-25°E and latitudes 18°45′N-33°N and bounded to the north by the Mediterranean Sea, to the east by Egypt, to the west by Tunisia and Algeria, and to the south by Sudan, Chad, and Niger. Its area is approximately 1,775,500 sq. km. and comprises four regions: Tarabulus (Tripolitania) in the northwest, Benghazi (Cyrenaica) in the northeast, Sabha (Fazzan) in the southwest, and Al Khalij (Syrtica) in the mid-North and Southeast. The capital city is Tarabulus (Tripoli). The Tarabulus region consists of a coastal plain area, the Jifarah, which rises southward to Jabal Nafusah, up to 300 m (984 ft) in height. The Sabha region consists of a series of east-west depressions and some large oases. In the extreme south, the Tibesti Mountains rise to a height of 3000 m (9843 ft). The Benghazi region consists of a small narrow coastal plain less than 30 km wide, and the terraces of the Al Jabal Al Akhdar (Green Mountains) (Fig. 11.1). The terraces rise to heights of 882 m (2894 ft) and slope southward into the Benghazi Pain; this area includes the Marmarica Plateau which extends into Egypt. Benghazi, the second largest city in the country, is located on the coast in that region. The Al Khalij region consists of a series of undulating plains and low-lying lands. There are a few oases, the Al Kufrah being the largest. Libya has an estimated population of about 6,500,000 (2011) (compared to 1,500,000 in 1964). Magnificent ancient Greek and Roman ruins are present in Leptis Magna (Labdah), Ptolemia (Tolmitha) and Cyrene (Shahat) in addition to many prehistoric cites. Tubruq is famous for World War II cemeteries. Prehistoric caves with well-preserved paintings are found in the Ghat area, in the extreme southwest corner of Libya. Circular features, interpreted as impact structures, occur in the Kufrah Basin in Souteast Libya (Schmieder et al., 2009, Ghoneim, 2009) and in the Gilf Kebir area in Southwest Egypt (Paillou et al., 2004) which provide opportunities for impact geology studies.

Recoverable oil reserves are estimated at 43.7 BBO (2008) with a daily production of 1.9 mmbopd (eia, 2009), and 581 mmcfgd. In additon to petroleum, large iron-ore deposits occur in the Wadi Ash Shati area in west-central Libya with estimated reserves of 3.5 billion tonnes at 35–55% Fe. In addition, appreciable non-metallic deposits of gypsum-anhydrite, building stones, and silica sand are being exploited.

Early geological reconnaissance work in Libya was conducted by Desio and his Italian colleagues who published the first geological map of Libya in 1933 at a scale of 1:30,000 and of the Fazzan area (west-central Libya) in 1936 (see Part I). Desio (1943)

*Figure 11.1* The ancient Greek city Cyrene, Al Jabal Al Akhdar, build on Eocene rocks.

discussed the phosphate and sabkha deposits of Libya in much detail. Christie (1955), under the sponsorship of the UN, carried out the most important work in Northwest Libya and divided the Cretaceous and Tertiary sequence in the Gharyan area into new stratigraphic units. French geologists did much work in the Jabal Al Qarqaf (Jabal Fazzan) area in the post-war years and in the early sixties, among them were Collomb, Jacqué, Burollet, and Massa.

Petroleum exploration has been carried out by oil companies since 1956, and the first commercial oil was discovered in Early Tertiary rocks by Esso Standard in January 1958 in the Oasis B1-32 well in the Dahra Field in the north-western Sirte Basin. Results of the work conducted by oil companies in the fifties were summarized by Burollet (1960) in the *Lexique Stratigraphique International* sponsored by the Commission of Stratigraphy of the International Geological Congress. That work included the definition of approximately 135 lithostratigraphic units.

Hecht et al. (1963) presented an overview of the geology of Libya which summarized the information available at that time. Conant & Goudarzi (1964) produced a geological map of Libya which was published by the U.S. Geological Survey, Washington, D.C. at a scale of 1:2,000,000 and discussed the tectonics of the country in 1967. Goudarzi (1970) also discussed the geology and mineral resources of Libya. The second edition of the Geological Map of Libya (1977) was compiled and updated by Maghrabi & Cheshitev of the Industrial Research Centre, Tripoli. Barr & Weegar (1972) carried out one of the most important works in the Sirte Basin and established at least 40 new formations and groups. Six symposiums were organized by the Geological Exploration Society of Libya. The first symposium was held in 1963 (published by the *Institut Francais du Pétrole*). Papers of the 1968 symposium were published in Geology of Libya in 1971. The results of the 1978 symposium were published in 1980 in volumes I to III of The Geology of Libya. Another symposium was held in 1987, and the papers were published in 1991 in volumes IV to VII of The Geology of Libya. These volumes also contained a number of papers on the offshore of Libya. A symposium in 1993 (published in three volumes in 1996) was dedicated entirely to the geology of the Sirte Basin. Two more conferences were held on Northwest Libya (2003) and Northeast Libya (2006).

Results of palynological research on the subsurface of Northeast Libya Libya were published in the Journal of Micropaleontology (volume 5) in 1985 under the auspices of AGOCO and a second volume was published by the Garyounis University, Benghazi, in 1988. An International symposium on the geology, flora and fauna of the Paleocene Sahabi Formation in the Sahabi area, one of the richest localities in fossil mammals in North Africa, was held at the *Naturhistorisches Museum*, Main, Germany in 1981. The results were published in 1982 as Special Issue No. 4 of the University of Garyounis, Benghazi. Tawadros (2001) and Hallett (2002) gave detailed accounts of the geology of Libya.

Hammuda et al. (1985) reviewed the stratigraphy of the offshore of Northwest Libya and introduced a number of new formations and groups that were correlated with those of the onshore Northwest Libya and Tunisia. From 1974 to 1984, the Industrial Research Centre commissioned different organizations to carry out geological mapping at the scale 1:250,000 in Libya. These organizations included the Czechoslovakian company Geoindustria, the Yugoslavian Geoinstitute (Institute for Geological and Mining Explorations), the General Egyptian Organization for Industrialization (work was carried out by the Geological Survey of Egypt), and the *Centro Ricerche Geologiche* (C.R.G.) S.p.A., Firenze, Italy. The result of this work was 61 maps covering the northwestern half of Libya accompanied by explanatory booklets and a geological map of Libya at a scale of 1:1,000,000 published in 1985. The results of most of this work were also summarized by Mamgain (1980), and Megerisi & Mamgain (1980a, b), and a new Lexicon was edited by Banerjee (1980) which included 455 entries or lithostratigraphic units. The authors of Geoindustria, however, changed the original spelling of formational names to agree with English transliteration of Arabic names. They also used a set of accents which are not known in the English language which makes typing and spelling of the new names very difficult. The present author concurs with Barr & Berggren (1980) that the original spelling of formational names should be maintained even if the spelling of geographical names has been changed. This practice is in agreement with the North American Commission on Stratigraphic Nomenclature (1983, Article 8d). In the following discussion, the original spelling of names of formations and other stratigraphic units have been used, since they have priority. However, the new spellings of geographic names, especially those of towns, have been used, except where they constitute well established geological names. For example, the name Gargaf Arch has been used although the name of the mountains has been changed to Jabal Al Qarqaf. Mapping of the southeast corner of Libya is currently underway; the remaining part of the country is covered by sand dunes of the Great Sand Sea.

# Chapter 12

# Paleozoic

Paleozoic rocks are widely distributed in Libya. They crop out in the Jabal az Zalmah (Jabal Dalma) and Jabal Arknu in Southeast Libya, over large areas of the Jabal Al Qarqaf (Jabal Gargaf), and in Southwest Libya. They are encountered in the subsurface of the Kufrah, Sirte, Ghadames, and Murzuq basins, and the Jifarah Plains. In the Sirte Basin, only a small record of the Paleozoic sediments remains after the Hercynian orogeny. As in Egypt, Early Cambrian sediments and igneous rocks originated during the late stage of the Pan-African orogeny and they are often included in the basement.

Most of the Paleozoic sediments in Libya are poorly fossiliferous, and the fauna recovered are usually poorly preserved and undiagnostic. The fossils are limited to ichnofossils *Tigillites (Skolithos), Harlania (Arthrophycus), Spirophyton (Zoophycus)*, and locally *Cruziana*, in addition to a few species of *Lingula*, brachiopods, and wood fragments that have a little or no stratigraphic value. Rare graptolite horizons were reported from Silurian shales, especially in West Libya. Bonnefous (1963) subdivided correlative Silurian strata in the subsurface of the Jifarah Plains of Southern Tunisia based on graptolites. In recent years, palynology has been of great advantage in dating many of the apparently unfossiliferous Paleozoic formations. Collomb (1962), Jacqué (1963), Massa & Moreau-Benoit (1976), Seidl & Röhlich (1984), Parizek et al. (1984), Jakovljevic (1984), and Grubic et al. (1991) gave an elaborate description of the Paleozoic sediments in West Libya. Paleozoic sediments were studied on the southwestern side of the Kufrah Basin, along the track from Faya Largeau to Tekro which straddles the Libyan-Chad border, by Wacrenier (1958), De Lestang (1965, 1968), and Klitzsch (1966), in southern Jabal Azbah (Jabal Asba) and west of the Jabal Arknu by Bellini & Massa (1980), and in the Jabal Awaynat (Gebel Aweinat) by Burollet (1963b) and Vittimberga & Cardello (1963). Other recent studies on the Paleozoic sediments of the Kufrah Basin were carried out by Turner (1980, 1991), Grignani et al. (1991), Lüning et al. (2000a), Seilacher et al. (2002), Ghanoush & Abubaker (2007), Le Heron et al. (2009), Schmieder et al. (2009), and Paillou et al. (2009). There is, however, little paleontological support for regional correlations and most of the Paleozoic lithostratigraphic units are dated based solely on their stratigraphic positions. Paleozoic rocks do not crop out in Northeast Libya Libya. El-Arnauti & Shelmani (1985, 1988) reviewed the stratigraphy of the Paleozoic section in the subsurface of Northeast Libya.

## 12.1  CAMBRIAN AND ORDOVICIAN

### 12.1.1  Cambrian and Ordovician of West Libya

Cambrian and Ordovician sediments in West Libya are represented by the Gargaf Group which includes in an ascending order the Hassaouna Formation, traditionally ascribed to the Cambrian, followed by the Ordovician Achebyat, Haouaz, Melez Chograne, Tasghart and Memouniat formations. In this present work, the name Gargaf Group is extended to include all the Cambro-Ordovician rocks in Libya, both in outcrop and in the subsurface.

The name Gargaf Sandstone (*Grès du Gargaf*) was first used by Lelubre (1946a) to describe a sandstone succession in the Fazzan area (Sabha region) which he ascribed to the Devonian. Burollet (1960) proposed the name Gargaf Group for the Cambro-Ordovician sandstone sequence in the Jabal Al Qarqaf area and divided it into four formations, from base to top: the Hassaouna Formation, Haouaz Formation, Melez Chograne Formation, and Memouniat Formation. The type section of the Gargaf Group extends from the igneous basement outcrops in Jabal Al Qarqaf (Jabal Al Hassawnah 28°15′N, 14°00′E) to the Silurian outcrops at Awaynat Wanin (28°0′N, 12°5′E), and to Bir el Qasr (27°00′N, 12°05′E). Jacqué (1963) divided the Gargaf Group in the Jabal Al Qarqaf area into four formations: Hassaouna, Haouaz, Emi Daoun, and Dor El Fatta formations. The last two formations are equivalent to the Melez-Chograne and Memouniat formations of Burollet (1960), respectively. However, the names Emi Daoun and Dor El Fatta formations were not accepted by subsequent authors, and justifiably fell into oblivion. Havlic ek & Massa (1973) recognized another unit within the Gargaf Group between the Hassaouna and Haouaz formations on the western flank of the Jabal Al Qarqaf for which they introduced the name Achebyat Formation.

The Hassaouna Formation (spelled Hasawinah on the Geological Map of Libya, 1985) is the most extensively developed Lower Paleozoic unit and has been recognized in western, central and eastern Libya. It was first separated from the overlying Haouaz Formation by Massa & Collomb (1960) and used by Burollet (1960) without reference to their work. It was named after Jabal Al Hassawnah. No type section was designated for this formation. In the Jabal Al Qarqaf area, the Hassaouna Formation unconformably overlies sandstones of the Mourizidie Formation, or the Precambrian basement. The lower contact of the Hassaouna Formation with the basement rocks is best exposed in Wadi ad Dharman (approx. 28°30′N, 13°55′E) (Jurak, 1978) where it overlies the basement with a basal conglomerate, 13 m (43 ft) in thickness. Collomb (1962) measured a thickness of 350–400 m (1148–1312 ft) for the Hassaouna Formation at Dur al Qussah on the eastern edge of the Murzuq Basin. Klitzsch (1963) postulated a thickness of 1700 m (5578 ft) in the same area. Jacqué (1963) reported a thickness of 1253 m (4111 ft) on the eastern flank of Jabal Mourizidie. On the western side of the Gargaf Arch, the Hassaouna Formation is overlain with an angular unconformity by the Ordovician "Achebyat", Haouaz, Melez Chograne, or Memouniat formations (Parizek et al., 1984). Toward the centre of the uplift, the Cambro-Ordovician section is progressively eroded and the Memouniat Formation rests unconformably on the Cambrian Hassaouna Formation (Seidl & Röhlich, 1984, Collomb, 1962, Massa & Collomb, 1960). The Hassaouna Formation was assigned a Cambrian age by Massa &

Collomb (1960), Jacqué (1963) and Burollet (1960), and it was treated as such by all subsequent workers. Palynological analysis of the Hassaouna Formation in the subsurface of the Ghadames Basin (well A1-69) by Albani et al. (1986) yielded only age undiagnostic Sphaeromorphs. However, a rich assemblage was recovered from its equivalent the Sidi Toui Formation in the A1-70 well.

The Hassaouna Formation in the Jabal Al Qarqaf area is composed of cross-bedded, medium to coarse- grained sandstones which are silica cemented with a kaolinite matrix and rare carbonate cement. The sandstones are mostly coarse-grained and feldspathic in the lower part, becoming medium-grained upwards with *Tigillites* (Collomb, 1962, Mennig & Vittimberga, 1962). Polymictic conglomerates at the base contain devitrified rhyolite fragments. Grain size decreases westwards and the Formation becomes more feldspathic away from the Gargaf Arch (Mennig & Vittimberga, 1962). The Formation is intruded by Tertiary phonolites, basalt dikes and cones (Jurak, 1978). Freulon (1964) divided the Hassaouna Formation in the central Sahara into lower, middle and upper units. The lower unit starts with several meters of basal conglomerates composed of quartz pebbles, quartzite and metamorphic rock fragments, followed by coarse-grained, quartzitic, cross-bedded sandstones. The middle unit is approximately 45 m (148 ft) in thickness and composed of very fine-grained sandstones and silty sandstones, thin-bedded, laminated and rippled with slump structures. The upper unit is approximately 100 m (328 ft) and consists of fine to medium-grained sandstones, partly coarse grained and conglomeratic, thick-bedded, cross-bedded, friable, and cemented with carbonates.

On the southern flank of the Jabal Al Qarqaf, the Hassaouna Formation is 50 m (164 ft) thick (Seidl & Röhlich, 1984) and represents a fining-upward transgressive cycle. The lower part with high-angle planar cross-stratification was probably deposited in fluvio-deltaic environments. The direction of transport was to the north and northwest. The upper part is composed of medium to coarse-grained sandstones with a few siltstone beds. These beds display planar and trough cross-bedding with bipolar orientation, tabular lamination, convolute bedding, ripple bedding and reactivation surfaces (Cepek, 1980).

On the southwestern side of the Murzuq Basin, the Hassaouna Formation unconformably overlies basement rhyolites, dacites and granites of the Hoggar Massif (Tassili Tafasasset) and is approximately 300–400 m (984–1312 ft) thick (Grubic, 1984, Galecic, 1984). The Formation in that area consists mostly of silica-cemented quartz pebbles and siliceous pelites, grading upward into cross-bedded quartzose sandstone with the trace fossils *Harlania, Bilobites, Rusophycus, Fraena,* and *Cruziana.* Protic (1984) recognized only the middle and upper units of the Hassaouna Formation of Freulon (1964) in the Tikiumit area. The middle unit is approximately 45 m (148 ft) in thickness and composed of very fine-grained sandstones and silty sandstones, thin-bedded, laminated and rippled with slump structures. The upper unit is approximately 100 m (328 ft) and consists of fine to medium-grained sandstones, partly coarse-grained and conglomeratic, thick-bedded, cross-bedded, friable, and cemented with carbonates. In the Ghat area, the upper boundary of the Formation with the overlying "Achebyat" Formation is gradational (Protic, 1984). Grubic et al. (1991) placed this boundary at the first upward appearance of *Tigillites* and the change of color from grey to white. The Hassaouna Formation is equivalent to the Sidi Toui Formation in the subsurface of the Jifarah Plains and the northern Ghadames Basin

(Deunff & Massa, 1975, Massa et al., 1977, Albani et al., 1986). The Hassaouna Formation is equivalent to the Ajjer Formation, the Hassi Massaoud Sandstone, and the Hassi Leila Formation, or Unit II in the Illizi Basin in Algeria (Beuf et al., 1971, Bennacef et al., 1971, Dubois, 1961, Corriger & Surcin, 1963).

The Haouaz Formation[1] was described by Burollet (1960) as composed of fine-grained sandstones, well-bedded, cross-bedded, silty and micaceous, and green and grey shales. It can be differentiated from the Hassaouna sandstones by the abundance of feldspars (Collomb, 1962, Jacqué, 1963), and by an abrupt change from coarse-grained, cross-bedded sandstones of the Hassaouna Formation to fine-grained, massive sandstones of the Haouaz Formation (Massa & Collomb, 1960). Diagenetic minerals include illite, dickite, and kaolinite (Abouessa & Morad, 2009). In the Jabal Al Qarqaf area, the thickness of the Haouaz Formation is 120–190 m (393–623 ft). Collomb (1962) divided the Formation into three units, lower *Tigillites*, middle sandstone (*grès intermédiaires*), and upper *Tigillites*. The latter is equivalent to the Haouaz Formation as defined by Burollet (1960). The lower *Tigillites* unit is 10–50 m (33–164 ft) thick and consists of massive very fine to fine-grained, quartzitic sandstones, interbedded with argillaceous siltstones rich in *Tigillites*. The middle sandstone unit is coarse-grained, lenticular, and contains a few *Tigillites*. The upper *Tigillites* unit is 90–120 m (295–393 ft) thick, fine-grained and feldspathic at the base. It was assigned an Arenig age by Jacqué (1963) and a Tremadoc-Arenig age by Collomb (1962). Havlicek & Massa (1973) suggested a Middle Ordovician (Llanvirnian-Llandeilian) age based on its stratigraphic position below the Melez Chograne Formation which is dated by brachiopods. It was also correlated with the Llanvirnian-Llandeilian Bir Ben Tartar Formation in the subsurface of the north Ghadames Basin by Deunff & Massa (1975) and Massa et al. (1977). The base of the Haouaz Formation in the Murzuq Basin is marked by an unconformity (Burollet, 1960, Klitzsch, 1963). The Haouaz Formation rests on the Cambrian with a strong unconformity at Dur al Qussah (Collomb, 1962) which is not visible in the Jabal Al Qarqaf (Burollet, 1960, Bellini & Massa, 1980). The Formation wedges out to the southeast between the Melez Chograne and the Hassaouna formations and it is missing in the eastern part of the Dur al Qussah area (Klitzsch, 1963). The Haouaz Formation is also absent in the southwestern corner of Libya, west of longitude 12°E and south of latitude 28°N, and the Memouniat Formation overlies the "Achebyat" Formation (Grubic et al., 1991). The sedimentology of the Haouaz Formation in the Jabal Al Qarqaf outcrops has been examined by Vos (1981b) and Ramos et al. (2006). The Haouaz Formation is equivalent to the Bir Ben Tartar Formation in the Ghadames Basin, and the Ourgala Sandstone and Azzel Shale and the In Tahouite Formation in Eastern Algeria.

The presence of *Tigillites* has been used by many workers to correlate and differentiate between the different Cambro-Ordovician formations and to erect new ones. Unfortunately, this has led to a lot of confusion and stratigraphic disorder. Havlicek & Massa (1973) introduced the Achebyat Formation (spelled Ashebyat on the Geological Map of Libya, 1985) which they considered to be transitional between the Cambrian and the basal Middle Ordovician on the western edge of the Jabal Al Qarqaf. The Formation overlies with a gradational contact the Hassaouna Formation and underlies the Haouaz Formation. Its thickness is approximately 65 m (213 ft),

---

1 Also spelled Hawaz.

and contains abundant *Tigillites*, *Cruziana* and *Harlania*. It was assigned a Tremadoc age by Massa et al. (1977) and Bellini & Massa (1980) based on regional correlations with the Sanrhar Formation (Tremadoc) in the subsurface of the northern Ghadames Basin. Havlicek & Massa (1973) selected the type area of the Formation in the eastern margin of Jabal Ash Shabyiat (longitude 13°05′E, latitude 28°00′N), but no type section has been designated. In the south-western part of the Jabal Al Qarqaf area, the Formation is composed of 17–40 m (56–131') of white to yellowish grey, fine, quartzitic sandstones with *Tigillites* in the lower part and 20–25 m (66–82') of light grey, medium to coarse-grained, thick-bedded sandstones in the upper part (*grès intermédiaires* of Collomb, 1962) (Parizek et al., 1984). On the western side of the Murzuq Basin, the Achebyat Formation conformably overlies the Hassaouna Formation. The Achebyat Formation is eroded in some places and the Hassaouna Formation is overlain directly either by the Melez Chograne Formation or the Tannezouft Formation. In the Ghat area where the Haouaz Formation is assumed to be missing (Radulovic, 1984), the Achebyat Formation is overlain directly by the Melez Chograne (Grubic et al., 1991). Radulovic (1984) identified the Achebyat Formation in the Ghat area solely by the presence of *Tigillites*, especially *Tigillites dufrency*, up to 0.5 m in length. He reported the acritarchs *Baltisphaera* cf. *ternata*, *Zonosphaeridium* sp. and *Trachypsophosphaera* sp., as well as indeterminate algae. Protic (1984) treated the lower and middle parts of Collomb's (1962) Haouaz Formation as the Achebyat Formation and restricted the Haouaz Formation to the upper *Tigillites* unit only.

The Achebyat Formation is probably equivalent to the upper part of the Hassaouna Formation defined by Massa & Collomb (1960) and the lower part of the Haouaz Formation described by Collomb (1962). It is considered in the present work to be a unit of the Haouaz Formation and should not be separated from it until its exact stratigraphic position has been clarified. The Achebyat Formation as defined is probably equivalent to the Sanrhar Formation (Tremadoc) in the subsurface of the northern Ghadames Basin and the In Kraf Formation, subunit 4b of Unit III, In Tahouite, or Edjeleh Formation, in the Illizi Basin.

The name Melez Chograne Formation was introduced by Massa & Collomb (1960) and used by Burollet (1960) for an argillaceous sequence with intercalations of greenish, yellow and grey micaceous sandstones on the western edge of the Jabal Al Qarqaf (28°20′N, 13°00′E). The type section has not been established. This unit is equivalent to the defunct Emi Daoun Formation of Jacqué (1963). The Formation is overlain by the Memouniat Formation and unconformably overlies the Haouaz Formation in the Jabal Al Qarqaf (Collomb, 1962), or the Hassaouna Formation with an angular unconformity, in the Dur al Qussah area (Klitzsch, 1963). The Formation forms lenticular bodies due to the infilling of the irregular topography and post-Melez Chograne erosion. Where it is absent, the Memouniat Formation rests unconformably on the older Ordovician units or the Cambrian Hassaouna Formation. The thickness of the Formation is very variable and varies between 10–60 m (33–197 ft) in the type area. In the Jabal Al Qarqaf area, the Formation is composed of thin-bedded, variegated green-brown, violet, and black, unfossiliferous shales and siltstones (Collomb, 1962, Parizek et al., 1984), sandy shales, diamictites (Mennig & Vittimberga, 1962), and thin sandstone and sandy fossiliferous carbonate beds. The carbonate rocks of the Melez-Chograne, where present, are richly fossiliferous. The fauna includes trilobites, brachiopods, especially *Lingula*, echinoderm fragments, bryozoans, hyalothids, and conularids. A ferruginous, phosphatic, oolitic conglomeratic horizon with bryozoans and the trilobites *Synhomalonotus tristani* and *Kloucekia* sp., are also present at the base of the Formation in the Jabal Al Qarqaf area (Collomb, 1962). There has been no general agreement on the age of the Melez Chograne Formation. It was assigned a Middle Ordovician (Llandeilian) age by Burollet (1960), a Llanvirnian-Llandeilian age by Collomb (1962), and a Caradocian age by Bergström & Massa (1991), Havlicek & Massa (1973) and Parizek et al. (1984).

In the subsurface of the Murzuq Basin, the Melez Chograne is made up of highly radioactive shale interbedded with siltstones and fine sandstones (Pierobon, 1991). Its thickness varies from 15–120 m (49–394 ft), and the unit is absent in places due to erosion. The contact with the underlying Haouaz Formation or Hassaouna Formation is unconformable and marked by ferruginous oolites. In the Ghat area, in the southeastern corner of Libya, the exposed Melez Chograne Formation is 20 m (66 ft) thick. It consists of greenish-grey shales with intercalations of thin layers and lenses of sandstones and pebbly shales to siltstones (diamictites) (Grubic et al., 1991). The upper part is composed of very hard, greenish-grey fine-grained sandstones, carbonate cemented with slump structures of folds and rolls. In the Tikiumit area to the north, the unit is also 20 m (66 ft) thick (Protic, 1984). It is composed of reddish brown, laminated, fine-grained sandstones with boulders of granites, crystalline schists, conglomerates, phosphate concretions, brachiopods, gastropods and cephalopods, followed upward by siltstones and sandy siltstones with slump structures, convolute bedding, flute casts, and fragments of granites, diorites, and crystalline schists (gravely mud, mixton). South of Ghat, the Memouniat is absent and the Melez Chograne is unconformably overlain by the Tannezouft Formation. Further to the south, in the Anay and South Anay areas, both the Memouniat and Melez Chograne formations are absent and the Silurian Tannezouft Formation rests directly on the Hassaouna Formation (Galecic, 1984, Grubic, 1984).

In Eastern Algeria, the equivalent of the Melez Chograne, Tasghart, Memouniat, and basal Tannezouft formations is known as Unit IV (Bubois, 1961, B.R.P. et al., 1964, Corriger & Surcin, 1963), Felar-Felar Formation (BRP et al., 1964), or the Tamadjert Formation (Bennacef et al., 1971). This unit ranges in thickness from 10 m (33 ft) to 450 m (1476 ft) and its age extends from Late Caradocian to Middle Llandoverian. It unconformably overlies the In Tahouite Formation (Edjeleh Formation, or Unit III), and it is overlain by the Imirhou Formation (Bennacef et al., 1971) (equivalent to the Tannezouft Formation). Variations in thickness and lithology are abrupt. Unit III (equivalent to the Haouaz and Achebyat formations) unconformably overlies Unit II (equivalent to the Hassaouna Formation), and it is believed to be Late Cambrian-Tremadocian in age. Unit II unconformably overlies the Precambrian basement. These lithostratigraphic subdivisions were extended into the subsurface of southeast Algeria by Corriger & Surcin (1963) where they show similar lithologies and facies relationships. These units can be correlated with some degree of confidence with the Cambro-Ordovician succession cropping out in Southwest Libya, and are believed to continue into the subsurface of Southwest Libya. Figure 12.1 shows a tentative correlation of the Cambro-Ordovician units between Southeast Algeria, Southwest Libya, and the Gargaf Arch area. It appears that the Hassaouna, Achebyat, and Haouaz formations form a sequence bounded on top and base by unconformities (although many unconformities may be present within the sequence, they are difficult to detect from the available information).

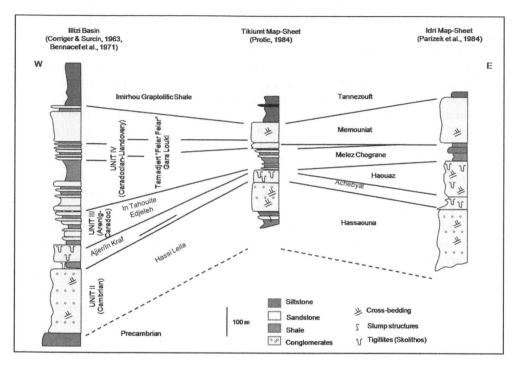

*Figure 12.1* Correlations of Cambro-Ordovician successions in West Libya (after Tawadros, 2001) (See color plate section).

The Melez Chograne Formation probably marks the beginning of a marine transgression (Mennig & Vittimberga, 1962, Massa et al., 1977) at the end of Ordovician glaciation. The depositional environment of the Melez Chograne Formation was probably neritic and glacio-marine extending over a large area of West Libya (Massa et al., 1977, Parizek et al., 1984, Grubic et al., 1991) and Eastern Algeria. Boulders and pebbles of granites, gneisses, quartzites, metamorphic schists and sandstones with polished and striated surfaces occur in the lower part of the Formation, and may represent glacial deposits (Collomb, 1962, Bennacef et al., 1971, Beuf et al., 1971, Radulovic, 1984, Bergström & Massa, 1991).

Another problematic unit is the Tasghart Formation. Radulovic (1984) proposed the name Tasghart Formation for a 70 m (230 ft) unit between the Melez Chograne Formation and the Memouniat Formation at Wadi Tanazuft and assigned it a Caradocian-Ashgillian age based on its stratigraphic position. The type area of the Formation is in the Wadi Anlalin (between latitude 25°54′32″N & 25°54′52″E, longitude 9°53′52″N & 9°59′56″E). The unit has a very limited areal distribution, mostly in the Wadi Tasghart area. Where the Melez Chograne Formation is absent, the Memouniat Formation rests with a gradational contact on the Tasghart Formation. The Formation starts at the base with sequences up to 5 m (16 ft) in thickness, composed of purple to grey, medium to coarse-grained sandstones with sole marks, contorted bedding, parallel lamination, and ferruginous and carbonate concretions at the base, followed by massive or distinctly-bedded sandstones in the middle, and then by fine to medium-grained, silty sandstones with horizontal lamination at the top. This sequence was interpreted to be a turbidite by Radulovic (1984), Protic (1984), and Grubic et al. (1991). It is distinguished from other units based mainly on the presence of slump structures and contorted bedding. These structures, however, are also present in the Memouniat Formation, Melez Chograne Formation, and the Haouaz Formation, and the Tashgart Formation can be confused with other units. The Tasghart Formation should be restricted to the equivalent of Unit IV-1d in the Illizi Basin (Fig. 12.1).

The name Memouniat Formation was introduced by Massa & Collomb (1960) and used by Burollet (1960) to describe a unit, often unconformably, overlying the Melez Chograne Formation, in the Jabal Al Qarqaf area. The Formation is composed of massive, cross-bedded, and medium to coarse-grained sandstones with *Tigillites*, slightly cemented with kaolinite, finely bedded fine-grained sandstones and very rare micaceous silty shales. The Formation is equivalent to the obsolete Dor El Fatta Formation of Jacqué (1963). At Jabal Douesa, the Formation is made up of feldspathic sandstones and siltstones. South of the Awaynat Wanin area, the unit consists of non-feldspathic, medium to coarse-grained sandstones (Mennig & Vittimberga, 1962). The middle part of the Formation contains a coquina bed made up of the brachiopod *Plectothyrella*. It was assigned an Ashgillian age by Havlicek & Massa (1973), Massa et al. (1977), and Bergström & Massa (1991). The Formation overlies the Melez-Chograne shales or the Haouaz sandstones. On the southern flank of the Gargaf Arch, the unit unconformably overlies the Hassaouna Formation and is unconformably overlain by the Aouinat Ouenine Group (Seidl & Röhlich, 1984, Parizek et al., 1984). Parizek et al. (1984) designated the section at Gara Antelope (27°43'25"N, 12°51'56"E), on the northeast side of the Idri-Awaynat Wanin track approximately 45 km north-northwest of Idri as the type section of the Memouniat Formation. In that section, the Formation is 118 m (387 ft) in thickness, base not exposed, and it is unconformably overlain by the Devonian Bir al Qasr Formation. It consists of jointed, greyish-white to greyish-violet, fine-grained sandstones, becoming medium-grained and cross-bedded in the upper 20 m (66 ft) with thin layers of micaceous siltstones near the top. In basin areas, the Formation is overlain by Tannezouft shales or younger formations. Its thickness varies from 150 m (492 ft) in the northern Jabal Al Qarqaf area to 20 m (66 ft) in the southern part of the area. The Formation becomes eroded to the east (Seidl & Röhlich, 1984). The Memouniat Formation in the subsurface of the Murzuq Basin has a thickness of 20–170 m (66–558 ft) (Pierobon, 1991). In the Ghat and Wadi Tanazuft areas (Fig. 12.3), the Memouniat Formation consists of thick-bedded to massive, whitish-grey, yellowish, or pink medium to coarse-grained sandstones with wavy and cross-lamination (Grubic et al., 1991) (Fig. 12.2). The lower contact with the Melez Chograne Formation or the Tasghart Formation is conformable. The upper contact with the Tannezouft Formation is unconformable and marked by a deep-red hematitic sandstone bed in the Ghat area. It was assigned an Ashgillian age by Grubic et al. (1991) and Bergström & Massa (1991) based on coquina beds in the middle part of the Formation composed of *Plectothyrella libyca* and *Hirnantia* aff. *sagittifera* near Ghat. Grubic et al. (1991) reported a palynomorph assemblage from the Memouniat Formation in the Wadi Tanazuft area which includes the acritarchs *Veryhachium* sp., *Leiosphaeridia* sp. and *Leiofusa* sp., the chitinozoans *Cymatiosphaera* sp. and *Sporites* sp., as well as the algae *Tortunehma* sp. This assemblage indicates an Upper Ordovician (Ashgillian)

*Figure 12.2* Memouniat Formation, North of Ghat, Southwest Libya (courtesy of Sebastian Lüning).

age. The Memouniat Formation probably represents a second glacio-marine episode in the Ordovician. The Memouniat Formation includes beach and tidal flat sediments in the Ghadames Basin (Bracaccia et al., 1991), and glacio-marine sediments in Southwest Libya (Parizek et al., 1984, Grubic et al., 1991). The Melez Chograne and Memouniat formations are equivalent to the Djeffara Formation in the subsurface of the northern Ghadames Basin, and the Rhazziane and Mkratta formations, most of the Tamadjert Formation, Gara Louki Formation, and Unit IV in Eastern Algeria.

Cambrian and Ordovician sediments occur widely in the subsurface of the Jifarah Plains of Northwest Libya and Southern Tunisia. The lithostratigraphic nomenclature established mostly in Southern Tunisia is applied equally in Libya (Fig. 12.3).

Cambrian sediments were encountered in the subsurface by a number of wells in the Jifarah Plains of NW Libya and Southern Tunisia (Fig. 12.3). The Cambrian is represented by the Sidi Toui Formation[2]. The Sidi Toui Formation was introduced by Deunff & Massa (1975) and Massa et al. (1977) for a Cambrian rock unit, named after the type well Sidi Toui (St1) about 45 km south of Ben Gardane in Tunisia (Ben Ferjani et al., 1990). The Formation consists of coarse arkosic sandstones at

---

2  The name Sidi Toui Limestone or Formation was used by Burollet (1963) and Burollet et al. (1978) for a
   Triassic unit equivalent to the Rehach and Azizia formations in Libya and Tunisia, respectively.

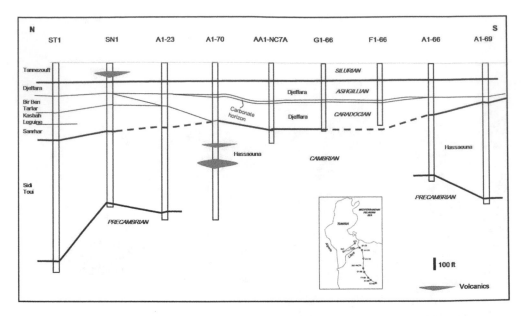

*Figure 12.3* N-S stratigraphic cross-section of Cambrian-Silurian lithostratigraphic units (Northwest Libya to Southern Tunisia (modified from Massa, 1988). Age determinations are based on Massa (1988), Albani et al. (1991), and Vecoli et al. (2009).

the base and grades upward into coarse clayey quartzitic sandstones. Its thickness is between 350 m (1148 ft) (Massa et al., 1977, Massa, 1988) and 1000 m (3281 ft) (Deunff & Massa, 1975). In the St1 well, it has a thickness of 1020 m (3347 ft) (Albani et al., 1991). It is conformably overlain by the Lower Ordovician Sanrhar Formation. In the A1-23 well, the Hassaouna/Sidi Toui Formation probably overlies the Infra-Cambrian Mourizidie Formation that fills the Pan-African topography (Letouzey et al., 2005). It was assifgned Middle to Late Cambrian age by Deunff & Massa (1975) based on the presence of algal spores and acritarchs such as *Staplinium*, *Leiosphaerida*, and *Baltisphaeridium*. Kalvacheva & Kazandiev (1978) reported the acritarchs *Acanthodiacrodium*, *Cristallinium*, *Cymatiosphaera*, *Leiosphaerida*, and *Micrhystridium* from the A1-100 well in the Ghadames Basin near the Tunisian border. Albani et al. (1991) assigned it a Middle-Upper Cambrian age based on acritarchs recovered from the A1-70 and Tt1 wells in northern Tripolitania and Southern Tunisia, respectively. The assemblage includes *Adara alea*, *Comasphaeridium* sp., *Cristallinium cambriense*, *Cristallinium* sp. cf. *C. ovillense*, *Eliasum microgranulatum*, *Multiplicisphaeridium martae*, *Stelliferidium* sp., *Strictosphaeridium brayense*, *Timofeevia lancarae* (reported as *Baltisphaeridium* n. sp. aff. *lancarae* by Deunff & Massa, 1975), *T. phosphoritica*, *Vulcanisphaera turbata*[3], and a number of species of *Micrhystridium*, in addition to abundant Sphaeromorphs. *Vulcanisphaera turbata* has been reported from the Upper Carboniferous Belhedan Formation in the Sirte Basin by Tawadros

3 Reported from the Sirte Basin by Tawadros et al. (2001).

et al. (2001). The formation grades from braided stream deposits in the south to tidal and shallow marine in the north (Bellini & Massa, 1980, Albani et al., 1991). The Sidi Toui Formation extends into the Jifarah Plains in Northwest Libya (Deunff & Massa, 1975, 1988, Massa et al., 1977, Albani et al., 1991) and is equivalent to the Hassaouna Formation to the south. It is also equivalent to the Ajjer Formation, the Hassi Massaoud Sandstone, the Hassi Leila Formation, and Unit II in the Illizi Basin in Algeria (Beuf et al., 1971, Bennacef et al., 1971, Dubois, 1961, Corriger & Surcin, 1963).

In the subsurface of the Jifarah Plains of NW Libya and Southern Tunisia, the Ordovician succession includes four formations, in an ascending order: Sanrahar Formation, Kasbah Leguine Formation, Bir Ben Tartar Formation, and Jeffara (Djeffara) Formation (Fig. 12.3).

The Sanrhar Formation was named after an area in Tunisia by the same name to characterize a section composed of 100 m (328 ft) of argillaceous sandstones. The type locality of the Sanrhar Formation is in the well Sanrhar 1 (SN-1) (Fig. 12.3), about 36 km south of Bordj. Bergström & Massa (1991) described the Formation as fossiliferous shales. It conformably overlies the Sidi-Toui Formation and is conformably overlain by the Kasbah-Leguine Formation. It is rich in acritarchs (Deunff & Massa, 1975) such as *Acanthodiacrodium petrovi*, *A. simplex*, *A. lineatum*, *Lophodiacrodium tuber*, *L.* cf. *salebrosum*, *Arbusculidium destombesii*, *Goniosphaeridium uncinatum*, *G.* cf. *dentatum*, *G. pungens*, *Filisphaeridium capillatum*, *Ploygonium gracilis*, *Vulcanisphaera africana*, *Stelliferidium cortinulum*, *Cymatiogalea bellicosa*, *C. cuvillieri*, *C. membranispina*, *C. multiclaustra*, *Saharidia downiei* and *Leiofusa squama*, suggesting a Tremadocian age. It also contains abundant inarticulate brachiopods (Massa et al., 1977) such as *Tunisiglossa tripolitanea*, *Palaeoglossa crumena* and *Siphonotreta* sp., supporting a Tremadocian age. This unit is probably equivalent to the "Achebyat" Formation in the Murzuq Basin, and the In Kraf Formation, El Gassi Shale and El Atchane Sandstone in Eastern Algeria.

The Kasbah-Leguine Formation was introduced for a Lower Ordovician (probably Arenigian) unit in the subsurface of the Ghadames Basin and named after an area in Tunisia. The Formation is 10–25 m (33–82 ft) in thickness and consists of fine sandstones with ferruginous oolites. It reaches a thickness of 56 m (184 ft) in its type locality in the Sidi Toui 1 (ST-1), but may reach up to 130 m southward (Ben Ferjani et al., 1990). It conformably overlies the Sanrhar Formation and is conformably overlain by the Bir Ben Tartar Formation. It contains the acritarchs *Priscotheca raia*, *Goniosphaeridium dentatum* and *Polygonium gladkovae*. This unit is probably equivalent to the basal Haouaz Formation. It is equivalent to the Hamra Quartzites in Eastern Algeria.

The name Bir Ben Tartar Formation is derived from its type locality in the Bir Ben Tartar well (Tt1), about 43 km east of Remada (Ben Ferjani et al., 1990) on the Sidi-Toui Uplift "*Môle de Sidi-Toui*" in Tunisia, which is the western continuation of the Nefusah Uplift of Northwestern Libya (Massa, 1988). The Formation consists of fine-grained sandstones with horizons of clays and oolites. It conformably overlies the Kasbah-Leguine Formation and is conformably overlain by the Djeffara Formation. It is rich in trilobites such as *Neseuretus* sp., the graptolite *Didymograptus*, and acritarchs such as *Veryhachium horridum*, *V. sartbernadensis*, *V. trisulum*, *V. angstum*, *V. valiente* and *V. splendens*, indicating a Llandovery-Llanvirnian age (Deunff & Massa, 1975, Massa et al., 1977). Adloff et al. (1985, 1986) suggested that its age may extend into the Arenig. The Bir Ben Tartar Formation is equivalent to the Haouaz Formation in West Libya, and the *Argilo grèseux of Hassi Touareg*, the Edjeleh Formation, Tahouite Formation, the combined Ouargla Sandstone-Azzel Shale, and the Oued Saret Formation in Eastern Algeria. The St-3 well on the Kasr Hadada Permit of Petroceltic, as well as the recently drilled (August 2010) St-4 well penetrated 364 m of the Bir Tartar Formation and encountered oil and gas shows in fractured reservoirs. The Oryx-1 well proved dry (Petroceltic News Release, 2010).

Recent palynostratigraphic studies of the Ordovician, glacial-related sediments in the northern Sahara Platform, such as the *Argiles Microconglomératiques* in Algeria, the Djeffara Formation in Tunisia and NW Libya, and the 2nd Bani Formation in Morocco confirm a Hirnantian (latest Ordovician) age (Paris et al., 1995, 2000, Bourahrouh et al., 2004, Ghienne et al., 2007, Vecoli, 2000, Vicoli et al., 2009).

The Djeffara (Jeffara, Jifarah) Formation was introduced by CFP (1964) for Upper Ordovician rocks in Tunisia. It is equivalent to the *Argiles Microconglomératiques* in Algeria. The unit consists of poorly stratified shales in the lower part and an upper unit composed of microconglomerates (diamictites) of glacio-marine origin. The upper and lower units are separated by a dolomitic limestone unit with mud-mounds, up to 100 m (328 ft) in thickness of early Ashgillian age (Buttler et al., 2007), especially in Libya (Fig. 12.3). Its thickness varies from 80 m (262 ft) to 150 m (492 ft). It conformably overlies the Bir Ben Tartar Formation, and is unconformably overlain by Silurian rocks of the Tannezouft Formation. Massa et al. (1977) reported the brachiopods *Orbiculothyris costellata, Onnizetina* sp., *Aegiromena* sp., *Torynelasma* sp., and *Lingulella* sp., and the ostracod species *Bollia henningsmoeni*. Jaeger et al. (1975) reported the acritarchs *Veryhachium irroratum, V. bromidense, V. trisulcum, V. setosapelliculum, Orthosphaeridium insculptum*, and *Stelliferidium* sp. Conodonts are abundant in this Formation (Bergström & Massa, 1991) and include *Istorinus erectus, Panderodus* sp., *Sagittodontina robusta, Scabbardella altipes, Dapsilodus mutatus*, and *Amorphognatus* sp. cf. *A. ordovicius*. The acritarchs *Veryhachium irroratum* and *V. Setosapelliculum* suggest a Caradocian-Ashgillian age (Molyneux & Paris, 1985). The Djeffara Formation was dated Ashgillian (with the absence of the Caradocian part) by Adloff et al. (1985, 1986) and correlated with the Memouniat Formation in Libya. However, in Libya, the Caradocian part is well developed (Massa, 1988, Molyneux & Paris, 1985, Ben Ferjani et al., 1990) (Fig. 12.3). Palynomorphs of the uppermost Djeffara Formation (Jaglin && Paris, 2002, Vecoli et al., 2009) are dominated by light-colored palynomorphs composed of chitinozoans, rare cryptospores, and badly preserved amorphous organic matter. The acritarchs include *Acanthodiacrodium crassus, Leprotolypa evexa, Neoveryhachium carminae, Ordovicidium* sp., *Orthosphaeridium insculptum, O. rectangulare*, and *Villosacapsula villosapellicula*, as well as various species of large *Baltisphaeridium*. In addition, the assemblage includes *Dactylofusa cucurbita, Deunffia* sp., *Hoegklintia* sp., *Saharidia munfarida which supports* a latest Ordovician (Hirnantian) age of these sediments (cf. Vecoli & Le Hérissé, 2004, Vecoli, 2008). Reworked acritarchs have been also observed and include Middle Ordovician species, such as *Frankea* and *Dicrodiacrodium*. Chitinozoans include *Angochitina* sp., *Belonechitina* sp., *Conochitina* sp., *Conochitina* cf. *elongata*, and *Lagenochitina* sp. Carbonates 10–100 m (33–328 ft) thick occur in the lower part of the Formation (Bergström & Massa, 1991) and consist of bryozoan biostromes with brachiopods, trilobites, echinoderm fragments, stromatoporoids, protospongia and ostracods. The development of these biostromes follows the east-west structural trend of the Tripolitania area. The Djeffara Formation was extended into the subsurface of the Ghadames Basin by Deunff & Massa (1975) and Jaeger et al. (1975). The Formation is equivalent to the Melez Chograne and Memouniat formations in West Libya, and the Tamadjert Formation, Gara Louki Formation, and the combined Rhazziane and Mkratta formations in Eastern Algeria. The Ashgillian mud-mounds are dominated by bryozoans, such as *Orbipora indenta*, *Dekayia* sp., *Hallopora* sp., *Monotrypa* sp., and *Jifarahpora libyensis* gen. et sp. Nov. (Buttler et al., 2007). The Djeffara Formation sediments are considered to be of a distal, peri-glacial origin (Jaeger et al., 1975, Massa, 1988, among others).

## 12.1.2 Cambrian and Ordovician of the Sirte Basin

Barr & Weegar (1972) extended the name Gargaf Group of Burollet (1960) to include undifferentiated Paleozoic rocks in the subsurface of the Sirte Basin and introduced two new correlative units of presumably Cambro-Ordovician age, the Hofra and Amal formations. The name Hofra Formation was named after the Hofra Field in the northwestern Sirte Basin. The type section is located in the A1-11 well (29°23′09″N, 17°59′06″E) at a drill depth of 4352–7977 ft. It consists of a 3625 ft (1105 m) sequence of relatively clean quartz sandstones with minor amounts of shales, siltstones, and conglomerates. The sandstones are white to light grey, occasionally red and hematitic, and firmly cemented with silica. Grey, brown-grey, and rare red micaceous shales and siltstones occur in thin beds. Olivine basalts and rhyolites occur as thin interbeds in the type well and some other wells in the area. In the type well, the Formation is unconformably overlain by the Satal Formation (Danian-Maastrichtian), and it unconformably overlies phyllites of uncertain age. In other areas, the Formation is overlain by Mesozoic Nubian sandstones, Bahi Formation, or younger formations. Differentiation between the different units in such cases is always problematic. In the D1-32 well, it is overlain by Silurian graptolitic shales. It was assigned a Cambro-Ordovician age based on its correlation with the Gargaf Group in West Libya. The Gargaf Group and Hofra Formation were assumed to be equivalent to the Amal Formation in the eastern Sirte Basin. The name Amal Formation (previously used informally by Roberts, 1970) was proposed by Barr & Weegar (1972) for a rock unit in the subsurface of the eastern Sirte Basin of an assumed Cambro-Ordovician age and composed predominantly of white to red, purple, tan and grey sandstones, very fine to cobbles in size, poorly sorted with frequent conglomerate beds. The sandstones are composed dominantly of quartz, cemented by clays, sericite and rarely dolomite. The sandstones are occasionally interbedded with grey silty clays, and grey, green and red micaceous shales. Volcanic rocks in the form of dikes, sills and flows are found, especially in the upper part of the Formation. Approximately 1461 ft (445 m) of the Formation have been penetrated in the type B1-12 well (29°26′38″N, 21°09′43″E) at a drill depth of 9829–11,290 ft, the base of the unit was not reached. In that well, the Formation is overlain by the Late Cretaceous Maragh Formation.

The Gargaf Group sandstones in the Sirte Basin are usually deeply silicified and intensely fractured (Hea, 1971). The fractures are healed by silica, protomylonite, marcasite, dolomite, kaolinite, and dead hydrocarbons. The matrix is mainly illite and chlorite.

Recent work suggests that the Gargaf Group, Amal Formation and Hofra Formation in the Sirte Basin include different ages such as Cambro-Ordovician, Silurian, Devonian, Carboniferous, Permian, Triassic, Jurassic, and Cretaceous (van Erve, 1993, Thusu & Mansouri, 1993, Hallett & El-Ghoul, 1996, Wennekers et al., 1996, Thusu, 1996, Sinha & Eland, 1996, Tawadros et al., 2001) (Fig. 12.4). The presence of sediments representing other ages within the Gargaf Group was first suggested by Hea (1971). Bonnefous (1972) pointed out, though based on erroneous data (most of the nannofossils and palynomorphs reported by him are probably contaminants), that the upper part of the quartzites in the D2-104A well in the Wadi Field which was previously thought to be Cambro-Ordovician in age, is Cretaceous in age. He dated the lower part Cambro-Ordovician based on the occurrence of acritarchs such as *Cymatiogalea cuvillieri*, *Pterospermopsis* sp. and *Attritasporites messaoudi*. In the meantime, Bonnefous (1972) reported the palynomorphs *Leiospheres*, *Veryhachium granulosum* and *Saharidia downiei* from the "Gargaf" succession in the V1-59 well and suggested a Cambro-Ordovician age. This part of the section was proven later to be of Early Carboniferous age based on palynological analysis (Wennekers et al., 1996). Tawadros et al. (2001) confirmed a Late Carboniferous age for the upper part of the section and determined the lower part to be of Devonian age. Recent palynological work on the type section of the Hofra Formation in the A1-11 well has yielded a rich Cenomanian dinocyst assemblage from the upper part of the Formation (Hallett & El-Ghoul, 1996). Mid-Triassic miospores and fish fossils were reported from interbedded lacustrine-lagoonal shales of the upper Amal Formation (Thusu & Mansouri, 1993, Thusu, 1996, Sinha & Eland, 1996) in the L4-51 and A1-96 wells in the eastern Sirte Basin, as well as from the Amal Type section in the B1-12 well. In the light of these arguments, it is apparent that the definition and stratigraphic relationships of the Gargaf Group and related formations in the Sirte Basin should be revised.

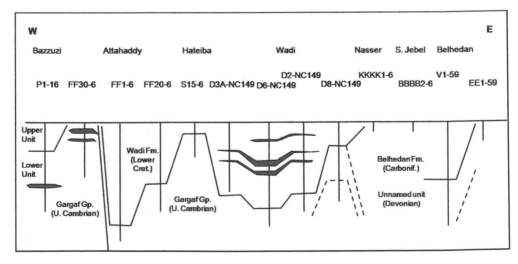

*Figure 12.4* Cross-section of pre- and syn-rift reservoirs (Cambrian-Lower Cretaceous) in the central part of the Sirte Basin (modified from Tawadros et al., 2001).

## 12.1.3    Cambrian and Ordovician of the Kufrah Basin

Cambro-Ordovician sediments of the Gargaf Group crop out in the Tibesti Mountains on the western side of the Kufrah Basin (De Lestang, 1965 & 1968, Klitzsch, 1966, Le Heron & Howard, 2010) which extends into the Chad Republic (Burollet, 1963b, Hesse et al., 1987) and at Jabal Azbah and Jabal Arknu in the northern part of the Basin (Bellini & Massa, 1980). However, in the latter two areas, Cambrian and Ordovician strata are undifferentiated (Le Heron & Howard, 2010). On the other hand, Lüning et al. (1999) and Seilacher et al. (2002) recognised ichnofauna of an Ordovician affinity in pre-glacial strata, to which Le Heron & Howard (2010) applied the name Haouaz Formation. Seilacher et al. (2002) identified the ichno-fauna *Skolithos*, *Cruziana goldfussi*, *C. furcifera*, and *Arthrophycus linearis* from the Haouaz Formation (Fig. 12.5). At Jabal Arknu, the Cambro-Ordovician succession is made up of fining-upward sequences (Turner, 1980) of medium to coarse-grained sandstones with large-scale planar cross-bedding, and fine to medium-grained with small-scale trough-cross-bedding, and abundant *Monocraterion* vertical burrows. The exposed thickness of the Cambro-Ordovician succession is up to 250 m (820 ft), the base is not exposed. On the northeastern side of the Jabal Arknu, the Hassaouna Formation is approximately 200 m (656 ft) in thickness and made up of quartz sand-stones with *Tigillites* and conglomerates at the top. It is separated from the overlying

*Figure 12.5 Arthrophycus*, Ordovician, Jebel Asba, Southeast Libya (courtesy of Sebastian Lüning).

Memouniat Formation by an erosional surface (Burollet, 1963). West of Jabal al Qardah, the Hassaouna Formation consists of thick-bedded sandstones and exhibits glacial morphology such as eroded "*moutonnée*" surfaces, and it is overlain by over 40 m (131 ft) of medium to coarse-grained, cross-bedded sandstones of the Memouniat Formation.

At Jabal Azbah, the Hassaouna Formation is made up of white, micaceous, very fine to medium-grained, quartz-rich sandstones forming fining-upward sequences 1–10 m (3–32 ft) in thickness (Turner, 1980, 1991). The overlying Memouniat Formation is made up of medium to coarse-grained, cross-bedded sandstones with horizontal branching burrows and *Planolites* and pebble conglomerates at the base. Igneous ring complexes ranging in composition from granitic to syenitic, as well as trachytic ring dykes of Upper Proterozoic to Cambrian and Ordovician age intrude the Formation.

Cambro-Ordovician sediments also form the major part of the eastern Tibesti mountain ranges. The thickness of the sequence is 550–600 m (1641–1969 ft) in the area of Jabal Nuqay and up to 1300 m (4265 ft) near the Libyan-Chad border (Bellini & Massa, 1980). On the southwestern side of the Kufrah Basin, Cambro-Ordovician sediments are exposed along a major fault plane trending northwest-southeast. The Hassaouna Formation is 1150 m (3773 ft) thick and the Memouniat is 150 m (492 ft). The two formations in that area are separated by a thin *Harlania* (*Arthrophycus*) zone (Grignani et al., 1991) (Fig. 12.5).

Dalloni (1934) introduced the name Eghei Sandstone[4] after Jabal Eghei (now Jabal Nuqay) in the Tibesti area for a unit of massive, cross-bedded pebbly sandstones with local conglomeratic sandstones and ferruginous beds at the top, and the ichnofossils *Tigillites*, *Harlania*, and *Cruziana*. This Formation was included in the Cambro-Ordovician rocks on the Geological Map of Libya (1977) and called Nuqay Sandstone, but it was not mentioned on the 1985 map.

The Eghei Formation unconformably overlies Precambrian basement rocks and it is unconformably overlain by Lower Paleozoic cross-bedded sandstones of the Lori Sandstone of uncertain age. The latter is considered by Burollet (1960) to be Cambrian to Devonian in age (if the Eghei Sandstone is Upper Ordovician, then the Lori Sandstone is probably Silurian to Devonian and most probably Devonian in age). Burollet (1960) considered the Eghei Sandstone to be equivalent to his Gargaf Group and of Cambro-Ordovician age, but since it overlies Lower Ordovician rhyolites and granites dated at 497–454 Ma by the K-Ar method (Hunting Geology & Geophysics, 1974, cited by Banerjee, 1980, p.108), the Formation is possibly Upper Ordovician in age and equivalent to the Melez Chograne and Memouniat formations.

De Lestang (1965, 1968) and Klitzsch (1966) used the name Largeau Formation for a Cambro-Ordovician unit in the southwestern Kufrah Basin near the Libya-Chad border, overlain by the Silurian Bedo Formation.

On the eastern side of the Kufrah Basin at Jabal Awaynat (Gebel Oweinat) in the Karkur Talh area near the Egyptian-Libyan border, Klitzsch (1990a) applied the name Karkur Talh Formation to a unit several tens of meters in thickness and composed of well-bedded shallow-marine sandstones. The sandstones contain *Cruziana rouaulti*,

---

4 The name Eghei Sandstone was also used for a Cretaceous unit in the subsurface of the Western Desert of Egypt (now obsolete).

believed to be of Ordovician age, and they are intercalated with coarse-grained con-glomeratic fluviatile sandstones. This Formation is probably equivalent to the Eghei Formation in the Jabal Arknu area and the Memouniat Formation.

The Cambro-Ordovician sequence in the subsurface of the Kufrah Basin is 1226 m (4023 ft) in the A1-NC43 well and 1430 m (4692 ft) in the B1-NC43 well. The sequence is made up quartz sandstones, fine to very coarse-grained, poorly cemented by silica, kaolinite, and argillaceous matrix with levels of gravel. The stratigraphy of the section has not yet been resolved, but the upper part of the Mem-ouniat Formation has been recognized based on acritarchs and chitinozoans of Late Ordovician (Caradocian-Ashgillian) age by Grignani et al. (1991). They reported Upper Ordovician acritarchs and chitinozoans from the A1- and B1-NC43 wells, including *Baltisphaeridium longispinosum delicatum, B. perclarum, Veryhachium subglobosum. V. lairdi, V. oklahomense, V. trispinosum, Navifusa* sp., *Villosacap-sula setosapellicula, Eupoikilofusa striatifera, E. parvuligranosa, Dactylofusa cucurbitaa, D. spinata, Leiofusa estrecha, Ancyrochitina merga, Armoricochitina nigerica, Cyathochitina* cf. *campanulaeformis, Desmochitina minor, Sphaerochi-tina lepta,* and *Conochitina baculata. Villosacapsula setosapellicula, Ancyrochitina merga,* and *Armoricochitina nigerica* were reported by Molyneux and Paris (1985) from Northeast Libya Libya.

Vittimberga & Cardello (1963) described the Hauaisc Formation in the Jabal Hawayish area in the northern Kufrah Basin. The Formation is composed of medi-um-grained sandstones, occasionally coarse-grained and quartzitic. It is unconform-ably overlain by the Tannezouft Formation, but its base is not exposed. It correlates with the Memouniat Formation. Banerjee (1980) recommended the name for local use only.

In Jabal Dohone in the Kufrah Basin, Vittimberga & Cardello (1963) subdivided the Cambro-Ordovician section, in an ascending order into the Zouma, Teda, and Munchar formations. The Cambrian Zouma Formation is more than 100 m (328 ft) thick and it consists of medium-grained, illitic sandstones. The Teda Formation is of probable Cambrian age, and it is composed of medium to coarse-grained, quartz-itic, and illitic sandstones. The Munchar Formation is approximately 100 m (328 ft) thick and composed of ferruginous microconglomerates at the base followed upward by fine-grained, feldspathic, and kaolinitic sandstones interbedded with some shales. The Formation contains *Cruziana, Harlania,* and *Tigillites* (Klitzsch, 1966) and it is assumed Ordovician in age. The Cambro-Ordovician section is overlain by the Tannezouft Formation. No type sections were designated for any of these forma-tions. Klitzsch (1966) correlated the Munchar Formation with the Haouaz, Melez Chograne, and Memouniat formations, and the Zouma and Teda formations with the Hassaouna Formation, and considered all three formations to be equivalent to the Gargaf Group.

## 12.1.4  Cambrian and Ordovician of Northeast Libya

Cambro-Ordovician rocks of the Gargaf Group are distributed extensively in the subsurface of Cyrenaica. El-Arnauti & Shelmani (1985) estimated thickness between 200 m (656 ft) to 2000 m (6562 ft) for the Paleozoic section and attributed these vari-ations to post-Cambrian erosion and tectonics. Upper Ordovician rocks have thicknesses of between 400 m (1312 ft) and 2000 m (6562 ft), the presence of Cambrian and Early Ordovician is still unconfirmed.

Two Late Ordovician units have been described by El-Arnauti & Shelmani (1988), the lower unit is 250–680 ft (76–207 m) thick, and consists of interbedded fine to medium-grained sandstones and shales with minor siltstones, overlying directly Proterozoic sediments. Three acritarch zones have been recognized within this interval (Hill & Molyeux, 1988) and three chitinozoan zones (Paris, 1988). The acritarchs and chitinozoans suggest a Caradocian-Ashgillian age and deposition in a shallow-marine, nearshore environment. Molyneux & Paris (1985) reported the acritarchs *Veryhachium setosapelliculum* and *V. irroratum*, and the Ashgillian chitinozoans assemblage composed of *Calpichitina lenticularis, Armorichitina nigerica, Plectochitina sylvanica, Tanuchitina bergstromi*, and *Ancyrochitina merga*, as well as *Belonechitina micracantha typica, B. capitata, Cyathochitina latipatagium, Desmochitina (pseudodesmochitina) minor* cf. *typica, Spinachitina bullmani, S. sp.*, and *Conochitina elegans*. Similar assemblages were reported from the Djeffara Formation by Jaegar et al. (1975) and from the subsurface of the Kufrah Basin by Grignani et al. (1991). The lower unit is probably equivalent to the Melez Chograne and Memouniat formations in the Kufrah Basin and West Libya. Paris (1988) suggested similarities to the Hirnantian (latest Ashgillian) glacio-marine environment reported from Morocco. The upper unit is made up of unfossiliferous, fine-grained sandstones with quartz pebbles in some wells. It varies in thickness between 70 ft (21 m) and 696 ft (212 m). El-Arnauti & Shelmani (1988) suggested a possible Ashgillian age and a peri-glacial origin for this unit. It is correlatable with the Memouniat Formation and the "Djeffara" Formation in Northwest Libya.

## 12.2 SILURIAN

Silurian sediments in Libya are exposed in the Tadrart ranges and Wadi Tanazuft on the southwestern side of the Murzuq Basin, on the western side of the Jabal Bin Ghanimah, on the eastern slopes of the Jabal Nuqay in the Tibesti area, and in the Jabal az Zalmah north of the Kufrah Basin. Silurian sediments represent the first major marine transgression after the Upper Ordovician period of glaciation and erosion. They comprise two units, the Tannezouft Formation composed of graptolitic shales and the Acacus Formation composed predominantly of sandstones. Locally, the basal Tannezouft Formation contains a silty shale unit known as the Iyadar Formation. In Southeast Libya and Northern Chad, Silurian sediments are undifferentiated and belong to the Bedo Formation. Source rock potential of the Silurian shales have been examined by Lüning et al. (2003c), Sikander et al. (2003), Fello et al. (2006), Belaid et al. (2009), and Vecoli et al. (2009).

### 12.2.1 Silurian of West Libya

Silurian rocks in West Libya are represented by the Tannezouft and Acacus formations. The two formations crop out along the eastern and western sides of the Murzuq Basin. The Tannezouft and Acacus formations are probably one continuous transgressive-regressive cycle (Burollet & Manderscheid, 1967, Elfigih, 1991, 2000, Le Heirisse, 2002), progradational northward and westward (Bonnefous, 1963, Ben Ferjani et al., 1990). The Tannezouft (Llandovery) thins northwards, while the Acacus (Wenlockian-Ludlovian) thins northward (Bonnefous, 1963). The Silurian progressively disappears from Tunisia in the west to the Gargaf Arch in the east, by either erosion or non-deposition. In the Awaynat section, the Lower Llandovery is absent (Bellini & Massa, 1980) and the Middle Silurian rests unconformably on the Memouniat Formation (Ashgillian).

Desio (1936) introduced the Tannezouft Formation (spelled Tanezzuft on the Geological Maps of Libya, 1977 & 1985) named after Wadi Tanazuft between Ghat and Al Awaynat for a series of occasionally sandy, dark grey and green with spots of red, thinly bedded shales with sandstone stringers. The shales contain graptolites, such *Monograptus nilssoni, M. dubius, Climacograptus libycus, C. scalaris scalaris*, and *C. medius*. The type section is at Wadi Tanazuft about 65 km north of Ghat (25°0'N, 10°10'E). Klitzsch (1965, 1969) considered the section described by Desio (1936) to be unsuitable because of poorly defined contacts and described a complete section of the Tannezouft shales, approximately 475 m (1558 ft) in thickness at Takarkhouri (24°8'N, 10°3'E) in Wadi Tanazuft about 35 km south of Ghat. Although this section was designated as a type section by Klitzsch (1965, 1969) it is in reality a reference section.

In the Ghat area, the Tannezouft Formation is 480 m (1575 ft) thick. The lower part of the Formation is composed of greenish-grey and reddish kaolinitic shales and contains intercalations of grey, pink and reddish brown fine-grained sandstones with ripples, sole marks, and trace fossils (Grubic et al., 1991). The amount of sandstones decreases in the upper part. It was dated Llandoverian in that area by Protic (1984), based on graptolites.

On the eastern side of the Murzuq Basin, in the Jabal Al Qarqaf area, the base of the unit is made up of graptolitic shales with siltstone intercalations. The siltstone content increases upwards. The basal shales yielded Llandovery graptolites (Jacqué, 1963). Klitzsch (1963, 1969) examined the Silurian section exposed in the Dur al Qussah area and reported a graptolites assemblage from the Tannezouft Formation composed of *Climacograptus medius, C. rectangularis, C. scalaris scalaris, C. scalaris normalis, Cryptograptus* sp. cf. *tamriscus*, and *Clyptograptus sinuatus* and assigned it Llandovery age. Thin coral reefs were reported from the Tannezouft Formation between Mourizidie and Jabal Oueda by Fürst & Klitzsch (1963).

Krumbeck (1906) called the Tannezouft Formation the "Taita Slate" after Wadi Taita east of the Akakus-Tadrart ranges. Ratschiller (1967) named it the "Idinine Formation" after Jabal Kaf Idinine (presently Kaf Janu) north of Ghat.

The Silurian sequence in the Ghadames Basin was divided by Burollet & Manderscheid (1967), using the Algerian nomenclature, into a lower unit (the Principal Shale Formation) and an upper unit (the Sandy/Shaly Formation). These two units are equivalent to the Tannezouft and Acacus formations, respectively. Bellini & Massa (1980) recommended the use of the latter names for both Murzuq and Ghadames basins. The Tannezouft in the Ghadames Basin was deposited over a regional unconformity (Bracaccia et al., 1991). The Silurian is truncated by an Early to Middle Devonian unconformity in the Dur al Qussah area and west of the Jabal Al Qarqaf area. This unconformity is basin-wide (Pierobon, 1991).

The Tannezouft Formation ranges from 45 m-320 m (148–1050 ft) in thickness in the Murzuq Basin and pinches out due to erosion, on the Brak-Bin Ghanimah paleohigh (Fürst & Klitzsch, 1963, Pierobon, 1991), and increases in thickness in the Dur al Qussah Basin where it is represented mainly by clastic facies (Fürst & Klitzsch, 1963). It conformably overlies the Memouniat Formation or unconformably older formations in the Dur al Qussah area, on the southeastern flank of the Murzuq Basin (Fürst & Klitzsch, 1963).

In the extreme Southwest Libya (Grubic, 1984, Galecic, 1984) the Tannezouft Formation overlies the Cambrian Hassaouna Formation. Tempestite deposits are common in the Silurian Tannezouft and Acacus formations (De Castro et al., 1991). Tempestites occur in coarsening and thickening-upward sequences up to 50 m (164 ft) in thickness with hummocky cross-stratification and they are associated with offshore bars. Al-Ameri (1983a & b) carried out a detailed palynofacies analysis of the Silurian sediments in the northern Ghadames Basin.

The Iyadhar Formation (spelled Ayadhar on the Geological map of Libya, 1985) was introduced by Massa & Jaeger (1971) for a basal unit of the Tannezouft Shale in the western margin of the Murzuq Basin and named after the type area in Wadi Iyadar where the Formation is 200 m (656 ft). It was also reported from the subsurface of the Ghadames Basin by Deunff & Massa (1975) and Jaeger et al. (1975). It consists of silty and hematitic shales rich in feldspars and rare intercalations of calcareous sandstone. It unconformably overlies the Memouniat Formation in the south (Massa & Jaeger, 1971) and the "Djeffara" Formation in the subsurface of the Ghadames Basin (Deunff & Massa, 1975), and passes gradually upward into the Tannezouft shales. It contains a rich graptolite fauna and a rich microfloral assemblage of acritarchs, spores and chitinozoans which indicate a Lower Silurian (Lower Llandovery) age. Radulovic (1984) rejected the Iyadar Formation because it is a chronostratigraphic unit defined by the graptolites and not a lithostratigraphic unit, although he recognized two parts in the Tannezouft Formation. In the northern Illizi Basin in Tunisia, the base of the Silurian is marked by a fine grained quartzitic sandstone horizon with pyrite known as "dalle de M'kratta" (Jaeger et al., 1975).

The name Acacus Formation was introduced by Desio (1936) as the "Acacus Sandstone" for a thick unit of quartzitic sandstones which forms the high parts of the topography of the Acacus and Tadrart ranges. This sandstone unit is a stratigraphic equivalent of the lower part of Kilian's (1928) "Grès supérieurs du Tassili" and to the term "Zone de passage" of the Algerian Sahara (Burollet, 1956), or the "Gothlandien argilo-gréseux" (Bonnefous, 1963). The type section in the Takarkhuri Pass was described by Klitzsch (1965) where the Formation is made up of brown, fine-grained sandstones, coarse-grained to conglomeratic in the upper part and quartzitic at the top with shallow marine trace fossils such as Harlania, Cruziana, and Tigillites. The Acacus Formation is the oldest unit exposed on the western rim of the Murzuq Basin (Jakovljevic, 1984). In the Ghat area, the Acacus Formation is about 220 m (722 ft) thick. The Formation is unconformably overlain by the Devonian Tadrart Formation. Bonnefous (1963) reported the graptolites Monograptus fritschi from the D1-23 well and M. dubius from the B1-61 well (both wells are in the northern Ghadames Basin, near the Tunisian border), and assigned the Acacus Formation a Lower Ludlow age.

The Acacus Formation is present in the northern part of the Dur al Qussah area where it has a thickness up to 465 m (1526 ft) (Klitzsch, 1963, Fürst & Klitzsch, 1963), and it is eroded to the south. It overlies the Tannezouft Formation, and is unconformably overlain by the Tadrart Formation. It was assigned an Upper Llandovery-Ludlow age in the Dur al Qussah area by Klitzsch (1963) based on its stratigraphic position and a Wenlochian-Ludlovian age by Jakovljevic (1984). It is composed of greyish yellow, pinkish grey, brownish red, very fine to medium grained, feldspathic sandstones with tabular cross-bedding, and the trace fossils *Tigillites* and *Harlania*. The upper part of the Formation is made up of alternating micaceous shales and fine sandstones with convolute bedding, often rich in *Harlania*. Grubic et al. (1991) identified the trace fossils *Tigillites dufrenoyi, Arthrophycus alleghaniensis, Corophioides* sp., and *Cruziana* sp. The Acacus Formation is absent in the Jabal Al Qarqaf area and the eastern side of the Murzuq Basin.

The boundary between the Silurian and Devonian is determined only based on sedimentological criteria. It is placed between interbedded shales and sandstones of the Acacus Formation and the coarse sandstones of the Tadrart Formation (Burollet & Manderscheid, 1967). In places, this boundary is marked by a ferruginous stromatolitic sandstone bed (*l'horizon d'Ikniouen* of Freulon, 1964) (Jakovljevic, 1984). In the Ghadames Basin, the Acacus Formation is separated from the overlying Tadrart Formation by iron oxide-rich sandstone layers (Bracaccia et al., 1991). In the absence of these layers, separation between the two formations is not possible. The boundary between the Tannezouft Formation and Acacus Formation is gradational and diachronous. Grubic et al. (1991) selected a marker bed of hard, fine-grained sandstone, 2–3 m thick, within this zone to mark the boundary between the two formations.

Silurian rocks are not exposed in the Sirte Basin and they have been reported only in a few drill holes such as the T-2D/35 water well at Al Fuqaha (Mamgain, 1980), in the exploratory D1-32 well (Barr & Weegar, 1972), and in the C1-44 in the Eastern Sirte Basin where Hill et al. (1985) reported the Silurian acritarchs *Veryhachium carminae*. Hill et al. (1985) reported the following acritarchs assemblage from the subsurface of Northeast Libya: *Veryhachium trispinosum, V. valiente, Diexallophasis caperoradiola, Oppilatala eoplanktonica, ?Dateriocradus monterrosae, Multiplicisphaeridium arbusculiferum,* and *Visbysphaera gotlandicum, Neoveryhachium carminae* is present sporadically. They also reported the chitinozoans *Conochitina edjelensis elongata, C. armillata, Plectochitina pseudoagglutinans, Cyanochitina* sp., and "*Sphaerochitina*" *vitrea*, and the miospores *Dyadospora murusdensa, D. murusattenuata, Nodosphaera* sp., *Rugosphaera* sp., and *Ambitisporites dilutus*. The differences between these assemblages and those reported from the Kufrah Basin have been attributed to differences in sedimentary facies.

## 12.2.2  Silurian of the Kufrah Basin

Silurian sediments crop out at Jabal Nuqay on the western side of the Kufrah Basin, at Jabal az Zalmah on the northern side, and in Chad. They also extend into the subsurface of the Basin.

In the Kufrah Basin, the Tannezouft Formation is composed of greenish grey, red, and black shales with coatings of manganese and interbeds of micaceous siltstones. The

Formation is up to 82 m (269 ft) thick. Graptolites reported from this Formation such as *Climacograptus medius*, and the chitinozoans assemblage of *Ancyrochitina ancyrea, A. fragilis brevis, A. saharica, A. tumida, A. aculeata, A.* cf. *corniculans, Lagenochitina* sp., *Angochitina* sp., and *Sphaerochitina* sp. indicate a Lower Llandovery age (Bellini & Massa, 1980). The Tannezouft Formation encountered in the A1- and B1-NC43 wells contains acritarchs and chitinozoans assemblages similar to those reported by Bellini & Massa (1980). In addition, Grignani et al. (1991) reported a chitinozoan assemblage composed of *Sphaerochitina lepta, Ancyrochitina ancyrea, A. laevaensis, A. tumida, Angochitina* sp., and *Spinachitina* sp. from the Tannezouft Formation, and assigned it an Early Silurian (Early Llandovery) age. Seilacher et al. (2002) identified the ichnospecies *Cruziana kufraensis* in transgressive sandstones at the base of the Tannezouft Formation.

The Acacus Formation consists of dark grey and greyish green siltstones with thin interbeds of fine-grained sandstones and micaceous shales. To the southeast of the Kufrah Basin, the Early Devonian Tadrart Formation is eroded and the Acacus is unconformably overlain by the Upper Devonian Binem Formation. The thickness of the Formation varies between 2 m (7 ft) and 25 m (82 ft). Ripple marks, flute casts, and the trace fossils *Cruziana furcifera* and *Arthrophycus* sp. are common near the top of the Formation near Jabal Azbah (Bellini & Massa, 1980). The Acacus Formation encountered in the subsurface of the Kufrah Basin in the A1- and B1-NC43 wells is composed of barren sandstones (Grignani et al., 1991).

Klitzsch (1966) and De Lestang (1965, 1968) described the Silurian Bedo Formation in the southwestern Kufrah Basin near the Libya-Chad border which unconformably overlies the Cambro-Ordovician Largeau Formation, and it is unconformably overlain by the Devonian Gouring Formation or *Spirophyton* Group. The Formation is composed mainly of greenish grey to purple red sandstones and siltstones. This Formation is equivalent to the Tannezouft and Acacus formations. The Silurian succession along the eastern flank of the Tibesti massif and the western side of the Kufrah Basin in northern Chad is represented by the Dohone Trail Formation (*Formation de la Piste du Dohone*). De Lestang (1965, 1968) subdivided the succession in ascending order into three members: a carbonate member, a shale member, and a silty-sandstone member. The first two members are equivalent to the Tannezouft Formation and the third member is equivalent to the Acacus Formation, or the lower part of the Gouring Formation.

## 12.3 DEVONIAN

The Devonian in Libya crops out on the western and southern edges of the Jabal Al Qarqaf, along the western and eastern flanks of the Murzuq Basin, in the Ghadames Basin, eastern Tripolitania, south-West Libya, the Kufrah Basin, Jabal Dohone (Eghi), Jabal Oweinat, and Jabal Hawayish. The Devonian is also known from wells in the Murzuq and Ghadames basins, Southern Tunisia and Eastern Algeria. In Cyrenaica, Early to Late Devonian sediments were reported by El-Arnauti & Shelmani (1985, 1988), Paris et al. (1985), and Paris (1988). Tawadros et al. (2001) described Devonian sediments in the Central Sirte Basin. Reworked Devonian palynomorphs have been reported in many parts of the subsurface of the Sirte Basin (Thusu, 1996, Wennekers et al., 1996) which suggests that these sediments are common in the area. Devonian rocks are generally poor in fossils. Only the Aouinet Ouenine Group which crops out on the western side of the Gargaf Arch contains abundant brachiopods. However, extensive palynological studies have contributed to our understanding of the Devonian stratigraphy in Libya. The boundary between the Silurian and Devonian is usually arbitrary (Burollet & Manderscheid, 1967). The basal Lower Devonian is missing in Libya (Bellini & Massa, 1980, Grignani et al., 1991, Massa & Moreau-Benoit, 1976) and in the Illizi Basin in Eastern Algeria (Jardiné & Yapaudjian, 1967, Jardiné et al, 1974).

## 12.3.1   Devonian of West Libya

In West Libya, the Devonian crops out on the western and southern edges of the Jabal Al Qarqaf, and along the western, eastern, and northern flanks of the Murzuq Basin. The Devonian is also encountered in the subsurface of the Ghadames and Murzuq basins. Wells show 350 m (1148 ft) of fine to medium-grained sandstones with silty shale interbeds (Burollet & Manderscheid, 1967). On the western side of the Murzuq Basin, Devonian sediments overlying Silurian shales crop out in the Akakus-Tadrart ranges over a distance of 300 km. Devonian sediments also crop out in the Tihemboka Arch area. The basal Lower Devonian is absent and the Middle Devonian is present only on the north and south borders of the Arch. The Upper Devonian is transgressive over the Arch and rests unconformably on the Silurian shales (Burollet & Manderscheid, 1967). Devonian and Carboniferous brachiopods and bivalves in the Djado and Murzuq Basins were reviewed by Mergl et al. (2001).

The Devonian in West Libya includes six major units: the Lower Devonian Tadrart, Ouan Kasa, and Emgayet formations, the Middle Devonian Aouinet-Ouenine Group, the Upper Devonian Ash Shati Group, and the Strunian Tahara Formation.

The Tadrart Formation was introduced by Burollet (1960) for a unit forming massive scarps in the Tadrart area and composed of massive cross-bedded sandstones with plant fragments near the top, and many ferruginous beds. The Formation overlies the Acacus Formation and is overlain by Emsian shales and sandstones of the Ouan Kasa Formation. Klitzsch (1965) selected the type section of the Tadrart Formation in the Takarkhouri Pass, approximately 35 km south of Ghat (the reference section for the Tannezouft Formation). In that section the Formation unconformably overlies the Acacus Formation and it is 317 m (1040 ft) thick and composed of brown to red brown, fine to coarse-grained, massive and cross-bedded with rare plant remains and abundant trace fossils.

Klitzsch (1963) and Fürst & Klitzsch (1963) designated it a Sieginian-Emsian age based on its stratigraphic position. It was assigned a Siegenian (Pragian) age by Burollet (1960) and Burollet & Manderscheid (1967) based on correlations with similar units in Algeria and Tunisia, an upper Siegenian age by Bellini & Massa (1980), a Sieginian-lower Emsian age by Jacovljevic (1984) and Grubic (1984), and a Gedinnian-Siegenian age by Galecic (1984) and Grubic et al. (1991), also based on its stratigraphic position. Massa & Moreau-Benoit (1976) reported the spores *Archaeozonotriletes divellomedium* and *A. chulus*, and the chitinozoans *Desmochitina elegans* (an index fossil), *Ancyrochitina ancyrochitina*, *A. fragilis* var. *brevis* and *Linochitina cingulata* from the Ghadames Basin and Jifarah Plains and assigned it a Pragian (Siegenian) age.

The Tadrart Formation varies in thickness from 40 m (131 ft) north of the Gargaf Arch, to 130 m (427 ft) in wells in Northwest Libya and 200 m (656 ft) in southern Tunisia. In Southwest Libya the unit varies between 0–415 m (0–1362 ft) in thickness. It has a thickness between 300 m (984 ft) and 350 m (1148 ft) in the Jabal Tadrart and between 375 m and 415 m (1230 ft and 1362 ft) in the Dur al Qussah area (Klitzsch, 1963, Fürst & Klitzsch, 1963). It is eroded on the southern flank of the Gargaf Arch.

The Tadrart Formation represents a transgressive phase which overlaps local unconformities near paleohighs (Burollet & Manderscheid, 1967). The Formation rests conformably on the Silurian Acacus Formation only in the Ghadames Basin. Everywhere else the lower contact is unconformable. On the northern and western sides of the Gargaf Arch, the Tadrart rests unconformably on Ordovician sediments, the Silurian being progressively eroded. On the eastern side of the Murzuq Basin, and the northwest side of the Tibesti high, the Formation rests unconformably on Cambro-Ordovician sediments, the Silurian is absent probably by non-deposition. The upper boundary of the Tadrart Formation with the overlying Ouan Kasa Formation is often conformable and gradational. The top of the Formation is placed between the last cross-bedded sandstones and the first quartzitic sandstones (Jacqué, 1963).

In Southwest Libya, the Tadrart Formation crops out along the Akakus-Tadrart range where it has a maximum thickness of 380 m (1247 ft) (Jacovljevic, 1984). The Formation is composed mainly of medium to coarse-grained, conglomeratic, feldspathic sandstones with ferruginous sandstone interbeds. The sandstones are plane-laminated, ripple cross-laminated in the lower part with *Tigillites, Cruziana*, and brachiopod fragments. The sandstones become finer grained and parallel-bedded with trace fossils in the upper part (Jacovljevic, 1984). The source area of the Tadrart sands was probably located on the south and southwestern of Libya with the Sidi Toui Arch in Tunisia and the Gargaf Arch in Libya contributing minor amounts to the basin (Burollet & Manderscheid, 1967). In the subsurface of the Murzuq Basin, the Tadrart Formation deposits start with festoon cross-bedded sandstones, and contain herringbone cross-beds at the top (Pierobon, 1991). In southern Tunisia, the upper part of the Formation grades into carbonates with bryozoans, echinoderms, and brachiopods. It is equivalent to the Oued Mehaiguene Formation in Eastern Algeria.

The name Ouan Kasa Formation (spelled Wan Kaza on the Geological Map of Libya, 1977 and Ouan Kasa in the 1985 edition) was introduced by Borghi & Chiesa (1940). The type locality of the Ouan Kasa Formation is at Gour Iduka in Wadi Wan Kaza, southeast of Ghat where it is made of 43 m (141 ft) of green and grey shales with siltstone intercalations. Klitzsch (1965) described the Ouan Kasa Formation in the Takarkhouri Pass in the reference section for the Tannezouft and Acacus formations where the Formation is 43 m (141 ft) thick and composed of grey silty shales with argillaceous siltstones and fine-grained sandstones, rich in brachiopods, tentaculites, cephalopods, and trilobites. He dated it Couvinian based on *Acrospirifer arduennensis* var. *mosellanus* and *Spirifer andaculus medialis*. In addition, Fürst & Klitzsch (1963) reported *Tropidoleptus carinatus* and *Chonetes orientalis* var. *admixta* and abundant *Spirophyton*. The Ouan Kasa Formation was assigned an Emsian age by Collomb (1962) in the Jabal Al Qarqaf area. Fossils recovered from the middle of the Ouan Kasa Formation such as *Acrospirifer mosellanus* and *Mediospirifer audaculus* indicate a basal Middle Devonian Eifelian (Couvinian) age, however, the unfossiliferous lower part may be Emsian in age (Burollet & Manderscheid, 1967). Massa & Moreau-Benoit (1976) assigned the Ouan Kasa Formation in the Ghadames Basin and Jiffarah Plains an Upper Emsian age. Fossils reported by Jakovljevic (1984) and Galecic (1984) from the Formation on the southwestern side of the Murzuq Basin are *Chonetes minuta, C. hardensis, Spirifer dombrowiensis, S. carinatus, S. mucronatus, Camarotoechia neapolitana, C. prolifica, Rhipidomella vanuxemi, C. prolifica,* and *Terebollites fezzanensis.* It was dated Couvinian by Grubic (1984) and Grubic et al. (1991) and Emsian-Lower Couvinian by Galecic (1984).

The distribution of the Tadrart and Ouan Kasa formations is controlled by the structural configuration of the basins formed during the Caledonian period. The separation of the Ouan Kasa from the underlying Tadrart Formation in the subsurface is not always evident and regional correlations are often difficult. Up to 427 m (1401 ft) of combined thickness of the two formations was penetrated in the deep parts of the Illizi Basin along the Tunisian-Libyan borders (Poyntz, 1995) where the two formations represent a complete cycle starting with conglomerates or coarse-grained sandstones and conglomerates, changing gradually into fine-grained sandstones, then into siltstones and shales of the Ouan Kasa Formation. On the other hand, the transition from the Tadrart to the Ouan Kasa is marked by a change from a continental palynological association with abundant spores into a marine association with abundant chitinozoans (Massa & Moreau-Benoit, 1976). In the Awaynat Wanin section, the Tadrart Formation is cross-bedded and the Ouan Kasa is finely laminated with abundant bioturbation. Basinward in the Ghadames Basin, the Ouan Kasa consists of three units (Massa & Moreau-Benoit, 1976), the lower member is composed of green or grey shales with dolomitic limestone beds and sandy bioclastic dolomite rich in molluscs and bryozoans, the upper two members are each made up predominantly of dark grey to brownish red shales and beds of fine to very fine grained sandstones with ferruginous oolites and beds of limestone. In Jabal Al Qarqaf area, the Ouan Kasa Formation is approximately 20 m (66 ft) thick and composed of fine-grained, nonfossiliferous, argillaceous sandstones. It conformably overlies the Tadrart Formation, and is unconformably overlain by the Aouinat Ouenine Formation. On the western side of the Murzuq Basin, the Ouan Kasa Formation crops out in a narrow belt, trending north-northwest along the eastern slopes of Jabal Tadrart (Jacovljevic, 1984). It conformably overlies the Tadrart Formation and it is unconformably overlain by the Carboniferous Mrar Formation (the Aouinat Ouenine Group is absent in that area). The Formation is composed mainly of light grey, grey, and violet, siltstones with minor kaolinitic shales, and feldspathic sandstones, oolitic, ferruginous, and gypsiferous in places with tabular cross-lamination The formation also contains limonitic intervals with ferruginous oolites (Adamson et al., 2000). Adamson et al. (2000) reported the trace fossils *Planolites, Tigillites, and Chondrites.*

Jacqué (1963) proposed the name Emi-Magri Formation for a succession exposed on the eastern side of the Murzuq Basin between the top of the Tadrart Formation and the pre-Upper Devonian unconformity. However, the name Emi-Magri Formation was discarded by subsequent authors in favour of the Ouan Kasa Formation. The Ouan Kasa and Emi-Magri formations on both sides of the Murzuq Basin have the same lithology and are equivalent (Jacovljevic, 1984), although the sand content increases near the Gargaf Arch. Both units conformably overlie the Tadrart Formation and are unconformably overlain by the Aouinet Ouenine Group. Burollet & Manderscheid (1967) restricted the name Ouan Kasa to the Devonian sediments exposed on the western side of the Murzuq Basin, and the Emi-Magri to sediments exposed on the eastern flank of the Basin. In the meantime, they proposed the name Emgayet Formation for Emsian sediments made up of silty shales with fine sandstone intercalations in the Ghadames Basin between the Tadrart Formation and the Aouinat Ouenine Group. They suggested that the Emgayet Formation is Emsian in age based on the presence of *Rhipidomella hamoni* and *Spinocytria subcuspidata*, while the Ouan Kasa Formation extends to the lower Middle Devonian. They mentioned that the type section of the Emgayet Formation is in the subsurface in a well which they did not identify in the Emgayet region in the south of the Ghadames Basin. Some authors restrict the Ouan Kasa Formation to the sandstones and the Emgayet Formation to the shales. The thickness of the unit varies from 15 m (49 ft) at the edge of the Ghadames Basin to 100 m (328 ft) in the centre of the Basin.

The Middle and Upper Devonian sediments in West Libya are included in the Aouinat Ouenine Group and the Ash Shati Group. The Strunian stage, represented by the Tahara Formation, is regarded by some workers to be basal Carboniferous, or a transitional zone between the Devonian and Carboniferous, and by others workers to be Devonian. In this present work, it is considered to be Devonian. On the Algerian side, two main units have been recognized, the Alrar Formation (Middle Devonian) and the Gazelle Formation (Upper Devonian).

The name Aouinat Ouenine Formation (spelled Awaynat Wanin on the Geological Map of Libya, 1977 and 1985) was introduced by Lelubre (1946a) in the Fazzan area. A few attempts have been made by different authors to correlate and subdivide this unit. Collomb (1962) restricted the name Aouinat Ouenine to the lower part of the unit and introduced the name Chati Formation for the upper part; however, this name was not accepted by later workers. Collomb (1962) was the first to realize the cyclic nature of the Devonian sequence and divided the Aouinat Ouenine and the Chati formations each into four cycles. In the Chatti Formation, each cycle starts with shales at the base followed by siltstones and ferruginous oolites at the top. Burollet & Manderscheid (1967) applied the name Aouinat Ouenine Formation to the Middle and Upper Devonian succession. They also noticed the cyclic nature of the Devonian sequence in northWest Libya and the western side of the Gargaf Arch. They described the unit as a cyclic repetition of coarsening-upward sequences, each cycle starts with shales at the base, grades upward into siltstones, and ends with sandstones. The amount of shale increases away from the Gargaf Arch toward the centre of the Ghadames Basin. Massa & Moreau-Benoit (1976) recommended abandoning the name Chatti Formation. On the other hand, Banerjee (1989, p.98) and Turk et al. (1980) proposed the name Wadi Ash Shatti Member of the Aouinat Ouenine Formation for the Upper Devonian iron-bearing beds. Mamgain (1980) proposed the name Wadi Ash Shatti Formation for this unit. Jacqué (1962) proposed the name Meerscheema Formation to include shales and sandstones of Middle to Upper Devonian which are exposed on the eastern side of the Murzuq Basin and disconformably rest on the Emi Magri (Ouan Kasa) Formation. He divided it into three members: a predominantly shale member at the base, a predominantly sandstone member in the middle, and a ferruginous sandstone member with *Bifungites fezzanensis* at the top. Klitzsch (1963) described the equivalent of the Aouinat Ouenine Formation in the Dur al Qussah area. The succession is made up of thin to thick-bedded, fine-grained, cross-bedded sandstones with abundant *Tigillites* and casts of *Lepidodendron*, intercalated by shale beds up to 10 m (33 ft) in thickness. Abundant *Bifungites fezzanensis* and *Tentaculites* occur in the upper part. The thickness of the exposed section at Dur al Qussah is between 140 m (459 ft) and 310 m (1017 ft). The thickness of the exposed section of the Aouinat Ouenine Formation at Gour Iduka described by Burollet & Manderscheid (1967) is 180–200 m (591–656 ft). However, the thickness of the unit ranges from 300 m (984 ft) at the edge of the Ghadames Basin to 750 m (2461 ft) in the centre of the Basin. The unit unconformably overlies the Ouan Kasa Formation in the Ghadames Basin and Tripolitania, but the contact becomes unconformable in south-West Libya and near tectonic arches. The base of the unit is diachronous from north to south and ranges from Middle to Upper Devonian (Burollet & Manderscheid, 1967). It overlies Cambro-Ordovician sediments in places.

Massa & Moreau-Benoit (1976) reviewed the stratigraphy and palynology of the Devonian system in the Ghadames Basin and Jifarah Plains. They raised the Aouinat Ouenine Formation to group status and divided it into four formations designated A.O.I to A.O.IV. The four formations thicken northward in the subsurface of the Ghadames and Jifarah Plains.

The Middle Devonian is represented by the A.O.I (Couvinian) and A.O.II (Givetian) formattions. The A.O.I (Couvinian) is made up of fine-grained sandstones, coarse-grained at the base with gypsum nodules and kaolinite clasts. It contains brachiopods and Tentaculites in the subsurface. In the central and northern parts of the Ghadames Basin the unit becomes predominantly shales with argillaceous sandstone intercalations. The A.O.I Formation belongs to Palynozone 4 (Couvinian) which is characterized by *Emphanisporites annulatus* and *Dibolisporites eifeliensis*, and abundant megaspores such as *Grandispora macrotuberculata*, *G. protea*, *G. megaformis*, *Apiculiretusispora brandtii*, *Diatomozontrolites* sp., and *Ancyrospora nettersheimensis*. Chitinozoans are abundant and include *Angochitina devonica*, *A.* cf. *ramusculosa*, and *Hoegisphaera glabra*. This unit is correlatable with unit VIII (Emsian-Eifelian) of Jardiné & Yapaudjian (1968) or unit L1 of Jardiné et al. (1974) in the Illizi Basin. The A.O.II (Givetian) is an alternation of finely laminated shales and fine to medium-grained sandstones with argillaceous and ferruginous cements with cross-bedding and oblique stratification rich in brachiopods, lamellibranchs, gastropods, tentaculites, arthrodires, bryozoans, and rare corals. Northward it changes into dark grey shales with thin beds of fine sandstone or sandy limestone. The lower part of the A.O.II Formation belongs to Palynozone 5 (Lower Givetian) which includes *Retusotriletes biarealis*, *R. goensis*, *Apiculiretusispora brandtii*, *Aneurospora goensis*, *Corystisporites multispinosus*, *Emphanisporites annulatus*, *Anapiculatisporites devonicus* var. *azonatus*, and *Rhabdosporites langi*. Chitinozoans include *Ancyrochitina* cf. *cornigera*, *Eisenachitina castor* and *Urochitina* sp. The upper part of the A.O.II Formation spans Palynozone 6 (Upper Givetian), characterized by *Grandispora echinata*, *G.* cf. *Spinosa*, *G. mamillata*, *Ancyrospora magnifica*, *Raistrickia aratra*, *Verrucosisporites inaequalis*, *Lophozonotriletes crassata*, *L. gibberulus*, and *L. media*. Chitinozoans are rare.

It is proposed herein to restrict the Aouinat Ouenine Group to the Middle Devonian and to raise the Ash Shatti Member of Banerjee (1980) and Turk et al. (1980), or the Chatti Formation of Collomb (1962), to group status, the Ash Shatti

Group, and restrict it to the Upper Devonian. The Aouinat Ouenine Group and the Ash Shati Group are absent in extreme southWest Libya.

The Upper Devonian Ash Shatti Group is represented by the A.O.III (Frasnian), A.O.IV (Famennian) and Tahara (top Famennian-Strunian) formations (Massa & Morreau-Benoit, 1976). The Middle-Upper Devonian boundary is very sharp and abrupt. On the Algerian side equivalent sediments were called the Gazelle Formation by Jardiné et al. (1974). In the subsurface of the Ghadames Basin, the A.O. III unit is composed of radioactive, finely laminated, grey shales, alternating with grey, very fine-grained sandstones or yellow siltstones with *Bifungites fezzanensis* and large phosphatic nodules at the base. In the subsurface of the Jifarah Plains it consists of intercalations of dark-colored microcrystalline limestones with Tentaculites, conodonts, and rare goniatites and molluscs. The thickness of the unit is 30–55 m (98–180 ft). In Tripolitania the lower part of the A.O.III Formation belongs to Palynozone 7 (Lower Frasnian) which contains an abundance of Tasmanites and small forms of *Lophosphaeridium* associated with the acritarch *Navifusa bacillum*, and rare spores and chitinozoans. In the Ghadames Basin the A.O.III Formation contains abundant *Grandispora cassidea*, as well as *Retusotriletes raisae*, *Reticulatisporites textilis*, *Contagisporites optivus* var. *vorobjenensis*, the megaspore *Enigmophytospora simplex*, and several species of *Lophozonotriletes*. The upper part of the A.O.III Formation belongs to Palynozone 8 (Upper Frasnian) which is characterized by the abundance of *Lophozonotriletes* such as *L. crassata*, *L. concessus*, *L. densdraconis*, *L. evlanensis*, *L. gibberulus* and *L. media*, as well as *Apiculiretusispora granulata*, *Raistrickia baculosa*, *Camarozonotriletes parvus* and *Ancyrospora langii*. Chitinozoans are represented by *Ancyrochitina pilosa*, *A.* sp. and *Fungochitina pilosa*. The basal 6 m (20 ft) of the A.O.IV Formation contains many large nodules or lenses of green and grey calcareous shales and pseudo-oolitic limestones with fragments of bryozoans, phosphatic debris, and mudstone clast conglomerate. The lower part of the A.O.IV Formation spans Palynozone 9 (Lower Famennian) which characterized by abundant *Geminospora lemurata* and the last occurrence of *Rhabdosporites parvulus*. Also present are *Retusotriletes pychovii* var. *major*, *Archaeozonotriletes famennensis* and *Ancyrospora langii*, as well as the first monolete spores *Laevigatosporites desmoinescensis*. Chitinozoans include *Lagenochitina* cf. *elegans*, *L. macrostoma*, *L.* sp., and *Sphaerochitina fenestrata*. The upper part of the A.O.IV unit (Palynozone 10, basal Lower Famennian) includes *Verrucosisporites bullatus*, *Archaeozonotriletes famennensis*, *Spelaeotriletes* (*Humenozonotriletes*) *granulatus*, *Retusotriletes planus*, *Hystricosporites multifurcatus*, *Grandispora* cf. *uncata*, *Ancyrospora langii*, and *A. furcula*. Chitinozoans (close to their extinction at that time) are few and include *Lagenochitina* sp. and *Sphaerochitina schwalbi*.

The Aouinat Ouenine Group (*sensu lato*) in the northeastern Murzuq Basin and southern Jabal Al Qarqaf area, the type area of the group, was divided into six new formations by Seidl & Röhlich (1984): Bir al Qasr, Idri, Quttah, Dabdab, Tarut, and Ashkidah. The Middle Devonian succession (the Aouinat Ouenine Group of this work) has been divided into the Bir al Qasr and Idri formations.

The Bir al Gasr Formation was named by Seidl & Röhlich (1984) after Bir al Gasr, 40 km west-northwest of Idri. This unit is equivalent to the Aouinat Ouenine I Formation of Massa & Moreau-Benoit (1976). The type section was described by Parizek et al. (1984). In the type section the Formation consists of two medium-grained sandstone intervals, 10 m (33 ft) thick, separated by a 22 m (72 ft) thick silty shale interval. In the Jabal Al Qarqaf area, argillaceous sediments prevail in the western part and ferruginous sandstones on the eastern side. The Formation pinches out to the east (Seidl & Röhlich, 1984). It overlies the Hassaouna Formation, or the Memouniat Formation to the west. In that area, the Formation is predominantly composed of variegated, greenish grey, brownish red silty shales interbedded with purple, medium grained, cross-bedded sandstones which become light grey and quartzitic upwards. It contains the brachiopods *Schizophoria* sp., *Iridistrophia* sp., *Rhynchospirifer* cf. *halleri*, and *Cupularostrum arenosum*. The trace fossils *Spirophyton (Zoophycus)*, *Chondrites*, *Cruziana*, *Planolites* and *Bifungites* are common. Palynological assemblages from quartzitic sandstones from the Bir al Qasr Formation are characterized by Middle Devonian terrestrial vegetation such as *Protolepidodendron helleri* and *P.* cf. *scharyanum*, in addition to brachiopods and bone fragments (Hlustik, 1991). Seidl & Röhlich (1984) and Hlustik (1991) assigned it a Couvinian (Eifelian) age based on palynology.

Seidl & Röhlich (1984) proposed the name Idri Formation after the village of Idri. The Formation corresponds to the Aouinat Ouenine II Formation of Massa & Moreau-Benoit (1976). The type section is 12 km east-northeast of Idri and was described by Parizek et al. (1984). In the type section the Formation is 65 m (213 ft) and consists of fine to medium-grained sandstones and silty shales. In the northeastern side of the Murzuq Basin (Seidl & Röhlich, 1984) the Formation consists of cyclic repetition of shales, planar cross-bedded, medium-grained quartzitic and ferruginous sandstones. Silty shales predominate to the west and in the Tripolitania area. Bellini & Massa (1980) reported a thickness of 80 m (262 ft) (for their unit II) of sandstones and coquina with brachiopods, trilobites, bryozoans, corals and arthrodires. Large tree trunks occur within layers of ferruginous, oolitic, fine-grained sandstones and siltstones with shale clasts, brachiopods and bone fragments (Hlustik, 1991). These fragments form the so-called "microconglomerate" (this term was also used to describe the diamictites in the Cambro-Ordovician sediments). The Formation was assigned a Givetian to early Frasnian age by Seidl & Röhlich (1984) and Hlustik (1991).

The Upper Devonian succession (Ash Shatti Group) includes the Quttah, Dabdab, and Tarut formations, as well as the Ashkidah Formation (Strunian).

The Quttah Formation was introduced by Seidl & Röhlich (1984) to replace the lower part of the Aouinat Ouenine III Formation of Massa & Moreau-Benoit (1976). It corresponds to cycle IV of Collomb's (1962) Aouinat Ouenine Formation or "*Le membre inferieur à Biofungites*". The type section is located 3 km northwest of Quttah. In the type section the unit is approximately 30 m (98 ft) thick and composed of shales and silty shales with thin sandstone intercalations in the lower part, and fine to coarse-grained purple to light brown sandstones with brachiopods and pelecypods in the upper part. *Bifungites fezzanensis* is common throughout. It rests unconformably on the Idri Formation. The unit is absent to the east. It was assigned a Frasnian age by Seidl & Röhlich (1984). In the subsurface of the Ghadames Basin the facies consists mainly of dark shales with dark limestone intercalations and contains abundant *Tasmanites* and *Lophozonotriletes* (Weyant & Massa, 1991).

The name Dabdab Formation was proposed by Seidl & Röhlich (1984) after the village of Dabdab near Brak. It is equivalent to cycle I of the Chatti Formation

(Collomb, 1962) and the upper part of the Aouinat Ouenine III of Massa & Moreau-Benoit (1976). The type section is 3.5 km north of Dabdab. The thickness at the type section is 13 m (43 ft). The Formation is made up mostly of variegated shales with a few beds of ferruginous fine-grained sandstones and siltstones. The uppermost part is composed of a 2 m (7 ft) interval of ferruginous oolites. This unit represents the first cycle of iron-ore sedimentation. Seidl & Röhlich (1984) and Collomb (1962) observed the unconformity at the base of this unit. The unit contains only poorly-preserved atrypidine brachiopods. It was assigned a Middle-Upper Frasnian by Collomb (1962) and Frasnian age by Seidl & Röhlich (1984).

The name Tarut Formation was proposed by Seidl & Röhlich (1984) after the village of Tarut. It corresponds to the Aouinat Ouenine IV Formation of Massa & Moreau-Benoit (1976) and the lower part of Cycle II of the Chatti Formation of Collomb (1962). It represents the second iron ore-bearing cycle of the Aouinat Ouenine Group. The type section is 5 Km southwest of Tarut. In the type section, the Formation is composed mostly of greenish grey, light grey and purple sandy shales. A thin light grey fine-grained sandstone bed with pelecypods occurs in the lower part. The upper 4 m (13 ft) consist of cross-bedded, rippled and burrowed oolitic iron ore unit. Ooids contain hematite, limonite, berthierine, and siderite. The top of the Formation is marked by a conglomerate which may indicate a minor interruption in sedimentation. However, this conglomerate is absent in the Aouinat Ouenine area and sedimentation may be continuous in that area. In the Jabal Al Qarqaf area the Formation thins eastward from 45 m (148 ft) to 10 m (33 ft), and the sand content increases. The Formation contains poorly preserved inarticulate brachiopods, pelecypods, and abundant plant fragments. It was assigned a Famennian age by Seidl & Röhlich (1984). Conodonts found at the base of the Tarut Formation represent transition between Frasnian and Famennian, while conodonts at the top represent transition between Famennian and Strunian (Bellini & Massa, 1980). The depositional environments were interpreted by Seidl & Röhlich (1984) to be bay, lagoonal, and shallow-marine swamps.

The upper part of the Ash Shatti Group in the Ghadames and Murzuq basins can always be detected on gamma-ray logs as a radioactive unit due to the high concentrations of uranium (Weyant & Massa, 1991) and it is believed to be a good source for hydrocarbons. It represents a transitional zone between the Tarut and Tahara formations. It was included in the A.O.IV Formation by Massa & Moreau-Benoit (1976) and in the Ashkidah Formation by Seidl & Röhlich (1984). This interval is also marked in places by a polymictic conglomerate with phosphatic pebbles and fragments of arthrodires. Its age is probably Famennian-Strunian. This unit was given different names such as the Cues Limestone Horizon (Bellini & Massa, 1980), *Horizon repère* (Massa & Moreau-Benoit, 1976), and the *Tornoceras* Limestone layer (Bracaccia et al., 1991). It is made up of dark grey shales and limestones. The limestones are wackestones and packstones. Fossils include goniatites such as *Tornoceras simplex*, *Manticoceras* and *Orthoceras*, fish debris, thin-shell pelecypods, ostracods, and Tentaculites. These fossils indicate deep-marine facies (Weyant & Massa, 1991).

The Tahara Formation was introduced by Burollet & Manderscheid (1967) as "*grès de Tahara*" for a sand-shale sequence underlying the marine Carboniferous and marking the transition between the Devonian and Carboniferous systems in the Ghadames Basin. The Tahara Formation was separated from the Aouinat Ouenine

Group by Massa et al. (1974) and Massa & Moreau-Benoit (1976), but was included in the group by Burollet & Manderscheid (1967) and Seidl & Röhlich (1984) (under the name Ashkidah Formation). The unit increases in thickness from 50 m (164 ft) at the Jabal Awaynat Wanin to 75 m (246 ft) in the subsurface of the Jifarah Plains. Massa & Moreau-Benoit (1976) selected the B1-49 well as the type section of the Tahara Formation. They reported the spores *Spelaeotriletes lepidophytus, Verrucosporites nitidus, Vallatisporites* cf. *pusillites, Dictyotriletes (Reticulatisporites) fimbriatus, D.f.* var. *spathulatus, Knoxisporites hederatus, K. literatus, Lophozonotriletes rarituberculatus, L. cristifer, Rugospora flexuosa (Hymenozonotriletes famennensis), Hystrichosporites* cf. *obscurus, H.* cf. *reflexus* and *Grandispora uncata*, and the chitinozoans *Lagenochitina macrostoma* and *Sphaerochitina fenestrata*. The boundary between the Strunian (Tahara Formation) and the Carboniferous (Mrar Formation) was taken by Massa & Moreau-Benoit (1976) at the first appearance of Tournaisian marine fauna. This boundary is also characterized by the disappearance of the spore *Spelaeotriletes lepidophytus* and Chitinozoans.

The name Ashkidah Formation was proposed by Seidl & Röhlich (1984) after the village of Ashkidah east of Brak. The type section is on the western side of Wadi Dabdab, 4 km west of Ashkidah. It is composed mainly of greenish grey silty shales. Light grey, fine-grained, quartzitic sandstones occur in the upper part of the Formation. The sandstones are cross-laminated, rippled with flaser bedding and *Tigillites*. A 2 m (7 ft) ferruginous oolite bed is present in the middle of the Formation. A thin conglomerate with fossils occurs in the lower part. The unit has its maximum thickness at al Mahruqah and thins easwards. The contact with the overlying Mrar Formation is conformable. The Formation contains poorly preserved inarticulate brachiopods and bone fragments of *Placodermi*. Plant fragments are common and indicate a Tournaisian age. Plant microfossils in the lower part of the Ashkidah Formation such as *Retispora lepidophyta, Tumulispora malevkensis, T. ratrituberculata* and *Vallatisporites verrucosus* indicate an uppermost Devonian age (Vavrdova, 1991). The age of the Ashkidah Formation may, therefore, extend from Late Devonian to early Tournaisian. Plant fragments mixed with pebbles and brachiopods form a "microconglomerate" similar to that in the Idri Formation. The freshwater algae *Botryococcus* was reported by Vavrdova (1991). The Ashkidah Formation corresponds to the Tahara Formation of Burollet & Manderscheid (1967), the upper part of cycles II, III, and IV of the Chatti Formation of Collomb (1962), or "*Le membre superieur à Bifungites*" and the Meerschema Formation of Jacqué (1963). The Tahara Formation, however, has priority and should be retained and the name Ashkidah Formation should be suppressed.

Iron ores occur in the Devonian sediments (Goudarzi, 1971, Turk et al., 1980) of the Aouinat Ouenine and Ash Shatti groups (Middle-Upper Devonian), especially on the southern flank of the Gargaf Arch. Detailed geological studies of the Wadi Ash Shati iron ore deposits were carried out by the U.S. Geological Survey between 1955–1958 (Goudarzi, 1971), and a consortium of French organizations (The French Group) between 1971–1976 under the sponsorship of the Industrial Research Center, Libya. The results of this work were summarized by Turk et al. (1980). Goudarzi (1970, 1971) believed that the iron-bearing beds are mostly Lower Carboniferous. It is for these beds that Banerjee (1980) and Turk et al. (1980) proposed the name Ash Shatti Member (Frasnian-Strunian) of the Aouinat Ouenine. Total indicated reserves of ore are about 3.5 billion tonnes with Fe content between 35 and 55% (Turk et al., 1980). At a locality near Tarut stems of *Leptophloem rhombicum*, originally up to 3 m (10 ft) long, are associated with the oolitic iron ores of the Wadi ash Shati area (Seidl & Röhlich, 1984).

The subdivision of the Middle-Upper Devonian succession into Units I–IV of Massa & Morreau-Benoit (1976) or the formations of Seidl & Röhlich (1984) is not visible in the subsurface of the Murzuq Basin (Pierobon, 1991). The succession ranges from 20 m (66 ft) to about 210 m (689 ft) in thickness and it is composed mostly of marine sediments with *Spirophyton* and coquina beds, but fluvial-deltaic deposits occur in the middle and upper parts. Middle Devonian sandstones (F3 Sand) form important reservoirs in the Al Wafa Field in West Libya (Tawadros et al., 1999) and the Alrar Field in Eastern Algeria (van de Weerd & Ware, 1994, Chaouchi et al., 1998). The Alrar Field has recoverable gas reserves of 4.6 TCF and condensate reserves of 277 MMB (van de Weerd & Ware, 1994).

### 12.3.2   Devonian of the Sirte Basin

Rocks of Devonian age occur locally in the central Sirte Basin. Devonian (Strunian) sandstones have been identified in the V1-59 and EE1-59 wells (Tawadros et al., 2001) (Fig. 12.4) based on the presence of the acritarch *Gorgonisphaeridium solidum* and the spore *Verruciretusisporites* cf. *Famenensis* in the former and the spores *Grandispora* sp. and *Calamospora* sp. in the latter. The sandstones are yellow, red and brown, very fine to very coarse-grained and pebbly, massive, laminated, or cross-bedded. Reworked Devonian palynomorphs have been recorded from several Cretaceous sections in the Basin (Wennekers et al., 1996).

### 12.3.3   Devonian of the Kufrah Basin

In southeastern Libya Devonian rocks crop out in Jabal Eghi (Jabal Nuqay) on the western side of the Kufrah Basin, in the Gebel Oweinat and Jabal Azbah on the eastern side of the Basin, and in the Jabal az Zalmah and Jabal al Qardabah on the northern side of the Kurfrah Basin. Devonian exposures also extend southwards along the eastern, western and southern sides of the Basin into Chad. They also occur in the subsurface of the Kufrah Basin. The Devonian sequence is composed entirely of clastic rocks, represented by the Tadrart, Ouan Kasa, Binem and Blita formations, and the lower part of the Dalma Formation. The basal Lower Devonian is absent in the Kufrah Basin (Grignani et al., 1991), as it is in the rest of Libya. In the Gebel Oweinat area the Tadrart Formation is overlain by the Kurkur Murr Formation.

The combined thickness of the Tadrart and Ouan Kasa formations (undifferentiated) in the Kufrah Basin is 90–130 m (295–427 ft). The two formations thin to the south. The succession is made up of conglomerates at the base, followed upward by medium to coarse-grained massive sandstones with irregular stratification. In Jabal Azbah it is composed of brown to greyish violet, fine to very coarse grained sandstones with irregular beds of conglomerate at the base (Bellini & Massa, 1980). Cross-bedding, slumping phenomena, and ironstone beds are common. To the southwest, the succession conformably overlies the Acacus Formation and is unconformably overlain by the Mesozoic "Nubian" sandstones. The combined thicknesses of the Tadrart and Ouan Kasa formations in the subsurface of the Kufrah Basin are 57 m (187 ft) and 42 m (138 ft) in the A1- and B1-NC43 wells, respectively (Grignani et al., 1991). The two formations were dated Middle Siegenian-Late Emsian by Grignani et al. (1991). These authors reported the following miospores from the Tadrart Formation in the A1-43 well: *Acinosporites lanceolatus, Camarozonotriletes* cf. *sextantii, Emphanisporites schultzii, Cirratriradites* sp., and *Retusotriletes* cf. *ocellatus perotriletes* sp. The Ouan Kasa Formation contains two palynomorphs assembalges representing the Lower and Upper Emsian. The Early Emsian assemblage is characterized by *Emphanisporites erraticus, E. spinaeformis, E. rotatus, E. obscurus, Dibolisporites variverucatus, Dictyotriletes emsiensis, Leiotriletes pagius, Camptozonotriletes aliquantus, Brochotriletes* sp., and the last appearance of *Retusotriletes* cf. *ocellatus*. The Late Emsian assemblage is represented by *Emphanisporites anulatus, Granulatisporites muninensis, Craspedispora craspeda, Grandispora douglastownense*, and *Camarozonotriletes* sp., and by the first appearance of *Grandispora medicona* and *Anapiculatisporites devonicus* var. *azonatus*, in addition to *Veryhachium trispinosum* and *Sphaerochitina* cf. *fenestrata*. These two units are probably equivalent to the Ouri Sao Formation of De Lestang (1965, 1968) on the western side of the Kufrah Basin in northern Chad.

Vittimberga & Cardello (1963) described the Binem Formation as a sequence of Devonian feldspathic sandstones and siltstones in the Jabal Dohone in the Tibesti area in southeastern Libya. It overlies the Tadrart sandstones and it is overlain by the Carboniferous Ounga Formation (equivalent to the Dalma Formation). It was assigned a Middle-Upper Devonian age based on its stratigraphic position. Bellini & Massa (1980) selected the type section at the Wadi Binem-Oudi Ounga in the Tibesti area where the Formation is 150 m (492 ft) thick and consists of white, grey, red and purple, fine-grained sandstones, sandy siltstones, argillaceous siltstones, and ferruginous fine to medium grained sandstones. *Spirophyton (Zoophycus)* is abundant; *Bifungites fezzanensis* is common in places. It thins southward from 150 m (492 ft) to 140 m (131 ft). North of Jabal Azbah, the Formation contains plant remains more than 1 m (3 ft) long (Bellini & Massa, 1980). In the subsurface of the Kufrah Basin the Binem Formation is 555 m (1821 ft) thick in the A1-NC43 well and 612 m (2008 ft) in the B1-NC43 well (Grignani et al., 1991). Grignani et al. (1991) recognized 8 palynological zones in the Binem Formation, correlatable with those of Massa & Moreau-Benoit (1976) and Jardiné et al. (1975). The Couvinian (zone 4) is dominated by miospores, but chitinozoans and acritarchs are also present. The assemblage includes *Grandispora medicona, Anapiculatisporites devonicus* var. *azonatus, Emphanisporites anulatus, E. rotatus, Diatomozontriletes* sp., in addition to *Neoraistrickia* sp., *Retusotriletes witneyanus, Evittia remota, Horologinella horlogia, Leiofusa* sp., and *Sphaerochitina complanata*. The Early Givetian (zone 5) is characterized by the presence of *Retusotriletes goensis, R. biarealis, Apiculiretusispora brandtii, A. plicata, Aneurospora goensis, Anapiculatisporites devonicus* var. *azonatus, Grandispora inculata, G. velata, Dictyotriletes* sp., *Cirratriradites dissutus*, and *Diatomozontriletes* sp. The zone is also characterized by the last appearance of *Emphanisporites anulatus* and *E. rotatus*, and by the first appearance of *Grandispora libyensis* and *Dibolisporites echinceus*. The Late Couvinian (zone 6) is distinguished by the appearance of large spores and the genus *Lophozonotriletes* such as *L. gibberulus* and *L. media*, in addition to *Acanthotriletes horridus*, *Raistrickia aratra*, *Grandispora* sp., *G. inculta*, *G. velata*, and *Apiculiretusispora brandtii*. These zones span the Aouinat Ouenine Group in West Libya. The Fransnian (zones 7 & 8) is characterized by a bloom of *Lophozontriletes* and rare acritarchs. The assemblage includes *Ancyrospora langii*, *A. capillata*, *Lophozonotriletes densdraconis*, *L. concessus*, *L. macrogrumosus*, *L. dentatus*, *L.* sp., *Hystrichosporites* cf. *obscurus*, *H. multifurcatus*, *H.* sp., *Smaratosporites triangulatus*, *Verrcosporites scurrus*, *Camarozontriletes parvus*, *Tetraletes* sp., *Stellinium octoaster*, and *Navifusa* sp. The Early Famennian (zone 9) is dominated by *Geminospora lemurata*, *Verrucosisporites bullatus*, *Hystricosporites gravis*, *H. porcatus*, *Retusotriletes pychovi major*, *Cymbosporites* cf. *famennensis*, *Apiculiretusispora granulata* cf. *Densosporites anulatus*, *Ancyrospora langii*, and very frequent *Horologinella horologia*. The Late Famennian (zone 10) is typified by the last occurrence of *Geminospora lemurata* and *Verrucosisporites bullatus*, by the presence of *Retusotriletes planus*, *Emphanisporites* sp., *Endosporites* sp., *Ancyrospora langii*, and *Hystrichosporites multifurcatus*, and by the first occurrence of *Auroraspora hyalina*, *Spelaeotriletes granulatus*, and *Retispora lepidophyta*. Zones 7–10 span the Ash Shatti Group in West Libya. The latest Famennian-Strunian (zone 11) is identified by *Retispora lepidophyta*, *Leiotriletes struiensis*, *Hystrichosporites* cf. *obscurus*, *Verrucosporites nitidus*, *Dictyotriletes fimbriatus*, *Raistrickia baculosa*, *Auroraspora hyalina*, *A. asperella*, *Endosporites micromanifestus*, *Spelaeotriletes granulatus*, *Knoxisporites literatus*, *K.* sp., *Lophozonotriletes rarituberculatus*, *L. cristifer*, *Grandispora uncata*, *Verruciretusispora famenensis*, *V. magnifica*, *Emphanisporites* sp., and *Pustulatisporites gibberosus*, and scarce acritarchs such as *Horologinella horologia* and *H. quadrispina*. The interval covered by this zone is equivalent to the Tahara Formation in West Libya. In the A1-NC43 well the Devonian is followed by Late Permian and in the B1-NC43 well by Early Carboniferous. The Binem Formation correlates with the Blita Formation on the northern side of the Kufrah Basin and the Aouinat Ouenine and Ash Shatti groups in the Ghadames Basin, and also probably with the Lori Sandstone (Goudarzi, 1970) in the Tibesti area.

The Blita Formation was described by Vittimberga & Cardello (1963) as a thick sequence of Devonian siltstones, shales, and fine-grained sandstones with kaolinitic and ferruginous cement exposed in the Jabal Hawaish on the northern side of the Kufrah Basin. The Formation overlies the Tadrart Formation and is overlain by the Carboniferous Dalma Formation. Vittimberga & Cardello (1963) correlated the Blita Formation with the Binem Formation in Jabal Dohone which they considered to be Middle-Upper Devonian. They differentiated between the Binem and Blita formations based on the presence of feldspars and more shale intercalations in the latter. Near Jabal az Zalmah the pelecypods *Kufrahlia* and *Pleoneilos* sp. are present (Burollet & Manderscheid, 1967). The exposure extends over a distance of 200 km. The thickness of the Blita Formation varies between 270 m (886 ft) and 290 m (951 ft) and it is composed of fine-grained sandstones, well bedded with ferruginous nodules, abundant *Spirophyton* and rare lamellibranchs and brachiopods. It was assigned a Middle-Upper Devonian age by Burollet & Manderscheid (1967) based on *Tropidoleptus carinatus*.

The Blita and Binem formations are probably equivalent to the Bideyat Formation, and the *Spirophytons* Formation (Kora Formation) of De Lestang (1965, 1968) and Klitzsch (1966) on the western side of the Kufrah Basin in northern Chad.

## 12.3.4   Devonian of Northeast Libya

In the subsurface of Northeast Libya Libya, Devonian sediments unconformably overlie the Silurian. No Gedinnian-Siegenian sediments are present in the area. El-Arnauti & Shelmani (1988) recognized seven shallow-marine lithological units (I-VII). Unit I (Emsian) is restricted to the eastern edge of the area near the Egyptian border, but units II-V (Eifelian "Couvinian"-Givetian) are present over most of the area. Units VI and VII (Frasnian-Famennian) are present in the central and northern parts of the area. The distribution of these units is related to N-S and E-W structural highs. Unit I is represented by dark grey shales, silty and pyritic in part, grading upward into fine grained sandstones which include miospores and marine microplanktons. A miospore assemblage was recovered from the A1-33 well (Paris et al., 1985) which includes? *Procoronaspora* sp., *Diatomozonotriletes* sp., *Craspedispora craspeds*, *Emphanisporites annulatus*, *E. erraticus*, *E. rotatus*, *E. obscurus*, *Dibolisporites eifeliensis*, *D. echinaceus*, and *Apiculiretusispora brandtii*. The Eifelian (Couvinian) represents a shallowing-upward marine sequence and consists of three units. The sequence starts at the base with 300–1500 ft (91–457 m) of alternating sands and shales followed upwards by 300–600 ft (91–183 m) of dark grey shales, and then by 300–800 ft (91–244 m) of fine-grained sandstones with rare shale interbeds, argillaceous limestones occur at the base. A miospore assemblage has been recovered from the A1-37 well (Paris et al., 1985) which is characterized by *Anapiculatisporites* sp., *Dibolisporites* cf. *Bullatus*, Hystricosporites sp., *Emphanisporites spinaeformis*, *Diatomozonotriletes* sp., Clayptosporites cf. *Velatus*, *C. spinosus*, *Grandispora* cf. Libyensis, Samarisporites sp., S. cf. *Megaformis*, *Hymenozonotriletes discors*, and *Ancryospora nettersheimensis*. The Givetian consists of interbedded reddish brown to yellow fine-grained, glauconitic sandstones and dark grey silty shales, 400–700 ft (122–213 m). The Givetian has been identified in the A1-37, B1-31, E1-82, and G1-82 wells. It is characterized by abundant miospores, acritarchs and chitinozoans (Paris et al., 1985, Streel et al., 1988) such *as Acinosporites acanthomammillatus*, *A. macrospinosus*, *Dibolisporites achinaeceus*, *Geminospora tuberculata*, *Grandispora inculta*, *G. libyensis*, *Rhabdosporites langii*, *Verrucosisporites premnus*, and *V. scurrus*. Chitinozoans are dominated by *Eisenackitina castor*, *Gotlanochitina* sp. (*Angochitina devonica*), *G. milanensis*, *Ancyrochitina* ?*aequoris*, and *Hoegisphaera glabra*. The Upper Devonian (Frasnian-Strunian) is restricted to the northern and central wells. The Frasnian is made up of shallow, open-marine, dark brown to black shales with a rich assemblage of miorspores and acritarchs which was recorded from the C1-125 and A1-115 wells (Paris et al., 1985). This assemblage includes the miospores *Verrucosiporites bullatus*, *Samarisporites triangulatus*, *Ancyrospora langi*, and *A. multifurcata*, and the acritarchs *Horologinella quadrispina*, *H. horlogia*, *Unellium winslowae*, and *Billosacapsula globosa*. The Frasnian section has a thickenss of 350 ft (107 m) in the north and wedges out to the south. The Famennian is represented by open-marine to marginal marine shales, siltstones, and sandstones. The sandstones become more abundant upwards and contain ferruginous oolites and glauconites. It varies in thickness between 100 ft (30 m) and 2200 ft (671 m). A rich assemblage of miospores have been recorded in the A1-NC92 well (Paris et al., 1985) which includes *Cryroapora cristifer*, *Crassiangulina tesselita*, *Umbellasphaeridium deflandrei*, *Unellium winslowae*, *Veryhachium pannuceum*, *Villosacapsula globosa*,

*Leiotriletes libyensis,* and *Rugospora flexuosa.* The Strunian is represented by sandstones, siltstones and shales. Strunian palynomorphs were recorded from the C1-125 and A1-37 wells (Paris et al., 1985) and include the *miospore Retispora lepidophyta, Knoxisporites literatus,* and *Vallatisporites pusillites.* The whole Devonian succession is correlatable with the Zeitoun Formation of Keeley (1989) in the subsurface of the northern Western Desert of Egypt.

## 12.4 CARBONIFEROUS

Carboniferous formations crop out on the western rim of the Murzuq Basin along north-northwest-trending belts in the Akakus-Tadrart range and the Messak-Mellet Escarpment (Jakovljevic, 1984). They overlie Silurian and Devonian sediments and are overlain by Mesozoic rocks. They also crop out along the southern flank of Jabal Al Qarqaf, on the eastern side of Jabal Bin Ghanimah and Jabal Ati on the western margin of the Tibesti Massif, at Jabal Nuqay (Eghi) on the eastern flank of the Tibesti Massif on the western side of the Kufrah Basin, at Jabal Oweinat on the eastern side of the Kurfrah Basin, and at Jabal az Zalmah and Jabal Azbah on the northern flank of the Basin. They also crop out in a small area in the Jifarah Plains in northWest Libya. Carboniferous sediments also occur in the subsurface of the Murzuq Basin, Ghadames Basin, the Jifarah Plains, the Kufrah Basin, and Northeast Libya Libya. They have been reported recently from the subsurface of the Sirte Basin (Wennekers et al., 1996, Tawadros et al., 2001).

### 12.4.1 Carboniferous of West Libya

The Carboniferous sequence in West Libya includes in ascending order, the Mrar Formation, *Collenia* Beds, Assedjefar Formation, Dembaba Formation and Tiguentourine Formation.

The name Mrar Formation was introduced by Lelubre (1948) for a Carboniferous unit composed predominantly of shales, siltstones and sandstones in the Garat el Marar area (28°20′N, 12°15′E) in the Ghadames Basin. Previously Desio (1936) gave the name Serdeles Sandstone to this unit and considered it to be of Devonian age. Menchikoff (1944) assigned it a Carboniferous age based on brachiopods. Burollet (1960) recommended abandoning the name Serdeles Sandstone in favor of the Mrar Formation. No type section has been designated for the Mrar Formation. Massa (1988) in his Ph.D. thesis used the A1-49 well as a type section of the Mrar Formation. However, this section has not been published. The sedimentology, diagenesis, and sequence stratigraphy of the Mrar Formation in the Ghadames Basin have been examined by Fröhlich et al. (2010).

Collomb (1962) described the Mrar Formation section exposed in the Awaynat Wanin-Dambabah area on the northwestern side of Jabal Al Qarqaf. In that area, the Mrar Formation is 750–850 m (2461–2789 ft) in thickness. Seidl & Röhlich (1984) reported a total thickness of 150 m (492 ft) for the Mrar Formation (including the *Collenia* Beds) on the southern flank of the Gargaf Arch. On the southwestern side of the Murzuq Basin (Jacovljevic, 1984) the Mrar Formation varies in thickness between 70–250 m (230–820 ft) and it unconformably overlies the Early Devonian Ouan Kasa

Formation and it is conformably overlain by the Assedjefar Formation. The lower part of the Mrar Formation is composed of shallowing-upward coarsening-upward cycles (Jacovljevic, 1984), within an incised valley system (Fröhlich et al., 2010). Each cycle starts at the base with grey to black shales followed by light and dark green silty and micaceous shales, and fine-grained sandstones and shales in the upper part. Whitebread & Kelling (1982) described 15 such cycles from outcrops of the Mrar Formation and the overlying Assedjefar Formation on the western side of the Gargaf Arch. The Mrar Formation contains ferruginous concretions, plant fragments of *Lepidodendropsis africanum*, and reddish brown siltstones and ferruginous oolitic lenses with stems of *Lepidosigillaria intermedia* (Hlustik, 1991). The top of the Formation is marked by ferruginous oolitic beds and coarse-grained sandstones which are rich in wood fragments (Seidl & Röhlich, 1984). The lower part of the Formation is composed of bioturbated and burrowed, medium to coarse-grained sandstones and kaolinitic shales with abundant ferruginous concretions. The bottom 1 m (3 ft) is characterized by abundance of *Bifungites fezzanensis*.

Klitzsch (1963) examined the stratigraphy of the Paleozoic outcrops in the Dur al Qussah area on the eastern flank of the Murzuq Basin. The equivalent of the Mrar Formtion is made up of a basal shale unit containing the brachiopods *Chonetes*, *Camarotoechia* and *Ligula*, crinoids, bivalves, and sporadic spirifers, and a thin limestone bed which contains *Syringothyris cuspidata*. The upper unit consists of rhythmic alternations of shales with brachiopods and sandstones with plant fragments. The Formation also contains *Orthoceras* sp., *Camarotoechia* sp., *Productus costatus*, *P. semireticulatus*, *Spirifer* sp., *Megalodus* sp., and *Chonetes* sp. Plant fragments include *Lepidodendron losseni*, *L.* cf. *volkmannianum* and *L.* cf. *veltheimi*. The contact between the Mrar Formation and the underlying Devonian strata is marked by a 1 m (3 ft) thick basal conglormeratic sandstone bed with numerous bone fragments. On the southwestern side of the Murzuq Basin coarsening-upward sequences start with plane-laminated silty sandstones and sandy shales with ferruginous and oolitic beds, and grade upward into fine-grained, gypsiferous sandstones, ripple cross-laminated with brachiopods in places (Jacovljevic, 1984). *Tigillites* are also common near the base of the Formation (Jacovljevic, 1984). The upper cycles of the Mrar Formation contain hummoky cross-stratification interpreted to be tempestites by De Castro et al. (1991).

The upper part of the Mrar Formation may contain locally stromatolitic beds known as the *Collenia* Beds. The name *Collenia* Beds was first introduced by Freulon (1953) for a stromatolitic limestone unit in the southern part of the Ghadames Basin, southeast of Dambabah and north of the Tihemboka High. This unit was treated as a separate unit by Collomb (1962), but was included in the Mrar Formation by Bellini & Massa (1980). The unit is composed of a cyclic repetition of dolomitic marls, argillaceous marls, limestone beds with *Collenia*, oolitic sandy limestones, dolomites, kaolinitic shales, siltstones, bioturbated and burrowed fine-grained sandstones, and a few conglomerates. *Collenia* oncolites are up to 40 cm in diameter, silicified and have algal structures (Seidl & Röhlich, 1984). Brachiopods, bryozoans, gastropods, crinoids and corals are found in the limestones and dolomites. Collomb (1962) reported the brachiopods *Productus punctatus*, *P. semireticulatus*, *Pustula interrupta*, *Tylothyris laminosa*, *Spirifer striatus*, among others, and assigned the unit an Upper Visean age. The *Collenia* Beds is recognized in outcrops in the southern

Gargaf Arch and southWest Libya and in boreholes in the northern and northwestern Murzuq Basin. In these areas it can reach thicknesses between 50–83 m (164–272 ft). On the southwestern flank of the Murzuq Basin the *Collenia* Beds is approximately 25 m (82 ft) in thickness (Jacovljevic, 1984). It is missing in the eastern, central and southern parts of the Basin (Pierobon, 1991).

The Mrar Formation was assigned a Visean age by Collomb (1962), an Upper Visean age by Massa et al. (1974) and an Upper Tournaisian-Visean (Dinantian) age by (Burollet, 1960), Jacqué (1963), Massa & Moreau-Benoit (1976), and Seidl & Röhlich (1984). It was assigned a Middle Tournaisian-uppermost Visean age by Bellini & Massa (1980). Jakovljevic (1984) also considered the Mrar Formation to be Lower Carboniferous (Tournaisian-Visean) in age based on the presence of *Lepidophytes* and other fossils in the *Collenia* Beds. However, all these authors included Collomb's (1962) cycle IV (the upper *Bifungites* unit) of the Ash Shatti Group in the Mrar Formation. The Mrar Formation in the southern Murzuq Basin in Southwest Libya and the Djado Basin in North Niger was assigned a late Tournaisian-Visean age by Mergl et al. (2001) based on the occurrence of *Syringothyris* cf. *ahnetensis*, a large rhynchonellid *Rhynchopora magnifica* sp. nov., *Saharonetes* aff. *saharensis*, *Schuchertella* sp., and *Syringothyris* p., as well as rare bivalves characterised by the giant globose rhynchonellid *Paurogastroderhynchus serdelesensis* and the productid *Fluctuaria undata*.

The name Assedjefar Formation was introduced by Lelubre (1952a). Collomb (1962) described the type section in the Adrar Uan Assedjefar Hill (28°30'N, 11°30'E). In the Jabal Al Qarqaf area the lower part is 80 m (262 ft) in thickness and composed of a thick continental sandstone sequence. The upper part is 40 m (131 ft) in thickness and composed mostly of limestones and anhydrites. In the Dur al Qussah area the Assedjefar Formation is 420 (1378 ft) and composed of shales and marls with rare sandstones (Klitzsch, 1963). Limestone beds contain *Trilobitae* ind., *Neoglyptioceras* sp., *Chonetes carbonifera*, *Linoproductus cora*, *Orthotetes* (*Schellwinella*) *crenistria*, *Productus constatus*, *P. semireticulata*, *P.* (*Gigantella*) *undatus*, *Spirifer conolutus*, *S. grandicostata*, *S. striatus*, *Bellerophon* sp., *Composita ambigua*, *Derba* cf. *robusta*, *Schizostoma catilla*, *Porcellia* sp., *Fenestrellina* cf. *ripisteria*, and *Crinoidea* ind. It was assigned a Visean-Namurian age by Burollet (1960), Klitzsch (1963), and Jacovljevic (1984). Therefore, Lower-Upper Carboniferous (Pennsylvanian-Mississippian) boundary is placed at the top of the Assedjefar Formation (Bellini & Massa, 1980). In the southern Murzuq Basin and north Djado Basin, Mergl et al. (2001) reported a weakly diversified brachiopod association near the base of the formation characterized by the large syringothyrid *Syringothyris jourdyi* and the terebratulid *Beecheria*, as well as *Streptorhynchus*, the productids *Antiquatonia*, *Flexaria*, *Ovatia*, *Rhipidomella*, *Composita*, *Syringothyris*, *Anthracospirifer* and small rhynchonellids. In Southwest Libya, the Formation is composed of greyish green, medium to coarse-grained sandstones and silty to sandy limestones, cross-bedded, plane-laminated, and ripple cross-laminated with occasional conglomerates, ferruginous concretions, gypsum, brachiopods, and gastropods (Jacovljevic, 1984). The amount of limestone increases upwards. A sandstone interval 7–15 m (23–49 ft) thick with large concretions known as "Cannon ball sandstone", "Mushroom sandstone", or "*Grès de champignon*" occurs within the Formation. The Assedjefar Formation conformably overlies the Mrar Formation and it is comformably overlain by the Dembaba Formation. It thins southward from 127 m (417 ft) to 60 m (197 ft).

The Dembaba Formation (spelled Dambabah on the Geological map of Libya, 1977) was introduced by Lelubre (1952a) for a succession in a well in Hasi Dembaba (28°30′N, 11°33′E). Collomb (1962) described the exposed section around Hasi Dembaba. In the Jabal Al Qarqaf area the Formation is made up of microcrystalline limestones and dolomitic marls. On the western side of the Murzuq Basin (Jacovljevic, 1984) the Formation is made up of silty limestones, interbedded with cross-laminated feldspathic sandstones with stromatolitic beds at the base. The limestones contain brachiopods, crinoids, corals, and molluscs in the lower part, and analcime (zeolites) oolites in the upper part. The lower contact of the Formation with the Assedjefar Formation is conformable. The upper contact with the Triassic Zarzaitine Formation is unconformable. In the Dur al Qussah area the Dembaba Formation thins southward from 86 m (282 ft) to 40 m (131 ft) (Klitzsch, 1963). The thickness of the Dembaba Formation in the subsurface of the Murzuq Basin is 35 m (115 ft) to 55 m (180 ft). Brachiopods reported by Jacovljevic (1984) are *Pungnax minutus, Cliothyris roussii, Spirifer subconvolutus, Schucherfella bituminosa, Schizophoria* sp., and *Camarophoria* sp. Klitzsch (1963) reported the fossils *Productus semireticulata, P. (Gigantella) undatus, Pustula pustulosa, Bellerophon* sp., *Schizostoma catilla, Zaphrentis* sp., *Porcellia* sp., *Fenestrellina* sp., and *Crinoidea* ind. from the Dembaba Formation in the Dur al Qussah area and assigned it an Upper Carboniferous (Namurian-Westphalian) age. It was assigned a Namurian-Moscovian age by Burollet (1960). It was assigned a Middle Carboniferous (Bashkirian-Moscovian) age based on brachiopods and fusilinids by Jacovljevic (1984) in southWest Libya and by Bellini & Massa (1980) in the Jabal al Qarqaf area. In the southern Murzuq Basin and north Djado Basin, the upper part of the marine Carboniferous (Moscovian) in the Djado sub-basin yielded an assemblage consisting of the chonetid *Rugosochonetes* cf. *chesterensis* associated with Moscovian foraminifera (Mergl et al., 2001).

Carboniferous sediments are also known from the subsurface of the Jifarah Plains. Massa et al. (1974) introduced the name Hebilia Formation for a subsurface unit, named after a well (32°24′N, 11°51′E) by that name. It overlies Silurian rocks and is overlain by Permian sediments. The Formation is about 183 m (600 ft) in thickness and it consists of: a) A lower dolomitic and anhydritic fine-grained sandstone unit, about 79 m (259 ft) thick. b) A 4 m (13 ft) thick bed composed of coarse conglomeratic sandstones. c) An interval made up of about 47 m (154 ft) of fine-grained sandstones with dolomitic and anhydritic cement, silty shales, and oolites. d) A 57 m (187 ft) inteval composed of wine red and grey yellow crystalline dolomites, microcrystalline limestones *Textularides* and *Endothyrides*, and thin beds of anhydrite. The Hebilia Formation was assigned an Upper Carboniferous (Namurian-Moscovian) age by Massa et al. (1974) based on regional correlations. It was considered to be equivalent to the Dembaba Formation of the Ghadames and Murzuq basins by Banerjee (1980), but it may include also a part of the Assedjefar Formation.

The name Tiguentourine Formation (Upper Carboniferous, probably Stephanian) was introduced by De Lapparent & Lelubre (1948) north of the Edjeleh area on the Algerian/Libyan border for the basal unit of the *Continental post-Tassilien* and was used by Burollet (1960) in the *Lexique*. The Formation crops out in southWest Libya. It was also encountered in wells in the southern Ghadames Basin and the Murzuq Basin. In the type section the unit is made up of red-brown dolomitic shales with shaly dolomites and anhydrite in the lower 20 m (66 ft) (Bellini & Massa, 1980).

In southWest Libya the unit is composed of varicolored siltstones with intercalations of calcareous siltstones, shales, gypsum, and anhydrite (Grubic et al., 1991). The Carboniferous sediments in southWest Libya are overlain by the Zarzaitine Formation (Triassic), Taouartine Formation (Jurassic), and Messak Formation (Upper Jurassic-Lower Cretaceous) or its equivalent the Ben Ghanima Formation. The thickness of the Tiguentourine Formation in the subsurface of the Murzuq Basin is 130–520 m (426–1706 ft). The Formation represents the last regressive phase of the Carboniferous in West Libya. It is absent in the Dur al Qussah area (Klitzsch, 1963) where the Dembaba Formation is overlain directly by the Upper Jurassic-Lower Cretaceous Ben Ghanima Formation (Korab, 1984). It is also absent in the southwestern corner of Libya where the Dembaba Formation is overlain by Triassic sediments (Jacovljevic, 1984, Galecic, 1984, Grubic, 1984).

## 12.4.2 Carboniferous of the Sirte Basin

The presence of Carboniferous sediments has been confirmed in the V1-59 well in the Belhedan Field and the C3-6 well in the Nasser Field in the central Sirte Basin (Wennekers et al., 1996). These sediments had been previously assigned to the Cambro-Ordovician or the Cretaceous (Bonnefous, 1972). Carboniferous sandstone occurrences have been confirmed also in the KKKK1-6 well in the Nasser Field and the BBBB2-6 well in the Jabal Field (Tawadros et al., 1999, Tawadros et al., 2001). The sandstones are red, green, brown, grey and yellow, fine to very coarse-grained and pebbly. They contain the palynomorphs *Aratrisporites saharaensis*, *Calamospora* sp., *Polycospora* sp., *Verrucosisporites* sp., *Punctatisporites* sp., *Cymatiosphaera* sp., *Veryhachium trispinosum*, *Granulatisporites* sp., *Raistrichia* sp., and *Gorgonisphaeridium* sp. These sandstones were named the Belhedan Formation by Tawadros (2001).

## 12.4.3 Carboniferous of Northeast Libya

In the subsurface of Northeast Libya the Tournaisian and the Upper Carboniferous are absent in the central part of Cyrenaica, but a complete section from Early Visean to Gezelian is present in the north. The Bashkirian (early Upper Carboniferous) is probably absent throughout the whole area.

El-Arnauti & Shelmani (1988) divided the Carboniferous succession into five units C-I to C-V. The Early Visean (unit C-I) in the northern wells is made up of 1260 ft (384 m) of deltaic interbedded sandstones, siltstones and shales. A miospore assemblage has been recorded from the A1-37 and A1-NC92 wells by Clayton & Loboziak (1985) which includes *Spelaeotriletes balteatus*, *S. owensi*, *S. pretiosus*, *Radiizonates genuinus* and *Vallatisporites agadesi*, and *V. vallatus*. The Late Visean (unit C-II) is 880–2240 ft (268–683 m) in thickness and shows shallowing upward and southward. Shallow marine shaly and silty limestones at the base become oolitic near the top in the northern wells, and change into limestones and dolomites southwards. A miospore assemblage was recorded from the A1-NC92, J1-81A, and A1-14 wells by Clayton & Loboziak (1985) which is distinguished by the appearance of the genus *Lycospora* and *Spelaeotriletes triangulatus*. Toward the south, the sandstones contain ferruginous pellets, glauconite, and beds of shales and lignite. Units C-I and C-II were correlated with the Mrar Formation in West Libya.

Unit C-III (Serpukhovian or Namurian) is made up of silty shaly limestones and minor sandstone intercalations with thicknesses of 900–1000 ft (274–304 m). Monosaccate pollen appear in that interval in the A1-NC92 and A1-14 wells (Clayton & Loboziak, 1985). It was correlated with the Assedjefar in West Libya. Unit C-IV (Moscovian) is composed of oolitic limestones in the lower part and grades upward into shaly limestones and sandstones. It is correlatable with the Dembaba Formation in West Libya. Unit C-V (Gezelian-Early Asselian) is a sequence of 400 ft (122 m) of limestones with shell debris, siltstones, shales, and minor coal laminations. This unit is characterized by the monosaccate pollen genera *Potonieisporites*, *Plicatis-pollenites*, *Cannanoropollis*, and *Barakarites*, the striate bisaccate pollen *Illinirwa*, *Protohaploxypinus*, *Strotersporites*, *Striatoabieites*, and *Distriatites* (Brugman et al., 1985b). This unit which straddles the Carboniferous-Permian boundary was correlated with the Tiguentourine Formation. The facies distribution and the absence of Upper Devonian and Upper Carboniferous in the south are attributed to uplift and erosion due to the Late Hercynian Orogeny.

## 12.4.4  Carboniferous of the Kufrah Basin

Only the Lower Carboniferous is present in southeastern Libya and it is partly represented by the Kurkur Murr, Dalma, and Ounga formations.

The term Kurkur Murr Formation (also spelled Karkur Murr) was used by Vittimberga & Cardello (1963) and Burollet (1963b) for an Upper Devonian-Lower Carboniferous quartzitic conglomeratic sandstones and siltstones overlying the Precambrian basement in the central part of Jabal Awaynat (Gebel Oweinat). The Formation is overlain by siltstones, shales, and sandstones of the Ouadi Ouaddan Group (Burollet, 1963b). It has a thickness of approximately 80 m (262 ft). To the east the basal 30 m (98 ft) of the siltstones and white and pink san dstones section contain *Archaeosigillaria* aff. *vauxemi* which indicates an Upper Devonian-Lower Carboniferous age. The Formation is intruded by rhyolitic veins (Menchikoff, 1927a, b, Burollet, 1963b) and overlain by 12 m (39 ft) of syenites. The Formation is equivalent to the Dalma Formation and partly to the Wadi Malik Formation. The latter also contains *Archaeosigillaria* aff. *vauxemi* in southwestern Egypt and was assigned a Tournaisian-Visean age by Klitzsch (1990a).

Burollet (1963b) subdivided the Carboniferous section overlying the Kurkur Murr Formation in the Jabal Oweinat (Awaynat) into three units and used the informal names Ouadi Oaddan Argillites, "cima" Sandstones, and Plateau Silts. The Ouadi Ouaddan Argillites is a Carboniferous unit exposed at the heart of the Jabal Awaynat. It is composed of steeply dipping alternating siltstones, shales, and sandstones overlain by quartzites. The Formation is approximately 280 m (919 ft) in thickness. Burollet (1963b) applied the name "cima" Sandstones to a very thick Carboniferous series of quartzitic sandstones which overlie the Ouadi Ouaddan Argillites. The thickness of this unit is 370 m (1214 ft) at Karkur Murr. It has undergone some contact metamorphism in the central part of the Awaynat Massif. The name Plateau Silts was given to a Carboniferous unit overlying the "cima" Sandstones. The unit is approximately 60 m (197 ft) thick and composed of black silts interbedded with mudstones and coarse-grained beds. The silts are intercalated with trachytes and they are overlain by 500 m (1641 ft) of purple to green volcanic breccias of andesites and rhyolites. These three units combined are equivalent to the Dalma Formation in southeastern Libya and the Wadi Malik Formation in southwestern Egypt.

The name Dalma Formation was applied by Vittimberga & Cardello (1963) to Carboniferous rocks exposed in the Jabal Hawaish. It consists of yellow, medium to coarse-grained quartzose, cross-bedded sandstones. *Lepidodendron* is abundant at the top. It was assigned an Upper Devonian-Lower Carboniferous age by Burollet & Manderscheid (1967). Its thickness is about 500 m (1641 ft). The Formation is unconformably overlain by Nubian sandstones. The Dalma Formation is absent in the A1-NC43 well and has a thickness of 541 m (1775 ft) in the B1-NC43 well (Grignani et al., 1991). The Dalma Formation in the B1 well consists of dark grey-black, locally silty, shales in the lower 100 m (328 ft), followed upward by a succession of alternating white-light brown, varicolored, fine to coarse-grained sandstones, varicolored siltstones and dark grey to varicolored shales. Its age is Tournaisian-Early Middle Visean based on the presence of the palynomorphs assemblage *Grandispora balteata, Densosporites variomarginatus,* and *Spelaeotriletes pretiosus,* the last appearance of *Vallatisporites vallatus, Raistirckia baculosa, Verrucosisporites nitidus, V. macrogrunosus, Diatomozontriletes fragilis, Apiculiretusispora multiseta, Knoxisporites pristinus, Savitrisporites nux, Radiizonates genuinus, Vallatisporites baffensis, V. ciliaris, Denosporites sp., Pustulatisporites gibberosus, Reticulatisporites canellatus, Spelaeotriletes owensi, S. triangulus, Brochotriletes diversifoveolatus, Apiculiretusispora sp., Krauselisporites ornatus, Cristatisporites sp., Denosporites sp.,* and *Convolutisporites* cf. *balmei.* It correlates with Mrar Formation in West Libya, and partly with units C-I and C-II in Cyrenaica.

The Ounga Sandstone was used by Vittimberga & Cardello (1963) for a sequence of Carboniferous rocks overlying the Binem Sandstones in Jabal Dohone (Eghi) in southeastern Libya. It consists of varicolored fine-grained sandstones at the base with *Licopodiale* plant remains, followed upward by marine ferruginous sandstones with *Sanguinolites variabilis, Chonetes* cf. *hardrensis, Camarotoechia* cf. *pleurodon* (?), *Schellwinella* (?), and *Syringopora* (?), then by thin-bedded sandstones and siltstones. The Formation is equivalent to the Dalma Formation of southeastern Libya and the Ouddi Kozen and Arkenouga Sandstone in northern Chad (De Lestang, 1965, 1968).

## 12.5  PERMIAN

Permian rocks do not crop out anywhere in Libya. Lower Permian sediments, 660–2520 ft (201–768 m) in thickness, have been recorded in the subsurface of Northeast Libya Libya in the northernmost wells in Cyrenaica (Brugman et al., 1985b, El-Arnauti & Shelmani, 1988). The Upper Permian is absent Northeast Libya Libya. Two Lower Permian units have been recognized: The lower unit is composed of limestones, shales, and sandy shales with dolomite near the top. The upper unit grades from continental to deltaic sandy shales at the base into sandstone at the top. In the Kufrah Basin the Permian is probably represented by the basal part of the Permo-Triassic Madadi Formation. The Madadi Formation was used by De Lestang (1965, 1968) for the Permo-Triassic rocks exposed in southwestern Kufrah Basin along the track from Faya Largeau to Tekro near the Libyan-Chad border. The Formation consists of basal varicolored silty shales, followed upward by red to brown and fine to medium-grained, partly quartzitic sandstones. Its thickness varies between 50 m (164 ft) and 100 m (328 ft). It unconformably overlies Carboniferous rocks and is overlain by the Soeka Formation. Permo-Triassic volcanics occur on the eastern side of the Kufrah Basin in the Jabal Oweinat area (Schandelmeier et al., 1987). Permian sediments were also recorded from the subsurface of the Kufrah Basin in the A1- and B1-NC43 wells by Grignani et al. (1991) where they have thicknesses of 109 m (358 ft) and 315 m (1034 ft), respectively. They were assigned to the Upper Permian based on diagnostic specimens such as *Lueckisporites virkkiae, Nuskoisporites dulhuntyi, Vittania ovalis, V. subsaccata, Hamiapollinites medius, H. insolitus, Gardenasporites heisseli, Gigantosporites hallstattensis,* and *Playfordiaspora crenulata.* The species *Lueckisporites virkkiae* and *Gigantosporites hallstattensis* were reported by Adloff et al. (1985, 1986) from the Uotia Formation in the K1-23 well in northWest Libya.

The distribution of the Permian sediments in the subsurface of northWest Libya is very limited. Their thickness is approximately 700 m (2297 ft), although Upper Permian sediments with thicknesses up to 4000 m (13,124 ft) were encountered in the subsurface in the Tebaga area in Tunisia (Bishop, 1975). Mennig et al. (1963) introduced the Permian Uotia (Watiah) Formation, named after the village Al Watyah in the western Jifarah Plains. The Formation is composed of fine to medium-grained, argillaceous, dolomitic, anhydritic, and ferruginous sandstones and thin intercalations of red shales with iron-bearing horizons, lignite debris, and sideritic shales. The upper part of the Formation is dominated by brown and red silty and dolomitic shales in the upper part (Adloff et al., 1985, 1986). It is overlain by the Bir el Jaja Formation. It was dated Lower Permian based on its stratigraphic position by Mennig et al. (1963), but was assigned a Middle-Upper Permian age by Adloff et al. (1985, 1986) based on the palynomorphs *Lueckisporites virkkiae, Paravesicaspora splendens, Klausipollenites schaubergeri* and *Gigantosporites hallstattensis.* The Uotia Formation was encountered only in the type K1-23 well where it has a thickness of 700 m (2297 ft).

# Chapter 13

# Mesozoic

## 13.1 TRIASSIC-EARLY CRETACEOUS

Differentiation between the Triassic, Lower and Middle Jurassic, and Upper Jurassic-Lower Cretaceous sequences is often too difficult due to the scarcity or lack of fossils, the gradual transition of lithologies which straddles the boundaries between the systems, and the prevalence of unfossiliferous sandstones, dolomites and evaporites in many sections in the north and continental sandstones in the south.

In Northwest Libya Triassic to Cretaceous sediments are exposed along the Jifarah escarpment for a distance of 350 km (Fig. 13.1). They also occur in the subsurface of the Jifarah Plains and northern Ghadames Basin and extend westwards into Tunisia. The Triassic-Early Cretaceous succession consists of the Bir el Jaja, Ouled Chebbi, Ras Hamia, Azizia, Bu Sceba, Bir el Ghnem, Bu Gheilan, Bu en Niran, Abreghs, Tocbal, Giosc, Khashm az Zarzur, Scuicsiuch, Cabao and Chicla formations, followed by the Late Cretaceous Sidi Asid, Garian, and Gasr Tigrinna formations. Most of these units extend into the offshore of northWest Libya (Hammuda et al., 1985). Assereto & Benelli (1971), Fatmi et al. (1980), Bellini & Massa (1980) reviewed the stratigraphy of the Jabal Nafusah area and Busson (1967) gave an excellent discussion on these sediments, especially in relation to similar sequences in Tunisia. Koehler (1982) carried out a detailed work on the Cenomanian Ain Tobi Formation Adloff et al. (1985, 1986) conducted an elaborate study of the Permian-Early Jurassic sediments in the subsurface of the Jifarah Plains. Their age datings will be used in this work.

In the following discussion the Triassic-Early Cretaceouus units will be divided arbitrarily into four successions: Triassic, Jurassic, Upper Jurassic-Lower Cretaceous (undifferentiated) and Aptian-Albian.

## 13.2 TRIASSIC

Triassic rocks are exposed in the Jabal Nafusah in Libya and the Tebaga of Medinine in Tunisia, the Ghadames Basin, the Murzuq Basin, and in the southwestern Kufrah Basin in Libya and Chad. They have been described from the Jabal Gharya by Brichant (1952). Sur la découverte du Trias au pie du Djebel Garian. They also occur in the subsurface of the Jifarah Plains, the Ghadames Basin, the Kufrah Basin, and

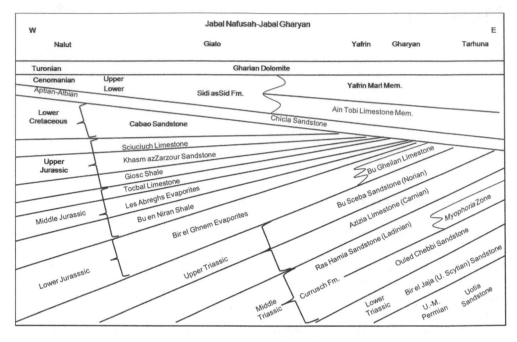

*Figure 13.1* Schematic stratigraphic cross-section of the Permian-early Upper Cretaceous in the Jabal Nafusah-Jabal Gharyan area: Based on Christie (11955), Burollet (1963a), Mennig et al. (1963), El-Hinnawy & Cheshitev (1975), and Adloff et al. (1986).

Northeast Libya Libya. Recently, Triassic sediments have been reported from the sub-surface of the Sirte Basin.

## 13.2.1   Triassic of Northwest Libya

The Triassic succession in northWest Libya consists of the Early Triassic Bir el Jaja and Ouled Chebbi formations, the Middle Triassic Ras Hamia (Currusc) Formation, and the Late Triassic Azizia and Bu Sceba formations (Fig. 12.6).

The Lower Triassic Bir el Jaja Formation (Scythian) was named by Mennig et al. (1963) after the Bir al Jaja well in the Jifarah Plains and it is composed of fine-grained, glauconitic and phosphatic sandstones with some clayey dolomitic beds. The Formationis overlain by the Ouled Chebbi Formation and unconformably overlies the Uotia Formation (Middle-Upper Permian). It was assigned an Upper Permian age based on microfossils by Mennig et al. (1963), but was later assigned a Lower Triassic (Scythian) age by Adloff et al. (1985, 1986) based on the foraminifera *Meandrospira pusilla* and the annelid *Spirorbis phlyctaena*. The species *Meandrospira pusilla* is considered to be characteristic exclusively of the Upper Scythian but was placed erroneously in the Permian by Mennig et al. (1963).

The Ouled Chebbi Formation was introduced by Mennig et al. (1963) for a subsurface Triassic unit in the Jifarah Plains composed of fine and medium-grained sandstones which are feldspathic, quartzitic, dolomitic, anhydritic, and pyritic with abundant quartzite and coal fragments. It overlies the Bir el Jaja Formation and is overlain by the Ras Hamia Formation. In the D1-23 well it is composed of medium to coarse-grained, cross-bedded sandstones. It has a thickness of 60 m (197 ft) in the type section in the A1-23 well. It varies in thickness from 15 m (49 ft) in the south of the Jifarah Plains to 735.5 m (2413 ft) in the north Adloff et al. (1985, 1986). The Formation is unfossiliferous, except for fossil wood. It was assigned a lower Triassic age based on its stratigraphic position by Mennig et al. (1963) and a lower Middle Triassic (Anisian) by Adloff et al. (1985, 1986). Kruesman and Floegel (1980) described the Ouled Chebbi Formation from the subsurface of the Jifarah Plains, and considered it to be Permo-Triassic in age. The Formation consists of red conglomeratic sandstones and red shales with a few sandstone and dolomitic beds at the top.

Christie (1955) introduced the name Boutoniere Formation for the oldest exposed rocks in the Gharyan area composed of "fine-grained, dark red, micaceous sandstone, red and green silty claystone and clay, and lesser narrow carbonate bands". The Formation is exposed in only four places in the area and all exposures are at or near the centres of small domes. Two exposures are just to the north of Gharyan and two are just to the east of Wadi Ghan near the Jabal front where 34 m (112 ft) of the section are exposed in the core of one of the domes. North of Gharyan in the bed of Wadi Sart Bu On Christie (1955) reported *Ostrea* from a yellow dolomitic limestone band, and Birchant (1952) reported fragments of *Nothosaurus* bones from the same locality. Desio & Rossi-Ronchetti (1960) and Desio et al. (1960) discarded the Boutoniere Formation and introduced the name Currusc Formation to replace it (this name is spelled Kurrush on the Geological Map of Libya, 1985). The latter was named after the type section at Jabal Kurrush. Mennig et al. (1963) reported echinoderms, gastropods, foraminifera, and pteropods, in addition to a few species of *Myophoria*, dwarfed *Lingulas* fromf the Currusc Formation. Burollet (1960) also rejected the Boutoniere Formation and introduced the Ras Hamia Formation[1] to substitute for it. He reported a thickness of 74 m (243 ft) for the exposed part of the Formation at Ras Hamia, although he included the top sandstone unit in the overlying Azizia Formation. The base of the Ras Hamia Formation is not exposed in outcrops, but the Formation overlies the Ouled Chebbi Formation in the subsurface of the Jifarah Plains. It is overlain by the Azizia Formation and the definition of the contact between the two units has been a subject of a long dispute (Busson, 1967a, b). Kruesman & Floegel (1980) reported thicknesses from 450–600 m (1476–1969 ft) from the subsurface of the Jifrarah Plains. It was assigned a Middle Triassic age based on its stratigraphic position by Burollet (1960), a Middle Triassic (Ladinian) age by El-Hinnawy & Cheshitev (1975) based on the faunas *Myophoria elegans, M. inaequicostata, M. orbicularis, Lingula* aff. *subelliptica, L. keuperea, Nucula elliptica* and *Avicula mediocalcis*. Adloff et al. (1985, 1986) also dated it Ladinian based on the above molluscs and the following palynological assemblage: The spores *Calamospora tener, Leiotriletes* sp., *Punctatisporites triassicus, Verrucosisporites* sp., *Palaeospongisporis europeus, Maculatasporites* sp., *Keuperisporites baculatus*, and *Aratrisporites granulatus*, the saccate pollen *Heliosaccus dimorphus, Tsugaepollenites* sp., *Ovalipollis pseudoalatus, Staurosaccites quadrifidus, Tridispora staplini, T. falcata, T. plicata, Vitreisporites*

---

1 Although the name Currusc Formationis favored by the authors of the Geological Map of Libya (1985), the name Ras Hamia Formation is also receiving wide acceptance.

*pallidus, Pityosporites neomundanus, Cuneatisporites radialis, Microcachryidites doubingeri, Protodiploxypinus gracilis, Sulcatisporites institatus, Alisporites magnus, A. catheutensis, A. grauvogeli,* and *Fuldaesporites* sp., and *Paracirculina scurrilis,* and the plicate pollen *Cycadopites* sp. Mennig et al. (1963) examined the petrography of the Ras Hamia Formation in the subsurface of the Jifarah Plains in detail. The Ras Hamia Formation is composed of green, red, and black sandy shales with sandstone intercalations, and occasional dolomitic limestones. The Formation consists of two units: an upper sandstone unit and a lower *Myophoria* Zone (Mennig et al., 1963, Adloff et al., 1985, 1986). The *Myophoria* Zone varies in thickness from 30 m (98 ft) in the south to 125 m (410 ft) in the north and marls and limestones with foraminifera in the north to sideritic dolomites in the south. The upper unit has thicknesses between 248 m (814 ft) in the south and 633 m (2077 ft) in the north (Adloff et al., 1985, 1986). The unit is composed of white to beige, fine-grained and glauconitic sandstones with common polycrystalline quartz grains, alternating with red and green silty, dolomitic shales with abundant pyrite and coal fragments. The Ras Hamia Formation is equivalent to the Kirchaou Sandstone in Tunisia which forms the main reservoir in the Giant El Borma Field.

Parona (1914) described the Triassic section in a quarry at Azizia and distinguished a Middle Triassic (Ladinian) lower unit of oolitic limestones and an upper unit of Upper Triassic cherty limestones. The name Azizia Formation (spelled Al Azizyah on the Geological Map of Libya, 1985) was formally introduced by Christie (1955) for a unit of Middle Triassic (Maschelkalk) age composed of limestones or dolomitic limestones in places containing narrow bands or nodules of chert. The lower part of the Formation consists of a richly fossiliferous light to dark grey limestones. This unit is exposed in isolated outcrops in the Jifarah Plains and in the Jabal Nafusah between the town of Azizia and Kaf Muebba (Desio et al., 1963). The type section of the Formation is on the south side of the Wadi Sart Bu On, a tributary of Wadi Bu Shayba. The thickness of the Formation at the type section is 110 m (361 ft). The Formation also occupies the core of the Gharyan anticline. The lower boundary of the Formation with the underlying Currusc (Ras Hamia) Formation was taken by Christie (1955) at the top of red sandy sediments, however, this contact is not visible at the type locality. The upper boundary with the Bu Sceba Formation was taken at the appearance of the first red sandstone beds. Burollet (1963a) subdivided the Formation into three units: The lower unit is composed of red sandstones and shales with beds of yellow or green dolomite, and marks the transition from the Ras Hamia Formation into the Azizia Formation. It was assigned to the upper Middle Triassic (Ladinian) based on the fauna *Enantiostreon difforme, Lingula tenuissima, Ostrea (Lopha)* cf. *montis caprilis* and bones of *Nothosaurus* (this unit was included in the Ras Hamia Formation by Busson, 1967). The middle unit consists of massive, finely crystalline, pseudo-oolitic dolomite and calcareous dolomite with several beds of chert, and spicules, small ostracods, crustacean coprolites, and dasycladacean algae such as *Acicularia*. Desio et al. (1960) considered this middle unit to be the Azizia Formation and dated it Carnian based on *Lopha calceoformis, Lopha montis-caprilis, Nucula strigillata* var. *aziziensis, Modiolus raiblianus, Pleuromya mactroides,* and *Spiriferina lipoldi* var. *tripolitana.* The upper unit is made up of dolomitic limestones with several red dolomitic sandstone and sandy to shaly limestone beds with *Myophoria inaequicostata, M.* cf. *vestita, M. goldfussi, Eniantostreon difforme, Lopha calceoformis, Entolium filosus, Lima* cf. *costata, Pleurotomya mactroides,* and *Trachyceras (Protrachyceras)* cf. *mandeslohi. Myophoria inaequicostata* and *Spiriferina lipoldi* were reported from the Mohilla Formation (Carnian-Nornian) in Sinai by Druckman (1974a). Assereto & Benelli (1971) defined six microfacies in the Azizia Formation in the Jabal Nafusah, planar and domal algal biolithites, burrowed and laminated micrites and dolomitic micrites, argillaceous biomicrites and dolomitized biomicrites, pelmicrites, oosparites, and medium to coarsely crystalline dolomites. El-Hinnawy & Cheshitev (1975) placed the upper boundary of the Azizia Formation at the undulating surface of the top limestone bed, and the lower boundary at the first appearance of limestone beds overlying the sandstones of the Currusc (Ras Hamia) Formation. They assigned it a Ladinian-Carnian age based on the Ladinian fossils *Lingula tenuissima, Enantiostreon difforme* and *Lopha montis caprilis,* and the Carnian fauna *Spiriferina lipoldi* var. *tripolitana, Nucula strigillata, Modiolus paroni, Dielasma wohrmannianum* and *Lopha calceoformis.* The Azizia Formation according to their definition includes the lower part of the Bu Sceba Formation defined by Christie (1955). In the subsurface of the Jifarah Plains the Azizia Formation is 390–500 m (1280–1641 ft) in thickness (Kruesman & Floegel, 1980) and composed of beige to dark grey dolomitic, partially silicified limestones, anhydrites, and minor shales and siltstones containing lignite fragments. The Formation extends westward to the Tunisian border (Burollet, 1963a, Busson, 1967b, Demaison, 1965). It was assigned a Carnian age by Adloff et al. (1985, 1986) based on the presence of the dasycladacean alga *Poikiloporella kuthani, Aulotortuspraegasclei, Mesodiscus eomesozoicus, Lamelliconus biconvexus* and "*Trocholina*" *procera,* and palynomorphs. The Lower Carnian is characterized by abundant *Circumpolles Classopollis),* bisaccate pollen such as *Samarapollenites speciousus, Elongatosaccus triassicus,* and *Allisporites* sp., as well as *Vallasporites ignacii, Classopollis classoides,* and *Brodispora striata.* The Upper Carnian microfloras are characterized by the abundance of *Circumpolles* such as *Praecirculina granifer* and *Camerosporites secatus,* the bisaccate pollen *Pityosporites neomundanus, Alisporites magnus,* and *A. cacheutensis,* in addition to the pollen *Camerosporites secatus, Alisporites magnus, A. cacheutensis* and *Patinasporites densus* which are also found in the Lower Carnian. Spores are rare and represented by *Ovalipollis pseudoalatus* and *O. minimus*

The succession between the Azizia Formation (Carnian) and the Tocbal Formation (Early Bathonian) in the Jabal Nafusah and the Jifarah Plains is made up of unfossiliferous evaporites, dolomites, limestones and sandstones of presumably Upper Triassic to Jurassic age. The lack of fossils from these units led to long debates and controversies (efficiently summarized by Busson, 1967b) which have not yet been resolved. Although the facies relationships and the stratigraphic positions of the different units appear to be simple and well understood, the boundaries between the different units and their ages are still arbitrary (including the ones in this work). The same is also true for the equivalent units in Tunisia, Egypt and Sinai.

The name Bu Sceba Group was introduced by Christie (1955) for a supposedly Jurassic unit composed mainly of cross-bedded red to brownish sandstones and small-pebble conglomerates, white sandstones, red and green clays, in part bentonitic, and minor amounts of gypsum. A brown fossiliferous marly limestone band, 5–15 m (16–49 ft) thick in thickness, occurs near the base. Burollet (1960) assigned the Bu Sceba to formation status. The best exposed section of this Formation is between Wadi Sart Bu On and Gharyan where the Formation reaches 162 m (532 ft) in thickness.

Desio et al. (1960, 1963) published a detailed description of the unit in the type section and also ranked it as a formation. They divided it into three members: a lower member of red sandstones, a middle member of fossiliferous limestones, and an upper member of continental sandstones. The lower beds of the Bu Sceba Formation (spelled Abu Shaybah on the Geological Map of Libya, 1977 & 1985) were identified as Carnian and the upper beds as Upper Triassic-Jurassic (Norian-Rhetian-Liassic). The Bu Sceba Formation is confined in distribution to western Tripolitania. The Formation changes facies from course to medium-grained, trough cross-bedded sandstones east of Wadi Ghan into sandstones and shales at Gharyan and west of Giado. White silicified wood fragments were found in the Formation at the Wadi Abu Shayba section. The Formation was divided by Assereto & Benelli (1971) into two members: the lower member consists of laminated, red siltstones with feldspathic, carbonate-cemented sandstones, fossiliferous sandy dolomite with crinoids, algae, ostracods and pelecypods, and mud-cracked marls and collapse breccia at the top. The upper member consists of trough and planar cross-stratified pebbly quartzarenites, feldspathic sandstones, conglomerates, plane-laminated and rippled cross-laminated shales and siltstones, and massive shales. The upper member grades laterally into evaporites and vertically into dolomitic deposits. The contact between the two members is marked by erosional channels filled with conglomerates and coarse sandstones. El-Hinnawy & Cheshitev (1975) restricted the Formation to the upper member of Desio et al. (1960, 1963) and placed the lower boundary of the Formation at an erosional surface at the base of the member. They included the two lower members of Desio et al. (1960, 1963) in the Azizia Formation. They assigned the Bu Sceba Formation an upper Carnian age. Adloff et al. (1985, 1986) assigned the Formation tentatively to the Norian. The Bu Sceba Formation is considered in this work to be Norian-Rhaetian and correlated with the Mohilla and Mish'hor formations in Sinai.

## 13.2.2　Triassic of Northeast Libya & the Sirte Basin

Triassic sediments occur in the subsurface of Northeast Libya Libya. A major hiatus is present between the Permian and Triassic in the subsurface of Cyrenaica (El-Arnauti & Shelmani, 1988). The Upper Permian and the Lower Triassic (Scythian) are absent. Triassic sediments occur sporadically in the northern wells. A non-marine section occurs in the A1-19 well in the northeastern Cyrenaica (Thusu et al., 1988). Anisian continental and lagoonal sediments about 8 ft (2.4 m) in thickness with fish remains of *Cleithrolepis major* sp. nov. were encountered in the L4-51 well (Gardner, 1988) and in the A1-96 well in the eastern Sirte Basin (Thusu, 1996, Kuehn, 1996, Sinha & Eland, 1996). Marine Middle-Upper Triassic (Anisian-Rhaetian) limestones are restricted to the north. Brugman et al. (1985a) reported a Middle Triassic palynological assemblage dominated by species of the monolete lycopodiophytic miospore *Aratrisporites* such as *A. centratus*, *A. parvispinosus*, *A. strigosus*, *A. saturni*, *A. paenulatus*, *A. tenuispinosus*, and *A. ovatus*, in addition to the species *Tiadispora crassa*, *T. plicata*, and *Lunatisporites moviaulensis*. The Anisian section is characterized by *Stellapollenites thiergartii*, *Stroterisporites* sp., and *Angustisulcites grandis*. The Ladinian section contains rare *Partitisporites novimundanus* and *Duplicisporites granulatus* and some triletes miospores such as *Keuperisporites baculatus* and *Paleospongisporis eurapaeus*. This association is different from that of the Triassic

Zarzaitine Formation of West Libya reported by Jackovljevic (1984) probably due to differences in the depositional facies.

## 13.2.3 Triassic of the Kufrah Basin

The Permo-Triassic Madadi Formation (Wacrenier, 1958, De Lestang, 1965, 1968) in the southwestern Kufrah Basin along the track from Faya Largeau to Tekro near the Libyan-Chad border is made up of basal varicolored silty shales, followed upward by a red to brown and fine to medium-grained partly quartzitic sandstones. The Formation uncoformably overlies Carboniferous rocks and is overlain by the Jurassic Soeka Formation.

In the subsurface of the Kufrah Basin the Triassic is included in the *Continental post-Tassilien* unit which ranges in age from Triassic or Permian to Jurassic (Bellini et al., 1991). It consists of more than 550 m (1,805 ft) of orange to whitish, fine to coarse-grained, unconsolidated sandstones with interbeds of shales which contain an impoverished palynological assemblage suggestive of a Carnian age.

## 13.3 JURASSIC

## 13.3.1 Jurassic of Northwest Libya

Overlying the Bu Sceba Formation in the Jabal Nafusah is a succession of poorly dated Jurassic rocks. This succession includes the Liassic (Lower Triassic) Bu Gheilan, Bir el Ghnem and Bu en Niran formations, the Middle Jurassic Abreghs, Tocbal, Giosc, and Khashm az Zarzur formations, and the Late Jurassic Scecsciuch Formation (Fig. 12.6).

The name Bu Gheilan Limestone (spelled Bu Ghaylan on the Geological Map of Libya, 1977, and Abu Ghaylan in the 1985 edition) was introduced by Christie (1955) for a formation in the Gharyan area composed "mainly of light colored white to buff to light grey to light brown limestone and dolomitic limestone". The unit conformably overlies the Bu Sceba Formation and is unconformably overlain by the Chicla Sandstone. The distribution of the Bu Gheilan Formation is restricted to the Jabal Nafusah escarpment in the Gharyan area. The unit is eroded to the east and it grades into the Bir el Ghnem Gypsum (dated Liassic by Adloff et al. (1985, 1986) to the west in the valley of Wadi Zagut, near Kaf Mantrus (32°11′N, 12°53′E). Assereto & Benelli (1971) divided the Bu Gheilan Formation into three lithologic units: the lower unit is composed of fine and coarse dolomite breccia and flat-pebble conglomerates, interbedded with thin bedded, laminated dolomicrites, stromatolitic dolomites, pellet dolomicrites, and laminated microgranular gypsum. The middle unit contains cross-laminated oolitic and intraclastic calcarenites, interbedded with massive and bedded dolomicrites, and subordinate stromatolitic dolomites. The upper unit is composed of massive stromatolitic dolomites, laminated dolomicrites and pellet dolomites with subordinate flat-pebble conglomerates and dolomitic crusts. The Bu Gheilan Formation extends into the offshore of Libya. Hammuda et al. (1985) selected a reference section for the Bu Gheilan Formation in the Shell A1-38 well at a drill depth of 2662–3836 ft (811–1169 m). The lithology is similar to that exposed in the Jabal Nafusah area and it is made up mainly of grey and light brown dolomites with relics of oolites and pellet fabrics and minor amounts of shale and anhydrite. It is barren of fossils. They assigned the Formation a Late Triassic-Early Jurassic age, based on its stratigraphic position. The Bu Gheilan Formation will be restricted in this work to the Liassic.

The Bir el Ghnem Gypsum was first described as the Bir el Ghnem Group by De Lapparent (1952) and named after the village of Bir el Ghanam (32°16′N, 13°02′E) between Al Aziziyah and Yafrin. Christie (1955) divided the Group into two laterally equivalent units: the Bu Gheilan Limestone and Bir el Ghnem Gypsum, a division which is accepted by most of the subsequent workers. Magnier (1963) divided the Bir el Ghnem Formation into three units: the Abreghs Gypsum, Bu en Niran, and Bir el Ghnem. El-Hinnawy & Cheshitev (1975) designated these three units as members of the "Bir al Ghanam" Formation. However, the use of the name Bir el Ghnem (or Bir al Ghanam) for a formation as well as a member is unacceptable according to the rules of stratigraphic nomenclatures. The Bir el Ghnem Formation consists mainly of gypsum and anhydrite with frequent dolomitic limestone and dolomite interbeds. It is unconformably overlain by the Chicla Formation or conformably by the Bu en Niran Formation. Desio et al. (1960, 1963) and Smetana (1975) reported *Nucula* sp., *Polypamina* cf. *gregaria*, and *Pleuromya uniformis*, among others. However, these fossils are not age diagnostic. The Bir el Ghnem Formation was given a post-Carnian-pre-Bajocian age by El-Hinnawy & Cheshitev (1975) based on its stratigraphic position. It was assigned a Liassic age by Adloff et al. (1985, 1986) based a paly-nomorphs assemblage composed of the spores *Concavisporites torus, Todisporites major, T. minor, Mantonisporites equiexinus, Trilites micorverrucosus, Ischyosporites mesofoveasolidus, Kraeuselisporites* sp., *Laevigatosporties* sp., the pollen *Patinasporites densus*, a number of species of *Classopollis* such as *Spiritisporites spirabilis, Circulina meyeriana* and *Classopollis classoides* (the latter group is dominant in the Liassic of Portugal and Luxembourg) in addition to *Cycadopites* sp., *Chasmatosporites* sp., *Inaperturopollenites orbicularis, Araucariacites* sp., and *Exesipollenites tumulus*, and the freshwater algae *Botryococcus*.

The Bir el Ghnem and Abreghs evaporites are separated by a limestone unit known as the Bu en Niran Formation. This unit forms a marker bed and it is equivalent to the "Horizon B" in the subsurface of Tunisia or the Zmilet Haber Formation in outcrops which separates the Bhir Evaporites from the Mestaoua Formation (Busson, 1967, Ben Ferjani et al., 1991). It forms a strong seismic reflector in the offshore area (Bobier et al., 1991). The term was introduced by Burollet (1963a) after the village of Bu en Niran at the foot of the scarp between Bir el Ghanam and Yafrin. The unit consists of a bed of massive dolomite about 2 m (6.5 ft) thick at the base, argillaceous shales and green shales approximately 15 m (49 ft), and a red shale bed about 6 m (20 ft) thick at the top (Magnier, 1963). The age of this unit is problematic. Burollet (1963a) assigned it a Middle Jurassic (Bajocian) age based on the presence of echinoid fragments and rare *Valvulinids* and *Ophthalimiids*. Desio et al. (1963) reported *Astarte douvillei, Avicula* sp., *Pseudomorphia* sp. and *Rhabdocolpus* sp. and assigned it to the Bajocian, an age favored also by El-Hinnawy & Cheshetiv (1975). Magnier (1963) believes the unit represents the top of the Liassic. On the other hand, Busson (1967) suggested a basal Liassic age for it. The equivalent Zmilet Haber or "Horizon B" in Tunisia has been dated Pliensbachian by Ben Ferjani et al. (1991) based on its stratigraphic position. A Bajocian age for the Bu en Niran Formation is most likely which is in line with equivalent units in Egypt.

The overlying Abreghs Formation was introduced by Burollet (1963a) after the Abregh Deba hill northwest of Nalut (31°52'N, 10°59'E). The Formation is composed of gypsum and anhydrite with dolomite and shale beds in the lower part and green to yellowish green shales with limey and gypsiferous beds in the upper part. It is separated from the Bir el Ghnem evaporites by the Bu en Niran Formation. In the absence of the latter the Les Abreghs and Bir el Ghnem evaporites form one continuous unit. The Abreghs Formation is conformably overlain by the Tocbal Formation. Magnier (1963) reported a thickness of 130 m (427 ft) for this unit. Burollet (1963b) assigned it a Bajocian age based on the occurrence of *Avicula constatula*. Magnier (1963) also assigned it a Bajocian age based on its stratigraphic position. This Formation is equivalent to the Mestaoua Formation in Tunisia (Busson, 1967, Ben Ferjani et al., 1990).

Desio et al. (1960) introduced the Tocbal Limestone (spelled Takbal in Banerjee, 1980) in the Jabal Nafusah area, named after the village of Takbal (32°08'N, 12°45'E). It is composed of fine-grained limestones, occasionally dolomitic with several beds of gypsum in the lower half and shale beds near the top. Burollet (1960) reported *Burmirhychia* cf. *parva*, *Palaeonucula waltoni*, *Modiolus imbricatus*, *Eomiodon fimbriatus*, *Anisocardia* cf. *loweana*, *Protocardia lycetti*, *Ceromya concentredia similis*, *Homonia* cf. *gibbosa*, *Nuculana (Dacromya) lacryma*, and *Procerithium variabile* and assigned the Tocbal Limestone a Bathonian age. Desio & Rossi-Ronchetti (1960) maintained that the limestones of the Tocbal Formation change facies eastwards into the Gorof Gypsum (obsolete) which changes in turn into dolomites of the Bu Gheilan Formation. The Tocbal Formation is equivalent to the Krachoua Formation in Tunisia. The two formations occupy the same stratigraphic level. The latter was dated Bathonian by Busson (1967) and Bajocian by Ben Ferjani et al. (1990).

The entire Upper Jurassic-Lower Cretaceous succession which crops out in the Jabal Nafusah escarpment (Fig. 12.6) above the Tocbal Formation (Bathonian) was called the Chicla Formation and assigned a Wealden (Early Cretaceous) and Albian age by Desio et al. (1963). Burollet (1963a) divided this succession in ascending order into the Giosc Shale, Chameau Mort Sandstone, Scecsciuch Limestone, Cabao Sands and Chicla Sands. Burollet (1963a) grouped the Tocbal Limestone, Giosc Shale and the Chameau Mort Sandstone into the Tigi Formation which was raised to group status (Tigi Group) by Magnier (1963). Burollet (1963a) changed the name of the "Chameau Mort Sandstone" to Gorria Formation, but this name did not find acceptance among the geologists in Libya. El-Hinnawy & Cheshitv (1975), following Desio et al. (1963), considered the whole sequence between the Sidi asSid and Tocbal formations (which they assigned a Bathonian age) west of Wadi al Mazayidah to belong to the Chicla Formation (which they spelled Kiklah). They divided the section into three members, from base to top: the Khashm az Zarzur, Shakshuk (Arabic transliteration of the Scecsciuch Formation of Burollet, 1963a), and Ar Rajban members. They combined the Giosc Shale and Chameau Mort Sandstone of Burollet (1963a) into one unit, the Khashm az Zarzur Member of the Kiklah (Chicla) Formation. The Member was named after the Khashm az Zarzur promontory (32°03'N, 12°23'E) and assigned an Oxfordian-Kimmeridgian age by Burollet (1963a). They recognized two units within this member: a lower lacustrine clay with minor sandstone interbeds and an upper unit continental sandstone sequence with clay intercalations. Fatmi et al. (1980) suggested raising the Khashm az Zarzur Member to formation status. It is recommeded here to apply the name Khashm az Zarzur Formation to the "Chameau Mort Sandstone" to conform with the rules of stratigraphic nomenclature, and to treat the Giosc Shale as a separate unit with formation status.

The name Giosc Shale was applied by Burollet (1963a) to a green shale unit, greyish and violet with several irregular sand beds which he assigned a Middle Jurassic (Dogger) age. Its thickness varies between 60 m (197 ft) and 90 m (295 ft) (Magnier, 1963). El-Hinnawy & Cheshitev (1975) and Smetana (1975) included this unit in the Khashm az Zarzur Member of the Kiklah (Chicla) Formation. The Giosc Shale was dated Bathonian by Magnier (1963) and Busson (1967). Novovic (1975) and Antonovic (1977) assigned it an Early Bathonian age as part of the Khashm az Zarzur Formation. The Giosc Shale overlies the Tocbal Limestone and underlies the Khashm az Zarzur Formation or the Chicla Formation. It is equivalent to the Techout Formation in Tunisia, dated Bathonian by Ben Ferjani et al. (1990).

The Giosc Shale is overlain with a gradational contact by the Khashm az Zarzur Formation (Chameau Mort Sandstone). The Khashm az Zarzur Formation is made up of a white or yellow sand unit with brown ferruginous sands and varicolored shales, and lignite beds at the base. Burollet (1963a) reported *Laccopteris saighanensis* and attributed the age of the unit to the Callovian-Oxfordian. The Khashm az Zarzur Formation is exposed in the scarp face of Jabal Nafusah in the vicinity of the village of Kiklah to the east to the village of Wazim to the west, near the Tunisian border (El-Zouki, 1980). The unit crops out best in the Yafren escarpment and in the wadi east of the village of Shakshuk. It is truncated by the Chicla Formation east of the village of Takbal (Hammuda, 1971). In the Yafren and Kiklah areas, the Formation is about 40 m (131 ft) thick. In these areas it is overlain by the Chicla Formation (Lower Cretaceous) and underlain by the Tocbal Formation/Giosc Shale (Bathonian). In the Jadu area (31°59′N, 12°01′E) and westward it is overlain by the Scecsciuch Formation (Upper Jurassic) (El-Zouki, 1980). The Formation contains lenses of clay with plant leaf impressions at the top. The fossil plants recovered from this unit are identified as *Neuropteris (Pectopteris) philipsi* and *Zamites* (Hammuda, 1971), *Dionites buchianus, Sphenolepidium kurrianum, Gingkoidium gracile*, and *Otozmites* n. sp. (Sassi, 1942, mentioned in El-Zouki, 1980), *Piazopteris branerii, Pagiophyllum* n.sp., *Brachyphyllum* sp., *Hirmerella* sp., *Otozmites sp.*, and *Samaropsis sp.* (El-Zouki, 1980). The Khashm az Zarzur Formation was assigned an Upper Jurassic (Callovian-Oxfordian) age by Burollet (1963a), Hammuda (1969), and El-Zouki (1980), and Bathonian-Callovian by Walley (1985). The lower part of the Formation consists of yellowish-white, fine-grained, well-sorted, micaceous and cross-bedded sandstones. The most common sedimentary structures in these sandstones are low-angle planar cross-bedding sets with erosional bases. Clay drapes are common. In two quarries south of the Shukshuk village, the Formation consists of 40 m (131 ft) of sandstones with clay and carbonate horizons. The upper part comprises multicolored, and gypsiferous shales with ferruginous, organic-rich laminae, and thin beds of marl containing abundant plant fossils (El-Zouki, 1980). Field observations suggest that there is a facies transition between the Scecsciuch limestones and the Chameau Mort sandstones with the section becoming more terrigenous to the east (Walley, 1985).

The Scecsciuch Formation (spelled Shakshuk on the Geological Map of Libya, 1985) was introduced by Burollet (1963a) after the village of Shakshuk (32°02′N,

11°57′E) for a unit composed of alternating limestones, dolomitic limestones, sandstones, and gypsiferous, green and sandy shales which form large outcrops at the foot of the cliff in the Shakshuk area. He identified Middle-Upper Jurassic fossils such as *Goniocora* sp., *Pleurotomaria* sp., and *Nerinea* sp., in addition to *Pseudocylammina jaccardi* and *Valvulinella jurassica* known from the Lower Kimmeridgian in the Middle East. Hammuda (1971) recorded the echinoid *Monodadema cotteaui* which he considered also to be Kimmeridgian. El-Hinnawy & Cheshitev (1975) reported *Nuculana ovum* and coalified plant debris of Jurassic age and assigned the Formation an Oxfordian-Kimmeridgian age. Burollet (1963a) and Magnier (1963) assigned it an Upper Oxfordian-Lower Kimmeridgian age, but Busson (1967) argued for a Callovian-Oxfordian age. The Formation is equivalent to the Tlalett Formation or Foum Tatahouine Formation in Tunisia. The Scecsciuch Formation overlies the Khashm az Zarzur Formation and has a thickness between 100 m (328 ft) and 110 m (361 ft).

## 13.4  MESOZOIC CONTINENTAL SEDIMENTS

The Mesozoic continental sediments in West Libya include the Cabao and Chicla formations in Northwest Libya (Fig. 12.6) and the Zarzaitine, Taouartine, Messak, and Bin Ghanimah formations in southWest Libya. For paleogeographic considerations, the Aptian-Albian succession (including the Chicla Formation) will be discussed under a separate heading.

On the Geological Map of Libya (1985) the Triassic and Jurassic continental sediments are grouped under three categories: undifferentiated Jurassic which includes the Jarmah Member of the Messak Formation and its equivalent in the Murzuq and southwestern Hammada al Hamra basins, undifferentiated Jurassic and Triassic which include the "*Continental post-Tassilien*", and undifferentiated Triassic. Only the Awbari (Ubari) Member of the Messak Formation and its equivalent the Bin Ghanimah Formation are mentioned as representatives of the Lower Cretaceous rocks in Libya.

The Cabao Formation (spelled Kabaw on the Geological Map of Libya, 1977) was first separated from the Chicla Formation of Christie (1955) in the western Jabal Nafusah by Burollet (1963a) who ascribed it to the Wealden. Magnier (1963) attributed the Formation to the Early Cretaceous (Wealden). Desio et al. (1963) did not recognize the Formation as a separate unit from the Chicla Formation, and assigned the latter partly to the Wealden and partly to the Albian. Burollet & Manderscheid (1965) designated the Cabao Formation as Wealden and the overlying Chicla Formation as Aptian-Albian. Busson (1967b) postulated a topmost Jurassic-Wealden (Purbecko-Wealden) age for the Cabao and a Barremian-Aptian age for the Chicla. El-Hinnawy & Cheshitev (1975) assigned an Upper Jurassic age for the Ar Rajban Member of their Kiklah Formation of which the Cabao forms the lower part. Smetana (1975), Novovic (1977) and Antonovic (1977) assigned the Ar Rajban Member an Oxfordian to Albian age. It is likely that the Cabao Formation is Upper Jurassic-Early Cretaceous in age, while the Chicla Formation is Aptian-Albian. The Cabao Formation crops out along the Jabal Nafusah escarpment under the Chicla Formation. At Giado, the Formation reaches a thickness of 80 m (262 ft). It thickens westward into Tunisia, and thins eastward until it disappears east of Gasr el-Haj (Hammuda,

1971). In the western Jabal Nafusah the Cabao Formation rests conformably on the Scecsciuch Formation and it is unconformably overlain by the Chicla Formation. The Cabao Formation is composed of yellowish-grey, friable, fine to medium-grained, planar and trough cross-bedded sandstones. At the base the sandstones are ferruginous and conglomeratic with fragments of silicified wood. At the top they become finer grained with shale beds (Hammuda, 1971). In the area between Jadu and Nalut, especially north of the village of Tandanmmirah, the upper part of the Formation contains cherty, brecciated, dolomitic horizons 2–7 m (7–23 ft) thick (El-Zouki, 1980). In a sand quarry at Jamnawan just below the town of Jadu the lower part of the Cabao Formation contains abundant marine and continental vertebrate fossil remains and silicified wood. These fossils include dinosaur vertebra of *Spinosaurus?*, shark teeth of the species *Priohybdus arambougi*, croccodile teeth and dermal plates and turtle shell fragments. The Cabao Formation is partly equivalent to the Messak Formation in Southwest Libya and probably spans the Asfar, Boudinar, Bouhedma, and Sidi Aich formations in Tunisia.

El-Hinnawy & Cheshitev (1975) combined the Cabao and Chicla formations of Burollet (1963a) into the Ar Rajban Member of their Kiklah Formation which they assigned a Upper Jurassic age. However, they did not present any new evidence for this revised dating or a convincing justification for deleting the name Cabao Formation (which Banerjee, 1980 also recommended). The "absence" of an unconformity between the Chicla and Cabao formations is not a good reason to justify the change in terminology. Furthermore, these authors were able to recognize two units within the Ar Rajbah Member: a lower conglomeratic sandstone unit with clay intercalations (probably equivalent to the Cabao Formation), and an upper faint red unit of alternating clays and sandstones (probably equivalent to the Chicla Formation). Therefore, it is recommended to retain the Chicla and Cabao formations as separate units.

Hammuda et al. (1985) designated the section in the Aquitaine L1-37 offshore well at a drill depth interval of 10,040–11,218 ft as the reference section for the Chicla Formation in the offshore area. It consists of fine to medium-grained sandstones, grey pyritic shales and thin beds of brown argillaceous dolomites in the lower part, dolomites and limestones in the middle part, and fine dolomitic sandstones and siltstones, shale, and subordinate dolomites in the upper part. The Formation is overlain by a volcanic (basaltic and andesitic) and sedimentary succession. Fossils are restricted to *Kurnubia*. The Formation was assigned a Middle Jurassic-Early Cretaceous age by these authors. However, this unit is probably equivalent to the Cabao Formation in onshore northWest Libya. In this work the Chicla is assigned an Aptian-Albian age following Fatmi et al. (1980).

### 13.4.1 Nubian Sandstones/*continental intercalcaire*

The name "Nubian Sandstone" was introduced by Russeger (1837) for continental sandstones in the Nubia region in Upper Egypt. The term was extended into Libya by Desio (1935) and Sandford (1935). In Libya the term Nubian Sandstone was applied to all continental beds between the marine Carboniferous and marine Cretaceous (Conant & Goudarzi, 1967), therefore it includes *the Continental post-Tassilien* and *Continental Intercalcaire* of the Central Sahara. Kilian (1931) divided the continental succession between the marine Carboniferous and marine Cenomanian in the Central Sahara into the *Continental post-Tassilien* and the *Continental Intercalcaire*. The latter overlies the former with an angular unconformity. Like the Nubian Sandstone, this term was used extensively and carelessly until it lost its original meaning. Subsequently the name "Nubian" became intermingled with Kilian's terminology. Kilian & Lelubre (1946) discussed the age of the Nubian Sandstone in the Messak section.

De Lapparent & Lelubre (1948) divided the *Continental Intercalcaire* (post-marine Carboniferous) of Kilian (1931, 1938) in ascending order into the Tiguentourine, Zarzaitine and Taouartine formations. De Lapparent (1954) recognized that the "Nubian Sandstones" in the Central Sahara may include Carboniferous and older sediments, overlain by Early Cretaceous and Cenomanian sandstones similar to the Wealden facies of Europe. He recognized two units in the *Continental Intercalcaire*: a lower unit 100 m (328 ft) thick containing horizons with Lamellibranchs such as Nuculides, Lucinides and Myides, as well as fossil wood and bone and reptile fragments at higher levels. The second unit is at least 400 m (1312 ft) and starts with a red sandstone bed overlain by cross-bedded sandstones and shales with abundant silicified wood at various levels. A ferruginous lense in the middle of this unit contains imprints of plants such as *Yuccites* sp., *Weichselia reticulata*, *Caldophlebis zaccagnai*, and *C. albertsii*. De Lapparent (1954) postulated that a hiatus which includes the Permian, Triassic, and Jurassic separates the two successions. Freulon & Lefranc (1954a, b) believed that the Nubian section at Messak is Turonian and Senonian in age. Freulon (1964) included the *Continental post-Tassilien* in the *Continental Intercalcaire*. Klitzsch (1963) found *Caldophlebis zaccagnai* in the middle of the Nubian section (Ben Ghanima Formation) at Dur al Qussah. Burollet (1963a) in a discussion during the First Saharan Symposium commented that this plant is Jurassic in age. This plant was also reported from the Jurassic part of the Nubian succession in southwestern Egypt (Klitzsch et al., 1979). *Caldophlebis albertsii* is also characteristic of the Middle and Upper Jurassic, and *Weichselia reticulata* is a common plant in the Early Cretaceous section in North Africa (Lefranc & Guiraud, 1990).

Burollet (1960), following Lefranc (1958) used the names Djoua Group and Djoua Formation, respectively, for the Early Cretaceous continental sediments between Algeria and Libya. The sequence is made up of a basal bed (Bone Bed) composed of sandy green marls with bones, teeth, and plant remains followed by red and green gypsiferous clays (Gypsum Member). The type section of this unit is at Wadi Djoua in the Hammadah de Tinrhert in Algeria.

Klitzsch (1963) applied the name Nubian Formation to the succession at Jabal Ben Ghanimah in the Dur al Qussah area. Klitzsch (1966) subdivided the Mesozoic continental sequence in the Murzuq Basin into three lithostratigraphic units: The Zarzaitine (Triassic), Taouartine (Jurassic) and Messak (Upper Jurassic-Lower Cretaceous) formations. These formations are separated by hiatuses, lateritic horizons, and erosional discordances.

In southWest Libya the Triassic is represented by a continental facies of the Zarzaitine Formation. The name Zarzaitine Formation was introduced by De Lapparent & Lelubre (1948) in the Zarzaitine area in Algeria. In the Murzuq Basin the Formation is about 130 m (427 ft) in thickness and thins to 90 m (295 ft) in the south due to erosion (Grubic et al., 1991). In Wadi Irawan and Al Awaynat areas in West Libya the Zarzaitine Formation unconformably overlies different levels of the Dembaba Formation. The upper contact with the overlying Jurassic Taouartine Formation is erosional and marked by a 0.5 to 5 m (2–16 ft) thick lateritic level which indicates an important break in sedimentation (Jakovljevic et al., 1991). The Formation is locally overlain by Quaternary deposits (Grubic et al., 1991). The Zarzaitine Formation consists of alternations of thick sets of red to reddish-brown siltstones and cross-bedded sandstones and lenses of conglomerate. The lower part of this Formation contains abundant freshwater shells and calcitized fossil logs (Jakovljevic et al., 1991, Grubic et al., 1991). The following palynomorphs were identified in the Zarzaitine Formation (Jakovljevic, 1984, Jakovljevic et al., 1991): *Punctatisporites* sp., *Deltoidospora neddeni, Osmundacidites* sp., *Cycadophytes* cf. *carpanteri, Circulina* sp., and *Vitreisporites* sp., indicating a Triassic age. In the Tamasah and Dur al Qussah areas the Zarzaitine and Taouartine formations are believed to be absent and the Messak (Ben Ghanima) Formation overlies directly the Carboniferous Mrar Formation (Korab, 1984).

The name Taouartine Formation was introduced by De Lapparent & Lelubre (1948) in the Taouartine area in Algeria. The Taouartine Formation on the western side of the Murzuq Basin has a thickness of 240 m (787 ft) (Jakovljevic, 1984) and consists of a number of bipartite sequences composed of cross-bedded sandstones with thin intercalations of siltstones, shales, and lenses of conglomerate (Grubic et al., 1991). Plant fragments are abundant and include *Weichselia reticulata, Laccopteries* cf. *polypodioides,* and *Sagenopteries* sp. (Lelubre, 1952a). The palynological assemblage reported by Stefek & Röhlich (1984) includes *Dictyophyllidites harrissi, Cyathidites minor, C. australis, Gleicheniidites feroneasis, G. senonicus, G.* sp., *Concavisporites infirmis, Osmundacites wellmanii, Ginkocycadophytes* sp., *Classopollis* sp., *Eucomiidites* cf. *troedsoni,* and *Sphaeripollenites subgranulosus* which suggests a Jurassic (probably Lower to Middle) age for the Formation. The upper contact with the overlying Messak Formation is unconformable and marked by a lateritic thin bed up to 1 m (3 ft) thick and composed of brownish-red to brownish-purple shales, siltstones and sandy siltstones (Jakovljevic, 1984, Jakovljevic et al., 1991).

Klitzsch (1970) replaced the names *Continental post-Tassilien* and Nubian Sandstone by the Tilemsin Formation and Messak Formation, respectively. Klitzsch (1970) introduced the name Tilemsin Fomation after Wadi Tilemsin for the undated non-marine red bed sequence in the wesern edge of Jabal Messak (24°20′N, 26°00′E) on the western side of the Murzuq Basin which unconformably overlies the Carboniferous beds and underlies the Messak Formation. The Tilemsin Formation is made up of fine to coarse-grained, partly conglomeratic, cross-bedded sandstones, siltstones, and variegated shales. It disconformably overlies the carboniferous Dembaba Formation and conformably underlies the Messak Formation. The Tilemsin Formation is equivalent to the Zarzaitine and Taouartine formations. However, Klitzsch (1972) erroneously correlated this Formation with the "Ubari Beds" and the "Germa Formation" of the Messak Formation, both of which subsequently became members of the Formation. Jakovljevic (1984) rejected the name Tilemsin Formation based on the poor definition of the Formation and substituted it with the better-defined Taouartine Formation. He also showed that the Tilemsin and Dembaba formations are in fault contacts, not stratigraphic contacts. Lefranc & Guiraud (1990) reviewed the stratigraphy of the *Continental Intercalcaire* of the northwestern Sahara. They found that the top and base of the succession are determined differently in various parts of North Africa and recommended that the name *Continental Intercalcaire* should be used only in basins around the Hoggar massif.

The Upper Jurassic-Neocomian interval in southWest Libya is represented mainly by the Messak Formation and its equivalent the Ben Ghanima Formation. The Messak Formation[2] was named by Klitzsch (1966) after Jabal Messak (24°20′N, 26°00′E). The type section is located at Wadi Tilemsin. The type section was divided by Klitzsch (1972) into a lower unit which he called the Germa beds and an upper unit which he called the Ubari sandstone. Kallenback (1972) also described the Formation in the same area, but both Klitzsch (1972) and Kallenback (1972) used the names Ubari and Germa for the same unit. The subdivision of the Messak Formation into two members is not always possible in other areas. The Messak Formation overlies the Lower-Middle Jurassic Taouartine Formation and is unconformably overlain by Quaternary rocks. The Messak Formation on the western side of Murzuq Basin has a thickness of 340 m (1116 ft) and it is composed of cross-bedded sandstones, conglomerates, shales and siltstones (Jakovljevic, 1984). The top of the Formation consists of an extremely hard silcrete up to 10 m (33 ft) in thickness (Jakovljevic et al., 1991). In the southern and eastern parts of the Basin the Messak Formation overlies Precambrian to Paleozoic strata, while to the west the Formation overlies with a gradational contact the Taouartine Formation (Jakovljevic, 1984, Grubic et al., 1991).

The type section of the Germa Member was described by Klitzsch & Baird (1969) in Jabal Zankakrah near the village of Jarmah (26°32′N, 12°59′E). It overlies the Early-Middle Jurassic Taouartine Formation. Its thickness in the type section is 89 m (292 ft), and ranges between zero in the east and 235 m (771 ft) in the west along the Messak escarpment (Stefek & Röhlich, 1984). In the southern Jabal Al Qarqaf the Germa Member is composed of at least two cycles (based on Seidl & Röhlich, 1984, Fig. 9), each cycle starts with trough cross-bedded, conglomeratic, coarse-grained sandstones, and passes upward into siltstones and shales with a few sandstone beds and abundant plant fragments. The clays are kaolinitic. The Germa Member was assigned a Jurassic age by Seidl & Röhlich (1984). However, palynological studies indicate that the Germa Member is Lower Cretaceous in age. In the type locality of the Messak Formation, the palynomorphs include *Ginkopollites* sp., *Araucarites* sp., *Sporites* sp., *Cicatricosisporites* sp. and *Podocarpidites* sp. (Galicic, 1984, Jakovljevic, 1984). The palynomorphs assemblage in the Germa (Jarmah) shales in the subsurface of the northern part of the Murzuq Basin, encountered in the S-192 and S-196 wells in the Sabha area, consists mainly of Mesozoic fern and gymnospermic vegetation (Konzalova, 1991). The spores and pollen are dominated by *Pteridophytes, Cyathidites, Dictyophellidites* and *Gleicheniidites*. Tekbali (1994) examined the palynology of the Germa Member from the S-192, S-194 and S-196 wells and assigned it a Late Berriasian age based on the spore assemblage *Cicatricosisporites* sp., *Pilosisporites trichopapillosus, Aequitriradites spinulosus, Concavissimisporites punctatus*, and *Trilobosporites apiverrucatus*. However, similar assemblages have been reported by Schrank (1987, 1992) and Ibrahim et al. (1995) from Egypt and Libya and indicate an age no older than Neocomian.

---

2 The name Messak Formation first appeared in Klitzsch (1963, Fig. 3) in the 10th Annual Field Conference, Tripoli, without any explanation. It was shown to overlie unconformably the "Post Tassilian Formation".

The Ubari Member was first described by Klitzsch & Baird (1969) as "Messak Sandstone". The name Ubari was given later to this unit by Klitzsch (1972) after the village of Awbari (26°36′N, 12°37′E), and the name Messak Sandstone was applied to a larger unit which included the Ubari Sandstone and Germa beds as members. In the south of the Gargaf Arch the Ubari Member consists mainly of cross-bedded, dark brown conglomerates in the lower part and brown to black, siliceous, ferruginous, conglomeratic sandstones in the upper part. The only fossils present in the Ubari Member are silicified tree trunks up to 20 m (66 ft) long (Stefek & Röhlich, 1984). The Member was assigned a possible Jurassic-Early Cretaceous by Seidl & Röhlich (1984). Stefek & Röhlich (1984) proposed the upper part of the Messak Formation at Jarmah (described by Klitzsch & Baird, 1969) as the type section of the Ubari Member (Awbari Member) where it attains approximately 55 m (180 ft) in thickness. The maximum thickness of the Ubari Member in the Awbari sheet area, on the northern flank of the Murzuq Basin is 103 m (338 ft). The Ubari Member in that area is made up of fine to coarse-grained sandstones and conglomeratic and pebbly sandstone interbeds with tabular and planar cross-bedding. The sandstones are cemented with silica, calcite, and ferruginous material. The matrix is usually kaolinite.

Along the southeastern edge of the Murzuq Basin, in the Dur el Qussah area, the Messak Formation grades into predominantly conglomeratic sediments (Klitzsch, 1963, 1966, Korab, 1984). This unit was called the Ben Ghanima Formation by Klitzsch (1972). It unconformably overlies the Mrar Formation and contains fossil plants such as *Cladophlebis zaccagni*, typical of the Middle and Upper Jurassic. The Formation is approximately 100 m (328 ft) in thickness and composed of rounded gravels and boulders of quartzitic Paleozoic sandstones, between 5–8 cm in size with a minor sand fraction (Korab, 1984).

### 13.4.2   Upper Jurassic-Lower Cretaceous (Neocomian) of the Sirte Basin

In the Sirte Basin the name Nubian has been used to include any continental sand overlying the Cambro-Ordovician Gargaf quartzitic sandstones and underlying Cretaceous marine sediments. However, confusion between the different units has always resulted due to the lack of age dating, silicification, and the similarity between the lithological make up of the units. No formal names have been proposed for the Nubian Sandstone equivalent in the Sirte Basin, although the name Sarir Sandstone was used by Sanford (1970) and the so-called "Calanscio[3] Formation" was used by Bonnefous (1972) in the eastern Sirte Basin.

Many discussions and debates about the name Nubian have taken place since the inception of the term (Tate, 1871, Pomel, 1877, De Lapparent 1954, Pomeyrol, 1968, 1969, Whiteman, 1970, Weissbrod, 1970, Rigassi, 1969, 1970, Ward & McDonald, 1979, Klitzsch et al., 1979, Wycisk, 1990, Tawadros, 2001, Tawadros et al., 2001). Most of these arguments were nonetheless premature for the lack of suitable means

---

3 It appears that there is some confusion with the use of this name. On one hand, Bonnefous (1972) and El-Hawat (1992) mention the similarities between the Nubian sandstones (Upper Jurassic-Lower Cretaceous) and those of the Calanscio Formation. On the other hand, Benfield and Wright (1980) mention that the Calanscio Formation overlies the Marada Formation (Lower and Middle Miocene).

of age dating the succession at that time such as palynology. And even now it is not always possible to differentiate between the various "Nubian" or "*Continental Intercalcaire*" succession based on lithology. Most of these successions are also barren of palynomorphs and their entity and stratigraphic relations will remain unknown for some time to come. For this reason, the author concurs with Barr & Weegar (1972) and the few others who see that this terminology is useful. They are indeed useful operational units, in places where the stratigraphy of the succession is in doubt. In the meantime, wherever possible, subdividing the succession into workable units should be attempted.

Nubian sandstones are widespread in the Sirte Basin, especially in the eastern part of the Basin. They form good reservoirs in the Magi A and C, Sarir, Bu Attiffel and Wadi fields. They unconformably overlie the Cambrian-Ordovician Gargaf Group quartzitic sandstones, or Precambrian basement rocks and they are unconformably overlain by predominantly marine Upper Cretaceous and Tertiary sediments. Much of the Nubian sequence in the Sirte Basin is unfossiliferous; however, fossil plants and spores have been recovered from a number of wells which indicate an Upper Jurassic-Lower Cretaceous age for the major part of the Nubian succession. Recent work also showed that some Nubian sections may contain Devonian, Carboniferous and/or Triassic rocks. The pre-Upper Cretaceous succession in the V1-59 well believed to be Cretaceous in age by Bonnefous (1972) proved to be Late Devonian and Early Carboniferous in age (Tawadros et al., 2001) (Fig. 12.4). The sandstones also contain reworked Devonian palynomorphs. Most of the Paleozoic sediments have been removed during uplift and erosion of the Sirte Arch prior to rifting in the Early Mesozoic. Only remnants of these sediments remain, but are difficult to identify without palynological work. In most cases, these sediments are included either in the Gargaf Group or in the Nubian sandstones.

During the Late Jurassic and Early Cretaceous the Kalanshiyu and Sirte arches which consisted primarily of exposed metamorphic and igneous basement rocks collapsed and resulted in the initiation of the Sirte Basin as an intracratonic rift basin. Subsidence was accompanied by continental sedimentation of the Nubian sandstones in the Sirte Basin (Tawadros et al., 2001). Some workers, for example Conant & Goudarzi (1967), Pomeyrol (1968, 1969), Gumati & Kanes (1985) believe that Nubian deposition was also widespread over the present horsts, while others (Hea, 1971) believe that deposition was limited to the troughs in which the greatest thickness of the Nubian is found today.

In the central part of the Sirte Basin the tripartite division of the Nubian sandstone section has not been recognized, probably due to the fact that very little work has been done or published on these successions, or because the section is heavily silicified and fractured in that part of the Basin. Unfortunately, the recent paper by El-Hawat et al. (1996) is littered with erroneous information and their data are not supported by reliable palynological dating. In fact, the P1-16 which was used by them as an example of the marine Nubian, is actually Cambrian in age, and the Middle Shale Member in the D6-NC149 well is composed of acidic volcanics with the virtual absence of shales (Tawadros et al., 2001). Bonnefous (1972) reported a Cretaceous palynological association which includes *Hystrichosphaeridium* sp., *Hystrichosphaera* sp., *Hystrichokolpoma* sp., *Pediastrum* sp., and dinoflagellates of the group *Deflandrea* sp. and *Gymmodinium* sp., some *Cysteracea* spores and microscopic unicellular algae, along with a large number of Cretaceous nannofossils from the Nubian section in the D2-104A. However, this assemblage was recovered from ditch cuttings samples and probably represents contaminants from higher intervals. He also reported a similar nannofossil assemblage from the V1-59 well and concluded that the two units represent Nubian sandstones. Later work by the Robertson Group (1991) and Tawadros et al. (2001) proved that the interval in the V1-59 well is Early Carboniferous in age and also showed the complete absence of nannofossils from these intervals (Wennekers et al., 1996). Palynomorphs are extremely rare in the Nubian sandstones in the Central Sirte Basin. Samples from the D3A-NC149 well yielded an impoverished assemblage of land-derived Tricolporate, Monocolpate and Trireticolpate pollen, in addition to the spore *Laevigatosporites* sp. (Tawadros et al., 2001). Tawadros (2001) proposed to introduce the name Wadi Formation for the Early Cretaceous "Nubian" quartzites and volcanics succession in the Central Sirte Basin. The type section is in the D6-NC149 well in the Wadi Field at a drill depth interval 10,200–12,030 ft. The sequence in that well consists of two quartzite units separated by a middle unit of quartzites and rhyolitic volcanics. The Formation conformably overlies the Late Cambrian Red Beds and is unconformably overlain by Late Cretaceous sediments.

Hea (1971) recognized two facies of the Nubian sandstones in the subsurface of the Sirte Basin, the Guilat facies composed of red beds of quartz arkoses and arkoses and the Amrha facies of feldspathic quartzites and orthoquartzites. The latter is often preserved in basement lows. Lithologically the Nubian sandstones are similar to the Gargaf sandstones and may also become quartzitic, in which case differentiation between the two sandstones become impossible, although the Gargaf sandstones appear to have a finer grain size and a better sorting than the Cretaceous sandstones (Bonnefous, 1972). Hea (1971) believes that the Nubian sandstones can be recognized by the presence of clays. Bonnefous (1972) on the other hand shows that the Cambrian-Ordovician sandstones may contain up to 11.6% clays, while the Cretaceous sandstones contain 4.5% clays. However, Hea (1971), Bonnefous (1972), and Tawadros et al. (2001) show that the clay minerals in the Nubian sandstones are dominated by kaolinite and in the Gargaf sandstones by illite and chlorite.

The Nubian Sandstone section in the eastern Sirte Basin can often be divided, especially near the edge of the troughs, into three units (Rossi et al., 1991, El-Hawat, 1992): a lower argillaceous sandstone, a middle variegated shale, and an upper argillaceous sandstone. The Middle Shale Unit changes facies near structural highs into clean sandstones and the whole sequence becomes fine-grained sandstones toward the center of the troughs. The Lower Argillaceous Sandstone Unit consists of a fining-upward sequence of variegated, blue-grey, purple and green, arkosic to subarkosic, fine to coarse-grained, conglomeratic sandstones, massive, planar, and trough cross-bedded. Sanford (1970) called this unit informally the Sarir Sandstone; no type section has been defined. The unit fills the irregular basement topography and varies in thickness from 15 m (50 ft) to more than 305 m (1000 ft). It corresponds to the Basal Sandstone unit of Gillespie & Sandford (1967), who believed that the unit is no older than Albian, based on the presence of angiosperm pollen. The unit directly overlies a basement complex of granite and metamorphic rocks. Thusu & van der Eem (1985) documented the following palynomorphs from the Sarir Sandstone: *Cerebropollenites mesozoicus, Concavisporites* sp., *Classopollis* sp., *Araucariacites* sp., and rare *Allisporites* sp. which suggest an Upper Jurassic-Lower Cretaceous age for the sandstones.

This unit is probably equivalent to the Ounianga and Soeka formations in the Kufrah Basin, the Six Hills/Gilf Kebir formations in southwestern Egypt, and the Sirual, Qahash, Mallegh, and Ghurab formations in the offshore of Cyrenaica. The Middle Variegated Shale Unit in east-central Libya consists of coarsening-upward sequences of cross-bedded and bioturabed shales and siltstones, and laminated and burrowed red and green shales interbedded with siltstones and bioclastic shales rich in brackish water ostracods (Rossi et al., 1991). Viterbo (1968) recorded the freshwater ostracods *Theriosynoecum*, ?*Darvinula*, and *Disulcocypris*, along with the Aptian *Charophyta oogonia*. It is devoid of foraminiferas, but contains a dinocyst assemblage dominated by *Phoberocysta neocomia*, *Muderongia simplex*, and *Oligosphaeridium performatum*, and abundant algae (Thusu et al., 1988). The greenish grey shales predominate to the north and contain glauconite, foraminiferal lining, fish teeth, and plant remains. The Middle Variegated Shale unit changes facies near structural highs into sandstones up to 260 m (853 ft) in thickness and consists of fining upward cycles of quartzarenites and minor shales with shale rip-up clasts, mud flasers, trough, planar, and herringbone cross-bedding. These sandstones grade upward into bioturbated, fine-grained sandstones, siltstones, and shales with horizontal and vertical burrows such as *Planolites* and *Skolithos* (El-Hawat, 1992). This unit forms the main reservoir in the Bu Attiffel Field. The middle variegated shale unit is probably equivalent to the Daryanah Formation in the subsurface of the offshore of Cyrenaica, the Chieun Formation in the Kufrah Basin, the Abu Ballas Formation in southwestern Egypt, and the Alamein Dolomite in the northern Western Desert of Egypt. The Upper Argillaceous Sandstone Unit consists of white to red-brown to grey, poorly cemented, fine to coarse-grained, subarkosic to quartzarenites with a kaolinite-rich matrix, quartz granules and shale rip-up clasts, and shaly sandstones, interbedded with red and green shales and sandy mudstones. The sandstones are massive, cross-stratified, rippled, and flat-bedded, grading upward into greyish green, hematitic, argillaceous, very fine to fine-grained sandstones with plant rootlets and burrows (El-Hawat, 1992). Only dinocysts are present in this unit (Thusu et al., 1988). Its thickness varies from 0–135 m (0–450 ft) and is absent over much of the Sarir Field due to either erosion or non-deposition over high areas. Several porous beds yielded prolific production (Gillespie & Sanford, 1967, Clifford et al., 1980, Gras & Thusu, 1998, 1990). This unit was informally called the "Calanscio Sandstone" of probable Albian-Cenomanian age (Bonnefous, 1972). It is probably equivalent to the Daryanah Formation in the subsurface of the offshore of Cyrenaica, the Tekro Formation in the Kufrah Basin and possibly the Bahi Formation in parts of the Sirte Basin.

## 13.4.3   Upper Jurassic-Lower Cretaceous (Neocomian) of Northeast Libya

Onshore Northeast Libya Libya the Upper Jurassic-Early Cretaceous sequence is made up of shallow-marine limestones, oolitic and detrital limestone, poorly fossiliferous dolomitic limestones and dolomites and fossiliferous shales, and sandstones, minor anhydrite, gypsum, coal and lignite occur in places. These sediments grade into argillaceous limestones interbedded with dark grey shales. The Upper Jurassic-Early Cretaceous sequence along the northern margin of northern Cyrenaica consists of a deeper open-marine facies composed of grey, brown and green claystones, interbedded with brown micritic limestones, siltstones and sandstones. Differentiation between Jurassic and Cretaceous sections can be made only in the northern part of Cyrenaica (Thusu et al., 1988).

Thusu et al. (1988) gave a detailed account of the facies variation of the Upper Jurassic-Early Cretaceous sequence in Northeast Libya Libya, and Batten & Uwins (1985), Uwins & Batten (1985, 1988), and Thusu & van der Eem (1985) reported on the palynological assemblages found in these sediments.

Undifferentiated Jurassic sediments occur widely in the subsurface of Northeast Libya Libya. Thusu & Vigran (1985) reported Middle to Late Jurassic (Bathonian-Tithonian) palynomorphs including rich pollen, miospores and dinoflagellates from a continental fluvial, lagoonal and lacustrine succession in the central and southern parts of the area. These sediments become shallow marine immediately to the north, deep-marine to the northeast and mixed marine and continental in the northernmost part. No Early Jurassic palynomorphs are present. Miospores are dominated by small gymnosperm pollen such as *Classopollis* sp., *Exesipollinites* sp., *Sphaeripollinites* sp., *Araucariacites* sp., *Concentrisporites* sp., *Perinopollenites* sp., *Callialasporites* sp., and *Inaperturopollenites* sp. This assemblage is similar to that of the Sahara, but different from that of Europe. This difference was attributed by these authors to rifting during that time (i.e. due to provincialism). Dinoflagellate cysts are dominant in the north and northeast such as *Systematophora penicillata*, *Adnatosphaeridium caulleri*, *Sentusidium echinatum*, *Escharisphaeridia pococki*, *Pareodinia ceratophora*, *Dimedidinium dangeardii*, *Ellipsoidictyum gochtii*, *E.* sp., *Korystocysta* cf. *kettonensis/gochtii*, *Ctenidodinium* cf. *tenellum*, *Bradleyella (Dichadogonyaulax)* sp., and *Gonyaulacysta filapicata*. They considered the last five species to be diagnostic for the Late Bathonian-Early Callovian rocks in the area. The Late Callovian-Early Oxfordian dinoflagellate cyst assemblage includes *Wanea digitata*, *Energlynia acollaris*, *Cribroperidinium granulatum*, *Korystocysta pachyderma*, and *Ganyaulcysta scarburghensis*. The Late Kimmeridgian-Tithonian is characterized by *Millioudinium globatum* and a few typical Late Jurassic such as *Muderognia* sp., *Ctenidodinium panneum*, *Leptodinium* cf. *aceras*, and *Lithodinia* cf. *jurassica*.

Duronio et al. (1991) studied in detail the stratigraphy of the Upper Jurassic-Lower Cretaceous sequence in the A1-NC120 and A1-NC128 wells in the offshore of Cyrenaica. They divided the section in the A1-NC120 well on the western side of Cyrenaica in ascending order into the Sirual, Qahash, and Daryanah formations. They divided the section in the A1-NC128 well on the eastern side of Cyrenaica into the Ghurab, Mallegh, Qahash and Daryanah formations. All these names, except the Daryanah Formationwere used informally because no type sections or depth intervals were given.

The Sirual Formation (Middle-Late Jurassic-Berriasian) in the Al-NC120 well consists of limestones composed of packstones, grainstones, and wackestones with ooids, large foraminifera, locally common dasycladacean algae, dolomite, intraformational breccia, and coal seams. It contains abundant fossils such as *Protopeneroplis striata*, *Trocholina* sp., *Clypeina jurassica*, *Kurnubia palastiniensis*, *Salpingoporella annulata*, *Pseudocyclammina* sp. and *Anchispirocyclina* sp. The dasycladacean algae *Clypeina jurassica*, as well as *Salpingoporella annulata* are typical of the Late Jurassic and has been reported from the Arab-D Formation in Arabia by Powers (1962) and Wilson (1975, p.264). The Formation is overlain by the Qahash Formation

(Valanginian-Barremian) and grades in the A1-NC128 well into the Ghurab and Mallegh formations.

The Middle-Late Jurassic Ghurab Formation (Callovian-Kimmeridgian) in the A1-NC128 well in the Gulf of Bambah is made up of brown, locally silty, marls with traces of sandstones, grading into intraclastic, sandy wackestones, sometimes oolitic, and grey-brown, silty and occasionally argillaceous mudstones-wackestones. These sediments contain sedimentary structures attributed to turbidity flows and several levels of resedimented allochthonous material composed of rounded bioclasts of fossil fragments, micritic pebbles, peloids, ooids, coarse rounded quartz grains, quartzitic sandstones, and sandstones with carbonate cement. The fragments were derived from reefs, carbonate platform material and emergent land masses. Fossils include *Calpionella alpina*, *C. elliptica*, *Calpionellites lata*, *Crassicollaria brevis*, *C. intermedia*, *C.* cf. *parvula*, *Stenosemellopsis involuta*, *Saccocoma* sp., *Stomiosphaera* sp., *Globochaete alpina*, Radiolaria, *Aptychus*, sponge spicules, pelagic pelecypods, *Spirillina* sp., Nodosariidae, Ophthalmidiidae, *Haplophragmoides* sp., miliolids, valvulinids, *Protopeneroplis striata*, *Acolisaccus* sp., and fragments of gastropods, echinoderms, brachiopods, bryozoans, *Cladocoropsis* cf. *mirabili*, serpulidae, dasycladacean algae, and reworked foraminifera. Thusu et al. (1988) reported a similar association from the Tithonian-Berriasian of the subsurface of the onshore of Cyrenaica (Jabal Akhdar facies), but Duronio et al. (1991) dated it Callovian-Kimmeridgian.

The Mallegh Formation (Tithonian) in the A1-NC128 well conformably overlies the Ghurab Formation and is conformably overlain by the Early Cretaceous Qahash Formation. It consists of alternations of grey to brown marls and silty, sandy, argillaceous mudstones and wackestones, micritic sandstones, quartzitic sandstones, and siltstones. Resedimented clastics include coarse rounded quartz grains, micritic pebbles, ooids, and rounded fragments of fossils. Duronio et al. (1991) identified *Crassicollaria intermedia*, *C.* sp., *C. brevis*, *C. parvula*, *Calpionella alpina*, *C. a. grandis*, *C. elliptica*, *Callpionellites lata*, *Tintinnospella colomi*, *Stenosemellopsis involuta*, *Cadosina* sp., *Globigerina helveto-jurassica*, *Spirillina* sp., Radiolaria, sponge spicules, *Nautiloculina oolitica*, *Verneuilina* cf. *tricarinata*, *Lenticulina* sp., ostracods, brachiopods, gastropods, serpulidae, *Macroporella* sp., *Cylindroporella* sp., and *Cladocoropsis* sp. Thusu et al. (1988) dated a similar assemblage as Tithonian-Berriasian.

From the above description it seems that there is not much difference between the Ghurab and Mallegh formations, and both units may represent two cycles within the Upper Jurassic-Lower Cretaceous sequence.

The Qahash Formation (Valanginian-Barremian in the A1-NC120 well and Berriasian-Barremian in the A1-NC128 well) consists of varicolored shale followed by whitish carbonaceous sandstones, grading into sandy mudstones and wackestones with common coal, and then by purple-brown siliceous shales including ooids. The Berriasian is represented by stylolitic breccia of wackestones, packstones, aggregate ooids and shales. The Formation also contains oolitic grainstones-packstones, oncoids, algal limestones, and probably include resedimented deposits. The Formation is unconformably overlain by the Daryanah Formation (Aptian-Albian). Fossils include *Pseudocyclammina litus*, *Trocholina* sp., *Orbitammina* sp., *Anchispirocyclina* cf. *lusitanica*, *Choffatella* sp., *Labyrinthia* sp., *Trochammina* sp., *Actinoporella podolica*, *Salpingoporella* sp., *Solenoporaceae*, and *Cladocoropsis mirabilis*, as well as pelecypods, ammonites and ostracods. There is a striking similarity between lithology and palynological assemblages of the Sirual Formation and the Qahash Formation.

### 13.4.4    Upper Jurassic-Lower Cretaceous (Neocomian) of the Kufrah Basin

In the Kufrah Basin the equivalent of the Zarzaitine, Taouartine, and Messak formations are represented by the so-called Nubian Sandstone. The Nubian succession was divided by Wacrenier (1958), Klitzsch (1966), and De Lestang (1965, 1968) in the southwestern Kufrah Basin in the Borkou area near the Libyan-Chad border in ascending order into the Soeka, Lakes (*formation de lacs*), Ounianga, Chieun, and Tekro formations. On the western side of the Kufrah Basin in northern Chad the Soeka Formation is absent and the Lakes and Ounianga formations are replaced by the Ouadi Mouro and Ehi Micha formations, respectively (De Lestang, 1965, 1968).

Burollet (1963b) divided the Nubian succession on the eastern side of the Kufrah Basin into a Lower Sandstone (Jurassic), Chieun Limestone (Wealden), and an Upper Sandstone (Lower Cretaceous).

The Nubian in the Kufrah Basin in Libya was also divided into three units by Grignani et al. (1991): 1) a lower *Continental post-Tassilien* (Permian-Jurassic), probably equivalent to the Soeka, Lakes, and Ounianga formations of Wacrenier (1958) and De Lestang (1965, 1968) or to the Lower Sandstone of Burollet (1963b). 2) A middle cherty limestone unit (Chieun Limestone) exposed east of Jabal Nuqay. 3) An upper unit of Early Cretaceous age which covers the central and eastern parts of the Kufrah Basin and is probably equivalent to the Tekro Formation of Wacrenier (1958) and De Lestang (1965, 1968) or the Upper Sandstone of Burollet (1963b).

The name Soeka Formation was applied by Wacrenier (1958) and De Lestang (1965, 1968) to the continental Mesozoic rocks overlying the Madadi Formation in the southwestern Kufrah Basin near the Libyan-Chad border. It consists of white and varicolored, fine to coarse-grained, kaolintic sandstones with varicolored silty shales at the base. Its thickness is approximately 300 m (984 ft). It passes gradually upward into the Lakes Formation (*Formation des Lacs*) which is considered to be a part of the Soeka Formation in this study. It was assigned a Jurassic age.

De Lestang (1965, 1968) applied the informal name Lakes Formation (*Formation des Lacs*) to Mesozoic continental deposits of Ounianga Chebir and Ounianga Sarir in the southwestern Kufrah Basin in northern Chad. It is composed of white to brown kaolinitic and well-bedded fine-grained sandstones. It overlies the Soeka Formation (Lower-Middle Jurassic) and is overlain by the Ounianga Formation. The Formation is 200–250 m (656–820 ft) thick. The lower 30 to 40 m (98 to 131 ft) of the Formation consists of red silty shales which dam the water to form large lakes in the area, hence the name "Lakes" Formation. Since the name of this Formation does not conform to the Code of Stratigraphic Nomenclature, Banerjee (1980) suggested discarding the name without recommending an alternative name. In the present work the Lakes Formation is considered to be the upper part of the Soeka Formation. The name Ounianga Formation was used by Wacrenier (1958) and De Lestang (1965, 1968) to describe continental Mesozoic sediments near the Libyan-Chad border. The Formation is composed of white to multicolor, massive to cross-bedded, fine to medium-grained sandstones, less frequently coarse-grained argillaceous and congolmeratic with thin red shale beds at the base, and abundant silicified wood fragments (Wacrenier, 1958). Its thickness is about 400 m (1312 ft). It overlies the Lakes Formation and it is overlain by the Chieun Formation of assumed Aptian age.

## 13.5   APTIAN-ALBIAN

Unlike the majority of the Lower Cretaceous succession, the Aptian-Albian sediments often contain fossils that enable their identification. In northWest Libya, the name Chicla Formation (spelled Kiklah on the Geological Map of Libya, 1977, 1985) was introduced by Christie (1955) for a sequence of clastic rocks underlying the Ain Tobi Formation which crop out along the Jabal escarpment to the west of Wadi Ghan. The Chicla Formation is widespread and it is composedo of more than 200 km from Gharyan to Tunisia (Assereto & Bellini, 1971). The Formation is 60 m (197 ft) at Bu Gheilan, 10 m (33 ft) at Ras at-Tahuna, and 40 m (131 ft) at Yafren and Giado (Hammuda, 1971). East of Gharyan, however, the Formation wedges out below the Ain Tobi Limestone. Fossils recovered from the type locality of the Chicla Formation include *Cyclas brougnarti*, *Corbicula* sp., and *Sphenolepidium kurrianum* forma *stonbergianum* sp. (Christie, 1955). In addition, several species of freshwater pelecypods have been identified (Desio et al., 1963). The age of the Chicla Formation is controversial. It was designated as Wealden (Lower Cretaceous) by Christie (1955), Albian by Burollet (1963), Upper Cretaceous by Klitzsch (1963), and Aptian-Albian by Fatmi et al. (1980). Ward et al. (1987) assumed that the Chicla Formation is of early Albian age based on the presence of *Classopollis*, ephedroids, *Callialasporites*, *Araucariacites*, and *Eucommiidites*, the spores *Aequitriradites*, *Triporoletes*, *Locopodiacidites*, *Klukisporties*, *Cicatricosisporites*, *Perotriletes pannuceus*, and *Afropollis*, and other Angiosperms and dinoflagellate cysts. However, this assemblage may represent an Aptian-Albian age (Rasul, pers. comm., 1999). Goudarzi (1970) believes that the Chicla Formation is a tongue of the Nubian sandstones of southwesterrn Libya. In the Gharyan area the Chicla Formation is composed of trough cross-bedded coarse-grained to pebbly sandstone. Lenticular beds (50–200 cm) and erosional channels filled with conglomerates are very common. Channels have depths of 40–150 cm. Fine-grained sandstones and mudstones with parallel laminations, ripples, and small-scale trough cross-lamination are rare and occur mainly at the top. Plant fragments and silicified wood with large tree trunks are fairly common at the base of the Formation (Hammuda, 1971). In the Jabal Nafusah the Formation varies in lithology from one area to another. In the eastern region it is characterized by a coarse conglomeratic facies, while the western facies is composed of sandstones in the lower part and muds in the upper part. Regional correlations suggest the presence of an angular unconformity between the Chicla Formation and the underlying Jurassic units (Hammuda, 1971). At Bu Gheilan the Chicla Formation unconformably lies on top of the Bu Gheilan Limestone. At Wadi Zaret (25 km to the west) it rests on the Giosc Shale (Middle Jurassic). At Yafren it overlies the Khashm az Zarzur (Upper Jurassic), and at Giado it overlaps the Cabao Sandstone (Upper Jurassic-Lower Cretaceous). At Takbal the Chicla Formation rests on the Tocbal Limestone (Assereto & Benelli, 1971, Fig. 3) and at Kaf Mantrus it rests on the Bir el Ghnem Evaporites.

In the offshore of Northwest Libya the Aptian and Albian are represented by the Masid and Turghat formations, both introduced by Hammuda et al. (1985).

The type section of the Masid Formation is in the Esso B1-NC35A offshore well at a drill depth of 10828 ft-12042 ft, the only well where the Formation was recognized. It is composed of cream to grey shales with white, buff, and tan argillaceous limestone beds. Foraminifera include *Planomalina buxtorfi* and *Rotalipora appenninica* suggesting an Albian age. The type section of the Turghat Formation is in the A1-NC35A offshore well at a drill depth of 12730 ft-16,097 ft. It is made up of interbedded tan and buff dolomites and dark grey argillaceous wackestones. Benthonic foraminifera are abundant and include *Cuneolina camosauri, C. laurenti, C.* sp., *Dictyoconus* sp., *Pseudocyclammina hedbergi, Ovalveolina reicheli, Coskinolina sunnilandensis*, and *Choffatella decipiens* indicating an Aptian-Albian age for the unit. These two formations combined are probably equivalent to the C hicla Formation in Northwest Libya.

The name Chieun Formation was applied by De Lestang (1965, 1968) to designate continental Mesozoic deposits exposed near the Libyan-Chad border in the southwestern Kufrah Basin. The Formation consists of variegated silty shales with lenses of limnic limestones. It overlies the Ounianga Formation and underlies the Tekro Formation. Its age is assumed to be Jurassic-Lower Cretaceous based on its stratigraphic position. However, the present writer assigns the Chieun Formation an Aptian age based on its correlation with the middle variegated shale unit of the Nubian sandstones in the Sirte Basin. Wacrenier (1958) and De Lestang (1965, 1968) described the Tekro Formation to designate the youngest unit in the southern Kufrah Basin in northern Chad. It consists of white, red, and brown medium-grained to conglomeratic, massive to cross-bedded sandstones. It contains abundant silicified wood such as *Dadoxylon*. Its thickness is more than 400 m (1312 ft). It overlies the Chieun Formation. The age of this unit was assumed to be Jurassic to Lower-Upper Cretaceous, however, it is probably equivalent to the upper argillaceous unit of the Nubian sandstones in the Sirte Basin, or the Sabaya Formation in southwestern Egypt, hence its age could be Albian to Cenomanian.

Duronio et al. (1991) introduced the name Daryanah Formation, named after the village of Daryanah, located along the coast about 33 km northeast of Benghazi and 20 km east of the A1-NC120 offshore type well. The unit was encountered at a drill depth of 2334–2572 m. The Daryanah Formation is a predominantly carbonate sequence. The lower part of the Daryanah Formation consists of thick greyish-grey shales with ferruginous ooids, and brownish packstones-grainstones, partly dolomitic, oolitic and sandy with bioclasts, coated rounded clasts, pyrite and glauconite, followed upwards by fine to medium-grained quartz sandstones with carbonate cement, grading into sandy intraclastic and oolitic packstones. It was dated Aptian-Albian by Duronio et al. (1991). Fossils are mainly benthonic foraminifera such as *Epistomina caracolla, E. spinulifera, Lenticulina* gr. *nodosa, Marginulina* sp., *Hedbergella* sp., *Schakonia cabri, Orbitolina* sp., *Choffatella decipiens, Haplofragmoides excavata, Trocholina* sp., in addition to ostracods, gastropods, pelecypods, brachiopods, echinoderms, bryozoans, and hydrozoans. The Daryanah Formation is also present in the A1-NC128 well, but in a deep-marine facies.

The upper part (Albian) of the Daryanah Formation consists of greyish packstones-wackestones, often recrystallized, slightly pyritic or silty with bands of greenish shales and pink dolomitic limestones in the lower part and greenish-grey shales, pyritic with ferruginous ooids and brick-red, hard siltstones, probably representing paleosols. Fossils are common (Duronio et al., 1991), especially in the lower part such as the foraminiferas *Epistomina cretacea, E. caracolla, E.* sp., *Lenticulina* sp., *Frondicularia inversa, Vaginulina gaultina, Valbulineria gracillima, Dorothia oxycona, Orbitolina* sp., *Ammodiscus gaultinus, Trocholina transversari, Nubeculinella tibia* and *Hedbergella* sp., in addition to ostracods, gastropods, pelecypods, brachiopods, bryozoans, serpulids, and algae.

The Daryanah Formation is unconformably overlain by the Eocene Derna Formation in the A1-NC128 well where the entire Upper Cretaceous and Paleocene sequences are absent, and by the Cenomanian Gasr al Abid Formation in the A1-NC120 well.

It is in the present writer's opinion that the Chieun Formation of the Kufrah Basin, the Abu Ballas Formation in Southwestern Egypt, the Middle Variegated Shale unit of the Nubian Sandstone in the Eastern Sirte Basin, the Daryanah Formation in the subsurface of Cyrenaica, and the Chicla Formation in Northwest Libya, as well as the Alamein Dolomite in the subsurface of the Western Desert of Egypt, are all equivalent and represent the Aptian marine transgression.

## 13.6 UPPER CRETACEOUS

Upper Cretaceous sediments are widespread in Libya. They are largely marine in the northern parts of the country, but interfinger with continental sediments southward. They crop out in the Al Hamada Al Hamra plateau in northWest Libya and extend into the subsurface of the offshore of Libya and Tunisia. Upper Cretaceous rocks also crop out in a small area in the Al Jabal Al Akhdar and along the coast of Northeast Libya Libya and extend into the offshore of Cyrenaica. They also occur extensively in the subsurface of the Sirte Basin.

### 13.6.1 Upper Cretaceous of Northeast Libya

Upper Cretaceous marine sediments are exposed in Northeast Libya Libya in a small area in the Al Jabal Al Akhdar and in limited areas along the coast in northern Cyrenaica; they were also reported from the subsurface of the offshore of Cyrenaica.

The surface exposures of Upper Cretaceous rocks in the Al Jabal Al Akhdar area have an average thickness of 980 m (3215 ft) and consist of the Gasr al Abid, Benia, Hilal, Majahir, Atrun, and Wadi Ducchan formations. The Hilal and Majahir formations crop out in very small areas along the coast of Cyrenaica. Upper Cretaceous sediments also extend into the subsurface of the offshore of northwestern Cyrenaica and into the subsurface of Cyrenaica, as far south as latitude 27°N. Another unit, the Tukrah Formation (Klen, 1974) still needs revision.

Kleinsmeide & van den Berg (1968), and Pietersz (1968) proposed the names Jardas Formation and Jardas el Abid Formation, respectively. Barr & Weegar (1972) suggested the use of the shorter name Jardas Formation for these rocks. The type locality of the Formation is near the village of Jardas al Abid (now Jardas Al Ahrar) (32°18′N, 20°58′E), about 15 miles south of Al Marj in northern Cyrenaica. Kleinsmeide & van den Berg (1968) subdivided the Jardas Formation into four members, from base to top: Gasr al Abid, Benia Limestone, Got Sas and Feitah Limestone. The stratigraphy of this section has been reviewed by Röhlich (1974) who divided it into the Qasr al Abid Formation and Benia Formation. An intra-Senonian unconformity was found by Klen (1974) and Röhlich (1974) within the Got Sas Member in the Ghawt Sas area (32°21'N, 20°56'E). Consequently, the lower part of the member was included in the Benia Formation (Coniancian) and the upper part in the Al Majahir Formation (Lower Campanian).

The Qasr al Abid Formation[4] (mentioned as Jardas al 'Abid Formation on the Geological Map of Libya, 1977) is the oldest unit exposed in the Al Jabal Al Akhdar. It was considered to be a member of the Jardas Formation by Kliensmeide and van den Berg (1968) and Barr & Weegar (1972), but was raised to formation status by Röhlich (1974). It consists of white to yellow brown marls and dark green calcareous shales with rare intercalations of marly limestones, 20–30 m (66–98 ft) in thickness. It has yielded a Cenomanian foraminiferal assemblage. Barr (1972) reported the following fossils: *Rotalipora cushmani*, *R. greenhornensis*, *Praeglobotruncana stephani*, *Hedbergella delrioensis*, *H. brittonensis*, and *Heterohelix moremani*, in addition to molluscs and echinoids.

The Upper Cretaceous is absent in the A1-NC128 well in the offshore of Cyrenaica. In the A1-NC120 well the Lower Cenomanian is represented by the Qasr al Abid Formation. It is overlain by the Benia Formation and it overlies the Daryanah Formation. It consists of shales with interbdded mudstones, wackestones, and packstones with *Praeglobotruncana* sp., *P. stephani*, *P. turbinata*, *Thalmanninella (Rotalipora) greenhornensis*, *T. appenninica*, *Favusella washitensis*, and very rare *Planomalina buxtorfi*. Uwins & Batten (1988) recorded numerous chorate and proximochorate dinoflagellate cysts such as *Cyclonephelium*, *Dinopterygium*, *Florentinia*, *Oligosphaeridium*, *Spiniferites*, and *Palaeohystrichophora infusorioides* from the Early Cenomanian of the subsurface of Cyrenaica. Thusu & van der Eem (1985) reported the dinocysts *Canningia baculata* and *Subtilisphaera* sp., and the marker miospore species *Classopollis brasiliensis*, in addition to the palynomorphs *Cyclonephelium vannophorum*, triporate angiosperms, and the pollen *Ephedripites* sp. and *Afropollis* sp. Batten & Uwins (1985) also reported *Elaterosporites klaszii* and *Elaterocolpites castelainii*.

The Benia Formation was considered to be a member of the Jardas Formation by Kleinsmeide & van den Berg (1968) and Barr & Weegar (1972), but was raised to formation status by Röhlich (1974). The Formation (Al Baniyah Formation) as defined by Röhlich (1974) includes a part of the Got Sas Formation of Kleinsmeide & van den Berg (1968). It is composed of well-bedded, microcrystalline limestones intercalated with marls and dolomitic limestones. Its thickness varies from 35–600 m (114–1969 ft) due to pre-Campanian erosion. Röhlich (1974) reported a thickness of 300–600 m (984–1969 ft) in the Darnah area (32°46'N, 22°38'E), and Klen (1974) reported a thickness of 500 m (1641 ft) in the Al Baydah area (32°46'N, 21°43'E). Kleinsmeide & van den Berg (1968) reported occasional beds rich in gastropods, pelecypods (including rudists and oysters), *Cuneolina pavonia parva* and *Rotalipora greenhornensis*. The foraminiferal assemblage indicates a Cenomanian to Coniacian age (Röhlich, 1974, Klen, 1974). The Benia (Al Baniyah) Formation conformably overlies the Qasr al Abid Formation in the A1-NC120 well in the offshore of

---

4  The new name of the town of Gasr al Abid is Qasr al Ahrar. Consequently, the new stratigraphic nomenclature chaged the name of Gasr al Abid Formation to Qasr al Ahrar Formation.

Cyrenaica (Duronio et al., 1991). It contains *Rotalipora cushmani* and *Oligosteginae* in the lower part, *Thomasinella punica* in the middle part (Upper Cenomanian), and *Nezzazata simplex, Dicyclina*, and *Cuneolina* in the upper part (Turonian).

The Upper Cretaceous sequence is exposed in a narrow belt along the coastline of Cyrenaica and consists of marine sediments ranging in age from Albian to Maastrichtian. The section was divided by Barr & Hammuda (1971) into two units: the Hilal Shale and Atrun Limestone. Barr & Hammuda (1971) proposed the name Hilal Shale (named Al Hilal Formation by Röhlich, 1974) for the Cretaceous shales exposed along the southwestern shore of the Marsa al Hilal embayment (32°52′N, 22°10′E) in Northern Cyrenaica. Th shales are mostly dark grey to greenish grey, often glauconitic with a thin limestone bed near the top. The base is not exposed, but Barr & Weegar (1972) estimated a thickness in excess of 328 m (1000 ft). The Formation is conformably overlain by the Upper Cretaceous limestones of the Atrun Formation (Lower Santonian-Upper Maastrichtian). The shales are rich in foraminifera with rare ostracods, calcispheres, echinoid fragments and *Inoceramus* prisms. Barr & Hammuda (1971) distinguished seven foraminiferal zones, from top to bottom: *Globotruncana concavata concavata, G. concavata, G. sigali, Praeglobtruncana helvetica, Rotalipora cushmani, R. appenninica,* and *Ticinella robeti.* This foraminiferal assemblage indicates an age of Albian to lower Santonian and a bathyal facies. The Hilal Formation is the deep water equivalent of the Early Cenomanian Qasr Al Abid and the Cenomanian-Turonian Al Benia formations (Röhlich, 1974). The Hilal Formation in the A1-NC120 well in the offshore of Cyrenaica was dated Coniacian-Santonian by Duronio et al. (1991). The Santonian part consists of light brown pelagic wackestone and argillaceous mudstone-wackestone with *Globotruncana lapperenti, Dicarinella concavata, Hedbergella* sp., *Gavelinella* sp., *Globorotalites* sp., *Rotalia algeriana,* and *Inoceramus* prisms. The Coniacian part is composed of green-grey marl and shale, pyritic and silty with *G. lapparenti, G. fornicata, D. concavata, Marginotruncana coronata, Sigalia deflaensis, H.* sp., and *Inoceramus* prisms. The Hilal Formation is equivalent to the Rachmat Formation in the subsurface of the Sirte Basin.

Crema (1922, in Desio, 1935) described a unit at the base of the section exposed in Wadi Bakur near Tukrah which he called the Wadi Baccur Limestone. The unit is composed of white and yellowish-white limestones with nodules of chert, hard to soft, occasionally floury, and contains lamellibranchs such as *Lucina dachelensis, L. calmoni, L. cusnumismalis, Cytherea rohlfi,* among others. He assigned it a Maastrichtian age. Desio (1935) examined the limestones exposed at Wadi Bakur and noticed that the nummulitic limestones contain fossils of older age and subsequently assigned the whole section an Eocene age. Burollet (1960) applied the name Tocra Limestone to this unit and similar limestones exposed at Tulmithah and Wadi Athrun. Along the coast between Tukrah and Tulmithah Klen (1974) described the Tukrah Formation (Campanian) as composed of light grey, well-bedded limestones with chert nodules and selected the type area on the Benghazi-Al Marj road. The exposed thickness is a few tens of meters. It is also exposed in excavations in the ancient Greek City of Tulmithah (Ptolmais) and on the beach to the northeast. Klen (1974) reported a rich benthonic formaminiferal assemblage with rare planktons such as *Heterohelix* sp., *Gabonella* ex gr. *hulsewei, G. rowei, Stensioeina* sp., *Bolivinoides* sp., *Rugoglobigerina* sp., *Gavelinopsis* sp., and *Globotruncana* ex gr.

*fornicata* and assigned the Formation a Senonian (Campanian) age. The Formation is overlain by the Appollonia Formation (Eocene) or by the Al Uwayliah Formation (Paleocene) in the subsurface. It occupies the same stratigraphic position as the Atrun Formation. The Tukrah and Appollonia formations are very similar in lithology and the bounda ry between the two units is not very clear. The Tukrah Formation sediments are also deep marine (Klen, 1974). It should be noted that the Wadi Baccur Limestone of Crema (1922) and Desio (1935) and the Tocra Limestone of Burollet (1960) are the same unit, Eocene in age. They are probably equivalent to the Appollonia Formation, whereas the Tukrah Formation of Klen (1974) represents a different unit which is Campanian in age and equivalent to the Al Majahir Formation. Although Burollet (1960) drew attention to the similarity between the limestones at Wadi Bakur, Tulmithah, and Wadi Athrun, he was probably referring to the Eocene limestones (Appollonia Formation) of those areas. Therefore, Burollet's Tocra Limestones is erroneous and should have been called Wadi Bakur Formation (Wadi Baccur Limestone) as Crema (1922) designated it. It is probably advisable to keep the name Wadi Bakur Formation confined to the section in Wadi Bakur until the problem has been resolved.

The name Atrun Limestone (called Athrun Formation by Röhlich, 1974, but herein the original spelling is retained and the unit is called Atrun Formation) was proposed by Barr & Hammuda (1971) for tan white microcrystalline limestones with marly intercalations and lenses of brown chert in the Wadi al Athrun (32°52'N, 22°17'E). The exposed thickness varies from 45–52 m (148–171 ft). In the Al Hilal-Al Atrun area the Paleocene is absent and the Formation is unconformably overlain by the Eocene Appollonia Limestone, and it conformably overlies the Upper Cretaceous Hilal Shale. The foraminifera were described by Barr (1968, 1972) and Barr & Hammuda (1971) who recognized the following foraminiferal zones, from top to bottom, *Abathomphalus mayaroensis, Globotruncana gansseri, G. tricarinata, G. elevata,* and *G. concavata concavata,* in addition to coccoliths, ostracods, and brachiopods. They suggested a lower Santonian-Upper Maastrichtian age. The Atrun Formation is the deep-water equivalent of the Wadi Ducchan Formation in the Jardas al Ahrar area (Röhlich, 1974). Convolute layers and slump structures, planktonic foraminifera, *Inoceramus* shells, and ichnofossils such as *Zoophycos* and *Chondrites* indicate deep-water, slope, syndepositional mass movements.

In the subsurface of the offshore of Cyrenaica the Atrun Formation (Maastrichtian) in the A1-NC120 well is composed of interbedded light grey to whitish chalky mudstones to wackestones and grey-brown marls with nodules of chert. It contains *Globotruncana stuarti, G. conica, G. contusa, G. gansseri, G. falsostuarti, Stensioeina exsculpta, Rugoglobigerina rugosa,* and *Bolivinoides draco,* and limestone clasts with *Orbitoides media* and *Sirtina* sp. It is unconformably overlain by the Late Paleocene Al Uwayliah Formation. The Atrun Formation is equivalent to the Kalash Formation in the subsurface of the Sirte Basin.

The Upper Cretaceous section overlying the Benia Formation was divided by Röhlich (1974) and Klen (1974) into the Al Majahir and Wadi Dukhan formations. The Al Majahir Formation was proposed by Röhlich (1974). The type section of the Al Majahir Formation is the eastern slope of the Valley Shatib Huluq al Jir (32°27'N, 21°41'E) about 8 km east-northeast of the old Majahir fortress (Röhlich, 1974). It rests unconformably on the Benia Formation. Its thickness varies from 70–200 m (230–656 ft). The

Formation is composed of light grey marls at the base, followed by marly limestones, thick-bedded and dolomitic in the upper part, and it contains abundant benthonic and planktonic foraminifera, molluscs, and ostracods. Barr (1968) reported *Globotruncana coronata* and *G. angusticarinata*, and Röhlich (1974) reported *Inoceramus balticus* and *Pycnodonta vesicularis* all of Campanian age. Tröger & Röhlich (1996) and Röhlich & Youshash (1996) studied the *Inoceramus* of the Al Majahir Formation and showed that *Inoceramus* is common in the Campanian. The Al Majahir Formation was divided by Duronio et al. (1991) in the A1-NC120 well offshore of Cyrenaica into three parts: the lower part consists of light brown packstones-grainstones grading upward into mudstones-wackestones and includes large pelecypods, locally common fragments of *Globorotalites michelinianus*, dasycladacean algae, and bryozoans, echinoderms, and *Inoceramus* prisms. The middle part is composed of chalky, locally recrystallized mudstones to packstones with *Globotruncana stuartiformis*, *Marginotruncana coronata*, *Dicarinella asymetrica*, *Globotruncana fornicata*, and *G. vetricosa*. The upper part consists of cream-whitish wackestones and packstones with dolomitic limestone intervals, and contains *Accordiella conica*, *Cuneolina pavonia parva*, *Valvulammina picardi*, and *Rhapydionina* sp. The fauna confirms a Campanian age for the Formation. The Al Majahir Formation is equivalent to the Sirte Shale in the subsurface of the Sirte Basin.

The name Wadi Ducchan Formation was introduced by Kleinsmeide & van den Berg (1968) and Pietersz (1968) after Wadi Dukhan about 10 km west of the village of Jardas al Ahrar. It consists of grey to brown dolomites and vuggy dolomitic limestones, massive, indistinctly bedded and mostly devoid of fossils. Pietersz (1968) estimated a thickness of about 610 m (2001 ft), but the exposed thickness varies between 40 m (131 ft) and 150 m (492 ft). The type locality is between Ghawt Sas and Wadi Dukhan. Röhlich (1974) assigned the Wadi Dukhan (Ducchan) Formation a Maastrichtian age based on its stratigraphic position. In the Al Uwayliah area (32°33′N, 20°59′E) the Formation is overlain unconformably by the Eocene Derna Formation. In the Shataatah area, the Wadi Ducchan Formation is overlain by the fossiliferous calcilutites of the Uwayliah Formation (Paleocene). In the Jardas Al Ahrar area the Wadi Ducchan Formation rests on limestones, shales and dolomites of the Al Majahir Formation (Campanian). The Wadi Ducchan Formation is correlatable with the Waha Formation in the subsurface of the Sirte Basin. The diagnostic Maastrichtian larger foraminiferal species *Omphalocyclus macroporus*, *Siderolites* cf. *calcitrapoides*, and *Orbitoides* cf. *media* have been reported for the first time by Tmalla (2007) from the Wadi Dukhan Formation in well B7-41 (Cyrenaica) and in well U2-6 (Northeast Libya Sirte Basin).

### 13.6.2   Upper Cretaceous of Northwest Libya

In Northwest Libya the Upper Cretaceous Sidi asSid, Garian, and Gasr Tigrinna formations of the Nafusah Group (Cenomanian-Santonian) are exposed in the Jabal Nafusah. In the Al Hamada Al Hamra Plateau south of the escarpment the Upper Cretaceous is represented by the Mizda and Zmam formations of the Al Hamada Al Hamra Group (Santonian-Maastrichtian). The thickness of surface exposures of the Upper Cretaceous in northWest Libya averages around 870 m (2854 ft) (Megerisi & Mamgain, 1980a, b). Burollet (1960) divided the section in the area near the village of Uazzen

(Wazun) (31°57′N, 10°39′E) in the western Jabal Nafusah near the Tunisian border into the Chicla Formation, Giado Dolomite, Uazzen Sandstone, and the Nefusah Group. The Giado Dolomite was considered by Busson (1967b) to be the basal part of the Ain Tobi Formation. He also argued that the Uezzan Sandstone changes facies into the Ain Tobi carbonates. El-Hinnawy & Cheshitev (1975), Novovic (1977), and Anotonovic (1977) included the Uazzen Sandstone, Giado Dolomite and Chicla Formation in the Ar Rajbah Member of the Kiklah Formation. Burollet (1960) proposed the name Nafusah Group to include most of the Upper Cretaceous rocks exposed along the northern escarpment of Jabal Nafusah such as the Garian Limestone, Jefren Marl and Ain Tobi Limestone, an d equated it with the Zebbag Group in Tunisia. Its type section is at Gharyan (32°39′N, 14°16′E). In the central and eastern Jifarah Plains all post-Triassic formations were eroded prior to the Miocene sedimentation, including parts of the Upper Triassic Bu Sceba Formation. Consequently, the Miocene immediately overlies Bu Sceba deposits. In the El Khums area the Turonian Garian Formation is unconformably overlain by the Miocene Al Khums Formation. El-Hinnawy & Cheshitev (1975) introduced the Sidi as-Sid Formation to include the Ain Tobi Limestone and Jefren Marl units of Christie (1955). It forms the base of the marine Upper Cretaceous succession exposed along the Jabal Nafusa escarpment and unconformably overlies the Chicla (Lower Cretaceous) and the Bu Sceba (Triassic) Formations, west and east of Wadi Ghan, respectively. It increases in thickness from 65 m (213 ft) in the east to 380 m (1247 ft) in the west. The sequence changes from predominantly marls in the west to predominantly limestones and dolomitic limestones in the east. The Formation was divided into the Ain Tobi Member and the Jefren Member. The Ain Tobi Member was originally proposed by Christie (1955) as Ain Tobi Limestone to describe a unit of light grey to white limestones, sandy at the base and containing narrow limestone bands rich in *Exogyra flabellata* or the rudist *Ichthyosarcolites* in the Gharyan area. The type section is located on the Tripoli-Gharyan Highway where the exposed thickness is 79 m (259 ft). The following fossils have been reported by Christie (1955): *Exogyra flabellata, E. columba, E. delettrei, Leda africana, Trigonia Beyrichi, T. ethra, Credita forgermoli, Roudairia* sp., *Cyprina (Venericadia) barroisi, Gervilleia* n. sp., *G.* cf. *bicostata, Avicula cenomaniensis, Corbula striatula, Ichthyosarcolites triangularis,* and *I. bicarinatus.* The Jefren Member (spelled Yifran Marl on the Geological Map of Libya, 1985) is characterized by a clay-marl-evaporite sequence. The name Jefren Marl was originaly proposed by Christie (1955) for an 80 m (262 ft) section of soft yellow marls, clays, silts and minor limestones with a 2 m (7 ft) band of gypsum. The unit is best exposed at El Mehla (30°01′N, 12°31′E) between Rumia and Yafrin (32°00′N, 12°30′E) which is considered to be the type section by Desio et al. (1963). The unit rests conformably on the Ain Tobi Limestone and is conformably overlain by the Garian Limestone. Christie (1955) reported the following gastropods from the Jefren Marls at Kaf Tobi: *Nerina gemmifera, Tylostoma* sp., and *Turritella* sp. The molluscan and foraminiferal assemblages suggest a Cenomanian age, probably Early Cenomanian.

The name Garian Limestone[5] or the Garian Formation (named Gharyan Formation on the Geological Map of Libya, 1977, and Nalut Formation, 1985) was proposed by

---

5 The replacement of the Garian Limestone by the Nalut Formation as suggested by El-Hinnawy & Cheshitev (1975) is not recommended. Although the Nalut Formation was named by Zaccagna (1911),

Christie (1955) for a light grey to cream, massive crystalline dolomitic limestone. It is equivalent to the Nalut Formation of Zaccagna (1919). However, the latter has never been used by subsequent workers until El-Hinnawy & Cheshitev (1975) proposed the resurrection of the name basing their argument on the rule of priority of the Code of Stratigraphic Nomenclature. It is recommended to use the name Garian Formation because it is well established in the geological literature (this is also in line with the recommendations of the Code of Stratigraphic Nomenclature). The type section is in Wadi Gharyan from a point just north of Qasr Tigrinnah towards Abu Ayad (32°07′N, 12°59′E). The thickness of the unit at its type locality is 55 m (180 ft), but varies from 40 m (131 ft) in the west to 200 m (656 ft) in the east. The unit is characterized by grey to cream, red near the top, crystalline dolomitic limestones and dolomites with common chert concretions in the upper part and large vugs filled with calcite or quartz crystals. It becomes more siliceous near the top. Fossils are rare, but Christie (1955) reported *Neitea showi*, *Lima iteriana*, and *Pecten* sp. Mann (1975a) reported from the Al Khums area the foraminifera *Praeglobotruncana* cf. *Stephani*, *Rotalipora* sp., *Hedbergella* sp., *Stensioeina* sp., as well as the radiolaria *Radiolarites* cf. *Zulfardi*, *Cadosina* sp., and *Globochaete* sp., and assigned it a Cenomanian-Turonian age. Burollet (1960, 1963) assigned the Garian Formation a Turonian age, but no reasons were given. Christie (1955), Desio et al. (1963), and El-Hinnawy & Cheshitev (1975) assigned it a Cenomanian age. Its age is probably Late Cenomanian.

The Gasr Tigrinna Formation (spelled Qasr Taghrannah on the Geological Map of Libya, 1977, and Qasr Tigrinnah on the Geological Map of Libya, 1985) is the highest and youngest of the pre-Quaternary sediments in the Gharyan area. The name was proposed by Christie (1955) for a succession of soft marls with red, pink, and yellow bands of limestones, and white, porous chalky limestones at the top. The measured section is 87 m (285 ft). The old Berber fort of Qasr Tigrinnah stands above the type section. The type section consists of 73 m (240 ft) at the base of buff to grey to greenish grey marls and shales, containing red and pink limestone bands in the upper 38 m (125 ft), followed by 15 m (49 ft) of white, yellow, and pink chalky limestones interbedded with marls and shales. To the east, in Wadi Ghan, the unit becomes mostly yellow and contains chert bands and nodules near the top. Christie (1955) reported the gastropods *Cerithium (Terebralia) sancti arromani*, *Nerinea (Ptymatis) requieniana*, *Turritella chaffati*, and *Turbo* sp., the lamellibranchs *Cardium (Trachicardium) productum*, *Sanwagesiav garianica*, and *Astarte* n. sp., and the echinoderm *Hemiaster* cf. *meslei* n. sp., and ascribed the Formation a Turonian-Cenomanian age. Burollet (1960) assigned it a Turonian age. In the Ghadames Basin the Formation contains anhydrite beds (Röhlich, 1979). Its thickness varies from 30–130 m (98–426 ft). Jordi & Lonfat (1963) included the Gasr Tigrinna as a member in the Mizda Formation. Further south the unit conformably overlies the Garian Limestone and underlies the Zmam Formation (Barr & Weegar, 1972). Novovic (1977) reported the gastropod *Cryptorhytis bleicheri* (Coniacian) from the Djeneien area, in addition to *Nerinea (Plesioptygmatis)* cf. *coutinhoi*, the algae *Marinella lugeoni*, *Halemida* and *Dasycladacea*, the foraminifera *Valvulammina picardi*, *Bulimina* sp., the ostracod *Ovocytheridea brevis*, *Acutostrea incurva*, and *Lopha destefanii*, the gastropod *Aporrhais* sp.,

___

the name Garian Limestone is preferred because it is a well established unit (see N. Am. Comm. Strat. Nom., 1983, Article 8c).

and the echinoid *Cyphosoma delaunayi*. Novovic (1977) and Zivanovic (1977) assigned it a Late Turonian-Coniacian. The fauna reported by Röhlich (1979) from the Gasr Tigrinna Formation on the western side of the Ghadames Basin are *Haplophragmoides glabra, H. eggeri, H. excarata, H. rugosa, H.* cf. *aequale, Ammobaculites* cf. *subcretaceus, Flabellammina* cf. *saratongensis, F. compressa, Gavellinella* ex. gr. *tumida,* and *Orbignya* cf. *ovata,* in addition to *Orbitoides* sp., *Clithocyridea* ex. gr. *senegali, Cibicidoides* ex. gr. *erickdalensis,* and *Lituola nautiloides.* Silicified wood is also present in the Formation. He gave the Formation a Late Turonian-Santonian age.

Burollet (1960) introduced the name El Hamra Group in the type section around Mizdah and to the southeast of Shuwayrit (30°00′N, 14°15′E) and divided it into three formations from base to top: the Mizda, Socna, and Gheriat formations. Jordie & Lonfat (1963) used the name Hamada Group for the same group. Banerjee (1980) recommended the use of the present English transcript of the plateau's name and named the group the Al Hammada al Hamra Group to include the Nefusa, Mizda, and Zmam formations (although he proposed to abrogate the name Nefusa).

The Mizda Formation (spelled Mizdah on the Geological Map of Libya, 1977) crops out south of Jabal Nafusah from Beni Walid and into southern Mizdah to Nalut in the west. Burollet (1960) divided the Mizda Formation (Turonian-Senonian) into an upper dolomite-shale member and a lower dolomite member.

Jordi & Lonfat (1963) divided the Mizda Formation into three members, the Tigrinna Member, Mazuza Limestone, and Thala Member. El-Hinnawy & Cheshitev (1975) reinstated the Gasr Tigrinna as a formation, and subdivided the Mizda Formation into the Mazuza and Thala members. The Formation was described by Cepek (1979), Salaj (1979), and Röhlich (1979). In the Ghadames Basin, the Mazuzah Member is composed of yellowish-white to pink crystalline limestones, occasionally dolomitic and silicified, often cross-bedded with intercalations of marly limestone near the base. Calcareous sandstones are rare. The unit is exposed only in a few places. The exposed thickness varies from 5–27 m (16–89 ft). Limestones contain abundant *Inoceramus* and some calcarenites are composed entirely of *Inoceramus* prisms (Röhlich, 1979). The Thala Member consists of alternations of chalky limestones, marls and green shales with abundant gypsum, especially in the lower part. The limestones are oolitic in places, often dolomitized and chertified. The top of the unit is marked in places by siliceous dolomitic limestone beds (Salaj, 1979). The fossils are limited to a few bivalves and gastropods, benthonic foraminifera, and algae. Silicified wood fragement are present in places. Planktonic foraminifera are rare and include *Globotruncana* cf. *fornicata, G.* gr. *tricarinata,* and *Hedbergella* sp., in addition to a few miliolids (Röhlich, 1979, Cepek, 1979). The unit varies in thickness from 15–45 m (49–148 ft). Most of the authors agree on a Campanian age for the unit, based on the presence of the bivalve *Lopha (Actinosteroan) morgani* and the foraminiferal assemblage, although Röhlich (1979) assigned it a Santonian-Campanian age based on the *Inoceramus* fauna. The Gheriat Formation[6] (Maastrichtian-Paleocene) is a light colored microcrystalline, dolomitic detrital limestone with cross-bedding and fragments of ammonites in the lower part. It overlies the Socna Formation and it is overlain by the "Chueref Lime-

---

6 Bezan et al. (1997) used the name Gheriat Group to include the Maastrichtian formations in the subsurface of the Sirte Basin.

stone" (Shurfa or Gelta Chalk of the "Uaddan" Formation) (Paleocene). The Gheriat Formation is equivalent to the Upper Tar Mar Member (Danian) and includes, where recognizable, the Socna Mollusc Beds of Jordi & Lonfat (1963). The Lower and Upper Tar members are equivalent to the obsolete name "Gebel Es-Soda series" of Desio (1935) in the Jabal AsSawda area.

Burollet (1960) applied the name Socna Formation to a unit of green shales and yellow shelly, detrital, fossiliferous limestones with echinoids, *Alectryonia*, *Exogyra*, and *Omphalocyclus* which he divided it into an upper shaly member and a lower detrital limestone member, and assigned it a Maastirchtian age. Banerjee (1980) recommended the deletion of the term Socna because of the presence of another homonym term, the "Socna Mollusc Bed" of Jordi & Lonfat (1963). However, the rule of priority of nomenclature was not followed in this case. The Lower Tar Marl Member of the Zmam Formation (Jordi & Lonfat, 1963) is equivalent to the Socna Formation of Burollet (1960). Nevertheless, the name Lower Tar Marl became well entrenched in the geologic literature of Libya.

The Zmam Formation was originally described by Jordi & Lonfat (1963) as a formation composed of shales, marls and limestones exposed on the southwest border of the Hun (Hon) Graben. The type section is located on an isolated hill near the entrance of Wadi Tar Al Kabir (29°22′N, 15°42′E). The Formation is Upper Cretaceous to Paleocene in age. Jordi & Lonfat (1963) subdivided it into four units (from base to top): the Lower Tar Member, Socna Mollusc Bed, Upper Tar Member, and Had Limestone Member. The Lower Tar Member corresponds to the Socna Formation as defined by Burollet (1960). The Cretaceous-Tertiary boundary falls within the Zmam Formation. In the Ghadames Basin the unit conformably overlies the Upper Cretaceous Mizda Formation and it is overlain by the Hagfa Formation (Barr & Weegar, 1972). It is equivalent to the Sirte Shale, Waha, and Kalash formations in the subsurface of the Sirte Basin.

The name Lower Tar Member was introduced by Jordi & Lonfat (1963) for a unit composed of green shales, marls and limestones which crops out from the Hun graben to Wadi Suf Ajjin to the Tunisian border. The Lower Tar Marl is composed of light to dark greyish-green gypsiferous shales at the base, brownish light green shales intercalated with yellowish-grey fossiliferous marls near the middle of the section, and greyish-yellow limestones and greyish yellow fossiliferous marls with thin beds of greyish-green shales at the top. Jordi & Lonfat (1963) assigned the unit a Late Campanian-Maastrichtian age based on foraminiferal (both planktonic and benthonic) assemblages. The base of this unit is not exposed at the Wadi Tar. The exposed thickness in northWest Libya varies from 40–190 m (131–623 ft) (Röhlich, 1979, Cepek, 1979). Barr & Weegar (1972) estimated 79 m (260 ft) for the thickness of the exposed section in the type section. Jordi & Lonfat (1963) estimated a thickness of 230–241 m (755–790 ft) for the unit in nearby wells, and Eliagoubi & Powell (1980) postulated a subsurface thickness of about 140 m (459 ft). The Lower Tar Marl is rich in both microfossils and macrofossils (Barr & Weegar, 1972, Cepek, 1979, Röhlich, 1979) such as pelecypods, gastropods, echinoids, and small button corals. The upper beds contain large specimens of the benthonic foraminifera *Omphalocyclus macroporus* and *Siderolites calcitrapoides*. These species are index fossils of the Maastrichtian. The lower beds contain abundant planktonic foraminifera. Barr (in Barr & Weegar, 1972, p.172) described the following planktonic foraminifera from the type section: *Globotruncana gansseri, G. stuarti, G. arca, G. contusa, G. mariei, Rugoglobigerina rugosa,* and *Heterohelix* sp. Salaj (1979) reported *Omphalocyclus macroporus, Inoceramus goldfussianus, Orbitoides media, Lopha dichotoma,* and *Pycnodonte vesicularis,* in addition to globotruncanids from the Ghadames Basin. In the Ghadames area at the conjuction of the Libyan-Tunisian-Algerian borders the unit is made up predominantly of calcarenites and dolomitic calcarenites rich in *Amphidonte overwegi* (Röhlich, 1979). The microfossils are mostly composed of miliolids and arenaceous foraminifera. Silicified wood fragments are occasionally found (Röhlich, 1979). Oyster banks made up mostly of *Rostelum serratum* occur in the Kaf Zarzum area (Cepek, 1979) and *Agerostrea ungulata* near Ghadames (Röhlich, 1979). *Inoceramus ianjonaensis* was reported from the Lower Tar Member in that region by Röhlich (1979). The Socna Mollusc Bed is a highly fossiliferous marly limestone unit which locally separates the Lower Tar Marl (Late Campanian-Maastrichtian) from the Upper Tar Marl (Danian). It is well developed in the Socna (Sawkna) area (29°05'N, 15°46'E) and has a thickness of 12 m (40 ft) (Gohrbandt, 1966b) and wedges out northward. Lefeld & Uberna (1991) reported the ammonites *Nostoceras magdadiae* sp. nov. and *Baculites* from the Lower Tar Member, associated with the foraminifera *Globotruncana gansseri, G. arca, G. fornicata, Bolivinoides draco,* and *B. miliaris.* They placed the Lower Tar Member in the Lower Massatrichtian, the Upper Tar Marl and the Scona Mollusc Bed in the Middle-Upper Maastrichtian, and the Had Member in the Paleocene. Nevertheless, the planktonic foraminiferal assemblage suggests an Upper Maastrichtian age for this unit. Luger & Gröschke (1989), and Hamama & Kassab (1990) reported a number of species of *Nostoceras* and *Baculites,* among others, from the Rakhiyat (Duwi) Formation in the Eastern Desert of Egypt which they assigned to the Campanian. None of these authors, however, consulted the work of the others and major differences exist between their interpretations.

The Lower Tar Marl Member changes facies south of Jabal Assawda into the sandstones of the Bin Affin Member (Woller, 1978, Seidl & Röhlich, 1984). The member was first recognized by Fürst (1964) which he called *Ben Afen Schichten* and described it as a 80 m (262 ft) to 100 m (328 ft) thick complex of fine- grained to conglomeratic sandstones with carbonate or clay cement, and rare intercalations of sandy carbonate rocks. It contains the benthonic foraminifera *Omphalocyclus* and *Siderolites,* in addition to *Lacazpsis termieri, Pseudorbitolina* sp., and pelecypod, gastropod, bryozoan, and echinoderm fragments, *Glyptoactis (Baluchicardia) ameliae* and *Cardium (Acantthocardium) pullatum.* In the Washikah map-sheet area it has a maximum thickness of 30 m (100 ft) and overlies directly the Cambrian Hassaouna Formation. It is overlain by a sandy dolomitic limestone bed which represents the basal part of the Upper Tar Marl Member (Woller, 1978). The Bin Afin interfingers with Lower Tar Marl southwest of the Jabal Assawda volcanics. It was assigned a Maastrichtian age by Woller (1978).

Nairn & Salaj (1991) proposed the Al Gharbiyah Formation for the sequence designated as the Lower Tar Member of the Zmam Formation by Cepek (1979). They divided the Formation into four members, from base to top: the Bir Bu al Ghurab Member, Lawdh Allaq Member, Bir az Zamilah Member, and the Lower Tar Member. They restricted the Zmam Formation to the Paleocene Upper Tar and Had members. The Al Gharbiyah Formation consists of a sequence of lagoonal to shallow-marine open-shelf limestones and marls (Salaj & Nairn, 1987). The age of the Formation ranges from Late Campanian to the end of Maastrichtian. The Formation includes three cycles of sedimentation (Salaj & Nairn, 1987, Nairn & Salaj, 1991) separated by phosphatic horizons. Only the youngest cycle correlates with the Lower Tar Member in its type section. There is also a conglomerate at the base of the third cycle (the Lower Tar Marl). According to this interpretation the Al Gharbiyah Formation is equivalent in part to the Kalash Formation, Waha Formation, and the upper part of the Sirte Shale in the subsurface of the Sirte Basin, and their redefined Zmam Formation is equivalent to the Paleocene Hagfa Shale and Lower Sabil Carbonates. It is recommended to retain the Zmam Group as originally defined by Jordi & Lonfat (1963) and to restrict the Al Gharbiyah Formation to the Bir Bu al Ghurab Member, Lawdh Allaq Member, and Bir az-Zamilah members only.

Hammuda et al. (1985) introduced and extensively described a large number of new units in a study of the geology of the offshore area of Northwest Libya. They divided the Upper Cretaceous section in ascending order, into: the Alalgah, Jamil, Bu Isa, Makhbaz, and Al Jurf formations. The Al Jurf Formation straddles the Cretaceous-Tertiary boundary. The type section of the Alalgah Formation is in the NOC-Aquitaine K1-137 offshore well at a drill depth of 9578–10,164 ft. It is made up of brown dolomites with black shales, grey argillaceous dolomites, and massive and nodular anhydrites in the lower part, and black to greyish brown dolomites interbedded with fossiliferous black shales in the upper part. It is equivalent to the Sidi Assid Formation in Jabal Nafusah and the lower member of the Zebbag Formation in Tunisia. It was assigned a Cenomanian age by these authors. The Jamil Formation in the type section in the NOC-Aquitaine K1-137 well at a drill depth of 7908–9150 ft is composed of grey to greyish brown, calcareous shales and a few interbeds of limestones. It reaches its maximum thickness of 1638 ft (499 m) in the Aquitaine L1-NC41 well. The unit is underlain by the Makhbaz Formation and overlain by the Bu Isa Formation. The fossils reported are of the *Globotruncana concavata* assemblage Zone of Santonian age. The overlying Bu Isa Formation is composed of cream to light brown, argillaceous and glauconitic micrites to biomicrites with the foraminifera *Globotruncana tricarinata*, *G. fornicata*, *G. stuartiformis*, *G. lapparenti*, and *G. elevata* indicating a Campanian age. El-Waer (1992) reported the ostracod genera *Krithe*, *Bairdia*, *Bythocypris*, *Pontocyprella*, and *Paracypris* from the Bu Isa Formation. The type section of the Bu Isa Formation is in the NOC-Aquitaine H1-137 well at a drill depth of 9150–10102 ft. The Formation is overlain by the Al Jurf Formation or the Farwah Group. It is equivalent to the Sirte Shale in the Sirte Basin, the Thala Member of the Mizda Formation in northWest Libya, and the Abiod Formation of Tunisia. The Makhbaz Formation was introduced for a unit in the type section in the NOC-Aquitaine I1-137 well at a drill depth of 8326–8928 ft. The Formation is composed of greyish-brown, dolomitic limestones and dark grey calcareous shales at the base, and white and light grey to tan, argillaceous and glauconitic micrites and biomicrites in the upper part. Hammuda et al. (1985) assigned it a Turonian-Coniacian age. It is equivalent to the Garian and Gasr Tigrinna formations in the Jabal Nafusah and the Bouleb Member of the Kef Formation in Tunisia. The Al Jurf Formation was introduced by Hammuda et al. (1985) in the offshore of northWest Libya for a unit of black shales with limestone interbeds, overlying the Bu Isa Formation and underlying the Lower Eocene Farwah Group. The Formation derives its name from the Arabic term "Al Jurf" which means continental shelf. The lithology of the Al Jurf Formation is very variable, but mainly consists of black shales and chalky limestones. Hammuda et al. (1985) designated the interval at a drill depth of 9208–10,565 ft in the NOC-AGIP C1-NC41 well as the type section. The thickness of the unit varies from 13,576 ft (4138 m) in the type well to 52 ft (16 m) southward. They assigned it a Maastrichtian-Paleocene age based on the presence of the foraminifera *Globotruncana stuarti, G. gansseri, G. contusa, G. gagnebini, G. conica, G. arca, G. falsostuarti, G. calciformis, Heterohelix consulata, H. pseudotessera, Bolivina incrassata, Bolivinoides draco,* and *Bolivinitella eleyi.* El-Waer (1992) recorded the ostracod genera *Kirthe, Bairdia, Paleocosta, Reticulina, Cristaleberis,* and *Buntonia* from the Al Jurf Formation. It should be noted, however, that no Paleocene fossils were reported. The Al Jurf Formation is equivalent to the Lower Tar Marl of the Zmam Formation in northWest Libya and in part to the Kalash and Sirte formations in the Sirte Basin, and the El Haria Formation in Tunisia (Fournié, 1978).

### 13.6.3  Upper Cretaceous of the Sirte Basin

Barr & Weegar (1972) reviewed the stratigraphic nomenclature of the Sirte Basin and introduced forty (40) new names of groups, formations, and members for the Upper Cretaceous and Tertiary. The Upper Cretaceous units include the Maragh, Bahi, Lidam formations, the Rakb Group which includes the Argub, Etel, Rachmat, Sirte Shale, Tagrifet Limestone formations, the Samah Dolomite, Ras Lanuf, Lower Satal, Waha, Kalash Limestone formations (Fig. 13.2). The Maastrichtian formations were grouped into the Gheriat Group by Bezan et al. (1993) and include the Lower Satal, Kalash, Waha, and Ras Lanuf formations which they considered to be co-eval. It should be noted that the name Gheriat Formation was used previously by Burollet (1960) for a Maastrichtian-Paleocene unit in northWest Libya.

The Maragh Formation was introduced by Barr & Weegar (1972) for a predominantly clastic unit in the subsurface of the Sirte Basin. The type section is in the Mobil N1-12 well (29°33′32′N, 21°09′04″E) at a drill depth of 3077–3120 m (10,096–10,237 ft). It unconformably overlies various units such as the Nubian sandstones, Gargaf Group, or volcanic and granitic basement rocks, and is conformably overlain by the Rachmat Shale, Tagrifet Limestone, or Sirte Shale formations. The unit is composed of quartz sandstones, argillaceous arkosic conglomerates, and dolomites. Glauconite and hematite are very common. The sandstones are cemented with calcite and dolomite. The Formation ranges in thickness from zero on the highs to 152 m (500 ft) in low areas. It forms a time-transgressive unit and ranges in age from Cenomanian to Danian. Roberts (1970) described this Formation in the Amal Field. At the top there are generally vugular and crystalline sandy dolomites which grade downward into dolomitic sandstones, and glauconitic sandstones with basal quartzite cobble conglomerate derived from the underlaying Gargaf (Amal) sandstones. Heselden et al. (1996) divided the Maragh Formation in the Masarb Field and surrounding area into a lower transgressive sandstone member and an upper shale/evaporite member which includes algal boundstones, anhydrites, shales and sandstones. The present author believes that the Maragh Formation is the shallow-water equivalent of the Sirte Shale, Rachmat Shale, and Bahi Formation. Its stratigraphy needs revision.

Barr & Weegar (1972) proposed the name Bahi Formation for a subsurface Cretaceous unit in the northwestern Sirte Basin. The type section is in the A3-32 well

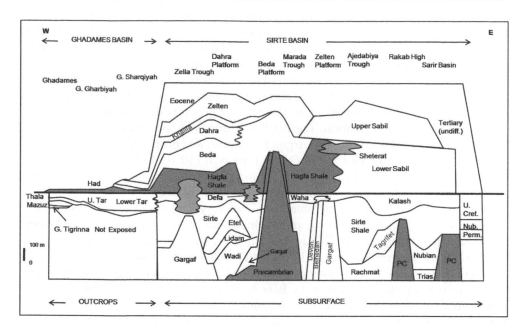

*Figure 13.2* Schematic Cretaceous-Tertiary cross-section across the Sirte Basin (modified from Tawadros, 2001).

(29°44′12″N, 17°35′22″E) at a drill depth of 6400–6763 ft in the Bahi Field where the Formation derives its name. It consists of interbedded sandstones, siltstones, conglomerates and shales. Medium to coarse-grained sandstones and pebbly sandstones are the most common lithologies in the Bahi Field area. Glauconite is common in the upper 10–20 ft (3–6 m). A 10–20 ft (3–6 m) thick conglomerate, composed of rounded quartzite in a sandstone matrix occurs in the basal part. The thickness of the Formation varies from a few feet to a maximum of 400 ft (122 m). It is overlain by the Cenomanian Lidam Formation or other Cretaceous formations, and it unconformably overlies the Paleozoic quartzites. According to these authors, the Formation may be partly equivalent to the Nubian sandstones.

The Lidam Formation was introduced by Barr & Weegar (1972) for a unit composed of light brown to light grey dolomite and pelletal limestone in the Mobil G1-57 well (28°34′18″N, 18°20′24″E) at a drill depth of 7012–7376 ft. The maximum thickness of the unit in the Sirte Basin is 600 ft (183 m). This unit represents the earliest marine deposits in the Basin. It is conformably overlain by the Etel Formation and it unconformably overlies various older sediments, including the Nubian sandstones, Gargaf Group, and Precambrian basement metamorphic rocks and granites. The unit is poorly fossiliferous, except for ostracods, miliolids, and the large benthonic foraminifera *Ovalveolina ovum* which indicates a Cenomanian age. Heselden et al. (1996) divided the Lidam Formation in the area west of the Masrab Field into four alternating lithostratigraphic units composed of limestone and dolomite-dominated lithologies. In the Masrab Field the Formation consists of interbedded massive dolomites and thin anhydrites.

The name Rakb Group was used informally by Williams (1972) in the northwestern Sirte Basin until it was introduced formally by Barr & Weegar (1972). The Group was named after the Rakb Field. The type section is located in the Oasis O2-59 well at a drill depth of 7702–9264 ft. Its thickness in the type section is 490 m (1562 ft). It is conformably overlain by the Satal Formation or the Kalash Limestone. Because of its transgressive nature it may rest conformably on the Lidam Formation or unconformably on older sediments. The Group includes, from bottom to top: the Argub carbonate, Rachmat Formation, and Sirte Shale. In the southern part of the Basin the Argub is replaced by the Etel Formation. To the east the group is divided into the Maragh Formation, Rachmat Formation, Tagrifet Limestone, and Sirte Shale.

The name Argub Carbonate or Argub Formation was proposed by Barr & Weegar (1972) for a subsurface unit in the D1-32 type well (30°01′14″N, 17°33′14″E) at a drill depth of 6672–6885 ft. In the type section the unit consists mainly of hard, tan, finely crystalline, slightly glauconitic dolomites with subordinate interbeds of white to grey, argillaceous, glauconitic limestones and very rare thin sandy stringers. The dolomites are occasionally vugular, and the limestones increase southward and to the northwest in the Bahi Field. Th Argub Formation overlies the Cenomanian Lidam Formation and is overlain by the Rachmat Formation (Santonian-Coniacian) with apparent conformity. It was assigned an Upper Cretaceous, probably Turonian age, based on its stratigraphic position by Barr & Weegar (1972). It is probably the lateral equivalent of the Etel Formation. Wennekers et al. (1996) designated the Argub Formation a Santonian-Coniacian age without giving the reasons for that dating.

The Etel Formation was introduced by Barr & Weegar (1972) for a unit composed predominantly of grey to brown dolomites, argillaceous silty limestones, abundant anhydrites, especially in the upper part, and grey to green shales with rare siltstones and fine-grained sandstones. The type section is in the Oasis O2-59 well (28°12′15″N, 19°295′43″E) at depth of 8730–9264 ft. Its thickness in the type section is 161 m (529 ft). It is absent on regionally high areas and thickens to over 396 m (1300 ft) in troughs. Barbieri (1996) reported a thickness more than 587 m (1926 ft) in the Hameimat Trough. It was assigned a Turonian age based on its stratigraphic position. The basal part of the Rachmat Formation contains evaporites similar to those in the Etel Formation. This led Wennekers et al. (1996) to erroneously date the Etel Formation Coniacian-Santonian. The Etel Formation is equivalent to the Argub Formation (Barr & Weegar, 1972). However, the latter is composed mostly of glauconitic dolomites.

The name Rachmat Formation was introduced by Barr & Weegar (1972) for a shale section that underlies the Sirte Shale. The type section is in the Oasis O2-59 well at a drill depth of 8272–8730 ft. The unit is composed of dark grey, glauconitic shales with light to medium grey limestones in the lower part. The limestones are argillaceous and fossiliferous and include biomicrites and biosparites with abundant rudists, *Inoceramus*, gastropods, crinoids, bryozoans, red algae, benthonic foraminifera, and ostracods. Dolomites are common in the basal part. In the Augila Field the Rachmat Shale ranges in thickness from zero to 2000 ft (610 m) and is composed of fine-grained glauconitic sandstones, glauconitic shales, very fine-grained sandstones, and silty shales (Williams, 1972). The Formation is overlain either conformably or unconformably by the Sirte Shale, and overlies various older sediments in different parts of the Sirte Basin. Where the Tagrifet Limestone overlies the Rachmat Shale the contact between the two units is gradational. The Rachmat Formation has a uniform thickness of more than 367 m (1204 ft) in the Hameimat Basin, but it is absent on regional highs (Barbieri, 1996). It was assigned a Coniacian-Santonian age by Barbieri (1996) who reported the following fossils: the agglutinated foraminifera *Haplophragmoides, Ammonomarginulina, Trochammina, Ammondiscus* and Glomospira, the ostracods *Cytherella, Ovocytheridea, Protobuntonia* cf. *numidica, Veenia, Acanthocythereis, Brachycythere,* and cf. *Linburgina.* The top of the Formation contains *Rotalia cayeuxi.* In the northern part of the Sirte Basin it contains the planktonic foraminifera *Globotruncana concavata, G. coronata,* and *G. angusticarinata,* indicating a Santonian to Coniacian age. However, Wennekers et al. (1996) extended its age to the Campanian, but their unit may include also the Sirte Shale.

The Sirte Shale was introduced by Barr & Weegar (1972). The type section is in the Oasis O2-59 well at a drill depth of 7702–8272 ft. The unit consists mainly of dark grey to dark brown shales, carbonaceous and calcareous, and grades into limestones, especially in the lower part. Sand, glauconite, and phosphatic grains are present in places. It is conformably overlain by the Satal Formation or Kalash Formation, and it overlies various older units depending on its location within the Basin. The unit is partly equivalent to the Thala Member (Campanian) of the Mizda Formation in northWest Libya and to the Atrun and Al Majahir formations in Northeast Libya Libya. Planktonic foraminifera are common throughout the Formation in the northern part of the Basin. In the southern part of the Basin planktonic foraminifera are common in the upper part of the Formation, and benthonic foraminifera are abundant in the lower part. The Formation contains *Globotruncana fornicata, G. arca, G. tricarinata, G. linneiana,* and *G. obliqua.* It was assigned a Santonian-Lower Maastrichtian age by Barr & Weegar (1972), Lower Maastrichtian by Barbieri (1996), and Campanian by Butt (1986). The latter identified the planktonic foraminifera *Globotruncana stuartiformis* and *G. elevata,* and the benthonic foraminifera *Siphogeneroides cretacea* and *Bulimina prolixa.* Jones (1996) assigned an Upper Maastrichtian age for an interval which he designated as Sirte Shale in the P4-6 well on the Zelten Platform based on the presence of planktonic foraminifera of the *Gansserina gansseri* Zone such as *Globotruncanita conica, Gansserina gansseri, Globotruncanita stuarti,* and *G. stuartiformis.* The upper part of this interval is made up of dark grey to black calcareous shales and very argillaceous lime wackestones. Although this lithology is consistent with that of the Sirte Shale, it may represent a facies variation of the Waha or Kalash formations. Sand bodies developed in the Sirte Shale in the central Sirte Basin such as in Concession 71 (Hammuda, 1980). These sands are very fine to fine-grained with scattered coarse grains. The matrix is made up of micrite with calcite, pyrite, glauconite, and bryozoan and pelecypod fragments. The sands grade into shales and are interbedded with sandy wackestones with bryozoan and pelecypod fragments. These facies are overlain and underlain by dark grey to grey, calcareous and pyritic shales. In the YYY1-6 well on the western edge of the Zelten Platform the Waha Formation overlies a basal 550 m (1804 ft) section of calcareous marine siliciclastics with thin limestones. The age of this thick succession is ill defined. The presence of radiolitid rudists in thin limestone beds within the interval indicates an age no older than Barremian at that level (Jones, 1996). This section is probably equivalent to the Sirte Shale and, therefore, of Campanian-Maastrichtian age.

On the southeastern margin of the Sirte Basin such as in the Augila and Amal fields the name Sirte Shale is restricted to the upper part of the unit only. The lower part grades into limestones of the Tagrifet Limestone (Barr & Weegar, 1972). The Tagrifet

Limestone is equivalent to the Rakb Carbonates of Williams (1972) in the Augila Field. The type section of this unit is the Oasis N1-59 well (29°08′11″N, 21°33′45″E) at a drill depth of 8390–8690 ft. The limestones contain globotruncanids, heterohelicids, ostracods, and *Inoceramus*, rudistid, pelecypod, gastropod, bryozoan, and algal fragments. Quartz sand is common at the base. In the Augila Field the basal sandstone is 3–10 m (10–33 ft) thick, medium-grained to granular, and it contains granitic fragments, rudistids, crinoids, and bryozoans, and grades upwards into sandy wackestones. Based on its age, stratigraphic position and lithology, the present author believes that the Tagrifet Limestone is equivalent to the lower part of the Sirte Shale unit in the Oasis O2-59 well and it should be treated as a member of the Sirte Shale.

Barr & Weegar (1972) proposed the name Samah Dolomite for an Upper Cretaceous unit in the subsurface of the Samah Field in the south-central Sirte Basin. The type section is located in the L3-59 well (28°10′45″N, 19°09′45″E) at a drill depth of 6382–6468 ft. The Formation has a thickness of 86 ft (26 m) in the type well and a maximum thickness of about 300 ft (91 m) in the Samah Field. It consists predominantly of a hard, light to dark brown or grey, mottled, limey dolomites, becoming dolomitic limestones in part. It contains pelecypod fragments, including rudistids, and rare foraminifera. The dolomites are often vugular with fractures filled with calcite or dolomite. Quartz grains are found in the basal few feet. According to these authors, the Samah Dolomite is restricted to the Samah Field. It overlies the Hofra quartzites, and it is overlain by the Kalash Formation. Barr & Weegar (1972) assigned it an Upper Cretaceous age. On their stratigraphic table the Samah Dolomite straddles the Campanian-Maastrichtian boundary. In that respect it may be a dolomitic equivalent of the Sirte Shale. The petrography of the Samah Dolomite has been examined by Barton (1979).

The Ras Lanuf Formation (Bezan et al., 1993) is made up of micritic limestones and it was deposited on a shallow platform adjacent to the Satal platform. It was assigned a Maastrichtian age and considered to be an equivalent of the Waha, Kalash, and Lower Satal formations by these authors.

The Satal Formation was proposed by Barr & Weegar (1972) for a unit in the subsurface of the Dahra-Hofra area of the northwestern Sirte Basin which forms the principal reservoir in the Bahi and Dahra 'B' oil fields. The type section is in the Oasis B2-32 well (29°28′09″N, 17°47′20″E) at a drill depth of 3782 ft to 4811 ft. The Formation is 314 m (1030 ft) in the type section. It was divided into a lower Maastrichtian member and an upper Danian member. The lower member consists of grey to white calcilutites grading into calcarenites near the top, a few thin dolomitic and thin shale intervals are present at the base. The lower member contains abundant pelagic foraminifera in the lower part, sparse pelagic and benthonic foraminifera in the middle part, and abundant benthonic foraminifera in the upper part, the latter include *Siderolites calcitrapoides*, *Orbitoides* sp., and *Omphalocyclus* sp. The upper member consists of light to white calcarenites and calcilutites with pellets, benthonic foraminifera with frequent miliolids and rotalids, molluscs, echinoids, algae, and corals. The sediments are often dolomitized. Anhydrite beds occur in higher shelf areas and shales are present along platform margins. The lithology of the lower part is strikingly similar to its equivalent the Waha Formation on the Zelten Platform. Both units are Maastrichtian in age and change facies into the Kalash Formation. The Satal Formation is also equivalent to the Lower Tar Member of the Zmam

Formation in outcrops. The Danian upper member grades into the Hagfa Shale and is equivalent to the Upper Tar Marl Member of the Zmam Formation in outcrops. Bezan et al. (1993) considered the Lower Satal Formation to be equivalent to the Kalash and Waha formations and to their Ras Lanuf Formation, all forming a part of their Gheriat Group. The sedimentology of the Satal Formation have been recently studied by Kardoes (1993).

The Waha Formation was proposed by Barr & Weegar (1972) for the tan to white skeletal calcarenite in the subsurface type section in the A29–59 well. It is characterized by the presence of the benthonic foraminifera *Omphalocyclus macroporus*, *Sidrolites calcitrapoides*, *Lenticulina muenstri*, and *Gaudryina faujasi*. These fossils indicate an Upper Maastrichtian age and they are characteristic of very shallow, high-energy shoals. On the Zelten Platform the Waha comprises a very variable succession of shelf limestones with localized siliciclastics occurring in the lower levels (Jones, 1996). The sands were derived from the erosion of pre-existing basement highs (Bezan et al., 1993). The Waha Limestone rests directly on upfaulted crystalline basement in the A29–59 well (Barr & Weegar, 1972) and the B2-59 well (Eliagoubi & Powell, 1980). The Waha Formation is restricted in distribution to the southern and central parts of the Sirte Basin, especially on platform areas. The thickness of the Formation varies from zero to 120 m (394 ft) as a result of deposition over an irregularly eroded surface and differential erosion at the top (Jones, 1996). The Waha Formation produces in the Ar Raqubah Field where it attains a thickness of 800 ft (244 m) and is made up of calcareous sandstones and skeletal sandy limestones with bioclasts, dasycladacian algae, and foraminifera (Broughton, 1996).

The Kalash Limestone was introduced by Barr & Weegar (1972) for a limestone composed of grey to dirty white argillaceous micritic limestones with planktonic foraminifera and intercalated with thin beds of dark grey calcareous shales. The type section is in the Mobil E1-57 well (28°18′29′N, 18°22′59″E) at a drill depth of 6890' to 7200'. The unit is very widespread in the Sirte Basin and maintains a relatively consistent lithology regionally. Its thickness ranges from a few feet to over 500 ft (152 m). This variation stems from a diachronous upper surface and basal transgression across a differentially eroded surface (Jones, 1996). Planktonic and benthonic foraminifera are abundant (Eliagoubi & Powell, 1980, Jones, 1996). The unit was assigned a Maastrichtian age by Eliagoubi & Powell (1980) and Butt (1986) based on the presence of *Globotruncana stuarti*, *G. conica*, *G. gansseri*, *G. havanensis*, *Rugoglobigerina*, *Bolivina Draco* and *Heterohelix*. These authors placed the Maastrichtian-Tertiary boundary between the Kalash and Hagfa formations. On the Zelten Platform, for example in the Lahib and Jabal areas, the Kalash Formation contains planktonic foraminifera of late Maastrichtian (*Gansserina gansseri* or *Globotruncana contusa* Subzone) to Early Danian age (Jones, 1996). In intervening areas the Kalash is exclusively Danian. In the Nasser Field the Kalash overlies the eroded upper surface of the Waha Formation and contains Danian species such as *Globoconusa daubjergensis*, *Globigerina triloculinoides*, and *Morozovella pseudobulloides*. Tmalla (1992, 1996) also recognized that the Maastrichtian-Tertiary boundary falls within the Kalash Limestone. In the Sirte Basin the Kalash Formation is conformably overlain by the Paleocene Hagfa Shale or the Lower Sabil Carbonates and it conformably overlies the Sirte Shale, the Samah Dolomite, or the Waha Limestone, or unconformably overlies the Wadi Formation, the Hofra Formation (Gargaf), or Precambrian basement.

The contact between the Kalash carbonates and the Sirte Shale is intra-Maastrichtian and unconformable (Jones, 1996). In the Jebel Field the Kalash Formation directly overlies orthoquartzites of the Gargaf Group in places. The presence of glauconite and phosphate at the base of the Kalash indicates a basal unconformity. The upper contact with the Hagfa Shale is gradational through interbedding.

Barr & Weegar (1972) and Eliagoubi & Powell (1980) suggested that the Waha Limestone, Kalash Limestone, and Lower Tar Marl are equivalent. However, Jones (1996) argued that the Waha is Maastrichtian in age and separated from the Maastrichtian-Danian Kalash by an unconformity. The unconformity surface is characterized by widespread karsting phenomena. Nevertheless, the Maastrichtian Kalash Formation is equivalent to the Waha Formation. Jordi & Lonfat (1963) observed no physical discontinuity between rocks of Maastrichtian and Danian ages in outcrops and placed the boundary arbitrarily at the base of the Socna Mollusc Bed in the middle of the Tar Marl of the Zmam Formation. Goudarzi (1970) on the other hand placed the Maastrichtian-Danian boundary at the top of the Socna Mollusc Bed.

# Chapter 14

# Tertiary

Tertiary rocks occur extensively in Libya. They crop out in northWest Libya in the Hammada Al Hamara Plateau south of Jabal Nafusah. Only the Middle Miocene is present in the Jifarah Plains, but thick Tertiary sequences have been reported from the subsurface of offshore northWest Libya and Tunisia. They also cover most of onshore and offshore Northeast Libya Libya, the Al Khalij Province (Syrtica), and the Tibesti area. They crop out in limited areas on the northern side of Jabal az Zalmah near the Libyan-Egyptian border and also near the Libya-Chad border. They are also widespread in the subsurface of the Sirte Basin. Tertiary volcanic rocks occur in Jabal Oweinat across the Libyan-Egyptian-Sudanese borders, in Jabal Nuqay, and in Jabal Bin Ghanimah on the eastern and western sides of the Tibesti Massif, respectively. They extend over large areas in northern Chad, the Al Haruj Al Aswad, Jabal As Sawda, and Jabal Hassawnah in central Libya, as well as in small areas in the Gharyan area.

Burollet (1960) reviewed the stratigraphy of Tertiary sediments in Libya based on published data and unpublished oil company reports. Barr & Weegar (1972) extended a number of the established Tertiary units into the subsurface of the Sirte Basin, revised other units, and established new ones. The Tertiary sediments were mapped extensively in Libya by geologists of the Industrial Research Center as a part of the 1:250,000 Geological Map of Libya project from 1974 to 1984. Some of the results were summarized briefly by Banerjee (1980) and Megerisi & Mamgain (1980a & b), the latter however also included information from unpublished oil company reports.

## 14.1  PALEOCENE

Paleocene rocks crop out over large areas in the Al Hammada Al Hamra Plateau in northWest Libya, south of Jabal Nafusah, and extend south to about latitude 29°N. They also occur in the subsurface of the offshore of northWest Libya and extend westward into Tunisia. They are also exposed along a narrow strip west of the Al Haruj Al Aswad and in the Hun Graben. They are widely distributed in the subsurface of the Sirte Basin. In eastern Libya they crop out in a limited area in the Al Jabal Al Akhdar and the offshore of Cyrenaica.

### 14.1.1   Paleocene of offshore Northwest Libya

The Paleocene in the offshore area of northWest Libya is represented by the Ehduz, Ajaylat, and Bouri formations, as well as the upper part of the Bilal Formation. All these formations were introduced and described by Hammuda et al. (1985). The Ehduz Formation is composed of limestones in the type section in the Aquitaine C1-137 well at a drill depth of 9529–9902 ft. The limestones are dolomitic, argillaceous, and anhydritic in part. The unit is underlain by the Al Jurf Formation and is overlain by the Bouri Formation. Hammuda et al. (1985) reported *Globorotalia trinidadensis* and *Globigerina triloculinoides* from the unit and assigned it a Danian age. The Ehduz Formation is probably a carbonate buildup equivalent to the middle part of the Al Jurf Formation. The type section of the Ajaylat Formation is in the NOC-Aquitaine G1-NC137 well at a drill depth of 7665–7920 ft where it is made up of non-fossiliferous variegated shales. It overlies the Ehduz Formation and underlies the Bouri Formation. It was assigned a Middle Paleocene age based on its stratigraphic position. It is probably equivalent to the Haria Formation in Tunisia. The Bouri Formation was established in the AGIP D2-NC41 type well at a drill depth of 8900–9308 ft. The Formation is made up of mudstones and wackestones with *Globorotalia mackannai, G. elongata, G. pseudomenardii, G. aequa, G. velascoensis,* and *Rotalina stellata,* indicating a Late Landenian.

### 14.1.2   Paleocene of Northwest Libya

Paleocene rocks crop out along the western and southwestern margin of the Sirte Basin and include the Upper Tar Marl and Had Limestone members of the Zmam Formation and the lower part of the Waddan Group which is represented by the Surfa Formation. The westernmost exposure of basal Paleocene occurs immediately west of latitude 12°N in the Mizdah map-sheet area (Novovic, 1977).

The Upper Tar Marl in the type section exposed in Wadi Tar is made up of marls and lime mudstones with thick shaly intercalations particularly in the middle part of the sequence (Jordi & Lonfat, 1963). The exposed thickness is 70 m (230 ft). Zivanovic (1977) showed that the thickness of the unit is variable in the Bani Walid area with thicknesses varying between 5–100 m (16–328 ft). In Northwest Libya the unit is 2–10 m (7–33 ft) in thickness (Cepek, 1979). Salaj (1979) reported brackish water species in the upper Tar Marl Member such as *Elphidiella, Protoelphidium,* and *Ammonia.* The lower part of the type section contains a very fossiliferous limy interval called the Socna Mollusc Bed (named Sukhnah Formation on the Geological Map of Libya, 1977, and Socna Beds on the Geological Map, 1985). Jordi & Lonfat (1963) and Conley (1971) observed no physical discontinuity of deposition between rocks of Maastrichtian and Danian ages and placed the boundary arbitrarily at the base of the "Socna Mollusc Bed".

The Had Limestone Member in its type section at Wadi Tar consists of three thick outstanding dolomite and dolomitic limestone layers separated by chalky marls (Jordi & Lonfat, 1963). The unit in the type section is 50 m (165 ft) thick. The Had Member is fossiliferous with a few gastropods, echinoids, and nautiloids, and abundant benthonic foraminifera such as miliolids and rotaliids, and corallinacean algae. The Had Member was assigned a Montian age by Cepek (1979) and Salaj (1979), and was considered as such on the Geological map of Libya (1985). The upper Tar Marl and Had members form a single cycle with dolomitic marls and green, gypsiferous shales at the base and limestones at the top (Cepek, 1979). This unit is correlatable with the Beda Formation of Barr & Weegar (1972) in the subsurface of the Sirte Basin.

The Jabal Waddan Group was introduced by Chiesa (1940) as the "Gebel Uaddan Series" for Eocene rocks exposed in the Giofra (Al Jufrah) area, in north-central Libya where he described the type section at Wadi Ammur (29°18′N–16°10′E). The base of the Group is not exposed in the type section, but is exposed at Wadi Zimam and Wadi Tar where it disconformably overlies the Upper Cretaceous El Hamra Group (Burollet, 1963, Jordi & Lonfat, 1963).

Burollet (1960) used the name Jabal Uaddan Formation to replace a few of the names current at that time such as the El Fogha Formation, Gara Gazalat Series, Wau El Kebir Series, and Bir Zidan Formation and divided it, from base to top, into Gelta Chalk, Operculinoides Limestone, Kheir Marl, Flosculina Limestone, Rouaga Chalk, Ben Isa Chalk, and the *Orbitolites* Limestone. The two latter units change facies southward into a more gypsiferous unit, the Gir Gypsum of Burollet (1960). Burollet (1963) assigned a Thanetian-Ypresian age to the Jabal Uaddan Formation[1] and divided it into the Chueref Limestone, Gelta Limestone, Operculinoides Limestone, Kheir Marl, *Flosculina (Alveolina)* Limestone, and Rouaga Chalk, overlain by the Wadi Tamet Formation (Lutetian-Priabonian). In the meantime, Jordi & Lonfat (1963) raised the Ueddan Formation to group status (the Waddan Group) in the exposed section on the western side of the Hun Graben near Wadi Tar and divided it into two formations: the Surfa Formation (Paleocene) and the Beshima Formation (Eocene). They divided the Surfa Formation into the Bu Ras Marlstone, Gelta Chalk, and *Operculina* Limestone members. Fürst (1964) divided the section (Surfa Formation equivalent) into Dor el Msid Formation (equivalent to Bu Ras Member) and *Operculinoides* Schichten (equivalent to the *Operculina* Limestone and Gelta Chalk members). Shakoor (1984) described the Surfa Formation in the Hun area and replaced the *Operculina* Limestone Member by the Ammur Member.

Desio (1935, 1943) applied the name El Fugha Series to a unit exposed in the eastern Fazzan on the slopes of Wadi El Fuqaha (27°50′N, 16°22′E) composed of green and yellow, argillaceous, sandy and gypsiferous limestones with foraminifera, echinoids, lamellibranchs, and nautiloids. Desio (1943) reported the fossils *Operculina hardiei, O. thouini, Echinolampas sp., Linthia desioi, L. navillei, Koilospatangus cleopatrae, Schizaster mokattamensis, Vulsella eymari, V. cf. crispata, V. cf. pseudocrispata, Gryphaea pharaonum* var. *aviculina,* and *Ostrea aviola.* Desio (1943, 1951) assigned it a Middle Eocene (Lutetian) age. Burollet (1960) considered the El Fugaha Formation to be a shaly equivalent of the basal part of the Wadi Tamit Formation (Middle-Upper Eocene) or the Gebel Uaddan Formation (Paleocene to Lower Eocene). Goudarzi (1970) considered the sediments of the El Fugha Formation to be nearshore or continental facies equivalent of the upper part of the Jabal Waddan Group includes the Beshima Formation and the Gir Gypsum.

Barsotti (1963) studied the Paleocene "Fugha series" ostracods of the El Fugaha Formation in the Al Fuqaha region and the A1-85 well in the Sirte Basin. He divided the sequene into a lower interval of marly limestones and bioclastic limestones, and

---

1 Burollet (1963b) simultaneously intorduced the name Ouadi Ouaddan (Wadi Waddan) for a Carboniferous unit in the Jabal Al Awaynat.

an upper interval of bioclastic limestones, marly limestones, occasionally dolomitic and gypsiferous, marls and intercalations of clays. He recorded an association of *Dahomeya alata* and *Leguminocythereis lockossaensis* at the top of the Paleocene interval in the A1-85 well similar to those of the coastal basins of Dahomey-Togo and the Ivory coast, as well as *Buntonia virgulata*, *B. tichittensis*, *B. pulvinata*, *Bairdoppilata magna*, *Krithe* cf. *perattica*, *Xestoleberis* sp. aff. *X. bumblei*, *Bradleya praecrassa*, *B. cultrata*, *B. teiskotensis*, *B. vesiculosa*, *Uroleberis glabella*, *U. teiskotensis*, *Ambocythere? tatteuliensis*, *Leguminocythereis teiskotensis*, *L. lokossaensis*, *L. lagaghiroboensis*, *Dahomeya alata*, *Actinocythereis modesta*, *A. teiskotensis*, *Isohabrocythere teiskotensis*, *Soudanella* cf. *laciniosa triangulata*, *S.* cf. *nebulosa*, and *Anticythereis bopaensis* from the Upper Paleocene basal part of the El Fogha Series in outcrops and the A-85 well with the exception of *L. lokosaensis* which was not found in outcrops. Salahi (1966) described the ostracod *species Buntonia bapaensis*, *B. vigulata*, *Costa dahomeyi*, *Acanthocythereis* n. sp.2, *Cythereis teiskotensis*, *Bradleya* aff. *B. cultrata* from the Paleocene interval in the C3-6 well in the Nasser Field. In addition to these ostracod species, Reyment & Reyment (1980) identified the following species from the Paleocene of the A5-32 well: *Cytherilla sylveterbradleyi*, *Paracypris nigeriensis*, and *Bairdia ilaroensis* which also occur in the Iullemedan Basin, *Mehesella biafrensis* (a synonym of *Leguminocythereis lokossaensis*), *Iorubaella ologuni* (also reported fromTogo Republic and the Ivory Coast), *Buntonia attitogonensis* (also occurs in the Togo Basin), *Protobuntonia ioruba*, and *Xestoleberis kekere* (Iullemeden Basin). Barsotti (1963) also reported the foraminifera *Operculina canalifera sidensis*, *Lochkartia haimei*, *Rotalia hensoni*, *R. trochidiformis*, *Cibicides* cf. *simplex*, and *Elphidium africanus*. The basal part of the El Fugaha Formation is probably equivalent to the Surfa Formation.

Bebout & Pendexter (1975) shows the Fogaha Group (Paleocene to basal Lower Eocene) to be composed of the Heira Shale and Ruaga Limestone; the latter includes the Zelten (Upper Paleocene) and Meghil (basal Eocene) members. Therefore, the name Fugha (or Fogaha) was used to cover all of the entire Paleocene and Eocene interval or the Waddan Group and the upper part of the Hamada Group which includes the Wadi Tamet Formation.

The Bu Ras Marlstone Member of the Surfa Formation was introduced and described by Jordi & Lonfat (1963). In the type section at Wadi Bu Ras (29°40′N, 16°28′E) the unit is 35 m (115 ft) thick and consists of gypsiferous marls, chalky limestones, and brecciated limestones. Its thickness is 10–15 m (3–49 ft) in the Al Qaryat Al Sharqiya area (Cepek, 1979), 15–24 m (49–79 ft) in the Al Qaryat Al Gharbiya area (Salaj, 1979), and 5 m (16 ft) in the Bani Walid area (Zivanovic, 1977). Strong recrystallization due to dolomitization is common. The unit is poorly fossiliferous with only bivalves (mainly oysters) present. It was assigned an Upper Paleocene (Thanetian) age by Cepek (1979) and Salaj (1979). This Formation is equivalent to the Dahra and Khalifa formations in the subsurface of the Sirte Basin (Barr & Weegar, 1972).

The Gelta Member was introduced by Burollet (1960), named after the Al Galta spring (29°48′N, 16°00′E). Jordi & Lonfat (1963) described the type section on the western rim of the Hun Graben where the unit is 50 m (164 ft) thick, and made of chalky marls, limestones and dolomites with occasional chert. He included it in his Gebel Uaddan Formation. It is the youngest unit exposed in the Ghadames Basin area (Salaj, 1979). In the Al Qaryat Al Sharqiyah area the unit is composed predominantly of intraformational limestone breccia. This brecciation was probably the result of karstification (Cepek, 1979). The limestones are peloidal, algal, oolitic, and contain gastropods and bivalves. The limestones are gypsiferous, recrystallized, and dolomitized. The Member was assigned a Thanetian age by Cepek (1979) and Salaj (1979).

Shakoor (1984) described the Surfa Formation in the Hun area and proposed the name Ammur Member to replace the *Operculinoides* Limestone of Burollet (1960, 1963) and Fürst (1964) and the *Operculina* Limestone Member of the Surfa Formation of Jordi & Lonfat (1963). The Ammur Member of the Surfa Formation in the type section is composed of 12 m (39 ft) of well-bedded limestonse, dolomites, marls, and marly chalks. In Al Qaryat Al Sharqiyah the unit does not contain *Operculina*, and it is made up of alternating silicified dolomites, limestones, and chalky limestones, composed mostly of peloidal, bioclastic grainstones, algal and oolitic in part with gastropods and miliolids. The Ammur Member was assigned a Thanetian age by Cepek (1979).

The Gelta and Ammur members are equivalent to the Upper Sabil Formation of Barr & Weegar (1972) in the subsurface of the Sirte Basin.

## 14.1.3 Paleocene of the Sirte Basin

Paleocene rocks are extensively developed in the subsurface of the Sirte Basin. The average thickness of the Paleocene sequence on platforms in the Sirte Basin is up to 2000 ft (610 m) and consists mainly of carbonates. It increases abruptly to more than 4000 ft (1219 m) of mainly clastic sediments in the structural low areas (Berggren, 1974). The Paleocene sequence becomes mostly dolomite southwards and predominantly shale in the northernmost part of the Sirte Basin (Conley, 1971, Berggren, 1974). Sandstones are common near the shoreline of the Basin. The stratigraphy of the Paleocene sequence was worked out by Barr & Weegar (1972). Although there are a few problems to be resolved such as the Cretaceous-Tertiary and Paleocence-Eocene boundaries, this work still stands as the most important and authoritative for the Sirte Basin. The Paleocene is represented by the Kalash Limestone, Hagfa Shale, Beda Formation, Lower Sabil Formation, Khalifa Shale, the upper Satal Formation, Zelten Formation, Dahra Formation, Harash Formation, and Kheir Formation.

The Kalash Limestone may contain planktonic foraminifera of Early Danian age (Jones, 1996) near its top in some areas on the Zelten Platform such as the Lahib and Jabal areas. In other areas such as the Nasser Field, the Kalash is entirely Danian and contains *Globoconusa daubjergensis*, *Globigerina triloculinoides*, and *Morozovella pseudobulloides*. The upper contact of the Danian Kalash limestones with the Hagfa Shale is gradational.

The name Hagfa Shale was introduced by Barr & Weegar (1972) for a Paleocene unit in the subsurface of the Sirte Basin, named after Wadi al Hagfa (28°15′N, 19°12′E), about 15 miles southeast of the Tagrifet Oasis. The type section is the Y1-59 well (28°09′14″N, 18°36′15″E) at a drill depth of 5710–6733 ft. The Formation consists mainly of grey, grey brown, grey green and black shales with occasional limestone beds. The limestones are more common in the upper part and are grey, tan, and brown, very fine-grained, fossiliferous, and rarely glauconitic. It conformably overlies the Kalash Formation and it is overlain by the Middle Paleocene Beda Formation of Barr & Weegar (1972). It contains the planktonic foraminifera

*Globoconusa daubjergensis, Globorotalia compressa, Globigerina pseudobulloides,* and *G. triloculinoides* which indicate a Danian age. The unit is thick in troughs and thin over paleotopographic highs. In the eastern Sirte Basin it passes into carbonates of the Lower Sabil Formation. The Hagfa Shale in the Ar Raqubah Field has a thickness of 1513 ft (461 m) to 1780 ft (543 m) and thickens southward.

The Hagfa Formation (Danian) in the western Sirte Basin is conformably overlain by the Paleocene Beda Formation[2] which was proposed by Barr & Weegar (1974). The type section is located in the BBB1-59 well (28°04′45″N, 19°22′45″E) at a drill depth of 5525 ft to 5950 ft. The Beda Formation consists of interbedded limestones with subordinate dolomites and calcareous shales. The limestones are argillaceous calcilutites and skeletal and oolitic calcarenites with abundant dasycladacean algae, especially in the lower part. For this reason the Formation was called previously the Dasycladacean Limestones by oil companies geologists. Fenestral fabrics are common in the dolomites. The Formation has a thickness of 425 ft (130 m) in the type section, but its thickness varies from a few feet to over 800 ft (244 m). In the Defa Field, the Beda Formation overlies the Defa Limestone, and in the Dahra area, it overlies the Satal Formation and conformably underlies the Khalifa Formation or the Dahra Formation. Foraminifera, especially miliolids, indicate a Middle Paleocene (Montian) age. Barr & Weegar (1972) divided the Beda Formation into a lower Thalith Member and an upper Rabia Member. Differentiation of the Beda Formation into these two members, however, is restricted to an area near the Dahra and Hofra fields. The type section of the Thalith Member is in the A1-11 well (29°23′09″N, 17°59′06″E) at a drill depth of 3502 ft to 3620 ft and composed predominantly of light to dark, microcrystalline, fossiliferous, argillaceous limestones with rare pyrite and glauconite and minor amounts of light to dark grey, grey-green to black, calcareous, fossiliferous shales. The type section of the Rabia Member is also in the A1-11 well at a drill depth of 3288 ft to 3502 ft. It is composed predominantly of dark grey to dark green, calcareous, fossiliferous shales with minor very argillaceous and marly limestone interbeds. The Member is 214 ft (65 m) thick in the type well. Other workers divide the Beda Formation in the southwestern and west-central Sirte Basin into three members: the Thalith Member, the Lower Beda Member, and the Upper Beda Member. According to Sinha & Mriheel (1996), all three members change facies laterally into the Hagfa Shale. Bezan et al. (1996) on the other hand believed that the Thalith Member grades laterally into the Hagfa Shale, and the Upper Beda Member into the Khalifa Shale. In the central Sirte Basin isolated carbonate and reefal buildups were developed in the Upper Beda Formation along major faults which form the margins of the Zelten Platform (Bezan et al., 1996). Berggren (1974) assigned the Beda Formation an Upper Danian age (Planktonic foraminifera Zone P2) and equated it with the Paleocene Middle Limestone Member of Conley (1971), Scedida Dolomite of Gohrbandt (1966b), and the Had Limestone of Jordi & Lonfat (1963).

The name Defa Limestone was proposed by Barr & Weegar (1972) for a Lower Paleocene unit in the subsurface of the Sirte Basin. The Formation unconformably overlies, either the Upper Cretaceous Waha Formation, or thin Hagfa Shale, and is

---

2 Röhlich (1974) introduced the Al Bayda Formation for an Oligocene unit in the Jabal Al Akhdar area, which may cause confusion between the two units. A new nomenclature is proposed later in this book for the Oligocene.

conformably overlain by the Paleocene Beda Formation. It changes facies laterally into the Hagfa Shale. The type section of the Defa Formation is in the B4-59 well (28°05′45′N, 19°52′45″E) at a drill depth of 5621 ft to 6184 ft. It is composed of chalky calcilutites with burrows and fenestral fabrics, and highly fossiliferous, pelletal calcarenites rich in molluscs, algae, benthonic foraminifera, and corals. The unit is 563 ft (171.5 m) in the type section with an overall average thickness of 400 ft (122 m) in the south-central Sirte Basin.

The Dahra Formation was introduced by Barr & Weegar (1972) for a unit composed of white to light grey, chalky, calcarenites and calcilutites, and subordinate tan to brown microcrystalline dolomites, and thin interbeds of dark shales. Minor anhydrite is occasionally present in the upper part of the Formation. The Formation is restricted to the western part of the Sirte Basin. The lower part becomes more shaly southward. The type section of the Dahra Formation is in the F1-32 well (29°28′19′N, 17°53′26″E) at a drill depth of 3040 ft to 3350 ft. It conformably overlies the Beda Formation and it is overlain also conformably by the Khalifa Formation (Landenian). Benthonic foraminifera such as miliolids and textularids are common. It was assigned a Lower Landenian age based on unspecified planktonic species.

The Khalifa Formation was proposed by Barr & Weegar (1972) for a Paleocene unit underlying the Beda Formation and overlying by the Zelten Limestone. In the type section in the AA1-59 well (28°23′15′N, 19°17′48″E) at a drill depth of 4804 ft to 5333 ft, the Formation can be divided into two units, an upper argillaceous limestone uni and a lower shale unit. The upper unit is 196 ft (60 m) thick and consists of dark grey, very argillaceous calcilutites with some shale interbeds, the lower unit is 333 ft (101 m) thick and made up of dark grey to black shales with occasional thin calcareous stringers. The Formation was assigned an upper Landenian age based on an unspecified planktonic foraminiferal assemblage. It grades laterally into the carbonates of the Dahra Formation and the Lower Sabil Formation.

Barr & Weegar (1972) introduced the names Lower Sabil Carbonates and Upper Sabil Carbonates for two Paleocene carbonate units in the subsurface of the eastern Sirte Basin. The type section of the two formations is in the C1-12 well (28°52′07′N, 21°07′24″E) at a drill depth 8630–9958 ft for the lower unit and 7700–8485 ft for the upper unit. The two units in the type section are separated by the Sheterat Formation, but where the Sheterat shales are not present, they may form one continuous unit. The Lower Sabil Carbonates is composed of tan to brown, very finely crystalline, occasionally sucrosic dolomites and lesser amounts of white to tan, fossiliferous, calcilutites, and occasional chalky limestones. Anhydrite occurs throughout the section. The unit is confomably overlain by the Sheterat Formation and underlain by the Upper Cretaceous Kalash Formation. The unit is 1328 ft (405 m) thick in the type section and thickens westward. It was assigned a Danian-lower Landenian age, but no reasons were given. The Upper Sabil Carbonates consists of predominantly limestones with lesser amounts of dolomites and some chalks. The limestones are pink, white and grey to tan, fossiliferous calcilutites, the dolomites are light to medium brown, fine to medium crystalline and anhydritic, the chalks are white to pink in color. The dolomites and anhydrites make up a larger percentage eastward. The Upper Sabil Carbonates is conformably overlain by the Kheir Formation and passes laterally into carbonates of the Zelten Formation. It has a thickness of 785 ft (239 m) in the type section but thins westward in the trough areas. The Upper Sabil Formation has a wider distribution than the Lower Sabil Formation, and contains some biohermal and reef buildups such as the Intisar "D" oil field. The buildups are composed of algal-foraminiferal biomicrites and biosparites, biomicrites, and coral biolithites (Brady et al., 1980). The sedimentology of the Upper Sabil carbonates has been reviewed by Spring & Hansen (1998).

The Lower and Upper Sabil formations are separated by a thin unit composed of light grey to tan, fossiliferous calcilutites and some chalks, and interbeds of grey, green to brown, fossiliferous, calcareous, and pyritic shales, for which Barr & Weegar (1972) proposed the name Sheterat Formation, named after Wadi Sheterat. It was assigned a Middle Paleocene age based on its stratigraphic position. The type section is in the Mobil C1-12 well at a drill depth of 8485 ft to 8630 ft where it has a thickness of 145 ft (44 m).

The name Zelten Limestone was introduced by Barr & Weegar (1972) for a unit composed predominantly of limestones in the central and western Sirte Basin. In the type section in the AA1-59 well at a drill depth of 4470 ft to 4804 ft the unit is 333 ft (101 m) thick and consists of cream, tan and grey, argillaceous to shaly, chalky, fossiliferous calcilutites and some calcarenites with thin stringers of grey-green, pyritic shales. Occasionally the unit becomes biohermal in character, and in other areas it consists of very fossiliferous, arenaceous, glauconitic, vuggy calcilutites and calcarenites with subordinate amounts of finely crystalline dolomites. In some wells such as the B1-13 well (29°02′N, 19°38′E) the unit is made up entirely of anhydritic dolomites. It conformably overlies, either the Khalifa Formation or the Lower Sabil Formation, and is overlain by the Harash Formation. In the south-central part of the Sirte Basin the Harash and Zelten formations can not be differentiated and they were both included in the Zelten Group by Barr & Weegar (1972) and Banerjee (1980), a practice which is not supported by the North American Code of Stratigraphic Nomenclature.

The Zelten Formation forms the principal reservoir in the Zelten Field (Bebout & Pendexter, 1975, Harding, 1984, Hallet, 2002). Bebout & Pendexter (1975) in a classic study of the Zelten reservoir treated the Zelten as a member of the Ruaga Limestone of the Fogaha Group. The Zelten Member is overlain by the Meghil Member of the Ruaga Limestone and underlain by the Heira Shale. The latter is equivalent to the Hagfa and Kheir formations of Barr & Weegar (1972).

### 14.1.4 Paleocene of Northeast Libya

In the Al Jabal Al Akhdar area the Paleocene is represented by the Al Uwayliah Formation, the upper and lower parts of which occur in two different localities. The first outcrop is along the Al Marj-Darnah highway just east of the village of Al Uwayliah (32°33′N, 20°59′E). The section is made up of 20 m (66 ft) thick chalky limestones of Landenian age with *Globorotalia pseudomenardii* Zone fauna, *Globorotalia chapmanii*, *G. angulata*, and *Globigerina triloculinoides*. This locality was first described by Barr (1968) and designated as the type locality of the Al Uwayliah Formation by Röhlich (1974). However, this section represents only the upper part of the Formation. The *Morozovella velascoensis* Zone is missing in outcrops. The second locality is 70 km to the east in the vicinity of the abandoned fortress of Jardas al Jarrari (32°32′N, 21°47′E) and represents the lower part of the Formation (Barr & Berggren, 1980). The Uwayliah Formation in this locality is 20 m (66 ft) thick and made up of chalky limestones, white chalks, and greenish marls. It was described by Röhlich (1974) who reported *Globorotalia perclara*, *Globigerina ex. gr. spiralis*, *G. cf. daubjergensis*, *Gumbelitria cretacea*, *Anomalinoides burlingtensis*, *Anomalina (?) ekblomi*, and *Cibides* sp., indicating a Danian age. In the offshore of Cyrenaica the Paleocene is represented in the A1-NC120 well by grey-green and brownish shales and argillaceous mudstones-wackestones with *Morozovella* (*Globorotalia*) *velascoensis*, *M. aequa*, *Acarinina mckannai*, *Gavelinella beccariiformis* of the Uwayliah Formation (Duronio et al., 1991). The Early Paleocene and Late Maastrichtian are missing. The sequence is probably Middle-Late Paleocene. It is conformably overlain by the Early Eocene Apollonia Formation, and it rests unconfromably on the Maastrichtian Atrun Formation. The stratigraphic position of the Al Uwayliah Formation has been reviewed by Tmalla (2007) who argued that the type section of the Al Uwayliah Formation should be considered a composite stratotype. According to Tmalla's (2007) interpretation, the section east of Al Uwayliah village is the holostratotype (upper part of the formation) and the Jardas al Jarrari section is the parastratotype (lower part of the formation), the middle part of this formation, between the holostratotype and the parastratotype (late Danian to early Selandian), has never been observed.

## 14.2    EOCENE

Eocene sediments crop out over large areas in the eastern Ghadames Basin, western Sirte Basin, the Al Jabal Al Akhdar in northern Cyrenaica, and extend southwards to the Tibesti Mountains where they overlie directly Precambrian basement rocks. They also occur widely in the subsurface of the Sirte Basin, the offshore of northWest Libya and Tunisia, and the offshore of northeastern Cyrenaica. Correlations, stratigraphic relations, and both lateral and vertical facies changes of the Ypresian to Lutetian units are not well documented, not only between outcrops, but also between outcrops and the subsurface of the Sirte Basin.

### 14.2.1    Eocene of Northwest Libya & the Sirte Basin

Eocene sediments are well exposed in the Jabal Waddan area in the western Sirte Basin. The Jabal Waddan is an arcuate chain of mountains with heights up to 550 m (1805 ft) above sea level which borders the eastern side of the Hun Graben and it is delimited on its eastern side by northwest-trending faults. The Jabal Waddan is dissected by a number of very long and deep valleys which provide excellent exposures of the Eocene succession, most important of which are the Wadi Thamit and Wadi al Kabir.

The most important pioneering studies of the Eocene rocks in the Jabal Waddan area were carried out by Chiesa (1940) and Desio (1943). Chiesa (1940) described a large number of outcrops of the "Gebel Uaddan Series" and divided them into units which he correlated over the whole area. Although he considered the entire succession to be Middle Eocene, his subdivisions provided the groundwork for all subsequent workers such as Burollet (1960), Jordi & Lonfat (1963), Fürst (1964), and eventually the team of the Geological Map of Libya 1:250,000 (1977–1985). Chiesa (1940) stated that the succession of the Wadi Amur Series (in outcrops) at Wadi Ammur (29°18′N, 16°10′E) represents the most typical of the limestone-marl succession of

Jabal Waddan. He also demonstrated that this succession is overlain at Jabal Margub to the north by a white limestone horizon with *Orbitolites complanatus* which he correlated with a similar horizon at Bir Zidan (31°03′N, 15°35′E) and also with the limestones with *Cardita* and *Callianassa* at Wadi Sheghega and Wadi Jir al Kabir. He dated the whole succession, including the latter horizons, Middle Lutetian and considered the beds with *Operculina alpina* to be the lowest horizons in the Middle Eocene succession. Desio (1943) considered this species to be Lower Eocene. It is now known that this succession includes different units ranging in age from Paleocene to Middle Eocene.

Desio (1943) divided the Eocene succession exposed between Wadi Thamat (31°00′N, 16°05′E) and Abu Ngem (Bunjim) (30°55′N, 15°24′E) into the Lower Eocene Garet Fologhi and Gara Gazalat series, the Middle Eocene Uadi Tofel, Sgbel, Dor Dogol and Bir Zidan, Roraiet Seteré, Garet el-Debeibat, and Gara el-Merbaa, and Gebel Besciascim series (equivalent to the Gara Gazalat and Uadi Tofel series), and the Upper Eocene Gara del Forte, Wadi Tamet, Guerat el-Melah, and Zella series. Chiesa (1954) dated the Wadi Tamet, Garet el-Merbaa, and Garet el-Debeibat series Lutetian, based on the presence of *Conulites* (*Dictyoconoides*) *cooki*, and reaffirmed the Upper Eocene age of the Garet el-Melah and Zella series.

Burollet (1960) divided his Jabal Uaddan Formation, from base to top, into the Gelta Chalk, Operculinoides Limestone, Kheir Marl, *Flosculina* Limestone, Rouaga Chalk, Ben Isa Chalk, and *Orbitolites* Limestone. He also indicated that the last two units change facies southwards into a more gypsiferous unit which he called the Gir Gypsum. A few years later, Burollet (1963a) assigned a Thanetian-Ypresian age to the Jabal Uaddan Formation[3] and amended it to include the Chueref Limestone at the base. In the meantime, Jordi & Lonfat (1963) raised the Uaddan Formation to group status, the Waddan Group, and divided it into two formations, the Surfa Formation (Paleocene) and the Beshima Formation (Eocene). Their stratigraphic nomenclature was adopted with minor modifications, by the Geological Map of Libya 1:250,000 project (1977–1985) geologists.

## 14.2.2 Lower Eocene (Ypresian)

Jordi & Lonfat (1963) in their classical paper on the Zmam Formation showed a composite stratigraphic section of the Zmam and Surfa formations in the Wadi Tar area, on the western side of the Hun Graben, and a schematic cross-section of the two formations in the Hamamda (Ghadames) Basin. Their stratigraphic section and cross-section also show a Beshima Formation which included the Kheir Marl, Rouaga Chalk, and Gir Gypsum members. That was the first mention of the Beshima Formation in the literature. No type secttion or locality were mentioned, and no reference was made to the origin of the name or any previous work. Banerjee (1980) assumed that the name was derived from Dor al Bashashim (29°30′N, 15°30′E) in the Abu Ngim area, and that the type section was located in Wadi Zimam, on the western side of the Hun Graben where Chiesa (1940) and Desio (1943) described the Besciascim Series. In reality, the reference (type?)

---

3 Burollet (1963b) simultaneously introduced the name Uuadi Ouaddan (Wadi Waddan) Formation for a Carboniferous unit in the Jabal Al Awaynat (Gebel Aweinat) in Southeast Libya.

section which appears on Jordi & Lonfat's (1963) cross-section is located in the Jabal Waddan area, on the eastern side of the Hun Graben. Chiesa (1940) previously gave a detailed description of the Dor al Bashashim section (Gebel Besciscim Series). The succession is similar to that of Wadi Amur, but thinner, and includes rocks ranging in age from Paleocene to Middle Eocene.

Goudarzi (1970) separated the Gir Gypsum from the Beshima Formation, and divided the latter following Fürst (1964), into the Kheir Marl, *Flosculina* Limestone, and Rouaga members. Woller (1978) described the three members of the Beshima Formation in the Al Washkah map-sheet NH33-15 area, and substituted the name Wadi Zakim Member for the *Flosculina* (*Alveolina*) Limestone.

The Beshima Formation was assigned a topmost Landenian-Ypresian age by Jordi & Lonfat (1963), Hecht et al. (1963), Woller (1978), and Shakoor (1984), a lower Landenian-lower Ypresian age by Goudarzi (1970), and an Ypresian age by Berggren (1974). In the following discussion, the Kheir Member has been separated from the Beshima Formation, for reasons discussed below, and it is raised to formation status following Barr & Weegar (1972). The Beshima Formation is divided into the Wadi Zakim Member and Rouaga Member.

The Kheir Formation was originally introduced by Burollet (1960) as the Kheir Member of the Uaddan Formation. Chiesa (1940) originally described this unit in the type section at Wadi Ammur as a 6 m (20 ft) unit of yellow, brecciated gypsiferous marls. In the Al Qaryat Ash Sharqiyah map-sheet NH33–6 area, the unit has a thickness between 20 (66 ft) and 35 m (115 ft) (Cepek, 1979). Two facies are present in that area: the first consists of greenish yellow and ochrous marls, the second is composed of dolomitic marlstones, green shales, and gypsum. Further to the south in the Tamasah map-sheet NG33-7 area the Kheir Formation is the youngest unit present and forms the upper part of the Dur al Masid escarpment. It is composed of limestones, dolomitic limestones, and dolomites, alternating with marls and marly limestones with a maximum exposed thickness of 12 m (39 ft) (Korab, 1984). Northwards in the Bunjim map-sheet NH33-7 area the Kheir Formation is made up of pale greenish yellow gypsiferous marls and marly limestones with dolomitized cross-bedded calcarenites and calcilutites (Said, 1979). The base of the unit is not exposed, but more than 25 m (82 ft) of section crops out in that area. The fossils are rare and include *Rotalia* sp., *Cibicides* cf. *libyca*, *Spiroloculina* sp., *Quinqueloculina* sp., *Textularia* sp., as well as echinoderm, gastropod and pelecypod fragments, spicules, and fish teeth. The Kheir Member may conformably or unconformably overlie the Paleocene Surfa Formation and is conformably overlain by the Wadi Zakim Member (Woller, 1978).

The Kheir Formation has been recognized and correlated in the subsurface of the Sirte Basin by Barr & Weegar (1972) who proposed the Oasis E1-59 well (28°41′13″N, 21°24′11″E) at a drill depth interval of 5998 ft to 6272 ft as a subsurface reference section. However, the unit was reassigned its member status by the members of the Geological Map of Libya 1250,000 (1977–1985), and Megerisi & Mamgain (1980a, b). The lithology of the Kheir Formation in the subsurface of the Sirte Basin is predominantly grey shales with some marls and limestones. Its thickness in the reference well is 274 ft (83.5 m). It conformably overlies the Paleocene Upper Sabil carbonates and it is overlain by the Lower Eocene Gir Formation of Barr & Weegar (1972). The unit was assigned an Upper Paleocene age by Goudarzi (1970) and Ypresian (Zone P6) age by Fürst (1964) and Berggren (1974). However, the unit contains *Operculina* sp., *Globorotalia velascoensis* (in the lower part), and *G. Subbotinae* (in the upper part) and most authors assign it a Landenian-Ypresian age (Jordi & Lonfat, 1963, Barr & Weegar, 1972, Woller, 1978, and Cepek, 1979). The Kheir Member occupies the same stratigraphic position as the Guss Abu Said Member of the Thebes Formation in Egypt. The unit contains *Globorotalia subbotinae* in the lower part, and most workers accept an Upper Paleocene-early Ypresian age for the unit.

The name Wadi Zakim Member of the Beshima Formation was introduced by Woller (1978) to replace the *Flosculina* (*Alveolina*) Limestone of Burollet (1960) and Fürst (1964). Chiesa (1940) originally described this member at Wadi Ammur as two separate units, a lower 1.5 m (5 ft) unit of dense, yellow-white, cherty limestones, and an upper 7 m (23 ft) unit of white, porous limestones with beds of chert and silicified *Alveolina*. The latter was called the Uadi Omur horizon (Chiesa, 1940, p.193). In the Al Qaryat Al Sharqiyah map-sheet NH33-6 area the Wadi Zakim Member is made up of light grey, silicified carbonates, dolomitic in part with common pellets and fine fragments of gastropods and bivalves, and display bird's-eye structures (Cepek, 1979). The unit is rich in planktonic foraminifera such as globorotaliids, as well as benthonic foraminifera such as miliolids. Its thickness varies from 6–15 m (20–49 ft). It is unconformably overlain by the Oligocene Maazul Naina Formation. Woller (1978) and Said (1979) reported the following foraminifera species, from the Al Washkah map-sheet NH33-15 and the Bunjim map-sheet NH33-6 areas, respectively: *Alveolina schwageri*, *A*. gr. *subpyrenaica* var. *Flosculina silvestri*, *A*. gr. *parva*, *A*. gr. *oblonga*, *A*. gr. *fornasinii*, *A*. gr. *decipiens*, and *A*. gr. *solida*, as well as *Quinqueloculina laevigata* and *Q. flectata*, among others. It was assigned an Early Eocene (Ypresian) age by earlier workers and later by Cepek (1979), Shakoor (1984), and Woller (1978).

The Rouaga Member was introduced by Burollet (1960) as a member of the Gebel Uaddan Formation which overlies the *Flosculina* Limestone (Wadi Zakim Member) and underlies the Ben Isa Chalk. Jordi & Lonfat (1963) included the *Flosculina* Limestone in this unit. Chiesa (1940) originally described this unit in the type area in the Wadi Rawagah as an 8 m (26 ft) interval of limestones with Fe-oxides impregnations, followed upward by a 15 m (49 ft) interval of porous, white limestones. Fürst (1964) described the unit in its type locality and estimated a thickness of 35–40 m (115–131 ft), but he apparently included the Ben Isa Chalk in that unit. Bebout & Pendexter (1975) divided the Ruaga Limestone of the Fogaha Group in the subsurface of the Sirte Basin into a lower Zelten Member (Upper Paleocene) and an upper Meghil Member (Lower Eocene), but the Ruaga Limestone used by them is probably a different unit. Shakoor (1984) designated the section in the type locality at Wadi Rawaghah (29°5'37"N, 16°13'22"E) as the type section. He described the Rouaga Member as a complex of dolomites, dolomitic limestones, and calcareous dolomites with lenticular beds of nodular and tabular chert. The cherty horizon is probably a part of the overlying Ben Isa Member of the "Gir Formation" of Burollet (1960). Fossils are rare and restricted to a few molds of pelecypods and foraminifera such as *Cibicides libycus*, *Miliola* sp., *Tritaxia* sp., *Quingueloculina* sp., *Triloculina* sp., *Spirolina* sp., and *Sigmolina* sp. In the Al Qaryat Al Sharqiyah map-sheet area the unit is mostly chalky and brecciated at the base. The upper part is cherty, dolomitic, and poorly fossiliferous except for agglutinated foraminifera at the top. The top of the Rouaga Member is erosional. The Rouaga Member was assigned an Ypresian age by

Cepek (1979), Woller (1978), Shakoor (1984). The thickness of the Rouaga Member ranges between 10 m (33 ft) and 30 m (98 ft) in the Zallah map-sheet NH33-16 area (Vesely, 1984), 25 m (82 ft) and in the type area in the Hun map-sheet NH33-6 area (Shakoor, 1984). It increases to 35 m (115 ft) northward and decreases to less than 10 m (33 ft) in the Al Wahskah map-sheet NH33–15 area (Woller, 1978). The Rouaga Member was assigned an Ypresian age by all workers.

It is in the writer's opinion that the Kheir Member of the Beshima Formation should be raised to formation status equivalent to the Kheir Formation defined by Barr & Weegar (1972), and assigned an Ypresian age following Berggren (1974). The Wadi Zakim (*Flosculina*) and Rouaga members become the only two members of the Beshima Formation, and accordingly the Beshima Formation becomes equivalent to the Gir Formation of Barr & Weegar (1972). The Gir Formation in the subsurface is divided into three members by Barr & Weegar (1972) and further studies may prove correlations between these members and the Wadi Zakim and Rouaga members. This is in variance with Lashab & West (1991) who suggested raising the Rouaga Member to formation status and restricting the Beshima Formation to the Kheir and Wadi Zakim members.

### 14.2.3   A review of the Gir Formation

The name Gir Formation has been used for different Eocene formations. The first was introduced by Burollet (1960) to denote a unit which crops out in the Jabal Waddan area for which he assumed an Ypresian age, but there is very little doubt that it is mostly of Lutetian age, an age accepted by most workers. The second unit was proposed by Barr & Weegar (1972) for a subsurface unit in the Sirte Basin, for which they assigned and Ypresian age. Recent works confirm this age (Wennekers et al., 1996). The Gir Formation of Barr & Weegar (1972) is, in all probability, the equivalent of the Beshima Formation (restricted) in outcrops. In the following discussion it will be recommended to discontinue the use of the name Gir Formation of Burollet (1960) (Al Jir Formation of the Geological Map of Libya 1:250,000 team) and to include its two members in the overlying Tamit Formation. This revised nomenclature will separate the Ypresian sequence from the Lutetian sequence, will help in relating the surface units to those in the subsurface, and facilitate the correlation between these units and those of Egypt and Tunisia.

Burollet (1960) introduced the name Gir Gypsum as a southern facies equivalent of the two upper membes of his Gebel Uaddan Formation, i.e. the Ben Isa Chalk and *Orbitolites* Limestone. However, no type section or description was provided for the Gir Gypsum. The Ben Isa Chalk Member and the *Orbitolites* Limestone Member were described earlier by Chiesa (1940) and Desio (1943) in the Jabal Waddan area, who included them in the Lutetian, based on the presence of *Orbitolites complanatus* and *Conulites (Dictyoconoides) cooki*. The Ben Isa Chalk Member was named after the well Bir Ben Isa (30°55′N, 15°25′E) and described as a widespread, white, compact limestone with nodules of chert and silicified algae with a thickness of up to 60 m (197 ft) at Wadi Ammur. The Bir Zidan Member was initially described by Chiesa (1940) and Desio (1943) as a white limestone with *Orbitolites*, molluscs and algae at Wadi Marghub and Bir Zidan, or with *Cardita* and *Callianassa* at Wadi Sheghega and Wadi Al Jir Al Kabir. Goudarzi (1970), following Desio (1943), replaced the name

*Orbitolites* Limestone with the name Bir Ziden Limestone Member, named after the well Zidan (31°00′N, 15°30′E) in the Wadi Bey el Kebir, between Gheddahia and Bunjim. The unit is 10 m (33 ft) in thickness at Bir Zidan.

The Ben Isa and Bir Zidan members are not always readily separable in the field. Mijalkovic (1977a) mapped the Gir (Al Jir) Formation in the Al Qaddahiyah map-sheet NH33-3 area where the type section of the Bir Zidan Member is located. He mapped the two members as one unit because of the insignificant thickness of the Ben Isa Member. In that area the "Al Jir Formation" (mostly the Bir Zidan Member) is made up of marly and sandy limestones, calcareous sandstones and claystones with gypsum interbeds in the lower part, and limestones, marly limestones, and brecciated limestones with clay and gypsum interbeds in the upper part. Mijalkovic (1977a) reported abundant micro- and macrofauna from the "Al Jir" Formation in outcrops. The microfauna includes the foraminifera *Orbitolites complanatus, Pararotalia* cf. *viennoti, Rotalia* cf. *tronchidiformis, Alveolinidae (Fascolites), Globigerina* sp., and *Globorotalia* sp., in addition to *Discorbis, miliolids,* and rare *Spirolina,* the marine algae *Thyrsoporella* cf. *silvestri,* and the freshwater Charophytes *raskyella pecki meridionale.* The macrofauna includes the pelecypods *Gryphaea exogyroides, G.* cf. *arabica, Ostrea malticostata strictiplacata, Cardita aegyptica,* the gastropods *Gisortia (Vicetia) gigantea, Turitella inbricataria,* and the echinoids *Echinocyamus cambonensis* and *Echinatus costeri.* Said (1979) stated that the two members in the adjacent Bunjim map-sheet NH33-7 to the south could be differentiated only in the northern part of the area. The two members decrease in thickness southward. East of the type section the Ben Isa Member consists of a sequence of white to yellow, hard limestones, marly limestones, and gypsum, capped by a light grey, siliceous limestone bed containing chert nodules and concretions. The Member is 24 m (79 ft) thick in the north. The overlying Bir Zidan Member is made up of 27 m (89 ft) of marly limestones and cavernous limestones. *Orbitolites* are concentrated in a 1 m (3 ft) thick bed at the base. In the Hun map-sheet NH33-11 to the south this marker bed is absent. The absence of this bed led Shakoor (1984) to map the two members as one unit, as did Vesely (1985) in the Zallah map-sheet NH33-16, Zikmund (1985) in the Abu Na'im map-sheet NH33-13, and Jurak (1985) in the Wadi Bu Ash Shaykh map-sheet NH33-12. Both Jurak (1985) and Vesely (1985) identified two laterally equivalent facies in the exposed part of the "Al Jir" Formation in their respective areas: a poorly fossiliferous dolomite and anhydrite facies to the west and south, and a bioclastic limestone facies to the east (on the western edge of the Az Zahara uplift). Zikmund (1985) recognized only the bioclastic limestone facies in the Abu Na'im map-sheet. This facies contains *Orbitolites commplanatus, Dictyoconoides cooki, Somalina stefanini,* and *Coskinolina (Fallotella) balsilliei.* The evaporite and dolomite facies is illustrated by a 44 m (144 ft) section exposed at Wadi Shuqayaqah (29°09′29″N, 16°32′28″E) at the southern end of the Jabal Al Jir Al Kabir. This facies probably represents the Gir Gypsum originally intended by Burollet (1960). The succession (Jurak, 1985) starts at the base with an 11 m (36 ft) bed of coarse-grained, wavy laminated anhydrite with two interbeds of bioclastic dolomite, followed upward by a 19 m (62 ft) interval of porous, chalky dolomite with a large amount of poikilitic gypsum and intercalations of anhydrite, a 6 m (20 ft) bed of anhydrite, and then by an 8 m (26 ft) section of gypsiferous dolomite at the top. The dolomites contain poorly preserved *Vulsella crispata* and *Gisortia (Vicetia) stefanini.* The stratigraphic relation between the two facies is not clear. Either the upper or the lower contact is visible only in very few and separate locations. Both facies appear to overlie the Rouaga Member of the Beshima Formation and they are overlain by the Al Gata Member of the Wadi Tamit Formation with gradational contacts which suggests that the two facies are lateral equivalents.

Mijalkovic (1977a) assigned the "Al Jir Formation" an Ypresian-Lutetian age. Shakoor (1984), Vesely (1985), Zikmund (1985), and Jurak (1985) ascribed the Gir Formation a Lutetian age based on its stratigraphic position and correlations with neighbouring areas.

## 14.2.4   The Gir Formation in the subsurface of the Sirte Basin

Since the Gir Gypsum of Burollet (1960) was poorly defined, Barr & Weegar (1972) considered the name to be invalid and reintroduced the name for a new Ypresian unit, the Gir Formation, in the subsurface of the Sirte Basin. This choice of name, although conforms by the rules of the North American and International codes of nomenclature, is unfortunate because the name is now needed for a member of the Wadi Tamit Formation.

The name Gir Formation was introduced by Barr & Weegar (1972) for a surface and subsuface unit composed of interbedded dolomites and anhydrites with subordinate amounts of limestones and shales. Halite is present in the southwestern part of the Sirte Basin. The Formation is 2003 ft (610 m) thick in the type section in the Oasis Y1-59 well (28°09'14"N, 18°36'15"E) at a drill depth of 1082 ft to 3085 ft. The Formation is conformably overlain by the Gedari or Gialo formations and underlain by the Upper Paleocene Harash Limestone or the Ypresian Kheir Formation. The unit was assigned a Lower Eocene age by Barr & Weegar (1972) based on the presence of unspecified species of *Nummulites*. Its Ypresian age has been recently confirmed in the subsurface (Wennekers et al., 1996).

Barr & Weegar (1972) divided the Gir Formation in the subsurface of the Sirte Basin into the Mesdar Limestone Member, the Hon Evaporite Member, and the Facha Dolomite Member. The Mesdar Limestone Member is made up of massive limestones, glauconitic in part with rare shale and dolomite beds. The type section of this unit is in the Mobil B1–13 well (29°00'24", 19°32'25"E) at a drill depth of 4183 ft to 5970 ft amounting to a thickness of 1787 ft (545 m). It is conformably overlain by the Gialo Formation and it conformably overlies the Kheir Formation. In the eastern and central parts of the Sirte Basin, the Mesdar Member becomes very thick and forms the entire Gir Formation. The Hon Evaporite Member consists of interbedded anhydrites and dolomites with minor shales. It contains halite in the western part of the Sirte Basin. The evaporites intertongue eastward with limestones of the Mesdar Member. The type section of the Hon Member is in the Oasis Y1-59 well (28°09'14"N, 18°36'15"E) at a drill depth of 1160 ft to 2774 ft giving the unit a thickness of 1614 ft (492 m). The Facha Dolomite Member is composed predominantly of tan, argillaceous, finely crystalline, vuggy dolomites with minor anhydrite and limestone beds. It conformably overlies the Harash or Kheir formations and is conformably overlain by the Hon Evaportie Member. The unit is restricted to the western part of the Sirte Basin. Its type section is in the Oasis Y1-59 well at a drill depth of 2774 ft to 3085 ft where it has a thickness of 311 ft (95 m).

The Lower Eocene hydrocarbon production in the Sirte Basin is mainly from the Facha dolomites, the Hon evaporites provide the seal. In the Zella Field the Facha Member consists of a variety of lithologies of argillaceous limestones with thin interbeds of terrigenous mudstones, passing up into clean limestones, overlain by variably anhydritic dolomites and anhydrites, and capped with "chicken-wire" or laminated anhydrites (Williams, 1995). The sedimentology and diagenesis of the "Jir" Formaion in West Libya has been examined by Lashhab & West (1991).

Although facies similarities between the Gir Gypsum of Burollet (1960) (Al Jir Formation of the Geological Map of Libya, 1974–1985, team) and the Gir Formation of Barr & Weegar (1972) are striking, the two units are not age equivalent, the former is Lutetian, the latter is Ypresian. It is recommended, therefore, to discontinue the use of the Gir (Al Jir) Formation for the outcrops and to place its two members, the Bu Isa and Bir Zidan, in the Wadi Tamet Formation (a review of this Formation follows). Wennekers et al. (1996) equated the Gir Formation of Barr & Weegar (1972) with the Beshima Formation in outcrops, the overlying Gir Formation (*sensu* Burollet, 1960) does not appear on their correlation chart.

### 14.2.5   Middle Eocene (Lutetian)

### 14.2.6   Revision of the Wadi Tamet Formation

In the present work, it is proposed to include in the Wadi Tamet Formation all the Lutetian units in outcrops as members, namely the Ben Isa, Bir Zidan, and Al Gata members. These units are mapped together as Lutetian on the Geological Map of Libya (1985).

Desio (1943) described the Wadi Tamet Series at Wadi Thamat (31°00′N, 16°05′15″E) as a succession of interbedded creamy, cherty limestones, grey sandstones with chert nodules, crystalline gypsum, yellow sandy gypsiferous marls, and yellow fossiliferous sandstones. The fossils reported in these sediments are *Ostrea multicostata, O. bouillei, Diplodonta cycloidea, Lucina pharaonis, L. polythele, Cytherea nitidula, Dentalium pagellai, Gisortia gigantea, Scutellina concava, Echinolampas crameri, E. africanus,* and *Brissoides figarii.* Chiesa (1954) assigned the Wadi Tamet Series a Lutetian age based on the foraminifera *Conulites (Dictyoconoides) cooki.*

Burollet (1960) designated the Wadi Tamet Series as the Wadi Tamet Formation and extended it to include Middle and Upper Eocene beds of Syrtica. The Formation was described as a succession of green shales interbedded with coquinoid limestones, chalk, and often gypsum with abundant fossils such as oysters, echinoids, and nummulites. It overlies the *Orbitolites* Limestone Member (Bir Zidan Member) of the Gebel Uaddan Formation and is overlain by the Upper Eocene-Oligocene Dor El Abd Limestone (presently Um ad Dahiy Formation). He selected the type locality at Wadi Thamat (31°00′N, 16°05′E) and divided the Formation into three members, Gara Gata, Tmed el Ksour, and Graret el Jifa. Defined as such, the Wadi Tamet Formation of Burollet (1960) includes the Wadi Tamet Series and Zella Series of Desio (1943).

Goudarzi (1970) raised the Wadi Tamet Formation to group status and designated the three members as formations. However, Shakoor (1984), Said (1979), Banerjee (1980), Srivastava (1980), Jurak (1985), Zikmund (1985), and Vesely (1985) reinstated the original designations of these units, and assigned the Wadi Tamet Formation a Lutetian-Priabonian (Bartonian).

Jurak (1985) described the three members of the Wadi Tamet Formation in the Wadi Bu Ash Ashaykh map-sheet NH33–12 area where their type areas are located. The Wadi Tamet Formation is about 250 m (820 ft) thick and rests conformably on the Lutetian "Al Jir" Formation.

In the type section of the Gata Member at Qarat al Gata (29°48′45″N, 16°36′30″E), only the upper 75 m (246 ft) of the member is exposed (Jurak, 1985), and it is overlain with a conformably sharp contact by the Tmed el Ksour Member. The exposed part consists of dolomites, argillaceous limestones, marlstones, calcareous claystones, and claystones. The limestones are very fossiliferous. The upper 18 m (59 ft) is made up of cross-bedded dolomites and contains an *Exogyra Exogyroides* coquinoid bed. The lower part is composed of dolomites with abundant quartz geodes, 5–20 cm in diameter. Celestite aggregates are common at the base. Jurak (1985) dated the Gara Member Lutetian, based on *Discorbis vesicularis* and *Dictyoconoides cooki*.

Northward, in the Bunjim map-sheet NH33–7 area (Said, 1979), the Wadi Tamet Formation outcrops in isolated hills on the western side of Wadi Thamat at Quwayrat az Zubayyib, Dur al Mrabi, and Ruus al Makharam. The Gata Member has an estimated thickness of 38 m (125 ft) and it is composed predominantly of marls interbedded with hard fossiliferous limestones, and becomes sandy to the south. The unit is eroded farther north in the Al Qaddahiyah map-sheet NH33–3 area (Mijalkovic, 1977a), and probably in its immediate offshore area.

In the immediate offshore of northWest Libya, near the Tunisian border, the Eocene sediments are predominantly clayey sandy wackestone-type limestones, poor in fossils with only ostracods and rare benthonic foraminifera (Fucek et al., 1998). A few *Nummulites* sp. with small benthonic forams occur near the top of the Eocene sequence. They probably represent the northern extension of the Middle Eocene (Wadi Tamet Formation) sediments along the western margin of the Sirte Basin.

South of the type area in the Zallah map-sheet NH33–16 (Vesely, 1985) the exposed part of the Gata Member at Wadi Wadi Falazlaz (28°50′40″N, 16°43′02″E) has a total thickness of 100 m (328 ft). It consists of green claystones and marls interbedded with bioclastic limestones or dolomites, followed upward by a rhythmic alternation of claystones and gypsum, and then by a cross-bedded or horizontally bedded calcarenite bed 4–6 m (13–20 ft) in thickness.

East of the type area and before the Eocene sequence disappears in the subsurface of the Sirte Basin the Gata Member in the Abu Na'im map-sheet NH34–13 is made up predominantly of bioclastic limestones, chalky limestones, clayey limestones, and marlstones with a few intercalations of dolomitized limestones (Zikmund, 1985). It has a thickness between 70 m (230 ft) and 80 m (262 ft). The unit is highly fossiliferous with abundant pelecypods, gastropods, echinoids, and foraminifera. Oysters form coquinoid limestone beds up to 2 m (7 ft) thick. In addition to the foraminifera *Orbitolites complanatus*, *Dictyoconoides cooki* (both are common in the Bir Zidan Member), *Nummulites discorbus*, *N. uranensis*, and *N.* aff. *gizehensis*, the unit contains the molluscs *Exogyra exogyroides*, *Carolia placunoides*, *C. cymalea*, and *Ostrea elegans*, and the echinoids *Echinolampas perrieri*, *E. fraasi*, *Sismondia saemanni*, and *Fabularia luciani*, and the nautiloid *Nautilus* cf. *leonei*.

## 14.2.7 Middle Eocene in the subsurface of the Sirte Basin

The Middle Eocene in the subsurface of the Sirte Basin is represented by the Gedari Formation and its lateral equivalent the Gialo Formation.

The name Gedari Formation was proposed by Barr & Weegar (1972) for a complex assemblage of shales, limestones, microcrystalline dolomites, fossiliferous sandstones, chalks, and gypsum. They stated that the Formation extends from outcrops into the subsurface of the western Sirte Basin, but it is not clear with which unit(s) they correlated it. The type section is located in the Mobil AA1-11 well (28°58′15″N, 17°23′31″E) at a drill depth of 586 ft to 3218 ft where it has a thickness of 3632 ft (1109 m). In the type section the Formation conformably overlies the Hon Evaporite Member of the Ypresian Gir Formation. The contact becomes unconformable in the extreme western part of the Sirte Basin. The faunal content includes foraminifera, pelecypods, gastropods, echinoids and bryozoans. It was assigned a Middle Eocene age by Barr & Weegar (1972), but it may contain Upper Eocene sediments in the upper part. The Gedari Formation is equivalent to the Wadi Tamet Formation (revised) in outcrops. The Gedari Formation grades eastwards into its lateral equivalent the Gialo Limestone.

Barr & Weegar (1972) proposed the name Gialo Formation for a thick subsurface sequence of grey or tan to brown, shallow marine limestones. The type section is in the Oasis E91–59 well (28°42′14″N, 21°22′11″E) at a drill depth of 2750 ft to 4337 ft where the unit attains a thickness of 1577 ft (484 m). The unit is highly fossiliferous and contains several species of *Nummulites* and molluscan, echinoid, and bryozoan fragments. Fossils include *Nummulites gizehensis*, *N. curvispira*, *N. discorbina*, *N. bullatus*, and *N. partschi tauricus*, in addition to *Coskinolina*, *Dictyoconoides* and miliolids. Arni (1967) described a number of species of *Nummulites* from the Gialo Formation. The unit conformably overlies the Lower Eocene Gir Formation and is unconformably overlain by the Bartonian Augila Formation. The wireline logs of the type well show three main units within the Gialo Formation, but Barr & Weegar's description does not provide a precise definition of these units. It was assigned a Middle Eocene age by Barr & Weegar (1972). The Gialo Formation is the main reservoir in the Gialo Field.

## 14.2.8 Upper Eocene (Bartonian)

It is proposed to introduce the new name Zallah Formation to include the Upper Eocene units Tmed el Ksour and Graret el Jifa members of the Wadi Tamet Formation of Burollet (1960) in outcrops. The two units are mapped together on the Geological Map of Libya (1985). The Formation is named after the Zallah Oasis (28°50′N, 17°50′E). Since there is no complete exposed section, the composite type section of the members may serve as a type section.

Desio (1943) previously described the Zella Series from base to top as: 1) marls and sandstones with gypsum intercalations with *Cardium*, *Ostrea multicostata*, wood fragments, fish bones, and crustacean shells, 2) sandy marls, green glauconitic marls with large *Ostrea praehongirostris*, *Carolia placunoides*, *C. cymbalea*, *Alectryonia (Ostrea) clot-beyi*, *Ostrea multicostata*, *O. reili* var. *abundans*, and *Nautilus leonei*, 3) sandy marls, brecciated, red with phosphatic nodules, fish teeth, turtle plates, vertebra, and crocodile teeth, and calcareous marl conglomerates, 4) a continental unit composed of white, siliceous limestones with pseudoconglomerate and breccia up to 30–40 m (98–131 ft) in thickness with green gypsiferous shales at the base, and 5) basalts. Chiesa (1954) assigned an Upper Eocene age for the lower two units and an Upper Miocene for the upper two units.

The Tmed el Ksour Member (Thmed al Qusur on the Geolgical Map of Libya, 1985) is a persistent unit and has a uniform lithology along the western margin of the Sirte Basin. It is composed predominantly of chalky limestones to chalky dolomites with chert beds, and celestite and anhydrite layers (Said, 1979, Jurak, 1985, Vesely, 1985, Zikmund, 1985). Celestite occurs as cavity and burrow-filling, or fossil replacement. The unit is very rich in fossils which include large gastropods such as *Gisortia* (*Vicentia*) *stefanini* and *Rostellaria* (*Amplogladius*) *turgida*, the mollusc *Vulsella crispata*, *Rosetellaria* (*Hippocerne*) *columbaria*, in addition to nautiloids, echinoids, and remains of crabs, it also contains ostracods and benthonic foraminifera. Jurak (1985) reported rare specimens of the coral *Placophyllia bartai* from the type area. It was assigned a Priabonian age based on the foraminifera *Globorotaloides* cf. *sutari*, *Subbotinae* cf. *eocaena*, *Turborotalia* cf. *increbenscens*, and *Gyoidina* cf. *complanata* (Zikmund, 1985) and the coral *Placophyllia bartai* which is known to occur in the Upper Eocene of Italy (Jurak, 1985). The Tmed el Ksour Member varies in thickness from 27 m (89 ft) to 40 m (131 ft) in the type area in the Wadi Ash Shaykh map-sheet area, from 10 m (33 ft) to 167 m (52 ft) in the Abu Na'im, and between 11 m (36 ft) and 30 m (98 ft) in the Zallah area. It varies in thickness from 9 m (30 ft) to 20 m (66 ft) in the Bunjim area north of the type area where it becomes more sandy (Said, 1979). The Tmed el Ksour Member is eroded further to the north in the Al Qaddahiya area, and probably in the immediate offshore area.

In many places such as the Hun map-sheet NH33-11 area (Shakoor, 1984) the Graret el Jifa Member is missing, and the Tmed el Ksour Member is unconformably overlain by the Oligocene Maazul Naina Formation. In the northeast corner of the Bunjim map-sheet area the Tmed el Ksour Member is overlain directly by the Miocene Al Khums Formation (Said, 1979).

The Graret el Jifa Member has its type section near Qararat al Jifa (29°58'45"N, 17°44'00"E). However, the succession is incomplete. Srivastava (1980) described a more complete section of the member in the An Nuwfaliyah map-sheet NH33-8 area, north of the type area and divided it into three units with a total thickness between 110 m (361 ft) and 160 m (525 ft). The lower unit is 78 m (256 ft) thick and composed mainly of bioclastic limestones, alternating with clays. The middle unit consists of 39 m (128 ft) of interbedded clays and gypsum. The upper 11 m (36 ft) unit is made up of dolomitized, bioclastic limestones and marls with sandy admixtures. Jurak (1985) recognized only the lower and middle units in the Wadi Bu Ash Shaykh map-sheet area where the type section is located. He estimated a thickness of 110 m (361 ft) for the Graret el Jifa Member in the type section and 160 m (525 ft) at Qarat ar Raqubah (29°07'24"N, 17°33'21"E). Vesely (1985) reported a comparable thickness of 150 m (492 ft) for the member in the section at Dur Bu Zanad (28°55'02"N, 17°40'25"E) in the Zallah map-sheet NH33–16 area south of the type locality, and he was able to recognize all three units in the Graret el Jifa Member. The lower 90 m (295 ft) interval is characterized by the predominance of bioclastic limestones and dolomites, usually glauconitic and containing gypsum interbeds. The middle unit is 38 m (125 ft) thick

and made up of green non-fossiliferous clays, interbedded with gypsum. The uppper unit is 22 m (72 ft) in thickness and composed of clays and gypsum with beds of
bioclastic dolomite. The Graret el Jifa Member is highly fossilifirous and has been
dated Priabonian by Burollet (1960), Goudarzi (1970), Jurak (1985), Vesely (1985),
and Zikmund (1985). Fossils reported by Zikmund (1985) from the Abu Na'im map-
sheet area include *Nummulites beaumonti, N. fabiani,* and *N. bouillei,* the pelecypod
*Exogyra exogyroides, E. maradensis Carolia placunoides, C. lefevrei, Ostrea reili,
O. elegans, Plicatula varioloia, P. bellardii,* and *Vulsella crispata,* the echinoids *Echinolampas chiesai messlei, E. perrrieri, E. Africanus, E. libycus, Eupatangus formosus,
Schizaster meslei, Kismondia saemanni,* and *Porocidaris Schmiedelli,* the sharks *Car-
caredon augustidens, Pycodus* sp., and *Chrysophyrs* sp., and the gastropods *Gisortia
(Vicetia) stefanini, Terebellum fusiforme,* and *Turbinella fegnens. Nummulites gizehensis,* a typical Lutetian species, reported by Jurak (1985) from this unit is probably
a misidentification.

The Graret el Jifa Member is either partially or completely eroded in many places.
It overlies with a sharp conformable contact the Tmed el Ksour Member and is uncon-
formably overlain by either the Oligocene Maazul Naina Formation or the Miocene
Al Khums Formation. It has not been recognized by Said (1979) in the Bunjim map-
sheet area, and is completely absent to the north in the Al Qaddahiyah map-sheet area
(Mijalkovic, 1977a) and the immediate offshore.

In the subsurface of the Sirte Basin the Upper Eocene is represented by the Augila
Formation (Bartonian). Barr & Weegar (1972) introduced the name Augila Formation
for an Upper Eocene unit in the subsurface of the central and eastern Sirte Basin. The
unit disconformably overlies the Gialo Formation and it is overlain by the Oligocene
Arida Formation. The type section is in the E1-59 well (28°41′13″N, 21°24′11″E) at
a drill depth of 2444 ft to 2707 ft. The Augila Formation in the type well has a thickness of 263 ft (80 m). It was divided into three units: the lower unit is made up of
grey to green shales with thin argillaceous limestone or dolomite interbeds, occasionally sandy and glauconitic at the base. Its thickness is 100–250 ft (30.5–76 m). The
unit contains unspecified Upper Eocene benthonic and planktonic foraminifera. The
middle unit is 100 ft (30.5 m) thick and composed of porous, glauconitic sandstones.
This unit is missing in the type section. The upper unit consists of sandy, slightly
glauconitic shales, occasionally very argillaceous with common molluscan debris. It
is 100 ft (30.5 m) thick in the type section. The unit contains *Nummulites fabianii*
and *N. striatus.* Along the eastern margin of the Sirte Basin the Formation includes a
limestone unit at the base named the Rashda Member. The type section of the Rashda
Member of the Augila Formation is in the D2-12 well (29°04′35″N, 21°14′18″E) at a
drill depth of 2522 ft to 2689 ft where it has a thickness of 167 ft (51 m). The unit is
composed of tan limestones with thin olive green shales at the top. The limestones are
fossiliferous, calcilutites and calcarenites, chalky to dolomitic, and glauconitic with
abundant algal and dolomitic beds near the base. It disconformably overlies the Gialo
Formation. The Augila Formation was assigned an Upper Eocene age based on the
occurrence of *N. fabianii* and other unspecified small and planktonic foraminifera.

The Augila Formation in the subsurface of the Sirte Basin is equivalent to the
Tmed el Ksour and Graret el Jifa members in outcrops along the western margin
of the Sirte Basin. The Rashda Member is probably equivalent to the Tmed el Ksor
Member.

In southcentral Libya at the southern end of the Sirte Basin Wight (1980) described a 100 ft (30 m) Paleogene section exposed at Dur at Talhah in the Sarir-Tibesti area (see also Savage, 1971) and noted the similarity and correlatability between the sections in that area and the Eocene-Oligocene Qasr el Sagha and Gebel Qatrani formations of the Fayium Depression in Egypt. Wight (1980) divided the sediments into three units, from base to top: the Evaporite Unit, the Idam Unit, and the Sarir Unit. The Evaporite Unit contains Upper Eocene oyster beds with *Ostrea clot-beyi* and *O. cubitus* at the base and grades laterally into chert and dolomitic limestones and upward into limestones and marls with gypsum. The unit directly overlies the Precambrian basement and it is overlain by the Lower Oligocene Idam Unit (Wight, 1980). This unit is included in the undifferentiated Eocene sediments on the Geological Map of Libya (1985).

## 14.2.9    Eocene of offshore Northwest Libya

The Eocene sequence in the subsurface of the offshore of northWest Libya has been studied by Hammuda et al. (1985) who introduced a number of new formations and groups for that area and correlated them with those of Tunisia and onshore Libya. The Eocene sequence includes the Bilal Formation, Jirani Dolomite, and Jdeir formations (Fig. 14.1), and their equivalent the Tajoura and Taljah formations of the Far-wah Group, the Ghalil Formation, Hallab Formation, and the Harsha, Dahman, and Samdun formations of the Tellil Group.

Hammuda et al. (1985) proposed the name Farwah Group for a Lower Eocene carbonate sequence in the subsurface of the central and southern Tripolitania Basin in the offshore of Libya. The type section is in the NOC-AGIP B2-NC41 well at a drill depth of 8084–9082 ft. The group is unconformably underlain by the Al Jurf Formation and overlain by the Tellil Group or Ghalil Formation. These authors

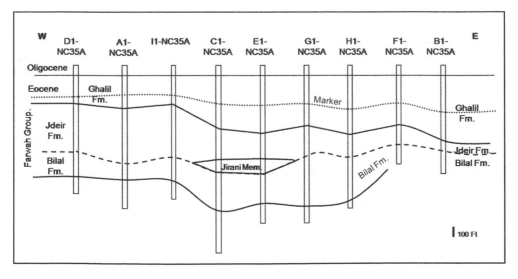

*Figure 14.1* Eocene stratigraphic cross-section (offshore NC35A area). Based (with modifications) on borehole logs in a cross-section in El Ghoul (1991).

divided the group in the type section, from oldest to youngest, into the Bilal Formation, Jirani Dolomite, and Jdeir Formation. The thickness of the Farwah Group is between 24–1334 ft (7–407 m). Southward, the Jdeir Formation and Jirani Dolomite are replaced by the Taljah Formation. The group passes into the deeper water facies of the Tajoura Formation to the east and northeast. The Farwah Group is equivalent to the well-established Metlaoui Formation of Burollet (1956) or the Metlaoui Group of Fournié (1978) in Tunisia. Therefore, the name Farwah is redundant and it should be discarded. The Metlaoui Group in Tunisia includes deep-water globigerinid facies of the Bou Dabbous, shallow-shelf nummulitid facies of the El-Garia, evaporitic anhydrite/dolomite facies of the Faid, the phosphate-bearing Chouabine Formation, and the evaporitic and dolomitic Tselja Formation (Fournié, 1978).

Hammuda et al. (1985) established the Bilal Formation in the type section in the NOC-AGIP B2-NC42 well at a drill depth of 8638–9082 ft. It is made up of interbedded dolomitic and glauconitic limestones, marls, shales, anhydrites, sandy dolomites, and limestones, sandstones, and conglomerates. The fauna reported includes *Orbitolites*, miliolids, bolivinoides, rotaliids, globorotaliids, alveolinids, algae, ostracods, and very rare *Nummulites* and locally globigerinids. It is overlain by the Jirani Dolomite in the north and center of the Tripolitania Basin and by the Taljah Formation in the south and unconformably underlain by the Al Jurf Formation. Its thickness is 600 ft (183 m) in the center of the Tripolitania Basin and it thins southward to 100 ft (328 m). It was assigned a late Late Paleocene-Early Eocene age. The type section of the Jirani Dolomite is in the NOC-AGIP B2-NC41 well at a drill depth of 8514–8638 ft. The dolomites are very finely crystalline, locally fossiliferous with anhydrite nodules, and include miliolids, rotaliids, textularids, peneroplids, algae, and ostracods. The Formation is conformably overlain by the Jdeir Formation and conformably underlain by the Bilal Formation. The Formation has a limited geographic distribution. It grades into the Tajoura Formation eastwards and into the Bilal Formation southward. It was assigned an Early Ypresian age by Hammuda et al. (1985). It is probably equivalent to the Facha Dolomite Member of the Gir Formation of Barr & Weegar (1972). The Jdeir Formation in the type section in the NOC-AGIP B2-NC41 well at a drill depth of 8084–8514 ft consists of light brown nummulitic wackestones and packstones. The upper part is made up of wackestones-packstones with abundant *Nummulites*, miliolids, ostracods, pelecypods, gastropods, and algal fragments and grades into dolomitic and cherty facies with globigerinids. The lower part is composed of packstones-grainstones with abundant *Nummulites*, echinoderms, pelecypods, *Operculina* sp., alveolinids, ostracods, gastropods, and algal fragments. The Formation is unconformably overlain by the Ghalil Formation and underlain with gradual contact by the Jirani Dolomite. In the center of the Tripolitania Basin it overlies the Bilal Formation. The Formation grades eastward into the bryozoan limestones of the Tajoura Formation. The thickness of the unit varies from 17–600 ft (5–182 m). It thins southward. Hammuda et al. (1985) reported the following fossils: *N. rollandi, N. irregularis, N. distans*, and *N. pomeli*, in addition to ostracods, pelecypods, and gastropods, indicating an Ypresian-early Lutetian age. The Formation is the principal reservoir in the offshore.

The Farwah Group changes facies into the Tajoura and Taljah formations. The Tajoura Formation was introduced by Hammuda et al. (1985) for an Early Eocene unit in the NOC-AGIP F1-NC41 well at a drill depth of 8666–9239 ft. The unit is made up of limestones, partly dolomitic, wackestones to packstones with bryozoa, ostracods, miliolids, rotaliids, nummulites, pelecypods, and algal fragments. It grades in places into oolitic grainstones and bryozoan boundstones with phosphatic and glauconitic grains. It was assigned an Early Eocene (Ypresian-Lutetian) age based on its stratigraphic position. This unit is correlated with the Farawah Group in the offshore of Libya, the Metlaoui Group and Tajoura Formation in Tunisia, and the Gir Formation in the Sirte Basin. The Early-Middle Eocene Taljah Formation was proposed by Hammuda et al. (1985) for a unit in the type section is in the NOC-Aquitaine H2-137 offshore well at a drill depth of 7923–8044 ft, composed mainly of dolomites, argillaceous in the upper part, anhydritic in the middle, and sandy in the lower part. It is underlain by the Bilal Formation.

The Ghalil Formation was introduced by Hammuda et al. (1985). The type section of the unit is in the NOC-AGIP D2-NC41 at a drill depth of 7680–8284 ft and consists of fossiliferous shales and marls with lenses of limestones, and abundant planktonic foraminifera. It conformably overlies either the Tajoura Formation or the Farwah Group and it is overlain by the Oligocene Ras Abd Jalil Formation (Sbeta, 1991, El-Ghoul, 1991). Its thickness varies from 236 ft (72 m) to 624 ft (190 m). In the Bouri Field the Ghalil Formation is 666 ft (203 m) in thickness (El-Waer, 1992). The fossils in the Ghalil Formation belong to three zones (Hammuda et al., 1985): the *Globorotalia aragonensis* and *G. rex* zone, the *Truncorotalia topilensis* zone, and the *Globorotalia ooki* zone which belong to the Middle-Upper Eocene. The Formation represents a pelagic facies in the east and northeast of the Tarabulus Basin. It is equivalent to the Gialo and Aguila formations (Barr & Weegar, 1972) in the Central and Eastern Sirte Basin, and the Souar Formation in Tunisia (Fournié, 1978). El-Waer (1992) reported the ostacod genera *Kirthe, Paleocosta, Soudanella, Buntonia, Bairdia,* and *Argilloecia* from the Ghalil Formation. The lower part of the Ghalil Formation was designated in the B1-NC35A well as the Hallab Formation by Hammuda et al. (1985), who considered it to be a deep-water equivalent of the Farwah Group. The type section is in the NOC-ESO B1-NC35A well at a drill depth of 8462–8620 ft. The unit consists of white limestones at the base with abundant planktonic forams, overlain by interbedded shales and limestones. The top of the unit is marked by a persistent limestone bed (El-Ghoul, 1991, Fig. 4). The fossils reported from this unit are *Globorotalia* cf. *acuta, G. lehneri,* and *G. aragonensis* and assigned it an Ypresian to Lutetian age. El-Ghoul (1991) indicated that the Hallab Formation is Bartonian in age and therefore it is not equivalent to the Farwah Group and recommended that the name should be abandoned. However, the planktonic foraminifera species indicate an Ypresian-Lutetian age and the present author recommends that the Hallab Formation should be retained. The Hallab Formation can be recognized and correlated over a large area. Only the top part of the Formation was encountered in the F1–NC35A well.

The Tellil Group was introduced by Hammuda et al. (1985) for a Middle-Late Eocene unit in the subsurface of the offshore of northWest Libya. The type section is in the NOC-Aquitaine H1-137 well at a drill depth of 6142–8020 ft. It consists mainly of alternating shales, marls, and limestones with anhydrites and sandstones in places. The Tellil Group includes the Harsha, Dahman, and Samdun formations. It grades northward into the pelagic shales and marls of the Ghalil Formation (Sbeta, 1991). It is equivalent to the Djebs, Cherahil, and Souar formations (Fournié, 1978) of Tunisia, and the Gedari, Gialo, and Augila formations of the Sirte Basin, and the Derna Formation in the Al Jabal Al Akhdar. El-Waer (1992) divided the Tellil Group into five ostracod zones based on the genus *Heptaloculites*, from the base to top, *H. harshae* (zone E), *H. semirugosa* (zone D), *H. minuta* (zone C), *H. cavernosa* (zone B), and *H.* aff. *gortanii* (zone A). Zones B-E are Middle Eocene, and zone A is Upper Eocene. He placed the Tellil-Ghalil contact at the base of zone B (*H. cavernosa*). The Harsha Formation derives its name after the Harsha area, 5 km west of azZawyah. It grades northward from shales with sandstone and anhydrite beds, to dolomitic limestones, limestones and shales, then into shales. It is rich in benthonic foraminifera, pelecypods, and ostracods with rare *Nummulites* (Sbeta (1991). It ranges in thickness from 700–1100 ft (213–335 m) in the center of the Basin and thins to the north and south. The Harsha Formation is 461 ft (141 m) in the C1-NC41 well and 1186 ft (361 m) in the K1-NC41 well (El-Waer, 1992). The type section is in the H1-137 well at a drill depth of 7323–8080 ft where it has a thickness of 760 ft (232 m) (Hammuda et al., 1985). El-Waer (1992) reported the ostracod genera *Asymmetricythere*, *Hermanites*, *Heptaloculites*, *Isobuntonia*, *Leguminocythereis*, *Schizocythere*, *Pokornyella*, *Urocythereis*, *Cytheropteron*, *Flexus*, *Loxochona*, and *Cytherella* from the Harsha Formation. The Harsha Formation is equivalent to the Cherahil Formation in Tunisia (Fournié, 1978). The Dahman Formation is a limestone sequence rich in molluscs and *Nummulites*. The Dahman Formation ranges in thickness from 42 ft (13 m) in the H1-NC41 well to 70 ft (21 m) in the B1a-NC41 well (El-Waer, 1992). It has a maximum thickness of 200 ft (61 m) (Sbeta, 1991). Its type section is in the H1-137 offshore well at a drill depth of 7247–7323 ft (Hammuda et al., 1985) where it has a thickness of 76 ft (23 m). It is consistent in thickness and lithological character over an extensive area. It changes facies to the north and northwest of the Tarabulus Basin into the pelagic facies of the Ghalil Formation. The Samdun Formation, named after Sabkhat Samdun, a few kilometers southeast of Bu Kammash, is made up of shales and limestones. It has a thickness of 1200 ft (366 m) in the center of the Basin and it thins both to the north and south (Sbeta, 1991). The Samdun Formation varies in thickness from 300 ft (91 m) in the B1-NC41 well in the Bouri Field to 1211 ft (369 m) in wel H1-NC41 (El–Waer, 1992). In the type section in the H1-137 well it was encountered at drill depth of 6142–7247 ft (1105 ft) where it is made up mostly of greenish grey, glauconitic shales (Hammuda et al., 1985). The Samdun Formation is unconformably overlain by the Oligocene *Nummulites vascus* marker bed of the Dirbal or Ras Abd Jalil formations. El-Waer (1992) reported the ostracod genera *Leguminocythereis*, *Heptaloculites*, *Asymmetricythere*, *Hermanites*, *Loxoconcha*, *Urocythereis*, *Pontocyprella*, *Paracypris*, and *Argilloecia* from the Samdun Formation.

## 14.2.10   Eocene of Northeast Libya

Gregory (1911) subdivided the Eocene rocks of northern Cyrenaica into three formations, from base to top: Apollonia, Derna, and Slonta (spelled Salantah on the Geological Map of Libya, 1977). The Apollonia Formation is well developed along the base of the northernmost coastal escarpment of Al Jabal Al Akhdar. Its type area is located south of Sousa (ancient Apollonia). The unit is composed of thin-bedded limestones with chert nodules. It unconformably overlies the Upper

Cretaceous Atrun Limestone and is conformably overlain by the Middle Eocene Derna Limestone in the Al Bayda area (Röhlich, 1974). It also grades laterally into the Derna Formation in the Benghazi area (Klen, 1974). It contains planktonic foraminifera such as *Globorotalia subbotinae* in the lower part indicating an Ypresian age, and *Nummulites, Truncorotaloides rohri* and *T. topilensis* in the upper part indicating a Middle Eocene age. It was assigned a Lutetian age by Röhlich (1974) who reported *Nummulites ex. N. gizehensis, N. aff. planulatus, Operculina pyramidum,* and *Discocyclina pratti.* It was assigned a Lutetian-Priabonian age by Klen (1974) who reported the Lutetian species *Nummulites gizehensis lyelli, Discocyclina pratti,* and *Actinocyclina* sp., and the Priabonian planktonic foraminifera *Globigerapsis mexicana, Globigerinita africana, Globigerinatheca lindiensis, Globigerina linaperta, G. tripartita,* and *G. galavisi,* in addition to numerous benthonic species. The Formation contains reworked limestone fragments of the Derna Formation, and displays graded bedding and slump structures, probably due to deposition by turbidity currents. The Formation interfingers with the Derna Formation as a result of transportation of detrital carbonate material of the Derna Formation into the deep-marine areas of the Apollonia Formation (Klen, 1974). The total thickness at Al Athrun is 380 m (1247 ft) (Röhlich, 1974), but may reach up to 800 m (2625 ft) in wells in the coastal region. The Formation wedges out toward the Eocene shoreline to the south.

The Apollonia Formation (Early-Late Eocene) in the A1–NC120 well in the offshore of Cyrenaica (Derna Formation in the A1–NC128 well) represents a complete Eocene section of open deep-marine facies composed of whitish to brownish argillaceous mudstones-wackestones grading into grey-green marls and shales, locally silty with chert nodules, pyrite, glauconite and traces of fine-grained quartz sandstones (Duronio et al., 1991). It conformably overlies the Late Paleocene Al Uwayliah Formation and is unconformably overlain by the Oligocene Beda Formation. The Early Eocene fossils in the Apollonia Formation include *Morozovella aragoensis, M. subbotinae, M. broedermanni, Acarinina pseudotopilensis, A. spinuloinflata, Cuvillerina eocenica, Discocyclina sp.,* and *Nummulites* sp. The Upper Eocene assemblage includes *Turborotalia cerrozulensis, T. cocoaensis, T. pomeroli, Hantkenina brevispina, H. alabamensis,* and *Discocyclina* sp.

The limestones overlying the Apollonia Formation were originally subdivided by Gregory (1911) into the Middle Eocene Derna Limestone and the Upper Eocene Slonta Limestone. This subdivision was also accepted by Pietersz (1968) and Kleinsmeide & van den Berg (1968). Gregory (1911) chose the type locality of the Derna Formation (spelled Darnah on the Geological Map of Libya, 1985) on the road between Dernah and Tubruq (32°44'50"N, 22°38'45"E) where the unit is composed of white to creamy limestones, algal and reefal in the upper part. It contains abundant *Nummulites gizehensis curvispira,* and algal, echinoid, and *Crassostrea* fragments. The Derna Formation was assigned a Lutetian age by Burollet (1960). Zert (1974) selected another type section (or more accurately a reference section) for the redefined Derna Formation 2 km south of Dernah where the unit reaches a total thickness of 140 m (459 ft) (Zert, 1974). The age of the Derna Formation is Upper Lutetian-Priabonian according to Zert (1974), Röhlich (1974) and Klen (1974). The Formation overlies the Upper Cretaceous Wadi Ducchan Formation and is progressively overlain by the Labrak (Oligocene) or Regima (Miocene) formations. It contains many *Nummulites* species and some specimens of nautiloids up to 75 cm in size. In the Dernah area reefs occur in the upper and middle parts of the Formation which contain abundant algal limestones, some layers are rich in solitary corals and large foraminifera such as *Nummulites* and *Orbitoides* (Zert, 1974). In the Al Hmidah road cut section, southwest of Tukrah, bioclasts are dominated by large-sized *Nummulites* sp., common operculinids, *Orbitolites complanatus* and rare fragmented discocyclinids (Abdulsamad et al., 2008). Some of the fossils reported from the Derna Formation are *Nummulites gizehensis, N. incrassata, N. straitus, N. beaumonti, N. chavannesi, N. fabianii, N. pulchllus, Heterostegina reticulata, Operculina pyramidum, Discocyclina varians,* and *D. pratti,* in addition to the nannofossils *Discoaster distinctus, D. kuepperi,* and *Lophodolithus reniformis.* This assemblage contains Lutetian and Bartonian elements. El-Hawat & Shelmani (1993) described the sedimentology of the Upper Eocene in the Al Marj Quarry.

The Derna Formation in the subsurface of the offshore of Cyrenaica unconformably overlies the Aptian-Albian Daryanah Formation and uncoformably underlies the Oligocene "Beda" Formation. Duronio et al. (1991) reported the following fossils from the Derna Formation (Middle-Late Eocene) in the A1–NC128 well in the offshore of Cyrenaica. From the lower part of the Middle Eocene section they reported *N. gr. gizehensis, N. millecaput, N. aturicus, N. laevigatus, N. striatus, Operculina* sp., *Heterostegina* sp., *Discocyclina* sp., and *Assilina* sp., as well as fragments of *Alveolina* sp., *Gypsina* sp., ostracods, anellids, broyozoans, pelecypods, and echinoids. From the upper part of the Middle Eocene sediments they recovered *N. cf. striatus, N. cf. variocolarius, N. cf. incrassata, N. cf. orbignyi, Valvulineria* sp., *Planulina marialana, Nonion granosum, Bolivina* sp., and *Bulimina* sp., in addition to planktonic foraminifera such as *Subbotinae yeguaensis, S. bolivariana, Turborotalia* cf. *bowerzi,* and *T.* cf. *wilsoni.* The Late Eocene assemblage includes *N. gr. striatus, Operculina* sp., *Rotalia trochidiformis, Turborotalia cocodensis, T. ampliapertura, T. boweri,* and encrusting algae.

The Apollonia andd Derna formations are coeval and both of them span the Lower/Upper Eocene interval. Oil and gas have been discovered recently in the Derna Formation in the offshore well A1-NC202, at depths of 4,442–4,484 ft, and at rates of 1,264 bopd and 0.58 mmcfgd[4]. Block NC202 was part of a package of exploration blocks including the offshore blocks NC201 and NC199 (Cyrenaica), NC200 (Murzuq Basin), and NC203 and NC204 (Kufrah Basin) which were awarded to Repsol and OMV in June 2003.

## 14.3  OLIGOCENE

Oligocene sediments crop out in northern Cyrenaica, the Sirte Basin, and the Hun Graben. They also occur in the subsurface of the Sirte Basin, the offshore of Cyrenaica, and the offshore northWest Libya. The stratigraphy of the Olig ocene sediments, especially in Northeast Libya Libya, is still in need of clarification.

---

4 "OMV confirms oil discovery in Sirte Basin Offshore Libya" (Rigzone News, Tuesday, April 21, 2009).

## 14.3.1   Oligocene of Northeast Libya

In Northeast Libya Libya the Oligocene is represeted by the Cyrene and Labrak formations, the upper part of the Khowaymat Formation (Upper Cretaceous-Lower Oligocene), and by the Ketatna Formation in the subsurface of the offshore of Cyrenaica.

Gregory (1911) introduced the name Slonta Limestone after the village of Salantah (Slonta) (32°36′N, 21°43′E) in the central Al Jabal Al Akhdar. The type locality (32°25′N, 21°45′E) extends for a distance of 19 km between Salantah. Pietersz (1968) selected another type locality at the western margin of the town of Salantah where the Eocene section is 60 m (197 ft). Burollet (1960) applied the name Slonta Formation to a part (probably the lower part) of Gregory (1911) Slonta Limestone. The Formation consists of beige-white, hard or porous limestones, in places chalky and cavernous with chalky marls (Burollet, 1960) and contains *N. striatus*, *N. fabianii*, peneroplids and gastropods. It was assigned an Upper Eocene age by Burollet (1960). Kleinsmeide & van den Berg (1968) included the Slonta Limestone as a member in their Akhdar Formation which also included the Apollonia and Derna limestones of Gregory (1911) as members. Burollet (1960) and Pietersz (1968) reported *N. gizehensis* and *Echinolampus* from the lower part of the Slonta and *N. fabianii* from the upper part and dated the unit Middle-Upper Eocene. Röhlich (1974) in his mapping of the Al Bayda map-sheet NH34-15 area believed that the Slonta and Derna formations are equivalent and do not superpose and recommended that the name Slonta to be dropped. On the other hand, Zert (1974) indicated that the Slonta Formation overlies the Derna Formation and that the two formations are, therefore, not equivalent. According to Zert (1974), the "yellowish horizon" of Gregory (1911) equivalent to the Slonta Formation is Oligocene in age with a high percentage of reworked Priabonian fauna. It appears therefore that the type section of the Slonta Formation chosen by Gregory (1911) is Oligocene in age and equivalent to his Cyrene Formation, but the type section designated by Pietersz (1968) is Eocene in age according to Röhlich (1974) and Zert (1974) and may be equivalent to the Derna Formation of Röhlich (1974). Accordingly, the Oligocene Slonta and Cyrene formations of Gregory (1911) and the Al Baydah Formation of Röhlich (1974) are all equivalent.

Gregory (1911) proposed the name Cyrene Limestone after the name of the ancient Greek city of Cyrene (presently Shahat) for soft, earthy, white and grey limestones and marls overlying his Slonta Limestone. He first placed these beds in the Aquitanian, then considered them to be Upper Oligocene. The limestones contain *Aequipecten zitteli*, *A. camaratensis*, *A. scabrellus*, and *Pecten vezzanensis*. Stefanini (1921) reassigned the unit to the Aquitanian, an age which was also accepted by Desio (1935) and Burollet (1960). Pietersz (1968) changed its name to Cyrene Formation and described its type section east of Shahat. He divided the Formation into three members, from bottom to top: the Marly Limestone Member, Algal Limestone Member, and Calcarenite Member. Kleinsmeide & van den Berg (1968) renamed the Marly Limestone and Calcarenite members the Shahat Marl Member and Labrak Calcarenite Member, respectively, and included them in their Oligocene Al Kuf Formation. Barr & Weegar (1972) divided the Cyrene Formation into Shahat Marl Member, Algal Limestone Member, and Labrak Calcarenite Member. Röhlich (1974), Klen (1974) and Zert (1974) divided the Oligocene section into a lower Al Bayda Formation (equivalent to the Shahat Marl and Algal Limestone members of Kleinsmeide & van den Berg, 1968) and an upper Al Abraq Formation (equivalent to the Labrak Calcarenite Member of Kleinsmeide & van den Berg, 1968), separated by an unconformity. Röhlich (1974) proposed the name Al Bayda Formation for an Oligocene unit in the Al Jabal Al Akhdar, Cyrenaica. However, the name Beda Formation (a homonym) has been used by Barr & Weegar (1972) for a Paleocene unit in the subsurface of the Sirte Basin which has priority and is still in current use. Therefore, it is recommeded to reinstate the name Cyrene Formation for this unit instead of the name Al Baydah Formation. The maximum surface thickness of the Al Bayda Formation is 50 m (164 ft) in the Wadi Azzad and Wadi Halag al Azayez sections (Abdulsamad et al., 2008). In the Wadi Ekhil and Wadi Zazah sections the formation is missing. Both contacts with the underlying Derna Formation and the overlying Al Abraq Formation are disconformable. Röhlich (1974) divided his "Al Baydah" Formation into two members: the Shahat Marl and Algal Limestone. The Shahat Member is developed only on the northern slope of the Al Jabal Al Akhdar but disappears to the south. Fossils in the Algal Limestone Member (it is proposed to call this unit the Salantah Member, now that the name is available) are the rock-forming algal species *Archaeolithothamnion maughini*, the foraminifera *Operculina discoidea, O. ex. gr. heterostegina*, N. cf. *hantkeni, Florilus scaphum, Textularia agglutinans*, echinoids, especially *Echinolampas cherichirensis*, pelecypods and gastropods, in addition to reworked Priabonian and Oligocene microfauna (Zert, 1974). This member represents the shallowest facies of the Cyrene Formation but grades into deeper-marine marly facies in the north. The Cyrene Formation also wedges out to the south. It is absent in the Benghazi area and the Labrak Formation rests directly on the Derna Formation (Klen, 1974).

Röhlich (1974) used the name Al Abraq Formation to replace the Labrak Calcarenite Member of the Kuf Formation of Kleinsmeide & van den Berg (1968) and the Calcarenite Member of the Cyrene Formation of Pietersz (1968). It consists of a carbonate unit transgressively overlying the Upper Cretaceous, Darnah and Al Bayda formations and unconformably overlain by the Al Fayidiyah Formation in the Al Jabal Al Akhdar area. The unit is named herein the Labrak Formation. The type section was designated by Röhlich (1974) at the head of a small Wadi (32°48′24″N, 21°59′54″E) north of Al Abraq settlement. It is composed of calcarenites with abundant reworked Eocene *Nummulites* species. It also contains *N. fichteli* and *N. vascus* and many echinoid species such as *Echinolampas cherichirensis*. The upper part contains *Operculina complanata* and *O. africana*. It contains dolomite beds in the Dernah area. It was dated Lower Oligocene to basal Upper Oligocene by Röhlich (1974), El Deftar & Issawi (1977), and Francis & Issawi (1977).

Mazhar & Issawi (1977), Swedan & Issawi (1977), El Deftar & Issawi (1977), and Megerisi & Mamgain (1980c) proposed the name Al Khowaymat Formation for a unit which crops out in the Zt. Msus, Bir Hacheim and Al Bardia map-sheet areas in northern Cyrenaica. The unit is named after the village of Al Khowaymat a few kilometers southeast of Bir Habas (32°04′N, 21°35′E). The unit is unconformably overlain by the Middle Miocene Qaret Meriem Formation, but its base is not exposed. The Formation consists of two members which are mapped separately. The lower member is made up of thick-bedded dolomites with limestone intercalations and minor silica intercalations. The unit contains *Nummulites intermedius*, *N. boucheri, Operculina complanata, Globigerina* sp., *Globorotalia* sp., echinoids and other shell remains. It was assigned an Upper Eocene to Lower Oligocene age by these authors. However, Röhlich (1974) assumed an Upper Cretaceous age for this part of the Formation. The upper member crops out in a narrow strip in the coastal area in the Zt. Msus map-sheet area. The type section was designated southeast of Bir el Giuma (31°47'40"N, 21°52'10"E) where the unit is approximately 6 m (20 ft) thick. It is composed of green clays with occasional oyster shells, and limestone interbeds with *Nummulites* in places. It contains *Nummulites intermedius, N. vascus, N. boucheri, N.* cf. *fichteli, Operculina complanata, Globorotalia* sp., *Globigerina* sp., the miliolid *Amphisorus, Peneroplis* sp., *Textularia* sp., and different species of ostracods. It was assigned an Upper Eocene to Lower Oligocene age.

In Offshore Cyrenaica the Oligocene sediments overlie with apparent conformity the Eocene sequence. The whole series belongs to the Early-Middle Oligocene Ketatna Formation and the Late Oligocene Labrak Formation. The Ketatna Formation (Duronio, 1993, 1996) is composed of wackestones and packstones grading into mudstones with *Nummulites* n. sp. *jamahiriae* (Duronio, 1996), *Nummulites vascus, N. fichteli-intermedius, Lepidocyclina* sp., *Turborotalia ampliampertuna*, bryozoans, hydrozoans, and pelecypods (Duronio et al., 1991). The Formation is equivalent to the Al Bayda Formation of Röhlich (1974) and the Cyrrene Formation of the present work. The Late Oligocene Labrak Formation consists of wackestones and packstones with *Eulepidina dilatata, Operculina complanata*, bryozoans, and the algae *Corallinaceae* (Duronio et al., 1991, Duronio, 1996). The Ketatna and Labrak formations are separated by an unconformity.

## 14.3.2   Oligocene of the Sirte Basin

Oligocene sediments are present in the central and eastern Sirte Basin. They rest conformably on the Upper Eocene Augila Formation. The latter is eroded to the west and Oligocene sediments rest unconformably on the Middle Eocene Gialo or Gidari formations. Barr & Weegar (1972) proposed the name Najah Group for the Oligocene-Miocene sequence in the surface and subsurface of central and eastern Sirte Basin. They divided the group into the Arida, Diba, and Marada formations. The first two formations are Oligocene in age and the latter is Miocene.

The type section of the Arida Formation is in the Oasis E3-59 well (22°44'15"N, 21°24'15"E) at a drill depth of 2178–2578 ft where it consists of fine-grained, glauconitic sandstones with shale interbeds in the lower part and green glauconitic shales with sparse planktonic and benthonic foraminifera in the upper part. It unconformably overlies the Augila Formation and is conformably overlain by the Diba Formation. The unit forms the reservoir in the Gialo (Jalu) Field and a few other wells in Concession 59 and also in well G7-51 in the An Nafurah Field (Bezan & Malak, 1996).

The name Diba Formation was proposed by Barr & Weegar for a unit in the subsurface of the Sirte Basin. The type section is in the Oasis E3-59 well (22°44'15"N, 21°24'15"E) at a drill depth of 1584–2178 ft where it is 594 ft (181 m) thick. In the type section the Formation consists of interbedded fine to coarse-grained, glauconitic sandstones and grey to green silty shales. The unit overlies the Arida Formation and is overlain by the Miocene Marada Formation. No diagnostic fossils have been reported from this unit. The Diba Formation produces oil only from the 2E1-12 well southwest of the Ar Rakb Field (Bezan & Malak, 1996).

In the southeastern Sirte Basin Oligocene sediments are overlain by varying thicknesses of the Marada Formation and younger sediments. The Oligocene sediments are made up of a lower fine-grained, fossiliferous, glauconitic and calcareous sandstones with shale interbeds and occasional carbonates and evaporites, and an upper unit of non-glauconitic, non-fossiliferous sandstones (Benfield & Wright, 1980). The lower unit contains marine fossils and represents marine deposits, and the upper unit represents non-marine facies. Shales increase in abundance in deep troughs. In the coastal area of the Sirte Basin Oligocene and Miocene sedimentation is represented by a thick section up to 3000 m (9843 ft) and composed mostly of shales (Fürst, 1993, Wennekers et al., 1996, Bezan & Malak, 1996).

The name Idam Unit was proposed by Wight (1980) for a unit which overlies the Upper Eocene Evaporite Unit in outcrops in south central Libya. It contains a rich verterbrate fauna of mammals, reptiles, and fish of a Lower Oligocene age. This unit is probably equivalent to the Augila and Zallah formations, and to the El Sherit Formation of Fürst (1964). The Sarir Unit[5] (Wight, 1980) is made up of unfossiliferous fluviatile sediments. The Idam and Sarir units are probably equivalent to the Arida and Diba formations in the Sirte Basin, respectively.

### 14.3.3 Oligocene of Northwest Libya

Oligocene sediments in northWest Libya are represented by the Dor el Abd, Greir Bu Hascisc, and Maazul Naina formations in the western Sirte Basin and eastern Ghadames Basin, and the Tarab Formation in the Gargaf Arch area.

Burollet (1960) introduced the Dor el Abd Formation for the lower part of the Oligocene deposits in the Sirte Basin. The type locality is at Dur el Abd cliffs (29°40′N, 17°20′E) where it consists of white, grey, and yellow dolomitic limestones with thin beds of light grey clays and gypsum in the middle part. The contacts with the overlying Greir Bu Hascisc Formation (Upper Oligocene) and the underlying Wadi Tamet Formation (Eocene) are gradational. Its thickness is 70–80 m (230–262 ft) (Hecht et al., 1963). It was assigned an Eocene-Lower Oligocene age based on the presence of *Crassostrea* cf. *gryphoides* and *Nummulites intermedius fichteli*. Srivastava (1980) restricted the Dor Al Abd Formation to the lower Oligocene and changed the name to Um ad Dahiy Formation after the Dur Umm ad Dahiy (the present-day name of Dur Al Abd). The upper part is marked by a 1.8 m (6 ft) bed of white chalky, fossiliferous and cross-bedded limestones. The thickness of the Formation in the An Nuwfaliyah sheet area is 8–13 m (26–43 ft) and probably thickens in the subsurface of the Sirte Basin. The Formation wedges out to the northwest. In the central and eastern parts of the area the Formation contains algal limestones and layers rich in solitary corals (described by Zuffardi-Comerci, 1940) and grades into coral limestones. Fossils include the pelecypods *Crassostrea* cf. *gryphoides*, *Cubitostres plicata*, molds of the gastropod *Turitella*, and the echinoids *Scutella chiesai*, *Mellitels* sp., and *Amphioppe duffi*. The foraminifera include *Operculinoides* cf. *africana*, *Operculina*

---

5 Note that the name Sarir is used for an Upper Jurassic-Lower Cretaceous (Nubian) unit in the Southeastern Sirte Basin.

cf. *complanata, Pararotalia* sp., *Nonion* sp., and *Tritaxia* sp. Hladil et al. (1991) described corals from the Dor el Abd (Um ad Dahiy) Formation in the Al Hofra Oil Field. The Formation conformably overlies the Upper Eocene Qarat al Jifah Member of Wadi Tamet Formation and is overlain by the Greir Bu Hascisc Formation. It consists of reefal, massive, bedded dolomites with abundant *Operculina* sp. and common glauconite, peloids, and bioclasts. Moldic porosity is up to 30%. The thickness of the Formation varies from 10 m (33 ft) to 23 m (75 ft). The Upper part is composed of gypsiferous silty clays with fossil plants. Carbonate nodules are common at the bottom of the unit. The main reef builders are *Stylophora parvistella, S. thirsiformis, Madracis decaphylla, Astraeopora decaphylla, Monticulastrea ex. gr. daedalea, Actinacis rollei, Agathiphyllia gregaria, Antiguastraea maradahensis* sp. nov., and *Athecastraea hufrahensis* sp. nov.

The Greir Bu Hascisc Formation (cited as Qurayyir Bu Hashish on the Geological Map of Libya, 1977, 1985) was introduced by Burollet (1960) in the type locality south of the Wadi Araua Valley. It is composed mainly of green shales interbedded with minor beds of fossiliferous limestones. The upper part contains white coquinoid dolomitic limestone. The Formation is 80 m (262 ft) in thickness and contains echinoids, oysters, bryozoans, and *Nummulites fichteli*. It was appointed an Upper Oligocene age. Srivastana (1980) changed the name according to the Arabic transcription to the Bu Hashish Formation and described two reference sections in the An Nuwfaliyah map-sheet NH33-8 area. In that area the Formation is made up of alternations of limestones, dolomites, chalks, and chalky marls with frequent intercalations of nummulitic coquina and clays beds. The basal beds are hard, cherty and dolomitic limestones. In addition to the foraminiferal assemblage reported from the Dor el Abd Formation, this Formation contains *N. fichteli, Echinolampas chericherensis, E. crameri, Crassostrea* cf. *gryphoides*, and silicified shells of *Ostrea (Giganostrea) crassostrea*, and *Clypeaster biarritzensis*, among others. Its thickness is 10–17 m (33–56 ft) in the western part of the area to 70–75 m (230–246 ft) in the eastern part.

The Maazul Naina Formation (spelled Maazul Neina in Banerjee, 1980, and M'azul Ninah in Megerisi & Mamgain, 1980a, b) was introduced by Jordi & Lonfat (1963) for the Oligocene sediments filling the Hun Graben. The type section is at Wadi Bay al Khayib (29°33′38″N, 15°56′25″E) (Shakoor, 1984). It rests with a sharp unconformity on Eocene sediments such as the Ammur Member, Kheir Member, or Wadi Zakim Member. The unit is composed of dark red brown, fine-grained sandstones and silty clays with abundant gypsum, and intercalations of medium to coarse-grained and conglomeratic calcareous sandstones at the top. Fürst (1964) described the Oligocene sediments in the Hun Graben without any reference to that name. The Maazul Naina Formation in the Hun Graben is assumed to unconformably overlie the Ypresian Rouaga Member of the Beshima Formation and is overlain by Miocene volcanics of the Jabal Assawda dated by the K/Ar method at 10.5–12.3 Ma (Ade-Hall et al., 1975). In the central part of the graben the exposed thickness of the Formation is 23 m (75 ft) and is composed of greenish claystones at the base with gypsum intercalations, overlain by calcarenites and sandy limestones, followed by gypsum, and capped by dolomitic limestones. Calcareous sandstones and conglomerates are common in the upper part of the Formation on the western side of the graben. It was assigned an Oligocene age based on its stratigraphic position by

Woller (1978), Cepek (1979), Shakoor (1984), Salaj (1979), Jurak (1985), Vesely (1985), and Zikmund (1985).

Jurak (1978) introduced the Tarab Formation for a tectonically wedged-in relics of mostly carbonate rocks occurring on the western and eastern margins of the Jabal Al Hassawnah. The type section is about 10 km northeast of the Trunzah hill on the western margin of the Jabal Al Hassawnah near Wadi Tarab (28°13'20"N, 13°49'50"E). The Formation is made up of green-yellow gypsiferous, soft claystones, followed upward by green-yellow gypsiferous marlstones with abundant intercalations of greyish yellow calcilutites including a 1–1.5 m (3–5 ft) thick bed composed of laminated calcilutites rich in fossil fish and other vertebrate remains. The top of the succession is made up of light brown, recrystallized calcarenites and calcilutites with abundant nodular chert. The Formation contains abundant remains of bones, a skull and a skeleton of *Titanohyrax paleotherioides*[6], bone remains of a frog assigned to *Libicus hasaunus* n.g., n.sp., fish remains, probably belonging to the the species *Paleochromis rousseleti*, silicified shells of the gastropods *Melania (Melanoides) fasciatus nysti* and an uncertain *Succinea* sp. He assinged the Formation an Early Oligocene age. The base of the Formation is not exposed, but it is assumed to overlie the Hassaouna Formation.

## 14.4 MIOCENE

### 14.4.1 Miocene of the Sirte Basin

The Miocene in the Sirte Basin is represented by the Marada and Regima formations of the Najah Group. The Marada Formation crops out in the central Sirte Basin and the Regima Formation crops out in northern Cyrenaica. The two formations were extended into the subsurface of the Sirte Basin by Barr & Weegar (1972).

The name Marada Formation was introduced by Desio (1935) for a sequence exposed along the southern escarpment of Dur Marada (29°14'N, 19°12'E) in the central Sirte Basin. The Formation consists of a large variety of lithofacies including limestones, shales, sandstones, and evaporites.

Selley (1969) restricted the term to those sediments lying beneath a prominent marine limestone, commonly cropping out around the tops of the jabals of the Maradah region which he suggested might correlate with the Marmarica Formation of the Western Desert of Egypt. Goudarzi (1970) applied the name Marada Group to the whole succession in the northern cliffs of the Marada Depression. Barr & Weegar (1972) included in the Marada Formation the whole of Lower and Middle Miocene sequence of the southeastern Sirte Basin. In the southeastern part of the Sirte Basin the Marada Formation disconformably overlies the Oligocene Diba Formation and is overlain by a thick sequence up to 213 m (700 ft) of unfossiliferous sands which Benfield & Wright (1980) designated as the Calanscio Formation.

The Marada Formation was divided in the Bir Zaltan area into two members (Domaci, 1985, Mastera, 1985), a lower Qarat Jahannam Member and an upper Ar Rahlah Member. The type section of the Qarat Jahannam Member is in Qarat

---

6 This species has been reported near the town of Zallah by Arambourg & Magnier (1961).

Jahannam. Its thickness is 123 m (404 ft). It increases to the south to over 160 m (525 ft) and decreases to the north to about 5 m (16 ft) (Mastera, 1985). The Member is made up mostly of green and yellow sandstones with minor green shales, calcareous sandstones, limestones, dolomites, and gypsum. The foraminifera reported from the unit include *Sorites* sp., *Discorbis discoides, Rotalia calcar, Ammonia beccarii, Pararotalia* sp., *Elphidium crispum, E. minutum,* and *Homotrema* sp. which confirm a Miocene age for the member. However, it was assigned an Aquitanian-Burdigalian age (Mastera, 1985, Domaci, 1985) based on the following fossils: *Placuna miocenica, Chlamys submalvinae, Echinolampas orlebari, Pecten cf. fuchsi, Euthecodon* sp., *Podocnemis aegyptica, Trionyx* sp., *Prolibytherium magnieri,* and *Mastodon cf. angustidens.*

The Ar Rahlah Member consists of carbonates in the lower part which change facies both laterally and vertically into terrigenous clastics (Domaci, 1985). The lower part (the carbonate facies) is made up of green and yellow calcareous sandstones, dolomites, and coquinoid limestones. The upper part (the clastic facies) is composed of white and green, bioturbated, calcareous sandstones and minor conglomerates. Its thickness is approximately 47 m (154 ft) (Domaci, 1985, Fig. 40). The carbonates are very rich in benthonic foraminifera, ostracods, and invertebrate fossils. Planktonic foraminifera include *Globigerina bulloides, Globigerinoides cf. quadrilobatus trillobus,* and *Borelis melo.* It was assigned a Lower-Middle Miocene ((Aquitanian-Serravallian) age by Domaci (1985) based on the presence of *Lopha plicatula var. virleti, Codakia leonina var. mediolaevis, Stombus coronatus, Pecten vezzanenzis,* and *Pecten (Janira) cristatocostatus.* Mastera (1985) reported *Porites colligniana gerudiensis* from the lower part of the member in the Maradah area and suggested that the base of the Ar Rahlah Member is Aquitanian in age.

The biostratigraphy of the vertebrate mammals of the Marada Formation sands at Jabal Zaltan (Gebel Zelten) has been discussed by Pickford (1991), Savage & White (1965), and Agrasar (2004). The latter described crocodile remains from that area. This fauna suggests an early Burdigalian age (Savage & White, 1965). The limestones contain *Borelis melo* (Langhian-Messinian) and *Borelis melo curdica* characteristic of the Burdigalian (?) of North Africa (Said, 1962a, Barr & Walker, 1973). The unit was described in detail by Selley (1969) in the central Sirte Basin.

The basal sequence of the Marada Formation at the central Jabals and Dur Maradah-Dur Zaqqut escarpment starts at the base with channel lag material dominated by reworked shells, bone fragments, shark teeth, silicified wood, boulders encrusted with calcareous algae and bored by bivalves, followed by channel-fill composed of fine to coarse-grained, tabular and trough cross-bedded, terrigenous and skeletal sandstones. This sequence grades upward into a calcareous shale facies which grades vertically and laterally into calcareous sandstones. Within these sands, oyster bioherms up to 700 m (2297 ft) in length and 6 m (20 ft) in thickness are formed by *Ostrea, Laufa,* and *Crassostrea,* as well as coralline red algae, corals, echinoids, bryozoans, and foraminifera (El-Hawat, 1980a). Cross-bedded sandy grainstones form fining and coarsening-upward cycles, containing reworked limestone cobbles and pebbles, oyster shells, coral debris, and bone and wood fragments. The cross-bedded grainstone facies is composed of wackestones, packstones and grainstones containing bryozoans, echinoids, molluscs, foraminifera, calcareous algae, and oncolites. These limestones are partly dolomitized and range from dolomites to dolomitic mudstones, mottled wackestones, pelletal mudstones, and skeletal grainstones. The dolomitic limestone and wackestone facies of El-Hawat (1980a) forms parts of the detrital limestone facies of Selley (1968). The wackestone facies contains bivalves, bryozoans, foraminifera, ostracods, worm tubes, coralline red algae, and echinoids, especially *Echinolampas* sp. This facies forms a regional marker horizon throughout the area. *Ostrea* sp., *Laufa* sp., and *Crassostrea* sp. form shell banks within this facies. Trace fossils include *Thalassinoides* and *Ophiomorpha* and worm tubes. The marl facies is composed of green to brownish grey marls containing bryozoans, echinoids, foraminifera, ostracods, *Archaeolithothamnium*, vertebrate bones, and silicified wood fragments.

In the Jabal Zaltan area the Marada Formation is locally overlain by a duricrust made up of limestones, dolomites, and calcareous sandstones with veinlets and concretions of chert. Its thickness is up to 10 m (33 ft). It includes altered marine sediments of the Marada Formation. This duricrust was probably formed in the Upper Miocene to Lower Pliocene (Domaci, 1985).

## 14.4.2   Miocene of Northeast Libya

In Northeast Libya Libya the Miocene is represented by the Faidia, Regima, Qaret Meriem, and Sahabi formations. The Miocene of Northeast Libya Libya is divided on the Geological Map of Libya (1977, 1985) into the Bu ad Dahhak and Fortino formations (Lower Miocene), the Mullaghah Formation, Binghazi Limestone, and non-marine calcareous marly, and sandy beds (Middle Miocene), and marine and continental marls, limestones, gypsiferous shales, clays, calcareous sands, and conglomerates (Upper Miocene).

Burollet (1960) divided the Miocene rocks in the Wadi Beddahach (Wadi Bu ad Dahhak) (32°37′N, 22°29′E) in Cyrenaica into two units: the Beddahach Formation (called Bu ad Dahhak Formation on the Geological Map of Libya, 1977, 1985) is of Aquitanian-Burdigalian age and forms the upper reaches of Wadi Dernah, and the Maalegh Formation which is of Helvetian age and occurs in Wadi Maalegh. Desio (1928) originally proposed the name "Porto Bardia Series" for marine sandy fossiliferous limestones with argillaceous and gypsiferous, cross-bedded sandstones which form the sea cliffs at Al Bardia (31°45′N, 25°05′E). Burollet (1960) introduced the name Beddahach Formation to replacce the name "Porto Bardia Series". Pietersz (1968) described the composite type section of the Bardia Formation at the base of the scarp along the road from the harbor to the village of Bardia just south of the Second World War cemetaries. He placed the Bardia Formation between the Faidia and Regima formations and considered it to be an equivalent to the Giarbub Series of Desio (1928). Goudarzi (1970) named the unit "Porto Bardia Formation" and placed it between the Beddahach Formation and the Maalegh Formation. The lower part (Aquitanian) of the Beddahach Formation is composed of massive, white, chalky, detrital limestones with *Miogypsina globulina*, *Miogypsinella complanata*, *Lepidocyclina (Eulepidina) dilatata*, *L. (Nephrolepidina) marginata*, and *Rotalia viennoti*, dolomite is developed locally. The upper part (Burdigalian) consists predominantly of detrital limestones, commonly marly or with algae and abundant mollusc casts and occasional echinoid fragments. The Beddahach Formation is equivalent to the Al Khums Formation (Mann, 1975a, b). The Maalegh Formation was described by Desio (1935) in the Wadi al Muallaq (Wadi Maalegh) (32°27′N, 22°51′E) which flows into the Gulf of Bomba. The Formation overlies the Beddahach Formation and consists of fine to medium-grained, detrital, organic, and slightly recrystallized limestones and contains *Ficus conditus* and *Pecten subarcuatus*. It was assigned a Helvetian age by Desio (1935) and Langhian-Helvetian by Burollet (1960) and Hecht et al. (1963). The Maalegh Formation is equivalent to the uppermost part of the Labrak Formation and the lower part of the Faidia Formation (Zert, 1974).

The name Faidia Formation was introduced by Pietersz (1968) for a shale and limestone sequence overlying the Cyrene Formation in Al Jabal Al Akhdar. The type section is near the entrance of the Village of Faidiyah (Faidia) (32°40′N, 21°55′E) about 10 miles south of Shahat. The Formation is composed of limestones in the upper part and shales and glauconitic marls in the lower part. The unit has a thickness of 143 ft (44 m). The lower contact with the Oligocene Cyrene Formation is controversial. Pietersz (1968) considers it disconformable. Kleinsmeide & van den Berg (1968) suggested that the contact is conformable and included the Cyrene and Faidia formations into a single unit. Barr & Weegar (1972) showed conformable upper and lower contacts on their correlation chart. It contains common *Pecten*, oysters, echinoids, and algae. It was assigned a Lower-Middle Miocene (Aquitanian-Helvetian) age by Pietersz (1968). The Faidia Formation is eroded in the Jardas al Ahrar area.

In the Soluq area the Faidia Formation is overlain by the Regima Formation and composed of hard, highly argillaceous limestones (Francis & Issawi, 1977) with *Spondylus crassiocosta, Flabellipecten schweinfurthi, Ostrea* sp., corals, calcareous algae, *Operculina complanata, Heterostegina heterostegina, Ammonia beccarii, Elphidium crispum, E. fichtellianum*, and *Amphistegina radiata*. This assemblage indicates an Upper Oligocene-Lower Miocene age for the Faidia Formation. Offshore Cyrenaica the Faidia Formation consists of wackestones and packstones containing *Miogypsina* sp., *Operculina* sp., and fragments of bryozoa, hydrozoa, echinoderms, and the algae *Corallinaceae* (Duronio et al., 1991).

The name Regima Formation was introduced by Desio (1935) in the Benghazi area in northern Cyrenaica. The type locality is in the town of Ar Rajmah (32°04′N, 20°22′E) 18 miles east of Benghazi. Kleinsmeide & van den Berg (1968), Pietersz (1968), and Barr & Weegar (1972) used the name Regima Formation for the whole Middle Miocene sequence in the Al Jabal Al Akhdar area. The Regima Formation consists of highly fossiliferous dolomitic limestones. Pietersz (1968) reported a thickness of 205 ft (62.5 m) in the type section and 285 ft (86.9 m) to the north. Along the western flank of the Al Jabal Al Akhdar the unit overlaps unconformably several formations including the Eocene Apollonia and Derna formations and the Oligocene Labrak or Cyrene formations. The fossils include echinoids, bryzoans, molluscs, corals, and foraminifera. The foraminifera *Borelis Melo* is common, indicating a Middle Miocene age for the unit. Klen (1974) applied the name Ar Rajmah Formation (the Arabic transcription of the Regima Formation) to this unit in the Benghazi area and divided it into the Benghazi and Wadi Al Qattarah members. Francis & Issawi (1977) raised the Regima Formation to group status and divided it in the Soluq area into the Benghazi, Al Sceleidima, and Msus formations. The latter is equivalent to the Wadi Al Qattarah Member and the name Msus Formation should be abandoned. Therefore, the Regima Formation in the Soluq area includes three members: the Benghazi Member, the Al Sceleidima Member, and the Al Qattarah Member.

Gregory (1911) first used the name Benghazi Limestone in Cyrenaica to describe a sequence of white fossiliferous limestones without specifying its stratigraphic position. The name Benghazi Member was introduced by Klen (1974) in the Benghazi area for the lower part of the Regima Formation. The type locality is at the dam site in Wadi Al Qattarah, approximately seven kilometers southeast of Ar Rajmah. It was named Benghazi Formation by Abdulsamad et al. (2008). The unit is approximately 60 m (197 ft) in thickness in the type locality. The lower part of the Member consists of highly fossiliferous algal limestones and the upper part consists of thick-bedded, fossiliferous limestones and marly dolomites. Klen (1974) reported the following foraminifera from the Benghazi Member in the Benghazi area: *Asterigenina planorbis*, *Borelis melo*, *Cibicides boueanus*, *C. praecinctus*, *C. pseudoungerianus*, *Elphidium advenum*, *E. craticulatum*, *E. crispum*, *Epinoides haidingeri*, *Heterostegina heterostegina*, and *Rotalia beccarii*. Miliolids appear in the upper part of the Member. The Benghazi Member is exposed in the Soluq area at the Al Sceleidima escarpment and composed of biosparites, biomicrites, and oosparites. The Member rests unconformably on the Faidia Formation and is overlain by the Al Sceleidima Member. Northward near Wadi Al Qattarah, the Al Sceleidima Member wedges out and the Benghazi Member is overlain directly by the limestones of the Al Qattarah Member. Fossils reported from the Benghazi Member in that area include *Turritella* sp., *Borelis melo*, *Amphistegina radiata*, *Textularia* sp., *Lithophyllus* sp., *Ostrea digitalina rholfsi*, *Lithodomus* sp., *Arca umbonata*, *Tellina luconosa*, *Lucina ornata*, *Elphidium macellum*, *E. vicoriense*, *Rotalia* sp., *Globigerinoides triloba immatura*, *Orbulina bilobata*, *Bolivina arta*, *B. spathulata*, *Nodosaria multicostata*, *Uvigerina pygmaea*, *U. bononiensis*, *Quinqueloculina poeyama*, *Miogypsina* sp., *Streblus beccarii*, *Heterostegina costata costata*, *H. heterostegina praecostata*, and *Lithothamnium* sp. The presence of *Borelis melo*, *Globigerinoides triloba immatura*, *Miogypsina* sp., and *Heterostegina costata costata* indicates a lower Middle Miocene (probably Langhian) age for this unit (Francis & Issawi, 1977).

The Al Sceleidima Member was introduced by Francis & Issawi (1977) in the Soluq area (31°35′N, 20°34′E). The type section of the Al Sceleidima Member is at the Al Sceleidima escarpment northwest of Burj al Sceleidima. The unit is 31.5 m (103 ft) thick and made up of gypsiferous, highly fossiliferous limestones, gypsum, and green claystones in the lower part, fine-grained gypsiferous sandstones which become calcareous upward in the middle part, and green claystones, fossiliferous limestones and cherty marls in the upper part. Francis & Issawi (1977) reported the following fossils: *Tellina lacunosa*, *Ostrea digitalina rohlfsi*, *Lithodomus* sp., *Arca umbonata*, *Lucina ornata*, *Flabellipecten schweinfurthi*, *Chlamys zitteli*, and *Pectunculus pilosus*. Northward near Wadi Al Qattarah the Al Sceleidima Member wedges out and the Al Qattarah Member (Msus Formation of Francis & Issawi, 1977) rests directly on the Benghazi Member.

The Wadi Al Qattarah Member was introduced by Klen (1974) in the Benghazi area for the upper part of the Regima Formation. It was designated as Wadi Al Qattarah Formation by Abdulsamad et al. (2008). The type locality is also at the dam site. The thickness of the unit varies between 20–60 m (66–197 ft). It is made up of greyish white, soft oolitic limestones with lenses of gypsum and thin green claystone intercalations in the upper part and grades into the Benghazi Member. The unit wedges out north of Wadi Zazah and the Wadi Al Qattarah Member rests directly on

Eocene sediment. In the Soluq area the Wadi Al Qattarah Member (Msus Formation of Francis & Issawi, 1977) conformably overlies the Al Sceleidima Member and is overlain by Quaternary sediments. The thickness of the member at the Al Sceleidima scarp is 12 m (39 ft), but increases to 40 m (131 ft) to the north. The unit is made up of hard fossiliferous dolomitic limestones in the lower part with a persistent 1 m (3 ft) thick oolitic limestone bed at the base. Francis & Issawi (1977) reported the following fossils from the Member in the Soluq area: *Borelis melo, Nonion* sp., *Elphidium* sp., *Ostrea frondosa, O. digitalina rohlfsi, Lucina multilamellata, Dosina orbicularis, Chlamys zitteli, Xenophora cumulans, Meretrix incrassata, Tellina planata, T. lacunosa, Tapes vetula*, and *Cytherea erycina*. This assemblage suggests a Middle Miocene (Serravallian-Tortonian) age for the Member.

The contact between the lower Middle Miocene Benghazi Member and the Serravallian-Tortonian Wadi Al Qattarah Member is marked by hardgrounds which represent the Langhian-Serravallian unconformity which is recognized almost everywhere in Egypt and Libya.

In the offshore area north of Benghazi the Middle Miocene Regima Formation in the A1-NC120 well (Duronio et al., 1991) consists of wackestones-packstones with common *Operculina* sp., *Heterostegina* sp., and *Amphistegina* sp. and locally abundant fragments of bryozoans, hydrozoans, and pelecypods. The Formation contains gypsum and gypsiferous limestones toward the top.

Swedan & Issawi (1977) introduced the name Qaret Meriem Formation for a unit which unconformably overlies the Al Khowaymat Formation in the Bir Hacheim map-sheet NH34-4 area. The type section is at the Qaret Meriem hill (31°47'50"N, 22°51'10"E) where the unit is 15 m (49 ft) thick. It is composed of highly fossiliferous reefal limestones, chalky limestones, marls, marly limestones, and clay intercalations. Fossils reported from the Formation include the foraminifera *Massilina tenuis, Quinqueloculina seminula, Peneroplis* cf. *planatus, Borelis* sp., *Puteolina proteus, Elphidium tota, Amphisorus* sp., *Textularia* sp., and *Amphistegina* cf. *radiata*, and the macrofossils *Bammosolen strigillatus, Cardita jananheti, Cardium gallicum, Chlamys zitteli, Cytherea erycina, Lucina multimellata, Meretrix incrassata, Ostrea verleti, Tellina lacunosa, T. planata, Natica* cf. *millepunctata, Strombus* cf. *bonelli, Turritella terebralis, Xenophora estigera*, and *Echinolampus amplus*, in addition to ostracods, bryozoans, the calcareous algae *Lithophyllum* sp., corals, and worm tubes. The unit interfigers and passes laterally into the Wadi Al Qattarah Member (Msus Formation) of the Regima Formation.

Desio (1935) first used the name "Sahabi Series" for a succession cropping out near Qasr As Sahabi. Petrocchi (1934, 1941, 1943) conducted the earliest systematic studies of the vertebrate fauna in the Sahabi area. Burollet (1960) described the lower part of the Sahabi Formation (named the Sahabi Beds on the Geological Map of Libya, 1977) as Upper Miocene lagoonal and shallow marine sediments and the upper part as Pliocene deltaic and continental deposits. Goudarzi (1970) used the name Sahabi Group and divided it into a lower gypsiferous clay and limestone unit of Miocene age and an upper unit of Pliocene age. De Heinzelin et al. (1980) and Boaz et al. (1982) divided the Sahabi sequence into nine units: M, P, Q, R, S, T, U, V and Z. Units M and P were ranked as formations and assigned a Middle Miocene age and units Q-V were treated as members and assigned a Late Miocene-Early Pliocene age (Boaz et al., 1982) or Pliocene age (De Heinzelin et al., 1980). Giglia (1984) restricted the Sahabi

Formation to units M-T and assinged units U and V to the Pliocene-Pleistocene Garet Uedda Formation. Unit Z was considered to represent a soil horizon.

In the Ajdabiya area the Sahabi Formation overlies Middle Miocene siliceous limestones of the Wadi Al Qattarah Member of the Regima Formation and is unconformably overlain by the Garet Uedda Formation. The Sahabi Formation is made up of lagoonal, coastal bar, and evaporitic deposits of Late Miocene age. Giglia (1984) divided the Formation into Sabkha Al Hamra Member, Sabkha Al Qunnayyin Member, and Wadi Al Farigh Member. The Sabkha Al Hamra Member corresponds to units M, P and Q of De Heinzelin et al. (1980) and Boaz et al. (1982). The unit is 24 m (79 ft) in the type section near Qarat Al Ush. It consists of transgressive shallow marine fossiliferous sandy limestones and calcareous sandstones in the lower part, and a regressive evaporitic facies of sands, green clays, siltstones, gypsum, and subordinate sandy limestones, as well as paleosoil horizons in the upper part. These sediments contain vertebrate remains, molluscs, corals, benthonic and planktonic foraminifera, and ostracods. Foraminifera reported from this Member include *Globorotalia humerosa, Ammonia tepida, Quingueloculina oblonga, Elphidium aculeatum, E. decipiens*, and *Borelis melo* (this fossil is taken by many workers to be indicative of the Middle Miocene). It also contains the hermatic corals *Tembellastraea reussiana* and *Siderastraea*, the echinoid *Echinolampas hemispaer*, as well as the brackish water ostracod *Cyprideis torosa*. These assemblages indicate a Late Tortonian-Early Messinian age. The depositional environments of the Messinian Sabkha Al Hamra Member were probably protected lagoons, hard-bottom, well-oxygenated and brackish water. The Sabkha Al Qunnayyin Member interfingers with the Wadi Al Farigh Member and is overlain with a sharp unconformable contact by the Garet Uedda Formation. The type section is near Bir Hasi Rasur where the unit is 8.5 m (28 ft) in thickness. The Sabkha Al Qunnayyin Member consists of transgressive facies of low-moderate energy, shallow marine, protected environment with local high energy of interbedded limestones, calcareous sandstones, siltstones, and clays. The unit is poorly fossiliferous. Foraminifera reported include *Ammonia beccarii, Astigerina planorbis, Borelis melo*, and *Dendritina rangi* (cf. the Miocene of the Gebel Gharra on the eastern side of the Gulf of Suez, Souaya, 1963b). It was assigned a Late Tortonian-Early Messinian age by Giglia (1984). The Wadi Al Farigh Member consists of high-energy lime sand shoal and oolitic bar deposits. It is unconformably overlain by the Pliocene-Pleistocene Garet Uedda Formation. The exposed thickness of the unit in its type section in Wadi Al Farigh is 23 m (75 ft). It consists of cross-bedded oosparites and calcarenites, minor interbeds of green shales, calcareous siltstones, silty clays, sandy marls, and dolomitized marls. Fossils reported include the foraminifera species *Borelis melo, Dendritina rangi* and *D. cornucopiae*, the ostracod species *Xestobeberis geometra, Callistocythere pallida*, and *Candites aff. calceolatus*. It was assigned a Messinian age by Giglia (1984).

### 14.4.3 Miocene of Northwest Libya & offshore

In Northwest Libya in the Jifarah Plains all post-Triassic formations were eroded prior to Miocene sedimentation (Kruesman & Floegel, 1980). The Miocene overlies immediately the Norian-Rhaetian Bu Sceba Formation. Lower Miocene deposits consist of shales and calcareous sediments. The Middle Miocene is composed of sandy limestones and sandy shales with mollusc fossils. In the eastern part of central Jifarah polygenetic conglomerates of Middle Miocene overlie the Mesozoic rocks.

In northWest Libya the Miocene is represented by the Al Khums Formation. The Formation consists chiefly of limestones, chalks, and minor sands and shales. The name was introduced by Mann (1975a) in the Al Khums area east of Tripoli who estimated a thickness of 100 m (328 ft) for the unit. Salem & Spreng (1980) mapped the Formation in the Al Khums area and divided it into a lower An Naggazah Member and an upper Ras al Mannubiyah Member. They estimated a thickness of 155–175 m (806–574 ft) west of the Al Khums area. The former overlies Mesozoic rocks and is missing in areas of high relief. It is composed of bioclastic and bioher-mal limestones, calcareous, fine to medium-grained conglomeratic sandstones with pelecypods, gastropods, and corals, brownish-yellow, argillaceous limestones, sandy detrital limestones, and green gypsiferous shales. Fossils are pectinids, gastropods, and echinoids. The Ras al Mannubiya Member consists of three limestone units: a lower unit of white to light grey sandy limestones with abundant *Borelis Melo*, and a conglomeratic bed with abundant corals and other fossils at the base. The second unit is made up of non-fossiliferous sandy chalks. The upper unit consists of bedded limestones with locally abundant gastropods and algae. The Al Khums Formation was assigned a Middle Miocene age by Salem & Spreng (1980), Mann (1975b), and Sherif (1991) based on the presence of the alveolinid *Borelis melo*, while a Late Miocene age was proposed by Innocenti & Pertusati (1984) and El-Waer (1991) based on ostracods. The Al Khums Formation was correlated with the uppermost Marada Formation, Regima Formation, and Al Jaghbub Formation of Libya, and the Marmarica Limestone and equivalent Middle Miocene rocks in the Nile Delta and Cairo-Suez regions (Salem & Spreng, 1980, Barr & Walker, 1973). Hladil et al. (1991, Fig. 2) showed the Marada Formation in the Hofra Oil Field to be over-lain by the Al Khums Formation (Wadi Yunis Member), and assigned the latter a Tortonian-Messinian age. Salem & Spreng (1980) reported the following foraminifera from the Formation: *Elphidium crispum, Rotalia beccarii, Cibicides boueanus, C. parecinctus, Quinqueloculina laevigata, Q. Lamarck,* and *Globigerinoides trilo-bus,* in addition to thin-shelled oysters, pelecypods, echinoids, and the coralline algae (*Archaeolithothamnion*). The upper unit of the Al Khums Formation contains frag-ments of corals, oysters, and other pelecypods, and the foraminifera *Borelis melo, Rotalia beccarii,* and *Elphidium* sp. Wood (1964) observed hollow spheroids (pseu-do-oolites) with no vestiges of nuclei which may have initiated growth within beds presumed to be of Upper Miocene age in outcrops near Nofilia on the northwestern margin of the Site Basin. These spheroids are associated with oolitic limestones, but it is not known to which formation they belong to.

In the offshore area of northWest Libya the Miocene is represented by the Al Maya, Sidi Bannour, Ras Abd Jalil, Dirbal, Tubtah, Bir Sharuf, and Marsa Zoua-ghah formations, all established by Hammuda et al. (1985). The Al Mayah Forma-tion was introduced in the G1-NC41 type well at a drill depth of 3079–5886 ft for a unit composed of alternating shales, sandstones, and carbonates. It overlies the Dirbal Formation and is overlain by the Sidi Bannour or Ras Abd Jalil formations. Fossils reported are *Globigudrina dehiscens, G. altispira, Globigerinoides bispher-ica, Uvigerina rutila,* and *Textularia carinata,* suggesting a Middle Miocene age. The Sidi Bannour Formation in the E1-NC41 type well at a drill depth of 2420–3868 ft is made up predominantly of shales with intervals of sandstones. The unit is overlain by the Tubtah or Bir Sharuf formations and overlies the Al Mayah Formation. Fossils reported include miliolids, *Ammonia beccardai, Rotalia* sp, *Lithothamnium*, ostracods, gastropods, bryozoans, and echinoderms. Hammuda et al. (1985) assigned it a Late Miocene age. The Ras Abd Jalil Formation in the type section is in the E1-NC41 well at a drill depth of 4071–7019 ft consists of light grey marls and sandy shales in the lower part, argillaceous wackestones-mudstones interbedded with marls in the middle part, and glauconitic and pyritic grey and green shales and white to beige wackestones and packstones in the upper part. The Formation overlies the Ghalil Formation and is overlain by the Al Mayah Formation. The Ras Abd Jalil Formation is equivalent to the Ketatna Formation in the Tunisian offshore. The fauna reported are *Catapsydrax dissimilis, Globorotalia opima,* and *Globigerina ampliapertura.* This assemblage suggests a Burdigalian age for the Formation. The Dirbal Formation is made up of two units, an upper unit of wackestones-packstones and a lower unit of grainstones-boundstones interbedded with dolomites. The Formation consists locally of biostromes. The Formation is unconformably overlain by the Al Mayah Formation and overlies also unconformably the Tellil Group. The base is represented by the marker bed with *Nummulites vascus.* The unit is equivalent to the deep-water facies of the Ras Abd Jalil Formation. The fauna reported includes *Operculina lepidocyclina* and *Nummulites* sp. The lower part of the Tubtah Formation in the type section in the NOC-AGIP C1-137 well at a drill depth of 1661–2838 ft is made up of wackestones-packestones, grey-green, poorly fossiliferous shales in the middle part, and mudstones-wackestones in the upper part. The Formation overlies the Sidi Bannour Formation and is overlain by the Marsa Zouagha Formation. The fossils reported include ostracods, miliolids, Elphidiiae, and *Ammonia beccarii* indicating a Late Miocene (Tortonian) age for the Formation. The Bir Sharuf Formation was introduced by Hammuda et al. (1985) for a deep-water facies equivalent of the Tubta Formation. The type section is in the B1A-137 well at a drill depth of 1410–2568 ft. The Formation is made up of biomicrites, biosparites, and pelletal micrites, interbedded with calcareous shales. The unit is sandy in the middle and evaporitic at the top. The age and stratigraphic correlations of the Formation are not defined. The authors assumed a Serravallian-Tortonian age based on its stratigraphic position. The Bir Sharuf overlies the Sidi Bannour Formation and is overlain by the Marsa Zouagha Formation. The Marsa Zouaghah Formation is a Late Miocene carbonate and evaporite sequence in the C1-137 type well at a drill depth of 1685–1997 ft. The lower part is dolomitic mudstones-wackestones, and the upper part is made up of alternating gypsum, shales, and marls. The Formation overlies the Sidi Bannour, Tubta, or Al Mayah formations and is unconformably overlain by the Sbabil or Assabria formations.

## 14.5  PLIOCENE-PLEISTOCENE

The stratigraphy of the Pliocene and Quaternary sediments in Libya, like those of Egypt, is poorly understood. Differentiation between Pliocene and Pleistocene deposits is almost impossible with our current state of knowledge. Lithostratigraphic units that are believed to span the Pliocene-Pliestocene interval include the Garet Uedda,

Calanscio, Al Mahruqa, Al Assah, and Al Hishash formations from the onshore, and the Sbabil and Assabria formations in the offshore of northWest Libya.

In the Jabal Zaltan area the Marada Formation is overlain by the Garet Uedda Formation. The Garet Uedda Formation was introduced by Di Cesare et al. (1963) for a unit exposed at Qarat Weddah (29°43′N, 24°36′E) in northeastern Cyrenaica. In its type area the Garet Uedda Formation is composed of 84 ft (25.6 m) of quartzitic eolian sands with interbedded sandy shales, locally gypsiferous which contain brackish-water pelecypods and gastropods, foraminifera, and ostracods (Domaci, 1985). In the Ajdabiya area the Garet Uedda Formation is equivalent to members U and V of the Sahabi Formation of De Heinzelin et al. (1980) and Boaz et al. (1982). It unconformably overlies the Sabkha Al Qunnayyin and Wadi Al Farigh members of the Sahabi Formation and is covered by a caliche or a paleosoil horizon, which probably formed during the Pliocene-Pleistocene time (Giglia, 1984). The caliche is made up of reddish, pisolitic, sandy limestones with remains of terrestrial gastropods. Quartz sands, calcareous sandstones, silts, clays, and gypsum of the Garet Uedda Formation fill erosional channels cut into Middle Miocene carbonates (e.g. the Sahabi Channel of Barr & Walker, 1973, Griffin,, 2006). These sediments change facies eastward into marine or coastal-sabkha cross-bedded calcareous sandstones and sandy oolitic limestones, and biopelmicrites (Giglia, 1984). The Formation is barren of fossils in the Jabal Zaltan and Ajdabiya areas. It was assigned a Pliocene-Pleistocene age by Domaci (1985) and Giglia (1984) based on its stratigraphic position.

The name Calanscio Formation was introduced by Benfield & Wright (1980) because of difference in facies and thickness between the Garet Uedda Formation in the type locality and the Marada area. The Formation is made up of a thick sequence of up to 213 m (700 ft) of unconsolidated, medium to coarse-grained sands with rare thin clays. The type section is in the S1-103 well. The configuration of the base of the Calanscio Formation was controlled by structure and incised valley-like features. The unit has a maximum thickness of 150 m (490 ft) in the Marada area. It is limited in distribution to the eastern part of the Sirte Basin but is eroded to the west. It unconformably overlies Oligocene rocks such as the Diba Formation. The upper contact is erosional and capped with a limestone bed. Abundant plant remains and terrestrial vertebrate fossils occur in the lower part of the unit at Jabal Zaltan. Shallow marine fossils are common in the upper part of the unit and to the north. The lower part of the unit was assigned a Lower Miocene (Burdigalian) age by Savage & White (1965). While the age of the upper part remains uncertain, it is probably Middle Miocene. The Calanscio Formation was considered to be equivalent to the Garet Uedda Formation by Domaci (1985) who indicated that the name is no longer in use. However, the Garet Uedda Formation is believed to be Pliocene-Pleistocene in age while the Calanscio Formation is probably Lower-Middle Miocene.

Tertiary-Quaternary sediments in the southern Gargaf Arch were mapped as Miocene Lacustrine deposits by Collomb (1962) and as Tertiary continental rocks by Conant & Goudarzi (1964). The Tertiary-Quaternary continental deposits in the Sabha and Idri areas (Seidl & Röhlich, 1984, Parizek et al., 1984) consist of conglomerates at the base, calcareous sandstones, sandstones, sandy limestones with fragments of ostracods and Characeae and bird's-eye structures, glauconite, bones and scales of fish, pelecypods, gastropods, superficial ooids, and laminated dolomites with sporadic gypsum intercalations.

The name Al Mahruqah Formation was introduced by Seidl & Röhlich (1984) for a predominantly carbonate rock unit unconformably overlying the pre-Tertiary formations at the northern margin of the Murzuq Basin. The name is derived from the village of Al Mahruqah 25 km west of Brak. The type section is on a table hill 5 km east of Al Mahruqah. In the type section the Formation is 12 m (39 ft) thick and decreases to 5 m (16 ft) both eastward and westward. It is composed of sandy dolomitic limestones at the base and passes upward into limestones with ostracods, gastropods, and pelecypods and represents deposition in an extensive lacustrine system (Seidl & Röhlich, 1984, Parizek et al., 1984). The unit unconformably overlies the Mrar Formation in the Sabha map-sheet NG33-7 area (Seidl & Röhlich, 1984) and Ordovician to Devonian sediments in the Idri map-sheet NG33-7 area to the west (Parizek et al., 1984). It overlies and interfingers with Late Pliocene-Pleistocene clastic continental deposits in the Tmassah map-sheet NG33-7 area (Korab, 1984, Seidl & Röhlich, 1984). The Al Mahruqah Formation probably grades southward into continental clastic and carbonate rocks. The Middle Pleistocene Al Mahruqah Formation consists of at least four limestone beds at four distinct elevations. The corresponding limestone beds are named (from oldest to youngest) the Antalkhata Member, the Brak Member, the Bi'r az Zallaf Member, and the Aqar Member, and were formed in humid episodes with duration of several tens of thousands of years as shown by $^{230}$Th/U dating (Geyh & Thiedig, 2008).

The term Murzuq Limestone was introduced by Desio (1936) after the village of Murzuq (25°54'N, 13°55'E) for a unit composed of greyish white, yellow and cream limestones. It frequently weathers to rubble. Cherty limestones and argillaceous and silty sandstones occur in the upper part. The unit overlies older sediments and is limited to morphological depressions. It has been assigned extremely varied ages ranging from Uppper Jurassic to Quaternary. The Murzuq Limestone has been discussed in detail by Desio (1936), Freulon et al. (1955), Banerjee (1980), and Seidl & Röhlich (1984). A review of the controversy of the Murzuq Limestone was given by Banerjee (1980) who recommended abandoning the term.

In Northwest Libya in the Jifarah Plains and the northeastern Ghadames Basin the Pliocene-Pleistocene consists of the Al Assah and Al Hishash formations introduced by Mijalkovic (1977) and considered older than the Gargaresh Formation. Al Assah Formation is made up to 8–250 m (26–820 ft) of lagoonal sediments of clayey sandstones, gypsum, sands, silts and gravels. The Al Hishash Formation consists of estuarine sediments of cross-bedded greyish, loosely cemented sands with gypsum, calcarenites, sandy limestones, and marly and gypsiferous beds, developed mostly near the mouth of major wadis. These two formations are equivalent to the Pliocene-Pleistocene Raf Raf and Porto Farina formations in Tunisia.

In the offshore of Northwest Libya the Pliocene is represented by the Assabria and Sbabil formations. The Sbabil Formation was proposed by Hammuda et al. (1985). The type section is in the F1-NC41 well. The Formation is composed of grey to green fossiliferous shales and sandstones. The Formation is restricted to the basinal areas and its thickness is more than 550 m (1805 ft). It unconformably overlies the Marsa Zouagha, Sidi Bannour, or Assabria formations. The top of the Formation is the seabed. The fauna reported is mainly *Globorotalia inflata* and *G. crassformis*, in addition to *Bulimina* sp. and *Nonion boueanum*. It was assigned an Early Pliocene (Zanclean) age, but its age could be Late Pliocene (Piacenzian) because of the presence of *G. inflata* and *G. crassformis*.

The Assabria Formation was introduced by Hammuda et al. (1985). The type section is in the NOC-AGIP B1-NC41 well at a drill depth of 885–990 ft. The unit is the highest in the offshore stratigraphic section and its top represents the seabed. The Formation is made up of white grey fossiliferous sands with a few limestone beds. It unconformably overlies the evaporites of the Marsa Zouagha Formation. The fauna reported include *Orbulina universa*, *Guttulina communis*, and *Elphidium* sp., *Discorbis* sp., *Epinoides* sp., gastropods, pelecypods, bryozoans, and ostracods. It was assigned a Pliocene age by Hammuda et al. (1985). The Assabria Formation is probably the lateral equivalent of the Sbabil Formation.

## 14.6 PLEISTOCENE-HOLOCENE

Alternating humid and dry periods took place in Libya and Egypt during the Quaternary. Pluvials correspond approximately to the four classical glacial phases of the Alpine region: Günz, Mindel, Riss, and Würm. The sea level underwent a series of oscillations which caused lateral shifts of the shorelines. The origin of the coastal terraces of Cyrenaica and Marmarica is controversial.

McBurney & Hey (1955) described three well-defined Quaternary deposits in Cyrenaica: 1) shoreline at 6 m above sea level with marine shells, 2) tuffaceous deposits with fossil remains of plants and mammals and paleolithic remains (Levalloiso-Mousterian affinities) (McBurney, 1967), and 3) young fossil dunes with *Helix melanstona* and *Rumina decollata* and associated gravels (Younger Gravels). The terraces are alternating with steep slopes and partly covered by calcreted fluviatile gravel (Hey, 1968). Patches of cemented marine calcarenites (Ajdabiya and Gargaresh formations) are often hidden beneath the fluviatile gravel. The terraces are of marine origin. The foot of each slope marks the position of an ancient shoreline. The lowest terrace maintains a constant level of six meters. The other terraces are at 15–25 m, 35–40 m, 44–55 m, 70–90 m, and 140–190 m (49–82 ft, 115–131 ft, 144–180 ft, 230–295 ft, 459–623 ft, respectively). To the southeast of Benghazi, the six-meter terrace is underlain by earlier Pleistocene calcareous silts with marine shells and brackish-water ostracod species (Hey, 1968). They contain plant remains and rootlets in the lower part and abundant pebbles of local Miocene limestones in the upper part. These deposits are believed to be estuarine deposits formed during a marine transgression that preceded the accumulation of the 6 m shoreline (Hey, 1968). The six-meters shoreline is also covered by the so-called Younger Gravels (McBurney & Hey, 1955). Their red color suggests the absence of perennial water and their coarseness indicates violent floods. The Younger Gravels yielded Levalloiso-Mounsterian artifacts. Near the Gulf of Bomba Moseley (1965, cited by Hey, 1968) described two successive series of ancient gravels (Fig. 14.3). The older is completely calcareted, the younger is only partially calcareted. The latter is probably equivalent to the Younger Gravels (Hey, 1968).

Vita-Finzi (1971) divided the wadi fills in the Al Jabal Al Akhdar area into two units: the Younger fill (the Bel Ghadir Alluvium), and the Older fill (the Kuf Alluvium). The Kuf Alluvium was named after Wadi al Kuf (32°43′N, 21°40′E). It is composed of gravels in terra-rossa matrix and forms alluvial fans at the foot of the coastal escarpment. Vita-Finzi (1971) correlated the Kuf Alluvium with the Younger

Gravels of Hey (1968). The Bel Ghadir Alluvium was named after Wadi Bel Ghadir Kuf (32°50′N, 21°44′E). It lies unconformably on the Kuf Alluvium. It has a maximum thickness of six meters (20 ft). It yielded Greek and Roman pottery. Vita-Finzi (1971) correlated the Bel Ghadir Alluvium with the Lebda Alluvium in Tripolitania. The name Lebda Alluvium was introduced by Vita-Finzi (1971) for the Younger fill in Tripolitania. It was named after Wadi al Lebda (32°37′N, 14°17′E) east of Al Khums. It is composed mainly of fine and coarse gravels, sands, and silts which fill the wadis. It has a maximum thickness of 10 m (33 ft). The Lebda Alluvium overlies with an erosive contact the Jefara Alluvium of Vita-Finzi (1971) which was included in the Jefarah Formation by El-Hinnawy & Cheshitev (1975).

Quaternary deposits along the Libyan coast are represented by the Ajdabiya and Gargaresh formations, sabkha sediments, eolian deposits, beach and coastal dune sands, and alluvium deposits. Desio (1971) attempted to correlate these Quaternary deposits in various parts of the Libyan coast.

Quaternary sediments in the Jifarah Plain consist of three formations: the Qasr al Haj Formation, the Jifarah Formation, and the Gargaresh Formation. These formations are equivalent and represent a south to north change in facies from alluvial fan gravels near the Jabal Nafusah to fluvio-aeolian sands and silts to aeolian calcarenites and littoral sands and gravels in the coastal areas (Anketell & Ghalleli, 1991b).

The Agedabia (Ajdabiya) Formation was first described by Gregory (1911) as the Ajdabiya Sandstone, and later by Desio (1935) as "*Panchina* with *Cardium*", that is marine terrace deposits. The Formation is composed of white calcareous sandstones essentially of skeletal debris and rounded limestone clasts. The type locality is in small quarries in the vicinity of Ajdabiya. Fauna reported from this Formation include *Acanthocardia* cf. *echinata, Arca noe, Chamalea gallina, Dosinia lupinus*, and *Mytilus edule* (Giglia, 1984). It was deposited during the Pleistocene-Lower Holocene time probably as a result of the Tyrrhenian marine transgression (Desio, 1935, Giglia, 1984). In the Ajdabiya area the Ajdabiya Formation is overlain by the Gargaresh Formation (Giglia, 1984). The Gargaresh Formation was introduced by Burollet (1960) as the "Gargaresc Sandstone" which is quarried at Qarqarish (32°49′N, 12°55′E) west of Tripoli. It forms steep cliffs along the Mediterranean coast in Libya and Tunisia. Burollet (1960) considered the Agedabia Formation to be equivalent to the Gargaresh Formation, a name which he preferred. The thickness of the Gargaresh Formation varies from 10–50 m (33–164 ft). It is composed of cross-bedded calcarenites and oolitic limestones with abundant reworked planktonic and benthonic foraminifera, echinoid fragments, calcareous algae, molluscs and bryozoans (Giglia, 1984) (Figs. 14.3a, b). It was described as eolinites or fossil dunes of Pleistocene-Holocene age (El-Hinnawy & Cheshitev, 1975, Francis & Issawi, 1977, Giglia, 1984). El-Hinnawy & Cheshitev (1974) selected the type section of the Gargaresh Formation along the shore to the west of Janzur village (32°49′N, 12°55′E) where the Formation is approximately 14 m (46 ft) in thickness, but the base is not exposed. They dated the exposed part as Late Pleistocene-Holocene.

El-Hinnawy & Cheshitev (1975) mapped the Ajdabiya and Gargaresh formations in the Tripoli area under the name Gargaresh Formation (called Qarqarish Limestone on the Geological Map of Libya, 1977, 1985). Francis & Issawi (1977), on the other hand, considered these two units in the Soluq area to be equivalent and mapped them together under the name Ajdabiya Formation. In other areas along the coast such as in Cyrenaica these units were mapped as marine and eolian calcarenites (Röhlich, 1974, Klen, 1974) (Fig. 14.2a, b).

The Qasr al Haj Formation was introduced by El-Hinnawy & Cheshetiv (1975) who divided it into two members: a lower member of cemented conglomerates, capped by a thick calcrete of possible Villafranchian age, and an upper member of unconsolidated gravels which correspond to the two series of ancient gravels of Moseley (1965). The Qasr al Haj Formation is composed of conglomeratic mudstones and siltstones with large pebbles up to 80 cm in diameter floating in a silty clay groundmass, conglomerates, and sandy conglomerates (Anketell & Ghellali, 1991b) (Fig. 14.3).

The name Jeffara Formation was introduced by Desio et al. (1963). El-Hinnawy & Cheshitev (1975) selected the type section of the Jifarah "Jeffara" Formation 6 km to the west of Sebratah along the sea cliffs where the Formation is more than 14 m (46 ft) in thickness, but the base is not exposed. Anketell & Ghellali (1991a) redefined the Jifarah Formation based on a composite stratotype. The redefined Jifarah Formation is described as white, buff, pink, and red to red-brown, fine sands and silts with subordinate gravel. The thickness of the Formation varies from 0–50 m (0–164 ft) and it is locally overlain by the Gargaresh Formation, Qasr al Haj Formation, or Holocene sabkha deposits. They also divided the Jifarah Formation into

*Figure 14.2a* A polished rock slab of the Gargaresh calcarenite, near Marsa El Brega.

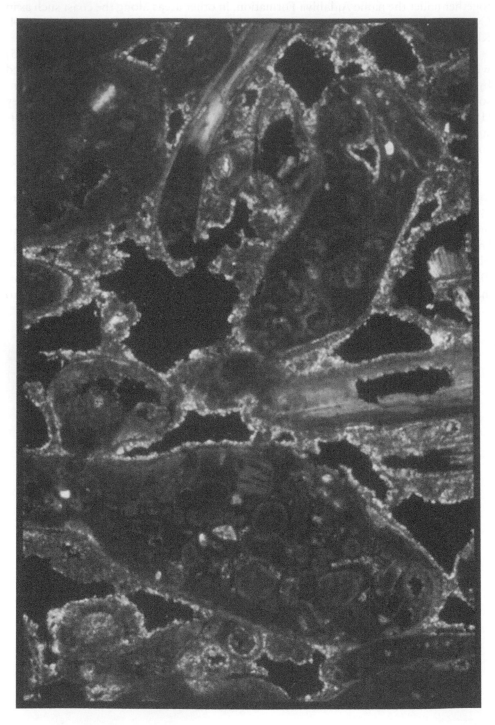

*Figure 14.2b* Thin section photomicrograph of the same sample.

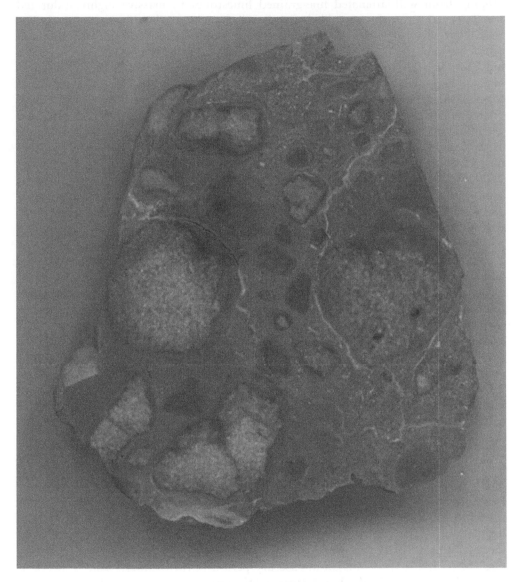

*Figure 14.3* A polished rock slab of a red conglomerate, probably equivalent to the "Old Conglomerate", composed of calcareous pisoids, Gulf of Bomba.

seven members: Arroman (Q1), Shuraydat (Q2), Karawah (Q3), Qarabulli (top Q4), Zabayin (Q4), Wadi Raml (Q5), and Wadi Megenin (Q6). The Arroman and Shuraydat members are separated by the Udul calcrete and the Shuraydat and Karawah members are separated by the Wadi Sariyah calcrete. The Arroman Member is composed of calcreted pink silts and conglomeratic fine-grained sands. Its age is probably late Pliocene-early Pleistocene. The Udul calcrete is made up of varied lithologies ranging from well-laminated fine-grained limestones to massive highly indurated conglomeratic crust. Its most likely age is Villafranchian (Pliocene-Pleistocene). The Shuraydat Member is red, pink fine sands and buff silts and calcretes. Its age is in doubt but could be Würm/Weichsel or Riss/Saale. The Wadi Sariyah Calcrete is a well-indurated white, pink, and brown calcrete with solution pipes filled with terra-rossa-like soil and a well formed root system. Its age is also doubtful, but it was probably formed during the Riss/Saale time. The Karawah Member is made up of white pure quatrz sands and buff loessic silts. Its probable age is Würm/Weichsel. The Qarabulli Member is grey to black weakly calcretized with freshwater gastropods. Archeological evidence suggests that its age is probably Pleistocene. The Wadi Raml Member is composed of buff to ochre, fluvial sands and gravels and dune sands. Its age is probably 1500–500 years BP. The youngest member, the Megenin Member is made up of fluvial sands and gravels and active sand dunes which are still forming at the present time. Anketell & Ghellali (1991a) indicated that the type section chosen by El-Hinnawy & Cheshitev (1975) from the coastal cliff west of Sebratah is composed only of the upper part of the Shuraydat Member, and that the name should be abandoned.

In the Jabal Zaltan area the Pleistocene deposits overlie Oligocene sandstones and form numerous small conspicuous hills. The sediments are composed mainly of lacustrine sandstones interbedded with green sandy claystones with Mn oxides, and thick laminated, whitish limestones with gypsum, and green silty and gypsiferous claystone laminae (Domaci, 1985). The thickness of the sediments in that area does not exceed 10 m (33 ft).

At the end of the Upper Pleistocene and during the Holocene three successive alluvial terrace deposits were present around the northwest Tibesti Mountains (Rognon, 1980): 1) the oldest terrace (*Oberterrasse*) is an early Pleistocene aggradation, also known as the *Hauptterrasse*. The Oberterrasse accumulated during an arid phase of climatic desiccation. It is composed of sands and gravels similar to the modern Wadi deposits. The age assigned to this terrace is late Acheulian. 2) The middle terrace (*Mittelterrasse*) formed during an interval of downcutting from several metres to 30 m above sea level The terrace is characterized by widespread accumulation of silts and clays. The brown or grey silts are often highly organic, suggesting overbank accumulation. The terrace is very discontinuous. Deposition of clays and silts is usually bracketed by intervals of greater fluvial discharge. The middle terrace was formed during the latter part of a pluvial during the Early Holocene, perhaps during the late Würm. It has been dated at 15,000–8000 yrs. BP. Following the *Mittelterrasse* aggradation there was an intense dry period in the southern Sahara at about 8000–7500 yrs. BP. 3) the lower terrace (*Niederterrasse*) consists of cross-bedded coarse sands and fine to medium gravels. This alluvial terrace was deposited under wetter conditions than those which were prevalent during the aggradation of the upper terrace. Deposition of the *Niederterrasse* alluvium probably took place toward the close of a pluvial phase which took place around 5700–4100 yrs. BP. Aggradation of the terrace continued until after 2690 ± 435 BP.

On the northern slope of the Tibesti Mountains, Hagedorn (1980) recognized three terraces which formed during the Late Pleistoncene-Holocene. 1) The oldest terrrace is the high terrace (*Hochterrasse*) of which only remnants have been preserved. It is a rock-cut terrace, aggradational in places. 2) The next terrace is the *Oberterrasse* or *Hauptterrasse* which is composed of alternating beds of gravels and sands. This terrace yielded a radiocarbon date of 15,000–8000 yrs. BP. Hagedorn (1980) believed that the *Mittelterrace* described by Rognon (1980) and the *Oberterrasse* belong to the same cycle. The youngest terrace is very similar to the high terrace (*Hochterrasse*). It is equivalent to the lower terrace (*Niederterrasse*) of Rognon (1980). This terrace is made up of alternating layers of gravel and sands which have been dated at 6000–3000 yrs. BP. A renewed phase of deposition took place around 2000 yrs. BP.

# Chapter 15

# Phanerozoic geology of Algeria

Algeria is bordered on the north by the Mediterranean Sea, with a 1200 km of shoreline, by Morocco to the west, by Tunisia and Libya to the east, Mauritania and Western Sahara to the south-west, and Mali and Niger to the south (Fig. 15.1). Algeria has an area of 2,381,741 km$^2$, which makes it the second largest country in Africa after Sudan. It has a population of 35.7 million (2010 estimates). The main mineral resources of Algeria are oil and natural gas, iron ore, phosphate, uranium, lead, and zinc. Proven oil reserves are estimated at 13.4 billion barrels (The World Factbook, 2010), with an average production rate of about 2.1 million barrels of oil/day (2009 estimates). Proven gas reserves are estimated at 159 TCF (Oil & Gas Journal, January 2010). It ranks tenth in the world's natural gas reserves and 14th in oil reserves (eia, 2010). Major fields include the Cambrian Hassi Messaoud Field and the Triassic Hassi Rmel Field. The former was discovered in 1957 and the latter was discovered in 1956 and put on production in 1961.

The geology of Algeria was discussed by Gautier (1922) and has been summarized more recently by Askri et al. (1995). Algeria's hydrocarbon potential is discussed by Askri et al. (1995), MacGregor (1998), Echikh (1998), Klett (2000a, b), and Attar & Hammat (2001).

Algeria consists, from south to north, of the following structural domains (Fig. 15.1): the Sahara Platform, the Atlas System, the External Tell System, the Flyschs Domain, and the Kabylides Domain. The Mesozoic-Tertiary tectonic history of Northern Algeria includes: A Triassic rifting episode, a Late Triassic-Early Jurassic post-rift phase dominated by thermal subsidence, and a Cenozoic compressional phase (Benaouali et al., 2006). The compressional phase resulted in a minimum value of 40 km (20%) of horizontal shortening. This phase is reflected in a Late Eocene deformation pulse, Early Miocene deposition of a thick flexural sequence, Middle-Late Miocene emplacement of the Tell nappes, and Late Miocene-Recent thrusting. In addition, the offshore continental shelf of Algeria is covered by Tertiary and Quaternary sediments, with 1000 m (3281 ft) to 3500 m (1148 ft) of Mio-Pliocene sediments, overlying a metamorphic basement (Askri et al., 1995). Little is known of the Pan-African basement of most Algerian basins.

The Sahara Platform consists of a Pan-African Precambrian basement, represented by the Touareg Shield, and a weakly deformed Paleozoic and Mesozoic sedimentary cover. The Shara Platform is separated from the Atlas System by the South-Atlas Front (SAF). The Saharan Desert is covered by large sand dunes (East and West Grand Erg)

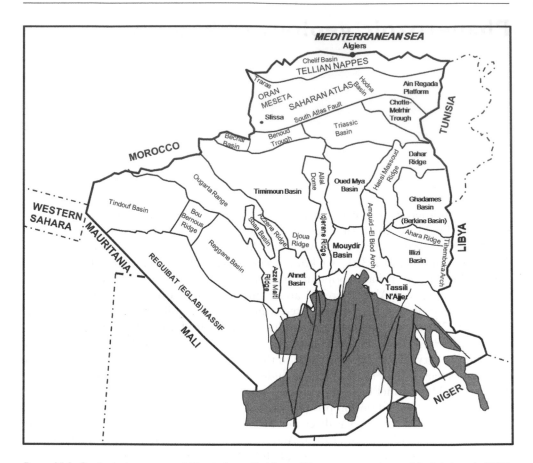

*Figure 15.1* Geological structure of Algeria (compiled from different sources, but mainly Askri et al., 1995).

and gravel plains (regs) with dispersed oases such as El-Oued, Ghardaia and Djanet. The hypothetical line linking the Eglab Range in the west and the Hoggar Mountains in the east marks the southern limit of the Algerian Sahara (Askri et al., 1995). The Sahara Platform includes a number of major Paleozoic basins, such as the Tindouf, Reggane, Bechar, Ahnet-Timimoun, Mouydir, Aguemour-Oued Maya, and Illizi-Ghadames, separated by tectonic highs, as well as the Tassili N'Ajjer (Askri et al., 1995) (Fig. 15.1):

The Tindouf Basin and Reggane Basin located on the northern and northeastern side of the Reguibat Shield. The sedimentary column reaches over 8000 m (26, 248 ft) in the Tindouf Basin and 6500 m (21, 327 ft) in the Reggane Basin.

The Bechar Basin is bordered to the north by the High Atlas and to the south and west by the Ougarta Range. Its sedimentary succession reaches 8000 m (26, 248 ft) in thickness.

The Ahnet-Timimoun Basin is limited to the north by the Oued Namous shoal, to the west by the Ougarta Range, to the south by the Touareg Shield, and to the east by the Idjerane-Mzab Ridges. Sediments thickness is up to 4000 m (13,124 ft). In the south, Ordovician and Devonian reservoirs are gas prone. In the north, oil has been discovered in all the Paleozoic formations in the Sbaa Trough.

The Mouydir Basin and Aguemour-Oued Mya Basin are bordered to the west by the Idjerane-Mzab Ridge and to the east by the Amguid-El Biod Ridge. Paleozoic sediments crop out in the Mouydir area. The Oued Mya Basin contains approximately 5000 m (16,405 ft) of Paleozoic-Cenozoic sediments. The Cambrian Hassi Messaoud gas field and the Triassic Hassi R'mel Field are located within this basin.

The Berkine/Ghadames Basin (Fig. 15.1) is the western extension of the pericratonic Ghadames Basin in Libya. It is bordered to the west by the Amguid-El Biod High and to the east by the Tihemboka Arch. It covers parts of Eastern Algeria, Southern Tunisia, and West Libya. It contains more than 6000 m (19,686 ft) of Paleozoic and Mesozoic sediments (Echikh, 1998). The area is covered by sand dunes of the Grand Erg Oriental more than 300 m in height. The Berkine Basin is bordered on the northeast by the Damah zone and on the northwest and west by the Hassi Messaoud Ridge. It is separated from the Illizi Basin on the south by the Mole D'Ahara. The Berkine Basin contains a number of giant oil and gas fields, including Rhourde El Baguel (Mitra & Leslie, 2002), Gassi Touil, Tin Fouye-Tabankort, and Alrar. El Borma Field, which straddles the Algerian-Tunisian border, is located in the northeast part of the Berkine Basin. The geology and petroleum geology of the Berkine Basin were reviewed by Gauthier et al. (1995), Askri et al. (1995), Boote et al. (1998), Echikh (1998), Guiraud (1998), Daniels & Emme (1995), Boote et al. (1998), Macgregor (1998), Echikh (1998), Cochran & Petersen (2001), Yahi et al. (2001), and Underdown & Redfern (2008), among others. Hydrocarbon accumulations are restricted to the west and north portions of the basin and are closely related to truncation edges of various Paleozoic strata (Cochran & Petersen, 2001).

The Illizi Basin is separated from the Mouydir Basin to the west by the Amguid-El Biod High and from the Murzuq Basin in Libya by the Tihemboka Arch (Fig. 15.1). Paleozoic sediments, up to 3000 m (9843 ft) crop out on the Southern fringes of the basin where they constitute the Tassilis (Askri et al., 1995). Mesozoic rocks crop out in the centre of the basin and Tertiary deposits are exposed on the northwestern flank. The petroleum systems and potential of the Illizi Basin have been treated by Askri et al. (1995), Boote et al. (1998), Chaouchi et al. (1998), Klett (2000a, b), Hirst et al. (2002), Quesada et al. (2003), Dixon et al. (2008), Wendt et al. (2009), and Galeazzi et al. (2010).

Tassili N'Ajjer in SE Algeria takes its name from the word Tassili which is the Tuareg name for the large plateaus where the Ajjer tribe lives. The area was first surveyed by Killian (1922). Some 3000 m (9843 ft) of Paleozoic sediments crop out in the Tassilis (Askri et al., 1995) and extend into the subsurface of the Illizi and Berkine basins (Eschard et al., 2005). It was subdivided by Kilian (1922) into Internal and External Tassili areas.

Northern Algeria is subdivided into four paleogeographical and structural units (Fig. 15.1); the Oran Meseta, the Saharan Atlas, the Southeastern Constantine area, and the Tellian Domain.

The Atlas system is comprised between the Tell and South-Atlas Fronts. The Atlas has been recognized as part of the Alpine System for a long time (Gautier, 1922). The System extends from the Atlantic Ocean in Southwest Morocco to Tunisia (Fig. 15.1) (see discussions of the Atlas System in respective parts). In Algeria, the Atlas System comprises the Atlas fold-thrust belt (Sahara Atlas and Aurès) and the High Plateau Domain with a thin and tabular Mesozoic sequence. The contact between the Tell and

Atlas systems is marked by the Pre-Atlas Domain, a region characterized by anticlines developed over Upper Triassic evaporites (cf. Tunisia). The Pre-Atlas and the Sahara Atlas are separated by the North Atlas Fault (NAF), a north-directed thrust fault. The North Atlas Fault is a composite feature involving shallow thin-skinned deformation within the sedimentary cover (Bracène & Frizon de Lamotte, 2002, Bracène et al., 1998). Inversion and related uplift of the Atlas System of Algeria resulted from two main tectonic events; the first led to major deformation and the development of large NE-SW-trending folds from the Middle Eocene to the Aquitanian; the second generated E-W-trending folds and thrusts during the Pleistocene (Benaouali et al., 2006, and references therein). The Miocene was a period of relative quiescence, characterized by the deposition of a relatively thick continental and marine sequence. The Miocene sequence is preserved in the Burdigalian-Tortonian M'sila Basin which rests unconformably on folded and faulted Cretaceous rocks of the Atlas System (Benaouali et al., 2006).

The External Tell System is located between the Flyschs Domain and the Atlas Chain. It is formed by an accretionary complex deriving from the African paleo-margin of the Maghrebian Tethys. The Triassic to Neogene sequence is composed predominantly of marls. The Tell System overthrusts the Atlas. It is overlain by the Miocene sequences of the M'sila, Chelif, and Hodna basins, which contain a thick pile of Upper Cretaceous to Lower Miocene allochthonous terranes of Tellian nappes (Askri et al., 1995, Benaouali et al., 2006). To the north, the Tell nappes rest upon Jurassic to Albian strata (Benaouali et al., 2006).

The Flyschs Domain is exposed in large thrust sheets or nappes, between the Bibans imbricated thrust stack and the Kabylides (Benaouali et al., 2006). In the Flyschs Domain is a major décollement situated between the Lower Cretaceous siliciclastic levels and the Jurassic substratum, the Achaiches Series. The Oligocene-Lower Miocene levels are also detached from their substratum, forming the Numidian thrust sheet (Moretti et al., 2007) composesd of up to 2000 m (6562 ft) thick turbidite sequence.

The Kabylides form an upper crustal wedge that was initially situated north of the Maghrebian Tethys (Bouillin, 1986), but was overthrust the North African margin along the Miocene suture zone during the Alpine Orogeny. It is divided into two geographical units; the Grande Kabylie and the Petite Kabylie (Fig. 15.1). It comprises different nappe complexes equivalent to the ones known in the Rif. The Kabylie Dorsale and the Kabylie Massifs consist of a pile of thrust sheets involving thick Lower Jurassic platform carbonates resting on Upper Paleozoic phyllites and Triassic sediments, followed by a thin cover of Jurassic-Cretaceous pelagic and Cenozoic siliciclastic sediments. The Kabylides are covered by Oligocene-Miocene molasses. The *Dorsale calcaire* is considered as the former northern passive margin of the Maghrebian Tethys (Bouillin, 1986).

## 15.1   LITHOSTRATIGRAPHY OF THE SAHARA PLATFORM

## 15.2   PALEOZOIC

Little is known about the Lower Paleozoic sediments in Northern Algeria. Gneisses, mica-schists, sandstones, and quartzites ranging in age from Ordovician to Permian have been reported from the Ghar-Rouban Mountains, Tiffrit, Traras, Chenoua, and

Kabylie (Askri et al., 1995). In the Little Kabylie metamorphic basement outcrops, up to 5000 m (16,405 ft), are covered by basal conglomerates followed by graptolitic shales and limestones with Tentaculites and Orthoceratid.

## 15.2.1    Cambrian/Ordovician of the Sahara Platform

Pioneering work on the Paleozoic of Algeria was carried out by Rolland (1890), followed by Flamand (1911, cited by Ouzegane et al., 2003) and Kilian (1925). The discovery of hydrocarbons in the Illizi Basin in the late 1950s provided invaluable stratigraphic and sedimentological data (Echikh, 1998, Traut et al., 1998), but most of the data remain unpublished (Dubois et al., 1968, Beuf et al., 1971, Eschard et al., 2005). Kilian (1922) differentiated three units in the Lower Paleozoic succession, from base to top: the Internal Tassili later identified as the Cambrian-Ordovician sandstones (Fabre 1988); the Intra-Tassilian characterised by the graptolitic Silurian shales, and the External Tassili, comprising the upper Silurian and Devonian shales and sandstones.

Lower Paleozoic succession exposed in the Tassili N'Ajjer area is the equivalent to the reservoir and source rocks in the Illizi and Berkine basins (Eschard et al., 2005). Large flexures induced regional unconformities. Fluvial and shallow marine sedimentation alternated in response to sea-level changes. The Ahara High, between Berkine (Ghadames) and Illizi, was also active during most of the Lower Paleozoic. Lower Paleozoic Fluvial sediments form continuous sand-sheets at a regional-scale. Stacked estuarine channels and bars form complex reservoirs, while wave and storm-dominated shoreface units make more homogeneous reservoirs. Many of the Lower Paleozoic reservoirs have been silicified and tightly cemented during diagenesis (Kaced, 2003). However, fractures may enhance the permeability of the reservoir, such as the case of the Hassi Messaoud Field in Algeria and the Attahaddy Field in the Sirte Basin (Tawadros et al., 2001).

Early Ordovician (Taconic) and Late Silurian-Early Devonian (Caledonian) tectonics caused uplift and erosion on the southwestern and southern flanks of the Berkine Basin (Echikh, 1998, Cochran & Petersen, 2001) which led to the deposition of lower Paleozoic reservoirs and Lower Silurian source rocks.

Late Ordovician (Ashgillian) glaciogenic sequences are important exploration targets and important producers of petroleum in North Africa and the Middle East (Beuf et al., 1971, Crossely & McDougall, 1998, Hirst et al., 2002). The glaciogenic sandstones of the Tamadjert Formation equivalent to the Memouniat Formation in West Libya, form a gas reservoir at the Tiguentourine Field, Illizi Basin, Algeria. A Late Ordovician glacial event led to the formation of incised valleys which were filled by glacio-marine sediments (Eschard et al., 2005). Outcrops in the Tassili N'Ajjer region show tunnel valleys incised into the subglacial surface and were attributed to the action of subglacial meltwater flows by Hirst et al. (2002).

The Ashgillian glaciogenic sequences typically have complex architectures that reflect multiple phases of ice advance and retreat and a variety of depositional processes, and glacio-tectonism (Dixon et al., 2008). The Late Ordovician outcrops of the Tassili N'Ajjer show the development of extensive erosional features including mega-valleys, paleovalleys, and intra-valley channels. The paleovalleys in outcrops comprise slightly sinuous valley segments, up to 50 km in length, with an early fill dominated by ice contact facies. Intra-valley surfaces are marked by large scale glacial lineations generated during ice advance. Subsequent ice retreat is marked by the deposition of high and low-density turbidites interbedded with debrites, with sheet-like architecture overstepping the paleovalley margins. They pass upward into sinuous and meandering channel/levee complexes. The sequences are interpreted as a large deltaic system fed by glacial outwash (Dixon et al., 2008).

The Cambrian-Ordovician (Internal Tassili) succession in the Ougarta Range consists of six transgressive-regressive (T-R) low-frequency (15–25 Ma) depositional sequences composed of inner-shelf siltstones to fluvial sandstones (Ghienne et al., 2007). These sequences have been correlated with the Anti-Atlas of Morocco and the Ahnet area in Southern Algeria. Sequence 1 (Lower Cambrian) comprises fluvial deposits. Sequence 2 (lower and middle Middle Cambrian) contains marine sediments, but fluvial conditions were maintained to the South. Sequence 3 (upper Middle Cambrian) is erosionally truncated. Sequence 4 (Lower Ordovician) inner-shelf siltstones extend to southern Ougarta. Sequences 5 and 6 (Middle-Upper Ordovician) were developed under inner-shelf conditions throughout the area and represent maximum flooding. The Hirnantian glacial sediments form a lowstand wedge in the upper part of Sequence 6 regressive system tract. Petroleum geologists consequently subdivided the Cambrian-Ordovician succession into four stratigraphic units; units I to IV of Beuf et al. (Debyser et al., 1965, 1971). Unit IV corresponds to the latest Ordovician (Hirnantian) glacially related succession. Pioneering work of Poueyto (1952), Arbey (1962), and Gomez Silva et al. (1963) proposed in the Ouagarta region led to the lithostratigraphic nomenclature that is still currently used (Legrand, 2002, Fekirine & Abdallah 1998).

The Cambrian-Ordovician succession in the Ougarta Range consists, in an ascending order of, the Sebkhet el Mellah, Ain Nechea, Kheneg el Aatene, Foum Zeidiya, Bou M'haoud, and Djebel Serraf formations (Fig. 16.4). It shows a variety of facies ranging from fluvial coarse-grained sandstones to offshore fine-grained sediments (Ghienne et al., 2007). They consist of pre-glacial sediments, braided fluvial, bioturbated sand-flat, mixed tidal and upper shoreface, upper shoreface, tidal ridges, storm-dominated shoreface, storm-dominated inner shelf deposits, and glacially-related deposits of the uppermost Ordovician (Hirnantian). The fluvial facies form the main part of the Sebkhet el Mellah Formation and part of the overlying Ain Nechea Formation and consists of course to very coarse arkosic, high-angle trough or planar cross-laminated sandstones. The Cambrian succession is made up of bioturbated, more mature, shallow-marine tidal, coastal plain and estuarine sediments. The latter are represented by very fine-grained, argillaceous and micaceous sandstones interfingering with cross-laminated *Skolithos*-bearing sandstones with occasional liquefaction structures. The Bioturbated sand-flat facies consists of sandstones with *Skolithos* in the Ain Nechea Formation and the Kheneg el Aatene Formation. The mixed tidal and shoreface facies is composed of a coarsening- and thickening-upward succession, with hummokey cross-stratification (HCS), wave ripples, and linguliform brachiopod shell beds at the top. This facies is common in the Ain Nechea Formation, the upper Foum Zeidiya Formation, and the Bou M'haoud Formation. *Daedalus* structures are relatively abundant in the Foum Zeidiya Formation. The upper shoreface facies comprise thick sandstone beds with horizontal to low-angle laminations and very coarse-grained lenses with rip-up clasts and occur in the middle part of the Kheneg el Aatene Formation. The Storm-dominated shoreface facies consist of large-scale amalgamated HCS, micaceous sandstone beds with frequent shell-rich limy concretions and siltstone clasts and disarticulated linguliform brachiopods, and occasional trilobites, bivalves, gastropods and graptolites. It has been recognized in the Kheneg el Aatene Formation, the upper part of the Foum Zeidiya Formation, the uppermost Ain Nechea Formation, and the Bou M'haoud Formation. The Tidal ridges facies occurs in the Bou M'haoud Formation and is composed of thickening- and coarsening-upward sandstone units with HCS, and medium to coarse-grained trough-cross-bedded sandstones. *Skolithos* and shell beds (echinoderm and brachiopod fragments) occur in both facies. Uppermost Ordovician (Hirnantian) glacially-related sediments constitute the Djebel Serraf Formation, which has been subdivided into three members by Legrand (1988). The middle El Kseib Member consists of glacial deposits, while the upper Ksar d'Ougarta Member is made up of post-glacial transgressive facies. A Fluvio-deltaic facies forms the main part of the glacial sediments of the lower member of the Djebel Serraf Formation. A Glaciomarine facies constitutes the El Kseib Member of the Djebel Serraf Formation. It consists of thin, <25 m (82 ft), and lenticular sandy diamictite bodies resting on an erosional surface. They contain abundant chitinozoans of the Hirnantian *elongata* Biozone (Paris et al. 1995, 2007). A post-glacial facies association comprises the bulk of the upper member or the Ksar d'Ougarta Member of the Djebel Serraf Formation (Legrand, 1988) and forms a widespread sandsheet that rests unconformably on underlying glacially-related strata of the lower and middle members. This facies is made up of medium- to coarse-grained, *Skolithos*-bearing, cross-bedded sandstones with abundant brachiopods and trilobites, and a few graptolites, grading upward in into wavy heterolithic fine-grained sandstones and then into siltstones and black shales of lowermost Silurian.

## 15.2.2 Silurian

The Silurian in Northwest Africa is characterized by exensive black graptolitic shales, sandy shales, and more rarely, carbonate layers especially in the north. They form a marker bed throughout the Sahara Platform. Silurian shales are very rich in organic matter and provide prolific source rocks for many the fields in Northwest Africa. Balduchi & Pommier (1970) reported two Silurian balck shale intervals with high organic carbon content and kerogen to the northwest of the Hassi Messaoud Field in Algeria. Silurian shales have been studied extensively by Lüning and others in a series of papers (for example Lüning et al., 2000c, 2003a, b, 2004, Vecoli et al., 2009).

The lithostratigraphy of the Silurian-Devonian succession in the outcrops was established by Dubois et al. (1967) and later by Beuf et al. (1971). Sequences defined in outcrops in the Tassili N'Ajjer area can be correlated with those in the subsurface of the Illizi Basin (Eschard et al., 2005). The facies changes gradually to more marine from south to north. However, rapid facies and thickness changes occur near the paleohighs which were active during sedimentation. Lower Devonian sediments completely pinch out over the Tihemboka high (Beuf et al., 1971).

Kriz (2008) described the bivalve *Algerina* gen. nov. and its type species *Algerina algena* sp. nov. from the Silurian (lower Ludlow) Oued Ali Clay Formation in the Ougarta Range of Algeria, the western Hoggar Mountains, and the Ahnet Massif. This genus is also known from the upper Wenlock of France, the lower Ludlow of Italy, and the Carnic Alps, as well as the lower and middle Ludlow of the Prague Basin in Bohemia and the lower Ludlow of the Tajmyr Basin, Russia.

### 15.2.3 Devonian

In the Illizi Basin, the lower Devonian sequence (Lochkovian-Emsian or Gedinnian-Siegenian) includes the F6 reservoir (equivalent to the Tadrart Formation), made of regressive deposits, marginal marine to fluvial, medium to coarse sandstones (Askri et al., 1995, Hallett, 2002). The F6 is characterized by porosities that may reach more than 20% (Askri et al., 1995). The F6 Sand offshore coastal bars in the northwestern Illizi Basin contain important hydrocarbon reservoirs (Alem et al., 1998). The F5 Sand (Emsian, equivqlent to the Ouan Kasa Formation) is a transgressive episode comprising marginal marine to shelf sandstone facies. The Middle Devonian (Eifelian-Couvinian) includes the F4 Sand (Couvinian, equivqlent to the AO I or Emgayet Formation) and the F3 Sand (Givetian, equivqlent to the AO II or Idri Formation). The F5 and F4 have average porosity values between 6% and 12%. The F6 covers most of the Saharan Platform while F5 and F4 are essentially confined to its northwestern and southern parts (Askri et al., 1995). The uppermost reservoir corresponds to the reservoir F2 (Strunian) equivalent to the Tahara Formation in West Libya, and consist of shaly-sandstones, siltstones, clays, and bioclastic and oolitic carbonates, with medium to low porosities and permeabilities of up to 1000 mD. The F2 reservoir is a producer in the Illizi Basin as well as in the Sbaa Trough (Askri et al., 1995).

In the southwestern part of the Oued Mya Basin, the Lochkovian is represented by sandy shales of up to 850 m (2789 ft) in thickness (Askri et al., 1995). It progressively pinches out to the northeast. It is absent in the Illizi Basin. The Pragian shows a greater extension than the Lochkovian and is made up essentially of sandstones. In the Illizi Basin, Pragian sediments rest unconformably on various Silurian levels. They have a thickness of up to 400 m (1312 ft) in the northern part of the Tindouf Basin and in the Ougarta area (Fig. 15.1). The Emsian is dominated by sandstones and marls. In the western regions it contains a limestone bed known as the "China Wall" at the base. In the eastern regions, the Emsian is represented chiefly by shaly sands and sandstones which include the F4 and F5 reservoirs of the Tadrart and Ouan Kasa formations. The F4 Sand is the principle reservoir in the Zarzaitine Field discovered in 1957.

The Middle and Upper Devonian are characterized by a number of sedimentary gaps due to Frasnian, Famennian, and other pre-Mesozoic erosional cycles (Askri et al., 1995). The Middle Devonian sequence is much thinner than the Lower Devonian and varies from 100 m (328 ft) to 250 m (820 ft) in thickness. The Frasnian shales of the western regions change progressively eastward into shales, marls, carbonates, and sandstones. The latter include the F3 reservoir (Givetian) of the Alrar Field in Eastern Algeria (Askri et al., 1995, Chaouchi et al., 1998) and the Al Wafa Field in West Libya (Tawadros et al., 1999, Hallett, 2002). The Upper Devonian sediments have the same geographical extent as those of the Middle Devonian. In the western Algeria it is dominated by calcareous shales up to 1700 m (5578 ft) in thickness, while in the Eastern areas it is thinner. Upper Devonian sediments include radioactive shales, limestones, and marls. Black shales deposited during the Frasnian Anoxic Event have high TOC values of up to 14% (Boote et al., 1998, Lüning et al., 2004c) and provide the source of hydrocarbons in Alrar and Al Wafa fields and other fields in West Libya and Algeria. The Frasnian source rocks in the Tindouf and Reggane basins are in the oil window; whereas in the Timimoun, Ahnet, Tindouf, Reggane, and Ghadames basins they are in the gas window (Askri et al., 1995).

In the central Illizi Basin, the Devonian is predominantly argillaceous and includes the Alrar Formation and the Gazelle Formation which pass into sandstones toward the top. Several palynological studies on chitinozoans and acritarchs were performed by Jardiné et al. (1974), Boumendjel (1985, 1987), Boumendjel et al. (1988), Coquel & Latreche (1989), Moreau-Benoit et al. (1993), Abdesselam-Rouighi & Coquel (1997), and Coquel & Abdesselam-Rouighi (2000). Clastic input from the Tihemboka Ridge led to the deposition of tidal bars of the Givetian F3 Sand which forms important hydrocarbon reservoirs in the northeastern Illizi Basin and West Libya (Chaouchi et al., 1997, 1998, Tawadros et al., 1999). Middle Devonian limestones and shales constitute a narrow band between the escarpment of the Lower Devonian sandstones and the Upper Devonian shales.

Jardiné & Yapaudjian (1968) proposed the name Alrar Formation (*Formation Argilo-Gréseuse d'Alrar*) for a Devonian subsurface unit in the Illizi Basin, Algeria. The type section is in the East-Alrar well. In that well the unit is 173 m (568 ft) in thickness. It thins westwards until it wedges out on the Tinrhert uplift. They divided the Formation into three members, from base to top: 1) a grey-black micaceous shale member with regular bedding, 2) a heterogeneous sandy-shaly member with ferruginous coarse-grained sandstones, changing into argillaceous sandstones with variable thickness upwards, a fine-grained fossiliferous sandstone bed is often present in its upper part, and 3) the Alrar massive sandstones member, composed of medium to coarse-grained sandstones with variable degrees of cementation and oblique stratification. To the south, the upper Alrar massive sandstones (F3 Sand) disappears abruptly between East Alrar and Ouen Tradjel while the two lower members, the heterogeneous argillaceous sandstone (F4) and the grey-black shale, including the F5 Sand, persist. Jardiné & Yapaudjian (1968) assigned an Emsian age to the grey-black shale member, a Couvinian age to the middle member, and a Givetian age to the massive Alrar sand, all based on palynology. The tops of depositional sequences are time lines, but within each sequence the lines are diachronous. Jardiné et al. (1974) divided the Alrar Formation into four palynological zones K and L1–L3. Zone K is Emsian with marine microplanktonic facies characterized by the appearance of *Triangulina alargada*, *Onondagella asymmetrica*, *Diexallophasis remotum*, and *Pterospermopsis circumstriata*, as well as the progressive appearance of *Navifusa bacillum*, *Crameria valentina*, *Stellinium octoaster*, and very rare *Polyedryxium*. This zone corresponds to the lower member of the Alrar Formation which correlates with the Emgayet Shale of Burollet & Manderscheid (1967) and the Ouan Kasa and upper Tadrart formations according to Massa & Moreau-Benoit (1976). Zones L1–L3 are Couvinian-Givetian in age. They span the Aouinat Ouenine Group. The middle member of the Alrar Formation (Zones L1–L2) contains the palynomorphs assemblage *Navifusa bacillum*, *N. brasiliensis*, *Crameria valentina*, *C. pharaonis*, *C. pyramis*, *Stellinium octoaster*, *Polyedryxium decorum*, ?*Veryhachium exasperatum* and ?*Verhachium libratum*, and the appearance of *Duvernaysphaera tessella*, *D. angelae*, *D. kraueseli* and *D. stellata*. It is equivalent to the Bir al Qasr Formation of Seidl & Röhlich (1984) or Aouinat Ouenine I of Massa & Moreau-Benoit (1976). The upper Alrar massive sandstone member (F3 Sand, or zone L3) is characterized by the frequent occurrence of *Duvernaysphaera* and the appearance of new forms of *Polyedryxium* cf. *decorum*, *P.* cf. *talus*, *Dailllydium quadridactylites*, *Veryhachium stelligerum*, *Duvernaysphaera radiata* and *Umbellasphaeridium deflandrei*. It correlates with the Idri Formation of Seidl & Röhlich (1984) and A.O.II of Massa & Moreau-Benoit (1976).

The Alrar Formation is overlain by the Upper Devonian Gazelle Formation (Frasnian-Strunian). This Formation is equivalent to the Ash Shatti Group and Tahara Formation in West Libya. Jardiné et al. (1974) divided the Formation into three units. The lower unit is a shaly sand with typical Frasnian macrofauna such as *Sphenotus* cf. *dolomus, Ptychopteria* e.g. *vanuxeni* and *Avicula bodana,* and includes a radioactive zone. It corresponds to their palynological zone L4 (Frasnian). A transitional zone between the Middle and Upper Devonian is characterized by the appearance of *Maranhites winslowae, M. brasiliensis, Unellium winslowae, Veryhachium distratum, Horologinella horologia* and the abundance of *Umbellasphaeridium deflandrei.* The middle unit is a shaly-sandy sequence with the macrofauna *Avicula* cf. *sociale* and *Posidonia venusta* var. *depressa* and corresponds to their zones L5–L6 (Frasnian-Famennian). It is characterized by the presence of *Umbellasphaeridium saharicum, U. deflandrei, Maranhite mosesiim M. brasiliensis, Crassiangulina tesselita, Gorgonisphaeridum solidum* and *Horologia quadrispina.* This unit is equivalent to the A.O.IV Formation and the Tarut Formation in West Libya. The upper unit is a regressive detrital sequence with *Spirifer verneuilli* and *Primoceras divisum* and corresponds to zone L7 (Strunian) which is distinguished by the disappearance of *Daillydium quadridactylites* and *Stellinium octoaster,* the last occurrence of *Navifusa,* and the appearance of *Schizocystia bicornuta* and the global spore *Hymenozonotriletes lepidophytus.* This unit is equivalent to the Tahara Formation in West Libya. The Gazelle Formation is equivalent to Qattara and Dabdab formations (L4), Tarut Formation (L5-L6) and Ashkida Formation (L7) of Seidl & Röhlich (1984), or Aouinat Ouenine III and IV and Tahara formations of Massa & Moreau-Benoit (1976).

The Middle Devonian (Givetian) reef belt of Mauritania and West Sahara, where six reef cycles separated by shales and sandstones have been recognized (Wendt & Kaufman, 2006), extends into the southern border of the Tindouf Basin and the northern margin of the Reguibat Massif and the Gourara-Ahnet area in Central Algeria (Kaufmann & Wendt, 2000). The reefs are similar to the Givetian mud mounds in the Maader Basin in the Eastern Anti-Atlas of Morocco.

In the Gara el Kahla section of the Gourara-Timimoun area (Fig. 15.1), 30 km SW of Timimoun, the Fammenian-Tournaisian Kahla Formation consists of three units (Conrad, 1985d); Kahla Mudstone, Lower Kahla Sandstone, and Upper Kahla Sandstone. Brice et al. (2007) recognized three successive faunas near the Devonian-Carboniferous boundary in the same sections, which they labeled as Lower, Middle, and Upper Fauna. The Lower Fauna is late Famennian in age, while the Upper Fauna is early Tournaisian in age. The Intermediate Fauna was identified only in Southern Morocco.

The Upper Devonian (Famennian-Strunican) is represented by the lower part of the Kahla Formation (Famennian-Tournaisian) (Conrad et al. 1985d, Brice et al., 2007). The Devonian part consists of the Kahla Mudstone with *Gonioclymenia* (Famennian, zone V), Lower Kahla Sandstone (Famennian), approximately 280 m (919 ft), with brachiopod coquinas, followed by unfossiliferous deltaic sandstones and shales. The Lower Kahla Sandstone is characterized by the Lower Fauna, which contains the productids *Hamlingella* sp., *Whidbornella* cf. *pauli radiata, Kahlella meyendorffi, Steinhagella* cf. *membranacea, Mesoplica praelonga,* and *M. nigeraeformis,* the rhynchonellids *Centrorhynchus* cf. *lucida, Paurogastroderhynchus lakahalensis, Megalopterorhynchus* sp., and *Gastrodetoechia* sp., and the spiriferids *Cyrtospirifer* aff. *oleanensis*, *Sphenospira* cf. *julii*, and *Parallelora* aff. *Subsuavis*. The correlation between the Lower Fauna from the Kheneg Lakahal secion in SW Morocco and the Gara el Kahla section in Algeria is achieved using upper Famennian brachiopod taxa common to both sections, such as the strophomenid *Leptagonia* cf. *analoga*, the productids *Whidbornella* cf. *pauli radiata*, *Mesoplica praelong*, and *Steinhagella* cf. *membranacea*, the rhynchonellids *Paurogastroderhynchus*, *Megalopterorhynchus* and *Gastrodetoechia*, and the spiriferids *Cyrtospirifer* aff. *oleanensis* and *Sphenospira* cf. *julii*. The last five species are considered as index fossils for the period Brice et al. (2007).

Devonian rocks crop out along the erosional southern border of the Illizi Basin and are divided into three major lithostratigraphic units: Lower Devonian sandstones (Orsine Formation or the upper part of the External Tassilis), Middle Devonian carbonates (Illizi Formation of Wendt et al., 2009b), and Upper Devonian shales (Tin Meras Formation). Middle Devonian carbonates were dated by conodonts and correlated over a W-E distance of about 350 km by Wendt et al. (2009b). The sequence shows a lateral transition from condensed series in the west deposited on a submarine high (Amguid Ridge) into a shallow basin (Illizi Basin) in the east. The eastern part of the Illizi Basin received clastic input from the Tihemboka Ridge, which separates the Illizi Basin from the Murzuq Basin in West Libya.

The Tin Meras Formation is Upper Devonian in age and is equivalent to the Argiles de Temertasset in the Ahnet and Mouydir basins (Wendt et al., 2006, 2009b). The Tin Meras Formation includes Middle and Upper Devonian sediments. Wendt et al. (2009b) separated the marine Middle Devonian part as a new formation and proposeed the term Illizi Formation for that interval, with a type section situated along the old trail from Illizi to Ohanet. Upper Silurian-Lower Devonian Faddoun Sandstone (Grès de Fadnoun) forms the rocky plateau of the External Tassilis, which overlies the Cambro-Ordovician Internal Tassilis and the Lower Silurian "*Depression Intratassilienne*" (Beuf et al., 1971). The sandstones of the External Tassilis have been subdivided into four formations (Dubois et al., 1968): The Oued Tifernine Formation at the base (subdivided into the Barre inférieur and the Talus à Tigillites), the Tamelrik Formation (subdivided into the Barre moyenne and Trottoirs), the Oued Samene Formation (also named Barre supérieure), and the Oursine Formation. The Tifernine Formation is probably Late Silurian (late Ludlow to Pridoli) in age, while the other formations are assigned to the Early Devonian (Eschard et al., 2005). Different formation names have been used for the same interval in wells drilled in the southern and northwestern Illizi Basin (Jardiné & Yapaudjian, 1968, Jardiné et al., 1974, Boumendjel et al., 1988).

The base of the Devonian is marked by breccias followed by a succession of shales and sandstones with patches of limestone reefs, such as in the Ghar-Rouban Mountains, while flysch deposits accumulated in adjacent basins (Askri et al., 1995).

## 15.2.4   Carboniferous

Carboniferous deposits are sporadically present in Northern Algeria. Continental Carboniferous sediments occur in the Djurdjura area. Shales, conglomerates, and volcanics are present in the Ghar-Rouban and Tlemcen areas (Askri et al., 1995). Western and Eastern Algeria were separated during the Carboniferous by the

Amguid-El Biod-Messaoud Ridge which was an area of non-deposition (Askri et al., 1995). The Carboniferous sediments in Algeria vary from deep marine to continental. Thicknesses of the Carboniferous sequence decrease from 2800 m (9187 ft) in the west to 900 m (2953 ft) in the east.

The most complete section of the Carboniferous is located in the Bechar Basin with a thickness of about 5000 m (1641 ft) thick (Askri et al., 1995). The western Bechar Basin contains up to 2000 m (6562 ft) of Mississippian carbonates composed of shoaling-upward sequences (Madi et al., 2000). The Visean part of the succession is made up of Waulsortian-type buildups overlain by oolitic and crinoidal grainstones. Most pores within the buildups were occluded by diagenetic cements, but the oolitic and crinoidal grainstones contain intergranular porosity and may have hydrocarbon potential.

Hercynian (Carboniferous-Middle Triassic) tectonic activities led once more to the truncation of Paleozoic sediments on the western and northern parts of the Berkine Basin (Cochran & Petersen, 2001, Figure 3).

The Djebel Bechar and the High Zousfana areas were surveyed by R. Capot-Rey in 1927 who recognized the folded Devonian sandstones, Carboniferous limestones, and the oil horizon of Kednasa and the unconformably overlying Cenomanian rocks.

The Carboniferous formations of the Bechar Basin have been divided into three groups by Lemosquet & Pareyn (1985):

I. The lower group (Strunian-early Serpukhovian) or the Zousfana Detrital and Peri-reefal Limestones Group (Les Calcaires détritiques et périrecifaux de la Zousfana) is composed of mixed carbonates, clastics, and local reef sequences. It includes, from base to top; the Ouarourout, Oulad Bou Hadid, Hassi Sguilma, El Hariga, Akacha-Mazzer, Boulmane, Harrez, Igli, Taouerta, Zousfana, El Gelmouna, Ain El Mizab, and Djanien formations.

The Ouarourout Formation (Strunian, here considered Devonian), at the base of the Zousfana Detrital and Perireefal Limestones Group, consists of about 4000 m (13,124 ft) of sandstones and overlies the Devonian Marhouma Sandstone in the Djebel Bechar-Zousfana Valley area, and is overlain by shales of the Olad Bou Hadid Formation (lower Tournaisian).

The Hassi Sguilma Formation (upper Tournaisian) is composed of sandstones and shales; it overlies the Olad Bou Hadid Formation and is overlain by the El Hariga Formation. The Hassi Sguilma Formation contains a latest Tournaisian conodont assemblage of *Scaliognathus anchoralis* (Nemyrovska et al., 2006). Korn et al. (2010c) described two new ammonite species *Ammonellipsites sguilmensis* and *Muensteroceras beniabbesense* from the Hassi Sguilma Formation of the Saoura Valley (Northwestern Algeria). The Formation is time equivalent of the Iridet Formation.

The El Hariga Formation (Visean) is composed mostly of shales with some limestones beds. The conodont assemblage of the El Hariga Formation (lower Visean) in the Saoura Valley, Algeria, contains the latest Tournaisian gnathodontids and the Visean *Pseudognathodus*, and an unornamented early Visean species of *Lochriea*, as well as *Lochriea saharae* (Nemyrovska et al., 2006). The El Hariga Formation overlies the Hassi Sguilma Formation.

The Akacha-Mazzer Formation (Visean), also called the "*Banc de Mazzer*", is a carbonate limestone bed with the goniatites *Merocanites ogivalis*, and the benthonic form *Neospirifer fascicostatus*. The upper part of the formation contains reef domes built up by the lithostrotionids *Siphonodendron pauciradiale* and *S. martini*. *Siphonodendron pauciradiale* in the Adarouch area of NE Morocco (Said & Rodriguez, 2007). The Akacha-Mazzer Formation also contains mid-Visean conodont fauna, such as *Gnathodus bilineatus* (Nemyrovska et al., 2006).

The Boulmane Formation, Harrez Formation, Igli Formation, and Taouerta Formation form a succession of limestones separated by sandstones and shales. They contain the conodont *Gnathodus bilinea*, several species of the foraminifera Archaediscidae, the rugose coral *Lithosstrotion decipiens;* reported from the Adarouch area of NE Morocco by Said & Rodriguez (2007), and the brachiopod *Neospirifer fascicostatus*.

The Zousfana Formation, El Guelmouna Formation, Ain El Mizab Formation, and Djenien Formation are well developed in the Cirque d'El Guelmouna area. The Zousfana and Ain El Mizab formations are made up of shales and limestones; the El Guelmouna and Djenien formations are massive limestones. Goniatites are common in the upper part of the Zousfana Formation and the lower part of the Djenien Formation, and include *Dombarites*, *Platygoniatites* and *Cravenoceras*, as well as *Girtyceras*, *Lyrogoniatites* and *Neoglyphioceras*. The El Guelmouna Formation is divided into two members; a lower Ain Mezerelt Member and an unnamed upper member. The Ain Mezerelt Member contains the serpukhovian foraminifera *Eosigoilina explicate*, *Eolasiodiscus priscus*, and *Eostaffella pseudostruvei*.

The Tagnana Formation (straddles the Serpukhovian-Bashkirian boundary). Lemosquet & Pareyn (1985) place the Visean-Serpukhovian boundary between the Oum el Graf Member of the Zousfana Formation and the Ain Mezerelt Member of the El Guelmouna Formation. The Ain El Mizab Formation consists of alternating dark shales and thin-bedded fossiliferous limestones (Manger & Pareyn, 1979) rich in rugose corals, such as *Koninckophyllum, Paleosmilia, Clisiophyllum,* and *Axophyllum, Latiproductus latissimus* and *Gigantoproductus meridionalis* (Lemosquet & Pareyn, 1985). The upper Mouizeb el Atchane Member of the formation contains *Astereoarchaediscus gregorii* and *Eolasiodiscus transitorius*.

On the northern side of the Saoura Valley (shown as Grand Erg Occidental section on Lemosquet & Pareyn, 1985, Fig. 3), the Tournaisian and the lower Visean are condensed and incomplete and consists of carbonate reefal faciess. The palynology was studied by Lanzoni & Magloire (1969), the microflora by Sebbar & Mamet (1996), and the foraminifera by Sebbar (1997). The reefs form the massifs between the Erg de Taghitand and the Grand Erg Occidental (Taoudraras, Ioucha, Goumriats, Meharez, Sameh, and El Oubeur formations). They are bioherms constructed by *Stromatactis*, sponges, rugose corals, and fenestillids. Crinoids predominate in the associated pre-reefal limestones. Reefs continue through the late Visean Ioucha and El Oubeur formations (Lemosquet & Pareyn, 1985), but diminish in the Serpukhovian El Aouidja, El Hamra, and lower Bent el Goumi formations. The latter is overlain by the Mezarif Formation (Bashkirian-Moscovian) (Sebbar, 1997). In the northern Djebel Antar and Djebel Horreit areas, the Tournaisian is represented by very thin micritic limestones of the basal Oued Souari Formation (Tournaisian-Late Serpukhovian), which rests conformably on top of Devonian strata and is overlain disconformably by lower Visean massive limestones with *Siphonophyllia*, 20–40 m (66–131 ft) thick. Late Visean reefs are made up of bioherms surrounded by reef boulder beds. The Oued Souari Formation is overlain by the Ain Antar Formation (Late Serpukhovian-Bashkirian). In the Ben Zireg area, the Ben Zireg Flysch Facies contains olistoliths ranging in age from Cambro-Ordovician to Tournaisian, embedded in an early Visean matrix. The Tournaisian blocks contain *Scaliognathus anchoralis* (Lemosquet & Pareyn, 1985). The Flysch sequence is overlain by the Lower Reefs, which is overlain in turn by the Ben Zireg Formation (Late Visean-early Serpukhovian). The Ben Zireg Formation is made up of shales equivalent to the lower part of the Oued Souari Formation at Djebel Antar, the Ioucha-El Aouidja succession in the grand Erg Occidental, and the Mazzer-El Guelmouna succession at Djebel Bechar. It is overlain by the La Gare Limestone Formation *"récif de la Gare"* (equivalent to the Ain el Mizab Formation). The *récif de la Gare* is of late Visean age (Lemosquet & Pareyn, 1985). At Djebel Grouz, the Carboniferous succession represents the basinal facies equivalent to the carbonate shelf and reefal facies. The Ain El Mizab Formation is overlain by the Djenien Formation (late Serpukhovian).

The Djenien Formation (late Serpukhovian) in the Bechar-Zousfana area is made up of cliff-forming, 100–150 m (492 ft) in height, peri-reefal crinoidal limestones, locally dolomitized, with chert layers. It contains an assemblage of the goniatites *Delepinoceras*, *Cravenoceras* and *Anthracoceras*, as well as the brachiopods *Gigantoproductus superbus*, *G.irregularis Latinoproductus* in the lower part and the genus *Beleutella* at the top. Its top is marked by paleokarsts filled with the overlying Tagnana Formation sediments. The Djenien Formation has been assigned to the Serpukhovian (Lemosquet & Pareyn, 1985).

II. The middle group (latest Serpukhovian-top Bashkirian) or the Djebel Bechar Limestone Group (Les Calcaires du Djebel Bechar) is characterized by intraformational channels and detrital deposits. The Djebel Bechar Limestone Group includes the Tagnana Formation (subdivided into three members), the Hassi Kerma Formation (consists of two members), and Oued El Hamar Formation (includes two members).

The Tagnana Formation (straddles the Serpukhovian-Bashkirian boundary) appears as cylindrical pillars of white sandstones, weathering brown, penetrating vertically into the limestones and dolomites of the Djenien Formation. These dikes are up to 15 m (49 ft) in depth and up to 3 m (10 ft) in width. The contact between the two formations is marked by erosional and dissolution features and ferruginous crusts associated with paleokarsting. The Tagnana Formation is divided into three unnamed members. Several fluviatile channels filled with plant-bearing, coarse-grained sandstones and conglomerates occur within the Tagnana Forrmation, especially in the lower and middle members (Lemosquet & Pareyn, 1985). The channels are capped by well stratified shales and limestones markers. Based on these markers, the lower member of the Tagnana Formation has been divided into nine units designated A to I (Lemosquet & Pareyn, 1985). Units A and B belong to the conodont *Adetognathus unicornis* Zone. Foraminifera, such as *Asteroarchaediscus postrugosus*, appear in unit C. Unit D contains *Titanaria Africana* and *Rhachistognathus muricatus*, and forms a regional marker in the Sahara. It represents the last marine deposits in the Tindouf Basin in Western Algeria. Unit E contains *Homoceras* and *Isohomoceras* and conodonts of the *Declinognathodus noduliferous* Zone. Units F to I consist of shales, sandstones and limestones, which contain *Asteroarchaediscus gregorii acutiformis*, *Eostaffella acuta* and *Millerella uralica*. The middle member is also an excellent regional marker composed of limestone ledges overlain by biostromes with primitive Durhaminidae and *Multithecopora*. Goniatites of the *Cancelloceras-Bilinguites* association make their first appearance. The upper member contains the Bashkirian foraminifera *Pseudostaffella antiqua*, which makes its first appearance, whereas the brachiopod genus *Choristites* becomes diversified (Lemosquet & Pareyn, 1985).

The Hassi Kerma Formation (upper Bashkirian) is made up of coarse-grained sandstones at the base, overlain by shales and limestones containing fauna of the *Bilinguites-Cancelloceras-Gaitherites* association, as well as *Wiedeyoceras, Diaboloceras, Anthracoceratites*, and *Phaneroceras*. Fluviatile channels filled with plant-bearing sandstones occur also within this formation. The limestones yield the foraminifera *Pseudostaffella antique grandis, P. Comressa, Novella evoluta*, and *N. pulchra*, and conodonts of the *Idiognathodus delicatus* Zone (Lemosquet & Pareyn, 1985).

The Oued el Hamar Formation (late Bashkirian) contains abundant foraminifera, such as *Profusulinella parva, Seminovella carbonica, Novella pulchra, Profusulinella extensa*, and *P. rhomboids*, and goniatites of the *Gastrioceras-Gaithrites-Donetzoceras* association. The Formation ends with the blue Djenien Limestone marker (Deleau, 1951, 1952). Fluviatile channels filled with plant-bearing sandstones occur in the upper member of the formation (Lemosquet & Pareyn, 1985). Saunders et al. (1979) described the species *Donetzoceras pareyni* and assigned it Westphalian age.

III. The upper group (Moscovian) (unnamed, but the name Chebket Mennouna Group seems appropriate when the need for naming it arises) includes the Kenadza Formation, the paralic coal measure unit, and the red shale unit. It is composed of terrigenous deltaic sediments grading into fluvial and lacustrine deposits; red shales occur near the top. It grades into the Oued Bel Groun Formation in the northeastern part of the Bechar Basin (Hassi Metired area). The upper group is well developed on both sides of the Chebket Mennouna Anticline between the Kendaza area in the north and the Abadla area in the south (Lemosquet & Pareyn, 1985). This group is referred to in here tentatively as the Chebket Mennouna Group. It includes the Kendaza Formation, the Paralic Coal Measure Unit, and the Red Shale Unit. Equivalents of this group are missing from the Grand Erg Occidental, Djebel Antar, and Ben Zireg sections.

The Kendaza Formation consists of several marine sequences, each starting with cross-bedded or rippled sandstones, followed by bioclastic limestones with brachiopods and crinoids, then by shales with goniatites. In the northeastern part of the Bechar Basin (Hassi Metired area), these sediments are replaced by carbonates with goniatites and foraminifera of the Oued Bel Groun Formation (early Moscovian). In the eastern part of the Basin (Nekheila area), the Chebket Mennouna Group is replaced by bioclastic limestones with algae and foraminifera (upper Moscovian), which grade upward into coal- bearing terrigenous sediments of Westphalian age. An Upper Moscovian succession has been reported from the Bordj Nili well N1–1 near Hassi R'Mel (Lemosquet & Pareyn, 1985).

The Paralic Coal Measure Unit is 400–1700 m (1312–5578 ft) in thickness and consists of rhythmically developed fluviatile deposits (Lemosquet & Pareyn, 1985) (cf. West Libya). The succession starts with coarse-grained sandstones with shale clasts and ferrigenous granules from reworked soils. The sandstones are overlain by sandy shales with plant fragments and coal seems, 0.1–0.4 m (4–16 in.) in thickness, and rootlet beds. The coals are overlain by carbonaceous shales wit plants remains, followed by thin-bedded shales with plant fragments of late Westphalian age (see Dealeau, 1951, 1961).

The Red Shale Unit (misnomer?) is composed of grey nodular limestones with ostracods and *Carbonicola*, overlain by fine-grained sandstones, greenish-grey shales, and red cross-bedded sandstones. The presence of *Estheria* of the *E. simonytenella* group suggests a Stephanian age (Lemosquet & Pareyn's personal communication with R. Feys).

The Carboniferous stratigraphic nomenclature of the Charouin-Timimoun & Gourara areas has been redefined by Conrad (1985a). The Carboniferous succession overlying the Strunian Kahla Sandstones and Shales (according to Askri et al., 1995) consists of the Kahla Sandstones, Timimoun Shales, Arhald Sandstones, and Bahmer Limestones.

In the Gara el Kahla section of the Gourara-Timimoun area, SW of Timimoun, the Upper Kahla Sandstone (Carboniferous), 130 m (427 ft) thick, is Tournaisian in age (Conrad, 1985d). contains, in an ascending order (Brice et al., 2007): a) basal unit of white quartzitic sandstones with abundant brachiopods (Conrad, 1985d) yielding abundant brachiopod faunas, Early Tournaisian in age, b) a 10 m (33 ft) bed of shales with *Gattendorfia* cf. *Crassa*, *Imitoceras*, *Cyathaxonia* fauna, and the brachiopods *Eobrachythyris strunianus*, *Eochoristites*, *Syringothyris ahnetensis*, and *Unispirifer*. Ebbighausen et al. (2004) described this horizon and identified *Kahlacanites* nov. gen., *Gattendorfia* cf. *crassa*, and *Acutimitoceras* and assigned it a probable late Early Tournaisian age, c) siltstones and sandstones, and d) the top of the Kahla Formation contains shales with *Pustula interrupta* and *Spirifer subcinctus* (Conrad, 19850, as well as *Acrocanites* (Brice et al., 2007) assigned to the Late Tournaisian. Korn et al. (2010a) identified nine new ammonoid species from the Upper Kahla Formation, which include *Imitoceras altilobatum*, *Triimitoceras amplisellatum*, *Kazakhstania inequalis*, *Acrocanites imperfectus*, *A. recurvus*, *Xinjiangites scalaris*, *Becanites canalifer*, *B. singularis*, and *B. inflateralis*.

The base of the Upper Kahla Sandstone is characterized by the strophomenid *Leptagonia* cf. *analoga*, the productids *Acanthatia* (?) sp., *Productina* sp., *Spinocarinifera* aff. *bulbosa*, *Spinocarinifera* aff. *arcuata*, the rhynchonellids *Macropotamorhynchus* n. sp. aff. *insolitus*, *Shumardella* aff. *fracta*, all of which occur in ferruginous sandstones, and the spiriferids *Prospira* sp.1, *Prospira* sp. 2, *Unispirifer unicus*, *Voiseyella* aff. *sergunkovae*, *Voiseyella* (?) aff. *tylothyriformis*, *Eochoristites platycosta*, *Eomartiniopsis lakahalensis*, and *Syringothyris* sp. The brachiopod succession may reflect the 'Hangenberg Event' characterized the sudden appearance of black shales just before the Devonian-Carboniferous boundary (Wendt & Kaufman, 2006), which has been recognized in the Maader-Tafilalt area near the Moroccan border.

The Timimoun Shales (lower-Upper Visean) consists of two shale units with limestone intercalations, separated by the Rhnet Sandstones (Conrad, 1985a). The Rhnet Sandstones is made up of 80 m (262 ft) of fine-grained sandstones and siltstones with the brachiopod *Fluctuaria* and a bioclastic limestone bed.

The Lower Timimoun Shales are 200 m (656 ft) in thickness. The lowermost layer is called "*dalle à Merocanites*" or "*dalle des Iridet*" in the northern and southern provinces, respectively. The two units contain the same ammonite fauna, such as *Merocanites applanatus*, *Ammonellipsites kochi*, and *Muensteroceras*, as well as zaphrentoid corals and condonts of the *Scaliognathus anchoralis* (Conrad, 1985a). Korn et al. (2010c) described four ammonoid species Iridet Formation "*dalle des Iridet*" in the Ahnet-Mouydir area and attributed them to the North African

*Ammonellipsites* -*Merocanites* Zone (Late Tournaisian-Early Viséan). They include the three new species *Eurites temertassetensis, Trimorphoceras teguentourense,* and *T. azzelmattiense.*

The Upper Timimoun Shale is 80 m (262 ft) thick. It consists of shales with *Beyrichoceras castletonense* and *Dimorphoceras gilbertsoni* of late Visean age, and *Neospirifer fascicostatus fascicostata.* The top of the shales contain either sandstones or limestones with zahrentoids corals and *Siphonophyllia* associated with *Neospirifer fasciococstatus gwinneriformis* and *Saharopteria.* Bockwinkel et al. (2010) described twenty-seven ammonoid species from the Timimoun Formation, and assigned them to the *Bollandites* -*Bollandoceras* Zone (Early and Middle Viséan). The taxa include *Rhnetites* n. gen., *R. rhnetensis, R. ouladallalensis, Parahammatocyclus mutaris, Bollandoceras nitens, B. subangulare, B. politum, B. aridum, B. zuhara, B. mirrih, Benimehlalites* n. gen., *B. benimehlalensis, Benimehlalites belkassemensis, B. brinkmanni, Pachybollandoceras* n. gen., *P. intraevolutum, P. repens,* Bollanditinae n. subfam., *Gourarites* n. gen., *G. hagaraswad, G. hagarkarim, G. mustari, G. zuhal, Semibollandites* n. gen., *S. kamil, S. pauculus, S. qawiy, Timimounia* n. gen., *Timimounia timimounensis, T. lunula,* Daaitidae n. fam., *Daaites* n. gen., *D. daaensis, Dimorphoceras lanceolobatum, Nomismoceras salim,* and *N. waltoni.*

The Arhald Sandstones is composed of fine-grained micaceous sandstones with ripples and plant fragments, intraformational conglomerates and fossiliferous limestones (Conrad, 1985a). It overlies the Timimoun Shales (Askri et al., 1995) or the *Dimorphoceras Clays* (Lefranc & Guiraud, 1990) and is overlain by the Bahmar limestones (Askri et al., 1995) or the Tala Limestones (Lefranc & Guiraud, 1990).

The exposed part of the Tala Limestones is 80 m (262 ft) thick and consists of bioclastic limestones separated by sandstone and green shale beds (Conrad, 1985a). The limestone bed at the base contains the brachiopod *Neospirifer fascicostatus gwinneriformis* and tabulate corals, followed by beds with corals and giganoproductids, such as *Gigantoproductus menchikoffi,* then by *G. dubokensis.* The corals include domes of *Siphonodendron* in growth position. The top of the formation is made up of massive limestones with *Giganoproductus* cf. *edelburgensis* and *Lithostrotion* (Conrad, 1985a). The Tala Limestones is overlain unconformably by the Neocomian Toubchirine Sand (Lefranc & Guiraud, 1990).

The area is located at the margin of the African continent during the Carboniferous. The area was studied by Fallot (1953). The Carboniferous succession comprises nine formations belonging to three main transgressive sequences (Conrad, 1985b). The first sequence is Famennian-Tournaisian; the second is Visean-Serpukhovian, and the top sequence is Serpukhovian-Bashkirian. The succession in the Mouydir, Ahnet, and Reggane basins is the same, with minor variations, except for the eroded section in the Mouydir Basin; the Upper Devonian-Carboniferous succession from bottom to top includes:

Khenig Shales (Famennian)
Khenig Sandstones (Strunian in lower part, Lower Tournaisian in upper part)
Teguentour Shales (lower Upper Tournaisian)
Tibaradine Sandstone (Upper Tournaisian)
*Dalle des Iridet* (at the Tournaisian-Visean boundary)
Kreb ad Douro Sandstones (Lower Visean)

Tirechoumine Shales (Lower-Upper Visean)
Garet-Dehb Sandstones (Upper Visean)
Djebel Berga Limestones (latest Visean-serpukhovian)
Gypsum of Hassi-Taibine (Serpukhovian)

Red Beds (Lower Moskovian-Bashkirian) (Red Beds of Ain ech Chebbi in the north of Reggane Basin and Red Beds of Azzel-Matti Formation in the south), and Red Beds of Hassi-Bachir Formation in the Ahnet Basin)

The Carboniferous along the northern border of the Hoggar Massif starts with a regressive facies of massive fine-grained clastics of the Khenig Formation (Strunian-early Tournaisian) (Conrad, 1985b). The average thickness of the formation is about 100 m (328 ft), but thickens southwards and may reach 250 m (820 ft) in places. The Lower Khenig Sandstones (Strunian) are strongly bioturbated and grade southward into continental sandstones. The top of the unit show evidence of emergence with paleosols, desiccation cracks, and wood fragments. It contains sparse bivalves and rhynchonellid brachiopods. The Upper Khenig Sandstones are more marine in the northern parts of the Mouydir, Ahnet, and Reggane basins than in the south and contain the brachiopods *Syringothyris ahnetensis*, *Verkhotomia* sp., *Unispirifer* ex. gr. *Pesasicus*, *Eochoristites* sp., and poorly preserved ammonoids, such as *Gattendorfia* and *Imitoceras subbilobatum*, indicating an early Tounaisian age.

The Teguentour Shales (lower Upper Tournaisian) contains ammonoids of the *Protocanites-Acrocanites* Zone, conodonts of the *Pseudopolygnathus dentilineatus* fauna and corals of the *Cyathaxonia* fauna and Tabulata. Its thickness is 100–200 m (328–656 ft) (Conrad, 1985b). The Teguentour Shales is absent in the Reggane area. Mottequin & Legrand-Blair (2010) confirmed the early Late Tournaisian age based on a small brachiopod assemblage of orthids, rhynconillids, and spiriferids, and goniatites. Korn et al. (2010b) described new families, genera and species from three successive assemblages within the Teguentour Shales, considered as the richest ammonoid faunas of the time interval worldwide, which include *Imitoceras dimidium, I. strictum, Triimitoceras tantulum, Acrocanites disparilis, Jdaidites cultellus, Pericyclus tortuosus, P. circulus, P. trochus, P. intercisus, Nodopericyclus* n. gen., *Nodopericyclus circumnodosus, N. deficerus, Ammonellipsites serus, Helicocyclus formosus, H. inornatus* n. sp., *H. laxaris, Ouaoufilalites creber*, family Temertassetiidae n. fam., *Temertassetia* n. gen., *T. temertassetensis, T. secunda, T. decorata, T. coarta, Jerania* n. gen., *J. jeranensis, J. sicilicula, J. pusillens, J. subvexa, J. persimilis, Kusinia* n. gen., *K. falcifera, Bouhamedites insalahensis, Muensteroceras subparallelum, M. multitudum, Follotites* n. gen., *F. folloti, F. stelus, F. flexus*, family Rotopericyclidae n. fam., *Eurites permutus, E. doliaris, Mouydiria* n. gen., *M. mouydirensis, M.scutula, Rotopericyclus kaufmanni, R. rathi, R. wendti, R. lubesederi*, subfamily Dzhaprakoceratinae n. subfam., *Dzhaprakoceras punctum, D. amplum, D. vergum, D. biconvexum*, Progoniatitinae n. subfam., *Progoniatites uncus, P. pilus, P. paenacutus, P. globulus*, Habadraitinae n. subfam., *Habadraites* n. gen., *H. weyeri, H. supralatus, Primogoniatites* n. gen., *Primogoniatites fundator*, Antegoniatitinae n. subfam., *Antegoniatites* n. gen., and *Antegoniatites anticiparis*.

The Tibaradine Sandstones (Upper Tournaisian) is about 100–200 m (328–656 ft) in thickness. It consists of neritic coquinoidal calcareous sandstones, becoming fluviatile southwards. Ammonoids are rare, but brachiopods are abundant and include *Syringothyris folloti*, *S. sefiatensis*, *Histsyrinx vautrini*, *Marginatia betainensis*, and *M. vaughani* (Conrad, 1985b). In the Reggane area, the Tibaradine Sandstones span the entire Upper Tournaisian.

The Kreb ed Douro Sandstones (early-late Visean) is separated from the Tibaradine Sandstones by the *dalle des Iridet*, which marks the Tournaisian-Visean boundary. This boundary also marks the limit between the first trangressive cycle and the following regressive sequence of the Kreb ed Douro Sandstones (Conrad, 1985b). The *dalle des Iridet* consists of thin oolitic, phosphatic, and pelletal limestones and exhibits marked faunal changes across the boundary, such as the appearance of *Merocanites* and *Spirifer* sp. and the extinction of *Pericyclus*, *Imitoceras*, and *Ammonellipsites*. It contains the early Visean foraminifera *Valvulinella angulata* and ammonoids and late Tounaisian conodonts of the *Scaliognathus anchoralis* Zone, such as *Pseudopolygnathus pinnatus*. The Kreb ed Douro Sandstones (absent in the Reggane Basin according to Fig. 6 of Conrad, 1985c, p.319) is made up of 100–300 m (328–984 ft) of fluviatile sandstones in the southern part of the Ahnet and Mouydir basins, changing into marine facies in the northern part.

The Tirechoumine Shales (late Visean) has an average thickness of 200 m (656 ft) (Conrad, 1985b). In the Mouydir Basin, the shales are of late Visean age as indicated by *Dimorphoceras gilbertsoni* and *Beyrichoceras castletonense*, in addition to *B. hodderense*, *B. redesdalene*, *B. micronotum,* and *B. obtusum*. It also contains specimens of *Marginatia* and *Fluctuaria,* foraminifera, and corals of the *Cyathonia* fauna and tabulate corals. The top of the formation includes *Neospirifer fasciocostatus gwinneriformis* and remains of *Saharapteria*.

The Garet Dehb Sandstones (late Visean) is made up of fluvial sediments rich in *Archaeocalamites* and *Lepidodendropsis* and has about 300–400 m (984–1312 ft) in thickness. The sandstones were deposited in N-S-oriented fluvial channels and were probably derived from the Hoggar Massif (Conrad, 1985b). Its age is constrained by the late Visean foraminifera which occur below and above the formation.

The Djebel Berga Limestones "*Calcaire du Jebel Berga*" (late Visean-Serpukhovian) signals the start of a new marine transgression in the region. The formation has a thickness of 300–400 m (984–1312 ft). The lower part of the Djebel Berga Limestones consists of oolitic clastic limestones with shales and fine-grained sandstone interbeds. The upper part is dominated by shaly-dolomitic rhythmic units. Ammonoids disappeared at that time from the area, and giganoproductids evolved (Conrad, 1985b). The formation has been dated late Visean-Serpukhovian by Conrad (1985b) based on foraminifera. Wendt et al. (2009a, 2010) assigned it a Late Visean-Lower Bashkirian age based on the conodont *Declinognathodus noduliferus* (see Legrand-Blain et al., 2010 and Wendt et al., 2010, for discussion). Corals are abundant and include *Lithostrotion decipiens* and *Solenodendron* at the base, followed upward by *Syringopora geniculata* and *Striatifera*, and finally by *Siphonodendron* and *Titanaria Africana* at the top. It also contains conodonts of the *Paragnathodus nodosus* Zone.

The Hassi Taibine Gypsum (Serpukhovian) occurs in the upper part of the Djebel Brega Limestones and is well developed in the Reggane Basin (Conrad, 1985b). It consists of shales with thick lenticular layers of gypsum and crinoidal dolomites, as well as the marker horizon *Titanaria africana- Siphonodendron* at the top.

The Hassi Bachir Red Formation (Bashkirian-early Moscovian) is made up of red sandstones and shales with a few calcareous-dolomitic interbeds (Conrad, 1985b). It is equivalent to the Ain ech Chebbi Formation, which crops out in the area between Ain ech Chebbi and Hassi Taibine, and the Azzel Matti Formation in the Reggane area.

The Tindouf Basin will be discussed again in the Chapter on the geology of Morocco. The Carboniferous sequence is exposed in four locations on the northern side of the Tindouf Basin in Western Algeria and Southeastern Morocco, each forming a topographic feature (Conrad, 1985d):

1    Sandstones and shale of the Djebel Tazout Sandstones in Djebel Tazout.
2    Shales and sandstones of the Betaina Formation in the Betaina Plain.
3    Carbonates of the Djebel Ouarkziz Formation in Djebel Ouarkziz.
4    Continental sediments of the Betana Beds[1] in the Betana Plain.

The Djebel Tazout Sandstones (Famennian-early Tournaisian) forms three cuestas corresponding to three units with a total thickness of 400 m (1312 ft) (Conrad, 1985d). They constitute the Tzout I, Tazout II, and Tazout III formations, separated from each others by shales with ironstone or phosphatic nodules. They are composed of fine-grained sandstones with ripples and tracks with intraformational conglomerates, shales, and siltstones. The Tazout I and Zazout II formations are Devonian in age and Tazout III is Carboniferous. The top of the latter marks the Devonian-Carboniferous boundary. The Tazout I (Famennian) sandstones contain *Mesoplica praelonga* and *Cystospirifer verneuili* and grade into shales and limestones with *Clymenia*. The Tazout II (Strunian) consists of shales with ammonoids and sandstones with *Eobrachythyris strunianus*. The Tazout III (early Tournaisian) consists of shales with *Gattendorfia*, and sandstones contain *Marginatia vaughani*. The red Bou Mgheirfa Conglomerate occurs at the base of the *Gattendorfia* shales in the Zemoul area.

The Betaina Formation (late Tournaisian-late Visean) is composed of shales with sandstone, thin carbonate, and phosphatic beds intercalations (Conrad, 1985d). It has 700–1100 m (2297–3609 ft) in thickness. The basal part contains the late Tournaisian fauna *Muensteroceras rotella* and *M. duponti*, whereas the upper sandstones have the late Visean *Beyrichoceras micronotum*, *Dimorphoceras*, and *Neospirifer fascicostatus*.

The Djebel Ouarkziz Formation (late Visean-Serpukhovian) consists of two limestone units separated by a 50–90 m (164–295 ft) thick shaly-evaporitic-dolomitic unit, which includes the "*gypses d'Oum el Achar*" of Fabre (1976) (Conrad, 1985d). The Lower Ouarkziz is 300 m (984 ft) thick and consists of massive bioclastic pelletic limestones separated by sandstone and siltstone beds. The basal part consists of sandstones, siltstones, and limestones with *Lithostrotion* biostromes and gigantoproductids, such as *Gigantoproductus africanus* and *G. menchikoffi*, as well as the cerioid *Lithostrotion vorticales*, as well as late Visean foraminifera. Shales increase in the upper part of the lower Ouarkziz Formation and limestones become dolomitic. Fossils include *Gigantoproductus meridionalis*, *Lonsdaleia duplicate* and Serpukhovian foraminifera. The Upper Ouarkziz is made up of nodular limestones, shales, siltstones,

---

1 Note the subtle difference in spelling of the names Betana and Betaina.

and bioclastic limestones with *Siphonodendron, Lonsdaleia, Titanaria Africana,* and *Brachythyrina libyca.*

The Betana Beds consists of two formations (hence it should be called the Betana Group), each starting with a paleosol or a conglomerate. The Djebel Reouina Formation is 500 (1640 ft) thick and consists of fluvial sandstones and red shales with wood fragments of Namurian age (Conrad, 1985d). The Merkala Formation is 500–700 m (1640–2297 ft) thick and consists of calcareous conglomeratic sandstones with Westphalian plant remains at the base followed upward by red beds composed of shales and sandstones with carbonate concretions and Stephanian plant remains.

Carboniferous sediments occur in the extreme southern part of Algeria and south of the border in the Iullemedden and Taoudenni basins. They are described briefly by Legrand-Blain (1985). The Iullemedden Basin is located south of the Hoggar and extends into Niger, Mali, Nigeria, and Algeria. The Carboniferous crops out in the northeastern part the basin (known as the Tim Mersoi Basin) west of Aïr. The Carboniferous above the Devonian Taberia Formation is represented by the Upper Tournaisian Tim Mersoi Formation and the Visean Tagora Formation (Legrand-Blain, 1985). The Taoudenni Basin is located mostly in Mali south of the Reguibat Shield and west of the Hoggar. Carboniferous sediments crop out in the northern part of the basin (Legrand-Blain, 1985), but they are concealed beneath the *"Continenal Intercalcaire"* in the south (Legrand-Blain, 1985). The succession includes, in an ascending order, the sandstones of the Bir en Naharat Formation (Lower Carboniferous), carbonates of the Hamada Safia Formation (upper Visean) and Hamad el Haricha Formation (Serpukhovian), and clastics of the Hamou Salah Formation (Namurian) and Jakania Formation (Westphalian)

The Illizi Basin is separated from the Mouydir Basin by the Amguid-El Biod High and from the Murzuq Basin in Libya by the Tihemboka Arch (Fig. 15.1). The Carboniferous sediments crop out on the northern and eastern sides of the Illizi Basin and extend into the subsurface. The Carboniferous succession consists of five formations (Legrand-Blain, 1985): The Hassi Issendjel, Assekaifaf, Oued Oubarakat, El Adeb Larache, and Tiguentourine formations.

The Hassi Issendjel Formation (Upper Tournaisian-middle Uppar Visean) overlies the Djebel Illirene Sandstones (late Devonian-Strunian) in the center of the basin. At Tihemboka, it rests unconformably on top of Silurian sediments. The lower part consists of silty shales and sandy limestones with upper Tournaisian brachiopods and microflora (Legrand-Blain, 1985). The Formation grades upward into siltstones and shales with sandy limestone intercalations. *Beyrichoceras* suggests a middle or upper Visean age. The Hassi Issendjel Formation ends with *"Grès à champignons inférieurs"*, named as such owing to their weathering in the form of domes. These domes are also characteristic of the Mrar Formation in West Libya.

The lower part of the Assekaifaf Formation (Upper Visean-early Serpukhovian age) is made up of marls and limestones with rich marine fauna. Foraminifera, such as *Asteroarchaeodiscus bashkircus, Globivalvulina, Eostaffella pseudostruvei* indicate a late Visean-early Serpukhovian age. Stromatolites within this formation are developed on the Tihemboka Arch in Eastern Algeria and West Libya, and are known as the *Dome à Collenias.* The top of the formation is marked by the *"Grès à champignons supérieurs"* with lycophyte stem remains (Legrand-Blain, 1985) (cf. Mrar Formation in West Libya).

The lower part of the Oued Oubarakat Formation (uppermost Serpukhovian-Bashkirian) is made up of silty shales with sandstone and sandy coquinoidal limestone intercalations with brachiopods (Legrand-Blain, 1985). Foraminifera, such as *Eosigmoilina* and *Neoarchaediscus angulata*, indicate a late Serpukhovian age for this part of the formation. The age of the upper part ranges from the uppermost Serpukhovian, based on the presence of *Palaeosmilia*, *Latinproductus*, and *Beleutella* in some coquinas, to Bashkirian, based on the foraminifera *Eostaella pseudostruvei chomatifera*, *Endothyra bashkirica*, and *Endothyranella* (Legrand-Blain, 1985).

The El Adeb Larache Formation (Serpukhovian-Bashkirian) starts with limestones with nautiloids and brachiopods (Legrand-Blain, 1985). The *Syringthyris* brachiopod suggests a Serpukhovian age. A Bashkirian age is indicated by choristitids and the foraminifera *Pseudostaffella antique*, *Endothyranella gracilis*, and *Eostaffella lepida* (Massa & Vachard, 1979). These limestones are followed upward by marls, gypsum, and sandstones. The upper part of the formation is composed of variegated limestones, dolomitic and oolitic in part, as well as coquinas with brachiopods and gastropods. The foraminifera include species of the *Aljutovella*, *Profusulinella*, and *Glomospirella* Zones, associated with the goniatite *Eoxaralegocera*.

Tiguentourine Formation (cf. West Libya) starts with thin layers of stromatolitic limestones with Euestheria (Bertrand-Sarfati & Fabre, 1972), followed upward by red and green shales and bioturbated siltstones with ripple marks and mud cracks and *Euestheria* and fish remains (Legrand-Blain, 1985). The formation ends with red shales and cross-bedded sandstones. Its equivalent in Libya is Upper Carboniferous, but may extend into the Permian.

In Northern Algeria, the Carboniferous is present in the High Plains and the coastal area of the Oranais (Oran Province), and in the Internal Massif of Chenoua and Kabylie (Legrand-Blain, 1985). However, the stratigraphy is not very clear because of tectonics. The Carboniferous is represented by marine and volcano-clastic sediments. In the coastal area, the Carboniferous consists of sandstones, red shales, and conglomerate.

## 15.3   MESOZOIC

### 15.3.1   Triassic

Triassic deposits include sandy shales and lacustrine deposits (anhydrites and salts) unconformably overlying the Paleozoic formations. Triassic sandstones of the *Trias Argilo-Gréseux Inférieur* (TAGI) form the main reservoir in the Hassi R'mel Field in the Oued Mya Basin, about 550 km south of Algiers, and the El Borma Field in Tunisia (Askri et al., 1995) with estimated recoverable reserves of 880 mmbo and 1.3 TCF of gas (Galeazzi et al., 2010).

The Hassi R'Mel (Rmel) Field was discovered in 1956 following a seismic campaign which led to the drilling of the first well south of Berriane in October 1952 (Legrand, 2002). The well showed a potential hydrocarbon reservoir in the Trias Argilo Grèseux (TAG). Consequently, the Hassi R'Mel Field was discovered by the drilling of the HR1 well, which was put on production in 1961 and proved to be one of the world's largest gas fields. The Field extends 70 km from north to south and 50 km from east to west (Bencherif, 2003). The Gassi Touil Field is another example of Triassic fields in Algeria. It was discovered in 1960 by the drilling of the GT-1 well based on seismic and gravity surveys. Oil occurs in the sandstones and limestones of the lower, middle, and upper Trias Argilo Grèseux Formation (TAG) (Askri et al., 1995, p.88). The trap is an N-S anticline bounded by normal faults.

Triassic sediments have been enounterd in some wells in Northern Algeria. They are composed of red sandstones at the base, followed by thick evaporites, carbonates, and dolomites with basic volcanic intercalations at the top (Askri et al., 1995). Lower Triassic red sandstones are followed upward by Middle Triasssic limestones and another red sandstone interval with some doleritic rocks. In the Babors area to the south, the Middle Triassic is overlain by Upper Triassic lagoonal, gypsum-bearing units. Upper Triassic evaporites occur over most of Northern Algeria, except in some local areas, such as Doui Zaccar and Beni Snassene (Askri et al., 1995). The Djurdjura Mountains in the Great Kabylie in Northern Algeria are divided into three tectono-morphological units; the internal Kouriet, the middle Haizer-Akouker, and the external Tikjida and Tamgout units, which are thrusted to the south on Cretaceous flysch (Kotanski et al., 2004). The whole Triassic sequence, from Anisian to Rhaetian is represented in the middle unit. Kotanski et al. (2004) identified trackmarks of *Rotodactylus* cf. *bessieri* in the late Anisian section in the Haizer-Akouker Unit and postulated the presence of a terrestrial sedimentary regime in the Maghrebides during that period.

In the Hodna Mountains in Northern Algeria (Fig. 15.1), numerous Triassic outcrops line the major faults of Ouled Tebben and Boutaleb (Bouchareb Haouchine & Boudoukha, 2009). The Triassic succession consists of a sandstone unit at the base and an evaporite unit with dolomitic limestone beds at the top. Basic volcanic intercalations are common.

Topographic relief created on the Hercynian unconformity also played a major role in the deposition of Late Triassic sediments in the Berkine Basin (Cochran & Petersen, 2001).

The Triassic reservoirs are the most prolific plays in Algeria (Eschard & Hamel, 2002). They are exemplified by the giant Hassi R'Mel and El Borma fields and the more recent oil discoveries in the Berkine Basin[2], such as the oil discovery by Sonatrach in 2006 by the drilling of the well AIIMN-1(Ait Hamouda North-1) in Block 405-a and the well GSM-1by the Sonatrach and First Calgary Petroleum Partnership in Block 405–b in the Ledjmet contractual area (Sonatrach Magazine, February 2007, p.7 &10). However, the reservoirs are heterogeneous and the prediction of the reservoir extension is difficult.

The Triassic succession is dominated by fluvial channel sediments interfingering and alternating with floodplain and sabkha sediments (Eschard & Hamel, 2002, Bourquin et al., 2010). Major thickness and facies variations occurred across normal faults. Many Triassic depositional sequences have been recognized between the Hercynian unconformity and the Rhaetian and correlated across the area (Eschard & Hamel, 2002). Triassic sedimentation started during Upper Ladinian times in the northern part of the Berkine Basin. During Carnian times, an extensional phase produced two major basins, the Berkine and the Oued Mya basins were separated by

2 See Gulf Oil & Gas online (11/12/2003).

the El Biod-Hassi-Messaoud High. Extension was accompanied by widespread lava flows which covered the Northern part of the Hassi-Messaoud High and the Hassi Rmel area. Sandy braided-plains and sand flats of the TAGI reservoir unit prevailed in the Oued-Mya Basin. Restricted marine carbonate sediments of the Trias Carbonaté Formation were deposited in the graben axis following the major marine transgression of the Paleo-Tethys (Eschard & Hamel, 2002).

In the Berkine Basin/Ghadames Basin in Algeria and Southern Tunisia, the Triassic succession consists mainly of fluvial deposits in the south, becoming paralic and evaporitic facies (Busson, 1967a, b, Boote et al., 1998, Eschard & Hamel, 2002, Turner et al, 2001, Sabaou et al., 2005, Turner & Sherif, 2007). The Triassic out-crops of the Zarzaitine area, south of In Amenas, correspond to the most proximal facies of the Triassic depositional systems preserved in Algeria (Bourquin et al., 2010). In the Berkine Basin, the Triassic succession has been divided by Busson (1971a, 1971b, 1974) and Busson & Comee (1989) into three units, from the base to the top; Trias Argilo-Gréseux Inférieur" (TAGI), Trias Carbonaté, and Trias Argilo-Gréseux Supérieur" (TAGS). The Early Triassic is missing in Algeria and the basal TAGI fluvial sequence was deposited on top of the Hercynian unconformity during the initial sag-ging phase of the Basin. Fluvial channels flowed northward toward the Tethys Domain (Rossi et al., 2002) with sediments supplied by the El Biod Hassi-Messaoud High. The TAGI sequence was dated palynologically as late Ladinian-late Carnian in the north-ern part of the Berkine Basin. Tectonic activities increased during the late Carnian and Norian with the deposition of the Upper TAGI, Trias Carbonaté, and TAGS. N-S and NE/SW normal faults inherited from the Pan-African basement formed on both sides of the El Biod-Hassi-Messaoud High (Fig. 15.1). Horsts, grabens, and tilted blocks were formed in the Berkine basin (Turner et al., 2001) during the Late Triassic and Liassic periods. During the Norian time, evaporitic sediments (named the S4 salt unit, Busson, 1971b) (shown as Rhaetian by Turner & Sherif, 2007 and Bourquin et al., 2010) were deposited in the north and a sand-rich delta (TAGS) formed in the south (Boote et al., 1998). During the Rhaetian a major transgression is marked in Algeria by a dolomitic marker known as the D2 Dolomite in the Oued Mya Basin (Fig. 15.1). The D2 Dolomite pinches out toward the Berkine Basin where it interfingers with the shales of the Trias Argileux. An extensive sabbkha extended into the northern part of the Berkine Basin (S4 Salt Unit). In the southern part of the basin, a large coarse-grained braided delta (the TAGS Unit) interfingered with the sabkhas (Eschard & Hamel, 2002). Upper Triassic sedimentation ended with a regional transgressive event depositing a thin dolomite bed with a regional extension (D2 Bed). A salt basin then was developed during the Lower Jurassic.

Late Triassic and Early Jurassic evaporites of the Berkine/Ghadames Basin in Eastern Algeria and West Libya form the regional seal to a number of hydrocar-bon reservoirs in the Triassic Argilo-Greseux Inferieur (TAGI). The evaporitic suc-cession (Norian to Late Liassic) has been divided into three units (S1, S2 and S3) (Turner & Sherif, 2007). S1 consists of five evaporitic cycles dominated by mud-stones and halite, which represent the main salt deposits of the Berkine Basin. At that time, the Berkine/Ghadames Basin was a restricted evaporitic basin with a bar-rier to the north (Medenine High) separating the basin from the peri-Tethys and the North Atlantic Ocean. The succession thickens to the west, thins around the Hassi

Messouad Ridge, and continues into the Western High Atlas of Morocco. The S2 is represented by the basin-wide development of a carbonate platform associated with the Pliensbachian relative sea-level rise and the opening of the Atlantic. The S3 comprises five cycles that are predominantly mudstones, fine-grained carbonates and anhydrites. Mound structures, evident on seismic sections, may represent the development of patch reefs and the whole succession has been interpreted as a shallow-marine carbonate-evaporite ramp (Turner & Sherif, 2007). Restricted circulation was caused by an east-west oriented sill to the north (Medenine High in Central Tunisia).

The sedimentary succession exposed in the Zarzaitine area includes both Triassic and Liassic deposits. The Triassic interval is made up of fluvial and floodplain sandstones alternating with reddish mudstones of the Lower Zarzaitine Formation (Bourquin et al., 2010). The thickness of the Triassic succession is about 200 m (656 ft), but thins out over a structural high. The Lower Zarzaitine Formation rests on top of red and purple shales of the Upper Permian Tiguentourine Formation. The top of the Lower Zarzaitine Formation is marked by a sandy dolomitic bed attributed to the D2 Marker (Bourquin et al., 2010). Busson (1971) subdivided the Lower Zarzaitine Formation into two sandstone units separated by thick red purple shales attributed to the Middle Unit, with a maximum thickness of 40 m (131 ft). The Lower Sandstone Unit has a maximum thickness of 70 m (230 ft) and is composed do fine-grained, through cross-bedded sandstones. The Upper Sandstone Unit is 100 m (328 ft) in thickness and consists of fine- to coarse-grained sandstones and red shales. The whole succession is dated as Late Triassic (Lehman 1957, Busson 1971b, 1974, Busson & Burollet, 1973) based on dinosaur remains. Jalil & Taquet (1994) identified remains of Tetrapod vertebrates in the Lower Unit of the Zarzaitine series giving a late Carnian to Norian age. Jalil (1999) confirmed the Triassic (Anisian-Norian) age of the lower unit.

## 15.3.2  Jurassic

Jurasssic Marine and lacustrine deposits are present over most of the Triassic Province (Fig. 15.1). The Jurassic sequence starts with dolomites of the Horizon B.

The Lower Liassic of Northern Algeria is dominated by dolomitic and oolitic limestones, changing progressively upward into Upper Liassic marls (Askri et al., 1995). Hettangian-Pliensbachian sediments consist of about 200 m (656 ft) of cherty and dolomitic limestones, red Ammonite limestones, and reefal and algal patches. The pelagic Ammonitico rosso facies, well developed in Sicily and the Alps during the Middle Jurassic-Lower Cretaceous, is found in the Djurdjura area. Kotañski et al. (reported reptile tracks of *Rotdactylus* in the Middle Triassic in the same area. The Jurassic crops out only in the central and eastern parts of the Hodna Mountains (Bouchareb Haouchine & Boudoukha, 2009). The Liassic is represented by massive dolomites and dolomitic limestones. The Toarcian consists of marls and marly limestones, up to 350 m (1148 ft) in thickness. The Dogger is composed of limestones and marly limestones, 80 m (262 ft).

The Middle Jurassic changes facies from thin carbonates in the Kabylie Domain into thick shaly-carbonates in the Tellian Zone, limestones in the High Plateau and the Constantine areas, and thick shales and sandstones of more than 2000 m (6562 ft) in thickness in the Atlas Basin (Askri et al., 1995). Fluvial and deltaic clastic sediments predominate in the southern part of the Oran plateau and the Western Saharan Atlas.

The Upper Jurassic marine sediments in the north change into deltaic and continental sediments southward (Askri et al., 1995). However, in certain areas, such as Sedrata-Laghouat at the southern edge of the Saharan Atlas, marine sediments predominate throughout the Upper Jurassic. The Upper Jurassic crops out in isolated mounts south of Chott el Hodna (Aissaoui, 1989). It consists of shallow platform limestones composed of peloidal mudstones and wackestones, with oncolites, oolites, and lithoclasts composed of the benthonic foraminifera *Anchispirocyclina lusitanica*, *Pseudocyclamina lituus*, *Trochlina* gr. *alpina*, and *T.* gr. *elongata*, in addition to gastropods, echionderms, dasyclads, and algal fragments, and Upper Tithonian calpionellids (Aissaoui, 1989). The top of the succession is often dolomitized. The Upper Jurassic-Lower Cretaceous (Kimmeridjian-Berriasian) succession occurs only in the eastern part of the Hodna Mountains and is made up of interbedded limestones and marls (Askri et al., 1995).

For biostratigraphical zonation of the Middle Jurassic Brachiopods in Western Algeria and Morocco, see Almeras & Faure (2008).

The Ksour Mountains (named after Berbers palaces) in Northern Algeria are located close to the Moroccan border and form a series of NE-SW-oriented folds between the North (NAF) and South Atlas (SAF) Faults (Mekahli et al., 2004). The Middle-Upper Jurassic Ksour Sandstones form a delta in the Algerian Atlas Domain (Piqué, 2001, p.100).

Sauropod dinosaur vertebrae were described by De Lapparant (1960) discovered in 1953 by J. Cassedanne in the Middle Jurassic hard calcareous sandstones at Rhar Rouban about15 km southeast of Oujda near the Algerian-Moroccan border. The discovery represented the first report of Jurassic dinosaurs in Algeria at the time. The remains were dated Middle Callovian by ammonites associated with these deposits. Mahammed et al. (2005) reported the discovery of a skeleton, including cranial material, of a new cetiosaurid sauropod *Chebsaurus algeriensis* n.g., n.sp. in the Middle Jurassic of the Algerian High Atlas, which represents the most complete Algerian sauropod. Sauropod remains were documented from the Early and Middle Jurassic strata of North Africa by Moroccan sauropods through the descriptions of "*Cetiosaurus*" *mogrebiensis* (De Lapparent, 1960), *Atlasaurus imelakei* (Monbaron et al., 1999) and more recently *Tazoudasaurus naimi* (Allain et al., 2004). Sauropod remains occur also at Sfissa (Fig. 15.1).

The Atlas Chain is oriented SW-NE in the Ksour area, whereas it trends roughly E-W in the High Atlas and NW-SE in the Aurès (Mekahli et al., 2004). In the Ksour Mountains, an initial carbonate platform probably formed as early as the Rhaetian-Hettangian, the marine Lower Hettangian and the Sinemurian are dated by rich ammonite fauna. The subsiding basins became hemipelagic during the Sinemurian, with radiolarian limestones and several biostratigraphical markers. Regional deepenings took place during the Late Pliensbachian, the Early Toarcian, and at the beginning of the Late Bajocian. The return to a platform environment began diachronously from the Late Bajocian in the SW to the Early Bathonian in the NE.

The Ksour Mountains consist of various sectors (Mekahli et al., 2004), each characterized by its own stratigraphy; from north to south: A) The Aïn Ben Khellil Basin area, B) the Melah-Souiga (or Mekalis) median shoal, C) the Ain Ouarka subbasin, and D) the southern border (Kerdacha).

In the northeastern part of the Ain Ben Khellil area, the basal Guettob Moulay Mohamed Carbonates (equivalent to the Koudiat el Beia Formation) (Late Sinemurian) contains *Orbitopsella dubari, Planisepta compressa (Labyrinthina recoarensis)*. The Oulad Amor Formation, 35–85 m (115–279 ft) in thickness, contains abundant *Palaeodasycladus mediterraneus* attributed to the Lower Pliensbachian (Carixian) and the brachiopod *Hesperithyris renierii*. The Toarcian Jebel Nador Formation is made up of thick purple marls, bioclastic limestones, and dolomites, 60–120 m (197–394 ft) in thickness. It is overlain by the Middle Jurassic shallow carbonates of the Antar Dolomites Formation. To the northwest (Fig. 15.2), the Lower Jurassic outcrops at the foot of Djebel Reha consist of three formations: The Guettob Moulay Mohamed Carbonates Formation at the base has a thickness of more than 150 m (492 ft), and begins with a Massive Dolomites Member, changing upwards into the Lithiotis Limestones Member rich in large bivalves and occasional corals, and the Gaaloul Cherty Formation, 95–100 m (312–328 ft) of grey cherty limestones and marls. The tops of these beds are hard ground with belemnites, oolites and bioclasts. The age of the succession ranges from Carixian (with *Tropidoceras* sp.) in the lower part, Middle Domerian (with *Reynesoceras* gr. *Indunense, Prodactylioceras* sp., *Arieticeras* sp.) in the middle part, to the Late Domerian (with *Emaciaticeras* sp. and *Tauromeniceras*) at the top. The following Reha Formation is about 150 m (492 ft) thick and represents an Early Toarcian cyclic sequence. It contains *Paltarpites* sp., *Dactylioceras (Eodactylites)* gr. *mirabile*, and *Dactylioceras (Orthodactylites)* sp., followed upward by marls with pyritic ammonites, such as *Pseudogrammoceras subregale* and *"Podagrosites"* gr. *Aratum*, and *Audaxlytoceras dorcadis*. The presence of Toarcian marls dwarf ammonites (Mekahli et al., 2004) and the variations in

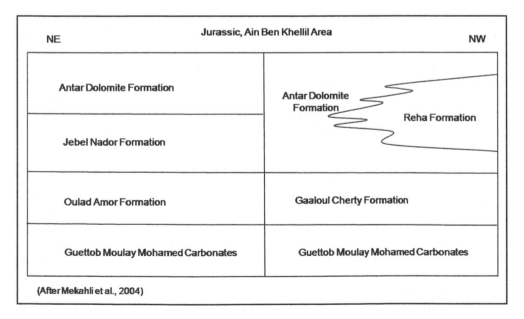

*Figure 15.2* Jurassic stratigraphy of the Ain Ben Khellil area, Ksour Mountains, NE Algeria (after Mekahli et al., 2004).

Nodosariid foraminifera (Sebane et al., 2007) indicate the isolation of the area and the development of anoxic conditions that prevailed globally during the Toarcian OAE (Jenkyns, 1988, 1991, Sebane et al., 2007). The return to a platform setting was marked by the deposition of limestones with *Zoophycus*, passing vertically and laterally into shallow carbonates of the Antar Dolomites Formation (not accurately dated, is probably Middle Jurassic). The Late Bajocian is characterized by alternating marls and neritic limestones with *Ermoceras* fauna, overlain by oolitic carbonates and reefs of the Guettaï Formation.

In the Melah-Souiga (or Mekalis) shoal area, the succession consists of the Souga Formation at the base, followed upward by the Ammonitico rosso, Aouinat es Siah Formation, Ain Beida Formation, Teniet el Klakh Formation, and Tifkirt Formation. The upper part of the Souiga Dolomites Formation contains a rich rhynchonellid fauna associated with *Zeilleria hierlazica*, (Sinemurian) (Mekahli et al., 2004). The Formation is overlain by calcareous red nodular bioturbated limestones of the Ammonitico rosso, then by the Aouinet es Siah Formation (Domerian) composed of marls at the base and cherty limestones at the top. The overlying Ain Beida Formation (Toarcian) is represented by marls in lower part, followed upward by *Zoophycos* limestones (Aalenian), changing upward into a marly Ammonitico rosso. The Early Bathonian is represented by distal turbidites, slumps, and brachiopod-rich limestones of the Teniet el Klakh Formation. The type locality of the the Teniet el Klakh Formation and Tifkirt Formation is situated between Djebel Souiga and Djebel Tifkirt. The Tifkirt Formation consists of 500 m (1640 ft) of limestones, marls, and sandstones. The lower part is made up of reef build ups and oolitic horizons (Mekahli et al., 2004). The boundary between the Tikfirt and Teniet el Klakh formations is placed at the base of the first channelized sands by Mekahli et al. (2004). Its age ranges from late Bajocian to early Callovian. The Tifkirt Formation consists of two member; Guettai Member and El-Gouachiche Memebr, which contain *Burmirhynchia termierae*, *Monsardithyris uniplicata*, and *Rugitela prebullata* (Bajocian *Niortense* Zone), in addition to the the ammonite *Ermoceras* (Almeras & Fauré, 2008).

The Callovian in the Ouarsenis Mountains, Tell of Algeria, includes the Encrinitic Limestones and Massive Oolitic Limestones Units in fault blocks at Bou Hadjar, northwest of Tiaret (Atrops & Alméras, 2005). The two units are rich in brachiopods that gave a Lower Callovian date, such as *Terebratula balinensis* (*Aromasithyris balinensis*), *Aromasithyris ameuri* nov. sp., *Wattonithyris* cf. *roettingensis*, *Linguithyris bifrons*, *Linguithyris vicaria*, *Antiptychina* cf. *teisenbergensis*, *Caucasella rectecostata*, *C. trigona*, and *Kallirynchia anglica*.

In the Ain Ouarka Subbasin, the Jurassic succession begins with shallow platform carbonates of the Chemarikh Dolomites Formation with brachiopods at the top, including *Zeilleria perforate* (= *Terebratula psilonoti*), followed unconformably by the Ouarka Pelagic Limestones Formation. The Ouarka Formation is 120–160 m (394–525 ft) thick and consists of seven members designated A to G (Mekahli et al., 2004): Member A consists of limestones with rare bioclasts and foraminifera, including Nodosariids and *Involutina liassica* and the ammonites *Arnioceras geometricum*, *A. miserabile*, *A.* cf. *speciosum*, and *A.* cf. *flavum*. Member B is composed of bioturbated limestones with *Asteroceras* beds or *Gleviceras* gr. *Doris*. Member C is made up of Radiolaria-rich cherty limestones with the ammonites *Plesechioceras* cf. *delicatum*, *Angulaticeras* gr. *dumortieri*, and *Paltechioceras* sp. Member D is dominated by bedded limestones and marls. Coquina with thin-shelled bivalves of *Diotis janus* and the small ammonite *Galaticeras* cf. *aegoceroides* occur in the upper part. Member E is made up of marls with *Tropidoceras calliplocum, Protogrammoceras* gr. *dilectum-pseudodilectum,* and *P.* gr. *Volubilepantanelii.*

Member F is a marly Ammonitico rosso rich in *Reynesoceras.* Member G is composed of green marls and limestones with *Pleuroceras solare.* The overlying Ain Rhezala Formation (Toarcian) is a thick marl unit changing to micritic limestones at the top, with *Pseudogrammoceras subregale* and *P. pinnai.* It is followed by reworked oolites and breccias of the Raknet el Kahla Megabreccias (Early Toarcian-Early Bajocian) with the ammonites *Leioceras (Cypholioceras)* cf. *comptum, Staufenia (=Ancolioceras) opalinoides, Ludwigella* cf. *arcitenens,* and *Euaptetoceras* sp. The Upper Bajocian-Lower Bathonian includes the Teniet el Klakh Formation, the Bou Lerfahd Reefs and Limestones, and the Tifkirt Formation with a total thickness of 500 m (1641 ft) (Mekahli et al., 2004). The Bou Lerfahd reefs have a maximum thickness of 14 m (46 ft) and start with stratified fossiliferous bioclastic limestones with the bivalves *Isognomon* and *Lopha,* solitary corals, crinoids, bryozoans, calcisponges, and rare Lenticulinids foraminifera. The core of the reef is made up of massive coral colonies of *Isastrea, Thamnastrea,* and stylinids in bioclastic micrites. The Tifkirt Formation (Late Bajocian) consists of alternating sandstones, shales, marls, and oolitic intervals rich in brachiopods. The Djara Formation consists of laminated sandstones, shales, and oolitic dolomites. It has a thickness of 450 m (1476 ft) at Ain Ouarka. The Jurassic Ksour Sandstones form a delta in the Algerian Atlas domain (Piqué, 2001, p.100), with delta -front sandstones and lagoonal green and red shales. The age of the formation is poorly defined, but may range from topmost Bathonian to Oxfordian (Mekahli et al., 2004). The highest Jurassic unit is the Aissa Formation overlying the Ksour Sandstones and is composed of 450–500 m (1476–1641 ft) of fluvial sandstones and variegated shales. It has been dated Latest Oxfordian-Early Kimmeridgian based on correlations with the Tlemcen Domain (Benest, 1985, cited by Mekahli et al., 2004). There are no Jurassic outcrops south of Ain Ouarka, except at Djebel Kerdacha (Mekahli et al., 2004). The Kerdacha Formation, with *Monsardithyris ventricosa* and *Rugitela prebullata* of the *Niortense* Zone, is equivalent to the Tifkirt Formation (Almeras & Fauré, 2008).

In the Ain Ouarka Subbasin in Northern Algeria, the Upper Jurassic-Lower Cretaceous succession is represented by the Tiloula Formation of probable Latest Jurassic-Earliest Cretaceous age and consists of variegated shales, channelized pebbly sandstones, and stromatolitic dolomites with bird's-eyes and halite. It overlies the Upper Jurassic Aissa Formation. The Tiloula Formation has a thickness of 500–650 m (1641–2133 ft) in the Ain Sefra area. It is equivalent therefore to to the "Nubian Sandstone" *s.s.* in Libya anad Egypt (Tawadros, 2001). The Tiout Formation is composed of pebbly sandstones, and has been assigned tentatively a Lower Cretaceous-Albian age by Mekahli et al. (2004). The Tiout Formation is overlain by the Upper Cretaceous Mdaouer Formation composed of shales and gypsum, 100–150 m (328–492 ft) thick in the Ain Sefra area, followed by about 130 m (427 ft) of carbonates and marls with the Cenomanian *Vascoceras* ammonite.

The Mellala section in the Traras Mounts of the Tlemcen Domain, Northwestern Algeria, displays limestones of the Tisseddoura Formation, about 9 m (30 ft) in thickness, overlain by the Bayada Formation (Late Pliensbachian-Early Toarcian) (Elmi et al., 2010). The Bayada Formation is made up of a deepening-upward sequence of biomicrites with brachiopods and belemnites, as well as *Zoophycos* and *Steinmania*. The Bayada Formation is about 50 m (164 ft) in thickness. It consists of two unnamed members, 23–25 m (75–82 ft), separated by a marker-bed composed of bioclastic quartz-rich limestones marks the limit between the two members of the formation. The base of the Bayada Formation contains *Dactylioceras* (*Orthodactylites*) *crosbeyi* and *Pleuroceras solare* (Upper Domerian/Pliensbachian) occurs in the lower part of the Formation, followed upward by *D.* (*Eodactylites*) *polymorphum*, *D. mirabile*, and *D. pseudocommune*, then by *Eodactylites*. The top of the Byada Formation yields *Eleganticeras* sp. The occurrence of the brachiopod *Koninckella* indicates a restricted environment within a deep basin durin the end of the Pliensbachian-Lower Toarcian, while *Lenticulina obonensis* indicates stressed conditions (Elmi et al., 2010). The Bayada Formation consists of three units (Hadji et al., 2006): 1. Lower Bayada Formation is about 22.3 m (ft) in thickness and consist of alternating marls and limestones rich in the ammonites *Hildoceras* sp., *Polyplectus pluricostatis*, *Harpoceras*, and *Hildoceras* cf. *sublevisoni*, changing upward into bioturbated limestones, then into bioclastic limestones with bivalves and the ammonites *Collinites*, *Collina*, and *Osperleioceras* of Middle Toarcian age. 2. Traras Ammonitico rosso, 14.8 m (49 ft) thick, is made up of interbedded micritic limestones and reddish marls, rich in small ammonites of Upper Toarcian age. 3. Upper Bayada beds, 3.2 m (10 ft), consists of interbedded marls and limestones with ammonites, such as *Hammatoceras* sp., *Dumortiera* sp., *Osperleioceras* gr. *Nadorense*, *Phylloceras* sp., *Leioceras* sp., and *Harpoceras costula* of Upper Toarcian age. Benthonic foraminifera are dominated by the Nodosariidae *Lenticulina*, associated with *Lenticulina chicheryi*, *Lenticulina toarcence*, and *Dentalina pseudocommunis*.

### 15.3.3 Cretaceous

Cretaceous sediments crop out mostly in Central, Western, and Northern parts of Algeria (Geological Map of Algeria). The stratigraphy of the Cretaceous in NE Algeria follows that of Tunisia (Askri et al., 1995, Fig. 14). In the Berkine Basin, the stratigraphic column includes the Lower Cretaceous "Contintental Intercalcaire" (Neocomian-Barremian), the Aptian Dolomite, the Cenomanian Lagunaire Argileux, and the Turonian Calcaire Argileux, Senonian Carbonate Anhydritique Lagunaire, overlain unconformably by Middle Pliocene sandstones (Yahi et al., 2001). However, because in the Berkine Basin, for example, all the reservoir and source rocks are pre-Liassic, stratigraphic studies of younger succession are almost lacking.

Cretaceous rocks are spread over the whole Sahara Platform. The sequence consists mainly of sandstones at the base, changing upwards into carbonates (Askri et al., 1995).

In Northeastern Algeria, the Lower Cretaceous consists of the Neocomian Sidi Kralif Formation, Meloussi Formation; the basal part of the Boudinar Formation (Neocomian-Aptian), Bouhedma Formation, Sidi Aich Formation (Aptian-Albian), Orbata Formation (Aptian-Albian), and the Lower Zebbag Formation (Albian-Cenommanian).

Most of Northern Algeria was subjected to a general regression during the Barremian. The Tellian Atlas, the Constantine Uplift, and the northeastern part of the Atlas Basin were totally emerged (Askri et al., 1995). Neocomian-Barremian sediments consist mainly of alternating sandy shales and carbonates. Clastic Cretaceous sediments are found over the whole Saharan Atlas (Askri et al., 1995), which reach a maximum thickness of 1200 m (3937 ft) in the Western Atlas.

The Maghrebian Orogeny extends for over 2000 km in the Western Mediterranean and is structurally characterized by nappes verging towards the external zones. The Flyschs Domain is a series of stacked-up allochthonous units containing of various Cretaceous-Tertiary sedimentary sequences.

The Lower Cretaceous occupies large areas in the Hodna Mountains. The succession includes, from base to top (Bouchareb Haouchine & Boudoukha, 2009): A marly and silty sandstone formation (Valanginian-upper Berriasian), about 250 m (820 ft) thick; a dolomitic sandstone formation (Hauterivian), a unit composed of limestones, dolomites, and sandstones (Barremian) up to 200 m (656 ft) in thickness. The Aptian-Albian succession include rudist limestones, sandstones, oyster limestones, and red marls (Aptian), calcareous marls with *Orbitolina*, and a calcareous sandstone unit (Albian) about 370 m (1214 ft) in thickness. The Aptian carbonates and reefal deposits in the Hodna and the Aures areas change facies in the south and west into fluvio-deltaic sandstones (Askri et al., 1995). In the Atlas Trough, the Aptian and Albian cannot be distinguished within the carbonate sequence. The Albian in the Saharan Atlas displays lithological variations from sandstones in the north to flysch facies in the south. In the Tell, the Albian is composed of sandy shales, whereas in the Constantine area limestones predominate. The sandstones and flysch deposits of the Tell and Hodna regions are overlain by black marls and limestones (Askri et al., 1995).

The sections of the Ouled Nail Massif in the western part of the Saharan Atlas on the northern edge of the Saharan Craton display three successive periods within the Late Cenomanian-Early Turonian interval 1 (Grosheny et al., 2008). The first (lower part of the upper Cenomanian) and the last (the whole of the Turonian) periods correspond to a ramp-type depositional system. The ramp is S-N oriented, from the Saharan craton to the deeper depositional environments of Algerian coastal range. During the intervening phase (latest Cenomanian-earliest Turonian); isolated carbonate platforms (rudists and/or oyster-bearing facies) developed. Cenomanian-Turonian black shales, probably related to the OAE, occurred between the platforms. The shales pinch out on the platform edges.

The Upper Cretaceous in Northeastern Algeria includes the Upper Zebbag Formation (Cenomanian); the Hassi Bouras, Ras Toumb, and Kemaken formations (Turonian), the Doulid Formation (Turonian-Coniacian), the Aleg Formation (Coniacian-Campanian), the Oued Mellah Formation (Santonian), and the Abiod Formation (Campanian-Maastrichtian). The Upper Cretaceous forms the bulk of the Mesozoic outcrops in the Western Hodna Mountains (Bouchareb Haouchine & Boudoukha, 2009). It is represented mainly by Lower-Middle Cenomanian marls and marly limestones, 200–300 m (656–984 ft) thick, with fossiliferous beds at the top; Upper Cenomanian-Turonian limestones and dolomites, 150–200 m (492–656 ft) in thickness; Senonian marls and marly limestones with *Globotruncana* and rudists; marls and marly limestones, about 200 m (656 ft), and Maastrichtian limestones about 100 m thick (328 ft).

The Upper Cretaceous-Paleocene succession in the El Kantara Syncline in the Western Aurès, Algeria, includes two Cretaceous units; the Djar ed Dechra Formation and the Evaporitic Green Marls of Tamarins, followed upward by the Upper Paleocene (Thanetian) El Kantara Red Marls and Oued El Haï Marly Limestones, and then by an Eocene-Oligocene terrigeneous succession in the center of the El Kantara Syncline (Belkhodja & Bigot, 2004). The Djar ed Dechra Formation (Maastrichtian) consists of of limestones with radiolite rudistid, inoceramid, and oyster fragments, as well as foraminifera, such as *Cuneolina* sp. and *Fleuryana adriatica* and the orbitoids *Orbitoides tissoti, Omphalocyclus* sp., *Orbitoides concavatus*, and the algue *Lithothamnion*, followed by a horizon rich in *Laffitteina boluensis, L. conica, L. mengaudi, L.* cf. *monody, L. oeztuerki,* and *L. bibensis*. It has been assigned Upper Maastrichtian age by (Belkhodja & Bignot, 2004) based on orbitoides. It overlies the Evaporitic Green Marls of Tamarins (Upper Cretaceous).

The geology of the Constantine region has been reviewed by Bensalah et al. (1991). From south to north, the following tectono-stratigraphic units are distinguished (Moretti et al., 2007): Constantine High, Peni-Tellian, Central Tellian, Ultra Tellian, 'Ain Kerma', and Numidian.

In the Constantine High Unit (Mesozoic), the Mesozoic rocks are represented by neritic limestones; changing from the Campanian onward into marls and greyish limestones (Lahondere & Magne, 1983). It is overlain by the Tell Zone succession which includes Paleocene marls, Lower Eocene greyish limestones, and Eocene-Oligocene marls and shales. South of Constantine area, for example in the Ain Mlila region, this succession extends up to the Early Miocene, whereas further east, in the Guelma region, it extends only to the Late Paleocene. In the Oran region, some hills, such as Djebels Mekaïdou, Kerbaya, and Ouazzene, are composed of thick continental conglomerates, marls, red shaales and silts of Eocene age. These sediments represent fluviatile and occasional lacustrine deposits and often show thick calcrete beds. They contain an abundant terrestrial Bulimes genera, such as *Romanella* and *Vicentinia* similar to those of Brezina in the southern Western Sahara Atlas, El-Kantara (Aures), Central and Southern Tunisia, and Morocco (Bensalah et al., 1991).

Peni-Tellian Unit 'Penitelliennes' (Upper Cretaceous-Paleocene) is applied to a group of formations of alternating Barremian-Aptian and Upper Cenomanian neritic limestones and marls. To the north and west of Constantine, the Penitelian succession may extend beyond the Maastrichtian. The uppermost units (Uppermost Senonian and Paleogene) occur in thrust sheets which further south form the Oued Zenati-Sigus Unit. The Penitellian Domain was situated on the northern slope of the Tell Basin.

Central Tellian Unit (Jurassic-Eocene) crops out mainly in the Constantine area (Dj. Ouach) and in the east, near the Algerian-Tunisian border (Ain Kerma region). The Central Tellian sediments are composed of Lower Eocene blackish limestones and Middle Eocene blackish-grey marls. The upper section of the marls contains abundant reworked microfauna and, locally, blocks of Jurassic-Cretaceous limestones.

The Ultra Tellian Unit is a group of formations characterized by marls and light-colored limestones containing rich pelagic microfauna (Durand-Delga, 1969a, b). These rocks were deposited on a bathymetic high which separated the Tell Basin from the Great Flysch Basin. This formation shows a decrease in terrigenous sediments southward.

The Ain Kerma Series *"Serie d'Ain Kerma"* (Jurassic-Cretaceous) is equivalent to the Ultra Tellian succession. It consists of a lower unit similar to the Djebel Guetar

Series in the north and ranges in age from Turonian to Eocene and an upper unit of Eocene-Oligocene age. The upper unit consists of greenish pelites with 'silexite' levels, extremely poor in microfossils. Moretti et al. (2007) suggested that the upper unit of the Ain Kerma Series originally constituted the Paleogene cover for the Massylian Units and was later detached and tectonically transported southward. The older units (Lower Cretaceous) overlie younger Paleocene-Middle Eocene rocks.

## 15.4   TERTIARY OF ALGERIA

### 15.4.1   Paleocene

The Upper Paleocene (Thanetian/Landenian) succession in the El Kantara Syncline in the Western Aurès overlying the Upper Cretaceous Djar ed Dechra Formation and the Evaporitic Green Marls of Tamarins, includes the El Kantara Red Marls and Oued El Hai Marly Limestones, about 50 m (164 ft) in total thickness, followed upward by an Eocene-Oligocene terrigeneous succession in the center of the El Kantara Syncline (Belkhodja & Bigot, 2004). The Danian is missing due to emersion during the early Paleocene (cf. Egypt).

The El Kantara Red Marls contain an assemblage of *Microcodium (Paronipora)*, freshwater Charophytes, and foraminifera (Belkhodja & Bigot, 2004). The Oued El Hai Marly Limestones overlie the Kantara Red Marls and contain a Thanetian assemblage of *Vania anatolica (Broeckinella arabica)*, *Idalina* aff. *sinjarica*, *Periloculina* aff. *Pseudolacazina alpani*, *Spirolina* sp., *Rotalia* sp., *Lockhartia tipper*, *Operculina* sp., *Verneuilina limbata*, *Valvulina angulosa*, *V. nammalensis*, *V. guillaumei*, *V. limbata*, *V. pupa*, *V. terquemi*, *V. triangularis*, *V. triedra*, and *Gyrovalvulina corrugata*, in addition to small agglutinated foraminifera and Miliolids (*Quinqueloculina*), *Glomalveolina primaeva*, *Laffitteina bibensis*, *Lockhartia* sp., and the Discorbid *Rosalina bractifera (R. elegans)*.

### 15.4.2   Eocene

In the Tell, Lower Eocene (Ypresian) sediments are represented by fossiliferous shales, marls, and carbonates. In the Aurès, they include clastics and red marls which pinch out under the Miocene unconformity in the Hodna Basin. South of the Hodna area, the Lower Eocene consists of gypsum-bearing marls followed upward by phosphatic and cherty limestones. The Lutetian is composed of nummulitic limestones and fossiliferous marls (Askri et al., 1995, Belkhodja & Bignot, 2004).

In Northern Algeria, Eocene-Oligocene sediments overlie the Oued El Hai Formation in the El Kantara Syncline in the Western Aurès Belkhodja & Bignot, 2004). The Ypresian is absent, but it crops out at Tighanimine and West Tinrhert areas to the south (Amard & Blondeau, 1979) and extends into the subsurface (Cornet & Gouskov, 1959).

Eocene phosphate deposits in Algeria form part of the phosphate belt of the southern Mediterranean. They occur in two belts; one that crosses Algeria from east to west; the other is more limited to Eastern Algeria (Villevain, 1924). They crop out at Medjerda and in the Guelma Basin, in the Kabylie of Babors, and the Middle Aurès.

Duffetaut (1982) described what was believed the first evidence of an African terrestrial mesosuchian crocodilian with a ziphodont dentition (compressed serrated teeth) from Middle Eocene sediments from Southern Algeria. Ziphodont mesosuchians are known mainly from South America and Europe and their presence in Algeria suggests faunal connection between South America and Europe via Africa.

The Glib Zegdou Formation is a continental series unconformably overlying Paleozoic sediments. The middle member of the formation contains several vertebrate species, with lungfish, actinopterygians, chelonians, and mammals, associated with Early to Middle Eocene charophyte. Adaci et al. (2007) reported a hyracoid (*Titanohyrax tantulus* ) and a macroscelidid (?*Chambius* sp.) from the formation. A small species of Simiiformes (Anthropoidea) was also reported from the same formation (Godinot & Mahboubi, 1992). The species is similar to *Aegyptoplthecus* and suggests a long and endemic African history for higher primates.

The Eocene in the northern High Lands of Oran, such as at Djebel Mekaïdou, Djebel Kerbaya and Djebel Ouazzene, is composed of thick fluviatile and lacustrine conglomerates, marls, red clays, siltstones, and calcretes. They contain an abundant terrestrial malacofauna of Bulime with *Romanella* and *Vicentinia* (Bensalah et al., 1991).

## 15.4.3  Oligocene/Miocene

Oligocene sediments in Northern Algeria are essentially clastics (Askri et al., 1995, Belkhodja & Bignot, 2004). They form part of the undifferentiated Eocene-Oligocene succession in the El Kantara Basin.

During the Early Miocene, marine transgression covered Northern Algeria along a more or less E-W line from the Tlemcen area to the Biskra depression and led to the deposition of thick blue marls over 1000 m (3281 ft) in thickness, changing laterally into marine shaly sandstones (Askri et al., 1995). The Tafna, Chelif, Hodna, and Sebaou basins were formed during that time. The end of this period is marked by a regression during the late Burdigalian. The Upper Miocene transgression over Northern Algeria led to the deposition of black and blue marls, sands and sandstones, limestones, diatomites, and gypsum in the Chelif Basin (Askri et al., 1995).

The Upper Oligocene-Lower Miocene (Lower Burdigalian) in the Atlas Domain in Northern Algeria is represented by the tectono-stratigraphic unit, the Numidian Sequence (Moretti et al., 2007). The Numidian rocks are dominated by poorly sorted quartzarenites with a siliciclastic matrix. The succession of the Constantine Mountains includes arenaceous and conglomeratic debris-flows, turbidites, and slump units in places. The *Continental Intercalaire* of the Hoggar, Tassili, and Fezzan) and the Series Pharusienne of the the Hoggar and Eglab are the most likely source of the Numidian quartz (Moretti et al., 2007).

In Northwest Africa, the Betic Cordillera and Southern Apennines, the term Numidian is used to indicate a rock sequence characterized by an alternation of light colored, generally very quartz-rich, and in places coarse-grained sandstones with greyish, brownish and multi-coloured pelites(Wezel, 1970, Rouvier, 1977, Hoyez, 1989, Moretti et al., 2007, Talbi et al., 2008, Sami et al., 2010). The thickness of the sequence varies from a few hundred meters to 3000 m (9843 ft), with an average of

1000 m (3281 ft). Numidian clays generally contain *Tubotomaculum,* and gradually change upward into quartzarenites.

The Numidian Sequence (Late Oligocene - Lower Burdigalian) crops out between Guelma and Constantine (Askri et al., 1995). It includes three units; the basal unit is represented by the 'sub-Numidian clays' with *Tubotornaculum.* The middle unit is characterized by an alternation of quartz sandstones and shales. The top unit is constituted by "supra Numidian marls" (Late Oligocene-Early Aquitanian) with associated volcaniclastics. The Numidian succession is always found on top of the Tell and Flysch thrust sheets.

In Greater Kabylie, Bossière & Peucat (1985) described a N70° dextral mylonitic belt, 3 km wide, cross-cutting metamorphic rocks and their crystalline basement (Piqué et al., 1993). Monié et al. (1984, 1988) attributed metamorphism to Alpine deformation. However, zircon from ounger peraluminous granite intrusions yielded a Rb/Sr age of 271 ± 12 Ma (Bossière & Peucat, 1985, 1986) and a U/Pb age of 273 ± 6 Ma (Peucat & Bossière, 1991). Accordingly, they postulated a major crustal melting episode, linked to deep seated mylonitic belts, occurred in this area around 270–280 Ma. Similar mylonitic zones have been described in the Edough Massif in NE Algeria.

The Edough Massif is the easternmost crystalline core of the Maghrebides and represents the Pan-African basement of the west Mediterranean Alpine Belt (Hammor et al., 1998, Caby et al., 2001). U-Pb zircon dating gave ages of 595 ± 51 Ma and 606 ± 55 Ma for an orthogneiss of the lower granite and the upper pelitic unit, respectively. These ages suggest emplacement of the two granitoids during the Pan-African Orogeny, with subsequent metamorphism during the Miocene. Monazites from a paragneiss gave a U-Pb age of 18 ± 5 Ma, which suggests a Miocene age for the high-temperature metamorphism (Hammor & Lancelot, 1998). Peridotites in the lower units of the Edough Massif, dated by U-Pb at 17.84 ± 0.12 Ma (Late Burdigalian), was probably responsible for melting and metamorphism of the surrounding Hercynian rocks (286–308 Ma) during the rifting phase that led to the opening of the Algerian Basin (Bruguier et al., 2009).

The Upper Miocene fauna from Bou Hanifia, Algeria include *Orycteropus mauritanicus, Ceratotherium primaevum, Hipparion africanum, "Palaeotragus" germaini, Damalavus boroccoi, Gazella praegaudryi, Allohyaena (Dinocrocuta) algeriensis, Progonomys cathalai, Testudo* sp., and *Struthio* Sp. (Geraads, 1989). Bernor & Scott (2003) reported *Hipparion africanum* from the Sahabi area in Libya.

## 15.4.4 Pliocene and Quaternary

During the Pliocene marine conditions prevailed in the Chelif and the Mitidja basins in Northern Algeria. Up to 1000 m (3281 ft) of blue marls and sandstones were deposited. In other areas of Algeria, Pliocene lagoonal sediments grade upwards into continental Quaternary deposits. Global climatic changes mark the period around 2.5 Ma accompanied by eustatic sea-level changes that led to global cooling and resulted in increased aridity in Africa (Lahr, 2010). The Saharan-Arabian Desert Belt was established in North Africa, and the alternating dry and wet conditions that followed led to faunal turnover. The late Early Pleistocene global climate transition caused a critical climate change in Africa probably triggered hominin migrations to southern Europe (Muttoni et al., 2010).

Pleistocene archeological sites in Northern Algeria related to that period include the Ain Bouchert, the Ain Hanech, and the El-Kherba (Sahnouni, 2006). The fauna of the Ain Hanech site are similar to those of the Thomas Quarry near the Atlantic Coast of Morocco; both contain *Connochaetes taurinus*, *Vulpes* sp., *Theropithecus*, *Ellobius*, *Gerbillus*, *Paraethomys*, *Ceratotherium*, *Hippopotamus*, *Crocuta*, *Ursus*, and *Homo erectus*. The Ternifine fauna have been dated near the Lower-Middle Pleistocene boundary (about 700,000 Ma) (Geraads et al., 1989).

The Ain Hanech site near Sétif in the High Plateau was discovered by Camille Arambourg in 1947, who described an Oldowan archeological assemblage. The site has been dated at 1.8 Ma based on paleomagnetic and biochronological data (Sahnouni & de Heinzelin, 1998, Sahnouni, 2006). Recent excavations yielded a Plio-Pleistocene savannah fauna associated with Oldowan assemblages deposited in a floodplain developed on top of the Beni Foudha sedimentary basin, with Upper Miocene-Holocene sediments. The nearby Oldowan El-Kherba site yielded *Mammuthus meridionalis*, *Equus tabeti*, *Sivatherium maurusium*, and *Kolpochoerus phacochoeroides* and was assigned to the Olduvai subchron (1.95–1.78 Ma) (Sahnouni, 2006). This dating has been disputed by Geraads et al. (2004), who believe that its age is around 1.2 Ma, since stone-tools are absent from the Ain Boucherit (2.3 Ma) and Ahl al Oughlam (2.4 Ma) and up to 1.0 Ma in North Africa (Lahr, 2010).

The Ain Hanech site is believed to be the oldest hominid occupation in North Africa. The oldest archeological occurrence is that of Thomas Quarry near Casablanca which yielded chopping tools, polyhedrons, and some cleavers, trihedrons and bifaces (Lahr, 2010). The Acheulean site of the Errayah locality, discovered in 1996, on the western coastal area of Algeria, consists of fine sandy deposits overlying Pliocene marls. It contains two Acheulean layers (Derradji, 2006). Other Oldowan later Acheulean sites in North Africa include Abbassiya in Cairo, Bir Kiseiba (Haynes et al. 1997, Haynes, 2001), the Fayum Depression, Wadi Midawara in the Kharga Oasis, and Bir Tarfawi (Wendorf et al. 1994, Hill, 2001, 2009, Smith et al. 2004), as well as Acheulean deposits in the Tadrart Acacus, Awbari, and Murzuq in West Libya (Lahr et al. 2010).

Major mammalian and hominin evolution occurred during episodes of a wetter, but highly variable climate in the Pliocene-Pleistocene interval (Trauth et al., 2009). Human migration into Europe took place during the late Early Pleistocene climate transition from the Southern Mediterranean, such as Ain Hanech in Northern Algeria (Sahnouni, 2006, Muttoni et al., 2010) and the Thomas Quarry in Morocco (Raynal et al., 2001, 2002). The role of the Gibraltar Strait in the process of migration has been rejected by Muttoni et al. (2010) who argued that human contacts across the Gibraltar were possible only in the terminal Paleolithic, ca.11 ka. The most important factor in migration in North Africa was probably the periodic expansion of the Sahara Desert (Aouraghe, 2006).

The succession in the Ain Boucherit Valley consists of the Oued Laatach, Oued Boucherit, and Ain Hanech formations, overlain by surfacial sediments (Sahnouni & de Heinzelin, 1998, Sahnouni, 2006).

The Oued Laatach Formation consists of over 100 m (328 ft) of poorly stratified, yellowish red clay-silt sediments with occasional lenses or interbeds of sands, conglomerates, and travertines. The Formation extends northward into the Beni Fouda Basin (Sahnouni & de Heinzelin, 1998).

The Oued Boucherit Formation was named after Arambourg's site of Ain Boucherit (Upper Pliocene, Lower Villafranchian). It consists of reddish yellow coarse sands, lenses or tongues of very coarse, conglomerates cemented by carbonates, silt-stones, and lacustrine limestones (Sahnouni & de Heinzelin, 1998). The Ain Hanech Formation is named after Arambourg's site of Ain Hanech on the right bank of Ain Boucherit. In the type section on the left bank of Ain Boucherit is 30 m (98 ft) thick (Sahnouni & de Heinzelin, 1998). The top Complex of Surfacial Formations in the Ain Hanech type section consist of pebbles and gravels calcrete, and paleosols. Later deposits consist of buttes and interfluves with Late Paleolithic or Epipaleolithic human occupations (Sahnouni & de Heinzelin, 1998).

# Chapter 16

# Phanerozoic geology of Tunisia

## 16.1 INTRODUCTION

Tunisia is bordered on the southeast by Libya, on the southwest by Algeria, and on the north and west by (the Mediterranean Pelagian) Sea (Fig. 16.1). It has a surface area of 163,610 km² and a population of about 10.5 millions (2009 estimates). Northern Tunisia forms part of the Alpine Chain (the Maghrebides) of the Western Mediterranean and is occupied by the easternmost extension of the Atlas Mountain Chain, including the Tell and the Dorsale domains. The highest point in Tunisia is at Jebel ech Chambe at 1544 m (5066 ft) in the Tell Domain. The Mejerda River (Wadi Majardah) is the largest and only permanent river in Tunisia. It originates in Northern Algeria and runs eastward for about 450 miles (720 km) in Tunisia before it empties in the Gulf of Tunis (Gulf of Utica). Phosphate deposits, with reserves of about 1 billion tons and an average annual production of 8 million tons constitute 5.2% of world's supply. Phosphate based fertilizers are the main contributions to the world mineral market (Tunisia Mineral Book, 2007). Tunisia produced 12 Mt of phosphates from 7 open pit quarries and one underground mine in 2007. The main mines are located at Metlaoui, Kef Eschfair, Jallabia, Kef Eddour, Redeyef, Moulares, and Mhdilla (Ben Charrada, 2004) (Fig. 16.1). The phosphate is being mined by the Compagnie des Phosphates de Gafsa (CPG). Tunisia is also a minor producer of oil and gas. Proven reserves are estimated at 425 million barrels of oil and 2.3 TCF of gas (eia, 2010). Daily production was about 95,000 bopd in 2010 (ETAP), coming mostly from the Gulf of Gabes and Southern Tunisia.

Tunisia is divided into five tectono-stratigraphic provinces (Fig. 16.1): Southern Tunisia (Sahara Platform), Western Tunisia, Central Tunisia, Eastern Tunisia (occupied mostly by the Pelagian Platform), and Northern Tunisia (including the Atlas and Tell domains). Western Tunisia belongs to the Atlas Domain. The Tunisian Sahara is a large monocline dipping south. It is separated from the northern and northeastern domains by the Jifarah-Gafsa fault system. The Mesozoic Dahar Platform or Uplift covers most of Central Tunisia. Eastern Tunisia and the High Atlas to the west are separated by the N-S Axis (NOSA) (Burollet, 1991). The E-W Tilamzane (Talemzane) Arch runs in the middle of the Saharan Domain. Each of these domains is characterized by a distinctive geological history and stratigraphy. Most of the onshore stratigraphic nomenclature of Tunisia and Libya extend to the offshore area. The Ghadames Basin in West Libya and Eastern Algeria extends into southernmost Tunisia.

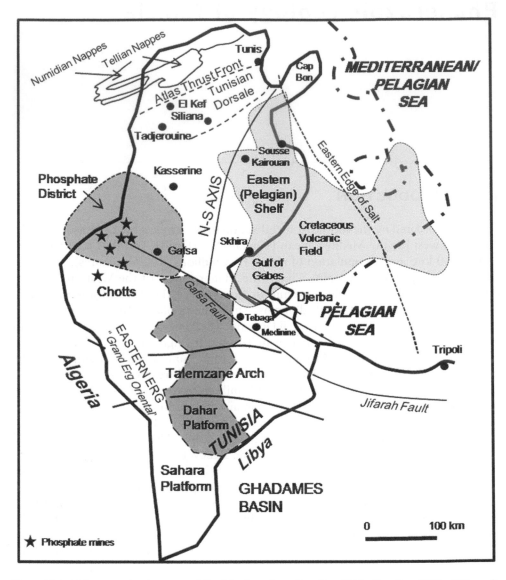

*Figure 16.1* Main tectonic elements of Tunisia (based on Burollet, 1978, Burollet, 1991, Ahlbrandt, 2001, and Mattoussi et al., 2009).

## 16.2   PHANEROZOIC GEOLOGY OF TUNISIA

### 16.2.1   Precambrian

The Precambrian basement does not crop out anywhere in Tunisia (Ben Ferjani et al., 1990) and was encountered in only one well in Tunisia.

## 16.2.2   Paleozoic of Tunisia

The Paleozoic crops out in Tunisia in a small exposure of Permian rocks at Jebel Tebaga near Medinine in the southeastern part of the country (Fig. 16.1), but other Paleozoic sediments are recorded in numerous wells in the subsurface. Therefore, the stratigraphic nomenclature comes essentially from subsurface units (Bishop, 1975, Ben Ferjani et al., 1990). The Early Paleozoic marine transgression in North Africa proceeded from the west and northwest starting in the Infracambrian (Furon, 1968). From Early Permian onward, marine transgressions came from the north and northeast (Hoffman-Rothe, 1966, among others). South of the Tilemzane (Talemzane) Arch, Paleozoic sequences are eroded or pinch out northward and Cambrian to Silurian sediments are overlain transgressively by Permian and Triassic deposits in the higher parts of the arch (Burollet, 1991).

### 16.2.2.1   Cambrian

Cambrian sediments, like most of the Paleozoic units, do not crop out in Tunisia. In the subsurface of Southern Tunisia, the Cambrian rests unconformably on top of the Precambrian basement. The Lower Cambrian is either absent or forms part of the Pan-African basement (Tawadros, 2001). Cambrian sediments were encountered in the subsurface by a number of wells, such as the Sidi-Toui (St1), Sanrahar (Sn1), and Bir Ben Tartar (Tt1) (Fig. 12.3). The Cambrian is represented by the Sidi Toui Formation (discussed in the chapter on Libya).

### 16.2.2.2   Ordovician

The Ordovician succession in the subsurface of Southern Tunisia includes four formations, in an ascending order; they are the Sanrahar Formation, Kasbah Leguine Formation, Bir Ben Tartar Formation, and Jeffara (Djeffara) Formation (Fig. 12.3). The Tt2 well drilled in 2009 by Chinook Energy of Calgary on the Remada Sud Permit flowed at approximately 225 bopd of 43°API oil from an Ordovician zone at a depth of 1422 m (4666 ft) (Chinook Energy News Release, Oct. 2010, World Oil, News Center, 19/08/2010).

### 16.2.2.3   Silurian

The Silurian is recorded only in the subsurface of the Ghadames Basin in Southern Tunisia, but it crops out in West Libya and Southern Algeria. It includes the Tannezouft Formation at the base overlain by the Acacus Formation. The thickness of the Silurian succession may exceed 1000 m (3280 ft), for example in the well OZ1, which has 1078 m (3537 ft) of Silurian sediments (Jaeger et al., 1975, Ben Ferjani et al., 1990, Vecoli et al., 2009).

The Tannezouft Formation (also spelled Tanezzuft) was introduced by Desio (1936) and named after Wadi Tanazuft between Ghat and Al Awaynat in West Libya. Jaeger et al. (1975) called it in Tunisia the *Argiles Principales* Formation to distinguish it from the Tannezouft Formation. The names Tannezouft Formation and *Argiles Principles* Formation are being used concurrently in the literature depending on the worker's preference. The formation consists of a series of occasionally sandy, dark grey and green with spots of red, thinly bedded shales with sandstone stringers. The age of the Tannezouft Formation is Lower Llandoverian to Ludlowian, based on the Middle-Upper Llandovery graptolites *Climatograptus rectangularis, C. torquisti; followed upward by C. extremus, Monograptus priodon, M. dubius, M. priodon flemmingii, Retiolites geinitzianus, Cryptograptus hamatus,* and *C. lundgreni,* as well as the Ludlow species *Monograptus vulgaris, M. nilssoni,* and *M. tumescens* (Ben Ferjani et al., 1990). The Tannezouft Formation contains black shale intervals that are often radioactive, expressed in a high Uranium peak on logs (Lüning & Kolonic, 2003, Lüning et al., 2004b) and contain a high TOC content between 5% and 13%, which can be correlated to high Gamma ray values on the wireline logs and stable isotope ($\delta^{13}C$) values (Lüning et al., 2000d, Vecoli et al., 2009). The Tannezouft is the source rock of the Giant Triassic El Borma Field,[1] as well as many other fields in Northwest Africa. In Tunisia, an organic-rich interval has been identified in well EB-1, near the southeastern flank of the Talemzane Arch. Other intervals with TOC values of about 7% were reported from the wells RA-1, BJA-1, and MER-1, as well as in the wells SN-1, Tt-1, Lg-3, Lg-1, ST-1, AMC-1, and OZ-1 (Lüning et al., 2000c).[2]

In Southern Tunisia and Northwest Libya, the *Argiles Principales* Formation (early Llandovery-Ludlow) overlies the Djeffara Formation (Jaeger et al., 1975, Massa, 1988). The base of the *Argiles Principales* Formation in the Tt1 well is characterized by fine-grained sandstones and quartzites with interbedded thin shale layers which yielded graptolites of the *vesiculosus* zone (Rhuddanian- early Llandovery) (Jaeger et al., 1975, Vecoli et al., 2009). The uppermost shaly interval of the formation in the Tt1 has yielded miospores and chitinozoans which indicate a late Ludlow age. The Silurian sequence is truncated by a major erosional surface above which sediments of Late Permian age occur (Jaeger et al., 1975, Vecoli et al., 2009). The palynomorphs at the base of the *Argiles Principales* Formation indicate a Rhuddanian-early Llandovery (Vecoli et al., 2009). Acritarchs include long-ranging species (Late Ordovician-Silurian), such as *Evittia remota, Neoveryhachium carminae,* and *Baltisphaeridium,* as well as some middle-late Silurian species *Visbysphaera, Oppilatala, Triangulina,* and *Tunisphaeridium, Deflandrastrum millepiedi,* and *D. leonardi, Ammonidium microcladum, Tylotopalla,* and *Leiofusa.* Chitinozoans[3] include *Spinachitina fragilis* (an index species for the basal Rhuddanian), *Spinachitina maennili,* and *Conochitina edjelensis edjelensis,* in addition to *Angochitina longicollis* (Vecoli et al., 2009), as well as *Angochitina echinata, Ancyrochitina primitiva, Cingulochitina* cf. *convexa,* and *Sphaerochitina acanthifera.* The acritarchs and chitinozoans assemblages suggest a late Ludlow age for the upper part of the formation. Land-derived organic debris, miospores, cryptospores, and trilete spores are represented by *Ambitisporites avitus, ?Archaeozonotriletes* sp. cf. *A. divellomedium, Laevolancis divellomedium, Retusotriletes warringtoni, Synorisporites verrucatus,* and *Emphanisporites rotatus.* Prasinophytes are rare and include *Tasmanites, Leiosphaeridia* and *Cymatiosphaera.*

---

1  El Borma Field straddles the Tunisian-Algerian border. It is a faulted anticline and has about 15 m of net pay, 18% porosity, 200 mD, and API of 40° (Bishop, 1975). The same interval produces gas at Hassi Keskessa to the northwest and the D-23 and F-90 fields in Northwest Libya.

2  See this reference for a review of Silurian black shales in North Africa.

3  For earlier work on chitinozoans of Silurian and Devonian samples in some Tunisian well samples, see Grignani (1967).

The Tannezouft/*Argiles Principales* Formation is overlain by the Acacus Formation (Desio, 1936) composed of dark grey shales and interbedded quartzitic sandstones in the Acacus and Tadrart ranges in West Libya. It is a stratigraphic equivalent of the lower part of "*Grès supérieurs du Tassili*" of Kilian (1928), the "*Zone de passage*" of Burollet (1956) of the Algerian Sahara, and the "*Gothlandien argilo-gréseux*" of Bonnefous (1963) (a name preferred by Bellini & Massa, 1980, for the Acacus Formation in Tunisia). It was dated Upper Ludlow in the lower part and Pridolian in the upper part, based on graptolites. However, the lower part of the formation was dated Llanovery by Klitzsch (1966). The thickness of the Acacus Formation reaches up to 800 m (2625 ft) (Ben Ferjani et al., 1990).

The Acacus Formation forms reservoir rocks in many fields in West Libya and Southern Tunisia (Fig. 12.3). The recently drilled well Jenein Centre-1 by Chinook Energy of Calgary on their Jenein Centre Permit in north-Central Ghadames Basin in Tunisia proved the Acacus sands to be dry (Chinook Energy News Release, Oct. 2010, World Oil, News Center, 19/08/2010).

### 16.2.2.4 Devonian

In Southern Tunisia, the Silurian is overlain by Devonian rocks. The Lower Devonian includes the Tadrart and Ouan Kasa formations. The Middle Devonian includes Units AO I and AO II of the Aouinat Ouenine Group, while the Upper Devonian includes Units AO III and AO IV, as well as the Strunian Tahara Formation. The entire Devonian section (Units I-IV) was encountered in the two wells SB1 and MG1, with up to 1245 m (4085 ft) in the latter (Ben Ferjani et al., 1990, p.32). For a discussion of these formations, see the chapter on Libya!

### 16.2.2.5 Carboniferous

The Carboniferous is represented by marine facies of sandstones, shales and carbonates, and includes in an ascending order: The M'Rar (Mrar) Formation, Assedjefar Formation, Dembaba Formation, and Tiguentourine Formation, all of which are described in detail in the part on Libya. The Carboniferous does not crop out in Tunisia, but it has been encountered in few wells in Southern Tunisia, such as Krachoua (KR1) and Kasbah Leguine (LG1, LG2). The most complete Carboniferous succession has been encountered in the KR1 well, but only the Moscovian deposits are present in the other wells (Lys, 1985). The Lower Carboniferous starts with an oolitic limestone horizon (Visean) with the algae *Koninckopora inflata* and the foraminifera *Tubispirodiscus cornuspiroides* and *Archaediscus covexus*, followed by a Serpukhovian (lower Namurian) interval with *Globivalvulina scaphoidea*. The Middle Carboniferous is made up of marly limestones, occasionally very sandy or oolitic, with a Lower Bashkirian association of *Archaediscus donetzianus, Millerella umbilicata, M. umbilicatula, Asterarchaediscus gregorii acutiformis*, and *Eostaffella pseudostruvei angusta*. The overlying Upper Bashkirian contains *Endothyranella gracilis, Schubertella obscura*, and *Pseudostafella* cf. *gorskyi*. The Moscovian is represented by greyish black marly, oolitic, and highly fossiliferous limestones, identified in the LG2 well by the presence of *Dvinella comata, Profusulinella rhombiformis nibelensis*, and *Aljutovella postaljutovica*. The Upper Carboniferous is made up of grey to brown, oolitic, crinoidal limestones and shale intercalations with several species of the *Protriticites* and *Triticites* zones. The highest Carboniferous limestone unit contains *Quasifusulina cayeuxi caspiensis* and *Pseudofusulina* cf. *firma* (Lys, 1985). Devonian and Carboniferous units are discussed in detail in the chapter on Libya.

### 16.2.2.6   Permian

The Permian strata were discovered in Tunisia in 1932 by Eugene Berkaloff and described by Douvillé et al. (1933) and Solignac & Berkaloff (1934). A detailed account of the Permian stratigraphy was also published by Solignac & Berkaloff (1934) who provided a detailed description of the Upper Permian succession at Tebaga which includes from base to top:

1   Siliceous limestones, 2–3 m (7–10 ft).
2   Varicolored sandstones with *Neoschwagerina* and *Mcdlicottia*, 135 m (443 ft).
3   Highly fossiliferous green marls and limestones, up to 180 m (590 ft).
4   Siliceous limestones, 60 m (197 ft).
5   Poorly exposed marls, sandstones, and limestones, with *Neoschwagerina* and bellerophonitids, 450 m (1476 ft).
6   Green marls and limestones, with calcareous sandstones with *Neoschwagerina*, 120 m (394 ft).
7   Siliceous, recrystallized limestone, 50 m (164 ft), which form the crest of the central Djebel Tebaga.
8   A conglomerate.
9   Varicolored, highly fossiliferous sandstones with bellerophonitids and Strigogoniatites.
10  Yellow and red fossiliferous sandstones and marls, 60 m (197 ft)
11  Brown cross-bedded sandstones, 80 m (262 ft).
12  Sandstones and marls, 60 m (197 ft).

The marls contain a relatively rich assemblage of isolated sponges, gastropods, smaller foraminifera, fusulinids, dasycladacean algae, and ostracods (Lethiers et al., 1990). Miller & Furnish (1957) reported the ammonites *Agathiceras*, *Popanoceras*, *Peritrochia*, *Stacheoceras*, and *Cibolites* from the Permian section in Djebel Tebaga and assigned the succession a late Middle Permian age. Newell et al. (1977) described ammonites and fusulinids from the same area and assigned the assemblages to the Guadalupian.

Crinoids were later described by Termier & Termier (1958). Lane (1979) described the assemblage includes *Tetrabrachiocrinus*, ?*Paragaricocrinus*, *Yakoviev* sp., *Tunisiacrinus imitator*, *Trinlicrinus tunisiensis*, and *Strobocrinus brachiatus* from Djebel Tebaga and assigned it Guadalupian age. The crinoids are associated with shallow water sponge-rich bioherms. Ostracods from the Marls of Merbah el Oussif, which forms part of the Permian succession of the Djebel Tebaga, suggest a high sedimentation rate, a well oxygenated, nearshore shallow marine environment (Lethiers et al., 1990), and connection with the Tethys. Lethiers et al. (1990) described a new ostracod species of *Bairdiacypris postrectiformis*.

Lower Permian sediments are absent in outcrops because they onlap on the Nefusa Arch. Lower and Middle Permian sediments are known only in wells in the Kirchaou, Sidi Toui, Kasbah Leguine, and Bir Ben Tartar areas (Burollet et al., 1978). Some of these sections were described by Glintzbockel & Rabate (1964, cited by Lys, 1985 and Toomey, 1991) from wells drilled in the Tunisian part of the Djeffara (Jifarah) Plains.

The Middle and Upper Permian sediments encountered in wells in the Kirchaou-Kasbah Leguine area were assigned to the Zoumit Formation by El Euchi et al. (1998). The Permian succession is overlain by Triassic to Cretaceous sediments. The Permian consists of bioclastic limestones with *Pseudoschwagerina*, *Schwagerina*, *Rugofusulina*, overlain by detrital limestones with *Hemigordius*, *Eoverbeekina*, *Calcitornella*, and *Mizzia*, and marls and anhydrites at the top (Burollet et al., 1978). The Upper Permian in the subsurface overlies eroded Paleozoic sediments and contains *Streblospira*. In the northern Djefara Plains, the Upper Permian is composed of shales with interbeds of sandstones and limestones. It has been divided by Burollet et al. (1978) into three units, based on their fossil content: 1) A 2900 m (9515 ft) unit with *Polydiexodina*, *Yangchiena*, *Bunbarula*, and *Parafusulina*. 2) A 1700 m (5578 ft) unit with *Codono-fusiella*, *Schwagerina*, *Bunbarula*, *Verbeekina*, and *Geinitzina*, in addition to crinoids and algae. 3) A 1500 m (4922 ft) unit with *Bunbarula*, *Neoschwagerina*, *Yabeina*, *Tetrataxis*, *Osagia*, *Mizzia*, *Epimastopora*, *Gymnocodium*, and *Permocalculus*.

Only the Upper Permian crops out at Tebaga in Southern Tunisia, and is the only marine Permian outcrop known in Africa (Kilani-Mazraoudi et al., 1990, Toomey, 1991), which comprises the western tip of the Late Permian Tethyan seaway (Toomey, 1991). The succession at Tebaga is 850 m (2789 ft) thick (Toomey, 1991), but the Upper Permian carbonates may reach up to 4000 m (13,124 ft) in thickness in the subsurface, such as in the TB1 well in the Djebel Tebaga area (Lys, 1985). The exposed succession has been named the Tebaga Formation (Termier & Termier, 1958). The Upper Permian Tebaga sediments are overlain by Lower and Middle Triassic Red beds (Burollet et al., 1978). The Tebaga Formation is equivalent to the Watia (Ouatia) Formation in Northwest Libya.

The Tunisian Permian outcrops at Tebaga represent a biohermal complex situated between a subsurface shale basin to the north and shelf sediments on the margin of the Sahara Shield to the south (Toomey, 1991). Two intervals of bioherms have been identified separated by a thick interval of marine shales with abundant calcisponges. The bioherms are made up of encrusting algae, a diverse calcisponge assemblage, corals, and encrusting bryozoans. They form small limestone bioherms less than 60 m (197 ft) in length, 6 m (20 ft) in height, and up to 32 m (105 ft) in diameter. The bioherms are encased in intraclastic, algal, and foraminiferal grainstones and fine-grained sands. They were deposited in a relatively low-energy, middle to outer-shelf to shelf-margin environment (Toomey, 1991). The reefoid facies is rich in fauna of the *Dunbarula* and *Afghandia* Zones (Burollet, 1978). Senowbari-Daryan & Rigby (1991) described the Upper Permian thalamid sponge species *Intrasporeocoelia hubeiensis*, *Rhabdactinia depressa*, and *Glomocystospongia gracilis*, in addition to *Tristratocoelia rhythmica* from the reefs of Djebel Tebaga.

Kilani-Mazraoudi et al. (1990) identified Upper Permian sediments in the subsurface in two wells south of the outcrops based on palynomorphs such as *Lueckisporites virkkiae*. They also postulated the presence of an unconformity at the Permo-Triassic boundary marked by the absence of parts of the Upper Permian and the Lower Scythian.

## 16.2.3   Mesozoic of Tunisia

The majority of the Tunisian Mesozoic stratigraphic nomenclature was established by Burollet in the 1950s and was later reviewed and amended by Fournié (1978) and the *Entreprise Tunisienne d'Activités Pétrolières* (ETAP) (El Euchi et al., 1998). Mesozoic stratigraphy was extended into the offshore of Tunisia by oil companies. Nevertheless, the Mesozoic regional stratigraphy changes rapidly across the Pelagian Province and the stratigraphic nomenclature become complicated and varied. The stratigraphy of the Libyan offshore has been introduced by Hammuda al. (1985), although Tunisian nomenclature is still preferred.

The Mesozoic paleogeography in Tunisia is classically divided into three main paleogeographic zones (Chihaoui et al., 2010) (Fig. 16.1); from south to north, they are:

1   The Saharan Platform, covered by thin, mainly clastic, continental to shallow marine deposits.
2   The Central Tunisian Platform (Dahar Platform), marked by shallow marine siliciclastic and calcareous deposits. The platform is bounded to the north by the present-day NE-SW trending thrust and fold belt of Northern Tunisia. Central Tunisia includes the southern Chott Depression that separates the Saharan and Central Tunisian platforms. Central Tunisia is part of the Eastern Atlas Domain. It consists of Mesozoic and Cenozoic sedimentary rocks deposited in several basins. Two major tectonic cycles affected this area, which is characterized by NE-SW trending folds, strike-slip faults and diapirism (Burollet, 1991, Bouaziz et al., 2002). The first tectonic cycle was related to the NE-SW to N-S opening of the Tethys in the Late Permian to Early Cretaceous and led to the formation of rifts, tilted blocks, horsts and grabens by extensional tectonics. The second tectonic cycle was characterized by NW-SE compression in the Cenozoic and caused inversion of normal faults, reactivating uplifts and intrusion of diapirs of Triassic evaporites (Ben Ferjani et al., 1990, Grasso, 1999, Bouaziz et al., 2002).
3   The North Tunisian Basin.

### 16.2.3.1   Triassic

The Triassic in Tunisia was first recognized by Thomas (1891, in Thomas, 1908, cited by Burollet, 1963). Bertrand (1896) and Blayac (1898, cited by Demaison, 1965) identified the Triassic deposits of the Constantine area in Northern Algeria. Pervinquière (1903, 1912) gave a detailed account of the Triassic sediments in Central Tunisia. These initial studies were followed by more detailed studies by Solignac (1927), Castani (1951), Sainfeld (1952), and Burollet (1956, 1963, 1967, 1991), Bobier et al. (1991), and more recently by El Euchi et al. (1998), Kamoun et al. (2001), Bédir et al. (2001), Mzoughi et al. (1994), Courel et al. (2003), Decree et al. (2008), and Bourquin et al. (2010).

The Lower and Middle Triassic formations Bir Mastoura, Bir El Jaja and Ouled Chebbi in the Kirchaou area are equivalent to the Trias Grèseux in Southern Tunisia.

The Makraneb, Touraeg, Rehach, Messaoudi, and Bhir formations are equivalent to the Trias Evaporitique (Adjadj) in Southern Tunisia (Ben Ferjani et al., 1990, p.36).

The Triassic succession crops out at Tebaga in the Jifarah Plains near Medenine (Fig. 16.1), Foum Tataouine, and the Libyan border, and the Dahar area in Central Tunisia, but in Southern Tunisia, it is known only in the subsurface from exploration wells in the Kirchaou area (Peybernès et al., 1993). The Triassic succession in Central Tunisia overlies conformably Permian sediments, while it rests unconformably on older Paleozoic rocks in the south.

The Lower Triassic (Scythian) consists of red sandstones and shales of the Bir Mastoura Formation,[4] dated as Lower Scythian by palynomorphs in well B1-23 in Libya (lower part of Bu Sceba Formation, see Burollet, 1963 and Demaison, 1965). It is overlain by the Bir El Jaja Formation dated by the foraminifera *Meandrospira pusilla*[5] as Upper Scythian (Adloff et al., 1985, 1986),[6] although its base probably belongs to the Upper Permian. Hammuda et al. (1985) designated the well B1-23 as the type section of the Bir El Jaja Formation. It has a thickness of 61 m (200 ft) in its type section, although in the well K1-23, slightly to the north, it reaches a thickness of 215 m (706 ft) (Hammuda et al., 1985). The Bir Al Jaja Formation overlies the Al Watyah (Watia, Ouatia) Formation in the K1-23 and the Silurian Tannezouft Shales in the B1-23.

In Southern, Central, and Eastern Tunisia, the lower clastic interval (Scythian-Ladinian) includes sandstones and mudstones of the Bir Mastoura, Bir El Jaja, Ouled Chebbi, and Kirchaou or Trias Argilo-Grèseux (TAG) Inferieur Formations. Depositional facies were continental in the southern part of the Pelagian Province and graded into shallow marine facies northward. The northern limit of the clastic interval is not known (Bishop, 1975). The Kirchaou or Trias Argilo-Grèseux Inferieur Formation provides major oil and gas reservoirs in the Saharan Platform basins of Libya, Tunisia, and Algeria. The middle carbonate interval (Carnian) includes the Mekraneb, Touareg, and Rehach formations, as well as the lower part of the Mhira Formation. It is equivalent to the Azizia, Trias Carbonaté, and Trias Argilo-Grèseux Superieur (TAGS) Formations. The upper interval (Norian-Rhetian) contains interbedded anhydrites, salts, and dolomites (Busson, 1967, Bishop, 1975, Kamoun et al., 2001) and includes the Mhira and Messaoudi formations, as well as the lower part of the Bhir Formation and its equivalent Trias Evaporitique (Adjadj). The upper unit may include the lower parts of the Adjadj Evaporites, Fkirine Formation, Nara Formation, and Oust Formation. These intervals were included in three depositional "megacycles" by Bourquin et al. (2010).

The Middle Triassic (Anisian and Ladinian) consists of the Ouled Chebbi (Al Guidr) Formation (Anisian) and the Kirchaou Formation (Ladinian). The name Ouled Chebbi Formation was introduced for a subsurface unit in NW Libya by Mennig et al. (1963), although is crops out near Medenine in Tunisia, but without providing any

---

4 The Bir Mastoura Formation is considered equivalent to the Bir Al Jaja Formation by El Euchi et al. (1998) and is treated as such by Klett (2001). It overlies the Watia (Ouatia) Formation and is overlain by the Anisian Ouled Chebbi Formation (also known as Al Guidr Formation in Hammuda et al., 1985 and Hallet, 2002).

5 Reported by Kilani-Mazrzoui et al. (1990) from the subsurface of Southern Tunisia.

6 Considered Permian-Lower Triassic by Hallet (2002, p.143).

information. Subsequently, Hammuda et al. (1985) designated the well B1-23 as the type well and gave a formal description. In the type well, the formation is composed of 244 m (800 ft) of fine- to medium grained lignitic sandstones with minor grey micaceous shales and rare dolomites. The Anisian-Carnian deposits consist of sliciclastic sediments with representatives of the *Alisporites* and *Densoisporites* palynological groups (Kilani-Mazraoudi et al., 1990). It overlies the Bir El Jaja/Mastoura Formation and the Kirchaou Formation (Ladinian). The Kirchaou Formation is exposed in the Jifarah Plains. They extend into the subsurface in the south. The Kirchaou Sandstone forms the main oil reservoir of the giant El Borma Field (Ben Ferjani et al., 1990) and gas in well A1-23 in Libya. The Kirchaou Formation includes a limestone horizon known the *Myophoria* Limestone in Libya.

In the Kirchaou area, the Upper Triassic (Carnian to Rhetian) succession includes, from base to top (Ben Ferjani et al., 1990, p.36): Mekraneb Dolomite, Touareg Sandstone, Rehach Dolomite, Mhira Formation, and Messaoudi Dolomite. The Mekraneb Dolomite (Lower Carnian) is an 8–10 m (26–33 ft) thick fossiliferous dolomite horizon, with interbedded clay and marl levels, overlying the Kirchaou Sandstone. Its top is marked by a ferruginous hard ground. The Touareg Sandstone[7] (Middle Carnian) is composed of conglomerates and cross-bedded sandstones, 15–20 m (49–66 ft) thick. The Rehach Dolomite (Carnian) is made up of pink or yellow dolomites and dolomitic limestones, well-bedded, about 80 m (262 ft) in thickness, and forms the hills at Jebel Rehach and Jebel Sidi Toui. The Mhira Formation (Upper Carnian-Lower Norian), dated by the palynomorphs *Precirculina granifer, Ganmiosporites secatus, Vallosporites ignacin, and Patinosporites desus*, is made up of 120–150 m (394–492 ft) of red and green clays interbedded with argillaceous sandstones and fibrous gypsum beds (Ben Ferjani et al., 1990). The Messaoudi Dolomite (Norian-Rhetian) consists of a 3 m (10 ft) polygenic conglomerate at the base followed by grayish dolomites, laminated, brecciated, and oolitic. The Triassic succession ends with the Bhir Formation (Late Triassic-Early Jurassic).

In Northern Tunisia, the Triassic was dominated by open-marine conditions, while neritic shelf conditions prevailed in the south (Burollet et al., 1978, Burollet, 1991). Triassic rocks consist of three main intervals: a lower clastic interval, a middle carbonate interval, and an upper evaporite interval. Triassic sandstones produce oil at El Borma, Makhrouga, Larich, and Chaouch Essaida Fields (Ben Ferjani et al., 1990).

In Northern Tunisia, the Triassic occurs in two large anticlinal domes at Jebel Hairech and Jebel Ichkeul. The Triassic Salt Basin (Fig. 16.1) continues northward under the Atlas Mountains of Eastern Algeria and Tunisia where numerous anticlines or diapirs have brought Triassic or Jurassic strata to the outcrops (Demaison, 1965).

The Triassic of the Tunisian Atlas is made up of siliciclastic and evaporitic deposits interbedded with carbonates (Burollet, 1956, Soussi et al., 1998, Courel et al., 2003). They crop out mainly in Western Tunisia at Jebel Rheouis (about 100 km northeast of Gafsa) and in Northern Tunisia in the diapir and nappe zones (Fig. 16.1).

The Triassic succession at Djebel Rheouis, between Sfax and Sidi Bou Zid, was divided into six units by Burollet (1963), from base to top: 1) A lower gypsum Unit composed of white massive gypsum, up to 400 m (1312 ft) in thickness, and contains

7 Note that there is an Ordovician Hassi Touareg Formation in Algeria.

intercalations of green, yellow, and purple shale in the upper part. The base of this unit is not visible in outcrops. 2) A black, grey, or blue dolomitic limestone unit with casts of undetermined fossils, 150 to 300 m (492–984 ft) thick, with sandstone intercalations, especially in the lower part. 3) A middle red or black gypsum unit, 400–800 m (1312–2625 ft) in thickness, with intercalations of shales and siltstones and dolomites, and large dolomite and quartz crystals. 4) A yellow limestone *"en dailles"*, dolomitic or ferruginous in places, alternating with grey and brown shales, yellow sandstones, purple and redish siltstones. It was assigned an Upper Triassic age by Burollet (1963) based on the presence of *Alectryonia spondyloides*, *Ostrea* (*Alectryonia*) cf. *calceiformis*, *O.* (*Lopha*) *montis-caprilis*, and *Myophoria goldfussi*, among others. 5) An upper siltstone and shale unit, 50–60 m (164–197 ft) thick, red, purple, or grey, with thin beds of gypsum and grey argillaceous limestones with undetermined bivalve molds. 6) An upper gypsum unit, 150–200 m (492–656 ft) in thickness, compact, white or red, with a few beds of red and green shales. Burollet (1963) correlated unit 1 with the Ras Hamia Formation in the D1-23 in NW Libya, units 2–4 with the Azizia Formation, unit 5 with the Bu Sceba Formation, and unit 6 with Bir El Ghnem Formation.

At Djebel Rheouis, they are represented by a Ladinian-Norian succession. Norian-Rhetian outcrops occur in the Nara-Sidi Khalif Range (25 km north of Djebel Rheouis) (Courel et al., 2003). Palynological studies on these sediments have been carried out by Soussi (2003), Soussi & Ben Ismail (2000), and Kamoun et al. (1994). The complete Triassic succession is more than a 1 km (3281 Ft) thick (Courel et al., 2003) and consists of four megasequences. The succession begins with Ladinian-Carnian evaporites and black carbonates (Peybernès et al., 1993) (megasequence 1), followed by Megasequence 2 composed of evaporites, siltstones, carbonates, and shales rich in organic matter (Soussi et al., 1997, in Courel et al., 2003). Megasequence 3 consists of a thick evaporite succession, 200 m (656 ft) in thickness and a 40 m (131 ft) carbonate. The evaporites consist of massive black gypsum and interbedded dolomites and black and brown shales. Megasequence 4 (Courel et al., 2003) begins at Jebel Rheouis with siltstones and red evaporites, more than 100 m (328 ft), dated Norian, and continues northward at Jebel Akrouta (Nara Range) with siltstones and evaporites alternating with Rhetian dolomites (Kamoun et al., 1994). The Triassic succession is overlain by Liassic shallow marine carbonates (Ben Ismail & M'Rabet, 1990, Soussi & Ben Ismail, 2000). Triassic outcrops of the Koudiat El Halfa and Rheouis Domes (Mehdi et al., 2009) (Fig. 16.1) are mainly composed diapirs of gypsum, red marls, clays, silt, and dolomites of the Rheouis Formation (Ladinian-Lower Norian, according to Courel et al., 2003). The outcropping Triassic succession was assigned a Middle Triassic (Ladinian-Carnian) age. However, palynological analysis by Mehdi et al. (2009) postulated an Early Carnian age for the Triassic sediments in the well KEA5 in the Koudiat El Halfa Dome. The palynomorphs consist of *Enzonalasporites vigens*, *Patinasporites densus*, *Vallasporites ignacii*, *Brodispora striata*, *Samaropollenites speciosus*, and *Ovalipollis pseudoalatus* in association with Circumpolles species such as *Praecirculina granifer*, *Camerosporites secatus* and *Duplicisporites granulatus*. In addition, *Lagenella martini* and *Aulisporites astigmosus* occur in the upper part of the succession. They correlated the section with the Rheouis Formation in outcrops along the North-South Axis of Central Tunisia. The core of the

NE-trending Djebel Debadib Anticline in the Tunisian Atlas, near El Kef (Fig. 16.1), consists of a deformed Triassic (Rheouis Formation), gypsum-matrix breccia containing clasts of supratidal dolomites, terrigenous clastic rocks, and ophites (metabasites) (Snoke et al., 1988). Locally, the evaporites contain clasts of low-grade metasedimentary rocks, which may represent fragments of basement rocks. The Triassic rocks are juxtaposed against Cretaceous or younger rocks due to faulting. Many studies suggest a diapirism to explain the internal structure of the Triassic rocks and their contact relations with the younger rocks. Triassic rocks are unconformably overlain by Aptian-Albian sediments.

### 16.2.3.2  Jurassic

The Jurassic succession crops out in Tunisia in three main areas (Fig. 16.2); in the Jurassic Range (Tunisian Dorsale) in the north, along the North-South Axis (NOSA) and the Dahar area in Central Tunisia, and locally in the Tebaga area. It has a thickness of more than 2900 m (9515 ft) in the subsurface and shows great variations in facies from north to south.

The earliest studies on the Jurassic of Northern Tunisia were carried out by Spath (1913) and Solignac (1927). Jurassic ammonites were described by Spath (1913) from Jebel Zaghouan. Stratigraphic studies of Jurassic outcrops were carried out by Castany (1951), Busson (1967), and Bonnefous (1972). Peybernès (1991) reviewed the Jurassic stratigraphy established by Rakus (1973) (cited by Peybernès, 1991), Fauré & Peybernès (1986), and Busson & Comee (1989) and introduced four new formations. Other studies were carried out by Peybernès et al. (1996), Soussi & Ben Ismail (2000), Soussi (2003), and Enay et al. (2002).

In the Tunisian Atlas Domain of Northern Tunisia, the Jurassic crops out along a NE-SW trend extending from the Gulf of Gabes to the Kairouan area (the Tunisian Jurassic Range) in fault blocks such as Djebel Bou Kornine, Ressas, Zaghouan, Staa, El Azeiz Kohol, Ben Saidane, Fikrine, and Zaress, as well as Djebel Oust, Beni Kleb, Raouas, Hammam Zriba, Garci, Mdeker, and Hammam Jedidi. It also crops out in Northern Tunisia at Djebel Ichkeul, Djebel Hairech, and Thuburnic.

In the Thuburnic (Oued Maaden) area in Northwestern Tunisia, the basal Liassic corresponds to the upper part the Chemtou Marble[8] (Ben Ferjani et al., 1990), which is probably equivalent to the Bu Gheilan Formation in West Libya.

The Jurassic succession of the Tunisian Dorsale has been described in detail by Peybernès (1991), who divided it into three major depositional "megacycles" and introduced a number of new formations. The Jurassic in that area is represented, in an ascending order, by the Oust, Zaghouan, Staa, Kef El Orma, Ben Saidane/Bou Kornine, Zaress, and Ressas/Bene Klab formations. The last two formations straddle the Late Jurassic-Early Cretaceous boundary.

Oust Formation was introduced by Rakus (1973) and later studied by Fauré & Peybernès (1986, in Peybernès, 1991) and named after Djebel Oust to the west of the Tunisian Dorsale. The type locality of the formation is at a quarry to the SE of Djebel Oust (Bonnefous, 1972). It has a thickness of more than 300 m (984 ft). It is

---

8  Named after the City of Chemtou (Roman Simitthu), where remains of ancient quarries and a basilica still exist.

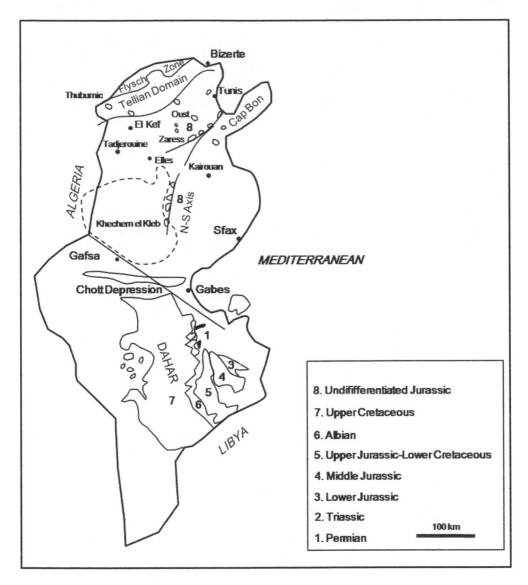

*Figure 16.2* Permian and Mesozoic outcrops in Tunisia (after Peybernès, 1991, Benton et al., 2000, Anderson et al., 2007, Tanfous Amri et al., 2008, and Boughdiri et al., 2009).

composed of massive limestones, occasionally dolomitic in the lower part, with onco-lites and bird's eye structures. Ammonites occur in the upper part with the Dasyclad *Palaeodasycladus mediterraneus* (Peybernès, 1991). In addition, Tanfous Amri et al. (2008) reported *Thaumantoporella parvovesiculifera, Cayeuxia Africana, Involutina liassica* gr. *Tenera, Lingulina* gr. *Prima, Dentalina* gr. *vestustissima, Nodosaria* gr. *Tenuissima*, as well as Verneuilinidae and Trochamminidae. It overlies unconformably the late Triassic Rheouis Formation, with a marked hardground. It was assigned a Rhaetian-upper Sinemurian by Peybernès (1991). Its thickness is 640 m (2100 ft) at Djebel Staa in the Zaghouan area.

Zaghouan Formation was introduced by Fauré & Peybernès (1986, in Peybernès, 1991). Its name is derived from Djebel Zaghouan. Its type locality is along a mule trail near the Village of Zaghouan. Its thickness is 30 m (98 ft) to 50 m (164 ft). It is composed of thin-bedded limestones with black flints (see Fauré et al., 2007). It overlies the Oust Formation with a subtle break and grades into the overlying Staa Formation. It was assigned a Lotharingian age (substage of the Sinemurian of the Jurassic) by Rakus & Biely (1970, in Peybernès, 1991) in its type locality and a Domerian age by Bonnefous (1972). It is probably equivalent to condensed glauconitic and highly fossiliferous sections, with the ammonites *Tropidoceras flandrini*, *T. mediterraneum*, *T. calliplocum*, among others, at Djebel Staa, Djebel Oust, and Djebel Azreg (Peybernès, 1991).

In Jebel Zaghouan, the Bou Gabrine Formation overlies the Stah Formation and is overlain by the Zaghouan Formation. It was considered previously as a member of the Stah Formation (Soussi (2003), but was raised to a formation status by Fauré & Peybernès (1986) and Fauré et al. (2007).

Staa (Stah) Formation was introduced by Fauré & Peybernès (1986, in Peybernès, 1991) and named after Djebel Staa (Stah) to the southwest of Djebel Zaghouan. Its type locality is an old fluorite mine in the southwest side of Djebel Staa. Its thickness varies between 10 m and 55 m (33–180 ft). Peybernès (1991) recognized two members within this formation; the lower member is made up of flaggy marly, partly glauconitic limestones; the upper member (Cabrine Formation) is composed of interbedded nodular marly limestones and yellow marls, rich in ammonites, such as *Emaciaticeras* gr. *emaciatum* (in the top layer), *Protogrammoceras* aff. *Dilectum*, *P. brevispiratum*, *P. celebratum* in the lower member, and *Protogrammoceras madagascariensis*, *Hildaites serpentium*, *Harpoceras subexaratum*, *Merlaites alticarinata*, and *Dumortieria latumbiliaca*, among others in the upper member. It was assigned a Late Carixian-late Toarcian age by Peybernès (1991). It grades into the overlying El Kef Orma Formation.

The Kef El Orma Formation was introduced by Fauré & Peybernès (1986, in Peybernès, 1991) and named after Djebel Kef El Orma. Its type locality is situated along the trail from Bou Gabrine to Zaghouan. Its thickness ranges between 10 m (33 ft) and 40 m (131 ft), and may reach up to 150 m (492 ft) in the Tunisian Dorsale (Sekatni et al., 2008). The Kef El Orma Formation consists of massive or flaggy, well-bedded, light-grey limestones, with flints. The upper boundary of the formation is marked by a regional hardground surface. It was assigned Aalenian to early Bajocian age by Peybernès (1991). Ammonites were reported from the formation include *Erucites* sp., *Leioceras* gr. *subglabrum*, *Haploceras* gr. *Mundum*, *Sonninia* sp., *Skirroceras* sp., and *Witchellia* sp., as well as belemnites near the top of the formation. At Jebel Bou Kornine of Hammam Lif (Northern Tunisia), the transition between the Lower Jurassic (Liassic) and Middle Jurassic (Dogger) is located within the carbonates of the Kef El Orma Formation (Upper Toarcian-Lower Bajocian). It separates two carbonate conglomerate horizons with slump structures of uppermost Toarcian (the ammonite *Aalensis* Zone) from another two conglomerate horizons of the Lower Aalenian ammonite *Opalinum* Zone. According to Sekatni et al. (2008), these conglomerates were formed during the formation of the Tunisian Trough and the Tunisian Ridge (Dorsale) during the Upper Toarcian.

The Ben Saidane Formation was proposed by Peybernès (1991) for a unit at Djebel Ben Saidane, between Djebel Fkirine and Djebel El Azeiz, north of which the type locality is located. Its thickness ranges from 50 m (164 ft) to 250 m (820 ft). The Ben Saidane Formation is composed of yellow-green marls alternating with marly limestones with "microfilaments". The formation overlies the erosional surface of the Kef El Orma Formation and grades upward into the Zaress Formation (Balusseau & Cariou, 1983). It is laterally and partly equivalent to the Bou Kornine Formation. It was assigned early Bajocian-Callovian age by Peybernès (1991), based on the successive occurrence of *Emileia* aff. *brocchii*, *Parkinsonia* sp., *Cadomites* sp., and *Macrocephalites* sp.

The Bou Kornine Formation was introduced by Peybernès (1991) for a Jurassic unit at Djebel Bou Kornine of Hammam-Lif in the northernmost part of the Tunisian Dorsale. Its type locality is along the trail from Hammam Lif to the summit of the Djebel. It corresponds to a flysch-type complex between the Kef El Orma Formation and the Zaress Formation. It consists of six units (from base to top); 1) a unit of interfingering pebbly mudstones and marly turbidites, 2) a "Lower Megabreccia" with micrite-supported clasts and olistoliths, 3) a distal carbonate turbidites unit with reworked ammonites, such as *Morphoceras multiforma*, 4) an "Upper Megabreccia", 5) a proximal turbidites unit, and 6) a distal turbidites unit. It was assigned to the Early Bajocian-early Callovian by Peybernès (1991).

The Zaress Formation was established by Peybernès (1991) for a unit at Djebel Zaress, in the southernmost Tunisian Dorsale. The type locality is in a hill on the western side of the Djebel, where the formation was described by Castany (1951, 1955) and Balusseau & Cariou (1983). The latter called it *Calcaires noduleux rouges* (Red Nodular Limestones) and reported the ammonites *Athleta*, *Plicatilis*, *Transversarium*, *Bifurcatus*, *Bimammatum*, and *Planula* zones. In addition, Bonnefous (1972) reported the foraminifera *Globouligerina oxfordiana* from the lower part of the formation. The Zaress Formation is composed of red-wine, nodular, amygdaloidal, and bioturbated marly limestones with abundant ammonites. The similarity between the Zaress Formation and the *Ammonitico Rosso* facies of Italy was emphasized by Peybernès (1991), Soussi et al. (2000), Enay et al. (2002), Boughdiri et al. (2005), and Fauré et al. (2007). The formation has been dated middle Callovian-early Later Kimmeridgian by Balusseau & Cariou (1983) and Peybernès (1991), with a gap equivalent to the Early Oxfordian. It has a thickness between 14 m (46 ft) and 20 m (66 ft).

The Ressas Formation was introduced by Rakus (1973, in Peybernès, 1991) and named after Djebel Ressas, about 35 km southeast of Tunis. The type section is located at Petit Ressas. The Ressas Formation has a thickness between 400 m (1312 ft) and 500 m (1641 ft). The formation is made up of massive grey limestones, with oolitic and bioclastic levels, which contain fragments of the rudistid *Heterodiceras*. It was assigned a Kimmeridgian-late Berriasian age and considered as the lateral equivalent of the Bene Kleb Formation by Peybernès (1991). It overlies the Zaress Formation and is overlain by the Bene Klab Formation. According to Enay et al. (2005), the Formation starts in the Oxfordian *Bimammatum* Zone.

The Bene Klab Formation was introduced by Peybernès (1991) for a unit at Djebel Bene Klab (also spelled Kleb) in the western part of the Tunisian Dorsale where its type section is located. The formation is composed of 100–150 m (328–492 ft) of monotonous alternation of grey-yellow marls and marly limestones, with tintinnids and ammonites, such as *Aspidoceras* gr. *acanthicum* and *Virgatosphinctes* gr. *eystettensis*.

It grades upward into Middle-Late Berriasian sandy marls. The Bene Klab Formation was assigned a Late Kimmeridgian-early middle Berriasian age by Peybernès (1991). It was appointed a Kimmeridgian-Tithonian age by Enay et al. (2005).

A Jurassic sequence was also described from the subsurface in the CF1, CF2, ZB1 boreholes in the Chott Range by Bonnefous (1972) and Peybernès (1991). The Jurassic succession is absent just south of the CF1 well due to erosion. In the CF1 well, the Jurassic sequence consists of four units (Peybernès (1991, p.1698): 1) Unit 1 consists of 445 m (1460 ft) limestones, dolomites, and anhydrites. 2) Unit 2 is made up of 190 m (623 ft) of micritic limestones. 3) Unit 3 is composed of 990 m (3248 ft) of marls with protoglebigerinids and fragments of *Hecticoceras* sp. 4) Unit 4 is up to 1005 m (3297 ft) in thickness and consists of limestones and dolomites with the dasyclads *Clypeina jurassica* and *cylindroporella* aff. *arabica*, and reworked foraminifera, such as *Alveosepta jaccardai, Kurnubia palastiniensis* and *Anchispirocyclina lustitanica*. These units were correlated with the Oust, Kef El Orma, Ben Saidane, Zaress, Bene Klab, and Ressas formations.

The Tunisian Fluorine Province contains abundant Pb, Zn, Ba, and F mineral deposits along the Zaghouan Fault in the Lower Triassic, basal Jurassic, and Uppermost Jurassic (Trabelsi et al., 2005), as well as Aptian-Albian sediments (Amouri, 1989), and have been attributed to the circulation of hydrothermal fluids along karsts and fractures. Mineral concentrations occur in fractures and karsts fillings (Trabelsi et al., 2005). The distribution of minerals was controlled by tectonics, geochemistry, hydrogeology, and diapirism.[9] The ore deposits are generally distributed around the salt bearing Triassic outcrops near the thermal springs (Trabelsi et al., 2005). Fluorite-celestite-barite ores, such as at Bou Kornine, Jebel Ressas, Jebel Mecella, Jebel Oust, Jebel Kohol, Hammam Zriba, and Hammam Jedidi.

Jurassic sediments crop out along the North-South Axis (NOSA) in Central Tunisia in Djebels El Haouareb, Nara, and Sidi Khalif. The succession includes the Nara Formation and the Sidi Khalif Formation.

The Nara Formation was introduced by Burollet (1956) as the *Formation carbonateé du Nara*. It has a thickness between 300 m (984 ft) and 400 m (1312 ft). It consists of well-bedded or massive dolomites. It overlies the Triassic Rheouis Formation with a tectonic contact. Ammonites from the Nara Formation were studied by Soussi (2003) and M'Rabet (1987). Rakus (in Bonnefous, 1972 & Peybernès, 1991) reported *Hildoceras bifrons* from Djebel Touila and *Witchella* sp., *Sonninia* sp., *Cadomites* sp., and *Sphaeroceras* sp. The formation was assigned a Liassic-Tithonian age by Peybernès (1991). It is overlain by the Upper Jurassic-Lower Cretaceous Sidi Khalif Formation (late Tithonian-late Berriasian).

The Sidi Khalif Formation (also spelled Kralif) (Tithonian-Berriasian according to Ben Ferjani et al., 1990) was introduced by Burollet (1956) as the *Formation argileuse de Sidi Kralif*, named after Djebel Sidi Khalif. The formation consists of interbedded marls and marly limestones rich in ammonites, becoming sandy in the upper part. In other places along the N-S Axis, such as at Jebel Bou Zer, the Sidi Khalif Formation consists of yellow limestones with belemnites, green marls with pyritized ammonites, and orange red dolomites at the base (Guibert et al., not dated). The formation has a thickness of up to 200 m (656 ft) in the type locality. In Northern Tunisia, the Sidi

9  For Miocene Pb-Zn mineralizations, see Decrée et al. (2008).

Khalif Formation is composed of grey shales with green weathering, interbedded with argillaceous or fine grained limestones in the lower part or siltstones and sandstones in the upper part. The Early Cretaceous is made up entirely of shales with minor sandy or calcareous levels. Previously, these dark shales were included in the Sidi Khalif Formation. However, Ben Ferjani et al. (1990) proposed to apply the name M'Cherga Formation to include the Valanginian-Aptian shales, and restrict the name Sidi Khalif Formation to the Tithonian-Berriasian shales. The formation overlies unconformably the upper Nara Formation and is overlain with a gradual contact by the Neocomian Meloussi Formation (Busnardo et al., 1980). The Sidi Khalif Formation is rich in fossils, which include ammonites, brachiopods, belemintes, crinoids, foraminifera, tintinnids, ostracods, and coccoliths (Ben Ferjani et al., 1990, p.43). It was assigned a Tithonian-Berriasian age by Ben Ferjani et al. (1990) and to the late Tithonian-late Berriasian by Peybernès (1991) based on ammonites. The assemblage contains *Paraulacosphenctes senex*, *Micracanthoceras* sp., *Tirnovella alobrogensis*, *T. suprajurensis*, *Calpionella alpine*, *Crassicollaria parvula*, and *Tintinnospella carpathica*. This assemblage has been attributed to the Tithonian. The Berriasian part of the Sidi Khalif Formation contains *Pseudosubplanotes grandis* and *P. lorioli*, the ostracod *Pontocyprella hodnensis* at the base, followed by an assemblage of the ammonite *Occitania*, the tintinnids *Protancyloceras punicum*, *P. depressum*, *Tiranovella occtanica*, *Berriasella subcallisto* and *B. privasensis*, among others (Ben Ferjani et al., 1990, p.43). Its top and bottom are diachronous (Ben Ferjani et al., 1990, Peybernès, 1991).

The Jurassic succession along the North-South Axis (NOSA) was studied recently by Tanfous Amri et al. (2008), who included the entire Jurassic and Lower Cretaceous succession in the Mesozoic Tataouine Basin, north of the Tebaga Uplift, in the Nara Group. In that area, the Triassic evaporites (Rheouis Formation & Fkirine Formation and the Lower Jurassic carbonates (Chaabet el Haris, Oust, and Stah formations, equivalent to the Lower Nara Formation/Group) occur in a series of SE-tilted fault blocks, which are bounded by NW-dipping normal faults (Tanfous Amri et al., 2008, Fig. 3), and capped by Middle Jurassic (Guemgouma and Khechm el Kelb formations) and Upper Jurassic (Kef el Hassene Formation, equivalent to the Upper Nara Formation) deposits, followed by shales of the Lower Cretaceous Sidi Khalif Formation and the overlying formations.

In Southern Tunisia, Jurassic outcrops form a long cliff of more than 200 km (125 miles) long on the eastern side of the Dahar Platform, from Tebaga de Medenine in the north to the Tunisian-Libyan border in the south (Ben Ismail & M'Rabet, 1990, Peybernès, 1991, Benton et al., 2000) (Fig. 16.2).

In the Tebaga area, Jurassic outcrops along the South Tunisian Fault are represented by a single condensed unit, the Remtsia Formation (Peybernès, 1991, p.1698). The Remtsia Formation (Middle Bathonian-Oxfordian) was introduced by Peybernès et al. (1985) for a carbonate series exposed at Djebel Remtsia and Djebel Ouget El Gabel. It overlies with an angular unconformity the Permian Seikra Formation, and is overlain unconformably by Lower Cretaceous sandstones (Peybernès, 1991, Fig. 10). It has been divided into three members in its type locality (Peybernès, 1991, p.1698): Member 1 is made up of 20 m (66 ft) of interbedded green shales, limestones, and dolomites. At the top, massive limestones contain corals and oncolites and in some sections, *Satorina apuliensis* (middle-late Bathonian). At Djebel Ouget El Gabel, it contains the foraminifera *Satorina apuliensis*, and *Protopeneroplis striata*, and the algae *Salpingoporella annulata*. Member 2 consists of 10 m (33 ft) of shales and white marly limestones with stromatoporoids. Member 3 is composed of 20–25 m (66–82 ft) of massive, cliff-forming dolomites with flints and reefal limestones with silicified coral biostromes. At Djebel Ouget El Gabel, it contains *Trocholina gigantea* at the base.

In the Dahar area, on the western side of the South Tunisian Fault, the Jurassic is represented by the upper part of the Bhir Formation (Upper Triassic-Liassic), the Zmilet Haber Formation, and Mestaoua Formation (Liassic), followed by the Middle Jurassic (Dogger) Krachoua Formation, Techout Formation, and Foum Tataouine Formation, then by the Upper Jurassic-Neocomian Merbah el Asfer Formation.

The Bhir Formation (Upper Triassic-Liassic) is made up of gypsum and anhydrite with thin interbedded varicolored clays and very thin layers of dolomite (Ben Ferjani et al., 1990). It overlies the Upper Triassic Mhira Formation at Dehebat and Remada, and the Messaoudi Formation to the north. It is equivalent to the Lower Evaporites of Busson (1967) and Bir El Ghnem Formation in Libya. The two formations are overlain by the Bu En Niran Horizon (Pliensbachian) in Libya or its equivalent the Zmilet Haber Limestone in Tunisia. The age of the Bhir Formation could not be determined with precision. It was assigned to the Norian-early Liassic by Peybernès (1991, p.1700). The Triassic-Jurassic boundary falls within the Bhir Formation.

The Zmilet Haber Formation (Pliensbachian) is also known as the "Horizon B" (Busson, 1967) in the subsurface of Tunisia and forms a strong seismic reflector in the offshore area (Bobier et al., 1991). The Zmilet Haber or "Horizon B" in Tunisia has been dated Pliensbachian by Ben Ferjani et al. (1991) based on its stratigraphic position and by Peybernès (1991), based on limited paleontological data. The formation consists of flaggy dolomites rich in the coprolites *Favreina* sp. and *Glomospira* sp., the dasyclad *Neomizzia* sp., as well as the echinoid *Pseudodiadema* cf. *prisciniacense* (Peybernès, 1991, p.1700). It has a thickness between 10 m (33 ft) and 35 m (115 ft) (Peybernès, 1991). The Zmilet Haber Marker is overlain by evaporites and dolomites of the Mestaoua Formation (Liassic. The Zmilet Haber Formation represents the first Jurassic marine transgression (Peybernès, 1991). It is equivalent to the Bu en Niran Formation in Libya.

The Mestaoua Formation (Upper Liassic) is equivalent to the Upper Evaporites of Busson (1967). It is about 500 m (1640 ft) (Ben Ismail & M'Rabet, 1990) in thickness and composed of thick beds of gypsum, interbedded with thin carbonates and shales, and capped by a dolomitic bed rich in dwarfed lamellibranchs (Peybernès, 1991). The Mestaoua Formation overlies the Zmilet Haber Formation and is overlain in turn by the Krachoua Formation.

The Krachoua Formation (Bajocian) was introduced by Busson (1967). It is about 100 m (328 ft) thick and consists of a number of sequences composed of shales at the base, followed upward by carbonates, and capped by algal mats with desiccation cracks. The Krachoua Formation contains three members (Peybernès et al., 1985, (Peybernès, 1991); they are from base to top, the Kazzani, Grimissa, and Djelidat. The Grimissa Member contains at the top a gypsiferous mudstone layer with the dasyclad *Selliporella donzellii* (Bajocian-Bathonian) and *Parurgonina* sp. (Peybernès, 1991). The Djelidat Member contains a yellow dolomitic limestone layer near the top with echinoids and brachiopods, such as *Rhynchonellidae* (Peybernès, 1991).

The Techout Formation (late Bathonian-Bajocian) was named by Busson (1967) for a complex of clays, dolomites, and sandstones, about 70–110 m (229–361 ft) in thickness. It has been divided into a Zahra Member at the base and a Mened Member at the top by (Peybernès et al., 1985, Peybernès, 1991). The two members represent transgressive and regressive cycles, respectively. The top of the Zahra Member contains a fossiliferous red marly limestone bed with brachiopods. The Mened Member grades upward into fluvial sandstones with wood fragments (Peybernès, 1991).

The Foum Tataouine Formation (or simply the Tetaouine Formation) is dated Callovian based on the foraminifera *Alzonella cuvileri*, *Praekurnubia crusei*, the ammonites *Erymnoceras* sp. and *Pachyerymnoceras philbyi*, as well as the brachiopods *Septirhynchaia numidiensis*, *Daganirhynchia daganiensis*, and *Bihenithyris barringtoni* (Ben Ferjani et al., 1990, p.41). The Oxfordian is characterized by a red rubble facies of the Ammonitico rosso or the Zaress Formation[10] (or Member) in the Jurassic Ranges, with a thickness of 56 m at Djebel Bou Kornine of Hammam Lif. The Kimmeridgian-Tithonian consists of three main facies: 1) A reefal facies at Djebel Ressas, Djebel Sidi Salem, and Mecella, made up of bioclastic limestones interbedded with shales, which contain *Ellipsactinia*, *Heterodiceras*, coelentrates, hydrozoans, rudistids, nerinids, crustaceans, serpulids, and algae. 2) A transitional facies at Djebel Staa characterized by limestones interbedded with marls with abundant ammonites. 3) A pelagic facies in the Jurassic Ranges composed of nodular limestones interbedded with marls, rich in the ammonites *Saccocoma* sp., and *Globochate* sp. The uppermost limestones contain abundant calpionnellids. Busson (1967) subdivided the Tataouine Formation (Callovian) into four units; units 1 and 2 were assigned to the Callovian and units 3 and 4 to the Oxfordian. Walley (1985) designated the units as members; namely the Bouret Member (Callovian), Smida Member (Oxfordian), Tlalett Member (Oxfordian), and M'Rabtine Member (Kimmeridgian-Tithonian). Enay et al. (2002) consequently assigned the members different names; Beni Oussid 1 (Beni Youssid I in Peybernès, 1991) (Callovian according to Enay et al., 2002, and mid-late Bathonian according to Peybernès, 1991), Beni Oussid 2 (Beni Youssid II in Peybernès, 1991) (Callovian in age), and equated them to Unit 1 of Busson (1967) and the lower two members of Walley (1985), Khechem[11] el Miit (Oxfordian), equivalent to Unit 2 and the Smida Member, the Ghomrassen Member (Unit 3) or the Tlalett Member (Lower Oxfordian), and the Ksar Haddada (equivalent to Unit 4 or the M'Rabtine Member) (probably of Oxfordian age). Enay et al. (2002) correlated the Tataouine Formation with the Masajid Formation and Zohra Formation in Egypt.

### 16.2.3.3 Upper Jurassic-Lower Cretaceous

The limit between the Jurassic and Cretaceous is not obvious in the field (Ben Ferjani et al., 1990). It is located in the marine shales of the Sidi Khalif Formation

---

10 Mentioned as member in some publications.
11 Also spelled Krechem. The difference in spelling is a common problem in the transliteration of Arabic names/words that contain the Arabic letter "kh" which has no equivalent in the the English or French languages. It sounds like the German "ch" in the word "Nacht". In English writings the "kh" is commonly used, while in French the "kr" is used instead. This also explains the differences in the spelling of the names Khalif/Kralif Formation.

(Upper Jurassic-Early Cretaceous) (already discussed) or in the continental sands of the Merbah el Asfer Formation (Ben Ferjani et al., 1990) or the Merbah el Asfar Group (Benton et al., 2000, Anderson et al., 2007) overlying the Tataouine Formation. The Upper Jurassic-Lower Cretaceous succession belongs to the Supersequence S1 of Bobier et al. (1991), which includes two smaller megasequences: A lower megasequence (Late Oxfordian-Early Barremian) and an upper megasequence (Upper Barremian-Lower Aptian).

Many recent attempts have been made to subdivide the Upper Jurassic-Lower Cretaceous successions in Tunisia. As a result, some formations have been raised to group status and other new groups were erected. However, the stratigraphic limits and regional correlations have not been established for these groups. They are not accepted in the present work, although they will be mentioned occasionally when reference is made to these units.

The Upper Jurassic-Lower Cretaceous succession of the North-South Axis (NOSA) of Central Tunisia is similar to that described from the Middle and Central Aurès Range (Eastern Algeria) (Boughdiri et al., 2006a & b, Boughdiri et al., 2009, Tanfous Amri et al., 2008). The Tunisian Dorsale succession in Northern Tunisia is similar to that of the Northern Aurès, the Hodna Mountains, and the South Tellian Border (Tiaret area) in Algeria (Atrops & Benest, 2007), and the Prerif and Mesorif Upper Jurassic facies in Morocco. On the other hand, the equivalent deeper facies of the Tunisian Dorsale and the Tunisian Trough are different from that of the Internal Rif and Kabylias' Dorsales Calcaires, and indicate different paleogeographic settings between these domains.

At Chott Fedjedj in West-Central Tunisia, the Upper Nara is overlain by the Asfer Formation or the Asfer Group of Anderson et al. (2007). It was assigned an Upper Tithonian-Lower Barremian age by Anderson et al. (2007). Busson (1967a, b) used the name Merbah El Asfer for the same formation, where the type section is located. The formation is made up of interbedded variegated shales and sandstones, with minor amount of limestones, dolomites, and evaporites. It includes a calcareous horizon with *Alveosepta jaccardi*, the algae-like *Tricocladus perplexus*, and *Salpingoporella annulata* (Ben Ferjani et al., 1990). It has been described as "Wealdian" or "*Continental Intercalcaire*". Its stratigraphic position is not clear. Lefranc & Guiraud (1990) show it overlain by the Orbata Formation and underlain by the Tatahouine Formation. The Merbah el Asfer Formation was raised to a group status, the Merbah el Asfer Group (Oxfordian-Aptian) (Benton et al., 2000 & Anderson et al., 2007) to include the succession overlying the Foum Tataouine Formation (Callovian) in the Dahar area. The group includes, from base to top, the Bir Miteur, Boulouha, and Douiret formations. The Merbah el Asfer Group varies from a maximum thickness of 220 m (722 ft) in the south at Merbah el Asfer to just a few meters in the north. The Merbah el Asfer Group is overlain by the Ain el Guettar Group (Lower Albian-Cenomanian).

The Bir Miteur Formation is a 70 m (230 ft) thick unit of late Oxfordian to Kimmeridgian age. The formation consists of alternating green shales, gypsum, silts, sands and dolomites, indicating lagoonal conditions (Benton et al., 2000). It is overlain by a marginal marine carbonate sequence of the Boulouha Formation, reaching a maximum thickness of 80 m (262 ft), and ranging from Tithonian to Neocomian in age. The uppermost Douiret Formation (early Aptian) (Benton et al., 2000) consists mainly of mudstones. The Douiret Formation is characterized by dominance of sharks, such as the hybodont *Priohybodus*, thought to be restricted to freshwater environments, while the ray *Rhinobatos* is considered mainly marine. The Douiret Formation has been therefore interpreted to represent marginal marine deposition, possibly within a large delta (Benton et al., 2000). At Jebel Boulouha, the Douiret Formation consists of intercalations of fine sands, sandy mudstones, and dolomites, overlain by a thick bed of green clays, and occasional bands of gypsum. The vertebrate remains are preserved at the base of each sandy unit, with abundant fish and reptile remains. The base of the Formation is a regional discordance marked by a conglomeratic unit with fossil wood and vertebrate debris.

The Ain el Guettar Group (Lower Albian-Cenomanian) (Benton et al., 2000 & Anderson et al., 2007) consists of two formations, the Chenini Formation (Lower Albian) and the Oum ed Diab Formation (Middle Albian to Cenomanian).

The Chenini Formation (Lower Albian) is 50 m (164 ft) thick and consists mainly of coarse sandstones, and interspersed conglomerates, breccias, and mudstones, with fossil plant fragments. The coarse sandstones yielded numerous fish, turtle, crocodile and dinosaur remains (Benton et al., 2000). At Oued el Khil, the formation consists of yellow cross-bedded sandstones and occasional breccias and conglomerates. The basal conglomerates are indurated and cemented by iron oxides and contain wood fragments, isolated bones and teeth, and coarse quartz grains. The base of the sands is marked by a channel, 3 m (10 ft) wide and 1 m (3.3 ft) deep, with gravels, flattened clay clasts, and rare wood and bone fragments. Sedimentary structures suggest deposition of the sands in fluvial point bars (Benton et al., 2000). The dinosaurs belong to the North African theropods *Spinosaurus* and *Carcharodontosaurus*, as well as a medium-sized sauropod and an iguanodontid (Anderson et al., 2007). The Chenini Formation also yielded a typically marine shark assemblage (Benton et al., 2000).

The overlying Oum ed Diab Formation is 15 m (49 ft) thick. It starts with alternating shales and sands at the base, marked by a clay layer resting on an erosive surface rich in debris of vertebrates, and dated as Albian-Cenomanian (Bouaziz et al., 1988, Benton et al., 2000). The remainder of the formation is composed of fine micaceous sands, in places conglomeratic. At Touil el Mra, the basal unit consists of 3 m (10 ft) of deep red, unconsolidated coarse sandstones, with numerous channels filled with cross-bedded sandstones with coarse basal lags. The lags are composed of mudstone and sandstone clasts with abundant bones and teeth. The Oum ed Diab Formation (Lower Albian-Cenomanian) at the Touil el Mra and Oum ed Diab was interpreted as a freshwater environment (Benton et al., 2000), because of the terrestrially-derived fossils and the absence of marine faunal remains (Anderson et al., 2007). The Radouane Member is 10–15 m (33–108 ft) thick and rests directly on the Permian in the Tebaga area. It is made up of calcareous marls and bioclastic limestones rich in ammonites such as *Knemiceras syriacum* and *K. gracilis* (reported also from Sinai, Egypt) of Middle Albian age (Benton et al., 2000). A carbonate-rich sandy member with ammonites, Lepidotes and pycnodonts represents the major Cenomanian transgression (Anderson et al., 2007).

Anderson et al. (2007) postulated that the vertebrate assemblage in the Douiret Formation (Aptian) was formed in a terrestrial carbonate-rich environment with relatively little mixing of previously buried bones. The Touil el Mra assemblage in the Oum ed Diab Formation (Lower Albian) indicated a marginal marine environment with some mixing. The Oued el Khil assemblage in the Chenini Formation (lower

Lower Albian) and the Oum ed Diab assemblage in the Oum ed Diab Formation (upper Lower Aptian) suggest mixed freshwater and marine influences. On the other hand, Cuny et al. (2004) and Gilles et al. (2010) described remains of sharks, actinopterygians, and crocodiles from the Douiret Formation and Ain el Guettar Formation and concluded that the vertebrate assemblage of the Douiret Formation is pre-Aptian. The presence of *Bernissartia* in the Ain el Guettar Formation confirms faunal exchange between Africa and Europe during the Early Cretaceous (Gilles et al., 2010).

In Central Tunisia, the Upper Jurassic- Early Cretaceous comprises the Sidi Khalif, Meloussi, and Boudinar formations. The Meloussi and Boudinar formations are included in the Sened Group (Tithonian-Lower Hauterivian). They are overlain by the Gafsa Group (Hauterivian-Albian according to Ben Ferjani et al., 1990 & Aptian according to Fournié, 1978, Fig. 1) which includes the Bou Hedma, Sidi Aich, and Orbata formations. It was originally defined as a formation by Burollet (1956), but was elevated to a group status by M'Rabet (1987). The Sened and Gafsa Groups constitute the Maknassy Supergroup (Hauterivian-lower Cenomanian) (Burollet, 1956).

Upper Jurassic- Lowermost Cretaceous formations in Northwestern Tunisia include the Zaress Formation (Upper Jurassic) (Peybernès, 1991) or the *Ammonitico Rosso* facies (Middle Callovian-Oxfordian) of the Tunisian Dorsale, and the Fahs Formation and the Jedidi Formation (radiolarian-bearing series of Upper Bajocian-Oxfordian age) of the nearby surrounding exposures and the Tunisian Trough, respectively. Boughdiri et al. (2006a, b, 2009) studied the Jurassic-Cretaceous calpionellid association from Jebel Amar and Jebel Jedidi in the Tunisian Trough and assigned the interbedded marls and limestones of the Beni Kleb Formation a Kimmeridgian-Middle Berriasian age. The age of the radiolarians from the siliceous series of the Jedidi Formation spans the Late Bathonian-Early Oxfordian time interval (Cordey et al., 2005), and is correlatable with those of the External Zone of the Maghrebides. It is therefore equivalent in age to the lower part of the Tataouine Formation (see Enay et al., 2002). Coeval reef facies in the Ressas Formation (Upper Oxfordian-Tithonian) were developed on the crests of tilted blocks (e.g. J. Ressas, eastern J. Zaghouan) (Boughdiri et al., 2009). Enay et al. (2002) placed the Kimmeridgian-Tithonian boundary in the pelagic and transitional facies of the Beni Kleb Formation of the Tunisian Dorsale based on the presence of the Lowermost Tithonian *Hybonotum-Lithographicum* Zone. They distinguished, within the formation, between the Kimmeridgian-lower Tithonian grey pseudonodular mudstones and the massive Tithonian limestones. The *Hybonotum-Lithographicum* Zone was not identified in the reefal facies of the Ressas Formation and the Kimmeridgian and Tithonian are not divided. The Ressas Formation starts within the Upper Oxfordian *Bimammatum* Zone.

The paleobiogeographical distribution of Oxfordian ammonites and coral reefs in Northern and Central Europe, the Mediterranean area, North and East Africa, and the Middle East and Central Asia suggests a Late Callovian-Early Oxfordian crisis in carbonate production as indicated by the widespread absence of Lower Oxfordian reefal formations (Cecca et al., 2005). There is a gap (hiatus) in deposition on epicontinental platforms, with Middle Oxfordian deposits resting unconformably on Upper Callovian, while shales accumulated in adjacent intracratonic basins. Simultaneously, in Mediterranean Tethys, radiolarites accumulated in deep troughs while Rosso Ammonitico facies formed on pelagic swells. However, deposition on swells was also discontinuous with numerous gaps (hiatuses) and sequences that are much reduced in thickness (Cecca et al., 2005). Middle Callovian deposits are generally overlain by Middle Oxfordian limestones.

At the end of the Jurassic the paleogeography of Northeastern Tunisia Dorsale (Fig. 16.2) was controlled by the presence of a series of positive areas aligned along the Zaress-Zaghouan axis which delimited two subsiding basins: the Tunisian Trough to the northwest and the Hammamet Trough to the northeast. These positive areas were sites of development of reefs which graded to open marine facies (Memmi et al., 1989).

The Sened Group (Tithonian-Lower Hauterivian) includes the Meloussi and Boudinar formations. It is the southern shallow marine equivalent of the Sidi Khalif Formation. The Meloussi Formation (Tithonian-Lower Hauterivian) was introduced by Burollet et al. (1954) and Burollet (1956), with the type locality at Jebel Meloussi. It is composed of fine to coarse grained sandstones with intervals of green shale, dolomite, or dolomitic limestones. The Boudinar Formation (Hauterivian) is composed of cross-bedded, fine- to coarse-grained sand with layers of gravel, and abundant silicified wood remains. It grades laterally into the Meloussi Formation. Its occurrence is limited to Southern Tunisia.

### 16.2.3.4  Lower Cretaceous of Tunisia

During the Late Barremian-early Late Aptian, Central Tunisia was dominated by shallow-marine sedimentation in intrashelf basins. A stable high; the Kasserine Island, was located in the eastern part (Lehman et al., 2009). Typical deposits of this shallow-marine domain are limestones and sandstones (M'Rabet, 1987, Ben Ferjani et al., 1990, Chaabani & Razgallah, 2006). The northern part of Tunisia or the Tunisian Trough was dominated by basinal shales, marls, and minor turbidites (Turki, 1985, Memmi, 1989).

In Central Tunisia, the Early Cretaceous formations crop out mainly along the North-South Axis (NOSA) (Fig. 16.1), and in the mountains east of Gafsa (Gallala et al., 2009). The area is bounded by two major dextral strike-slip fault zones; an E-W system at Kasserine and a NW-SE-trending zone near Gafsa. Early Cretaceous deposits are dominated by siliciclastic sediments and carbonates. Fluvio-deltaic sandstones, marly sandstones and rare carbonates of the lower part of the Early Cretaceous succession form a prograding megasequence which grades laterally into a deeper-water marly facies towards the north (Burollet, 1956, M'Rabet, 1987, Ben Ferjani et al., 1990, Chaabani & Rezgalla, 2006). In the Central Tunisian Platform, some halokinetic vertical movements were responsible for the generation of several highs during the latest Barremian (Chaabani & Rezgalla, 2006).

Sedimentation became mainly continental during the Barremian in the Western and Central Atlas and argillaceous to marly and thicker towards the Aurès Mountains in Algeria (Bracene et al., 2003).

The Upper Jurassic-Lower Cretaceous of Tunisia includes the Merbah el Asfer, Boudinar, Meloussi, Sidi Khalif, M'Cherga Bouhedma (Bou Hedma) (and its equivalent Limaguess and Berada formations), Mrhila, Sidi Aich, Orbata, Serdj, Hameima, and the Lower parts of the Fahdene and Zebbag formations (Upper Albian-Cenomanian according to Ben Ferjani et al., 1990). The Jurassic-Lower Cretaceous boundary falls within the Merbah el Asfer (*Continental Intercalcaire* or Nubian Sandstone equivalent), Meloussi, and M'Cherga formations. Other Lower Cretaceous formations are the Douiret and Ain el Guettar formations (Gilles et al., 2010). The Bouhedma (Hauterivian-Barremian) (equivalent to Limaguess or Berada Formation and Mrhila Formation), Sidi Aich (Barremian), and Orbata (Aptian) formations were included in the Gafsa Group (Hauterivian-Albian) by Burollet (1956). However, the Orbata Formation is Aptian and should be separated from the Gafsa Group to facilitate correlations with the rest of North Africa. The Group overlies the Boudinar and Meloussi formations.

The Bouhedma (Bou Hedma) Formation (Hauterivian-Barremian) was introduced by Burollet et al. (1954) and Burollet (1956) in the type locality of Jebel Bou Hedma, east of Gafsa. It is made up of oolitic, fine grained or bioclastic limestones, laminated dolomites, variegated shales, gypsum, anhydrite, and fine grained sandstones (Burollet, 1954, 1956, M'Rabit 1987). It overlies the Boudinar Formation or the Meloussi Formation and is overlain by the Sidi Aich Formation. It contains *Chofatella decipiens*, *Ammobaculites whitneyi*, *A.* aff. *Obscures*, *Daxia cenomana*, *Campanellula capuensis*, *Lenticulina* aff. *Roemeri*, *Lingulina* sp., *Discorbis* aff. *Minima*, and *Spirillina minima*, in addition to ostracods, mollusks, serpulids, bryozoans, and algae. It was assigned a Barremian-Late Hauterivian by Damotte et al. (1987) and Ben Ferjani et al. (1990). Its thickness varies between 60 m (197 ft) at Jebel Boudinar and 840 m (2756 ft) in the ZB1 well. The formation is missing in the Tunisian Sahara and in various places along the N-S Axis (NOSA). In the Chott Fedjej, the Bouhedma Formation is replaced by the Limaguess or Berada Formation (Ben Ferjani et al., 1990, p.49) composed of shales, dolomites, and gypsum. Pervinquière (1902, 1912) named these two units the *"marnes á Saptangues"* and described their fauna. The Mrhila Formation is described at Jebel Mrhila where it has a thickness of 290 m (951 ft). It is a lateral equivalent of the Bouhedma Formation where the facies becomes shaly.

The M'Cherga Formation (Valanginian-Barremian) was named and dated by Ben Ferjani et al. (1990) and El Euchi et al. (1998). It is composed predominantly of black shales. The formation is named after the M'Cherga Village near Djebel Oust where the unit crops out. The M'Cherga Formation consists of three unnamed members (Ben Ferjani et al., 1990, p.60). The lower member is made up of dark shales with interbedded limestones and sandstones, and contains the Valanginian *Thaurmanniceras* and the Tinitinnid *Calpionellites darderi* at the base. The upper part of the member contains the Late Valanginian *Phylloceras thetys*, *Bochianites* sp., *Neolissoceras grasi*, *Olcostephanus sayni*, and *Teschenites* cf. *paralesius*. The middle member is composed of shales, marls, and minor argillaceous limestones. It has been dated Hauterivian based on the presence of *Distoloceras* sp., *Oosterella vidali*, *O.* gr. *Gaudryi*, *Olcostephanus hispanicus*, *Phylloceras thetys*, *P. rouyi*, *Apthycus didayi*, *Neolissoceras grasi*, *P. baborense*, *Barremites difficile*, and *B. strettostoma*, among others. The upper member is about 500 m (1640 ft) in thickness and consists of dark shales and marls, with pyritized Aptian ammonites, which belong to the *Deshayesites deshayesi*, *Aconoceras nisum*, and *Diadochocers subnodocostatum* zones. The M'Cherga black shale are considered as seals and primary source rocks for hydrocarbons as they contain type II kerogen with TOC content of 0.2–3% (Klett, 2001).

Sidi Aich Formation (also written as Aïch) (Barremian) was named after the type locality on the SE flank of Jebel Sidi Aich, north of Gafsa in Southwestern Tunisia (Burollet, 1956, Ben Ferjani et al., 1990). The Sidi Aich Formation crops out continuously between Central and Southwestern Tunisia (Gallala et al., 2009), but is missing on the Sahara Platform and in some places along the N-S Axis (NOSA) (Ben Ferjani et al., 1990). A Barremian age has been suggested for the Sidi Aich Formation (Ben Ferjani et al., 1990, Gallala et al., 2009). It grades into the shales of the M'Cherga Formation in Northeastern Tunisia. The Sidi Aich Formation overlies the Bouhedma Formation or M'rhila (Mrhila) Formation and is overlain by either the Orbata Formation or its equivalent the Serdj Formation. It is made up of varicolored coarsening-up fine to medium grained sandstones with occasional large silicified wood fragments (Marzouk & Ben Youssef, 2008, Gallala et al., 2009), interbedded siltstones and shales, and minor carbonate beds with pelecypods and gastropods. It has a thickness between 50–250 m (163–820 ft). The formation consists of coarsening-upward sequences, within which, cross-bedding and ripples indicate a provenance from the southwest (Marzouk & Ben Youssef, 2008). Various facies have been recognized in the Barremian (Sidi Aich Formation and its equivalent) of Central and Southern Tunisia and include alluvial fan deposits in the south, changing northward into braided stream facies and laterally into deltaic facies, and then into shallow marine facies (Gallala et al., 2009). The presence of *Skolithos* burrows suggests deposition in a shallow intertidal marine environment. The distal offshore marine facies is represented by marls (M'rabet, 1987).

### 16.2.3.5  Aptian-Albian of Central Tunisia

The paleogeography of Tunisia during the Aptian-Albian consisted of the Saharan Platform in the south, the Chott Depression in the center, and the Northern Tunisian Basin in Northern Tunisia, including the Cap Bon Peninsula (Chaabani & Rezgalla, 2006, Chihaoui et al., 2010) (Fig. 16.2). The Central Tunisian Platform during the Aptian consisted of inner and external platforms separated by the highs of the Central Tunisia. Carbonate-dominated platform deposits change northward into pelagic marl-dominated sediments. The SW part of the platform is characterized by tilted blocks, bounded by NW-SE-to east-west-oriented faults, which had an important control on sedimentation (Chaabani & Rezgalla, 2006). Aptian and Albian rocks in Tunisia include limestones, dolomites, sandstones, mudstones, marls, and some evaporites of the Orbata, Serdj, Hameima, and Lower Fahdene formations (Burollet et al., 1978, Salaj, 1978, Chihaoui et al., 2010). The Aptian is present in the Gulf of Gabes and Eastern Tunisia and thickens toward the northwest (Bishop, 1975). The Nafusa High in Libya and the Tebaga Uplift in Tunisia were exposed at that time. The Albian sequence in Eastern Tunisia is locally intruded by volcanics (Mattoussi Kort et al., 2009).

Pioneering work emphasized the scarcity of ammonites in the pelagic facies of the Aptian deposits of Tunisia (Krenkel, 1911). However, an ammonite zonal scheme for the region was developed by Stranik et al. (1970, 1974), Biely et al. (1972), and Memmi (1999). Pervinquière (1903) first described the Djebel Serdj section and mentioned the occurrence of the cephalopods '*Douvilleiceras' martini*, '*Hoplites' fissicostatus*, as well as undetermined belemnites. In addition, Pervinquière (1903) reported the *Belemnites* (*Hibolites*) *Smicanaliculatus blainville* and *Parahoplites uhligi* from Djebel Bargou, north of Djebel Serdj. For a review of Pervinquière's (1903) identification of these ammonites, see Lehman et al. (2009).

The Aptian-Albian of the Gafsa Basin includes the Orbata Formation (Aptian, but Albian according to Burollet et al., 1978), or the Orbata Member of the Gafsa Formation in earlier publications, is composed of limestones and dolomites (Bishop, 1975, Ben Ferjani et al., 1990, Marzouk & Ben Youssef, 2008); the Bou Hedma (Aptian); the Sidi Aich Formation (Aptian), the Hamada Formation, the Gafsa Formation (Burollet et al., 1978) (Aptian-Albian), or the Gafsa Group (Aptian) (Fournié, 1978, Fig. 1), which includes the Bouhedma, Sidi Aich, and Orbata formations. The Orbata are lateral equivalents of Serdj Formation. The Aptian-Albian sequence also includes the basal Zebbag Formation (latest Albian-top Cenomanian) and the basal Fahdene Formation. They also include the Chenini Formation and Oum ed Dib Formation of Ain el Guettar Group (Lower Albian-Cenomanian) (Benton et al., 2000).

Orbata Formation (Aptian) overlies the Sidi Aich Formation in Central Tunisia and is overlain unconformably by the Zebbag Formation. Its type locality is the southern cliff of Jebel Orbata. It consists of three members (Burollet, 1956, M'Rabet, 1987, Marzouk & Ben Youssef, 2008): A lower member of cliff-forming dolomites, silty or phosphatic in places, with *Choffatella, Paleorbitolina lenticularis*, and pelecypod fragments (Burollet, 1956, M'Rabet, 1987, Marzouk & Ben Youssef, 2008). The lower member (the Barrani Dolomite) is composed of massive dolomites (Busson, 1967, Kamoun, 1988). The dolomites contain echinoids and oolites, and abundant *Orbitolina lenticularis* (Marzouk & Ben Youssef, 2008). A middle member composed of bioclastic limestones, with *Choffatella decipiens, Cuneolina* sp., *Nautiloculina* sp., *Quinqueloculina* gr. *Antique, Patellina subcretacea, Textularias*, the green algae *Pianella dinarica*, in addition to ostracods, serpulids, and fragments of echinoids, crinoids, pelecypods, gastropods and bryozoans (Burollet, 1956, M'Rabet, 1987, Marzouk & Ben Youssef, 2008). The middle member grades upward and laterally into evaporites. *Choffatella decipiens, Orbitolina parva*, and *Cylindroporella sugdeni*, suggest a Bedoulian to Gargasian (Early to Middle Aptian) age (Marzouk & Ben Youssef, 2008). An upper member (Ben Ferjani et al., 1990, p.56) composed of interbedded bioclastic limestones; argillaceous limestones and marls with oysters, ostracods, algae, and rudistid fragments; dolomitic sandstones with phosphatic and glauconitic oolites and rudistid fragments, and bioclastic limestones with oysters, rudistids, serpulids, ostracods, and the foraminifera *Mesorbitolina texana* (Burollet, 1956, M'Rabet, 1987, Marzouk & Ben Youssef, 2008). The upper member is made up of marls, sands and calcarenites, capped by limestones containing rudists and foraminifera, such as *Orbitolina parva* and *Mesorbitolina texana*, the ammonites include *Deshaysites callidiscus, D.* cf. *planus, D. weissi, D. furcata*, and *Valdorsella* sp., and fragments of chelonicertidae. This fauna is middle Bedoulian to Gargasian in age (Marzouk & Ben Youssef, 2008). The top of the Orbata Formation is marked by a conglomerate, which is dated as Gargasian based on the presence of *Orbitolina minuta, Orbitolina texana*, and *Paracoskinolina tunesiana*. The upper member is eroded south of the Chott Fedjedj, at Foum El Argoub section.

The Orbata Formation correlates with Aptian formations at Jebel Nefusa in Northwest Libya (Burollet, 1963, Magnier, 1963) and the Alamein Dolomite in Egypt (Tawadros, 2001).

In the late Aptian, carbonates were deposited on an expanded Central Tunisian Platform under shallow marine conditions (Ben Ferjani et al., 1990), and contain large and medium-size parahoplitids, such as *P. maximus, P.* cf. *nutfieldiensis,* and

*P. laticostatus*. At the base of the upper Aptian section, the nautiloid *Cymatoceras neckerianum* occurs with *Parahoplites laticostatus* and *Parahoplites* cf. *nutfieldiensis*. *Cymatoceras neckerianum* is also common in bioclastic to peloidal wackestones and packstones in the lower part of the Serdj Formation.

The Hamada Formation is 305 m (1000 ft) thick and consists of three members; a lower member of 82 m (269 ft) of limestones and marls, a middle member composed of 154 m (505 ft) of marls, and an upper member of marls, 70 m (230 ft) thick. The Hamada Formation is overlain by limestones of the Serdj Formation. Tlatli (1980, cited by Heldt et al., 2008) proposed an Early to Late Aptian age for the Hamada Formation and an early Late Aptian age for the base of the Serdj Formation, based on planktonic and benthonic foraminifera. Heldt et al. (2008) assigned the Hamada Formation at Djebel Serdj a late Late Barremian-early Late Aptian in age based on planktonic foraminifera and $\delta^{13}$ C stratigraphy.

The Upper Barremian-lower Upper Aptian hemipelagic deposits of the Hamada Formation in the Djebel Serdj consist eldt et al., 2008). Based on planktonic foraminifera and $\delta^{13}$C stratigraphy, Heldt et al. (2008) recognized four genetic intervals: 1) a pre-Oceanic Anoxic Event (OAE 1a) interval, 2) a Lower Aptian OAE 1a interval, 3) a platform-drowning-equivalent interval, and 4) a post-platform-drowning interval. Deposits of the OAE 1a include bioclastic wackestones and packstones with abundant poorly preserved radiolarians and moderately to well-preserved planktonic foraminifera. Mudstones of the platform-drowning interval directly overlying the OAE 1 are characterized by a pronounced drop in carbonate content and scarcity of macrofossils. The Lower/Upper Aptian boundary was formerly placed by Tlatli (1980) at the base of the *Globigerinelloides ferrolensis* Zone, or the base of the upper member of the Hamada Formation. Recently, Heldt et al. (2008) placed it on top of the OAE 1a in the lower part of the *L. cabri* Zone, or in the lower part of the middle member of the Hamada Formation. The lowermost member of the Hamada Formation is latest Barremian, while its top is early late Aptian (Heldt et al., 2008); the other two members extend from the lower Aptian to the beginning of the upper Aptian.

Tlatli (1980) interpreted the deposits of the Hamada Formation as basinal deposits. On the other hand, Heldt et al. (2008) suggested deposition in middle to outer-ramp setting. Thick lower Aptian hemipelagic sediments reflect a sea level rise during the OAE 1a event, whereas the upper Aptian shallow marine sediments were deposited on a carbonate platform (Lehmann et al., 2009). Ammonites include the lower Aptian *Deshayesites*, *Dufrenoyia*, *Pseudohaploceras*, *Toxoceratoides* and *?Ancyloceras*, and the upper Aptian *Zuercherella*, *Riedelites* and *Parahoplites* (Lehmann et al., 2009).

The Aptian-Albian interval in the Tadjerouine area, south of El Kef in Northwestern Tunisia (Fig. 16.1) includes the Orbata, Serdj Formation, Hameima Formation, and Fahdene Formation (Burollet, 1956). These units were recently studied and described in detail by Chihaoui et al. (2010).

The Serdj Formation (Aptian-Albian) was originally defined by Burollet (1956) as the "Serdj Limestones" and described it as massive limestones and dolomites, rich in reefal to peri-reefal faunas such as rudistids, corals and benthic foraminifera. The Serdj Formation was assigned an Aptian age by Burollet (1956) who also realized that the top of the formation may reach the Albian (Burollet, 1956). The latter has been confirmed by Tlatli (1980) who assigned the Formation an Upper Aptian-Lower Albian age in its type locality at Jebel Serdj, Northeastern Tunisia. The Serdj

Formation was dated Aptian by M'Rabet (1981) who proposed that the formation is diachronous. Chihaoui et al. (2010) reported *Acanthohoplitinae* and *Epicheloniceras* sp. of Late Aptian age, about 30 m (303 ft) below the top of the Serdj Formation at Jebel Harraba, southwest of El Kef. The Serdj Formation is a reef limestone equivalent to the Orbata Formation and overlain by clastic sediments of either the deeper marine Hameima Formation or Lower Fahdene Formation (Burollet et al., 1978).

The Hameima Formation (Late Aptian) was defined by Burollet (1956) as "shales with sandy interbeds of "Hameima" and described it as a black to dark green shale succession, with numerous thick interbeds of sandstones, dolomites and biogenic limestones". The unit was later assigned a formation status by M'Rabet et al. (1995). In the Tadjerouine area, southwest of El Kef (Fig. 16.1). The Hameima Formation is 150–230 m (492–755 ft) thick and overlies the Serdj Formation. The Hameima Formation is made up of massive limestones, locally rich in orbitolinids, with quartzarenite intercalations, separated by thick marl intervals. The base of the Hameima Formation is placed at the top of the uppermost thick, massive, rudistids and chondrodonts-bearing limestones or dolomites of the Serdj Formation (Chihaoui et al., 2010). The upper contact is placed either at the top of the uppermost thick sandstone or limestone bed, or at the base of the lowermost thick argillaceous marly interval of the Fahdene Formation (Chihaoui et al., 2010). The Hameima Formation is subdivided into three members (Chihaoui et al., 2010): The Lower Member is 70–120 m (230–394 ft) thick and rests on top of the Serdj Formation. The top of this unit is marked by deep karsts, locally filled with iron-rich minerals. The unit consists of marls and limestones, rich in orbitolinids, pectinids, and oysters, with few beds of calcareous sandstones in the lower and upper parts. It contains the ammonites *Knemiceras compressum* and *Douvilleiceras* gr. *albense-monile*, as well as *Acanthohoplitinae*, *Parengonoceras* sp., and *Hypacanthoplites paucicostatus* and *Parengonoceras* sp. The Middle Member ("*Clansayes* Sandstones" of Vila et al., 1994) is 35–70 m (115–230 ft) thick. It consists of marls, sandstones, and limestones in the lower part, grading upward into cross-laminated sandstones, and then into orbitolinid-rich limestones, with *Orbitolina (Mesorbitolina) parva* and *O. minuta* and *Praeorbitolina wienandsi* at the top. The macrofossils include oysters, pectinids, echinoids, annelids, trigonids, corals, gastropods, and plant remains. The Upper Member is made up of up to 55 m (180 ft) of marls, limestones, and glauconitic, phosphatic, and clast-rich beds. The fauna consists of oysters, pectinids, brachiopods, and scarce ammonites, such as *Douvilleiceras* sp., *Acanthohoplitinae* and "*Hypacanthoplites*" sp. (Chihaoui et al., 2010).

The Fahdene Formation (Albian-Cenomanian) was originally defined by Burollet (1956) as the "Fahdene marls and shales" for a very thick series of marine, grey to black marls and shales, with a few beds of limestones or marly limestones above the Serdj limestones. It was later designated as the Fahdene Formation (Burollet et al., 1983). The type locality of the Fahdene Formation is in the core of the Oued Bahloul anticline, southeast of Makthar (Burollet, 1956). The Fahdene Formation has been assigned an Albian-Cenomanian. The Fahdene is the lateral equivalent of the Zebbag Formation. The boundary between the Hameima and Fahdene Formations is believed to be diachronous, being older in the northwest (Chihaoui et al., 2010). The Fahdene Formation has been divided into five informal units by Burollet (1956), in an ascending order; they are the Lower Shale, Allam limestones, Middle Shale, Mouelha limestones, and Upper Shale.

The Lower Shale Member is 180–400 m (590–1312 ft) thick (Chihaoui et al., 2010) and overlies massive sandstones and limestones of the Hameima Formation. The base of the unit consists of dark green shales with few thin calcareous interbeds and phosphatic nodules. The Lower Shale Member includes an ammonite-rich horizon ascribed to the "*Clansayes* zone", which was considered as earliest Albian (Burollet, 1956), but is now assigned to the latest Aptian (Burollet et al., 1983, Chihaoui et al., 2010). Ammonites include "*Hypacanthoplites*" *paucicostatus*, *Parengonoceras* sp., and *Acanthohoplitinae* indet., followed upward by "*Hypacanthoplites*" *ouenzaensis*, *Douvilleiceras* sp. nov., *Parengonoceras* sp., and *Prolyelliceras gevreyi* (Chihaoui et al., 2010). A second carbonate interval contains belemnites, echinoids, ammonites, pectinids, brachiopods, oysters, and gastropods, in addition to the ammonites *P. gevreyi*, *Mirapelia* cf. *alticarinata*, *Beudanticeras* sp., and *Desmoceras latidorsatum* (Chihaoui et al., 2010). The upper part of the Lower Shale Member is dominated by pyritized ammonites, belemnites, brachiopods, pectinids, and a few echinoids. The ammonites include *Prolyelliceras radenaci*, *D. mammillatum*, *Beudanticeras revoili*, *B. dupinianum africanum*, *Phylloceras* (*Hypophylloceras*) *velledae*, *Ptychoceras hamaimensis*, *Pictetia astieriana*, *Silesitoides thos*, *Desmoceras* sp., *Protanisoceras* cf. *acteon*, and *Tegoceras mosense* (Chihaoui et al., 2010).

The Allam Limestone Member was described by Burollet (1956) as the Allam Limestones composed of hard, fine-grained black limestones alternating with dark shales and marls, with belemnites and poorly preserved ammonites, such as *Puzosia* sp. Burollet (1956) ascribed the Allam Limestones a Middle Albian age. The Allam Member is 110–180 m (361–590 ft) thick and contains locally minor oil seeps (Chihaoui et al., 2010). The base of the unit is locally marked by a limestone bed containing phosphatized clasts, while the upper part is commonly silicified with large limestone nodules. The fauna consists of belemnites with scarce ammonites and fish remains. The lower part of the unit contains the ammonites *Beudanticeras* sp., *Ulighella* sp., *Douvilleiceras* sp., *Oxytropidoceras* sp., *Pusozia* sp. and *Desmoceratids*, *Tegoceras camatteanum*, and *L. pseudolyelli* (Chihaoui et al., 2010).

The Middle Shale is composed of 280 m (919 ft) of black shales and marls, dated as Late Albian (Chihaoui et al., 2010). The fauna consists of numerous crushed ammonites, associated with belemnites and scarce echinoids and bivalves. The base of the unit is locally glauconitic and contains *Hamites* sp., *Kosmatella* sp. and *Mirapelia* sp., *Venezoliceras* sp., and *Oxytropidoceras* sp., followed upward by an assemblage of *Mortoniceras* (*Deiradoceras*) sp., *Mortoniceras* (*Mortoniceras*) sp., *Oxytropidoceras* sp., *Venezoliceras* sp., *Hysteroceras* sp., *Neokentroceras* sp., *Puzosia* sp. (*Mortoniceras pricei* Zone) (Chihaoui et al., 2010).

The Mouelha Limestone Member consists of black laminated limestones, about 50 m (164 ft) thick, assigned to the Late Albian age.

Upper Shale Member is made up of black shales with some marly intercalations, more than 1200 m (3937 ft) in thickness, and ranges in age from uppermost Albian (Vraconian) to Cenomanian.

During the Upper Cretaceous, Tunisia was differentiated into platforms, basins, and exposed areas (Marie et al., 1984, Burollet, 1991, and Ben Ferjani et al., 1991, Klett, 2001, Bensalem et al., 2002, Chihaoui et al., 2010). The Cenomanian-Turonian succession in Eastern Tunisia contains basaltic flows, mafic intrusions, pyroclastics, tuffs, and volcano-sedimentary rocks, while the Senonian contains altered basalts and pyroclastics (Mattoussi Kort et al., 2009). Geological, geophysical, and geochemical data from the onshore and offshore of Eastern Tunisia indicate a crustal thinning produced by the Tethyan rifting (Mattoussi Kort et al., 2009). This thinning is responsible for the subsequent evolution of the North African passive margin during the Late Cretaceous and the development of the fold-thrust belt and associated foreland deformations, as well as mantle upwelling which led to high heat flow, basaltic magma activities, and the circulation of hydrothermal fluids. From the mid-Cretaceous to the late Eocene, the continental margin of Northern Tunisia was a regional basin in which a thick sequence of carbonate rocks was deposited (Snoke et al., 1988).

The Upper Cretaceous lithostratigraphic units in Tunisia include the Bahloul (lateral equivalent of Zebbag and Fahdene Formations), Aleg, Abiod, base of El Haria, Annaba (member of the El Kef or Bahloul formations), Bireno, Miskr, and Bir Aouina formations. All the Upper Cretaceous formations were included in the Sidi Mansour Group by Ben Ferjani et al. (1990).

The Zebbag Formation (Albian-Turonian) overlies unconformably the Aptian Orbata Formation. The Zebbag sediments were deposited on a shallow marine carbonate platform in Central Tunisia, the Gulf of Gabes, and the offshore (Bishop, 1988). Subtidal conditions existed in Central Tunisia and in the Gulf of Gabes, where lagoonal mudstones, dolomites, and anhydrites were deposited (Bishop, 1975, Bishop, 1988). Reefs, reefoid facies, and rudist banks are also present (Bishop, 1975, Burollet et al., 1978). In the Pelagian Province, the Zebbag Formation is made up of limestones, dolomites, and bioclastic rocks, and is laterally equivalent to basinal mudstones and argillaceous limestones of the Fahdene and Bahloul formations (Bishop, 1988). The Cenomanian fractured Zebbag dolomites form the main reservoir in the Ezzaouia Field discovered by Marathon in 1986 (currently operated by Maretap, a joint venture company between Candax and ETAP), about 12 km north of Zarzis (Rodgers et al., 1990) and produces at about 2000 bopd. The Fahdene Formation is made up of dark-grey mudstones and the Bahloul Formation is composed of dark-colored, laminated, euxinic, argillaceous limestones (Burollet et al., 1978, Bishop, 1988).

Burollet (1956) introduced the Aleg Formation and subdivided it into the Annaba and Bireno members. It has a thickness of more than 208 m (682 ft) at its type section at Djebel Om El Aleg on the western side of Djebel Orbata. Ben Ferjani et al. (1990) assigned the formation a Turonian-Lower Campanian age. They also included the Beida Evaporites and Bou Douaou Carbonates in the internal shelf facies of the lower part of the Aleg Formation, and the Annaba and Bireno members in the laterally equivalent transitional facies. The Santonian Jamil Formation and the Campanian-Maastrichtian Bu Isa Formation are laterally equivalent in part to the Aleg and Abiod Formations (Klett, 2001).

The Cenomanian-Turonian Bahloul Formation in Tunisia consists of well-bedded dark grey limestones, with local intercalations of marls and argillaceous limestones. The Bahloul Formation is well exposed on the northwestern flank of the Oued Bahloul Anticline, in Central Tunisia about 100 km west of Kairouan City (Fig. 16.1), and 10 km to the southeast of the city of Makthar (Burollet, 1956). The core of the anticline is occupied by Cenomanian marls of the Fahdene Formation. In the Oued Bahloul section, the Bahloul Formation is about 28 m (92 ft) in thickness and overlies a 4 m (13 ft) massive conglomeratic sandy limestone beds with an erosional base.

In the Oued Bahloul section the Formation is composed of deepening-upward cycles (Zagrarni et al., 2008). Each cycle starts with laminated dark grey limestones, changing gradually into argillaceous limestones, and terminates with bioturbated marls containing microfauna of the *Whiteinella archaeocretacea* biozone. The upper boundary of the Bahloul Formation represents a maximum flooding surface (mfs) which is directly overlain by marls of the Annaba Member of the El Kef Formation. The Annaba Marl contains the planktonic foraminifera *Dicarinella imbricate*, *Whiteinella praehelvetica*, *W. archaeocretacea*, as well as *Helvetoglobotruncana helvetica* biozone, indicating an early Turonian age. It also includes the ammonites *Pseudaspidoceras flexuosum* and *Mammites nodosoides* (Robaszynski et al., 1990, Camoin, 1993). Toward the SW, in Jebel Bireno, the Bahloul Formation is about 6 m (20 ft) and is made up entirely of shallow-marine carbonates. At Jebel Asker, limestones, more than 27 m (89 ft) in thickness, highly bioturbated and dolomitized in the upper part. The upper 6 m (20 ft) contain *Vascoceras* gr. *kossmati* and *Spathites* gr. *subconciliatus*, indicating an early Turonian age (Zagrarni et al., 2008). In the Gafsa Basin and in Southern Tunisia, the Bahloul Formation equivalent is capped by rudist-rich carbonates of the Gattar Member. The Bahloul sediments were deposited during the late Cenomanian-early Turonian Oceanic Anoxic Event (OAE) identified in numerous regions in Tunisia (Burollet, 1956, Razgallah et al., 1994, Heldt et al., 2008), North Africa (Kuhnt et al., 2004, Kolonic et al., 2005, Keller et al., 2008), and the Middle East (Lüning et al., 2004, Zagrani et al., 2008, Rodriguez-Rover et al., 2009).

The Turonian-Lower Campanian Aleg Formation includes mudstones, limestones, and marls (Bishop, 1975, Salaj, 1978) and overlies the Fahdene, Bahloul, and Zebbag Formations. The Bireno, Miskar, and Douleb formations are equivalent to some intervals of the Aleg Formation (Klett, 2000). Both the Bireno and Douleb formations are, in part, laterally equivalent to or are members of the Miskar Formation. These formations consist of limestones, dolomites, and marls (Salaj, 1978, El Euchi et al., 1998).

The Miskar Formation represents a rudist bank or banks on the shelf edge or slope (Bishop, 1988, Knott et al., 1995). Oil is produced from Coniacian limestones in the Djebel Onk Field (Bishop, 1975). The Bireno Limestone contains benthonic foraminifera, such as *Cuneolina pavonia*, *Pseudolituonella reicheli*, *Montcharmontia appenninica?* and *Nazzazatinella picardi*, the rudists *Hippurites requieni*, *H. praetouscasi*, and *Biradiolites* sp. aff. *B. lumbricalis*, and the ammonites *Kamerunoceras turoniense* and *Neoptychites cephalotus*. The assemblage suggests a Middle Turonian age (Robaszynski et al., 1990, Camoin, 1993). The Douleb Limestone contains the benthonic foraminifera *Rotalia algeriana*, *Cuneolina* sp. gr. *C. pavonia*, *Nezzazatinella picardi*, *Pseudocyclammina sphaeroidaea*, *Dicyclina schlumbergeri*, *Montcharmontia appenninica*, and *Minouxia* sp. cf. *M. lobata*, in addition to the rudists *Hippurites variabilis* and *Vaccinites taburni*, and the ammonite *Barroisiceras haberfellneri*. The foraminiferal assemblage and the ammonite give the Douleb Limestone a Coniacian age (Robaszynski et al., 1990, Camoin, 1993).

During the Campanian to Maastrichtian, chalky limestones, micrites, and marls of the Abiod-Formation were deposited on top of the Aleg Formation (Burollet et al., 1978, Salaj, 1978, Bishop, 1975).

The Abiod Chalk Formation (Campanian-Maastrichtian) is exposed at El Kef, Kalaat Senan, and Elles in Northwestern Tunisia. It was originally defined as

*Les calcaires de l'Oued El Abiod* by Burollet (1956). It derives its name from its type section at Oued El Abiod,[12] east of Maknassi. The Abiod Formation has a thickness of about 400 m (1312 ft) in Western Tunisia and 100 m (328 ft) in the offshore of the Pelagian Sea (Fournié, 1978, Davies & Bel Haiza, 1990). It has been subdivided into three members (Burollet, 1956, Burollet et al., 1954, Ben Ferjani et al., 1990): a lower chalk member and an upper chalk member, separated by a middle shale member. Robaszynski et al. (2000) subdivided the Abiod Formation into seven units, in an ascending order; the Assila, Haraoua, Mahdi, Akhdar, Gourbeuj, Ncham, and Gouss members. Jarvis et al. (2002) retained the original tripartite division of the Abiod Formation, and assigned the three units to the Haraoua, Akhdar, and Nachm members, and considered the others as transitional lithologies at the top or bottom of the other members. The Haraoua Member has a thickness of about 20 m (66 m) and is made up of interbedded white chalks and light grey marly chalks, with globigerinids, globotruncanids, and orbitoids. It contains a bed with abundant inoceramid bivalves and another with reworked orbitoids. Its lower part is rich in *Pseudotextularia* and *Neoflabellina*. It has a thickness of 31 m (102 ft). The Akhdar Member is made up mainly of medium to dark blue-green to greenish grey marls with interbeds of light grey marly chalks and pale yellow limestones. Its thickness is 194 m (637 ft). The limestones contain the inoceramid bivalves *Endocostea* cf. *ghadamesensis* and the ammonite *Nostoceras (Bostrychoceras) polyplocum*. The upper Nachm Member is about 143 m (469 ft) in thickness and is composed predominantly of thin rhythmically interbedded marls and chalks. The lower chalks contain very large *Zoophycos* burrows, up to 1 m (3 ft), and abundant *Planolites* burrows, and scattered crinoids. A bed near the contact with the overlying El Haria Formation contains abundant very large inoceramids of *Trochoceramus nahorianensis*. Although this tripartite subdivision is recognizable in the majority of sections, some areas show continuous carbonate sedimentation (Fournié, 1978). In many places, the Abiod Formation reflects tectonic instability with turbidites and mud flows mixed with deeper pelagic calcareous mud (Ben Ferjani et al., 1990). Along the N-S Axis (NOSA), the formation becomes sandy due to reworking of Lower Cretaceous sands (Ben Ferjani et al., 1990). The formation overlies unconformably units of various ages, such as Neocomian at Bou Gobrine, Jurassic at Hammam Zriba, and Aptian at Khanguet El Hajej, and is overlain by the El Haria Formation (Ben Ferjani et al., 1990). The top of the Abiod Formation corresponds to a strong reflector called Horizon H6, which is very distinct on seismic profiles (Bobier et al., 1991).

In Central Tunisia, at Wadi Ed Dam near the Village of Elles, the Abiod Formation crops out in an almost uninterrupted exposure about 2860 m (9384 ft) in thickness (Robaszynski & Mzoughi, 2010). It is underlain by the Kef Marls and capped by the El Haria Marls. Fossils include foraminifera and ammonites. The Campanian-Maastrichtian boundary at Elles is established by Robaszynski & Mzoughi (2010) by comparison with the Kalaat-Senan section and in the international stratotype of Tercis, France, in the lower third of the upper indurated limestones of the Abiod Formation, namely the Ncham Member, which is in agreement with Li et al. (1999). The location of the boundary is based on the upper limit of the occurrence of *Nostoceras hyatti*

---

12 The area was visited and its topograpghy was mapped by E. de Larminant in 1896.

and *Pseudokossmaticeras brandti* and the first appearance of *Nostoceras magdadiae* (a Maastrichtian ammonite).

The Abiod fractured carbonates form gas reservoirs at Tazeka, Maamoura, and Miskar fields (Hallett, 2002, p.408). They form the main reservoir in the Maamoura Oil Field in the Gulf of Hammamat, with reserves of 50 mmbbls of oil and a flow rate of 3300 bopd (Davies & Bel Haiza, 1990).

In the Tunisian Atlas, such as at Ousselet, Bou Dabbous, and Bou Hajar areas, the Abiod Carbonates contain synsedimentary breccias and slumps (Dhahri & Boucadi, 2007).

The Abiod Formation (Campanian-Maastrichtian) is equivalent to the Berda Formation and Merfeg Formation (Fournié, 1978, Marie et al., 1984).

The Berda Formation was named after its type section at Djebel Berda (described in detail by Fournié, 1978, p.131–133). It consists of three members; the upper and lower members are composed of bioclastic limestones, while the middle member is made up of marls, marly limestones, and minor gypsum, with pelecypods, oysters, echinoids, inoceramids, and benthonic foraminifera, such as *Orbitoides* and *Sidrolites*. It was dated Campanian-Maastrichtian by Fournié (1978). It is equivalent in part to the Lower Tar Formation in West Libya and the Waha Formation in the Sirte Basin.

At Jebel Merfeg, the Abiod Formation changes facies into rudistid and coral mud mounds, with breccias and chalky limestones rich in calcispheres of the Merfeg Formation. The Merfeg Formation was originally named *Formation des récifs du Merfeg* by Fournié (1978). Its type section is at Djebel Merfeg on the southern flank of Djebel Kebar (described in detail by Fournié, 1978, 134–135). The Merfeg Formation rests on shales and marls of the Kef Formation or its equivalent the Aleg Formation and is unconformably overlain by Miocene and Pliocene shales. In the Gafsa Basin, lagoonal Paleocene sediments rest directly on top of the Campanian part of the Abiod limestones; the Maastrichtian and Danian are missing (Bensalem, 2002).

## 16.2.4  Tertiary

### 16.2.4.1  *Paleocene*

The Paleocene in Tunisia includes El Haria Formation and the lower part of the Tselja Formation of the Metlaoui Group.

El Haria Formation was introduced by Burollet (1956) as El Haria Shales *"Formation des Argiles El Haria"* along the track of Hammam Mellegue near El Kef. Burollet (1956) described the following succession in the El Haria type section, with a total thickness of 700 m (2297 ft):

a  Glauconitic and phosphatic marls and marly limestones, 90 m (295 ft).
b  Brownish to dark clays with *Globorotalia angulata* and *Globorotalia velascoensis* of Paleocene age, 310 m (1017 ft).
c  Grey marls with a limestone interbed with *Globorotalia trinidadensis* and *Globorotalia uncinata* of Danian age, 65 m (213 ft).
d  Dark clays and marls with *Globotruncana*, 200 m (656 ft).
e  Interbedded marls and limestones at the base.

The El Haria Formation is widely distributed in Tunisia and extends into the offshore with little variations in lithology. The El Haria Formation is made up of fissile black, dark grey or brown shales rich in microfossils, with minor limestone intercalations, especially in the middle part of the formation. It overlies the Abiod Formation and is overlain by the Lower Eocene Metlaoui Group, usually with a contact marked by phosphatic beds. It was appointed a Maastrichtian-Paleocene age by Burollet (1956), Fournié (1978), and Ben Ferjani et al. (1990). In Northern Tunisia, the carbonate beds in the middle of the formation contain Danian brachiopods and corals. In the Tellian Domain it contains large boulders of black and yellow weathered limestones (Ben Ferjani et al., 1978, p.77). Its thickness may attain 700–1000 m (2297–3281 ft), with an average thickness of 400 m (1312 ft) (Bensalem, 2002, Dhahri & Boukadi, 2007). However, in the stable shelf area of Eastern Tunisia and in some positive areas, such as parts of the N-S Axis (NOSA), the formation is missing, especially the Danian part (Bishop, 1975, Ben Ferjani et al., 1990). In some parts of the Gulf of Hammamat, the entire El Haria Formation is absent. In Northwestern and Central Tunisia, eroded shales and marls of the El Haria Formation form wide valleys and plains between the Cretaceous Abiod carbonates and the Eocene Metlaoui phosphatic limestones (Bensalem, 2002). The El Haria Formation extends over much of the Pelagian Province but is absent in the southern and southwestern parts, as a result of either non-deposition or removal by erosion (Bishop, 1975, Klett, 2001). The Al Jurf Formation, Ehduz Formation, and part of the Bilal Formation are laterally equivalent to the El Haria Formation (Klett, 2001). The El Haria Formation is equivalent to the Upper Tar Formation in West Libya, the Hagfa Shale in the Sirte Basin in Libya, and the combined Dakhla and Esna Shales of Egypt.

In the Tellian Zone in Northwestern Tunisia, the El Haria Formation forms a continuous and very thick succession characterized by yellow dolomitic concretions, "*marnes à boules jaunes*" (Bensalem, 2002). Some thin interbedded limestones appear near the Upper Maastrichtian-Danian boundary. In Jebel Ed Diss of Nefza (Tamera railway station), where the total thickness of the El Haria Formation is of about 900 m (2953 ft), Rouvier (1985, in Bensalem, 2002) described, the following succession below the Ypresian limestones and above a tectonic shear plane:

a    Grey to dark clays with dolomitic concretions, 100 m (328 ft).

b    Marls with marly limestones, 100 m (328 ft).

c    Grey to dark clays with dolomitic concretions, 300 m (984 ft). The clays contain *Globotruncana angulata* and *Globotruncana pseudomenardii* which indicate a Thanetian (Landenian) age.

d    Marly limestones, 150 m (492 ft). *Globigerina pseudobulloides* occurs in a thin yellow limestone level.

e    Grey to dark clays, 100 m (328 ft), with dolomitic concretions. The presence of *Globotruncana trinidadensis* indicates a Danian age.

f    Grey to dark clays with dolomitic concretions and *Globotruncana contusa* and *Globotruncana gansseri* of Late Maastrichtian age.

The Tselja Formation (also spelled Selja) was introduced by Burollet (1956) as the *Formation des marnes et calcaires du Tselja*. The formation has a thickness of 90 m (295 ft) in its type section at Djebel Alima at the gorge of Oued Tselja (Fournié (1978) and is made up of fossiliferous limestones with abundant *Ostrea* at the base followed by evaporites with dolomitic shales, dolomites, and gypsum. It overlies the Paleocene El Haria Formation and is overlain by Ypresian black shales and phosphates of the Chouabine Formation. The age of the Tselja Formation is Upper Paleocene. Equivalent shales and marls with *Globorotalia velascoensis* and *G. pseudomenardii* have been encountered in wells in the Gulf of Gabes (Fournié (1978). They reach a thickness of 116 m (381 ft) in the well Ashtart Nord No.1. The Tselja Formation is equivalent to Had Formation in West Libya and the Esna Shale in Egypt.

### 16.2.4.2  Cretaceous-Tertiary boundary

The Latest Cretaceous-Paleocene paleogeography of Tunisia is characterized by two subsiding troughs in the north, the NW Tunisian Trough and the NE Tunisian Basin (Bensalem, 2002). The Saharan Platform is separated from the northern basins by the Kasserine Island. In the northern basins, the Cretaceous-Paleocene boundary lies within the marine El Haria Formation (Maastrichtian-Landenian), which attains a thickness of 700–1000 m (2297–3281 ft). The GSSP (stratotype) of the K/T boundary at El Kef lies in the NW Tunisian Trough. In the Gafsa-Metlaoui area, the El Haria Formation was deposited in an intracratonic basin and consists of about 100 m (328 ft) of lagoonal sediments. In the eastern part of the Gafsa -Metlaoui Basin, the Upper Maastrichtian and Danian are missing. The El Haria Formation covers much of the Pelagian Province but is absent in the southern and southwestern portions, as a result of either non-deposition or removal by erosion (Bishop, 1975). The El Haria Formation is the lateral equivalent of the Al Jurf and Ehduz formations and part of the Bilal Formation (Fig. 13.1). It overlies the Abiod and Bu Isa formations (Burollet, 1967, Bishop, 1975).

A large number of paleontological studies on the K/T stratotype at El Kef has been published by Perch-Nielsen (1979, 1981a, 1981b), Pospichal (1994), Verbeek (1977), Brinkhuis & Zachariasse (1988), Keller (1988), Keller et al. (1995, 2002, 2008), Gardin (2002), Karoui-Yaakoub et al. (2002).

Planktonic foraminiferal fauna across the K/T transition at the El Kef stratotype, the Ain Settara, and Elles and El Melah (Keller, 1988, 1996, Keller et al., 1995, Molina et al., 1998, Luciani, 2002), in northwestern and Northeastern Tunisia show similar patterns of species extinctions and survivorship, with more than 2/3 of the species disappearing at or before the K/T boundary and slightly less than 1/3 surviving into the Danian (Karoui-Yaakoub et al., 2002). At Elles, they show major changes in the Tethyan marine ecosystem during the upper Maastrichtian (Abramovich & Keller, 2002). A progressive cooling trend resulted in the decline of globotruncanid species, followed by a further decline at the climax of a warm event about 300 ka before the K/T boundary. Long-ranging species fluctuated in abundance considerably reflecting high-stress environmental conditions. At the end of the Maastrichtian, rapid cooling associated with accelerated species extinctions followed by the extinction of all tropical and subtropical species at the K/T boundary.

Calcareous nannofossils of the Elles section across the K/T boundary by Gardin (2002) show that the late Maastrichtian assemblages are rich and well preserved. Pulses of surface water cooling increased the abundance of cool water taxa, such as *Kamptnerius magnificus*, *Nephrolithus frequens*, and *Gartnerago* sp., and a decrease of the warm water species *Watznaueria barnesae*, as well as *Arkhangelskiella cymbiformis*. On the other hand, the Danian at El Kef and Elles is characterized by newly evolved species.

The late Maastrichtian-early Danian $^{87}Sr/^{86}Sr$ is characterized by maxima at 0.3–0.4 Ma around the K/T boundary. The latter was attributed to Sr release by soil leaching combined with increased rainfall associated with the K/T.

Li et al. (1999, 2000) postulated that cooler and arid climates generally accompanied low sea levels across the Cretaceous/Tertiary (K/T) in Tunisia. Adatte et al. (2002) also observed that the Late Campanian to early Danian succession at El Kef and Elles indicates an increasingly more humid climate associated with sea-level fluctuations and increased detrital influx that culminates at the K/T transition, associated with middle and high latitude cooling. Short-term changes across the K/T boundary indicate a sea-level lowstand in the latest Maastrichtian about 25/100 ka below the K/T boundary marked by increased detrital influx at El Kef and Elles and a short hiatus at Ain Settara. A rising sea-level at the end of the Maastrichtian is expressed at Elles and El Kef by deposition of a foraminiferal packstone. A flooding surface and condensed sedimentation highlighted by clays that are rich in terrestrial organic matter. Adatte et al. (2002) suggest that in Tunisia, long-term environmental stresses during the last 500 ka before the K/T boundary and continued into the early Danian are primarily related to climate and sea-level.

### 16.2.4.3   Eocene

The Eocene in Tunisia includes the Ypresian Metlaoui Group, which comprises the uppermost Paleocene Tselja Formation and the Ypresian Chouabine, Faid, El Garia, and Bou Dabbous formations, and the Lutetian-Priabonian (Bartonian) Djebs (Jebs), Cherahil, and Souar formations, as well as the Tanit and Bou Loufa formations. The stratigraphy of the Ypresian Metlaoui Group in Tunisia has been studied extensively by Fournié (1975, 1978), Bishop (1975, 1988), Comte & Dufaure (1973), Comte & Lehmann (1974), Fournié (1975), Blondeau (1980), Burollet & Odin (1980), Winnock (1980), Bismuth & Bonnefous (1981), Moody (1987), Moody et al. (1998), Reali & Ronchi (1998), Loucks et al. (1998), Jamoussi et al. (2001, 2003), Jorry et al. (2001, 2003), Reali et al. (2003), Jorry (2004), Racey (1994, 2001), Racey et al. (2001), and Tawadros (2001).

The Metlaoui Group was originally introduced by Burollet (1956) as the *Formation des calcaires Metlaoui* (Metlaoui Formation) in the Oued Seldja Canyon, west of Metlaoui, where Philippe Thomas discovered the phosphates in 1885 (Ben Ferjani et al., 1990, Burollet, 1995). The formation was later raised to a group status by Fournié (1978)[13] to include all the Ypresian facies variations. Six different facies have been recognized in the Metlaoui Group and each was designated as a formation by Fournié (1978). They include the *Formation lumachelles du Cherahil* (Cherahil Formation), *Formation des calcaires d'El Garia* (El Garia Formation), *Formation de phosphates de Chouabine* (Chouabine Formation), *Formation de évaporites du Tselja* (Tselja Formation), Formation des évaporites du Faid (Faid Formation), and

---

13 Fournié (1978), however, continued to use the name Metlaoui Formation, inspite of the fact that he included six formations in it. Ben Ferjani et al. (1990) followed the same concept.

*Formation des calcaires du Bou Dabbous* (Bou Dabbous Formation). The age of the Metlaoui Group is Ypresian, although it may include the uppermost Paleocene and exend into the Paleocene (Selja or TseljaFormation). The Libyan Jirani Dolomite and Jdeir Formation are lateral equivalents of the Metlaoui Group.

The Chouabine Formation was first described by Burollet (1956). It consists of two units with a total thickness of 35 m (115 ft). The lower unit consists of interbedded black shales and white micritic limestones rich in gastropods and contains thin phosphate beds. The upper unit is made up of phosphatic sands with *Ostrea* (Burollet, 1956, Fournié, 1978). The type section is located on the northern side of Djebels Chouabine and Alima, about 4 km to the SE of Redeyef (Fournié, 1978, p.116). The formation was dated Lower Ypresian by Fournié (1978) based on the presence of *Globorotalia wilcoxensis* at the base, as well as the nannofossils *Toweius craticulus*, *T. eminens*, *T. tovae*, *Fasciculithus involutus*, and *Neochiastozygus concinnus*. The Chouabine Formation also forms the base of the cliffs delimiting the Kesra Plateau in Central Tunisia, where two units were recognized within the formation (Jorry et al., 2003, Jorry, 2004); a glauconitic marl unit at the base and a bioclastic limestone bed at the top. The basal unit is strongly bioturbated and contains abundant iron oxide concretions. The foraminifera include *Morozovella subbotinae*, *Morozovella aragonensis*, *Morozovella formosa formosa*, *Morozovella* aff. *caucasica*, *Morozovella lensiformis*, *Morozovella quetra*, *Acarinina primitiva*, *Acarinina pentacamerata*, *Acarinina broedermami*, *Turborotalia* aff. *frontosa*, *Globigerina linaperta*, *Globigerina inaequispira*, which confirm the late Ypresian age. The bioclastic limestone bed is marked by concentrations of phosphatic granules at its base, followed by bioclastic wackestones with phosphatic nodules and small gastropods, bivalves, nummulites and nautiloids. The nautiloids may reach more than 30 cm in diameter. The top of the unit is marked by a bioturbated horizon with *Thalassinoides*-type burrows. In the Gulf of Gabes, equivalent radioactive limestones, marls, and shales with phosphatic matrix and nodules were encountered in wells, such as the Ashtart Nord 1. This unit was described and named by Fournié (1978, p.118) *"Formation radioactive"*. The Chouabine Formation is overlain by the Ypresian Faid Formation (a continental facies with evaporites), the Ain Merhotta Formation (lagoonal gastropod-rich carbonates), the El Garia Formation (nummulitic limestones), and the basinal Bou Dabbous Formation (lime mudstones) (Jorry et al., 2003, Jorry, 2004).

The Faid Formation was defined by Fournié (1978, p.123) as *Formation des evaporites du Faid*; a dolomitic unit between the Chouabine and Djebs formations, locally evaporitic, rich in mollusks and gastropods, with gypsum and phosphatic nodules. The type section, where the formation is 59 m (194 ft) thick, is situated at Djebel Faid. The formation did not yield any fossils, but it was assigned a Ypresian age based on its stratigraphic position. It is a lateral equivalent of the Bou Dabbous and Ain Merhotta formations.

The Tselja Formation is overlain by the El Garia Formation which shows marked lateral facies variations from nummulitic packstones-grainstones in the SW to thick nummulitic accumulations in the NE. Nummulites in this formation include *Nummulites perplexus*, *Nummulites formosus* (*Nummulites operculiniforme*), *Nummulites* cf. *pomeli*, *Nummulites* aff. *pomeli*, *Nummulites rollandi (caillaudi)*, and *Nummulites tenuilamellatus*, which corresponds to the boundary between the late Ypresian and the early Lutetian. At Kesra, they are associated with *N. rollandi*, *N.* aff. *rollandi*,

*N. perplexus, N. tenuilamellatus*). *N. perplexus* (Jorry et al., 2003, Jorry, 2004). The El Garia Formation is composed of two shallowing-up depositional sequences separated by a deepening-up sequence (*ibid.*). The Ypresian nummulitic limestones of the El Garia Formation form important carbonate reservoirs in North Africa. In Tunisia, they form reservoirs at the Ashtart Field and the Sidi El Itayem Field (discovered in 1971) and in numerous small pools, as well as the Bouri Field in Libya (Vennin et al., 2003) and the Hasdrubal Field (Macaulay et al., 2001). The reservoir consists of shallow-water carbonates composed of *Nummulites* and *Discocyclina*. The basinal facies of the Bou Dabbous Formation represents the source rock, which also interfingers with the carbonates reservoir rocks (Bishop 1988). The Bou Dabbous Formation which contains abundant globigerinid planktonic Foraminifera and is rich in organic matter (Bishop 1988). The organic matter varies between 1% and 3.1% TOC and of marine Type II origin (Bishop, 1988, Racey et al., 2001, Vennin et al., 2003). The Ousselat Member (Moody et al., 1989) represents clastic nummulitic accumulations in the El Garia Formation.

The El Garia Formation limestones are locally capped by the Lutetian Souar Formation composed of yellowish marls interbedded with massive oyster beds and rare thin beds of argillaceous or glauconitic limestone beds (Fournié, 1978). This formation is only exposed in the centre of the Kesra Plateau and near the village of El Garia (Jorry et al., 2003, Jorry, 2004). In Northeastern and Eastern Tunisia, the Souar Formation contains a horizon with large Nummulites and Discocyclinas, which suggest a basal Upper Lutetian age. This horizon was named the Reineche Limestone by Burollet (1956). In the Tellian Nappes in Northwestern Tunisia, the Souar Formation contains large yellow concretionary boulders of black limestones Ben Ferjani et al. (1990). To the northeast, from Cap Bon to the Gulf of Hammemat, the Souar Formation is replaced by bioclastic platform carbonates of the Halk El Menzel Formation (Bismuth & Bonnefous, 1981). In Central Tunisia, the Souar shales contain oyster coquinas and bioclastic limestones, which were encountered in several wells.

The Souar Formation grades laterally into the Cherahil Formation (Comte & Dufaure, 1973) named after Djebel Cherahil. In its type section, it consists of nodular limestones with lamellibranchs. It consists of two members separated by the coquinoidal Siouf Limestone with abundant echinoids (Ben Ferjani et al., 1990). It has been assigned middle to Upper Eocene age. On the northeastern border of the Kasserine Island, the Cherahil Formation is replaced by its equivalent the Djebs Formation composed mostly of white gypsum, up to 900 m (2953 ft) in thickness, with some intercalations of green and red shales and dolomites.

In Central and large parts of Southern Tunisia, the Metlaoui Group is absent. However, in places around the Kasserine Island, it is replaced by continental Eocene sediments of the Bou Loufa and Tanit formations (Ben Ferjani et al., 1990, Jamoussi et al., 2001, 2003). The Tanit Formation is made up of varicolored shales with Eocene palynomorphs. It has been encountered in several wells in the southern part of the Gulf of Gabes (Ben Ferjani et al., 1990, Jamoussi et al., 2003). The Bou Loufa (Bouloufa) Formation is made up of conglomerates, red shales, lacustrine limestones, and caliche, with the continental gastropod *Bulimes* (Ben Ferjani et al., 1990). At Chebket Bouloufa, west of Gabes, the Bou Loufa Formation overlies dolomites of the Aleg Formation (Coniacian) and is made up of 87 m (285 ft) of reddish and grayish calcareous and silty shales interbedded with carbonate crusts (Jamoussi et al., 2001).

At Jebel Rheouis, the Formation is made up of 82 m (269 ft) of varicolored shales rich in palygorskite with *Bulimes*. In Jebel Boudinar, the Bou Loufa Formation is 140 m (259 ft) and overlies El Haria Shale and is overlain by Miocene sandstones (Jamoussi et al., 2001, 2003). Sassi et al. (1984) reported the continental gastropods *Romanella hopii* and *Vidaliella* and the helicids *Paleocyclotus* and assigned the Bou Loufa Formation a Lutetian-Bartonian age.

Rouvier (1977, cited by Talbi et al., 2008 and Samir et al., 2010) recognized three units in the Tunisia Atlas; the Ed Diss Carbonates and Marls (Cretaceous), the Adissa and Ain Draham Shales and Breccias (latest Cretaceous-Eocene), and the Kasseb Unit (Paleogene-Eocene) composed of Paleocene shales, Ypresian limestones with globigerinids, and Lutetian shales with yellow carbonate concretion (Fig. 16.3).

### 16.2.4.4 Oligocene-Miocene of Tunisia

Tectonic activity during the late Oligocene and early Miocene resulted in erosion and non-deposition. The Miocene in Tunisia shows a complex distribution of sedimentary sequences as a result of the interplay of tectonic, eustatic and climatic factors (Mannaï-Tayech, 2009). The dating and correlation of certain formations are not yet resolved (Mannai-Tayech, 2009), due to the lack of a confirmed biostratigraphic framework. Moreover, there are marked differences in nomenclature between Gafsa-Metlaoui Basin in Southwestern Tunisia, Central Tunisia, and Northwestern Tunisia.

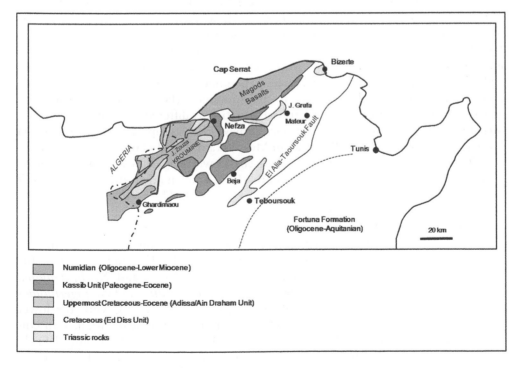

*Figure 16.3* Map of Numidian Flysch (after Talbi et al., 2008 and Sami et al., 2010) (See color plate section).

The Oligocene-Miocene sediments overlie unconformably the Upper Eocene units, such as the Souar, Djebs, and Cherahil formations. The Oligocene unconformity is marked by a shale horizon with interbedded fine-grained sandstones and some nummulitic limestones called the *Nummulites vascus* Horizon (Bishop, 1975, Burollet et al., 1978, Salaj, 1978, Fournié, 1978). The *Nummulites vascus* marker bed has been recognized in the Dirbal and Ras Abd Jalil formations in Libya (Hammuda et al., 1991), the Labrak Formation in NE Libya (Tawadros, 2001), and the Ketatna Formation in the offshore of Cyrenaica, NE Libya (Duronio, 1993, 1996). The *Nummulites vascus* Horizon is overlain by a succession of Oligocene-Lower Miocene terrigenous to marine sediments of the Fortuna Formation, Ketatna Formation, and the Salammbo Formation, in addition to the Numidian Flysch in Northern Tunisia. The lithostratigraphy of Miocene sediments in Tunisia has been reviewed recently by Mannaï-Tayech (2009).

Oligocene-Lower Miocene facies vary from shallow-marine to basinal. The most landward unit is the Fortuna Formation (Klett, 2001). Nearshore marine rocks are represented by limestone of the Ketatna Formation, which interfingers with offshore mudstones of the Salammbo Formation (Schwab, 1995). The ages of all these formations range from Oligocene (Rupelian) to Miocene (Burdigalian) (El Euchi et al., 1998). The Fortuna, Ketatna and Salammbo Formations are equivalent to the Dirbal and Ras Abd Jalil Formations in Libya (Hammuda et al., 1991). Basal Miocene sediments are commonly conglomeratic and unconformably overlie older sediments (Burollet, 1967).

The Fortuna Formation (Oligocene-Lower Miocene) was introduced by Burollet (1956) as the *Formation des grès de Fortuna*. It was assigned an Oligocene age by Burollet (1956). It was assigned an Oligocene-Aquitanian age by Fournié (1978, 105–106), and an Upper Oligocene (Rupelian)-Lower Miocene (Aquitanian) by El Euchi et al. (1998). It consists of a lower part of interbedded shales and sandstones, and an upper part of massive sandstones. The upper part of the formation is made up of continental red beds attributed to the Aquitanian (Fournié, 1978, p.106). The geographical distribution of the Fortuna Formation is different from earlier Tertiary formations and reflects a new tectonic framework (Bishop, 1975). The Fortuna Formation is overlain by the Burdigalian Messiouta or Oued Hammam formations.

The Ketatna Formation (Upper Oligocene-Lower Miocene) was introduced by (Fournié, 1978, p.106–109, after Bismuth, 1973) and named after the type well Ketatna-1 (10°55′28″E, 34°55′33″N), drilled in 1964. The type section has been described in detail by Fournié (1978, p.108). It was the first well to encounter a carbonate Oligocene-Miocene section unknown from outcrops. It is the carbonate equivalent of the Fortuna Formation sandstones. The carbonates are rich in bryozoans, amphistegenes, operculinids, and lepidocyclinids. The middle part of the Formation contains the lepidocyclinids *Eulepidina dilatata* and *Nephrolepidina tournoueri*. The shaly parts yielded specimens of *Preoperculina* and *Globigerinoides sicanus* attributed to the Lower Langhian (Fournié, 1978, p.109), while the lower part contains Oligocene Nummulites. The Ketatna Formation was assigned an Oligocene-Lower Langhian age by Fournié (1978, p.109). On the other hand, El Euchi et al. (1998) restrict it to the Rupelian-Aquitanian interval. The Ketatna Formation is overlain by the Burdigalian Ramla Formation. It is overlain by the Ain Grab Formation (Fournié, 1978, p.106).

The Salammbo Formation (Oligocene-Langhian) was introduced and named by Fournié (1978, p.103) as "the Salammbo Clays and Limestones Formation" after the type well Salammbo-1 (SLB-1) (11°41′12″E, 33°55′53″N) drilled in 1972 in the Gulf of Gabes. This formation is not known in outcrops. It consists of two units of shales. The lower unit is made up of greenish grey to beige shales, associated with fine argillaceous micritic limestones. The upper unit is composed of sandy shales. The two units contain planktonic foraminifera and represent open-marine sediments. The formation has a thickness of 671 m (2202 ft) in its type well. It was assigned an Oligocene-Langhian age based on the fossils *Nummulites vascus* at the base, followed upward by *Globorotalia opima opima*, *Globigerina* cf. *ampliapertura*, *Globorotalia kugleri*, *Cyclammina* sp., *Marginulinopsis* sp., *Globigerina ciperoensis*, *G. angulisaturalis*, *G. dehiscens*, and *G. dissimilis* at the top (Fournié, 1978, p.109–111). It overlies the Eocene Cherahil Formation and is overlain by the shales of the Mahmoud Formation.

The Middle Miocene (Langhian-Serravallian) is represented by regionally transgressive limestones of the Ain Grab Formation and marls of the Mahmoud Formation, the Messiouta Formation, Oued Hammam Formation, and Ramla Formation. Other formations include the Grijma Formation, Oued Hajel Formation, and the Halluf Formation.

The Ain Grab Formation (Burdigalian) was introduced by Burollet (1956) for a highly fossiliferous limestones, with conglomerates at the base. In the well Ketatna-1, it overlies the Ketatna Formation. It marks the Miocene transgression on top of the continental Oligocene sediments. It was placed in the Lower Langhian by El Euchi et al. (1998). The Ain Grab Formation rests directly on the Salammbo Formation. The Fortuna Formation is separated from the Ain Grab Formation by the Messiouta or Oued Hammam formation, and the Ketatna Formation is separated from it by the Ramla Formation. In the offshore of Tunisia, it is identified as Horizon H2 (Langhian) on seismic sections (Bédir et al., 1991, Bobier et al., 1991). The Ain Grab and Mahmoud formations are included in the Cap Bon Group by Fournié (1978, p.106).

The type section of the Mahmoud Formation is located at Djebel Derhafla (Marabout of Sidi Mahmoud), where it consists of 40 m (131 ft) of marls and grayish green shales, locally gypsiferous with sandy and glauconitic intercalations in the lower part and fine argillaceous sandstones with bryozoans, mollusks, and gastropods in the upper part (Fournié, 1978, p.106). The Mahmoud Formation has been dated Langhian-Serravallian by Salaj & Stranik (1970, in Fournié, 1978, p.106), based on the planktonic foraminifera *Orbulina suturalis*, *O. universa*, *Preorbulina glomerosa*, and *Globigerinoides sicanus*.

The Ain Grab and Mahmoud formations have included in the Cap Bon Group. The Group spans the Burdigalian-basal Tortonian age according to Biely et al. (1972, in Fournié, 1978, p.106) and the upper Burdigalian-lower Serravallian according to Robinson & Wiman (1976, cited by Wiman, 1980).

Messiouta Formation (Burdigalian) overlies the Fortuna Formation and is overlain, probably unconformably, by the Ain Grab Formation. The Messiouta Formation was considered Aquitanian in age by Biely et al. (1972), Robinson & Wiman (1976) and Wiman (1980, Fig. 2). It is partly equivalent to the Sehib Formation (Aquitanian-Burdigalian).

Oued Hammam Formation (Burdigalian) overlies the Fortuna Formation and is equivalent to the Messiouta Formation.

Ramla Formation (Burdigalian) is equivalent to the Messiouta and Oued Hammam formations. It overlies the Ketatna Formation.

The Late Miocene to Early Pliocene interval was a time of widespread tectonic activities in Northwestern Tunisia (Wiman, 1980, Burollet, 1991). According to Wiman (1980), there is no evidence of active erosion and channel cutting as it was in Libya and Egypt. The uppermost Miocene rocks include the lower part of the Segui Formation and lagoonal and brackish-water sediments, including anhydrites (Jongsma et al., 1985) of the Messinian Oued Bel Khedim Formation in Tunisia and the Marsa Zouaghah Formation in Libya (Burollet et al., 1978, Salaj, 1978, Hammuda et al., 1985). The youngest known Miocene deposits in the southern Dorsale in Northeastern Tunisia are the marls of the Saouaf Formation (Middle Miocene) exposed near El Haouaria on the Cap Bon Peninsula (Wiman, 1980). Biely et al. (1972) extended the top of the Saouaf Formation to the latest Miocene time. Robinson & Wiman (1976) placed the top of the formation in the upper Tortonian on the bases of planktonic foraminifera.

The Upper Middle (Serravallian) and Lower Upper (Tortonian) Miocene sediments show a more complex stratigraphy. The succession includes the Beglia, Birsa, Zelfa, Saouaf, Sehib, Kef Ettir, Oum Douil, Somaa, Melqart, and Oum Douil formations.

The Melqart (Melquart) Formation (Tortonian) was introduced by Fournié (1978, p.103–105) as the *Formation des calcaires de Melqart* for a subsurface carbonate unit in the Gulf of Gabes. It is made up of seven units (12–56 m in thickness) composed of bioclastic limestones rich corals, bryozoans, mollusks, and algues, intercalated with grey gypsiferous shales. The type well is the Melqart-1(11°38′05″E, 34°13′18″) drilled in 1972. In the type well, the Melqart Formation overlies the Beglia Formation and is overlain by the Messinian Oued Bel Khadem Formation, and has a thickness of 242 m (794 ft). It was assigned a Tortonian age by Fournié (1978) based on the faunal association *Ammonia* gr. *Beccardii*, *Borelis melo*, *celanthus*, *Pararotalia*, *Elphidiella*, *Hemicyrhere defomis*, and *Parakarithe*. Bryozoans were described by Moisette (1997) from wells in the Gulf of Gabes. It is equivalent to the more marine Saouaf Formation. The Melqart Formation was also described in the Cap Bon area (Amari & Bédir, 1989, Bédir et al., 1996). The Melqart Formation was assigned a Messinian age by Moissette (1997), Mannaï-Tayech (2009) and Bismuth et al. (2009) in the Ashtart-28 well in the Gulf of Gabes.

In the Phosphate and Metlaoui-Gafsa areas, the Neogene succession is approximately 600 m (1969 ft) in thickness and consists of sands, shales, silts, conglomeratic lenses, and rare lacustrine limestone and gypsum beds (Mannaï-Tayech, 2009). The succession overlies unconformably Eocene and Cretaceous sediments. It includes three formations; the Sehib, Beglia and Segui (Burollet, 1956).

Sehib Formation (Aquitanian-Burdigalian) is made up of siltstones, red shales, and conglomerates, and bounded by two major erosional surfaces. It has a thickness of up to 100 m (328 ft) (Mannaï-Tayech, 2009). It can be correlated with the Messiouta Formation in Central Tunisia. Its age may extend from Oligocene to Lower Serravallian (Mannaï-Tayech, 2006).

Beglia Formation is largely exposed in the Metlaoui-Gafsa area in Southwestern Tunisia (Fournié, 1978, p.103, Boujamaoui et al., 2000). It has a thickness of up to 600 m (1969 ft). It rests unconformably on the Sehib Formation and is overlain by the Kef Ettir Formation (Mannaï-Tayech, 2009). The Formation consists of fluvial coarse-grained, yellow sandstones with quartz pebbles, with conglomerates at the base. It contains remains of mammals in the lower part. The upper part of the Beglia Formation changes eastwards into thin lacustrine limestone beds. Pyrite fills the pore space of the sandstones and jarosite occurs between the quartz grains and their silica overgrowths. Mineralization probably took place in the Upper Miocene (Serravallian-Tortonian) (Decrée et al., 2008). Bruyère et al. (2010) attributed mineralization to downward-moving undersaturated meteoric fluids along the faults which carried with it pre-Miocene Miocene base metal sulphides. Vertebrate fauna from the Beglia Formation at Oued Thelja (Metlaoui), Sehib and Sidi Bouhlel in the Gafsa Metlaoui Basin occur in sands with shale clasts. The assemblage contains sparse sirenians, Crocodylus, remains of Palaeotragus (Giraffidae), Gazella (Bovidae), Hipparion (Rhenoceratidae), Perissodactyla, Miotragocerus (caprine antelope), Proboscidea (angulidae), Nile perch (Teleostidae), catfish, and antelopes, such as *Pachytragus solignaci*. It has been assigned an Upper Serravallian-Lower Tortonian based on vertebrates, especially the caprine antelope *Miotragocerus* (Mannai-Tayech, 2009).

The Beglia and Saouaf formations were grouped in the Oum Douil Group by Biely et al. (1972, 1974, cited by Fournié, 1978, p.103). Originally, Burollet (1956) introduced the Oum Douil Formation and divided it, in an ascending order, into the Mahmoud Shales, Beglia Sandstones, and Saouaf Lignitic Shales and Sandstones.

The Segui Formation (Miocene-Pliocene) was introduced by Burollet (1956) for a unit composed of yellow or ochre sandstones at the base, followed upward by interbedded sandstones and shales with conglomerates. The formation becomes more conglomeratic upwards. The sandstones show southward-oriented paleocurrents and form fining-upward channel sequences (Mannaï-Tayech, 2009). The are interpreted as brackish water deposits that accumulated on the south side of the Atlas Mountains during the collision of the African and European plates. At least part of the Segui Formation may be of marine origin (Swezey, 2009). The formation was assigned a Miocene-Pliocene age by Fournié (1978). However, the foraminifera suggest a Tortonian and/or Messinian age. In the Gafsa-Metlaoui Basin in Southwestern Tunisia, the Segui Formation overlies the Beglia Formation and is overlain by Villafranchian conglomerates (Mannaï-Tayech, 2009). In the Kef Ettir's (Et Tir) area in Northern Tunisia, the upper part of this Formation is made up of ochre-colored, unfossiliferous siltstones, with conglomerate, sandstone and gypsum intercalations. These deposits in the Gafsa-Metlaoui region are always organized in decreasing graded bedded sequences. The Segui Formation overlies the Beglia Formation and is capped by Pleistocene conglomerates. It was correlated with the sandstones of the Somaa Formation in the Cap Bon Peninsula and assigned a Lower Tortonian-Upper Messinian age by Mannaï-Tayech (2009). Mallouli et al. (1987) reported vertebrate fragments of skulls, dorsal plates, vertebras, and teeth of crocodiles, the mammifer Perissodactyles (Rhinoceras), *Hipparion* cf. *primigenium*, *Merycopotamus* sp., Giraffides, *Pachytragus* sp., *?Prostrepsiceros* sp., bovides, and the Proboscidien *Tetralophodon* cf. *longirostris*. In Central Tunisia the formation yielded vertebrate fossils of Upper Miocene age, including numerous fishes, several reptiles, a Sirenian, a large *Tetralophodon*, Hipparion, a horned Rhino, Bovids, a Giraffid, an Anthracothere, a possible *Nyanzachoerus*, in addition to *Choerolophodon* and *Brachypotherium* (Geraads, 1989). The Segui Formation of Central and Southern Tunisia is probably the equivalent of both the Hergla and Melqart Formations encountered in the HGA-1 well in the Gulf of

Hammamet (Mannaï-Tayech, 2009), which in turn are the lateral equivalents of the Somaa Formation and Beni Khiar Formation of the Cap Bon Peninsula. The Melqart and Beni Khiar Formations are confined to the present coastline.

The Kef Ettir Formation is made up of green clays, usually sandy, gypsiferous, with carbonate intercalations. It crops out at Kef Ettir "Et Tir", Oued Hachana, Bled Douarah, and Henchir Souatir in Southwestern Tunisia (Mannaï-Tayech, 2009). These facies have been assigned a Tortonian and at least partly a Messinian age since the underlying Beglia sands were dated Serravallian to Lower Tortonian based on their vertebrate faunal content (Robinson & Black, 1974, Mannaï-Tayech and Otéro, 2005). Locally, the clay-rich lithofacies bears shark teeth, coprolites, Helicidae, *Elphidium hauerinum* and *Ammonia beccarii* (Biely et al., 1972, Mannaï-Tayech, 2006). The clay-rich facie was assigned to the Saouaf Formation by Biely et al. (1972). Mannaï-Tayech (2009) argued that neither the stratigraphic position nor the lithology of this clay-rich unit support its the inclusion in the Saouaf Formation and proposed a new formation, the Kef Ettir Formation, named after the locality.

North of the Dorsale in Northeastern Tunisia, the Tortonian is represented by the Ketchabta Formation (equivalent to the Beglia Formation) (Wiman, 1980).

The Messinian in Tunisia is represented by two formations (Burollet et al., 1978, Salaj, 1978); the Oued Bel Khedim and part of Segui Formation and the Ketchabta Formation in Northern Tunisia.

The Oued Bel Khedim Formation (Messinian) was introduced by Burollet (1951) as the *Formation des marnes et gypses de l'Oued Bel Khedim* for a group of lagunal and lacustrine sediments of Uppermost Miocene (Messinian) age (Fournié, 1978, p.103). Equivalent sediments in at Sidi Driss and Douahria in the Nefza mining district, Northern Tunisia, contain sulfide ore deposits (Decree et al., 2008). The Oued Bel Khedim Formation overlies the Ketchabta Formation. It is composed of bedded evaporites, freshwater limestones, diatomites, and marls with shallow-water benthonic foraminifera, ostracod fragments, and reworked planktonic foraminifera (Wiman, 1980). The Oued Bel Khedim Formation is overlain by marls of the Raf Raf Formation. The Oued Bel Khedim Formation is equivalent to the Marsa Zouaghah Formation in the offshore of Libya (Hammuda et al., 1985).

The Atlas Phase was initiated in response to movements of the African plate against the Eurasian plate, which induced collision in Northwest Africa during the Early Neogene (Cohen et al., 1980, Guiraud et al., 2005). It has been dated from Upper Miocene to basal Pliocene (Cohen et al., 1980, Talbi et al., 2008) and is manifested by thrusts and deformations in Northwestern Tunisia. During the Middle Miocene Atlas Orogeny, the rocks of the Tunisian Trough were deformed into a series of NE-SW folds deformed. These folds were extensively faulted during the Late Miocene and Pliocene (Snoke et al., 1988). At the end of this phase, magmatic activities changed from acidic calc-alkaline to basic alkaline. Post-orogenic extension was accompanied by the intrusion of basaltic rocks with alkaline affinity, such as those in the Nefza and Mogods areas (Talbi et al., 2008) (Fig. 16.1). Late Oligocene sands accumulated unconformably on the older carbonate rocks in the Tunisian Basin (Snoke et al., 1988).

The Sidi Driss and Douahria sulfide ore deposits of the Nefza mining district, Northern Tunisia, are hosted in Upper Miocene (Messinian) deposits equivalent to Oued Bel Khedim Formation (Decrée et al., 2008). They occur as void-filling within carbonate lenses of Fe-Mn-enriched dedolomites, partially or totally replaced by barite and celestite. The ores consist mainly of galena and spherulitic/colloform sphalerite and partially replaced by later Fe sulfides in the form of disseminated or banded ores. They are believed to have formed through a process of dissolution and replacement. The remaining voids are filled by late calcite and barite.

In Northern Tunisia, the nappe pile is composed of the Kasseb and Ed Diss thrust sheets (Upper Cretaceous-Eocene) and is overlain by the Numidian Nappes (Rouvier, 1977). The latter consists of more than 1000 m (3280 ft) of siliciclastic sediments composed of sandstones and shales of Oligocene to Burdigalian age. The nappe pile is cross-cut by felsic plugs and mafic dikes, sills, and basaltic flows of the Late Miocene and Pliocene Nefza magmatic province (Jallouli et al., 2003). The small Sidi Driss-Tamra and Douahria basins superimpose the eroded nappe pile. The Sidi Driss Basin is a syncline with NE-SW-oriented syn-sedimentary faults, with a sedimentary fill dominated by marls, with several intercalations of conglomerates, limestones, and evaporites. A volcaniclastic horizon in the lower part contains rounded pebbles of calcitized rhyodacite (Dermech, 1990) and clasts from the nappe substratum (Numidian Sandstones) and the Oued Belif metamorphosed Triassic shales. The top of the pile is a supratidal limestone horizon of variable thickness of stromatolitic and micritic facies. Lignite also occurs in the upper part of the succession. The formations of the Sidi Driss Basin are unconformably overlain by the ferruginous formations of the Mio-Pliocene Tamra Basin (Decrée et al., 2008). The Douahria Basin is a small graben filled with marls with numerous volcaniclastic intercalations, and shales and lenticular limestone intercalations. The shales contain remnants of a terrestrial mammal fauna dated from the latest Messinian (Roman & Solignac, 1934, Jaeger, 1977). The upper part of the volcaniclastic unit contains silicified ashes and tuff levels and ferruginous impregnations. The latter is known as the "ferruginous formation" (Negra, 1987), which was later called the Tamra Ferruginous Formation by Decrée et al. (2008). The base of this "ferruginous formation" contains a thin lenticular shallow marine limestone horizon composed of micritic and anhydrite-bearing stromatolitic limestone which hosts the Douahria Pb-Zn ores (Decrée et al., 2008).

The Numidian Flysch Domain is composed of a thick sequence of Oligocene-Lower Miocene turbidites, with a thickness of more than 2000 m (6562 ft) (Talbi et al., 2008), deposited in a subsiding basin bounded by Oligocene rift faults (Frizon de Lamotte et al., 2004). The Numidian Formation has been considered as either a complex of allochtonous Numidian units with a tangential contact with the underlying units, or autochtonous with a sedimentary contact. The allochthonous nature of these units has been challenged by Ould Bagga et al. (2006). Talbi et al. (2008) argued against a tangential tectonic contact at the base of Numidian series, whereas Sami et al. (2010) supported the allochthonous position of the Numidian complex and its displacement southward.

The Oligocene-Miocene Numidian Flysch of Northern Tunisia (Sami et al., 2010, Talbi et al., 2008) (Fig. 16.3) has been divided into three distinct lithostratigraphic units considered as vertically superimposed. The Numidian Flysch was sub-divided into three members by Rouvier (1977, Sami et al., 2010); the Zouza Shales, 1500–2000 m (4922–6560 ft) thick at the base, the Kroumirie Sandstones, 1000–1200 m (3281–3937 ft) thick in the middle, and the Babouch Shales, 300 m (984 ft) thick at the top. The first two members are mostly Oligocene in age and the Babouch

Member is Lower Miocene (Sami et al., 2010). Palynological studies by Torricelli & Biffi (2001) suggested that Zouza and Kroumirie members are age equivalent rather than stratigraphically superimposed. Planktonic foraminiferal analysis by Sami et al. (2010) agrees with the palynological results. These three units represent the principle facies of the Tunisian Tell (Talbi et al., 2008).

### 16.2.4.5  Pliocene of Tunisia

Lower Pliocene sediments were deposited during the marine transgression that followed the Messinian Salinity Crisis and the opening of Mediterranean. The top of the Miocene is marked by an angular unconformity (Bishop, 1975, Burollet et al., 1978). In the meantime, local subsidence occurred in troughs and basins, in which thick accumulations of terrigenous and marine clastic sediments were deposited (Bishop, 1975, Burollet et al., 1978). Pliocene Formations include the Raf Raf Formation (Lower Pliocene) and Porto Farina Formation (Upper Pliocene (Fournié, 1978, p.101), both are lateral equivalents to the upper part of the Segui Formation (Miocene-Pliocene) in Tunisia and the Sbabil Formation in Libya.

The Raf Raf Formation (Burollet, 1954) is a thick unit of grey marls, occasionally brown or green. It starts with a transgressive facies marked by conglomerates at the base. It grades upward into the sandstones of the Porto Farina Formation. The oldest fossiliferous Lower Pliocene marine deposits that crop out in the Nabeul quarry on Cap Bon, south of the Dorsale; belong to the Raf Raf Formation. The benthonic and planktonic foraminifera indicate an Early Pliocene (Zanclean) to lower Pliocene (Piacenzian) (Wiman, 1980). The benthonic forms include *Bulimina aculeate, Cibicidoides robertsonianus, Planulina ariminensis, Pullenia bulloides*, and *Sphaeroidina bulloides* (see Wiman, 1980, Fig. 3, for a list of fossils from the Nabeul section). Early-Middle Pliocene mammal localities in Tunisia were studied by Arambourg at Lake Ichkeul and Ain Brimba.

Porto Farina Formation (Burollet, 1954) is a thick yellow or reddish brown sandstone unit rich in pectinids and ostreids, grading into eolian and continental sediments of the basal Quaternary.

The six Mediterranean biozones, previously established by Cita (1975), have been identified in the Nabeul-Hammamet unit by Derbel-Damak & Zaghbib-Turki (2002). They are the *Sphaeroidinellopsis subdehiscens, Globorotalia margaritae margaritae, Globorotalia margaritae, Globorotalia puncticulata, Sphaeroidinellopsis subdehiscens, Globigerinoides elongates*, and *Globorotalia inflata* biozones. They were identified in argillaceous and yellow sandy lithological units in Northeastern Tunisia, such as the Potier Shale (*Argiles des Potiers*), Nabeul Yellow Sands (*Sables Jaunes de Nabeul*), and the Sidi Barka Shales (*Argiles de Sidi Barka*). These units combined are equivalent to the Raf-Raf Formation developed in the Bizerte area toward the North. The last two biozones belong to the sands and sandstones of the Hammamet Unit, equivalent to the Porto-Farina Formation.

### 16.2.4.6  Quaternary

Quaternary deposits include clastic terrigenous and marine sediments (Burollet et al., 1978), as well as red beds and caliche. Middle to late Pleistocene marine deposits discontinuously crop out in the Cap Bon Peninsula. They consist of raised terrace deposits and coastal dunes. They have been investigated recently by Elmejdoub & Jedoui (2009) and Elmejdoub et al. (2011). These terraces have been attributed to the interplay between the Pleistocene sea-level changes and tectonic uplift during the Quaternary (Elmejdoub & Jedoui, 2009). They are composed of mixed carbonate and siliciclastic bodies and unconformably overlie a Miocene-Pliocene and Quaternary slope succession. The sedimentary sequences represented by mixed carbonate and siliciclastic bodies are related to the succession of highstand shorelines and contrasting climate conditions. The intercalated soils and calcretes occurred during episodes of relative sea-level fall. They occur at elevations of about 100 m, 60 m, and 40 m (328, 197, and 131 ft). The coastal dunes (eolianites) along the western littoral of the Cap Bon Peninsula (Northeastern Tunisia) probably formed under a northwesterly wind regime and supplied by sand from the shore in association with a sea level lower than today. The ages for these eolianites range from $112 \pm 10$ to $53 \pm 2$ ka (Elmejdoub et al., 2011). In the Chott Basin, Quaternary sediments of about 12,000 years old rest unconformably on Miocene strata (Swezey, 2009). Marine deposits 35,000–25,000 years old are recorded in the Chott el Djerid (Richard & Vita-Finzi, 1982). These deposits suggest a marine connection during that period, and in variance to previous ideas and imply uplift by some 80 m (262 ft) in an area long regarded as either stable or subsiding throughout the Quaternary.

Chapter 17

# Phanerozoic geology of Morocco

## 17.1 INTRODUCTION

Morocco is located in the northwest corner of Africa. It is bounded by the Western Sahara to the south, Algeria to the east, the Mediterranean (Alboran Sea) to the north, and the Atlantic Ocean to the west. It has an area of 458,730 km². Morocco is separated from Spain by the Strait of Gibraltar, a narrow water passage with a width of 13 km in its narrowest point and bordered by the Pillars of Hercules (Fig. 17.1).

Morocco is the world's third leading exporter of phosphate rock after China and the United States (Jasinski, 2008, Jasinki et al., 2008), although exports have declined since the early 1990s. Estimated phosphate reserves are 84,000 billion tons, and represent nearly 40% of global reserves (Vaccari, 2009), and annual production of 19 million tons (Bakkali, 2006). In addition, the country produces a wide variety of minerals, which include silver (Morocco's second most significant mineral product[1]), barite, clays, coal, cobalt, copper, fluorspar, gold, iron ore, lead, nickel, salt, talc, and zinc. Morocco has oil reserves of about 1 million barrels and gas reserves of about 60 billion cubic feet (Bermúdez-Lugo, 2005), and is the only North African country with no significant oil or gas production. Small amounts of oil and natural gas are produced from the Essaouira Basin and small amounts of natural gas are produced from the Gharb (Rharb) Basin (Fig. 17.1).

The geology of Morocco has been treated extensively by Michard (1976), Petters (1991), Piqué (2001), and more recently by a large group of experts in the field whose work was compiled and edited by Michard et al. (2008). A brief, but useful, introduction can be found in the recent book by Schlüter (2008). The amount of geoscientific information on Morocco is tremendous and extremely complicated; first because of the presence of the Atlas Mountains, and secondly due to the scarcity of an established stratigraphic nomenclature over large areas. There is no lexicon of Moroccan stratigraphic nomenclature. The natural recommendation is then to form a stratigraphic committee for Morocco whose main objective is to compile a comprehensive lexicon of stratigraphic nomenclature similar to those of Banerjee (1980) for Libya and Fournié (1978) for Tunisia.

The physiography of Morocco is dominated by the Atlas System (Fig. 17.1, Table 17.1). The Atlas Mountains belt trends in an ENE direction and extends from the Atlantic coast of Morocco to Tunisia. The chain consists of a series of mountainous belts, namely the Western High Atlas, Central High Atlas, Middle Atlas, Eastern

---

1 Silver deposits at Zgounder are reviewed by Petruk (1975).

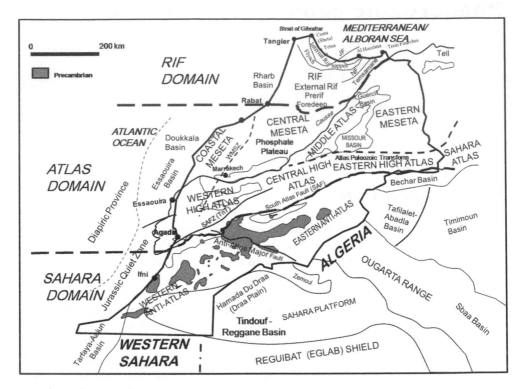

*Figure 17.1* Tectonic map of Morocco. Note the direction of the Ougarta Range and the Tafilalt-Timimoun basins, almost perpendicular to the Anti-Atlas Mountains direction or at high angle to the Atlas Mountains. Compiled and modified from Michard et al. (2008), Frizon de Lamotte et al. (2008), Piqué (2003), and Askri et al. (1990). Atlantic margin provinces are after Holik & Rabinowitz (1992).

High Atlas, and the Saharan Atlas. The Saharan Atlas continues into the Aurès Mountains, which straddle the Algerian-Tunisian border. The Middle Atlas branches out in a NE direction from the Central Atlas Mountains. The Atlas System also includes a number of plateaus, such as the Coastal Meseta, Central Meseta, and Eastern Meseta (also called the High Plateau or Oran Meseta), as well as the Middle Atlas "Causse" (the name is derived from a similar group of limestone plateaus, between 700–1200 m (2297–3937 ft) of elevation in the *Massif Central* of southern France). The highest peak of the Atlas Mountains of 4167 m (13,671 ft) is located in Morocco at Jbel Toubkal[2], about 63 km south of Marrakech (Marrakesh). Djebel Aissa is the highest peak in the Algerian Atlas at 2236 m (7336 ft), while Djebel Chelia is the highest point in the Aurès Mountains with an elevation of 2328 m (7638 ft). To the north, the Rif Range and the Tellian-Kabylian belt (Maghrebides), with their allochthonous terrains, extend along the Mediterranean coast (Alboran Sea). From the tectonic point of view, Morocco is located at the triple junction between the African continent, the

2 Jbel is the preferred spelling in Morocco, although Djebel and Jebel are occasionally used.

*Table 17.1* Tectono-stratigraphic units of Morocco.

| Age | Tectonic domain | Sudivisions | |
|---|---|---|---|
| I | West African Craton (WAC), Reguibat Shield & Pan-African Belt | Reguibat Shield: Two domains; Archean (west, Ghallaman region) & Eburian (east, Sfariat region) | Zag Basin<br>Reggane Basin<br>Taoudenni Basin |
| II | Saharan Platform (Stable Shelf) | | Tindouf Basin<br>Tafilalt Basin<br>Timimoun Basin, Bechar Basin & Ougarta Massif (mostly in Algeria) |
| III | Anti-Atlas | Western Anti-Atlas | Souss Basin<br>Ouarzazate Basin |
| | | Central Anti-Atlas | |
| | | Eastern Anti-Atlas | Draa Plain |
| IV | Mesetas | High Plateau or Oran Meseta | |
| | | Western (Moroccan) Meseta | |
| | | Central Meseta, Phosphate Plateau & "Causse" | Saiss Basin<br>Phosphate Plateau |
| | | Eastern Meseta | Guercif Basin<br>Missour Basin<br>Moulouya |
| | | Coastal Meseta | Doukkala Basin<br>Rharb Basin<br>Argana Basin<br>Oued Zat Basin |
| V | Atlas System | Western High Atlas | Essaouira Basin<br>Agadir Basin<br>Souss Basin<br>Tizi n'Test Basin |
| | | Central High Atlas | Bahira-Tadla Basin<br>Haouz Basin |
| | | Middle Atlas | |
| | | Eastern High Atlas | |
| | | Saharan High Atlas & Oujda Mountains or Tell Atlas | |
| | | Western Moroccan Arch (WMA) | |
| VI | Rif | Internal Rif | |
| | | External Rif | Prerif nappes<br>Mesorif<br>Intrarif |
| | | Flysch | Messylian nappes<br>Mauritanian nappes |
| | | Foredeep | Rharb (Gharb) Basin |
| VII | Atlantic passive margin | | Essaouira-Agadir Basin<br>Tarfaya-Aaiun Basin<br>Bas Draa |
| VIII | Mediterranean margin (Tethyan Domain) | Alboran Sea<br>Gibraltar Strait | |
| IX | Cenozoic volcanics | | |

Atlantic Ocean, and the Alpine subduction-collision zone (Michard et al., 2008b). It shares a common tectonic history with southern Spain for the most part.

This particular location makes Morocco vulnerable to earthquakes, some with devastating results, such as the Aghadir (Agadir) Earthquake (1960) (Mridekh et al., 2009), Rissani Earthquake (1992), and Al Hoceima Earthquake (2004) (magnitude 6.3) (Tahayt et al., 2009). Morocco still exhibits significant tectonic activities, especially in the Rif and the Alboran Sea, which are located in the contact zone between two converging plates, Eurasia and Africa. The active zone follows the Algerian coast, crosses obliquely the Alboran Basin from the eastern Rif to the Eastern Betic Cordilleras (Trans-Alboran Seismic Zone, TASZ), and continues to the west in the Atlantic Ocean (Michard et al, 2008).

## 17.2  TECTONIC SUBDIVISIONS OF MOROCCO

## 17.3  SAHARA PLATFORM

The Sahara Platform includes the northern part of the Mauritinides, such as the Ouled Dlim and the Zemmour areas, as well as the Anti-Atlas and the Ougarta Massif. The latter has been interpreted as an aulacogen by Donzeau (1974). The Sahara Platform also encompasses a number of intracratonic basins, such as the Bechar Basin, Tafilalt-Abadla Basin, Timimoun Basin, Tindouf Basin, Zag Basin, Reggane Basin, Taoudenni Basin, Sbaa Subbasin, and Ahnet Basin. These basins are located mainly in Algeria and/or Mauritania, but extend into Morocco. The thickness of the sedimentary successions in these intracratonic basins reaches about 8 km (5 miles) in the Taoudenni Basin and up to 10 km (6.3 miles) in the Tindouf Basin. Detailed description of these basins can be found in the chapter on Algeria.

## 17.4  ANTI-ATLAS

Reviews of the geology of the Anti-Atlas have been presented by Michard (1976), Soulaimani et al. (1997), Piqué (2001), Ennih & Liégeois (2001), Thomas et al. (2002, 2004), Burkhard et al. (2006), Soulaimani & Burkhard (2008), and Frizon de Lamotte et al (2008). The Anti-Atlas Basin is an intracratonic basin similar to other basins in the stable platform of North Africa, such as the Tindouf, Reggane, Bechar, Ahnet, Ghadames, Illizi, Hamra, Murzuq (Boote et al., 1998), and the Nile Basin (Tawadros, 2001).

The Anti-Atlas is a SW-NE trending mountain chain, about1000 km long and 200 km wide, located on the northern border of the West African Craton (WAC) in Southern Morocco (Fig. 17.1). The chain is bordered in the north by the High Atlas and in the south by the Moroccan and Algerian Sahara. The South Atlas Fault (SAF) and the Tizi n'Test Fault mark the limit between the Anti-Atlas and the High Atlas domains. The nature of the Tizi n'Test Fault is controversial. Mattauer et al. (1972, 1977) postulate strike-slip movements along the fault during the early Mesozoic. Laville & Piqué (1991) suggest the formation of a series of Variscan pull-apart depocenters bounded to the south by the Tizi n'Test Fault. On the other hand, Jenny (1984), Schaer (1987), and El Kochri & Chorowicz (1996) suggest limited strike-slip movements. Frizon de Lamotte et al. (2008) and Toto et al. (2008) did not identify transfer faults and emphasize the role of tensional tectonics in the formation of the Middle to Late Jurassic depocenters.

The closure of the Paleotethys Ocean (Stampfli & Borel 2002) during the Variscan Orogeny produced the collision between Laurussia and Gondwana (Beauchamps et al., 1997). This compression led to the inversion of normal faults, uplift of the Precambrian basement, and folding of the overlying cover. However, the Anti-Atlas was mildly deformed during the Variscan Orogeny. Metamorphism is extremely weak to non-existent (Soulaimani & Burkhard, 2008). The Anti-Atlas is considered as the southern limit of the Variscan chain (Piqué et al., 1993). The Anti-Atlas Basin have been interpreted as the southeastern half of an oblique back-arc basin that evolved into a passive margin in Silurian-Devonian times, following a right lateral strike-slip departure of the Hunic terrains.

The Western Anti-Atlas is dominated by highly cylindrical, upright fold trends. There is no evidence for any thin-skinned thrusting with the exception of the western-most parts of the Anti-Atlas along the Atlantic Coast (Belfoul et al., 2002) and some blind thrusting below the Ouarkziz Ridge. Helg et al. (2004) interpret the Ouarkziz ridge, slightly curved monocline of more than 400 km in length, as the surface expression of a triangular structure. Burkhard et al. (2006) postulate that the blind thrust ends below the Tindouf Basin (Fig. 17.1) and a major NNW-verging backthrust re-emerges within a thick series of shales above the Devonian Rich.

The transition between the Anti-Atlas and the Bechar and Reggane basins is obscured by a thick blanket of Upper Cretaceous sediments (Hamada) (Arthur et al., 2003). The Anti-Atlas appears to be the northern half (or the northern extension) of the Tindouf Basin (Fig. 17.1). However, the structure of the Tindouf Basin, the Timimoun Basin, and the Reggane Basin is nearly perpendicular to the Anti-Atlas trend. The Ouarkziz Chain (Mamet et al., 1966) represents the deformation front of the Anti-Atlas belt and the northern border of the Tindouf Basin. The Anti-Atlas belt changes gradually into the Ougarta Chain (Coward & Ries, 2003). The same structural trend is also present in neighboring basins of Reggane and Bechar, which was probably inherited from the Pan-African Orogeny. East of Tata, the Anti-Atlas folds decrease in number, amplitude, and tightness. There is also a change in the general orientation from SW-NE to E-W (Coward & Ries, 2003).

The Eastern Anti-Atlas differs structurally from the Western Anti-Atlas (Helg et al., 2004), regional change in the deformation style, intensity, and orientation is also observed from north (thick-skinned inversion style with ENE-WSW fold axis) to south (thin-skinned detachment folding with NW-SE fold axis) (Robert-Charrue & Burkhard, 2005). In the Eastern Anti-Atlas (Ougnat and Tafilalt areas), Cretaceous and Neogene sediments are tilted northward (Robert-Charrue 2006). The Alpine uplift (Pliocene-Pleistocene) of the Anti-Atlas Belt (Frizon de Lamotte et al., 20008 was partly due to a Neogene thermal uplift (Teixell et al., 2005, Missenard et al., 2006).

Many authors have interpreted the Anti-Atlas fold belt as a product of strike-slip movements (Mattauer et al., 1972, 1977, Piqué et al., 1991), while others depict the Western Anti-Atlas as a series of highly cylindrical frontal folds related to collision in a NNW-SSE direction (Helg et al., 2004, Soulaimani et al., 1997). Folding of the cover is dominated by doming and a complete absence of any thrusting and duplex structures (Helg et al., 2004).

The Anti-Atlas Domain probably formed by up-doming of marginal parts of the West African Craton (WAC). The Precambrian basement crops out in a series of inliers, called *boutonnières* (buttonholes) by Choubert (1963), such as the Bas Draa, Kerdous, Irherm, Zenaga, Bou Azzer-El Graara, Saghro (Figs. 5.14, 17.1). The Precambrian basement is represented by crystalline, metamorphic and sedimentary rocks (Ennih & Liégeois, 2001). Carbonates of Early to Middle Cambrian age are exposed around these Precambrian basement inliers. Because Cambrian carbonates are more resistant to erosion than the crystalline basement rocks, the Precambrian rocks form topographic depressions between Cambrian highs. Soulaimani & Piqué (2004) propose that the Anti-Atlas *boutonnières* are reactivated metamorphic core complexes. Burkhard et al. (2006) infer that they were associated with a series of major crustal reverse faults that resulted in 12 to 15% crustal shortening, which amounts to some 17 km. The boutonnières were uplifted during the Hercynian compressive phase in the Late Carboniferous-Permian (Beauchamps et al., 1997, Soulaimani & Burkhard, 2008). The Paleozoic cover is gently folded and was affected only by low grade metamorphism (Robert-Charrue & Burkhard, 2005).

## 17.4.1   Phanerozoic stratigraphy of the Atlas Mountains

The Anti-Atlas of Southern Morocco has an almost continuous exposure of Paleozoic rocks (Bourcart, 1927, Choubert & Faure-Muret, 1969, Lubeseder et al., 2003, 2009, 2010). Interestingly, Thomson (1899) thought that the Cretaceous sediments overlie the Precambrian with the complete absence of Paleozoic, Triassic and Jurassic rocks. The origin of the sediments is under debate (Burkhard et al., 2004). Their origin has been attributed to either Late Pan-African molasses deposition during plate collision, or to post-orogenic extension and collapse within tilted half-grabens (Leblanc et al., 1980, Soulaimani et al., 1997, Soulaimani et al., 2004). Tectonic studies support both hypotheses.

There were at least two Early Paleozoic rifting phases that took place along the northern border of Gondwana (Stampfli & Borel, 2002). Burkhard et al. (2006) proposed that the future Avalonian and Hunic terrains represent former active subduction margins off the Northwestern African Craton. Both terrains drifted away during two consequent episodes of back-arc rifting, which generated the Rheic (Cambrian) and Paleo-Tethys (Lower Silurian) oceans. Burkhard et al. (2006) assume that no rifting took place in the Tindouf Basin or in the Tafilalt Subbasin from the Middle Cambrian times onward.

The Paleozoic cover of the Anti-Atlas reaches more than 10 km (33,000 ft) in thickness in the Westernmost Anti-Atlas near Tiznit and decreases to less than 6 km (19,700 ft) in the easternmost Anti-Atlas of the Tafilalt (Burkhard et al., 2006, 2008). It is dominated by shallow marine fine-grained clastic sediments (Cavalazzi, 2006). Detrital input was from the WAC during Cambrian to Silurian times. The Lower to Middle Cambrian succession of the Eastern Anti-Atlas consists of sandstones, conglomerates, shales, and volcanic rocks (Destombes et al., 1985). Upper Cambrian deposits, which are common in the rest of North Africa, have not been found so far in Morocco. Ordovician strata are dominated by argillaceous rocks, which alternate with several hundred meters thick sandstones. Carbonate sedimentation took place from the end of Silurian to the end of Devonian (Döring & Kazmierczack, 2001).

### 17.4.1.1   Cambrian

The Adoudounian or the Ouarzazate Supergroup in the Anti-Atlas Domain is unconformably overlain by a Lower Cambrian post-rift sedimentary sequence composed of a thick sequence of carbonate-siliciclastic rocks with locally developed volcanic rocks of the Tata and Taroudant groups (Álvaro et al., 2008). The Precambrian-Cambrian boundary falls within the lowermost part of the sequence (Landing et al., 1998). These rocks developed within a gradually subsiding extensional foreland basin, the Tindouf Basin. The Tata and Taroudant groups are not considered to be part of the Anti-Atlas Supergroup, but as the lowermost units of the next supergroup cycle (Tindouf Supergroup).

The Bas Draa Valley in the westernmost part of the Anti-Atlas is located between the Bas Draa Inlier to the east and the Atlantic Coast to the west (Burkhard et al., 2006) (Fig. 17.1). Deformed Cambrian rocks form NNE-SSW-trending ridges (Choubert 1963, Soulaimani 1998, Belfoul et al. 2001). In outcrops, the main structure in the Cambrian rocks is a west-dipping penetrative cleavage. Along the Atlantic coast, Lower Cambrian strata form tight recumbent folds, trending NNE-SSW and overturned eastward. Their axial planes are oriented W-NW, for example, at the Plage Blanche[3]. Regionally, the deformation intensity decreases eastward, where the folds are cut by east-dipping thrust faults (Burkhard et al., 2006). For example, in the Bou-Jerif area, the Cambrian strata are folded with a westward steeply-dipping cleavage. Farther to the east, in the Guelmine area, the Middle Cambrian strata are only slightly tilted and lack cleavage.

The Middle Cambrian to Lower Devonian sediments in the Anti-Atlas Domain consist of shallow-marine fine-grained detrital deposits eroded from the West African Craton, followed by Middle Devonian platform carbonates (Soulaimani & Burkhard, 2008). The Early Carboniferous of the Eastern Anti-Atlas includes olistostromes that were probably deposited in an intramontane trough. The Late Visean, Namurian and Westphalian sequences of the Ouarkziz, Betana and Bechar areas in Morocco and Algeria represent a regressive megasequence (Soulaimani & Burkhard, 2008).

In the Anti-Atlas, the sedimentary cover and crystalline basement reacted differently to the Late Paleozoic Variscan/Hercynian Orogeny; the Paleozoic cover was affected by the reactivation of basement faults, while the crystalline basement remained relatively undeformed. The sedimentary cover, including the Late Precambrian volcaniclastic rocks of the Ouarzazate Supergroup, shows intense deformation at its base. On the other hand, the crystalline basement shows only reverse faults along the borders of the inlier (Burkhard et al., 2006, Beauchamps et al., 1997, Soulaimani & Burkhard, 2008). Many Anti-Atlas rocks have experienced a Variscan lower-greenschist facies metamorphism at 150–300° C (Buggisch, 1988, Soulaimani, 1998, Burkhard et al., 2001, Soulaimani & Burkhard, 2008) related to deep burial beneath 10 km of sediments. The thin-skinned tectonic style observed along the Atlantic Coast differs considerably from the thick-skinned tectonics throughout the Anti-Atlas Chain, where the involvement of the basement plays a critical role during the Variscan compressive deformations.

---

3 The Plage Blanche is an alluvial plain at the mouth of the Oued Draa River. The latter is an intermittent stream.

The Pan-African extensional tectonics continued into the Earliest Cambrian in the Western Anti-Atlas region (Algouti et al., 2001, Piqué, 2003, Benssau & Hamoumi, 2003, 2004, Buggisch & Flügel, 2006, Burkhard et al., 2006, Soulaimani & Burkhard, 2008).

The late Neoproterozoic Ouarzazate Group is overlain by the Lower Cambrian Adoudounian Formation (Group), composed of conglomerates at the base, followed upward by carbonates, marls, and sandstones (Algouti et al., 2001, Belfoul et al., 2001, Faik & Belfoul, 2001, Arboleya et al., 2008) (Fig. 17.2). The presence of the calcareous algae *Kundatia composita* (Buggisch & Flügel, 2006) indicates an Early Cambrian age for the Lower Adoudounian Formation. It is overlain by the Lie de Vin Formation.

The Lower Cambrian succession in the Western Anti-Atlas, such as in the Guelmine area (Belfoul et al., 2001), consists of carbonates at the base, which include the Upper Archeocyathids and Trilobites Limestones *"calcaires superieurs à archeocyathes"* and the Schists and Limestones Series *"Serie schistocalcaire"*, and an upper succession of brecciated schists and sandstones *"Schistes et grès terminaux"*. The Lower Cambrian sequence has been divided into the Taroudant and Tata groups. The Taroudant Group includes the Adoudounian and Lie de Vin formations. The overlying Tata Group consists of the Igoudine, Amouslek, and Issafene formations (Maloof et al., 2005, Álvaro et al., 2008) (Fig. 17.2). The first marine transgression across the rifted Pan-African margin marks the base of the Taroudant Group (Maloof et al., 2005).

The Precambrian-Cambrian boundary (Nemakit-Daldyan (ND)-Tommotian (Tm)) boundary at 525.5 Ma is characterized by an abrupt increase in diversity and abundance of calcifying organisms, such as the reef-building *Archaeocyatha* and the shelly fauna, accompanied by the appearance of vertical burrows, such as *Skolithos* and *Diplocraterion* (Maloof et al., 2006) (See also Sdzuy & Geyer, 1988 for discussion of the boundary). The first marine transgression across the rifted Pan-African margin

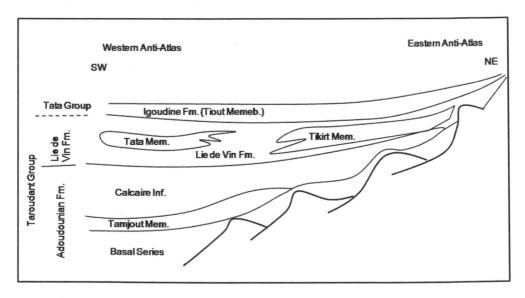

*Figure 17.2* Schematic section showing Lower Cambrian stratigraphy in the Anti-Atlas Domain (after Álvaro et al., 2008).

marks the base of the Taroudant Group (Maloof et al., 2005). This boundary coincides with the base of the Lie de Vin Formation (525–521 Ma) (Belfoul et al., 2001, Baidder et al, 2008). Based on the comparison of the Cambrian record in Morocco and Siberia, Maloof et al. (2005) placed the Precambrian-Cambrian (Ediacaran-Cambrian) boundary at the base of the Tifnout Member of the Adoudounian Formation, at around 525.5 Ma[4]. The contact between the basement and the sedimentary cover is defined by the first carbonate-bearing units of the lowermost Adoudounian, which overlies the PIII conglomerates of the Ouarzazate Supergroup (Buggisch & Flügel, 1988). This contact is also marked by a major color contrast in the field and it is easily visible in satellite images (Burkhard et al., 2006, Soulaimani & Burkhard, 2008). The PIII behaves as a competent unit, while the overlying carbonates are detached and folded. This boundary is difficult to recognize in Egypt and Libya and falls within the latest Pan-African successions.

The Adoudounian (Adoudou) Formation (Ediacaran-Lower Cambrian) (Maloof et al., 2006) was traditionally divided into a basal unit known as the *Série de Base* and an upper unit called the *Calcaires Supérieurs*. Maloof et al. (2006) elevated the two units to member status, and renamed them the Tabia Member and Tifnout Member, respectively. The Tabia Member (named after a town on the Oulili River) is equivalent to the *Série de Base*. It is up to 250 m (820 ft) thick unit of peritidal carbonates, evaporites, mud-cracked siltstones, fluvial sandstones, and coarse siliciclastic debris that were deposited in half grabens bounded by steep normal faults (Destombes et al., 1985, Benziane et al., 1983, Azizi Samir et al., 1990, Chbani et al., 1999, Algouti et al., 2000, 2001, Benssaou & Hamoumi, 2001, 2003). The upper Tabia Member contains discoid structures of unknown affinity that resemble the Ediacaran *Aspidella* (Houzay, 1979). The Tifnout Member, named after a town on the Sdas River, is equivalent to the *Calcaires Supérieurs*. It is composed of peritidal dolomites, with siltstones and marls, at the base, followed by wavy-laminated micrites and grainstones, and capped by microbiolaminites, occasionally with stromatolites and signs of subaerial exposure. In the El Graara Inlier, andesitic and trachytic flows and breccias are interbedded with the Tifnout Member. The Jbel Boho Syenite plug has been dated at 534 ± 10 Ma (Ducrot & Lancelot, 1977) and provides a minimum age for the Tifnout Member. In the southwestern corner of the El Graara Inlier, Buggisch & Flügel (1988) reported the algae *Kundatia composita* from stromatolites in the lower part of the Tifnout Member, indicating an Early Cambrian age.

Important platform carbonate build-ups formed during the Lower Cambrian in the Western Anti-Atlas (Drebenne & Drebenne, 1978, James & Drebenne, 1980, Soulaimani & Burkhard, 2008, Maloof et al., 2005). Lower Cambrian bioherms similar to those of the Lower and Middle Cambrian of Spain and Sardinia occur in Morocco at Jbel Taissa, SE of Guelmine (Guelmim). Mounds of up 100 m (328 ft) in height are composed mostly of algae with Archaeocyaths and lime mudstones, surrounded by nodular limestones also rich in archaeocyathids. Bioherms, with algae and the archaeocyathids *Agastrocyathus* and *Halysicyathus*, also extend along a trend from Draa to Jbel Irhoud and are well developed at Issafen (James & Drebenne, 1980). The oldest archaeocyathid fauna from Tiout (Anti-Atlas, Morocco) contains

4  For comparison, the Ediacaran-Cambrian boundary in Oman is placed at 542 ± 0.6 Ma by Amthor et al. (2003).

the following species (Debrenne & Debrenne, 1978): *Aldanocyathus cribratus, Neoloculicyathus magnus, Coscinocyathusprimus, C. secundus, C. fasciola, Tumulifungia marocana,* and *Gordonifungia cowieiare.*

The Lie de Vin Formation (Maloof et al., 2006) is made up of peritidal dolomites and purple argillite-biohermal limestone couplets (Destombes et al., 1985). It thins to the East and South from 950 m (3117 ft) at Tiout to less than 150 m (492 ft) at El Graara and Bas Draa. An ash layer from the upper Lie de Vin at Tiout gave a U-Pb age of 522.4 ± 2.0 Ma (Compston et al., 1992). The limestone bioherms are stromatolitic or thrombolitic and commonly contain calcified algae such as *Tarthinia, Renalcis* and *Kordephyton* (Latham and Riding, 1990, Bertrand-Safarti, 1981). The vertical burrow *Diplocraterion* and horizontal traces of *Diplichnites*-type are found in the upper Tikirt Member of the Lie-de-Vin Formation in the Siroua area (Geyer & Landing, 1995, Álvaro et al., 2008).

The Tata Group overlies the Taroudant Group and includes the Igoudine, Amouslek, and Issafene Formations, and their correlatives (Geyer & Landing, 1995). The Group consists of shoaling-upward parasequences, which start with green or grey shales, followed by black limestones with wavy laminations, and capped by trilobite hash packstones, cross-bedded oolitic grainstones, or archaeocyathan-algal bioherms. Desiccation cracks and vadose gypsum occur at the top of some parasequences (Maloof et al., 2006). The Trilobites *Antatlasia hollardi, Antatlasia gutta-pluviae,* and *Sectigena* occur in the Issafene Formation. An ash bed from the equivalent to the lower Issafen Formation in the Lemdad syncline of the High Atlas is dated at 517 ± 1.5 Ma (Landing et al., 1998).

The Igoudine Formation contains the oldest known skeletal fossils from Morocco (Choubert & Hupé, 1953, Choubert et al., 1975, Boudda et al., 1979, Maloof et al., 2006). Trilobites of the *Eofallotaspis* zone first appear in the Tiout Member of the upper Igoudine Formation (Geyer & Landing, 1995). Trilobites in the uppermost Tiout Member are transitional into the *Fallotaspis tazemmourtensis* Zone.

The Amouslek Formation is composed of shales and limestones with rich fauna of trilobites, and archaeocyathids, as well as brachiopods, such as *Brevipelta* in the middle part of the formation, the so called *"Schist de Timoulay"*, which consists of yellowish to greenish fine sandstones (Boudd et al., 1979). The fauna indicate a Lower Cambrian. It contains trilobites of the *Choubertella* and *Daguinaspis* zones (Geyer & Landing, 1995) and archaeocyathids of the *Erismacoscinus marocanus* Zone (Debrenne & Debrenne, 1995).

The Middle Cambrian in the Western Atlas consists mainly of *Paradoxides* Schists *"schistes à paradoxides"* with a few beds of quartzitic sandstones *"barre de Guelmine"* (Boudda et al., 1979, Choubert & Faure-Muret, 1983), or the Tabanite Sandstones and Schists (Belfoul et al., 2001, Faik & Belfoul, 2001). The *Paradoxides* Schists are equivalent to the *"Schistes de Feijas"* of Belfoul et al. (2001) and Faik & Belfoul (2001). These sediments were deposited on a proximal carbonate platform which transgressed over an aborted Precambrian basement rift associated with the opening of the Rheic Ocean (Jeannette & Piqué, 1981, Michard et al., 1982, Wendt, 1985, Belfoul et al., 2001).

From the Middle Cambrian to Late Silurian times, sedimentation was dominated by detrital input from the West African Craton (Buggisch & Siegert, 1988). The Upper Cambrian is not exposed in the Anti-Atlas.

In the Western and West-Central Anti-Atlas, the Middle Cambrian is represented by the Goulimine Quartzitic Series (also called Group or Bar) containing various tri-lobites, or the Tabanite Sandstones and Shales in the Tata area. In the eastern Ougnat area, the Middle Cambrian is represented by the Tabanit Sandstones with the *Conocoryphe* trilobite and *Lingula*, overlain by Trachy-basalts (Baidder et al., 2008) and by Lower Ordovician sandy schists in the Tata area (Belfoul et al., 2001, Faik & Belfoul, 2001). They overlie the Lower Cambrian Tata Group.

There is a major unconformity at the end of the Upper Cambrian. This hiatus is recognizable everywhere in North Africa from Morocco through Tunisia, Algeria, Libya, and Egypt (Crossley & McDougall, 1998), and marks the base of the Ordovician, a time of incised valleys and low sea-level (Tawadros, 2001). This break has been attributed to the Sardic Orogeny by Michard (1964) or to epeirogenic uplift by Destombes et al. (1985). Burkhard et al. (2006) ascribed the break to a new phase of rifting and erosion.

The Paleozoic High Atlas is a horst of Paleozoic terrains intruded by granites with thin cover of Mesozoic strata. The Precambrian High Atlas is a horst of slightly deformed Infra-Cambrian and Precambrian rocks with very thin Mesozoic cover.

Two Precambrian granitoid intrusions, the Tighardine and Takoucht, crop out near the village of Wirgane along the NE-SW-trending N'Fis Fault in the Western High Atlas (Fig. 17.1). The country rocks of the intrusions are basalts to andesites, rhyolitic ignimbrites, lapilli and ash tuffs, volcano-clastics, such as pyroclastic breccias and tuffaceous sandstones, and schists, quartzites, greywackes, and carbonates (Eddif et al., 2007). The rocks were subjected to a regional greenschist facies metamorphism and locally to a low pressure andalusite-cordierite thermal metamorphism. The two Wirgane intrusions are similar and consist of granodiorites, porphyritic monzodiorites, and monzogranites with a porphyritic to granophyric texture. U-Pb geochronological dating on single zircon grains indicates a Neoproterozoic age ($625 \pm 5$ Ma) for the two intrusions, contrary to the previously published Variscan age. Therefore, Eddif et al. (2007) view the Wirgane intrusions as remnants of the Neoproterozoic basement. However, no other Neoproterozoic intrusions are known to occur in the area (Pouclet et al., 2008).

Early Cambrian rocks crop out in the southeastern part of the central part of the Western High Atlas along the Tizi-n'Test Fault. Two Cambrian structural and stratigraphic formations can be distinguished (Pouclet et al., 2008): the Ifri-Azegour Formation and the Ouzaga-Tizzirt Formation to the north and south of the Middle Western High Atlas Fault (MWHAF) (Fig. 17.1), respectively. This fault system consists of reverse faults dipping 50–60° to the south.

The Ifri-Azegour Formation (Lower Cambrian) crops out in two areas, Ifri and Azegour, separated by a Cretaceous graben (Pouclet et al., 2008). It consists of fine-grained sandstones, shales, and siltstones, alternating with limestones and volcanic basaltic flows. In the Ifri area, the Formation is 1400 m (4593 ft) thick and consists of three successive units of shallow-water sediments. The lower and middle units are made up of interbedded shales and siltstones with a few limestone beds and lava flows or sills. The top unit is characterized by a decrease in volcanics and an increase in argillaceous sediments. In the Azegour area, two units are recognized within the formation; a lower unit of carbonates and volcanics similar to those of the Ifri area, and an upper shale unit. The thickness of the whole sequence is about

1000 m (328 ft). A dacite intrusion in the upper part of the formation gave a U-Pb date of 533 ± 4 Ma (Pouclet et al., 2008).

The Ouzaga-Tizzirt Formation (Lower-Middle Cambrian) crops out between the Middle Western High Atlas Fault and the Tizi-n'Test Fault, to the NW of the Tichka Igneous Massif (Pouclet et al., 2008). The total sequence is about 3100 m (10,171 ft) and consists of siliciclastic and carbonate sediments, including turbidites. The Formation has been dated Early-Middle Cambrian by correlations with other areas. Interbedded flows at the base are of continental tholeiite composition and suggest a continental rift tectonic setting (Pouclet et al., 2008).

### 17.4.1.2  Ordovician

Ordovician sediments in Morocco crop out mainly in the Anti-Atlas (Destombes et al., 1985). In the Eastern Anti-Atlas, the oldest Ordovician rocks belong to the Caradocian First Bani Formation; considered transitional between Middle and Upper Ordovician "Tremadocian-Caradocian" (Nardine, 2007), followed by the Lower Ktaoua Formation (Caradocian). On the southern border of the Tindouf Basin, the Paleozoic succession starts with Upper Ordovician sandstones (Cavaroc et al., 1976, Malla et al., 2000).

Ordovician sediments crop out on the eastern edge of the Tafilalt Basin in Eastern Morocco (El Maazouz & Hamoumi, 2007). During the Lower and Middle Ordovician, the Tafilalt area was covered by a tide-dominated epeiric shelf (Maazouz & Hamoumi, 2007). Two eocrinoid genera have been described from the Ordovician of Morocco; *Rhopalocystis* from the upper Tremadocian in the Zagora region (Central Anti-Atlas) and *Balantiocystis* from the upper Arenig in the Tan Tan region (Western Anti-Atlas) (Nardin, 2007). Upper Ordovician (Caradocian-Ashgillian) extensional tectonics divided the Tafilalt shelf area into two subbasins (Maazouz & Hamoumi, 2007): The Khabt el Hejar (Maader[5]) Subbasin to the northeast and the Western Tafilalt Subbasin to the southwest. The Khabt el Hejar Subbasin was occupied by a carbonate platform with bryozoan mounds and high-energy peritidal mixed siliciclastic-carbonate deposits. On the other hand, in the Western Tafilalt Subbasin, half-grabens were filled with fan-delta deposits with thick conglomerates supplied by the Sahara glaciers and the carbonate platform to the northeast.

The Middle Ordovician north of Zagora includes the Upper Fezouata Formation (shales and lenticular sandstones) (Van Roy et al., 2008, 2010) and the Tachila Formation shales (Nardin, 2007). The Fezouata Formation reaches a thickness of 1100 m (3609 ft) north of Zagora (Destombes et al., 1985) and yields numerous soft-bodied fossils comparable to the Burgess Shale in Canada (Van Roy et al., 2008). The assemblages are dominated by benthonic organisms. The shelly fossils include conulariids trilobites (asaphids, harpetids, odontopleurids, phacopids, proetids, ptychopariids and agnostids), articulated hyolithoids and other mollusks (helcionelloids, bivalves, gastropods, nautiloids), brachiopods, occasional bryozoans, and echinoderms (homalozoans, asterozoans, various eocrinoids, cystoids, rare crinoids), in addition to planktonic and benthonic graptolites (Van Roy et al., 2010).

---

5  Also spelled Ma'der or Maâder, but the form Maader is used here for its simplicity.

The Caradocian includes the First Bani Formation (Group) and the Lower Ktaoua Formation (Group) (Fig. 17.3). In the Khabt el Hegar Subbasin (northeastern Tafilalt), the Lower Caradocian deposits form two sedimentary sequences (Maazouz & Hamoumi, 2007). The basal sequence is composed of shales, siltstones, calcareous sandstones, and bivalve calcareous lenses and represents wave- and tide-dominated subtidal sediments. The second sequence exhibits subtidal- intertidal and wave- and tide-dominated subtidal facies (Maazouz & Hamoumi, 2007). In the Western Tafilat Subbasin, the Lower Caradocian sediments consist of coarsening-upward, thickening-upward sequences of shales, siltstones, and minor carbonates. In the Khabt el Hegar Subbasin, the Middle and Upper Caradocian sediments of the First Bani and Lower Ktaoua formations (groups) consist of bryozoan siltstones, argillites, limestones, and mixed siliciclastic-carbonates, capped by microconglomeratic shales *"argile microconglomératique"*, and coarse sandstones (Maazouz & Hamoumi, 2007). The coarse sandstones display shattering, friction cracks, and soft sediments deformations, which indicate sedimentation in a periglacial estuarine channel and

| | | MOROCCO | | LIBYA |
|---|---|---|---|---|
| UPPER ORDOVICIAN | Ashgill | Second Bani Grp. | Upper Second Bani / Lower Second Bani | Memouniat |
| | | Ktaoua Group | Upper Ktaoua | |
| | | | Rouid-Assa/ Upper Tiouririne | |
| | Caradoc | | Lower Ktaoua | Melez Chograne |
| | | First Bani Group | Izegguirene | |
| MID. ORD. | Llanvirn | | Ouine-Inime | |

Figure 17.3 Upper Ordovician lithostratigraphic units of the Anti-Atlas and their correlation with Libya (after Nardin, 2007 and Hunter et al., 2010).

sandy to muddy intertidal flat environments. In the Western Tafilalt Subbasin, the Middle and Upper Caradocian crop out in the Erfourd and Imzizoui areas and consist of conglomerates with slump structures, microconglomerates, and coarse sandstones. Nardin (2007) reported the echinoderm *Cardiocysties bohemicus*, as well as asteroids, crinoids, eocrinoids, rhombiferans, and solutes from the Lower Ktaoua Formation at Oued El Caid Rami, about 41 km southwest of Erfourd. Hunter et al. (2010) documented *Ascocystites, Cardiocystites, Asterocystis, Anatifopsis, Aspidocarpus,* and *Eumitrocystella* from Upper Ordovician of the Anti-Atlas. The First Bani Group comprises the Middle Ordovician Ouine-Inirne Formation (Llandeilian) and Izegguirene Formation (lowermost Caradocian), overlain by the Upper Ordovician Ktaoua Group composed of the Lower Ktaoua (lower Caradocian-lowermost Ashgill), Upper Tiouririne (lower Ashgillian), and Upper Ktaoua formations (lower Ashgillian) (Lefebvre et al., 2010, Hunter et al., 2010) (Fig. 17.3).

The Ashgillian consists, in an ascending order, of the Rouid-Assa Formation, Upper Ktaoua Formation, and the Second Bani Formation (Figs. 17.3, 17.5). In the Khabt el Hejar area, the Upper Ashgillian sediments are similar to those of the Middle and Upper Caradocian, except in northwestern Merzane, where the succession contains carbonate mounds in the lower part (Maazouz & Hamoumi, 2007), composed of bryozoans associated with brachiopods and echinoderms. Mound initiation and growth occurred below fair-weather wave-base in a subtidal environment. The Ashgillian clastics were derived from the Precambrian Shield and its sedimentary cover. In the Western Tafilalt Subbasin, Upper Ashgillian sediments crop out at Imzizoui and Tabhet el Khir and consist mainly of red coarse sandstones (Maazouz & Hamoumi, 2007).

The Ashgillian carbonate ramp and the mixed siliciclastic-carbonate peritidal system of the Eastern Anti-Atlas can be correlated with the southern and northern Ahaggar, West Libya, Tripolitania, and Tunisia where contemporaneous carbonate deposits are associated with glacial sediments (Maazouz & Hamoumi, 2007).

The Second Bani Formation is made up of sandstones, micaceous clays and occasional limestones, with graptolites and trilobites. Ashgillian glaciation is manifested by tillites (microconglomerates/diamictites) at Djebel Serraf (Ghienne, 2003, Legrand, 2003, Sutcliffe et al., 2000, El Maazouz & Hamoumi, 2007, Le Heron, 2007, Le Heron & Craig, 2008, Soulaimani & Burkhard, 2008) (Fig. 17.4) It is divided into a Lower and Upper units. The Lower Second Bani Formation consists of wave-dominated, transgressive shoreface to offshore sediments. The thickness variations of the shell beds within these deposits were controlled by the paleotopography (Destombes et al., 1971, 1985) that existed prior to the advance of the ice sheets in the Hirnantian (Le Heron, 2007). The presence of ice sheets is shown by soft-sediment striations, streamlined bedforms, downward-injected sedimentary dikes, shear zones, and chaotic fold zones, capped by diamictites. High-energy meltwater channels occur in the High Atlas and a deep-marine system occurs in the Tazzeka Massif the Eastern Meseta (Le Heron, 2007).

The Upper Second Bani Formation represents the Late Ordovician glaciation phase. It contains five facies associations (Le Heron, 2007, Fig. 4): 1) Tabular sandstones (shallow-marine/shoreface deposits). 2) Massive sandstones and conglomerates (ice contact debrites). 3) Meander sandstones (ice proximal sandur). 4) Stratified diamictites (ice-rafted debris). 5) Sigmoidally bedded sandstones (intertidal sandstones).

In the Central High Atlas, Soufiane & Achab (1993) identified five chitinozoan assemblages in Ordovician samples from the Bj wells Bj103, 106, and 109 in the

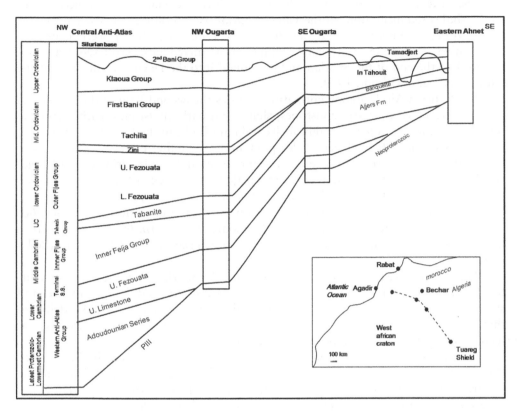

*Figure 17.4* Cross-section depicting the correlations of Cambro-Ordovician units between Morocco amd Algeria (after Ghienne et al., 2007).

Boujad Permit, as well as the KAT1 well in the Kasba Tadla Permit located in the eastern part of the Tadla Basin (Fig. 17.1). The successions are composed mostly of grey-black and black shales with minor siltstones and fine grained quartzitic sandstones. The black shale section in the KAT1 well consists of 17 m (56 ft) of microconglomeratic shales (diamictites). The five assemblages are of upper Arenig, latest Arenig, early Llandeilian, early Caradocian, and late Ashgillian ages and include *Laufeldochitina protolardeuxi* (Arenig), *Lagenochitina spinatostoma*, *Sagenachitina oblonga* (at the boundary of the Middle and Upper Arenig), *Desmochitina bulla*, *Armoricochitina nigerica*, and *Calpichitina lenticularis* (Upper Arenig), *Kalochitina inflata* (Lower Llandeilian), *L. deunffi* (Lower Caradocian), and *Jenkinochitina lepta* (Upper Ashgillian).

### 17.4.1.3 Silurian

Silurian sediments have been described in the Anti-Atlas and the Meseta domains. Correlations of Silurian lithostratigraphic units based on graptolites in Morocco were presented by Legrand (1989) and Petters (1991).

The transgression, related to the Early Silurian deglaciation, induced graptolitic shales sedimentation. In the Upper Silurian two prominent marker horizons, the *Orthoceras* Limestone and *Scyphocrinites* Limestone (Beds) were deposited (Destombes et al., 1985). Carbonate sedimentation, mixed with clastic sedimentation resumed at the end of the Silurian and continued into the Devonian throughout the Anti-Atlas region (Wendt, 1989).

The Silurian Ain Delouine Formation overlies the Second Bani in the Anti-Atlas and is represented by platy sandstones, shales and dark mudstones, with carbonate nodules, graptolites, a few lamellibranchs and nautiloids. In the Eastern (edge of the) Anti-Atlas in the Tafilalt, black shales predominate (Lüning et al., 2000d, Kolonic et al., 20002). They are equivalent to the Tannezouft and Argiles Principales of Algeria, Tunisia, and West Libya and represent a major source rock in North-Africa (Boote et al., 1998, Morabet et al., 1998, Lüning, 2000a-d, Tawadros, 2001).

The Lower Devonian Lmhaifid Formation at Ain-Delouine overlies the *Scyphocrinites* Beds near the Silurian-Devonian boundary. *Scyphocrinites* Aff. *elegan* from Djebel Issoumour, Alnif, is assumed to be Upper Silurian or basal Devonian (Belka et al., 1999).

The Moroccan Meseta was dominated during the Silurian by a tide-dominated littoral, differentiated, from NW to SE, into: 1) a storm-dominated shelf, 2) a wave- and storm-dominated delta, and 3) a tide-dominated littoral. Attou et al. (2004) attributed this paleogeography to the interplay between the Infracambrian-Silurian extensional tectonics and the glacio-eustatic sea rise during the Silurian.

Silurian conodonts have been described from the Western Meseta by Benfrika & Raji (2003) and Benfrika et al. (2007). In the area between Rabat and Tiflet, in the Central Meseta, they include the two guide species of the European Silurian zones *Ozarkodina sagitta sagitta* and *Ozarkodina remscheidensis eosteinhornensis*, and two guide fossils of the North American zones *Kockelella stauros* et *Kockelella variabilis* which indicate the presence of the Wenlock on top of the Ludlow-Pridoli in Morocco (Benfrika & Raji, 2003).

### 17.4.1.4   Devonian

Pioneering studies on the Devonian of the Moroccan Sahara were carried out by Le Maître (1952), Choubert et al. (1952), Hollard (1967), Massa et al. (1965), and Michard (1976).

During the Devonian, a number of basins were established on the northern margin of the Sahara Craton, such as the Draa, Tindouf, Bechar, and Tafilalt, and Maader basins. In the Early Devonian, a carbonate ramp developed in the Eastern Anti-Atlas, while in the Southwestern Anti-Atlas (Draa Plain) (Fig. 17.5), a mixed siliciclastic-carbonate shelf prevailed. During the Middle Devonian, Northwest Africa was dominated by widespread carbonate deposition. Shale deposition prevailed once more in the Late Devonian until a late Famennian sea-level fall led to rapid progradation of deltaic sandstones (Lubeseder et al., 2009, 2010).

In Morocco, the Lower Devonian includes the Rich Group, Tafilalt Limestones, Megsem-Mdarsal Group, and Draa Plain Clays (Soulaimani & Burkhard, 2008, Fig. 2, Aitken et al., 2002, Lubeseder et al., 2003 & 2009, Jansen et al., 2007). It displays two types of facies: In the Western Anti-Atlas, sandy mudstones and limestones

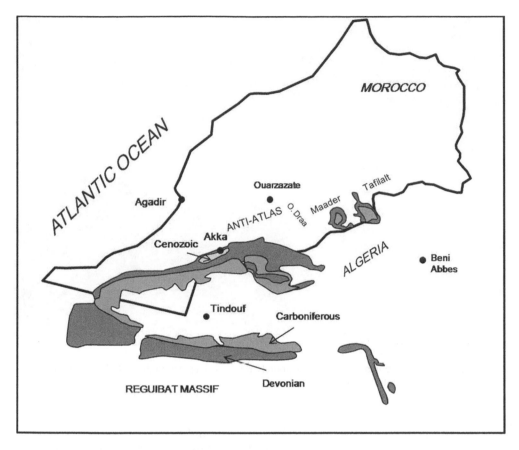

*Figure 17.5* Devonian outcroups in Draa Valley and Tindouf Basin (after Cavaroc et al., 1976 and Brice et al., 2007).

of the Gedinnian-Siegenian with conodonts and tentaculites predominate. In the Eastern Anti-Atlas, for example in the Tafilalt Basin, Devonian mud mounds and basaltic rocks occur at Hamar Laghdad (Aitken et al., 2002) (Fig. 17.6, 17.8). The Middle Devonian and the Frasnian are represented in the Western Anti-Atlas mostly by black limestones, whereas to the east, the Upper Tafilalt Limestones are rich in goniatites and tentaculites. Plant fragments of *Callixylon henkei* occur in Devonian (Frasnian) marine deposits exposed in the Draa Valley of the Southern and Eastern Anti-Atlas of Morocco associated with conodonts and tentaculites (Meyer-Berthaud et al, 2004). The Famennian of the Western Anti-Atlas exhibits a clayey facies at the base, followed by detrital beds and then by a calcareous horizon. In the Eastern Anti-Atlas, platforms were uplifted in the Maader and Tafilalt and resulted in a condensed limestone section with cephalopods (Cephalopod Ridge or Tafilat Platform). Equivalents of these sediments form prolific hydrocarbon reservoirs and source rocks in Algeria and Libya.

Early stratigraphic studies of the Draa Valley (Webster & Becker, 2009) began in the 1920s by Bourcart (1927) and Gentil (1929) and continued by Descossy & Roch

*Figure 17.6* Lower Devonian (Emsian) mud mounds at Hamar Laghdad in the Tafilalt region, Anti-Atlas (Courtesy of Steven Aitken, Calgary).

(1934), and Bondon & Clariond (1934). Detailed and systematic studies were later carried out by the famous French geologist Henry Hollard (d.1981), whose publications from 1960 to 1986[6] established the stratigraphic framework that form the basis of all subsequent works.

Hollard (1981) assigned the Rich Group a Late Siegenian-Early Eifelian age. It includes neritic and pelagic units divided into four main cycles called '*Riches*' (Choubert 1956, Hollard, 1967, Jansen et al., 2007). A 'Rich' generally consists of limestones at the base, silty and sandy shales in the middle, and sandstones at the top (Hollard 1967). Each of these cycles has been designated a formation status and named 'Rich 1' to 'Rich 4' (Hollard, 1967). Significant lateral facies changes from SW to NE are present within these units, which led to different nomenclatures for units of the same age. The neritic sandstones contain Rhenish brachiopod fauna (Jansen et al., 2007); whereas the pelagic facies contains conodonts, tentaculitids, and goniatites of the global biozones (Bultynck & Hollard 1980, Lazreq & Jansen et al., 2007). Jansen et al. (2007) reviewed the Devonian stratigraphy and the Rich Group and assigned formational names to the four units in the Draa Valley.

The Draa Valley of Southwestern Morocco extends over a distance of about 600 km and runs parallel to the south side of the Western Anti-Atlas Mountains from near

6 His papers of 1981 and 1986 were published posthumously.

Zagora in the NE to Tan-Tan near the Atlantic Ocean in the SW (Webster & Becker, 2009). The Draa Basin forms an ENE-WSW channel with two internal high blocks, surrounded by highs and regional faults (Ouanaimi & Lazreq, 2008).

Lower Devonian sediments are widely exposed in the Draa Valley in the Southwestern Anti-Atlas (Hollard, 1967, Jansen et al., 2007) and in the Tafilalt area in the Northeastern Anti-Atlas (Wendt et al., 1984, Döring & Kazmierczack, 2001, Aitken et al., 2002, Lubeseder et al., 2003, 2009, 2010).

The Lower Devonian successions in the Anti-Atlas show two main facies; a thick neritic clastic 'Rhenish Facies' in the Western Anti-Atlas (Draa Valley) and a thin calcareous pelagic and hemipelagic 'Hercynian Facies' in the Eastern Anti-Atlas and the High Atlas (Jansen et al., 2007, Ouanaimi & Lazreq, 2008). In the Rhenish Facies, ages are determined mainly by brachiopods, whereas in the Global (Hercynian) Facies they are based mainly on conodonts (Jansen et al., 2007).

In the Southwestern Anti-Atlas-Draa Valley, the Lower Devonian (Rich Group) in (Lubeseder et al., 2003, 2009, 2010): a) A carbonate ramp system is dominated by offshore mud grading into an outer carbonate ramp facies of nodular limestones with orthocone nautiloids, ammonites, large bivalves, trilobites and small rugose corals, overlain by a regressive sequence of tentaculite-packstones, crinoid-packstones to grainstones or bioclastic-grainstones. b) A regressive siliciclastic-carbonate system composed of offshore mud that grade into a siliciclastic sequence dominated by storm deposits.

The Lower Devonian sequence of the Draa Basin is represented by the Rich Group, Lmhaifid Formation, and Oued el Mdaouer Formation.

The Lmhaifid Formation (lower-middle Lochkovian in the global subdivision or lower-upper Lower Gedinnian in the Rhenish subdivision) (Hollard, 1967, Jansen et al., 2007) or the 'Série de Passage' of Hollard (1963) consists of about 600 m (1969 ft) of sandy shales, sandy limestones and limestones. Hollard (1963, 1981, in Jansen et al, 2007) subdivided the Lmhaifid Formation into Ain Deliouine, Lower Ouadi Hara, Upper Ouadi Hara, Sidi MBark, and Talmadert members. However, no type section has been designated (Becker et al., 2004a, b). It is known only from the section at el Ayoun south of Tata in the NE Draa Valley. Conodonts indicate a Lower- to lower Middle Lochkovian age. Brachiopods from the same section indicate Lower Gedinnian age. The Lmhaifid Formation contains the index fossils *Latericriodus woschmidti*, *Acastella heberi*, and *A. tiro*, *Eoglossinotoechia temieri*, and *Lanceomyonia occidentalis*. Conodonts, brachiopods, and trilobites confirm a Gedinnian age. The Lmhaifid Formation at Ain-Delouine in the SW overlies the lowermost Devonian *Scyphocrinites* Beds (Hollard, 1967, Belka et al., 1999).

The Oued el Mdaouer Formation (Middle Lochkovian-Upper Emsian) (Jansen et al., 2007) overlies the Lmhaifid Formation in the NE Draa Valley. It consists of crinoidal limestones with corals, bryozoans, conodonts and marly limestones, with thicknesses of 30–80 m (98–262 ft) near Tata (Jansen 2001). The formation is equivalent to the interval that spans the upper part of the Lmhaifid Formation and the Assa Formation in the SW Draa Valley. The Oued-el-Mdaouer Formation is probably equivalent to the *Orthoceras* Limestone in the Achguig region to the NE.

The four formations of the Rich Group (Pragian-late Eifelian); Rich 1 (Assa Formation), Rich 2 (Merza-Akhsai Formation), Rich 3 (El Annsar Formation or Mdaouer El Kbir Formation, and Lower Timrhanrhart Formation), and Rich 4 (Nkheila Formation or Khebchia Formation) in the Draa Valley (Lubeseder et al., 2009) represent four third-order depositional sequences, each containing a transgressive systems tract and a highstand systems tract deposited on a storm- and tide-dominated shelf (Ouanaimi & Lazreq (2008). The depositional sequences are related to global sea-level changes, tectonic subsidence, and sediment supply. Deposition probably took place in a subsiding passive margin on the northern edge of the WAC, where early Paleozoic and Precambrian faults are reactivated in an extensional or transtensional regime.

The Assa Formation (Rich 1) (Hollard 1967, Jansen et al., 2007) in the SW Draa Valley consists of: 1) Basal limestones composed of coarse grained limestones and reddish sandstones. 2) Middle shales with siltstone and sandstone beds (Becker et al., 2004a, b). 3) Upper massive sandstones with brachiopod coquinas, oolitic beds, and phosphate nodules at the top. In its type area near the town of Assa, the formation reaches a thickness of approximately 300 m (984 ft). It overlies the Lmhaifid Formation. Becker et al. (2004a, b) separated the limestones above the sandstones from the Assa Formation and included them in the overlying Merza Akhsai Formation. The enigmatic *Nowakia acuaria* occurs in the basal part. Conodonts include *Latericroidus steinachensis*, *Platyorthis hollardi*, and *Dixonella assaensis* (Becker et al., 2004a, b). It has been dated uppermost Lochkovian-upper Lower Siegenian by Jansen et al. (2007) and Upper Lochkovian-Pragian by Becker et al. (2004a, b).

The Merza Akhsai Formation (Rich 2) (Hollard, 1981, Jansen et al., 2007) consists of two units (Jansen et al., 2007): 1) Basal bluish-grey, crinoidal limestones with trilobites, brachiopods, and conodonts, such as *Latericroidus steinachensis* and *Caudicriodus curvicauda* (Jansen et al., 2007). 2) Middle shales. 3) Upper sandstones. Conodonts in the limestones are of middle to upper Pragian age, whereas middle and upper parts of the formation are dated Middle to Late Siegenian (Jansen et al., 2007). It is equivalent to a limestone unit in the Achguig region (cf. Bultynck & Walliser, 2000). The Merza Akhsai Formation has been dated Pragian (Middle-Late Siegenian) (Jansen et al., 2007) and Upper Pragian to lowermost Emsian (Becker et al., 2004a, b). The ammonites of the Merza Akhsai Formation and Mdaouer el Kebir Formation have been discussed recently by De Baets et al. (2010), based on which, they correlated the early Emsian (Zlichovian) strata in the Tafilalt and the Draa Valley regions.

The Mdaouer el Kebir Formation (Rich 3) (Hollard 1967, Jansen et al., 2007) is equivalent to the 'Rich d'El Annsar'. It is developed in the NE Draa Valley between Akka and Foum Zguid. Its type locality is near Foum Zguid where its thickness is 295 m (968 ft). The abrupt change from sandstones of the Merza-Akhsai Formation to the basal limestones of the Mdaouer-el-Kbir Formation (equivalent to the Oui-n'-Mesdour Formation) (see also Becker et al., 2004a, b) has been referred to the Basal Zlichov Event (Chlupac & Kukal, 1988, Jansen et al., 2007). The lowermost part of the Formation contains the conodonts *Eucostapolygnathus excavatus*, *Caudicriodus sigmoidalis*, and *Latericriodus celtibericus*. The overlying black limestones contain *L. multicostatus*, *L. bilatericrescens bilatericrescens*, and *L. celtibericus*; the tentaculitids *Viriatellina pseudogeinitziana* and *V. hercynica*, as well as the primitive goniatite *Erbenoceras advolvens* (Bultynck & Hollard, 1980). The Mdaouer el Kebir Formation has been assigned to the Lower-Middle Emsian by Becker et al. (2004a, b) and the Lower-Upper Emsian by Jansen et al. (2007). The massive sandstones of the Rich 3 Sandstone Member, at the top of the Mdaour el Kbir Formation, form a widely recognizable marker unit that contains the Lower-Upper Emsian transition (Jansen et al., 2004, Webster & Becker, 2009).

The Oui n'Mesdour Formation (Hollard 1978) is restricted to the SW Draa Valley and correlates with the Mdaouer el Kebır Formation in the NE (Jansen et al., 2007). It consists of limestones and marly limestones with thicknesses of up to 45 m (ft). Its type locality is situated near Aouinet Torkoz. The Formation has been subdivided into the Akhal Tergoua and Black Marl members (Jansen et al. 2004, 2007). The fauna include the conodonts *Caudicriodus sigmoidali*, *Latericriodus beckmanni*, *L. bilatericrescens*, *C.* cf. *ultimus*, and *L. beckmanni sinuatus*, in addition to the Tentaculitid *Viriatellina* cf. *pseudogeinitziana* and the goniatite *Mimagoniatites* and indicate a Lower Emsian (Zlichovian) age for the Formation (Jansen et al., 2007).

The Khebchia Formation (Rich 4) (Jansen et al., 2007) consists of limestones and marlstones at the base, followed upwards by shales and sandstones. It overlies the Oui n'Mesdour Formation in the SW Draa Valley. Its thickness reaches up to 300 m (984 ft). The formation has been subdivided into five members: *Hollardops* Limestone Member, Brachiopod Marl Member, *Sellanarcestes wenkenbachi* Limestone Member, Bou Tserfine Member, and 'Rich 4' Sandstone Member (Jansen et al. 2004, Becker et al., 2004). The age of the *Hollardops* Limestone Member at Bou Tserfine is Upper Emsian. The presence of the icriodid fauna *Caudicriodus culicellus*, *Icriodus corniger ancestralis*, and *I. fusiformis* indicate early Upper Emsian (Weddige, 2004, Becker et al., 2004). The conodont and goniatite fauna also indicate a lower Upper Emsian age for the formation. The Kebchia Formation is correlatable with the Timranrhart Formation in the NE Draa Valley (Jansen et al., 2007). The 'Rich 4' sandstones have been dated by brachiopods as latest Emsian-early Eifelian ((Jansen et al., 2007, Becker et al., 2004, Webster & Becker, 2009). The boundary between the Khebchia Formation and the underlying Oui n'Mesdour Formation is gradual, and is placed between massive trilobite limestones and nodular limestones with abundant goniatites and trilobites (Becker et al., 2004). The Bou Tserfine Member is made up of unfossiliferous green shales and siltstones with siltstone concretions in the lower part. A Cone-in-Cone Marker Bed is well developed within the member. The Bou Tserfine Member grades upward into the sandstones of the "Rich 4" Sandstone Member. The boundary between the two members is usually placed at the base of the first massive sandstone bed (Becker et al., 2004).

The Timrhanrhart Formation (Rich 4) (Upper Emsian-Eifelian) is well exposed in the upper part of the Foum Zguid section where it overlies the Mdaouer el Kbir Formation in the NE Draa Valley (Becker et al. 2004, Jansen et al., 2007, Webster & Becker, 2009). The formation starts with trilobite-bearing limestones, overlain by a massive *Sellanarcestes* Limestones, and then by nodular limestones, marls and greenish shales. It also contains pyritized cephalopods (Lubeseder et al., 2009). The formation is overlain by dolerites. At the type locality, the formation has a total thickness of 37 m (121ft) (Hollard 1978). The Emsian-Eifelian boundary falls within the upper part of the Formation as indicated by conodonts, tentaculitids, trilobites, and ostracods (Jansen et al., 2007, Webster & Becker, 2009).

Givetian table reefs, biostromes and patch reefs occur in a narrow belt along the northern margin of the West African Craton (Reguibat Massif) extending for 1500 km from the Zemmour (Mauritania) in the SW to the Taout-Gourara area in Central Algeria (Wendt & Kaufmann, 2006) in the Uein Terguet, Gor Loutad, and

NW Semara (Smara) areas in the Tata area, West Sahara (former Spanish Sahara). Reefs cycles are separated by shales and sandstones. They consist of stromatactis mud mounds, large stromatoporoids, associated with tabulate as well as solitary and colonial rugose corals, and reef-dwelling organisms. The reefs at Smara are equivalent to the 6th reef cycle of Wendt & Kaufmann (2006). The cyclicity have been attributes either to relative sealevel changes or of oscillating siliciclastic input from the West African. May (2008) described the three tabulate coral species *Caliapora robusta*, *Pachyfavosites tumulosus*, and *Thamnopora major*, the rugose coral *Phillipsastrea* ex gr. *irregularis*, and the chaetetid *Rhaphidopora crinalis*, from the Givetian (Eifelian)/ Frasnian collection from the Semara area (Morocco, former Spanish Sahara). May (2008) conclude that these species have close biogeographic relationships to Central and Eastern Europe, as well as to Western Siberia.

The Upper Frasnian of the Eastern Draa Valley is represented by the Anou Smaira Formation (Late Frasnian) (Webster & Becker, 2009). The Formation consists of interbedded calcareous marly layers with pyritized fauna, such as goniatites, thin siltstones, nodular limestones, and irregular large concretions of marls and limestones. The so-called Upper Red Griotte unit with the conodont *Avignathus decorosus* forms a local marker bed and helps divide the formation into lower and upper members (Webster & Becker (2009). Pyrite-rich sediments with pyritic goniatites are present in the section. The Givetian succession at Oued Mzerreb, SE of Tata (Fig. 17.1) has been assigned to the Ahrerouch Formation and subdivided into the Tigusselt Member and the Oued Mzerreb Member. The formation consists mostly of pelagic deeply weathered marls, often with pyritic faunas, and thin limestone and siltstone beds. Reddish nodular limestones have been called Red Griotte, in reference to those of the Montagne Noire in France (Webster & Becker (2009). The Lower Red Griotte contains neritic trilobites, gastropods, bivalves, brachiopods, rugose and thamnoporid corals, as well as pelagic fauna, such as styliolinids and goniatites. The assemblage has been dated Givetian age based on the marker species *Maenioceras decheni*. Crinoids include *Hexacrinites chenae*. A thin siltstone bed contains common *Agoniatites* and has been called the *Agoniatites* Bed by Abousallam & Becker (2004) and Webster & Becker (2009). Late Middle-lower Upper Devonian crinoids have been also described from pelagic facies of the Eastern Draa Valley (Tata area) by Webster & Becker (2009). The crinoids show only minor relationship with the crinoids from the Tafilalt and Maader areas of the Eastern Anti-Atlas, which are attributed to differences in the environments between the two domains. The Draa Valley hexacrinids show greater affinity with the European faunas, whereas the amabilicrinids show more affinity with the North American taxa.

The Upper Devonian (Frasnian-Famennian) is represented in the Western Anti-Atlas by black shales and limestones (see Meyer-Berthaud et al., 2004, for Frasnian land plants in the Draa Valley, and Lüning et al., 2003a, 2004c, Wendt & Belka, 1991, Belka et al., 1999). In the Eastern Anti-Atlas, they are represented by the Upper Tafilalt Limestone rich in goniatites and tentaculites. The Late Devonian Kellwasser Extinction Event at the Frasnian/Famennian boundary (Buggisch, 1991, Wendt & Belka, 1991) was probably caused by a meteorite impact, although sea-level rise (Wendt & Belka, 1991), anoxia, and plant evolution have been considered as alternative causes. The so-called Kellwasser Facies consists of black bituminous limestones and shales, which were deposited on pelagic platforms and in adjacent shallow basins (Wendt & Belka, 1991). The Moroccan Kellwasser sediments are extremely fossiliferous and contain pelagic and benthonic organisms. Two Kellwasser episodes are recognized (Cavalazzi, 2008) in the lowermost Frasnian and another late Frasnian-early Famennian successions. The latter consists of coquinas made up of cephalopods, tentaculites, and styliolinids, which grade basinward into shales with lenses of fine-grained sandstones and limestones. Similar black shales, marls, and limestones occur in the Cues Limestone in West Libya and its equivalent in Eastern Algeria (Lüning et al., 2004c).

The Devonian succession of the Tafilalt area (Eastern Anti-Atlas) is highly fossiliferous and contains a diverse assemblage of goniatites, trilobites, crinoids, and tentaculites and formed the subject of numerous studies (Bondon & Clariond, 1934, Alberti 1981, Wendt et al., 1984, 2006, Brachert et al., 1992, Belka et al., 1999, Döring & Kazmierczack, 2001, Aitken et al., 2002, Lubeseder et al., 2003, 2009, 2010, Berkowski, 2008, Preat et al., 2008, Toto et al., 2008). Devonian reefs occur in Morocco in two main areas, the Eastern Anti-Atlas and the Western Ant-Atlas. They also occur in the Meseta Domain. Two reef-building episodes during the Lower and Middle Devonian have been recognized in the Tafilalt Basin.

In the Eastern Anti-Atlas, an Early Devonian carbonate platform prevailed (Wendt et al., 2006). However, by the Middle and Upper Devonian, the platform was divided into two subbasins as a result of early Variscan tectonic movements and differential subsidence (Wendt et al., 1984): The Tafilalt Subbasin in the east and the Maader Subbasin in the west (reminiscent of the Late Ordovician time). The two subbasins were separated by the narrow N-S condensed Cephalopod Ridge (Tafilalt Platform) from the Eifelian until the late Famennian (Lubeseder ct al., 2003, Toto et al., 2008). The Maader Basin was filled with siliciclastic and calcareous sediments up to 3000 m (9843 ft) thick (Döring & Kazmierczack, 2001). Lower Famennian is represented by quartz-rich brachiopod coquinas, crinoidal limestones, thick-bedded cephalopod limestones, and nodular limestones. In the Late Famennian, marls and nodular limestones were deposited in slightly deeper environments, and debris flows and slumps on the platform margins. In the Tafilalt Subbasin, marls and nodular limestones were deposited in shallow environments. On the other hand, in the Maader Basin, sandy and calcareous turbidites accumulated in deeper waters. During the Strunian-Tournaisian the whole area was covered by a thick deltaic sequence (Wendt et al., 1984). Lubeseder et al. (2003, 2009, 2010) recognized six third-order transgressive-regressive cycles in the Eifelian to Tournaisian succession. These cycles were correlated with those in the Devonian outcrops (Ougarta, Ahnet, Mouydir, and Tassili areas in southern Algeria) and in the subsurface of the Illizi and Ghadames basins.

The Lower Devonian (Lochkovian-Emsian) deposits in the Tafilalt region in the Eastern Anti-Atlas have been subdivided by Hollard into three groups: Merzane Group, Seheb el Rhassel Group (formerly called Kess Kess Formation and Amerboh Group (Aitken et al., 2002) (Fig. 17.6). The rugose corals in that succession have been described by Berkowski (2006, 2008).

The Merzane Group (Aitken et al., 2002) has a maximum exposed thickness of 85 m (279 ft), but may reach up to 135 m (443 ft) (Aitken et al., 2002). It thins abruptly to the east and west. The Merzane Group is dominated by pyroclastic and volcaniclastic sediments composed of tuffs with pumice pyroclasts in a groundmass of devitrified glass and calcite cement. Channels filled with immature volcaniclastic deposits with angular clasts up to 30 cm in diameter occur in the upper part of the group. The uppermost part of the Merzane Group consists of 3 m of fine- to medium-grained tuffaceous sandstones. The Group overlies the lowermost Devonian *Scyphocrinites* Beds.

The Seheb el Rhassel Group (Aitken et al., 2002) is equivalent to the Kess Kess Formation of Brachert et al. (1992). It has a thickness of 35–140 m (115–459 ft). Its contact with the underlying Merzane Group is sharp and wavy and marked by a discontinuous and condensed bed. The group is made up mostly of crinoidal and trilobite wackestones, with favosites, thamnoporids, bivalves, brachiopods, gastropods, ostracods, and foraminifera, among others. Sedimentary structures include large hummocky cross-bedding, burrows of *Thalassinoide* and *Cruziana*, and graded bedding. It has been assigned an Emsian age (Klug, 2001, Berkowski, 2006, Kröger et al., 2005).

The Amerboh Group (Pragian-early Emsian) (Aitken et al., 2002) is 20–125 m (66–410 ft) in thickness and thins abruptly to the east and west. It consists of green to black, laminated marlstones with scattered limestone nodules up to 25 cm in diameter and thin nodular beds. Fauna include styliolinids, tentaculites, bivalves, rugose corals (see Berkowski, 2008), trilobites, and brachiopods. The upper contact with the Tchrafine Group (Middle Devonian) is placed at the first appearance of nodular limestones. The latter are made up of platform limestones with goniatites, rugose corals, bivalves, gastropods, crinoids, styliolinids, wavy lamination, and bioturbations (Aitken et al., 2002, Fig. 4).

The Kess Kess mud mounds (Brachert et al., 1992, Aitken et al., 2002, Cavalazzi, 2006) are exposed in the Eastern Anti-Atlas in the Hamar Laghdad Ridge, a small mountain range about 18 km southeast of Erfoud in the Tafilalt region, Morocco (Fig. 17.5). Biostratigraphical, sedimentological, and paleontological studies of the mounds have been carried out by Gendrot (1975), Alberti (1981), Brachert at al. (1992), Peckmann et al. (1999, 2005), and Aitken et al. (2002). The Kess Kess mud mounds have circular to sub-elliptical bases ranging in diameter from few meters to about 100 m (328 ft) (Cavalazzi, 2006). Their elevation is between 20–55 m (66–180 ft) with steep (20°–65°) asymmetrical flanks. The Kess Kess Mound and the intermound deposits, as well as the underlying crinoidal limestones, are cut by a large number of veins. In the Early Devonian, mud mounds formation started with the accumulation of bioclastic piles on top volcanic highs, such as the Hamar Laghdad Ridge, followed by the deposition of more than 140 m (459 ft) of Lower Devonian crinoidal and tabulate coral limestones of the Kess-Kess Formation (Brachert et al., 1992, Belka, 1998, Cavalazzi, 2006), or the Seheb el Rhassel Group of Aitken et al. (2002). Pyroclastic volcaniclastics were derived from calc-alkaline basalts (Belka, 1998, Mounji et al., 1998). The mounds occur mostly in the topmost part of this formation/Group (Brachert et al., 1992) (Fig. 17.7). The origin of the Devonian mud mounds is controversial; Belka (1998) and Mounji et al. (1998) postulated a hydrothermal origin for the micrite that form the mounds. On the other hand, Brachert et al. (1992) proposed two major factors that controlled the development of the Kess Kess mud mounds: 1) External factors, such as the existence of a paleohigh formed by the volcanics, the dominance of unidirectional currents, storms, and sea-level fluctuations. 2) Internal factors, such as the preferred growth position of the organisms; for example, on the flanks of the mound or on top of it, and synsedimentary cementation. Aitken et al. (2002) rejected

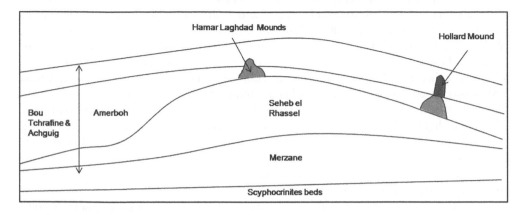

*Figure 17.7* Hamar Laghdad reef model (simplified from Aitken et al., 2002).

the role of hydrodynamic piling, baffling, and biotic production in the formation of the mud mounds and opted for a hydrothermal origin.

Extensive Lower Devonian shallow-water ooidal ironstones, contemporaneous with the mud mounds, occur in the Upper Djebilt Formation (Pragian) at Gara Djebilt along the southern margin of the Tindouf Basin (Guerrak et al., 1988, 1989, Aitken et al., 2002) and are economically exploited. They were deposited closer to the shoreline than the mud mounds. The source of the iron is believed to be the Precambrian basement. Aitken et al. (2002) postulate a seep processes for their origin.

Middle Devonian mud buildups are common in NW Africa. They have been described from Hamar Laghdad in the Tafilalt Basin, the Maader Basin (Belka, 1998, Aitken et al., 2002), and the Tindouf Basin, as well as the Ahnet Basin in Algeria (Wendt et al., 1993). Givetian mud mounds also occur in the Tata (Semara) area in the West Sahara, but unlike the other mud mounds, the reefs, biostromes and patch reefs are made up stromatoporoids (Wendt & Kaufmann, 2006, May, 2008).

During the Middle Devonian (Couvinian-Givetian), the Tafilalt area in the Eastern Anti-Atlas was completely covered by Middle Devonian shales, nodular limestones and marls, and mud mounds (Belka, 1998, Aitken et al., 2002). The Middle Devonian Hollard Mound on the eastern margin of the Hamar Laghdad Ridge (south of the Kess Kess Mound) is the largest mud mound in the area (Brachert et al., 1992) (Fig. 17.7). Conodont biostratigraphy indicates that the Hollard Mound was initiated in the earliest Eifelian (Couvinian) (Brachert et al., 1992, Peckmann et al., 1999, Aitken et al., 2002). Similar to the Kess Kess mud mounds, the Hollard Mound consists of fossiliferous micrite crosscut by numerous veins *"neptunian dykes"*. Most of these veins are filled with black-grey and yellowish-red laminated, microcrystalline calcite. The infills have negative carbon and oxygen stable isotope values of -19.5% PDB (Cavalazzi & Barbieri, 2006, Cavalazzi in press, Cavalazzi et al., in press). The core of the Hollard Mound contains a dense cluster of worm tubes and bivalves enclosed in microbial carbonates (Peckman et al., 2005). Since these organisms flourished on the location of an Eifelian hydrocarbon seep, Peckmann et al. (1999, 2005) attributed that the origin of the mounds to hydrocarbon venting and oxidation.

The Bou Tchra-Wne section is highly condensed and composed of 32 m (105 ft) of bedded to nodular limestones (Termier, 1936, Bultynck & Walliser, 1991, Becker & House, 2000, El Hassani & Benfrika, 2000, Kaiser et al., 2007, Preat et al., 2008, Webster & Becker, 2009). Conodonts from the same section were examined by Bultynck & Hollard (1980). The succession contains numerous red carbonate intervals of the so-called Red Griotte (Preat et al., 2008) with conodonts and sporadoceratids (Aboussalam & Becker, 2004). Bioclastic tempestites form a transition between an outer shelf and a hemipelagic basin. The paleoenvironment was characterized by the absence of light as indicated by the lack of phototropic algae and corals, low oxygen recorded by low content of $Fe_2O_3$, and a low sedimentation rate indicated by numerous ferruginous hardgrounds and condensed series. Preat et al. (2008) postulated that the red pigmentation was not caused by iron oxides, but by the action of iron bacteria and fungi in dysaerobic microenvironments.

Gouwy et al. (2007) carried out graphic correlations of several upper Givetian and Frasnian pelagic and neritic successions from the Tafilalt and the Maader areas (Eastern Atlas), in order to establish a regional reference section for the Frasnian conodonts. The Frasnian section corresponds to the Bouia Formation of the Achguig Group. The Middle Famennian succession in the Amessoui Syncline in the Eastern Anti-Atlas contains the following faunal succession, from top to bottom: *Kosmoclymenia lamellose, Endosyphonites muesteri, Protoxclymenia wendti*, and *Sporadoceras orbiculare*. Between the *Maeneceras biferum* and *Goniclymenia speciosa* Fauna, three other biozones have been recognized by Korn et al. (2000), namely *Playclymenia annulata, Sporadoceras orbiculare*, and *Endosiphonites muensteri*. Although most of the North African Famennian fauna are similar to their counterparts in Europe, the *Endosiphonites muensteri* Fauna show remarkable differences (Korn et al., 2000).

The Upper Devonian (Fammenian-Frasnian) is marked by subsidence and a return to detrital sediments until the Carboniferous (Hollard, 1971, Brice et al., 2007, Soulaimani & Burkhard, 2008).

The most complete Devonian-Carboniferous stratigraphic sections are exposed in the Oued Zemoul area in the Western Anti-Atlas Mountains (Choubert et al. 1971, Kaiser et al. 2004, Brice et al., 2007) (Fig. 17.1). They are represented by the Tazout Group. The top of the Tazgout 2 Formation is erosional and marked by conglomerates and red sandstones. The Tazout 3 Formation is Tournaisian in age. The lower part of the formation contains a *Gattendorfia* cf. *crassa* marker bed.

In the Western Anti-Atlas, for examples in the Assa and Kheneg Aftes areas in the Draa Valley, the sediments become sandier and form three successive sandstone ridges of the Tazout 1, 2 and 3 formations (Brice et al., 2007). However, *Gattendorfia* is absent in these sands. The thickness of the Tazout Sandstones decreases westwards (Cavaroc et al. 1976, Choubert et al. 1969). The Tazout 1, 2 and 3 formations were assigned to the Tazout Group as component members with new names: The Maader Talmout Member is 130 m (427 ft) in thickness and consists of sandstones and siltstones with brachiopod coquinas (Becker et al. 2004, Kaiser et al. 2004). It is equivalent to Hollard's Tazout 1–2 (Brice et al., 2007). The Kheneg Lakahal Member is made up of unfossiliferous siltstones. The third member is unnamed and corresponds to the upper Tazout 3 ridge.

Brice et al. (2007) recognize three brachiopod associations which they labeled as Lower, Middle, and Upper Fauna. The Lower Fauna of late Famennian age occurs in all sections, except in the Tafilalt-Maader.

In Southwestern Morocco, the Lower Fauna occurs at Hassi Rharouar and Kheneg Aftes sections in the Tazout 1 Sandstone Formation (Maader Talmout Member). Fossils include *Leptagonia* cf. *analoga*, *Schuchertella* sp., the productids *Hamlingella talmouti* and *Mesoplica praelonga*; the rhynchonellids *Paurogastroderhynchus lakahalensis*; the spiriferids *Cyrtospirifer pseudorigauxia*, *Dichospirifer zemoulensis*, *Prospira struniana*, and *Eobrachythyris hollardi*. The Upper Fauna of early Tournaisian age are present in the Assa, Akka, and Zemoul areas and in the Timimoun area in Algeria. In the Tafilalt-Maader basins, the Upper Fauna occurs above a 'Hangenberg Black Shale' equivalent.

The Middle-Late Devonian paleogeography of the Western Meseta consisted of a western carbonate platform and an eastern fault-bounded turbidite basin (Piqué, 1987, Bouabdelli & Piqué, 1996, Baidder et al., 2008). There are faunal similarities between the platform deposits of the Western Meseta and the Anti-Atlas, and both belong to the "Bohemian facies" according to Hollard (1981). These similarities support the view that the Meseta Domain was in close proximity to the Anti-Atlas during the Devonian (Piqué & Michard, 1989, Hoepffner et al., 2006, Baidder et al., 2008). On the other hand, the tectonic setting of the Meseta Domain and the Anti-Atlas showed significant differences during the Late Devonian. The Doukkala and other Western Meseta basins were dominated by a compression-transtensional regime (Piqué, 1987, Bouabdelli & Piqué, 1996, Echarfaoui et al., 2002) and the Eastern Meseta by the Variscan folding phase (Piqué & Michard, 1989, Hoepffner et al., 2006). These differences led to the notion that the Meseta Domain and the Anti-Atlas were not attached during the Late Devonian Stampfli & Borel, 2002, Burkhard et al. (2006).

The Devonian in the Azrou-Khenifra area is represented by two units; an auto-chthonous platform unit (Upper Devonian) and an allochthonous unit of thick calcareous and siliciclastic turbidites with provenance from the east. A few intermediate and basic veins of dolerites and ophetic gabbros are associated with these deposits (Bamoumen et al., 2008). The eastern limit of the Azrou-Khenifra Basin coincides with the Hercynian front (Piqué & Michard, 1989, Hoepffner et al., 2006).

El Hassani & Benfrika (1995) described five Devonian formations in the Khatouat and Mdakra zone in the Meseta; they are the Fouizir, Chabet El Baya, Bir En Nasr, Souk Jemaa, and M'Garb formations.

Various studies of the Paleozoic succession of the Doukkala Basin have been carried out since the mid-1960s by Hollard (1967), Destombes (1971), and El Attari (2001), Backer et al. (1965), Benfrika (1994); the tectonics by Michard (1976), and the sedimentology by Ben Bouziane (1995).

The Silurian-Lower Devonian succession on the northeastern side of the Doukkala Basin in the Western Coastal Meseta is represented by interbedded carbonates, shales, and sandstones succession of the Oued Cheguigua Formation (Cherradi et al., 2007). The succession consists of three depositional sequences: SD1, SD2, and SD3, dated Silurian-Lochkovian, Pragian, and Emsian, respectively. These sequences were developed on a platform-type ramp, in response to sea level changes. Dolomitization and recrystallization of the Devonian carbonates occluded all pore space and reduced their potential as reservoir rocks for hydrocarbons (Cherradi et al., 2007).

In the Rabat area, in the Central Meseta, the oldest rocks exposed along the left bank belong to the Ordovician and Silurian. They are overlain by a Lower Devonian succession composed of the following formations (Fig. 17.8): Hossei Formation (Lochkovian), Bou Regreg Formation (Pragian), Oued Akrech Formation (Pragian-Emsian), and Rahal Formation (Eifelian) (Benfrika et al., 2007).

| | | Formations | |
| --- | --- | --- | --- |
| | Stages | RABAT AREA | TIFLET AREA |
| Middle Devonian | Lower Givetian | *Not recognized* | Formation B |
| | Eifelian | Rahal | Formation A |
| Lower Devonian | Emsian | Oued Akrech | Safsaf |
| | Pragian | Bou Regreg | Tiflet |
| | Lochkovian | Hosei | Zemmour |
| Lower Silurian | Pridoli | | |

*Figure 17.8* Devonian stratigraphy of the Rabat-Tiflet area (from Benfrika et al. 2007).

At Bou Regreg, In the Rabat area, the Devonian succession includes in an ascending order, the Hosseia, Bou Regreg, Oued Akrech, and Rahal formations (Benfrika et al., 2007) (Fig. 17.8).

Only the upper part of the Hossei Formation crops out at Bou Regreg and consists of interbedded dark limestones and thin black graptolitic shales. The limestones are locally dolomitized, with tentaculites, dacryoconarids, orthocones, ostracods, bryozoans, and crinoids. The Lochkovian index fossil *Monograptus uniformis* occurs in the lower part of the unit. Conodonts are represented by *Caudicriodus* cf. *woschmidti*, *Ozarkodina remscheidensis remscheidensis*, and *Belodella devonica* (Benfrika et al., 2007).

At Bou Regreg, the Bou Regreg Formation has a thickness of about 30 m (98 ft). The lower part of the formation is made up of pale grey nodular bioclastic wackestones and packstones with tentaculites, ostracods, and the conodonts *Eognathodus* cf. *sulcatus eosulcatus*, *Ozarkodina pandora*, and *Ozarkodina* cf. *repetitor* (Benfrika et al., 2007). The upper part is composed of micritic greyish limestones with tentaculites, bivalves, bryozoans, crinoids, fish scales, *Nowakia acuaria*, and *Guerichina strangulate*. The Bou Regreg Formation has been assigned a Pragian age by Benfrika et al. (2007).

At Bou Regreg, the Oued Akrech Formation has a total thickness of about 50 m (164 ft). It is made up of thick-bedded dolomitic limestones with tentaculites, ostracods, bryozoans, brachiopods, and crinoid fragments at the base, followed upward by massive dolomitic limestones and thick-bedded dolomites, with crinoids, brachiopods, gastropods, tentaculites, and fish bioclasts, and then by bedded wackestones-packstones. Benfrika et al. (2007) reported the following conodonts from the Oued Akrech Formation: *Ozarkodina steinhornensis miae* and *Caudicriodus celtibericus* near the base and *Latericriodus bilatericrescens bilatericrescens* in the lower part, *Polygnatus dehiscens* vel. *kitabicus*, *Latericriodus beckmanni*, and *O. steinhornensis steinhornensis* in the middle part, and *Icriodus fusiformis*, *I. corniger ancestralis*, *I. culicellus culicellus*, *I. rectirostratus*, and *Polygnathus laticostatus* in the upper part.

The Rahal Formation is only partly exposed at Bou Regreg and consists of shales with the conodont *Belodella devonica* (Benfrika et al., 2007).

At Al Khaloua, SW of Tiflet, on the left bank of the Oued Bou Regreg, the Rechoua Formation is about 38 m (125 ft) thick and consists of interbedded limestones and dolomitic limestones. It is overlain by Safsaf Formation. The Rechoua Formation is a lateral equivalent of the Bou Regreg Formation (Pragian) (Michard, 1976). The Safsaf Formation consists of grainstones with bioclasts of crinoids and brachiopods, dolomitic limestones, and dolomites. Conodonts include *Caudicriodus celtibericus*, *Belodella devonica*, *Polygnathus dehiscens*, *Ozarkodina steinhornensis miae*, *Latericriodus bilatericrescens bilatericrescens*, and *Pelekysgnathus serratus serratus*. The Safsaf Formation has been dated Pragian-Emsian (Benfrika et al., 2007).

In the Oued Tiflet area, east of Rabat, the Devonian succession is made up of the following units in an ascending order (Fig. 17.8):

Zemmour Formation (Lochkovian)
Tiflet Formation (uppermost Lochkovian-Emsian)
Safsaf Formation (Emsian)
Formation A (uppermost Emsian-basal Givetian)
Formation B (Lower Givetian)

At Oued Tiflet, the Devonian is represented by the Zemmour, Tiflet and Safsaf formations, ranging in age from Wenlock to the upper Emsian.

The Zemmour Formation starts with an 18 m (59 ft) thick interval of interbedded black shales and dark grey limestones rich in bivalves, gastropods, orthocones, crinoids, and ostracods, followed by 34 m (112 ft) of black shales with intercalations of nodular limestones or thin-bedded limestones, and then by 40 m (131 ft) of micaceous platy limestones and shales in the upper part. Conodonts include *Belodella devonica*, which first occurs at the base of the unit, as well as *B. devonica, Caudicriodus* cf. *C. postwoschmidti*, and *Ozarkodina remscheidensis remscheidensis*. The Zemmour Formation has been assigned a Pridoli-Wenlock, based on the presence of *Monograptus lochkovensis* and the conodonts *Kockelella absidta* and *Ozarkodina bohemica* (Benfrika et al., 2007).

The Tiflet Formation has a total thickness of 48 m (157 ft) and consists of pale grey nodular limestones with thin shale interbeds, with tentaculites, trilobites, and crinoids. It contains the conodonts *Ozarkodina pandora* and *Latericriodus steinachensis* in the lower part, followed by *Caudicriodus celtibericus* and *O. steinhornensismiae* in the upper part (Benfrika et al., 2007).

The Tiflet Formation is overlain by a unit equivalent to the Oued Akrech Formation about 45 m (148 ft) in thickness and composed of dolomitic limestones and thin shales with tentaculites, ostracods, crinoids, and the conodonts *Latericriodus bilatericrescens bilatericrescen, O. steinhornensis steinhornensis*, and *L. beckmanni* (Benfrika et al., 2007). It has been assigned to the Emsian based on comparison with the Oued Akrech Formation (Benfrika et al., 2007).

Devonian rocks exposed in an abandoned quarry south of Oued Tiflet was divided into two units designated as Formations A and B. Formation A overlies 11 m (36 ft) of the Safsaf Formation and consists of dark grey shales and thin limestones with tentaculites, bivalves, and brachiopods. Gouwy et al., (2007) described a complete section of Formation A, about 60 m (197 ft) in thickness, where they identified the conodonts *Polygnathus linguiformis bultyncki* and *P. serotinus* in the lower part of the section, as well as *P. hemiansatus* in the upper part. They assigned the formation a Givetian age and correlated it with the Rahal Formation (upper Emsian-lower Givetian, according to Benfrika et al., 2007) in the Oued Bou Regreg area to the east. The overlying Formation B is 35 m (115 ft) thick and consists of reefal limestones with tentaculites, ostracods, crinoids, and corals, in addition to the conodonts *Polygnathus timorensis, P. rhenanus*, and *P. varcus*, in addition to various species of *Ozarkodina, Tortodus, Icriodus*, and *Eognathodus*.

The Ahrerouch Formation (Givetian) south of the Tiguisselt Village in Oued Tata in the Draa Valley is about 70 m (230 ft) in thickness and consists shales, marls, with thin layers of limestones and siltstones (Aboussalam & Becker, 2004). The formation consists of the Tigusselt Member and Oued Mzerreb Member. The Ahrerouch Formation has been dated Givetian based on ammonites (Aboussalam & Becker, 2004). The succession consists, from top to base, of:

*Agonoatites* and *Afromaenoceras* Beds (equivalent to Tully Limestone at Mzerreb and Oufrane)
Juvenocostatus Beds
Grey Marker Limestone (equivalent to the Tigusselt Member)

Lower Red Griotte (equivalent to the Oued Mzerreb Member)
Upper *Pumilio* Bed
Coral Marl and Upper *Maenioceras* Beds
Lower *Pumilio* Bed

The *Pumilio* Bed is made up of 15 cm (6 in.) of platy to laminated, bioclastic, dark grey limestones with styliolinids and small brachiopods. The same interval occurs at Oufrane (Ifrane). The Coral Marl and Upper *Maenioceras* Beds are composed of shales with the hematitic gastropods *Sobolewia virginiana* and *Maenioceras* sp., among others, marls, nodular limestones and marls, massive limestone, silty limestones, and styliolinitic beds. The Grey Marker Limestone overlies the Oued Mzerreb Member (Middle Givetian) (equivalent to the Lower Red Griotte) of the Ahrerouch Formation and consists of massive limestones and marls, styliolitinid wackestone with ostracods and some brachiopods. The *Juvenocostatus* Beds contain *Mzerrebites*, *Atlantoceras*, rare *Maenioceras* n. sp.II, and *Afromaenioceras crassum*. The *Agoniatites* and *Afromaenioceras* Beds consist of three shale, marl, and limestone cycles, with abundant hematitic *Tervoneites*, *Wedekindella lata*, and *Agoniatites meridionalis*. It was correlated with the Middle and Upper Tully Limestone at Mzerreb and Oufrane.

Upper Givetian to lower Frasnian conodont fauna from the Tafilalt of the Eastern Anti-Atlas have been described by Bultynck & Hollard (1980), Bultynck & Jacobs (1981), Bensaid et al., (1985), Gouwy et al., (2007), Aboussalam & Becker (2007), and Webster & Becker (2009). The Givetian of the Tafilalt is mostly characterized by condensed pelagic limestones rich in ammonites, nautiloids, tentaculitoids, and ostracods, as well as rare trilobites, crinoids, small solitary rugose corals, rare tabulate corals (thamnoporids and cladochonids), brachiopods (rhynchonellids and lingulids), bivalves, such as the large-sized genus *Panenka*, and gastropods. Coral biostromes with *Phillipsastrea* and some stromatoporoids are developed only in the southern Tafilalt Amessoui Syncline (Massa, 1965, Aboussalam & Becker, 2007). The upper Givetian-basal Frasnian Bouia Formation (Unit K, Bultynck & Walliser, 2000) overlies the *Pharciceras* aff. *amplexum* Bed, (of the Upper Tully Event Interval), (topmost middle Givetian) (Aboussalam & Becker, 2007), from base to top:

1 The *Mzerrebites Erraticus* Beds consist of greenish-grey, thin-bedded, nodular limestones, interbedded with marls or thin crinoidal beds with compressed *Mzerrebites Erraticus* goniatites and rare *Pharciceras*.
2 The Red *Lunupharicceras* Beds is made up of reddish-grey, hematitic, nodular or thin-bedded limestones with *Lunupharicceras*.
3 The Lower Marker Bed consists of massive limestones, which Ebert (1993) named the "Unterer *Pharciceras*-Horizont". It represents a shallowing-upward cycle with a hardground at the top. In the basinal Hassi Nebech area, the massive limestones change into fossiliferous grey nodular limestones with *Synpharciceras*. In the Amessoui Syncline (Jebel Ouaoufilal) the Lower Marker Bed is replaced by shales with small goethitic goniatite.
4 The *Taouzites* Beds consist of nodular and thin-bedded limestones with ammonite-rich levels with very large *Taouzites taouzensis* and small *Pseudoprobeloceras pernai*.

5   The Upper Marker Bed was named the "Oberer *Pharciceras*-Horizont" (Ebert,
    1993). It represents a shallowing upward cycle. The top contains many goniatites
    and large nautiloids.
6   The *Petterroceras* Beds consist of reddish iron-rich micritic limestones crusts
    on top of Upper Marker Bed and contains various species of *multilobed*, *Pette-
    roceras*, and *Ponticeras*.
7   The *Rotundiloba Pristina* Bed is a red, thin, unfossiliferous, micritic limestone
    bed.

The Frasnian Event (Lower Styliolinites) (Aboussalam & Becker, 2007) is represented
by the widespread occurrence of black marls with lower Frasnian styliolinids (Gouwy
et al., 2007). They have been reviewed by Bensaid et al. (1985). They occur within
the Bouia Formation. Wendt & Belka (1991) applied the term "Lower Kellwasser
Member" to the succession but was later correlated with the Frasnian Event Interval
by Belka et al. (1999) and Lüning et al. (2004c).

The Moroccan Variscan belt is represented north of the Atlas Belt by the Meseta
Domain (Piqué, 1989, Piqué & Michard, 1989, Hoepffner et al., 2005) (Fig. 17.1).
The Mesetas are separated from the Atlas Domain by the Atlas Paleozoic Trans-
form Zone (APTZ) (Piqué et al, 1998). They are separated from the Anti-Atlas
by the western extension of the APTZ, which includes the Agadir lineament, the
Western South Atlas, and the Tizi-n'Test (TTFZ) faults and from the Bechar-Oujda
corridor by the Bsabis-Tazeka-Melilla Fault. The Mesetas comprise a set of NE-SW-
trending Paleozoic structures (anticlinoria), which include the High Plateau, Jebilet,
Rehamna, and the Central Massif (Ziar-Azrou and Khouribga-Oulmes). The NE-
SW-trending Middle-Atlas fold belt runs in the middle of the Meseta Domain. The
Meseta Domain is separated into two parts by (or the Middle Atlas axis). The Mese-
tas are further subdivided into: High Plateau or Oran Meseta, Western (Moroccan)
Meseta, Central Meseta, Phosphate Plateau & the "Causse", Eastern Meseta, and
Coastal Meseta.

Devonian sequences occur in the SW-NE Khouribga-Oulmes anticline in the Central Meseta (Piqué & Michard, 1981, Kaiser et al., 2007). Termier (1938) reported
the presence of Lower Famennian ammonites fauna from outcrops at Tabourit, Bou
Gzem, Sidi Bou Sif, and Ain Djema.

The Middle and Upper Devonian succession at Bou Gzem (Bou Keziam) (Cogney,
1967, in Kaiser et al., 2007) starts at the base with the Slimane Formation. The Slimane
Formation consists of Eifelian trilobite-bearing dark limestones (Ain Jemaa Member), and Givetian reefal limestones (Bou Sif Member), then by Frasnian greenish-grey
shales, Lower Famennian nodular limestones with pyritic or hematitic tornoceratids
and cheiloceratids (Becker 1993), sandy shales, dark limestones with Middle Famennian trilobites (Franconicabole), grey shales, and a thick quartzite unit at the top.
Alberti (1970) reported early Famennian conodonts *Palmatolepis glabra glabra*,
*P. glabra pectinata*, and *P. quadrantinodosa inflexoidea*).

Two pelagic successions occur near Ain Jemaa dated by conodonts and ammonites as Middle-Upper Famennian (Kaiser et al., 2007). The Bou Gzem Formation
(Upper Devonian) was erected by Kaiser et al. (2007). It is made up of shales and
nodular limestones and consists of three members. The Upper Member consists of
black shales correlatable with the black shales of the Hangenberg Event. The Bou

Gzem Formation is overlain by quartzites the Ta'arraft Formation (also introduced by Kaiser et al., 2007) that probably correlates with the major regressive phase of the Hangenberg Event (Kaiser et al., 2007). Strunian coarse siliciclastics end the succession. In the meantime, other Meseta areas show contemporaneous reworking and deposition of mass flows, conglomerates, and turbidites in the adjacent pelagic basins. In other Meseta regions the Middle and Upper Famennian succession is interrupted by transgressive pulses of the global *Annulata* and Dasberg Events (Kaiser et al., 2007) represented by thin pelagic limestone beds. A different sequence occurs east of the Oulmes Fault in the Moulay Hassani section which belongs to a more basinal setting. The Moulay Hassane Formation consists of black shales and limestones with Upper Frasnian and Lower Famennian goniatites and brachiopods. The Kellwasser beds occur in the section. The upper member of the Moulay Hassane Formation is a thick sandstone unit correlatable with the Ta'arraft Formation at Ain Jemaa. The Bou Gzem Formation is missing above the reefal limestones. In the Sidi Bettache Basin, south of Rabat, Famennian conglomerates occur in the Ain Hallaouf Formation within a basinal facies with thick mass flow deposits (Piqué & Kharbouch 1983, Piqué, 1984, Kaiser et al., 2007).

### 17.4.1.5   Carboniferous

In the Eastern Anti-Atlas, in the southern Tafilalt area (Fig. 17.9), the Lower Carboniferous consists of shallow-marine siltstones and sandstones. The presence of deep water sediments in the Anti-Atlas is disputable (Michard et al., 1982). Burkhard et al. (2006) could not find any convincing flysch sequences that previously thought to have existed near the High Atlas Front (Michard, et al., 1982). On the other hand, the transition of this sequence into slope and basinal facies is probably located in the northern Tafilalt region, as indicated by debris flow and turbidite deposits (Wendt et al., 1984).

The Devonian-Carboniferous succession in the Tafilalt-Maader area shows considerable variations of sedimentary facies and thickness (Hollard 1967, Korn 1999, 2000, Brice et al., 2007). Two regressive siliciclastic units occur at the base: the Aguelmous-nou-Fezzou Sandstone (*Grès de l'Aguelmous-nou-Fezzou*) at Maader, and the Ouaoufilal Sandstone (*Grès d'Ouaoufilal*) at Tafilalt. Basal Tournaisian spiriferids occur immediately above the Aguelmous Sandstones in the Tafilat area (Kaiser et al. 2004). The lower Ouaoufilal Formation consists of a "Hangenberg Black Shale" equivalent, just below the Devonian-Carboniferous boundary, followed by a thick succession of unfossiliferous shales and turbidites. The upper Ouaoufilal Formation is a 55 m (180 ft) thick and consists of brachiopod-bearing sandstones with the rhynchonellid fauna *Centrorhynchus* (?) sp. (Famennian) near its base, and *Hemiplethorhynchus* (?) sp. (Tournaisian) in the upper sands (Brice et al., 2007). The basal shales of the overlying Oued Znaigui Formation are unfossiliferous, but middle Tournaisian ammonites occur higher in the formation (Korn et al., 2000). The Devonian-Carboniferous boundary has been placed at the occurrence of the *Acutimitoceras* fauna by Brice et al. (2007). Two equivalent Carboniferous ammonic horizons occur in the Tafilalt-Gourara area (Ebbighausen et al., 2004, Bockwinkel & Ebbighausen, 2006, Korn et al. 2010a-c). The *Gattendorfia-Eocanites and Gattendorfia-Kahlacanites* horizons were assigned a middle-Early Tournaisian age by Brice et al. (2007).

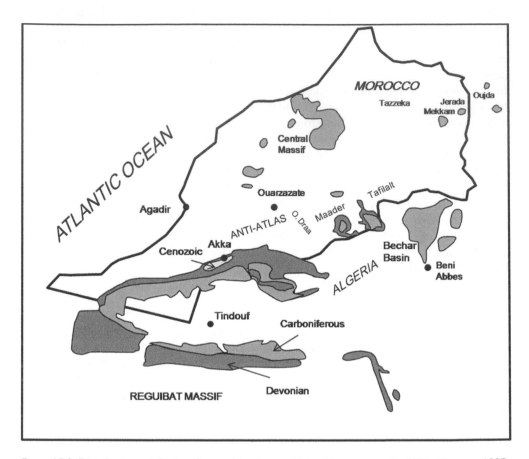

*Figure 17.9* Distribution of Carboniferous Northwest Africa (Cavaroc et al., 1976, Weynat, 1985, Brice et al., 2007).

No Permian-Jurassic sediments are found in the Anti-Atlas and Carboniferous sediments are followed by Cretaceous conglomerates (Soulaimani & Burkhard, 2008).

The Lower Carboniferous in the Western Anti-Atlas is represented in a series of cuestas in the Draa Valley which forms the northern side of the Tindouf Basin, which includes the Tazout Group, and the Betaina and Jbel Reouina Formations (Soulaimani & Burkhard, 2008). Deposition was continuous in that area from uppermost Devonian into the Carboniferous. Strunian (latest Devonian in this book) deposition was dominated by sandstones and arkoses with brachiopods and goniatites, such as *Gattendorfia*, of the Tazout Formation (Conrad, 1984)[7]. The upper member of the formation has been dated Tournaisian. The lower Visean is represented by the overlying Betaina Formation composed of shales and sandstones. The upper Visean marks

---

7 The Tazout Formation was raised to group status "Tazout Group" by Becker et al. (2004) and divided it into two members; a lower Maader Talmout Member and an upper Kheneg Lakhal Member. The latter could not be dated (Kaiser et al., 2004).

the most important transgression in the Paleozoic, accompanied by subsidence, which led to the deposition of more than 2500 m (8203 ft) of shales and sandstones (in the Tazout region) and up to 650 m (2133 ft) of massive limestones with biostromes with brachiopods and foraminifera (Djebel Ouarkziz area) of the Ouarkziz Formation. The formation is overlain by the Betana Formation (Namurian-early Stephanian) with a thickness of 1400 m (4593 ft) of continental fine-grained sandstones and shales with plant fragments (Conrad, 1985).

The Meseta Domain constitutes the principal orogenic zone of the Variscan Chain in Morocco (Bamoumen et al., 2008) and has been subdivided into numerous structural zones (Michard, 1976, Piqué & Michard, 1989, Hoepffner et al., 2006). The Variscan cycle started with a phase of extension which led to the formation of Devonian and Carboniferous basins.

The Carboniferous Jerada (Djerada) Basin (Fig. 17.9) is located southwest of Oujda in the Eastern Meseta. The Visean basin-fill sequence starts with rhyolitic, andesitic, and dacitic volcano-sedimentary rocks of the Oued Defla Formation (Herbig & Aretz, 2007). The Oued Defla Formation is unconformably overlain by the Namurian-Westphalian succession of the Cafcaf (Safsaf) Formation composed of interbedded cherts and pyroclastics, followed by the Oued El Koriche Formation (Herbig & Aretz, 2007, Aretz et al., 2007). Goniatite-bearing, predominantly shaly slope deposits are represented by the Oued Es-Sassi Formation. The basin fill ends with a prograding shallow-marine carbonate facies with various reef facies (Aretz et al., 2007, see below) and siliciclastic sediment of the Koudiat Es-Senn Formation. Emersion took place at the end of the Visean. The geodynamic setting of the Jerada Basin is controversial. The Jerada Basin is believed to be part of a larger basin comprising parts of the Eastern Meseta, and probably represents a continental margin back-arc basin, dissected by a strike-slip fault system associated with an oblique, south-directed subduction. The Visean basin fill sequence contains well developed late Visean reefs (Herbig & Aretz, 2007, Aretz et al., 2007). Three distinctive reef intervals can be recognized on the southern edge of the Jerada Basin. The oldest reefs are limestone olistoliths tens of meters in size (Aretz et al., 2007) within the Oued el Koriche Formation, which represents toe of slope facies (Herbig & Aretz, 2007, Aretz et al., 2007). They might attain sizes of several thousands of cubic meters. The reefs are made up essentially of siliceous sponges, microbialites, and subordinate bryozoans. Complex cavity systems are widespread. Similar reefs occur within a shallowing-upward cycle of carbonate-siliciclastics of the overlying Koudiat es-Senn Formation (Aretz et al., 2007). The reefs flank facies consists of bryozoans and rugose corals. Reef formation ended with the deposition of cross-bedded crinoidal and oncolitic grainstones-rudstones with *Girvanella* patch reefs. On the northern flank of the basin, a single coral patch reef is recorded within oolitic shoal deposits from the upper part of a second shallowing-upward cycle of the Koudiat es-Senn Formation (Aretz et al., 2007).

The Djerada coal was discovered in 1928 by Andre Brichant who also discovered the coal in Algeria (Missenard et al., 2008). The Coal Basin of Djerada (Delépine, 1939, 1941, Owodenko, 1946, Bouckaert & Owodenko, 1965) yielded Late Visean goniatites and Namurian and Westphalian faunas similar to those known from Central Europe. The Djerada Coal Measures contain also limnic and land plants dated Westphalian C (Bensaid et al., 1985).

Fossils from two formations in the Jerada (Djerada) Basin, the Mekam Shales and the Tafechna Limestone have been described by Bensaid et al. (1985a).

The Mekam Shales (Visean) contain *Delepinea comoides*, cf. *Merocanites planorbis, Asterocalamites?*, followed unconformably by a transgressive sequence composed of shales with limestone clasts with *Palaeosmilia murchisoni, Plectogyra arcuata, P. prisca, Archaediscus krestovinkovi, Ungdarella,* and *Anthracoporellopsis machaevi, the* goniatites *Beyrichoceras obtusum, B. micronotum, B. truncatum, Paraprolecanites mixolobus,* and *Goniatites falcatus* (Bensaid et al., 1985b).

The Tafechna Limestone yielded Lingula elongata, Spirifer striatus, Coelonautilus, and Eumorphoceras bisulcatum. At Oued Agaia, the succession contains shales with Goniatites crenistria, G. straiatus, Posidonia becheri, as well as Antiquatonia, Gigantoproductus, and Spirifer bisulcatus. These shales are overlain by Namurian marine sediments, followed upward by the Westphalian A and B, with plants appearing for first time. A major regression is indicated at the beginning of Westphalian B (Bensaid et al., 1985b).

Two volcaniclastics intervals occur within the Lower Carboniferous shales and limestones sequence (Late Visean-Serpukhovian) in the Eastern Meseta (Berkhli et al., 2000, 2001). The first interval (Visean) accumulated in horsts and grabens during the opening of the Beni-Snassene, Jerada, and Mekam basins; the second spans the Visean-Serpukhovian boundary and accompanied the opening of the Debdou-Tazekka Basin.

The Lower Carboniferous (Mamet et al., 1966) in the Central Meseta is marked by a renewed increase in detrital input from the east, the south and the north (Michard et al., 1988, Piqué & Michard, 1989). Regional facies and thickness changes point toward an open ocean to the west and northwest throughout the Paleozoic. The Visean transgression in northern Morocco was synchronous with faulting (Huvelin & Mamet, 1997). Pre-Visean blocks in the Khenifra area in the Central Meseta were covered by shallow seas and Visean deposits overlie unconformably the pre-Visean sediments. Four transgressive phases can be recognized (Huvelin & Mamet, 1997): V1 overlies Ordovician sediments and is made up of limestones followed by flysch deposits. V2a consists of platform carbonates and rests on Devonian sediments. V2b-V3a is made up of detrital platform sediments and overlies Cambro-Ordovician strata. V3b is made up mainly of partly carbonated detrital sediments (see Bensaid et al., 1985, Herbig & Aretz, 2007).

During the Visean, the sea transgressed over the entire Meseta Domain. Synsedimentary tectonics, for example the Jebilet and Azrou-Khenifra basins of the Central Meseta received large thicknesses and lithological variations of Carboniferous sediments (Bensaid et al., 1985, Bamoumen et al., 2008). The Upper Visean basins of the Jebilet and Azrou-Khenifra form part of the Central Meseta, between the Eastern and Western Mesetas (Piqué & Michard, 1981, Piqué et al., 1993, Bouabdelli & Piqué, 1996, Bamoumen et al., 2008). The development of the two basins located between the Eastern and Western Mesetas was controlled by major strike-slip faults (Bamoumen et al., 2008). This phase was accompanied by the emplacement of synorogenic and late orogenic granites; for example in the Sidi Bettache Basin, which is separated from the Azrou-Khenifra Basin by the Zaer-Oulmes Ridge (see Hoepffener et al., 2006, Fig. 3).

Three structural units can be recognized the Azrou-Khenifra Basin in the Eastern Meseta (Bouabdelli, 1994):

1  An eastern allochthonous terrain which contains the nappes of Khenifra, Mrirt, and Azrou. The Azrou nappes consists of Upper Ordovician shale and sandstones, Silurian black graptolitic shales, Lower Devonian clastic rocks, Middle and Upper Devonian carbonates, upper Tournaisian conglomerates and greywackes, and Lower and Middle Visean limestones and shales at the top. The Mrirt nappe is located south of the Azrou nappe and consists of Upper Ordovician sediments between 500–700 m (1641–2297 ft) in thickness. The Khenifra nappe occurs SW of the Mrirt nappe and is made up of Ordovician sandy schists.

2  A central unit composed of Upper Tournaisian-Upper Visean autochthonous transgressive sediments which overlie Cambro-Ordovician rocks. They include the Bou Khadra-Afoud Oulgham, Mouchenkour, and Ain Ichou formations. Calcareous sandstones and olistoliths in the middle part of the Mouchenkour Formation contain calcareous microfossil assemblages contain the problematic algae *Ungdarella*, *Fasciella*, and *Koninckopora* and the foraminifera *Archaediscus*, *Pseudoendothyra*, and *Eostaffella* (Vachard et al., 2006).

3  A western unit with folded Upper Carboniferous (Namurian) autochthonous sediments bordered to the west by the Fourhal Basin in the Central Massif.

The Jebilet Massif is a chain of Paleozoic rocks between the Haouz and Bahira areas. Two Visean formations, juxtaposed along the Sidi Bou Othmane strike-slip Fault can be recognized (Bensaid et al., 1985, Bamoumen et al., 2008): The Kharrouba (Flysch) Formation to the east and the Sarhlef Schists to the west. The Eastern Jebilet is characterized by the presence of Ordovician and Devonian allochthonous terranes resting directly on the sandy shales of the Kharrouba Formation.

The Kharrouba (Flysch) Formation is made up of flysch deposited during the Visean. Its base has not been recognized. The flysch contain chaotic olistostrome blocks composed of Ordovician sandstones and quartzites, Devonian limestones with Tentaculites, and sandstones, and Upper Visean black crinoidal limestones (Bamoumen et al., 2008). The Kharrouba Formation has been dated Visean based on *Posidomya becheri* and *Goniatites crenistria* (Huvelin, 1977; cited by Bamoumen et al., 2008). The Kharrouba sediments show synsedimentary folds cut by vertical faults. The central part of Jebilet contains bimodal volcanic composed of tholeiitic and calc-alkaline suites. U-Pb zircon dating yielded an age of 330.5 Ma (Essaifi et al., 2003). The Eastern Jebilet, on the other hand, contains basaltic volcanics and volcaniclastics with pillow-lavas and pyroclastic breccias (Bamoumen et al., 2008).

The Sarhlef Schists (Visean according to Bensaid et al., 1985) are also associated with abundant volcanics with calc-alkaline affinities and probably formed in an intracontinental anorogenic setting (Bamoumen et al., 2008). In the M'rirt region, there are two Carboniferous formations associated with volcanic; the Tanwalt Formation "*Serie de Tanwalt*" (Upper Visean) and the Talgarat Formation (Namurian). The Tanwalt Formation which is equivalent to correlates with the Kharrouba Formation was intruded by volcanics with alkaline affinities, while the Talgarat Formation contains calc-alkaline volcanics. The latter are similar to the volcanics of the Sarhlef

Schists. Carboniferous siliciclastic sediments in the eastern part of the Azrou-Khenifra Basin are also intruded by basic igneous rocks "*Serie de Tanwalt*".

The affinity and origin of the basic volcanics in the Fourhal Basin in the Central Meseta is controversial. Kharbouch (1994), Remmal et al. (1997), and Remmal (2000) maintain that the basic volcanics have an alkaline to tholeiitic affinity. On the other hand, Ben Abbou (2001) and Roddaz et al. (2002) advocate a calc-alkaline affinity. Roddaz et al. (2002) describe basic volcanics of pillow lava flows and gabbros and dolerite sills of calc-alkaline affinities associated with turbidites. According to Driouch et al. (2010), trace element patterns of the parental melt inferred from clinopyroxene analyses include high LILE and low HFSE concentrations compatible with magmas from a subduction-related geotectonic setting. Accordingly, two different geodynamic settings have been proposed: The first proposes an intraplate extension setting (Kharbouch, 1994, Remmal et al., 1997, Remmal, 2000); the other assumes an oceanic subduction zone (Roddaz et al., 2002, Driouch et al., 2010). These views were challenged by Hoepffner et al. (2005, 2006) who pointed out the difficulty of assigning these volcanics to either intracontinental extension or back-arc basin setting.

The basal Carboniferous of the Eastern Meseta lies conformably on Upper Devonian sediments (Bensaid et al., 1985a). The succession in the Eastern Meseta consists of Bou Mia Shales, Djebel Goulib Formation, Tizi ben Tizuit Formation, Titar Oumjel Formation, Tizi n'Salihine Formation, Djebel Hadid Conglomerates, Koudiat Ain Jemaa Formation, Dchar Ait Merrouane Formation, Igrou Hammoud Formation (consists of *Neoglyphioceras subcirculare* Horizon, Mohammed ben Si Mohammed Conglomerates, and Ich ou Mellal Shales and sandstones), ochreous nodules shales, and Sidi Kassem Formation and Zeiliga coal measures. The Koudiat Ain Jemaa Formation, Dchar Ait Merrouane Formation, and the *Neoglyphioceras subcirculare* Horizon are equivalent to the Afroug Formation. The succession starts with Strunian deposits which contain fauna of Devonian and Tournaisian affinities. The Strunian (latest Devonian) succession starts with ferruginous sandstones with *Lepidodendron* bark fragments and microfauna of late Devonian affinity, such as *Evlania transversa*, *Girvanella media*, *Issinella devonica*, *Anthracoporellopsis* index, *Septaglomospiranella nana*, the ostracods *Cryptophyllus*, and the macrofauna *Prospira greenockensis*, *Plicochonetes elegans*, *Rugosochonetes hardrensis*, *Marginatia burlingtonensis*, *Syringothyris hannibalensis*, *Cyrtospirifer* cf. *verneuili*, *Sphenospira julii*, *Orbiculoidea*, *Proetus* (*Pudeproetus*), *Aganaster*, and *Conularia*. Shales contain bryozoans, bivalves, and eggs containing *Atactotoechus* (see Bensaid et al., 1985a, p. 336, for more fossils).

Marine Tournaisian outcrops are very rare. In the Rabat area, a narrow gulf was filled with 2000 m of flysch with *Syringothyris* and goniatites. In the Eastern Meseta, the *Fenestella* Shales of Bou Mia Formation yielded Tournaisian fossils (citing Dedok & Hollard, 1980). The lower Visean (V1) is missing. The Bou Mia Shales "Tournaisian" are overlain unconformably by the Djebel Goulib Formation (V2a), according to Bensaid et al. (1985a). The middle and upper Visean rest unconformably on rocks of various ages. The Visean succession consists of shales, quartzites, and sandstones with brachiopods and large caniniid corals, and rests disconformably on the Strunian (Visean V2b according to Bensaid et al., 1985). The Djebel Goulib Formation (V2a) (Bensaid et al., 1985) is a sandy, muddy, and ferruginous wackestone, rich in debris of sponges, corals, fenestellid bryozoans, productid brachiopods, bivalves, trilobites, ostracods, and echinoderms. Microfossils include *Girvanella wetheredi*, *Nanopora anglica*, *Polysphaerinella bulla*, and *Pachysphaerina pachysphaerica*, among others (see Bensaid et al., 1985a, p.338, for a list of fossils). The formation overlies unconformably the Strunian Bou Mia Formation (according to Bensaid et al., 1985a) and is overlain by conglomerates of the Zizi Ben Zizuit Formation. Tizi ben Zizuit Formation (V2b according to Bensaid et al., 1985) composed of conglomerates at the base, followed by oolitic fossiliferous limestones (Bensaid et al., 1985a, p.338 for a long list of fossils). It contains, among many others, *Koninckopora sahariensis*, *Aoujgalia variabilis*, *Pachysphaerina pachysphaerica*, *Stacheoides tenuis*, *Endothyranopsis compressa*, *Archaediscus chernoussovensis*, among others (see Bensaid et al., 1985a, p.338 for a list of fossils). The overlying Titar Oumjel Formation at Tizi n'Salihine starts with conglomerates at the base followed by 600–700 m (1969–2297 ft) of shales, sandstones, and limestones. It contains *Girvanella ducii*, *Endothyropsis compress*, *Earlandia minor*, *Pseudotaxis eominima*, *Planoarchaediscus*, *Archaediscus chernoussovensis*, *Hexaphyllia*, *Striatifera striata*, *Kannsuella maxima*, *Linoprotonia hemisphaerica*, and *Caldenocrinus curtus*. The Titar Oumjel Formation is overlain by the Djebel Hadid Conglomerate with abundant microfauna, such as *Aoujgalia variabilis*, *Fasciella kizilica*, *Epistacheodes* cf. *nephriformis*, *Anthracoporellopsis index*, *Pachysphaerina pachysphaeric*, etc. Accumulations of knoll-like white buildups occur at Tabainut, Tizra Sghrira, Arfoug, Akerchi, and Tizra Kebira. These limestones contain *Goniatites crenistria*, *Latiproductus latissimus*, *Productus productus*, and *Pachysphaerina pachysphaerica*, etc. (see Bensaid et al., 1985a, p.339). The shales around these buildups, such as the Koudiat Ain Djemaa Formation, contain *Goniatites crenistri*, *Sagittoceras complicatum*, *Dimorphoceras* cf. *gilbertsoni,* and *Posidonia becheri*. The limestones and shales continue into the Visean/Namurian passage beds, for example the Akerchi buildup contains lower Namurian fossils, such as *Latiproductus edelburgensis*, *Brachythris gracilis*, and *Martinia minima*, in addition to Visean fossils (citing Dedok & Hollard, 1980). The lower Dchar Ait Merrouane Member is a flysch deposit with thin sandstone layers and a few limestone lenses with *Posidnia becheri*, *Goniatites crenistria mediterraneus*, *G. striatus*, and *Asterocalamites*. According to Bensaid et al. (1985, Fig. 12), the Dchar Ait Merrouane Formation is of lower Visean age, and is overlain by the *Neoglyphioceras subcirculare* Passage Beds of lower Serpukhovian.

The Fourhal Group, equivalent to the Arfoug Formation and up to Migoumes Formation interval, is a non-calcareous flysch. However, at the Igr ou Hammoud ridge, the group contains *Lusitanites subcircularis*, *Lyrogoniatites* sp., *Sagittoceras* sp., etc., as well as crinoids and plant fragments. Its age is late Visean-early Westphalian (Bensaid et al., 1985a, Fig. 12). The Mohamed ben Sidi Mohamed Conglomerates of the Fourhal Group contains pebbles of ferruginous carbonates that contain *Diaschophyllum chevalieri*, *Lonsdaleia duplicate*, *Ivanovia*, *Girvanella*, *Pachysphaerina pachysphaerica*, etc. (see Bensaid et al., 1985a, p.339–340). Bensaid et al. (1985) postulated that these fossils were derived from younger buildups. The conglomerates are overlain by shales and sandstones of the Ich ou Mellal Formation, followed by Migoumes Formation or the "ochreous nodules shales", and then by the Khenifra Formation (Westphalian) and the Bou Achouch Formation (Stephanian).

In the western part of the Middle Moroccan (Bensaid et al., 1985, Fig. 12), the Carboniferous consists of Et Touijinne Formation, flysch of the Khorifla Formation,

Rabat Formation, Ouedzemrine Formation, unnamed interval with *Delpines comoides* and *Fluctuaria undata,* and Si Mohammed Abdallah Formation. At Rehamna, the flysch of the Mechra Ben Abbou Formation (Visean-Serpukhovian) is overlain unconformably by red conglomerates of Westphalian age. At Jebilet, the Kharrouba Flysch is overlain by Sarhlef Formation; both are of Visean age.

The Lower Carboniferous (Mississippian) of the Adarouch region, in the northeastern part of the Central Meseta (part of the Azrou-Khenifra Region of Piqué, 1979) consists successively of two shallow platform carbonate intervals (Visean), separated by turbidites, shales, and olistostromes of Devonian and Visean age, followed upward by Serpukhovian sandstones and carbonates (Berkhli et al., 2001). The eastern and western parts of the region display different successions.

The Devonian (Givetian-Strunian) beds with *Spelaeotriletes lepidophytus* (Berkhli et al., 2001) are followed unconformably by the Middle Visean in the Agourai area and by Upper Visean in the Imouzzer area. The uppermost Visean is represented by an oolitic carbonate marker bed. The Lower Serpukhovian in the Akerchi Mountain (in the Meknes-Tafilalt area in the Eastern Anti-Atlas) is marked by the disappearance of algae *Koninckopora* sp., a low diversity of foraminifera, and the early appearance of the brachiopod *Titanaria*. The Lower Carboniferous beds crop out in two areas oriented NE-SW, separated by the Namurian M'taoutoult Formation (Berkhli et al., 2001).

In the eastern Adarouch area, the Lower Carboniferous consists of four formations: Oued Amhares, Tizra, Mouarhaz, and Akerchi. Farther to the east, another unit rich in rugose corals, the Idmarrach Formation, crops out near El Hajeb City (Said et al., 2010).

The Oued Amhares Formation (Berkhli et al., 2001) is about 500 m (1641 ft) in thickness and consists of three members: The AKN1 Member is about 140 m (459 ft) in thickness and is made up of sandstone and limestone olistostromes, beds of calcareous sandstones, and quartzitic sandstones rich in plant debris intercalated with black shales with nodules. The succession is capped by a channelized conglomeratic horizon with pebbles of limestones, sandstones, siltstones, and quartz. The Middle Carboniferous boundary (the H stage) crops out on the southern flank of the Jerada Basin in the Agaia area. The AKN2 Member is about 180 m (591 ft) thick and made up of alternating black shales and black bioclastic limestones, with crinoids, brachiopods, and mollusks. The AKN3 Member is 130–150 m (427–492 ft) thick and composed of a succession of black shales, black bioclastic limestones, and sandy limestones with plant fragments, capped by a calcareous sandstone bed.

The Oued Amhares Formation was dated previously as Middle-Upper Visean by Archaediscidae (Verset, 1983). Berkhli et al. (2001) suggest a Visean V3b age based on the occurrence of *Stacheoides tenuis, Mametella skimoensis, Ungdarella uralica, Pseudoendothyra* sp. *Koninckopora inflata, Aoujgalia* sp., *Earlandia vulgaris, Forschia parvula Pseudolituotubagravata, Endothyra* sp., *Omphalotis samarica, Globoendothyra* sp., *Endothyranopsis crassa, Koskinobigenerinaprisca, Eostaffella* sp., *Paraarchaediscus* sp., *Koninckopora tenuiramosa, Stacheoides* sp., *Earlandia minor, Endothyra* sp., *Endothyranopsis* sp., *Koskinotextularia bradyi, Mediocris mediocris, Pseudoendothyra struvei, Paraarchaediscus* sp., *Koninckopora inflate, Luteotubulus licis, Fourstonela* sp., *Ungdarella uralica, Plectogyranopsis regularis, Endoth yranopsis compressa, Endothyranopsis crassa, Valvulinella youngi, Koskinotextulafia strica,*

*Mediocris mediocris, Eostaffella* sp., *Pseudoendothyra* sp., and *Paraarchaediscus* cf. *redita*.

The Tizra Formation (Berkhli et al., 2001) is about 160 m (525 ft) in thickness and includes three members: The TZ1 Member is made up of 90 m (295 ft) of poly-genetic conglomerates with pebble of limestones and sandstones at the base, followed by bioclastic limestones, and green shales. The TZ2 Member is about 44 m (144 ft) in thickness and consists of bioclastic spongistrome buildups followed upward by shales and black bioclastic limestones, then by black shales. The TZ3 Member is a 130 m (427 ft) succession of black shales, black bioclastic limestones, and spongis-trome buildups, and sandy limestones, grading into green shales and thin siltstones. Said et al. (2010) reported the rugose coral genus and species *Tizraia berkhlii* in the Tizra and Akrechi formations. The Tizra Formation was dated Late Visean by Dedok & Hollard (1980) and suggested that its age may extend into the Namurian, based on brachiopods *Schizophoria resupinata, Krotovia spinulosa, Cancrrnella undata, Productusproductus, Spiriferbisulcatus.* Ouarhache et al. (1991) confirmed a Late Visean age based on foraminifera and algae, such as *Biseriella parva, Endostaffella* cf. *parva, Neoarchaediscus incertus, and Ungdarella uralica.* Foraminifera in the Tizra Formation also include *Windsoporella* sp., *Kamaena* sp., *Earlandia minor, Endothyra* ex. gr. *similis, Endostaffella* sp., *Eostaffella* sp., *Biseriella* sp., *Paraarchaediscus* sp., *Endothyranopsis crassa, Tetrataxis* sp., *Koskinotextulariinae* sp., *Archaediscus karreri*, and *Haplophragmiina*s sp. (Berkhli et al., 2001).

The Mouarhaz Formation is 500 m (1641 ft) in thickness and consists of two members (Berkhli et al., 2001): The MZ1 Member is 95 m (312 ft) succession of black bioclastic limestones, oolitic limestones, black shales, black bioclastic limestones with crinoids and mollusks, and channelized, calcite-cemented polygenetic conglomerates with pebble of limestones, sandstones, and Sandy limestones. The MZ2 Member is 300–400 m (984–1312 ft) in thickness and composed of black shales, interbedded with siltstones and limestone olistostromes of Devonian and Visean age, capped by cross-bedded ferruginous quartzitic sandstones with shales interbeds. The Mouarhaz Formation has been dated by Verset (1983) as Upper Visean based on foraminifera and *Gigantoproductus.* Berkhli et al. (2001) extended its age into the Early Visean based on an exhaustive assemblage of foraminifera, algae, and pseudo-algae including *Koninckopora tenuiramosa, Aoujgalia* sp., *Ungdarella uralica*, and *Neoachaediscus gregorii*, among others. Two horizons of olistostromes are recognized in this forma-tion, both are Visean in age. The first horizon contains an assemblage of *Earlandia* sp, and *Paraarchaediscus grandiculus*, and the second horizon includes *Konincko-pora inflata, Pseudostacheoides* sp., *Omphalotis* sp., *Endothyranopsis compressa*, and *Archaediscus karreri.*

The Akerchi Formation has a thickness of up to 140 m (459 ft) and includes two members (Berkhli et al., 2001): The lower AK1 Member is a 16 m (52 ft) inter-val of channelized polygenetic calcite-cemented conglomerates with beds of cross-bedded, bioturbated, bioclastic limestones into oolitic limestones. The upper part of AK1 member contains a biostrome, 2–5 m (6–16 ft) thick, composed mainly of rugose corals, dominated by *Siphonodendron junceum*, and gigantoproductid brachiopods embedded in marly limestones (Said et al., 2010). The upper AK2 Member is 80 m (262 ft) thick and consists of calcite-cemented conglomerates with subrounded peb-bles, black limestones with *Gigantoproductus*, microconglomerates with pebbles of limestones, quartz, calcite-cemented sandstones, and sandy limestones with crinoids and brachiopods. The age of the Akerchi Formation is uncertain. Dedok & Hollard (1980) assigned it an Early Namurian age based on unpublished assemblage of the foraminifera Archaediscidae, and subdivided it into two parts; the lower part has been attributed to the Upper Visean and contains *Gigantoproductus superior, Productus* cf. *Redesdalensis, 'Camarotaechia'* sp., *Syringothyris* sp., *Martinia* sp., *Punctospirifer* sp., *Dielasma* sp., and *Athyris* sp., whereas the upper part was assigned a Lower Namurian age based on *Gigantoproductus edelburgensis, Brachythyris gracilis,* and *Martinia minima.* Vachard & Tahiri (I 992) assigned the formation an Early Serpukhovian age.

In the western part of the Adarouch area (Eastern Meseta), the Lower Carboniferous is made up of two formations:

The Goulib Formation (Berkhli et al., 2001) has a thickness of 300–350 m (984–1148 ft) and is divided into three members: The OBH1 Member is made up of 100 m (328 ft) of black shales. The OBH2 Member consists of 40 m (131 ft) of siltstones and shales with quartzitic sandstones with flute casts oriented NNW-SSE. The OBH3 Member is 150–200 m (492–656 ft) thick and is composed of interbedded siltstones and shales. Laterally, the formation changes into olistostromes which form the Goulib Hill. The olistostromes are made of interbedded black bioclastic and oolitic limestones and shales.

The Oued Bournhares Formation (Berkhli et al., 2001) has a thickness of about 350–450 m (1148–1476 ft) and consists of two members: The OBH4 Member is made up essentially of shales. The OBH5 Member is between 150–200 m (492–656 ft) in thickness and consists of interbedded green shales and quartzitic sandstones with limestone pebbles. The unit is overlain by basal Namurian conglomerates, with sandstone, limestone, siltstone fragments.

The Carboniferous (Stephanian) Ida Ou Zal and Ida Ou Ziki (Western High Atlas) and the Ida Ou Zal basin (Western Meseta) are extensional grabens bounded by faults (Saber et al., 2007). The Oued Zat and Ida Ou Zal basins are similar in that they both contain a Stephanian sequence unconformably overlain by Upper Triassic sediments. The Stephanian succession in the Ida Ou Zal Basin is over 1,600 m (5250 ft) (Saber & El Wartiti, 1996). In the Oued Zat Basin, the succession decreases in thickness to 50–450 m (164–1476 ft) and is dominated by fluvial deposits of the Tighadwiyn and Mçtour formations (Saber et al., 2007).

In the Tighadwiyn Formation (Stephanian) in the Oued Zat Basin is composed of purple-red conglomerates composed of rounded quartzite pebbles, with an erosional base, overlain by red sandy and grey or yellowish shales. In the Ida Ou Zal Basin, the Lower Conglomerate Formation (Tighadwiyn Formation) is 400–600 m (1312–1969 ft) in thickness. It crops out in the Ikhourba Valley and in the Houria Valley. It consists of conglomerates composed of quartzite, sandstone, and carbonate fragments (Saber et al., 2007).

In the Oued Zat Basin, the Mçtour Formation (Stephanian) consists of shales and siltstones with carbonaceous lenses, native sulfur and plant remains, such as the macroflora *Odontopteris dufresnoyi, O. subcrenulata, Pecopteris densifolia, P. rarinervosa, P. polypodioides, Scolecopteris, Cordaites lingulatus, Sphenophyllum costa, S. thoni, Sphenopteris* sp., *Aulacopteris, Calamites suckowi,* and *Lebachia parvifolia,* and are assigned a Stephanian age (Doubinger & Roy-Dias, 1985, Saber et al., 2007). In the Ida Ou Zal Basin, the Upper Clay-Sandstone Formation (Mçtour Formation) is more than 1000 m (3281 ft) in thickness and is made up interbedded fine to conglomeratic sandstones and shales, dolomitic in places.

Uppermost Carboniferous (Late Pennsylvanian) sediments were deposited on top of deformed and metamorphosed basement rocks (Saber et al., 2007). The succession starts with a basal coarsening-upward megasequence composed of conglomerates and sandstones of about 400–600 m (1312–1969 ft) in thickness in both basins, followed upward by up to 1200 m (3937 ft) of grey, alluvial plain and fluvial channel sandstones, lacustrine black shales, and coal seams. The Ikhourba (in Ida Ouzal Basin) and Tajgaline (in Ida On Ziki Basin) conglomeratic formations are overlain by greenish-grey fluvio-lacustrine beds (of El Menizla Formation and Oued Issene Formation, respectively) with abundant imprints of leaves, stems, and trunks (Saber et al., 2007). The type section of the El Menizla Formation is located along the El Menizla River near the El Menizla Village on the southern flank of the Ida Ou Zal Subbasin. It contains *Calamites, Asterophyllites equisetiformis, Annularia, Macrostachya, Sphenophyllum oblongifolium, Lepidodendron, Pecopteris arborescens, P. hemitelioides, P. unita, P. paleacea, Asterotheca, Odontopteris,* and rare *Walchia*. Plant remains in the lacustrine laminated and varved black shales contain conifers, such as *Otovicia hypnoides, Ernestiodendron filiciforme,* and cf. *Lepidostrobophyllum,* and rare *Odontopteris subcrenulata* and *Autunia* cf. *conferta*. A rich Stephanian assemblage of fossil insects occurs near the towns of Taghzout, Agadir-Ou-Anzizen, and El Menizla, and considered the oldest in Africa (Hmich et al., 2006). It includes representatives of the Families Spiloblattinidae, Mylacridae, Phyloblattidae, and Poroblattinidae of the Order Blattodea (coackroaches). The Mylacridae family is represented only by *Opsiomylacris thevenini,* which is the most abundant. Palynomorphs from the middle part of the Oued Issene Formation are dominated by trilete and monolete spores, such as *Laevigatosporites, Calamospora, Lycospora,* and *Reticulatisporites,* among others, as well as rare Monosaccate and bisaccate pollen (Aassoumi et al., 2003). The top of the Oued Issene Formation contains *Reticulatisporites, Lycospora, Thymospora,* and *Leiotriletes* among others; the monosaccate pollen *Potonieisporites* is less abundant (Saber et al., 2007). The Oued Issene Formation in the Ida On Ziki Basin is overlain by siltstones of the Tirkou Formation. Plant remains from these outcrops (collectively) and the siltstones of the Tirkou Formation consist of *Annularia stellata, A. sphenophylloides, Pecopteris candolleana, P. hemitelioides, P. cyathea, P. monyi, P. pinnatifida, Sphenopteris matheti, Odontopteris obtusa, Mixoneura neuropteroides,* roots of *Stigmaria, Sphenophyllum oblongifolium, Alethopteris subelegans, Odontopteris* cf. *subcrenulata, Taeniopteris* gr. *jejunata, Neuropteris cordata, N. neuropteroides, Poacordaites, Walchia piniformis, Otovicia hypnoides,* and some *Culmitzschia* species, as well as *Sphenobaiera* sp. (Clariond, 1932, Broutin et al., 1989).

## 17.4.2  Atlas Mountains

The Atlas Mountains System extends from Agadir on the Atlantic Ocean in the west to eastern Tunisia in the east (Figs. 2.1, 17.1). It is a Cenozoic intra-continental, autochthonous system, developed over a continental basement (Frizon de Lamotte et al., 2008). It is limited to the west by the Central Atlantic continental stable margin, to the north by the largely allochthonous Rif Domain (Negro et al., 2007), to the north-west by the Western Meseta Domain, and to the southeast by the Anti-Atlas Chain. The Cap Tafelney fold-belt marks the northern offshore front of the Atlas System (Frizon de Lamotte et al., 2008). The Mesozoic-Cenozoic South Atlas Fault (SAF) and the Tizi n'Test Fault (TTFZ) define the boundary between the Atlas Belt and the Anti-Atlas Domain to the south. The Atlas Domain is composed of four major units: The Western High Atlas, the Central High Atlas, the Middle Atlas, and the Eastern High Alas. The Central and Eastern High Atlas are also known as the Calcareous High Atlas because of the thick Mesozoic carbonates that extend into a deep rift in Algeria (Barbero et al., 2007). The sedimentary succession in the High Atlas extends from Precambrian to Recent.

Two main rifting episodes have been recognized in that Atlas Domain (Frizon de Lamotte et al., 2008): The first took place in Late Permian-Early Triassic time; the second occurred in Early Middle Jurassic. Mesozoic structures of the Middle and Central Atlas were controlled by pre-existing faults, which were reactivated again during the Oligocene compressive phase. Three sets of faults have been identi-fied (Brede et al., 1992); each reacted differently to the compressive stress. The first set oriented 35–45° dominates in the Middle Atlas and was reactivated as sinistral oblique slip reverse faults. The second set runs at 70° and is common in the southern part of the Central High Atlas; it was reactivated as upthrust. The third set occurs at the northern border of the Central High Atlas and was reactivated as dextral oblique slip reverse faults.

### 17.4.2.1  Permian-Jurassic (Central High Atlas)

The Tizi n'Test Basin, along the Tizi n'Test Fault Zone (Fig. 17.1) in the Central High Atlas, is a Triassic rift basin (presently inverted) (Qarbous et al., 2003, Corel et al., 2003) bounded by ENE-WSW-trending faults dipping to the NNW. Strike-slip movements along the faults appear to have been very small during the Triassic. The Permo-Triassic succession comprises six formations defined by Biron & Courtinat (1982, cited by Qarbous et al., 2003) (see Corel et al., 2003, for correlations with the Argana Basin):

> The F1 Formation or the Anrar Conglomerates is probably of Upper Permian age.
> The F2 Formation or the Cham-El-Houa Siltstones, is probably also of Upper Permian age.
> The F3 Formation or the Timalizene Conglomerates is of Middle Triassic age.
> The F4 Formation or the Anouffig Siltstones is also Middle Triassic in age.
> The F5 Formation or the Oukaimeden Sandstones is Carnian in age.

The F6 Formation or the Tafilalt Formation (Upper Siltstone Formation in Corel et al., 2003) is probably of Norian age based on correlations with neighboring areas. It includes basalt flows and is overlain by Jurassic sediments. The succession includes thick tholeiitic basalt flows dated at about 200 Ma (Sebai et al., 1991). Only the upper four formations are present in the Tizi n'Test area. The thicknesses of these formations vary according to their relative position in the basin.

### 17.4.2.2 Jurassic

In the beginning of the Jurassic (Liassic), the paleogeography of the Atlas Mountains was dominated by two domains (Frizon de Lamotte et al., 2008): The Tethyan Domain was connected to the Tethys Ocean and covered the areas now occupied by the Central High Atlas, Eastern High Atlas, and Middle Atlas. The Atlantic Domain extended from the Central Atlantic Ocean and covered the area of the Western High Atlas, as well as the Essaouira and Agadir basins in the Atlantic passive margin. The West Moroccan Arch (WMA) (the "*Terre des Almohades*" of Choubert & Faure-Mauret (1960) formed a shallow platform during the late Middle Jurassic-Early Cretaceous (Saddiqui et al., 2008). Coarse clastics were deposited near the Arch, while silts and evaporites were deposited away from it (Courel et al., 2003, Teixell et al., 2007).

Ibouh et al. (2001) identified eight Jurassic formations in the Ait Bou Guemez area (Central High Atlas). All the formations are dated by foraminifera:

The Ait Bou Oulli Formation is a basal massive carbonate unit with a thickness of 30–400 m (98–1312 ft). The formation is overlain by the Agnane Formation. In the R'bat area[8], it contains *Lituosepta recoarensis*, which indicates an Upper Sinemurian age (Brede et al., 1992).

The Aganane Formation is best developed in the R'bat and Agouti area. It consists of three units, designated as Ag1, Ag2, and Ag3, in an ascending order (Brede et al., 1992). The Ag1 Unit is made up of a rhythmic succession of fine foraminifera limestones and laminated and marly dolomites. It was assigned an Upper Sinemurian to Carixian (equivalent Pliensbachian) age, based on the foraminifera *Orbitopsella primavera* and *Pseudopfeinderina butterlini* (Ibouh et al., 2001, El Bchari & Souhel, 2008). The Ag2 Unit corresponds to the marls and dolomites of the Ah Bazzi Formation on the 100,000 map of Zaouit Ahncal area (Jossen, 1988, cited by Ibouh et al., 2001). In the R'bat area, it starts with 8–10 m red marls with paleosoil horizons, followed by unfossiliferous yellow marls with brecciated dolomite intercalations. In the Agouti area, the base of the formation made up of 15–20 m (49–66 ft) of dolomitic breccias with slump structures. The unit is assigned to the Middle-Upper Carixian based on its stratigraphic position. The Ag3 Unit is similar to Unit Ag1, with an increase in dark bioclastic, conglomeratic limestones, rich in the algae *Cayeuxia* and foraminifera. The top of this unit shows numerous dinosaur footprints. It was assigned a Liassic (Domerian) age based on the benthonic foraminifera *Lituosepta compressa*, *Pseudocyclamia liassica*, and *Haurania* sp. In the Tizi n'Tirghist area, it includes channelized limestone conglomerates derived from the underlying units. This unit is shown as Pliensbachian on Jossen's (1988) map. It rests unconformably on top of the Lower Liassic with the absence of the Carixian units (Ibouh et al., 2001, El Bchari & Souhel, 2008).

The Jbel Choucht Formation is a lateral facies equivalent of the Aganane Formation (Brede et al., 1992). It was introduced by Septfontaine (1986, in Ibouh et al., 2001) in its type locality of Jbel Choucht, NW of Zaouit Ahancal, where it is made up of massive limestones. The formation straddles the Pliensbachian Carixian-Domerian boundary. Ibouh et al. (2001) disagree with Septfontaine's age designation and environmental interpretation for that formation. They describe it as a bedded, very thick

---

8 Not to be confused with the Rabat.

unit rich in Megalodonts bivalves, and assigned it a Domerian age. The section at Jbel Azourki starts with fine limestones with the benthonic foraminifera *Mayncina termieri* and *Lituosepta compressa*, followed upward by calcareous wackestones to packstones with Megalodont fragments (El Bachri & Souhel, 2008).

The Amezrai Formation was dated latest Domerian-Lower Toarcian based on brachiopods (Brede et al., 1992, Bouchouata et al., 1995). It is best developed at Jebel Amzourlai where it is composed of carbonates, sandstones, and marls, associated with conglomerates derived from the Paleozoic basement (Ibouh et al., 2001, El Bachri & Souhel, 2008).

The Assemouk Formation is limited to a small basin in the Assemouk area where it is made up of marls with resedimented blocks and slump structures (Brede et al., 1992). It has been assigned Middle-Upper Domerian age by Jossen (1988).

The Tamadout Formation crops out in the Assemouk area and consists of greenish papery marls (Brede et al., 1992). It has been assigned a Lower Toarcian age by Jossen (1988). It rests directly on top of the Assemouk Formation.

The Wazzant Formation represents a western lateral facies partly equivalent of the Amezrai Formation (Brede et al., 1992). It was assigned a latest Domerian-Aalenian age by Jossen (1988) based on its stratigraphic position. It is made up of complex alluvial sedimentary associations. The lower part of the formation is made up of channel conglomerates with quartz and Paleozoic basement fragments, lenticular sandstones, and red shales (Ibouh et al., 2001, El Bachri & Souhel, 2008).

The Tafraoute Formation[9] is Toarcian in age according to Bouchouata et al. (1995) and represents a lateral facies equivalent of the uppermost part of the Wazzant Formation and the Amerzai Formation. The formation crops out only on the northern side of Jbel Azourki (Brede et al., 1992). It consists of nodular oolitic limestones, condensed ferruginous surfaces with brachiopod fossil fragments, and green marls and sandy limestones with bioclastic fragments (Ibouh et al., 2001, El Bachri & Souhel, 2008).

### 17.4.2.3  Upper Jurassic-Lower Cretaceous

During the Late Jurassic-Early Cretaceous (Nubian time), the whole of Morocco, like the rest of North Africa was emerged, except for the coastal areas. The *Terre des Idrissides* of Choubert & Faure-Mauret (1960)), or the *Terre Sud-Rifaine* of Michard (1976), believed to be an emergent land in the Upper Jurassic-Early Cretaceous, did not exist at that time according to Frizon de Lamotte et al. (2008), because it was formed during the Senonian inversion episode. They refer to this highland as the North Moroccan Bulge[10]. On the other hand, Frizon de Lamotte et al. (2008) suggested the presence of two narrow gulfs during the Early Cretaceous (Valanginian-Aptian): The North Middle Atlas Gulf (NMAG) (also called the Boulmane Gulf) with connection to the Tethys, and the North High Atlas Gulf (NHAG), connected with the Atlantic margin and the Essaouira Basin. The two gulfs reached their maximum extension during the Aptian (cf. Western Desert of Egypt and Sarir Basin & Trans-Africa Seaway).

---

9 Not related to the Tafraoute Granite.
10 The present author suggests that the priority rule for stratigraphic nomenclature should be applied to naming geological and paleogeographic features.

The regional Cenomanian marine transgression associated with the global sea-level rise led to shallow marine conditions in the High Atlas, as well as the rest of North Africa. Cenomanian-Turonian carbonates were deposited unconformably on the Jurassic rift structures. (See Rodriguez-Rovar et al., 2009, Cavin et al., 2010).

The Bathonian-Berriasian Red Beds are about 900 m (2953 ft) in thickness in the Ouaouizarht Syncline, south of Bani Mellal in the Central High Atlas (Fig. 17.1). The Red Beds are divided, from base to top (Charrière et al., 2005), into the Guettioua Formation (Bathonian), Iouaridene Formation (Upper Jurassic-Barremian), and the Jbel Sidal Formation (Barremian). The Red Beds are overlain by the Ait Tafelt Formation (Aptian), and then by post-Aptian strata (see also Haddoumi et al., 2002). The Red Beds contain, in addition to ostracods and charophytes, numerous bones and footprints of dinosaurs, as well as a complete skeleton of a giant Sauropod with legs longer than 3.5 m (Monbaron et al., 1999). Volcanic gabbro intrusions and basalt lava flows occur in the Bathonian-early Upper Jurassic sediments (see also Laville & Piqué, 1992). Frizon de Lamotte et al. (2008) attributed the volcanic flows to the previous extensional regime.

The Guettioua Formation (Bathonian) consists of fluvial conglomeratic siltstones and sandstones, separated by shales with occasional paleosoils. In the Tilogguit Syncline, it yielded an almost complete skeleton of *Atlasaurus imelakei* (Monbaron et al., 1999). The top of the formation contains a number of basaltic lava flows (Haddoumi et al., 2002).

The Iouaridene Formation consists of two members; a lower shale member and an upper evaporite member. Charophytes from the top of the Iouardene Formation and the base of the Jbel Sidal Formation gave an Upper Hauterivian-Lower Barremian age (Haddoumi et al., 2002). It was assigned an Oxfordian-Kimmeridjian age by Charrière et al. (2005).

The Jbel Sidal Formation (defined by Jenny et al., 1981) is formed of lenticular channelized sandstones, siltstones, and conglomerates separated by shales, and occasionally associated with limestones or dolomites in places (Haddoumi et al., 2002). Charophytes in the Jbel Sidal Formation are dominated by *Globator trochilicoides* in addition to *Feistiella* sp. It is overlain by limestones of the Ait Tafelt Formation.

In the beginning of the Jurassic, the Central and Eastern High Atlas were occupied by two seaways open to the Tethys and the Atlantic. The two seaways were separated by the WMA. Limestones and marls of up to 9 km (29,530 ft) in thickness were deposited in the High Moulouya Trough (El Kamar et al., 1998). The succession includes, from base to top; the Ouchbis Formation, Tagoudite Formation, Agoudim Formation, Flilo Beds, *Calcaires Corniches*, and Tazegzaout Formation (El Kamar et al., 1998).

In the Talghemt area on the southern border of the High Moulouya in the Central Atlas, the Liassic (Upper Domerian-Lower Toarcian) succession consists of two formations; the Ouchbis Formation and the Tagoudite Formation (El Kamar et al., 1998).

The Ouchbis Formation, overlies the Aberdouz Formation and is overlain by the Tagoudite Formation (Chafiki et al., 2004), is composed of interbedded marls and limestones rich in large ammonites, such as *Canavaria (Naxensiceras) depravatum*, *C. zancleana*, *Emaciaticeras imitator*, *Emaciaticeras emaciatum*, *Emaciaticeras* sp., *Emaciaticeras gracile?*, *Phylloceras* sp., *Tauromeniceras nerinum*, *Tauromeniceras* sp.,

*Tauromeniceras occidentale, Tauromeniceras* cf. *nerinum, Atractites* sp., *Lytoceras* sp., *Neolioceratoides* gr. *hoffmanni, Lioceratoides lorioli, Lioceratoides* sp., *Paltarpites?* sp., *Eodactylites mirabile, E. pseudocommune, Paltarpites* sp., *Hildaites (Murleyiceras)* aff. *gyrale,* and *Juraphyllites libertus,* in addition to foraminiferas and ostracods (El Kamar et al., 1998).

The Liassic succession in the Midlet-Errachidia area west of the Talghemt area in the Central High Atlas consists of two massive intervals separated by a regional sedimentary break (Chafiki et al., 2004). The lower interval includes the Idikel, Aberdouz, Ouchbis, and Tagoudite formations Chafiki et al., 2004). Mehdi et al. (2003) introduced the Foum Zidet Formation to replace the upper three units in the upper part of the Idikel Formation, which includes bioclastic limestones with biostromes, fine-grained argillaceous limestones, and sponge mounds. Mud mounds in the succession were described by Dubar (1960, 1962, Neuweiler et al., 2001, Mehdi et al., 2003, Chafiki et al., 2004). The first mounds are 30–130 m (98–427 ft) in thickness and occur at the top of the Idikel Formation; the second is about 80 m (262 ft) thick and straddles the Aberdouz-Ouchbis formations contact (Chafiki et al., 2004). The mud-mound horizons reach their maximum development of about 280 m (919 ft) at Foum Tillicht. In that area, the mounds of the Idikel Formation overlie a Lower-Upper Sinemurian transitional horizon with *Asteroceras* and *Arnioceras* (Chafiki et al., 2004). The cores of the mounds are made up of thrombolites, stromatolites, leiolites[11], the sponge Hexactinellids, Lithistides, the annelids *Serpula* and *Terebella,* bryozoans, brachiopods, bivalves, and gastropods set in a micritic matrix.

Jurassic rocks crop out between Midelt and Errachidia in the Central High Atlas Mountains and are represented by two sedimentary facies (Ait Addi, 2006); a basinal facies in the north and a carbonate platform facies in the south. They were deposited on top of continental Triassic rocks. In places, the Upper Jurassic is eroded and Middle Jurassic rocks are overlain directly by Lower Cretaceous continental red beds. The basinal facies includes the Tagoudite Formation (Lower Toarcian), the Agoudim Formation (Toarcian-Upper Bajocian), and Tazigzaout Formation (Bathonian-Upper Bajocian).

The Tagoudite Formation (Lower Toarcian), on the southern border of the Moulouya Valley, the Tagoudite Formation is made up of green marls interbedded with thin graded oolitic and glauconitic beds and turbidites, with poorly preserved nodosariids, ostracods, echinoderms, and lamellibranchs. Its top is marked by silty marls and thin sandstones. It has been attributed to the Lower Toarcian based on *Eodactylites simplex* (El Kamar et al., 1998) is made up of calcarenites and sandy marls. On the southern border of the Moulouya Valley, the Tagoudite Formation is made up of green marls interbedded with thin graded oolitic and glauconitic beds and turbidites, with poorly preserved nodosariids, ostracods, echinoderms, and lamellibranchs. Its top is marked by silty marls and thin sandstones. It has been attributed to the Lower Toarcian based on *Eodactylites simplex* (El Kamar et al., 1998). The Toarcian represents an important stage in the evolution of the Atlas Domain developed during the rifting on the West Tethyan margin during the Early Jurassic. In the Todrha-Dades area, Tagoudite Formation contains a rich fauna of foraminifera and ostracods in the marls and marly sandstones particularly in the lower part (Ettaki & Chellai, 2006). In the Todrha-Dades

11 Dense micrites with abundant microborings.

area, it increases in thickness from SW to NE. The extensional tectonics were responsible for isolation of the basin and led to the development of anoxic conditions that controlled the replacements of the fauna (Ettaki & Chellai, 2006).

The Agoudim Formation (Toarcian-Upper Bajocian) consists of four members (I–IV) (Ait Addi, 2006), in an ascending order: Member I is made up of grey-green marls with ammonites, Member II consists of shaly marls and dark grey oolitic-peloidal limestones, Member III composed of dark blue shaly marls with *Posidonomyes*, and Member IV is equivalent to the *Calcaire Corniche*. On the southern border of the Moulouya Valley, the Agoudim Formation (Member I of Ait Addi et al., 2006) is composed on interbedded marls and limestones with ammonites, nodosariids, verneulinoids, Protoglobigerinas, and ostracods. It has been dated Middle-Upper Toarcian based on ammonites (El Kamar et al., 1998). The Ait Daoud Formation (or Member according to Brechbühler, 1984) (Member II of Ait Addi et al., 2006) consists of oolitic limestone breccias of Lower Bajocian age (El Kamar et al., 1998). The Central Atlas gastropod fauna show close affinity to those of Europe, but show no relation with the faunas of the inner area of the Western Tethys (Costi & Monari, 2001). The Flilo Beds (Member III of Ait Addi et al., 2006) are made up of marls and limestones rich in benthonic foraminifera and poorly preserved ostracods (El Kamar et al., 1998). The *Calcaire Corniche* (Member IV of Ait Addi et al., 2006) is composed of bioclastic limestones with bivalves, gastropods, bryozoans, echinoderms, and brachiopods. Its date is controversial (El Kamar et al., 1998). Southwards, the Calcaire Corniche consists of two limestones beds with reefal fauna, separated by marls with rich in benthonic foraminifera, ostracods, echinoderm fragments, and coprolites.

The Tazigzaout Formation (Upper Bajocian-Bathonian) is composed of limestones and brown green marls with brachiopods dominated by *Zeilleria* and *Monsardithyris* and the ammonites *Cadomites* sp. and *Phylloceras* sp. (Ait Addi, 2006). On the southern border of the Moulouya Valley, the Tazegzaout Formation is made up marls separated by nodular limestone beds with brachiopods, foraminifera, ostracods, and coprolites (El Kamar et al., 1998). It has been dated Upper Bajocian-Bathonian (Ait Addi, 2006).

The platform facies is represented by the Bin Elouidane Group (Toarcian-Bathonian) composed of two informal units designated as Formation 1 and Formation 2 (Ait Addi, 2006, Table 1). The limit between the two formations is not clear. The group consists of interbedded grey blue marls with ammonites and bioclastic limestones, followed upward by reefal and oolitic limestones and dolomites, then by interbedded dolomites, limestones, marls, and red shale, and then dolomitic limestone buildups, and finally by dolomitic and oolitic limestones. The group is overlain by the Tilougguitr Formation, followed by the Guettiou Formation. Conti & Monari (2001) identified a new genus of *Sadkia* (family Eucyclidae), and two new species, *Sadkia richensis* and *Pirper ouchenensis* among the rich gastropod fauna in the Jurassic Upper Aalenian-Lower Bajocian sediments (equivalent to Member II of the Agoudim Formation) of the Central High Atlas.

The Middle Atlas is a NE-trending mountain belt bordered on the north by the Rif Domain, on the east by the Eastern Meseta (Missour and Moulouya basins), on the west by the Western Meseta, and on the south by the Central High Atlas. It belongs to the Tethyan Domain of Frizon de Lamotte et al. (2008).

The Middle Atlas comprises three structural zones corresponding to distinctive paleogeographic domains during the Mesozoic (Frizon de Lamotte et al., 2008): 1) The Tabular Middle Atlas (Middle Atlas "Causse") extends over the Western Meseta and Tazekka basement and consists of tabular Triassic-Liassic sequences, and 2) the Folded Middle Atlas, bordered by two NE-SW-trending faults; the North Middle Atlas Fault (NMAF) and the South Middle Atlas Fault (SMAF) and its extension, the Ait Oufella Fault (AOF) to the SW. The structural development of the Middle Atlas and the Central High Atlas was controlled by a pre-existing fault patterns (Brede, 1992).

The Missour-High Moulouya Basin is the sedimentary cover of the Eastern Meseta and exhibit tabular Liassic and Middle Jurassic carbonates "Dalle des Hauts Plateau".

The NE-trending Middle Atlas and the E-W-trending Central High Atlas were probably derived from inversion of an orthogonal and an oblique Jurassic rifts, respectively (Teixell et al., 2007). Deformation probably started in the Middle Eocene and continued into the Quaternary. The mean elevation exceeds 2000 m (6562 ft) over large areas. The Middle Atlas and the Central High Atlas are separated by the Jurassic Ridge. This ridge separates the Moulouya Valley (at the junction of the Central High Atlas and Middle Atlas)[12] from the Tadla Basin[13] (in Central Atlas) (Charrière et al., 2005). The southern boundaries of the Tadla and Bahira Basins are marked by reverse faults (toward the south). The Mesozoic units in the Tadla Plains are very thin. Jurassic beds are thin or missing, but reach 2000 m in the hanging wall of the fault. However, the Cretaceous succession consists of 200 m (656 ft) of Lower Cretaceous continental beds and 400 m (1312 ft) of Upper Cretaceous (Cenomanian-Turonian and Senonian) marine lagoonal sediments, overlain by 400 m (1312 ft) of Tertiary beds.

Unconformable Molasses are found in the core of the Central High Atlas Chain, such as Zawyat Ahansal[14] or the "Cathedral Conglomerate"[15] (Miocene-Pliocene) in the Amesfrane Cliff near Imilchil in the High Atlas, similar to the Skoura Conglomerate in the Middle Atlas.

The Jurassic to Tertiary succession at the junction of the High and Middle Atlas of Morocco is as follows:

a    Carbonates (Early-Middle Liassic) on top of terrigenous-evaporitic-basaltic Triassic unit. Two formations have been recognized (Löwner et al., 2002, El Hammichi et al. (2008).
b    Lower Liassic limestones and massive dolomites.
c    Middle Liassic limestones with flint nodules "Calcaire à Silex".
d    Upper Liassic green marls.
e    Dogger carbonates (Upper Aalenian-Middle Bajocian), oolitic-oncolitic-bioclastic (brachiopod fragments).
f    Red Beds (Jurassic-Cretaceous) filling small trough along the fault (Formation des Couches Rouges) (Charrière et al., 2005, Jenny et al., 1981).

---

12  Bouabdli et al. (2005) & Ensslin (1992) for Moulouya Basin.
13  See Soufiane & Achab (1993) for Ordovician in Tadla Basin.
14  Also spelled "Ahancal" or "Ahançal".
15  The Amesfrane Cliff in the High Atlas was named by early travellers "La Cathédrale" (Monbaron, 1985, Jossen, 1990).

g   Dogger basalts (mainly gabbroic dikes) (N30°–N45° and N130°–N140°).
h   Cretaceous carbonates, subdivided into:

1   "Barre aptienne" (Aptian)
2   Red marls, sandstones, and anhydrites (Albian-Cenomanian).
3   Limestones rich in planktonic foraminifera "*barre turonienne*" (Cenomanian-Turonian).

i   Conglomerate (Miocene-Quaternary).

The Southwestern Middle Atlas contains many occurrences of Toarcian and Aalenian ammonites, bivalves, and brachiopods (El Hammichi et al., 2008). The first study of the ammonites collected Termier (1927) from Toarcian marls from the Middle Atlas was carried out by Monestier (1930) followed by Termier (1936). The ammonites in the Middle Atlas belong to the Western Tethys Domain (El Hammichi et al., 2008). However, differences within the Middle Atlas Domain exist due to differences in tectonic settings. Bajjaji et al. (2010) recognized Lower Jurassic (Toarcian) benthonic foraminifera biozones in the Middle Atlas and correlated them with those of the Western Tethys area. They occur in hemi-pelagic sediments deposited in platform to basinal environments.

The Toarcian lithostratigraphy of the Middle Atlas was reviewed by El Arabi et al. (2001) and more recently by El Hammichi et al. (2008) and Bajjaji et al. (2010). Overlying the limestones with large bivalves of the Tizi Nehass Beds and unnamed Pliensbachian argillaceous marls, the Toarcian starts with red marls similar to the lower part of the Mibladene Beds. Typical facies of the Mibladene Beds are known from the Bou Angar Syncline and occur along the southwestern Middle Atlas (El Hammichi et al., 2008). They consist of fossiliferous marls with ferruginous oolites and bioclastic coquinas in the lower part changing into cherty limestones of the *calcaires à silex* (Aalenian-Bajocian). El Hammichi et al. (2002) proposed the name Al Yabes Beds for the beds between the Pliensbachian platform carbonates and the Bajocian marly limestones with *Zoophycos* of the Iwansitn Beds.

## 17.4.3   Western High Atlas

The Western High Atlas is a mountainous region composed of folded Neoproterozoic and Paleozoic formations. Mesozoic sediments are preserved in a number of basins, such as the Argana Basin, Oued Zat Basin, Souss Basin (Ida Ou Ziki and Ida Ou Zal Subbasins), and Haouz Basin.

The succession in the southwesternmost part of the High Atlas is divided into five tectono-sedimentary units (Saber et al., 2007):

TS1 (Late Permian = Lowest Argana sequence) made of fluviatile red beds.
TS II (dated Anisian based on palynology by El Arabi et al., 2001) is composed of fluvial-lacustrine and palustrine beds.
TS III (= Oukaimeden Formation, dated Carnian (Fabuel-Perez et al., 2009), composed of deltaic conglomeratic sandstones with a thickness of 600 m (1969 ft) in the centre of the basin and thins to 20–50 m on top of highs.
TS IV (dated palynologically Carnian-Norian-Rhetian by Marzoli et al., 2004), made up of eolianites and lacustrine clays and silts.

TS V (Hettangian = Earliest Jurassic), composed of interbedded volcanics and silts and marls. The volcanics yielded $^{39}$Ar-Ar$^{40}$ dates of 196 ± 1.9 to 199.0 Ma and are believed to have caused a biological crisis at the Triassic-Jurassic boundary (Frizon de Lamotte et al., 2008).

The Permian to Early Jurassic continental basin fill of the Argana Basin starts with the Ikakern Formation (Brown, 1980, El Wartiti et al., 1990, Saber et al., 2007, Hmich et al., 2006), which includes Units T1 and T2 of Tixeront (1974). The two units were named the Ait Driss Member) (Driss River Conglomerates) and the Tourbihine (Tourbiain) Member, respectively by Saber et al. (2007). The Ikakern Formation is restricted to the Argana Graben, in the central part of the Argana Valley (Brown, 1980). Maximum thickness of about 2,000 m (6562 ft) is found in the SE. The basal conglomerates of the Ait Driss Member (Unit T1 of Tixeront, 1974) rest unconformably on Cambrian to Devonian folded metasediments and locally with an angular unconformity on Late Carboniferous sediments (Hmich et al., 2006). Tetrapod skeleton remains have been discovered in the 1970s by Dutuit (1988) in the Argana Basin in the Tourbihine Member (T2) (also of the Argana Formation). They were determined as *Diplocaulid nectrideans (Diplocaulus minimus)*, the captorhinid *Acrodonta,* and a moradisaurine (Jalil & Dutuit, 1996). Studies on vertebrate fossils carried out by Jalil & Dutuit (1996) and Jalil (2001) in the T2 Unit, indicate an Upper Permian (Tatarian) age (Saber et al., 2007). The relation between the Ikakern Formation and the Argana Formation is not clear. Ikakern Formation is the former "Lower Series" of the Argana Formation, but has been used as a synonym of the latter.

The Ikakern Formation (Saber et al., 2007) overlies older sediments with an angular unconformity. An Early Triassic NS to NNW-SSE extension phase accompanied by the intrusion of alkaline granites, such as the Azegour granites (Pouclet eta al., 2008) in the High Atlas and the Sebt Labrikiine granites (Brikiine Granite in Amenzou & Badra, 1996) in western Rehamna.(in the Western Meseta. The origin and age of the Azegour and Brikiine granites is controversial. Based on zircon typology, Amenzou & Badra (1996) concluded that the Azegour Granite and Medinet Quartzdiorite (in the Western High Atlas) and Brikiine Granite (in the Western Meseta) belong to calc-alkaline granites associated with late-orogenic Hercynian phase. The Azegour Granite has been dated (Permian) 271 ± 3 Ma, the Birkiine granites 273 ± 2 Ma by Mrini et al. (1992). The Azegour Granite is composed of k-feldspars, quartz, albite, and biotite, and intrudes Lower Cambrian folded volcanic and sedimentary rocks (Amenzou & Badra, 1996). The Medinet Quartz-diorite is composed of plagioclase, quartz, biotite, amphibole, and sericite. The Brikiine Granite intrudes Lower Paleozoic schists and sandstones in the Meseta. The Ikakern Formation was dated Upper Triassic by Brown (1980), Upper Permian by Jalil & Dutuit (1996), and Tatarian (uppermost Permian) by Jalil (2001). The Argana Formation is dated Permian (Jalil & Dutuit, 1996). Voigt et al. (2009) assigned it Middle to Upper Permian age.

Ikakern Formation (Permian) (Brown, 1980) consists of two members. The Ait Driss River Conglomerates Member is up to 1500 m (4922 ft) in thickness and confined mainly to the central Argana Graben (Basin). The conglomerates lay with angular unconformity on Cambrian, Ordovician, and Silurian quartzites, phyllitic schists, undifferentiated Devonian and Upper Carboniferous (De Koning, 1957, Van Houten, 1976). Jones (1975) described the conglomerates as consisting of moderately to poorly sorted cobbles and pebbles up to 90 cm (3 ft) in diameter in a matrix of coarse sand, silt, and clay. In the Argana Graben the conglomerates grade upward and laterally into the overlying Tourbihine Member. Matrix sand grains are coated with hematite and cemented by calcite. Beds of coarse pebbly sandstones contain trough and planar cross-bedding, imbricate pebbles, and asymmetric ripple laminate. The source of the conglomerates is the lower Paleozoic metamorphic rocks and the latest Carboniferous deposits. The conglomerate is overlain by up to 300 m (984 ft) of coarse sandstones interpreted as meandering stream deposits.

The Tourbihine Conglomerate Member in up to 1000 m (3281 ft) in thickness (Brown, 1980): The Tourbiain Conglomerate. It is similar in composition to the underlying Ait Driss River Member from which it grades. Beds display evidence of exposure such as caliche horizons, root casts, reptile tracks of *Rhynchosauroides* sp. (Jones, 1975, Hmich et al., 2006), plant remains of *Vohzia heterophyl*, mud cracks, and raindrop imprints. Jones (1975) assigned the unit a Triassic age based on the reptile tracks *Rhynchosauroides*. Voigt et al. (2009) described tetrapod foot prints in the uppermost layers of the red-bed sequence of the Ikakern Formation in the Argana Basin. They associated the ichnofossil *Hyloidicus* and *Tachypes* with the body fossils found in the formation and assigned the formation a Middle to Late Permian age. Medina (2009) attributed the Ikakern Formation to the early Triassic.

The Argana Formation crops out in the area between Marrakech and Agadir, known as "*Couloir d'Argana*". It overlies the Paleozoic Massif of Ida-ou-Mahmoud in the east and is overlain by the Jurassic tablelands of Ida-ou-Tanan in the west (Jalil & Dutuit, 1996). The formation was subdivided into eight lithostratigraphic units, namely T1-T8 by Tixeront (in Brown, 1980). These units are described in detail by Brown (1980). The formation has been assigned an Upper Permian age based on the presence of the tetrapod *Diplocaulus minimus* (Jalil & Dutuit, 1996). Other Permian Captorhinid reptiles from the Argana Formation (Unit T2) have been described by Jalil & Dutuit, 1996). They include *Acrodonta irerhi*, *Moradisaurinae* Gen. et sp. indet.

The Timegadiouine Formation (Liassic according to Brown, 1980; Carnian according to Olsen et al., 2000) has been divided into three members by Brown (1980): the Tanameurt Volcaniclastic Member, the Aglegal Sandstone Member and the Irohalene Member.

The *Tanameurt Volcaniclastic Member* is up 10 m (33 ft) thick and consists of conglomerate. It lies with an angular unconformity on eroded beds of the Tourbihine Member. In the vicinity of Iferd and 4 km south of Timesgadiouine, it lies on truncated beds of the Driss River conglomerates. The Tanameurt Member is conformably overlain by the Aglegal Sandstone. The conglomerates of the Tanamert Member contain abundant volcanic rock fragments. Angular porphyritic volcanic clasts mainly of rhyolitic composition constitute as much as 25% of the cobbles and pebbles, and 10% of the matrix grains. Potassium-Argon dates of three igneous rock fragments gave ages of $247 \pm 5$ Ma, $280 \pm 15$ Ma, and $300 \pm 20$ Ma (Jones, 1975, p. 101).

The Aglegal Sandstone Member (Brown (1980) has a thickness of 800–1500 m (2625–4922 ft) and overlies unconformably deformed Paleozoic basement rocks or conformably on local deposits of the Ikakern Conglomerate. The unit grades into homogeneous mudstones of the Irohalene Member. The Aglegal Member is made up of coarse-grained sandstones within thick sequences of mudstones, siltstones, and fine-grained sandstones. Bedding is locally lenticular with basal scour surfaces, marked by lag deposits grading upward into trough cross-laminated sandstones, laminated, and rippled. Trails and burrows are common along upper bedding planes. Mudstones are usually burrowed, with occasional roots and mud cracks.

The Irohalene Member (Brown (1980) is conformable on the Aglegal Member and is conformably overlain by the Bigoudine Formation. It thins from 500 m (1641 ft) in the grabens to 200 m (656 ft) over intervening horsts. On top of the Tirkou horst, carbonized plant debris of *Horizon de Boulba* and minor copper mineralization occur (Saadi, 1973, Tixeront, 1971). At Jebel Armrhouz, Estheriids occur at the top of the section. They suggest brackish water conditions. Fossil vertebrates from this member (DuTuit, 1966) are of amphibious and terrestrial origin.

The Bigoudine Formation comprises three members, from base to top: The Tadrart Ouadou Sandstone Member, the Sidi Mansour Mudstone Member, and the Hasseine Mudstone Member. Its age is Liassic according to Brown (1980) and Norian according to Olsen et al. (2000).

The Tadrart Ouadou sandstone Member is up to 150 m (492 ft) in thickness and overlies the Irohalene Member. It consists of bedded sandstones at the base followed by lens-shaped sandstone bodies. Clay chips and quartzite cobbles and pebbles form basal lag deposits. Parallel lamination, low-angle festoon cross-bedding, climbing ripples, and convolute structures are abundant in lower and middle parts. In the upper part, bedding becomes more lenticular and steep cross-bed sets predominate. At Jebel Amrhouz, thin beds of halite *(Salines de Tazelmat)* are interbedded with fine white sandstone and red siltstones. Minor amounts of gypsum and anhydrite and thin veins of carbonized plant debris, malachite, and azurite (Pouit & Saadi, 1964) are present.

The Sidi Mansour Member (Brown (1980) is up to 200 m (656 ft) in thickness and forms a fining-upward sequence composed of chocolate-brown mudstones and siltstones. It reaches a maximum thickness in the north-central area of the Argana Graben, but is absent on the Tirkou horst. The sandstones pass vertically into alternating siltstones and mudstones, with asymmetric ripple marks and mottling.

The Hasseine Mudstone Member (Brown (1980) is 300–1100 m (984–3609 ft) in thickness and marked lateral variations in thickness. It is capped by basalts, 10–20 m (33–66 ft) thick. The Hasseine Member is made up of mudstones interbedded with siltstones and fine sandstones with small trough cross-beds, symmetric and asymmetric ripple marks, and parallel and flaser laminations. Bioturbation is common.

The Hercynian basement of the Marrakech region hosts a large number of massive Cu and Pb-Zn sulfide deposits which often occur in rhyolites and rhyodacites intrusions emplaced during the Carboniferous around 330 Ma (Essafi et al., 2003, Essaifi & Hibti, 2008). Their mineralogical and chemical characteristics indicate a strongly reducing environment which led to the formation of pyrrhotite in the Guemassa-Jebilets metallogenic province. They occur along northern lineaments.

Two groups of Cretaceous-Tertiary plateaus occur north and south the Atlas Mountains in Central Morocco (Zouhri et al., 2008); the northern group includes the Oulad Abdoun Plateau (Phosphate Plateau), the Gannour Plateau, and the Meskala Plateau. The southern group includes the Coastal Laayoune-Bou Craa Plateau.

The phosphate deposits of Morocco, known since 1908, have been exploited since the 1920s (Office Chérifien des Phosphates, 1989, Bardet et al., 2005), and form part of the Mediterranean Tethyan phosphate province (Lucas & Prévôt-Lucas, 1996). Outcrops are situated in several basins, from northeast to southwest: Oulad Abdoun, Ganntour, Meskala, Souss and Bu-Graa in the Sahara (Bardet et al., 2005) (58). Phosphates occur in three basins (Charles & Charles, 1924): 1) The Oued Zem El Borouj, 2) Ganntour Basin, and 3) Chichaouna Basin. The first is the largest and richest with an area of 3000 km² and reserves of about 25–30 million tons (1924 estimates). It extends westward into the Tadla, but richness is reduced in that area. Exploitation (in 1924) is concentrated in the Ouled Abdoun area, with most production coming from Bou Jniba. Production was transferred to Khouribga in 1924. Stratigraphically, these phosphate strata range from uppermost Cretaceous (Maastrichtian) to middle Eocene (Lutetian), spanning the longest time interval of all Tethyan phosphates (Lucas & Prévôt-Lucas, 1996). The Maastrichtian phosphate series in the Oulad Abdoun Basin near Khouribga is only about 2–5 m (7–16 ft) thick. It consists of grey limestones and yellow and brown phosphates and marls, with many extremely hard lenses locally known as *dérangements* or disturbances (Bakkali, 2006). The series is Maastrichtian in age (Bardet et al., 2005). The phosphates in the Ganntour Basin (Ben Guerir area) reach about 45 m (148 ft) in thickness and are separated by barren calcareous and marly levels. The phosphates of the Oulad Abdoun and Ganntour basins were probably deposited in a long and narrow gulf which opened to the Atlantic Ocean (Herbig, 1986). In the northeastern part of the Oulad Abdoun Basin deposition probably took place in a high-energy nearshore environment; whereas in the southwestern part of the Ganntour Basin it occurred in a deeper open marine setting (Office Chérifien des Phosphates, 1989, Trappe, 1991, 1992, Lucas & Prévôt-Lucas, 1996).

The Cretaceous-Eocene beds rest unconformably on Hercynian and Middle Liassic limestones (the Foum Kheneg Unconformity). Folded Eocene rocks overlie tilted Senonian strata along the Azegour (Azgour) Fault (Bakkali, 2006, Fig. 3, Frizon de Lamotte et al., 2008). The Atlas Domain was still submerged during the Middle Eocene, which is the actual beginning of the Atlas Orogeny. The Middle Cretaceous Kem Kem beds contain rich terrestrial faunas (Gheerbrant & Rage, 2006), and the Eocene and Maastrichtian phosphate beds of the Ouled Abdoun Basin contain abundant mammals remains (Gheerbrant et al., 2001, 2007). The Maastrichtian phosphate beds contain the oldest known crocodilian from Africa (Jouve et al., 2008).

## 17.4.4   Atlantic Passive Margin

The Essaouira Basin is currently the most intensively studied part of the continental margin, both at outcrop and in subsurface, because of its potential wealth in hydrocarbons (Broughton & Trepanier 1993, Medina 1995, Le Roy et al., 1998).

The Essaouira and Haha basins are the most important oil producing basins in Morocco (Davison, 2005). Seven fields have been discovered onshore, with six producing from Jurassic, and one from Triassic reservoirs. The basins extend from the Atlantic margin eastward into the High Atlas. The Essaouira Basin is separated from the Doukkala Basin to the north by the Safi Strike-slip Fault and from the Souss Basin to the south by the Agadir Canyon and the South Atlas Fault (SAF) (Le Roy & Piqué, 2001, Davison, 2005). The continental margin is wide in this area and the free air gravity indicates the presence of a shallow basement (Ras Tafelney Plateau) (Davison, 2005, Meissenard et al., 2007). The sedimentary succession in the Essaouira and Haha basins reaches up to 8 km (26,248 ft) in thickness and ranges from Triassic to Recent in age near the coastline (Davison, 2005). The eastern part of the Haha Basin is undeformed while the western part is. From Late Jurassic to Early Cretaceous, the Essaouira Basin and the Western High Atlas formed a shallow platform (Soulaimani & Burkhard, 2008). From the Late Cretaceous onward, the region was subjected to a general NNW-SSE compression which led to folding and reverse-faulting.

The stratigraphy of the Essaouira Basin (Fig. 17.1) remains limited with the exception of the work of Le Roy et al. (1998). The Mesozoic and Cenozoic cover of the Essaouira Basin lies over a Paleozoic basement and consists of the following successions:

The Lower Triassic-Lower Liassic succession consists of terrigenous sediments. It lies with an angular unconformity on the Hercynian basement. It consists of red silty sediments with thick evaporitic beds and doleritic sills and lava flows. The $^{39}$Ar/$^{40}$Ar dates of the basal dolerites yielded ages near the Triassic-Jurassic boundary between 210–196 Ma (Fiechtner et al. 1992, Le Roy et al., 1998).

The Lower Liassic-Upper Jurassic succession is over 2500 m (8,202 ft) thick in the ESSI well near Essaouira (Fig. 17.1) and includes four units (Du Dresnay 1988, Le Roy et al., 1998):

A Sinemurian-Toarcian unit composed of carbonates at the base, is followed upward by shaly sandstones. The lower carbonates have been correlated with the oncolitic limestones of the Arich Ouzla Formation of Jebel Amsitten and the upper part with the Amsitten Formation (Domerian-Toarcian).

Limestones and dolomites of the Anklout-Tamarout Formation (Toarcian-Bathonian) about 700 m (2297 ft) thick in the ESSI well. The Formation is overlain by the Ameskroud Formation (lower Bathonian).

Marly limestones of the Si Rhalem Formation (Callovian-middle Oxfordian) rich in microfauna, such as the ammonites *Reineckia anceps* and *Aspidoceras phoenicicum*, brachiopods, and foraminifera, about 150 m (492 ft) thick in the ESSI well.

Evaporites of about 900 m (2952 ft) in thickness in the ESSI well, equivalent to the Imouzer Formation (lower Kimmeridgian) which yielded *Plesiocidaris durandi* and *Terebratula humeralis* and the Ihchech Formation and Timsilline Formation (upper Kimmeridgian-Berriasian) with *Pseudocyclammina lituus* and *Salpingoporella annulata*.

The middle Berriasian is represented by shales and has been correlated with the marls of the Si Lhousseine Formation and the limestones of the Agroud Ouadar Formation in outcrops (Le Roy et al., 1998).

The lower Valanginian-upper Hauterivian succession in the Agadir-Essaouira region is rich in belemnites with Mediterranean affinities. Three belemnite assemblages have been recognized (Mutterlose & Wiedenroth, 2008): 1) A *Hibolithes-Duvalia* assemblage which is typical of inner to middle shelf settings. 2) A *Duvalia-Hibolithes* assemblage of the outer shelf setting. 3) A hemipelagic assemblage of *Duvalia-Pseudobelus*.

The overlying transgressive Barremian-Aptian sequence consists of marls equivalent to the Aptian ammonites-bearing marls of the Oued Tidsi Formation (Medina 1995). The Barremian Taboulouart Formation is made up of interbedded marls, marly limestones, and coquina limestones (Medina 1995).

The Aptian-Albian succession of the Essaouira area is represented by marly limestones of the Tamzegrout Formation (marls and marly limestones with ammonites of Rey et al. (1988), the Tahdrat Formation (marls and limestones with ammonites of Rey et al., 1988), the Lemgo Formation, and the Oued Tidsi Formation.

The Albian succession of the Agadir-Essaouira Basins contains rich and well pre-served continental and marine microfloras (Herngreen et al., 1996, Bettar & Meon, 2006), such as Gymnosperms, Angiosperms, spores of Pteridophytes, ephedroids, and elaterates spores. Continental palynomorphs include *Cornetipollis herngreenii, Sergipea agadirensis,* and *Steevesipollenites stoverii*. The pollen are dominated by the characteristic forms of the Albian-Cenomanian Elaterates Equatorial Province (Hern-green et al., 1996). This indicates that during the Albian, the area was under humid climate. Elaterates, found in many contemporaneous Albian-Cenomanian sections worldwide (Herngreen et al., 1996), are absent from Egyptian sediments near Aswan (Schrank, 1992, Mahmoud & Essa, 2007). Their absence has been attributed to pale-oecological factors (Schrank, 1992) and the absence of marine influence in that area (Mahmoud & Essa, 2007).

The Upper Cretaceous succession of the Essaouira Basin is dominated by marine sediments. The Cenomanian high sea level rise and transgression led to the deposition of marls and marly limestones of the Ait Lamine Formation over the entire Essaouira Basin (Le Roy et al., 1998). The Turonian is represented by marly limestones containing interbeds of black shales, which are common in North Africa and the continental margin of the Atlantic (Kolonic et al., 2002, Lüning & Kolonic, 2003). The Senonian is composed of limestones, dolomitic limestones, and marls.

The Cenozoic in the Essaouira Basin is represented by Paleocene and Eocene deposits composed of phosphatic sands, limestones, and marly limestones with oys-ters, such as *Ostrea multicostata* and *O. strictiplicata*. In the offshore, they are rep-resented at the foot of the shelf slope by thick interbedded shales and limestones, probably of Eocene and Miocene ages, such as those encountered at the Deep Sea Drilling Site DSDP-416 (Lancelot & Winterer 1980).

The Agadir Basin is filled with Permian to Eocene continental shelf deposits (Zühl-lke et al., 2004). The basin shows an evolution related to that of the Central Atlantic rifting history (Medina, 1995). The synrift period (Early Triassic-Early Jurassic) is dominated by deposition of red clastic sediments and evaporites, which are overlain by or intercalated with basalt flows. Two extensional phases have been recognized. The first, probably in Middle Triassic times, is oriented NW-SE and resulted in the formation of NE-SW-oriented grabens and half-grabens. The second, a widespread NW-SE extension (Ladinian-earliest Liassic), and induced the formation of westward-dipping half-grabens. Lower Aptian strata exposed in the Agadir Basin represent a fluvial valley fill that was incised into the underlying Late Barremian marginal-marine deposits during the Early Aptian sea-level fall. It starts with fluvial conglomerates and pebbly sandstones, followed by fluvial, tidal flat, estuarine point-bar, and bay-fill deposits (Nouidar & El Chellaï, 2001).

The Cenomanian/Turonian organic-rich strata in North Africa have been most intensively studied in Morocco (Tarfaya, Rif) and Tunisia (northern and central parts) where the highest organic richness are reached (Kolonic et al., 2002, Lüning et al., 2004a). In contrast, very little data is available for the other parts of North Africa, such as Algeria, Libya, and Egypt (Tawadros, 2001).

On the northern side of the Central High Atlas, the Upper Cenomanian-Turonian boundary falls within the Ben Cherrou Formation, which Ettachfini et al. (2005) suggested to replace the incomplete Ait Attab Formation. Ichnological and facies across the Cenomanian-Turonian boundary in the westernmost Tethys, on the Spanish side, show that the Oceanic Anoxic Event (OAE-2) was interrupted by short aerobic or dysaerobic sub-events with slight differences in facies and ichnofabrics (Rodriguez-Rovar et al., 2009). Higher diversity and abundance of trace fossils can be caused by a higher abundance of food available for trace makers. Trace fossils (dominated by *Chondrites, Palaeophycus, Planolites, Taenidium, Thalassinoides,* and *Trichichnus*) show that the OAE-2 event consisted of a series of anoxic sub-events interrupted by short dysaerobic or aerobic sub-events (Rodriguez-Rovar et al., 2009). The development of anoxic conditions and deposition of black shales in the Western Tethys during Cenomanian-Turonian has been related with a sluggish circulation due to the narrow strait of Gibraltar (Lüning et al., 2004a). On the NW Africa Atlantic Margin (Fig. 17.1), the OAE-2 is associated with an expansion of the oxygen minimum zone (OMZ) towards the shelf, probably associated with a redeposition of the organic matter from shallower waters to deep-sea environments (Lüning et al., 2004a).

The stratigraphy, structure and petroleum geology of the Central Atlantic margin of NW Africa, from Morocco to Guinea are summarized by Davison (2005). Rifting of the margin began in Late Triassic (Carnian) times and clastic red bed sequences were deposited on both sides of the Atlantic. Red beds were followed by early Jurassic evaporite deposition in three separate salt basins developed. A major magmatic event with dikes, lavas and plutons occurred along the Central Atlantic margin at 200 Ma (early Jurassic) during salt deposition. A carbonate platform developed along the margin in Jurassic to Early Cretaceous times. A carbonate ramp and rimmed-shelf carbonate platforms developed in Senegal. The deepwater sections of the margin consist of predominantly deep-marine clastic sedimentation from the Jurassic to Recent. Deltas built out at Tan Tan, Cape Boudjour (Early Cretaceous), Nouakchott (Tertiary) and Casamance (Late Cretaceous). The delta deposits are important for oil exploration, because of the presence of potential rich Cenomanian-Turonian mature source rocks for hydrocarbons in these areas (Davison, 2005). Seismic reflection data indicate the Moroccan salt basin extends to the Cap Boudjour area in the Aaiun Basin (Davison & Daily, 2010). Salt diapir structures and collapsed strata indicate salt removal at the shelf edge. The presence of salt along the Atlantic margin may generate traps for hydrocarbons around the salt diapirs. The presence of deep early Cretaceous and Jurassic potential source rocks increases the hydrocarbon potential of unexplored areas (Davison & Daily, 2010).

The southern offshore part of the Doukkala Basin has also been termed the Safi Basin by Tari et al. (2000, 2003). The northern offshore area is known as the Casablanca Offshore Basin (Morabet et al., 1998). The Mazaghan Plateau in the north, a shallow basement capped by Jurassic carbonates, extends to the edge of the salt basin. The Jurassic carbonate bank in the Doukkala Basin contains many salt diapirs and salt-cored anticlines. The Jurassic to Pliocene succession is 4–5 km (13,124–16,405 ft) in thickness. Most of the salt basin lies under more than 2000 m (6562 ft) of water. No wells have been drilled so far in the deep offshore Doukkala Basin. Salt tectonics is also prominent in the Aaiun Basin (Davison & Daily, 2010).

## 17.5 TERTIARY

### 17.5.1 Tertiary of the Central High Atlas

#### 17.5.1.1 Oligocene-Pleistocene

A series of basins occur between the High Atlas and the Anti-Atlas/Sahara Platform, along the South Atlas Fault, also known as the Sub-Atlas Zone. These are the Souss Basin, Ouarzazate Basin, and Boudenib-Errachidia Basin (Fig. 17.1).

The Ouarzazate Basin is a Mio-Pliocene foreland Basin with an area of about 5400 km$^2$ (ONHYM) on the southeastern boundary of the Middle High Atlas and the northern flank of the Eastern Anti-Atlas (J. Saghro) (Missenard et al., 2007). Only two stratigraphic wells have been drilled in the basin. In the southern margin, Neogene sediments overlie the Precambrian basement. The northern margin consists of folded Jurassic to Neogene sediments. In the central part of the Basin, the Mesozoic sequence pinches out against the Paleozoic. Jurassic sediments were detached and thrust southward over the Ouarzazate Basin, for example in the Ait Kandoula nappes and the Ait Seddrat nappes (El Harfi et al., 2001). The South Atlas Front (SAF) bounds the northern border of the Basin and corresponds to a major thrust-fault zone. The structural style varies considerably within the basin along strike from east to west (Missenard et al., 2007), with décollement within the Senonian succession on top of the flat Cenomanian-Turonian carbonates and the underlying sandstones.

The Tertiary succession in the eastern part of the Ouarzazate Basin is Lower Paleocene (Danian) to Oligocene (Dragastan & Herbig, 2007). It includes, in an ascending order, the Upper Red Series (Danian), Asseghmou, Jbel Guercif, and Jbel Talouit formations Upper Paleocene), Ait Ouarhtane Formation (Ypresian), Jbel Tagount Formation (Lutetian), and Hadida Formation (Priabonian-Oligocene). The Paleocene-Oligocene succession has been assigned to the marine Subatlas Supergroup (Dragastan & Herbig, 2007).

The Asseghmou Formation is made up of supratidal green siltstones, dolomites, and lacustrine micritic limestones and has been attributed a basal Thanetian age by Herbig (1991).

The overlying Jbel Guercif Formation consists of a shallowing-upward sequence of shallow marine bioclastic limestones. The top of the formation is marked by a sedimentary break with borings. Herbig (1991) assigned a late early- mid Thanetian age to the Jbel Guercif Formation; while Dragastan & Herbig extended its age to the upper Thanetian. Dragastan & Herbig (2007) identified and amply illustrated the following *Halimeda* species from the Jbel Guercif Formation: *Halimeda nana*, *H. incrassata*, *H. cylindracea*, *H. opuntia*, *H. monile*, *H. erikfluegeli*, *H. lacunosa*, *H. barbata* n. sp., *H. marcconradi*, *H. praetaenicola*, and *H. unica*.

The Jbel Talouit Formation is subdivided into three unnamed members; the lower and upper members consist of greyish to greenish-grey siltstones with a few sandstone interbeds. The middle member is composed of marginal-marine, sandy, oolitic and oyster-bearing limestones. The Jbel Talouit Formation is latest Thanetian to basal Ypresian according to Herbig (1991).

The Ait Ouarhitane Formation is made up of a shallowing-upward sequence composed of bioclastic limestones, with Ypresian vertebrate fauna (Herbig, 1991). The Formation contains *Halimeda praeopuntia*, *H. praemacroloba*, *H. incrassata*, *H. cylindracea*, *H. opuntia*, *H. monile*, *H. opuntia* f. *triloba*, *H. simulans*, *H. tuna*, *H. tuna* cf. *platydisca*, *H. gracilis*, and *H. copiosa* (Dragastan & Herbig, 2007). *Halimeda tuna* was reported from the Ypresian of Egypt. Herbig & Geyer (1988) also reported abundant *Ovulites margaritula* from the uppermost Jbel Tagount Formation. *Ovulites margaritula* also occurs in the Anmiter Formation and Thersitea Formation in the southern Central High Atlas (Trappe, 1992), as well as the Jbel Tagount Formation (Herbig, 1991).

The Jbel Tagount Formation is a mixed carbonate-siliciclastic succession of shallow to marginal marine greenish-grey siltstones, sandstones, and bioclastic limestones. Palynomorphs indicate a latest Ypresian age for the basal part of the formation (Mohr & Fechner, 1986), and the top of the formation may extend into the Bartonian (Geyer & Herbig, 1988). The Subatlas Group in the southern Central High Atlas is conformably overlain by the continental sebkha-type red beds of the Hadida Formation (Middle Eocene-Oligocene) (Herbig, 1991, Tesón & Teixell, 2006).

The Imerhane Group (Oligocene-Pleistocene) crops out in the Ouarzazate Basin. The Group encompasses all the continental lithostratigraphic units from the Oligocene to Pleistocene (El Harfi et al., 2001, Fig. 4) (Fig. 17.1). It overlies the Neogene marine Sub-Atlas Group.

The Ait Kandoula Basin is located on the Southern slope of the High Atlas, northeast of the Toundout Village, with a surface area of about 330 km$^2$ (Benammi et al., 1996). There have been many controversies about the ages of the Neogene (Oligocene to Pliocene) continental formations.

The Ait Arbi Formation, in the Ouarzazate Basin, is Late Eocene and includes conglomerates at its base (El Harfi et al., 2001). It is probably equivalent in part to the Hamada Boudenib Formation in the Boudenib-Errachidia Basin.

The Ait Kandoula Formation is made up syndeformational alluvial fan conglomerates (Tesón & Teixell, 2006). It was assigned a Middle Miocene-Pliocene age based on micromammals and magnetostratigraphy (Benammi et al., 1996). The Formation consists of two members: The lower Ait Ibrim Member is composed of shales and lacustrine limestones with some conglomeratic intercalations. Its thickness is 400 m (1312 ft). The upper Ait Seddart Member has a thickness of about 300–400 m (984–1312 ft) and consists of conglomerates made up of Jurassic clasts. It was assigned a Pliocene age. Ankaratrite lava at Foum el Kous area gave an upper limit of the Ait Kandoula Formation at 29 Ma (Tesón & Teixell, 2006, citing Schmidt, 1992). The Ait Kandoula Formation is marked at the top by a level of conglomerates of possible Late Pliocene or Pleistocene age. Benammi et al. (1996) proposed a Middle Miocene-Upper Pliocene age for the Ait Kandoula Formation. An ash layer yielded an Ar/Ar age of 5.9 ± 0.5 Ma (Benammi et al., 1996). The formation contains the rodents *Occitanomys* and *Prolagus* cf. *michauxi* andothers with Western European affinities, indicating a trans-Mediterranean terrestrial faunal exchange between Africa and Europe during the Late Miocene (cf. Libya and Egypt).

The Ait Ouglif Formation (see also El Harfi et al., 2001) consists of polygenic conglomerates with rounded pebbles of Precambrian to Paleogene rocks deposited in an alluvial fan environment. It has an average thickness of 30–40 m (98–131 ft). It overlies older sediments with an angular unconformity. It was assigned a lower to middle Miocene by Tesón & Teixell (2006).

The Hadida Formation of the Imerhane Group (Herbig, 1991) is composed of red shales with gypsum and passes laterally into sandstones and microconglomerates of the Ait Arbi Formation with a thickness of about 700 m (2297 ft). Frizon de Lamotte et al. (2008) assigned the Ait Arbi Formation to the Upper Eocene. The two formations were assigned a Middle Eocene to Oligocene age (Tesón & Teixell, 2006). Tesón & Teixell (2006) postulate that the siliceous pebbles in the Ait Arbi Formation were derived from the erosion of the Lower Cretaceous conglomerates of the High Atlas. El-Harfi et al. (2001) suggest that the conglomerates of the Ait Arbi and Hadida formations were derived from the Anti-Atlas.

In the Boudenib-Errachidia Basin, the Hamada Boudenib Formation (Eocene) consists of two units (Armenteros et al., 2003); the Lower Hamada de Boudenib I (Lower Eocene) overlain unconformably by the Upper Hamada de Boudenib II (Middle/Upper Eocene). The two units dip and thicken to the south. Their lithology is dominated by mudstones and sandy mudstones, with frequent duricrusts. Fining-upward sequences of siltstones, sandstones and conglomerates occur at the base of the sequences. Limestones and dolomites increase in frequency upwards and show evidence of dedolomitization. The sands are fine to medium-grained and contain carbonate concretions. The carbonates contain *Microcodium*, bioclasts of gastropods and ostracods, in addition to calcareous and dolomitic lithoclasts. The two units are exposed in the southern scarps at Oued Guir. In the Tichniouine and Gara Titratine sections, they have total thicknesses of 25 m (82 ft) and 55 m (180 ft), respectively.

The history of stratigraphic nomenclature in the Guercif Basin (Fig. 17.10) has been summarized by Bernini et al. (1999). The pre-Neogene succession consists of a Paleozoic section overlain unconformably by a Permo-Triassic continental sequence of red marls and basalts up to 500 m (1641 ft) in thickness. Red marls and salt diapirs occur along faults in the basin. The overlying Jurassic succession is made up of dolomites, limestones and marls.

The Neogene (Miocene) basin-fill reaches about 1500 m (4922 ft) in thickness (Bernini et al., 1999). The succession includes the continental Draa Sidi Saada Formation at the base (contradiction with Gomez et al.), overlain successively by the Ras el Kasr Formation, Melloulou Formation, and the continental Kef ed Deba and Bou Irhardainene Formations (Bernini et al., 1999).

The Draa Sidi Saada Formation (Bernini et al., 1999) consists of lenticular bodies of fluviatile conglomerates and sandstones interbedded with grey and pink mudstones (*marnes rouge-briques* of Benzaquen, 1965). It crops out in the western part of the basin with a maximum of 100 m (328 ft). The Formation rests unconformably on the Jurassic basement is overlain by the Ras el Kasr Formation. Benzaquen (1965) assigned the marls a Tortonian age.

The Ras el Kasr Formation was introduced by Benzaquen (1965). It unconformably overlies the Jurassic basement and locally the Draa Sidi Saada Formation. It shows significant facies and thickness variations from shallow-marine to outer shelf (Bernini et al., 1999). Its thickness reaches up to 500 m (1641 ft). The basal part of the formation contains corals, sponges and algae or cross-laminated arenites.

The Melloulou Formation (Bernini et al., 1999) (*Marnes bleues* of Benzaquen, 1965) consists of four units, from base to top: 1) The Blue Marl (Tortonian), more than 500 m (1641 ft) in thickness. 2) The Rhirane Turbidite is exposed at the meeting

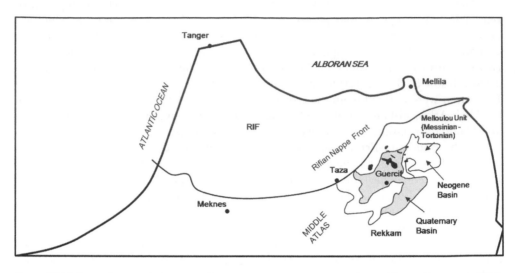

*Figure 17.10* Isopach map of Neogene-Quaternary succession in the Guercif Basin (Sani et al., 2000, Gomez et al., 2000).

point of the two rivers and composed of thin-bedded (about 10 cm) mudstones and sandstones. 3) The Tachrift Turbidite is made up of several turbidite units separated by mudstone intervals. 4) The Gypsiferous Marl consists of variegated gypsum-bearing mudstones. The upper part contains *Ostrea*-bearing lenses, laminated mudstones, and red wavy-bedded siltstones. It was dated Messinian based on the occurrence of *Globorotalia miotumida* gr. (Krijgsman et al., 1998). The last three units are absent in the Bab Stout area and the Safsafat Anticline. It overlies either the Ras el Ksar Formation or the Jurassic basement.

The above formations are disconcordantly overlain by the Messinian Kef ed Deba Formation (equivalent of the Laguno-lacustre of Kandek el Ourish of Benzaquen, 1965). Its thickness varies between 100 m and several hundred meters (Sani et al., 2000). It is composed of yellowish sandstones, calcilutites and thin layers limey layers. The Irhardaiene Formation (Pliocene) (equivalent to El Mongar Formation of Benzaquen, 1965 and the Grès et Conglomerats Continentaux of Colletta, 1977) overlies the Kef ed Deba Formation with an angular unconformity. The Grès et Conglomerats Continentaux (Colletta, 1977) is made up of continental conglomerates, calcarenites, and lacustrine calcilutites, more than 400 m (1312 ft) in thickness and is disconcordantly overlain by Plio-Quaternary alluvial conglomerates, sandstones, and occasional lake bed deposits.

All the major Neogene unconformities at the base of the Tortonian, the base of Messinian (Kef ed Deba) and the base of the Pliocene (Bou Irhardaiene) sequences can be observed on the surface and in the subsurface can be recognized in seismic sections (Sani et al., 2000).

Planktonic and benthic foraminifera from Upper Miocene-Lower Pliocene in the Eastern Rabat area (Wadi Regreg) in the Central Meseta suggest three different environments during that period: an *Ammorda beccarii-assemblage* of the inner shelf, a

*Hanzawaia boueanum* assemblage of the middle shelf, and a *Planulina ariminensis* assemblage of the upper slope (Gerbhardt, 1993). The succession at Ain Allal Ben Mehdi and Oued Akrech includes the following units (Wernli, 1988, in Gerbhardt, 1993):

1   Yellow bioclastic sands is composed of a basal conglomerate, followed upward by interbedded orange to yellow, calcareous marls and marly bioclastic limestones, with marls increasing upward. It rests unconformably upon the Paleozoic rocks.
2   Conglomeratic-coral bank is a 0.4 m (1.3 ft) thick conglomeratic bed between the Yellow bioclastic sands and the overlying "Marnes de Salé" or Blue Marls.
3   Blue marls (*Marnes de Salé*) is made up of laminated marls, changing upward into sandy marls. They have a thickness of 1.4 m (4.5 ft) at Ain Allal Ben Mehdi.
4   Dalle Moghrebienne (Pliocene-Pleistocene), at Oued Akrech, consists of cross-bedded conglomeratic limestones.
5   Grey sandy marls (*Marnes grises sableuses*) is made up of laminated, strongly bioturbated marls, with some hard calcareous horizons. It has a thickness of 4.6 m (15 ft) at Ain Allal Ben Mehdi where it overlies the Marnes de Salé.
6   Green glauconitic sands (*Sables verts glauconieux*) is composed of glauconitic bioturbated sandy marls with a thickness of 5 m (17.4 ft). It overlies the Grey sandy marls at Ain Allal Ben Mehdi.

The Yellow bioclastic sands and the Blue marls are of Messinian-Zanclean age. The Grey sandy marls and the Green glauconitic sands correspond to the Zanclean stage. The Miocene-Zanclean boundary falls within the Salted Marls Unit (Gerbhardt, 1993).

Miocene-Pliocene galena mineralization occurs in the Oued Mekta area, Eastern Morocco. The galena ore body is located in dolomitized algal limestones. Sulfur isotopes of galena show a mean $\delta^{34}S$ value of –0.32 per mil, a mean $\delta^{13}C$ value of 1.302 + or –0.51 per mil, and a mean $\delta^{18}O$ value of oxygen is –5.645 + or – 0.63 per mil. The mineralization is independent of the stratigraphic level of the host rock. It has been attributed to the final stage of Miocene-Pliocene tectonics accompanied by later basalt extrusions (Rajlich, 1983).

The Rif is an arcuate fold-thrust belt in Northern Morocco (Bernini et al, 1999), bordered on the north by the Alboran Sea, and on the south by the Mesetas, on the west by the Atlantic Ocean. The Rif Belt is part of the Betic-Rif-Tell belt which is part of the Mediterranean Alpine Chain (Fig. 17.11). It forms the westernmost part of the Maghrebide Belt, which extends along the North African coast and continues eastward to Sicily and Calabria in southern Italy (Chalouan et al., 2006, 2008). It also forms the southern part of the Gibraltar Arc. Jebel Moussa is the counterpart of the Gibraltar Rock on the opposite side of the Strait of Gibraltar (Durand-Delga & Villiaumey, 1963) (Figs. 17.11, 17.12), and together they form the ancient Pillars of Hercules. The Rif is divided into three Tectono-stratigraphic domains: The Internal Rif in the north, the Flyschs Zone, and External Rif in the south (Fig. 17.11). Each domain consists of tectonic complexes of stacked units or nappes (Chalouan et al., 2006, 2008). Deformation in the Internal Rif started during the early Alpine phase and was progressively thrust southward onto the Flyschs Domain. The External Rif was deformed in the Neogene, with the final thrusting onto the foreland during the

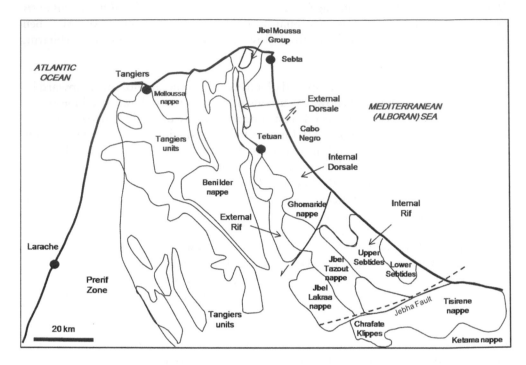

*Figure 17.11* Simplified geological map of the Rif Domain (after Kadiri et al., 2006, Chalouan et al., 2006, El Adraoui et al., 2007).

*Figure 17.12* Gibraltar Rock, view from the Spanish side.

Tortonian-Messinian. The *Rides prérifaines* (Rifian Ridge) is a Jurassic basement that crops out north of Mekness and was back trust during the Miocene and Pliocene (Bernini et al., 1999). The Jebha Fault is a sinistral shear zone that marks the boundary between the Internal Rif and the Flysch Zone (Fig. 17.11). The Rif is separated from the Middle Atlas and the Eastern Meseta by the Middle Atlas Shear.

The geology of the Rif Domain has been reviewed by El Hatimi et al. (1988), Barbieri & Ori (2000), Kadiri et al. (2006), Negro et al. (2007), and Chalouan et al. (2006, 2008). Holocene deposits were discussed by Barathon et al. (2000).

The stratigraphic succession (Durand-Delga et al., 2007, Chalouan et al., 2008) starts with Triassic-Lower Liassic sequence in the Internal Domain. The Aelenian-Bajocian is made up of nodular, ammonite-rich limestones. The sequence continues with Late Jurassic similar to the distal part of the External Domain. The Early Cretaceous sequence is similar to the Dorsal and Beni Ider Flysch succession, the "Aptychus Complex". The Upper Maastrichtian-Paleocene consists of Red beds "Couches rouge", and the Upper Eocene-Oligocene by colored marls and clastic-nummulitic mud flows. The sequence ends with Aquitanian pebbly sandstones of the "Numidian Facies" and Lower Burdigalian brown pelites similar to the Dorsal pre-nappe and the Ghomeride post-nappe successions.

Triassic rifting in Eastern Meseta has been also described in the Oujda Mountain (north of the Jerada Basin) by Oujidi & Elmi (2000) and Oujidi et al. (2006). According to Oujidi et al. (2006), the Ladinian-Carnian (middle Triassic) tectonic instability is recorded within the Carbonate Formation of the Oujda Mountain. ENE-WSW (direction of the Oujda Mountain) & NNW-SSE extension resulted in the formation of the Triassic Basin, which opened to the Western Tethys Domain (associated with the early episode of Tethyan rifting. Structural development was controlled by the NNE-trending Hercynian faults. The Triassic succession crops out in NE Morocco. The Triassic Series lies unconformably over a deformed Paleozoic basement composed of Ordovician-Silurian sandstones and schists and weathered Visean granodiorites. According to Rakus (1979), the complete Triassic succession includes:

> A silicified stromatolitic level (Oujidi & Elmi, 1992)
> Upper Red Beds
> Upper Dolerites
> Middle Carbonate Formation
> Lower Dolerites
> Lower Red Beds
> Lower Carbonate Formation

The Lower Carbonate Formation (Late Ladinian-Early Carnian) is made up of dolomites, limestones, and marls. It forms a regional marker and has a thickness of 3–13 m. It displays many synsedimentary deformation structures, such as breccias, slumps, and normal faults (Oujidi et al., 2006).

The sedimentary klippes (in the Flyschs Domain), composed of Jurassic and Cretaceous rocks, show that: The Ammonitico Rosso facies was deposited in the Liassic, radiolarites formed during the Dogger, and redeposition of the Dorsal Calcaires units in the form of nappes during the Domerian to Cenomanian (El Hatimi et al., 1988). During the Mesozoic, the continental margin subsided and the Flyschs Basin opened as a result of movements along a transform zone between the Central Atlantic and the Ligurian Ocean (El Hatimi et al., 1988), followed by a wrench compressive phase during the Cenozoic.

Three structural domains form the Rif (Chalouan et al., 2006) (Fig. 17.11):

1   The Internal Domain consists of the Dorsal, Ghomarides, and Sebtides nappes, which are made up of Mesozoic carbonate rocks, Variscan-derived Paleozoic terranes, and high-grade metamorphic and mantle rocks, respectively. The Internal Domain was a part of the AlKaPeCa microplate (Bouillin, 1986) located between Africa and Eurasia. The Flysch Trough, which extended between Sicily and the Gibraltar area, represented the deepest part of this paleogeographical setting and could have an oceanic to transitional crust (Durand-Delga et al., 2000).
2   The Flyschs Domain consists of allochthonous units ranging in age from lower Cretaceous at the base to Lower Miocene (Durand-Delga et al., 2000). The Flyschs units are grouped into two main stratigraphic successions (Bouillin et al. 1970, Durand-Delga et al., 2000, Chalouan et al., 2006):

   a   The Mauretanian Flyschs Series on the southern side of the Gibraltar Strait consists of the Tisirene Flysch (late Jurassic-early Cretaceous), and the Beni Ider Flysch (Paleogene-Middle Burdigalian), and
   b   The Massylian Series located close to the southern paleomargin and consists of the Massylian Flysch (early Cretaceous) and the Numidian Flysch (Oligocene-Miocene). They overlie the Melloussa and Intrarif nappes units, and are overlain in turn by the Predorsalian and Numidian nappes. The main structures were produced by a compression phase, during which N-S to NW-SE compression phases (produced by the Africa-Iberia convergence and resulted in small-scale thrusts and brittle deformation) alternated with ENE-WSW to E-W compression stresses caused by the westward drifting of the Alboran plate and resulted in folding and thrusting). The emplacement of the Mauretanian nappe over the Melloussa and Intrarif units was caused by a combination of compression and gravitational forces, which led to the stacking of the nappes in an out-of-sequence succession. The precise age of the tectonic phases remains uncertain because of paucity or absence of direct stratigraphic data (Chalouan et al., 2006), but it can be limited between the Middle Burdigalian when sedimentation ceased in the Flysch Basin and the Langhian-late Tortonian, the age of the overlapping sediments.

3   The External Domain consists of (a) Late Cretaceous, pelite-dominated parautochthonous terranes (Tangiers-Loukous units) and Triassic to Tertiary carbonate and siliciclastic parautochthonous to autochthonous terranes (Mesorif-Prerif). The External Domain derived from the African paleomargin of the Maghrebian-Apennine chain.

The Tisirene (Late Jurassic-Early Cretaceous) and Beni Ider (Paleogene-Middle Burdigalian) nappes correspond to the Flysch de Guerrouch in the Algerian Kabylies. The Late Cretaceous transitional strata between the Tisirene sandstones and the Beni Ider turbidites consist of incompetent thin-bedded calci-turbidites and red shales, which acted as a large-scale detachment level Chalouan et al. (2006).

The Eastern Rif is subdivided into the Mesorif and Intrarif nappes. The Mesorif is dominated by the Temsamane Massif, which is separated from the Intrarif in the north by the Nekor Fault (Fig. 17.11) and shows green-schist metamorphic grade and ductile deformation (Negro et al., 2007). The contact between the Eastern Rif and the Intrarif represents an ancient suture zone marked along the Nekor Fault by the Beni Malek serpentinites interpreted as a sliver of the Jurassic-Cretaceous oceanic crust (Michard et al., 2007).

The Temsamane Massif displays a complete lithological sequence from Paleozoic to Albian (Azdimousa et al., 2007). The Paleozoic metamorphic rocks of the Ras Afraou Unit are overlain by Triassic volcanic rocks, Jurassic marbles, Neocomian schists and marbles, and Aptian-Albian brown schists and quartzites. The Temsamane units represent the sedimentary cover and basement of the Mesozoic Maghrebian paleomargin, which was inverted and metamorphosed during the Lower Miocene. Black basic volcanic rocks occur at Ras Afraou. Triassic dolerites overlie the Paleozoic rocks with an angular unconformity and are overlain in turn by black marbles of probable Jurassic age (Choubert et al., 1984). The latter are covered conformably by Lower Cretaceous rocks. Belemnites have been found in Lower Cretaceous (Neocomian) sediments by Frizon de Lamotte et al. (2008), as well as *Phylloceras* sp. at the top of the succession. They are followed upward by brown schists and quartzites attributed to the Aptian-Albian. The Boudinar Basin is of Late Miocene and Pliocene ages and consists of detrital sediments and marls, with volcanic intercalations near the top (Azdimousa et al., 2006). The succession ends with Pliocene-Quaternary fluviomarine and continental deposits. The Temsamane units are characterized by intense metamorphism and penetrative ductile deformation (Negro et al., 2007). The Temsamane units underwent medium-pressure low-temperature metamorphism possibly during the Oligocene, and were later exhumed during Middle to Late Miocene, during which they were subjected to east-west stretching. The metamorphism of the Temsamane units has been attributed to subduction in the External Rif parallel (Alboran Domain) or the External Rif (Negro et al., 2007).

The Calcareous Dorsal is subdivided into Eastern, Central, and Western Dorsals (El Hatimi et al., 1988). The Central Dorsal of Hauz is located in the Internal Rif Domain and belongs to the Calcareous Dorsal. It occupied by discontinuous Jurassic, Cretaceous, and Tertiary sections. The presence of radiolarias suggests that during the Dogger-Malm the area was occupied by a deep open sea similar to that of the External and Internal Dorsal (Hlila et al., 1994). The Hauz Chain belongs to the Calcareous Dorsal and extends from Tetuan in the south to the Strait of Gibraltar in the north. The chain is subdivided (as the rest of the Dorsal) into Internal, Central, and External domains.

The Upper Rhetian-Jurassic dolomitic succession in the Central Dorsal (Hlila et al., 1994) consists in an ascending order of:

a   Thin red and green radiolarites (Dogger-Malm) unconformably overlies Rhaetian sediments.

b   Thin Upper Eocene grey marls with abundant planktonic foraminiferas, such as *Globorotalia cerroazulensis, G. cocaensis, Globigerinatheka index, Globigerina praebulloides, G. eocaena, G. galavisi,* and *Catapsydrax pera.*

c   Limestone breccias of variable composition.

d   A 40 m (131 ft) interval of yellow limonitic marls with sandstones of Eocene or the Oligocene age. It also includes Thanetian and Berriasian marly limestones, as well as Paleocene detrital limestones with *Microcodium*.

The Eocene succession of the Talembote upper Ghomaride Unit ranges from Ypresian (Ilerdian) to Early Bartonian, interrupted by surfaces of discontinuity (Hlila et al., 2007). The lower sediments correspond to shallow platform carbonate facies with benthic foraminifera of Upper Lutetian age, which was fragmented during the Early Bartonian.

El Kadiri et al. (2006) made regional correlations across the Internal-External Rif Front of the Cretaceous-Early Burdgalian succession in six areas:

The Predorsal Zone, defined by Didon et al. (1973), is a narrow transitional zone between the Internal and External Rif domains. It contains the Jbel Moussa Group and the Chrafate klippes. The Predorsal Zone was separated from the Flysch Trough during the Late Cretaceous-Eocene by a ridge, which collapsed in latest Eocene-early Oligocene, at the same time as the formation of the Chrafate Klippe Olistostromes, the *Breccias Nummuliticas*, and the Talembote Olistostromes.

The Internal Rif Domain consists of a complex nappe pile resulting from the westward stacking of three distinct units; the Sebtides, Ghomarides, and the Dorsal Calcaire (Calcareous Dorsal) (El Kadiri et al., 2006).

a   The Sebtides consists of metamorphic nappes including a granulitic-ultramafic unit, overthrust by detached high pressure-low temperature metamorphic shales of late Paleozoic-Triassic age, The Sebtides nappes were placed during the Alpine Orogeny.

b   The Ghomarides nappes thrust over the Sebtides and consist of Paleozoic schists, which formed the westernmost part of the AlKaPaCa paleoplate of Bouillin (1986).

c   The Dorsale Calcaire consists of Triassic-earliest Jurassic thick-bedded shallow-water limestones. Two zones are recognized in the Dorsale Calcaire; the Internal Dorsal and the External Dorsal. The limestones are marked by a paleokarst horizon in the Internal Dorsal, and by Fe-encrusted hardgrounds in the External Dorsal. Two formations have been recognized:

The Fnideq Formation in the Ghomeride Zone is composed of a large variety of facies, including marls and sandstones which become coarser toward the south with chaotic and limestone blocks derived from Liassic and Eocene sediments of the Ghomarides and the Dorsal (El Hatimi et al., 1988, Ouazani-Touhami & Chalouan, 1994). It overlies, with a tectonic contact, faulted Upper Oligocene red radiolites, which rest on top of Paleozoic sediments, and is overlain by the Sidi Abdesalam Formation (Lower Burdigalian) (El Hatimi et al., 1988). The Fnideq Formation is considered Aquitanian-Burdigalian by El Hatimi et al. (1988) and Upper Oligocene-Aquitanian by Ouazani-Touhami & Chalouan (1994).

The Sidi Abdesalam Formation (Lower Burdigalian) is composed of conglomerates rich in Paleozoic debris associated with synsedimentary faults and slump structures (Ouazani-Touhami & Chalouan, 1994).

In the Internal Dorsal, the paleokarst surfaces are overlain by an Ammonitico rosso facies (Domerian-Toarcian) and *Saccocoma* rich and/or *Calpionella*-rich mudstones (Tithonian-Berriasian).

The External Dorsal contains two Jurassic facies; the Ammonitico rosso and red and green radiolarites. The External Dorsal Calcaire consists of two terranes; a paraautochthonous terrane which comprises the Tangiers, Ketama, and Loukous units, and the overriding sandstone flysch nappes. Two main flysch successions in the External Dorsal, which are also recognized in the Kabylie Domain: a) The Mauritanian Series, composed of the Tisirine Sandstones (early Cretaceous) and the Beni Ider Sandstones and Calcirudites (late Cretaceous-Aquitanian). b) The Massylian Series composed of Barremian-Albian Sandstones and Numidian Sandstones.

In the two Dorsal zones, the Cretaceous displays two hiatuses; one during the Hauterivian-Albian; the other within the Campanian-Maastrichtian *Globotruncana* mudstones.

The Paleogene succession consists of black nodular shales, followed by Eocene-Middle Oligocene variegated marls and nummulitic sand interbeds, then by rust-colored, micaceous sandstones of late Oligocene-early Aquitanian age.

Durand-Delga & Villaumey (1963) defined the Beliounis Unit west of Sebta, which forms a narrow zone bordering the western side of the Dorsale Calcaire. It is composed of limestones with *Microcodium* fragments and Numidian Facies sandstones.

The northern part of the Dorsale Calcaire is occupied by the Jbel Moussa Group, composed of four tilted blocks of Triassic-early Cretaceous successions.

The Tariquide Ridge is used in a restricted sense to designate a paleogeographic high composed of shallow-water carbonates that existed during the late Cretaceous-early Oligocene.

The paleogeography of the area during the Late Miocene-Pliocene consisted of (El Hammichi et al., 2006):

1   The Eastern-Rifian Foreland composed of the Gareb chaotic unit and a deformed tectonic foreland (Terni Masgout, Beni Bou Yahi, Beni Mahiou and Beni Snassen), which extends eastward into the Traras Mountains in Algeria.
2   The Guercif and Taourirt-Oujda basins (in the Eastern Meseta) with Middle Miocene-Quaternary sediments, connected to the Tafna Basin in Algeria.
3   The Taourirt-Oujda Mountains (in the Eastern Meseta) with subhorizontal Jurassic sediments, and extends eastward into the Tlemcen Pre-Tellian foreland (Oujidi & Elmi, 2000, Oujidi et al., 2006).

In Northeastern Morocco, the Paleozoic basement crops out in the Terni-Masgout, Beni Snassen, and Taourirt-Oujda Mountains (El Hammichi et al., (2006), which form part of the Hercynian Belt. Deformation was concentrated within shear zones trending approximately NE-SW to E-W (Torbi & Gelard, 1994). Fault reactivation was initiated in the Late Triassic as a result of extensional tectonic. Upper Triassic red mudstones, carbonates, and tholeiitic basalts accompanied the first marine transgression of the Tethyan Ocean over the Northwestern African Margin. In the Early Liassic, a carbonate platform developed. NW-SE extension from the Upper Liassic through the Upper Jurassic-Lower Cretaceous resulted in extensional faults bounding NE-SW-oriented horsts and grabens with tilted fault blocks. The whole area was emerged in the Late Cretaceous and dominated by strike-slip tectonics contemporaneous with the dextral displacement of Africa relative to Iberia. From the Eocene to the Middle Miocene, a compressive event created NE-SW and NW-SE strike-slip faults and NE-SW to E-W folds (Chotin et al., 2000). During the Late Miocene, a NE-SW stress associated with metamorphism, led to the formation of the Guercif, Taourirt-Oujda, Triffa, Oued Hai basins, separated by emerged areas of the Gareb, Beni Snassen, and Taourirt-Oujda Mountains (El Hammichi et al., 2006). In Late Miocene-Quaternary, the region was subjected to compressive strike-slip movements associated with calc-alkaline volcanic rocks.

The Miocene-Pliocene stratigraphy of NE Morocco is as follows (El Hammichi et al., 2006): Lower Miocene (Burdigalian to Aquitanian) deposits are composed of conglomerates and sandy marls and overlie directly the Late Jurassic deposits of the Beni Bou Yahi, Beni Mahiou, and the Beni Snassen. Upper Miocene (Tortonian and Messinian) marls and limestones were deposited in the Guercif, Taourirt-Oujda, Triffa, and Gareb basins. Conglomerates, gypseous marls, and sandy marls of Middle Miocene-Pliocene (Aquitanian-Pliocene) crop out in the Oued Hai Basin. Upper Miocene-Pliocene shoshonitic volcanics of the Guilliz and Koudiat Hamra (near Taourirt), yielded ages of 8–4.5 Ma (Hernández & Bellon 1985); whereas Calc-alkaline rocks in the Oujda Mountains gave ages between 6.2–4.8 Ma (Tisserant et al., 1976 in El Hammichi et al., 2006).

The Melilla Basin (Fig. 17.11) is a post-orogenic basin in which deposition began in the Serravallian or the Tortonian (Guillemin & Houzay, 1982). At the base of the sedimentary succession of the Melilla Basin is a siliciclastic-rich unit composed of basal conglomerates, overlain by marine silty marls and minor siltstones and sandstones. The succession rests unconformably on Paleozoic metamorphic rocks, and the upper surface is an angular unconformity produced by tectonic tilting during the Tortonian (Guillemin & Houzay, 1982). These metamorphic and clastic rocks form the Tarjat High. The overlying Tortonian-Messinian carbonate complex includes both the ramp and bioclastic platform sequences.

The major drawdown of the Mediterranean Sea during the Messinian Salinity Crisis has been dated at 5.78 Ma based on stratigraphy, paleontology, magnetostratigraphy, and Argon dating of the Late Miocene sedimentary succession in the Melilla Basin, NE Morocco (Cunningham et al., 1997, Cunningham & Collins, 2002). In the Melilla Basin, this drawdown was associated with an increase in benthonic foraminifera in the deep-marine section in the Bou Regreg Valley, NW Morocco (Hodell et al., 1994). The Tortonian-lower Messinian section at Bou Regreg (Hodell et al., 1994) consists of: 1) an onlapping ramp, 2) a prograding bioclastic platform, 3) a prograding and, locally, downstepping *Porites*-reef complex, and 4) a topography-draping sequence composed of grainstones, *Porites* reefs, and stromatolites. A major fall in relative sea level occurred near the demise of the reef complex at 5.95 Ma and signals the initiation of drawdown and changing environmental conditions in the Melilla Basin (Cunningham et al., 1997, Cunningham & Collins, 2002). A megabreccia interpreted as forming by solution collapse of evaporites on the basin margin of the reef complex occurs at the base of the last carbonate complex. A major subaerial unconformity separates the reef complex and last carbonate complex. Evaporites, restricted to the Melilla Basin, were probably deposited at approximately 5.82 Ma during the initial stage of the drawdown of the Paleo-Mediterranean Sea (Cunningham et al., 1997, Cunningham & Collins, 2002).

### 17.5.1.2 Quaternary

The Atlas Mountains of North Africa (Fig. 2.1) show evidence of glaciations (Awad, 1963, Hughes et al., 2004, Mark & Osmaston, 2008). Glacial and periglacial features, formed during the Quaternary, are present throughout the Atlas Mountains in Morocco and Algeria and include moraines, U-shaped troughs, roche moutonée, riegels and cirques (Hughes et al., 2004). Glacial and periglacial processes have contributed to shaping the contemporary landscape of the Atlas Mountains, and periglacial processes are still active today (Hughes et al., 2004). At least three discrete Pleistocene glaciation episodes have been identified during the Soltanian, the Tensiftian, and Saletian stages. The Saletian glaciation may have occurred during the Younger Dryas Chronozone. Offshore marine pollen records between Portugal and the Canary Islands indicate a progressive aridity in the Atlas Mountains over the past two glacial periods (185–130 ka and 70–10 ka), along with gradual northward progression of the Sahara (Hooghiemstra et al., 2006). About 18 ka, a narrow cold/arid coastal belt prevailed because of meltwater from the Northern Hemisphere ice sheets (Rognon, 1987). In the period between 40–20 ka, the environment was wetter in the Northern Sahara (Rognon, 1987). The rainfall was probably regular and abundant as a consequence of the shifting of middle-latitude cyclone paths towards the South in both summer and winter. From about 20–18 ka, rainfall became more irregular. Between 12–10 ka, aridity prevailed in the South Maghreb and the Northern Sahara. After 6000 yr B.P., the climate became wetter again and led to the formation of Late Pleistocene red paleosoils (Rognon, 1987).

In the Eastern Rif, pollen analysis indicates a humid grassland environment in the Early Holocene (Early Rharbian) (ca. 10,000 yr BP), which was later replaced by arid steppes. The Middle Holocene (Middle Rharbian) (7000–5500 yr BP) is represented by silty deposits, and during the Recent (Late Rharbian) (<3000 yr BP) low sandy terraces developed (Barathon et al., 2000). The Upper Pleistocene was a period of incision and formation of valleys. In the Upper Pleistocene Soltanian terraces, two different facies can be observed. The older facies, composed of coarse alluvia probably dates back to the Middle Paleolithic (Rognon, 1987). The Mousterian sites are located on the surface or in caves. The second facies is reddish sand and silt and overlaps the coarse alluvia. This change in facies occurred before 40 ka. Three arbitrary paleoclimatic changes probably took places at 40–20 ka, 20–6 ka, and after 6000 yr BP (Rognon, 1987).

# Petroleum geology & petroleum systems of North Africa

# Chapter 18

# Petroleum geology & petroleum systems of North Africa

## 18.1  INTRODUCTION

It is not known what the long term effect of the events and turmoil of 2011 in Egypt, Libya, Tunisia, and the rest of the Middle East will have on oil exploration and production in the area. According to Fugro Robertson, Egypt was ranked first by companies who were interested in new ventures (AAPG Explorer, Oct. 2008). The changes in the political scene in Libya prior to the "revolutions" promised a more active exploration for oil with some successes. There is no doubt that this picture will change for the next while until stability returns to the area.

The combined recoverable hydrocarbon reserves of North Africa are estimated at 64 billion barrels of oil (2007 estimates) and 8.2 million cu m (290 TCF) of gas (Fig. 18.1). These hydrocarbon accumulations resulted from favorable tectonic and depositional settings, which led to the formation of numerous reservoirs and traps fed by extensive and rich source rocks. The dominant traps are structural, but stratigraphic and combination traps are also present. Deposition of source rocks, reservoirs, and seals took place in continental rift basins, cratonic basins, and passive continental margin basins. Reservoirs range in age from Precambrian to Pliocene and include fractured basement igneous rocks, fractured quartzites, peri-glacial deposits, turbidites, continental, deltaic, and shallow marine sandstones and carbonate build-

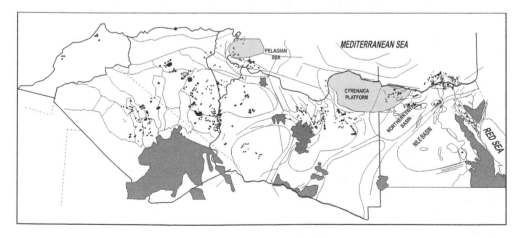

*Figure 18.1*  Oil and gas fields in North Africa.

| PHASES | | EVENTS | OROGENIES | SEA LEVEL H L |
|---|---|---|---|---|
| PLEISTOCENE | Nile | Nile & Nile Delta | | |
| PLIOCENE | | | | |
| MIOCENE | Gulf of Suez | Incised valleys, Nile, Sahabi, Abu Madi Channel<br>Gulf of Aqaba<br>Gulf of Suez<br>Souther Red Sea rifting | | |
| OLIGOCENE | Alpine | Western Mediterranean<br>Jabal Akhdar uplift<br>Flextural subsidence | | |
| EOCENE | | | | |
| PALEOCENE | | | | |
| CRETACEOUS | | Plate collision, basin inversion, collapse of arches | Alpine s.l. | |
| | Atlantic | North Atlantic<br>South Atlantic | | |
| JURASSIC | | Central Atlantic | | |
| | Mediterranean | Neo-Tethys | | |
| TRIASSIC | | Collapse of arches<br>Paleotethys | | |
| PERMIAN | Tectonic-Arches | Inversion of basins | Hercynian | |
| CARBONIFEROUS | | | | |
| DEVONIAN | | Uplift of North Africa & exhumation of Precambrian basement | | |
| SILURIAN | Glacial | Inversion of basins Glacial-eustatic sea-level rise | | |
| ORDOVICIAN | | Glaciation<br>Incised valleys | Caledonian | |
| CAMBRIAN | Graben | Grabens & pull-apart basins | | |
| PRECAMBRIAN | Magmatica-Arc | Magmatic arcs, consolidation of Gondwana | Pan-African | |

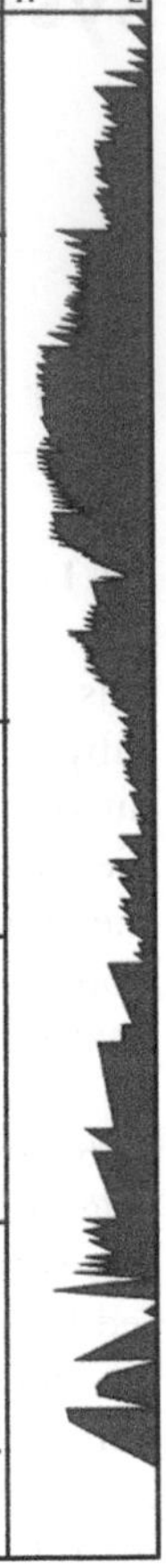

*Figure 18.2* Tectono-depositional phases (modified from Tawadros, 2001, 2003).

ups. Frequent Oceanic Anoxic Events (OAE) took place throughout the Phanerozoic (Fig. 18.2) and resulted in the deposition of extensive black shales and radioactive marls with TOC values exceeding 7% and reaching up to 17% in places (Lüning, 2000d), although the universality of the predominance of anoxic conditions have been questioned (Pancoast et al., 2004). Many Paleozoic oils in North Africa are similar and show typical Lower Paleozoic characteristics of high tricyclic terpane contents and a dominance of $C_{29}$ sterane over $C_{27}$ and $C_{28}$ (Quesada et al., 2003). Deposition in platform areas during these OAE episodes led to large commercial phosphate deposits. Campanian dark marls and black shales accompany the phosphate beds in the southeastern Desert in the Safaga-Qoseir region in Egypt that have the potential of being used for conventional power plants (Awad et al., 2001).

The petroleum geology and hydrocarbon systems of North Africa have been reviewed in numerous works (Boote et al., 1998, Keeley & Massoud, 1998, Macgregor et al., 1998, Macgregor & Moody, 1998, Moody & Clark-Lowes, 1998, Makhous,

2001, Badalini et al., 2002, Dolson et al, 2002, Casati et al., 2003, Lüning, 2004a, b, c, Craig et al., 2004, 2009). Precambrian basement reservoirs in North Africa are discussed by Craig et al. (2004), Aziz & Ghneia (2009), Lottaroli et al. (2009), and Lüning et al. (2009).

## 18.2  TECTONO-DEPOSITIONAL PHASES

The Precambrian-Recent history of North Africa has been divided into nine tectonic/depositional phases, based on sedimentary facies, relative sea-level changes, and tectonics (Tawadros, 2003) (Fig. 18.1). Each of these phases has distinct tectonic and depositional styles and hydrocarbon systems. These phases include, from older to younger:

1  The Magmatic-Arc Phase (Precambrian-Lower Cambrian) was dominated by the formation and break-up of supercontinents and the opening and closure of oceans. The stratigraphic record includes Archean to Early Cambrian igneous, volcanic, and sedimentary rocks, which form the foundation of North Africa. The tectonic structures of the basement, especially those of the Pan-African Orogeny, controlled the Phanerozoic geology and morphology of that part of the African Continent. The end of the Pan-African Orogeny produced a set of NW-trending faults (Klitzsch, 1971, 1990a). From the economic point of view, this phase led the formation of many ore deposits, such as the Imiter Ag-Hg deposits in Morocco (Bensaid & Kutina, 1986, Gasquet et al., 2008) and the precious metals in Egypt, many of which have been exploited since ancient times (Hussein & El Sharkawi, 1990, Harrell, 2004). Fractured Precambrian basement rocks form hydrocarbon reservoirs in the Hurghada, Zeit Bay, and Shoab Ali fields (Khalil & Meshrif, 1988, Rohrback, 1983, Salah, & Alsharhan, 1998, Sircar, 2004) in the Gulf of Suez, and the Maragh Field in the Sirte Basin (Williams, 1968). The Neoproterozoic-Early Cambrian (Infracambrian) successions in Libya and other parts of North and West Africa are discussed by Craig et al. (2009), Aziz & Ghneia (2009), Lottaroli et al. (2009), and Lüning et al. (2009).

2  The Graben Phase (Middle-Upper Cambrian) was dominated by extensional tectonics and the deposition in rift basins of shallow-marine sandstones and associated volcanics, many of which were converted later into fractured quartzites, and minor shales and carbonates. The fractured quartzites are hosts of giant oil and gas reserves, especially in Libya (Attahhady Field) and Algeria (Hassi Messaoud Field).

3  The Glacial Phase (Ordovician-Silurian): The South Pole was located in NW Africa during that phase (Scotese et al., 1999, Carr, 2002, Le Heron, 2007, Le Heron et al., 2007). North Africa was dominated by extensional tectonics and glaciation during the Ordovician. Ordovician continental, shallow-marine and peri-glacial deposits accumulated in half-grabens and incised valleys. Ordovician sands form good reservoirs, especially in West Libya and Eastern Algeria. The sea-level rise at the beginning of the Silurian led to the deposition of extensive blankets of organic-rich black shales "Hot Shales" (Lüning et al., 2000c, d) that represent one of the earliest Oceanic Anoxic Events (OAE) in the Phanerozoic. These black shales constitute source rocks and seals for hydrocarbons over most of North Africa.

4    The Tectonic Arches Phase (Devonian-Permian) was dominated by continental plate collisions and basin inversion during the Variscan/Hercynian Orogeny, which led to the formation of most of the prominent tectonic arches in the area. Thick continental and shallow-marine sandstones and shales, as well as patch reef buildups (Wendt & Kaufmann, 2006, Aitken et al., 2002, Brachert et al., 1992), were deposited. Early Devonian volcanics developed in the Anti-Atlas and Late Devonian granitoids were intruded during the Late Devonian. Frasnian radioactive limestones, shales and marls provided source rock for hydrocarbons, especially in Libya and Algeria. Sandstones which belong to that phase also form oil reservoirs in the Gulf of Suez. Lower Frasnian black shales have been deposited in many parts of North Africa and form an important hydrocarbon source rock in this region with total organic carbon (TOC) values of up to 14% (Boote et al., 1998). The maximum TOC values coincide with the U-enriched interval (Lüning et al., 2004b). A synthesis was compiled by (Lüning et al., 2004c) that allowed explain the mechanism that led to the accumulation of organic matter during the early Frasnian in North Africa. The organically richest and thickest Frasnian black shales are developed in central North Africa (Algeria, Western Libya). Vertical TOC trends have been interpreted as being controlled by the rising and falling of an oxygen minimum zone (OMZ) (Lüning et al., 2004c, Ghori & Mohammed, 1996) during the Early Frasnian eustatic transgression. In addition, iron ore deposits accumulated in Libya and Manganese deposits (mostly diagenetic) were formed in the Gulf of Suez/Sinai area of Egypt. Visean bimodal volcanic and pillow lavas and flysch deposits accumulated in the Meseta Domain in Morocco and volcaniclastics were deposited in the Argana Basin.

5    The Mediterranean Phase (Triassic-Lower Jurassic) was dominated by numerous rifting episodes which accompanied the opening of the Mediterranean Sea (Neotethys), the Atlantic Ocean, and the Atlas Basin in Morocco. Continental, redbeds, shallow-marine carbonates, and evaporites accumulated in rift basins in the Central Atlantic and coastal basins, such as the Aaiun (Laayoune), Essaouira, and Argana basins (Brown, 1980, Medina, 1988, Olsen, 2000, Hafid et al., 2000, Bouatmani et al., 2004, Et-Touhami et al., 2008). Structural and stratigraphic hydrocarbon traps were formed in Libya, Tunisia, and Algeria. Triassic sandstones form reservoirs at the giant Hassi R'Mel and Ourhoud fields in Algeria and the El Borma Field across the Algerian-Tunisian border. Small Triassic and Jurassic fields are present in Morocco. Volcanic rocks associated with rifting are common in North Africa. The most widespread magmatic types in the Sahara are dolerites, with ages varying from 166 Ma in the Reggane Basin, 166–170 M in Bechar Basin, 189–195 Ma in the Timimoun Basin, 180 Ma in Tindouf Basin, 180–200 Ma in the Anti-Atlas of Morocco, and 230–200 Ma at the Algeria-Mali border (Makhous & Galushkin, 2003).

6    The Atlantic Phase (Upper Jurassic-Lower Cretaceous): was dominated by the opening of the Central, North, and South Atlantic Oceans, in that order, and the associated rifting in North Africa. The Sirte Basin rift in Libya and tilted fault blocks in the Western Desert of Egypt were filled with sandstones (converted to fractured quartzites in parts of the Basin) that form large oil reservoirs. In the Western Desert and the Sirte Basin, sediments of this phase form source rocks, reservoirs, and seals. The whole of North Africa tilted northwards during that phase and led to vast continental and minor shallow-marine sandstone deposits

over North Africa and beyond. The Syrian Arc is an accretion belt of the Meso-zoic terranes, which was formed in the Early Cretaceous (Eppelbaum & Katz, 2011). Many of the oil and gas fields in North Africa belong to that phase. Early Cretaceous erosional channels with hydrocarbon potential occur in the Eastern Mediterranean (Eppelbaum & Katz, 2011). Commercial coal deposits accumulated in Sinai. The age equivalent sediments of the Jurassic coals form prolific source rocks in the Western Desert of Egypt.

7    The Alpine Phase (Upper Cretaceous-Eocene) started with an extension and the collapse of most of the arches during the Cenomanian. It was then dominated by the Alpine Orogeny, the formation of the Atlas Mountains in Morocco, Northern Algeria, and Tunisia, the Green Mountain (Jebel Akhdar) in Northern Libya, and the Syrian Arc in Egypt. Two major basin inversion episodes took place during that phase; one during the Santonian (Guiraud & Bosworth, 1997, Guiraud, 1998, Guiraud et al., 2001, 2005, Ziegler et al., 2001, Samuel et al., 2009); the other during the Oligocene (Moustafa & Khalil, 1990, Kuss et al., 2000a, b). Deposition was dominated by limestones, shales, continental sandstones, and shallow-marine carbonates. The Tertiary NNW to SSE compression led to the inversion of NE and NNE structures and the formation of salt anticlines in the Essaouira Basin in the Atlantic margin and the High Atlas Mountains (Medina, 1988, Zizi, 1996, Gomez et al., 2000, Zühlke et al., 2004). Volcanic activities took place during the Eocene in the High Atlas (Marks et al., 2008). Some of the monzonites from the Eocene Tamazegt Complex in the High Atlas gave radiometric ages of $44 \pm 4$ Ma (Rb/Sr) and $42 \pm 3$ Ma (K/Ar) (Tisserant et al., 1976). An OAE, known as OAE2 took place around the Cenomanian-Turonian boundary (Keller et al., 2008, Nederbragt et al., 2004). Cretaceous phosphate deposits were mined near the Red Sea in Egypt and now are being exploited in the Western Desert. The phosphates in Egypt are Campanian in age, and in Tunisia and Algeria, they are Ypresian in age. The equivalents of the Ypresian phosphates form good source rocks of hydrocarbons in Tunisia and its offshore area. The equivalent of the Campanian phosphates in Egypt also forms vast source rocks in the Gulf of Suez and the Sirte Basin. In Morocco, the phosphates range in age from Late Maastrichtian to Lutetian (Middle Eocene). This phase also saw the appearance of mammals in North Africa (Savage & White, 1965, Gingerich, 1992). The Paleocene-Eocene Thermal Maximum (PETM) represents a period of extreme global warming (Zachos et al., 1993), associated with a major extinction of deep-sea benthic foraminifera (Tjalsma & Lohmann, 1983) and evolutionary rejuvenations among planktonic foraminifera (Kelly et al., 1996), mammals (Clyde & Gingerich, 1998, Alegret et al., 2009), calcareous nannofossils (Aubry, 1995), diatoms and larger foraminifera (Scheibner et al., 2003). The PETM boundary is also marked by a negative carbon isotope excursion (CIE) (Speijer et al., 2003, Morsi et al., 2008, Morsi & Scheibner, 2009, Bornemann et al., 2009, Scheibner & Speijer, 2009). A sea-level fall immediately preceding the onset of the PETM, followed by a sea-level rise and enhanced upwelling during the PETM is postulated in Egypt (Speijer & Wagner, 2002).

8    The Gulf of Suez Phase (Upper Oligocene-Messinian) is characterized by the opening of the Red Sea and the Gulf of Suez in Egypt and widespread volcanic activities over most of North Africa (Montenat et al., 1988, Meshref, 1990, Bohannon &

Eittreim, 1991, Makris & Rihm, 1991, Girdler, 1991a & b, Bosworth & McClay, 2001, Khalil & McClay, 2002). Oligocene and Miocene source rocks, sandstone and carbonate reservoirs, and shale, carbonate, and evaporite seals were deposited in the Gulf of Suez rift where tilted fault blocks provided ideal traps for the hydrocarbons. Mud volcanoes and diapirs developed in the Alboran Sea (Sautkin et al., 2003). Desiccation (Hsü et al., 1972) and/or tectonic movements (Roveri & Menzi, 2006) led to deposition of extensive evaporites in the Mediterranean Basin. Inland major incised valleys were formed, such as the Nile Valley, the Abu Madi Channel in the Nile Delta in Egypt and the Sahabi Channel in Eastern Libya.

9   The Nile Phase (Pliocene-Recent) was dominated by the development of the Nile and the Delta. In the Early Pliocene, the Mediterranean Sea was invaded by oceanic waters, which led to the deposition of Pliocene marine sediments. The incised valleys were filled with early Pliocene sands. In the Late Pliocene, the Mediterranean became deeper and turbidites were deposited in the offshore of the Delta. The Early Pliocene channel sands and the Late Pliocene and Upper Miocene turbidite sands form reservoirs for the gas, condensate, and oil in the Nile Delta and the Mediterranean offshore. The source for the hydrocarbons is believed to be Jurassic onshore and mostly Miocene offshore. Sapropels, which are still forming today in the Mediterranean Basin, generate biogenic gas for the Pliocene reservoirs. This phase saw the appearance of man in North Africa and the Nile Valley.

## 18.3 TECTONO-STRATIGRAPHIC PROVINCES IN NORTH AFRICA

North Africa is made up of a number of tectono-stratigraphic provinces:

1   The Pan-African Shield and the West African Craton which form the backbone of the North African continent. The Pan-African Shield includes the Southern Sinai Massif, the Red Sea Mountains, Tibesti Massif, the Hoggar Massif, and the Anti-Atlas boutonnières includes

2   Intracratonic basins that include the Nile Basin, Kufra Basin, Murzuq Basin, Ghadames Basin, Illizi Basin, Oued Mya Basin, Gouara Basin, Bechar Basin, Tafilalt-Abadla Basin, Timimoun Basin, Tindouf Basin, Zag Basin, Reggane Basin, Taoudenni Basin, Sbaa Subbasin, and Ahnet Basin, as well as the Anti-Atlas Basin. The majority of these basins originated during the Hercynian compressional phase, followed by an extended thermal-subsidence phase. These basins are separated from each other by paleohigh or tectonic arches.

3   Rifted marginal basins, such as the Northern Egypt Basin, Cyrenaica Basin, Jifarah Basin, and the Central Atlantic Passive Margin.

4   Rift Basins, which include the Red Sea and Gulf of Suez basins, the Sirte Basin, and the Tethys/Mediterranean Basin, and associated volcanics.

5   Tertiary mountain chains and intermountain basins associated with the Alpine compression and thrusting, such as the Atlas Mountains, which developed on the site of Mesozoic wrench basins, and the Rif Domain.

6   Cenozoic volcanic provinces.

# Chapter 19

# Hydrocarbon systems in North Africa

Throughout the geological history, a number of Oceanic Anoxic events (OAE) took place. These were times of maximum flooding which led to the formation of oxygen-minimum zones and the deposition of organic rich black shales "Hot Shales" (Lüning et al., 2000c, 2004c), limestones, and glauconitic shales, characterized by a $\sigma^{13}C$ isotope levels and TOC values, which form excellent source rocks in the region (for a detailed discussion of the OAEs, see Summary). The source rocks which were deposited during these events extend from Silurian to Pliocene; the most important ones are:

1. Infracambrian black shales, especially in NW Africa.
2. The Silurian Tannezouft Shales in West Libya, Tunisia, and Algeria.
3. The Late Devonian Cues Limestone in Libya and its equivalent the Gazelle Formation in Eastern Algeria.
4. The Oxfordian Kidod Shale in Sinai, Jurassic coals in Central Sinai and equivalent coaly sediments form prolific source rocks in the Western Desert of Egypt. Potential Jurassic source occur in the Middle Atlas of Morocco.
5. The Campanian Sirte Shale in the Sirte Basin and the Brown Limestone in the Gulf of Suez. Extensive commercial phosphate deposits occur in Egypt and Morocco.
6. The Neocomian Matruh Shale in the Western Desert.
7. Aptian-Albian/Cenomanian/Turonian source rocks in the Western Atlas and the Central Atlantic basins, such as the Essaouira, Tarfaya, and Aaiun basins.
8. The Cenomanian-Turonian Etel Formation in the Sirte Basin, the Baharia Formation in the Western Desert, the Bahloul and Zebbag formations in Tunisia and the offshore, and the Abu Qada and Wata formations in the Gulf of Suez and Sinai.
9. The Santonian-Coniacian Abu Roash Formation in the Western Desert, the Matulla Formation in the Gulf of Suez,, and the Alalgah and Makhbaz formations in Tunisia and the offshore.
10. The Paleocene El Haria and Hagfa formations in the Sirte Basin and the Esna Shale in Egypt, the Eocene Bou Babbous Formation in Tunisia and the offshore and to a minor extent in the Gulf of Suez.
11. Pliocene Sapropels in the Mediterranean Basin.

## 19.1   NEOPROTERZOIC PETROLEUM SYSTEMS IN NORTH AFRICA

The 'Infracambrian' succession in North Africa (Lelubre, 1946b) is emerging as a hydrocarbon exploration target with considerable potential and proven petroleum systems in different areas (Lottaroli et al., 2009, Lüning et al. (2009).

Precambrian and "Infra-Cambrian" stromatolitic carbonate units of potential reservoir facies are widespread in North Africa. They have been identified in outcrop and in the subsurface of the Murzuq, Al Kufra and Sirte, and Cyrenaica basins (Aziz & Ghnia, 2009, Benshati et al., 2009, and Le Heron et al., 2009). Their hydrocarbon potential is discussed by Lottaroli et al. (2009), Lüning et al. (2009), and Craig et al. (2009). Basal Cambrian stromatolitic limestones have been recognized in outcrops in the Gulf of Suez (Omara, 1972). Fractured basement rocks host hydrocarbons in some fields in the Gulf of Suez in Egypt, such as the Hurghada, Zeit Bay, and Shoab Ali fields. The Hurghada Field is a shallow, granite-buried hill. Some wells penetrated weathered granites at depths of 1670–2000 ft (510–610 m) (Sircar, 2004). In the Zeit Bay Field, the well QQ 89-1 (1981) recovered gas from Precambrian porphyritic granites and metamorphic rocks, which contains approximately one-third of the total oil in place in the field (Zahran & Askary, 1988). The QQ 89-2well encountered a 253 m (830 ft) oil leg in Cretaceous and basement rocks, with flow rates of 700 to 10,000 bopd. The field is located in a NW-SE trending structure in the southwest corner of the Gulf of Suez (Khalil & Pigaht, 1991). Out of the 36 production wells drilled in the Zeit Bay field, 24 wells penetrated the fractured basement, and 14 of them were completed as open-hole wells. Precambrian fractured granites also produce at the Shoab Ali Field in the Gulf of Suez (Khalil & Meshrif, 1988, Rohrback, 1983) and the Maragh Field in the Sirte Basin (Williams, 1968).

Potential stromatolitic carbonate reservoirs have been identified in the Taoudenni Basin, the Anti-Atlas region of Morocco, and the Kufra Basin of Libya (Underdown & Redfern, 2008, Lottaroli et al., 2009, Lüning et al., 2009, Wenzhe, 2009, Craig et al., 2009). Palynological data gave a Late Riphean age for the stromatolitic limestone reservoir in the Abolag-1 well in the Taoudenni Basin. Only two wells have been drilled in the basin in 1974; the Abolag-1 (by Texaco) and Ouasa-1 (by Agip). The Abolag-1 well produced gas at about 13,600 m$^3$ per day (0.48 TCF) (Wenzhe et al., 2009). Hydrocarbon migration and accumulation took place at the end of the Carboniferous. However, the whole Taoudini Basin subsequently suffered denudation which limits its exploration prospectivity. The source rock is believed to be the Infracambrian black shales which occur in half-grabens and pull-apart basins (Lüning et al., 2009).

Infracambrian strata may also occur in the Tindouf Basin. However, their deep burial and maturation history may be unfavorable for the preservation of Infracambrian-sourced hydrocarbons (Lüning et al., 2009). Infracambrian source rocks may also occur in the Reggane, Ahnet, Mouydir and Iullemeden basins, as indicated by the presence of black shales in wells MKRN-1 and MKRS-1 in the Ahnet Basin.

## 19.2   PHANEROZOIC PETROLEUM SYSTEMS

In North Africa, the Lower Paleozoic succession is dominated by marine to marginal marine-fluvial sandstones interbedded with marine mudstones. Pinchouts, carbonate buildups, and unconformity truncations offer potential for stratigraphic traps. Faults and rotated fault blocks and folds provide structural traps. Quartz cementation and other diagenetic products in some areas, such as Libya and Tunisia, led to the virtual elimination of porosity in many types of sandstones, yet fracturing and dissolution generated excellent porosity and permeability in others. Primary porosity, however, has been preserved in some instances such as the Devonian sandstones in the Sirte Basin (Tawadros et al., 2001). Potential and actual reef reservoirs occur in the Jifarah Plains (Cambrian), Morocco (Devonian), Sinai (Jurassic), the offshore of Libya and Tunisia (Albian-Aptian), and the Gulf of Suez (Miocene).

By far, the richest and most widespread of these source rocks are the Silurian Shales, the Late Devonian Shales and Marls, the Campanian Sirte Shale and its time equivalent, the Brown Limestone in the Gulf of Suez, and the Eocene Shales in Tunisia and its offshore area.

The Jurassic Khatatba coaly source rocks in the Western Desert of Egypt, the Miocene *Globigerina* Shale in the Gulf of Suez, and the Miocene shales in the Nile Delta and the offshore area were deposited under different conditions.

Shales and evaporites, and in other cases, faults and fault smears have provided seals.

## 19.2.1    Egypt's hydrocarbon systems

Egypt's proven oil reserves are estimated at 4.4 billion barrels (2009 est.), with a daily production of 675,000 bopd (CIA Yearbook 2010). Proven gas reserves are 2.19 trillion cu. m. (77.3 TCF), with a daily gas production of 62.7 billion cu. m. (2.2 TCF). The local Egyptian oil consumption is amounted to 712,700 bopd, while oil exports were only 89,300 bopd. Gas consumption in 2009 was 42.5 billion cu m and gas exports were 8.55 billion cu m. (0.3 TCF) Egypt's oil production comes mainly from the Western Desert and the Gulf of Suez (Fig. 19.1). Gas production is mostly from the Western Desert and the Nile Delta and their offshore areas.

The first oil discovery in the Western Desert was the Alamein Field in 1966 (Fig. 19.2). Commercial gas was later discovered in the Abu Gharadig Field in 1969 and Khalda Field in 1970. The main reservoir in the first is the Aptian Alamein Dolomite and in the second is the Lower Cenomanian Baharia Formation. Oil also occurs in the Santonian-Coniacian Abu Roash 'E' Formation. Many other gas and condensate pools and fields were subsequently discovered in the area, especially in the 1980s. In 2000, for example, Apache discovered the Akik IX on its Khalda Concession, which tested at 49.4 mmcfgd and 4051 bcd, the Karama 1X with flow rates of 1520 bopd of 43°API oil and 500 mcfgd, and the Neath South 1X in the Khalda Offset Concession at 2778 bopd and 4.5 mmcfgd (World Oil, August 2000).

Presently, major oil and gas fields include the Alamein, Abu Gharadig, Razzak, Badr El Din, Umbarka, and Khalda, in addition to the recent discoveries of the Tanzanite, Syrah, Qasr, Abu Sir, Ozoris, Tut, and Qarun. The Qasr discovery was made by Apache in 2003 on its Khalda Concession with estimated reserves of 2 TCF of gas and up to 50 million barrels of condensate (OGJ., 2004). The discovery well Qasr-1X flowed at 51.8 mmcfgd and 2688 bcpd from the Lower Safa Formation (Fig. 19.3). The Abu Sir discovery was also made by Apache in 2002 with the well Abu Sir 1-X. In the Oziris Field, the Oziris-4 well tested 1775 bcpd from the Upper

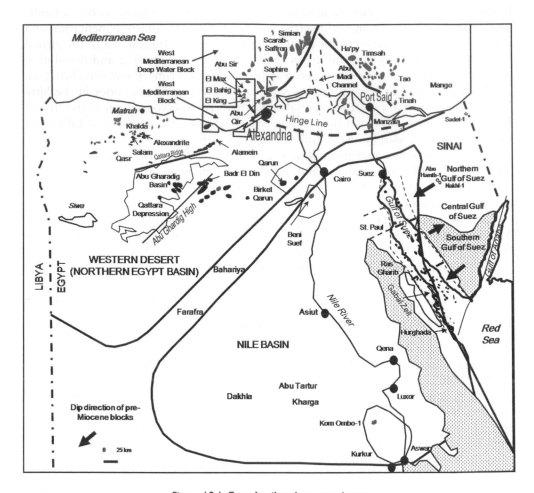

*Figure 19.1* Egypt's oil and gas provinces.

Safa Sand, in addition to oil and gas from the Kharita and AEB sands on the western side of the Khalda Ridge. The Mihos 1-X on Apache's Matruh Concession encountered gas and condensate in the Jurassic Lower Safa sands, and the Imhotep 1X discovery well on the Khalda Offset Concession tested gas and condensate from the same formation (OGJ., 2004).

The Khatatba Formation is composed mainly of deltaic shales and sandstones with coal seams and minor limestones which become more abundant northwards. It is overlain by the Late Callovian shallow-marine carbonates of the Masajid Formation, which were deposited during the maximum Jurassic transgression. The Masajid Formation was either eroded or not deposited in parts of the north Qattara Ridge and Umbarka Platform, although continuous marine sedimentation is postulated in the Matruh Basin and the Sidi Barrani area (Sultan & Abdel Halim, 1988). A major unconformity separates the Masajid Formation from the overlying Alam El Bueib Formation, which starts with Early Cretaceous shallow-marine sandstones and carbonates,

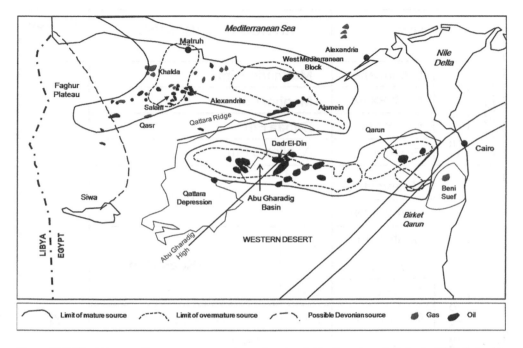

*Figure 19.2* The Western Desert source rocks and major fields (based on Rohrback, 1983, Sultan & Abdel Halim, 1988, Taher et al., 1988, El Ayouti, 1990, Fahmy, 1998, Younes, 2003c, Abdine, 1993, Alsharhan & Abdel Gawad, 2008).

followed by marine shale and a Neocomian succession of massive fluviatile sandstones. The sands are overlain by sands, shales, and carbonates of which ends with the Aptian Alamein Dolomite, which represent the Aptian marine transgression. The Dahab Shale (Albian) marks the end of this cycle. An unconformity separates the Dahab Shale from the Kharita Formation at the base of the third cycle, which extends from the Middle Albian to the latest Cretaceous. The continental and shoreline sandstones of the Kharita Formation are overlain by the shallow-marine and nearshore deposits of the Baharia Formation (Lower Cenomanian). Widespread transgression occurred during the deposition of the Abu Roash 'F' and 'G' and continued during the Senonian with the deposition of carbonates of the Abu Roash 'B' to 'E' during the Turonian and the Coniacian-Santonian Abu Roash 'A'. The Santonian compressional phase causes uplift and erosion in parts of the Western Desert, and the Abu Roash sediments were covered unconformably by the Khoman Chalk Formation (Campanian-Maastrichtian). The cycle is terminated by an unconformity at the base of the Eocene Apollonia Formation, followed upward by the Oligocene Dabaa Formation and the Miocene and Moghra Formations, and capped by the Marmarica Limestone.

Carbonate reservoirs include the Aptian Alamein Dolomite and the Upper Cretaceous carbonates of Abu Roash 'D', 'F', and 'G'. Sandstone reservoirs are numerous and include the Middle Jurassic Khatatba/Safa sands, the Barremian Alam el-Bueb (AEB), the Albian Kharita, the Upper Aptian Dahab, the Lower Cenomanian Baharia,

| AGE | | | LITHOSTRATIGRAPHY | | Source rock |
|---|---|---|---|---|---|
| CENOZOIC | TERTIARY | L. Miocene | Marmarica Fm. | | |
| | | E. Miocene | Moghra Fm. | | |
| | | Oligocene | Dabaa Fm. | | |
| | | Eocene | Apollonia Fm. | | |
| MESOZOIC | CRETACEOUS | Maastrichtian | Khoman Fm. | | |
| | | Campanian | | | |
| | | Santonian | A | Abu Roash Fm. | |
| | | Coniacian | | | |
| | | Turonian | B-E | | |
| | | Cenomanian | F-G | Baharia Fm. | ● |
| | | Abian | Kharita Fm. | | |
| | | Aptian | Dahab Fm. | | |
| | | | Alamein Fm. | | |
| | | Neocomian | Alam El Bueb Fm. | | |
| | | Barremian | | | |
| | JURASSIC | Late | Masajid Fm. | | |
| | | Middle | Khatatba Fm. | | ● |
| | | Early | Ras Qattara Fm. | | |
| | TRIASSIC | | Eghi Group | | |
| PALEO-ZOIC | | | | | |
| PRECAMBRIAN | | | | | |

Figure 19.3 Stratigraphic column of the Western Desert and major source rocks, Egypt.

and the Turonian Abu Roash 'C' and 'E'. The Lower Cenomanian Baharia Formation in the Khalda Field is made up of sandstones that were deposited in tidally dominated environments with mud-rich tidal flats, tidal channels and grade into tidal shelf sands (Conway et al., 1988a, b). Facies and diagenesis are the main controls on reservoir qualities in the Baharia sandstones.

The Western Desert oil province belongs to the Northern Egypt Basin of Tawadros (2001) or the Unstable Shelf of Said (1962a). It is a Mesozoic rifted margin basin, which consists of a number of subbasins dominated by half grabens and tilted fault blocks. Most of the traps are structural three and four-way closers or tilted fault block structures, but stratigraphic traps are also present. Accumulation took place mostly in early Tertiary time (El Ayouti, 1990).

The Western Desert oils belong to two groups (Wever, 2000, Bakr, 2009a, b) reflecting the two major source rocks; the first is represented by a high gravity ≥30°, paraffinic, low sulphur oils/condensates (common in the Sushan/Matruh Basin), probably originated from the coal-rich sediments of the Jurassic Khataba Formation. The second type is of intermediate maturity, has a moderate API gravity around 30°, and low Pr/Ph ratios, probably derived from the Upper Cretaceous Abu Roash 'F' and 'G'. The Abu Roash 'F' Member has an organic content of 1.5–3.5% TOC values and type I and II kerogen (Shahin & Shehab, 1988). Abu Roash 'F' was also considered as a source rock for the oil in the Badr El Din Field (El Ghoneimy & El Gohary, 1988). The Khatatba source rock is characterized by high Pr/Ph ratios and deposition under oxic conditions; whereas the Abu Roash source has low Pr/Ph ratios and was deposited under anoxic to suboxic conditions.

The Shushan Basin is dominated by Jurassic and Cretaceous extensional and strike slip faults (Sultan & Abdel Halim, 1988) inherited from the pre-existing north-south trending Paleozoic framework and resulted in the formation of a series of block-faulted horsts and half-grabens. The Jurassic to Turonian section generally dips and thickens to the north and east. The main potential source rocks in the Matruh-Shushan Basin of the Western Desert are the Lower Cretaceous Alam El Bueib (AEB) and the Jurassic Khataba formations (Metwalli & Pigott, 2005); although Taher et al. (1988) found that the kerogen in the AEB was of moderate quantities but of low quality with minor capacity to generate gas. The two formations share mixed kerogen type II/III. The AEB source rocks entered the oil window in the Late Cretaceous, while the Khataba entered the oil window in the Turonian. Geochemical characteristics of the crudes from the Meleiha Field also suggest a Khatatba/AEB mixed source (El Nady et al., 2003).

Alsharhan & Abd El-Gawad (2008) examined source rock potential of the Middle Jurassic- Upper Cretaceous succession (Khatatba, Masajid, Alam El-Bueib, Alamein, Kharita, Baharia and Abu Roash Formations) in wells Ja 27-2, Tarek-1 and Jb 26-1 in the central, structurally low part of the Shushan Basin and from well Lotus-1 in the structurally-elevated western part of the basin. The results indicate that the thermal maturity of the samples can be correlated closely with burial depth. Samples from the central part of the basin are more mature than those from the west. Samples from the central part of the basin (except those from the Albian Kharita Formation) have reached thermal maturities sufficient to generate and expel crude oils. Source rocks from the Middle Jurassic Khataba Formation and Early Cretaceous Alam El-Bueib Formation can be correlated with crude oil from well Ja 27-2. The well Lotus-1 in the western part of the basin, show a sufficiently high TOC content to act as a source rock. Thermal maturities range from immature to peak oil generation, and the top of the oil window occurs at approximately 2438 m (8000 ft).

Other source rocks include Lower Cretaceous Alam El Bueib (AEB) Formation (Metwalli & Pigott, 2005), the Kharita Formation (Albian), the Matruh Shale (Neocomian), the Abu Roash 'E' Member, Baharia Formation (Cenomanian) (Wever, 2000), the Turonian Abu Roash 'E' Formation (oil prone), and the Cenomanian Baharia Shale (oil prone). Some workers proposed the Ypresian Thebes (Apollonia) and the Maastrichtian carbonates and shales as source rocks.

Two crude oil families have been recognized in the Shushan and Abu Gharadig basins (Bakr, 2009a, and b). The Sushan Basin oil family was generated from clay-rich terrigenous source rocks with high plant input deposited under oxic conditions (Bakr, 2009a, and b). The Abu Gharadig Basin crude oil family was derived from a marine source rich in clay minerals and deposited under suboxic conditions. Biomarkers in this basin show peak oil maturity.

Barakat et al. (1997) pointed out that the high Pr/Ph ratio of 3.08–3.73 in the crude oils from the Umbaraka, Khalda, and Meleiha fields suggests deposition in an oxic environment. However, the crudes from the Yidma Field have Pr/Ph ratio of 0.89–1.53 and medium to low sulphur content, which suggest that the source rock was deposited in suboxic conditions.

The AEB, Abu Roash, and Thebes (Apollonia) source rocks in the south Alamein area were found to be thermally immature (Moretti et al., 2010), while the Khatatba source did not occur below 40°C and vitrinite reflectance of 0.9%. The main source kitchen is located north of the area.

Based on biomarkers distribution, Higazi et al. (2003) concluded that the oils from Badr El Din, Aman, Qarun, and Faras, Wadi Rayan, and El Raml fields were derived from a variety of source rocks, deposited in normal saline environments and suggest upper Cretaceous sources. Oils from the Aman and Qarun fields have high Pr/Ph ratios, 3.64 and 3.13, respectively, suggesting a highly oxic environment. For the other fields, lower Pr/Ph ratio values, 1.05–1.24 (similar to the Abu Roash Formation, according to Wever, 2000); indicate an anoxic to suboxic depositional environment. The abundant diastrane in the Badr el-Din Field indicates a marine deltaic source rich in clay minerals and terrigenous organic matter consistent with the Khatatba Formation. The geochemical characteristics of the oils from the Faras and SW Raml fields also suggest a marine deltaic origin with influx of terrigenous organic matter. The Wadi Rayan Field crudes with high sulphur content indicate a marine carbonate/evaporite depositional environment (Higazi et al., 2003).

Oil occurs in the Razzak Field in seven separate reservoirs (Jurassic-Upper Cenomanian) in a northeast plunging anticlinal nose in the Qattara Ridge (El Ayouty, 1990). The structure is cut by a number of faults without large displacement. Oil was found in the Baharia Formation sands and the Abu Roash 'G' dolomites. Terrestrial organic components dominate the Baharia Formation and the basal Abu Roash "G" Member; whereas amorphous organic matter and marine palynomorphs are more abundant in the upper part of the Abu Roash "G" Member (El Ayouty, 1990). Palynofacies analysis shows a predominance of type III kerogen, and palynomorph colors indicates that these sediments are mature (Zobaa et al., 2008). The porosity in the Abu Roash 'G' in the Rezzak Field ranges from 26% to 36% (Abdine et al., 1993). In the Razzak Field, the Jurassic Khatatba Formation may have reached peak generation and expulsion took place in the Late Cretaceous (Abdine et al., 1993). On the other hand, in the Alamein Basin, generation and expulsion stage in the Early Miocene The source in the Alamein Basin and the Razzak Field are derived from the Jurassic Khataba Formation (Abdine et al., 1993).

Sedimentary organic matters from Cretaceous to Middle Eocene source rocks in the Badr El Din Concession in the Abu Gharadig Basin, including the Alam El-Bueib, Kharita, Baharia, Abu Roash, and Khoman, and Apollonia formations, show different kerogen types and maturity stages (Maky & Saad, 2009). These source rocks were deposited during a transgressive stage (Alam El-Bueib Formation) and a high stand system tract (Kharita Formation). These two units are characterized by high terrestrial input and nearshore deltaic environment, whereas that of Baharia Formation (Cenomanian) is characterized by high content of oil-prone kerogen (sapropelic maceral). In Abu Roash Formation (*s.l.*), type II kerogen is dominant. In Khoman Formation (Campanian-Maastrichtian), types I, II and III kerogens. Kerogen from the Apollonia Formation (Lower Eocene) is composed of oil-prone organic matter (mixture of marine and terrestrial sapropelic maceral).

Bitumen and kerogen analyses of various formations in the well Jb26-1 in the central part of the Shushan Basin (Sultan & Abdel Halim, 1988) show that Type I/II kerogen is present in the upper part of the Khatatba Formation and Type II/III kerogen occurs in the Kharita and Alam EI-Bueib Formations and possibly in the topmost part of the Khatatba Formation. Type III kerogen is also present including the coaly shales and thin coals in the Khatatba Formation. Type IV kerogen was recorded in parts of the Alam EI-Bueib Formation. The Abu Roash Formation has an S2/S3 ratio of <2.3 indicating the presence of gas-prone kerogen, whereas S2/S3 ratio in the Baharia, Alam El-Bueib, and Khatatba formations is between 2.3 and 5.0 indicating that both oil- and gas-prone kerogen is present (Alsharhan & Abd El-Gawad, 2008).

In the Northeastern part of the Western Desert, open-hole well logs from the wells Harun-1, Qarun E-1, and Wadi Rayan-1 in the Gindi Basin and the Wadi Rayan Platform indicate that the Abu Roash "F" and "G" Members have the highest source rock potential in the area (Aly et al., 2003). Calculated TOC values derived from parameters such as $\Delta t/Rt$, $\varphi N/Rt$ or $\rho b/Rt$ range from 0.5% to more than 6.0% for the Abu Roash "F" Member. The Abu Roash 'G' Member may also contain average to very high organic contents.

Two oil-prone basins occur in the Nile Basin (Tawadros, 2001) (Fig. 19.1); the Komombo and Beni Suef basins, at the southern and northern borders of the basin, respectively. Only one petroleum system is believed to exist in the Nile Basin. It consists of a possible Jurassic source rock, probably equivalent to the Khataba Formation and Jurassic/Cretaceous reservoirs. Information is still scanty and contradicting.

The Komombo Basin is a rift Basin (Dolson et al., 2001). The Komombo Field (renamed Al-Baraka by the Petroleum Minister in 2007) was discovered in 1998 by the drilling of the Komombo-1 well by Repsol on the Ganope Block2. The well tested 37–39° API oil at a rate of 6976 bopd from Jurassic and/or Lower Cretaceous reservoirs (Dolson et al., 2001). The trap is fault controlled. The source rock is believed to be lacustrine Jurassic type II-III, probably equivalent to the Khatatba Formation in the Northern Western Desert.

The Beni Suef Field was discovered by Apache in partnership with Seagull in 1997 in the eastern part of the Beni Suef Concession by the drilling of the Beni

Suef-1X. The discovery well tested 40°API oil at 5200–6976 bopd from the Kharita Formation (Albian) and the Cenomanian Baharia Formation (Nemec & Colley, 1998, Dolson et al., 2001, Tawfik et al., 2005). The Yusuf Field was discovered by Dana Petroleum in 2008 in the same part of the Concession. The source rock is probably of Jurassic age.

The geology and hydrocarbon potential of the Gulf of Suez (Fig. 19.4) have been reviewed extensively in the literature (Hume, 1911b, 1917, Tromp, 1950, 1951, Robson, 1971, Moustafa, 1976, Garfunkel & Bartov, 1977, Brown, 1980, Elzarka & Moustafa, 1988, Evans, 1988, Jackson et al., 1988. Richardson & Arthur, 1988, Tewfik, 1988, Zahran & Meshrif, 1988, El Ayouty, 1990, Helmy, 1990, Abdine et al., 1992, Hughes et al., 1992, EGPC, 1996, Schütz, 1996, Patton et al., 1994, Sellwood & Netherwood, 1984, Alsharhan & Salah, 1994, 1995, 1997a, b, Bosworth, 1995, Bosworth & McClay, 2001, Lindquist, 1998, Khalil & McClay, 2002, Younes & McClay, 2002, Alsharhan, 2003). The history of oil exploration in Egypt and Libya is reviewed in detail by Tawadros (in preparation).

The Gulf of Suez contains 70% of Egypt's oil, but already more than 4.5 billion barrels have been produced. To date, more than 1900 wells have been drilled in the Gulf of Suez, which led to the discovery of more than 60 fields in addition to numerous one-well pools, for a total of 120 fields (Lindquist, 1998) (Fig. 19.4).

Oil seeps occur at Gebel Zeit at the southern end of the Gulf of Suez (Fig. 19.4) and have been exploited by the Ancient Egyptians since the Romans times (Harrell & Lewy, 2002). The first well in Egypt was drilled in 1868 which led to the discovery of oil at Gebel Zeit.

Perhaps the first attempt to speculate on the source of oil in North Africa was made by Captain Ardagh in 1886, who postulated that the oil seeps found in fractured basement granites at Gebel Zeit originated from Miocene reservoirs down dip, although his explanation would relate more to up dip migration rather than source rocks as we understand it today. His ideas were confirmed by recent geochemical studies (Alsharhan & Salah, 1997).

The Gulf of Suez Basin (Fig. 19.4) is a Miocene rift dominated by NW-SE-trending half-grabens and horst blocks that are crosscut by left-lateral transform faults related to the Gulf of Aqaba-Dead Sea left-lateral wrench faulting. The Gulf of Suez Basin is approximately 300 km in length and up to 80 km in width. Water depth reaches 60 m (197 ft) in its deepest part, but there are many islands, especially in the southern part. Its borders extend onshore beyond the actual shorelines of the Gulf, which led Hume (1911b) to apply the term Clysmic Gulf instead of the Gulf of Suez. The Gulf of Suez is bordered to the east and west by Precambrian highlands and NW-SE-trending faults.

The Gulf of Suez is divided into three provinces (Moustafa, 1976) (Fig. 19.4), separated by transform faults or "hinge zones", and characterized by a change in the general structural dip; a Northern Province, a Central Province, and a Southern Province. The northern and southern provinces dip to the southwest, whereas the central province dips to the northeast. The Northern and Central Provinces are separated by the Zaafarana Hinge Zone (Accommodation Zone of Younes & McClay, 2002). The Morgan Hinge Zone (Younes & McClay, 2002) separates the Central Province from the Southern Province. The Southern Province is separated from the Red Sea by the Aqaba Transform Fault.

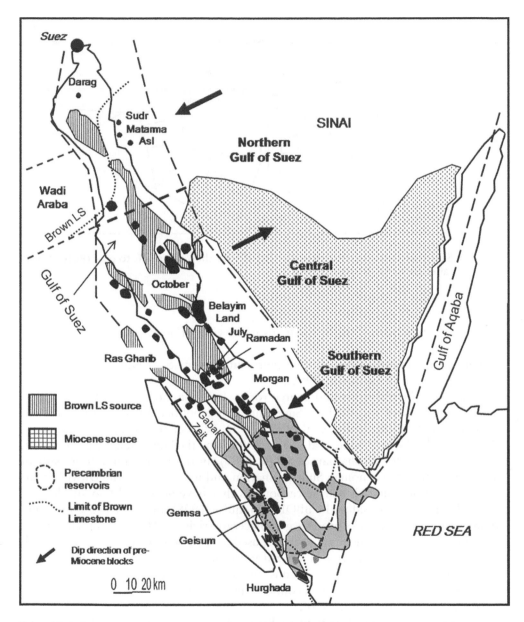

*Figure 19.4* Gulf of Suez oil (black) and gas (grey) fields and distribution of source rocks (after Hughes et al., 1992, Alsharhan & Salah, 1994, 1995).

The stratigraphic succession in the Gulf of Suez ranges from Precambrian to Recent and has been divided into three sequences (Garfunkel & Bartov, 1977) (Fig. 19.5); pre-rift, syn-rift and post-rift. Complex tectonics and facies changes provided various reservoirs, source rocks, seals, and traps. Reservoirs in the Gulf of Suez range from Precambrian to Upper Miocene.

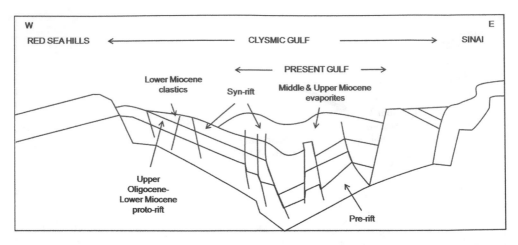

*Figure 19.5* Schematic cross-section across the Gulf of Suez showing pre-rift to post rift sequences (after Garfunkel & Bartov, 1977).

There are two main petroleum systems in the Gulf of Suez:

I    The Brown Limestone-pre-Miocene petroleum system (Sudr-Nubian Total Hydrocarbon System of Lindquist, 1998).

It includes pre-rift reservoirs and source rocks. It is common in the Northern and Central provinces of the Gulf of Suez and is exemplified by the October Field (El Ghamri et al., 2002, Abdine et al., 1992, 1994, Lelek et al., 1992, Sercombe et al., 1997), Ramadan Field (Abdine et al., 1992, 1994), and the Hurghada Field (El Ayouti, 1990, Kostandi, 1961, Alsharhan & Salah, 1997). In the Brown Limestone-Nubian petroleum system, the Senonian Sudr Formation and its equivalent the Duwi Formation consist of organic-rich marine carbonates that have generated known ultimate recoverable reserves exceeding 11 BBOE (Lindquist, 1998). In the Brown Limestone-Nubian system, source rocks produce gammacerane-rich oil. The reservoirs belong to the Nubian Sandstone which in this area is used to include sandstone reservoirs with a wide range of poorly constrained, pre-rift units ranging in age from Early Paleozoic to Early Cretaceous. The "Nubian" Sandstone is the primary reservoir in the B-Trend in the Southern Gulf of Suez, which contains several fields, such as the GS365, Sidki, East Zeit, and Hilala (Helmy, 1990).

The Late Cretaceous (Campanian) Brown Limestone (equivalent to the phosphatic Duwi Formation or the lower Sudr Formation) has been considered to be the main source for the oil and gas in the Gulf of Suez (Chowdhary & Taha, 1987, Atef, 1988, Shahin, 1988, Mostafa et al., 1993, Robison, 1994). It belongs to the pre-rift sequence of (Garfunkel & Bartov, 1977). Nearshore facies of the Brown Limestone is associated with phosphorites, glauconites, and abundant organic matter, and indicates deposition in a zone of high productivity and low oxygen (Robison, 1994). The equivalent basinal facies is also rich in organic matter, but lacks the phosphate. The phosphates of the Duwi Formation are exploited in the Eastern and Western

Desert, and contain oil shales in places. The Brown Limestone is equivalent to the Sirte Shale, a major source rock in the Sirte Basin of Libya.

The Brown Limestone contains high concentrations of liptinitic material, and kerogen type II (Atef, 1988). It contains an average TOC of 2.6% (maximum 21%) and thicknesses ranging from 25 to 75 m (82 to 246 ft) (Chowdry et al., 1987, Lelek et al., 1992, El-Ghamri et al., 2002).

The Paleozoic-Lower Cretaceous "Nubian" Sandstone in the Gulf of Suez area is a thick, continental to shallow marine, sequence of up to 1200 m (3937 ft) of sandstones, shales, and minor carbonates, deposited in continental and fluviomarine to marine settings (Alsharhan & Salah, 1997). The Nubian Sandstone is subdivided, in ascending order into the following groups and formations: the Qebliat Group (Araba and Naqus formations), Umm Bogma Formation, Ataqa Group (Abu Durba and Rod El Hamal formations) and El-Tih Group (Qiseib and Malha formations). The pre-Cenomanian Nubian Sandstone is one of the most prolific reservoirs in the Gulf of Suez oil province, with net pay thickness of up to 450 m (1476 ft), and net sand ratios ranging from 60% to 90%. Porosity varies from 10% to 29%, and permeability up to 850 mD (Alsharhan & Salah, 1997).

The Qiseib Formation (Permo-Triassic) consists mainly of a reddish, fine- to coarse-grained, cross-bedded sandstone with thin interbeds of shale. Petrographically, these sandstones are quartzarenites. The Qiseib Formation tested oil from an 18% porosity sandstone in the North Darag discovery in the northern Gulf of Suez (Alsharhan & Salah, 1997). The Ras Budran Field produces from the Carboniferous Nubian Sandstones, with subordinate reservoirs in the Raha and Matulla Formations (EGPC, 1996, Bakr & Wilkes, 2002). In the Abu Rudeis-Sidri Field, hydrocarbons are produced mainly from a conglomeratic sandstone unit in the Nukhul Formation, with secondary production from wells in lower Senonian and Nubian Sandstones (EGPC, 1996).

II    The Miocene-post-Miocene petroleum system (Maqna Total Petroleum System of Lindquist, 1998).

It consists of syn-rift Lower Miocene source rocks and syn- and post-rift Oligocene to Recent reservoirs. It is dominant in the Southern Province of the Gulf of Suez, such as the Morgan Field (El Ayouti, 1990, Rohrback, 1983). In many cases, however, both derivations from Brown Limestone and the Miocene sources, as in the Zeit East Field (Younes, 2001, Bakr & Wilkes, 2002, Khalil & Pigaht, 1991) or from "mixed sources" have been attested in some places, such as the Ashrafi Field (Younes, 2003b, Elewa et al., 2002) and the Gamma Field (Barakat et al., 1997). Other fields produce from Permo-Triassic reservoirs, such as the Darag North Field in the northernmost part of the Gulf of Suez, while others produce from Precambrian fractured and weathered granites, such as the Zeit Bay Field (El Ayouti, 1990, Khalil & Pighat, 1991, Bakr & Wilkes, 2002) and the Sidki Field (Helmy, 1990). The Miocene Kareem Formation is the major reservoir in the Miocene petroleum system (Lindquist, 1998). Middle Miocene shale source rocks produce oleanane-rich oils and wet gas trapped primarily in Miocene reservoirs. It is the predominant system in the Southern Gulf of Suez and the Red Sea basin. The Middle Miocene coralline algal facies "Nullipore carbonates" forms reservoir in the Ras Fanar Field, discovered in 1978, 2 km offshore Ras Gharib, Central Province. High porosity in this case is associated with erosional surfaces, karstification and collapse breccias (Vaughan et al., 2003). The Nullipore carbonates also produce in the Ras Bakr, South Ras Amer, and Ras Gharib fields.

The Rudeis Formation is generally considered the principal Miocene source rocks. In addition to the Campanian and Miocene source rocks, several others have been suggested, such as the Carboniferous Ataqa Formation, the Triassic Qiseib Formation, the Turonian Wata Formation, the basal Miocene Nukhul Formation, the Upper Miocene Kareem and Belayim formations, and the Lower Eocene Thebes Formation. Many workers postulated mixing of two or more source rocks, especially in the Northern and Southern provinces of the Gulf. Alsharhan & Salah (1997) found that the Hammam Faraun Member of the Belayim Formation is mature only in the Gemsa Trough; the Kareem and Rudeis formations are mature in the Gemsa, West Hurghada, and West Shadwan troughs, while the Rudeis Formation is marginally mature in East Dabaa and East Ras Abu Soma troughs. Mostafa (1999) observed that the Cenomanian-Turonian section in the Bakr area to be condensed and lack good source rock potential. Source-dependent biomarker properties from crude oils in the Southern Province of the Gulf of Suez suggest multiple sources (Barakat et al., 1997). Oils from the Ras Fanar and East Zeit fields show abundance of gammacerane indicating a marine saline marine depositional environment for the source rock (Brown Limestone or Lower Eocene Thebes Formation carbonates). The oils from the Gamma and Amal-9 wells show oleanane indices over 20%, indicating an angiosperm-rich Tertiary source (Lower Miocene Rudeis Formation). The oils from Amal-10 wells show geochemical characteristics that indicate mixing of oil from both sources. Organic geochemical analysis of crude oil from the Ashrafi Field, also in the Southern Province, show that oils in pre-Miocene reservoirs were derived from the Brown Limestone with gammacerane index over than 30% and oleanane index less than 10%., while the oils from Miocene crudes are rich in tricyclic terpanes and extended hoptanes indicating a source rock with angiosperm land plants (Rudeis Formation) (Younes, 2003b).

Elzarka & Mostafa (1988) concluded that in the Rahmi area in the Central Gulf of Suez the basal Miocene Nukhul Formation, the Eocene, and Senonian (Brown Limestone) all have good potential for generating hydrocarbons, with the Eocene having the best potential with TOC content of 2.1–3.58%. On the other hand, Rohrback (1983) based on carbon and sulphur isotope analysis of a large number of crude samples across the entire length of the Gulf of Suez from Ras Sudr in the north to Shoab Ali in the south emphasizes that all the oils analyzed are part of the same genetic family derived from a single marine source rock, regardless of the age of the reservoir rock. The low content of $C_{29}$ steranes, characteristic of higher land plants, supports a marine origin for the oils. Variations in oil composition parameters, API gravity and sulphur content are attributed to variations in the degree of maturity (Rohrback, 1983). The pristine/phytane ratio observed ranges from 0.6–1.0 and increased with the maturity level (API gravity), and suggest an extremely reducing environment. However, Rohrback (1983) does not speculate on the age or lithology of the source rock. Higazi et al. (2003) based on biomarkers distribution concluded that the source of the crude oils from Belayim Land and Belayim Marine fields (Central Province) were deposited in a suboxic, normal salinity, marine carbonate or evaporite environment as indicated by low Pr/Ph ratios (0.61–0.74), a low gammacerane index (0.08–0.09). The source rock of the Gemsa and Geisum (Southern Province) was deposited in a highly reducing hypersaline environment, based on the abundance of gammacerane (1.01–1.18) and dominance of C27 steranes. The two areas show an oleanane index of (0.03–0.08). They suggested a source from the Upper Cretaceous Brown Limestone or Lower Eocene Thebes Formation for both areas[1]. Chowdhary & Taha (1987) pointed out that the Brown Limestone and the Thebes Formation in the Ras Budran Field have the richest TOC values with an average of 2–8%. According to Wever (2000), based on C7 Star Plot, there are two source rocks in the Gulf of Suez; a carbonate-rich Cretaceous-Eocene source and a Miocene silicate-rich source rock. In the Northern Red Sea, in the "Safaga Concession", the pre-Miocene pre-rift sequence is absent, and therefore, the Miocene sediments are the only potential source rocks (and reservoirs) (Miller & Barakat, 1988). In that area, Miocene sediments overlay directly on top of the Precambrian basement. Bakr & Wilkes (2002) examined crude samples ranging from the Sudr Field in the north to South Hurghada Field in the south of the Gulf of Suez. However, out of the 27 samples Bakr & Wilkes (2002) analyzed, 24 samples came from Miocene reservoirs.

Oils from the Central Province show the lowest Pr/Ph ratio (pristine/phytane ratio) of 0.59–0.86 and indicate a carbonate source rock deposited in an anoxic environment consistent with a Brown Limestone or a Lower Eocene Thebes Formation source (Bakr and Wilkes, 2002). Based on organic geochemical and biomarker parameters, Barakat et al. (1997) determined that the crudes from the Central Gulf of Suez fields Belayim Land, Bakr, Gharib, and July show a low Pr/Ph ratio, between 0.7–0.84, and a high sulphur content, which suggest deposition under strongly reducing conditions. Roushdy et al. (2010) obtained Pr/Ph ratios of 0.14–0.36 and waxiness values of 0.68–0.94 from samples from the Central Gulf of Suez and suggested a marine organic source deposited under suboxic conditions.

Other potential source rocks include the Jurassic marine rocks of the northern-most Gulf of Suez, such as in the Abu Darag Basin (Tewfik, 1988) (Fig. 19.6) and the Eocene in the Rahmi Field (Elzarka & Mostafa, 1988, Bakr & Wilkes, 2002). Oil samples from the Northern Province of the Gulf of Suez, which include the Sudr and Zaafarana fields, fields, have a Pr/Ph values of 0.67–1.10, and were derived from a shale-rich, oxic source, probably from the Lower Cenomanian Raha Formation (Bakr & Wilkes, 2002) (suggested also by Shahin et al., 1994).

On the other hand, crudes from the South-Central and Southern Gulf of Suez fields Ras Fanar, Shark El-Zeit, Amal-9, Amal-10, and Gamma have a Pr/Ph ratio of 0.89–1.53 and medium to low sulphur content, which indicate the source rock was deposited under suboxic conditions. Miocene source include relatively low concentrations of gammacerane and 17α, 21β(H)-29-pentakishomohopanes and high concentrations of diasteranes, 18α(H)-30-neonorhopane, and oleanane (Harrel & Lewan, 2002). The Southern Province crude samples show Pr/Ph ratios of 0.79–1.29 (Shahin et al., 1994, obtained values of 1.44–2.47 and suggested Lower Senonian and Rudeis sources), which suggest that the Brown Limestone and Thebes Formation contributed oil to the area (Bakr & Wilkes, 2002).

Traps are predominantly tilted fault blocks of Miocene age; stratigraphic and combination traps are also common. Seals are mostly post-rift and composed of very thick Upper Miocene salt, anhydrite, and shale intervals. In other cases, seals are formed by syn-depositional facies changes from porous carbonate and/or sandstone

---

1 Most studies cannot differentiate between the Upper Cretaceous Brown Limestone and Ypresian carbonate sources.

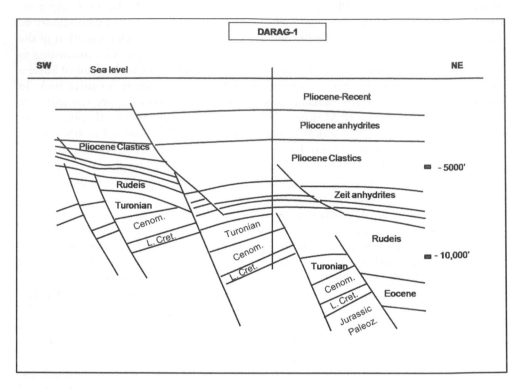

*Figure 19.6* Cross-section across the Abu Darag Basin, Northern Gulf of Suez (modified from Tewfik, 1988).

reservoirs into tight shales or carbonates. Structural traps can be found in the Belayim Land, Kareem, Morgan, Geisum, Shoab Ali, Hilal, East Zeit, Ramadan, Ras Budran, among many others. Combination traps are found in the Ras Fanar, Zeit Bay, East Zeit, Gemsa, and Hurghada fields (Alsharhan, 2003).

The Abu Durba Field discovered in 1916 in proximity of oil seeps in Wadi Araba and depleted in the early 1940s, is an open field with no apparent trap or seal (van der Ploeg, 1953). It produced oil from Nubian Sandstones composed of sands with igneous, limestones, sandstone, and shale pebbles (van der Ploeg, 1953).

The October Field is located in the Central Gulf of Suez. It contains both light and heavy oils. It consists of a series of elongated, pre-Miocene NW-SE trending and NE-dipping rotated faulted blocks (Zahran, 1986, EGPC, 1996, El-Ghamri et al., 2002, Lelek et al., 2002). The Field is bounded westwards by a series of large normal faults down throwing to the west. The field is divided into a main southern block and a subordinate northern block, with several additional smaller tilted fault-block compartments. Miocene and post-Miocene sediments drape across the pre-Miocene structure. Drilling and production are complicated by major Clysmic (NE-SW) and transverse faults, as well as by the shallow salt sections (Lelek et al., 1992).

Various reservoir intervals produce in the October Field; the Carboniferous-Lower Cretaceous Nubian Sandstones, the Upper Cretaceous Raha, Wata and Matulla sandstones and dolomites of the Nezzazat Group, the Lower Miocene Nukhul Formation, and the Lower Miocene Asl Member of the Upper Rudeis Formation (Alsharhan, 2003, Alsharhan & Salah, 1997). The Nubian sandstones are the main reservoirs. The hydrocarbon column is 330 m (1083 ft) (Zahran, 1986). A common oil-water contact is present across the field in the Nubian Sandstones at - 3,557m.

The Campanian Brown Limestone (Duwi, Sudr formations) is the dominant source rock for the oil in the October Field (Abdine et al., 1992, El Ghamri, 2002). The Brown Limestone is 25–75 m (82–246 ft) in thickness across the Gulf of Suez, with an average TOC of 2.6% (Lelek et al., 1992). Other source rocks include shales of the Eocene Thebes Formation, the Lower Miocene Rudeis Formation (Wever, 2000, Zahran, 1986), and the Cretaceous Nezzazat Group (Alsharhan, 2003, Alsharhan & Salah, 1997). The oil in the October Field may have been generated from adjacent basins (Lelek et al., 1992), such as the October South Trough immediately to the south and west (Zahran, 1986), the Baba Trough to the east (Abdine et al., 1992), and the Lagia Basin to the northeast. The main source kitchen was probably the Baba Basin east of the field (Zahran, 1986, El Ghamri et al., 2002), where hydrocarbons were expelled and migrated westward up-dip into the reservoirs. Up-dip vertical migration along fault planes is the most common mechanism in the Gulf of Suez (Schlumberger, 1995). However, migration path appears to be complicated and multi-phased in places. Younger Mesozoic sandstones are charged by vertical migration through seals and along faults (El Ghamri et al., 2002).

The July oil field is a normal fault-bounded structural block adjacent to a major structural transfer zone, which has controlled sediment influx to the rift basin (Pivnik et al., 2003). The Abu Rudeis-Sidri Field, about 10 km to the east of the October Field, consists of a similar but smaller asymmetric NW-SE trending faulted anticline (EGPC, 1996, El Ghamri et al., 2002). By contrast, the Ras Budran Field, about 15 km to the NE of the October Field, forms a unique structural feature in the Gulf of Suez in that it consists of NE-SW trending and NE dipping faulted blocks (Chowdhary & Taha, 1987). Both of these fields are up-dip of the Baba Trough (Abdine et al., 1992).

An example of stratigraphic traps is the Ras El Bahar Field, where Miocene carbonate reservoirs are sealed, both vertically and laterally, by tight carbonates (Alsharhan, 2003), the Belayim Land Field where sandstone reservoirs are sealed both vertically and laterally by evaporites, the Ras Gharib and Ras Fanar Fields where Miocene coralline algal reservoirs of the "Nullipore Carbonates" are sealed by evaporite-clastic-carbonate units of the South Gharib Formation (Vaughan et al., 2003). Other stratigraphic traps were formed as a result of truncation below Miocene or intra-Miocene unconformities, as in the Ras Gharib Field, or the pinchout of the sandstones of the Nukhul Formation on the flanks of tilted fault blocks (Alsharhan, 2003).

Although North African oils are known to be relatively light with low sulfur content, heavy oils were encountered in many places in North Africa. For example:

Heavy oil occurs in various fields in the Gulf of Suez of Egypt. The Kareem Field discovered in 1958 by GPC, started production of 17°–18° API oil from the Lower Miocene and Eocene reservoirs (EGPC, 1992?). The Ras Bakr Field, discovered in 1958 and put on production in 1960, had an estimated OOIP of 106 mmbbls of 10° oil. The remaining reserves are estimated at 12.9 mmbbls (HIS, 2005). The Bakr West Field, discovered by EGPC in 1978, produces 18° API oil since it was put on production by EPEDEPCO in 1980. The Ras Budran Field, discovered in 1978, has an 18°–25°API oil (with an average of 26.4°) in Cretaceous and Nubian reservoirs with about 14% porosity (El Ayouty, 1990). The Fadl-1 well (T.D. 5612 ft) was drilled by Tanganyika Oil Co. in April 2005, encountered a 40 ft pay in the Lower Rudeis Sand with 21% porosity, which produced at a rate of 250 bopd with no water (Source Tanganyika Oil Co., 2005). The Hoshia well drilled by Tanganyika in 2005, encountered 16.5°API oil in the Lower Miocene Rudeis Sand, with a net pay of 32 ft and an average porosity of 25%. Production was stabilized at 620 bopd. The West Hoshia well had an initial rate of 300 bopd with 40% water cut from 10° API oil (Source: Tanganyika Oil Co., 2005). The recent discovery of Al Amir-IX by Sea Dragon Energy of Calgary in their West Gemsa Concession tested 300 bopd of 18° API oil from the South Gharib Formation at a depth of 4750 ft (Sea Dragon Press Release, August 10, 2010).

The Issaran (Assaran) Field (Fig. 19.7) was discovered in 1981 and began production in 1998 under a joint venture between Scimitar and EGPC. The field contains heavy oil (API 10°–14°) in the fractured Upper Dolomite with about 140 ft in thickness (north part of the field) and the Lower Dolomite (in the southern part) (The Dolomite Units are equivalent to the South Gharib and Belayim formations), with net pay of 200–475 ft, highly fractured limestones of the Nukhul Formation (100 ft net pay, 30% porosity, and up to 300 mD permeability) (Samir et al., 2010), and Zeit Sandstone (net pay 10–30 ft). Current production is at 4500 bopd (Samir et al., 2010). Rally Energy (2007) estimated the oil-in-place volumes in the Issaran Field at 713 million barrels with an estimated 91.7 million barrels of recoverable proved and probable reserves. A large part of the recoverable reserves was attributed to thermal (SAGD) recovery.

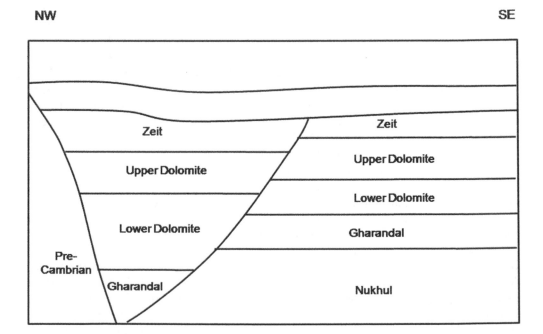

*Figure 19.7* Schematic cross-section of Issaran Field (based on MacKay & Lamb, 2010).

In the Assaran (Issaran) Field, the reservoir consists of about 229 m (750 ft) of dolomites (equivalent to the Gharib and Belayim formations). The oil is heavy with 10–18° API (Higazi & El-Gayar, 2009). The oil in-place is estimated to be total 670 million barrels. The oil in the Assaran Field has a sulphur content of 6.38% (Barakat et al., 1997a, b, Hegazi et al., 2003, Younes et al., 2007). The high sulphur content and oil gravity in the field have been attributed to biodegradation (Hegazi & El-Gayar, 2009). A low Pr/Ph ratio (0.59) and high sulphur content in the Assaran Field suggest generation in a strongly reducing environment (cf. Barakat et al., 1997). Biodegradation is probably due to sulphate-reducing anaerobic bacteria, which removed most of the n-alkanes although other hydrocarbon molecules are intact. The Ts/Tm ratio of the Assaran oil was 0.32 indicating a marine carbonate and evaporite depositional environment. The oil samples have elevated 22S + 22R $C3517\alpha(H),21\beta(H)$-pentahomohopane concentrations (homohopane index = 0.14), indicating that the oil was derived from a marine source rock deposited under low Eh conditions (Hegazi & El-Gayar, 2009). An abundance of $C_{35}$ homohopanes is commonly linked to large-scale bacterial activity in highly saline environments. A distinguishing feature of the Assaran oil was the presence of measurable amounts of gammacerane with a gammacerane index of 0.12, which indicates salinity and water column stratification in the source rock depositional environment. The Assaran oil has an oleanane index < 0.2, consistent with generation from a Cretaceous or Early Tertiary source rock (Hegazi & El-Gayar, 2009). The geochemical character-istics of the oil from the Assaran Field indicate that oil was derived from a marine carbonate source deposited in a highly reducing saline environment with a high bacte-rial contribution, consistent with the Upper Cretaceous Brown Limestone or Lower Eocene Thebes Formation Type II kerogen. The oil is marginally mature (Hegazi & El-Gayar, 2009).

In Libya, heavy oil shows with 3% sulfur content were encountered near Beda-Haram area, between the Intisar and Amal fields in the Sirte Basin (Ghori, 1996). In the Sarir Field, gravity segregation of the oil in the trap led to the formation a 10–30 ft thick tar mat at the OWC contact. The Tar has 24.4° API and a pour point of 160° F (Gillespie & Sanford, 1967, Ghori & Mohammed, 1996, Ahlbrandt, 2001). Heavy oil zones were also encountered in the Messla Field.

The first onshore gas field discovery in the Nile Delta, the Abu Madi gas Field, was made in Pliocene sands by Eni in 1967. This discovery was followed by a series of discoveries in various reservoirs in the prolific Abu Madi Channel, striking approxi-mately N-S in the middle of the Nile Delta (Fig. 19.8). In 1969, Philips discovered the shallow-water offshore Abu Qir gas field in the Pliocene Abu Madi and Miocene Sidi Salim formations in an E-W anticlinal closure cut by faults west of the Delta, opening a new phase of exploration in the offshore area. During the 1980s, other fields were discovered which established the Nile Delta and the offshore as a prolific gas and condensate province. The Abu Qir discovery was followed by the discovery of gas in the Pliocene Abu Madi and Kafr el Sheikh formations in the N. Abu Qir Field by Aquitaine in 1987 and in the Abu Madi Formation in the W. Abu Qir Field in 1989.

In general the Nile Delta is a gas province. The USGS estimates of the undis-covered recoverable oil and gas resources in the Nile Delta Basin Province at about 1.8 million barrels (mmb) of oil, 223 trillion cubic feet (TCF) of gas, and 6 million barrels (mmb) of natural gas liquids (Kirschbaum et al., 2010). The traps in the Nile Delta are predominantly structural, including gentle anticlines, large faulted anticline bounded by growth faults.

The Nile Delta and its offshore area are divided into three tectono-sedimentary zones; central, eastern, and western (Barsoum et al., 1998) (Fig. 19.8). The central zone is delimited on the south by the Hinge Line, which a faulted zone that separates a stable shelf in the south from a subsiding basin in the north. It includes the northern Delta and its offshore area, as well as the Abu Madi Channel. The eastern offshore zone is separated from the central zone by the NW-SE-oriented Bardawil Line, a zone of growth faults. The western zone is separated from the central zone by the Rosetta Line, a NE-SW left shear zone. There are three types of plays in the Nile Delta/offshore province (Provinces 2 to 4): 1) Early Pliocene incised channels in shallow water of less than 500 m (1640 ft). 2) Late Pliocene turbidites in deep water. 3) Upper Miocene deep-water channels (Abdel Aal et al., 2000).

The development of Upper Pliocene channels in the Western Nile Delta started initially with the formation of submarine valleys, up to 6 km wide and 200 m (656 ft) deep, that were cut by slumping and submarine erosion during a sea level lowstand (Samuel et al., 2004). These valleys were completely filled with debris flow deposits, sands, and shales, and then re-incised during similar processes. Amalgamated sheet units occur both at the base and top of the channel systems.

The Nile Delta and its offshore area consist of four main petroleum provinces (Levorsen, 1954, p. 30) (Fig). Province 1 covers most of the onshore Delta and is limited to the south by the Hinge Line. In addition to the group of fields along the Abu Madi Channel trend, it includes the Desouq, El Wastani, Sherbean, Abu Monkar, Gelgel and Qantara fields. Province 2 occupies the central part of the Delta and is dominated by the Abu Madi Channel, which extends northward into the offshore. Province 3 is located to the Northwest of the Delta, west of the Rosetta Line or Fault. It includes the Abu Qir, El King, El Max, El Bahig, Sapphire, Simian, and Scarab/Saffron fields. Province 4 lies to the Northeast of the Delta, east of the Bardawil Line, and includes the offshore area of Sinai. It includes the Ha'py, Timsah, Tao, Port Fouad, Kamose, and Tineh fields, among others. Reservoirs include Jurassic carbonates and Early Cretaceous turbidite sand reservoirs within Cretaceous inversion play (Syrian Arc) (Peck & Horscroft, 2005). Each of these provinces is characterized by various petroleum systems and hydrocarbon plays (Fig. 19.9). Although available information is still scanty, each of the provinces has different source rocks with different degrees of maturity.

The majority of oil and condensate was derived from Miocene shale-rich source rocks, but some may have originated from a Cretaceous-Jurassic source (EGPC, 1994, Wever, 2000, Sharaf, 2003). According to EGPC (1994) (Fig. 19.8), Jurassic and Lower Cretaceous source rocks are mature in the onshore Province 1; Province 2 contains mature Miocene rocks; Province 3 is dominated by mature Lower Miocene and Oligocene source rocks, but Pliocene and Pleistocene sapropels may have also contributes some of the gas (Vandre et al, 2007), and Province 4 contains Lower Miocene and Oligocene shales. Oil condensates typically have high API gravity and low sulfur content. According to Wever (2000), there are two types of oils in the onshore Nile Delta. The fist type has a lesser $\delta13C$ value and higher oleanane content than the second type and suggests derivation from Jurassic-Cretaceous source.

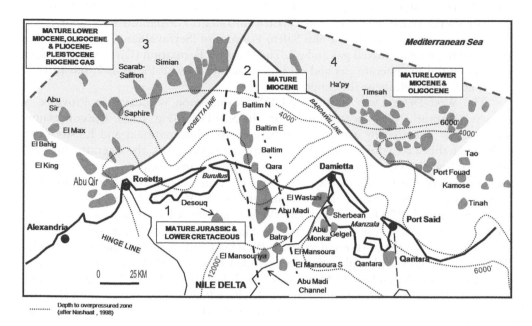

*Figure 19.8* Source rock potential in the Nile Delta (based on EGPC, 1994, and Sharaf, 2003). Depth to the top of overpressured zones (in ft) is after Nashaaat (1998). Note the Port Fouad Field is now called North Port Said Field.

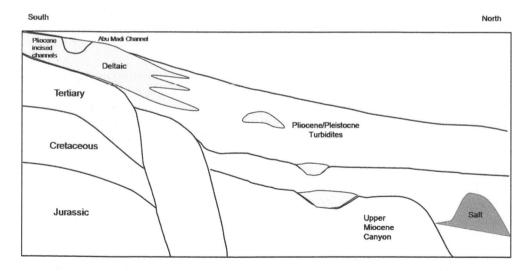

*Figure 19.9* Model of Nile Delta Plays (after Abdel Aal et al., 2000 and Samuel et al., 2003).

The second type was probably derived from Late Cretaceous or post-Cretaceous source.

Source rocks in the wells Port Fouad Marine Southeast-1 (PFM SE-1), Temsah-4 (T-4), and Barracuda Deep-1 (BD-1) in the offshore of the NE Nile Delta (Province 4) have the TOC values ranging from 0.37–2.87% (Sharaf, 2003, Table 1). Shales in the Kafr El Sheikh (Early Pliocene) and Wakar (Tortonian) Formations have poor capability to generate gas or oil; the Sidi Salem Formation (Serravallian) has poor to fair potential to generate mixed gas and oil. The Qantara Formation (Early Miocene) has poor potential to generate gas and oil in the vicinity of well PFM SE-1, but further north near well T-4, its quality improves.

The West Delta Deep Marine (WDDM) Concession is located in Province 3 (Sharp & Samuel, 2004) located in offshore the Nile Delta (Fig. 19.8). The concession was initially operated by a partnership between BG International and Edison Gas. Following the successful exploration and appraisal program, the operatorship has passed to Burullus Gas, a Joint Venture operating company composed of the Egyptian General Petroleum Company, BG International, and Edison Gas. Exploration has been focused on the Pliocene slope channel play in the WDDM (Samuel et al., 2003, Sharp & Samuel, 2004). Although the Delta Cone thins basinward (Sestini, 1989, Samuel et al., 2003), the thickness of the Pliocene-Pleistocene succession remains considerable up to the northern part of the WDDM Concession (Samuel et al., 2003). A total of 13 successful exploration wells drilled between 1999 and 2004. Six wells were drilled by the end of 1999; Scarab-1 and 2, Saffron-1 and 2, Serpent-1, and Simian-1. The simian-1 tested at 44 mmcfgd. There had been no instances of shallow water flows or over-pressured shallow gas anomalies being encountered in any of the six wells drilled.

Even prior to the events of 2011, participation in the March 2009 bid-round issued by Egypt in the deepwater Nile region was moderate. Exploration and foreign companies' involvement have been affected adversely by sectarian violence and political pressure (Feller, 2009). In November 2010, BP announced a new discovery by the drilling of the well WMDW-7 well in the West Mediterranean Deep Water Concession. The discovery is the first to recover hydrocarbons from Oligocene reservoirs in the area.

The sedimentary sequence consists of thick Miocene-Pleistocene siliciclastics (Sharaf, 2003) (Fig. 19.10). The following stratigraphic units are recognized (EGPC, 1994, El Heiny & Enani, 1996). At the base of the Miocene succession is the Aquitanian-Langhian Qantara Formation, up to 300 m (984 ft) thick, composed of calcareous marine shales and sandstones (Omar et al., 1988). This is overlain unconformably by the Sidi Salem Formation (Langhian-Serravallian) which is dominated by shales with some submarine fan sandstones. The formation is about 450m (1477 ft) thick. The Tortonian Wakar Formation overlies it unconformably, and comprises a thick sequence of deep-marine to shelf shales and sandstones, 500–600 m (1641–1969 ft) thick. Above is the Messinian Rosetta Formation, 100–300 m (328–984 ft) thick in the northern Nile Delta, which is composed of anhydrite with interbedded sandstones and shales. The formation is overlain unconformably by the shales and sandstones of the Kafr El Sheikh Formation (Early Pliocene); this unconformity marks the widespread Early Pliocene transgression which followed the Messinian Crisis in the Mediterranean region. The thickness of the Kafr El Sheikh Formation increases towards offshore areas; up to 1,800 m (5906 ft). The Kafr El Sheikh Formation passes up into the El Wastani Formation of about 400 m (1312 ft) of shales and sandstones, which is a transitional facies between the shelf facies of the Kafr El Sheikh Formation and the fluvio-deltaic facies of the overlying Mit Ghamr Formation, up to 600 m (1969 ft) thick.

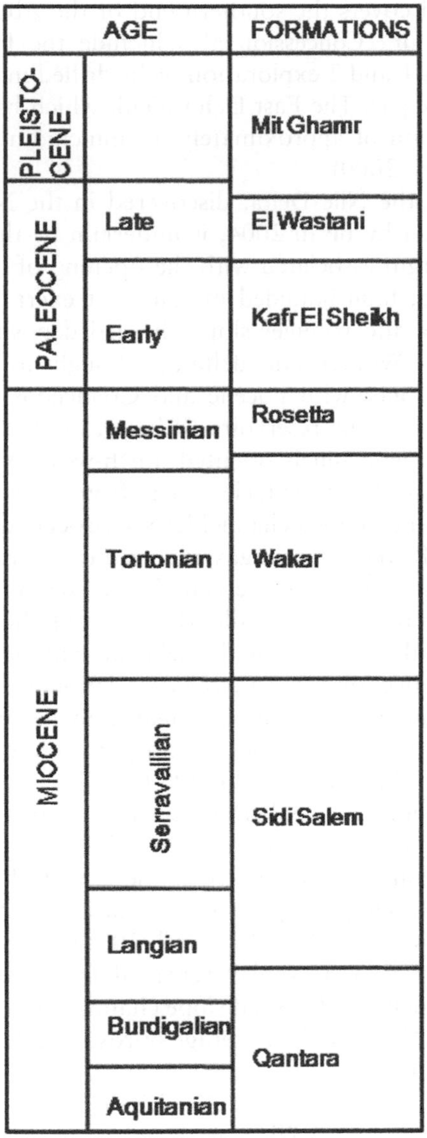

| AGE | | FORMATIONS |
|---|---|---|
| PLEISTO-CENE | | Mit Ghamr |
| PALEOCENE | Late | El Wastani |
| | Early | Kafr El Sheikh |
| MIOCENE | Messinian | Rosetta |
| | Tortonian | Wakar |
| | Serravallian | Sidi Salem |
| | Langian | |
| | Burdigalian | Qantara |
| | Aquitanian | |

*Figure 19.10* Stratigraphy of Nile Delta (after EGPC, 1994).

The Qantara Concession is situated in the Eastern Nile Delta (Melrose Resources, 2000). The concession includes the Qantara Field which was discovered with the drilling of the Qantara-1 well in 1976. The well tested at 19.2 mmcfd and 2,730 bopd of 55°API condensate/d from the Qantara Formation (Early Miocene-Late Oligocene). The gross proven reserves of the field have been assessed at 17 bcf of gas and 2.7 mmbbls of condensate.

The El Mansoura Concession, also operated by Melrose, is located west of the Qantara concession and covers the southern end of the Abu Madi Channel. Other fields on the El Mansoura Concession also include the Halawa and East Delta fields. The El Mansoura-1 and 2 exploration wells drilled in 2000 encountered non-commercial quantities of gas. The East Delta Field, which has been producing since 1996, averages production of approximately 35 mmcfgd and 570 bopd from two wells (Melrose Resources, 2000).

The Raven Field in the Nile Delta, discovered in the North Alexandria Block in the Western Nile Delta by BP in 2004, is underlain by the outer part of a rifted Jurassic continental margin associated with the opening of the Mediterranean. It is characterized by a steep, fault-bounded margin that exerts a fundamental control on the overlying Eocene and younger slope channel deposits (Whaley, 2008). The main source rocks in the Western Nile delta are thought to be Pliocene mudstones, which generate biogenic gas, with Eocene and Cretaceous sediments sourcing the recently discovered Pre-Pliocene reservoirs (Whaley, 2008). The Raven Field reservoirs are in turbidite slope channels oriented north-west to south-east draped over the Raven structural high. Reservoir facies range from coarse-grained gravels at the base of the channels to fine-grained channel levees associated with the final stages of channel abandonment. Raven-1 tested gas at a rate of approximately 36.6 mmcfgd (1.04 million $m^3$ gas per day) and 23 barrels condensate per million standard cubic feet with additional shows in the Upper Middle Miocene and Pliocene.

The Messinian Eo-Nile canyon could be delineated in the eastern part of the Nile Delta by using seismic reflection data. A close correlation is evident between the Late Quaternary Nile River channels and the Messinian canyons in the Eastern Nile Delta. This indicates that the Messinian canyons were a persistent topographic feature until historical branches (El Mahmoudi & Gabr, 2009).

The Miocene succession in the Nile Delta consists of the Qawasim and Sidi Salem formations. The Qawasim Formation (Early Messinian) is made up of shales with very coarse to medium-grained sand and conglomerate interbeds which form channel fills. In the Abu Madi gas field, the Qawasim Formation is about 70 m (230 ft) thick and has an average porosity of 17% (Mohamed, 2004).

The Sidi Salem Formation (Tortonian-Serravallian) in the Temsah Field forms a fining upward sequence deposited in shelf slope channels or a prograding delta. It has a net pay of about 16 m (52 ft) and an average porosity of 22%, an average $S_w$ of up to 47% (Mohamed, 2004).

Pliocene reservoirs are represented by Abu Madi, Kafr El Sheikh, and El Wastani formations, followed by the Upper Pliocene-Quaternary Mit Ghamr Formation. The Abu Madi formation (Messinian) in Baltim gas fields has an average porosity of 22%, water saturation ($S_w$) of about 30%, and a net pay of 31 m (102 ft). Permeability is about 120 mD in the Abu Madi gas field. It forms elongated channel fill (Baltim channel), or a prograding delta/delta plain coarsening-upward sequence in the Abu Qir and Nidoco fields. It was deposited in a fluvio-marine environment on shelf/ shelf slope (delta plain/ delta front), which is indicated by brackish water benthonic foraminifera and Dinoflagellates (Mohamed, 2004).

The Kafr El Sheikh Formation (Lower-Middle Pliocene) is the primary target in Mediterranean Sea and Nile delta. It forms a turbidite sequence deposited on the down thrown side of east-west growth faults in the Seth, Ha'py, King, and Max gas discoveries. Its thickness is up to 200 m (656 ft). Listric growth faults are common in Kafr El Sheikh Sequence. It forms reservoir in the Seth, Ha'py, King, Abu Sir, Scarab & Simian gas fields. It has porosity between 26–31%, an $S_w$ of 16–25%, and a net pay of 100 m (328 ft). In Abu Sir gas discovery, it has an average porosity of about 29%, an $S_w$ about 24%, a net pay of 120 m (394 ft), and an average permeability of more than 500 mD (Mohamed, 2004).

The Mediterranean offshore area of North Sinai is dominated by the northern extension of the Syrian Arc inversion structures (Peck & Horscroft, 2005). The Syrian Arc in the Eastern Mediterranean region formed during the Late Cretaceous to Miocene Alpine Orogeny and the inversion of a Late Triassic-Early Jurassic half-graben system (Moustafa & Khalil, 1990, Druckman et al., 1995). Seismic, gravity, and magnetic data from the offshore North Sinai show large buried NE-trending asymmetrical anticlines which formed as a result of the Late Mesozoic and Cenozoic inversion events. Three different phases of deformation have been recognized in offshore North Sinai (Yousef et al., 2010): 1) A Jurassic-Early Cretaceous extensional phase, which formed NE trending normal faults bounding asymmetrical half-grabens, 2) Post-Santonian-Middle Miocene positive inversion of these faults and half-grabens and 3) Post-Middle Miocene subsidence.

From 1967 until 1984, Sinai was under the Israeli occupation. Numerous studies were undertaken by Israeli geologists, the results of which were published in the form of a collection of reprints by the Geological Survey of Israel (1980–1984). In the period 1967–1971, work in Sinai concentrated on production activities of the previously discovered fields in the Gulf of Suez. Extensive exploratory work in Sinai took place in the period 1972–1979. Exploration in Sinai and the Gulf of Suez by the Israelis ended with the signing of the Egypt-Israel peace treaty in 1979.

Drilling onshore and offshore Sinai led to the discovery of a number of small oil and gas fields (Fig. 19.8) in NE-SW-oriented structural traps (Syrian Arc folds). The Sadot/Raad fields were discovered by the Israelis in 1975 in Cenomanian sands. IEOC made a series of discoveries in deep turbidite sands of the middle Miocene Wakar (Qawasim) Formation in the Port Fouad Field (later named North Port Said) in 1981, the Wakar Field in 1983, the Kersh Field in 1983, and Abu Zakn Field in 1986. The Tineh Field was also discovered by IEOC in 1981 in isolated Oligocene sands and Bougaz Field in 2006. The Mango Field was discovered by Total in 1986 in lower Cretaceous sands.

The only onshore field in Sinai is the small Sadot Gas Field (Fig. 19.1). Oil shows were recorded in wells Nakhl-1 and Abu Hamth-1 in Central Sinai (El Ayouty, 1990). Reservoirs in both the Negev and Sinai are mainly of Jurassic and Cretaceous age (Lüning, 2011). The traps are controlled by the Syrian Arc anticlines. The Mango-1 well tested 10,000 bopd plus gas from thin Albian-Aptian turbidite sands. Nevertheless, the field turned out to be too small to justify development (Peck, 2008, Peck & Horscroft, 2005). The Tineh-1 well, drilled on the crest of the structure, encountered significant oil indications in some sand reservoirs, interbedded with shales, within the Late Oligocene section (Yousef et al., 2010).

The Early Cretaceous sandstones are up to 200 m (656 ft) in thickness and are fine- to coarse-grained, cross-bedded, highly jointed and of high porosity and permeability (Ammar & Afifi, 1992). These sandstones belong to the Rakeiba and Risan Aneiza formations. The sandstone tested oil in the Mango-l discovery. The net sand thickness in the Mango discovery reaches 35 m (115 ft) with an average porosity of 25% (Alsharhan & Salah, 1996). The Sadot Gas Field produces from the Cretaceous sands. Potential source rocks include Jurassic shales, Early Cretaceous carbonates, and the Oligo-Miocene fine clastics. In addition to the Jurassic and Cretaceous carbonates, the sandstones of the Miocene, Oligocene, Cretaceous and Jurassic rocks form the major reservoirs in the north Sinai. The intraformational Mesozoic and Cenozoic shales and dense carbonates and the middle Miocene anhydrite form the seals for hydrocarbons.

Trap types include structural, stratigraphic and combination traps (Alsharhan & Salah, 1996). Mesozoic and Cenozoic reservoirs produce oil from fault traps in the Wakar, Kersh and Abu Zakin discoveries (Alsharhan & Salah, 1996) (Fig. 19.8). Other structural traps include four-way dip closure fold traps (Alsharhan & Salah, 1996), such as the Sadot Field, or hanging wall anticlinal reservoirs, as is the case of the Tineh discovery.

## 19.2.2  Libya's hydrocarbon systems

Oil was first discovered in Libya with the B2–1 well in the Atshan Field (Fig. 19.11). The Field is located on the Atshan Saddle, which separates the Murzuq and Ghadames basins. It encountered 47° API oil in Upper Ordovician to Carboniferous sands (OGJ, March, 1995).

The Cambrian to Silurian sequences of Libya, Algeria, and Tunisia form one of the world's greatest concentrations of Lower Paleozoic reservoirs (Crossley & McDougall, 1998). The Hassi Messaoud and Rhourde el Baguel fields in Algeria (Crossley & McDougall, 1998, Mitra & Leslie, 2003) and the Attahaddy Field in the Sirte Basin in Libya (Tawadros et al., 2001) are the largest fields in this system. The Ordovician reservoirs include the Hamra Quartzite in Algeria and the Haouaz and Memouniat formations in West Libya (Beuf et al., 1971, Burollet & Byramjee, 1964, 1969, Bellini & Massa, 1980, Turner, 1980, Klitzch, 1981, Vos, 1981b, Clark-Lowes & Ward, 1991). Through much of the Silurian, marine conditions prevailed over most of North Africa as relative sea level rose on top of the latest Ordovician erosional surface. The Sharara and Elephant fields in West Libya and the Tiguentourine, Brides (Berkine Basin), and Tin Fouye-Tabankort (Illizi Basin) fields in Algeria produce from Ordovician peri-glacial sediments (Cochran & Petersen, 2001). Equivalent sequences in Egypt and Morocco might also have hydrocarbon potential, although in the case of Egypt their potential might be restricted by the lack of Silurian source rocks and the limited distribution of Cambro-Ordovician reservoirs.

Thick graptolitic, locally organic-rich shales of the Tannezouft Formation in Libya and its equivalent in Northwest Africa formed a regional seal and source rock unit (Berry & Boucot 1973, Lüning, 2000, Lüning et al., 2004, Sikander et al., 2003). However, Silurian shales change facies in Egypt into coarse-grained facies with limited source rock potential (Keeley, 1989, 1994).

The basin fill in the Kufra Basin extends from the Cambrian to the Cretaceous. The basin is on trend with the hydrocarbon-bearing Benue Basin of Nigeria. The Kufra Basin is the least explored basin in Northeast Africa with only two dry holes drilled to date and there are no known hydrocarbon occurrences or source rocks. However, Lüning et al. (1999) and Le Heron & Howard (2010) postulated the presence of Ordo-

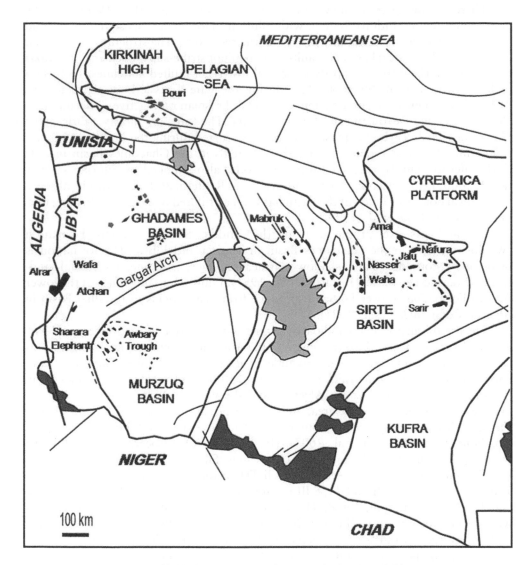

*Figure 19.11* Structural map of Libya and oil and gas fields.

vician reservoir rocks and Silurian source rocks in the Kufra Basin, while Ghanoush & Abubaker (2007) suggested the presence of structural traps in the area. On the other hand, the proximity of the basin to the Silurian shoreline and the dominance of course-grained material and doubts about the maturity of the Silurian shales limit the potential of this basin. Schmieder et al. (2009) suggested that the Jebel Hadid impact structure in the Southern Kufra Basin may have provided local secondary pathways and traps for hydrocarbon migration and accumulation in the Nubian Sandstones and therefore may represent a potential area for future hydrocarbon exploration feature.

There are two petroleum systems in the Ghadames Basin (Fig. 19.11): 1) The Tannezouft-Acacus/Tadrart System, with Silurian "Hot Shales" as source rocks and the Silurian/Devonian sandstones providing the reservoirs. The structure in the basin is dominated by vertical basement faulting with minor strike-slip movements (Galeazzi et al., 2010). The traps are mostly structural low-relief faulted anticlines. Expulsion of hydrocarbons probably took place during the Late Triassic-Lower Cretaceous. 2) The Cues-Aouinat Ouenine System consists of Frasnian radioactive shale and marl source rocks and Devonian sandstone reservoirs. The traps are mostly stratigraphic pinch-outs. Hydrocarbon expulsion probably occurred during the Early-Late Cretaceous time. Other secondary hydrocarbon systems in the Ghadames Basin include Carboniferous and Triassic sandstone reservoirs, also supplied by Silurian sources (Klett, 2000, Letouzey et al., 2005, Underwood & Redfern, 2007, 2008). The absence of hydrocarbons over much of the basin has been attributed to extensive flushing during Cenozoic (Boote et al., 1998).

Oil and gas are produced from the Middle Silurian Acacus Formation (Wenlock-Pridoli) in the Ghadames Basin in Tunisia and Libya. The formation is made up of storm-dominated shallow marine, shoreface, estuarine, and fluvial sandstones (Bracaccia et al., 1991, Castro et al., 1991). Equivalent sandstones of the lower Acacus Formation (Lower F6 Sand) produce from many fields in the Illizi Basin in Algeria (Cochran & Leslie, 2001).

The organic matter in the radioactive zones of the Silurian lower Tannezouft Shale and the Devonian Cues Marls (Frasnian) of the Aouinat Ouenine Group in West Libya is predominantly amorphous algal (Sikander et al., 2003). The Lower Tannezouft radioactive shales attain a total thickness of 46 m (150 ft) in the central Ghadames Basin (Libya), the eastern Illizi Basin (Algeria), and the western Atshan Saddle. They are mature over a large part of the area, but over mature in the western Ghadames Basin. The Cues Marls rapidly thicken from 3 m (10 ft) to over 61 m (200 ft) in the western Ghadames Basin (Sikander et al., 2003). The Tannezouft Hydrocarbons in West Libya were expelled during the Permian-Late Cretaceous (Sikander et al., 2003).

The subsurface Silurian section in well Tt1 in the Ghadames Basin in Southern Tunisia contains Type II and Type III organic matter with TOC values between 0.2 and 13.2% (Vecoli et al., 2009). The vertical distribution of TOC values correlates with the Gamma-ray curve (Lüning, 2000, Vecoli et al., 2009). Very low TOC values with an average of 0.4% are observed in the Hirnantian sediments of the Djeffara Formation. An increase in TOC values to about 5.6% is noted at the transition with the Argiles Principales Formation (Tannezouft equivalent) (Vecoli et al., 2009). The overlying Ludlow silty shales give low TOC values with an average of average 1.4%. Average $T_{max}$ value is 437 °C, which indicates that the organic matter is in the beginning of the oil window.

In April 2009, the NOC/Sonatrach partnership made the first discovery on Block 65 acquired in 2005 in the Ghadames Basin in West Libya with the wildcat A1-65 well, about 230 km south of Tripoli. The well encountered both oil and gas in the Upper Ordovician Memouniat Formation at intervals of 8532–8560 ft and tested 48.8° API oil at a rate of 1344 bopd and 1.88 mmcfd of gas[2].

2 www.gulfoilandgas.com.

In 1991, Sirte Oil discovered Al Wafa Field in Middle Devonian F3 Sand (Givetian) sands on the Libyan-Algerian border near the Algerian Alrar Gas Field, discovered by Sonatrach in 1980, with estimated reserves of 4.7 TCF of gas. Production started in 2004 at 23,000 bopd, 450 mmcfgd, and 15,000 bcd. Arguments and controversies followed regarding the continuity of the Devonian reservoirs between the two fields. An agreement was signed between NOC and Sonatrach in September 2006 to study the link between Al Wafa and Alrar fields.

The Alrar Field is located on the Ahara Ridge in the Illizi Basin in Eastern Algeria and the Al Wafa Field on the Atshan Saddle in West Libya (Fig. 19.11, 19.16). Chaouchi et al. (1998) and Tawadros et al. (1999) interpret the F3 Sand at Alrar as a series of tidally influenced barrier bars. The reservoir has porosities between 7% and 15%, and permeability of 5–43 mD. The Givetian F3 Sand is the main reservoir in the Al Wafa Field in West Libya (Tawadros et al., 1999) and the Alrar Field in Eastern Algeria (van de Weerd & Ware, 1994, Chaouchi et al., 1998). The Alrar Field has recoverable gas reserves of 4.6 TCF and condensate reserves of 277 MMB (van de Weerd & Ware, 1994).

The Lower Carboniferous Mrar Formation in West Libya contains potential fluvial hydrocarbon reservoir rocks deposited within incised valley systems (Fröhlich et al., 2010) (Fig. 19.12). In the Tinedhan Anticline on the southern margin of the Ghadames Basin, the formation consists of fluvial channels filled with sandstones up to 50 m (164 ft) in thickness. Diagenesis influenced the reservoir quality of the sandstones, which include kaolinitization of plagioclase grains, calcite cementation, and partial chloritization of kaolinite. Despite these diagenetic effects, the sandstones maintained an average porosity of 12%, and good permeability.

The regional petroleum geology of the Murzuq Basin is has been reviewed by Aziz (1998), Davidson et al. (2000), Echikh & Sola (2000), Hallett (2002), Sikander et al. (2003), and Belaid et al. (2009). Oil is restricted to the central and western parts of the Basin (Fig. 19.11). The Murzuq Basin consists of seven tectonic elements (Echikh & Sola, 2000) from west to east: Tihemboka Arch, Al Awaynat Trough, Tiririne High, Awbari Trough, Idhan Depression, Brak-Bin Ghanimah Uplift, and Dur al Qussah Trough.

Between 1984 and 1986 a total of 29 wells were drilled by Rompetrol in Concession NC115 in the Awbary Trough in the Murzuq Basin (Fig. 19.11), which resulted in the discovery of six oil fields, named collectively as the Sharara Field. The Field is a giant with OOIP of more than one billion barrels (Bertello et al., 2003). In 1997 Lasmo made the discovery of the Elephant Field with the well F1-NC174, which tested about 3000 bopd of 38° API oil in the Upper Ordovician Memouniat Formation. The primary petroleum system consists of the Upper Ordovician Memouniat and Haouaz Formation sandstones as reservoir and of the Lower Silurian Tannezouft Shales as source rock and seal. Hot-shales have been identified in the H and A Fields of the NC115 Concession and in Tannezouft outcrops at Ghat by Fello et al. (2006). A maximum TOC value of 22% was measured in well H29-NC115. Uranium content of the shales remained largely unaltered by the weathering processes and could therefore be used as a valid proxy parameter to distinguish between pre-weathering organically rich and lean shales (Fello et al., 2006). In the Murzuq Basin in Libya, more than 2 billion barrels of reserves have been proven in Ordovician reservoirs (Quesada

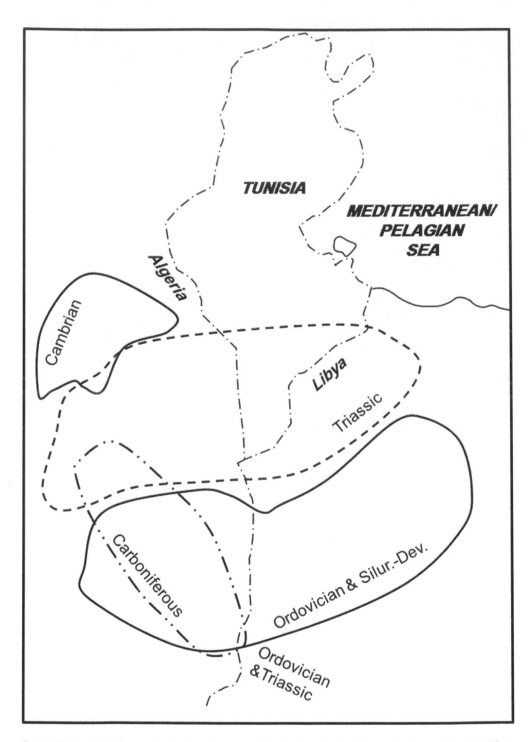

*Figure 19.12* Oil and gas plays in the Ghadames-Illizi-Berkine Basin (Libya, Algeria, and Tunisia) (after Dardour et al., 2004, Galeazzi et al., 2010).

et al., 2003). Ordovician reservoirs in the Sharara Field have porosity of 12–20% and permeability of 100–1000 mD.

There is only one major petroleum system in the Murzuq Basin, the Tannezouft-Memouniat System, which include a Silurian Hot Shale source and Ordovician peri-glacial sandstones, Upper Silurian sandstones (Acacus), and Lower Devonian sandstones (Tadrart) reservoirs. The traps are mostly structural. The hydrocarbon system in the Murzuq Basin is relatively complex and is controlled by a number of different factors including structural evolution and timing of oil generation relative to the structure formation (Bertello et al., 2003). The Hot Shales in the Murzuq Basin consist of anoxic radioactive shales with TOC content of up to 17% and a type I-II kerogen (Lüning et al., 2000, Bertello et al., 2003). Hot shales in wells A1-NC186, H29-NC115, F3-NC174, and A1-NC190 contain mainly Type II kerogen with TOC values from 3% to 23% (Belaid et al., 2010). Type II kerogen with over 80% liptinite suggest deposition in a restricted distal marine environment. The hot shales are immature to early mature in the southern part of the Murzuq Basin and mature in the northern part. The highest level of maturity is present in the Awbari Trough in NC115 (Belaid et al., 2010). The traps are structural and mainly consist of gentle 4-way-dip folds and fault-bounded folds (Bertello et al., 2003). The Tannezouft shales also provide the seal. Oil generation and charge are considered to have been active since the pre-Hercynian time although the main phase occurred during the Jurassic to Early Tertiary (Bertello et al., 2003).

The Sirte (Sirt) Basin is a three-arm failed rift. The basin consists of a series of platform oriented approximately north-south in the central part of the basin, east-west in the eastern part, and northeast-southwest in the southwestern part. The platforms are separated by elongated basins and troughs. Paleozoic to Triassic pre-rift fields occur mostly in the troughs, while syn- and post-rift Lower Cretaceous to Tertiary plays occur on the platforms.

The Sirte Basin ranks 13th among the world's petroleum provinces and contains reserves of about 36.7 billion barrels of oil, 37.7 trillion cubic feet (TCF) of gas, and 0.1 billion barrels of natural gas liquids (Ahlbrandt, 2001). Among its 16 major fields, the Amal Field and the Gialo Field have approximately 4.0 billion barrels of recoverable reserves, each.

Traps in the Sirte basin are mainly tilted fault blocks, but sandstone and reef stratigraphic and combination traps are also common. Reservoirs range in age from Precambrian to Eocene (Figs. 12.5, 13.1). They include a large spectrum of lithologies that provide good quality reservoirs. They include Precambrian fractured basement rocks, such as in the Maragh Field (Williams, 1968), Cambrian fractured quartzites, as is the case of the giant Attahhady Gas Field (Tawadros et al., 2001), Devonian-Carboniferous sandstones (Tawadros et al., 2001, Winnekers et al., 1996), Triassic-Lower Cretaceous sandstone and fractured reservoirs (Winnekers et al., 1996, Thusu, 1996, Thusu & Mansouri, 1998), Upper Cretaceous (limestones, dolomites, sandstones), e.g. Waha (Barr & Weeger, 1972, El-Ghoul, 1996), Jabal, and Nasser fields, Paleocene carbonates and reefs, such as the Paleocene pinnacle reefs of the Intisar, Sahabi, Mheirigah, and Shatirah fields (Terry & Williams, 1969, Brady et al., 1980), and the Nasser (Zelten) Field (Bebout & Pendexter, 1975), and Eocene sandstones and carbonates, for example the Gialo (Jalu) Field.

The Sirte Shale (Campanian) is the dominant source rock in the Sirte Basin (Parsons et al., 1980, Gumati & Schamel, 1988, El-Alami, 1996, Ghori & Mohammed, 1996, Macgregor & Moody, 1998a, b, Ambrose, 2000, Ahlbrandt, 2001). The thickness of the Sirte Shale ranges from a few hundred meters to as much as 900 m (2952 ft) in the troughs. These rocks are within the oil-generating window between depths of 2,700 and 3,400 m (8859–11,155 ft) in the central and eastern Sirte Basin (Futyan & Jawzi, 1996). Parsons et al. (1980) reported total organic content (TOC) values ranging from 0.5% to 4.0% with an average of 1.9% TOC for 129 samples from the Upper Cretaceous source rocks. Baric et al. (1996) estimated TOC values of up to 3.63% and type II amorphous kerogen from the Sirte Shale in wells on the Zelten Platform. Other source rocks in the Sirte Basin include. Maturation took place during the Paleocene-Eocene time.

Six types of quartzites and sandstones of Late Cambrian, Late Devonian, Devonian-Carboniferous, and Early Cretaceous age have been identified in the Central Basin (Tawadros et al., 2001) (Fig. 12.4). The Upper Cambrian reservoirs comprise two shallow-marine quartzite units separated by a *Skolithos*-rich, varicolored argillaceous feldspathic sandstone unit (Red Beds). The two quartzite units form fractured reservoirs in the giant Attahaddy Gas Field. A Late Devonian (Strunian) shallow marine to continental sandstone unit has been identified in wells V1-59 in the southwest of the study area and EE1–59 in the southeast. A Carboniferous (Visean) sandstone unit has been identified in wells C3-6, KKKK1-6, BBBB2-6 and V1-59. Triassic reservoir rocks have been identified in the As Sarah Field in the Eastern Sirte Basin. The Early Cretaceous (Wadi Formation) continental fractured quartzite unit dated by tricolpate pollen produces oil in the Wadi Field. The Upper Cambrian lower quartzite unit and the Wadi quartzites contain frequent rhyolitic volcanic horizons. The study reveals that the Devonian and Carboniferous sandstones (Belhedan Formation), with good primary porosity, previously considered Nubian and Cambro-Ordovician, respectively, form an attractive hydrocarbon bearing trend in the southern part of the study area.

Major Maastrichtian and Paleocene oil fields occur on the Zelten Platform edge (Fig. 19.13). The Maastrichtian Waha nearshore carbonate and sand fields include the Nasser, Ralah, Jabal, Ar Raquba (Broughton, 1996), and Waha (El-Ghoul, 1996). Paleocene fields include the reefal carbonate buildups of the Zelten (Nasser) Field (Landenian, equivalent to the Lower Sabil Formation) (Bebout & Pendexter, 1975), Shatirah Field, Harash Field, and the Defa Field (Danian, equivalent to the Lower Sabil Formation). Eastward, in the Ajdabia/Wadayat Trough, the Upper Sabil pinnacle reefs form a number of prolific oil fields, designated A to L, including two giants; the Intisar "A" and "D" fields (Figs.), with recoverable oil reserves of 1.0 and 1.4 million barrels of oil, respectively. The Intisar pinnacle reefs consist of algal-foraminifera limestones in the lower part and coral reefs in the upper part (Terry & Williams, 1969). The average porosity in these fields ranges from 15% to 27% and permeability is up to 490 mD, with a net pay of 16–640 ft (53–210 ft).

The Dahra Platform (Fig. 19.13) is characterized by high gravity anomalies on the eastern and western sides, while the middle part is characterized by low gravity anomalies (Elakkari, 2005). The high gravity anomalies are related to the Messlah High and Gatar Ridge at both edges of the platform. Most of the oil fields within the Dahra Platform are located on these highs, attributed to high density fringing reefs

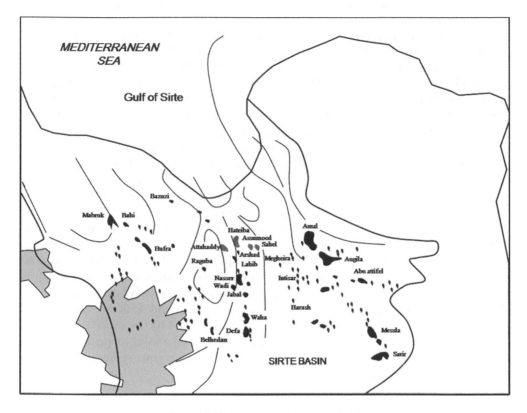

*Figure 19.13* Sirte Basin oil and gas fields.

or carbonate rocks. The low gravity anomaly within the middle part of the Dahra platform is related to the low density source rocks.

Intisar 103L Field is located approximately 370 km to the South of Benghazi in the Sirte Basin of Concession 103 in Libya. It was discovered in 1968 by drilling the first discovery well, L1-103 down to a total depth of 10,295 ft (Senturk et al., 2007). The field comprises two separate carbonate reservoirs; the Eocene Gir and the Paleocene Upper Sabil (Zelten). The traps are four way dipping anticlines with less than 15 m (50 ft) closures. The average porosity ranges from 15% to 28%, and permeability from 1 mD to 15 mD. The field has OOIP of 19.6 million barrels, of which 14.9 million barrels are in the Gir and 4.7 million barrels in the Upper Sabil. The field production started in 1976 from the two productive wells L1-103 and L2-103.

Upper Eocene carbonates are oil-bearing in the Maragh Trough in wells D1 and F1-96 (Gruenwald, 2001). Hydrocarbons probably migrated along graben-bounding faults from deeply-buried source rocks to platform and sub-platform areas. Traps are of combined structural and stratigraphic type.

The Cyrenaica Basin is a Mesozoic marginal rifted basin similar to the Northern Egypt Basin. It is overlain by the Miocene Cyrenaica-Marmarica Platform. The succession extends from Cambrian to Recent. The Santonian inversion led to the uplift of the Jabal Akhdar (Green Mountain) in the northwest, which was reactivated during the Eocene-Oligocene Syrian Arc compression.

There are only two offshore oil wells; the A1-NC120 discovery by Agip/NOC in 1983, which tested 36°API oil at a rate of 5263 bopd at 2436 m (7993 ft) from Lower Cretaceous and Turonian carbonates (Duronio et al., 1990) and the 2009 new discovery in A1-NC202 by Repsol/OMV, which flowed 1264 bopd and 0.6 mmcfgd from the Eocene Dernah Formation in the interval 4442–4484 ft (1354–1367 m) (World Oil, May 2009). The source rock is believed to be the Campanian Sirte Shale. The trap is a structural fault block (Gruenwald et al., 2010).

Ghori (1991) recognized three major petroleum provinces in Cyrenaica (Fig. 19.14):

Province 1 lies south of lat. 32°00'N and east of long. 21°30'E. The Devonian and Carboniferous contain good gas source rocks. The Devonian shales show the best hydrocarbon generating capacity. The present-day geothermal gradient is 24°C/km. The oil generative window is between 2250 m and 4000 m (13,124 ft). Hydrocarbon generation started during Late Cretaceous and reached its peak during the Tertiary.

Province 2 lies north of lat. 32°00'N. The Jurassic and Lower Cretaceous contain good oil and gas source rocks. The rocks of the western coastal region show the best hydrocarbon-generating capacity. The present-day geothermal gradient is 26°C/km. The oil generative window is between 2500 m and 4500 m, except in Al Jabal Al Akhdar uplift where it is between 1200 m and 4500 m (3937–14,765 ft). Hydrocarbon-generation started during the Paleocene and was at its peak during the Oligocene and Miocene. The Devonian and Jurassic-Cretaceous sediments form the most suitable source rocks for gas and liquid hydrocarbons.

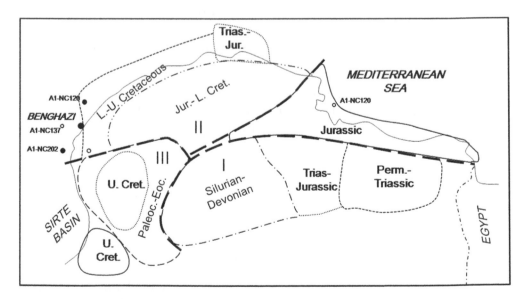

*Figure 19.14* Potential source rocks of Cyrenaica (Based on Thusu et al., 1988, Ghori, 1991, Hallett, 2002, and Burwood et al., 2003). The source of oil in the offshore well A1-120 is presumed to be Jurassic in age.

Province 3 is located south of 32°00'N and west of 21°30'E. The Upper Cretaceous, Paleocene and Lower Eocene sediments contain good oil and gas source rocks. The Upper Cretaceous shales show the best hydrocarbon-generating capacity. The present-day geothermal gradient is 31°C/km. The oil generative window is between 2500 m and 4200 m (8203–13,780 ft). Hydrocarbon generation started during the Oligocene and reached its peak during the Miocene-Pliocene time.

### 19.2.3  Algeria's hydrocarbon systems

Crude oil was first extracted from dug tunnels and oil seeps in 1875 in the Chelif Basin in Northernmost Algeria. Two small oil discoveries were made in Miocene sands in the Chelif Basin in the Ain Zeft and Tliouane fields in 1892. However, the first commercial discovery was made in 1949 in Nummulitic Eocene reservoirs in the Oued Gueterini Field in the Tellian nappes in Northern Algeria (Askri et al., 1995). Other discoveries were made at Djebel Foua and Djebel Onk.

The Greater Ghadames Basin covers an area of approximately 250,000 km² in Algeria, Tunisia, and Libya. In Algeria, the sedimentary succession reaches up to 7000 m (22,967 ft) of mixed clastic and carbonate rocks (Moffat & Johns, 2001). The Algerian portion of the Ghadames Basin has recently been renamed the Berkine Basin. The petroleum geology and basin evolution of the Ghadames/Berkine Basin have been summarized by Tissot et al. (1973), Gauthier et al. (1995), Askri et al. (1995), Boote et al. (1998), Daniels & Emme (1995), Echikh (1998), Fekirine & Abdallah (1998), Guiraud (1998), Macgregor (1998), Malla et al. (1998), Lüning et al. (2000), Klett (2000), Cochran & Peterson (2001), Moffat & Johns (2001), Yahi et al. (2001), Makhous & Galushkin (2003), Sikander et al. (2003), Underdown & Redfern (2008), and Galeazzi et al. (2010).

The Ghadames/Illizi/Berkine Basin is one of the cratonic basins of the Algerian Sahara which extend along the northern margins of the Precambrian Reguibat and Hoggar massifs. Devonian rocks crop out along the erosional southern border of the basin and have been divided into three lithostratigraphic units (Wendt et al., 2009b); Lower Devonian sandstones (upper part of the Tassilis Externes), Middle Devonian carbonates (Illizi Formation of Wendt et al., 2009b), and Upper Devonian shales (Tin Meras Formation). The sandstones of the Tassilis Externes have been subdivided, from base to top, into (Dubois et al., 1968) the Oued Tifernine (includes the Barre inférieur and the Talus à *Tigillites*), Tamelrik (subdivided into the Barre moyenne and Trottoirs), Barre supérieure (also named Oued Samene), and the Oursine Formation. The eastern Illizi Basin received clastic input from the Tihemboka Ridge which separates the Ghadames/Illizi Basin in Algeria from the Murzuq Basin in West Libya.

The Illizi Basin has about 4 billion barrels of oil reserves, primarily in Devonian reservoirs, with secondary Cambro-Ordovician and Carboniferous reservoirs (Quesada et al., 2003), and more than 20 TCF of gas have been found in Ordovician reservoirs. The sedimentary fill is similar in the Illizi and Murzuq basins, but the trap style is somewhat different (Quesada et al., 2003). In the Illizi Basin the traps are mainly structural; whereas in the Murzuq Basin it is a combination structural-morphological as a result of Late Ordovician glaciation. Structural traps are associated with Paleozoic tectonics, mainly Caledonian, Acadian, and Hercynian. In the Illizi Basin, a relatively short lateral up-dip migration during the Mesozoic is the most likely migration mechanism; while long distance migration is possible from Ghadames (Berkine) Basin in the north (Quesada et al., 2003).

The Rhourde el Baguel Field was discovered by Sinclair Oil Company in 1962 in the Ghadames Basin, about 100 km east of the giant Hassi Messaoud Field (Fig. 19.15). It is one of the largest oil fields in Algeria. It produces oil from fractured Cambrian sandstones at subsea depths of 2300–3070 m (7546–10,072 ft) (Mitra & Leslie, 2003). The RB 01 discovery well penetrated 550 m (1805 ft) of fractured marine Cambrian-Ordovician sandstone reservoir on the eastern flank of the structure. Most of the oil is produced from the upper 170 m (558 ft) of the Cambrian-Ordovician sandstones (Units Ri and Ra) (Fig. 19.15). The sandstones are silica-cemented marine sandstones with a primary porosity of 6–9% and permeability averaging about 20 mD (Mitra & Leslie, 2003). The lower zone is made up of course-grained braided-stream Cambrian-Ordovician reservoirs (R2 and R3 units). The trap consists of a three-way dip closure and a lateral seal on the western flank provided by a Triassic normal fault. The top seal for the reservoir includes the Ordovician Zone des Alternances, Argile d'El Gassi, and Larroque Shales. The source

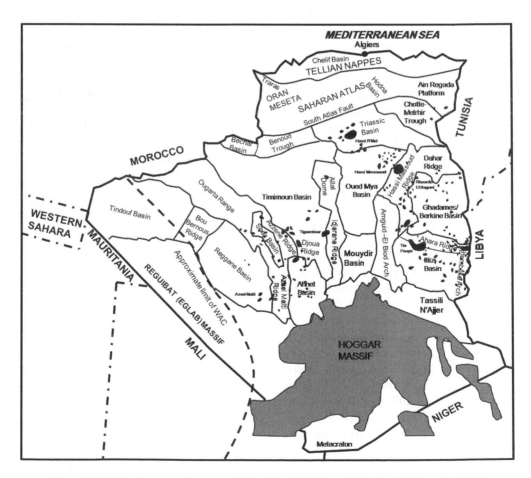

*Figure 19.15* Algeria tectonic map and oil and gas fields (modified after Askri et al., 1995).

rock is the mature Silurian-Devonian black shales. The Cambrian-Ordovician sandstones, which constitute the reservoir in the Hassi Messaoud and Rhourde el Baguel fields, were deposited as braided-stream, deltaic, and shallow-marine deposits over the eroded basement. Paleocurrent measurements show northward transport of sediments derived from the Hoggar Shield (Beuf et al., 1971). Tertiary recovery using miscible gas injection commenced in late 1999 by Sonarco, a joint-venture company between Sonatrach and ARCO.

The Ghadames/Illizi Basin/Berkine Basin shows two main cycles of subsidence. The first cycle spanned the whole Paleozoic and ended in the Upper Carboniferous with the Hercynian unconformity. The Ghadames/Illizi Basin was subjected to moderate erosion (Makhous & Galushkin, 2003). Burial of the source rocks of exceeded 2000 m (6562 ft) in the north of the basin and reached 2,600 m (8530 ft) in the southwest (Tissot et al., 1973). In the central part of the basin, the Paleozoic Tiguentourine shales are preserved under the Triassic, but in the other parts of the basin, erosion brought the source rocks to a shallower depth. The second cycle took place from the Triassic until the Cretaceous. Triassic sedimentation followed the Late Paleozoic uplift and peneplanation with fluvial sandstones and playa-lake mudstones interbedded with thin lacustrine and shallow marine shales and dolomites. Triassic deposition was accompanied by extensional faulting and basic volcanism. Post-rift halite of Late Triassic and Early Jurassic age provide a thick effective seal for the northern Saharan basins (Galeazzi et al., 2010). To the west of the Ghadames/Berkine Basin, the Late Carboniferous-Permian Hercynian uplift in the Dahar and Oued el-Mya basins caused erosion of about 2000–3000 m (6562–9843 ft) of Ordovician to Lower Carboniferous strata (Makhous & Galushkin, 2003). In Triassic-Cretaceous times, maximum subsidence occurred in the northern areas.

The same petroleum systems that operated in West Libya and Southern Tunisia also exist in Algeria. Other petroleum systems that generated most of the hydrocarbons in the Paleozoic have shown limited production potential and prove difficult to explore and exploit elsewhere in the Saharan Platform (Boote et al., 1998). The Silurian Oued Ali Shales and the Upper Devonian Meden Yaha Formation (F4) in the Ghadames/Berkine Basin (Mofffat & Johns, 2001), Reggane, and Ahnet basins in Algeria, equivalent to the Tannezouft and Cues marls in West Libya, respectively, are the major source rocks of hydrocarbons. Both were associated with eustatic highs and global anoxic events (Lüning, 2000, Lüning et al., 2000, Tawadros, 2001, Lüning et al., 2005, Fello et al., 2006, Galeazzi et al., 2010). Devonian source rocks are believed to be absent in the Zemlet Field and the Hassi Messaoud Field (Peters & Creaney, 2004). The oils originated from thermally mature Silurian source rocks and subsequently migrated updip to the fractured Cambro-Ordovician quartzite reservoirs. However, in the Assekaifaf, Oued Zenani, Zarzaitine, and Dome fields to the south, both Silurian and Devonian source rocks exist, but input was mainly from Devonian source rock (Peters & Creaney, 2004).

Silurian and Frasnian source rocks contain a type I-II oil-prone kerogen, and show TOC values of 2–14% for the Frasnian and 2–17% for the Silurian (Cochran & Peterson, 2001, Moffat & Johns, 2001). The Upper Devonian F4 source has up to 100 m (328 ft) (Moffat & Johns, 2001). The main Silurian kitchen is located in center of the Berkine (Ghadames) Basin and the southwestern portion of the Illizi basin, with a less mature area around the Ahara High. They are characterized by high organic carbon content and consist of predominantly amorphous and lipid-rich organic matter that was preserved under anoxic depositional conditions (Yahi et al., 2001). Bulk geochemical properties and biomarker correlations show insignificant differences between the two source rocks and the oils. Maturity data indicate that the Frasnian source rock is at the oil generation stage or the late oil/gas generation stage depending upon position within the basin (Yahi et al., 2001). The Llandoverian-Wenlockian source rock, on the other hand, is over mature and is in the gas generation stage (Yahi et al., 2001).

In the southwestern Berkine Basin the main trapping mechanism for the Triassic reservoirs is extensional faults with offsets which range from tens to hundreds of meters (Moffat & Johns, 2001). These faults appear to have originated from a Late Triassic-Jurassic reactivation of earlier faults. The trapping mechanism for Strunian reservoirs appears to be a result of a combination of structural and stratigraphic elements. The primary reservoir is the Triassic Argilo-Grèseux Inferieur (TAGI), described as fluvial to alluvial and eolian in origin. TAGI sandstones typically have porosities ranging from 13% to 18% and permeability from 40 to 90 mD, with thickness of 50–100 m (164–328 ft). Secondary reservoirs exist in the Strunian F1 and F2 transitional marine and Carboniferous Rhourde El Krouf (RKF) fluvial-deltaic sandstones. The F1 and F2 reservoir sandstones are relatively thin, about 10 m (33 ft), with porosity of 18–25% and permeability of more than 200 mD. The Rhourde El Krouf Field has approximately 500 MMB of oil in place. Hydrocarbon migration for the Strunian reservoirs is believed to be initially vertical from the F4 and then laterally towards the western up-dip pinch-out or truncation of the reservoirs.

The Lower Silurian Tannezouft source rock underwent two main phases of hydrocarbon generation (Underdown & Redfern, 2008): The first phase occurred during the Carboniferous, and the second started during the Cretaceous, generating most hydrocarbons in the East Libya. The Frasnian shales underwent an initial minor generative phase in the central depression during the Carboniferous; followed by the main generation during the Late Jurassic-Cenozoic (Underdown & Redfern, 2008). The Frasnian shales are currently only marginally mature in the eastern part of the Ghadames Basin (Underdown & Redfern, 2008).

Hercynian erosion could be partly responsible for the maturation of source rocks in the Saharan basins; on the other hand, Triassic and later volcanic activity and associated hydrothermal heat may explain the variations in the maturation profile (Tissot et al., 1973, Makhous & Galushkin, 2003). The observed maturity profiles in the Ghadames/Illizi Basin reflect the Late Cretaceous-Tertiary heat flow patterns. The Berkine Basin can be divided into a region of low-heat flow in the central depression and a region of higher heat flow near the anticlines, such as the El Biod-Hassi Messaoud Ridge. Peak oil generation for the Frasnian source rock occurred during the Late Cretaceous-Tertiary (Yahi et al., 2001). Because of the high maturity, it becomes difficult to distinguish between the various types of oil derived from different source rocks (Tissot et al., 1973).

Silurian source rocks also provided hydrocarbons to the Triassic TAGI reservoirs several fields, such as the El Borma Field, El Merk, Hassi Berkine, and Oughroud fields.

Triassic sands in Algeria, Tunisia and Libya are one of Africa's most significant plays, with an estimated 31 BBOE of reserves, which constitute about 14% of Africa's petroleum reserves (Boote et al., 1998). Over half of these gas reserves are in the giant Hassi R'Mel Field (Purdy & Macgregor, 2003). The majority of reserves lie in fluvial sandstones below the Triassic salt. Major Triassic reserves occur in the Southeastern Sirte Basin which originated from Cretaceous source rocks (Burwood et al., 2003).

The Triassic sediments form oil and gas reservoirs in the Berkine Basin (SE of Algeria) (Figs. 19.12, 19.17). In the Illizi Basin, the Triassic succession consists essentially of fluvial deposits becoming paralic and evaporitic northward close to the Paleo-Tethys margin (Busson, 1971a, 1974, Hamel, 1988, Aït Salem et al, 1998, Boote et al., 1998, Eschard et al, 1999a, 1999b, Turner et al, 2001, Sabaou et al., 2005, Bourquin et al., 2010). Late Triassic crops out at Zarzaitine in the Illizi Basin. Oil and gas were discovered in Triassic reservoirs in the late 1950s and early 1960s at the Hamra Field and Gassi Toui Field in Algeria (Ali, 1975). In 1994, the Anadarko group discovered Ourhoud Field (formerly known as Qoubba) (Fig. 19.16), with estimated reserves of more than 1.1 billion barrels of oil (OGJ, 1999). Giant Triassic fields include the Hassi

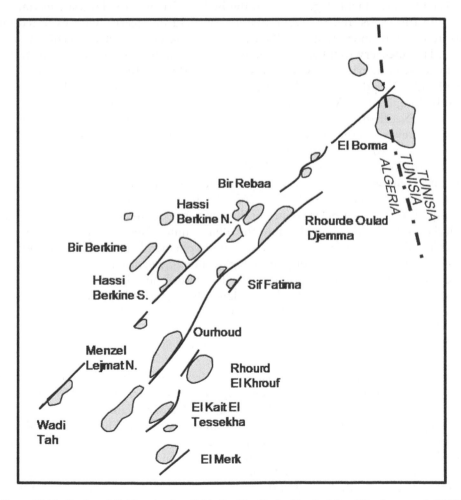

*Figure 19.16* Ourhoud Field and other fields, Berkine Basin, Algeria (after Needham et al., 2008).

R'Mel Field in the in Algeria and El Borma Field in Tunisia. These constitute about 14% of Africa's petroleum reserves (Boote et al., 1998). The majority of reserves lies in fluvial sands below Triassic salt sourced from (Silurian) source rocks. There are also major reserves at this level in the Southeastern Sirte Basin, sourced stratigraphically downward from Cretaceous source rocks (Burwood et al. 2003).

The Hassi R'Mel gas field is located about 550 km South from Algiers. It lies on top of the Talemzane Arch and flanked by the Benoud Trough to the north and the Oued Mya Basin to the southeast (Magloire, 1970). Hassi R'Mel Field was discovered in 1956 by the drilling of HR1 and came on production in 1961 (Bencherif, 2003). The Hassi R'Mel Field extends for about 80 km in a north-south direction and is about 60 km wide (Courel et al., 2000). The reservoir consists of three basal Triassic sands, A, B and C of the Trias Argilo Grèseux (TAG), overlain by Upper Triassic and Liassic evaporites (Hamel, 1990). Porosity is up to 22% and permeability may reach several Darcies (Magloire 1970, Boote et al., 1998), but reservoir quality deteriorates toward the North-East in wells Lg1, Ph1 and Pg1 (Baouche et al., 2003).

The Ourhoud Field (Fig. 19.16) in the Berkine Basin is the largest Algerian discovery in the 1990s and the second largest oil field in Algeria, with expected ultimate recovery of more than 1 billion barrels (Lunn et al., 2001, Needham et al., 2008). The discovery well BKE-1 encountered a 284 m (932 ft) (oil column in Triassic Argilo-Grèseux Inferieur (TAGI) sandstones and flowed 40° API oil at an average rate of 15,037 bopd (Lunn et al., 2001). The structure of the Ourhoud Field is dominated by long NNE-SSW-trending normal faults of relatively low throw on the eastern flank (Needham et al., 2008). The structure formed by late-Triassic divergent wrenching related to the break-up of Pangaea. The sandstones are fluvial, eolian, and deltaic, associated shales deposited in lacustrine and desiccated Chott basins. The Triassic Argilo-Gréseux Inferieur (TAGI Formation) reservoir of the Berkine Basin shows different oil-water contacts, suggesting fault compartmentalization. The TAGI Formation in the field is 80–120 m (262–394 ft) in thickness and has an average porosity of 16% and a permeability of 200 mD (Needham et al., 2008). The TAGI in Ourhoud has been subdivided into three basic units (Needham et al., 2008): The Lower TAGI is characterized by a succession of ephemeral channels, sheet flood and overbank mudrocks, which display a highly heterogeneous reservoir architecture and correspondingly complex connectivity. The Middle and lower Upper TAGI units are composed primarily of vertically and laterally amalgamated low-sinuosity channels with subordinate crevasse-splay and overbank deposits (Needham et al., 2008). The Silurian and Devonian shales are the principal source rocks (Lunn et al., 2001, Needham et al., 2008). Triassic-Jurassic salts provide a regional seal (Lunn et al., 2001, Needham et al., 2008).

In 1992, Cepsa discovered the Rhourde El Krouf Field which tested oil at rates in excess of 2500 bopd from Carboniferous sandstone reservoirs, followed by the discoveries of the El Merk, and Hassi Berkine fields by Anadarko and partners and of the giant Oughroud Field by Cepsa and partners, all of which tested high quality oils from Triassic sandstone reservoirs (Moffat & Johns, 2001). In 1995, the MLE-1 well drilled by Burlington Resources and Talisman Energy flowed oil from the Triassic and Strunian reservoirs. In 1996, the MLN-1 well tested 15,850 bopd from the Triassic and in 1997, the MLN-4 tested a total of 22,700 bopd from the Strunian reservoir. The MLSE-3 tested a rate of 21,857 bopd from the Triassic, Carboniferous and Strunian reservoirs.

More recently, a number of Devonian discoveries were made by Sonatrach and partners BG and Gulf Keystone on the Hassi Ba Hamou Permit, awarded in the 6th International Bid Round on April 2005. The discovery well HBH-1 spudded in 2007 tested Devonian reservoirs at 10.8 mmcfgd through an 88/64 inch choke[3]. Another discovery was made with the well RM-1[4].

The Ahnet Basin is a Variscan anticlinal structure (Klitzsch, 1970, Beuf et al., 1971, Beekman et al., 2000). The basin includes five stratigraphic units of Paleozoic age (Beekman et al., 2000), which overlie the Precambrian basement and are covered in turn by Mesozoic carbonates. The Paleozoic units comprise, from bottom to top; Cambro-Ordovician sandstones, Silurian shales, Lower and Middle Devonian sandstones and shales with a carbonate unit at the top, Upper Devonian Frasnian sandstones, and Famennian sandstones and shales. The source rocks are mainly the Silurian "Hot Shale" and Frasnian organic-rich shales and limestones (Tissot et al., 1973, Lüning et al., 2004). Cambro-Ordovician sandstones are the main reservoir rocks (Macgregor, 1998, Arthur et al., 2003).

The Reggane Basin is situated to the south of the Ougarta Ridge. It contains more than 6000 m (19,686 ft) of sediments in the central part of the basin. Dolerite sills lie upon the upper Paleozoic series.

The Melrhir Basin, located north of the Talemzane-Gafsa Arch in Northeastern Algeria, is a foreland trough that resulted from the inversion and tilting of the Triassic Basin (Boote et al., 1998, Klett, 2002). Potential reservoirs include clastic sediments of Cambrian, Ordovician, Carboniferous, and Triassic age in the southern part of the Melrhir Basin (Madi & Damte, 2002). Triassic clastic units on the northern flank of the Talemzane Arch are expected to be of fluvial-estuarine origin as they are in the Oued Mya and Berkine basins. Jurassic reefs outcropping in the central part of the Saharan Atlas have been seismically mapped in the subsurface, and form an east-west trend across the basin. In the north, several carbonate reservoirs, including rudist-bearing buildups of Cretaceous age, produce or have oil shows. Source rocks in the northern part of the Melrhir Basin are the Cenomanian-Turonian Bahloul Formation, with TOC values of more than 14% and the Cretaceous section with TOC values of more than 2% (Madi & Damte, 2002). The principal source rock in the Melrhir Basin is the Silurian Tannezouft Formation, with TOC content of 1–5% (Cunningham, 1988, Hammill & Robinson, 1992). Oil may have been generated as early as Carboniferous or Permian time. Reservoir rocks are predominately fluvial to marine sandstones of the Ordovician Hamra Formation and fluvial sandstone of the Triassic Kirchaou Formation, the lateral equivalent of the Trias Argilo-Grèseux (TAGI). Triassic to Jurassic evaporites, shales, and carbonate rocks provide a regional top seal, and the Tannezouft shales. Accumulations occur within tilted fault blocks and stratigraphic trap (Hammill & Robinson, 1992, Rigo, 1996, Klett, 2002).

## 19.2.4  Tunisia's hydrocarbon systems

Tunisia is divided into five structural provinces: Southern Tunisia covered by the Sahara Platform, Western Tunisia (part of the Atlas Domain), Central Tunisia,

---

3  www.scandoil.com.
4  www.bgara.blacksunplc.com.

Eastern Tunisia which include the Pelagian Sea, and Northern Tunisia (including the Atlas and Tell domains).

Oil seeps were recognized in Triassic and Cretaceous outcrops since the 1930s, especially in faulted folds and diapirs in Northern Tunisia (Tinthonian, 1939). A few boreholes were drilled in the mid-1930s to confirm the presence of hydrocarbons in the subsurface.

Tunisia had recoverable reserves of 700 mmbbls of oil (Ben Ferjani et al., 1990, Boote et al., 1998) and produced about 55,000 bopd in 1990 (Ben Ferjani et al., 1990), but decreased to 28,000 bopd by 1998 (Michalski, 1998). About 70% of Tunisia's production comes from six fields; El Borma, Ashtart, Oued Zar, Adam, and Miskar (Petroleum News, May, 2011)[5]. The largest and principal oil field in Tunisia is the Triassic El Borma Field discovered in the Ghadames Basin by Agip (Eni) in 1964 near the Tunisian-Algerian border. In addition, there are many fields in the Mediterranean offshore.

Oil and gas in the Tunisian Ghadames Basin originated from two main source rocks; the Middle-Upper Devonian marls and basal Silurian hot shales, with average TOC values of 3% and 5% of type II kerogen, respectively. In general, maturity decreases from the southern part of the basin to the northern part (Ferjaoui et al., 2001).

Four oil and gas condensate fields, including Sabri and E1 Franig, have been discovered in Southern Tunisia, on the northern flanks of the Talemzane Arch (Fig. 19.17) since the late 1980s (Cunningham 1988, Rigo 1996). One is in basal Triassic sands and the others are in fractured Ordovician sandstones. All four accumulations were sourced from the Tannezouft shales (Boote et al., 1998). Four hydrocarbon fields have been discovered since 2006 in the Jenein Sud Permit in Southern Tunisia. They are the Warda, Nawara, Ahlem, and Sourour. The recent discoveries of oil in the Acacus Formation in Southern Tunisia by a number of oil companies attest to the good potential of the Silurian reservoir sandstones. Synrift Permian clastics and reefal carbonates were deposited north of the Talemzane-Djeffara Arch (Rigo 1995, Rigby et al. 1979, Boote et al., 1998). These rocks include Lower Permian pelagic limestones and mudstones and Upper Permian bioherms, carbonates, and clastic rocks (Rigo, 1995). The Upper Permian Bir Jaja mudstones serve as a seal (Boote et al., 1998). Triassic reservoirs include sandstones of the Kirchaou Formation and its lateral equivalent the Trias Argilo-Grèseux (TAG) which produce from the El Borma Field and represent the most important reservoir in the Triassic Basin of Northeast Algeria and Southern Tunisia (Boote et al., 1998). Triassic and Jurassic evaporites, mudstones, and carbonates provide the seal rocks.

The Pelagian Province contains more than 2,300 million barrels of oil and approximately 17.2 TCF of gas (Klett, 2001). Two petroleum systems have been identified in the Pelagian Basin; the Jurassic-Cretaceous systems and the Bou Dabbous-Eocene (Ahlbrandt, 2001, Klett, 2001). Most of the reserves occur in the Bou Dabbous-Tertiary Petroleum System.

In the Jurassic-Cretaceous System, the primary source rocks are mudstones of the Jurassic Nara Formation, the Lower Cretaceous M'Cherga Formation, the Albian Lower Fahdene Formation, and the Cenomanian-Turonian Bahloul Formation (Klett, 2001). The Nara Formation source rocks are black mudstones interbedded with limestones, approximately 80 m (262 ft) thick, contain up to 2% TOC, and are mature

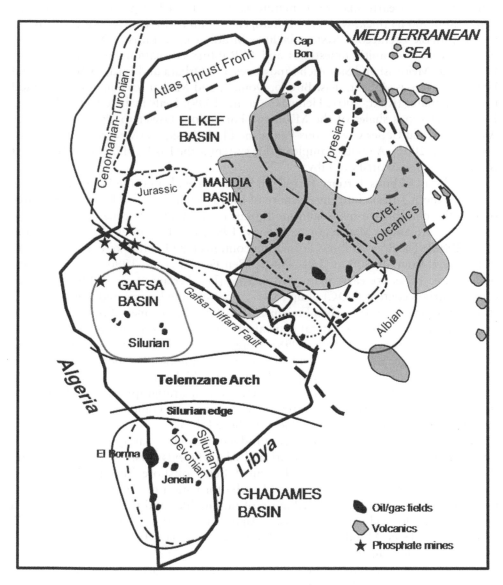

*Figure 19.17* Oil source rocks in Tunisia (based on Lüning et al. (2004). Large shaded area represents area where Cretaceous volcanics were reported (after Mattoussi et al., 2009). Other volcanic occurrences are from Klett (2002). Basin names after Jarvis et al., 2002).

to late mature (Ben Ferjani et al., 1990). The M'Cherga Formation source rocks are up to 100 m (328 ft) and consist of light- to dark-gray calcareous and dolomitic mudstones and contain 0.2–3% TOC and type II kerogen, and they are mature to late mature. The Lower Fahdene Formation source rocks are dark pelagic marls with interbedded limestones, up to 150 m (492 ft) thick, containing type II and III kerogen and 0.5–3% TOC, and they are early mature to mature. The Bahloul Formation source rocks are early mature to mature laminated black argillaceous limestones, about 20 m (66 ft) thick, with type II kerogen and up to 14% TOC. Reservoir rocks in the Jurassic-Cretaceous System include Middle-Upper Jurassic Nara Formation dolomites and dolomitic limestones and M'Rabtine Formation sandstones, Lower Cretaceous Meloussi and Sidi Aïch sandstones and Orbata and Serdj carbonates, and Upper Cretaceous Zebbag, Isis, Bireno, Douleb, and Miskar carbonates, and Abiod fractured chalks (Klett, 2001, Hennebert et al., 2009). Seals include Upper Jurassic-Lower Cretaceous shales of the M'Cherga Formation, Lower to Upper Cretaceous Aleg shales, and Upper Cretaceous-Paleocene El Haria shales (Macgregor & Moody, 1998, Klett, 2001). A good example of this system is the Isis Field offshore Tunisia. It is not yet proven in offshore Libya.

The Bou Dabbous-Tertiary Petroleum System extends into the Pelagian Province from Northern Tunisia into the offshore of Libya (Klett, 2001). The primary source rock is the dark-brown marls and mudstones of the Lower Eocene (Ypresian) Bou Dabbous Formation (Macgregor & Moody, 1998, Ben Ferjani et al., 1990, Vennin et al., 2003). The Bou Dabbous Formation contains type I and II kerogen and ranges in thickness from 50–300 m (164–984 ft) (Ben Ferjani et al., 1990). The Bou Dabbous source rocks are early mature to mature (Ben Ferjani et al., 1990) with total organic carbon content (TOC) of up to 4%. Seals include Eocene and Miocene mudstones and carbonate rocks (Ben Ferjani et al., 1990). Mudstones of the Cherahil and Souar formations and their lateral equivalents provide seals for many of the Eocene reservoirs (Macgregor & Moody, 1998). Traps are horst blocks and folds (Bédir et al., 1992). Most of the oil and gas fields that belong to the Bou Dabbous-Tertiary Total Petroleum System occur in low-amplitude anticlines, high-amplitude anticlines associated with reverse faults, wrench fault structures, and stratigraphic traps. Accumulations in combination structural-stratigraphic traps are also common (Klett, 2001). An example of this system is the Bouri Field offshore Libya.

In Tunisia and Libya, oil and gas production comes mainly from nummulitic limestone reservoirs, such as the Bouri and Al Jurf fields offshore NW Libya, and the Sidi el Itayem, Ashtart, and Hasdrubal fields offshore Tunisia (Fig. 19.18). Numerous models have been proposed for the formation of nummulitic accumulations (Tawadros, 2001). The distribution of nummulites and their size is controlled mainly by the hydrodynamic processes (Aigner, 1982, 1983, 1984, Moody, 1987, Moody et al., 1998, Moody, 1998a, b, Moody & Sandman, 1998, Braithwaite et al., 1998, Reali & Ronchi, 1998, Reali et al., 2003, Loucks et al., 1998a, b). The distinction between the various depositional modes is essential for determining the geometry, the spatial distribution, the internal heterogeneities, and the petrophysical properties of the nummulites reservoirs (Jorry & Davaud, 2004).

The porosity in the El Garia Formation of offshore Tunisia is up to 35% (Loucks et al., 1998). Intra-skeletal porosity of up to 40% in large Nummulites has been observed in some outcrops in Tunisia (Jorry et al., 2003). However, intra-particle porosity is generally isolated and ineffective, and permeability is relatively low, although it may reach one Darcy in places (Loucks et al., 1998). Reservoir quality is also controlled by diagenesis which led to the formation of dissolution porosity and dolomitization (Beavington-Penney et al., 2008). Dissolution porosity within the matrix of the nummulitic wackestones and packstones has typically been attributed to early meteoric processes in the El Garia and Jdeir Formation reservoirs in Tunisia and

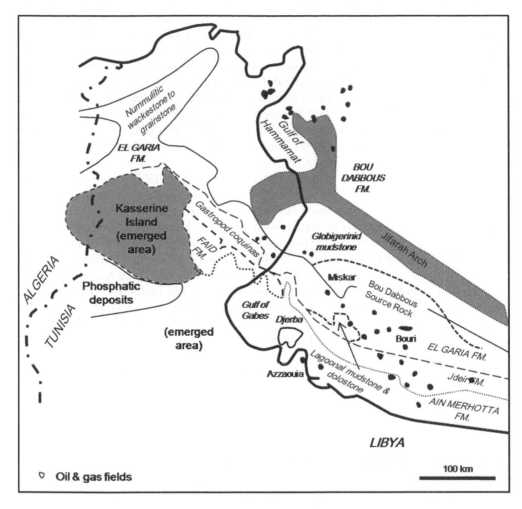

*Figure 19.18* Early Eocene facies and petroleum system, Pelagian Sea (after Bishop, 1988, Beavington-Penney et al., 2008, and Jamoussi et al., 2008).

Libya, respectively (Bernasconi et al., 1991, Bishop, 1988, Anketell & Mriheel, 2000, Reali et al., 2003). On the other hand, Racey et al. (2001) suggested that dissolution porosity in nummulites reservoirs was caused by acidic waters moving in advance of migrating hydrocarbons.

Other reservoirs include the Cretaceous Abiod Chalk and the Miocene Birsa Formation. Marathon discovered in 1986 the 20 mmbbls Ezzaouia Field near Djerba, and AGIP SpA discovered the Maamoura Oilfield in the Gulf of Gabes in 1988 with reserves of about 50 mmbbls of oil. The main reservoir is the Campanian-Maastrichtian Abiod Chalk (Davies & Bel Haiza, 1990). The Maamoura discovery was followed by onshore discoveries of Kupec's Sidi el Kilani-1 well, which tested a rate of 2600 bopd, Shell's Zinnia-1 at 1300 bcpd with gas, Tiref-1, and Miskar-6 with flow of more than 50 mmcfgd. The Abiod consists of three members; the lower member is 70 ft

(21 m) thick, the argillaceous middle is 100 ft (30 m) thick, and the upper carbonate member which is the most important economically, has a porosity of more than 20% and permeability of 1–3 Darcies. It is a massive chalk with slumps and turbidite fan deposits; it can exceed 600 ft (183 m) in thickness. El Haria Shale forms an excellent seal. The source rocks are Bahloul Formation (late Cenomanian-early Turonian) with TOC values of up to 8% or the Fahdene Formation (Albian-Cenomanian). The Mouelha Formation (Upper Albian) consists of dark grey to black shales up to 300 m (914 ft) thick and thin beds of black limestones, with Upper Albian microfauna such as *Rotalia breggiensis* and *Ticinella primula*. It has TOC values of 0.65–3.6%, but may reach up to 14% in places. At Shell's Zinnia-1 well, the Abiod flowed at 50 mmcfgd and 1300 bcd (Davies & Bel Haiza, 1990).

The Cap Bon Peninsula in Northeastern Tunisia (Fig. 19.17) is formed by the large Sidi Abderahman Anticline with thick sequences of shales and two main carbonate reservoir sequences of the Abiod Formation and the Metlaoui Formation (Group) (Ypresian). The Birsa Field is located in water depth is 140 m (459 ft) in the Cap Bon Marine permit, which was granted to Shell in 1973. Agip acquired 50% of the concession in 1975. The field is an anticline divided into two main compartments by SW-NE faults (Portolano et al., 2000). The Birsa Formation contains three Middle Miocene (Serravallian) sandstone reservoirs, separated by shales. The sands were deposited in a shoreface to foreshore marine environment. The Birsa Formation is overlain by shales of the Saouaf Formation (Serravallian-Tortonian). Another reservoir exists in the Ain Grab Limestone Formation (Langhian), which is separated from the Birsa Formation by the Mahmoud Formation (Upper Langhian). In 1976, the discovery well BRS-1 tested 480 bopd of 31°API oil from the Birsa Sands with GOR ratio of about 300 scf/bbl. Consequently, the well BRS-2 tested 2489 bopd and 8.95 mcfgd and the BRS-3 encountered oil in the upper and Middle Sands at rates of 4560 and 589 bopd, respectively. However, the BRS-4, 5 & 6 wells were dry.

## 19.2.5  Morocco's hydrocarbon systems

According to the U.S. Energy Information Administration (2006), Morocco has proven oil reserves of about 1 million barrels and natural gas reserves of about 60 bcf. The country produced small volumes of natural gas and oil from the Essaouira Basin and small amounts of gas from the Gharb (Rharb) Basin (Pratsch, 1996) (Fig. 19.19) and was North Africa's leading importer of oil and natural gas. In addition, Morocco has large oil shale deposits, estimated at 50 billion barrels, which are subject to extensive exploration and processing research by the *Office National de Recherches et d'Exploitations Pétrolières* (ONAREP). The Moroccan oil shale deposits include the Guir, Ganntour, Mescala, Oulad Abdoun, Tanger, Tarfaya, and Timahdit. The most important deposits were discovered during the 1960s at Timahdit in the Middle Atlas and Tarfaya along the Atlantic Coast, with estimated oil in place of 16 and 23 billion barrels, respectively (World Energy Council, 2007).

Oil exploration in Morocco started in the 1920s and remained active until the 1970s. In 1923, the Ain Hamra pool was discovered. Three other fields were discovered during that period, the Haricha Field, the Tselfat Field, and the Bou Draa Field. Each field produced oil or gas from the 1920s to the 1970s with limited production until the 1990s. In 1928, the BRPM (*Bureau de Recherches et d'Exploitations Minières*) was

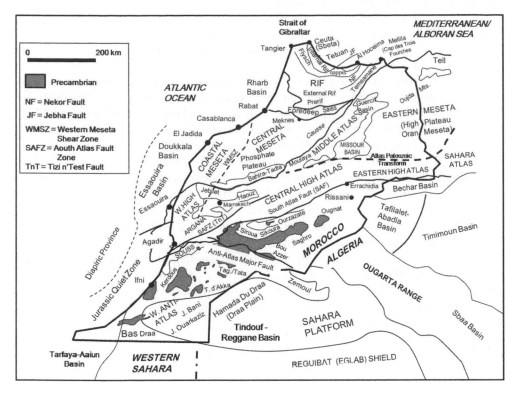

Figure 19.19 Tectonic map of Morocco.

founded, followed by the creation of the SCP (*Societé Cherifienne de Pétrole*) in 1929. Seismic reflection techniques were introduced in 1935 in the Gharb, Pre-Rif, Essaouira, Souss, and Guercif basins. In 1949, BRP (Bureau des Recherches Pétrolières) announced the discovery of the Mellaha Wadi Field in Morocco and at Cap Bon in Tunisia. The Wells are now either shut-in or abandoned. Exploration between 1958 and 1981 led to the discovery of oil in the Essaouira and Gharb Basins. Oulad Youssef and Oulad Bendich fields were discovered in 1980 by a consortium of BRPM/Elf Aquitaine/PCS/KPC (Kuwait Petroleum Corporation) in the Gharb Basin. In 1981, ONAREP (Office National de Recherches et d'Exploitations Pétrolières) was established, and ONHYM was established in 2003 through the merger of ONAREP and BRPM. The Hydrocarbon Law was amended in 1992 and 2000. In 2000, Lone Star Energy announced the discovery of what was supposed to be a major discovery in recent years. The controversial Talsint Field is located about 500 miles SE of Rabat. The field was assumed to have oil reserves between 50–100 million barrels, based on one well[6].

According to Onhym, source rocks in Morocco include Triassic (2.5% TOC), Jurassic (10% TOC), Cretaceous (Aptian-Cenomanian, 20% TOC), Neogene (7% TOC). Pliensbachian-earliest Toarcian marls and argillaceous limestones exposed at

6   Source: www.onhym.com.

the surface near Ait Moussa (Boulemane Province, Middle Atlas) include the only examples of effective petroleum source rocks so far known in the Moroccan Atlas rift basins (Rachidi et al., 2009). The outcrop interval includes hemipelagic, peri-Tethyan low latitude source rocks with Type II kerogen (total thickness of 2.5 m) with mean TOC of around 3.2%.

The Zag Basin occupies parts of Algeria, Morocco and Mauritania. It remains one of the largest prospective and sparsely explored frontier basins in North Africa (Gillhespy & Exton, 2008). Exploration drilling took place between 1959 and 1971, but drilling density is extremely low with an average of only one well in 13,000 km². The Zag Basin is an intra-cratonic basin developed on the West African Craton similar to the Reggane and Taoudeni basins. Sediments in the basin range from Infra-Cambrian to Late Paleozoic overlain by thin Cretaceous sediments. Over 60% of the wells drilled reported oil or gas shows, and several recovered hydrocarbons from open-hole tests (Gillhespy & Exton, 2007). Silurian and Late Devonian shales which crop out in the northern part of the basin have proven source rock potential. Ordovician, Devonian, and the Carboniferous sediments form potential reservoirs and are capped by extensive shale seals (Gillhespy & Exton, 2008); potential traps are low-amplitude folds. There are three petroleum systems in the Zag Basin (Thompson & Smith, 2009); 1) Upper Ordovician sandstone reservoirs-Middle Ordovician source rocks, 2) Silurian carbonate reservoirs-Silurian shale source, and 3) Upper Devonian-Lower Carboniferous reservoirs-Frasnian source rocks.

The High Atlas Mountains of Morocco are an intracontinental mountain belt formed by the Cenozoic inversion of a pre-existing Mesozoic rift system (Beauchamps et al., 1997). The Atlas rift system contains rich organic source rocks, thick evaporites and volcanic rocks providing seals, and clastic submarine fans and/or carbonate buildups forming reservoir rocks. The Atlas Mountain Belt provides complex and prospective exploration targets with similar potential to the underlying rift basins.

The Gharb (Rharb), Saiss, and Guercif basins constitute the main foreland basins of the Rif folded belt (Pratsch, 1996) (Fig. 19.19). A number of commercial gas discoveries have been made in Neogene sandstones. The Blue Marls (or Tortonian Marls) acted as both source and seal for hydrocarbons reservoir in the sandstones (Bernini et al., 1999). In the Gharb Basin (which was in continuity with the Taza-Guercif Basin during the Tortonian in the so-called South Rifian Corridor) (Morabet et al., 1998, Bernini et al., 1999). The sedimentary succession consists of about 6000 m (1829 ft) of Mesozoic and Cenozoic sediments, covered by a complex of nappes which are overlain in turn by Upper Miocene clastic sediments (Dakki et al., 2007). Potential Jurassic carbonates reservoirs also exist, but no commercial oil has been found to date (Bernini et al., 1999).

Petroleum exploration started as early as 1910 around surface oil seeps which led to the discovery of small oil and gas fields at Ain Hamra, along Sidi Fili Fault and in the Prerif Ridges of Haricha, Tselfat, and Bou Draa. Exploration activities between the early 1960s and the early 1970s resulted in the discovery of biogenic gas fields in the Gharb Basin. Mature source rocks in the Prerif area include Lower Jurassic marls and shales (Domerian-Toarcian) and Upper Cretaceous shales which are considered the main source rocks in the area. Lower Jurassic shales show TOC values of up to 2%. Upper Cretaceous formations in outcrops have TOC values of up to 15%. Maturation of Lower Jurassic source rocks may have started as early as Jurassic-Lower

Cretaceous in the deeper troughs probably after the emplacement of nappes during the Mio-Pliocene time. Porosities in the Miocene sands reach up to 35% (Dakki et al., 2007). A discovery was made by Dana Petroleum in 2009 south of Tanger in the Tanger-Larache Licence about 40 km from the Atlantic Coast of Morocco. The discovery well Anchois-1 was drilled to a depth of 2435 m (7989 ft) and encountered two sand intervals with net pays of 90 m (295 ft) and 40 m (131 ft) (Dana Petroleum). However, no information is available about the reservoir units. The Serica KSR-9 tested gas in onshore Sabou Licence in 2009. Circle Oil made another gas discovery on its Sabou Permit with the Add-1 well in the Rharb Basin in the main Hoot and Guebbas targets (World Oil, June 26, 2011). The KSR-1 exploration well tested gas at 10.6 mmcfgd on a 26/64 in. choke from the Main Hoot horizon at 1736.6–1728.2 m and at 2.4 mmcfgd in the interval 1650.5–1649 m, with a net pay of 1.6 m (5 ft). In the meantime, TransAltlantic Petroleum Ltd. started producing gas in January 2011 from the HR-33 well on its Tselfat Permit in the Gharb Basin.

Source rock sedimentation in the Southern Rif Jurassic Basin was controlled by tectonic, paleogeographic and physico-chemical factors, which led to the establishment of two source rock areas (Assaoud et al., 2007). The first area is located in the western part of the basin, dominated by terrigenous minerals, lignites, and organic carbon content not exceeding 1%, and low values of oil potential and S2/S3 ratio. However, the index of production (IP) is high and indicates the presence of important accumulation of hydrocarbons. The second area is located in the eastern part of the basin, characterized by fine sedimentary and shale deposits, with TOC content of more than 2.5% and high S2/S3 ratios and a hydrogen index up to 716 mg HC/G Org. Carb. Deposition in the eastern part of the Southern Rif probably took place during the Toarcian anoxic event.

The Central Atlantic Margin includes the Essaouira, Doukkala, Tarfaya, and Aaiun basins (Fig. 19.20). Several hydrocarbon discoveries have been made in the Central Atlantic margin at Cap Juby, Chinguetti, Banda Pelican, Tiof (Mauritania), Dome Flore, and Dome Gea (Senegal) (Davison, 2005). The first major discovery of hydrocarbons in the region off Morocco was made when in 1970 Exxon encountered 20°API heavy oil in the offshore part of the Tarfaya Basin. Gas and condensate were also discovered in the Essaouira Basin and oil in the Sidi Fili trend. The hydrocarbons are trapped in Hercynian structures cut by Triassic normal faults (Zizi et al., 2003). Further hydrocarbon discoveries were made in the Aptian Salt Basin between Nigeria and Angola (Coward et al., 1999).

Most traps are anticlinal, related to salt tectonics (Tari et al., 2003). A number of petroleum systems in these areas are now well documented (Tiesserenc et al., 1989, Katz, 2000, Schoellkopf & Patterson, 2000). Two potential petroleum systems separated by salts are present in the region (Purdy & Macgregor, 2003); a pre-salt system of syn-rift Early Cretaceous source rocks and reservoirs and post-rift carbonates and clastic reservoirs, and a post-salt system that includes Albian-Cenomanian and Turonian source rocks and clastic turbidite reservoirs.

The development of the Central Atlantic margin of NW Africa started with the deposition of continental Triassic red beds during the initial rifting stage in Late Triassic-Early Jurassic (around 225–200 Ma) with thicknesses of up to 5 km (16,400 ft) in the Moroccan onshore basins, such as the Argana Basin (Brown, 1980, Le Roy, 1997, Le Roy et al., 1997, Olsen et al., 2000, Et-Tahoumi et al., 2008), followed by

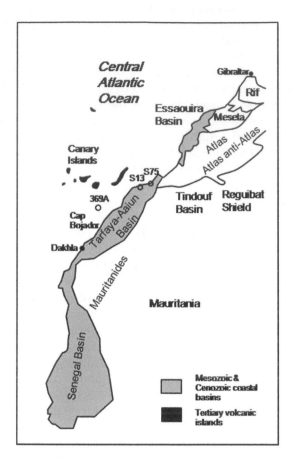

*Figure 19.20* The Moroccan Atlantic margin (after Kolonic et al., 2002, Cool et al., 2008, Cameron, 2007, Davison, 2005, Davison & Dailly, 2010).

Lower Jurassic salts and basaltic magmatism at 200 Ma (CAMP volcanics), oceanic spreading around 180–170 Ma, Jurassic-Early Cretaceous platform carbonates, and Cretaceous-Tertiary marine clastics (Davison, 2005). The presence of thick Triassic-Early Jurassic salts in the Cap Boujdour area of the Aaiun Basin has been confirmed on new seismic sections by Davison & Dailly (2010). The Essaouira Basin (Fig. 19.20) was formed during the Triassic on a Paleozoic platform (Broughton & Trepanier, 1993). Several kilometers of Upper Triassic and Lower Jurassic salts accumulated during the rifting episode. Sediments were deposited during the post-rift phase until the latest Cretaceous when it was uplifted and tilted seaward following the Atlas Orogeny. A gentle fold belt separates the Essaouira Basin from the Doukkala Basin to the north. The basin is divided into two subbasins by the Meskala-Zelten Horst; the Argana Buried Valley to the east and the Neknafa Syncline to the west. These salts formed diapirs, two of which reached the surface. A Cenozoic collapsed graben formed over the diapirs. The removal of salts also led to faulting in the Cretaceous and Cenozoic strata (Davison & Dailly, 2010).

The Essaouira Basin is an important hydrocarbon basin in Western Morocco. There are seven producing or shut-in oil and gas fields in the basin (Broughton & Trepanier, 1993), including the onshore fields of Meskala, Sidi Rhalem, Toukimt, Zelten, N'Dark, Ketchoula, and Jeer. Hydrocarbons in the Essaouira Basin were generated from two sources (Broughton & Trepanier, 1993): Jurassic (Oxfordian) shales, with up to 4.3% TOC of type II/III kerogen, in the Nektafa Syncline, and deep Carboniferous coals. The reservoirs include Jurassic fractured carbonates, associated with structurally closed drapes above deeply buried salt domes (Toukmit, Jeer, N'Dark, and Kechoula fields), and basal Liassic clastics (Zelten Field).

A number of major basins are located south of Essaouira Basin, such as the Ad Dakhla, Laayoune, and Tarfaya Basins (Davison, 2005) (Fig. 19.20). Two DSDP wells (site 369 and 397) were drilled NW of Cape Boujdour. Syn-rift deposits have been penetrated in several wells in the central onshore Aaiun (Laayoune) and Tarfaya basins. They consist of conglomerates, red shales, sandstones, evaporites, and volcanics (Ranke et al., 1982). Lower to Middle Jurassic strata are only present in the northern part near Tarfaya. Jurassic-Early Cretaceous bank carbonates are sporadically developed. Lower Cretaceous continental to marine deltaic sediments were deposited on the shelf (Von Rad & Wissmann, 1982, Ranke et al., 1982), followed by Upper Cretaceous and Paleocene Carbonates. Eocene and Oligocene sediments dominated by continental sandstones and conglomerates; however, the Oligocene is missing in the north due to erosion. The Paleogene succession reaches a maximum thickness of 1000 m (3280 ft) in the onshore part of the Aaiun and Tarfaya basins and Ad Dakhla Basin and decreases seaward to 200 m (656 ft) at the DSDP Site 369 (Ranke et al., 1982). Neogene succession is generally less than 100 m (328 ft) in thickness and composed mainly of sandy limestones and oyster beds.

The Aaiun Basin extends approximately 800 km from Northern Mauritania northward to the Canary Islands along the Atlantic margin of NW Africa. Only five wells have been drilled in the shallow shelf in the offshore of the Aaiun Basin. Salts of up to 2 km in thickness were deposited around the Triassic-Jurassic boundary along this margin. These salts were contemporaneous with volcanic activities of the CAMP at 200 ± 1 Ma around that boundary. Rifting continued after the deposition of the salts which were limited to the half-graben depocenters. However, no halite has been documented in the Aaiun Basin (Davison & Dailly, 2010).

The area contains numerous reservoir and source rocks. There are a few proven and potential source rocks in the Central Atlantic Margin. Triassic source rocks with 2.5% TOC occur in the Doukkala Basin (ONHYM website). Lower Jurassic carbonate source rocks are thought to be present in the area between Ad Dakhla and Pointe Noir (Davison, 2005). The DSDP boreholes suggest the presence of mid-Cretaceous carbonate oil-prone sources associated with the Jurassic shelf edge bank (Cameron et al., 2007). The Cenomanian-Turonian interval contains oil source rocks in Mauritania, but in the Western Sahara it probably has minor organic content (Tissot et al., 1980). However, this may be due to dilution of organic material by clastics in the deltaic areas. Type II/III kerogen has been recognized in the Cretaceous sediments of the Atlantic basins (Tissot et al., 1980). In Site 370 (Fig. 19.20) offshore Morocco, Lower Cretaceous shales and turbidites contain mainly terrestrial organic matter, whereas several Aptian and Albian beds show a predominance of planktonic material (type II). A widespread period of oceanic anoxia and high TOC accumulation in black shales

(Fig. 19.21) occurred during the late Cenomanian-early Turonian (known as OAE2), when sea levels reached their Phanerozoic maximum (Haq et al., 1987, Hallam, 1992). This phase was accompanied by a global positive δ13C and high organic carbon production in ocean basins (Schlanger et al., 1987, Jarvis et al., 2002, 2006, Jenkyns et al., 1988).

The Tarfaya Cenomanian-early Turonian sequences (OAE2) exposed along the Atlantic Coast contain thick black shales rich in organic carbon (Kolonic et al., 2002, 2003, 2005, Kuhnt et al., 1997, 2004) and is one of the most complete worldwide (Keller et al., 2008) (Fig. 19.20). The OAE2 interval spans a series of biotic and oceanic events. Planktonic foraminifera show species extinctions, evolutionary diversification, morphological adaptations, and changes in the relative abundance of species populations (Keller et al., 2008). Cenomanian-Turonian black shales from the Tarfaya Basin (Fig. 19.21) have high up to 18% TOC content of type I-II kerogen, and hydrogen indices between 400 and 800 mgHC/g TOC, which indicate that the black shales are oil-prone source rocks (Kolonic et al., 2002). Low Tmax values obtained from Rock-Eval pyrolysis (404–425 MC) indicate an immature to early mature level of thermal maturation. In the onshore S13 and S75 wells east of Tarfaya, the Late Cenomanian- early Turonian succession shows total organic carbon (TOC) contents between 10% and 20% (Kuhnt et al., 1990, 1997, 2001, 2009).

*Figure 19.21* Black Shales, Mafatma Section, Tarfaya Basin, Morocco (courtesy of Sebastian Lüning) (See color plate section).

The Late Albian-Campanian succession in outcrops and the exploration wells S13 and S75 displays two depositional environments (El Albani et al., 1999); the first is late Albian-early/middle Cenomanian in age and dominated by terrigenous siliciclastic and bioclastic sediments with terrestrial organic matter and a low organic content. The second comprises Late Cenomanian-Campanian sediments characterized by pelagic marls and limestones with a high organic content and abundant pelagic macro- and microfauna. The Albian-Campanian interval in the Deep Sea Drilling Project Site 369A, in the deepwater of the Aaiun-Tarfaya Basin, contains marine and oil-prone organic matter with a TOC content of 0.5–12% (Cool et al., 2008). An unconformity separates the Late Albian/Early Cenomanian and the early Campanian succession.

The lack of good traps and seals may limit the potential of the area. Structures in the shallow shelf area are rare, but some Albian roll-over structures generated by listric faulting may constitute exploration targets (Morabet et al., 1998, Heyman, 1989). The Jurassic carbonate bank forms a possible reservoir. However, no structural closures have been documented at this level (Davison, 2005). Lower Cretaceous sandstones probably form deep-water fans off Cape Boujdour (Ranke et al., 1982). Seals will be a problem in the northern part of the basin as the Lower Cretaceous sequence is sand rich (Davison, 2005).

The Tarfaya Basin is situated southwest of the Anti-Atlas Mountains and extends into the Tindouf Basin to the east and the Senegal Basin to the south and is limited to the west by the East Canary Ridge. The basement is composed of folded Precambrian and Paleozoic rocks, uncomfortably overlain by Mesozoic sediments more than 12 km (39,370 ft) in thickness (Kolonic et al., 2005). The basin hosts one of the largest oil-shale deposits in the world (Amblés et al., 1994) with up to 500 m (1640 ft) in thickness. The Timahdit oil shales in the Middle Atlas about 240 km east of Rabat, occur in an eroded Cretaceous anticline (Leine, 1986, Amblés et al., 1994, ) and are covered by basalt flows (Nummendal et al., 2009). They are estimated to have 22 Billion barrels of oil in place (ONHYM, 2010). The deposits occur in two basins, the El Koubbat Basin to the northwest and the Angueur Syncline to the southeast. The two basins are separated by the Jbel Hayane Anticline. In the Koubbat Basin, the Oil Shales are Maastrichtian in age and are 250 m (820 ft) thick. The Timahdit oil shales contain type I/II kerogen (Amblés et al., 1994).

The Mesozoic Tarfaya Basin extends along the Atlantic Coast of southern Morocco and is part of the western margin of the Sahara platform (El Albani et al., 1999). During the late Cenomanian and early Turonian, deposition occurred in an open shelf that became shallower toward the paleo-shoreline to the northeast (Gebhardt et al., 2004). Sediments consist of dark-light laminated marly shales, marlstones, and bioclastic limestones with foraminifera and calcareous nannofossils and occasional silicified nodules. The limestones have erosional bases and display hummocky cross-stratification.

Two main transgressive cycles occur in the Cenomanian succession, separated by a major regression at the Early-Middle Cenomanian transition in the Tarfaya Basin (Kuhnt et al., 2009). The regressive interval is characterized by lagoonal lowstand deposits, indicating an overall sea-level fall of more than 30 m (98 ft). Superimposed on the two main transgressive cycles, there are 11 third-order depositional sequences that correlate to globally recognized sea level fluctuations.

# PART 6

## Phanerozoic geological history

# Chapter 20

# Phanerozoic geological history

## 20.1 INTRODUCTION

This part of the book deals with the regional stratigraphic correlations and paleogeographic evolution of North Africa set in the context of local and global tectonics.

The geological evolution of North Africa can be discussed in terms of three megasequences spanning three time-stratigraphic units: Paleozoic-Albian, Upper Cretaceous-Upper Eocene, and Oligocene-Recent (Tawadros, 2001, Tawadros et al., 2006). These megasequences are subdivided further into sequences which represent genetically related rock bodies bounded by regional unconformities. Each sequence corresponds approximately to a major time-stratigraphic unit. This subdivision, however, is rather artificial and arbitrary. For example the Precambrian events, such as the Pan-African Orogeny, actually extend into the Lower Cambrian. In addition, the Mesozoic-Tertiary boundary is still controversial in many parts of North Africa. In addition, the stratigraphic nomenclature has changed drastically in the last three decades, especially in Egypt and Libya, which made correlations cumbersome; whereas the majority of the lithostratigraphic units in NW Africa are still unnamed or loosely defined.

## 20.2 PHANEROZOIC

The paleogeographic and tectonic evolution of North Africa have been described in numerous papers (Guiraud et al., 1987, 2000, 2001, 2005, Guiraud & Maurin, 1991, Guiraud & Bosworth 1999, Guiraud, 2001, Piqué, 2001, Piqué et al., 2002, Hoepffner et al., 2005, Hoepffner et al., 2006, Craig et al., 2009).

Since the stratigraphy of different areas was worked out by many workers, and due to the difficulty in assigning accurate ages to the different rock units, some of the stratigraphic interpretations have been modified to maintain consistency in correlations. Other interpretations have been extrapolated to certain areas because of the lack of adequate information. Some of the difficulties encountered in correlating the successions in the marginal areas of North Africa are due to differences in differential subsidence and the amount of sediment supply between different basins. For example, in some areas, the northern margin is abrupt, while in others; there is a transitional zone between the North African Craton and the Mediterranean. These differences produced different depositional cycles and facies. Moreover, the stratigraphy of the Atlas Domain in NW Africa has not been totally resolved. In addition, there is a tendency among the workers in NW Africa to use detailed lithostratigraphic description rather than lithostratigraphic nomenclature.

A series of time-slice paleogeographic maps has also been compiled from published data. These maps were used to deduce the sedimentary and tectonic evolution of North Africa from Cambrian through Quaternary times.

The Phanerozoic sedimentary column in North Africa can be divided into lower, middle and upper megasequences (Fig. 20.1). They were established initially by Tawadros (2001) in Egypt and Libya, but are expanded here to include all of North Africa. Tawadros (2007) called for the formalization of these sequences as lithostratigraphic units based on the fact that they have the same attributes of lithostratigraphic units. These subdivisions and the distribution of sediments and facies throughout the geologic column were controlled by global and local tectonics and sea level fluctuations. Local variations in lithofacies between different areas in North Africa were governed by relative movements and subsidence of the different tectonic plates (Fig. 2.1). The main characteristics of the three megasequences are:

1   The Lower Megasequence includes the post-Pan-African (approximately Middle Cambrian) to Albian sediments composed predominantly of siliciclastics interfingering with carbonate sediments and in places evaporites, especially in the northern parts of Tunisia, Algeria, Libya, and Egypt. The time interval is

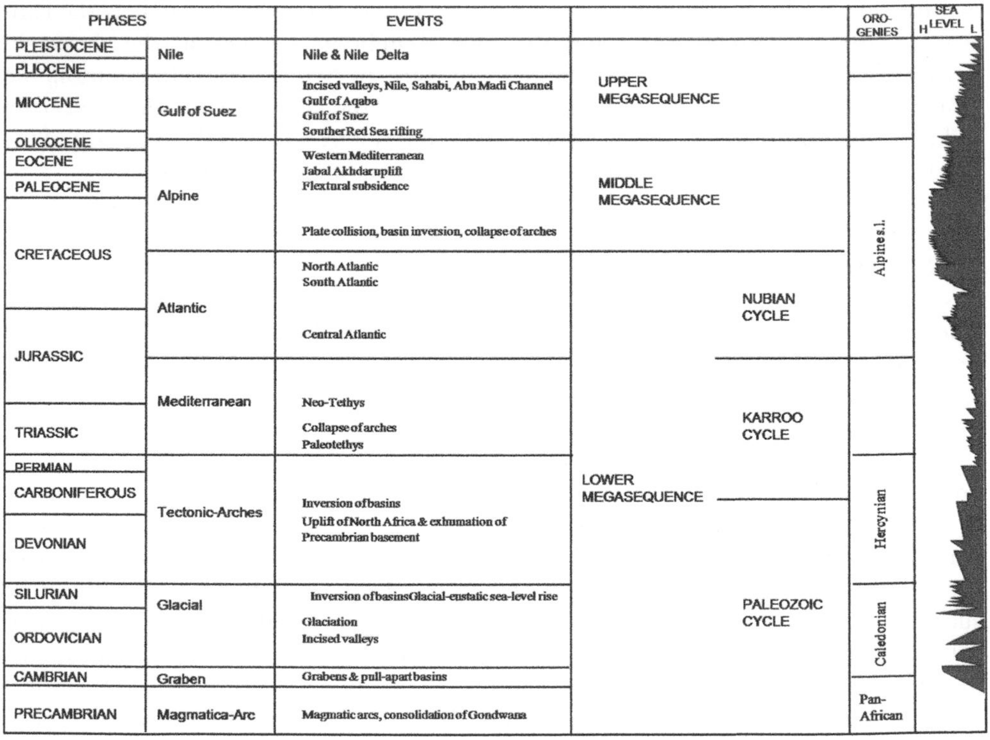

| PHASES | | EVENTS | | ORO-GENIES | SEA LEVEL |
|---|---|---|---|---|---|
| PLEISTOCENE | Nile | Nile & Nile Delta | | | |
| PLIOCENE | | | UPPER MEGASEQUENCE | | |
| MIOCENE | Gulf of Suez | Incised valleys, Nile, Sahabi, Abu Madi Channel, Gulf of Aqaba, Gulf of Suez, Souther Red Sea rifting | | | |
| OLIGOCENE | | | | | |
| EOCENE | Alpine | Western Mediterranean, Jabal Akhdar uplift, Flexural subsidence | MIDDLE MEGASEQUENCE | | |
| PALEOCENE | | | | | |
| CRETACEOUS | | Plate collision, basin inversion, collapse of arches | | Alpine s.l. | |
| | Atlantic | North Atlantic, South Atlantic | NUBIAN CYCLE | | |
| | | Central Atlantic | | | |
| JURASSIC | | | | | |
| TRIASSIC | Mediterranean | Neo-Tethys | KARROO CYCLE | | |
| | | Collapse of arches, Paleotethys | | | |
| PERMIAN | | | LOWER MEGASEQUENCE | | |
| CARBONIFEROUS | Tectonic-Arches | Inversion of basins | | Hercynian | |
| DEVONIAN | | Uplift of North Africa & exhumation of Precambrian basement | | | |
| SILURIAN | Glacial | Inversion of basins Glacial-eustatic sea-level rise | PALEOZOIC CYCLE | | |
| ORDOVICIAN | | Glaciation, Incised valleys | | Caledonian | |
| CAMBRIAN | Graben | Grabens & pull-apart basins | | | |
| PRECAMBRIAN | Magmatica-Arc | Magmatic arcs, consolidation of Gondwana | | Pan-African | |

*Figure 20.1* Megasequences and tectono-sedimenatry units of North Africa.

characterized by the formation of tectonic arches as a result of compression stresses which acted during the Hercynian Orogeny, rifting between Gondwana and Eurasia and the opening of the Mediterranean Sea and the Atlantic Ocean. This Megasequence is regressive and represents deposition on a passive margin with epeirogenic sag interrupted by a number of orogenies, such as the Taconian, Caledonian and Hercynian. It was accompanied by a gradual fall in sea level (Haq et al., 1987) (Fig. 20.1). It corresponds to the lower clastic unit of Said (1961, 1962a) in Egypt, the pre-graben or pre-rift cycle of Harding (1984), Baird et al. (1996), and Van der Meer & Cloetingh (1996) in the Sirte Basin, the Gondwana Supercycle of Galeazzi et al. (2010) in Algeria, and the Pre-rift Cycle of Beauchamps et al. (1996) in the Missour Basin in Morocco. This Megasequence can be subdivided into three broad cycles (Klitzsch & Squyres, 1990): a) A Paleozoic cycle (Cambrian-Early Carboniferous) characterized mainly by shallow marine with minor peri-glacial sedimentation during the Late Ordovician. The cycle started with the development of pull-apart basins over most of North Africa, followed by periods of tectonic uplift and erosion associated with the Taconian, Caledonian, and Hercynian orogenies. This cycle is partly equivalent to the *Continental de Base* (Cambrian), *Couverture Tassilienne* (Ordovician-Lower Devonian), and *Series post-Tassilienne* (Frasnian-Namurian) of Kilian (1931) in the Central and Western Sahara. It ended with the onset of the Hercynian Orogeny which led to mountain building, uplift, basin conversion, and erosion. Devonian iron ores were deposited in Ash Shati area in West Libya and reefal mud mounds were formed in the Anti-Atlas Domain in Morocco. b) A Late Carboniferous-Early Jurassic Karroo cycle composed predominantly of continental sediments. Evaporites dominated in NW Libya, Tunisia, Algeria, and Morocco (Aitken et al., 2002, Cavalazzi, 2006). This cycle witnessed the collision between Gondwana and Eurasia. Minor peri-glacial activities occurred during the Late Carboniferous. c) An Upper Jurassic-Lower Cretaceous Nubian cycle dominated by continental deposits in the south and marginal-marine sediments in the north. It is equivalent to the syn-rift cycle of Beauchamps et al. (1996) in the Missour Basin in Morocco. This cycle is related to the disintegration of Pangaea, rifting and hot spot activities. The last two cycles correspond roughly to the *Continental Intercalcaire* (Moscovian-Lower Cenomanian) of Kilian (1931) and Lefranc & Guiraud (1990). Upper Carboniferous manganese minerals occurred in Western Sinai, Jurassic coal deposits accumulated in Northern Sinai and Coaly source rocks were deposited over large areas in the Western Desert of Egypt.

2    The Middle Megasequence encompasses sediments from the base of Cenomanian to the top of the Eocene. It corresponds to Said's (1962a) middle carbonate unit in Egypt, Harding's (1984) graben-fill cycle, the graben subsidence and filling tectonic phase (Cenomanian-Santonian) and macrobasinal subsidence phase (Maastrichtian-Upper Eocene) of Baird et al. (1996) in the Sirte Basin, and the *Series Hammadienne* (Upper Cenomanian-Danian) of Kilian (1931) in the Central and Western Sahara. This cycle is dominated by carbonates on platforms, fine marine clastic sediments in deeper parts, and minor fluvial to shallow-marine sandstones. Evaporites were deposited in the western Sirte Basin. This transgressive Megasequence was deposited during a period of active tectonics

and rifting and a high global sea-level (Haq et al., 1987) (Fig. 20.1), interrupted by a major episode of basin conversion in Cyrenaica and the northern Western Desert during the Santonian due to a change in the relative motion of Africa and Eurasia (Guiraud et al., 2005). The time period also saw the final collapse of many of the tectonic arches, rapid basin subsidence, widespread deposition of carbonate sediments, drowning of platforms and frequent Oceanic Anoxic Events (Jarvis et al., 2002, 2006, Jenkyns, 1991, Kuhnt et al., 1997, El Albani et al., 1999, Kolonic et al., 2001, Lüning et al., 2004, Heldt et al., 2008, Rodriguez-Rovar et al., 2009). A thermal event (a period of global warming) took place around the Paleocene-Eocene boundary with significant effects on the ecosystem (Speijer et al., 2000, Schulte et al., 2009, Scheibner & Speijer, 2008, Morsi & Scheibner, 2009). This period is characterized by the flourishing of Rudistids in Sinai, Northern Egypt (Jenkins, 1990, De Castro & Sirna, 1996, Zakhera, 2010), and offshore Libya, Tunisia, and Morocco, the accumulation of vast Cretaceous and Eocene commercial phosphate deposits and equivalent hydrocarbon source rocks and iron ores. A Trans-Saharan Seaway was established in NW Africa and acted as a communication channel between Northwest and Northeast Africa from the Cenomanian to the Eocene.

3     The Upper Megasequence includes Oligocene to Recent sediments dominated by clastics. Limestones and evaporites were deposited in the Mediterranean Basin, the Gulf of Suez and the Sirte Basin. The Megasequence was deposited during a gradual fall in sea level (Haq et al., 1987) (Fig. 20.1). This Megasequence was initiated during the main phase of the Alpine Orogeny and convergence between Africa and Eurasia. These events led to rifting in East Africa and the Red Sea, the opening of the Gulf of Suez and the Gulf of Aqaba, and the building of the Atlas Mountains in NW Africa and the Syrian Arc System in the Northeast Africa. Closure of the Mediterranean and its desiccation took place during the Messinian, and turbidites accumulated in the offshore of the Nile Delta. It corresponds to the upper clastic unit of Said (1962a), the interior-sag or post-graben cycle of Harding (1984) and the dominance of the Ajdabiya Trough, Sarir and Abu Attiffel Basin tectonic phase of Baird et al. (1996) in the Sirte Basin, the *Continental Terminal* (post-Middle Eocene) of Kilian (1931) and Lang et al. (1990) in the Sahara, and the Maghrebian orogenic phase of Boote et al. (1998). The Megasequence terminates with the formation of the Nile River, the Nile Delta, and the advent of the human race.

The three megasequences have been further subdivided into nine tectono- depositional phases (Fig. 20.1), described in the chapter on the petroleum systems in North Africa.

## 20.3 LOWER MEGASEQUENCE

### 20.3.1 Paleozoic

Thick Paleozoic sediments with more than 5000 m (16,400 ft) occur in cratonic basins in North Africa (Guiraud et al., 2000, Avigad et al. 2003) (Fig. 20.2). In the Anti-Atlas of Morocco, the Paleozoic cover exceeds 10 km (6 miles) (Burkhard et al., 2006, Soulaimani & Burkhard, 2008).

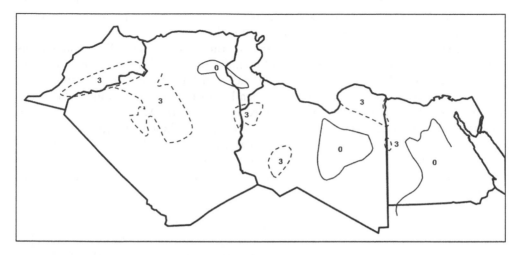

*Figure 20.2* Paleozoic isopachs, North Africa (simplified and modified from Avigad et al. 2003, Figure 1).
0 = Paleozoic absent, 3 = Paleozoic sediments >3.0 km in thickness.

The Paleozoic of North Africa started during the final stages of the Pan-African Orogeny with the accretion of island arcs to the continental nucleus and the emplacement of younger granites and associated volcanics. These events were followed by rifting and extensional tectonics in the Middle to Late Cambrian times (Klizsch, 1971, Schandelmeier, 1988) and by relative tectonic stability throughout the Paleozoic era until the commencement of the Hercynian Orogeny in the Upper Carboniferous. Two rifting episodes took place during the Early Paleozoic; the first led to the opening of the Rheic Ocean during the Cambrian; the second initiated the Paleotethys during the Silurian (Burkhard et al., 2006). Deposition during the Palcozoic was dominated by continental and shallow-marine clastic sediments with minor amounts of carbonates. Deposition was controlled by global eustatic sea-level changes, climate, and epeirogenic movements. Epeirogenic movements during the Paleozoic led to the formation of a number of topographic highs and subsidence of intra-cratonic basins, such as the Northern Egypt, Nile, Kufrah, Murzuq, Ghadames, Sirte, Illizi, Timmimoun, Bechar, and Anti-Atlas, which were filled with thick Paleozoic sediments. Erosion of the Pan-African Mountains led to the deposition of Cambrian clastics over vast areas of North Africa. Late Ordovician glaciation resulted in the deposition of peri-glacial deposits; best documented in Southwest Libya and Eastern Algeria (Beuf et al., 1971, Ghienne & Deynoux, 1998, Craig et al., 2002, Ghienne, 2003, Le Heron & Howard, 2010). Sporadic magmatic activities associated with extensional tectonics continued during the Early Paleozoic, especially in the Sirte Basin. Toward the end of the Paleozoic, magmatic activities were associated with minor tectonic movements related to the Hercynian Orogeny and the global re-organization of the continents (Morgan, 1990, Westphal et al., 1979, Saber et al., 2007). These movements led to the formation of new arches, the accentuation of old ones, and the stripping of the Paleozoic sediments over large areas. A number of prominent uplifts was present by the end of this orogeny, such as the Nafusah Uplift, Gharyan Dome, and Gargaf

Arch in Northwest Libya, the Tihemboka Arch in Southwest Libya, the Tibesti High in Southern Libya, the Sirte and Kalanshiyu arches in Central and Eastern Libya, the Jebel Oweinat, and the Helez Geanticline in Northern Sinai.

Paleozoic rocks occur in North Africa in three major NW-SE trending areas (Fig. 20.2); the first is in West Libya and includes the Jifarah Plains, the Ghadames Basin which extends into Tunisia and Algeria (Illizi Basin), and the Murzuq Basin, the second includes the Kufrah and Cyrenaica basins in East Libya and the Ghazalat Basin in Northwestern Egypt, and the third is in Western Algeria. A fourth trend, oriented NE-SW, occurs in the Anti-Atlas in Morocco. Sediments ranging in age from Cambrian to Carboniferous occur also in isolated areas along the Nile Valley, on both sides of the Gulf of Suez, and in the Sirte Basin. In the Sirte Basin, only a small record of the Paleozoic deposits with the exception of Cambrian sediments remained after the Hercynian Orogeny. Thick Paleozoic sediments also occur in Western Algeria and Eastern Morocco (Fig. 20.2).

The Paleozoic and Mesozoic successions in the Berkine and Illizi (Ghadames) basins have been divided into a lower Gondwana and upper Tethys supercycles, respectively, separated by the Hercynian Unconformity (Galeazzi et al., 2010). The two supercycles are first-order cycles and follow the same history of extensional basin formation, followed by regional subsidence, interrupted by minor transpressional tectonics, and ending with basin inversion and regional uplift. Although the two supercycles have similar tectonic histories, there are marked differences in their sedimentary successions, mostly due to major variations in relative sea-level changes and climate through time (Fig. 20.1).

The presence of reworked Devonian-Early Carboniferous miospores in Early and Late Cretaceous sediments in the Sirte Basin (Thusu, 1996, Wennekers et al., 1996) indicates that Early Paleozoic sediments are common and acted as source of sediments in the area. The recent discovery of Devonian and Carboniferous sediments in the Central and Eastern Sirte Basin (Tawadros et al., 2001, Thusu, 1996, Wennekers et al., 1996, van Erve, 1993, Hallett & El-Ghoul, 1996) shows that Lower and Upper Paleozoic and Early Mesozoic sediments were deposited in the Sirte Basin and were subsequently eroded during the Taconian, Caledonian, and Hercynian orogenies. This large-scale tectonics also resulted in uplift of large areas of Egypt, the removal of most of the Paleozoic sediments and the development of regional unconformities.

### 20.3.1.1  Cambrian period

Cambrian sediments are limited in distribution in Egypt, but are widespread in Libya, Algeria, and Morocco (Fig. 20.3). They are represented by the lower part of the Shifa Formation in the subsurface of the Northern Western Desert of Egypt, and localized occurrences of the Araba Formation in the Bahariya Oasis, West Wadi Qena, and Western Sinai. They also include the Hassaouna Formation in West Libya, its equivalent the Sidi Toui Formation in the subsurface of the Jifarah Plains and Tunisia, the Hassi Leila Formation in Algeria, and parts of the Hassaouna, Hofra, and Amal formations of the Gargaf Group in the subsurface of the Sirte Basin in Libya. Cambrian sediments are absent on the Tihemboka Arch and the eastern and northeastern parts of the Illizi Basin in Algeria (van de Weerd & Ware, 1994) and NE Libya (El-Arnauti & Shelmani, 1988) and from most of Egypt. In Algeria,

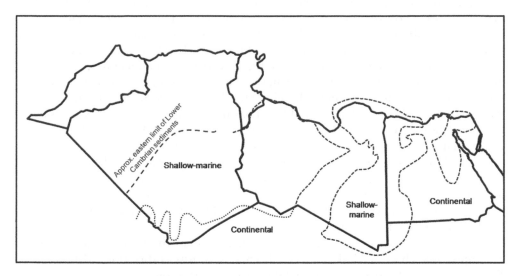

*Figure 20.3* Paleogeography of North Africa during the Cambrian (modified from Tawadros, 2001).

Cambro-Ordovician sediments have been included previously in the Internal Tassili (Chanut & Nyssen, 1958, Borocco & Nyssen, 1959, Fabre, 1988). Cambrian rocks crop out in the Tassilis (Eschard et al., 2005). In Morocco, Cambrian sediments belong to the Adoudounian Group in the Anti-Atlas Domain, which lie unconformably on top of Ouarzazate Supergroup. The Lower Cambrian post-rift sedimentary sequence is composed of a thick sequence of carbonate-siliciclastic rocks with locally developed volcanic rocks of the Tata and Taroudant Groups (Thomas et al., 2004, Álvaro et al., 2008). The Precambrian-Cambrian boundary falls within the lowermost part of the sequence (Thomas et al., 2004). These rocks developed within gradually subsiding extensional basins. In the Anti-Atlas and Atlas domains, the Early Cambrian is characterized in Morocco by the appearance of the reef-building Archaeocythids, the Paradoxides trilobites, and the appearance of the vertical burrows *Skolithos* and *Diplocraterion* (Drebenne & Drebenne, 1978, James & Drebenne, 1980, Geyer & Landing, 1995, Maloof et al., 2006, Soulaimani & Burkhard, 2008).

Most of the Early Cambrian sediments in North Africa form part of the metamorphosed and deformed Pan-African basement complex. Middle and Late Cambrian sediments were deposited during the rifting phase, dominated by extensional tectonics and volcanic activities which followed the Pan-African Orogeny (Tawadros et al., 2001). Albani et al. (1991) reported dolerite (diabase or microgabbro) from the Sidi Toui Formation in the Jifarah Plains. Barr & Weegar (1972) reported monzonites from the Amal Formation type well B1-12, olivine basalts from the Hofra Formation type well A1-11 and rhyolite tuffs from well I1-13. Tawadros et al. (2001) identified many horizons of volcanic rocks intercalated with the Late Cambrian quartzites in the Central Sirte Basin. Polymictic conglomerates at the base of the Hassaouna Formation in the Gargaf Arch area also contain rhyolite fragments (Mennig & Vittimberga, 1962) and the Formation is intruded by phonolites and basalts (Jurak, 1978). Most of these volcanics are believed to be subaerial volcanic eruptions. Weissbrod (1969) and Allam (1989) reported volcanic rock fragments from the Araba Formation in Sinai.

In Libya, Middle and Late Cambrian deposition was dominated by fining-upward transgressive cycles (Jurak, 1978, Cepek, 1980, Turner, 1980, Seidl & Röhlich, 1984). The Hassaouna Formation is made up of braided stream conglomerates, sandstones, and siltstones in the lower part grading upward into nearshore, subtidal and intertidal sediments in the Murzuq Basin (Collomb, 1962, Jurak, 1978, Seidl & Röhlich, 1984, Parizek et al., 1984), Ghadames and Jifarah basins (Albani et al., 1991), Kufrah Basin (Turner, 1980) and the Sirte Basin (Hea, 1971, Tawadros et al., 2001). *Tigillites* are common near the top of the section in outcrops in West Libya (Collomb, 1962, Jurak, 1978) and in the subsurface of the Jifarah Plains and the Ghadames Basin (Sidi Toui Formation) (Albani et al., 1991). These sediments also commonly contain abundant acritarchs, especially sphaeromorphs (Deunff & Massa, 1975, Massa, 1980, Kalvacheva & Kazandjiev, 1987, Albani et al., 1991, Tawadros et al., 2001) suggesting a shallow marine environment.

In the Jebel Gargaf area, paleocurrent directions in the Hassaouna Formation are approximately SSW-NNE. This consistency of transport direction led Burollet & Byramjee (1964, 1969) to conclude that the Gargaf uplift is younger than the Hassaouna deposition and it could not have had any influence on its sedimentation. On the other hand, the grain size and percentage of feldspars in the Hassaouna sandstones decrease away from the Gargaf arch. These phenomena led Mennig & Vittimberga (1962) to suggest that the Arch was already a high area and acted as a source for these sandstones in that area. The Hoggar and Tibesti uplifts probably did not act as a source of sediments at that time.

So far, no Lower Cambrian sediments have been identified in Northwest Libya, the Sirte Basin, Northeast Libya, or Egypt. The only exception is probably the Araba Formation near the Gulf of Suez. The initial Upper Cambrian sediments in the central Sirte Basin were deposited during an extensional tectonic and rifting phase (Tawadros et al., 2001). Deposition started with sandstones with tidal bedding and acritarchs in a subtidal environment, accompanied by subaerial volcanic activities. A drop in sea level led to the deposition of a thick succession of variegated sandstones and shales, rich in *Skolithos* burrows, in extensive intertidal flats. Late Cambrian deposition ended with a subtidal quartzarenites.

More than 1000 m (3281 ft) of shallow-marine Middle-Late Cambrian sediments with acritarchs (Gueinn & Rasul, 1986), brachiopods, and trilobites were deposited in the Ghazalat Basin in the Northern Western Desert (Keeley, 1989). Concentric facies belts can be recognized around the basin. Coarse alluvial fan deposits interfinger with tidal flat, shoreface and open-marine, fine to medium-grained sands, silts and clays (Keeley, 1994).

At Um Bogma and Wadi Feiran in Sinai, and Abu Durba and Somr el Qaa in the northern Eastern Desert, deposition of the early Cambrian Araba Formation arkosic sandstones and conglomerates started in fluvial environments. Consequently, fluctuating sea levels led to the deposition of interbedded fluvial, cross-bedded sandstones and shallow-marine nearshore sandstones with groove casts and load casts and abundant *Skolithos* and *Cruziana aegyptiaca* (Allam, 1989, Klitzsch, 1990a, Klitzsch et al., 1990a, Jenkins, 1990). Locally stromatolites and small Archaeocyathids were deposited (Omara, 1972).

Cambrian sediments are heavily silicified in places, especially in the Central Sirte Basin and the Hassi Messaoud in Algeria and the Tassilis (Eschard et al., 2005). However, the reasons for and extent of silicification are unknown. Silicification of the Gargaf Group sandstones was attributed to close proximity to unconformity surfaces where the sediments were exposed (Hea, 1971), or to hydrothermal activities in close proximity to phonolite volcanic bodies (Jurak, 1978). However, a combination of early freshwater diagenesis, circulation of hydrothermal fluids along faults and remobilization of silica due to tectonics could be responsible for the silicification of these sediments (Tawadros et al., 2001). These processes were probably responsible also for the destruction of kaolinite. Cambrian sediments in North Africa are characterized by the presence of chlorite and illite, with the virtual absence of kaolinite in many formations which becomes the dominant clay mineral in younger sediments (Hea, 1971, Bonnefous, 1972, Weissbrod & Perath, 1990, Tawadros et al., 2001). In the Anti-Atlas Domain, they are mildly metamorphosed (Thomas et al., 2002, 2004, Burkhard et al., 2006, Soulaimani & Burkhard, 2008). Sedimentation in the Atlas and Anti-Atlas domains was dominated by detrital input from the West African Craton (Buggisch & Siegert, 1988). In the Western and West-Central Anti-Atlas, the Middle Cambrian is represented by the Goulimine Quartzitic Group or the Tabanite Sandstones and Shales, with trilobites, such as *Conocoryphe* trilobite, as well as *Lingula*, overlain by Trachy-basalts (Baidder et al., 2008) or by Lower Ordovician sandy schists in the Tata area (Belfoul et al., 2001, Faik & Belfoul, 2001). In the Western High Atlas, the Early-Middle Cambrian includes fine-grained sandstones, shales, and siltstones, limestones, and volcanic basaltic flows of the Ifri-Azegour Formation and the Ouzaga-Tizzirt Formation (Pouclet et al., 2008).

### 20.3.1.2  *Ordovician period*

The Ordovician succession is bounded at the top by the Caledonian/Taconian unconformity and at the base by the Cambrian unconformity. Ordovician sediments are widespread in North Africa (Fig. 20.4). In West Libya, they include the Achebyat, Haouaz, Melez Chograne, Tasghart and Memouniat formations. In the subsurface of the Jifarah Plains and Tunisia, they are represented by the Sanrhar, Kasbah Leguine, Bir Ben Tartar and "Djeffara" formations (Vecoli et al., 2009). In the subsurface of the Sirte Basin, Ordovician sediments probably form parts of the Gargaf Group quartzitic sandstones. Ordovician sediments (Caradocian) have been palynologically identified in well D1-32 and a several thousand feet thick Early Ordovician section is present in well A1-10 (Wennekers et al., 1996). Movements on the Gargaf Arch were continuous from the Lower Ordovician through the Lower Devonian (Seidl & Röhlich, 1984, Collomb, 1960, Massa & Collomb, 1960). Ordovician sediments are of limited distribution in Egypt. They are represented by the Naqus Formation exposed locally in the Gulf of Suez (Issawi & Jux, 1982) and the upper part of the Shifa Formation (Keeley, 1989) in the subsurface of the Ghazalat Basin in the northern Western Desert and the Karkur Talh Formation (Klitzsch, 1990a, b) in the Gebel Oweinat area. Uplift of most of Egypt and Libya during the Caledonian/Taconian orogeny, combined with glacial erosion resulted in either the erosion or non-deposition of Ordovician sediments east of the Oweinat-Bahariya Arch.

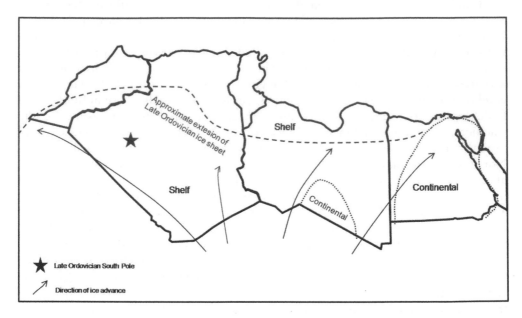

*Figure 20.4* Paleogeography of North Africa during the Ordovician (after Tawadros, 2001, Craig et al., 2002, Stampfli & Borel, 2002, Guiraud et al., 2005, Ghienne, 2003, Ghienne et al., 2007, Le Heron et al. (2009): reconstruction of Gondwana at ~440 Ma showing the distribution of Hirnantian ice sheets. The distribution of ice sheets is after Ghienne (2003).

In Libya, two high areas probably existed at the end of Ordovician time in Central Libya; one striking northwest-southeast; the other northeast-southwest. These two areas probably represent the nuclei of the Sirte Arch (Tripoli-Tibesti Arch of Klitzsch, 1970) and Tibesti-Kalanshiyo Arch, respectively. The Sirte Arch, the Gargaf Arch, and the Ben Ghanima-Brak High in the Murzuq Basin and its extension in the Ghadames Basin were probably established at that time. On the southern flank of the Gargaf Arch and the northern part of the Murzuq Basin most of the Ordovician section has been removed by erosion and the Memouniat Formation rests unconformably on the Cambrian Hassaouna Formation. Eroded Cambrian sandstones probably supplied most of these sediments.

The Early Ordovician (Tremadocian) of the Illizi Basin, Southeast Algeria, is represented by fluvial sediments of the In Kraf Formation (Maache, 1995). *Tigillites*-rich sandstones of the Achebyat Formation were deposited in the Murzuq Basin (De Castro et al., 1991, Parizek et al., 1984) and the Sanrhar Formation in the Jifarah Plains (Deunf & Massa, 1975, Jaeger et al., 1975, Massa et al., 1977, Massa, 1988). These sediments were deposited in shallow-marine, probably intertidal, environments. The Achebyat Formation is followed upwards by the shallow-marine foreshore to shoreface sandstones of the Haouaz Formation (Arenig-Llandeilo). The Sanrhar was followed by the Kasbah Leguine and Bir Ben Tartar formations in the Jifarah Plains. The In Kraf Formation was followed by the Hamra Formation, In Tahouite and Edjeleh formations in Southern Algeria. The In-Tahouite Formation displays shoreface and tidal prograding sandstones (Les Castelets Member) made up of siltstones and fine-grained sandstones with *Skolithos* and *Cruziana* ichnofacies (Eschard et al., 2005). The presence of *Tigillites* (*Skolithos*) in these sediments, in addition to a few acritarchs (Radulovic, 1984) suggests that sedimentary conditions were similar to the time of the Achebyat and Sanrhar. However, closer to the source of sands, such as the Gargaf Arch area, the Haouaz sandstones were deposited predominantly in alluvial fans (Vos, 1981b). The Achebyat and Haouaz formations appear to have been deposited during a period of tectonic stability which led to uniform deposition over vast areas.

A post-Llandeilo epeirogenic uplift took place during the Taconian orogenic phase which resulted in partial erosion of the Hassaouna, Achebyat, and Haouaz formations (Units II & III in SE Algeria). These activities are manifested by the absence of these units over the main uplifts, such as the Tihembouka Arch (Echikh, 1998).

This phase was followed by a period of glaciation which affected the northern Gondwana region during the Late Ordovician (Caradocian-Ashgillian) (Eyles, 1993, Deynoux, 1998). North Africa was positioned during that time near the South Pole and an extensive ice cap developed (Beuf et al., 1971, Bennacef et al., 1971, Paris et al., 1995, Ghienne et al., 2007, Konaté et al., 2006). The Hoggar and Tibesti massifs, as well as the Kandi Basin in the Niger (Konaté et al., 2006), were probably covered by glaciers at that time. This glaciation resulted in a global drop in sea level accompanied by the formation of incised glacial valleys (Dubois, 1961, Beuf et al., 1971, Bennacef et al., 1971) which were filled with glacial and periglacial sediments. Melting of the ice sheets when climate was much more temperate (Eyles, 1993) resulted in a marine transgression and the transformation of the area into an epicontinental sea. Extensive Late Ordovician glacial sediments were deposited over large areas in North Africa. Glacial valleys and heterogeneous valley-fills have been recognized in outcrop and the subsurface of the Murzuq and Illizi basins (Beuf et al., 1971). Glacio-marine diamictites and dropstones are common in the Taoudine Basin in West Africa, the Hoggar Massif, western Libya, the Tibesti Massif, west central Saudi Arabia and South America (Caputo & Crowell, 1985). Although these diamictites can originate in other non-glacial environments, they are the most typical glacial deposits (Deynoux, 1998), especially when considered with other glacial criteria.

The front of the ice is customarily determined by the areas dominated by glaciomarine shales. However, Rubino et al. (2003) suggested that the mapping of the stratigraphic gap at the base of the Ashgillian series, which is related to the erosion of the grounded shelf ice, is a key point in determining that front. Based on that assumption, Rubino et al. (2003) postulated that the ice cap reaches an east-west line joining the Morocco Anti-Atlas to Djerba in Tunisia.

Glaciation and deglaciation caused a frequent change in the volume of sea water which resulted in third-order or second-order sedimentary cycles (Williams, 1993, Craig et al., 2004). These cycles had durations between approximately 2 Ma (Paris, 1990) and 500,000 years (Fitches, 1998). Application of sequence stratigraphy to the Late Ordovician succession, however, is not very beneficial in resolving stratigraphic problems, for a number of reasons: 1) although the succession is made up of genetically related strata and bounded at its top and base by unconformities (*cf.* Mitchum et al., 1977), it contains a number of unconformities of unknown magnitudes. 2) The formation of incised valleys is not always related to a drop in sea level, but also to changes in climate. 3) The lack of paleontological control may result in miscorrelation between units. 4) The presence of slump or gravity flow units, resulting from the slope failure during the onlap of a slope (Cartwright et al., 1993), may lead to erroneous interpretations of the genetic relationships between units. These factors are combined with the difficulties in distinguishing diamictites from shales and/or sandstones on seismic sections and wireline logs, and during hasty fieldwork.

Two glacial episodes probably took place during the Late Ordovician. The first episode led to the deposition of the Melez Chograne periglacial diamictites and slump deposits during the Caradocian (Havlicek & Massa, 1973). The other took place in the Late Ashgillian (Pierobon, 1991, Paris et al., 1995) and preceded the deposition of the peri-glacial Memouniat sandstones.

The older Cambro-Ordovician succession (Hassaouna-Haouaz formations) in western Libya was truncated by incised valleys filled with variegated shales, siltstones, diamictites and coarse fluvio-glacial and glacio-marine sandstones of the Melez Chograne and Memouniat formations. These peri-glacial deposits were recognized in outcrops (Mennig & Vittimberga, 1962, Seidl & Röhlich, 1984, Grubic et al., 1991), and in the subsurface (Pierobon, 1991, Paris et al., 1995, Abuessa & Morad, 2009). They show complex facies associations which suggest deposition in a variety of environments, including streams, delta front, delta slope, and storm-wave influenced shoreface settings (McDougall & Martin, 1998). These facies formed a succession of proglacial deltas (Blanpied et al., 1998) associated with turbidites. These sediments are comparable to those described from outcrops in Algeria (Bubois, 1961, Beuf et al., 1971, Bennacef et al., 1971), and the subsurface of the Illizi in SE Algeria (Corriger & Surcin, 1963). They bear a striking resemblance to the Carboniferous glacial deposits of Oman (Levell et al., 1988), and the Late Ordovician glacial deposits of the Arabian Peninsula (Vaslet et al., 1998). The bottoms of these incised valleys in southeastern Algeria cut down to the basement rocks in places and may reach a depth of 248 m (814 ft) over a distance of only 3 km (Corriger & Surcin, 1963) (Fig. 20.5). The basal Caradocian-Ashgillian TST sediments of the Melez Chograne Formation transgressed over an eroded Llandeilo surface of the Haouaz Formation, marked by ferruginous oolites, phosphate nodules, trilobites and brachiopods (Collomb, 1961, Pierobon,

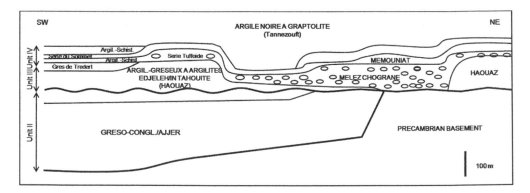

*Figure 20.5* Correlation of Lower Paleozoic successions in West Libya and the subsurface of the Illizi Basin: Based on Corriger & Surcin (1963), Bennacef et al. (1971), Protic (1984), and Parizek et al. (1984).

1991). In the Wadi Tanazuft area, the Formation contains fine-grained ferruginous sandstones with iron and phosphate concretions, boulders of granite, diorite, and schist, and conglomerates with remains of brachiopods and cephalopods. This surface resulted from subaerial exposure following the epeirogenic uplift and drop of sea level at the end of Llandeilo. In outcrops, on the southwestern side of the Murzuq Basin, the Melez Chograne is made up of shales, sandstones and lenses of diamictites (Grubic et al., 1991). In southeastern Algeria, the Tamadjert Formation (Unit IV) (equivalent to the Melez Chograne and Memouniat formations in Libya) is characterized by numerous channelled sandstone bodies forming a close network interpreted as proglacial or periglacial streams (Bennacef et al., 1971, El-Hawat & Bezan, 1998, McDougall et al., 2003). The deposition of the Melez Chograne in Southwest Libya, as well as its equivalent the Gara Louki Formation or Hassi Leila Formation in the Illizi Basin, was described as glacio-marine by Lelubre (1952a, b), Beuf et al. (1971), van de Weerd & Ware (1994), Maache (1995), and Fikirine & Abdallah (1998). The various materials which make up these diamictites were probably released by melting of floating icebergs and were mixed with argillaceous sediments (Radulovic, 1984). Glacio-marine diamictites of the Melez Chograne Formation and its equivalent were considered to represent a high-stand tract (HST). In the subsurface of the Murzuq Basin, the Melez Chograne Formation contains beds of highly radioactive shales. These shales probably represent the maximum flooding during Melez Chograne time (Pierobon, 1991).

The peri-glacial deposits of the Melez Chograne Formation in western Libya are associated with turbidites of the Tasghart Formation (Caradocian-Ashgillian) (Radulovic, 1984, Grubic et al., 1991). These turbidites are made up of interbedded coarse-grained sandstones, siltstones and shales. They display sole marks, flute casts and contorted bedding, and contain ferruginous and carbonate concretions. The Tasghart Formation turbidites appear to have been deposited near the end of the Melez Chograne glacial cycle (Fig. 20.5).

Late Ashgillian glaciation caused another global drop in sea level. The Memouniat Formation may represent the product of this second glacio-marine cycle (Pierobon, 1991, Boote et al., 1998) which probably took place during the latest Ordovician (Hirnantian) (Paris et al., 1995, 2000), the time of maximum extension of the South Pole ice cap. Shallow marine cross-bedded, conglomeratic, kaolinitic sandstones with *Tigillites* (Collomb, 1962) were deposited in the Murzuq Basin. The environment became shallower eastward in proximity of the Gargaf Arch (Parizek et al., 1984). In the Ghadames Basin to the north, the Memouniat Formation is characterized by beach deposits and tidal flat sediments (Bracaccia et al., 1991). In Southwest Libya and Algeria, peri-glacial sandstones of the Memouniat Formation infill north-south-aligned incised valleys. In the type section of the Memouniat Formation, a marked incised channel occurs in the middle of the section. The Late Ordovician of the Tassili N'Ajjer outcrop displays the development of extensive erosional features at the base in the form of paleo-valleys and channels (Hirst et al., 2002, Dixon et al., 2008). To the north, in the Jifarah Plains, open marine and glacial sediments of the "Djeffara" (Melez Chograne/Memouniat) Formation consist of diamictites and shales rich in conodonts (Bergström & Massa, 1991). Carbonates occur in the lower part of the Formation and consist of bryozoan biostromes and mud-mounds with brachiopods, trilobites, echinoderm fragments, stromatoporoids, protospongia, and ostracods

(Butler et al., 2007, Vecoli et al., 2009). The development of these biostromes follows the east-west structural trend of the Tripolitania area and may reflect warm water conditions preceding the onset of glaciation. In southwest Libya and Algeria, good hydrocarbon reservoirs are found in the periglacial Memouniat sandstones. Several discoveries have been made in the basin, among which are the two giants Sharara Field (NC115) and Elephant fields (Bortello et al., 2003). In the NC115 area, the trap is formed by fault structures (Boote et al., 1998).

In the subsurface of Northeast Libya, the Ordovician succession starts with Caradocian-Ashgillian interbedded fine to medium-grained sandstones and shales (El-Arnauti & Shelmani, 1988) correlatable with the Melez Chograne Formation in West Libya, followed upward by and Ashgillian fine-grained sandstones and coarse to pebbly sandstones "microconglomerates". This unit is correlatable with the Memouniat Formation. Ordovician acritarchs and chitinozoans, similar to those reported from the uppermost Ashgillian glacial sediments in North Africa (Paris, 1988, 1990); indicate an offshore marine shelf environment (Hill & Molyneux, 1988).

Eastward, in the Ghazalat Basin in Northwestern Egypt, the early Ordovician succession is made up of shallow marine sandstones and conglomeratic fluviatile sandstones of the Shifa Formation (Keeley, 1989, 1994). The Upper Ordovician (Llandeilo-Ashgillian), equivalent of the Melez Chograne and Memouniat formations, is missing (Gueinn & Rasul, 1986), probably due to glacial erosion, tectonic uplift, or a combination of both.

In the Kufrah Basin, for example in the area west of Jabal al Qardah (Turner, 1980, 1991, Seilacher et al., 2002, Le Heron & Howard, 2010), the exposed shallow-marine fine to coarse-grained, conglomeratic and cross-bedded sandstones of the Memouniat Formation were deposited unconformably on a glacially shaped morphology, or the eroded *"moutonée"* surfaces, of the Hassaouna Formation. The undifferentiated Cambro-Ordovician sediments in the subsurface of the Kufrah Basin form a fine to very coarse-grained, pebbly sandstone sequence which contains Caradocian-Ashgillian acritarchs and chitinozoans in the upper part (Grignani et al., 1991). Evidence for grounded ice sheets has been observed by Le Heron & Howard (2010) at the southeastern (Jabal Azbah) and northern (Jabal az-Zalmah) margins of the Kufrah Basin. Characteristic soft-sediment deformation structures, including soft-sediment folds, small-scale faults and striated pavements indicate subglacial shearing and the formation of glacial erosion surfaces.

In Southwest Egypt, the Ordovician shallow-marine sandstones with *Cruziana rouaulti* of the Kurkur Talh Formation are interbedded with fluviatile conglomeratic sandstones (Klitzsch, 1990a, b). Ordovician sediments are not common in the Eastern Desert of Egypt and Sinai. Pebbly, medium to coarse-grained sandstones of the Naqus Formation in the Red Sea area, may represent fluvio-glacial sediments (Issawi & Jux, 1982). Ordovician sediments crop out in the Anti-Atlas Domain of Morocco.

The Ashgillian glaciogenic deposits of the Upper Second Bani Formation in the Anti-Atlas Domain of Morocco are marked by paleovalleys incised into the underlying shallow marine clastic deposits (Le Heron, 2007) of the Lower Second Bani Formation, which include wave-dominated, transgressive shoreface to offshore sediments. The glaciogenic deposits are dominated by tillites (microconglomerates or diamictites) (Ghienne, 2003, Legrand, 2003, Sutcliffe et al., 2000, 2001, El Maazouz & Hamoumi, 2007, Le Heron, 2007, Le Heron & Craig, 2008, Soulaimani & Burkhard, 2008).

### 20.3.1.3  Silurian period

The Silurian succession includes silty and hematitic shales and siltstones of the Iyadhar Formation, black graptolitic shales of the Tannezouft, and sandstones of the Acacus Formation in Libya and equivalent units in Tunisia, the Kohla and Basur formations in Northwestern Egypt and the Um Ras Formation in Southwest Egypt, Ouled Ali Formation and Feguagira Shale in Western Algeria and the Silurian Shale, Imirhou Formation, and F6 Sand in Eastern Algeria, and the Mokattam, Kheiala, and Ain Delouine Formation and equivalent sediments in the Anti-Atlas and the Meseta domains.

The Silurian succession is predominantly marine. The sea covered smaller areas of North Africa during the Silurian (Fig. 20.6) but with a noticeable deepening of the depocenters in the Early Silurian followed by a regressive phase in the Late Silurian. These two phases led to the deposition of the transgressive system tract (TST) parasequence of the black graptolitic shales of the Tannezouft Formation (Llandovery), and feldspathic silty and hematitic shales of the Iyadhar Formation in places, and the low-stand (LST) parasequence of silty hematitic shales and siltstones of the Acacus Formation (Wenlockian-Ludlowian), respectively. The Silurian succession is progradational from south to north and shows a marked shallowing upward. The change in water depth may be attributed to global fall in sea level accompanied by epeirogenic tectonic movements. Silurian shales overlie Cambro-Ordovician sediments with a flat contact and almost without transitional lithologies, suggesting a rapid transgression (Beuf et al., 1971, van de Weerd & Ware, 1994). The Murzuq and Ghadames basins were probably connected through shallow to deep oceanic channels during that time (Al-Ameri, 1983b). The TST parasequence of the Tannezouft Formation in the Murzuq and Ghadames basins starts with open-marine nearshore littoral to outer neritic facies, followed upward by offshore sandstone bars and fluvial plain sediments. It terminates with stromatolitic beds deposited in subtidal and supratidal flats, marshes, and beach environments (Pierobon, 1991, Bracaccia et al., 1991, Vecoli et al., 2009).

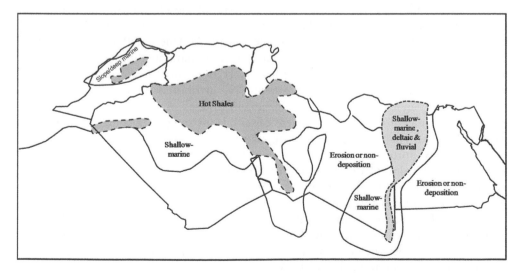

*Figure 20.6* Silurian paleogeography (after Tawadros, 2001, Lüning et al., 2003).

Minor turbiditic flows of very fine sandstones occurred in places (Pierobon, 1991). Tempestites associated with offshore bars were produced by storm waves (De Castro et al., 1991). These tempestites consist of sequences up to 50 m (164 ft) in thickness, forming coarsening-upward cycles of lenticular and wavy shales and sandstones, capped by hummocky cross-bedding with *Skolithos*. The Tannezouft graptolitic and radioactive shales (occasionally referred to as "Hot Shales") are known to be a good source rock for hydrocarbons with TOC content between 2–17% (Boote et al., 1998, Lüning, 2000, Lüning et al., 2004). These shales probably represent the maximum flooding of the Silurian sea, and an Oceanic Anoxic Event (OAE) (Schlanger & Jenkyns, 1976). Epeirogenic uplift at the end of the Llandovery (*Phase Tassilien* of Freulon, 1953) exposed most of North Africa and led to the development of a regional unconformity on top of the Tannezouft. Silurian sediments pinch out on the Brak-Ben Ghanimah paleohigh (Pierobon, 1991) suggesting that this paleohigh existed at that time. The presence of reworked Cambrian and Ordovician acritarchs in the Silurian sediments suggests that land-derived sediments were eroded from the uplifted areas, such as the Gargaf Arch, the Hoggar Massif and the Tibesti Massif (Al-Ameri, 1983b).

The overlying LST parasequence of the Acacus Formation reflects continued epeirogenic uplift and shallowing-upward. These movements are indicated by the upward change of the palynological assemblages in the Acacus sediments from acritarchs to spores (Al-Ameri, 1983b). In the Ghadames and Murzuq basins, the Acacus Formation is made up of shallow marine sediments including fan-delta, estuarine tidal flat, tidal channel and fluvial plain deposits. The lower Acacus Formation in the Ghadames Basin displays at 14 progradational deltaic packages which change from fluvial channel sandstones in the south to coastal deltaic sandstones, and eventually to offshore marine sandstones and shales (Elfigih, 1991, 2000). The upper contact of the Acacus Formation with the Tadrart Formation is marked by a ferruginous stromatolitic sandstone bed (*l'horizon d'Ikniouen* of Freulon, 1964) deposited probably in a subtidal to supratidal environment with extensive beaches (Jakovljevic, 1984). Iron oxide-rich sediments in the upper part of the Formation mark subaerial conditions (Bracaccia et al., 1991, Pierobon, 1991). Based on a study of palynomorphs of the Silurian (Ludlowian) in the Ghadames Basin, Al-Ameri (1983a) observed that Scolecodonts, Tasmanites and chitinozoans are abundant in the northern parts of the Ghadames Basin while in the southern part only rare chitinozoans and a less diversified acritarch assemblage are present. He attributed these variations to the occurrence of lagoonal and intermediate environments in the north and an open-marine environment in the south. These observations suggest that the northern part of the basin was probably uplifted during that time.

Secondary porosity, the dominant type in the Lower Acacus sandstones, was derived mainly from the dissolution of carbonate cements. The best reservoir quality is found in the proximal delta front facies with an average porosity of 20.3% and an average permeability of 921 mD (Elfigih, 1991, 2000).

In the subsurface of Northeast Libya, the Silurian period starts with a Llandoverian TST parasequence, equivalent to the Tannezouft Formation, made up of dark grey, pyritic and graptolitic shales and occasional intercalations of micaceous, glauconitic siltstones, up to 1600 ft (488 m) in thickness (El-Arnauti & Shelmani, 1988). This parasequence was deposited in a shallow, marginal-marine environment. A Wenlockian-Ludlowian parasequence, equivalent to the Acacus Formation, composed of non-fossiliferous; fine to very fine, partly micaceous sandstones was deposited during the low-stand.

Silurian sediments in the Kufrah Basin also form a thick, coarsening-upward progradational sequence. The TST parasequence of the Tannezouft Formation is made up of a deltaic sequence of marine and shoreface sediments, followed upward by the LST parasequence of the Acacus Formation, composed predominantly of a braided delta plain facies. The two parasequences are separated by a regional erosional surface (Turner, 1991), probably equivalent to the Tannezouft unconformity in western Libya (Massa & Moreau-Benoit, 1976, Lüning et al., 2003).

Eastwards, in the subsurface of the northern Western Desert of Egypt, the Silurian also starts with a transgressive (TST) parasequence of the marine Kohla Formation, equivalent to the Tannezouft Formation. However, graptolites are absent (Keeley, 1994), probably due to the shallow water conditions in proximity to the Silurian shoreline. Siltstones and shales were deposited in fluvial, tidal flat, and shoreface environments. These sediments were followed by a regressive phase (LST) of the Basur Formation, equivalent to the Acacus Formation, deposited in alluvial fans and braided streams (Keeley, 1989, 1994). Southwards, in the Gilf Kebir area, sandstones with *Cruziana*, *Harlania*, and *Skolithos* of the Silurian Um Ras Formation were deposited in fluvial, deltaic, and shallow marine environments (Klitzsch, 1990a). Differentiation between the Llandovery TST and Wenlockian-Ludlowian LST parasequences has not been made in that area. The Silurian was dominated by regional uplift and non-deposition over most of eastern Egypt and Sinai (Klitzsch & Smetner, 1993).

In the Anti-Atlas Domain, the Upper Silurian shales and clastic sedimentation was interrupted by the deposition of two prominent marker horizons, the *Orthoceras* Limestone and *Scyphocrinites* Limestone (Destombes et al., 1985).

### 20.3.1.4  Devonian period

Devonian sediments are present in the Murzuq, Ghadames, Kufrah and Cyrenaica basins in Libya, the Ghazalat Basin, Gebel Oweinat in Egypt, in Algeria, and the Anti-Atlas Domain in Morocco (Fig. 20.7). Only remnants of the latest Devonian (Strunian) exist in the subsurface of the Sirte Basin following the Hercynian uplift and erosion (Tawadros et al., 2001). The Devonian in West Libya is represented by the Tadrart Formation, Ouan Kasa Formation, the Aouinat Ouenine Group and the Ash Shatti Group. These sediments extend westwards into Algeria and include the Oued Mehaiguene, Alrar and Gazelle formations. The Aouinat Ouenine and Ash Shatti groups in northeastern Libya are replaced by the Binem and Blita formations in Southeast Libya, and by the Zeitoun Formation in the subsurface of the Ghazalat Basin in northwestern Egypt. In the Gebel Oweinat and Gilf Kebir areas, the latest Devonian Kurkur Murr Formation is the only record of this period. The Anti-Atlas area of southern Morocco lay within the continental passive margin of Gondwana during the Devonian Period (Belka et al. 1999, Toto et al., 2008). In Morocco, the Lower Devonian includes the Rich Group, Tafilalt Limestones, Megsem-Mdarsal Group, and Draa Plain Clays (Soulaimani & Burkhard, 2008, Aitken et al., 2002, Lubeseder et al., 2003 & 2009, 210, Jansen et al., 2007).

The Devonian succession is bounded by the Bretonian/Acadian unconformity at the top and the Taconian/Ardennian unconformity at the base. Near the end of the

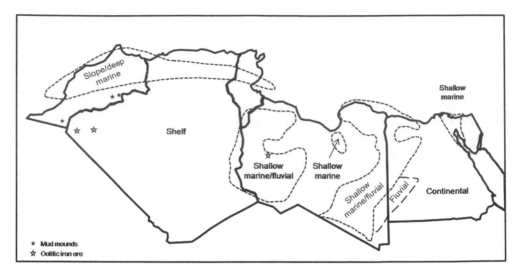

*Figure 20.7* Devonian Paleogeography (modified from Tawadros, 2001; NW Africa after Guiraud et al., 2005). Location of mud mounds is after Aitken et al. (2002).

Silurian the sea retreated from most of North Africa. Most of Central Libya and Egypt remained a high area during the earliest Devonian. The Gedinnian is absent almost everywhere in Libya and Egypt. The distribution of Lower Devonian sediments was controlled by the Caledonian structural configuration of the basins. The Lower Devonian is absent on of the Tihemboka Arch and the Gargaf Arch.

Devonian sediments represent a thick transgressive sequence starting with mostly continental deposition during the Lower Devonian and grading upward into marine sedimentation during the Middle and Upper Devonian. In general, three depositional parasequences which correspond to the Lower, Middle and Upper Devonian can be recognized. Jardiné et al. (1974) noticed that the tops of the depositional sequences in the Illizi Basin, Algeria, are time lines, but within each sequence the lines are diachronous.

Early Devonian sedimentation in West Libya forms a transgressive parasequence which starts with the Segienian low-stand tract (LST), dominated by continental and mixed sedimentation. It includes the cross-bedded sandstones with plant fragments of the Tadrart Formation in West Libya and Tunisia and the Oued Mehaigene Formation or the Hesi Tabankort Formation in the subsurface of the Illizi Basin in Algeria. In Morocco, the Lower Devonian displays two types of facies; sandy mudstones and limestones of the Gedinnian-Siegenian with conodonts and tentaculites predominate in the Western Anti-Atlas; in the Eastern Anti-Atlas, for example in the Tafilalt Basin, Devonian mud mounds and basaltic rocks occur at Hamar Laghdad (Aitken et al., 2002). They are followed by the deposition of more than 140 m (459 ft) of Lower Devonian crinoidal and tabulate coral limestones of the Kess-Kess Formation (Brachert et al., 1992, Belka, 1998, Cavalazzi, 2006), or the Seheb el Rhassel Group of Aitken et al. (2002). Because they occur in a cluster of mounds on top of a volcanic massif, they have been attributed to hydrothermal activities (Belka, 1998, Mounji et al., 1998). A hydrothermal origin of the Early Devonian Kess Kess mounds, exposed in the Hamar

Laghdad Ridge, Tafilalt region, eastern Anti-Atlas, SE Morocco, has been envisaged. However, their origin and growth are disputed (Massa et al., 1965, Brachert et al., 1992, Bełka, 1998, Mounji et al., 1998, Joachimski et al., 1999, Peckmann et al., 1999, 2005, Aitken et al., 2002, Jansen et al., 2007, Becker et al., 2004a, b, c, 2005, Webster & Becker, 2009).

The upper part of the Tadrart Formation grades westward in Tunisia into open-marine carbonates with bryozoans, echinoderms, and brachiopods (Burollet & Manderscheid, 1967, Massa & Morreau-Benoit, 1976). On the southwestern side of the Murzuq Basin, medium to coarse-grained, conglomeratic and feldspathic sandstones of the Tadrart Formation were probably derived from a source area located to the south or southwest (Jacovljevic, 1984) with minor contributions from the Gargaf Arch (Burollet & Manderscheid, 1967).

The Ouan Kasa Formation consists of rhythmic fining-upward sequences, and minor coarsening upward sequences, composed of sandstones, siltstones and shales with erosional contacts (Rizzini, 1975). It also contains thin beds or nodules of microcrystalline siderite. The presence of fine shell conglomerates with brachiopod fragments and mud-pebble conglormerates, as well as sedimentary structures such as planar and small trough cross-bedding, flaser and lenticular lamination, burrowings, and mud cracks, point to a storm-dominated tidal environment.

Tectonic uplift during the Caledonian Orogeny and the Middle Devonian resulted in the formation of the Gargaf Arch in central Libya and the Tihemboka Arch in southwest Libya (Burollet & Manderscheid, 1967). It also caused the removal or incision of the Ordovician, Silurian and Lower Devonian successions in the Gargaf Arch area and the northern part of the Murzuq Basin (Seidl & Röhlich, 1984, Parizek et al., 1984). During the Devonian, the Murzuq Basin was dipping at a low angle to the northwest and the changes in relative sea level transgressed and regressed large areas (Adamson et al., 1998).

The Caledonian Orogeny was initiated during the Pridoli-Gedinnian (sediments of these two periods are missing in many places) as a result of the collision between West Africa and North America, which caused uplift and erosion (Echikh, 1998). The Early Devonian (Gedinnian/Sieginian-Emsian) was dominated by erosion and continental sedimentation. This time interval is represented by a transgressive parasequence. The Gedinnian is absent. The Sieginian sediments form a low-stand tract (LST). They include the Tadrart Formation in Libya and southwest Egypt, the Gouring Formation in the Tibesti Massif, and the basal part of the Zeitoun Formation in the subsurface of the Ghazalat Basin in northwestern Egypt. The following Emsian transgressive tract (TST) includes the Ouan Kasa and Emgayet Shale formations in Libya, Unit I of El-Arnauti & Shelmani (1985) in the subsurface of Northeast Libya, the upper Gouring Formation in the Tibesti Massif, and the lower part of the Zeitoun Formation in the subsurface of northwestern Egypt. The distribution of the Tadrart and Ouan Kasa formations was controlled by the Caledonian structural configuration of the basins. The Lower Devonian is absent on top of the Tihemboka Arch, and the Middle Devonian is present only on the north and south borders of the Arch.

In the subsurface of the Illizi Basin, Algeria, the Lower Devonian (Gedinnian/ Sieginian) is represented by the Oued Mehaiguene Formation or the Hasi Tabankort Formation. The sequence includes the F6 reservoir (equivalent to the Tadrart Formation), made up of regressive deposits, marginal marine to fluvial, medium to coarse sandstones (Askri et al., 1995, Hallett, 2002), which form important hydrocarbon reservoirs (Alem et al., 1998). The F5 Sand (Emsian, equivalent to the Ouan Kasa Formation) is a transgressive episode comprising marginal marine to shelf sandstone facies. The Emsian is typified by grey-black micaceous shales of the lowermost member of the Alrar Formation (palynological zone K, including the F5 Sand) (Jardiné & Yapaudjian, 1968, Jardiné et al., 1974) deposited in a mixed continental-marine environment. In the Ghadames, Murzuq and Kufrah basins, the Emsian Ouan Kasa ferruginous sandstones and shales with spores and chitinozoans (Massa & Moreau-Benoit, 1976) and abundant brachiopods (Jakovljevic, 1984, Galecic, 1984, Grubic, 1984, Grubic et al., 1991) were deposited. They grade into the Emgayet Shale in western Libya and grey-black micaceous shales of the lowermost member of the Alrar Formation (palynological zone K, including the F5 Sand) (Jardiné & Yapaudjian, 1968, Jardiné et al., 1974) in the Illizi Basin in Algeria.

Devonian sediments in Northeast Libya and Northwestern Egypt, like most of the Devonian successions in West Libya and eastern Algeria, display at least three shallowing-upward cycles, corresponding to the Lower, Middle and Upper Devonian. In the subsurface of Northeast Libya (Cyrenaica), the Devonian rests unconformably on the Silurian due to the Caledonian uplift of the area and the drop in sea level at the end of the Silurian and the beginning of the Devonian. No Gedinnian or Sieginian sediments (equivalent to the Tadrart Formation) are present in the area (Streel et al., 1988) and the Emsian (Ouan Kasa Formation) (Unit I of El-Arnauti & Shelmani, 1988) is restricted to the eastern edge of the area near the Egyptian border. The Emsian is represented by shallow-marine, dark grey, pyritic siltstones, grading upward into fine-grained sandstones which include miospores and marine microplanctons, such as acritarchs and chitinozoans. Eastwards, in the subsurface of the northern Western Desert, the Early Devonian sandstones and conglomerates with acritarchs and miospores of the Zeitoun Formation (Keeley, 1989) transgressed unconformably over the regressive sequence of the Silurian Basur Formation. Southwards, continental fluvial sedimentation of fine-grained, tabular cross-bedded sandstones and conglomerates with convolute bedding of the Tadrart and Ouan Kasa formations in the Kufrah Basin, took place on top of the ferruginous and lateritic flat surface of the Tannezouft marine sediments (Bellini & Massa, 1980). Northeast of the Gebel Oweinat and the western Abu Ras Plateau areas in Egypt, Devonian sediments correlatable with the Tadrart Formation (Klitzsch, 1990a) were deposited on top of Ordovician and Silurian sediments of the Karkur Talh and Um Ras formations.

Tectonic uplift during the Middle Devonian caused the removal or incision of the Ordovician, Silurian, and Lower Devonian successions in the Gargaf Arch and the northern part of the Murzuq Basin (Seidl & Röhlich, 1984, Parizek et al., 1984), and in most of the Sirte Basin and Egypt. The lowermost Middle Devonian is absent almost everywhere in Libya and Egypt because of uplift and erosion during the Caledonian movement. During the Middle and Upper Devonian, sedimentation was controlled by eustasy, sedimentary supply, tectonism and subsidence (Adamson et al., 1998).

The Middle Devonian (Couvinian-Givetian) Aouinat Ouenine Group (s.s.) in western Libya represents the second major Devonian cycle of deposition with an overall shallowing-upward trend (Collomb, 1962, Seidl & Röhlich, 1984, Parizek et al., 1984, Pierobon, 1991). It starts with the deep marine sediments of the Bir al Qasr (A.O. I Formation of Massa & Moreau-Benoit, 1976). The sediments consist of medium-grained sandstones, ferruginous sandstones, and variegated silty shales with abundant brachiopods and bone fragments (Hlustik, 1991), and the trace fossils *Spirophyton, Chondrites, Cruziana, Planolites, Bifungites* (Seidl & Röhlich, 1984), *Harlania*, and *Lycophytes*, as well as *Tentaculites* (Bellini & Massa, 1980). Abundant land-derived palynomorphs (Loboziak & Streel, 1989) suggest the presence of a landmass. They grade into argillaceous sediments westward, and pinch out eastward toward the Gargaf Arch (Seidl & Röhlich, 1984, Parizek et al., 1984). The upper part of the cycle is made up of the Idri Formation (Givetian-Early Frasnian) (A.O. II Formation of Massa & Moreau-Benoit, 1976). The Formation consists of cyclic repetition of shales, cross-bedded sandstones, and ferruginous sandstones with coquinas, composed of brachiopods, trilobites, bryozoans, corals, and arthrodire fish (Bellini & Massa, 1980), and large tree trunks (Hlustik, 1991). The sandstones represent a deltaic complex containing progradational and transgressive facies (Vos, 1981a, Blanpied & Rubino, 1998). Delta-front and fluvial distributary channel sediments were reworked by tides and storms into tidal channels, bars and tempestites (Pierobon, 1991) which form the HST tract (Blanpied & Rubino, 1998). These sediments change gradually northward in the Ghadames Basin and Tripolitania into an open-marine, prodelta silty shale facies (Vos, 1981a, Weyant & Massa, 1991).

In the subsurface of the Illizi Basin, Algeria, the Middle Devonian (Couvinian-Givetian) is represented by the upper part of the Alrar Formation (*Formation Argilo-Gréseuse d'Alrar*) (Jardiné & Yapaudjian, 1968, Jardiné et al., 1974). Overlying the Emsian basal unit of the Alrar Formation is a heterogeneous sandy-shaly (F4 Sand) member (Couvinian) with ferruginous, coarse-grained sandstones changing upward into marine argillaceous sandstones with acritarchs, spores and chitinozoans, a fine-grained, fossiliferous sandstone bed caps the succession. The uppermost Alrar Member (F3 Sand) (Givetian) is hydrocarbon-bearing. It is composed of massive, cross-bedded, medium-grained sandstones. The F3 Sand in the West Alrar Field, Algeria, was deposited as tidally influenced littoral bars elongated in a northwest-southeast direction with interbar areas (Chaouchi et al., 1998). The succession forms a progradational sequence (Tawadros et al., 1999) which starts with a lower shoreface facies composed of bioturbated shales and fine-grained sandstones followed upward by a middle shoreface bioturbated sandstone facies. The sequence ends with an upper shoreface sandy bar facies. The sandy bar facies is cut by shale channels. Oil and gas were also discovered by Sirte Oil in 1991 in a combination structural-stratigraphic trap in the Devonian F3 Sand (Givetian) in the Al Wafa Field, about 15 km northeast of Alrar (Traut et al., 1998), with 4–7 TCF of gas reserves (World Oil, 1996). The reservoir is found mainly in the clean sands of the upper shoreface and middle shoreface facies. A silica-cemented horizon forms the lateral seal and the lower boundary of the reservoir (Tawadros et al., 1999).

Spectacular Givetian table reefs, biostromes and patch reefs occur in two areas (Uein Terguet and Gor Loutad) of West Sahara which are part of a narrow belt along the northern margin of the West African Craton (Reguibat Massif) extending for 1500 km from the Zemmour (Mauritania) in the SW to southern Algeria in the East (Wendt & Kaufmann, 2006). They are composed of stromatactis and stromatoporoid mud mounds. West of the Hamar Laghdad Frasnian-Famennian mud mounds (Hollard, 1967, Brachert et al., 1992, Montenat et al., 1996, Mounji et al., 1998, Baidder et al., 2008), a northern east-dipping normal fault was active during the

Emsian, and partly controlled the mud mound formation and preservation in the downthrown block (Montenat et al., 1996, Baidder et al., 2008).

In Northeast Libya, more than 3600 ft (1097 m) of the Middle Devonian sediments (Couvinian-Givetian) (units II–IV of El-Arnauti & Shelmani, 1988) (partly equivalent to the Zeitoun Formation) were deposited in a shallow to marginal marine environment. These sediments are composed of alternations of sandstones and dark grey shales with rare shale and limestone interbeds followed upward by interbedded reddish fine-grained sandstones, glauconitic sandstones, and dark grey shales.

The Ghazalat Basin in Northwestern Egypt was dominated by carbonate shoal, tidal-flat, and fluvial channel deposits of the Zeitoun Formation. Alluvial fan conglomerates were deposited at the edge of the basin (Keeley, 1989, 1994). The presence of acritarchs and large spores suggests marine influence in a very shallow nearshore marine environment (Schrank, 1987). Southwards, in Egypt and Southeast Libya, predominantly fluvial conglomeratic sandstones and siltstones with plant fragments of the Kurkur Murr Formation in the Gebel Oweinat area (Burollet, 1963b, Vittimberga & Cardello, 1963) and the Gilf Kebir area (Klitzsch, 1990a) were deposited directly over the Precambrian basement.

In the Tibesti area and the Kufrah Basin (Bellini & Massa, 1980) Middle Devonian variegated and ferruginous sandstones and siltstones in the lower part of the Binem Formation and its equivalent the Blita Formation contain the trace fossils *Spirophyton* and *Bifungites fezzanensis*, plant remains more than 1 m (3 ft) long and the pelecypods *Kufrahlia* and *Pleoneilos* sp. These facies point to deposition in lagoonal and deltaic environments. The Middle Devonian appears to be missing in Southwest Egypt.

During the Middle Devonian, Northwest Africa was dominated by widespread carbonate deposition. Shale deposition prevailed once more in the Late Devonian until a late Famennian sea-level fall led to rapid progradation of deltaic sandstones (Lubeseder et al., 2009, 2010). Middle Devonian (Givetian) reefs occur in Mauritania and West Sahara, where six reef cycles separated by shales and sandstones have been recognized (Wendt & Kaufman, 2006). They extend into the southern border of the Tindouf Basin and the northern margin of the Reguibat Massif and the Gourara-Ahnet area in Central Algeria (Kaufmann & Wendt, 2000).

The Late Devonian upper depositional cycle in western Libya (Ash Shatti Group of the present work) (Frasnian-Famennian) includes the Quttah, Dabdab, and Tarut formations. The Quttah Formation (Frasnian) (the lower part of the A.O. III Formation of Massa & Moreau-Benoit, 1976) was deposited unconformably on the Idri Formation (Collomb, 1962, Seidl & Röhlich, 1984). On the northern flank of the Murzuq Basin, deposition started with shales and thin sandstones, followed by fine to coarse-grained purple to light brown sandstones with brachiopods, pelecypods, and *Bifungites fezzanensis*. The depositional environments were probably fluvial channels (Blanpied & Rubino, 1998), grading upward into restricted marine to subtidal environments (Seidl & Röhlich, 1984). In the subsurface of the Ghadames Basin, the Quttah Formation is made up mainly of dark shales with dark limestone intercalations with abundant specimens of the palynomorphs *Tasmanites* and *Lophozonotriletes* (Weyant & Massa, 1991). The Dabdab Formation (Frasnian) (the upper part of A.O. III Formation of Massa & Moreau-Benoit, 1976) represents the first cycle of iron-ore sedimentation (Seidl & Röhlich, 1984) of the Ash Shatti Group.

Variegated shales, ferruginous fine-grained sandstones, siltstones, and ferruginous oolites with atripid brachiopods, were deposited in very shallow, restricted marine, and lagoonal environments. The ironstones, which occur in this group, are stacked storm-dominated parasequences with hummocky cross-stratification (Blanpied & Rubino, 1998). The Tarut Formation (Famennian) (A.O. IV Formation of Massa & Moreau-Benoit, 1976) represents the second iron ore-bearing cycle of the Ash Shatti Group and it is very similar to that of the Dabdab Formation. The top of the Formation is marked by a conglomerate which may indicate a minor interruption in sedimentation before the deposition of the Strunian Tahara Formation. This conglomerate is absent in the Awaynat Wanin area. The Formation thins eastwards and the sand content increases toward the Gargaf Arch, which suggest a derivation of the sandstones from the Arch. The depositional environments probably included bays, lagoonal, and shallow-marine swamps (Seidl & Röhlich, 1984). The ironstones in the Dabdab Formation are primary beach and upper shoreface deposits (Blanpied & Rubino, 1998). Stems of *Leptophloem rhombicum* up to 3 m (10 ft) long associated with oolitic iron ores of the Wadi Ash Shatti area (Seidl & Röhlich, 1984) may represent remnants of a forest-like development in tidal flats on the Gargaf uplift (Hlustik, 1991).

The origin of the iron ores in the Wadi Ash Shatti, central Libya, has been discussed by Wood (1964), Goudarzi (1971), Van Houten & Karasek (1981), and Turk et al. (1980). The ore occurs in the Aouinat Ouenine and Ash Shatti groups (Middle-Upper Devonian). They are sedimentary stratiform deposits of Miette- or Loraine-type, and contain thin beds of phosphate composed of collophane, worn fragments of bones and other organic debris and clastic material (Seidl & Röhlich, 1984). They were deposited in an oscillating (cyclic) shallow marine, continental-pelagic environment (Collomb, 1962, Goudarzi, 1971, Turk et al., 1980). Oolites and matrix are made predominantly of bertheriene (iron chlorite), limonite, magnetite and siderite. Iron was leached from soils, probably as colloidal solutions of ferrous hydroxides. On arrival at the sea, any ferrous iron was oxidized and iron was deposited (Goudarzi, 1971, Turk et al., 1980).

Deposition at the close of the Devonian succession is represented by the Tahara Formation. The Tahara Formation (Strunian) is dominated by silty shales, cross-laminated and rippled feldspathic sandstones with *Tigillites*, ferruginous oolite beds, and thin beds of conglomerate with brachiopods, bone fragments, and plant fragments (microconglomerate) similar to those of the Idri Formation (Seidl & Röhlich, 1984). These sediments form two coarsening-upward sequences deposited in a shoreface, tide and storm-dominated shelf environment. The lower sequence is formed of tempestites; the upper sequence is made up of tide-dominated facies and tempestites (De Castro et al., 1991). However, fossil plants and the freshwater algae *Botryococcus* indicate the presence of a brackish-water environment (Vavrdova, 1991). Similar cycles have been recognized in the Tahara Formation in the Illizi Basin (Hasi, 1995).

In the Ghadames Basin, the topmost Famennian unit and the lower part of the Tahara Formation (Strunian) form a radioactive unit (the Cues Limestone Horizon or the *Tornoceras* unit of Bracaccia et al., 1991). The dark grey shales and wackestone to packstone-limestones with goniatites, fish debris, thin-shell pelecypods, ostracods, and *Tentaculites*, which make up the unit, indicate deep-marine facies (Weynat & Massa, 1991). This unit probably represents the maximum flooding at the end of the Devonian and the beginning of the Carboniferous. In the Illizi Basin, Algeria, the Late Devonian Gazelle Formation (Frasnian-Strunian) (Jardiné et al., 1974) was deposited unconformably on top of the Alrar Formation. The lowermost unit is made up of shaly sandstones containing a radioactive zone (Frasnian palynological zone L4) equivalent to the Qattara and Dabdab formations (Seidl & Röhlich, 1984) in western Libya. The radioactive zone is probably diachronous, being younger in southeastern Algeria than in western Libya. This unit has proven to be a good source rock for hydrocarbons. Correlation between this zone and the Cues Limestone in southwest Libya, which is a high possibility, has not been established. In the Eastern Anti-Atlas of Morocco, the Frasnian-Famennian anoxia has been correlated with the Kellwasser Facies or Event (Wendt & Belka, 1991, Wendt & Kaufmann, Wendt et al., 2006, 2009b, Lüning et al., 2004, Bond & Wingall, 2008), which consists of highly fossiliferous black shales and limestones.

Shallow marine to continental, Late Devonian (Strunian) sandstones with *Gorgonisphaeridium solidum* and *Verruciretusisporites* cf. *famenensis* were deposited in the Central Sirte Basin on top of Late Cambrian quartzites (Tawadros et al., 2001). They were probably deposited during a quiet period during the Hercynian tectonic event. These sandstones have good primary porosity and contain hydrocarbons in the EEEE1-59 well.

In Northeast Libya, the Upper Devonian (Frasnian-Strunian) shallowing-upward cycle (Units VI & VII of El-Arnauti & Shelmani, 1988) of the Ash Shatti Group is restricted to the northern and central wells. The Frasnian black shales with acritarchs and chitinozoans were deposited in a shallow, open marine environment, and they were followed by Famennian shales, siltstones and sandstones which contain ferruginous oolites and glauconites deposited in a shallow, marginal marine environment. The Strunian (equivalent to the Tahara Formation and the Late Devonian sandstone in the central Sirte Basin) is represented by delta-front sandstones, siltstones and shales (El-Arnauti & Shelmani, 1988).

In the Ghazalat Basin, deposition of the Zeitoun Formation continued in a very shallow, nearshore, marine environment (Schrank, 1987) with carbonate shoals, tidal flats, and fluvial channels. Alluvial fan conglomerates continued to accumulate around the edge of the basin (Keeley, 1989, 1994). Southwards, in Southeast Libya and southwest Egypt, fluvial conglomeratic sandstones and siltstones with plant fragments of the Kurkur Murr Formation (Burollet, 1963b, Vittimberga & Cardello, 1963, Klitzsch, 1990a) were deposited unconformably on the Precambrian basement rocks.

### 20.3.1.5 Carboniferous period

The Carboniferous Epeiric Sea advanced over the eroded Devonian surface in North Africa. The Carboniferous sediments were probably widespread but were extensively removed during the Hercynian Orogeny. Deposition of sandstones and shales followed approximately the same trend as the Devonian, but also covered southwest the Murzuq Basin (Fig. 20.8). A few areas in Egypt also received sediments for the first time since Lower Silurian time. Deposition was probably related to rapid subsidence and/or sea-level rise, which started at the end of Devonian and continued at the beginning of the Carboniferous. The Devonian-Carboniferous boundary is often difficult to determine on a lithological basis. However, the boundary is characterized by the extinction of

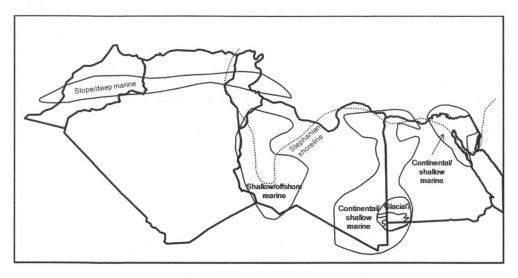

*Figure 20.8* Carboniferous paleogeography (modified from Tawadros, 2001). NW Africa Stephanian shoreline is after Guiraud et al. (2005). Possible glacial deposits may have accumulated in the southwest corner of Egypt (after Klitzsch, 1990a).

the palynomorph *Retispora lepidophyta* (Clayton, 1996), an event which can be recognized worldwide Massa & Moreau-Benoit, 1976).

Africa was dominated during the Middle and Upper Carboniferous Variscan (Hercynian) Orogeny by a compressional regime caused by the collision between Africa and Eurasia and plate organization along the margin of North Africa on one hand, and the subduction along the Paleo-Pacific margin of Gondwana on the other (Visser & Praekelt, 1996). This tectonic phase resulted in the development of a number of continental dome-shaped megastructures which may have reached several hundred kilometers in diameter (Gvirtzman & Weissbrod, 1984). In northern Sinai, the Helez Geanticline was truncated in two phases, the first in the Late Devonian to Early Carboniferous (Famennian-Tournaisian), the second in the Late Carboniferous or Early Permian. In Libya, the Sirte Arch was uplifted and resulted in the removal of Paleozoic sediments down to the Cambrian on the crest, or even down to the Precambrian basement in some areas. The Gebel Oweinat, Jabal Arknu, Tibesti and Hoggar mountains in southern Libya and Egypt were also uplifted during that time (Goudarzi, 1980, Lefranc & Guiraud, 1990, Guiraud et al., 2005). Paleozoic sediments are preserved only on their flanks. In Northwest Libya, the Gharyan and Jabal Nafusah highs were formed as a result of the Hercynian tectonics (Burollet et al., 1971, Koehler, 1982). The Middle Carboniferous compressional regimes resulted also in the formation of mega-shear systems, such as the Falkland-East Africa-Tethys Shear (Visser & Praekelt, 1996), and probably in movements along the Trans-Africa Lineament (Nagy et al., 1976).

The Hercynian tectonic uplift resulted in the erosion of the Carboniferous sediments and the development of two major unconformities within the Carboniferous succession (Table): the first at the end of the Visean, the second at the end of the

Carboniferous and marking the end of the Paleozoic Era. Local uplifts caused local unconformities at the end of Westphalian, Namurian, and Tournaisian. Recent work suggests the presence of remnants of these Carboniferous sediments in the central Sirte Basin (Hallett & El-Ghoul, 1996, Wennekers et al., 1996, Tawadros et al., 2001).

Lower Carboniferous mud mounds occur in the Tafilalt and Maader sub-basins in the Eastern Anti-Atlas of Morocco in several parallel, WNW-ESE-trending belts, within a 4000 m (13,124 ft) thick succession of shales with intercalated bedded limestones, sandstones, and siltstones of Visean stage. Individual mounds are up to 30 m (98 ft) high and have bases of up to 300 m (984 ft) in diameter (Wendt et al., 2001). Four types of mounds can be distinguished: (1) massive crinoidal wacke- or packstones without stromatactis; (2) massive crinoidal wacke- or packstones with rare stromatactis; (3) similar to (2), but allochthonous; and (4) bioclastic (skeletal) grainstone mounds. While carbonate deposition in types (1) to (3) was probably the result of microbial precipitation, type (4) is formed of skeletal debris. Rich assemblages of rugose corals occur in the Upper Carboniferous Tizra, Akerchi and Idmarrach formations in Central Morocco (Said et al., 2010). The biostromes are composed mainly of rugose corals and gigantoproductid brachiopods in a marly limestone matrix. In the Central Atlas, Permian sediments occur in Triassic rift basins, such as the Tizi n'Test Basin. They include the Anrar Conglomerates (F1) and the Cham-El-Houa Siltstones (F2), both of Upper Permian age (Qarbous et al., 2003, Corel et al., 2003).

In West Libya, the Carboniferous is represented by the Mrar Formation and the *Collenia* Beds, and the Assedjefar, Dembaba, and Tiguentourine formations. The Pennsylvanian-Mississippian boundary is placed at the top of the Assedjefar Formation (Bellini & Massa, 1980). All the Carboniferous formations display coarsening-upward shoaling-upward cycles. More than 15 cycles of fluvial-dominated deltas associated with lagoonal, tidal flats, beaches, and tidal and marine bars have been described from the Mrar and Assedjefar formations on the western flank of the Gargaf Arch (Whitebread & Kelling, 1982). These cycles were probably controlled by sediment supply, tectonics, and eustatic sea-level fluctuations and reflect the rapid modifications of the paleogeography. Sea-level lowstand led to the exposure of most of the west Libyan shelf three times during the Visean and once during the early Serpukhovian (Fröhlich et al., 2010).

The initial Carboniferous marine transgression is marked in the Gargaf Arch area by the deposition of the lower part of the Mrar Formation composed of bioturbated and burrowed, medium to coarse sandstones and kaolinitic shales with *Bifungites fezzanensis* and *Tigillites* and abundant ferruginous oolites and concretions (Seidl & Röhlich, 1984). In the Murzuq Basin, alternating intervals of sandstones with ferruginous and oolitic beds and gypsiferous siltstones with brachiopods (Jacovljevic, 1984) were probably deposited in a shallow-marine intertidal environment, tidal flats, offshore bars, and swamps (Jacovljevic, 1984, Seidl & Röhlich, 1984). The upper part of the Formation is represented by the *Collenia* Beds composed of shales, bioturbated fine sandstones and siltstones, pink dolomites and stromatolitic and oncolitic limestone beds which were probably deposited in intertidal environments (Jacovljevic, 1984, Pierobon, 1991). This unit is recognized at the top of the Formation in outcrops and in boreholes in the northern and western parts of the Murzuq Basin, but it wedges out to the south and east. Episodic events, such as tempestites with hummocky cross-stratification, were also common during the deposition of the Mrar Formation (De Castro et al., 1991).

Deposition of the Assedjefar Formation continued conformably on top of the Mrar Formation with no major changes in the depositional environments. The Assedjefar Formation is a transgressive sequence composed of continental, medium to coarse-grained cross-bedded sandstones, grading upward into silty and sandy limestones with occasional conglomerates, large ferruginous concretions, gypsum, brachiopods, and gastropods (Bellini & Massa, 1980, Jacovljevic, 1984). The Formation contains large concretions (*Grès de champignon*) of unknown origin. Deposition of the Assedjefar Formation marks the end of the Lower Carboniferous. The Murzuq Basin was probably interconnected with the Ghadames Basin at this time (Fröhlich et al., 2010). Both the transgressive (TST) and highstand (HST) systems tracts of the Lower Carboniferous are dominantly made up of shallow marine, siliciclastic storm and shoreline deposits alternating with offshore shales (Fröhlich et al., 2010). Deposition of carbonate sediments gradually increased during the Serpukhovian. Bashkirian and Moscovian strata are dominated by shallow marine, calcareous storm and shoreline deposits.

In the Jifarah Plains, Late Carboniferous sediments consist of conglomeratic, coarse sandstones, fine-grained sandstones with dolomitic and anhydritic cements, shales, and beds of limestone and wine red dolomite of the Hebilia Formation (Numurian-Moscovian). These sediments transgressively overlie Silurian sediments (Massa et al., 1974). Rare foraminifera such as Textularides and Endothyrides indicate a shallow marine environment. Upper Carboniferous sediments grade in the Ghadames Basin into carbonates with abundant marine fauna of the Dembaba Formation (Gundobin, 1985). Further south, in the Murzuq Basin, the Dembaba Formation is made up of interbedded silty limestones and cross-laminated feldspathic sandstones with stromatolitic beds at the base (Jakovljevic, 1984). The limestones contain brachiopods, crinoids, corals, and molluscs in the lower part and analcime (zeolites) oolites in the upper part. These sediments suggest deposition in shallow marine and nearshore environments. Jakovljevic (1984) postulated that interaction between brackish water and volcanic ash probably caused the formation of the analcime. However, there is no evidence of volcanic activities in Libya during that time and another mechanism should be sought. The sediments of the Dembaba Formation grade in the subsurface of the Murzuq Basin into fluvio-alluvial, brackish-water lagoonal and evaporitic sediments (Pierobon, 1991). They grade westward in Algeria into the El Adeb Larache limestones with abundant marine fauna (Lefranc & Guiraud, 1990). It appears, therefore, that western Libya was dominated during the deposition of the Dembaba Formation by continental and restricted environments changing into more open-marine facies both northward and westward.

At the end of the Carboniferous, evaporitic and restricted seawater conditions prevailed. This period is represented by varicolored siltstones of the Tiguentourine Formation with intercalations of calcareous siltstones, shales, gypsum, and anhydrite (Grubic et al., 1991). The Tiguentourine Formation is absent on the southwestern flank of the Murzuq Basin and the Dembaba Formation is overlain unconformably by the Triassic Zarzaitine Formation (Jakovljevic, 1984, Galecic, 1984, Grubic, 1984).

In Southeast Libya, the Lower Carboniferous is represented by the Dalma Formation (Tournaisian-Middle Visean) exposed in the Jabal Hawaysh (Vittimberga & Cardello, 1963, Burollet & Manderscheid, 1967) and consists of more than 500 m (1641 ft) of variegated fine to coarse-grained quartzose, cross-bedded sandstones with abundant

*Lepidodenderon* and spores (Grignani et al., 1991). Turner (1980) described the sedimentary facies of the Dalma Formation as braided stream at the base changing upwards into subtidal, lower shoreface and shelf-shoreface environments.

In Northeast Libya, the Tournaisian is missing, but a complete section from Early Visean to Gzelian (Stephanian) is present in the northern part of the area (El-Arnauti & Shelmani, 1988). The Visean succession is up to 2240 ft (683 m) in thickness and shows a shallowing trend both upwards and southwards. The Early Visean in the northern wells is represented by delta-front interbedded sandstones, siltstones and shales. The remainder of the Visean section includes shallow marine shaly and silty limestones at the base which become oolitic near the top in the north and change into limestones and dolomites southwards. Toward the south, the lower sandstones contain ferruginous pellets and glauconite and beds of shales and lignite. The Upper Carboniferous is present only in the north and is composed of limestones, sandstones, siltstones and shales with minor coal laminations. The succession was deposited in a high energy, near-shore to shoal, and marginal marine to lagoonal environments. The change in facies reflects the gradual withdrawal of the sea at the end of Carboniferous. The change in sedimentation patterns and the absence of the Upper Devonian and Upper Carboniferous in the south are attributed to uplift and erosion due to the Late Hercynian Orogeny.

In the Northern Western Desert of Egypt, deposition in the Ghazalat Basin ended in the Strunian with uplift and erosion. On the other hand, the Tehenue Basin started to subside (Keeley, 1989) and received thick sediments. The Carboniferous succession is made up of fluvial sediments at the base and gives way upward to fluvio-marine and marine sediments. High-energy fluvial sandstones, shoreface-delta front sandstones and siltstones and prodelta shales of the Desouky Formation (Tournaisian-Early Visean) (Keeley, 1989) were deposited unconformably on the Devonian Zeitoun Formation. The following Dhiffah Formation (Early Visean-Late Namurian) is composed of calcareous and carbonaceous shales, sandstones, and oolitic and bioclastic limestones deposited in alluvial-plains, tidal-flats and delta-front, prodelta shales and carbonate shoals. The overlying Safi Formation (Namurian-Early Permian) (Keeley, 1989, 1994) is made up of sandstones and siltstones changing facies into shales and oolitic limestones northwards. The Formation is present only in the westernmost part of Egypt due to erosion of the Carboniferous succession to the east.

In Southwest Egypt, Carboniferous sediments were deposited in fluvial environments, which gave way upward to marine environments. The Early Carboniferous Wadi Malik Formation fining-upward sandstones represent braided river deposits at the base, high-sinuosity river point bars in the middle, and shallow marine tidal channel-fill at the top (Wycisk, 1990). This sequence is capped with heavily bioturbated sandstones with brachiopods and the trace fossil *Bifungites*. Similar deposits occur in the Wadi Qena area (Somr el Qaa Formation). The Late Carboniferous North Wadi Malik Formation rests unconformably on the Wadi Malik Formation and it is characterized by glacio-fluvial diamictites, braided channel sediments and lake deposits (Wycisk, 1990, Klitzsch, 1983, 1990a, Klitzsch et al., 1989). These deposits are the only record of glacial deposits in the Carboniferous of Egypt and Libya. Jux & Issawi (1983) postulated that Carboniferous sediments in the Wadi Qena were deposited under glacial condition, but Bandel et al. (1987) refuted this idea. Late Carboniferous glacial deposits similar to those recorded from the Late

Ordovician of North Africa are widespread in Gondwana (Levell et al., 1988, Love, 1994). Paleomagnetic reconstruction of Gondwana (Scotese et al., 1979) indicates that during the Late Carboniferous-Early Permian, southern Egypt lay at a paleolatitude of about 40°S. The North Wadi Malik deposits may represent the edge of the Late Carboniferous periglacial deposits of Gondwana.

In Sinai, Lower Carboniferous fluvial and marine sediments were deposited over Cambrian sandstones (at Um Bogma) or the Precambrian basement (at Gebel Nukhul). The Carboniferous sediments consist of sandstones with *Skolithos* and *Cruziana*-like tracks, sandy dolomites and limestones of the Um Bogma Formation with manganese and iron ores and some coal seams at the base (Brenckle & Marchant, 1987). These sediments are overlain in turn by sandstones, siltstones, and shales of the Carboniferous Abu Thora Formation (Weissbrod & Perath, 1990) or the Ataqa Formation. At Wadi Araba, on the western side of the Gulf of Suez, the Um Bogma Formation ("Nubia A" in the subsurface of the Gulf of Suez) shales, sandstones, and crinoidal and dolomitic limestones are overlain by marls and limestones of the Ataqa Formation. Southward, the Abu Durba Formation consists of cross-bedded sandstones with *Lepidodenderon* and *Calamites* tree trunks in the lower parts, alternating shales and sandstones in the upper parts, and a coquina with bryozoans and spiriferids at the top. The lower part was considered as a fluvio-marine equivalent of the Um Bogma Formation (Issawi & Jux, 1982). These sediments continue northward into the Rod el Hamal Formation, Abu Darag Formation, and Ahmeir Formation. These three formations are composed of variegated shales, sandstones, and crinoidal, bryozoan, and/or oolitic limestones at the top. The microfaunal association and the presence of reef-building corals, bryozoans, and calcareous algae suggest deposition in shallow open-marine environments and carbonate shoals and mounds. In Wadi Qena, coarse, angular material was reworked from the granitic basement which it overlies, followed by reddish sandstones. It was followed by shallow-marine, fine-grained, laminated sandstones, and then by sandstones intercalated with thin conglomeratic beds. Iron-stained *Skolithos* burrows (pipe-rocks) and *Cruziana*-like trails are abundant in the upper part of the section (Bandel et al., 1987). At Wadi Qiseib, north of Abu Darag along the Red Sea coast, continental red shales and sandstones contain *Cordites* megaflora and microflora of latest Carboniferous-earliest Permian age (Schürmann et al., 1963). The Abu Darag Formation (Westphalian) sandstones, shales, and limestones with arenaceous foraminifera and basalt flows were deposited in a nearshore marine environment (Omara & Kenawy, 1966, Omara et al., 1966).

The Carboniferous fossil associations in Sinai indicate that the Carboniferous was deposited in a subtropical epicontinental sea which covered a great area in North Africa (Kora, 1995) (Fig. 20.8). The Lower Carboniferous paleoflora of Egypt and Libya are very similar and suggest a hot climate during that period, but the climate was probably not as dry in Egypt as in Libya, as indicated by the abundance of the vascular plants *Pteridophylla*. The abundance of these faunas in Sinai suggests that the climate in that area was humid and not as warm as in other areas (Lejal-Nicol, 1990). For a review of global diversity of Carboniferous phytoplanktons, see Mulling & Servais (2008).

The manganese deposits of Um Bogma form lenses in dolomitic beds at the base of the Um Bogma Formation. The ore lenses are 1 m (3 ft) to 200 (656 ft) in diameter and up to 6 m (20 ft) in thickness. The larger lenses are differentiated into Mn-rich cores and Fe-rich margins. The manganese minerals include psilomelane, pyrolusite, plianite, manganite, hausmannite and pyrochlorite, hematite and goethite are the only iron-bearing minerals. The average Mn content is 22% and the average iron is 35% (Mart & Sass, 1972). Two major theories have been put forward to explain the ore genesis at Um Bogma: 1) epigenetic or hydrothermal origin (Attia, 1956, Weissbrod, 1969), and 2) syngenetic or sedimentary origin (Mart & Sass, 1972). According to Attia (1956) the process leading to the formation of the manganese and iron ores is related to the Miocene faults and volcanic activities, resulting in hydrothermal fluids rich in manganese and iron. The hydrothermal fluids also resulted in the dolomitization of the limestones. The solutions were stopped by the dolomites at the base of the Um Bogma Formation which were replaced by the ores to form lenticular bodies. Mart & Sass (1972) rejected Attia's idea because most of the ore bodies are unaffected by faults or dikes. They proposed that the ore is of sedimentary origin formed syngenetically in tidal pools and that the differentiation of the lenses was post-depositional, probably due to Eh-pH differences between the cores and margins.

In Morocco, the Lower Carboniferous in the Draa Valley in the Western Anti-Atlas includes the Tazout Group, the Betaina Formation, the Jbel Reouina Formation, and the Ouarkziz Formation (Soulaimani & Burkhard, 2008) dominated by sandstones and arkoses with brachiopods and goniatites, such as *Gattendorfia*. In the Eastern Anti-Atlas, in the southern Tafilalt area, the Lower Carboniferous consists of shallow-marine siltstones and sandstones (Burkhard et al., 2006). The Early Carboniferous of the Eastern Anti-Atlas includes olistostromes that were probably deposited in an intramontane trough. Two equivalent Carboniferous ammonite horizons in the Tafilalt-Gourara area equivalent to the *Gattendorfia-Eocanites and Gattendorfia-Kahlacanites* Zones have been assigned a middle-Early Tournaisian age (Ebbighausen et al., 2004, Bockwinkel & Ebbighausen, 2006, Korn et al., 2000, Korn et al. 2010a-c, Brice et al., (2007). The Late Visean, Namurian and Westphalian sequences of the Ouarkziz, Betana and Bechar areas in Morocco and Algeria represent a regressive megasequence (Soulaimani & Burkhard, 2008). The succession is overlain by the Betana Formation (Namurian-early Stephanian) composed essentially of continental fine-grained sandstones and shales with plant fragments (Conrad, 1985).

The Carboniferous of the Jerada (Djerada) Basin in the Eastern Meseta contains a Visean basin-fill sequence which starts with rhyolitic, andesitic, and dacitic volcano-sedimentary rocks of the Oued Defla Formation (Herbig & Aretz, 2007), followed unconformably by the Namurian-Westphalian succession of the Cafcaf (Safsaf) Formation composed of interbedded cherts and pyroclastics, then by the Oued El Koriche Formation (Herbig & Aretz, 2007, Aretz et al., 2007). Shaly slope deposits of the Oued Es-Sassi Formation include Goniatite-bearing horizons. The basin fill ends with a prograding shallow-marine carbonate facies with various reef facies (Aretz et al., 2007) and siliciclastic sediment of the Koudiat Es-Senn Formation. Emersion took place at the end of the Visean. The Lower Carboniferous in the Central Meseta is marked by a renewed increase in detrital input from the east, the south and the north (Mamet et al., 1966, Michard et al., 1988, Piqué & Michard, 1989). Carboniferous siliciclastic sediments in the eastern part of the Azrou-Khenifra Basin are also intruded by basic igneous rocks "*Serie de Tanwalt*" (Bensaid et al., 1985, Bamoumen et al., 2008). Visean Waulsortian mudmounds occur in the Anti-Atlas (Wendt et al., 2001) and carboniferous rugose corals form biostromes, such as in the Akerchi Formation (Said & Rodríguez, 2007, Said et al., 2010). Fossil assemblages suggest a temperate to warm environment (Said et al., 2010).

At the end of the Carboniferous (Visean), the Djeffara Basin began subsiding and received thick Carboniferous and Permian sediments (Burollet et al., 1978). This Tethyan embayment retreated eastward from Morocco after Early Carboniferous time and from Tunisia after Late Permian time leaving Africa (Toomey, 1991).

### 20.3.1.6   Permian period

The release of heat in the Late Carboniferous-Early Permian resulted in subsidence and widespread basin formation in Africa (Visser & Praekelt, 1996). Sea levels reached their lowest level in the Late Permian (Haq et al., 1987) (Fig. 20.9). The Late Permian period was also dominated by rifting. Magmatic activity associated with Permo-Triassic rifting episodes occur son the eastern side of the Kufrah Basin in the Gebel Oweinat area (Schandelmeier et al., 1987). Trachytes are interbedded with black siltstones, mudstones and sandstones (*Silts des Plateau*) of Burollet (1963b). Spilitic basaltic and doleritic flows are also known from Permo-Triassic rocks in the northern part of the Oued Mya Basin in Algeria (Bossière, 1971, in Lefranc & Guiraud, 1990), and from the Sirte Basin and Northern Egypt (Massa & Delort, 1984). The Pelagian Block, the Jifarah Plains, Cyrenaica Basin, the Northern Egypt Basin and northern Sinai probably subsided at that time and received thick sediments. Shallow-marine carbonate deposition took place on the Pelagian Platform (Van Houten & Brown, 1977) and in the northern parts of Egypt (Keeley, 1989, El-Dakkak, 1988) and Libya (Brugman et al., 1985, Grignani et al., 1991, Adloff et al., 1985, 1986, Mennig et al., 1963, El-Arnauti & Shelmani, 11988). Thick Carboniferous-Permian shallow-marine detrital and minor carbonate sediments were deposited in the Jifarah Basin in

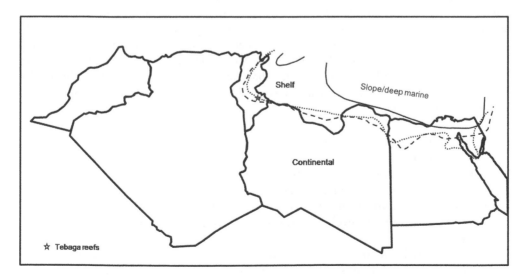

*Figure 20.9* Permian paleogeography (after Guiraud et al., 2005). Shorelines: dotted, Early Permian, dashed, Late Permian; solid, slope/deep water.

Northwest Libya north of the Azizia Fault (Koehler, 1982) and in the Permian Basin in Tunisia (Burollet et al., 1971, Bishop, 1975). However, deposition was dominated by continental and transitional facies southwards. A major hiatus exists between the Permian and Triassic. The Late Permian and Early Triassic are missing in many places in North Africa.

In Northwest Libya, the Lower Permian is absent and the area probably remained high during that time. Deposition during the following period in Middle-Upper Permian times was dominated by unfossiliferous sandstones and shales with iron-bearing horizons and abundant lignite debris of the Watiah (Uotia) Formation (Mennig et al., 1963, Adloff et al., 1985, 1986). The palynomorphs assemblage, such as with the palynomorphs *Lueckisporites virkkiae* and *Gigantosporites hallstattensis*, is mainly of continental origin and deposition probably took place in a brackish-water non-marine environment (Adloff et al., 1985, 1986), or marginal marine. An Early Permian shallowing-upward sequence, up to 2520 ft (768 m) in thickness, was deposited in Northeast Libya (El-Arnauti & Shelmani, 1988). The sequence starts with limestones, shales, sandy shales, and dolomites, deposited in a shallow marine environment, and grades upward into deltaic and continental sandy shales and sandstones. The Upper Permian is absent from that area. Equivalent sediments in the northern Western Desert of Egypt include sandstones and siltstones with the pollen *Vittania* sp. and the monosaccate *Cannanoropollis janakii* of the upper part of the Safi Formation, which change into shales and oolitic limestones northward (Keeley, 1989).

Continental Permian sediments constitute the upper part of the Permo-Triassic Madadi Formation (De Lestang, 1965) in the southwestern Kufrah Basin and they are made of variegated silty shales and red to brown fine to medium-grained sandstones and disconformably overlie the continental Carboniferous beds of the Ounga Formation (Vittimberga & Cardello, 1963). The Upper Permian in the subsurface of the Kufrah Basin is dominated by continental sandstones with the palynomorphs *Lueckisporites virkkiae* and *Gigantosporites hallstattensis* (Grignani et al., 1991). The two species have been reported from NW Libya by Mennig et al. (1963) and Adloff et al. (1985, 1986). In Tunisia, Lower and Middle Permian sediments were recognized in a few wells in Southern Tunisia (Burollet et al., 1978). The only marine Upper Permian sediments crops out only at Tebaga. Thick shales were deposited in basins to the north. In the Djeffara Basin in Western Tunisia, reefoid facies were developed (Burollet at al., 1978, Toomey, 1991).

In Northern Sinai, early Paleozoic sediments (Cambrian to Devonian) were uplifted and truncated in Permian times (Gvirtzman & Weissbrod, 1984). Permian to Triassic sediments overlie Precambrian basement on the crest of the Helez Geanticline and Paleozoic sediments (Cambrian to Carboniferous) wedge out northward toward the high. The crust of the geanticline is of typical continental composition. In the Late Permian the area began to subside in response to rifting (Garfunkel & Derin, 1984), but because of the low sea levels, deposition was dominated by continental variegated sandstones.

Along the Red Sea coast, deposition during the Early Permian was dominated by red, continental, ferruginous sandstones, siltstones and shales with *Cordites* megaflora and microflora (Schürmann et al., 1963) of the Qiseib Formation. These sediments were deposited by northward-flowing braided channels (Allam, 1989).

## 20.3.2 Mesozoic

### 20.3.2.1 Triassic period

Triassic sediments are exposed along the Jifarah escarpment in Northwest Libya and Tunisia for a distance of 350 km and extend into the subsurface of the Jifarah Plains. They also occur in the Ghadames Basin and Murzuq Basin in West Libya, and in the Kufrah Basin in Southeast Libya. Only remnants of unknown extent of Triassic sediments occur in the subsurface of the Eastern Sirte Basin. In Egypt, Triassic sediments occur in the subsurface of the Ghazalat Basin and the Nile Delta region, at Wadi Araba along the Red Sea coast, and in Northern and Central Sinai.

With the beginning of the Mesozoic the paleogeography of North Africa became differentiated into three major facies belts, an open-marine belt to the north, a shallow carbonate belt in the middle and a continental belt to the south. These facies belts are most visible in Tunisia (Marie et al., 1984), but they are also apparent in West Libya and Egypt, Algeria, Tunisia, and Morocco. The paleogeography during the Triassic was controlled essentially by the rifting events, sea-level fluctuations, and the opening of the Tethys and Atlantic oceans. The Triassic was a time of a low sea level (Haq et al., 1987), and the Tethys covered only small areas in Libya and Egypt (Fig. 20.10). Carbonate deposition prevailed over most of Northwest Africa and Northern Sinai, but continental clastic deposition of fluvial and deltaic sediments dominated in the south. To the east and northeast and in the Pelagian Sea (including the Gharyan High in Libya), carbonate rocks were replaced by evaporitic facies (Fig. 20.10).

The Triassic succession in North Africa is similar to the standard tripartite Germanic Triassic with detrital lower layers (Bundsandstein), middle carbonates (Muschelkalk) and upper evaporites (Keuper). The sequence corresponds to the supercycles UAA1 to UAA4 of Haq et al. (1987) and "Supersequence S" of Bobier et al. (1991). The Early Triassic (Scythian) started with a low-stand tract (LST), followed by a transgressive

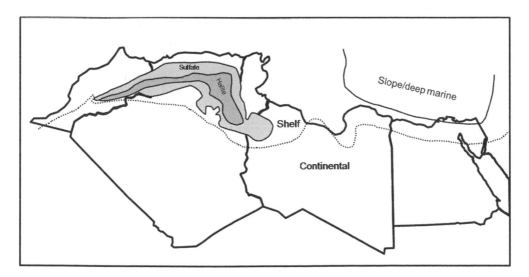

*Figure 20.10* Triassic Paleogeography (after Guiraud et al., 2005). Upper Triassic-Lower Jurassic salt basin is after Turner & Sherif (2007).

tract (TST) during the Middle Triassic (Anisian-Ladinian), which culminated in the early Upper Triassic (Carnian) with a high-stand tract (HST). The late Upper Triassic (Norian-Rhetian) was a time of gradual regression and uplift.

The opening and closing of the paleo-Tethys during the Late Permian-Middle Triassic dextral (clockwise) motion of Eurasia relative to Africa (Robertson & Dixon, 1984) was followed by the opening of the Neo-Tethys and the Eastern Mediterranean in the Middle Triassic, accompanied by the final closure of the Paleo-Tethys during the "Cimmeride Event" of Sengör et al. (1984). The Early-Middle Triassic extensional tectonics associated with the rifting stage resulted in the partial collapse of a number of paleotectonic highs such as the western margin of the Hercynian Helez Geanticline in northern Sinai (Gvirtzman & Weissbrod, 1984, Weissbrod & Horowitz, 1989), the Jifarah High, and the eastern margin of the Sirte Arch in Libya. Thermal subsidence, rifting and shear motion led to the formation of pull-apart basins along the margin of the North African craton in the Mediterranean Sea during that stage. Shallow-marine carbonate sediments accumulated in the offshore in the Ionian Sea, the Pelagian Platform and the Levant Sea and continental clastic sediments and evaporites accumulated inland. Clastic deposition in alluvial fans and braided and meandering river channels dominated in the Tripolitania Basin, the Jifarah Plains (Del Ben & Finetti, 1991), over most of Libya, northern Egypt, and Sinai. Restriction of the western Tethys basin during the Middle and Upper Triassic led to the deposition of a thick series of Triassic evaporites (Bishop, 1975, Demaison, 1965). Evaporites were deposited in the Gulf of Gabes and inland in Tunisia, especially in the Atlas Mountains, where halite forms prominent diapirs (Gill, 1965, Burollet, 1963c, 1979, Bishop, 1988, Ben Ferjani et al., 1990). These evaporites contain salt up to 300 m (984 ft) in thickness near Hassi Massoud (Messaoud) in Algeria (Demaison, 1965). Shallow marine, lagoonal and intertidal Triassic sediments with some evaporites were deposited unconformably on the truncated Precambrian basement in northern Sinai and on Paleozoic sediments southwards (Druckman, 1974, Gvirtzman & Weissbrod, 1984, Derin & Garfunkel, 1984). Lateral gradation from Triassic evaporites to carbonates was observed in the field along Jebel Nafusah (Magnier, 1963, Burollet et al., 1978, Burollet, 1991).

In the subsurface of Tunisia south Tebaga, the Triassic succession overlies unconformably the Permian sediments, where parts of the Upper Permian and Lower Scythian are absent (Kilani-Mazraoudi et al., 1990). In Central Tunisia, it overlies conformably Permian sediments, but rests unconformably on older Paleozoic rocks in the south (Peybernès, 1991).

Fragmentation of central Pangaea probably occurred in three major phases from the late Ladinian (230 Ma) to the Tithonian (147.7 Ma) (Schettino & Turco, 2009). In the first phase, from the late Ladinian (230 Ma) to the Rhetian (200 Ma), rifting proceeded along the Northwest African margin and the Atlas Domain. During the second phase, from the latest Rhetian (200 Ma) to the Pliensbachian (185 Ma), the Moroccan Meseta continued to drift away from North America and the Atlas rift reached its maximum extent. In the third phase, from the late Pliensbachian (185 Ma) to 147.7 Ma, rifting in the Atlas zone ceased and plate motion started along complex fault systems between Morocco and Iberia. The inversion of the Atlas rift and the subsequent formation of the Atlas Mountain Belt occurred during the Oligocene-early Miocene time interval, while in the Central Atlantic; higher spreading rates took place (Schettino & Turco, 2009).

This rifting stage is manifested by the presence of volcanic rocks and the formation of polymictic conglomerates at the southern rifted margin of the Tethys, such as those encountered in the Helez Deep Borehole in the coastal plains of the Levant Sea. The conglomerates are up to 1236 m (4055 ft) in thickness and composed of different carbonate rock types (Druckman, 1984). In the northern Ghadames Basin they are associated with the Azizia Fault (Mokaddem, 1995) and in the Jifarah Plains of Tunisia they occur at the base of the Norian-Rhetian dolomites (Ben Ferjani et al., 1990). Olivine basalt dikes intrude the lower part of the Arif en Naga Formation (Saharonim Formation) (Carnian) (Bartov et al., 1980). Andesite and dolerite (diabase or microgabbro) flows occur in the Early Triassic sediments in Tunisia along a major erosional unconformity (Demaison, 1965). These volcanics were probably the result of extensional tectonics associated with rifting. In the Argana Basin in Morocco, Lower Triassic sediments are cut by intrusions of alkaline granites (Pouclet eta al., 2008) in the High Atlas and in western Rehamna in the Western Meseta (Amenzou & Badra, 1996). The origin and age of the Azegour and Brikiine granites is controversial.

## 20.3.3   Early Triassic (Scythian)

The Early Triassic consists of the Bir el Jaja Formation (Scythian) in NW Libya and Tunisia, the lower parts of the Zarzaitine Formation in Southwest Libya, the Madadi Formation in Southeast Libya, and the Zafir Formation in the subsurface of Northern and Central Sinai. Triassic sediments occur in the subsurface of Northern Algeria where they are composed of red sandstones at the base, followed by thick evaporites, carbonates, and dolomites with basic volcanic intercalations at the top, followed upward by Middle Triassic limestones and red sandstones with doleritic rocks (Askri et al., 1995). In the Babors area to the south, the Middle Triassic is overlain by Upper Triassic lagoonal, gypsum-bearing units. Upper Triassic evaporites occur over most of Northern Algeria. In the Hodna Mountains in Northern Algeria, the Triassic succession consists of sandstones at the base and evaporites with dolomitic limestone beds at the top (Bouchareb Haouchine & Boudoukha, 2009). In Morocco, the Lower Triassic is represented by the Ikakern/Argana Formation in the Argana Basin.

Shallow-water carbonates were deposited on the Pelagian Shelf in Sicily and Northern Tunisia (Del Ben & Finetti, 1991). Southward in Tunisia, the Early Triassic low-stand tract (LST) is dominated by continental red sandstones and shales of the Bir Mastoura Formation (Lower Scythian) (Ben Ferjani et al., 1990), followed by reddish-brown shales, siltstones, and fine-grained sandstones of the Bir el Jaja Formation in the Jifarah Plains in NW Libya and Tunisia (Demaison, 1965). The Bir Mastoura Formation has not been reported from the Libyan Jifarah Plains. The Basal Triassic unit in the northwestern faulted margin of the Ghadames Basin consists mainly of conglomerates interfingering with sandstones and mudstones, and unconformably overlies Ordovician sedimentary rocks. These sediments were deposited in alluvial fans as debris flows, sheet floods, channel sandstones, lag conglomerates and interchannel deposits (Mokaddem, 1995). Similar Triassic deposits occur in the Oued Mya Basin in Algeria (Bossiere, 1971, Lefranc & Guiraud, 1990, Bourquin et al., 2010). In the Jifarah Plains, the Triassic conformably overlies the Permian succession, but further south there is a major unconformity at the base of the Triassic which overlies various Paleozoic formations (Ben Ferjani et al., 1990). A major hiatus is also present between the Permian and Triassic in northeastern Libya (El-Arnauti & Shelmani, 1988). The Upper Permian and the Lower Triassic (Scythian) are absent. The earliest Triassic marine transgression in Northern Sinai led to the deposition of shallow marine shales, fossiliferous limestones and sandstones of the Zafir Formation (Scythian) (represented only in the subsurface) (Druckman, 1974b, Bartov et al., 1980). Conformable relations between the Permian and Lower Triassic successions exist over Northern and Central Sinai, but southwards, Triassic sediments overlap eroded Late Paleozoic surfaces where deposition of fluvial and deltaic clastic sediments took place (Bartov et al., 1980).

In the Argana Basin in Southwest Morocco, continental sediments of the Ikakern/ Argana Formation (Brown, 1980, El Wartiti et al., 1990, Saber et al., 2007, Hmich et al., 2006), up to 2,000 m (6562 ft) in thickness, start with the Driss River Conglomerates followed by fluvial and lacustrine sediments of the Tourbihine Member (Saber et al., 2007). They contain Tetrapod skeleton remains (Jalil & Dutuit, 1996).

The Middle Triassic (Anisian-Ladinian) sedimentation was mostly nearshore marine with continental influence. In the offshore of Tunisia and Northwest Libya starts with a transgressive system tract facies (TST) made up of shales, dolomites and limestones of the Ouled Chebbi (Al Guidr) (Anisian) followed by another transgressive system tract facies composed of sandstones and minor limestones with species of *Myphoria*, dwarfed *Lingula*, and abundant pollen and spores (Adloff et al., 1985, 1986) of the Ras Hamia Formation (Ladinian) (Hammuda et al., 1985). The Ras Hamia Formation grades southwards into glauconitic and phosphatic sandstones, shales, red ferruginous shales, and minor limestones with dwarfed *Lingula* and crab coprolites, deposited in a nearshore, deltaic environment (Adloff et al., 1985, 1986) in the Jifarah Plains and the Jebel Nafusah area (Mennig et al., 1963) and westwards into the Kirchaou Formation sandstones (Ladinian) in Tunisia. These sediments (e.g. TAGI Sandstones) form oil reservoirs in the El Borma Oil Field in Southern Tunisia and Northeast Algeria (Ben Ferjani et al., 1990, Boote et al., 1998) and a gas reservoir in well A1-23 in Northwest Libya.

The Triassic succession at Djebel Rheouis, west of Sfax, is made up of a succession of gypsum, black dolomitic limestones, and minor sandstone intercalations equivalent to the Ras Hamia, Azizia, Bu Sceba, and Bir El Ghnem formations in Libya (Burollet, 1963). They are equivalent to a Ladinian-Norian succession at Djebel Rheouis and a Norian-Rhetian section that occur in the Nara-Sidi Khalif Range north of Djebel Rheouis (Soussi et al., 2000, Kamoun et al., 1994, and Courel et al., 2003). However, Triassic outcrops of the Koudiat El Halfa and Rheouis Domes are mainly composed of diapirs of gypsum, red marls, clays, silt, and dolomites of the Rheouis Formation (Courel et al., 2003, Mehdi et al. (2009). In the Djebel Debadib Anticline in the Tunisian Atlas near El Kef, the Rheouis Formation consists of a deformed breccia containing clasts of supratidal dolomites, terrigenous clastic rocks, and fragments of basement rocks set in a gypsum-matrix (Snoke et al., 1988).

In the Central Atlas, Triassic rift basins, such as the Tizi n'Test Basin were formed (Qarbous et al., 2003, Corel et al., 2003) bounded by ENE-WSW-trending faults dipping to the NNW. Overlying the Upper Permian sediments, the succession consists of the Timalizene Conglomerates (Middle Triassic), the F4 Formation or the Anouffig Siltstones (Middle Triassic), the F5 Formation or the Oukaimeden Sandstones (Carnian), and the F6 Formation or the Tafilalt Formation (Upper Siltstone Formation)

(Corel et al., 2003). In the extensional grabens of the Ida Ou Zal and Ida Ou Ziki basins (Western High Atlas) and the Ida Ou Zal Basin (Western Meseta), Upper Triassic succession overlies unconformably Upper Permian sediments (Saber et al., 2007).

The Middle Triassic in the anticlinal structure at Arif el Naga in Sinai begins with the Raaf Formation (Anisian) composed of fossiliferous limestones with echinoids, molluscs and foraminifera fragments (Druckman, 1974b, Bartov et al., 1980), followed upward by a transgressive system tract (TST) of the Gevanim Formation (Ladinian) (lower Arif el Naga Formation), composed of fine to medium-grained, cross-bedded sandstones and shales with plant fragments, petrified wood, poorly preserved fossils and a few limonitic limestone beds with ammonites (Druckman, 1974b, Bartov et al., 1980). Deposition took place in terrestrial, shallow coastal to shallow-marine environments (Bartov et al., 1980). Paleocurrent analysis shows a south-easterly source for the sediments (Karcz & Zak, 1968, Jenkins, 1990).

## 20.3.4 Upper Triassic (Carnian-Rhetian)

The Upper Triassic represents a shallowing-upward regressive cycle which started with a high-stand tract (HST) sequence during the Carnian and ended with uplift and erosion at the end of the Triassic. Maximum marine transgression of the Triassic Tethys took place during the Carnian and led to widespread deposition of a HST sequence of shallow marine carbonates on the Pelagian Platform, in Northern Tunisia, Libya, Egypt and Sinai, changing southward into tidal flat deposits and then into continental clastics. Dasycladacean algal and oolitic limestones, dolomites, dolomitic sandstones and shales of the Azizia Formation in the offshore of Northwest Libya (Hammuda et al., 1985, Del Ben & Finetti, 1991) and in the subsurface of the Jifarah Plains (Asserto & Benelli, 1971, Adloff et al., 1985, 1986, Kruesman & Floegel, 1980) were deposited in low energy, shallow subtidal-intertidal environments with estuaries, marshes and sabkha plains. The abundance of bisaccate pollen in the Azizia sediments (Adloff et al., 1985, 1986) indicates proximity to land. The entire Triassic succession changes facies southward into the continental clastic sediments of the Zarzaitine Formation. Westwards, the Carnian in Tunisia consists of three units equivalent to the Azizia Formation, the Mekraneb Formation dolomites, the Touareg Formation sandstones and the Rehach Formation dolomites (Marie et al., 1984, Ben Ferjani et al., 1990). Epeirogenic tectonic uplift took place at the end of the Middle Carnian. The top of the Touareg Sandstones is marked by a ferruginous hardground (Ben Ferjani et al., 1990). In Sinai, the Carnian marine transgression led to the deposition of a high-stand tract sequence (HST) of limestones and marls with ammonites of the Saharonim Formation (upper Arif el Naga Formation). The environment was dominated by a shallow subtidal, open-marine shelf in the north, changing southward into tidal flats (Jenkins, 1990). The whole Triassic succession changes facies in Central Sinai into continental clastics of the Qiseib and Budra formations.

The Carnian high-stand episode was followed by a global sea-level fall (Haq et al., 1987) and gradual regression during Norian-Rhetian times, which led to the formation of extensive tidal flats and the deposition of supratidal sabkhas and lagoonal sediments. Widespread dolomitization of carbonate sediments, associated with the restricted marine conditions, took place over most of North Africa and the Pelagian Platform. The Norian-Rhetian in Tunisia, the Pelagian Sea, and the Malta

Platform is dominated by supratidal-intertidal dolomites with scattered anhydrite nodules and a few basalt layers (Bishop & Debono, 1996). Deep exploration wells in southern Sicily and Malta penetrated more than 3000 m (9843 ft) of Late Triassic dolomites (Gill, 1965). In the Jifarah Plains, the facies of the Bu Sceba Formation grade upward from shallow open-marine feldspathic sandstones and siltstones and sandy dolomites with crinoids, algae, ostracods and pelecypods, into intertidal and shallow subtidal marls with mud-cracks and collapse breccia, then into non-marine alluvial delta channelized cross-bedded sandstones, conglomerates, shales, and siltstones, and finally into evaporites and dolomites deposited in restricted arid lagoons or coastal lakes (Assereto & Benelli, 1971). The sandstones were transported westwards from a granitic, gneissic, metamorphic and clastic source area. These sediments extend into Tunisia, and include clays and gypsum of the Mhira Formation, dolomites with polymictic conglomerates of the Messaoudi Formation, and evaporites of the Bhir Formation (Ben Ferjani et al., 1990).

In Northeast Libya, marine Late Triassic limestones are restricted to the north (El-Arnauti & Shelmani, 1985, Brugman et al., 1985). In the subsurface of the northern Western Desert, for example in the Matruh Basin, the Late Triassic is represented by carbonates, evaporites and sands of the Behir Formation (Early to Late Triassic), and dolomites and limestones and minor sandstones, shales, and evaporites of the Fadda Formation of Norian-Rhetian age (Keeley et al., 1990, Keeley & Massoud, 1998). In Northern Sinai, restricted marine conditions led to the deposition of dolomites, dolomitic shales, dolomitic limestones, algal stromatolites, and anhydrites (Druckman, 1974a, Bartov et al., 1980, Jenkins, 1990) with *Spiriferina lipoldi* and *Myophoria inequicostata* (Druckman, 1974b) of the Mohilla Formation, present only in the subsurface (Druckman, 1974b).

Tectonic uplift at the end of the Triassic affected most of the Pelagian Platform, Tunisia, Algeria, Libya, and Egypt and led to subaerial exposure, erosion and the formation of lateritic soils (Druckman, 1974b, Hirsch, 1984). The top of the continental Triassic Zarzaitine Formation in SW Libya is marked by lateritic levels which indicate an important break in sedimentation (Jakovljevic et al., 1991). The Ragusa Platform in Southern Italy and the Malta Platform were uplifted and remained positive areas since that time (Bishop & Debono, 1996). The Jifarah Plains were also uplifted and remained a high area until the Miocene. Most of Egypt remained a highland (Fig. 20.10).

## 20.3.5  Continental Triassic

The open marine Triassic facies prevailed along the margin of North Africa and graded southwards, through a sandy and evaporitic facies, into exclusively continental deposits in the south. Late Triassic-Early Jurassic evaporites accumulated in a giant evaporite basin extending through the Berkine/Ghadames Basin in Eastern Algeria and West Libya (Turner & Sherif, 2007) (Fig.). The total thickness of the evaporite sequence reaches 1250 m (4100 ft) and shows significant thickness variations which are related to differential subsidence along basement lineaments. These evaporites form seals for the TAGI/Kirchaou sandstones hydrocarbon reservoirs. The continental Triassic lithostratigraphic units include the Zarzaitine Formation in southwest Libya, Madadi Formation in the Tibesti area and the Kufrah Basin, the Behir Formation

(Ras Qattara Formation) in the subsurface of the Northern Western Desert and the Qiseib and Budra formations in Wadi Araba and Sinai. No attempt has been made to subdivide these continental sequences into Lower, Middle and Upper Triassic units, and it is doubtful if this is possible with the data available at the present time.

In Southwest Libya, the Triassic succession is dominated by continental sediments of the Zarzaitine Formation deposited unconformably on top of different levels of the Carboniferous Dembaba Formation. Thick red siltstones, cross-bedded sandstones, and conglomerates with freshwater shells and calcitized fossil logs (Jakovljevic et al., 1991, Grubic et al., 1991) were deposited in alluvial plains, lakes, deltas, and meandering river channels (Grubic et al., 1991). The Zarzaitine continental facies in Southwest Libya is separated from the open-marine facies in the Jifarah Plains in the north by the Gargaf Arch (Burollet, 1963b). Continental Triassic sediments outcrop in the Jifarah Basin of Southern Tunisia. The sandstones are channelized and encased in shales. The channels become braided in Southern Tunisia and Algeria (Grant, 1995, Eschard & Hamel, 2002).

Several hundred meters of Upper Triassic clastics occur in the Triassic Basin in Algeria (van de Weerd, 1995, Busson & Burollet, 1973, Bourquin et al., 2010). The Upper Triassic succession of the Zarzaitine outcrops display facies of braided river, alluvial plain, low sinuosity rivers, lake deposits and marginal sabkha (Bourquin et al., 2010). In Algeria, The Upper Triassic succession corresponds to a major transgressive regressive cycle (Eschard & Hamel, 2002). A Carnian extensional phase led to the formation of NE-SW normal faults, partly inherited from Pan-African structures. Two subbasins, the Berkine and the Oued Mya basins, were formed and separated by the El Biod-Hassi Messaoud High (Eschard & Hamel, 2002).

In Northeast Libya, Middle Triassic (Anisian-Ladinian) sandstones and shales were deposited in a non-marine lagoonal environment with some marine influence near the top (El-Arnauti & Shelmani, 1985, Brugman et al., 1985). These facies suggest that the Middle Triassic rifting was subaerial and that the sea did not reach northeastern Libya before the Upper Triassic. Southwards in the Hameimat Trough in the eastern Sirte Basin a thick Middle Triassic section of quartzarenites and minor shales has been reported recently from the Amal Formation in wells L4-51 and A1-96 (Thusu, 1996, Sinha & Eland, 1996). These sediments overlie Precambrian-Cambrian deposits and they are overlain by Late Cretaceous sandstones. The presence of miospores, fish fossils and coal fragments within the shales suggests a lacustrine to lagoonal environment. In the Sarir Field, these sediments consist of reddish-brown, hematitic sandstones (Thusu, 1996), believed to be deposited during the Triassic rifting phase (Gras & Thusu, 1998). Triassic sediments in the As Sarah and Tuama oil fields consist of cross-bedded, medium to coarse-grained, meandering and braided stream deposits, containing soil horizons with root traces up to 60 cm (2 ft) long (Kuehn, 1996).

In the Northern Western Desert, marine Triassic sediments in the north grade southward into predominantly clastic sediment of the Behir Formation (Ras Qattara Formation). In the southwestern Kufrah Basin, deposition of the Permo-Triassic Madadi Formation (De Lestang, 1968) variegated silty shales and red to brown fine to medium-grained sandstones (Vittimberga & Cardello, 1963) probably took place in extensional rift basins, associated with volcanic rocks in the Gebel Oweinat area (Schandelmeier et al., 1987, Burollet, 1963b).

During the entire Triassic, Southwestern Sinai was dominated by continental fluvial sediments of the Qiseib and Budra formations (Weissbrod & Perath, 1990) composed of variegated sandstones and shales with ripple marks, mud cracks, cross-bedding and silicified tree trunks several meters in length, representing channel and bank deposits. These sediments grade northwards in central Sinai and westwards in the Wadi Araba into the Qiseib Formation. The Qiseib Formation in Sinai and Wadi Araba appears to be transitional between the continental Budra Formation to the south and the marine Arif el Naga Formation to the north. The Upper Triassic in Morocco is represented by a thick red shaly detrital sediments at the base, grading upward into carbonates. Doleritic volcanics are common (Westphal et al., 1979).

### 20.3.5.1 Jurassic period

The Jurassic succession corresponds to the LZ1-3 and UAB3-4 supercycles of Haq et al. (1987). The Early Jurassic period was a time of major lithospheric extension and rifting in the Mediterranean, Central Atlantic, and the Atlas Domain, accompanied by extensive tholeiitic magmatism and alkaline intra-plate magmatism (Le Pichon, 1988, Wilson & Guiraud, 1998). Faulting and magmatism associated with this Liassic phase of rifting produced a passive continental margin along the Eastern Mediterranean (Garfunkel & Derin, 1984). This phase also produced faulting, sea-channels, and crustal thinning in the Sirte Rise and the Ionian Sea (Del Ben & Finetti, 1991). Carbonates were deposited on the Pelagian Platform (Fig. 20.11). Deep-water (pelagic) facies were deposited in a trough immediately north of the Jabal Nafusah (Bishop 1975, 1988). Shallow marine sands and carbonates accumulated in the Jarrafa and Tripolitania troughs offshore of Libya. Evaporites accumulated in Northern Tunisia, Northwest Libya, and the Nile Delta region. Several hundred meters of Lower Jurassic evaporites, including halite, occur in the Triassic Basin in Algeria (van de Weerd,

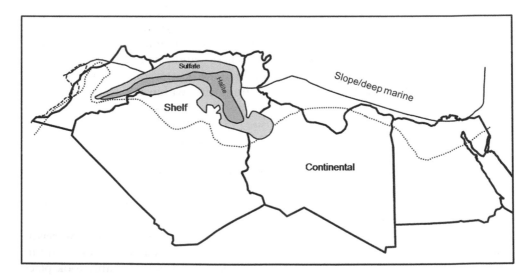

*Figure 20.11* Early Jurassic paleogeography (after Guiraud et al., 2005). Later Triassic-Early Jurassic salt basin after Turner & Sherif (2007).

1995) and the Atlas Mountains. In the coastal areas of Egypt, Jurassic sediments are composed of carbonates, sandstones and evaporites (Prior, 1976, Sestini, 1984). During the Jurassic the shallow sea penetrated the northern part of the Gulf of Suez as far south as Wadi Araba. Sauropod remains were documented from the Early and Middle Jurassic sediments in Morocco (Monbaron et al., 1999, Allain et al., 2004) and Algeria (Mekahli et al., 2004, Mahammed et al., 2005).

Restriction of the basin in Northwest Libya and the offshore area which started during the uppermost Triassic, continued into the Liassic (Assereto & Benelli, 1971, Hammuda et al., 1985) and led to the deposition of evaporites, dolomites and sandstones of the upper part of the Bu Sceba Formation, stromatolitic dolomites, oolitic limestones and evaporites of the Bu Gheilan Formation and evaporites of the Bir el Ghanam Formation. These sediments range from shallow supratidal to subtidal.

Early Jurassic sediments of Northeast Libya and the northern Western Desert are dominated by clastic sediments. Most of the early Jurassic, however, might be absent (Keeley et al., 1990). In North Africa, there is a marked hiatus spanning the late Hettangian, Sinemurian and early Pliensbachian (Keeley et al., 1990, Keeley & Massoud, 1998). In Sinai, sands of the late Pliensbachian-earlyToarcian Mashabba Formation (Al Far, 1966) were carried by northerly-flowing braided streams (Jenkins, 1990). The influx of these sandstones alternated with periods of shallow-marine limestone deposition. The rise in sea level during the Toarcian that led to the Oceanic Anoxic Event in deeper basins in North Africa resulted in the deposition of shallow-marine coralline algal limestones of the Rajabiah Formation. This unit becomes sandy and marly upwards, signalling the drop in sea level during the Aalenian. Fine to coarse-grained sandstones and shales with coaly material and abundant plant remains of the Shusha Formation (Keeley et al., 1990) were probably deposited in a fluviomarine environment during the Aalenian low-stand. The Mashabba, Rajabiah and Shusha formations grade southward into continental clastics of the Amir Formation.

In Northern Algeria, critical environmental changes occurred during the Lower Toarcian. In the Ksour Mountains in the Western Saharan Atlas near Morocco, Sebane et al. (2007) recognized several stages of Nodosariids evolution; the first started during the Lower Toarcian and characterized by the population of Nodosariids (stage of population), when the foraminifera lived under normal conditions, but during the following stage (stage of survival), isolation of the environment led the reduction in population and the development of atypical forms. During maximum deepening of the sea as a result of relative eustatic sea rise in the beginning of the Middle Toarcian, led to lethal conditions for the Early Toarcian benthonic (stage of extinction). The extinction stage coincided with the Toarcian OAE in the Western Mediterranean. These two stages were recognized within the Benia Marls Formation near Tiaret (Djebel Nador) in the Middle Atlas in Morocco and Beni Snassen (Boudchiche et al., 1987, Bejjaji et al., 2010).

The Lower Jurassic succession in the Ain Ben Khellil area in the Ksour Mountains (Mekahli et al., 2004) starts with the Guettob Moulay Mohamed Carbonates Formation (equivalent to the Koudiat el Beia Formation) (Late Sinemurian) which contains *Orbitopsella dubari* and *Planisepta compressa*, followed upward by the Oulad Amor Formation with abundant *Palaeodasycladus mediterraneus* attributed to the Lower Pliensbachian (Carixian) and the brachiopod *Hesperithyris renierii*. The Toarcian Jebel Nador Formation is made up of thick purple marls, bioclastic limestones, and dolomites and is overlain by the Middle Jurassic shallow carbonates of the Antar Dolomites Formation. The Lower Jurassic at Djebel Reha to the north consists of three formations: The Guettob Moulay Mohamed Carbonates Formation at the base composed of a Massive Dolomites Member, changing upwards into the Lithiotis Limestones Member rich in large bivalves and occasional corals, and the Gaaloul Cherty Formation of grey cherty limestones and marls. The tops of these beds are marked by hard ground with belemnites, oolites and bioclasts. The following Reha Formation represents an Early Toarcian cyclic sequence of marls with pyritized ammonites, such as *Pseudogrammoceras subregale* and *Audaxlytoceras dorcadis*. The presence of Toarcian marls dwarf ammonites (Mekahli et al., 2004) and the variations in Nodosariid foraminifera (Sebane et al., 2007) indicate the isolation of the area and the development of anoxic conditions that prevailed globally during the Toarcian OAE (Jenkyns, 1988, 1991, Sebane et al., 2007). The return to a platform setting was marked by the deposition of limestones with *Zoophycus*, passing vertically and laterally into shallow carbonates of the Antar Dolomites Formation (not accurately dated, is probably Middle Jurassic). The Late Bajocian is characterized by alternating marls and neritic limestones with *Ermoceras* fauna, overlain by oolitic carbonates and reefs of the Guettaï Formation.

The Liassic (Pliensbachian) succession at Ait Bou Guemmez in the Mellila region in the Central High Atlas in Morocco consists of the Ait Bou Oulli Formation at the base, the Agnane Formation and its equivalent Jbel Choucht Formation, and the Wazzat Formation and its lateral equivalents the Amezrane and Tafraout formations. The sequence is dominated by platform carbonates, generally dolomitized and display frequent emergence surfaces marked by mud cracks (El Bchari & Souhel, 2008). The Agnane Formation contains *Orbitopsella primavera*. The Pliensbachian-Toarcian succession in that area was deposited in a restricted marine environment within a deep subsiding basin (Elmi et al., 2010). The Toarcian Oceanic Anoxic Event is indicated by abnormal variability in the foraminifera assemblages. Stressed environmental conditions are suggested by the presence of *Lenticulina obonensis* (Elmi et al., 2010).

The Toarcian sediments of the Middle Atlas in Morocco show hemipelagic facies deposited in platform to basinal environments (Blomeier & Reijmer, 1999, Bejjaji et al., 2010). They are represented by marly thick sections that accumulated in troughs under restricted dysoxic conditions (Toarcian OAE), and condensed sections of marly limestones on high ridges. Benthonic foraminifera assemblages include species of *Lingulina, Marginulina, Lenticulina, Ichtyolaria*, and *Nodosaria*.

## 20.3.6  Middle Jurassic

A renewed phase of rifting took place in the Middle Jurassic in the Mediterranean and the Atlantic Ocean, which caused faulting, accompanied by magmatic activities. The Middle Jurassic of the eastern Pelagian Sea and along the Sicily-Malta escarpment is composed mostly of basaltic layers (Del Ben & Finetti, 1991). Shallow open marine sediments were deposited on the Pelagian Platform and in Northwest Libya, Northern Egypt, Sinai, Tunisia, Algeria, and Morocco (Fig. 20.12). These sediments graded southwards into fluvio-marine sediments, then into predominantly continental sediments. The Sirte Basin and Central and Upper Egypt probably were areas of non-deposition or erosion during that time (Keeley et al., 1990, Keeley, 1994).

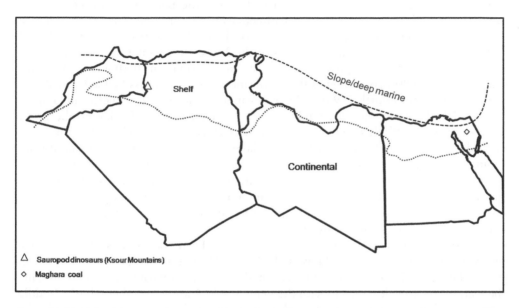

*Figure 20.12* Middle-Upper Jurassic paleogeography (modified after Guiraud et al, 2005, Aissaoui, 1989).

The Bajocian started with a widespread deposition of shallow marine carbonates along the southern margin of the Tethys. The Bu en Niran shallow marine limestones in Northwest Libya and its equivalent the Nara dolomites and Zmilet Haber limestones in Tunisia and the offshore area were probably deposited at that time. These carbonates represent the maximum marine transgression during the Bajocian and mark the beginning of a shallowing upward cycle. The Bu en Niran and Zmilet Haber limestones and the Nara dolomites yield a strong seismic reflector (Bobier et al., 1991). In NW Libya, Bathonian-Callovian deposition started with the deltaic Giosc Shale (Bathonian), and it was followed by the deposition of the Khashm azZarzur (Chameau Mort) (Callovian) fine cross-bedded sandstones, gypsiferous shales, ferruginous marls with abundant plant fossils and minor carbonates (El-Zouki, 1980). Direction of transport was to the north and northwest. The Khashm azZarzur deltaic deposits (Walley, 1985) grade into marine sediments to the north and into continental deposits of the Taouartine Formation to the south.

In the Northern Western Desert, the Bajocian-Bathonian shallow marine carbonates, alluvial plain shales, sandstones, and swamp coal deposits of the Khatatba Formation were deposited mainly in the Delta region and along the coastal area of the Western Desert (Hantar, 1990). The Khatatba coals contain large volumes of humic material in addition to minor vitrinite and algal kerogen. These coaly sediments provide a source for the Western Desert oils (Keeley, 1994, Bagge & Keeley, 1994, Alsharhan & Abdel Gawad, 2008, Baoumi, 2009).

Rifting began in earnest during the Early-Middle Jurassic, when East and West Gondwana separated, and the Central Atlantic Ocean began to open (Bumby & Guiraud, 2005). E-W trending half-grabens were formed during the Middle Jurassic (Dogger) along the Eastern Mediterranean margin (Guiraud & Bosworth, 1999,

Guiraud et al., 2005, Yousef et al., 2010). In Sinai the Bajocian-Bathonian succession started with the shallow marine sedimentation of coralline limestones (Jenkins, 1990) of the lower Mahl Member (Bajocian) of the Bir Maghara Formation, followed upward by shales, marls and sandstones of the Mowerib Member (Bathonian). This cycle was followed by a regressive cycle of the Safa Formation (Bathonian-Callovian) (Keeley et al., 1990), composed of fluvio-marine or tidal flat cross-bedded sandstones (Abdel Gawad & Gamiel, 1998), followed by marine shales with economically important coal beds of the Kabrit Member of the Safa Formation (Keeley et al., 1990, Ghandour et al., 2003, Baoumi, 2009), such as those at Ayun Musa. On the western side of the Gulf of Suez, north of the Galala Plateau, Bathonian low-stand tract sediments are made up of shallow-marine fossiliferous limestones and sandstones (Nakkady, 1955). Ghandour et al. (2003) attributed the dominance of kaolinite in Bajocian-Bathonian shales of the Bir Maghara and Safa formations; associated with coal deposits of the latter, at Gebel Maghara to humid climatic conditions and strong chemical weathering of the source rock.

In Southern Tunisia, sandstones of the Techout Formation represent a regressive interval during the Bathonian (Enay et al., 2002). The overlying Tataouine Formation marks the renewal of marine condition in shallow marine environments and represents the transgressive cycle, starting with the Beni Oussid 1 and 2 members (Early-Middle Callovian). The following Krechem el Miit Member shows a deepening, while the Ghomrassene Member represents a high stand. The Callovian transgression on the Saharan Platform in Central Tunisia (North-South axis) led to renewed sedimentation after a long hiatus (Middle Bathonian-Middle Callovian). In the Tunisian Dorsale, the Ammonitico Rosso facies was deposited from the Middle-Late Callovian (Enay et al., 2002). Sauropod remains, such as *Atlasaurus imelakei,* were documented from the Early and Middle Jurassic sediments of in Morocco (Monbaron et al., 1999, Allain et al., 2004) and from the Middle Jurassic of the Ksour Mountains in Algeria (Mekahli et al., 2004, Mahammed et al., 2005).

In the Ouarsenis Mountains in the Algerian Tell, the Encrinitic Limestones and Massive Oolitic Limestones Units, northwest of Tiaret (Atrops & Alméras, 2005) indicate a shallow marine environment.

Uplift accompanied by a drop in sea level at the end of the Callovian resulted in a complete regression and the development of a regional unconformity in North Africa. The top of the Callovian (Safa Formation) in Northern Sinai is marked by a hardground encrusted by ammonites (Keeley et al., 1990). The upper contact of the Taouartine Formation (Lower-Middle Jurassic) with the overlying Msak Formation (Upper Jurassic-Lower Cretaceous) in Southwest Libya is unconformable and marked by a lateritic bed of brownish-red to brownish-purple shales, siltstones and sandy siltstones (Jakovljevic, 1984, Jakovljevic et al., 1991, Grubic et al., 1991). Local basin inversion and folding occurred along the Moroccan Middle and High Atlas during the Middle Jurassic (Laville et al., 2004) that was probably associated with initiation of drifting along the Central Atlantic Domain (Guiraud et al., 2005).

### 20.3.7  Upper Jurassic

This succession is associated with the opening of the Central Atlantic Ocean and the Neo-Tethys. Sea levels reached their acme in the early Oxfordian. The sea transgressed over hardgrounds following the Callovian drop in sea level. In the northern regions of Egypt this event is marked by the deposition of ammonite-bearing shales of the Kidod Shale Member of the Masajid Formation (Lipson-Benitah et al., 1988, Keeley et al., 1990, Keeley & Massoud, 1998). A carbonate shelf was established along the subsiding margin. In the northwestern part of the Western Desert, Cyrenaica, and Northwest Libya, deposition was dominated by limestones.

Shallow open-marine conditions returned in Northwest Libya with the deposition of the Sciucsciuch limestones (Oxfordian-Kimmeridgian) and its equivalent the M'Rabtine Formation in Tunisia. Deposition was dominated by alternating limestones, dolomitic limestones, sandstones and green gypsiferous and sandy shales which contain *Goniocora* sp., *Pleurotomaria* sp., *Nerinea* sp., *Pseudocyclammina jaccardi* and *Valvulinella jurassica* (Burollet, 1963a), in addition to the echinoid *Monodadema cotteaui* (Hammuda, 1971) and coaly plant debris (El-Hinnawy & Cheshitev, 1975). These carbonate ramp sediments become more terrigenous both eastward and southward (McWilliams & Harbury, 1995). Deposition was dominated in the Murzuq Basin to the south by undifferentiated Jurassic continental sediments of the Taouartine Formation. These sediments are composed of cross-bedded sandstones, conglomerates, siltstones and shales with abundant plant fragments and they were probably deposited in braided rivers and lacustrine environments (Jakovljevic, 1984, Grubic et al., 1991). At the northern edge of the basin, shales were deposited in large lakes and swamps. The Tibesti Massif and its cover of Paleozoic rocks is the most likely source of these sands (Lorenz, 1980). Alluvial-fan deposits are predominant near that massive.

Upper Jurassic sediments in Cyrenaica, Northeast Libya, consist of a marginal-deep marine facies (Jabal Akhdar facies of Thusu et al., 1988). These sediments change facies southward into a transitional-lagoonal-marginal marine facies (Cyrenaica facies), then into the dominantly non-marine Nubian/Sarir facies.

The Jabal Akhdar facies in Northeast Libya includes the Ghurab Formation (Callovian-Kimmeridgian) and the Mallegh Formation (Tithonian) of Duronio et al. (1991). These two formations are made up of shales interbedded with siltstones, sandstones and micritic limestones. *Calpionella elliptica, C. alpina, Calpionellites darderi* and *Crassicollaria brevis* in the micritic limestones. This facies was deposited in marine paralic, sublittoral, probably inner shelf environments with occasional deeper marine influence. The two formations grade into the transitional lagoonal-marginal marine Cyrenaica Platform facies of the Sirual Formation (Duronio et al., 1991) which consists of interbedded sandstones, limestones, shales and oolitic and dolomitic limestones with occasional anhydrites, coals and lignites. The microfauna include arenaceous foraminifera, such as *Kurnubia jurassica, K. palastiniensis, Pseudocyclammina jaccardi* and *P. litulus*. Similar assemblages were reported from the Masajid Formation in Egypt (Said, 1962a, Al Far, 1966), the Northern Western Desert (Hantar, 1990), Sciucsciuch Formation in Northwest Libya (Burollet, 1963a), in Northeast Libya, and the Jurassic of Arabia (De Matos, 1994) and they probably thrived in table reefs on submerged highs in the Jurassic seas (Said, 1962a).

In the Northern Western Desert, more than 450 m (1476 ft) of Late Jurassic fossiliferous shales, limestones, sandy and silty shales, oolitic marly limestones, and glauconitic limestones with abundant brachiopods, ostriids and corals of the Masajid Formation (Callovian-Kimmeridgian) were deposited in a mid-neritic to nearshore environment (Hantar, 1990). Table reefs with abundant reef-forming foraminifera, such as *Kurnubia palastiniensis, Valvulinella (Kurnubia) jurassica, Pseudocyclammina virguliana* and *P. personata* were developed. More restricted conditions prevailed in the northwestern part of the Western Desert. Dolomites with a few interbeds of sandstones, shales and anhydrites of the Sidi Barrani Formation (equivalent to the Sirual Formation in Cyrenaica) were deposited conformably on top of the Khatatba Formation (Hantar, 1990).

The facies of the Masajid Formation in Sinai reflect the unstable sea level during the Late Jurassic which resulted in the deposition of shallow-marine limestones, sandstones and shales from the Callovian to the Kimmeridgian. The lower Kehailia Member (Callovian) is composed of fossiliferous shales with fossiliferous limestone beds rich in *Ostrea* and rare corals (Hegab, 1989), sandy and silty shales, cherty limestones, oolitic marly limestones and glauconitic fossiliferous limestones with brachiopods suggesting a mid-neritic to nearshore environment of deposition. At Gebel Mineshirah, the Kehailia Member represents intertidal-subtidal deposits (Abdel Gawad & Gamiel, 1998). Maximum marine transgression took place in late Oxfordian (Kidod Shale). The upper Arousiah Member (Oxfordian-Kimmeridgian) is made up of stylolitic, coralline and/or algal limestones with a few marl and clay interbeds. Ostracods in the Jurassic succession at Gebel Maghara suggest a shallow-marine warm environment of deposition (Rosenfeld et al., 1987). Tectonic movements at the end of the Jurassic are manifested in places, such as the Arif el Naga outcrops, by the presence of Early Cretaceous conglomerates (Arod Conglomerate) composed of quartzite pebbles up to 30 cm (12 in) in size, which overlie the Jurassic Inmar or Ardon formations with an angular unconformity (Bartov et al., 1980).

In the Ksour Mountains in Northern Algeria, a carbonate platform was established during the Jurassic (Early Hettangian-Rhetian) with the deposition of the Chemarikh Dolomites and Tiout Bridge Formation (Mekahli et al., 2004). Pelagic sediments prevailed during the Early Sinemurian. In the Southern Ain Ouarka Subbasin, radiolarias occurred at that time in lower member of the Ain Ouarka Pelagic Limestones. In the Figuig Subbasin to the west, basinal facies with radiolarias appeared in the Late Sinemurian and continued until the Late Bajocian. It was followed by turbidites mixed with basinal marls of the Teniet el Klakh Formation. The carbonate platform was established again with the development of reefs in the lower part of the Tifkirt Formation. Siliciclastic turbidites continued until the Early Bathonian. During the Late Jurassic, the area was invaded by deltaic and prodeltaic sediments of the Ksour Sandstones.

At the end of the Jurassic the paleogeography of Northeast Tunisia was dominated by a series of positive areas aligned along the Zaress-Zaghouan axis which delimited two subsiding basins; the Tunisian Trough to the northwest and the Hammamet Trough to the northeast (Memmi et al., 1989). These positive areas marked the transition between a shallow platform and a deep basin and were the sites of reef development.

### 20.3.7.1  Continental Upper Jurassic-Early Cretaceous

The Upper Jurassic-Lower Cretaceous sedimentary succession corresponds to super-cycles LZB1, LZB2 and LZB3 of Haq et al. (1987). It is bounded by unconformities.

The unconformity at the base (Cimmerian unconformity of Keeley et al., 1990) is associated with the Tithonian eustatic fall in sea level (Bobiér et al., 1991).

The Upper Jurassic-Early Cretaceous period is represented by the Cabao Formation in Northwest Libya and its offshore area, the Msak Formation in the Murzuq Basin, the Wadi Formation in the Central Sirte Basin, the Sarir Sandstone in the Eastern Sirte Basin, the Soeka and Ounianga formations in the Kufrah Basin, the "Nubian" and *"Continental Post-Tassilian"* sandstones in the Central and Western Sahara (Lefranc & Guiraud, 1990) (Fig. 20.13). The Upper Jurassic is absent in the Jifarah Plains. In the northern Western Desert, continental sediments are known under different names, such as the Bahrein Formation, Alam el Bueib Formation, Shaltut Formation, or Eghi Formation. This period is represented in Central Sinai by the continental sandstones of the Amir Formation (Early-Middle Jurassic) and the Hatira Formation (Early Cretaceous).

In the Murzuq Basin, continental cross-bedded sandstones, conglomerates, shales and siltstones, of the Msak Formation with Upper Jurassic-Lower Cretaceous pollen and spores (Konzalova, 1991, Galecic, 1984, Jacovljevic, 1984) were deposited. These sediments were probably transported by a north-flowing braided river system (Lorenz, 1980, Jacovljevic, 1984, Seidl & Röhlich, 1984), such as the Oued Tassili-Hamada River (Lefranc & Guiraud, 1990). These rivers carried sediments from the Tassili Plateau, the Hoggar, and Tibesti massifs, across the Hamada al Hamra (Ghadames Basin), and formed a delta near the Jifarah Plains and the Jabal Nafusah area in Northwest Libya. These sediments constitute the Cabao Formation (Upper Jurassic-Lower Cretaceous). They are composed of cross-bedded and conglomeratic sandstones in the lower part with abundant marine and continental vertebrate remains (El-Zouky, 1980), such as dinosaur vertebra, shark teeth, crocodile teeth and dermal plates, turtle shell fragments and silicified wood. Similar assemblages also occur in

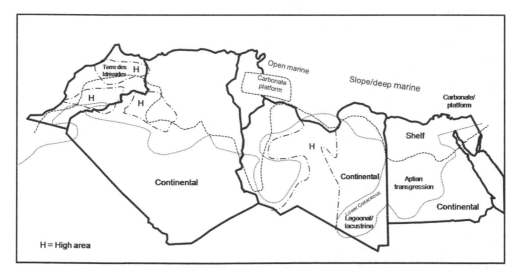

*Figure 20.13*  Upper Jurassic-Lower Cretaceous paleogeography (after Lefranc & Guiraud, 1990, Tawadros, 2001, Guiraud et al., 2005). Dashed line: Marine Jurassic; dotted line: Lower Cretaceous. Note: Marine Jurassic may reach the Kom Ombo Basin in Upper Egypt.

equivalent beds in Algeria, Tunisia, and Morocco. These sandstones change upward into fine sandstones, shales, and cherty, brecciated dolomitic horizons (Hammuda, 1971). Other local highs supplied sediments to the basins, for example in the Dur al Qussah area in the eastern Murzuq Basin, alluvial-fan conglomeratic deposits of the Bin Ghanema Formation comprise most of the succession (Woller, 1978, Lorenz, 1980). Paleocurrent measurements show directions of transport toward the northwest, east, and northeast (Lorenz, 1980, Stefek & Röhlich, 1984, Grubic et al., 1991). At the northern and western edges of the Murzuq Basin, large lakes and swamps collected massive clay deposits up to 20 m (66 ft) in thickness.

During the Upper Jurassic and Early Cretaceous times, the coastal area of the Western Desert and Cyrenaica was open marine (Prior, 1976, Sestini, 1984, Duronio et al., 1991). However, deposition was dominated by continental sandstones (the so-called Nubian) over most of Libya and Egypt as a result of the intense erosion of high areas. Following the drop of sea level during the Kimmeridgian-early Tithonian, restricted shelf conditions resulted in the deposition of evaporites and the development of syndepositional karstic surfaces (Keeley, 1994). Jurassic-Early Cretaceous deposition in Egypt and Libya was also influenced by the tectonic-sedimentary history of the region. Arches and basins became active. In Late Jurassic, northward tilting of North Africa and rejuvenation of faults changed the drainage pattern in Egypt to the north (Klitzsch, 1990b, Beauchamp et al., 1990). To the south, repeated uplift and erosion of high areas resulted in the supply of large amounts of sediments which were transported northward by meandering and braided channels. These processes produced widespread non-marine to marginal marine deposits in the subsiding basins in the southern margin of the Tethys (Van Houten, 1980).

The Northern Egypt and Nile basins subsided and were filled with non-marine, fluvial to lacustrine sands and shales (Hendriks et al., 1990). In the Northern Western Desert, shallow-marine and restricted-marine Jurassic sediments equivalent to the Masajid Formation grade southward into continental sandstones of the Bahrein Formation (Callovian-Kimmeridgian) (Hantar, 1990). The sediments of the Bahrein Formation are composed of red, fine to coarse-grained quartzose sandstones with interbeds of conglomerates, siltstones, shales and occasional anhydrites. The Formation was deposited unconformably on different Paleozoic units or the Precambrian basement. Westward it pinches out against a Paleozoic high which separates the northern Egypt Basin from the Cyrenaica Basin.

The opening of the South Atlantic during the Lower Cretaceous caused another major drop in sea-level and a regression in North Africa. Anticlockwise rotation of Africa and early fracturing of the Sirte Basin and northwestern Egypt began in the early Cretaceous, contemporaneously with rifting in Central Africa (Burollet, 1993, Guiraud & Maurin, 1993) and the opening of the Eastern Mediterranean (Guiraud, 1998). The triple-junction pattern of faults in the Sirte Basin probably owes its origin to a combination of strike-slip faulting, basement faults, and extension of the region above a fixed mantle hotspot over which Africa rotated (Van Houten, 1983, Anketell, 1996). The Sirte and Kalanshiyo arches in the Sirte Basin collapsed due to N-S extension (Clifford et al., 1980). The "Nubian" sandstones are considered pre-rift by some authors and syn-rift by others. In the Central Sirte Basin, thick syn-rift continental Early Cretaceous sandstones of the Wadi Formation were deposited in the rift basin (Tawadros et al., 2001).

They were later silicified and fractured. The fractured quartzites form oil reservoirs in the Wadi Field, which is located on a fault block in the Marada Trough (Wennekers et al., 1996). These quartzites are interbedded with acidic volcanic rocks. As the deposition of "Nubian" sandstones in the Sirte Basin was fault-controlled, preservation of the thickest section of the sandstones can be found in structural lows, while the thinnest sections are developed on the structural highs (Barsoum, 1995). These rifts were probably subaerial and remained exposed until the Upper Cretaceous marine transgression (Tawadros et al., 1999). Only during the Aptian time the sea advanced southward into Egypt, Libya and northern Sudan (Klitzsch, 1990b, Hendriks, 1985) (Fig.). However, the Central Sirte Basin was a high during the Aptian and had not been invaded by the sea at that time. That time was also a period of volcanic activities. Hot spot volcanic activities associated with extensional tectonics and rifting during the Early Cretaceous led to widespread occurrences of volcanic flows. Extensive basaltic and andesitic volcanic rocks occur in the offshore of Northwest Libya (Hammuda et al., 1985, Del Ben & Finetti, 1991, Tawadros et al., 2001). Rhyolites and basalts occur in the Sirte Basin (Barr & Weegar, 1972, Massa & Delort, 1984, Tawadros et al., 2001), at Wadi Araba and in Sinai (Said, 1962a, Bartov et al., 1980). In Central Sinai, variegated, cross-bedded sandstones of the Hatira Formation contain olivine basalt flows.

In the Southeastern Sirte Basin, the Late Jurassic-Early Cretaceous rift was filled with more than 3000 ft (914 m) of continental sandstones of the Sarir Sandstone (Upper Jurassic-Neocomian). The Sarir sandstones contain abundant pollen and spores (Thusu & van der Eem, 1985) and were deposited in braided and meandering channels and fan-deltas bordering the graben margins (Rossi et al., 1991, El-Hawat, 1992, Thusu et al., 1988). The Sarir Sandstone contains 24 BBOIP in the Sirte Basin. Hydrocarbon traps include traps associated with structural highs, structural-stratigraphic combination traps and relay ramp faults (Gras & Thusu, 1998). Seal and source rock are provided by the Late Cretaceous Rakb Shale or the Middle Variegated Shale. The organic matter in these shales is predominantly type II kerogen and has generated waxy oils.

Early Cretaceous sediments in the coastal area of Cyrenaica grade into marginal, shallow, open-marine (Marmarica facies) of the Qahash Formation (Neocomian) (Duronio et al., 1991), which consists dominantly of carbonate sequences of oolitic and detrital limestones with shale and sandy interbeds, becoming more sandy toward the margins. It contains a diversified assemblage of dinocysts, such *Microdinium* and miospores with abundant terrestrial (humic) fragments. Shales contain dinocysts, rare benthonic foraminifera and common calcareous nannofossils, such as *Nannoconus steinmannii, Polycostella senaria, P. beckmannii* and *Rhagodiscus eboracensis*, among others. The continuous detrital and humic kerogen input in all facies suggests renewed uplift of the source area (Duronio et al., 1991).

In Southwest Egypt, the Upper Jurassic-Lower Cretaceous succession is represented by medium to coarse-grained, cross-bedded sandstones which constitute soil, fluvial braided, tidal channel and shoreface deposits of the Gilf Kebir and Six Hills formations. The upper part of the Gilf Kebir Formation lies directly on the Precambrian surface (Wycisk, 1990, Klitzsch et al., 1979). The depression south of Assiut (in the Kharga-Baris area) was filled by fluvial and flood plain deposits of the Six Hills Formation which were deposited unconformably on the Precambrian basement (Klitzsch et al., 1979).

The Upper Jurassic-Lower Cretaceous (Neocomian) continental sediments of the Soeka Formation in the Kufrah Basin consist of variegated, fine to coarse-grained, cross-bedded sandstones, changing upward into fine sandstones and red silty shales, followed by conglomeratic fine to medium grained, cross-bedded sandstones of the Ounianga Formation (Wacrenier, 1958, De Lestang, 1965).

In the Northern Western Desert, the Upper Jurassic in the well Foram-1 contains purely terrestrial pollen and spores assemblages (Schrank, 1984, 1987, 1992), but the assemblage becomes partly marine in wells Betty-1 and Ghazalat-1 (Abdelmalik et al., 1981). The early Neocomian in all three wells contains a terrestrial pollen and spore assemblage with rare algal remains. A thick Early Cretaceous fluvio-deltaic sandstone sequence with minor shales (Metwalli & Abd el-Hady, 1975) of the Alam El Bueib Formation was deposited unconformably on the Bahrein sandstones. The presence of rare brackish-water fossils (Said, 1962a) suggests fluvio-marine conditions. The coals associated with these sands are vitrinitic and gas prone (Keeley & Massoud, 1998). The Formation grades northward in the Mamura-Sallum area into shallow marine carbonates with common grainstones and minor sandstones. These sediments grade in the Matruh-1 well into a section of more than 1000 m (3281 ft) of basinal and delta-front shales (Matruh Shale) (Prior, 1976). These shales can reach up to 2133 m (7000 ft) in places and form a good source rock (Amine, 1961, El-Ayouti, 1990). In the northern Eastern Desert, Early Cretaceous sandstones grade into shallow-marine sandstones with dinoflagellate cysts and miospores (Aboul Ela et al., 1998, Ibrahim et al., 1998).

Regionally, the Cretaceous changes facies throughout Tunisia from continental domain in the south to fully marine domain in the north (in the Tunisian Trough, passing through a marine carbonate platform (known as the Pelagian Platform, the Central Tunisia High, or the Kasserine Platform) in the center of the country, which extends eastwards into the Pelagian Sea (Marie et al., 1984).

The Upper Jurassic-Lower Cretaceous successions contain mostly carbonate rocks, representing increasingly deeper marine to pelagic deposition in Northern Tunisia (Burollet et al., 1978, Klett, 2001). Along the North-South Axis (NOSA) lagoonal, deltaic, and terrigenous facies in the south and southwest changed into carbonate shoal facies and some reefs during the Jurassic and Early Cretaceous (Bishop, 1975, Salaj, 1978, Burollet et al., 1978, M'Rabet, 1984, Klett, 2001). Calpionellids are common in the Upper Jurassic-Lower Cretaceous of Northern Tunisia (Boughdiri et al., 2006a, b, Boughdiri et al., 2009, Haddoumi et al., 2008). Farther east and to the south, pelagic sediments were deposited in a subsiding depositional trough in the Gulf of Gabes (Salaj, 1978, Burollet et al., 1978).

The Upper Jurassic-Lower Cretaceous succession of the Tataouine Basin on the Dahar Plateau in Southern Tunisia (Fig. 20.13) is represented by two groups, the Asfer Group (Bir Miteur, Boulouha, and Douiret formations) and the Ain el Guettar Group (Chenini and Oum ed Diab formations) (Anderson et al., 2007). The Asfer Group varies from a maximum thickness of 220 m (722 ft) in the south at Merbah el Asfer to just a few meters in the north. The Bir Miteur Formation (late Oxfordian-Kimmeridgian), at the base of this group, is 70 m (230 ft) thick and consists of alternating green shales, gypsum, silts, sands and dolomites, indicating lagoonal conditions (Benton et al., 2000). It is overlain by a marginal marine carbonate sequence of the Boulouha Formation (Tithonian-Neocomian), up to 80 m (262 ft). The Douiret, Chenini, and Oum ed Diab formations contain microvertebrate assemblages which indicate depositional environments that included terrestrial carbonate-rich, marginal marine, and mixed freshwater and marine (Anderson et al., 2007). The Douiret Formation (Lower Aptian) consists mainly of mudstones (Ben Ismaïl, 1991). The Douiret Formation is made up of interbedded fine sands, sandy mudstones, and dolomites overlain by a thick bed of green clay. The vertebrate remains include abundant fish and reptile fossils. The Chenini Formation is 50 m (164 ft) thick and consists mainly of fluvial coarse cross-bedded sandstones and minor conglomerates, breccias, and mudstones with fossil plants, fish, turtle, crocodile, and dinosaur remains (Benton et al., 2000). These assemblages are similar to those found in the Cabao Formation in Northwest Libya (El-Zouky, 1980). The overlying Oum ed Diab Formation (Albian-Cenomanian) is 25 m (82 ft) thick and marked by a transition to alternating shales and sands (Bouaziz et al., 1988). The Radouane Member of the Oum ed Diab Formation represents the major Cenomanian transgression, as indicated by a carbonate-rich sandy offshore bar with ammonites, Lepidotes and pycnodonts.

In Central and Southern Tunisia, the Lower Cretaceous comprises complex deltaic deposits, derived from the Northern Sahara, and shallow marine facies near the Chotts Chain (Marzouk & Ben Youssef, 2008). The Barremian-Aptian is characterized by carbonate and detrital deposits. A maximum flooding event occurred during the Barremian to Aptian (Bishop, 1975, Salaj, 1978). Barremian rocks consist of limestones, marls, and interbedded sandstones and shales of the Bouhedma and M'Cherga Formations, and the Sidi Aich Sandstones (Bishop, 1975, Salaj, 1978, El Euchi et al., 1998).

The Lower Cretaceous rocks of Southeastern Tunisia consist of four transgressive-regressive sedimentary sequences represented by the Bou Hedma, Sidi Aich, Orbata, and the Lower Zebbag formations, respectively (Marzouk & Ben Youssef, 2008), bounded by emersion surfaces. The Bouhedma Formation (Upper Hauterivian according to Busson, 1967 and Peybernès et al., 1985) consists of an alternating sequence of anhydrite, limestones, laminated and stromatolitic dolomites, and shales. The Sidi Aich Formation (Barremian) is composed of a coarsening-upward fluvial sequence of shaly sands and lenticular, cross-bedded, ferruginous, course-grained sandstones with occasional large silicified wood fragments (Marzouk & Ben Youssef, 2008). Based on variations in the sedimentary organic matter and clay minerals content of the Berriasian sediments of the Sidi Kralif (Khalif) Formation, at Jebel Meloussi in Central Tunisia, Schnyder et al. (2005) found that clay minerals display a shift in the kaolinite content at the early-middle Berriasian boundary, a time of high sea-level. A Contemporaneous change in clay mineral assemblages, documented in Southern Morocco (Agadir area) in the Atlantic Domain, has been attributed to a climatic change. A late Tithonian to early Berriasian dry and cooler phase (represented by the lower part of the Sidi Khalif Formation) is replaced by a middle to late Berriasian more humid phase, indicated by a general increase in the amount of kaolinite in the upper part of the formation. Tawadros et al. (2001) also noticed the presence of kaolinite as the dominant clay mineral in the Lower Cretaceous quartzitic sandstones in the Central Sirte Basin.

In the Ain Ouarka Subbasin in Northern Algeria, the Upper Jurassic-Lower Cretaceous succession is represented by the Tiloula Formation of probable Latest Jurassic-Earliest Cretaceous age and consists of variegated shales, channelized pebbly sandstones, and stromatolitic dolomites with bird's-eyes and halite. It overlies the Upper Jurassic Aissa Formation. It is equivalent therefore to the "Nubian Sandstone" *s.s.* in Libya and Egypt (Tawadros, 2001). The Tiout Formation is composed of pebbly sandstones, and has been assigned a tentative Lower Cretaceous-Albian age by Mekahli et al. (2004). The Tiout Formation is overlain by the Upper Cretaceous Mdaouer Formation composed of shales and gypsum, followed by carbonates and marls with the Cenomanian *Vascoceras* ammonite.

In Northern Algeria, Jurassic to Albian rocks crop out below the Tellian nappes in the Bibans window are affected by a synfolding cleavage, which is moreover expressed everywhere in the 'external massifs' of the Tell-Rif system (Benaouali-Mbarek et al., 2006).

In the Central High Atlas of Morocco, the Bathonian-Berriasian Red Beds occur in the Ouaouizarht Basin south of Bani Mellal (Charrière et al., 2005). The Red Beds consists of the Guettioua Formation (Bathonian), the Iouaridene Formation (Upper Jurassic-Barremian), and the Jbel Sidal Formation (Barremian). The Red Beds contain, in addition to ostracods and charophytes, numerous bones and footprints of dinosaurs (Regagba et al., 2007), as well as a complete skeleton of a giant Sauropod (Monbaron et al., 1999). The Jurassic-Lower Cretaceous contact probably falls within the Iouaridene Formation (Charrière et al., 2005).Volcanic gabbro intrusions and basalt lava flows occur in the area (Laville & Piqué, 1992, Monbaron et al., 1999), associated with an extensional regime (Frizon de Lamotte et al., 2008, 2009).

In the Eastern High Atlas of Morocco, continental "Red Beds" (Bathonian-Cenomanian) overlie the marine deposits of Jurassic and consist of three units; the Anoual Formation, the Ksar Metlili Formation, and the Dekkar Group (Albian-Cenomanian), bounded by two sharp sedimentary discontinuities (Haddoumi et al., 2002, 2008). The Anoual Formation (Lower Bathonian) consists of fluvio-deltaic plain deposits. The Ksar Metlili Formation (Tithonian-Berriasian) represents continuation of the fluvio-deltaic sediments with Late Tithonian-Early Berriasian charophytes.

### 20.3.7.2   *Aptian-Albian*

### 20.3.7.3   *Aptian*

On a global scale, the Aptian was a time of a major marine transgression (Matsumoto, 1980, Haq et al., 1988) (Fig. 20.14). On the Pelagian Platform in the offshore of Northwest Libya, Aptian interbedded dolomites and dark grey argillaceous wackestones of the Turghat Formationwith abundant benthonic foraminifera, such as *Cuneolina camosauri*, *C. laurenti*, *Dictyoconus* sp., *Pseudocyclammina hedbergi*, *Ovalveolina reicheli*, *Coskinolina sunnilandensis* and *Choffatella decipiens* (Hammuda et al., 1985) were deposited in a shallow-marine, partly restricted platform. This shallow platform extends into Tunisia and is bordered to the north by open deep-marine facies (Marie et al., 1984) (Fig. 20.14). Similar assemblages were reported from the Aptian of Arabia (De Matos, 1994). Aptian platform sediments in Tunisia are dominated by shallow-marine red dolomites and shales with *Orbitolina* and *Choffatella* of the Orbata Formation (Marzouk & Ben Youssef, 2008).

In Central and Southern Tunisia, Aptian and Albian rocks include limestones, dolomites, sandstones, mudstones, marls, and some evaporites of the Orbata, Serdj,

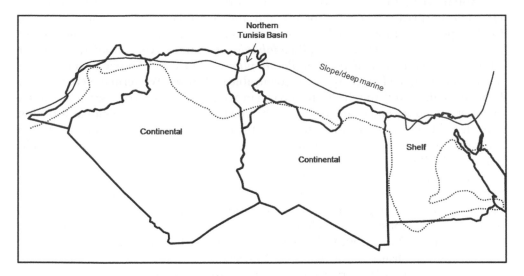

Northern
Tunisia Basin

Slope/deep marine

Continental

Continental

Shelf

*Figure 20.14* Aptian-Albian paleogeography (Tawadros, 2001, Guiraud et al., 2005).

Hameima, and Lower Fahdene Formations (Burollet et al., 1978, Salaj, 1978, Chihaoui et al., 2010). Facies and thickness variation are controlled by structure and halokinesis. During the early Albian stage, a large part of Tunisia including the Saharan Platform, Jifarah, and Central and Eastern Tunisia, was emergent. Only the northern part of Tunisia, the Chotts Chain in the south, and the Gafsa Basin have continuous sedimentary records during the Albian (Marzouk & Ben Youssef, 2008). In the Tadjerouine area of Northwestern Central Tunisia, south of El Kef, the Albian transgression is characterized by deposition of alternating marls, limestones and sandstones of the Hameima Formation (earliest Albian) that overly massive platform carbonate rocks of the Serdj Formation (late Aptian-early Albian) (Chihaoui et al., 2010). Continuing transgression resulted in the deposition of a thick series of marls and shales with subordinate carbonate beds of the Fahdene Formation (early-late Albian) (Chihaoui et al., 2010).

Upper Barremian-lower Upper Aptian hemipelagic deposits of the Hamada (Hameima?) Formation in the Djebel Serdj area, North-Central Tunisia, show evidence of the Lower Aptian Oceanic Anoxic Event 1a (OAE 1a) characterized by a short-lived negative $\delta^{13}C$ excursion, as well as time-equivalent deposits of shallow marine carbonate-platform drowning (Heldt et al., 2008). It is partly contemporaneous with the Continental sediments of the Malha Formation (Barremian-Aptian) and the green shales and marly limestones with cephalopods and planktonic foraminifera of the Resian-Aneiza Formation (Barremian-Albian) in Sinai, Egypt (Wanas, 2008, Abu Zied, 2008). Marl and limestone beds within the Halal Formation contain gastropods, rudists (*Eoradiolites liratus*) and orbitolinids (*Orbitolina parva, O. texana,* and *O. subconcava*), as well as planktonic foraminifera, such as *Favusella washitensis, Rotalipora subticinensis* and *Ticinella praeticinensis*) (Abu Zied, 2008).

Aptian cephalopods are common in North-Central Tunisia, such as in the Hamada Formation (Lower Aptian) and the Serdj Formation (Upper Aptian). Cephalopods are rare throughout the succession. Ammonite genera include lower Aptian *Deshayesites*, *Dufrenoyia*, *Pseudohaploceras*, and *Toxoceratoides*, and upper Aptian *Zuercherella*, *Riedelites* and *Parahoplites*. An Oceanic Anoxic Event (OAE1) occurred during the Lower Aptian, recognizable in Tunisia (Westermann et al., 2007, Lehman et al., 2009) and the Tarfaya Basin in Southern Morocco (Negri et al., 2009, Heldt et al., 2008) and characterized by black shales. A late Aptian global warming episode is indicated by the migration of Tethyan faunas toward the Boreal realm, a positive $\delta^{13}C$ excursion and a negative $\delta^{18}O$ excursion in deep-sea carbonates (Lehmann et al., 2009).

Africa became isolated after the breakup of Gondwana. The last connection between Africa and other continents was severed by Albian-Aptian times until the Early Miocene. During that period of isolation, an African endemic mammalian province was established and included freshwater fishes, reptiles, and rare mammals (Gheerbrant & Rage, 2006). A continuous terrestrial route between Africa and Eurasia was established during the Early Miocene.

In the Northern Western Desert, the Phanerozoic succession has been divided into Pre-rift, Syn-rift, and Post-rift sequence (El Sisi et al., 2002). The Pre-rift stage is restricted to the pre-Jurassic and older strata. The Syn-rift sequence encompasses Upper Jurassic-Lower Cretaceous sediments and was developed during the early extension phase of rifting and includes the Jurassic Masajid carbonates and the Lower Cretaceous Alam El Bueb Formation. The Post rift sequence extended from late Lower Cretaceous to early Upper Cretaceous and includes the Alamein Dolomite, Kharita Formation, and Baharia Formation. Numerous oil reservoirs and traps occur in all three sequences.

The earliest Aptian deposition started with sandstones and shales with a few interbeds of lignites, anhydrites and limestones of the Alam El Bueb Formation. Deposition took place in a near-shore, inner shelf, marine environment (Abdel Kireem et al., 1996). Deposition continued in the later part of the Aptian with the limestones containing *Orbitolina discoidea* of the Alamein Dolomite (Metwalli & Abd el-Hady, 1975, Chatelier & Slevin, 1988), which were later dolomitized. They were followed by the Dahab shales with interbeds of siltstones and sandstones. These sediments were deposited in a coastal shelf environment (Abdel-Kireem et al., 1996) either over thick Early Cretaceous shales of the Matruh Shale or the Burg el Arab Formation (Chatelier & Slevin, 1988). These sediments change facies northward into the Matruh Shale which constitute a source rock for the hydrocarbons in the area (Amine, 1961, El-Ayouti, 1990). The Alamein Dolomite form reservoirs in the Alamein, North Alamein, Yidma, and Horus fields (Younes, 2003c).

The Aptian of Northern Sinai is made up of oolitic and sandy limestones and non-fossiliferous variegated sandstones, alternating with fossiliferous marly beds of the Risan Aneiza Formation (Said, 1971), rich in *Orbitolina lenticularis* among other shallow-water foraminifera (Said & Barakat, 1957), as well as corals, pelecypods, gastropods, and ammonites. The Malha, the Risan Aneiza, and the Halal formations include rich assemblages of cephalopod horizons ranging in age from Barremian to late Albian (Abu Zied, 2008). These sediments grade southward into continental sandstones of the Hatira Formation.

The basal mudstones from the El-Nom borehole in the Gebel Abraq area near Aswan in Upper Egypt have yielded a diverse and well preserved terrestrial palynoflora that includes *Balmeisporites holodictyus*, *Crybelosporites pannuceus*, *Foveotricolpites gigantoreticulatus*, *Nyssapollenites albertensis*, *Retimonocolpites variplicatus* and *Rousea delicipollis*, which suggest an Albian-Cenomanian age and deposition in a fluvio-deltaic environment (Mahmoud & Essa, 2007).

The Aptian transgression in Southern Egypt (Fig. 20.14) is represented by coarsening-upward sequences composed of purple and green shales, siltstones and fine-grained, cross-bedded sandstones with wave ripples and burrows of the Abu Ballas Formation (Barthel & Boettcher, 1978, Klitzsch, 1979, 1990b). Each sequence is topped by thick paleosol horizons, which indicate oscillations of sea level. The unit contains abundant brachiopods (*Lingula* sp.), gastropods, pelecypods, plant remains, sea urchins, insects, *Archaeoniscus* and remains of fruits. The distribution of these sediments suggests a large embayment that extended between the Gilf el Kebir and the Kharga Oasis (Barthel & Boettcher, 1978). The increase of plant fossils toward the top suggests increasing freshwater influence.

In Cyrenaica, NE Libya, grey shales with ferruginous ooids, oolitic and sandy packstones-wackestones, and fine-grained sandstones with brachiopods, gastropods, ostracods, bryozoans, and echinoderms of the Aptian part of the Daryanah Formation were deposited in shallow-marine environment (Duronio et al., 1991). They grade locally into deep-marine sediments. Foraminifera in the Aptian part of the Daryanah Formation include *Nautiloculina oolithica, Ammobaculites* sp. and *Choffatella decipiens*. The latter species is characteristic of the Aptian in North Africa.

The Aptian sediments in Northeast Libya and Western Egypt are characterized by the appearance of angiosperm pollen. These pollen were reported from the Abu Ballas Formation (Aptian) in well Mawhoub West-2 located to the northwest of the Kharga Oasis by Schrank (1984, 1987, 1992), and from Libya, along with terrestrial miospores and freshwater ostracods (Ibrahim et al., 1995) and dinocysts (Thusu et al., 1988). West of the Abu Tartur Plateau in the southern Western Desert, the upper part of the Six Hills Formation contains exclusively land-derived palynomorphs such as spores, gymnosperms, angiosperms, and a few fungal spores. This part of the Formation is Aptian in age and was deposited in a continental environment (Mahmoud & Soliman, 1994). The presence of dinocysts in the middle variegated shales in the eastern Sirte Basin, in addition to glauconite, foraminiferal lining and fish teeth, indicates marginal marine influence during the Aptian (Thusu et al., 1988). This unit in the Sarir and Messlah fields contains also abundant algae, freshwater ostracods, and miospores, which were deposited in lacustrine, tidal flat, subtidal and marginal marine environments (Clifford et al., 1980, Thusu et al., 1988, Rossi et al., 1991, El-Hawat, 1992, Viterbo, 1969). The freshwater ostracod *Theryosyonecum* was reported from lacustrine and marginal marine shales and siltstones from the Eastern Sirte Basin and the Aptian shales and siltstones of the Bima Formation in the Benue Trough of northeastern Nigeria (Allix et al., 1981). Although the information is scanty, it is probable that rivers, bordered by freshwater lakes, were flowing in a southwest-northeast direction from the Benue Basin across the Kufrah Basin and the Eastern Sirte Basin into Egypt along the Trans-African Lineament during the Aptian. The Northern Egypt Basin and the Nile Basin were probably connected to the Kufrah Basin during that time.

Aptian sedimentation was probably associated with a new rifting and subsidence episode (Guiraud, 1998, Guiraud & Bellion, 1995). During that period, NW-SE-trending normal faults are often relayed by east-west dextral strike-slip faults (Guiraud, 1998). These east-west faults are now oriented northeast-southwest due to rotation of Africa. This rifting was followed by a tectonically (and magnetically) quiet period until the Santonian when inversion of basins occurred (Guiraud, 1998, Bosworth et al., 1999). In Southwest Egypt, the tripartite subdivision of the Nubian succession is evident and includes in ascending order the Gilf Kebir, Abu Ballas and Sabaya formations. The presence of *Skolithos* burrows in the Six Hills Formation and brachiopods in the Abu Ballas Formation suggests that the sea occasionally reached the area at that time. In the southern Eastern Desert, the equivalent of the Abu Ballas Formation is made up of shale-sandstone cycles with thin red-brown oolitic ironstone layers (Ward & McDonald, 1979). The sediments contain specimens of the freshwater bivalve *Unio*, fossil woods, and marine bivalves, such as *Inoceramus balli*, *Cyprina humei* and *Isocardia aegyptiaca* (Abbas, 1962). These fossils suggest deposition in coastal-plain to marginal marine environments. This unit overlies a thin unit composed of trough cross-bedded, fine to coarse-grained sandstones and conglomerates which is probably equivalent to the Six Hills Formation in the Gilf Kebir area.

Aptian sediments of the Chieun Formation (De Lestang, 1965) near the Libyan-Chad border in the southwestern Kufrah Basin consist of continental variegated silty shales with lenses of lacustrine limestones which were deposited unconformably on top of the Ounianga Formation. and have been correlated with the Middle Variegated Shale of the Nubian Sandstone in the Eastern Sirte Basin and the Abu Ballas Formation in Southwest Egypt (Tawadros, 2001) (Table). However, the age of the Chieun Formation is poorly constrained and has been assigned a possible Jurassic-Early Cretaceous age (Bellini et al., 1991, Lüning et al., 2000).

In spite of the global Aptian sea-level rise, the sea did not invade the Central Sirte Basin rift, with the exception of possible Aptian Variegated Shale in the Southeastern Sirte Basin. The absence of Aptian sediments in that area was due to either the possibility that the Central Sirte Arch was a high during that time or due to the lack of sediment supply (Posmantier & James, 1993).

A global drop in sea level took place at the end of the Aptian (Fig. 20.1) and resulted in an extensive unconformity which has been reported from Tunisia (Marie et al., 1984), the offshore of Northwest Libya (Bishop & Debono, 1996), Northwest Libya, Egypt and the Arabian Gulf region (Murris, 1980, Bulot et al., 2006). In Tunisia, this unconformity is marked either by karstification (Marie et al., 1984) or the deposition of continental sediments which contain ferrugenized woods and vertebrate fossils (Lefranc & Guiraud, 1990). In the northern Eastern Desert, at Gebel Shabrawet, the top of the Aptian succession is marked by emergence and the formation of paleosols (Mounir, 1998).

In the northern part of the Chott Fedjedj area (Fig. 18.18), the Orbata Formation (Aptian) consists of three members (Marzouk & Youssef, 2008). The lower member (the Barrani Dolomite) is composed massive dolomites (Busson, 1967, Ben Youssef et al., 1984, Kamoun, 1988), and limestones and marls rich echinoids, rudists, and abundant *Orbitolina lenticularis* (Marzouk & Ben Youssef, 2008). The middle member, made up of interbedded dolomitic limestones and marls, grades upward and laterally into evaporites. It contains *Choffatella decipiens*, *Orbitolina parva*, and *Cylindroporella sugdeni* (Marzouk & Ben Youssef, 2008). The upper member is made up of marls, sands and calcarenites, capped by limestones containing rudists and foraminifera, such as *Orbitolina parva* and *Mesorbitolina texana*, the ammonites *Deshaysites callidiscus*, *D.cf. planus*, *D.* weissi, *D. furcata*, and *Valdorsella* sp., and fragments of chelonicertidae (Marzouk & Ben Youssef, 2008). The top of the Orbata Formation is marked by a conglomerate with *Orbitolina minuta*, *Orbitolina texana*, and *Paracoskinolina tunesiana*.

### 20.3.7.4 Albian

This phase of global karstification and the formation of incised valleys during the sea-level lowstand was followed by drowning of carbonate platforms[1] as a result of a sea-level rise in the Upper Albian (*Rotalipora appenninica* Zone) (Gröschke et al., 1993). The top of these carbonates gives a good seismic reflector (Bobier et al., 1990). During that phase, the sea invaded most of North Africa (Lefranc & Guiraud, 1990).

Albian carbonates (*Albien carbonaté*) crops out near the Algerian-Libyan border (Lefranc & Guiraud, 1990). Albian basinal shales and argillaceous limestones with *Planomalina buxtorfi* and *Rotalipora appenninica* of the Masid Formation (Hammuda et al., 1985) were deposited in the offshore of Libya.

The sediments of the Turghat (Aptian) and Masid (Albian) formations grade in the offshore area of Northwest Libya into shallow-marine sandstones, shales, dolomites and limestones with the reef-forming foraminifera *Kurnubia* of the Chicla Formation (Hammuda et al., 1985, Bishop & Debono, 1996), and into the lower parts of the Zebbag and Fahdene formations and the Isis limestones in Tunisia (Bishop, 1988, Marie et al., 1984). These sediments grade in turn southward in Northwest Libya into coarse conglomeratic braided channel facies and meander channel fine-grained sandstones with plant fragments, silicified wood, and large tree trunks at the base of the Chicla Formation (Hammuda, 1971, El-Zouki, 1980). The Chicla sandstones were probably supplied by a source located to the south. Further south these sediments grade into the continental sandstones of the Msak Formation in western Libya.

In the northern part of the Chott Fedjedj area, Central Tunisia, the Lower Zebbag Formation (Albian) averages 90 m (295 ft) in thickness (Busson 1967). It starts with hard limestones with an erosional base which locally marked by reworked micritic clasts from the underlying sequence III. The limestones contain the ammonites *Knemiceras cf. gracile* and *Knemiceras* gr. *Compressum*, nautiloids, echinoids, and brachiopods, indicating a late Albian age. These carbonates are intercalated with fossiliferous marls with abundant oolites, lithoclasts, foraminifera, gastropods, and echinoids. This sequence reflects mixed marl-carbonate ramp sedimentation. The limestones with ammonites, nautiloids, and echinoids represent the maximum transgression during the Late Albian (Marzouk & Ben Youssef, 2008).

---

1 Schlager (1981) defines drowning of platforms as "an event in which combined effects of crustal subsidence and eustatic sea-level rise oupace the accumulation of carbonate sediments, resulting in the drop of the platform or reef top below the euphotic zone.

In the Eastern Sirte Basin, cross-bedded, argillaceous sandstones with plant rootlets and burrows (El-Hawat, 1992), and dinocysts (Thusu et al., 1988) of the upper "Nubian" sandstone unit (Albian-Cenomanian) were deposited in a nearshore environment. It is absent in much of the Messla Field and over high areas due to post-depositional erosion and/or non deposition (Clifford et al., 1980, Koscec & Gherryo, 1996, Gras & Thusu, 1998). Where it is present it forms prolific hydrocarbon reservoirs.

In the offshore of Northeast Libya deposition of the Albian platform sediments of the upper part of the Daryanah Formation (Duronio et al., 1991) started with oolitic grainstones with glauconitic and pyritic shales, followed by wackestones and packstones with scattered oolites and *Orbitolina* and *Neoaraqia*. These sediments contain ferruginous ooids and brick-red siltstones which probably represent paleosols.

In Southeast Libya, Albian sediments probably include white, red, and brown medium-grained to conglomeratic, massive to cross-bedded sandstones with abundant silicified wood, such as *Dadoxylon* of the Tekro Formation (Wacrenier, 1958, De Lestang, 1968). Eastwards the facies changes into braided fluvial channel, medium to coarse-grained, cross-bedded sandstones and paleosols with silicified and hematitized wood fragments of the Sabaya Formation (Wycisk, 1990, Klitzsch, 1990b, Klitzsch et al., 1979, Hendriks et al., 1990).

Albian deltaic and nearshore sandstones and shales of the Kharita Formation, which contain the land-derived spores *Cretacaciporites scabratus* and *Trilobosporites laevigatus* (El-Beialy, 1994), were deposited in the northern Western Desert and the Nile Delta region. The sandstones form reservoirs in the Qarun Field in the Gindi Basin, SW of Cairo (OGJ, 1996). Albian sediments become entirely limestones and dolomites in well Mamura-1 to the north (Said, 1962a), forming carbonate banks in that area (Sestini, 1984). At Gebel Shabrawet (between Ismailia and Suez) and in Central Sinai, Aptian-Albian sediments may include the upper part of the Risan Aneiza Formation (Said, 1971). Late Albian variegated shales and marls with *Hemiaster cubicus* and *Knemiceras syriacum* were deposited on top of Aptian sediments. The sediments consist of a lower retrogradational carbonate/siliciclastic parasequence and an upper aggradational carbonate facies (Mounir, 1998). The latter is composed of oolitic and sandy limestones and it was probably deposited during the *appenninica* event. The whole succession grades southwards into the Hatira sandstones in Sinai, and into shallow-marine sandstones with species of the dinocysts *Subtilisphaera senegalensis-Dinopterygium cladoides* (Aboul Ela et al., 1998), and the *miospores Crybelosporites pannuceus-Afropollis jardimis-Tricoporopollenites* assemblages (Ibrahim et al., 1998). In the Gebel El Minshera area, Northern Sinai, Albian-Cenomanian interval is represented by the Malha Formation and the overlying Galala Formation. The Malha Formation is made up of a clastic facies with a spore and pollen association that indicate a continental fluvial environment close to a vegetational source (El-Beialy et al., 2010).

The Albian-Cenomanian succession equivalent to the Timsah Formation encountered in borehole southeast of Aswan in Upper Egypt yielded terrestrial palynoflra that included the sporomorphs *Balmeisporites holodictyus*, *C. pannuceus*, *F. gigantoreticulatus*, *Nyssapollenites albertensis*, *Retimonocolpites variplicatus* and *R. delicipollis* (Mahmoud & Essa, 2007).

## 20.4   MIDDLE MEGASEQUENCE

### 20.4.1   Cretaceous

#### 20.4.1.1   Upper Cretaceous period

The Upper Cretaceous was a period of major eustatic sea-level fluctuations (Fig. 20.1) that led to a number of marine transgressions and regressions on a global scale (Vail et al., 1977, Matsumoto, 1980, Crux, 1982, Haq et al., 1987). The first marine transgression of the Upper Cretaceous submerged large areas of land, thus reducing the supply of detrital material. Consequently, thick sequences composed of limestones, dolomites and marine shales were deposited, especially in the northern parts of North Africa. Southward, large areas remained exposed and sandstones were predominant (Fig. 20.15). The rate of subsidence was low from the Cenomanian through the Late Campanian, followed by rapid subsidence rates during the Maastrichtian and the Paleocene (van der Meer & Cloetingh, 1996). Maximum subsidence took place in the Al Jabal Al Akhdar, Abu Gharadig and the proto-Clysmic[2] (Gulf of Suez) basins. Inversion of basins took place from the late Cretaceous through the Late Oligocene, as the sense of strike-slip motion of Africa relative to Europe was reversed from dextral to sinistral (Keeley, 1995, Keeley & Massoud, 1998, Guiraud, 1998, Guiraud et al., 2005). The resulting paleogeography in combination with sea level changes controlled the facies distribution throughout the Upper Cretaceous. A major change in the facies trends took place in the Upper

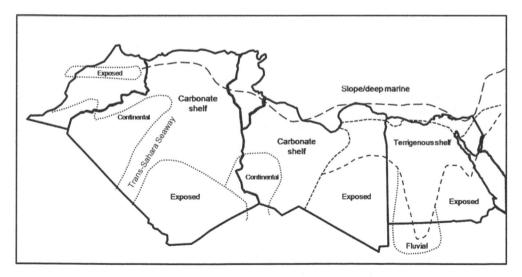

*Figure 20.15* Cenomanian paleogeography (modified after Philip et al., 2000, Stampfli et al., 2001, Bauer et al., 2002, 2003).

---

2   Clysma is the ancient Roman name of Suez. Hume (1910) used the term Clysmic Gulf to designate the ancestor Gulf of Suez which extended beyond its current limits.

Cretaceous in North Africa. Facies belts adopted east-west trends (Fig. 20.15) instead of the north-south trend that persisted during the Paleozoic.

The Upper Cretaceous was also a time of tectonic activities which led to the collapse of many of the tectonic arches, such as the Sirte Arch in Libya and the Dakhla Arch in Egypt. The precise timing and mechanism of the collapse of these arches are still not clear.

The Upper Cretaceous succession is equivalent to "Supersequence $S_2$ of Bobier et al. (1991) in Tunisia and the offshore of Northwest Libya. The succession is bounded by unconformities and it consists of sedimentary sequence that accumulated during a period of a relatively stable sea level highstand (supercycles UZA1 to UZA4 of Haq et al., 1987).

### 20.4.1.2  Cenomanian-Turonian

During the Cenomanian, the Tethys advanced from the north and the South Atlantic Ocean from the south, probably simultaneously, into the African continent through the epicontinental Trans-Saharan Seaway (Figs.). This seaway appears to have shifted eastward through time, starting in the Jurassic in the Atlas Domain to Algeria in the Late Cretaceous-Paleocene time (Figs.), probably as a result of the Atlas uplift in the west. The evidence for this seaway during the Late Cretaceous and Early Tertiary has been given by Barsotti (1963), Reyment (1963, 1966), Reyment & Reyment (1980), Dufaure et al. (1984), Keen et al. (1994), Carbonnel et al. (1990), and Bassiouni & Luger (1990). This evidence is based on similarities between ammonite, ostracod and pelecypod species which occur in North Africa and West Africa. Collignon & Lefranc (1974) suggested that communication took place via the Trans-Sahara seaway during the Upper Cenomanian, the basal Turonian and Lower Turonian. Reyment (1971) argued that communication between the two seas took place during the Upper Cenomanian but was interrupted due to a regression in basal Turonian and was resumed during the Maastrichtian. The similarity of the oysters and ammonites in West Central Sinai to those of North Africa, the Middle East, and western and southern Africa (Kassab & Ismael, 1994) supports a marine connection between the Mediterranean and West Africa during the Cenomanian. Migration routes could have been from the north (southern margins of the Tethys) or the south (eastern margins of the South Atlantic, via the Benue Trough (Courville, 2007).

On the other hand, some workers, such as Desio (1971) and Busson (1967b) opposed the notion of a seaway connecting the two bodies of water, based on paleo-geographic considerations.

Barsotti (1963) found that the majority of the ostracod species in the Paleocene "Fogaha Series" (equivalent to the Uaddan Group) (*G. velascoensis* Zone) in the Sirte Basin are also found in the West African basins, such as Senegal, Ivory Coast, Dahomey, Togo, and Mali. The genera suggest a carbonate-rich shallow shelf with warm-water conditions. Based on the similarities between the ostracod fauna of the Sirte Basin and those of the West African basins, such as *Buntonia pulvinata* and *Dahomeya alata*, he suggested a connection between these basins during the Paleocene. Reyment (1966) based on the study of ostracods, ammonites and molluscan faunas of West and North Africa during the Late Campanian to Paleocene concluded that migration toward

North Africa took place via the "Trans-Saharan Seaway" (Fig. 20.15). Reyment & Reyment (1980) studied some ostracod species from the Late Paleocene of the Sirte Basin in Libya and found that the vast majority of the species are identical to forms from the Paleocene of West Africa. Late Paleocene strata are much more common than earlier Paleocene strata in the region from West Africa to North Africa. It is believed that the maximum extent of the "Trans-Saharan Seaway" coincided with the Late Paleocene maximum transgression in many parts of the world (Keen et al., 1994). Bassiouni & Luger (1990) suggested that there was a shallow, warm, transcontinental connection between the southern Tethys and the Guinean province through Mali and Northwest Nigeria during the Late Paleocene transgression. Studies of ostracods at Gebel Nukhul and Wadi Feiran in West Central Sinai (Morsi et al., 2008) and around the P/E transition in the Southern Galala Plateau (Morsi & Scheibner, 2009) show faunal exchange between the Southern Tethys and the West African basins during the Maastrichtian-Early Eocene (via the Trans-Saharan Seaway). The ostracod fauna of Egypt described by Bassiouni & Luger (1990) from the Maastrichtian to Early Eocene of Egypt shows great similarities with those of the Sirte Basin. Migration of species was both to the south and to the north. According to Reyment (1981), the Paleocene Kalambina Formation of Northwest Nigeria contains an ostracod assemblage with roughly equal proportions of Southern Atlantic and Tethyan elements, thus indicating a marine connection between the Tethys and the Nigerian basin. El-Waer (1992) envisaged a connection between the Tethys and West Africa through the Trans-Saharan Seaway during the Paleocene based on the presence of an ostracod assemblage in the offshore of Northwest Libya, composed of *Dahomeya alata*, *Buntonia beninensis*, *Leguminocythereis lokossaensis*, *L. lagaghiroboensis* and *Cytheropteron lekense*. Barsotti (1963) suggested that separation between the two areas took place in the Lower Eocene as a result of tectonic uplift. The Paleocene-early Eocene boundary in the Sidi Nasseur-Wadi Mezaz area in Northwest Tunisia provided rich ostracod and foraminifera assemblages (Morsi et al., 2011) which reflect deposition in a coastal to inner neritic environment. Many of the taxa have a wide geographic distribution throughout the Middle East and North Africa, as well as West Africa. The fauna seem to have originated in West Africa during the Paleocene and migrated northwards during the late Paleocene to early Eocene. Sea-level change and decrease in oxygenation associated with the Paleocene-Eocene thermal maximum (PETM) caused the local disappearance of the South Tethyan Paleocene fauna and the appearance of a new but poorly diverse Afro-Tethyan fauna. After the PETM, diversity increased again and various ostracod taxa, such as *Bairdia aegyptiaca*, *Reticulina lamellata* and *Aegyptiana duwiensis* were re-established.

For the effect of the connection of the Gulf of Suez and the Mediterranean Sea on fauna, see Galil & Zenetos (1992). Their observations can be of interest to paleontologists and environmental geologists, and may shed the light on some of the faunal extinction in the region throughout time. For example, during the Cretaceous, the Tethys Ocean was connected periodically to the Pacific Ocean through the Trans-Saharan Seaway; the Mediterranean Sea was also connected intermittently to the Gulf of Suez and the Red Sea during the Miocene. According to these authors, the opening of the Suez Canal in 1869 allowed the entry Indo-Pacific and Erythean biota. Exotic macrophytes, invertebrates and fish are outcompeted or replaced native species locally, some are considered pests of cause nuisance, whereas other invaders are of commercial value. The invasion into the Eastern Mediterranean has increased the regions's biodiversity.

Many Upper Cenomanian to Lower Turonian sediments display anomalous lithological, geochemical, and faunal characteristics worldwide associated with the Cenomanian eustatic seal-level rise, which led to the major transgressive pulse during that period. This event was termed the "Oceanic Anoxic Event" (OAE2) by Schlanger & Jenkyns (1976) and has been studied extensively by Jarvis et al. (2002, 2006), Kolonic et al. (2005), Lüning et al. (2005), Marzouk & Lüning (2005), Kuhnt et al. (1990, 1997, 2004, 2009), Heldt et al. (2008), and Rodríguez-Rovar et al., 2009).

Major geological and biological events characterize the Cenomanian-Turonian period, such as a global high sea-level stand and warm sea-surface temperatures that allowed the development of large carbonate platforms; the radiation of marine fauna, and the OAE2. These events were subsequently reversed with a drop in sea-surface temperature, marine extinctions, and the drowning of carbonate platforms (Bardet et al., 2008).

Oceanic and marginal basins may contain dark-colored and black shales, associated with mass extinction of many species and the appearance of new species across the Cenomanian-Turonian boundary. The shales display anomalously high $\delta^{13}$ C values (Jarvis et al., 1988, 2006, Gasinski, 1988, Peryt & Wyrwicka, 1991, Keller et al., 2008). The OAE is more apparent in basinal sediments where black shales and marls are common. Shelf carbonates may display anomalous carbon stable isotope values (Jarvis et al., 1988,). In the Middle East, this Oceanic Anoxic Event has been recognized in outcrops and the subsurface (Lispon-Bentah et al., 1988).

The Cenomanian/Turonian organic-rich strata in North Africa have been most intensively studied in Morocco (Tarfaya, Rif) and Tunisia (northern and central parts) where the highest organic richness is reached (Kolonic et al., 2002, Lüning et al., 2004). In contrast, very little data is available for the other parts of North Africa, such as Algeria, Libya, and Egypt (Tawadros, 2001).

The Upper Cretaceous in Northeastern Algeria includes the Upper Zebbag Formation (Cenomanian), and the Hassi Bouras, Ras Toumb, and Kemaken formations (Turonian), the Doulid Formation (Turonian-Coniacian). In the Hodna Mountains, Lower-Middle Cenomanian marls and marly limestones are followed upward by Upper Cenomanian-Turonian limestones and dolomites (Bouchareb Haouchine & Boudoukha, 2009). In the Ain Ouarka Subbasin in Northern Algeria, carbonates and marls with the Cenomanian *Vascoceras* ammonite (Mekahli et al., 2004). Southward, in the Berkine Basin for example, the Cenomanian is represented by argillaceous lagoonal sediments (*Lagunaire Argileux*) and the Turonian by argillaceous limestones (*Calcaire Argileux*), overlain by Senonian lagoonal anhydritic limestones (*Carbonate Anhydritique Lagunaire*) (Yahi et al., 2001).

In the western Saharan Atlas in Algeria, the Cenomanian-Turonian was dominated by N-S oriented carbonate ramp, changing into deeper depositional environments in the Algerian coastal range. During the latest Cenomanian-earliest Turonian, isolated carbonate platforms with rudists and oyster-bearing sediments developed. Cenomanian-Turonian black shales, probably related to the OAE, occurred in the deeper areas between the platforms (Grosheny et al., 2008). Upper Cenomanian neritic limestones and marls also form part of the Peni-Tellian Unit 'Penitelliennes' (Upper Cretaceous-Paleocene) in the Constantine region (Moretti et al., 2007).

The late Cenomanian-early Turonian Oceanic Anoxic Event (OAE2) sedimentary record in the Tarfaya Basin along the Atlantic Coast of southern Morocco is one of the most complete worldwide (Keller et al., 2008). The OAE2 interval spans a series of biotic and oceanic events. Planktonic foraminifera are sensitive to changes in OAE2 and show species extinctions, evolutionary diversification, morphological adaptations, and changes in the relative abundance of species populations (Keller et al., 2008). However most species survived and returned when conditions improved (Keller et al., 2008).

Cenomanian and Turonian black shales and marls deposited during these Oceanic Anoxic Events form hydrocarbon source rocks. Cenomanian-Turonian black shales from the Tarfaya Basin (SW Morocco) have high up to 18% TOC content of type I-II kerogen, and hydrogen indices between 400 and 800 (mgHC/gTOC), which indicate that the black shales are excellent oil-prone source rocks. Low $T_{max}$ values obtained from Rock-Eval pyrolysis (404–425MC) indicate an immature to early mature level of thermal maturation (Kolonic et al., 2002, 2005). Kuhnt et al. (2009) recognized two main transgressive cycles that include eleven third-order depositional sequences within the lower and middle-upper Cenomanian sequence in the Tarfaya Basin. A major regression took place around the Lower-Middle Cenomanian boundary, characterized by lagoonal lowstand deposits, in response to sea-level drop. Kuhnt et al. (2009) also identified two positive carbon isotope $^{13}$C excursions in the middle and uppermost Cenomanian at 96.0 Ma and 94.0 Ma, respectively.

In the Abu Attifel Basin of the Eastern Sirte Basin, shales of the Cenomanian-Turonian Etel Formation contain TOC values in the range of 0.63% to 6.54% with an average of 2.85%. These values identify good source rock potential (El-Alami, 1996). They contain predominantly amorphous marine algae and type II kerogen, which were deposited in isolated basins under anoxic conditions (Gras & Thusu, 1996).

In Tunisia, the Cenomanian-Turonian Bahloul Formation consists of well-bedded dark grey limestones, with marls and argillaceous limestones intercalations which directly overlie a massive bed of conglomeratic sandy limestone with an incised base (Abdallah et al., 2000, Zagrani et al., 2008). The organic-rich facies of the Bahloul Formation change southward into open platform facies with planktonic and benthonic foraminifera and rudists and northward into basinal marls rich in ammonites and planktonic foraminifera. The Bahloul Formation developed during an anoxic event (Caron et al., 1999), as shown by the organic-rich black shales; the extinction of the foraminifera *Rotalipora* accompanied by the proliferation of *Heterohelix* and *Whiteinella*, and positive δ$^{13}$C excursions (Zagrani et al., 2008). The basal Turonian Bahloul Formation, believed to be the source of oil in the Upper Cretaceous Zebbag Formation and the Eocene Metlaoui Formation (Bishop, 1988), consists of thin, dark, laminated limestones with large globigerinids, also deposited under euxinic conditions (Marie et al., 1984, Bishop, 1988). The Bahloul Formation corresponds to the highest Cenomanian-Turonian mean sea level during the *Rotalipora cushmani, Globotruncana sigali* and *G. helvetica* zones. In the offshore of Northwest Libya the Turonian-Coniacian Makhbaz Formation contains black and dark grey calcareous shales in addition to dolomitic limestones, argillaceous and glauconitic micrites and biomicrites (Hammuda et al., 1985). The Cenomanian Alalgah Formation also contains black shales and black to greyish brown dolomites interbedded with fossiliferous black shales, argillaceous dolomites, and massive and nodular anhydrites

(Hammuda et al., 1985). These black shales were probably the result of the Oceanic Anoxic Event2 (OAE2).

In the offshore of Tunisia and Northwest Libya, Cenomanian carbonates of the Zebbag Formation and the Alalgah Formation were deposited, respectively. The Zebbag carbonates were deposited in platform, back-reef and lagoonal environments where wackestones and packstones with foraminifera and rudists occur in narrow belts and grade into deep water shales and micrites with planktonic foraminifera (Bishop, 1988). The presence of black to brown dolomites, black fossiliferous shales, and anhydrites of the Alalgah Formation suggests a restricted environment of deposition, ranging from a shallow shelf to deep anoxic basin.

Onshore Northwest Libya, the Cenomanian transgression resulted in the deposition of shallow platform limestones of the Early Cenomanian Ain Tobi Formation and marls of the Late Cenomanian Jefren Formation (the two members of the Sidi AsSid Formation of El-Hinnawy & Cheshitev, 1974) in the Jebel Nafusah area and the northern Ghadames Basin. These two formations were deposited unconformably on top of the eroded surface of the Chicla Formation (Aptian-Albian) and the Bu Sceba Formation (Triassic), west and east of Wadi Ghan, respectively. The succession changes from predominantly marly limestones in the west to predominantly limestones and dolomitic limestones in the east. The Early Cenomanian Ain Tobi Formation (Christie, 1955) is sandy at the base and rich in *Exogyra flabellata* and *E. columba* (These two species were reported from the Upper Cenomanian El-Heiz Formation in Egypt) and rudists. The Late Cenomanian Jefren Formation is characterized by a succession of clays, marls, limestones and evaporites with gastropods, such as *Nerinea gemmifera* and *Turritella* sp. (Christie, 1955). Restriction of the water circulation resulted in the deposition of evaporites in places.

The Cenomanian transgression initiated marine sedimentation in the Sirte Basin (Barr, 1972, Barr & Weegar, 1972). The sea transgressed over an eroded surface of Paleozoic and Early Cretaceous rocks. The Early Cenomanian transgression was associated with the collapse of the Sirte Arch. Structures of the Sirte Basin probably formed in response to dextral strike-slip movements along the Jifarah-Cyrenaica shear zone associated with the opening of the Mediterranean. Rapid erosion of the high relief topography at the end of the rifting phase led to the formation of coarse, conglomeratic basal sands of the Bahi Formation in many areas. The advancing sea reworked the Paleozoic and Lower Cretaceous sands. The Bahi sandstones contain rounded cobbles and boulders of Gargaf-type lithology. Deposition was initially continental fluvial, but changed into marine deposition. Glauconite is found only in the upper part of the Bahi Formation, indicating the initiation of a marine cycle (Barr & Weegar, 1972, Sghair, El Alami, 1996). Fracturing and brecciation of the pre-Bahi quartzites probably took place before the Cenomanian transgression. The Gargaf and Nubian quartzites were reworked and rounded by current action. The Maragh Formation is composed of pebbly sandstones, glauconitic sandstones, dolomites, evaporites and shales. It was probably deposited in alluvial fans, which graded into shallow-marine sandstones and evaporites (Roberts, 1970, Heselden et al., 1996, Sghair & El-Alami, 1996). The glauconitic sandstones represent transgressive sand bodies. The evaporites characterize a prograding sabkha coastline and an intertidal-supratidal environment (Heselden et al., 1996). The Bahi sandstones and the Maragh Formation were followed by the Cenomanian Lidam Formation composed of shallow-water marine dolomites and calcarenites that contain molluscs, miliolids, *Ovalvulina ovum*, algae and echinoids (Heselden et al., 1996, El-Bakai, 1991, 1996). These sediments form oil reservoirs in the Masrab Field in the Southeastern Sirte Basin and the Dur Mansur Field in the western part of the basin. Silicified Cenomanian limestones with marine microfauna occur on top of the Precambrian granites near the northern Tibesti foothills (Dufaure et al., 1984) and indicate the southern limits of the Cenomanian transgression.

In Northeast Libya, deposition continued uninterrupted from the Albian to the Cenomanian. Shallow-water, Late Albian-Early Cenomanian marls and shales of Qasr Al Abid Formation (Kleinsmeide & van den Berg, 1968, Röhlich, 1974) with miliolids, orbitolinids, *Thomasinella punica*, *Rotalipora appenninica*, *R. reicheli*, *R. cushmani*, and *R. greenhornensis* (Barr, 1972, Barr & Weegar, 1972), and species of dinoflagellates and miospores (Uwins & Batten, 1988, Thusu & van der Eem, 1985), Duronio et al., 1991) were deposited offshore and onshore northern Cyrenaica. Klen (1974) also shows the Gasr al Abid Formation, encountered in a water-well in the Benghazi sheet area, in a pelagic facies. Late Cenomanian-Turonian limestones, marls, and dolomitic limestones with gastropods, rudistids, oysters, *Cuneolina pavonia parva* and *Rotalipora greenhornensis*, *R. cushmani*, *Thomasinella punica*, and *Nezzazata simplex*, were deposited conformably on top of the Qasr Al Abid Formation (Röhlich, 1974, Klen, 1974, Duronio et al., 1991).

In the Kufrah Basin, the Cenomanian is represented by continental cross-bedded, medium-grained to conglomeratic sandstones with silicified wood of *Dadoxylon* of the Tekro Formation (Wacrenier, 1958, Burollet, 1963b, De Lestang, 1968, Hesse et al., 1987, Grignanni et al., 1991) equivalent to the Cenomanian Sabaya Formation in southwest Egypt.

The Cenomanian in Egypt shows a facies change from fully marine in the north to fluvio-marine to fluvial in the south. In the central Western Desert, Early Cenomanian deposition was dominated by fluvio-marine sandstones, glauconitic sandstones, sandy argillaceous limestones and bioturbated gypsiferous green shales of the Baharia Formation. The sediments contain a mixture of fresh and marine fossils, such as pelecypods, fishes, plesiosaurs, crocodiles, turtles and the snake *Symoliophis* in the Bahariya Oasis (Said, 1962a, Dominik, 1985, Soliman & Sultan, 1976). This area probably represents the transition from a marine environment in the north to continental in the south. Marine glauconitic sandstones in the Abu Roash; units F (Mansour Member) and G (Abyad Member, or the Baharia Formation) of the Abu Roash Formation, respectively (Jux, 1954) grade northward into open-marine limestones, sandstones and minor shales with *Thomasinella punica* (Metwalli & Abd El-Hady, 1975) (Medeiwar Formation of Schlumberger, 1984, Chatelier & Slevin, 1988). Baharia Formation and Abu Roash "G" Member sediments in the Razzak-7 well, North Western Desert, Egypt, contain abundant kerogen and fossil palynomorphs (Zobaa et al., 2008). Terrestrial organic components dominate in the Cenomanian Baharia Formation and the basal Abu Roash "G" Member. Dinoflagellate cysts in these units dominantly peridinioids such as *Subtilisphaera* suggest deposition under nearshore, moderate to high-energy conditions. In contrast, amorphous organic matter and marine palynomorphs are more abundant in the upper part of the Abu Roash "G" Member, suggesting deeper depositional conditions. In the Abu Gharadig Field these shales contain high TOC values of up to 1.41% and are considered the principal zone of oil generation (Khaled, 1999). Southward, the facies changes into braided fluvial channel, medium to coarse-grained, cross-bedded sandstones with paleosols of the Sabaya Formation. These sediments grade upward into ripple-laminated fine-grained sandstones and siltstones with silicified and hematitized wood fragments (Wycisk, 1990, Klitzsch, 1990, Klitzsch et al., 1979, Hendriks et al., 1990). Eastward, for example in Wadi Qena, Cenomanian sediments continue into massive fluviatile and deltaic sandstones with paleosol and root horizons of the Wadi Qena Formation. Direction of flow was northwards in the southern Western Desert, but it became variable with a predominantly southwest direction in the Southern Eastern Desert (Hendriks et al., 1990, Van Houten et al., 1985, Ward & McDonald, 1979). In the Aswan area, the unit interfingers with basaltic lava flows (Van Houten et al., 1985, Ward & McDonald, 1979).

The Late Cenomanian follows a similar trend. The El-Heiz Formation changes facies from sandy limestones with *Thomasinella punica* (also identified from the Ain Tobi Formation in Libya by Gohrbandt, 1966a) (Metwalli & Abd El-Hady, 1975), marine shales and limestones (Conway et al., 1988a, b) of the Abu Roash 'G' Formation in the subsurface of the Northern Western Desert into thick-bedded, white, foraminiferal limestones, highly argillaceous, cherty, and glauconitic limestones in the Abu Roash area. Oyster beds with *Cyphosoma abatei, Ostrea flabellata* and *O. columba* are common in that area.

In the Southern Western Desert, the Late Cenomanian transitional facies is dominated by fine-grained sandstones and shales with plant remains of the Maghrabi Formation. Deposition took place in fluvial, nearshore, supratidal-subtidal and mixed estuarine environments (Hendriks, 1986, Klitzsch et al., 1979). In the Abu Tartur area, the Maghrabi Formation contains the brachiopod *Lingula* sp., lamellibranchs, and remains of fish (Klitzsch, 1990b, Klitzsch et al., 1979). Eastward in the Wadi Qena area it changes facies into shallow-marine lagoonal deposits composed of sandy limestones, marls and limestones with coral and ammonite horizons of the Galala Formation (Luger & Gröschke,1989). These sediments interfinger with high-energy beach deposits and grade upward into shallow inner-shelf sediments (Klitzsch, 1990b, Hendriks et al., 1990). These sediments grade southward into mainly continental fluvial sediments with paleosols south of Gebel Kamil and parts of the Gilf Kebir Plateau (Hendriks, 1986). In the southern Eastern Desert, the Maghrabi Formation is made up of several cycles of variegated shales, sandstones and ripple-laminated siltstones. These cycles contain horizons of chamositic-hematitic oolitic ironstones, and phosphatic and conglomeratic lag deposits, which contain both freshwater and marine fossils (Van Houten et al., 1985, Ward & McDonald, 1979). The direction of flow was to the southwest and southeast in the lower part of the Formation changing to a northerly direction in its upper part (Van Houten et al., 1985).

In Northern and Central Sinai, a shallow platform, hundreds of kilometers wide existed during the Cenomanian (Kuss et al., 2000a, b, Bauer et al., 2001, 2002, 2003, 2004). Shallow-marine sediments with thicknesses of more than 450 m (1476 ft) were deposited. These sediments show a regional increase in thickness to the north (Bartov & Steinitz, 1977). Limestones, dolomites, fossiliferous marls, sandstones, and minor amounts of chert of the Raha Formation (Hazera Formation Bartov et al., 1980), equivalent to the Ain Tobi Formation in NW Libya were deposited on the platform. The Raha Formation represents a complete regressive-transgressive-regressive cycle (Cherif et al., 1989), which spans the Cenomanian. The cycle starts with Early Cenomanian shallow-marine glauconitic sandstones, limestones with *Thomasinella aegyptia* (Ghorab, 1961, Said, 1971) of the Abu Had Member, followed by fluvio-marine shales, limestones and marls with oysters, echinoids, ostracods and agglutinated benthonic foraminifera (Mukattab Member of Cherif et al., 1989) (Kora et al., 1994). A reefal biofacies containing orbitolinids and miliolids is succeeded upward by oyster banks, dominated by *Ostrea flabellata*, *O. africana*, and *Exogyra columba* (Said, 1962a, Bartov et al., 1980). This facies occurs at Gebel Shabrawet near Ismailia (Cuvillier, 1930), Gebel Maghara, Wadi Sudr, and Wadi Matulla in Central Sinai (El Shazly, 1998, Gendi, 1998). These carbonates were deposited during the maximum phase of the Cenomanian marine transgression. A regressive phase took place in the Upper Cenomanian, which resulted in the deposition of the sands of the Mellaha Member. The sands were deposited by rivers flowing from the east and southeast. These sediments grade northward into resistant dolomitic and recrystallized limestones with *Manelliceras* fauna and *Neolobites fourtaui* of the Halal Formation (Said, 1971), then into variegated shales and marls with thin limestone interbeds (Lewy, 1975). In the Gebel El Minshera area, Northern Sinai, the Galala Formation signals the Cenomanian marine transgression, with the establishment of proximal carbonate ramp sedimentary facies (El-Beialy et al., 2010). The spore-pollen assemblage near the base of the Galala is similar to that of the Malha Formation, in addition to *Classopollis* pollen, which suggessts coastal vegetation. The dinoflagellate cyst *Subtilisphaera senegalensis* in the upper Galala Formation reflects ecologically stressed conditions in a restricted marine environment (El-Beialy et al., 2010).

At Ain Sukhna on the western side of the Gulf of Suez, the Cenomanian Galala Formation consists of yellowish and greyish calcareous marls underlain by the Permo-Triassic Qiseib Formation and overlain by Senonian limestones and dolomites. The Cenomanian sediments contain a rich ostracod fauna typical of marine shelf setting of the Southern Tethys (Boukhary et al., 2009).

In the High Atlas, Cenomanian-Turonian carbonates were deposited unconformably on the Jurassic rift structures (Ensslin, 1992, Rodriguez-Rovar et al., 2009, Cavin et al., 2010). Cenomanian-Turonian marine lagoonal sediments were deposited in the Central High Atlas and the Middle Atlas. Albian-Cenomanian red marls, sandstones, and anhydrites were followed upward by Cenomanian-Turonian limestones rich in planktonic foraminifera "*barre turonienne*". The Cenomanian transgression also led to the deposition of marls and marly limestones of the Ait Lamine Formation over the entire Essaouira Basin (Le Roy et al., 1998), followed by Turonian marly limestones and black shale interbeds, which are common in the continental margin of the Atlantic (Kolonic et al., 2002, Lüning & Kolonic, 2003).

Differences in facies and ichnofabrics across the Cenomanian-Turonian boundary in the Western Tethys show that the Oceanic Anoxic Event (OAE-2) was interrupted by short dysaerobic conditions. Trace fossils, dominated by *Chondrites*, *Palaeophycus*, *Planolites*, *Taenidium*, *Thalassinoides*, and *Trichichnus*, show that the OAE-2 Event consisted of a series of anoxic sub-events interrupted by short dysaerobic or aerobic sub-events (Rodriguez-Rovar et al., 2009). The anoxic conditions and deposition of black shales in the Western Tethys during Cenomanian-Turonian has been related with a sluggish circulation due to the narrow strait of Gibraltar (Lüning et al., 2004). On the Northwest Africa Atlantic Margin, the OAE-2 is associated with an expansion of the oxygen minimum zone (OMZ) towards the shelf, probably associated with a redeposition of the organic matter from shallower waters to deep-sea environments (Lüning et al., 2004).

Continuing tectonic activity during the Turonian led to the formation of a number of shallow basins within the Sirte Basin in which evaporites were developed, such as the Hameimat and Maradah troughs, Al Kotlah Graben, and a salt basin to the west of the Jalu Field (Hallet & El-Ghoul, 1996, Hallet, 2002). In the Hameimat Trough, the Turonian is represented by the dolomites, anhydrites, and shales of the Etel Formation and its equivalent the Argub Formation (Barr & Weegar, 1972, Barbieri, 1996).

The Turonian in the subsurface of the Northern Western Desert, includes shallow-marine sediments of the Abu Qada and Wata formations (Rammak, Abu Sennan, Meleiha, and Miswag members of the Abu Roash Formation *sensu lato),* composed of very fine, calcareous glauconitic sandstones, oolitic- glauconitic limestones and shales, interbedded with sandstones and anhydrites (Metwalli & Abdel-Hady, 1975). In the Abu Roash area (see Jux, 1954) the Turonian is represented by white chalky, highly argillaceous, cherty, gypsiferous, sandy, and highly fossiliferous limestones, flint bands, bryozoan limestones, dolomitic limestones, sandstones with *Durania* biostromes, reddish brown sandstones, and yellowish brown shales (the lower part of the Abu Roash Formation "units B-E" as originally defined). These shales contain an average TOC value of 1.01% with sapropelic oil-prone Type II kerogen and are considered as a potential source rock for the oil in the Abu Gharadig Field (Khaled, 1999).

In Northern Sinai, Turonian sediments contain pelecypods, rudists, gastropods, and echinoderms (Said, 1962a, Lipson-Benitah et al., 1988) and grade southward in west-central Sinai into dolomitic limestones of the Wata Formation. In Upper Egypt, the Turonian is separated from the Upper Cenomanian by an unconformity. The Turonian in the Dakhla and Gilf Kebir areas is represented by the Taref, Abu Agag, and Umm Omeiyad formations. Medium to coarse cross-laminated sandstones with paleosols, roots and lacustrine intervals of the Taref Formation in the Kharga Oasis were deposited disconformably on the Upper Cenomanian Maghrabi Formation in braided stream channels (Wycisk, 1990, Hendriks et al., 1990). The direction of sediment transport is variable. Paleocurrents indicate an eastward sediment transport (Wycisk, 1990), or a northward direction (Klitzsch et al., 1979). In the Eastern Desert, the sandstones range from quartzarenites to subarkoses, rippled and burrowed at the top, and contain petrified wood and plant fragments, shale rip-up clasts, teeth and bone fragments, and oolitic ironstones (Van Houten et al., 1985, Ward & McDonald, 1979). The Abu Agag Formation overlies basement in the area between Qena and Abu Simbel. Basal conglomerates fill up the irregular topography of the basement and grade upward into coarse-grained, cross-bedded fluvial braided river sandstones, topped by paleosols and channel sandstones. The Umm Omeiyad Formation was deposited on the erosional surface of the Galala Formation and it is composed of fluvial sandstones in the lower part, followed upward by shallow-marine glauconitic sandstones, and then by marly limestones. Uplift of the area west of the Nile, probably during the Upper Campanian, resulted in the erosion of the Taref and Maghrabi formations (Hendriks, 1986). In that area, Late Turonian to Middle Campanian strata are missing.

Early Turonian deposition in Sinai started with a transgressive phase dominated by the deposition of shallow marine reefal facies with orbitolinids, miliolids, and oolitic shoals of the Abu Qada Formation. Deepening of the depositional basin led to the deposition of black marls and shales with *Heterohelix* and *Ostrea africana*, dolomitic limestones, and marls rich in ammonites of the Ora Shale (Bartov et al., 1980, Cherif et al., 1989, Kassab & Ismael, 1994). In Late Turonian time, expansion of marine waters led to the deposition of shallow platform carbonates of the Wata Formation. Reefal limestones with *Discorbinus* and rudistids, dolomites and cherts (Gerofit Formation) were deposited (Bartov et al., 1980, Cherif et al., 1989). Open-marine conditions returned at the end of the Late Turonian and resulted in the deposition of marls and shales with planktonic foraminifera, oysters and ammonites. Turonian sediments attain their maximum thickness in central Sinai. The Arif el Naga high was probably a positive area during the earliest Turonian time since deposits representing that time span are absent (Bartov et al., 1980). The Wadi Araba and St. Paul Monastery areas were probably high areas during the entire Turonian time and Turonian sediments are probably absent.

### 20.4.1.3 Coniacian-Santonian

Sinistral strike-slip shear movements took place in North Africa during the Santonian as a result of a northwest-southeast compression. This phase corresponds to the initial stages of collision between Africa and Europe and the initiation of the Syrian Arc fold belt (Neev & Ben-Avraham, 1977, Moustafa & Khalil 1989, Garfunkel 1998, Guiraud et al., 2005, Yousef et al., 2010). Many basins were inverted and some intraplate fault zones were rejuvenated (Lüning et al., 1998, El-Azabi & El-Araby, 2007). The change of sense of movement along the Trans-Africa Lineament resulted in the inversion of basins situated on either side. For example, subsidence of the Abu Gharadig Basin located on the western side of the lineament reached its maximum during the latest Cretaceous when the Gindi Basin, situated on the eastern side of the lineament, was a high (Keeley & Massoud, 1998). This zone separates the Northern Egypt Basin from the Nile Basin (Fig. 20.16). A reverse process took place during the Paleocene-Eocene. The Al Jabal Al Akhdar Basin which was inverted in Middle and Late Santonian (Röhlich, 1980) subsided rapidly during the Late Cretaceous-Early Paleocene, and was again inverted in Late Paleocene-Early Eocene.

Following the Turonian AOE, deposition during the Coniacian-Santonian period was dominated in the offshore area of Northwest Libya by dolomitic limestones, and argillaceous and glauconitic micrites and biomicrites of the Makhbaz Formation (Hammuda et al., 1985). The latest Santonian Jamil Formation deep-marine calcareous shales and limestone interbeds, with an assemblage of the *Globotruncana concavata* Zone were deposited on top of the Makhbaz Formation. Inland, shallow-water carbonates of the Gasr Tigrinna Formation were deposited in the Jebel Nefusah area and the Douleb Member of the Kef Formation in Tunisia.

In the Tunisian Trough in Northwestern Tunisia, the Coniacian-Santonian corresponds to the middle part of the Kef Formation. The section is rich in the inoceramid species *Platyceramus cycloides* associated with the foraminifera *Globotruncana manaurensis* overlain by a marker bed rich in inoceramid *Platyceramus cycloides,* which is an index taxon for the Coniacian-Santonian boundary (El Amri & Zaghbib-Turki, 2004). This part of the Kef Formation is correlatable with the Matulla Formation in West Central Sinai and the Gulf of Suez.

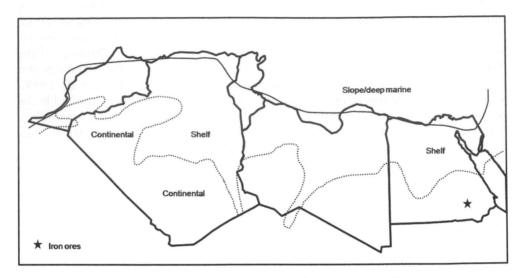

*Figure 20.16* Coniacian-Maastrichtian paleogeography (modified from Tawadros, 2001).

In the Sirte Basin, progressive onlap continued during Coniacian-Santonian time. Extensive land areas still existed during that period, particularly on the Az Zahrah (Dahra), Al Jahamah and Cyrenaica Platforms. That period was also dominated by erosion over most of central Libya. The Rachmat shales, calcarenites and dolomites were deposited in the troughs. The Rachmat Formation was capped in the Awjilah and An Nafurah areas by shallow-water calcarenites of the Tagrifet Limestone. The equivalent of this section in the Zallah, Maradah and Ajdabiya troughs consists predominantly of shale sequences with thin limestone and sandstone stringers. Deposition of the Rachmat Shale (Coniacian-Santonian) and Sirte Shale (Campanian) took place in a protected rift graben which prevented the oxygenation of organic matter and allowed for the preservation of source rock shales (Schröter, 1996). The oldest rocks encountered in the Al Wadayat Trough are of Coniacian-Santonian age (Hallett & El-Ghoul, 1996).

Coniacian-Santonian basinal sediments north of the Al Jabal Al Akhdar are represented by deep-water (bathyal) shales of the Al Hilal Formation. The shales are dark grey to greenish grey, glauconitic with thin limestone beds (Barr & Hammuda, 1971). They are rich in planktonic foraminifera and dynocysts (Thusu & van der Eem, 1985, Uwins & Batten, 1988) with rare ostracods, calcispheres, echinoid fragments and *Inoceramus* prisms. They grade into the offshore of Cyrenaica into wackestones and mudstones (Duronio et al., 1991). The Gasr al Abid and Benia formations change facies northward into pelagic shales, and distinction between them and the Al Hilal Formation is difficult.

Inversion of the Al Jabal Al Akhdar Basin took place during the Middle and Late Santonian (Röhlich, 1980). In the Ghawt Sas area, the Campanian Al Majahir Formation rests unconformably on the Late Cenomanian-Turonian Benia Formation (Klen, 1974, Röhlich, 1974). The equivalent of the Al Hilal Formation is absent in that area. The Campanian Al Majahir Formation contains a mixed Campanian-Lower Santonian fauna. Röhlich et al. (1996) attributed this mixing to reworking of Lower Santonian elements from older strata. The lack of Middle and Upper Santonian forms suggests that no sediments were deposited during that time (Röhlich et al., 1996), which was a time of erosion following folding and uplift of the area. Deposition was probably continuous in the offshore during Senonian (Duronio et al., 1991).

Sediments of the Abu Roash Formation (*sensu stricto*) (Coniacian-Santonian) were deposited over most of the platform areas in Egypt. They are absent in central Egypt (Dakhla and Wadi Qena areas), probably as a result of the Santonian compressive events. The white chalky, sandy, and glauconitic limestones with bryozoans of the Abu Roash Formation in the Bahariya Oasis were deposited in a shallow open-marine shelf environment. A few kilometers north of the pyramids marly limestones contain echinoids, ammonites and other molluscs (Cuvillier, 1930, Jux, 1954). Correlatable rocks in the subsurface of the Northern Western Desert form a sequence of cyclic shallow to deep-marine carbonates, shales, and sandstones rich in organic content (Awad, 1984, Zobaa et al., 2008). Southward, the Timsah and Um Barmil sediments consist of deltaic, beach, distributary channel, lagoon, and levee and pond sediments with intensely mottled paleosols (Hendriks et al., 1990). South of Aswan, they are replaced by backshore and shoreface sediments and contain oolitic iron-bearing shales (Attia, 1955, El-Naggar, 1970, Bhattacharyya, 1989) (Fig. 20.16).

From the Late Coniacian onwards, the shelf retreated southwards and was replaced by deeper water sediments in Central and North Sinai, and Israel (Lewy, 1975, Bauer et al., 2001, 2003, Kuss & Bachmann, 1996, Lüning et al., 1998a, Bauer et al., 2003, El-Azabi & El-Araby, 2007).

Two transgressive phases took place during deposition of the Matulla Formation (Coniacian-Santonian) (El-Dawy, 1994) in Sinai and the Gulf of Suez region, following the low sea-level period at the end of the Upper Turonian (Wata Formation). This low sea-level period was followed by a relative sea-level high stand during the Coniacian (Lower Matulla or Zihor Forrmation). The early Santonian was marked by a minor fall in sea level which led to a regression and an increased clastic detritus in the formation (Bosworth et al., 1999). Another transgressive phase started by a sea-level rise in the middle Santonian was followed by a major regression at the end of Santonian. The Matulla Formation in Nezzazat and Ekma areas on the eastern side of the Gulf of Suez is composed of various lithologies that have been deposited in lower intertidal to subtidal environment. Depositional facies include beach foreshore laminated sands, upper shoreface cross-bedded sandstones, lower shoreface massive-bedded, densely bioturbated and wave-rippled sandstones, shallow subtidal siltstones, deep subtidal shales, lower intertidal lime-mudstones to floatstones, dolomites, and oolitic ironstones, shallow subtidal shoals of rudstones and grainstones, and shoal margin packstones (El-Azabi & El-Araby, 2007).

The lower part of the Matulla Formation (Zihor Formation) (Coniacian), composed mainly of chalky bioclastic limestones and marls with ammonites, fish teeth, and *Pycnodonte vesicularis*, was deposited in a shallowing-upward open-marine environment in two depocenters in western and central Sinai (Bartov et al., 1980). An oolitic shoal with dascycladacean algae, hermatic corals and stromatoporoids separated an outer shelf to the north from a stable shelf to the south (Lewy, 1975). Tectonic activities took place during the deposition of the Matulla (Zihor) Formation in the Late Turonian-Late Coniacian (Lewy, 1975). In the Gebel Musabba Salama area, southeast of Abu Zneima, Sinai, the Upper Turonian-Coniacian is absent due to tectonics (Kassab & Ismael, 1994). The Santonian part of the Matulla Formation rests unconformably on top of the Lower Turonian Wata Formation. Deposition of the upper part of the Matulla Formation (Santonian) in that area formed a regressive sequence dominated by subtidal sediments with the gryphoid *Pycnodonte*, followed upwards by intertidal cross-bedded sandstones with gypsum bands.

Relative shallowing of the sea took place during Late Santonian-Campanian time. Deposition was dominated by massive chalks and marls containing phosphatic grains and corroded oysters, fish teeth and *Spinaptychus spinosus* of the upper part of the Matulla Formation (Menuha Formation) (Bartov et al., 1980) or the Campanian-Maastrichtian St. Anthony (Gebel Thelmet) Formation (Abdallah & Eissa, 1971, Kuss et al., 2000a, b) with *Dicarinella asymetrica* (El-Dawy, 1994). Matulla and Thelmet formations exposed at Wadi El-Ghaib in Southeastern Sinai, revealed the presence of six microfacies types namely: quartz arenite, sublithic arenite, sparite, dolomites, ferruginous dolomitized sparite and sandy micrite (Elhariri et al., 2007). Conglomerates, bioclastic limestones up to a few meters thick with reworked Cenomanian and Turonian limestone and dolomite pebbles were deposited close to the southern flank of the depositional basin.

### 20.4.1.4  *Campanian*

Campanian sediments correspond to the TST cycle of the Upper Zuni UZA3 super-cycle and the HST cycle of the UZA4 supercycle of Haq et al. (1987). Relative shallowing of the sea, which started during the Late Santonian, continued into the early part of the Campanian. The sea transgressed during the Campanian probably from the north and the south into North Africa. Deposition continued following the same pattern which existed during the Santonian (Fig. 20.17). Rifting during the Early Campanian-Late Maastrichtian took place in the Nile Basin. The Abu Gharadig Basin in the Northern Western Desert was rejuvenated as a pull-apart basin along dextral strike-slip faults (Guiraud, 1998). *Inoceramus* thrived in the Campanian seas in Egypt and Libya during that time. Economic phosphate deposits were formed during the Campanian low sea stand. Anoxic oceanic events during the rapid sea-level rise led to deposition of large volumes of hydrocarbon source rocks.

Campanian sediments in the offshore area of Tunisia and Northwest Libya are represented by deep-marine carbonates and marls. Argillaceous and glauconitic micrites to biomicrites with globotruncanids and deep-marine ostracods such as *Krithe* and *Bairdia* (El-Waer, 1992) of the Bu Isa Formation in the offshore of Northwest Libya (Hammuda et al., 1985) and the Abiod Formation in the offshore of Tunisia (Bishop, 1985, 1988) were deposited in a basinal to lower slope marine environment (Wilson, 1975, Microfacies MSF-3).

The Campanian-Maastrichtian succession near El Kef, Northwestern Tunisia is 500 m (1640 ft) thick. It includes the upper beds of the Kef Shale (Aleg Shale), the Abiod Chalk, and the El Haria Shale. A positive carbon isotope event of 0.3% $\sigma^{13}$C occur at the Santonian-Campanian boundary (at 83.7 Ma) and another one of 0.2%

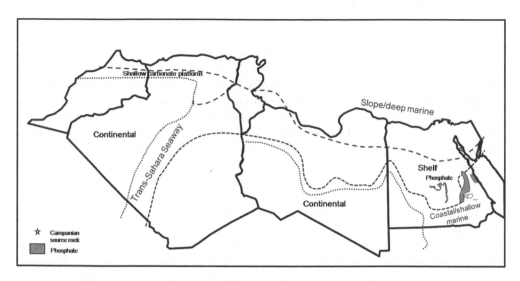

*Figure 20.17* Campanian-Maastrichtian paleogeography of North Africa (after Camoin et al., 1993, Kuss et al., 2000a, b, Scheibner et al., 2001, Tawadros, 2001).

in the mid-Campanian (at 78.7 Ma) (Jarvis et al., 2002). A negative excursion of −0.4% occurs in the Upper Campanian (at 74.8 Ma). Jarvis et al. (2002) attributed these events to rapid rise and fall in sea-level.

In the Internal and External Dorsale in the Rif Domain, the Campanian and Maastrichtian *Globotruncana*-rich marly mudstones are referred to as "Capas Rojas" and "Capas Blancas", respectively. A hiatus occurs within the Campanian-Maastrichtian *Globotruncana* mudstones (El Kadiri et al., 2006).

West Libya and the western Sirte Basin continued to receive shallow-water sediments of the Mizda Formation. Early Campanian deposition started with cross-bedded limestones and dolomites and rare calcareous sandstones of the Mazuzah Member. Some calcarenites are composed entirely of *Inoceramus* prisms (Röhlich, 1979). The depositional environment was probably intertidal. Deepening of the basin followed in the later Campanian with the deposition of chalky, dolomitic, siliceous and oolitic limestones, marls, green shales, and gypsum of the Thala Member. These sediments contain gastropods, miliolids, rare planktonic foraminifera, algae, and silicified wood fragments (Röhlich, 1979, Cepek, 1979) and were deposited in restricted lagoonal, shallow-marine open shelf, and deep marine environments (Salaj, 1979, Salaj & Nairn, 1987). Two cycles of sedimentation, separated by phosphatic horizons, can be recognized (Salaj & Nairn, 1987, Nairn & Salaj, 1991). It is interesting to see the striking similarity between this Formation in western Libya and its equivalent the Duwi Formation in Egypt.

In the subsurface of the Sirte Basin, deep-marine conditions became established in the trough areas during the Campanian (Fig. 20.17). Penecontemporaneous faulting of at least 1200 m (3937 ft) can be demonstrated in the Maradah Trough. The final collapse of the Sirte Arch took place at that time, leading to the formation of the Ajdabiya,

Hameimat, Maradah, Zallah and Tumayyim troughs (Hallett & El-Ghoul, 1996). The Sirte Shale sediments were deposited in the trough areas during the Campanian. The Sirte Shale glauconitic sediments reflect deep-marine and slow deposition under anaerobic conditions. These conditions were necessary for the formation of glauconite (Odin & Matter, 1981) and resulted in the deposition of dark, carbon-rich sediments. These anaerobic conditions probably resulted from a rapid sea-level rise during the Campanian (Fig. 20.17), and may reflect another OAE (Schlanger & Jenkyns, 1976), which may have occurred during the Campanian-early Maastrichtian times. The Sirte shales were deposited homogeneously in the Hammamat Trough and contain abundant planktonic foraminifera of the *Globotruncana linneiana* Zone and the lower part of the *Gansserina gansseri* Zone (Barbieri, 1996). The Sirte Shale in the Zaltan Platform area has good source rock potential with type II amorphous kerogen and TOC values up to 3.63% (Baric et al., 1996).

In the meantime, oxygenated-water dominated over the platform areas and deposition of carbonates took place, such as the Tagrifet Limestone (Barr & Weegar, 1972, Butt, 1984) and the Samah Dolomite (Barton, 1979). The Tagrifet Limestone (Campanian) shows remarkable lateral and vertical facies variations (Barbieri, 1996), the lower portion is made up of poorly sorted bioclastic detritus with abundant *Rotalia cayeuxi* deposited in a high energy environment, grading upward into planktonic foraminiferal, argillaceous micrites with silt-grade fossil fragments, inoceramid fragments, and fine grained, sandy limestones, suggesting active transport and increase in water depth.

Thick shallow-marine siliciclastic wedges formed along the edges of platforms during the Campanian (Fig. 20.17). On the western edge of the Zelten (Zaltan) Platform, such as in well YYY1-6, a section of more than 550 m (1805 ft) in thickness, composed of calcareous marine siliciclastics and thin limestones with radiolitid rudists (Jones, 1996) was deposited. Similar sand bodies formed long offshore bars along the Ad Daffah-Al Waha Ridge in the central Sirte Basin (Hammuda, 1980). These sands are very fine to fine-grained with scattered coarse grains, pyrite, glauconite, and bryozoan and pelecypod fragments. Sandy wackestones were deposited in the quiet shallow-water marine environment behind the bars. These facies are encased in dark grey, calcareous and pyritic basinal shales of the Sirte Shale.

In Cyrenaica, up to 200 m (230–656 ft) of Campanian neritic to deep-marine facies of the Al Majahir Formation were deposited unconformably over the Benia Formation (Röhlich, 1974, Klen, 1974). The succession consists of light grey marls, marly limestones, and dolomitic limestones with abundant benthonic and planktonic foraminifera, such as *Globotruncana coronata* and *G. angusticarinata* (Barr, 1968), molluscs, dominated by *Inoceramus balticus* and *Pycnodonta vesicularis* (Röhlich, 1974, Tröger & Röhlich, 1996, Röhlich & Youshash, 1996), and ostracods. In the offshore of Cyrenaica, the Al Majahir Formation is dominated by a shallowing upward sequence composed of fossiliferous mudstones-grainstones (Duronio et al., 1991) with large pelecypods, fragments of *Globorotalites michelinianus*, dasycladacean algae, bryozoans, echinoderms, *Inoceramus* prisms, *Globotruncana stuartiformis*, *Marginotruncana coronata*, *Dicarinella asymetrica*, *Globotruncana fornicata*, and *G. vetricosa*, *Accordiella conica*, *Cuneolina pavonia parva*, *Valvulammina picardi*, and *Rhapydionina* sp. southward, it changes into the Tukrah Formation, composed of deep-marine, cherty limestones, rich in benthonic foraminifera.

Central Egypt, between Wadi Qena and Um el Ghanayem, Kharga Oasis, was a high area until the Turonian (Fig. 20.15). The area subsided in the Campanian and Maastrichtian and received thick sediments. Campanian sediments include the Early Campanian high-stand tract (HST) of the Qoseir variegated shales, limestones of the lower part of the Khoman Formation in the subsurface of the Northern Western Desert (Samir, 1994, Metwalli & Abd el-Hady, 1975), and of the Sudr Formation (St. Paul Formation of Kuss, 1986) along the Gulf of Suez Coast and in north central Sinai (Samuel et al., 2009) (Fig. 20.17).

The Campanian platform carbonates occur near St. Anthony Monastery in the Western Galala Plateau Wadi Araba (Fig. 7.4). The carbonates belong to the Gebel Thelmet (St. Anthony) Formation and Sudr Chalk and contain the Campanian-Maastrichtian larger foraminifera *Orbitoides media* and *Omphalocyclus macropora* (Kuss et al., 2000a, b, Ismail et al., 2007). The Upper Cretaceous-Paleogene succession of the Galala plateaus in the Eastern Desert was controlled by a Syrian Arc Uplift that defined the subsequent basin morphology. During the Late Campanian, the Northern Galala/Wadi Araba Uplift was formed, as indicated by southward-prograding slope deposits of a Transition Zone that links the Uplift with the Southern Galala Subbasin (Kuss et al., 2000a, b). This phase was followed during the Late Campanian by a transgressive-system tract (TST), during which phosphate-bearing sediments were deposited. The period ended with a progradational low-stand tract (LST), where oyster beds were formed. These sediments are probably equivalent to the Sirte Shale in the Sirte Basin and the Bu Isa Formation in the offshore of Northwest Libya. The Early Campanian Qoseir Variegated Shales were deposited in low areas in the Nile Basin. They thin or wedge out on paleohighs In the Qoseir area, near the Red Sea coast. The Qoseir Formation is composed of variegated shales, ripple-laminated siltstones and fine-grained sandstones, and contains oyster-rich lenses (Youssef, 1957) and minor limestone sequences with pelecypods and gastropods (Kallenbach & Hendriks, 1986). Northwards in the Bir Umm Omeiyad-Bir Timimit el Shifa area, the Qoseir Formation is composed of coastal alluvial plain sediments. Southwards in the area between Qena and Abu Simbel, the Formation grades into back-shore to shoreface sediments and grades upward into shoreface to inner-shelf sediments (Hendriks et al., 1987). These sediments grade in turn southward into prodeltaic to inner shelf sediments of the Shagir Member of the Kiseiba Formation. In the Dakhla-Farafra area, the Qoseir Formation (Umm Barmil Formation of El-Naggar, 1970, and the Mut Formation of Barthel & Herrmann-Degen, 1981) is composed of red and green shales and white, very fine sandstones, and rare current-rippled sandstones. The lower part contains faunal and floral remains indicating terrestrial, lacustrine, brackish, and marine influence. These sediments were probably deposited in fluvial channels, coastal plains, lakes, and bays. These facies grade upward into tidal-flat, then into shelf facies made up of coarsening-upward sequences (Hendriks et al., 1990). Direction of transport was northward.

In Egypt, the transition from the Late Campanian to the Lower Maastrichtian was marked by the widespread deposition of phosphate-bearing formations which extend from the Red Sea coast in the Abu Tartur area east to the Bahariya Oasis in the west and the Kurkur-Dungul area in the south (Dominik & Schaal, 1984, Awadalla, 2009). The Duwi Formation is made up of a number of highly fossiliferous phosphatic, ammonite-rich sequences, separated by thick unfossiliferous shales. They were probably deposited in shallow offshore to shoreface, alternating high- and low-energy environments, on tectonically originated swells and depressions with reducing condition (Hendriks et al., 1990, Hendriks & Luger, 1987). Soliman et al. (1986) suggested deposition in deltaic and coastal bays for the phosphatic beds at Gebel Abu Had and Wadi Hamama. The main phosphatic components of the Duwi phosphates are diagenetic peloids and skeletal grains (Glenn, 1990). They represent transgressive-stand tract (TST) deposits, formed as a result of current winnowing and concentration of authigenic phosphatic minerals and their precipitation with organic-carbon-rich sediments, such as black shales and porcellanites. In the southern Western Desert, such as in the Abu Tartur area, the phosphate beds have formed as giant sand waves produced by tidal and storm reworking of sediments during the marine transgression accompanied by fluvial discharge (Glenn, 1990). Campanian deposition ended by a progradational low-stand tract (LST) during which oyster banks was formed.

Dinocysts are abundant in the shaly intervals of the Duwi Formation; acritarchs are more common in the phosphatic sandstone facies (Schrank, 1987, 1991), which indicate that the shales were deposited in a deeper marine environment than the phosphatic facies (Baudu & Paris, 1995). The high percentage of the peridinoids dinocyst assemblage, such as *Senegalinium* and *Andalusiella*, in the shales, along with the lack of planktonic foraminifera and calcareous nannofossils, suggests a low paleosalinity (Scharnk, 1987, 1991). On the other hand, the presence of the freshwater green algae *Pediastrum* and common pollen and spores in the sands indicates a marine depositional environment with continental input or a brackish water environment. The thin phosphatic interval on top of the Abu Roash Formation in the Bahariya Oasis and the phosphatic lenses near the base of the Sharawna Shale in the Kurkur-Dungul area are probably equivalent to the Duwi Formation.

South and west of Aswan, in the Bir Dungul-Bir Kiseiba area, the Campanian-Maastrichtian Shagir Member of the Kiseiba Formation (Hendriks et al., 1984, Hendriks, 1987), equivalent to the Qoseir and Duwi formations, comprises prodeltaic to inner shelf sediments.

In Sinai, Coniacian-Santonian sediments of the Mutalla Formation were followed during the Campanian by the deposition of a sequence composed of chalks, cherts, and bioclastic, phosphatic and glauconitic limestones of the Duwi Formation (Mishash Formation) with oysters, such as *Pycnodonte vesicularis*, *Lophas villei*, *L. morgani*, and fish teeth (Bartov et al., 1980). Campanian chalks of the St. Paul Formation were deposited in the Wadi Araba area unconformably on Cenomanian-Turonian sediments (Kuss, 1986). The base is marked by a sandy horizon with black phosphate grains, pellets, fish teeth, and phosphate layers followed by a layer with *Pycnodonte vesicularis*. The various lithological units of the Cenomanian-Santonian stratigraphic sequence reflect fluctuations in sea level. Campanian sediments are absent due to the major regressive phase during the Campanian (Haq et al., 1987) and the late Santonian inversion and tectonics of the Syrian arc system (Samuel et al., 2009).

In the Algerian Tell, the Campanian of the Constantine High Unit consists of marls and greyish limestones, overlain by the Tertiary Tell zone succession (Lahondere & Magne, 1983). In the Aaiun-Tarfaya Basin offshore Morocco, the Albian-Campanian interval is composed predominantly of black shale and marls wit *Zoophycos* and *Skolithos* burrows. The organic matters are marine in origin and oil-prone, with total organic carbon (TOC) contents ranging up to 12% (Cool et al., 2008).

### 20.4.1.5 Maastrichtian

Maastrichtian sediments correspond to the upper TST and HST cycles of the super-cycle UZA4 of Haq et al. (1987). The Early Campanian-Late Maastrichtian rifting phase was followed by a compressional event during the latest Maastrichtian-early Paleocene times (Guiraud, 1998).

During the Maastrichtian, the Tethys was reconnected to the South Atlantic Ocean through the Trans-Saharan Seaway. The similarity of ammonites, especially the species *Libycoceras afikoense*, large foraminifera and ostracods in the two oceans, suggests the existence of this seaway. The seaway was shallow and extended from Nigeria, through Niger and Mali to Algeria and Libya (Reyment, 1966, Kogbe, 1980, Zaborski, 1983). The connection between the Tethys and the South Atlantic is also supported by the presence of the ammonite species *Libycoceras ismaeli* in Egypt (Luger & Gröschke, 1987) and Algeria (Amard, 1996), which resembles *L. afikpoense* in Nigeria (Zaborsky, 1983) and also by the similarity of the ostracod assemblages (Bassiouni & Luger, 1990). The presence of *Inoceramus ianjoaensis* in the Lower Tar Member in the Ghadames region (Röhlich, 1979) also reinforces the connection between the Tethys and the southern African basins.

In the Western Sirte Basin, Maastrichtian deposition started with the Lower Tar Member of the Zmam Formation. The sediments consist of gypsiferous shales intercalated with fossiliferous marls and limestones (Jordi & Lonfat, 1963) rich in both microfossils and macrofossils, such as pelecypods, gastropods, echinoids, small button corals, globotruncanids (Barr & Weegar, 1972, Cepek, 1979, Röhlich, 1979, Salaj, 1979, Salaj & Nairn, 1987), and occasional silicified wood fragments (Röhlich, 1979). They form a shallowing- upward sequence grading from outer shelf-upper bathyal sediments with the pelagic foraminifera *Globotruncana gansseri*, into a high-energy neritic facies with *Omphalocyclus macroporus* and *Siderolites calcitrapoides*, then into deltaic and prodeltaic sediments with *Haplophragmoides desertorum* (Woller, 1978). Eastward in Tunisia, calcarenites and dolomitic calcarenites (Röhlich, 1979) rich in *Amphidonte overwegi* and oyster banks formed mostly of *Rostelum serratum* (Cepek, 1979) or *Agerostrea ungulata* (Röhlich, 1979) were developed. Southwards, miliolids and arenaceous foraminifera become more abundant, probably reflecting a more restricted environment. South of the Jabal Assawda, the Upper Cretaceous transgression onlapped over the eroded surface of the Cambrian Hassaouna Formation, and led to the deposition of shallow marine, fine-grained, cross-bedded sandstones, conglomeratic sandstones, and sandy limestone beds with *Omphalocyclus* and *Siderolites* of the Bin Affin Member (Fürst, 1964, Klitzsch, 1970, Woller, 1978).

Marine transgression reached its peak in the Sirte Basin during the Maastrichtian. Thick basinal Kalash limestones (Maastrichtian-Danian) with planktonic foraminifera of the *Gansserina gansseri* or *Globotruncana contusa* assemblages (Jones, 1996) typical of the lower slope marine facies (SMF3 of Wislon, 1975), accumulated in most of the Sirte Basin. Only the Al Jahamah Platform, a few isolated areas of the Az Zahrah Platform and the crest of the Cyrenaican ridge remained as islands during the Maastrichtian. During later Maastrichtian times, a major transgression from the south drowned all platform areas in the Sirte Basin with the exception of the Satal shallow marine and tidal flat area which persisted through the Paleocene (Bezan et al., 1996).

The Kalash Formation becomes chalky and dolomitic toward the east (Hallett & El-Ghoul, 1996) and shaly toward the south where it eventually pinches out.

The Zaltan (Zelten) and Al Bayda platforms were the sites of the shallow-marine carbonates and siliciclastics of the Waha Formation with Maastrichtian large benthonic foraminifera, such as *Siderolites calcitrapoides* and *Omphalocyclus macroporus* (Jones, 1996, Broughton, 1996, Hallett, 2002). In the Central Sirte Basin, the Waha comprises a very variable succession of shelf limestones with localized siliciclastics occurring in the lower levels. The thickness of the Formation is very variable because of deposition over an irregularly eroded surface and differential erosion at the top. Shallow-marine carbonates and quartzose sandstones of the Waha Formation form the main oil reservoir many fields, such as Nasser, Ralah, Jabal, Ar Raquba, and Waha. The Raguba Field, which still produces at a rate of 32,000 bopd of 42°API oil after 30 years of production (Broughton, 1996). The Waha Carbonates graded laterally into the basinal Kalash limestones and were overlain by the Kalash carbonates as the sea transgressed unconformably over the platform in Lower Paleocene time. In places, the Waha beds are absent and the Kalash Formation directly overlies quartzites of the Gargaf Group. With continued sea-level rise during the Maastrichtian, anoxic water lapped onto the Zaltan Platform and deposited carbon-rich shales and limestones of the Kalash Formation (believed to be equivalent to the Sirte Shale by many workers). These sediments are made up of dark grey to black calcareous shales and very argillaceous lime-wackestones. Skeletal content includes planktonic foraminifera and filamentous shells consistent with deep marine conditions (Jones, 1996). However, these shales might be a deeper facies variation of the shallow Waha facies.

The drop in sea-level at the end of the Maastrichtian is manifested by the presence of karsting and phosphatic and glauconitic grains at the base of the Danian Kalash Formation (Jones, 1996) and the presence of a thin radioactive bed at the boundary between the Waha and Kalash formations on the Zaltan Platform in the Central Sirte Basin.

In Cyrenaica, Maastrichtian shallow-marine dolomitic limestones of the Wadi Ducchan Formation and deep-marine sediments of the Atrun Formation were deposited. The Wadi Ducchan Formation (Kleinsmeide & van den Berg, 1968, Pietersz, 1968, Röhlich, 1974, Klen, 1974) consists of grey to brown, unfossiliferous dolomites and vuggy dolomitic limestones which were deposited on limestones, shales and dolomites of the Campanian Al Majahir Formation. These shallow-marine sediments grade into open-marine, neritic facies of the Atrun Limestone, which consists of tan-white, microcrystalline limestones, and marly intercalations and lenses of brown chert (Röhlich, 1974, Barr & Hammuda, 1971). They were deposited conformably on top of the Coniacian-Santonian Hilal Shale or conformably on the Campanian Tukrah Formation. Convolute layers and slump structures, planktonic foraminifera such as *Abathomphalus mayaroensis, Globotruncana gansseri, G. tricarinata, G. elevata,* and *G. concavata concavata,* in addition to coccoliths, ostracods, and brachiopods (Barr, 1968, 1972, Barr & Hammuda, 1971), *Inoceramus* shells, and ichnofossils, such as *Zoophycos* and *Chondrites* indicate deep-water slope syndepositional mass movements (Röhlich, 1974) during the Maastrichtian. These mass movements are also manifested in the subsurface of the offshore of Cyrenaica. Interbedded light grey to whitish chalky mudstones to wackestones and grey-brown marls with chert nodules of the Atrun Formation in well A1-NC120 contain a mixture of shallow-water limestone clasts with *Orbitoides media* and *Sirtina* sp. enclosed in deep-marine sediments with Globotruncana *stuarti, G. conica, G. contusa, G. gansseri, G. falsostuarti, Stensioeina exsculpta, Rugoglobigerina rugosa,* and *Bolivinoides draco* (Duronio et al., 1991).

In Egypt, continued subsidence during the Maastrichtian led to the deposition of lower slope to basinal carbonates of the Khoman/Sudr carbonates in the Northern Egypt and Nile basins. In Sinai, the Maastrichtian was dominated by the deposition of chalks of the Sudr Formation (Ghareb Formation) which accumulated in a NE-trending basin in Central and Eastern Sinai (Bartov et al., 1980). On the northern flank of the Arif el Naga Uplift, solitary corals, bivalves, gastropods, echinoids and crinoids suggest deposition in a low-energy marine environment. In Sinai, the Maastrichtian is represented by the chalks of the Sudr Formation in the deeper Nekhl section while it is represented by shales (Sharawana Shale) in the shallower Giddi section (Said & Kenawy, 1956).

Tectonic movements resulted in drastic variations in thickness of the Upper Cretaceous and Paleocene sediments. The Sudr Formation (Thelmet/St. Anthony Formation) (Abdallah & Eissa, 1971) in the Wadi Araba area, consists of marly, sandy, and *Orbitoides*-bearing limestones with baculites, ammonites, echinoids and oysters, changing upward into sandstones intercalated with clastic-dominated, dolomitic limestones with planktonic foraminifera of the *Globotruncana* group (Kuss, 1986). Upper Maastrichtian and Danian are absent in places in that area (Kuss et al., 2000a, b, Ismail et al., 2007). The Khoman Formation in the Bahariya Oasis (Dominik, 1985) consists of highly fossiliferous chalky limestones which were deposited unconformably on the Coniacian-Santonian Abu Roash Formation. The Khoman carbonates continue into the subsurface of the northern Western Desert where they are composed of pelagic limestones and shales characterized by the presence of *Globotruncana stuarti* (Girgis, 1989, Metwalli & Abd el-Hady, 1975). The top is characterized by a lithologic break caused by the drop in sea level. Planktonic foraminifera from the Khoman Formation between the Farafra and Bahariya oases are attributed to the Maastrichtian *Globotruncana aegyptiaca* and *Gansserina gansseri* zones (Samir, 1994). The diversified *Globotruncana* assemblage in the Khoman Chalk unit indicates deposition in an open-sea outer-shelf environment with gradual shallowing upward (Said & Kerdany, 1961). The lower part of the Khoman Formation (Khoman B) (assumed to be Campanian-Maastrichtian by El-Beialy, 1994) in well Bed 1-1, northwest of the Nile Delta is made up of limestones and shales which contain large amounts of land-derived plant debris and dinocysts, such as *Dinogymium acuminatum, D. euclaensis, D.* cf. *westralium* and *I. (Isabelidinium) cooksoniae,* in addition to freshwater algae (El-Beialy, 1994). The occurrence of land-derived plant debris and freshwater algae in these sediments (Hufhuf Formation) indicates the presence of a nearby landmass. The paleoclimate of the Late Cretaceous in the Farafra Oasis is suggested to be tropical or subtropical as deduced from the study of fossil wood by Kedves et al. (2004), Youssef & El-Saadawi (2004), and Kamal El-Din (2003).

The Khoman carbonates grade in other parts of Central and Southern Egypt and Sinai into the Sharawna Shale (El-Naggar, 1966). The Sharawna Shale is composed of shallow marine deposits with some intercalations of deep-marine sediments (Nakkady, 1957) which may represent prodelta sediments. A thick section of the Sharawna Shale was deposited in the Luxor Basin. Maastrichtian fauna of the Sharawna shales contain fewer planktonics (Nakkady, 1955) and seems to have been deposited in a near-shore muddy environment (Said & Kerdany, 1961). The change to more open-marine conditions during the deposition of the Sharawna Shale is evident by the presence of more cosmopolitan *gonyaulacoids* dinoflagellates, such as *Spiniferites*, *Florentinaia* and *Cordosphaeridium* (Schrank, 1987). An increase in terrestrial palynomorphs and peridinoids, such as *Deflandrea* and *Paleocystodininium* reported from the upper part of the Formation in well Younis-1 (Schrank, 1987) may suggest shallowing upward, parallel to that of the Khoman carbonates (Samir, 1994). Further south, in the Aswan-Dungul area, carbonates, shales, and sandstones of the Shab Member of the Kiseiba Formation, equivalent to the Sharawna Shale, were deposited in distributary mouth-bars, marshes, and delta front and delta plain environments (Hendriks et al., 1984, Hendriks, 1987).

The Upper Cretaceous in Northeastern Algeria includes the Upper Zebbag Formation (Cenomanian), Hassi Bouras Formation, Ras Toumb, and Kemaken formations (Turonian), Doulid Formation (Turonian-Coniacian), Aleg Formation (Coniacian-Campanian), Oued Mellah Formation (Santonian), and the Abiod Formation (Campanian-Maastrichtian).

In the Hodna Mountains (Askri et al., 1995, Bouchareb Haouchine & Boudoukha, 2009), the Lower-Middle Cenomanian succession consists of marls and fossiliferous marly limestones and thick marls with a pelagic fauna in the Tellian Atlas. The Upper Cenomanian-Turonian interval is represented by limestones, dolomites, marls and marly limestones with *Globotruncana* and rudists, followed by Senonian marls and marly limestones, and then by Maastrichtian limestones. Turonian marls with bivalves and echinoderms cover most of Northern Algeria and extend over the Sahara Platform, with the exception of some high areas, such as the High Plateaus and the Constantine Uplift. Senonian deposits in Northern Algeria are dominated by shelly marls and carbonates (Askri et al., 1995).

The Upper Cretaceous-Paleocene succession in the El Kantara Syncline in the Western Aurès is represented by limestones of the Cretaceous Evaporitic Green Marls of Tamarins and the Djar ed Dechra Formation, followed upward by the Upper Paleocene El Kantara Red Marls and Oued El Haï Marly Limestones, and then by an Eocene-Oligocene terrigenous succession in the center of the El Kantara Syncline (Belkhodja & Bigot, 2004). The Upper Maastrichtian Djar ed Dechra Formation consists of limestones with radiolite rudistid, inoceramid, oyster fragments, the foraminifera *Cuneolina sp.*, *Fleuryana adriatica*, *Orbitoides tissoti*, *Omphalocyclus* sp., *Orbitoides concavatus*, and the algae *Lithothamnion*, followed by a horizon rich in *Laffitteina boluensis*, *L. conica*, *L. mengaudi*, *L.* cf. *monody*, *L. oeztuerki*, and *L. bibensis* (Belkhodja & Bignot, 2004).

In Morocco, the phosphates of the Oulad Abdoun and Ganntour basins were deposited in a long and narrow gulf which opened to the Atlantic Ocean (Herbig, 1986). Vetebrate fossils from the phosphate deposits were described by Arambourg (1952). Deposition in the northern part of the Oulad Abdoun Basin probably took place in a high-energy nearshore environment; whereas in the southern part of the Basin deeper open marine conditions prevailed (Office Chérifien des Phosphates, 1989, Trappe, 1991, 1992, Lucas & Prévôt-Lucas, 1996). In the Rif Domain, the Upper Maastrichtian-Paleocene interval was dominated by the deposition of red beds (Chalouan et al., 2008, Durand-Delga et al., 2007). In the onshore part of the

Essaouira Basin, Eocene deposits are composed of phosphatic sands, limestones, and marly limestones with oysters, such as *Ostrea multicostata* and *O. strictiplicata*. In the offshore, thick interbedded shales and limestones were deposited at the foot of the shelf slope (Lancelot & Winterer 1980). In the Ouarzazate Basin, the Eocene is represented by a shallowing-upward sequence composed of bioclastic limestones, with vertebrate fauna (Herbig, 1991) of the Ait Ouarhtane Formation (Ypresian), followed upward by shallow to marginal marine greenish-grey siltstones, sandstones, and bioclastic limestones of the Jbel Tagount Formation (Lutetian) (Mohr & Fechner, 1986, Geyer & Herbig, 1988, Dragastan & Herbig, 2007). In the southern Central High Atlas, the Ypresian is conformably overlain by the continental sebkha-type red beds of the Hadida Formation (Middle Eocene-Oligocene) (Herbig, 1991, Tesón & Teixell, 2006).

## 20.4.2 Tertiary

Tertiary sediments form a shallowing-upward allocycle caused by the gradual global sea-level fall and include a number of prominent unconformities. The Tertiary includes four main sequences in Tunisia (Bobier et al., 1991): 1) A Paleogene to Lower Oligocene sequence. It starts with the Paleocene rise in sea level. Its upper boundary coincides with the LST associated with the Chattian eustatic fall in mean sea level. This sequence corresponds to "Supersequence $S_3$" of Bobier et al. (1991) and supercycles TA1 to TA4 of Haq et al. (1987). 2) A sequence which starts with the TST of the Upper Chattian and terminates with the HST of the Langhian. 3) A sequence which starts with the Serravallian HST and ends with the Messinian drop in sea level. It corresponds to the upper part of supercycle TB2 and the lower part of supercycle TB3 of Haq et al. (1987). 4) A Pliocene-Recent sequence, which corresponds to the upper part of supercycle TB3 of Haq et al. (1987).

### 20.4.2.1 Cretaceous-Tertiary boundary

The Cretaceous-Tertiary boundary is characterized by the extinction of many Cretaceous forms, such as the globotruncanids and the appearance of new forms in the Paleocene, such as the globorotalids. Large complex deeper dwelling forms were eliminated at the Maastrichtian-Tertiary boundary and only simple surface dwellers survived across this boundary (Canudo et al., 1991). These extinctions were attributed by many researchers to impact of asteroids or comets (Alvarez et al., 1980). However, Loper (1991) suggested that mass extinction could have been caused by environmental effects of volcanism associated with mantle plumes. In support of this hypothesis is the common presence of volcanic rocks at this boundary. A middle approach was taken by Sutherland (1994) who suggested that extinctions at the K/T boundary were caused by both impact-triggered volcanism and mantle plume events. This boundary is also marked by an interval of non-deposition or a hiatus from the uppermost Maastrichtian through the lower Danian which occurs in nearly all deep-sea Cretaceous-Tertiary sequences worldwide. This hiatus was probably the result of sediment starvation in deep ocean basins. The Upper Cretaceous fall in sea-level culminated just before the Cretaceous-Tertiary boundary (Haq et al., 1988) and was followed by a slow sea-level rise. In contrast, many continental shelf and upper slope sequences appear to contain a temporally complete record of sediment accumulation throughout this same interval, such as the Kef section in Tunisia (MacLeod & Keller, 1991, Keller et al., 1995). However, the Kef section in Tunisia (the stratotype of the Maastrichtian-Paleocene boundary), thought to be complete, is actually marked by extinctions, an $O_2$-minimum zone and regression at the end of the Maastrichtian (Donze et al., 1982). Another complete section which is also marked by strained environmental conditions is apparently present near the St. Paul Monastery on the western coast of the Gulf of Suez (Girgis, 1989). In that area the latest Maastrichtian nannofossil *Nephrolithus frequens* Zone and the earliest Danian *Biantholisus sparsus* Zone are present. In Egypt and Libya, the K/T boundary is marked either the absence of the uppermost Maastrichtian *Abathomphalus mayaroensis* Zone and one or more of the Danian *Globigerina-Globorotalia* assemblages (Said, 1962a, Said & Kerdany, 19661, Youssef & Abdel Aziz, 1971, Issawi, 1972), the absence of the nannofossil zones of the uppermost Maastrichtian *Micula prinsii* and Danian *Markalius inversus* and *Cruciplacolithus tenuis* zones (Faris, 1984), or by a thin conglomerate in many sections (El-Naggar, 1966, Issawi, 1972, Bartov et al., 1980). The Cretaceous-Tertiary boundary in the Farafra and Dakhla oases in Egypt is marked by a 1.5 m (5 ft) horizon composed of greyish to yellowish sandy marls, with glauconite and ferruginous ooids, phosphatized fossils and large gastropods (Barthel & Herrmann-Degen, 1981, Hermina, 1990). Karsting phenomena around the K/T boundary in the Sirte Basin (Jones, 1996) were probably caused by the drop in sea level at the end of the Maastrichtian. They were also observed in the Tilemsi Valley in Mali (Bellion et al., 1989), which falls within the Cretaceous Trans-Saharan Seaway. The unconformity at the Cretaceous-Tertiary boundary was also affected by tectonic activities at that time. The amount of unconformity at the K/T boundary is determined by the relative position on the structural highs (Shukri, 1954, Said & Kerdany, 1956, Kostandi, 1963, Salem, 1976). In the Southern Galala Plateau, the Sudr and Gebel Thelmet/St. Anthony formations represent deposition in an open marine environment in its lower part, and coastal marine to nearshore shelf in its upper part (Ismail et al., 2007). The Southern Galala Formation represents deposition in a shallow marine environment. The shallow water carbonates of the Southern Galala Plateau contain several large foraminifera and a few planktonic foraminifera, such as *Glomalveolina dachlensis* of Early-Middle Paleocene. The Paleocene/Eocene boundary is difficult to determine in that area because of the poor reservation state of the planktonic foraminifera (Ismail et al., 2007). The Early Eocene rocks contain *Nummulites* cf. *subramondi, Alveolina pasticellata,* and *Fabularia zitteli,* and the algae *Ethelia alba.*

Major extinction events took place at the K/T boundary. The majority of the calcareous planktonic foraminifera and calcareous nannoplaktons were extinct. However, land plants, crocodiles, snakes, and mammals were not affected. Many hypotheses have been forwarded to explain this extinction. For example extinction was attributed to the impact of a large asteroid (Alvarez et al., 1980). Planktonic foraminifera at Elles, Tunisia, show major changes in the structure of the Tethyan marine ecosystem during the upper Maastrichtian. The late Maastrichtian was dominated in the beginning by relatively stable environmental conditions and cool temperatures as indicated by diverse planktonic foraminifera populations with abundant intermediate and surface dwellers (Abramovich & Keller, 2002). A progressive cooling trend resulted in the decline of globotruncanid species, which experienced a further decline at the climax of a warm event before the K/T boundary. The abundance of long-ranging species fluctuated considerably, suggesting high stress environmental conditions. By the end of the Maastrichtian, rapid cooling led to accelerated species extinctions, followed by the extinction of all tropical and subtropical species at the K/T boundary (Abramovich & Keller, 2002).

The Latest Cretaceous-Paleocene paleogeography of Tunisia is characterized by two subsiding troughs in the north, the NW Tunisian Trough and the NE Tunisian Basin, and the Gafsa Gulf in the southwest (Ben Salem, 2002). The Gafsa Gulf, adjacent to the Saharan Platform, is separated from the northern basins by the Kasserine Island. In the northern basins, the Cretaceous/Tertiary (K/T) boundary lies within the continuously deposited marine El Haria Formation, which attains a thickness of 700–1000 m (2297–3280 ft). The GSSP (stratotype) of the K/T boundary at El Kef lies in the NW Tunisian Trough. The lateral facies and thickness variations of the El Haria shale are intimately related to the Late Cretaceous paleogeography (Fig. 20.17). Thus, in central Tunisia, and during Late Maastrichtian-Paleocene times, the large emerged zone of the Kasserine Island separated three paleogeographically distinct depositional zones (Burollet, 1956, 1967, Salaj, 1978): The Northwest Tunisian Trough, the Northeast Tunisian Basin, and the Southwest Gafsa-Metlaoui Gulf. During the Late Maastrichtian-Paleocene period, the northwestern part of Tunisia was occupied by a deep marine trough characterized by abundant pelagic microfauna (Tellian Zone and Tunisian Trough). During the Late Maastrichtian, northeastern Tunisia corresponded to small neritic and littoral basins with interbedded limestones and marls (Hammam Zriba Zone) (Ben Salem, 2002). The Danian is generally missing (Ben Salem, 2002). The Paleocene transgression covers larger areas than the Maastrichtian. To the southwest, in the Gafsa-Metlaoui Zone, the El Haria Formation changes into a euxinic and lagoonal basin with dark shales and very thick beds of gypsum (Gafsa Metlaoui Gulf) (Ben Salem, 2002). In the Northeastern Atlas, some small neritic littoral basins extend to the northeast of the Kasserine Island (Ben Salem, 2002).

In the Moroccan Atlantic margin, the Cretaceous-Tertiary boundary is marked by a prominent erosional surface on top of a chaotic seismic unit, which may be caused by reefs, gravity slides, or volcanics (Lancelot & Winterer, 1980, Hinz et al., 1982, Holik & Rabinowitz, 1992).

### 20.4.2.2  Paleocene period

The end of the Maastrichtian was a period of regression marked by a regional unconformity (the uppermost Maastrichtian and Lower Paleocene are missing in many areas), followed by a widespread marine transgression starting in the Danian with the deposition of shales over most of North Africa. Only the high areas received carbonates in West Libya, the Eastern Sirte Basin, and Cyrenaica. The sea advanced from the north into Egypt and Sudan, and reached as far south as the 24°N parallel in Libya during the Late Paleocene (Fig. 20.18). The Trans-Saharan Seaway crossed Algeria diagonally and connected with the Mediterranean Sea, the Sirte Basin, and Northern Egypt. The Paleocene contains about one-third of the oil discovered in the Sirte Basin.

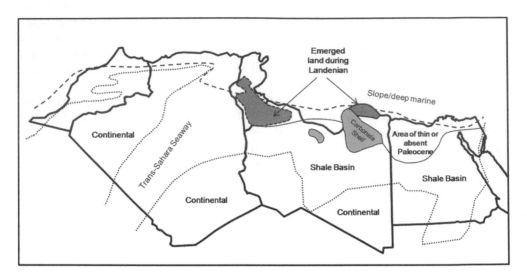

*Figure 20.18* Paleocene paleogeography (after Tawadros, 2001, Belkhodja & Bignot, 2004, Guiraud et al., 2005).

The regression in the Early Paleocene (Danian) led to the emergence of many parts of North Africa. In Northern Algeria, the Aures were emerged at that time (Belkhodja & Bignot, 2004). Like many parts in North Africa, the Danian is missing in that area. The return of the sea in the Landenian (Thanetian) was marked by the deposition of red marls with *Microcodium* and abundant Charophytes, followed by the establishment of a vast platform in Northern Algeria. In Egypt, a marked relative sea-level fluctuation occurred at the Danian-Landenian transition at ca.60.5 Ma (Speijer, 2003).

In the offshore of Northwest Libya the Danian is represented by the Ehduz Formation made up of dolomitic, argillaceous, and anhydritic limestones with *Globorotalia trinidadensis* and *Globigerina triloculinoides* (Hammuda et al., 1985, 1991). The unit is probably a deep-marine carbonate facies equivalent of the Al Jurf Formation black shales with limestone interbeds and abundant globotruncanids (Hammuda et al., 1985). The black shales of the Al Jurf Formation probably reflect an Anoxic Oceanic Event (AOE) across the Maastrichtian-Tertiary boundary. This event was paralleled by karsting and erosion in the platform areas. It was followed in the early Landenian by non-fossiliferous variegated shales of the Ajaylat Formation (Hammuda et al., 1985, 1991) probably equivalent to the El Haria Formation in Tunisia and the Kheir Shale in the central Sirte Basin. The Landenian was dominated by mudstones and wackestones of the Bouri Formation (Hammuda et al., 1985, 1991) with *Globorotalia mackannai, G. elongata, G. pseudomenardi, G. aequa, G. velascoensis,* and *Rotalina stellata.*

Paleocene rocks crop out along the western and southwestern margin of the Sirte Basin. The Paleocene in the western Sirte Basin includes three units, in an ascending order, the Upper Tar Marl and Socna Beds, and Had Limestone members of the Zmam Formation and the Bu Ras Marlstone and Gelta members of the Surfa Formation.

Deposition started in a shallow-marine environment with brackish water influence during the Upper Tarl Member (Salaj, 1979) followed by the deposition of fossiliferous limy sediments of the Socna Mollusc Bed. The depositional environment of the Had Limestone Member is intertidal to subtidal shallow marine and ranges from restricted to high-energy conditions (Jordi & Lonfat, 1963, Cepek, 1979, Salaj, 1979). The environment of deposition of the Bu Ras Marl Member was shallow subtidal to intertidal (Salaj, 1979, Zivanovic, 1977). The Gelta Member was deposited in a shallow low-energy marine environment (Cepek, 1979, Salaj, 1979). The depositional environment of the *Operculina* limestones, dolomites, marls, and chalks of the Ammur Member was shallow low-energy marine (Cepek, 1979).

In the Sirte Basin, carbonate platforms such as the azZahrah-Al Hufrah Carbonate Platform in the west, and the As Sababil Platform in the east were established early in the Danian. The location of the Paleocene carbonate-shelf edge is significantly different from the location of the Upper Cretaceous platform margins (Hallett & El-Ghoul, 1996). Troughs such as the Ajdabiya, Maradah, and Zallah, were principally filled with calcareous marine shales with thin stringers of shaly limestones. The Paleocene section is up to 2000 ft (610 m) in thickness on the platforms and increases to 4000 ft (1219 m) in the troughs. In the offshore of north-central Libya the whole Paleocene section decreases to less than 100 ft (30 m) with no Lower Paleocene (Bezan, 1996). The entire Paleocene succession grades southward into a continuous interval of carbonate sediments near the Paleocene shoreline and into shales in the northernmost part of the basin (Conley, 1971, Berggren, 1974). Sandstones accumulated along the southern shoreline of the basin.

The Paleocene succession in the subsurface of the Sirte Basin shows two main shallowing-upward cycles (Fig. 12.7), each cycle starts with shales and ends with carbonates. The first cycle started during the Danian with the deposition of the Hagfa Shale in the troughs, followed by Landenian carbonates and evaporites of the Thalith, Farrud, and Mabruk members of the Beda Formation. On the platforms, carbonate deposition of the Kalash, Defa, Upper Satal, and Lower Sabil carbonates dominated during the Danian. Danian deep-marine sediments of the Hagfa Shale were deposited on the eroded surface of the Waha Formation on the Zaltan Platform (Jones, 1996), and on top of the Kalash Formation on the azZahrah Platform (Kardoes, 1993). The second cycle started with the deposition of the shales of the Khalifa, Harash, and Sheterat formations in the Landenian, followed by the carbonates of the Dahra, Zelten, and Upper Sabil formations. The cycle ended by the deposition of shales with *Globorotalia velascoensis* of the lower part of the Kheir Formation during the latest Landenian. The entire cycle grades southward into a continuous interval of carbonate rocks and becomes dolomitic further south (Conely, 1971). Sandstones also accumulated along the southern shoreline of the basin.

The base of the Kalash-Lower Satal carbonate bank complex (Maastrichtian) on the azZahrah (Dahra) Platform is represented by the Kalash transgressive open-marine sequence on basement highs with syntectonic activity (Kardoes, 1993). The Lower Satal starts with a regressive restrictive facies overlain by transgressive, open-marine and high-energy deposits, and ends with a carbonate bank-building episode. The Upper Satal carbonates grade into the Hagfa Shale away from the Satal bank (Kardoes, 1993). The Upper Satal was also developed on the Al Bayda Platform

(Sinha & Mriheel, 1996). The Upper Satal was terminated by the deposition of thin Hagfa Shale over the entire bank.

The Hagfa Shale-Beda Formation cycle represents shallowing and clearing-upward marine conditions (Johnson & Nicoud, 1996). It started with argillaceous limestones and calcareous shales of the Thalith Member of the Beda Formation, followed by calcarenites, oolitic grainstones, and argillaceous calcilutites of the Farrud Member which were deposited in shoreface to offshore environments. The cycle ended with shales, limestones, anhydrites, and dolomites of the Mabruk Member. The Farrud and Mabruk Members grade into the shales of the Rabia Shale (Johnson & Nicoud, 1996).

The Paleocene Beda Formation sediments in the southwestern part of the Sirte Basin were deposited in a supratidal to intertidal, shallow marine lagoonal setting with high-energy oolite barriers (Grea, 1995). The carbonates change facies into the Rabia and Khalifa shales. Paleocene shoal carbonates of the Beda Formation were developed preferentially over a horst block which connected the Al Bayda Platform to the Southern Shelf (Sinha & Mriheel, 1996). Deposition of oolites, calcarenites and pellet al carbonate shoals of the Beda Formation spread over the south-central Hagfa Trough but the north-central part was dominated by the deposition of the Khalifa shales which continued on top of the Hagfa shales (Bezan, 1996).

Deposition in the Ajdabiya Trough during the Lower Paleocene was predominantly low-energy shallow-marine, where extensive shelf deposits of the Lower Sabil Formation were accumulated. Open-marine, high-energy conditions prevailed at the edges of the shelf areas (Brady et al., 1980, Kardoes, 1993). Deposition of the Lower Sabil was interrupted by an influx of the Sheterat Shale. Unlike Terry & Williams (1969) who suggested that deposition of the shales was due to subsidence and marine transgression, Brady et al. (1980) interpreted the Sheterat as an influx of shales from a distant provenance that supplied argillaceous material to the clear water environment.

The Lower Sabil carbonates and Sheterat shales were followed by deposition of shelf carbonates of the Upper Sabil Formation. On platform areas the Upper Sabil Formation has a wider distribution than the Lower Sabil. The Upper Sabil Formation consists of a prograding carbonate shelf with rimmed platform morphology and slopes. The overlying Harash Formation is characterized by its cyclic character and relatively high terrigenous content. The Harash Shales fill the Upper Sabil irregular surface. The morphology of the Upper Sabil platform coupled with sea level fluctuations controlled the distribution of depositional environments and thickness of the Harash Formation (Spring & Hansen, 1998).

In the southern Ajdabiya Trough, the Upper Sabil carbonates were deposited in various environments, such as platforms, bank margins, slopes, protected embayments, shoals, and isolated reefs (Meehan et al., 1993). Isolated organic buildups, such as the Intisar A to L fields were developed (Fig. 18.13). In the Intisar "D" Field pinnacle reefs are up to 5 km in width and 385 m (1263 ft) in height (Fig. 20.19). The main lithofacies of the Upper Sabil reefs are made up of algal-foraminiferal biomicrites, biosparites, and biolithites, coral debris, echinoderm fragments with calcareous red algae, bryzoans, pelecypods, crinoids, nummulites, dicyclinids, rotalids, miliolids, and alveolinids (Brady et al., 1980). The Intisar "D" Field (Fig. 20.19) contains more than 1 billion barrels of oil recoverable. Deposition of the reefs was contemporaneous with

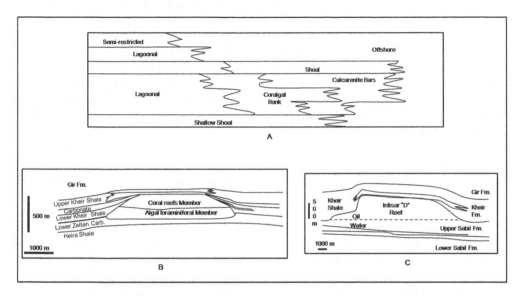

*Figure 20.19* Reef models in the Central Sirte Basin. A) Nasser Field (after Bebout & Pendexter, 1975). B) Intisar (Idris "A") Field in northeastern Sirte Basin (modified from Terry & Williams, 1969, Ahlbrandt, 2001). C) Intisar "D" reef (after Brady et al., 1980).

the Harash shales and was later covered by either the Harash Shales (Spring & Hansen, 1998), or the Kheir Shales (Landenian-Ypresian) which encased them. The Harash and Kheir formations are absent in the NE Ajdabiya and Suluq troughs (Bezan, 1996).

In the Zaltan (Nasser) Field (Fig. 20.19), the Zelten Formation (upper Landenian) is made up of micrites and calcarenites with miliolids, bryozoans, echinoids, molluscs, coralgals, corals, and discocyclinids in the upper part and dolomites in the lower part (Bebout & Pendexter, 1975). Foraminiferal calcarenites and micrites formed a broad shallow shoal upon which calcarenite bars developed along a northwest-southeast trend, and coralgal banks grew between the bars. Lagoonal facies developed to the southwest. All these facies graded laterally into offshore sediments with planktonic foraminifera which, as a result of continuous deepening, covered the whole area. Limestone beds within the dolomites in the lower part of the Zelten Formation suggest that these carbonates were deposited in a broad shelf-lagoon and were later dolomitized. Freshwater diagenesis of the carbonates resulted in the generation of secondary porosity, which reaches up to 40%.

In Northeast Libya, in the Al Jabal Al Akhdar area, the Paleocene is represented by chalky limestones and marls with planktonic foraminifera of the Al Uwayliah Formation (Danian-Landenian). Offshore Cyrenaica, the Paleocene is represented by shales and argillaceous mudstones-wackestones with planktonic foraminifera, such as *Globorotalia velascoensis*, *G. pseudomenardii*, and *G. triloculinoides*, indicating an open deep marine environment of deposition (Röhlich, 1974, Duronio et al., 1991). The earliest Paleocene and latest Maastrichtian are also missing from that area. The Formation was deposited unconformably on the Maastrichtian Atrun Formation in the offshore area (Duronio et al., 1991), or conformably on the Wadi Ducchan

Formation (Maastrichtian) in the onshore area (Röhlich, 1974). The Formation was deposited during deepening of the Al Jabal Al Akhdar area during the Paleocene but was uplifted and eroded during the Eocene (Röhlich, 1974).

Paleocene sediments are found in Egypt in the Farafra, Dakhla, Kharga, Kurkur, and Dungul oases in the Western Desert, in the Nile Valley near Esna and north of Thebes, in the Eastern Desert north of Qena, along the Red Sea coast, and in Sinai. Paleocene shales were encountered in a few wells in the northern Western Desert, such as the Natrun No.1 and Rabat No.1. They are generally absent north of a line passing approximately south of Maghara, Suez, Cairo, and Bahariya Oasis (this line represents the boundary between the stable and unstable shelf areas) (Fig. 20.18). Where Paleocene deposits are absent Cretaceous sediments are overlain unconformably by younger sediments. Absence or erosion of Paleocene is due to inversion of part of Egypt and northeastern Libya during the Eocene.

The Dakhla shales (Danian) were deposited unconformably over the Sharawna shales or the Khoman/Sudr carbonates (Fig. 20.20). Thick shale sections were deposited in the Mut Basin. These sediments were followed by the Upper Paleocene Tarawan Formation and then by the Esna Shale. Thick Esna shales were deposited in the Farafra Basin. They grade southward into the Kurkur sandstones and carbonates in the south Aswan/Dungul area. In Sinai, the Sudr Formation in the Nekhl section was followed by the Dakhla Shale of nearshore facies.

During the Maastrichtian-early Paleocene (C/T) a shallow sea covered the Western Desert of Egypt (Fig.) and clastic sediments were derived mainly from the Gulf El

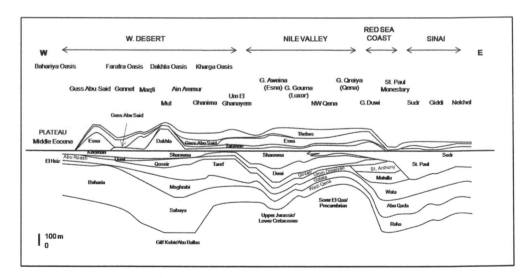

*Figure 20.20* Cretaceous-Tertiary cross-section A-A' across the Nile Basin (after Tawadros, 200; originally compiled from Said, 1960, 1962a), Said & Kerdany, 1961, El-Naggar, 1966, Said & Kenawy, 1956, Said & Sabry, 1964, Youssef & Abdel Aziz, 1971, Nakkady, 1950, 1957, Faris, 1984, Girgis, 1989, Hendriks et al., 1990, and Luger & Gröschke, 1989. Same scale as Figure. The Duwi phosphates (Campanian) are equivalent to the Brown Limestone in the Gulf of Suez and the Sirte Shale in the Sirte Basin. The Upper Jurassic/Lower Cretaceous succession is correlatable with the Khatatba and AEB formations in the Northern Western Desert and Hatira Formation in Sinai.

Kebir spur to the southwest of Dakhla and the Bahariya Arch. Uplift in the region resulted in major hiatuses in the late Maastrichtian-early Paleocene with increased erosion to the southwest (Tantawy et al., 2001).

Transgressions and regressions that occurred during the Thanetian (Landenian) (58.7–55.8 Ma) may provide an important constraint on the global paleoenvironment. This transgressive-regressive (T-R) cyclicity has been observed in various tectonically stable regions, such as the Eastern Russian Platform, Northwestern Europe, Northwestern Africa, Northeastern Africa, the Arabian Platform, the northern Gulf of Mexico, and Southern Australia (Ruban et al., 2010). However, except for a generally regressive trend occurring in the late Thanetian, no common T-R cycles can be delineated, which indicates an absence of global-scale T-R cyclicity during the Thanetian. Ruban et al. (2010) found no clear correspondence between documented T-R patterns and eustatic sea level changes. The reason for these discrepancies has been attributed to regional subsidence or uplift which governed transgressions and regressions locally and resulted in an inconsistency between T-R cycles in different parts of the globe (Ruban et al., 2010).

In the Late Paleocene section behind the St. Paul Monastery at Wadi Araba, the P4 Zone (*Planorotalites pseudomenardii*) and P5-P6a zones (*Morozovella velascoensis-M. edgari*) interval consists of a lower marly unit with wavy bedding and an upper unit of boulder-size carbonate clasts floating in a marly matrix, associated with lenses of cross-laminated sandstones, interpreted as mass flow deposits. This interval is marked by important changes of the depositional setting (*velascoensis* Event of Strougo, 1986) in a large number of sections from the Western Desert, the Gulf of Suez-Red Sea area, and the Nile Valley. Carbonate platforms (Abu Tartur and Galala) represent elevated structures on the deeper shelf (Speijer, 2003). The platform-basin transition at the Paleocene/Eocene (P/E) boundary can be seen in the Galala Plateau (Kuss et al., 2000a, b, Speijer & Scheibner, 2009). Late Cretaceous rimmed platform gradually changed to a Paleogene ramp-basin transition (Kuss et al., 2000a, b). The distally steepened Paleocene carbonate ramp evolved into an Early Eocene homoclinal ramp. Shallow-water limestones of the Southern Galala Formation were deposited on a gently south-dipping carbonate ramp. They crop out on the Northern Galala/Wadi Araba Uplift and interfinger with slope deposits in the Transition Zone and with basinal sediments (Esna Formation) of the Southern Galala Sub-basin. Evidence for these transitions is found in planktonic foraminiferal zone P4-P5.

The boundary between the Eocene and the Paleocene or older strata is mostly unconformable, but it might be continuous with the Paleocene in places, especially in basinal areas. Faris et al. (1998) observed no lithological change across the Paleocene-Eocene boundary in Central Egypt and suggested that the succession is complete across that boundary. The sediments across the Paleocene/Eocene (P/E) boundary at Gabal el Qreiya in the Nile Valley deposited during the PETM interval consist of three thin units of phosphatic shales characterized by high anomalies in chalcophiles and REEs, especially in the basal clay layers and bone-bearing beds (Soliman, 2003). The three phosphatic shale units reflect euxinic marine environments and deposition in anoxic $H_2S$-containing bottom waters rich in organic matter in a relatively deep water environment (Soliman, 2003). In the subsurface of the Sirte Basin, the Kheir Formation straddles the Paleocene-Ypresian boundary and often indicates continuous sedimentation. The boundary between the Eocene and the Paleocene successions in Tunisia is marked by phosphatic and glauconitic beds in Tunisia and Wadi Qena in Egypt (Comte & Lehman, 1974, Fournié, 1975, Loucks et al., 1998a, b).

During the Paleocene and Eocene Epochs (65.5–33.9 Ma) the Earth experienced the warmest conditions of the Cenozoic (Bornemann et al., 2009). This Paleogene greenhouse episode is characterized by several short-lived negative carbon isotope ($\delta^{13}C$) excursions, which are usually interpreted as transient warming events ('hyperthermals') as indicated by rising temperatures of surface and bottom waters, and accompanied carbonate dissolution in deep-sea settings. Among these events, the Paleocene-Eocene Thermal Maximum (PETM, c. 55.5 Ma) is globally the best documented and most prominent (Bornemann et al., 2009). The Paleocene-Eocene Thermal Maximum (PETM), around 55.5 Ma, marked by a 5–8°C warming, a perturbation of the hydrological cycle, and major biotic response on land and in the oceans, including radiations, extinctions and migrations. In addition, the PETM is associated with a pronounced negative carbon isotope excursion (CIE), recorded as a >2.5‰ decrease in the stable carbon isotope composition ($\delta13C$) of sedimentary components (Sluijs, 2008, and references therein).

The Paleocene-Eocene boundary is marked by a thermal maximum (the Paleocene-Eocene Thermal Maximum "PETM"), which was associated with biotic, climatic, and environmental events associated with negative carbon isotope excursion (CIE) and an increase in temperatures of about 2–8° C during the Landenian (Thanetian) (Zachos et al., 2001, Koch et al., 2003). This event has been recognized in many parts of the world including Egypt (Speijer et al., 1996, Speijer, 2003, Speijer & Wagner, 2002, Soliman, 2003, Guasti & Speijer, 2007, Morsi et al., 2008, Sluijs, 2008, Speijer & Scheibner, 2009, Morsi & Scheibner, 2009, Bornemann et al., 2009).

The gradual trends of warming and cooling since about 65 Ma until the PETM have been attributed to tectonics and periodic cycles driven by orbital processes (Zachos et al., 2001).

The global boundary stratotype section and point (GSSP) for the basal Eocene (Ypresian) has been established in the Dababiya section in the Nile Valley in Egypt (Aubry & Ouda, 2003, Schulte et al., 2009), which has been recommended as reference to the P/E boundary in marine and non-marine sequences. The onset of the carbon isotopic excursion (CIE) associated with the Paleocene/Eocene Thermal Maximum (PETM) Event acts as a prime criterion for delineating this boundary in sections around the world (Speijer & Scheibner, 2009).

The lower part of the Galala sequences corresponds to Zones P4-P5 (Speijer et al., 2000, Scheibner & Speijer, 2009). It is generally not possible in Upper Paleocene Egyptian sequences to reliably differentiate between Zones P4 and P5 on the basis of the disappearance of *G. pseudomenardii* or the lowest occurrence of *M. velascoensis*, which is a very common species in the Upper Paleocene of Egypt (Speijer, 2003). The P/E boundary apparently falls within Zone P5 according to Scheibner & Speijer (2008, Fig. 4). Therefore, the Paleocene/Eocene boundary has been placed at the base of the carbon isotopic excursion (CIE), which correlates with various evolutionary events in pelagic and deep marine microbiota, as well as a turnover in shallow platform biota (Aubry & Ouda, 2003).

In the Atlas Chain in NW Africa, magmatic rocks were emplaced during the Paleocene-Eocene period. They were probably derived from a giant asthenospheric plume which ascended below Northwest Africa and Western Europe during the Early Tertiary (Frizon de Lamotte et al., 2008, 2009).

### 20.4.2.3  Eocene period

The Eocene is divided herein into Ypresian (Lower Eocene), Lutetian (Middle Eocene), and Bartonian (Upper Eocene). The Priabonian units in the literature are included in the Bartonian. The Priabonian is the name preferred by the geologists working in Tunisia and Libya while the Bartonian is favored by Egyptian geologists. In spite of the suggestion by Harland et al. (1990) to include the Bartonian, along with the Lutetian, in the Middle Eocene and to restrict the Priabonian to the Upper Eocene, distinction between the two stages is not possible from the available data (but see Guiraud et al., 2005 and Bitner & Boukhary, 2009). The Bartonian-Priabonian transition is believed to be marked by major compression event (Guiraud et al., 1987) in the Northwest African Maghrebian/Alpine Belt (Fig. 20.21).

Eocene sediments are widespread and crop out over large areas in North Africa. The upper boundary of the Eocene succession with the Oligocene succession is often marked by a regional unconformity related to the global sea-level fall at the end of the Eocene.

The Eocene paleogeographic map (Fig. 20.21) shows a retreating sea from the Lower Eocene through the Upper Eocene. The Eocene shows three east-west facies belts representing progradational allocycles caused by global tectonics and sea-level

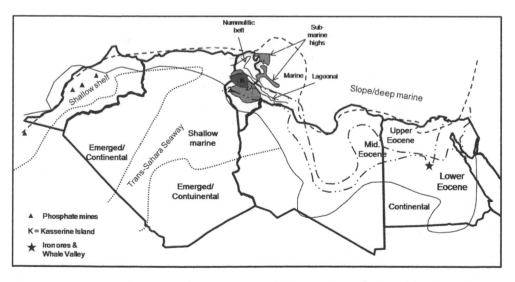

Figure 20.21  Eocene paleogeographic map (compiled from Klitzsch, 1990b, Sbeta, 1991, Bernasconi et al., 199, Röhlich, 1974, Duronio et al., 1991, Salem, 1976, Said, 1962a, Kogbi & Burollet, 1990, Tawadros, 2001, Jamoussi et al., 2003, Vennin et al., 2003, Guiraud et al., 2005, Beavington-Penney et al., 2008).

changes related to the collision between Africa and Eurasia. Autocycles caused by biological factors basin configuration and paleotopography occur within each belt. In the Sirte Basin the Eocene succession represents a post-rift sequence (Harding, 1984, van der Meer & Cloetingh, 1996, Schröter, 1996) deposited in a thermally sagging basin (Gumati & Kanes, 1985, Gumati & Nairn, 1991), but it was dominated by shallowing upward due to the gradual global fall in sea level. In the Gulf of Suez on the other hand Eocene sediments represent the pre-rift sequence (Garfunkel & Bartov, 1977, Orszag-Sperber et al., 1990). Eocene lithologies are dominated by carbonates, marls, and shales. Siliciclastics are common in the Upper Eocene. In the offshore of Northwest Libya, Tunisia, and the west-central Sirte Basin extensive evaporite basins developed during the Eocene. Extensive nummulitic banks which host hydrocarbons were developed in the offshore of Tunisia, Northwest Libya, and the Sirte Basin. Eocene sediments also host commercially exploited iron-ore deposits in the Western Desert of Egypt and phosphate deposits in Tunisia and Morocco.

The post-Eocene regression left vast areas which were subjected to continental erosion and led to the formation of depressions and escarpments in West and South Libya. These depressions were later filled with Tertiary and Quaternary lacustrine deposits ands conglomerates.

NW Africa was connected to NE Africa and Europe through the Trans-Saharan Seaway (Fig. 20.21). The Atlas Domain was occupied by the "Atlas Gulf" from the Maastrichtian until the Middle Eocene (the beginning of the Atlas Orogeny) which led to the deposition of phosphatic, carbonate, and mixed carbonate-siliciclastic facies during the interval (Frizon de Lamotte et al., 2008, Fig. 4.19). In the Late Eocene, the Saharan Atlas-Aurès Domain was subjected to major folding (Piqué et al., 2002, Guiraud et al., 2005), whereas the internal Rifian-Tellian Domain (Alboran-Kabylies) underwent thrusting and slight metamorphism probably related to the subduction of the Tethys underneath the Iberian Balearic margin (Frizon de Lamotte et al., 2000, Guiraud et al., 2005).

With the advent of the Eocene, Nummulites flourished and formed extensive bank deposits (Fig. 20.22) which form hydrocarbon reservoirs in a number of fields in Tunisia and Libya. Tertiary nummulitic deposits in Egypt, Libya, and Tunisia have been the subject of a number of studies by Arni (1963), Compte & Lohman (1974), Fournié (1975), Decrouez & Lanterno (1979), Aigner (1982, 1983,1984), Moody et al. (1998), Moody (1998a & b), Moody & Sandman (1998), Braithwaite et al. (1998), Reali & Ronchi (1998), Loucks et al. (1998a & b), Abdelsamad (1998), Elwarfalli & Stow (1998), Vennin et al. (2003, Jorry et al. (2001, 2003). Arni's (1963) sedimentological model for the formation of Nummulites accumulations was based on nummulitic deposits in the Sirte Basin of Libya. Figure 20.23 is a depositional model based on these works.

In this model, the ramp environment has been divided into three zones, which correspond to the X, Y and Z zones of Irwin (1965). The X Zone is deep open-marine (below storm wave base). It comprises three facies: 1–3. These three facies correspond to the basin, lower shoreface, and middle shoreface, respectively. The Y Zone represents the outer platform. It includes the nummulitic bank (Facies 4) and algal shoals (Facies 5) above fair-weather base. The Z Zone is equivalent to the inner platform environment. It includes four facies: the back-bank (Facies 6), lagoonal (Facies 7 and 8), and lagoonal-dolomitic-evaporitic (Facies 9).

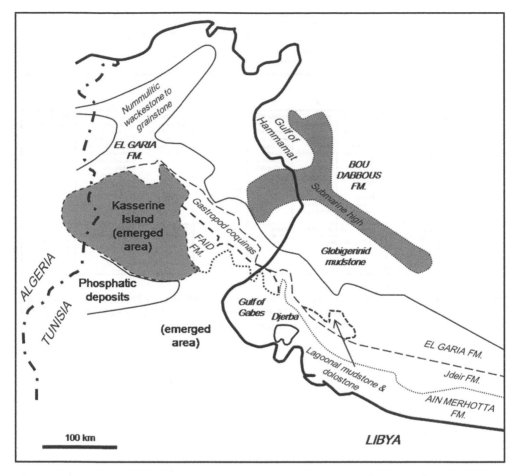

*Figure 20.22* Lithofacies map of the Metlaoui/Farwah Group (after Bishop, 1988, Beavington et al., 2008, and Jamoussi et al., 2008).

### 20.4.2.4   Basinal facies (facies 1–2)

The basinal facies is composed of mudstones-wackestones with pelagic foraminifera such as Globigerinids (Arni, 1963). Two subfacies may be recognized. The lower subfacies is characterized by *Globigerina* mudstones. The upper basinal facies is dominated by *Globigerina* and other small foraminifera, sponge spicules and radiolaria (Comte & Lehman, 1974).

Nummulites accumulations differ greatly throughout the Eocene-Oligocene. Productivity varies on a local and regional basis and in association with hydrodynamic and biological activities (Moody & Sandman, 1998). The Reinche Member of the Souar or Cherahill Formation (Middle-Upper Eocene) shows wide variations in sequence thickness and facies association between Tunisia and offshore Northwest Libya (Moody & Sandman, 1998). These variations are due to differences in

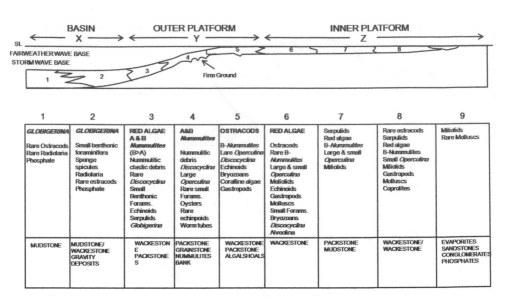

*Figure 20.23* Depositional model for nummulitic deposits (modified from Tawadros, 2001. Originally based on Comte & Lehman, 1974, Arni, 1963, Decrouez & Lanterno, 1979, Moody et al., 1998, Irwin, 1965, Wilson, 1975, and Aigner, 1982).

paleogeography, paleoenvironmental setting, substrata variation, and syn-sedimentary tectonics.

### 20.4.2.5   Fore-bank facies (facies 3)

The fore-bank facies is dominated by open-marine marls and marly limestone facies with planktonic microfauna. The fore-bank facies is a buildup of shell fragments derived from the bank facies. Closer to the bank edge in quiet waters the facies is made up of biomicrites with echinoderm fragments, molluscs, buliminoids, globigerinids, lenticulinas, rotaliids, miliolids, textularids, bryozoans, and red algae, or argillaceous biomicrites with lamellibranchs, echinoderms, nummulites, operculinids, rotaliids, globigerinids, *Elphidium* and miliolids (Fournié, 1975). The nummulitic forms are dominated by very thin Nummulites with many whorls (Decrouez & Lanterno, 1979), such as *N. planulatus*, and frequent *Discocyclina* and *Assilina* (Arni, 1963). Species of *Operculina* and benthonic foraminifera might be present.

### 20.4.2.6   Bank facies (facies 4)

The Nummulites bank facies is composed of pure limestones. Nummulites are often dominated by one or two whole large forms, such as *Nummulites gizehensis* or *N. perforatus* (Arni, 1963, Comte & Lehman, 1974). Nummulites may be associated with red algae and *Discocyclina*, depending on water depth (Decrouez & Lanterno, 1979).

### 20.4.2.7    Back-bank facies (facies 5–6)

The back-bank facies is composed of micritic-nummulitic limestones rich in pelecypod shells and small and medium size and robust Nummulites such as *N. deserti* and *N. beaumonti*, in addition *Gypsina*, small foraminifera and Dictyoconoides are common (Arni, 1963, Decrouez and Lanterno, 1979). A transitional facies grades into lagoonal facies, dominated by *Orbitolites* and *Alveolina* (Arni, 1963) may also be present. This transitional facies is made up of argillaceous, pyritic, micritic limestones with echinoids, gastropods, lamellibranchs, isolated nummulites, rare algal laminites, and traces of phosphate. It also includes algae and Nummulites shoals composed of grainstones with red melobesian algae (lives in restricted lagoons according to Flügel, 1982), bryozoans, small foraminifera, rare *Discocyclina*, *Alveolina* and *Nummulites* (Comte & Lehman, 1974).

### 20.4.2.8    Lagoonal-littoral facies (facies 7–8)

The lagoonal-littoral facies is composed of bioclastic limestone facies, and it consists of dolomitized, fine tests of molluscs and gastropods, miliolids, ostracods, and coprolites, associated with algal thalles (Arni, 1963, Comte & Lehman, 1974). Peneroplid shoals may locally develop (Arni, 1963).

### 20.4.2.9    Lagoonal-evaporitic facies (facies 9)

The lagoonal-evaporitic facies may form on the landside (Arni, 1963). It is dominated by evaporites and dolomites with gypsiferous and green and yellow shale intercalations. Phosphatic horizons, gypsiferous shales, sandstones, conglomerates with chert, and varicolored marls may occur (Comte & Lehman, 1974). The absence of evaporites and dolomites can be caused by the lowering of salinities in the back-bank zone by freshwater (Fournié, 1975). Evaporites can be replaced by coal and lignite (Decrouez & Lanterno, 1979).

The development of nummulitic banks within a basin can be attributed to morphological changes of the sea-bottom (Arni, 1963) such as may be caused by pre-existing tectonic structures (Said, 1961, 1962, Kostandi, 1963, Fournié, 1975, Aigner, 1983, Bernasconi et al., 1991). The presence of pre-existing tectonic structures was probably the major factor in the development of nummulitic banks in the Eocene of the offshore of Northwest Libya such as the nummulitic banks in the Bouri Field (Bernasconi et al., 1991), and the nummulitic bank facies of the Ypresian in the subsurface of the Western Desert (Said, 1961, 1962, Kostandi, 1963). Undulations in the relief of the Eocene depositional basin, along north-south and E–W direction in the Nile Valley and the Gulf of Suez areas, resulted in shallow and deep facies alternations (Boukhary & Abdelmalik, 1983).

Nummulites are large foraminifera characterized by dimorphic reproduction (Blondeau, 1972), the small A-forms (megalospheric) reproduced asexually, while the large B-form (microspheric) reproduced sexually. Variations in the ratios of the two forms in nummulitic deposits were attributed to either in-situ biological factors (Loucks et al., 1998a & b) or later physical processes (Aigner, 1982, Moody et al., 1998). The distinction between these depositional modes is essential for determining the geometry, the spatial distribution, the internal heterogeneities, and the petro-physical properties of the nummulites reservoirs (Jorry & Davaud, 2004). The ratio of the smaller megalospheric (A-type) to the larger microspheric (B-type) in the original assemblage is believed to be 10:1 (Blondeau, 1972). Loucks et al. (1998a, b) believe that variations in the A:B ratios were mainly the product of biological reproduction in response to environmental conditions rather than a product of hydrodynamic sorting. Nummulites matured rapidly and reproduced asexually under favorable conditions, and grew to larger sizes and reproduced sexually under stressed conditions (Wells, 1986, Hallock & Glenn, 1986).

The A:B ratio was probably also controlled by biological reproductive strategy in response to changing environmental conditions (Loucks et al., 1998b). Braithwaite et al. (1998) observed that units rich in the large B-form Nummulites alternate with units dominated by the small A-form nummulites. They suggested that an increase in water depth favored the large B-forms sexual reproductions while in a more stable shallow water the small A-forms reproduce asexually. This model is similar to that proposed by Loucks et al. (1998), but is in contradiction to Aigner (1984).

Once the Nummulites were deposited, they were subjected to sedimentary processes in the depositional environment. Physical processes involved in the Nummulites banks were episodic, storm-related winnowing of mud and the smaller A-forms, forming lag deposits of the larger B-forms. Aigner (1982) postulated that larger Nummulites would behave like coarse sand grains. Storms can result in the mixing of large foraminifera, such as Nummulites, Assilinas, Discocyclinas, and Operculinas (Blondeau, 1972). Both Nummulites banks of the Middle Eocene Mokattam Formation at Giza, Mokattam and Fayium and the shell banks of the Upper Eocene in Egypt, such as the shell beds with *Carolia*, *Plicatula* and *Ostrea,* were formed on scoured firm surfaces, followed by erosion and then inhabited by other organisms, such as *Ophiomorpha* (Aigner, 1982).

The Lower Eocene succession corresponds approximately to the upper part of supercycle TA2 of Haq et al. (1987) and consists of sediments that were deposited during a high-stand. The sea transgressed from the north and covered the Pelagian Block, Central and Northeast Libya, and most of Egypt. During that time, the "Pale-osirtic" Gulf (Desio, 1971) appeared (Fig. 20.21) and extended to the Tibesti Moun-tains, as far south as 23°N. This transgression was accompanied by uplift of most of western Libya. The sea transgressed over northern and central Egypt and southwards along the Nile Valley into Sudan, forming the "Paleonilotic" Gulf (Desio, 1971).

Early Eocene sediments are dominated by carbonates and shales. Evaporites occur in the west central Sirte Basin and the offshore of Northwest Libya and Tunisia. Siliciclastics are locally common. Facies changes are attributed to paleotopography which was controlled mainly by pre-existing tectonic structures.

The Lower Eocene succession in the offshore of Northwest Libya was deposited on the Pelagian Platform. This platform was separated from a shallow platform along the North African craton by the deep Tripolitania/Gafsa Basin which was dominated by the deposition of shales and marls. Three facies belts can be recognized in the offshore of Libya and Tunisia, from southwest to northeast: 1) A restricted evaporitic platform facies bordering the emerged land to the south. 2) A shallow-water nummulitic facies in the middle. 3) Open marine micrites and marls with abundant planktonic foraminifera to the north (Fournié, 1975, Bernasconi et al.,

1991, Bishop, 1975, 1988, Bishop & Debono, 1996). These facies belts were not recognized in Egypt.

The whole Ypresian succession belongs to a regressive sedimentary sequence delimited by unconformities and includes four regressive cycles (parasequences). This sequence is represented by the Metlaoui Group (Farwah Group of Hammuda et al., 1985) and includes the Chouabine Formation, El Garia Formation, Ousselat Formation, Faid Formation, and the Ain Merhotta Formation (Fournié, 1979), each representing a depositional facies. The sequence starts unconformably on top of the Cretaceous Abiod Formation or the Paleocene El Haria Formation with glauconitic and phosphatic transgressive units of the Chouabine Formation. These phosphatic deposits are exploited in Tunisia. The Lutetian unconformity at the top is also marked by glauconite, phosphate and pyrite. These minerals represent breaks in sedimentation (Fournié, 1975).

The Metlaoui Group was deposited on a broad carbonate ramp that deepened to the northeast (Loucks et al., 1998a & b, Reali & Ronchi, 1998, Reali et al., 2003, Braithwaite et al., 1998). The El Garia Formation (Jdeir Formation of Hammuda et al., 1985) consists of two minor sequences, separated by a transitional surface (Reali & Ronchi, 1998, Reali et al., 2003). These sequences are formed of high-energy shoal complexes and banks composed of nummulitic limestones with rare echinoids and bivalve fragments, and minor *Discocyclina* (Braithwaite et al., 1998, Loucks et al., 1998a & b). They occur as low-relief nummulitic banks or sheet-like bodies, lacking the distinct, high relief bank topography (Moody et al., 1998a), due to the successive erosion (ravinement) of the deposits (Fournié, 1975). These accumulations form elongate carbonate shoals parallel to the shelf margin, cut by major embayments (Moody, 1995). The shoals are locally cross-bedded and composed of A-*Nummulites*-rich grainstones and packstones with red algae, and discocyclinids (Comte & Lehman, 1974, Decrouez & Lanterno, 1979, Loucks et al., 1998a & b). They form prograding nummulitic marine bar sequences, deposited below fair-weather wave base and above storm wave base (Loucks et al., 1998a & b, Reali & Ronchi, 1998, Reali et al., 2003). These prograding sequences start with bioclastic packstones containing benthonic foraminifera, such as alveolinids, orbitolites, and miliolids, deposited in a restricted lagoonal environment. They grade upward into a coarsening-upward section of wackestones to rudstones, showing an increase in the size and sorting of Nummulites (Reali & Ronchi, 1998).

Nummulitic accumulations of the El Garia Formation (Jdeir Formation) form commercial hydrocarbon reservoirs at the Ashtart Field in the Gulf of Gabes, at the Sidi El Itayem Field in the Sfax area, Eastern Tunisia, and at the El Bouri Field, the major producer in the Mediterranean Sea, approximately 125 km northwest of Tripoli. Porosity in these fields is up to 16% and permeability varies from 40 to 100 mD (Ben Ferjani et al., 1990). Most of the porosity is secondary solution-enlargement (Bishop, 1988, Jorry et al., 2003).

Nummulitic banks of the El Garia Formation grade into an inner-ramp complex. This complex is composed of the restricted shallow lagoonal facies of the Ain Merhotta Formation and the sabkha facies of the Faid Formation in Tunisia (Fournié, 1979, Loucks et al., 1998a & b) and the Jirani Dolomite in the offshore of Northwest Libya (Hammuda et al., 1985). The latter is composed of burrowed packstones-grainstones with anhydrite nodules and gastropods, algae, miliolids, alveolinids, peneroplids, molluscs, echinoderm fragments, and ostracods. The Jirani Formation has limited distribution and is missing north of the Bouri Field (Bernasconi et al., 1991). Southward the Farwah/Metlaoui Group grades into the Taljah Formation made up mainly of sandy, anhydritic, and argillaceous dolomites (Hammuda et al., 1985, 1991).

The nummulitic bank facies grade basinward into the outer ramp Ousselat nummuliclastic packstones and grainstones deposited in a deep neritic environment then into the Bou Babbous basinal chalks composed of globigerinid packstones and wackestones (Loucks et al., 1998a & b, Vennin et al., 2003). The latter constitutes the source for oil at the Ashtart and Sidi el Itayem fields, with TOC values between 0.5 and 2.5% (Bishop, 1988, Ben Ferjani et al., 1990). Deposition of the source rock of the Bu Dabbous Formation (Bilal Formation of Hammuda et al., 1985) in the offshore of Northwest Libya and Tunisia was probably related to the Oceanic Anoxic Event (OAE) which took place during the Lower Eocene (Jarvis et al., 1988). Simultaneous deposition of phosphate deposits of the Chouabine Formation may coincide with the beginning of this anoxic event (Jenkyns, 1990). The Bou Dabbous (Bilal) Formation is missing north of the Bouri Field (Bernasconi et al., 1991).

The earliest Ypresian deposition in the western Sirte Basin started with the poorly fossiliferous limestones and dolomitic limestones, alternating with marls and marly limestones of the Kheir Formation under restricted, shallow-marine conditions (Burollet, 1960, Jordi & Lonfat, 1963, Cepek, 1979, Korab, 1984). In the subsurface of the Sirte Basin these sediments change facies eastwards into shales and provide the cap rock for the Late Paleocene Zelten reservoirs in some areas. In the Al Qaddahiyah map-sheet area (Mijalkovic, 1977a) the Kheir Formation changes facies into very fine calcareous sandstones, sandy limestones, marly limestones, and white limestones with *Globigerina triloculinoides* and the charophyte *Raskyella peckia meridionale* which suggest deposition in a brackish water environment. The presence of sands and charophytes suggests the existence of a nearby landmass.

Ypresian deposition continued with the deposition of the Wadi Zakim and Rouaga members of the Beshima Formation in the eastern Sirte Basin and the Gir Formation in the subsurface. Deposition of silicified and dolomitic carbonates with bird's-eye structures, pellets and fragments of gastropods and bivalves, globorotaliids, and miliolids of the Wadi Zakim Member (Woller, 1978, Cepek, 1979) probably took place in a shallow restricted lagoonal to supratidal environment. Deposition of the overlying Rouaga Member composed of poorly fossiliferous dolomites and dolomitic limestones with nodules and beds of chert (Jordi & Lonfat, 1963, Shakoor, 1984) probably occurred under similar conditions.

In the subsurface of the Sirte Basin, the total Eocene succession averages 900 m (2953 ft) in thickness on the platform areas but increases up to 2000 m (6562 ft) in the Ajdabiya, Maradah, Zallah trough, and Tumayyim troughs (Hallett & El-Ghoul, 1996, Hallett, 2002). An extensive evaporite basin was developed in the Zallah Trough during the Eocene with evaporite thicknesses over 300 m (984 ft) thick. In the Maradah trough, the Eocene is represented mainly by carbonates. A variety of environments were represented during the Ypresian. The Gir Formation changes from deep-water shales in the north to shallow-water carbonates then into evaporites southward. In the central and eastern part of the basin, the Ypresian is dominated by carbonates of the Mesdar Limestone Member. The Facha Member of the Gir Formation (Ypresian) (equivalent to the Kheir Formation of Barr & Weegar, 1972) in the Zallah Field, Sirte Basin, consists of a variety of lithologies of argillaceous limestones with thin interbeds of terrigenous mudstone, passing up into clean limestones overlain by variably anhydritic dolomites and anhydrites, and capped with anhydrites (Williams, 1995).

The Ypresian is poorly represented in Cyrenaica. If present, it may form the basal part of the Apollonia Formation. Apollonia Formation (Lower to Upper Eocene) cherty limestones were deposited unconformably on the Upper Cretaceous Atrun Limestone. The Formation contains reworked limestone fragments with a few species of *Nummulites*, *Operculina pyramidum*, and *Discocyclina pratti* (Röhlich, 1974), graded bedding, and slump structures were probably due to deposition by turbidity currents. It grades in the offshore of Cyrenaica into cherty and glauconitic argillaceous mudstones-wackestones, grey-green marls, and shales, deposited in an open deep-marine environment (Duronio et al., 1991).

The paleoenvironmental distribution of foraminifera in the Early Eocene-Middle Miocene rocks of northern Cyrenaica, Northeast Libya, indicates an overall shallowing-upward trend from open-marine (Apollonia Formation) to shallow-marine restricted conditions (Abdelsamad, 1998). This cycle consists of high to low-stand system tracts, approximately coincident with eustatic changes of sea level (Elwarfalli & Stow, 1998). Expansion of the nummulitic bank facies occurred during relative high-stand deposition. The bank facies include abundant shallow marine fauna such as Nummulites, algae, molluscs, miliolids, and echinoderms, forming grainstone to packstone textures and minor bioclastic wackestones with common red algae, deposited in the lower photic zone and above the storm wave base. The fore-bank facies is dominated by chalks, shales, and bioclastic packstones, including pelagic sediments and downslope resedimentation (El-Hawat & Shelmani, 1993). The back-bank facies comprises a restricted fauna, such as miliolids and molluscs, and interbedded dolomites and evaporites, indicative of hypersaline lagoonal environments.

Eocene sediments lie unconformably on older units in the Al Jabal Al Akhdar (Klen, 1974, Röhlich, 1974, Abdulsamad et al., 2008). The hiatus between the Paleocene and Eocene becomes less distinct toward the Mediterranean (Klen, 1974). Inland the Eocene Derna Formation lies unconformably on the Maastrichtian Atrun Formation or on the Wadi Ducchan Formation while seaward the Eocene Apollonia Formation lies unconformably on the Paleocene Uwayliah Formation in the Tulmetha area (Klen, 1974), or the Atrun Formation in the Marsa al Hilal area (Röhlich, 1974). The contact between the Maastrichtian Wadi Ducchan Formation and the overlying Eocene Derna or Apollonia formations is erosional (Klen, 19784, Röhlich, 1974). However, the contact between the Uwayliah Formation (Danian-Landenian) and the Wadi Ducchan Formation (Maastrichtian) seems to be conformable (Röhlich, 1974).

Northern Egypt was divided during the Eocene into platforms, talus and basins (Salem, 1976). On the platforms a few islands emerged. Under favorable conditions Nummulites banks developed (Decourez & Lanterno, 1979). Nummulitic bank development was interrupted frequently by vertical tectonic movements and regression toward the north. Emergent islands supplied clastics and breccias into the nummulitic banks.

In the Northern Western Desert, Upper Cretaceous tectonics controlled the facies, the thickness of Eocene sediments and the nature of the contact between the Mesozoic and Tertiary systems. The extent of the unconformity depends of the relative position on the structural high. Lower Eocene rocks are variable in thickness and missing on the crests of major subsurface structural highs such as Bahariya, Abu Rawash, Khatatba, Burg el Arab, Marsa Matruh and Mamura (Salem, 1976, Kostandi, 1963). They display current distribution due to Upper Cretaceous-Lower Cretaceous tectonics. In the Rabat-1 well, no Mesozoic rocks were encountered and a thick Eocene section directly overlies Paleozoic sediments (Salem, 1976, Awad, 1984). Eocene sediments consist dominantly of carbonates, including cherty, dolomitized, and argillaceous limestones on the highs and shales in the lows. On the flanks of the highs Lower Eocene sediments were deposited and preserved such as in the Wadi Natrun-1 and Faghur-1 wells. More than 229 m (751 ft) of Lower Eocene Nummulitic limestones were penetrated in the Wadi Natrun well. Due to northward tilting of North Africa tectonic highs, such as the Burg el Arab and Abu Rawash were submerged during the Middle Eocene while southern Egypt remained exposed. In the Upper Eocene, the tops of the Khatatba and Abu Roash highs rose above sea level (Kostandi, 1963).

At the beginning of the Ypresian large areas of Central and Southern Egypt formed a very shallow platform with emergent sections in the east. Boukhary & Abdelmalik (1983) believe that undulations in the relief of the depositional basin along N-S and E-W directions resulted in shallow and deep facies alternations throughout the Eocene. Deep-marine shales with thin limestone and marly beds intercalations of the Guss Abu Said Member of the Thebes Formation (Farafra Shale of Kostandi, 1963) with Nummulites, *Operculina libyca*, *Globorotalia wilcoxensis*, *Discocyclina nudimargo* in the Farafra Oasis (Youssef & Abdel-Aziz, 1971), as well as *Morozovella subbotinae* and *N. praecursor* in the Ghanima section (Faris, 1984) were deposited. This succession is absent in many places either due to non-deposition or most likely due to intra-Ypresian erosion. Deposition continued in the later part of the Ypresian with shales, marls, and limestones. Limestones were deposited with intense algal production by mainly udotacean forms (*Ovulites* and *Halimeda*) (Strougo et al., 1990). Nummulites accumulated in the deeper part of the platform, probably during catastrophic storm events (Aigner, 1983).

Limestones with flint concretions of the Thebes Formation (Said, 1960) near Luxor represent an open sea facies with pelecypods, Nummulites, and planktonic foraminifera dominated by globorotaliids and globogerinids. In the Farafra Oasis, the Thebes Formation (Ypresian) at El Guss Abu Said is composed of middle and inner-shelf limestones with alveolinids, especially *Alveolina schwager* and Nummulites. This Formation probably represents table-reefal facies of the Lower Eocene (Said, 1962a). This facies grades in the North Gunna section into deeper-water limestones facies. The latter is devoid of alveolinids, and contains a few Operculinas and Nummulites (Youssef & Abdel Aziz, 1971).

In the Kharga Oasis to the south marls, marly limestones, and thick-bedded chert limestones of the Rufruf Formation with globorotaliids and Nummulites represent a deep-marine facies (Faris, 1984, Faris et al., 1998) similar to the Thebes Formation.

In the area between Samalut and Assiut chalks with flints, devoid of large foraminifera, of the Assiuti Formation (Bishai, 1961) were deposited during the early part of the Ypresian. They were followed by shallow-marine, fossiliferous limestones of the Manafalut Formation with abundant Nummulites and discocyclinids. The two formations grade northward in the area between Assiut and Minia into thick-bedded, reefal, and lagoonal limestones of the Drunka Formation (El-Naggar, 1970).

The Thebes Formation section along the El Sheikh Fadl-Ras Gharib road in the Eastern Desert is composed of limestones with Nummulites and flint bands, deposited in a deep-shelf muddy environment, as suggested by the presence of the ostracod *Loxoconcha saharaensis* (Elewa et al., 1995). Southward in the south Aswan-Dungul area, the Dungul Formation is probably the nearshore equivalent of these units. In Sinai the Lower Eocene transgression covered all the structures with limestone facies having moderately deep-water faunas (Thebes or Safra Formation) (Said & Kenawy, 1956).

The sea retreated rapidly at the end of the Ypresian because of a drop in sea level accompanied by tectonics, which resulted in a regional unconformity over most of Egypt and over large areas in Libya. This unconformity is marked in places by hard-grounds or limestone conglomerate (Strougo et al., 1990). The marine connection between the Tethys and the South Atlantic (the Trans-Saharan Seaway) was cut off during that time (Barsotti, 1963, Reyment & Reyment, 1980). Deposition of the Lower Eocene sequence was terminated by the post-Early Eocene regression which was induced by the opening of the Gulf of Suez (Hendriks et al., 1990).

The Lutetian a period of rapid regression does not extend south of 28° parallel in Egypt and 25° parallel in Libya (Fig. 20.21). A notable increase in deposition of clastics took place this time. North Africa was in the tropical zone and covered with shallow waters (Decrouez & Lanterno, 1979).

In the offshore of Northwest Libya the Ypresian-Lutetian boundary is marked by a break in sedimentation evident from the local occurrence of phosphate, glauconite, dolomite, and anhydrite on top of the Ypresian Farwah Group (Sbeta, 1991). This break is due to the reactivation of the Jifarah Arch at the end of the Ypresian (El-Ghoul, 1991), which is also reflected by the absence of the Jdeir Formation, Jirani Dolomite, and Bilal Formation in places.

Middle and Upper Eocene deposits in the offshore of Northwest and Northeast Libya and Tunisia are also arranged in facies belts oriented northwest-southeast (Sbeta, 1991, Bernasconi et al., 1991, Duronio et al., 1991, Bishop, 1975, Bismuth & Bonnefous, 1981), representing a gradual increase in the paleobathymetry to the north. In Tunisia lagoonal deposits of gypsum and dolomites of the Djebs Formation (Middle-Upper Eocene) in the south, grade northward into thick neritic shales and coquinoid and bioclastic limestones of the Cherahil Formation, and shales and marls with pelagic fauna of the Souar Formation (Fournié, 1979). In the Gulf of Hammamat offshore of Tunisia argillaceous and pelagic deposits of the Souar Formation are abruptly substituted by a restricted carbonate platform facies of dolomites and limestones of the Halk el Menzel Formation (Bismuth & Bonnefous, 1981).

In the offshore of Northwest Libya the Tellil Group (Lutetian) consists mainly of alternating shales, marls, and limestones with anhydrites and sandstones in places (Hammuda et al., 1985, 1991). The Harash Formation grades northward from shales with sandstone and anhydrite beds into limestones and shales, then into shales rich in benthonic foraminifera, pelecypods, and ostracods, and rare Nummulites (Sbeta, 1991). It grades upward into limestones rich in molluscs and Nummulites of the Dahman Formation (Sbeta, 1991), and then into glauconitic shales and minor limestones of the Samdun Formation. All three formations are thick in the center of the basin and thin both to the north and south (Sbeta, 1991, Hammuda et al., 1985). These shallow water facies grade northward into the pelagic shales and marls of the Ghalil Formation, equivalent to the Souar Formation, with many species of *Globorotalia* (Sbeta, 1991).

In the northwestern Sirte Basin, open-marine conditions prevailed during the Lutetian, leading to the deposition of white chalks with miliolids, peneroplis, ostracods, marine algae and globorotaliids of the Ben Isa Member of the Wadi Tamet Formation. Shallowing of the sea led to deposition of dolomitized chalky limestones with *Orbitolites complanatus* and *Dictyoconoides cooki*, oysters, and echinoids of the Bir Zidan Member. Northward, in the Al Qaddahiyah map-sheet area (Mijalkovic, 1977a) the Wadi Tamet Formation contains marine algae and the freshwater charophytes *Raskyella peckia meridionale* in addition to *Orbitolites complanatus*, pelecypods and gastropods, suggesting transitional marine to brackish water sedimentation during the early Lutetian.

In the subsurface of the Sirte Basin the Gialo shallow marine limestones pass southward into shales, anhydrites, and dolomites of the Gedari Formation. Decrouez & Lanterno (1979) interpreted the Gedari Formation in northwest Sirte Basin as a transitional, "miliolid and evaporite-dolomite facies" of Arni (1963). In the Ajdabiya Trough, the Middle Eocene is represented by shales.

In Northeast Libya the Ypresian-Lutetian boundary falls within the Apollonia Formation. El-Hawat (1985, 1987), and El-Hawat & Shelmani (1993) placed this boundary at the top of a slump and debris flow unit. However, the present author concurs with Barr & Berggren (1980) and places the boundary at the base of the slump unit. This boundary is marked by hardgrounds and glauconite beds, which suggest a sea-level drop and interruption in deposition at the end of the Lower Eocene.

The upper Apollonia Formation interfingers with the shallow-water Derna Formation and wedges out toward the Eocene shoreline to the south. The Derna Formation (Upper Lutetian-Bartonian) was deposited in shallow neritic and littoral environment (Klen, 1974, Röhlich, 1974). It contains many Nummulites species and some specimens of nautiloids up to 75 cm in size. In the Dernah area, reefs occur in the upper and middle parts of the Formation which contain abundant algal limestones, some layers rich in solitary corals, and large foraminifers, such as *Nummulites* and *Orbitoides* (Zert, 1974). It represents a Nummulitic bank or "reef complex" (Decrouez & Lanterno, 1979).

In northwestern Cyrenaica the upper Apollonia and Derna formations grade into the lower Khowaymat (Mazhar & Issawi, 1977, Swedan & Issawi, 1977, El Deftar & Issawi, 1977). The latter is composed of thick-bedded dolomites with limestone intercalations and minor silica intercalations. It contains *Nummulites intermedius*, *N. boucheri*, *Operculina complanata*, *Globigerina* sp., and *Globorotalia* sp., echinoids, and other shell remains (Röhlich, 1974). The Dernah Formation extends into the offshore of Cyrenaica where it forms reservoirs, for example in the new discovery by Repsol in well A1-NC202 (WO May 2009).

In Egypt the Early Lutetian deposition started unconformably on top of the Thebes Formation and its equivalents. The Lower Lutetian Minia Formation is made up of limestones very rich in Nummulites, echinoids, molluscs and gastropods with the characteristic fossils *Nummulites atacicus*, *Orbitolites complanatus*, *Alveolina frumentiformis*, *Operculina canalifera*, as well as the small echinoids *Sismondia logotheti*, the crustacean *Callianassa* (Said, 1960, Cuvillier, 1930). Lutetian sediments along the El Sheikh Fadl-Ras Gharib road consist of chalky limestones and marls of the Maghagha Formation (Bishay, 1966) probably deposited in intermittent shallow waters with low salinity as suggested by the presence of the ostracods

*Loxoconcha mataiensis* and *Xestoleberis? kenawyi* (Elewa et al., 1995). These sediments grade laterally into calcareous sandstones and shales with macrofossils and Nummulites of the Qarara Formation. The presence of the ostracods *Loxoconcha mataiensis, Xestoleberis? kenawyi, Kirthe bartonensis, Bairdoppilata crebra* and *Leguminocythereis africana* suggests deposition in an open deep-marine environment. The overlying El Fashn Formation (Bishay, 1966) is composed of white limestones with flint band intercalations, is rich in *Nummulites* and bryozoans, and reflects the shallowing of the sea in the Late Lutetian.

The Upper Lutetian Mokattam Formation at Gebel Mokattam (Cuvillier, 1930, Said, 1960) was deposited unconformably on top of the Minia Formation. It starts with sandstones and conglomerates, approximately 3 m (10 ft) thick with ferruginous nodules followed by limestones rich in *N. gizehensis*, echinoids, molluscs, gastropods, and a few beds rich in *Lithothamnium*, corals, crustacea, *Operculina pyramidum*, orbitoides, a number of species of *Pecten* and *Cardita*, and large specimens of *Nautilus* sp. Within the Mokattam Formation *Nummulites gizehensis* forms a regressive succession of bank, fore-bank, and back-bank deposits around Cairo (Aigner, 1985). Bank deposits are made up of packstones in which larger Nummulites forms are dominant over the smaller forms (Fig. 20.23). Back-bank deposits are mudstones and wackestones and the smaller forms dominate over the larger forms. Bitner & Boukhary (2009) reported the first occurrence of the brachiopod species *Terebratulina tenuistriata*, associated with *Nummulites farisi, N. praestriatus*, and *N. bullatu*, from the Middle Eocene (Bartonian?) nummulitic limestone of the Upper Building Stone Member of the Mokattam Formation at El Basatin of Gebel Mokattam, Cairo, Egypt. The abandoned Mokattam Quarry at Cairo is believed to be one of the main sources of rock used in the Ancient Egyptian constructions around Cairo and Coptic limestone sculptures. The stones belong to the lower to middle part of the Mokattam Formation (Said, 1990e, Tawadros, 2001) (Park & Shin, 2008, Harrell, 2004).

The Mokattam Formation in the area between Assiut and Minia is composed of white limestones with abundant Nummulites, echinoids, and the molluscs *Carolia placunoides* and *Turitella pharaonica* (Cuvillier, 1930, Strougo et al., 1990). In the northeastern and northwestern areas of the Bahariya Oasis, thick nummulitic limestones of the Qazzun Formation (Said & Issawi, 1964, Said, 1971) were deposited unconformably on top of the Naqb Formation. Limestones and marls with Nummulites and molluscs of the Hamra Formation (Said & Issawi, 1963) were deposited on top of the Qazzun Formation. At Helwan, the Gebel Hof Formation (Farag & Ismail, 1959, Geological Map of Egypt, 1987, Said, 1990e) non-fossiliferous limestones are heavily burrowed and rich in *Nummulites gizehensis*. The Formation was successively followed by limestones and chalky limestones of the Observatory Formation, marly and chalky limestones, shales, sandy marls, and shell banks rich in *Ostrea reili, Nummulites beaumonti, N. striatus*, and *Operculina pyramidum* of the Qurn Formation, and then by poorly fossiliferous sandy shales and a hard, highly fossiliferous middle bed with *Plicatula polymorpha* and *Nicaisolopha clotbeyi* of the Wadi Garawi Formation. The Middle Eocene Observatory and Qurn Formations exposed in the Qattamiya area, northwestern part of the Eastern Desert, were deposited in a shallow reefal carbonate platform and lagoonal environments. The paleogeographic distribution of ostracods reveals that there was a direct connection between the Southern Tethyan Real and the Western Europe Tethyan Real through which the migration of the ostracods had occurred (Shahin et al., 2008).

In the Fayium district the Upper Lutetian consists of the Wadi Rayan and Gehannam formations. The Wadi Rayan Formation is composed of glauconitic limestones with *N. gizehensis, N. curvispira, N. atacicus,* and abundant corals, marls, shales and sandy shales, argillaceous sands with *Ostrea* sp. and *Carolia placunoides,* and white limestone with *N. gizehensis* and abundant gastropods (Beadnell, 1905b). The Gehannam Formation is composed of shales and marls with marine vertebrates in the lower part and sandy limestones and sandstones in the upper part (Said, 1962a, Said, 199d), *Nummulites contortus-striatus* and *Operculina pyramidum* (*Nummulites* aff. *Pulchellus,* Strougo, 1986).

Southwest of Gebel Shabraweet near Ismailia Early Lutetian sediments contain *Orbitolites complanatus, Dictyoconus aegyptiensis, Alveolina,* miliolids, Textularias, and *Echinolampas,* and may represent a shallow-marine and more restricted environment (back-bank). At Gebel Krar in Sinai on the eastern side of the Gulf of Suez the Lutetian marly limestones equivalent to the Mokattam Formation are deeper marine and contain abundant planktonic foraminifera (Cuvillier, 1930) (offshore, open-marine) in addition to many species of *Nummulites,* including *N. gizehensis.* On the eastern side of the Gulf of Suez, green, glauconitic marls of the Darat Formation with fish teeth such as *Carchardon auriculatus,* molluscs, and *Nummulites beaumonti* (Hume et al., 1920, Moon & Sadek, 1922, Said, 1962a, Viotti & El-Demerdash, 1969) were deposited in a shallow-marine environment. They were followed by locally fossiliferous shales, argillaceous limestones, and gypsiferous marls of the Khaboba Formation. Near the city of Suez the marls of the Khaboba Formation grade into dolomitic limestones and marls of the Suez Formation (El-Akkad & Abdallah, 1971) and clastics and marls of the El Ramiya Formation which indicate a very shallow marine environment in that area.

The Middle Eocene in the subsurface of the Northern Western Desert is more terrigenous than the Lower Eocene and it is composed of thick sand, silt, and shale beds (Salem, 1976, Kostandi, 1963).

The Upper Eocene (Bartonian) was a period of further regression, and larger areas became emergent in the western Sirte Basin, the southern Cyrenaica Shelf, Upper Egypt, and Western Sinai (Fig. 20.21). In the Sirte Basin the Augila carbonates and shales were deposited in the central and eastern parts of the basin, and limestones, dolomitic limestones, and marls of the Zallah Formation in the western part. In the Al Jabal Al Akhdar and Cyrenaica, the conditions which prevailed during the Lutetian persisted and deposition of the Apollonia and Derna carbonates continued. These carbonates graded into the Khowaymat Formation. The Dahman carbonates and Samdun shales were deposited in the offshore of Libya and the Souar, Cherahill and Djebes formations in Tunisia.

Late Eocene deposition along the western margin of the Sirte Basin started with shallow-marine, chalky limestones and chalky dolomites with chert beds, celestites and anhydrites (Said, 1979, Jurak, 1985, Vesely, 1985, Zikmund, 1985) of the Tmed el Ksour Member of the Zallah Formation. These sediments are rich in large gastropods, such as *Gisortia stefanini,* in addition to the mollusc *Vulsella crispata,* nautiloids, echinoids, remains of crabs, ostracods, benthonic foraminifera, and the coral *Placophyllia bartai,* known to occur in the Upper Eocene of Italy (Jurak, 1985). Deposition of the Upper Lutetian Graret el Jifa Member started unconformably on top of the Tmed el Ksour Member with bioclastic limestones and clays (Srivastava,

1980, Jurak, 1985, Vesely, 1985), followed by interbedded clays and gypsum, then by dolomitized, bioclastic limestones and marls with sandy admixtures with Nummulites, pelecypods, echinoids, sharks, and gastropods (Zikmund, 1985). These sediments contain a faunal assemblage very similar to that reported from the Maadi Formation in Egypt, which suggests that the same environmental conditions prevailed in Egypt and West Libya.

In the Central and Eastern Sirte Basin the Upper Eocene (Bartonian) Augila Formation (Barr & Weegar, 1972) is equivalent to the Tmed el Ksour and Graret el Jifa members exposed in outcrops along the western margin of the Sirte Basin. The Augila Formation was deposited disconformably on the Gialo Formation. Deposition started with grey to green shales and thin argillaceous limestone or dolomite interbeds. The shales are occasionally sandy and glauconitic at the base and contain benthonic and planktonic foraminifera. They were followed by porous, glauconitic sandstones, and then by sandy, slightly glauconitic shales with common molluscan debris, *Nummulites fabianii* and *N. striatus*. Decrouez & Lanterno (1979) interpreted the Gialo Formation as a "reef complexes". Along the eastern margin of the Sirte Basin the Formation includes fossiliferous, calcilutites and calcarenites, chalky to dolomitic, and glauconitic with abundant algal and dolomitic beds near the base of the Rashda Member, probably equivalent to the Tmed el Ksor Member. These carbonates were deposited disconformably on top of the Gialo Formation. In the Sarir-Tibesti Platform the Paleogene succession is similar and correlatable with the Eocene-Oligocene Qasr el Sagha and Gebel Qatrani formations of the Fayium Depression in Egypt (Wight, 1980). In that area Upper Eocene deposition was dominated by evaporites (the Evaporite Unit) which contain oyster beds with *Ostrea clotbeyi* and *O. cubitus* at the base. The evaporites grade laterally into cherty and dolomitic limestones and upwards into limestones and marls with gypsum, limonite, and manganese nodules. Vertebrate fauna includes whales, turtles, fish, and reptiles. These sediments probably represent lagoonal/deltaic conditions.

In Egypt, the Upper Eocene is marked by the appearance of *Nummulites contortus* and *N. striatus*. Horizons rich in bryozoans are common at the base of the Upper Eocene (Cuvillier, 1930). The Upper Eocene in Egypt is dominated by limestones, marls, and shales. Sandstones are common in the Fayium area. South of Cairo shallow-marine sandy limestones of the Maadi Formation (Bartonian) with Nummulites graded upward into shales and sandy shales with banks of *Carolia placunoides* and limestones rich in echinoids at the top (Ain Musa Bed) (Said, 1962a). These banks were probably deposited during storm events (Aigner, 1982). At Gebel Mokattam the shallow-marine Maadi Formation is made up of bryozoan and siliceous limestones and gypsiferous shales rich in molluscs, gastropods, Nummulites, coral fragments, annelids, fish remains, and coprolites, coquinoid limestones with *Carolia placunoides*, siliceous limestones with echinoids, and calcareous sandstones with abundant *Callianassa* at the top (Cuvillier, 1930). East of Bani Mazar in the Nile Valley the deep-marine facies of the El Mereir Formation (Omara et al., 1977, Cronin & Khalifa, 1979) are composed of limestones, arenaceous marls, marly limestones, and shales with globigerinids, globorotaliids, and a few Nummulites. In the Fayium district, the Upper Eocene Birket Qurun and Qasr el Sagha formations are made up of sandstones, shales, and a few limestone beds rich in Nummulites, a large number of oyster species, and abundant vertebrate fossils. Vertebrate fauna are similar to those reported from the Tibesti area (Wight, 1980). These sediments were deposited in fluvio-marine, tidal and subtidal environments (Said, 1962a) close to the Upper Eocene shoreline. Upper Eocene sequence maintains the same character on the eastern and western sides of the Red Sea area. Deposition of the Tanka Formation in Sinai started with non-fossiliferous marls and limestones, followed by red beds composed of sandstones, conglomerates, marls, and shales with Nummulites and molluscs (Cuvillier, 1930, Viotti & Mansour, 1969, Viotti & El-Demerdash, 1969).

The Eocene-Oligocene compressive episode resulted in the inversion of the High Atlas, as well as the generation of unconformities in the Anti-Atlas of Morocco (Frizon de Lamotte et al., 2008). In the Tell Domain of Algeria, Eocene sediments occur in allochthonous blocks and are represented by shales, highly fossiliferous marls, and carbonates. South of the Hodna, the lower Eocene consists of gypsum-bearing marls followed by phosphatic limestones and cherty limestones. The Lutetian is a shelly and nummulitic marl facies (Askri et al., 1995). The Upper Eocene in the Tell and the Atlas domains was dominated by a compressive tectonic phase, which led to the formation of NE-SW folds in the Aures and the Atlas. This phase was followed by an Oligocene extensional episode that formed small grabens oriented perpendicular to the Atlas structures (Askri et al., 1995).

## 20.5   UPPER MEGASEQUENCE

### 20.5.1   Oligocene period

The Oligocene signaled a major fall in sea level and maximum emergence of North Africa (Figs. 20.2 & 20.24) related to the Alpine Orogeny and the East-African and the Red Sea rifting. This period is marked by a major change in the depositional domain from mainly carbonate to mainly clastics (Said, 1962a, Desio, 1971) and widespread volcanic activities. Alpine orogenic movements also influenced the Cyrenaica uplift (Goudarzi, 1980) which was a part of the Mediterranean Tethys zone until the Oligocene (Klitzsch, 1970). The Fayium Basin was formed due to dextral strike-slip movements along the "Trans-Africa Lineament" (Guiraud et al., 2000) and the Abu Gharadig Basin was inverted (Bayoumi & Lotfi, 1989, Keeley, 1994). Tertiary volcanics are common in North Africa (Meneisy & Kreuzer, 1974a, b, Meneisy, 1990, Wilson & Guiraud, 1998, Mattoussi Kort et al., 2009). They occur in the Gebel Oweinat-Arknu Complex, Jabal Hassawnah, Jabal asSawda, and Jabal Al Haruj al Asswad in Libya (Ade-Hall et al., 1974a, b and 1975, Woller & Fediuk, 1980). Volcanic rocks are reported in Egypt from Gebel Oweinat, east of Kom Ombo, Bahariya Oasis, Wadi Araba, and Gebel Qatrani north of Fayium, the Nile Delta, Gebel Khesheb west of Cairo, the Gulf of Suez, and west-Central Sinai. Volcanic activities probably started in the Oligocene and continued to Recent times. The activities were probably concurrent with movements along deep-seated fractures related to the Alpine Orogeny (Conant & Goudarzi, 1967). The Sirte Basin was tilted northward, accompanied by the upheaval of the Al Haruj al Aswad during the Oligocene which led to reversal of the direction of the drainage system. In the meantime drainage into the paleo-Sirte gulf from the Murzuq Basin was cut off and the Murzuq Basin became a closed basin.

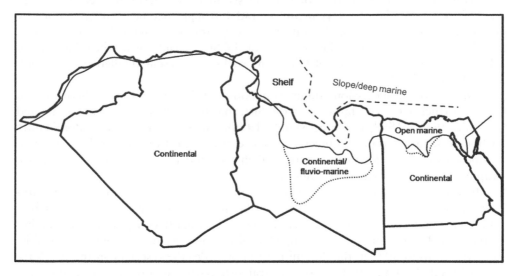

*Figure 20.24* Oligocene paleogeographic map: Compiled from Benfield & Wright (1980), Bishop (1975), Barr & Weegar (1972), and Said (1962a).

From the Oligocene to the Quaternary, reactivation of pre-Quaternary faults and subsidence of the Ionian Sea took place due to extensional tectonics (Biju-Duval et al., 1974). Late Paleocene- Pliocene tectonics acted along WNW-ESE and ENE-WSW axes but during the Quaternary tectonic phase, they acted along WNW-ESE trends. The western part of the Western Mediterranean was a land area in the Tethys in lower-middle Cenozoic time and it was subjected to strong volcanic and seismic activities (Blanpied & Bellaiche, 1983). The Western Mediterranean basins were formed during the Neogene.

In the offshore of Northwest Libya the Oligocene deep-marine marls and sandy shales, argillaceous wackestones-mudstones, and glauconitic and pyritic shales, wackestones and packstones of the Ras Abd Jalil Formation unconformably transgressed the Tellil Group. This deep-water facies grade upward into grainstones-boundstones interbedded with dolomites with *Operculina lepidocyclina* and *Nummulites vascus* at the base, and then into wackestones-packstones. Locally the Formation consists of biostromes. In the Early Oligocene, in the immediate offshore of Northwest Libya near the Tunisian border packstones with abundant *Nummulites vascus* and a few small benthonic foraminifera, coralline algae, and echinoid debris were deposited in a restricted shallow shelf (Fucek et al., 1998). Upper Oligocene deeper-marine sediments were also deposited on a shallow shelf (Kucenjak et al., 1998) with *Globigerina ciperoensis* and *Globorotalia nana*, and benthonic and larger foraminifera.

By the end of the Eocene the major tectonic events which formed the Sirte Basin had virtually ceased. The basin was transformed into an interior sag basin which represents the post-rift depositional phase of the basin development (Harding, 1984, Baird et al., 1996, Abadi et al., 2008). The rate of subsidence in the western troughs was reduced, and the distribution of Neogene sediments was less influenced by trough subsidence. The Zallah, Tumayyim and Maradah troughs had been largely infilled by the end of Eocene time. Rapid deposition; however, continued in the Ajdabiya Trough, which represents the principal depocenter of the Sirte Basin during the Oligocene in which over 1200 m (3937 ft) of Oligocene rocks are present (Hallett & El-Ghoul, 1996, Hallett, 2002).

The Oligocene succession attains its maximum thickness in the northern trough. In the northernmost coastal area of the Sirte Basin, sedimentation was dominated by shales (Fürst, 1993, Bezan & Malak, 1996). The shales pass into a dominantly carbonate sequence on the submerged highs east and west of the Northern Trough, composed of interbedded calcarenites, calcilutites and rare calcirudites, dolomites, evaporites and limestones in the inner zone.

In the western Sirte Basin, Oligocene dolomitic limestones with *Nummulites intermedius fichteli* and *Crassostrea* cf. *gryphoides* of the Dor el Abd (Um ad Dahiy) and Greir Bu Hashish formations were deposited in tidal flats, lagoons, and marshes (Srivastava, 1980). Coral reefs developed locally in mid-neritic environments. Moldic porosity of up to 30% in the dolomitic reefs provides the reservoir in the Al Hofra Oil Field (Hladil et al., 1991). In the Central Sirte Basin, shallow-marine, glauconitic sandstones of the Arida and Diba sandstones provide reservoirs in the Jalu, An Nafurah and Ar Rakb fields (Barr & Weegar, 1972, Bezan & Malak, 1996). In the Hun Graben, shallow-marine sandstones and conglomerates of the Maazul-Nina Formation (Jordi & Lonfat, 1963) were deposited around the edges of the graben (Woller, 1978, Salaj, 1979, Cepek, 1979), where they rest with a sharp unconformity on Eocene sediments. They grade in the central part of the Hun Graben into shales, calcarenites, sandy limestones, gypsum, and dolomitic limestones. Calcareous sandstones and conglomerates are common in the upper part of the Formation on the western side of the graben (Woller, 1978, Cepek, 1979).

The Oligocene at Wadi Bu Mras, in the northern Sirte Basin is composed of limestones and dolomites. Silty calcarenites composed of detrital bioclastic debris of lamellibranchs, gastropods, bryozoans, echinoids and foraminifera, are locally mixed with small quantities of quartz sand, silt, and glauconite, and capped by dolomites with numerous *N. interfichteli*. Trough and planar cross-bedding shows easterly and westerly directions (Selley, 1968). The base of the limestones is marked by erosional surfaces and bored limestone cobbles. The limestones are interbedded with unfossiliferous, gypsiferous mudstones. The limestones were deposited in shoal and barrier beach environments, and the mudstones in restricted lagoons behind them. To the southeast, west of Raguba, cross-bedded sandy limestones and calcareous sandstones are abundant. They overlie erosional surfaces and fill channels 2–3 m (6.6–10 ft) deep. Cross-beds dip to the north and south. They are intercalated with burrowed, unfossiliferous mudstones and fine sandstones with shark and crocodile teeth. Polygonal desiccation cracks filled with gypsum occur at the top. These mudstones and sandstones are overlain by quartzitic, cross-bedded sandstones with molds of vertical trees with their roots penetrating the Oligocene limestones below. These sediments were deposited in estuarine channels and tidal flats.

On the western and eastern margins of the Jabal Hasawnah Early Oligocene sedimentation started directly on top of the Cambrian Hassaouna Formation with a thick bed composed of calcilutites of the Tarab Formation (Jurak, 1978).

The calcilutites are rich in fossil fish and other vertebrate remains which include a skeleton of *Titanohyrax paleotherioides*, the frog *Libicus hasaunus*, and silicified shells of the gastropods *Melania fasciatus nysti*. The calcilutites were followed by gypsiferous shales and marlstones with intercalations of calcilutites. In the Sarir-Tibesti area these sediments represent a regressive sequence starting with lagoonal/deltaic sediments of the Evaporite Unit, changing into channel, interchannel, levee, and overbank sandstones and siltstones of the Idam Unit which is rich in vertebrate fauna (mammals, reptiles, and fish) of Lower Oligocene age similar to the Qasr el Sagha and Qatrani sediments in the Fayium Depression in Egypt, then into fluviatile sediments of the Sarir Unit (Wight, 1980). The sediments of the Tarab Formation, the Evaporite Unit, and the Idam Unit represent the southern and western shorelines of the Oligocene sea.

Carbonate deposition persisted on the Cyrenaica Platform, the Al Jabal Al Akhdar area, and the offshore of Cyrenaica. In Cyrenaica, the Oligocene is represented by shallow-marine limestones and marls of the Cyrene Formation (Pietersz, 1968, Kleinsmeide & van den Berg, 1968). The marls of the Shahat Member of the Cyrene Formation were developed only on the northern slope of the Al Jabal Al Akhdar, but they disappear to the south (Röhlich, 1974, Klen, 1974, Zert, 1974). They were followed by the deposition of the Salantah Member (Algal Limestone) limestones with the rock-forming algal species *Archaeolithothamnion maughini*, foraminifera, echinoids, pelecypods, and gastropods, in addition to reworked Priabonian and Oligocene microfauna (Zert, 1974). This member represents the shallowest facies of the Cyrene Formation but it grades into deeper-marine marly facies in the north. During the Upper Oligocene, calcarenites of the Labrak Formation with abundant reworked Eocene *Nummulites* species, *Nummulites fichteli, N. vascus,* and many echinoid species were deposited unconformably on the Cyrene Formation. The upper part contains *Operculina complanata* and *O. africana*. Dolomite beds were developed in the Dernah area (Röhlich, 1974, El Deftar & Issawi, 1977, Francis & Issawi, 1977). In the Benghazi area the Upper Oligocene Labrak Formation rests directly on the Derna Formation (Klen, 1974). In northwestern Cyrenaica, the Oligocene sediments grade into the upper part of the Khowaymat Formation (Mazhar & Issawi, 1977, Swedan & Issawi, 1977, El Deftar & Issawi, 1977) composed of green clays with occasional oyster shells, and limestone interbeds with *Nummulites intermedius, N. vascus, N. boucheri, N.* cf. *fichteli,* and *Operculina complanata* (Röhlich, 1974).

The Cyrene Formation grades in the offshore of Cyrenaica into shallow-marine sediments of the Early Oligocene Ketatna Formation (Duronio et al., 1991) composed of wackestones and packstones, grading upward into mudstones with many Nummulites species, bryozoans, hydrozoans, and pelecypods (Duronio et al., 1991, Duronio, 1996). Sediments of the Ketatna Formation (Duronio, 1996) were deposited with apparent conformity on top of the Eocene succession. The Late Oligocene Labrak Formation wackestones and packstones extended unconformably on top of the Ketatna Formation in the offshore area.

Up to 828 m (2717 ft) of prodelta and outer shelf shales and shelf limestones of the Dabaa Formation were deposited in the northern Western Desert and the Nile Delta region (Norton, 1967, Salem, 1976, Keeley, 1994, Hantar, 1990). More than 1800 m (5906 ft) of Oligocene sediments which probably represent a northerly pro-grading deltaic and submarine fan system were deposited in Northern Sinai and its offshore area (Jenkins, 1990).

Oligocene sediments thin southward and grade into variegated sandstones and shales of the Tayiba Red Beds (Said, 1971, 1990d, Jackson et al., 2008, Kuss & Boukhary, 2008) in Sinai where they are injected and capped by basalts, Upper Oligocene (Chattian) limestones, marls and dolomites of the Wadi Arish Formation at Risan Aneiza (Boukhary et al., 2008), and into reddish continental sandstones of the Gebel Ahmar east of Cairo, the Qatrani in the Fayium District, the Radwan in the Bahariya Oasis, and the Nakheil Formation (Akkad & Dardir, 1966) in the Red Sea area.

Oligocene rocks comprise more than 340 m (1116 ft) of cross-bedded sandstones and gravels with interbeds of shales and limestones rich in vertebrate fauna and fossil wood of the Qatrani Formation in the Fayium Depression southwest of Cairo, (Simons & Rasmussen, 1990), and variegated sandstones and shales with numerous marl beds with fossil wood and vertebrate remains in the southwestern part of the Qattara Depression (Albritton et al., 1990, 1991). The Qatrani sediments were deposited in nearshore marine, estuarine, bay, deltaic, point bar, and overbank environments (Said, 1962a). The sediments were carried by westerly-flowing rivers bordered by forests, into a large embayment of the Tethys (Bown, 1982, Bown & Kraus, 1988, Simon & Rasmussen, 1990). To the southwest in the Bahariya Oasis non-fossiliferous, ferruginous quartzites and coarse-grained sandstones of the Oligocene Radwan Formation (El-Akkad & Issawi, 1963) were deposited unconformably on top of both the Cenomanian Baharia Formation or Upper Eocene sediments. Deposition of the Qatrani Formation ended with an episode of Oligocene basalts dated at 31 ± 1 Ma (Fleagle et al., 1986). Oligocene quartzitic and reddish sandstones in the Gebel Ahmar (Red Mountain) east of Cairo and in the Lake Timsah area (Cuvillier, 1930) are associated with petrified wood and chert conglomerates. They contain miniature volcanoes (Rennebaum volcano) (Tosson, 1953) and hydrothermal pipes (Fig. 8.5) which have been attributed to silica-bearing hydrothermal fluids which were probably associated with Oligocene volcanic activities, based on the presence of silicified tubes cutting across the sandstones (Said, 1990e). In central Egypt gravel mounds are considered to be part of Oligocene deposits interpreted as erosional remnants of abandoned drainage systems (Said, 1990e).

During the Oligocene, most of the Gulf of Suez area was subaerially exposed and eroded (Schütz, 1988). On the Tiran Island at the southern end of the Gulf of Aqaba up to 100 m (328 ft) of Oligocene polymicitic conglomerates composed of pink granite, rhyolite, anhydrite, and dolomitic sandstone pebbles with arkosic sandstone lenses, were deposited unconformably on Precambrian volcanics (Goldberg & Beyth, 1991). In the Qoseir-Safaga area the Ypresian Thebes Formation is overlain unconformably by the Oligocene Nakheil Formation (Akkad & Dardir, 1966) which is composed of coarse breccia beds and variegated calcareous shales lacustrine.

Petrified wood is found in the Gebel Ahmar (Red Mountain) near Cairo, Gebel Khashab and Gebel Kibli el Ahram near the Pyramids, east of Maadi, and in the area between Cairo and Suez, the Siwa Oasis, the Qattara Depression, and in Northeast Libya. At Maadi, as well as in the Qattara Depression, tree trunks up to 40 m (131 ft) in height are oriented southwest-northeast. They were described by Buist (1846), Newbold (1848b), Unger (1859), Ibrahim (1943), and Shukri (1944). The origin of the petrified wood which range in age from Upper Eocene to Pliocene is controversial. Cuvillier (1930) and Said (1962a) believed that the woods were probably transported by a river from the south of Egypt. The absence of twigs, branches, and soft parts attest to long distance transport before the trunks were silicified. In northeastern Libya silicified trunks and quartzitic sandstone boulders occur along certain directions represented by extinct streams and at the base of hills (Di Cesare et al., 1963). Lefranc (1975) postulated that silicification occurred while the plants were still living in wet, tropical forests with alternating hot and cold seasons.

Iron ore deposits in the northern part of the Bahariya Oasis form layered lenticular bodies within the Lutetian Naqb Formation (Said & Issawi, 1963). Pisolitic and pseudo-pisolitic iron ore beds cap the Gebel Ghorabi. These ores are exploited in the El Gedida Mine, Bahariya Oasis. Basta & Amer (1969) postulated that the iron ore in the Bahariya Oasis was formed by metasomatic replacement. On the other hand the origin of iron-ores was attributed to fluctuations in sea level by Masaed & Surour (1998). Glauconitic sediments were deposited initially during a transgressive-stand tract (TST) to be altered during the following low-stand tract (LST). According to Masaed & Surour (1998), the ironstones of the El Gedida Mine were formed in three stages: 1) Stratiform diagenetic glauconitic ironstone pockets and concretions were related to dewatering and alteration of precursor greensands, and they occur at the base of small-scale depositional cycles. 2) Thin lateritic stratabound muddy and sandy ironstones terminated the small-scale depositional cycles. 3) Thick lateritic stratabound sandy ironstones were formed at the end of large-scale depositional cycles during periods of sea regression and laterization.

## 20.5.2   Miocene period

Marine and continental Miocene sediments covered the northern part of North Africa (Fig. 20.25) and rest unconformably on older rocks ranging in age from Precambrian to Oligocene. Two arms of the sea extended southward, one in the Gulf of Suez, the other in the western Sirte Basin. Miocene rocks crop out over much of the Northern Western Desert, along the Cairo-Suez road, and along the western and eastern sides of the Gulf of Suez, as well as in a small area in north-central Sinai. Miocene sediments also occur in the subsurface of the Gulf of Suez, the Nile delta, and the northern Western Desert. The sea transgressed over the Jifarah Plains and the northeastern Ghadames Basin during the Langhian-Serravallian for the first time since the end of Triassic (Fig. 20.25). The area was exposed one more time during the Upper Miocene. The sea also covered large areas of the Sirte Basin. Marine rocks however, are restricted to the northeastern part of the Basin with non-marine rocks extending southwestward to the basin margin.

In the Northern Western Desert Miocene deposits overlap Oligocene basalt flows or Oligocene sediments in the east and Middle and Upper Eocene sediments or Turonian dolomites in the west (Said, 1962a). Two facies can be recognized in the Lower Miocene (Said, 1962a, Salem, 1976), a fluvio-marine facies of sandstones and shales of the Moghra Formation connected to a river system that drained the North African continent and a marine facies of the Mamura Formation (Marzuk, 1970) consisting of reefal, open-bay, and open-marine sediments. The Middle Miocene started the deposition of richly fossiliferous limestones with shale intercalations (Said, 1962a, b, Gindy & El Askary, 1969) of the Marmarica Formation containing neritic and reefal assemblages of invertebrate fossils and foraminifera followed by white limestones. The Marmarica Formation in the Matruh area has yielded a rich

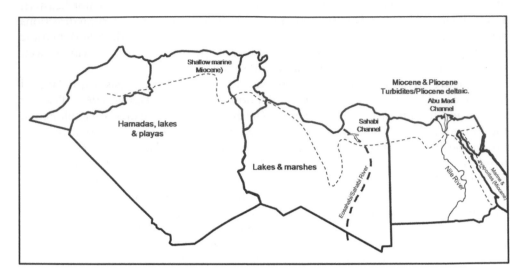

*Figure 20.25* Miocene-Pliocene paleogeographic map (compiled from Said, 1962a, Barr & Weegar, 1972, Salem, 1976, Goudarzi, 1980, Kogbe & Burollet, 1990). The Eosahabi/Sahabi River is after Griffin, 2006).

bryozoan assemblage which accumulated on a hard ground in bathyal during low rates of sedimentation (El Safori et al., 1998). The Sallum area west of Alamein was uplifted during the Middle Miocene time (Gindy & El-Askary, 1969) but the sediments thicken to the north. An unconformity which occurs between the Langhian and Serravallian can be recognized almost everywhere in Egypt and Libya (Said, 1962a, b).

Tectonics did not play an important role in the Nile Delta region during the Miocene (Rizzini et al., 1978). More than 3100 m (10,171 ft) of Miocene shales, sandstones, evaporites, and dolomites, were deposited in the Nile Delta region (Salem, 1976). Deposition during the Lower Miocene was dominated by sandstones and carbonates of the Gabal Khashab Formation, which grade westward into sandstones and limestones of the Moghra Formation. The overlying Middle-Upper Miocene succession forms a shallowing-upward cycle composed of deep-marine shales of the Sidi Salem Formation, fluvio-deltaic sediments of the Qawasim Formation, and Messinian evaporites of the Rosetta Formation (Rizzini et al., 1976). Said (1981) believed that the Eo-Nile was excavated during that time. The Sidi Salem Formation was followed by a period of uplift and erosion. Middle Miocene sediments offshore of the Nile Delta are composed of marine shales and sandstones of the Sidi Salem Formation which are subject to shale diapirism in places (Kilenyi et al., 1988). The Sidi Salem sandstones form reservoirs in the Delta region. Upper Miocene channel sediments of the Qawasim Formation, Messinian evaporites and channel deposits of the Rosetta Formation, and Early Pliocene Abu Madi Formation channel sandstones, form a seismic supersequence in the offshore of the Nile Delta (Kilenyi et al., 1988). The shales of the Sidi Salem Formation were deposited in an outer shelf environment followed by fluvial-deltaic sands and conglomerates of the Qawasim Formation with abundant slumps and slides. During this phase of emersion the area was subjected to subaerial erosion. Regression was accompanied by restriction of circulation in the basin during the Messinian and the deposition of the Rosetta evaporites. Open-marine fauna disappeared and were replaced by benthonic groups typical of the Messinian. Transgression during the Lower Pliocene led to deposition of fluvio-deltaic sediments followed by open-marine deposits of Abu Madi Formation.

Mesozoic and Lower Eocene shelf carbonates in the Nile Delta are overlain by a northward-thickening Middle Eocene-Holocene prograding fluvio-deltaic complex (El Mahmoudi & Gabr, 2009). The Mesozoic carbonates were folded and eroded during the Syrian Arc epeirogenic episode (Said 1962a, Salem 1976). Extensional subsidence then led to a marine transgression over the peneplained Mesozoic carbonates, with deposition of an unconformable layer of Lower Eocene shelf carbonates. In late Eocene times, clastic deltaic sedimentation took place (Barber, 1981). Volcanic activities during the Oligocene formed the Abu Zaabal and Bahariya Basalts. During Late Miocene (Tortonian), the Proto-Nile had established a delta complex represented by the Sidi Salem Formation, which was severely eroded during the evaporitic drawdown and subsequent desiccation of the Mediterranean Sea (Ryan & Cita, 1978, Hsü et al., 1978, Rizzini et al., 1978). Following the Oligocene emergence, the delta area underwent marked subsidence accommodated by a series of en echelon faults parallel to the coast.

Cenozoic tectonics led to the uplift of the east-west-trending Nubian Swell in Southern Egypt and Northern Sudan, (Thurmond et al., 2004); a structural high of Neoproterozoic crystalline basement and Paleozoic sediments. The Swell is still active and was probably responsible for the earth quakes in Upper Egypt during the 1980s (Thurmond et al., 2004).

The Nile Delta area became dry land during the Messinian Salinity Crisis (Harms & Wray, 1990). The Eo-Nile valley was excavated in Egypt (Said, 1981, 1993). A thick Pliocene to Pleistocene deltaic wedge was deposited as sea level was restored to near its present-day position. Isostatic loading and subsidence of the delta area has led to the deposition of up to 3,500 m (11,483 ft) of sediments (Harms & Wray 1990). Anhydrite beds were deposited in many areas at this stratigraphic level, which act as seismic reflectors. Griffin (1999) suggested that the Messinian was a time of marked high monsoonal rainfall activity and high sediment yield rates, which he named the "Zeit Wet Phase" in the Gulf of Suez area to mark the contrast to the arid conditions of the preceding Tortonian Stage.

Along the continental margin of the Southeastern Mediterranean off of Sinai the Miocene sediments of the Maviq'im Formation of late Tortonian-Messinian age (Ben Avraham & Mart, 1982, Mart & Ben Gai, 1982) form an evaporitic sequence up to 1400 m (4593 ft) in thickness. Two facies can be recognized in the Formation (Mart & Ben Gai, 1982), a shelf facies composed mainly of evaporites, and a basinal facies made up of gypsum, anhydrite and halite with interbeds of shales, sands and marls. The base and the top of the Formation are marked by the seismic reflectors "N" and "M" reflectors (Ryan et al., 1973), respectively. The "M" reflector marks the top of the evaporites or, where the evaporites are absent, an unconformity between the Maviq'im Formation and the Pliocene-Pleistocene Yafo Formation.

In the Cairo-Suez region Lower Miocene marine fossiliferous, sandstones, gravels, and limestones with oyster and *Pecten* banks, gastropods, pelecypods, corals, echinoids, and bryozoans, were deposited unconformably over basalts or gravels and sands. These sediments are equivalent to the Mamura Formation of the Western Desert (Said (1962a, b, Marzuk, 1970). They were followed during the Middle-Upper Miocene by non-marine fluviatile sediments equivalent to the Marmarica Limestone. These sediments represent an epicontinental Miocene sea facies which bordered the Gulf of Suez basin (Said, 1962a, b). Based on bryozoans, Ziko & El Safori (1998) suggested that the Early-Middle Miocene sediments cropping out on the western side of the Gulf of Suez, was deposited in a shallow marine and low-turbulent environment. Bryozoans indicate a free connection between the Clysmic area (Gulf of Suez) and the northern Western Desert during that time. Whole-rock analysis of the basalts yielded an Early Miocene age (Lotfy et al., 1995). Early Miocene section in the Gebel Gharra section, Eastern Desert of Egypt displays a transgression/regression cycle with siliciclastics at the base and shallow-water carbonates at the top. Echinoid fauna contains a succession of echinoid assemblages which reflect large-scale patterns of environmental changes. The assemblages range from high-energy, shoreface sediments with clypeasteroids and shallow-water carbonates, as well as storm deposits of disarticulated and fragmented echinoids (Kroh & Nebelsick, 2003).

Two distinct episodes of basalt formation occurred. The first is represented by very fine-grained porphyritic olivine-bearing basalts; the other consists of medium to coarse-grained crystalline, non-porphyritic, doleritic basalts. These two groups of basalts show two different paleomagnetic poles. Based on this information Lotfy et al. (1995) argued that a northward movement of Africa took place during the Tertiary followed by a clockwise rotation. This movement was associated with the opening of the Red Sea and the Gulf of Suez and the closing of the Mediterranean. At Gebel Ghara northwest of Suez Burdigalian shallow marine marls and shales of the Gharra Formation (Said, 1971) with *Miogypsina intermedia, Operculina complanata*, and *Heterostegina heterostegina* (Souaya, 1963) were deposited on top of the Abu Zaabal basalts. They were followed by Serravallian-Tortonian limestones, marly limestones, and arenaceous limestones with benthonic foraminifera (Souaya, 1963) of the Genefe and Hommath formations (Said, 1990e), deposited in shallow marine, semi-restricted environments.

Miocene sediments along the Red Sea coast are littoral in character (Said, 1990e). Miocene sediments are represented by the Ranga, Um Mahara, Abu Dabbab, Um Gheig, Samh, and Gabir formations. Polymictic conglomerates derived mainly from the basement with boulders up to 1 m (3 ft) in diameter embedded in a red matrix comprise the Burdigallian Ranga Formation (Samuel & Saleeb-Roufaiel, 1977). These conglomerates were deposited on top of sediments ranging in age from Precambrian in the south to Cretaceous in the north. They were followed upward by unfossiliferous, fine to medium-grained sandstones and shales. The Langhian sandstones of the Um Mahara Formation were deposited unconformably with a basal conglomerate on the Ranga Formation. They were followed by gypsiferous, fossiliferous, partly dolomitized limestones with coral reefs. The Ranga and Um Mahara formations are equivalent to the Gebel El Rusas Formation of El-Akkad & Dardir (1966), and to the Markha and Shagir members of the Kareem Formation in the Gulf of Suez. Serravallian deposition was dominated by gypsum with dolomitic limestones masses of the Abu Dabbab Formation (Samuel & Saleeb-Roufaiel, 1977) equivalent to the Belayim Formation in the Gulf of Suez, followed by dolomites of the Um Gheig

Formation (Samuel & Saleeb-Roufaiel, 1977). The Samh Formation (El-Akkad & Dardir, 1966) marls, sandstones, shales, intraformational conglomerates, and occasional limestone beds with casts of the freshwater species *Melania* (Said, 1990e) were deposited in a fluviatile environment. Deposition of the Middle Miocene sequence in the Safaga-Qoseir region, on the Red Sea coast, started with a succession composed of interfingering limestones and conglomeratic sandstones, with pebbles of flint, metamorphic and igneous rocks at the base, and thick evaporites at the top (Said, 1962a). The overlying succession consists of shales with brackish water fossils and sandstones at the base, and limestones rich in *Metis payracea* and *Ostrea gryphoides* at the top, indicating marine transgression during the upper part of the succession.

Approximately 1400 m (4593 ft) of Miocene sediments on Tiran Island at the southern end of the Gulf of Aqaba were deposited unconformably over Oligocene and Eocene polymicitic conglomerates (Goldberg & Beyth, 1991) which overlie a peneplained surface of Precambrian volcanics. The Early Miocene succession consists of fine-grained sandstones with igneous rock pebbles and a few sandy dolomitic limestone beds (Aquitanian-Burdigalian). They are followed by dolomites, shales, silts, and laminated and cross-bedded sandstones (Burdigalian-Langhian) followed upward by white massive anhydrite and gypsum with lenses of dolomite (Serravallian-Tortonian), a section of dolomitic shales with gypsum and halite veins, intraformational conglomerates, and fine-grained sandstones and conglomerates with igneous rock pebbles (Tortonian), overlain unconformably by dolomitic, organogenic limestones with echinoderms, pelecypods, gastropods, and benthonic foraminifera, such as miliolids. The dolomites and limestones grade into limy dolomites and a few beds of shales and conglomerates with igneous rock pebbles.

In the St. John's (Zabargad) Island, in the Red Sea, Middle-Upper Miocene evaporites composed of nodular, recrystallized gypsum with lenses of selenite and levels of reddish Fe hydroxides (Bonatti et al., 1983) of the Gypsum Valley Formation (El Shazly & Saleeb, 1977) were deposited.

The Miocene sediments in the Gulf of Suez (Figs. 18.6 & 20.24) include the syn-rift Lower Miocene open-marine clastics and Middle-Upper Miocene evaporites (Garfunkel & Bartov, 1977, Orszag-Sperber et al., 1990). The succession reaches up to 3657 m (12,000 ft) in structurally low areas (Abdine, 1981). The Miocene succession is cyclical, reflecting uplift, erosion, and gradual infilling of the intermittently subsiding basin and the development of restricted conditions. Shallow marine reefal carbonates predominate on submarine highs, sands are common in marginal seas, and fine-grained clastics are dominant in the deeper troughs (Hughes et al., 1992). Alluvial fans were present on both margins of the Gulf of Suez, where areas of high relief supplied abundant arkosic sediments from the Precambrian and pre-Miocene rocks (Salah & Alsharhan, 1997).

Highly bioturbated sediments of the Nukhul Formation (Aquitanian-early Burdigalian) form the basal marine Miocene beds in Sinai and the Gulf of Suez region (Winn et al., 2001, Malpas et al., 2005) and represent the initial early rift (proto-rift) sequence. The Formation consists of carbonates, coarse sands, and anhydrites. Facies distribution was controlled by the paleorelief of the basin. Anhydrite is common in the central and southern parts of the region. Reefs and platform carbonates were deposited on submerged paleohighs. The Formation contains reworked pre-Miocene fossils. Deep-marine carbonates, marls, shales, and sandstones of the Rudeis Formation (Burdigalian) were deposited during the main rifting phase. Deposition of the Rudeis Formation was interrupted by the "Mid-Rudeis Event" which resulted in the structural reorganization of the rift (Patton et al., 1994). This event correlates roughly with the renewed movements on the Dead Sea faults. The Upper Rudeis Formation is part of the synrift sequenc and was deposited as turbidites in a submarine-fan system sourced from the western rift shoulder (Pivnik et al., 2003).

Episodic hypersaline conditions during the Middle-Upper Miocene produced alternating fine-grained clastics and evaporites of the Kareem (Langhian-Serravallian) and Belayim (Tortonian) formations. The Kareem Formation includes thin anhydrite beds intercalated with shales and marls of the Markha Member (Rahmi Member). This episode was followed by the deposition of fossiliferous shales and marls with occasional limestone and fine to coarse-grained sandstone beds of the Shagar Member in an upper bathyal to shallow subtidal environment (Hughes et al., 1992). Contemporaneous faulting led to rapid facies change or absence of these sediments. Sandstones of the Kareem Formation were deposited in alluvial and submarine fans building out from the rift shoulders (Salah & Alsharhan, 1997). The carbonates and evaporites were deposited in marginal lagoons.

The main Miocene evaporite cycle started during the Middle Miocene (Serravallian) with deposition of the Baba and Feiran members of the Belayim Formation. These sediments were followed by shales with minor carbonates and sands of the Sidri Member and clastics or carbonates of the Hammam Faraun Member. In the marginal areas of the Gulf of Suez the Belayim Formation is composed predominantly of sandstones while reefal carbonates and oolitic shoals were deposited on submerged paleohighs. Deposition of carbonates predominated in places (the Gemsa Formation). A shallow to deep inner subtidal environment has been suggested for the Belayim Formation (Hughes et al., 1992).

A final phase of hypersalinity produced the thick, massive, halite and anhydrite sequence of the South Gharib and Zeit formations. The South Gharib Formation (late Serravallian-Tortonian) consists of massive halite and anhydrite and minor shales and very fine to fine sandstone beds. Anhydrite is common in the marginal areas, and halite in the depocentres. Restricted hypersaline, moderately deep to shallow-marine environments were suggested for the Formation (Hughes et al., 1992). Tortonian-Messinian deposition was dominated by anhydrites and shales with minor sandstones of the Zeit Formation in a shallow marine environment with alternating restricted and open marine conditions (Hughes et al., 1992). Sandstones predominate in marginal areas. Erosion resulted in the removal of the two formations in some areas.

An example of the Middle Miocene carbonate platforms buried under the Gulf of Suez is exposed along the Red Sea coast at the Esh el Mellaha (Rouchy et al., 1983, Coniglio et al., 1988, James et al., 1988, El-Haddad et al., 1984). These rocks are probably equivalent to the Rudeis Formation (James et al., 1988). Dolomitized platform carbonates of the Kharasa Member were deposited on horst blocks of Precambrian volcanics. High-energy platform margin reefs grade westward into bioclastic carbonates. Fringing reefs and associated reef-front and fore-reef sediments of the Mellaha Member developed on the seaward margin. Stromatolites were deposited at the top of the Mellaha Member. Coral biostromes developed in deep water at west

Gemsa. A period of subaerial exposure and meteoric water diagenesis resulted in karsting on top of the Mellaha Member. Where the Mellaha Member is absent karsting occurred on top of the Kharasa Member. Shallowing-upward sequences of peritidal and stromatolitic carbonates and evaporites of the Bali'h Member were deposited in arid tidal flats bordering the basin. The southern end of the Esh el Mellaha range shed alluvial conglomerates and sands. Coniglio et al. (1988) attributed dolomitization to either mixing-zone dolomitization or to marine to hypersaline fluids in a shallow subtidal to sabkha environment.

In west Central Sinai the Early Miocene Nukhul and Rudeis formations were followed by the Sarbut El-Gamal Formation and the Hammam Faraun Member of the Belayim Formation. The undifferentiated sequence of the Middle Miocene Sarbut El-Gamal Formation (Philip et al., 1997) equivalent to the Mreir Member of the Rudies Formation, Kareem Formation, and the Baba, Sidri, and Feiran members of the Belayim Formation, consists of up to 150 m (492 ft) of polymictic conglomerates composed of chert, limestone, and reefal-algal limestone clasts embedded in a sandy calcareous matrix, and interbedded with limestone beds with large benthonic foraminifera, such as *Borelis melo*, *Heterostegina* sp., and *Operculina* sp. These conglomerates were deposited unconformably on the Rudeis Formation. Polymictic conglomerates, a few hundreds of metres thick, interlayered with shallow marine limestones of the Abu Alaqa Group along the eastern margin of the Gulf of Suez were interpreted as representing the main Middle Miocene rifting phase (Garfunkel & Bartov, 1977). This unconformity was due to the Mid-Clysmic event which started at the beginning of the Middle Miocene and was responsible for the main rifting phase in the Gulf of Suez and the uplifting of Sinai (Garfunkel & Bartov, 1977). The Middle Miocene Hammam Faraun sediments of the Ras Gharib Formation change facies along the western coast of the Gulf of Suez into the Nullipore (algal *Lithothamnion*) fringing reefs. The Nullipore Limestones (Moon & Sadek, 1923) were deposited on a carbonate ramp in the Gulf of Suez and are reserved in fault blocks (Vaughan et al., 2003). They form good reservoir in a number of fields, including Ras Gharib, Ras Fanar, Ras Bakr, and Gemsa fields.

Miocene deposition in Central and East Libya was dominated by sediments of the Regima Formation in the Eastern Sirte Basin, and the Marada, Faidia, and Regima formations in the Cyrenaica Platform area. Deposition of the Marada and Faidia formations continued with no apparent interruption in sedimentation in northeastern Libya. However, the Regima carbonates (Langhian-Serravallian) were deposited unconformably (intra-Miocene) on the Marada and Faidia formations. Lower Miocene (Aquitanian-Burdigalian) limestones, shales, and glauconitic marls with oysters, echinoids, and algae of the Faidia Formation (Barr & Weegar, 1972), in northeastern Libya were deposited in a warm, nearshore, subtidal, and protected marine environment (Francis & Issawi, 1977). Northwards, offshore of Cyrenaica, the Faidia Formation changes into open, shallow-platform wackestones and packstones containing benthonic foraminifera, and fragments of bryozoans, hydrozoans, echinoderms, and the algae *Corallinaceae* (Duronio et al., 1991).

The depositional environments of the Benghazi Member (Langhian) of the Regima Formation (Langhian-Serravallian) were probably subtidal, protected, or lagoonal with oolitic shoals (Francis & Issawi, 1977). Deposition of the Al Sceleidima Member took place in brackish water lake and sabkha environments.

Terrigenous sediments were supplied by streams from the south and southwest. The Benghazi Member was followed by the Wadi Al Qattarah Member (Serravallian) (Barr & Weegar, 1972). Three main facies associations can be recognized in the Wadi al Qattarah Member (El-Hawat, 1980b): Coarsely crystalline gypsum facies, pelletal and mudstone facies, and cross-bedded conglomeratic oolitic grainstone facies. Domal algal stromatolites occur in the pelletal and mudstone facies. These facies represent deposition in storm-formed barrier beaches, tidal channels, tidal flats, shallow ephemeral ponds, and hypersaline pools. The presence of *Borelis melo*, miliolids, pelecypods, and gastropods indicate shallow marine, subtidal to neritic, protected environments of deposition. The Wadi Al Qattarah Member continues northward into the offshore area with wackestones-packstones with foraminifera, bryozoans, pelecypods, gypsum, and gypsiferous limestones deposited in open, shallow-platform environment (Duronio et al., 1991).

Miocene sediments of the Marada Formation in the Ajdabiya Trough reach 2100 m (6890 ft) in thickness in the region of Marsa al Burayqah and they consist mainly of shallow-water limestones. Deposition of sandstones and minor shales, limestones, dolomites, and gypsum of the Qarat Jahannam Member (Aquitanian-Burdigalian) of the Marada Formation took place in an inner to deep shelf environment, including fluviatile and estuarine environments in the south, and lagoonal, offshore bar and barrier beach environments in the north (Domaci, 1985). Ar Rahlah Member (Aquitanian-Serravallian) consists of calcareous sandstones, dolomites, and coquinoid limestones, very rich in benthonic foraminifera, ostracods, invertebrate fossils, and planktonic foraminifera. It changes facies, both laterally and vertically, into bioturbated, calcareous sandstones, and minor conglomerates.

Selley (1968, 1969, 1971) recognized five sedimentary facies, four of which trend parallel to the shoreline and locally are crossed by the fifth. These facies are from north to south, 1) detrital limestones (barrier bar and beaches), 2) laminated shales (lagoon), 3) interlaminated shales and sands (intertidal), 4) cross-bedded sands and shales (fluviatile), and 5) calcareous sandstone channels (estuarine). According to El-Hawat (1980a) the Marada facies can be grouped into three depositional sequences, from base to top, estuarine calcareous sandstones, a lagoonal barrier complex of interbedded cross-bedded grainstones, sandy grainstones and shales, and an open-marine complex of wackestone banks and deltaic marls. Southwards the first two sequences grade into undifferentiated estuarine and tidal flat deposits and lagoonal shales and the third sequence grades into fluviatile sands. Savage (1968, 1971) reported vertebrate fauna from these channels, such as crocodiles, tortoises, and land fauna including proboscideans (Mastodon), rhinocerotids, and giraffids. The mammalian fauna existed at a time of dramatic changes in environment from humid, forested, and wooded to semi-arid and steppes following the closure of the Mediterranean Sea (Pickford, 1991).

In offshore Libya and Tunisia the Miocene succession is bounded by unconformities commonly marked by erosion, incision, and local paleokarsts (Bédir et al., 1996). Supersequence $S_4$ of Bobier et al. (1991) includes sediments accumulated from the late Oligocene to the Present. The sequence starts with transgressive Upper Oligocene to Langhian sedimentary succession and includes the Oued Hammam, Behara, Ain Grab and Mahmoud formations. The Ain Grab Formation (Horizon H2) accumulated during the Langhian and corresponds to the TST and HST cycle 2.3 in the supercycle

TB2 defined by Vail et al. (1987). The Ain Grab Formation forms a transgressive systems tract (S1), composed of sandy limestones and sandy shales with molluscs. It overlaps Oligocene-early Miocene paleohighs and coincides with sea level rise Tejas B2.3 of Haq et al. (1987) (Bédir et al., 1996). The Langhian limestones of the Ain Grab Formation in the offshore area and their equivalents in Egypt and Libya produce a good seismic reflector representing the expansion of the platform setting during that period of high sea level (Bobier et al., 1991). The overlying seismic sequence includes the paralic fill of the Saouaf Formation. Marine sediments developed in the bordering Pelagian Sea and continental sediments accumulated in tectonic traps during the major tectonic phases, which took place during the Tortonian to Villafranchian. Hydrocarbon reservoirs in the Miocene of the Gulf of Hammamat and the Pelagian Block are principally the Langhian-Serravallian Ain Grab limestones and Beglia and Birsa sandstones. The Lower Miocene in the immediate offshore area of Northwest Libya was deposited on a shallow shelf (Fucek et al., 1998, Kucenjak et al., 1998). The sediments contain planktonic foraminifera, such as *Globoquadrina dehinscens, Globigerinoides primordius, G. altiaperturus* and larger foraminifera, such as *Operculina*, and small benthonic foraminifera, and fragments of echinoids, bryozoans, and coralline algae.

In the offshore of Northwest Libya the Miocene is represented by a shallowing-upward regressive sequence, which also shows a regressive trend from west to east (Doeven & Abdullah, 1995). During the Late Oligocene-Lower Miocene, deposition was dominated by the deep-water facies of the Ras Abd Jalil Formation composed of marls, sandy shales, argillaceous wackestones-mudstones, and glauconitic and pyritic shales with *Globorotalia opima* and *Globigerina ampliapertura* (Hammuda et al., 1985). This facies change into deep-water biostromal carbonate facies of the Dirbal Formation in the B1A-137 and A1-NC41 wells. These sediments are composed of wackestones-packstones and grainstones-boundstones interbedded with dolomites with *Operculina lepidocyclina* and *Nummulites* sp. The Mediterranean margin of Libya can be divided into three physiographic provinces, the Pelagian Shelf, the Sirte Embayment, and Offshore Cyrenaica. Cenozoic sediments along the Libyan margin display progradational character punctuated by erosional surfaces (Fiduk, 2009). In the Sirte Embayment a retrogradational surface coincides with the top of the Messinian unconformity on seismic sections (Fiduk, 2009). The retrogradation has been attributed to margin erosion during the Messinian Salinity Crisis or a cataclysmic failure of the Libyan margin during the late Miocene.

During the Late Langhian-Early Serravallian shallow marine, alternating shales, sandstones, and carbonates with benthonic foraminifera of the Al Maya Formation were deposited. These sediments grade into the Sidi Bannour Formation shales and minor sandstones with miliolids, *Ammonia beccardai, Rotalia* sp., *Lithothamnium*, ostracods, gastropods, bryozoans, and echinoderms deposited in shallow open to semi-restricted marine environment. The Sidi Bannour Formation equivalent to the Mahmoud Formation in offshore of Tunisia was deposited during a widespread marine transgression that took place during the Late Langhian-Early Serravallian. The two formations were followed during the Late Serravallian-Early Tortonian by the deep-water facies of the Bir Sharuf Formation composed of biomicrites, biosparites, calcareous shales, and sandstones, and evaporites at the top. The Tubtah Formation (Tortonian) is made up of poorly fossiliferous calcareous shales with ostracods, miliolids, Elphidiiae, and *Ammonia beccarii*, grading upward into mudstones-packstones. These were deposited in a shallow, semi-restricted marine environment on top of the Bir Sharuf Formation. Locally on the southern edge of the Jarrafa Trough, possible reefs developed during late Miocene time (Blanpied & Bellaiche, 1983).

The Late Miocene carbonate and evaporitic sequence of the Marsa Zouaghah Formation (Messinian) was deposited unconformably on the Sidi Bannour, Tubta, or Al Maya formations. Deposition started with dolomitic mudstones and wackestones, followed by alternating gypsum, shales, and marls. These sediments are probably equivalent to the Oued bel Khedim Formation in the offshore of northeastern Tunisia. The Messinian Oued bel Khedim Formation includes facies of bedded anhydrite, freshwater limestones, silty marls, and diatomites. Microfossils from the marls yield shallow-water benthonic foraminifera, fragmented ostracods, and reworked planktonic foraminifera (Wiman, 1980).

In Northwest Libya, the Middle Miocene (Langhian-Serravallian) sea transgressed over an eroded surface. Carbonate deposition returned with the Al Khums Formation in the Jifarah Plains and in the western Sirte Basin. In the central Jifarah Plains Middle Miocene polygenetic conglomerates overlie the Mesozoic rocks (Kruesman & Floegel, 1980). The Upper Miocene is absent. Limestones, chalks, and minor sandstones with foraminifera, oysters, pelecypods, echinoderms, and coralline algae of the Al Khums Formation were deposited over an irregular Cretaceous surface, under shallow water marine conditions. Biohermal reef patches formed near the shoreline (Hladil et al., 1991).

Southeast of the coast of the Gulf of Sirte, the Late Tortonian-Early Messinian Sahabi Formation (Pliocene according to Boaz, 1996) is made up of transgressive, nearshore bioturbated marine sands (Boaz, 1996) and lagoonal, coastal bar, and evaporitic deposits (Giglia, 1984) which overlie Middle Miocene siliceous limestones of the Msus Member of the Regima Formation. The Sabkha Al Hamra Member consists of shallow marine fossiliferous sandy limestones and calcareous sandstones, grading upward into evaporitic facies of sands, green clays, siltstones, gypsum, and sandy limestones with paleosoil horizons. These sediments contain mixed brackish water and shallow marine fauna. The overlying Sabkha Al Qunnayyin Member consists of a transgressive facies of shallow marine, protected environment limestones, calcareous sandstones, siltstones, and clays. These sediments interfinger with the Wadi Al Farigh Member (early Messinian) composed of high-energy lime-sand shoals and oolitic bars.

The Upper Miocene (Tortonian-Messinian) was a period of erosion over most of North Africa. The Messinian was a time of major global change in climate, evolution, and oceanic reorganization. The passageway (Strait of Gibraltar) between the Atlantic and the Mediterranean became constricted as a result of the relative movement of the African and European plates. This restriction led to isolation of the Mediterranean from the global oceans during the Messinian, evaporation of seawater, and the desiccation of the basin. The effect of this widespread regression is the Messinian Salinity Crisis (Nestroff et al., 1972, Ryan et al., 1973, Hsü, 1973, Cita, 1980, Hilgen et al., 2000, Roveri & Manzi, 2006, Roveri et al., 2006, Rouchy et al., 2007). During that period, deposition was dominated by the Marsa Zouaghah evaporites in the offshore of Libya, by sandstones and shales in the Jifarah Plains, and by the Rosetta evaporites in Northern Egypt. This Messinian crisis was terminated with restoration of open-marine connections with the Atlantic Ocean through the Straits of Gibraltar, marking the beginning of the Pliocene (Zanclean) (Wiman, 1980).

Marine Messinian deposits in Tunisia occur between the northern Bizerte and Cap Bon areas and the central-eastern Sahel region (Moissette et al., 2010). The lower Messinian deposits are dominated by siliciclastics and oolitic-bioclastic limestones deposited on an eastward facing ramp. The ramp changes westward into coastal lagoons with occasional evaporites and eastward into a reefal Pelagian Platform facies. The early Messinian succession starts with siliciclastics followed by oolitic-bioclastic carbonates of the Beni Khiar Formation and lower Oued bel Khedim Formation. The late Messinian consists of brackish to continental lagoonal sediments of the Oued el Bir Formation (probably equivalent to the Melqart Formation) and upper Oued bel Khedim Formation (Moissette et al., 2010).

The origin of the evaporites in the Mediterranean Sea and the Gulf of Suez is controversial (Said & El-Heiny, 1967, Friedman, 1972, Ryan et al., 1973, Cita, 1982, Sonnenfeld, 1975, Fabricius, 1984, Richardson et al., 1988, Hardie & Lowenstein, 2004, Roveri & Manzi, 2006, Roveri et al., 2006). One of the most puzzling problems is the occurrence of thick Miocene evaporites interbedded with deep-water sediments of Messinian age in the Mediterranean Sea, and Langhian-Messinian age in the Gulf of Suez.

Said & El-Heiny (1967) and Said (1962a) suggested that the evaporites in the Gulf of Suez were deposited in lagoons separated by phases of open marine conditions in which the intra-evaporite marls were deposited. Friedman (1972) agrees with Heybroek (1965) that no evaporites accumulate under deep-water conditions. Limestones formed under hypersaline conditions persist in deep basins. Sulphates cannot survive in deep-basin environments, and they are converted to calcite. Evaporitic sequences formed on sea-marginal flats (sabkhas), marginal to the Red Sea. Their presence in deeper parts of the basin is attributed to later subsidence. Based on the study of evaporite textures from coastal outcrops and subsurface cores, Richardson et al. (1988) suggested deposition in a shallow subaqueous hypersaline environment. Evaporites commonly interfinger with carbonates and clastics along the basin margins. Periodic influx of freshwater probably led to the formation of reefal carbonate deposits within the predominantly evaporitic series. Evaporites in coastal outcrops were probably deposited in a shallow subaqueous hypersaline environment (Richardson et al., 1988). Evaporites of the Markha Member of the Kareem Formation are interpreted as outer neritic sediments, based on the presence of planktonic and benthonic foraminifera in the intra-evaporite facies (Hughes et al., 1992).

During Leg 13 of the DSDP of the Glomar Challenger in the Mediterranean, three units were recognized in the Neogene-Quaternary section (Nesteroff et al., 1972): 1) A lower unit of poorly fossiliferous marls of Langhian, Serravallian, and Tortonian ages. 2) An evaporitic series of halite, anhydrite, gypsum, and dolomitic marls of Messinian age. Seismic reflection surveys suggest a thickness up to 3–4 km for these evaporitic series. Evaporites are very thick in the center of the basin and thin toward the margins. Brackish and lagoonal fauna, stromatolitic structures, and isotopic analysis suggest that the evaporites were deposited in shallow water. 3) An upper unit of hemipelagic marl oozes of Pliocene-Quaternary age.

The origin of shallow-water evaporites encased within deep-marine sediments in the Mediterranean Basin has been explained by one of three models, brine reflux in an inland basin (salina model), post-Messinian foundering of an originally shallow basin (sabkha model), or deep-basin desiccation (evaporative drawdown model). The first model was rejected by Cita (1982) because of the shallow facies of the anhydrite and the second because the Mediterranean is made up of a number of basins with different ages and tectonic histories. The deep-basin desiccation model suggests that the Mediterranean was deep before the onset of evaporitic conditions, that it was very shallow during the onset of evaporitic conditions, but was deep again in the Pliocene immediately after the termination of evaporitic conditions (Ryan et al., 1973, Cita, 1982).

According to Fabricius (1984) on the other hand, the facies distribution of the Upper Miocene deposits in the Mediterranean indicates shallow-water sedimentation within a shallow basin setting. The thickness of Messinian evaporites,  more than 1000 m (3280 ft) in the Ionian Sea, suggests basinal subsidence at an average rate of 1.0 m/1000 yrs. Basin subsidence was compensated by the deposition of evaporites, accompanied by episodic uplift of marginal areas. During the Pliocene-Quaternary time, there was little compensation of basin subsidence by sediment in-fill resulting in a starved basin.

Norman & Chase (1986) argued that the removal of the Mediterranean seawater would have led to a regionally compensated crustal warping, which would have caused uplift of the basin and shoreline bulges along its coast. They estimated a maximum uplift at the centre of the basin of 1.9 km (1.2 miles), and about 0.45 km (0.3 miles) at the shelf edge. These uplifts would have caused a reversal of the drainage pattern, but a competent river such as the Nile, would have excavated through the bulge as it formed.

Clauzon et al. (1995, 1996) proposed a two-stage process for the formation of the Mediterranean Miocene evaporites: First, marginal evaporites were deposited beginning at 5.80 Ma. This phase corresponds to a phase of global sea level fall, caused by the Antarctic glaciation. Second, a major and rapid tectonically induced sea level fall, which led to the desiccation of the basin, and the formation of deep-basin evaporites and strong erosion at the margins around 5.60–5.32 Ma.

Many factors probably interacted to lead to the deposition of evaporites in deep marine basins, such as the Gulf of Suez and the Mediterranean Sea. Rouchy (1986) drew the attention to the complex processes, which led to the formation of these evaporites. These processes include a balance between the influx of oceanic water, tectonics, eustatic sea level changes, climate, and geomorphology of the basin.

### 20.5.3   Pliocene period

The rapid sea-level fall during the Messinian led to widespread evaporitic conditions, karsting in carbonate terrains, and formation of incised valleys, depressions, and duricrusts on land. Substantial subaerial denudation of the North African continental margin took place during the Late Miocene salinity crisis. The Pliocene transgression was abrupt, with deposition of deep-sea sediments immediately after the flooding of the Mediterranean (Cita, 1982, Benson, 1972). The boundary between the Miocene and Pliocene, known as the "M" marker ("S" marker in the Red Sea), in the Mediterranean Basin is always sharp and erosional (Nestroff et al., 1972, Rizzini et al., 1978). The "M" marker beneath the continental shelf of Israel was interpreted as an erosional and incised surface on top of the Messinian with local relief of several hundreds of meters (Ryan, 1978). The basal Messinian unconformity ("M" marker) determined by seismic analysis and borehole correlation extends from sea level in the southern Delta area to a depth of at least 5 km in the Nile cone (Barber, 1981). During the Upper Miocene and Lower Pliocene the Mediterranean was subject to environmental cyclicity. This cyclicity was probably due to chemical variations in water composition, as well as variations in water-depth (Schreiber & Ryan, 1995) and climate. Approximately 10 depositional cycles are known from the subsurface of the Mediterranean coast of Egypt and from outcrops in Tunisia. The climate in the Mediterranean during the Early Pliocene alternated between relatively wet and dry intervals. The Early Pliocene is recognized as a very warm interval, as well as a period of glaciation. Recurring warm periods resulted in a very unstable ice sheet leading to high frequency changes in sea level (Burckle, 1995). These fluctuations in climate and sea level resulted in regular alternations between carbonate-rich and carbonate-poor marls with different oxygen and carbon isotope signatures (Gudjonsson, 1987).

An intense debate rose in the middle of the 1990's on how the Messinian Salinity Crisis ended and how marine conditions were restored at the onset of Pliocene in the Mediterranean basins (Rouchy et al., 2007). This debate examined whether the late Messinian conditions changed to lacustrine settings before the abrupt marine reflooding in the earliest Zanclean, or whether marine conditions were restored earlier, either episodically or definitely, during the upper Messinian. The Messinian Salinity Crisis is believed by some to have ended by generalized low salinity conditions, the so-called Lago-Mare that was suddenly interrupted in the earliest Zanclean by a fast marine transgression (Rouchy et al., 2007). How did the Messinian Salinity Crisis end is a matter of intense debate between two opposite concepts i.e., the generalized dilution event, the so-called Lago-Mare, followed by the sudden restoration of the marine conditions at the base of the Zanclean, or the early partial or complete marine refill that would have happened earlier during the upper Messinian. The Chelif Basin of Northwestern Algeria, one of the Messinian marginal basins of the Mediterranean, provides a record of the Miocene-Pliocene boundary. The late Messinian deposits exhibit a great lithological variability with predominant clastic deposits (sandstones, siltstones, conglomerates, carbonates); as well as large slide masses of lower Messinian (Tripoli Unit) in some marginal areas. The Messinian-Pliocene transition was marked by an abrupt change of the $\delta^{18}O$ values of carbonates from negative values characteristic of low salinity in the upper Messinian to higher values in the Zanclean marine conditions (Rouchy et al., 2007). Marginal areas were dominated by conglomerates and erosional features. The Messinian fall in sea level enhanced the incision of the major rivers in the Mediterranean region, such as the Nile River in Egypt and the Rhone River in southern France (Cita, 1982, and references herein). The Eo-Nile excavated a canyon about 1300 km long with a maximum depth of about 2500 m (8202 ft) (Harms & Wray, 1990). This event resulted in a valley that was filled by Pliocene estuarine sediments (Ryan, 1978, Said, 1981) and the development of a dendritic drainage system on the former Tortonian delta slope (Barber, 1981).

Late Tortonian-Messinian littoral carbonate complexes in Algeria and Morocco display two major biological sedimentary cycles and two regional surfaces (Cornée et al., 2004). The first represents a maximum flooding surface during around 6.7 Ma; the other is a regional marine paniplanation surface at around 5.95 Ma. Cornée et al. (2004) believe that the onset of the Messinian Salinity Crisis appears to be related to major environmental paleo-oceanographic changes in the Mediterranean, rather than to a major sea-level drop or to climatic change.

The Messinian salinity crisis was not probably related to drop in sea level around the around 7–5.6 Ma time-span, but to restriction of the Mediterranean. This idea is also supported by the evaporites of marginal basins which were coeval with littoral shallow water deposits including coral reefs Cornée et al. (2004).

During the Early Pliocene shallow marine and continental sands were deposited southeast of the shore of the Gulf of Sirte, as far south as the Al Jaghbub Oasis (Carmignani et al., 1990). The Nile Valley canyon was transformed into a narrow gulf reaching as far south as Aswan (Desio, 1971, Said, 1962a, Said, 1981). The Gulf of Suez rift opened to the Indian Ocean and connection with the Mediterranean ceased. Opening of the Bab el-Mandab Strait caused the Indo-Pacific fauna to invade the Red Sea.

In the Late Pliocene the Nile Gulf was converted to a river (the Paleo-Nile phase of Said, 1981). The Delta started to fan-out as far west as Wadi Natrun where mixed fluviomarine fauna are found. By the end of the Paleo-Nile phase the canyon was completely filled up and the delta surface became flat and sloping to the north. Sediments of the Paleo-Nile are dominated by fine sediments with montmorillonite and organic matter. Sediments were derived from local sources within Egypt such as the highlands in the Eastern Desert.

The Paleo-Nile was generated during the Upper Pliocene and includes deposits of the Gar el Muluk, Helwan, and Madamud formations in the Nile Valley, and Kafr el Sheikh Formation in the Nile Delta region.

On the periphery of the Nile canyon and in the Nile Valley the Lower Pliocene marine sedimentation started with coquinoid limestones, marls, and fossiliferous sandstones with *Pecten benedictus*, *Clypeaster aegyptiacus*, *Chlamys scabrella*, and *Strombus coronaturs*, in addition to a rich foraminifera and ostracod fauna (Said, 1981, Bassiouni, 1965) in the lower part of the Kom el Shelul Formation (Said, 1971) followed by non-fossiliferous sandstones with flint pebbles in the upper part. The fauna is of a Mediterranean affinity. They were deposited unconformably on top of the Late Eocene Maadi Formation. Fluvio-marine sediments of the earliest deltas of the Paleo-Nile are represented by the Late Pliocene Gar El Muluk Formation (Said, 1981, Blanckenhorn, 1902b), composed of black carbonaceous shales with plant fragments and *Ostrea cucullata* and fish bones, calcareous, cross-bedded sandstones with ostracods and remains of *Hipparion*, *Hippopotomus* sp., and crocodiles. South of Cairo the Upper Pliocene Paleo-Nile fluvial sediments (Said, 1981) include sands, sandstones, shales and a few marl interbeds with *Melanopsis* sp. of the Helwan Formation (Said, 1971). In Wadi Qena the Pliocene is made up mainly of gravels at the edge of the valley and grade into shales and marls at the center. In the upper parts of the Nile canyon Pliocene sediments grade into sands and clays with thin lenses of fine polymictic sands and sandy loams rich in authigenic minerals, such as glauconite, pyrite and siderite (Said, 1990f). On the east bank of the Nile in the Qena Province, the Paleo-Nile rhythmically banded fine sands and silts of the Madamud Formation (Said, 1981) abut against the sides of many of the wadis draining into the Nile from the Eastern Desert.

The Quaternary subsurface section of the Nile Delta is up to 1,000 m (3280 ft) thick and has been subdivided into the Bilqas Formation and Mit Ghamr Formation (Rizzini et al., 1978). The Bilqas Formation (Holocene) extends over the entire delta and its marginal zones. This formation has a maximum thickness of 77 m (253 ft) and is made up of alternating sand, silt, and clays, which may include peat layers. The Mit Ghamr Formation (Pleistocene) attains a maximum thickness of about 970 m (3183 ft); represented mainly by coarse sands with minor gravel and clay interbeds. During the Early Pliocene, marine sediments were deposited in the Nile canyon as the sea level rose after the reopening of the Gibraltar straits. These sediments were partly eroded during the Early Pleistocene. In the Middle Pleistocene, a vigorous river, called the Pre-Nile (Said, 1981), deposited thick layers of course to medium sands, locally more than 1,000 m (3280 ft) in thickness. The Pre-Nile sediments were partly eroded during the Late Pleistocene and during the times of the Pre-Nile/Neo-Nile and the Neo-Nile. The deeper Nile Delta distributaries and the flood plain appear to have been filled up with graded sand and gravel at this time. During the Holocene (since 10,000 BP), the Neo-Nile river was divided into a number of branches that shifted course through time, gradually building up a delta plain. The sediments of the Neo-Nile consist of fine sand, silt, and clay. Westward tilting led to the shifting of the main Nile branches to the west.

The Pliocene succession in the offshore of the Nile Delta is characterized by progradation, slumping and faulting (Kilenyi et al., 1988). The Lower Pliocene sediments are separated from the Miocene succession by a period of uplift and erosion (Rizzini et al., 1978). An angular unconformity separates the Tortonian Qawasim Formation from the Abu Madi Formation. The Pliocene-Quaternary succession forms a shallowing-up cycle (Rizzini et al., 1978) composed of the Lower Pliocene sands of the Abu Madi Formation, open-marine shales of the Kafr el Sheikh Formation, and Pleistocene fluvio-deltaic sediments of the El Wastani and Mit Ghamr formations. The cycle was followed by the present-day coastal and lagoonal sediments. Thick, rippled and bioturbated sands, rare conglomeratic sands, and clay layers of the Early Pliocene Abu Madi Formation (Rizzini et al., 1978) were deposited contemporaneously with the Kom El Shelul Formation. Foraminifera belong to the *Sphaeroidinellopsis* sp. Zone which is typical of the Mediterranean fauna. The basal Lower Pliocene Abu Madi channel sandstones (late Messinin according to Dalla et al., 1997) form the main reservoir in the offshore Abu Qir Gas Field northeast of Alexandria and in the onshore Abu Madi Field west of Port Said in the Delta region. In the Abu Qir Field gas is trapped in fault bounded structures sealed by shales of the Pliocene Kafr el Sheikh Formation. Sandstones of the Sidi Salem and Moghra formations provide secondary reservoirs. These sandstones are interpreted as gravity flow deposits whose distribution is controlled by syn-depositional tectonics (Ince et al., 1988). The Sidi Salem sandstones were followed by soft clays with some interbeds of sands of Lower-Middle Pliocene Kafr el Sheikh Formation (Rizzini et al., 1978), deposited unconformably on Middle Miocene limestones. The clays and shales contain abundant planktonic foraminifera (Omara & Ouda, 1972) of the Lower and Middle Pliocene *Globorotalia puncticulata, G. margaritae, G. aemiliana,* and *G. crassaformis* zones" (Rizzini et al., 1978). The development of Upper Pliocene channels in the Western Nile Delta started initially with the formation of submarine valleys, up to 6 km wide and 200 m (656 ft) deep, that were cut by slumping and submarine erosion during a sea level lowstand (Samuel et al., 2003, Sharp & Samuel., 2004). These valleys were completely filled with debris flow deposits, sands, and shales, and then reincised during similar processes. Amalgamated sheet units occur both at the base and top of the channel systems. In the offshore of the Nile Delta, they form hydrocarbon reservoirs in the King, El Bahig, Al

Max, Abu Qir, Abu Sir, Scarab-Saffron, and Simian fields. Hydrocarbons also occur in structural and combination traps in deltaic and turbidite sediments (Dolson et al., 2001, Abdel Aal et al., 2000, Samuel et al., 2003, Sharp & Samuel, 2004).

In the Qattara Depression a basal conglomerate composed of cobbles and pebbles of the Marmarica Limestone with scattered quartz pebbles and some oyster banks of the Kalakh Formation (Pliocene-Pleistocene) was deposited unconformably on the Middle Miocene Marmarica Formation (Albritton et al., 1990). In the northern part of the Farafra Oasis in the Ain el Wadi Depression, cross-bedded fine-grained sandstones, siltstones, silty shales and shales, capped by a thin allochthonous, fossiliferous limestone bed of the Ain Wadi Formation (Youssef & Abdel Aziz, 1971) fill ancient depressions. A mixed Danian-Landenian planktonic foraminiferal assemblage suggests that these beds are reworked Paleocene rocks.

Along the coast of the Northern Western Desert the Pliocene is represented by a shallow marine, pink dolomitic limestone sequence with a few gypsum beds of the Hagif Formation (Geological Map of Egypt, 1987) with a few benthonic foraminifera and abundant stromatolitic calcareous algae, which was deposited unconformably on the Middle Miocene Marmarica Formation. At Salum the Formation grades into sandy limestones (Mansour et al., 1969) and into open-marine Pliocene clays and shales with abundant planktonic foraminifers (Omara & Ouda, 1972) in the Burg el Arab area.

In Northern Sinai the Pliocene Saqia Marls (Said, 1962a) was deposited in a continental to littoral environment (Jenkins, 1990), unconformably over Upper Eocene rocks Said (1971). Said (1990d) correlated it with the Hamzi and Helwan formations. In the continental margin of the southeast Mediterranean off Sinai up to 2000 m (6562 ft) of the Pliocene-Pleistocene marls, shales and sandstones of the Yafo Formation (Ben Avraham & Mart, 1982, Mart & Ben Gai, 1982) derived predominantly from the Nile River and some wadis in Northern Sinai form a continental prism that lies unconformably on the Tortonian-Messinian Maviq'im Formation along the continental margin of the Southeast Mediterranean.

In the Cairo-Suez district, Late Pliocene fluviatile deposits of the Hagul Formation contain rolled fragments of silicified tree trunks (Said, 1990e). The Formation grades at Wadi Hagul into sandy limestones with flint of the Hamzi Formation. These sediments were followed unconformably by a sequence of porcellaneous limestones with the freshwater snail *Pirenella conica* suggesting deposition in freshwater lagoons, and numerous beds of flint (Said, 1971). The base is marked by a conglomerate consisting of pink limestone fragments.

Along the Red Sea coastal area in the Safaga-Qoseir area Pliocene coarse-grained sandstones, grits, and reefal limestones and sandstones of the Gabir Formation (El-Akkad & Dardir, 1966) containing the Indo-Pacific forms *Ostrea cucullata* and the echinoids *Clypeaster scutiformis* and *Laganum depressum* (Said, 1962a) were deposited unconformably on the Samh Formation. The overlying Ras Shagra Formation consists of sandstones, gravels, and marls, with reefal limestones near the top (Said, 1990e).

On the Tiran Island the Pliocene-Recent tectonic uplift led to the formation of raised marine terraces which extend from sea level up to 500 m (1640 ft) above sea level (Goldberg & Beyth, 1991). They are built of detrital limestones similar to the recent fringing reefs. In St. John's (Zabargad) Island, the Pleistocene fossiliferous biocalcarenites, reef limestones, and sandstones of the Old Reef Limestones

(Bonatti et al., 1983) were deposited directly on peridotites of the igneous basement in back reef, lagoonal, reef flat and reef edge environments. The limestones have been subjected to post-tectonics which resulted in tilting, faulting and folding, and north-south microfracturing. These sediments were followed by conglomerates and breccias, composed of large clasts of the older formations. These conglomerates were followed in turn by the Young Reef Limestones of typical coastal reefs with lagoonal, reef-flat, and reef-slope deposits. They are believed to have been deposited during the last interglacial.

The Pliocene sedimentation near the eastern edge of the Fayium Depression, started with channeled and sheet-like debris flow deposits of the Seila Formation (Hamblin, 1987), composed of subangular blocks of Middle and Upper Eocene calcareous sandstones, set in a muddy matrix. They were deposited unconformably on Eocene sediments. Unfossiliferous, fine-grained, well-sorted sands and muds draped over the breccia. The diatomites in the Seila Formation are dominated by freshwater assemblages that suggest that the taxa were introduced to the area with sediments of the Paleo-Nile River during the Late Pliocene (Zalat, 1995a, b).

The Qattara Depression was excavated in sedimentary rocks of Tertiary age and mantled by saline crusts, sabkha deposits, and salt marshes of Quaternary age (Albritton et al., 1990). The origin of the depression is still in dispute. Albritton et al. (1990, 1991) believe it originated as a stream valley (the Gilf River, or the Radar River) (Burke & Wells, 1989) draining into the Mediterranean that was subsequently dismembered by karsting processes during the late Miocene and it was deepened afterward by deflation and mass wasting. On the other hand, Gindy (1991) stressed the role of tectonism in the formation of the Qattara Depression, as well as other depressions, such as the Siwa and Kharga. The Siwa and Qattara depressions were initiated by tectonically controlled surface depressions. Tectonism and subsequent water erosion are primarily responsible for the Siwa Depression (Gindy & El-Askary, 1969). It is possible that erosional processes during the Messinian sea-level fall followed structural trends and led to the formation of these features.

In the Jabal Zaltan area and northeastern Cyrenaica, the Pliocene-Pleistocene (Domaci, 1985, Giglia, 1984) Garet Uedda Formation (the Calanscio Formation of Benfield & Wright, 1980) composed of eolian quartz sands, calcareous sandstones and beds of sandy shales, locally gypsiferous and which contain brackish-water pelecypods and gastropods, foraminifera, and ostracods was deposited. These sediments change facies eastward into marine or coastal-sabkha, cross-bedded, calcareous sandstones, and sandy oolitic and bioclastic limestones (Giglia, 1984). The Garet Uedda Formation represents interfingering facies of continental alluvial channel sandstones, shallow ephemeral lake limestones, and tidal-flat and sabkha limestones (Selley, 1969, Benfield & Wright, 1980, El-Hawat, 1980a).

The drop in sea level during the Messinian led to the incision of the Sahabi Channel (115) in Northeast Libya to a depth of more than 450 m (1476 ft) (Barr & Walker, 1973) cut into Middle Miocene carbonates of the Sahabi and Marada formations (Domaci, 1985, Giglia, 1985, Benfield & Wright, 1980). The channel is filled with unconsolidated quartz sands, silts, and clays of the Garet Uedda Formation (Pliocene-Pleistocene) (Barr & Walker, 1973). In the Jabal Zaltan area, sedimentation of the Garet Uedda Formation took place mostly in a fluviatile environment, rich in mammals, crocodiles and silicified trees. The sands were derived from the Marada Formation and duricrusts to the south and southwest (Domaci, 1985). In the Jabal Zaltan area, the Marada Formation is locally overlain by a 10 m (33 ft) duricrust made up of limestones, dolomites, and calcareous sandstones with veinlets and concretions of chert and includes altered marine sediments of the Marada Formation. This duricrust was probably formed in the Upper Miocene-Lower Pliocene time (Domaci, 1985).

A major drainage system developed ca.7.5–4.6 Ma in Northeastern Chad and adjacent areas of Libya. It started during the Messinian and early Pliocene with the formation of a large lake in the Chad Basin (Neogene Lake Chad of Griffin, 1999, 2002, 2006). It fluctuated in size in response to the precessional cycle and at times overflowed to the east, NE or north, giving rise to the Sahabi River. The Eosahabi River (late Messinian) (Fig. 20.25) flowed during the drawdown of the Mediterranean. Its paths may be represented by some part of the Radar Rivers (McHugh et al., 1988, McCauley et al., 1998). McCauley et al. (1982, 1986, 1996) identified three kinds of "Radar River" drainage channels (revealed by space shuttle imaging radar) buried at a depth of several meters, beneath the windblown sand deposits of the Arba'in Desert of the eastern Sahara in Upper Egypt and Northwestern Sudan. They are distinguished by their radar reflectance patterns and by characteristics of channel and valley forms. They proposed a westward-flowing, mid-Tertiary trans-African drainage system (TADS) that developed during the late Eocene and Oligocene from headwaters in the Red Sea Hills of eastern Egypt and Sudan. They postulated that a 4,500 km river system became integrated during the late Eocene, achieved maximum development during the early Miocene, and was disrupted by a late Miocene occurrence of intraplate doming and volcanism. McCauley et al. (1982, 1986, 1996) believed that this river system was independent of the evolution of the Nile. Landsat imagery shows ancient drainage networks, extending northward under the Sand Sea from the Gilf Kebir highlands (Haynes, 1982). The presence of a south-flowing river system implies former steeper gradients to the south and west than to the north in these areas (McCauley et al., 1986). Later during the Quaternary pluvial, parts of these valleys were reused by intermittent and then by ephemeral streams. By Holocene time, only scattered play as remained. The Quaternary eolian activities produced an extensive sand-sheet surface that obscures the underlying geologic and topographic features. Bown et al. (1982), based on the study of the Oligocene Jebel Qatrani Formation concluded that there is no evidence for the existence of a Tertiary "Proto-Nile". Bown & Kraus (1988) showed that deposition of the fluvial sediments of the Jebel Qatrani Formation was by meandering streams originating from the Red Sea Range. Their conclusion was supported by dominantly westward paleocurrent directions and the presence of diagnostic clasts of felsic feldspars and chert grains. Burke & Welles (1989), on the other hand, argued that the radar river valleys may have been parts of an old, north-flowing, Nile River system during the late Paleogene and early Neogene, rather than courses cut by a postulated westward-flowing, trans-African drainage system. According to their interpretation, the north-flowing river would have acted as a barrier to the westward drainage from the Red Sea Range across the Proto-Nile.

The Late Miocene-Early Pliocene was a time of widespread tectonic activity in the Pelagian Sea (Burollet, 1978, Wiman, 1980, Jongsman et al., 1985). Evaporitic sediments of the Messinian Marsa Zouaghah Formation were followed unconformably by the Early Pliocene Assabria and the Late Pliocene Sbabil formations (Hammuda et al., 1985). Fossiliferous sands and limestones of the Early Pliocene Assabria

Formation with benthonic foraminifera, gastropods, pelecypods, bryozoans, and ostracods, were deposited in a shallow marine environment. They grade in the offshore of Tunisia into middle to lower shelf unfossiliferous marls with sandy conglomerates at the base, of the Raf Raf Formation (Fournié, 1978). The Late Pliocene Sbabil Formation composed of more than 550 m (1805 ft) of grey to green fossiliferous shale and sandstone with *Globorotalia inflata*, *G. crassformis*, *Bulimina* sp., and *Nonion boueanum* was deposited unconformably in basinal areas on the Assabria, Marsa Zouaghah, or Sidi Bannour formations. They grade eastward, in the offshore of Tunisia, into sandstones with pictinids and oysters of the Porto Farina Formation (Fournié, 1978).

## 20.5.4   Quaternary (Pleistocene-Holocene)

The Pleistocene is marked by dramatic changes and tectonic activities associated with the Red Sea and Mediterranean rifting Said (1981). A world-wide lowering of sea level occurred between ca. 20 ka and 18 ka, followed by a rapid rise to 7 ka, and then by a more gradual rise to the present (Stanley & Warne, 1993).

Paleogeographic and climatic changes in the Eastern Mediterranean during the Neogene/Quaternary led to extended mammalian migrations and faunal exchanges between Eurasia and Africa (Koufos et al., 2005). Early Pliocene migrations were mainly controlled by climatic changes. Africa and Eurasia were connected after a long separation and a great number of faunal elements moved between Africa and Eurasia. During the Pleistocene, oscillation of glacial and inter-glacial periods caused an alternation of cold-steppe faunas with temperate ones in the Eastern Mediterranean. Endemic late Pleistocene mammalian faunas developed in the Mediterranean islands after their isolation, such as the dwarf elephants, cervids, and hippos occurred, as well as giant rodents (Koufos et al., 2005).

The most complete record of the Quaternary in Egypt is in the Nile trough (Said, 1983), in the other areas, the record is either thin or incomplete. The climate during the Quaternary in Egypt was arid in the north and wet in the south (Said, 1983). Alternating humid and dry periods took place in Libya and Egypt during the Quaternary. Pluvials correspond approximately to the four classical glacial phases on the alpine region: Günz, Mindel, Riss, and Würm. The Nile receives sediments from three major tributaries: the White Nile, Blue Nile, and Atbara River. Differences in sediment load record seasonal variations in rainfall and vegetal cover in the various drainage basins (Foucault & Stanley, 1989). Differences in composition of the three tributaries during the late Pleistocene to Recent are due to markedly different geological formations eroded in the Ethiopian highlands and central African plateau. Variations of the relative proportions of amphibole and pyroxene were attributed to time-related changes in the relative sediment contributions from the Nile tributaries and climatic oscillations over the Ethiopian plateau. The Blue Nile and Atbara are primary sources of pyroxene. The decreased proportions of pyroxene in sediments of the Main Nile and in the Delta would indicate an increased vegetation cover in the drainage basins of the Blue Nile and the Atbara. Tributaries from the Eastern Desert hills yielded more amphibole than pyroxene. These interpretations can be evaluated in the light of the late Quaternary paleoclimatological studies of lakes, such as measurements of their levels.

The Paleo-Nile/Proto-Nile, pre-Nile, and Neo-Nile Rivers were active successively during the Pleistocene. Their deposits are separated from one another by unconformities and long periods of recession (Said, 1981). During the Early Pleistocene, tectonism led to accumulation of thick breccia at the foot of cliffs and wadis. The Paleo-Nile/Proto-Nile was a highly competent river (Said, 1981) which carried cobble and gravel-sized sediments made up mainly of quartz and quartzites, of the Armant and Issawia formations. The Armant and Issawia formations were deposited during a pluvial that followed a brief period of aridity at the beginning of the Early Pleistocene. Travertine and tuffaceous limestones formed at the end of this pluvial. The Proto-Nile was contemporaneous with the earliest Pleistocene glaciation. The climate in Egypt was wet. The river carried cobbles and gravel-sized sediments of quartz and quartzites of the Idfu Formation, derived from distal and local sources.

Older Pleistocene deposits; 34 m and 44 m (112 ft and 144 ft) above sea level, in the Fayium Depression are possibly of fluvial origin. They are composed of a mixture of pebbles and sand containing *Corcibula*, rare *Unio*, and water-warn artifacts which include Levallois flakes and cores (Wendorf & Schild, 1976). Said (1981) estimated a period of about 130,000 years separated the deposition of the sediments from the succeeding Holocene lacustrine deposits. Near the eastern edge of the Fayium Depression the channel and debris flow sediments of the Seila Formation were followed during the Pleistocene by sands and gravels composed of poorly lithified, lenticular, gravelly sandstones with oysters, snails, silicified wood fragments, and silicified burrows (Hamblin, 1987).

In the Nile Delta region Early Pleistocene sediments are composed of thick sand beds interbedded with thin clays of the El Wastani Formation (Rizzini et al., 1978). These sediments were probably contemporaneous with the early Pleistocene Proto-Nile Idfu gravels (Said, 1981). They were followed by the Mit Ghamr Formation (Rizzini et al., 1978, Harms & Wray, 1990) composed of thick layers of sand with pebbles of quartzite, chert, and dolomite, and sporadic shells and interbeds of coquina and peat levels. The Mit Ghamr Formation forms the isolated mounds (turtle-backs) visible in the agricultural fields (Sandford & Arkell, 1939). The Mit Ghamr sediments correlate with the pre-Nile Qena Formation Said (1981). The Quaternary pelagic succession offshore the Nile Delta is interbedded with turbidites carried down from the Nile (Nestroff et al., 1972).

The sediments of the Middle Pleistocene pre-Nile are composed mainly of massive cross-bedded sands of the Qena Formation. The Pre-Nile was a competent river with large floods and an expanding delta. The river reached the Fayium Depression at that time. The Middle Pleistocene terminated with a pluvial, during which the Pre-Nile/Neo-Nile deposited the Abbassia Formation. Heavy rains led to the deposition of locally derived polygenetic gravels with abundant crystalline rock fragments and feldspathic sands. These gravels originated from the Eastern Desert. The pluvial probably corresponds to Riss/Würm interval (Said, 1981). The Abbassia Formation contains late Acheulian hand axes and Levallois (Middle and early Late Paleolithic) artifacts (Wendorf et al., 1970).

The Late Pleistocene is represented by the Neo-Nile (Said, 1981) aggradational deposits such as the Dandara, Masmas-Ballana (Dibeira-Jer of de Heinzelin, 1968), Sahaba-Darau, and Arkin aggradations which are separated by recessional deposits of the Gerza-Ikhtiariya/Korosko-Makhadma, Deir El-Fakhuri, and Dishna-Ineiba recessions. These recessional deposits are dominated by pond sediments, erosion, soil development, and wadi action (Wendorf et al., 1970). Wind-blown sands interfinger with the neo-Nile deposits. The neo-Nile deposits are indistinguishable from those of the present-day river.

The Dandara Formation deposits represent the basal aggradational sediments of the Neo-Nile (Wendorf et al., 1976, Said, 1981). Sediments of the Dandara and Qena (Pre-Nile) formations were supplied from the Ethiopian highlands. Mineralogy of sediments is different from the previous sediments. Fine sandy silts, brown silts, thin carbonate interbeds and occasional lenses of gravel of the Dandara Formation and the thick beds of sands and gravels of the Qena Formation record an interval of heavy rainfall in Egypt accompanied by a markedly different Nilotic regime from that which prevailed previously, when the bulk of the water was derived from the Abyssinian Plateau (Wendorf & Schild, 1976). Heavy mineral analysis of the Dandara Formation and the Qena Formation indicates a regimen similar to that of today with considerable water flow from the area of the Blue Nile and Atbara Basin (Wendorf et al., 1970, Wendorf & Schild, 1976). The Dandara aggradation was followed by a recession. Recessional deposits of that period at Dibeira west in the Sudanese Nubia are represented by eolian, well-sorted, massive, dune sands of the Gerza-Ikhtiariya Formation (Said, 1981) with Middle Paleolithic implements and a few mammal bones. These deposits are contemporaneous with the Mousterian-Aterian pluvial dated at 78,000–38,000 B.C. South of the Kom Ombo area sheet floods caused by torrential rains at the mouth of wadis which drained the Eastern Desert led to the deposition of subangular pebbles of local derivation, embedded in a matrix of coarse sands followed by poorly-sorted, medium-grained sands to sandy marls of the Korosko Formation (Butzer & Hansen, 1968, Said, 1981). These sediments contain numerous artifacts of Middle Paleolithic aspect at the base and extensive settlement of the early Late Paleolithic at the top (Wenford & Schild, 1976). They were dated at around 25,250–20,000 B.C. (Wendorf & Schild, 1976). Contemporaneous with these deposits were sheet-wash deposits of the Makhadma Formation (Said, 1981) composed of gravels, pebbles, and boulders with Middle Paleolithic tools (Wendorf & Schild, 1976).

The cold dry phase associated with the last glaciation maximum, from 18,000–11,000 B.C. resulted in a reduction in vegetation cover at lower altitudes and an increase in erosion. The Nile was a highly seasonal braided river which brought mixed coarse and fine sediments down to Egypt (Adamson et al., 1980). The Korosko-Makhadma recession was followed by a major Nile aggradation of the Masmas-Ballana Formation (Said, 1971, 1981, Wendorf and Schild, 1976) probably between 17,000–15,000 B.C. Flood-plain silts with a rich molluscan fauna (Butzer & Hansen, 1968) were contemporaneous with fine-grained, well-sorted dune sands (De Heinzelin, 1968) rich in late Paleolithic implements at the top and were topped by a podzol soil. The level of the Nile fell considerably between 15,000–12,000 B.C. (Wendorf et al., 1970) and the Masmas-Ballana aggradation was followed by a recessional period of the Deir El-Fakhuri Formation (Wendorf et al., 1970, Said et al., 1970). Permanent ponds developed in the peripheral swales abandoned by the river and in topographic lows between the dunes. Diatomites and pond sediments interrupted by silt units were formed in ponds that developed on the Ballana dunes in an arid grassland environment (Wendorf & Schild, 1976).

The next Neo-Nile aggradational phase led to the deposition of fluviatile silts of the Sahaba-Darau Formation (Said, 1981, Butzer & Hansen, 1968) on top of the recessional pond deposits of the Deir El-Fakhuri Formation, between 12,000–10,000 B.C. (Wendorf & Schild, 1976). They contain abundant Levallois artifacts (Wendorf et al., 1970). The Nile fell significantly between 10,000–7000 B.C. (Wendorf et al., 1970). From 8,500 B.C. overflow from Lake Victoria and higher rainfall in Ethiopia sent extraordinary floods to the main Nile (Adamson et al., 1980). The Nile flow became permanent and more regular and bed load moved along channels of higher sinuosity. The Holocene recessional phase that followed the Sahaba-Darau aggradation comprise a succession of playa deposits and Nile silt interbeds with gravels and pebble sheets embedded in fine sand of the Dishna-Ineiba Formation (Said, 1981, Said et al, 1970). The Dishna-Ineiba playa deposits accumulated behind the natural levees and in abandoned channels of the Sahaba-Darau aggradation (Said, 1981). Recessional features of this interval were recorded at El Kilh near Idfu. Neolithic occupation of the eastern Sahara is commonly associated with playa lakes and weak to moderate paleosols which are commonly truncated by wind erosion (Haynes, 1982, 2001).

The Arkin aggradation (de Heinzelin, 1968) followed the Dishhna-Ineiba recession. Silts and fine-grained micaceous sands of the Arkin Formation (Said, 1971) were deposited around 7000 B.C. The Arkin Formation is coeval with the Arminna Member of the Gebel Silsila Formation of Butzer & Hansen (1968) in the Kom Ombo area. The El-Kab Formation (Vermeersch, 1970, cited by Wendorf & Schild, 1976, p.310) forms a series of Nilotic sediments which are now under cultivation at El-Kilh, about 15 km north of Idfu on the east bank of the Nile River. Radiocarbon dates on charcoals yielded ages between 6400–5980 B.C.

The most detailed information on the Holocene comes from the Fayium Depression. The Fayium Depression is a circular basin in the Western Desert of Egypt, situated west of the Nile. A divide, ranging from 8 to 14 km (5–9 miles) in width and 30 to 90 m (98–295 ft) in height above sea level, separates the depression from the Nile Valley. A modern lake, Lake Qarun, occupies the lowest part of the depression. A canal, Bahr Youssef, connects the depression with the modern Nile. Holocene lacustrine deposits occur as embankments in the Fayium Depression (Said, 1981). They occur as a sequence of deep and shallow water sediments which appear to record at least three succeeding episodes of lake-aggradation and recession (Wendorf & Schild, 1976). Connections between the Nile and the depression occurred at 7000 B.C., 6150 B.C., 5190 B.C., and 4910 B.C. The oldest lacustrine deposits are in the form of two diatomites, each about 3 m (10 ft) in thickness. The first access of the river to the depression coincided with the Arkin aggradation (Wendorf & Schild, 1976, Said, 1981). The lacustrine deposits in the depression have been named the Fayum Formation by Wendorf et al. (1970) and Said et al. (1970). The oldest Paleomoeris Member is made up of two diatomites separated by a layer of sand with aquatic snails. Unconformably above this unit, silts, sands, and diatomites of the Premoeris Member occur. This unit was deposited in a nearshore and swampy environment between 6300 and 5000 B.C. It contains Late Paleolithic artefacts. These sediments are overlain by fine- to medium-grained lacustrine sands of the Moeris Member which contains Terminal Paleolithic artefacts at the base and Predynastic and Old Kingdom materials near the top.

In the Nile Cone three cyclothems dated 58,000–28,000, 28,000–17,000 and 17,000–0 yr BP were recognized (Stanley & Maldonado, 1979). Each cyclothem comprises a sapropel unit, muds and turbidites, and a calcareous ooze unit. Sapropels are brownish to black beds enriched in organic carbon and often laminated (Rohling & Hilgen, 1991, Capozzi & Negri et al., 2009). Sapropels occur in the transgressive system tracts sequences (Capozzi & Negri, 2009). The three sapropels coincide in time with high lake levels in the Nile basin and the deposition of Ethiopian silts in Egypt. The sapropel reflects stagnation of deep-ocean water rising from a major flux of freshwater in the eastern Mediterranean, such as from the Nile. The periods of calcareous ooze accumulation correspond to times of aridity in the Nile basin. The Mediterranean sapropels have been compared to the Cretaceous black shales (Meyers, 2006).

Sapropel formation is triggered by a vertical change in the water density, initiated by highly saline bottom waters and completed by a low salinity surface layer (Rossignol-Strick et al., 1982). This change in density results in stagnation of bottom waters which are rapidly depleted of oxygen. The resulting reducing environment preserves the pelagic organic matter. Isotopic and palynological analysis of the most recent sapropel suggests that the spreading over the sea surface of a low density, low salinity layer of Nile floods triggered total deep-water stagnation. The high-nutrient discharge into the Mediterranean reflects a very wet period in equatorial Africa. Two sapropel layers were dated with $^{14}$C at 11,800–10,400 and 9000–8000 BP. These dates are approximately equivalent to the Arkin aggradation of the Nile. Sapropel layers alternated with oxygenated nano-ooze layers in the Eastern Mediterranean (Kroon et al., 1998). The oxygenated layers were deposited during the Younger Dryas-Preboreal period, or the Neo-Nile recessions. During the last glacial maximum (19,000–13,000 yr BP), extreme aridity spread in the tropics. This period was immediately followed during the late Glacial-early Holocene by an equatorial heavy rainfall period of global scale. The Nile River carried enormous amounts of freshwater and nutrients from equatorial Africa into the Mediterranean. Rohling & Hilgen (1991), on the other hand, argued that the Nile flood did not account for humidity (freshwater) in Eastern Mediterranean as proposed by Rossignol-Strick et al. (1982). They suggested that sapropel formation took place during periods of intensified monsoons activities in the Mediterranean which resulted in increased discharge of freshwater and nutrients from the Nile. Sapropel provide source for biogenic gas in some Fields in the offshore of the Nile Delta.

The Nile flows across climate belts ranging from tropical to hyperarid. As these belts shifted in time there were changes in the relative proportion of sediments contributed to the Delta by headlands of the Central African Plateau (White Nile), Ethiopian Highlands (Blue Nile and Atbara River), and the Red Sea Hills (Said, 1981, Stanley & Warne, 1993).

Following the closure of the Aswan High Dam in 1964 and the increase of human activities the Nile Delta is no longer an active delta but, rather, a completely wave-dominated coastal plain along the Mediterranean coast, and no freshwater reaches the sea (Stanley & Warne, 1993). The evolution of the Nile Delta was controlled by a number of factors including sea-level changes, subsidence, climate, and sedimentary processes. Late Quaternary subsurface stratigraphy of the northern Delta consists of two major units, from bottom to top:

1   Late Pleistocene alluvial sand and stif mud from 35,000 to 12,000 yr BP. It is dominated by nonmarine, medium to coarse, iron-stained, quartz-rich sand, deposited on a subaerially exposed, partially vegetated, alluvial plain during a period

of low sea level low relief. The interfingering muds accumulated during floods in localized ephemeral depressions bordering the channels. In the west, quartz-rich sand and mud interfinger with carbonate-rich deposits. Four distinct stiff mud sequences were recognized by Chen & Stanley (1993) which were deposited between 34,000 and 10,000 yr BP: a) in seasonally flooded inland sabkhas proximal to Nile channels and were then covered by wind-blown sands, b) in central areas of the play as, c) in permanent salt lakes, and d) in freshwater-low salinity marshes. Deposition of these muds took place in ephemeral depressions bordering channels in a subaerially exposed alluvial plain (Stanley & Warne, 1993). These muds were correlated with several late Pleistocene aggradational phases of the Nile in Upper Egypt.

2    These sediments are unconformably overlain by late Pleistocene-Holocene shallow marine coastal transgressive sands from 12,000–8000 yr BP which were deposited during a rapid rise in sea level. The sands contain abundant shallow marine fauna and shell debris. A depositional hiatus occurs between 12,000–10,000 and 8000 yr BP (Foucault & Stanley, 1989). 3) This sand is unconformably overlain by a variable sequence of mid.-late Holocene deltaic sand, silt, and mud, approximately 7500 yr BP. These sediments were deposited in variable settings, from inner shelf to lower deltaic alluvial plain.

The distributaries of the Delta of the Neo-Nile were more numerous during the Holocene. Seven major branches of the Delta existed and were shown on historical maps (e.g. Herodotus, 484–425 B.C., Strabo, 63 B.C., Serapion, 350 A.D., and El-Idrisi, 1099–1154 A.D.). These branches included, from west to east, the Canobic, Bolbitine, Sebennitic, Fatmetic, Mendisy, Siatic (Tanitic), and Pelusiac branches. Five of these branches degenerated and silted up, and two did not, the present-day Damietta and Rosetta branches. Sneh & Weissbrod (1973), and recently Goodfriend & Stanley (1999) studied the Pelusiac branch. The silting up of the lower reaches of the Pelusiac branch was due to prograding beach accretion which occurred around 25 A.D. (Sneh & Weissbrod, 1973). The strand line which borders the deltaic plain in the north consists of numerous bundles of very low, accretional ridges. These beach ridges are composed of fine and coarse-grained sands, shell fragments, and accumulation of unbroken shells. The sands were deposited by the prevailing west-east longshore current in front of the deltaic plain along ancient shorelines, silting up the mouths of the distributary channels. The age of the earliest beach ridge (the one farthest inland) marks the stage of stream degeneration. Mollusks from this ridge gave a $^{14}$C date of 1925 ± 90 yr BP (25 A.D.) Goodfriend & Stanley (1999) obtained a $^{14}$C age of 1460 ± 25 yr BP of a *Donax* shell in a core cut from a low beach ridge located at the landward edge of the strand line. This age should represent the initial progradational phase of the plain. However, this date was calibrated and corrected to a calendar date of 800–1100 A.D. Furthermore, based on historical accounts and Nilometer records (which show three great floods during 813–820 A.D.); Goodfriend & Stanley (1999) concluded that the timing of the strand-plain accretion in the northeastern Delta has occurred in the first half of the ninth century A.D. This accretion led to the silting up of the Pelusiac branch. During that time the Nile broke through a new distributary to the west which probably represents the origin of the Damietta branch.

Evaluation of major deltaic processes indicates that deceleration in sea level rise was the essential factor in Holocene Delta formation between 8500 and 6500 yr BP

(Stanley & Warne, 1994). This deceleration led to progressive landward migration of near mean sea level depositional environments during the lower to mid.-Holocene. Radiocarbon dates confirm that basal Delta deposits range in age from about 8000 to 6000 yr BP (Stanley & Warne, 1993). The late Pleistocene to Holocene stratigraphic succession is composed, from base to top, of: 1) A basal Sequence I of late Pleistocene fluvial deposits. 2) Sequence II of late Pleistocene to early Holocene shallow marine transgressive sandy deposits. 3) Sequence III of Holocene deltaic aggradational to progradational deposits of variable lithologies (Stanley & Warne, 1993, 1994).

Sea level changes are the only processes that could induce early Holocene Delta development (Stanley & Warne, 1994). Sea level changes explain the tendency of the Delta deposits to be younger in a landward direction. Eustatic sea-level rise (~1 mm/year) and subsidence of the sediment substrate during the late Holocene resulted in submergence and/or burial of many ancient sites along the Nile Delta (Stanley, 2005).

In the Northeastern Delta region a coarsening-upward sequence characterizes the major Holocene depocenters along the northern margin of the present Delta. The coastal margin shifted northward by as much as 50 km 31 miles) during the past 5000 years. Four lobes of marine pro-delta and delta front deposits interfinger with each others (Coutellier & Stanley, 1987). Differential subsidence across the northern Delta resulted in a general northeastern tilting of the delta plain surface and consequent thickening of the Holocene section (Stanley & Warne, 1993). The section deepens and thickens from east to west. The highest subsidence coincides with the modern Manzala lagoon and is related to major eastern Mediterranean fault systems. In the northeastern Delta, the Holocene section coarsens upward from prograding open marine, prodelta and delta front deposits to coastal sand, silt, and mud. In the north-central Delta beach ridge, fluvial and coastal dune sand and lagoon mud dominate. In the northwest, marsh and lagoon deposits and floodplain silt and mud prevail. In the extreme northwestern part of the Delta thin; less than 10 m (33 ft), carbonate-rich lagoon and sabkha sands, silts, and muds are common. The Sebennitic branch was a major Holocene distributary channel delivering coarse sand to the north-central Delta coast from 7000–4500 yr BP (Stanley & Warne, 1993). The dominant wind direction is to the southeast which drives coastal currents to the east. These wind and wave regimes have been generally consistent since at least early Holocene time. These coastal currents transport large volumes of delta sediments eastward toward Sinai, Gaza, and the Levant.

The Northeastern Nile Delta has been actively sinking relative to the sea during the recent Holocene epoch (Stanley & Goodfriend, 1997). The main cause of subsidence is ongoing faulting, as well as downwarping, of the underlying 3000 m of Late Miocene to Quaternary succession. The subsidence rate is 3.98 mm/year from 8540 BP to the present, and the sea level is rising at roughly 1.0–1.5 mm per year, so the current relative sea-level rise in this area is at least 5 mm per year. There is a marked unconformity between the top of the transgressive sands and the immediately overlying marine prodelta muds. The sands are dated to 8540 ± 150 BP.

In the Qattara Depression, the Quaternary sediments are mainly fine-grained wind-blown sands and sabkha deposits of fine sands, silts, and clays, crusts of salt and gypsiferous silts. A 3 m (10 ft) thick bed of halite cut by numerous gypsum veins cap mesas of the Moghra Formation on the western part of the depression. They consist of about 2 m (6 ft) of laminated gypsum, rich in algal filaments at the base, reddish muddy sands in the middle, and nodular gypsum at the top (Hemdan & Aref, 1998).

Albritton et al. (1990) interpreted it as an erosional remnant of a sabkha deposit. Pliocene-Pleistocene sediments of the Kalakh Formation in the northeastern Qattara Depression were deposited unconformably on the Middle Miocene Marmarica Formation (Albritton et al., 1990). Deposition started with a basal conglomerate 1–2 m (2–7 ft) in thickness composed of cobbles and pebbles of the underlying Marmarica limestones with scattered quartz pebbles and some oyster banks.

In the northern part of the Farafra Oasis karsting of the Farafra Chalk during Pleistocene time resulted in the formation of depressions. These depressions were filled by cross-bedded fine sandstones, siltstones, silty shales and shales, capped by a thin allochthonous, fossiliferous limestone beds with mixed Danian-Landenian planktonic foraminifera. These sediments constitute the Ain Wadi Formation (Youssef & Abdel Aziz, 1971).

During the Quaternary the sea level underwent a series of oscillations that caused lateral shifts of the shorelines. In many Mediterranean areas elevated coastal terraces, show that the highest Quaternary terraces are the oldest (Fabricius, 1984). The origin of the coastal terraces of Cyrenaica and Marmarica is controversial. Fabricius (1984) explained this as due to uplift of coastal regions associated with basin subsidence during the Quaternary. Quaternary deposits in northernmost Egypt, such as in the Gulf of Arabs to the west of Alexandria, are represented by elevated offshore bars, lagoonal beds, evaporites, and marls (Said, 1983). Desio (1971) attempted to correlate the Quaternary successions in various parts of the Libyan coast. McBurney & Hey (1955) described three well defined Quaternary deposits in Cyrenaica: 1) shoreline at 6 m (20 ft) sea level with marine shells, 2) tuffaceous deposits with fossil remains of plants and mammals and paleolithic remains (Levalloiso-Mousterian affinities), and Young fossil dunes with *Helix melanstona* and *Rumina decollata* and associated gravels (Younger Gravels). Hey (1968), McBurney & Hey (1955), and (Desio, 1971) reported the following terraces in Cyrenaica, Northeast Libya: Calabrian (140–190 m) (459–623 ft), Sicilian (70–90 m) (230–295 ft), Milazian (44–55 m) (144–180 ft), Tyrrhenian (35–40 m) (115–135 ft), Monasterian (15–25 m) (49–82 ft), and Würm (6 m) (20 ft).

Fluctuations in sea level during the Quaternary combined with epeirogenic movements have caused the development of karstification of Eocene and Miocene calcarenites and chalky limestones in the Benghazi plain down to a depth of 130–150 m (427–492 ft) below sea level (Guerre, 1980). The deepest karstic features were attributed to one of the main sea-level regressions of the Pleistocene.

Lacustrine environments were widespread and river systems with intervening lakes are typical of early to middle Holocene times in Libya and Egypt (Pachur, 1980, 1996, Pachur & Hoelzmann, 2000, Pachur & Rottinger, 1997, Pachur & Wunnermann, 2002). Higher groundwater level in the west resulted in precipitation of lake-carbonates, whereas in the east only suspended load was deposited as a result of runoff in the depressions. Pleistocene lake carbonates have also been found in Egypt. The same river systems were probably active during the Aptian and resulted in the deposition of the freshwater variegated shales, Chieun limestones, and the continental sandstones of the Abu Ballas Formation. These channels were in close proximity to the Aptian shoreline and were occasionally invaded by seawaters. In the subsurface of the eastern Sirte Basin, the Pliocene-Pleistocene succession changes facies from thin sandstones and shales in the south to sandstones, shales, limestones, and dolomites in the centre, then into shales in the north. A northwest-southeast trending marine shale channel runs from the Sarir Field to the Ajdabiya (Wright & Edmunds, 1971).

In the Ajdabiya area the Garet Uedda Formation is covered by a caliche or a paleosoil horizon, probably formed during the Pliocene-Pleistocene time (Giglia, 1984). The caliche is made up of reddish, pisolitic, sandy limestone with remains of terrestrial gastropods. In the Jabal Zaltan area 10 m (33 ft) thick Pleistocene deposits overlie Oligocene sandstones. The sediments are composed mainly of lacustrine sandstones interbedded with green sandy claystone with Mn oxides in the lower part and a thick laminated, whitish, limestone, gypsum, and green silty and gypsiferous claystone laminae (Domaci, 1985). The Villafranchian was a new emersion phase dominated by the formation of caliche and silcrete (Carmignani et al., 1990). The sea transgressed again during the Tyrrhenian, and another regression took place in the Holocene.

Quaternary deposits along the Libyan coast are represented by the Ajdabiya and Gargaresh formations, sabkha sediments, eolian deposits, beach and coastal dune sands, and alluvium deposits. In other areas along the coast such as in Cyrenaica these units were mapped as marine and eolian calcarenites (Röhlich, 1974, Klen, 1974). The Ajdabia Formation is composed of white calcareous sandstones essentially of skeletal debris and rounded limestone clasts. It was deposited during the Pleistocene-Lower Holocene time probably during the Tyrrhenian marine transgression (Desio, 1935, Giglia, 1984). The Gargaresh Formation (Pleistocene-Holocene) is composed of cross-bedded calcarenites and oolitic limestones with abundant reworked planktonic and benthonic foraminifera, echinoid fragments, calcareous algae, molluscs and bryozoans. It was described as eolianites or fossil dunes (Hoque, 1975, El-Hinnawy & Cheshitev, 1975, Francis & Issawi, 1977, Giglia, 1984). Carbonate beach deposits were carried landward during the Würm glacial stage to form dunes (Desio, 1971). Cross-stratification measurements (Hoque, 1975) show bimodal modes in the Tripoli area. The morphology and orientation of these dunes were largely controlled by the prevailing wind at the time dominantly blowing from the northwest with balancing effects of strong bidirectional winds from the southwest and northeast. Carbon-14 dating of oolites from the Gulf of Gabes, Southern Tunisia yielded an age of 20,000 to 30,000 yr BP for the inland oolitic belts (Fabricius et al., 1970). The seaward parts near the present shoreline gave a Holocene age. These ages led them to conclude that the formation of the oolites took place during a period younger than the last main glaciation (Würm, Weichsel, Wisconsin).

In Northwest Libya in the Jifarah Plains up to 250 m (820 ft) of lagoonal sediments of the Pliocene-Pleistocene Al Assah Formation of clayey sandstones, gypsum, sands, silts and gravels (Mijalkovic, 1977a, b, Megerisi & Mamgain, 1980a) were deposited. Estuarine sediments of Al Hishash Formation composed of cross-bedded greyish; loosely cemented sands with gypsum, calcarenites, sandy limestones, marly and gypsiferous beds were developed mostly near the mouth of major wadis. These two formations are equivalent to the Pliocene-Pleistocene Raf Raf and Porto Farina formations in Tunisia.

Quaternary sediments of the Jifarah Plains display a south to north change in facies from alluvial fan gravels (Qasr el Haj Formation) through fluvio-eolian sands and silts (Jifarah Formation) to calcarenitic of marine origin associated with littoral sands and gravels (Gargaresh Formation) (Anketell & Ghellali, 1991b, c). Each formation has been divided into six members separated from each other by surfaces of unconformity associated with soil formation (Anketell & Ghellali, 1991b). The succession was interpreted in terms of response to changing sea level and climate associated with the advance and retreat of glaciers in northern Europe (Anketell & Ghellali, 1991b). The development of glaciers led to a sea-level fall, erosion, and fluvial deposition. Glacial retreat caused a transgression and an improvement in climate, accompanied by the formation of calcrete crusts and soil horizons.

On the northern margin of the Murzuq Basin Tertiary-Quaternary sediments of the Al Mahruqah Formation are composed of sandy dolomitic limestones and limestones with ostracods, gastropods and pelecypods were deposited in an extensive lacustrine system on top of older sediments ranging from Ordovician to late Pliocene-Pleistocene (Seidl & Röhlich, 1984, Parizek et al., 1984, Korab, 1984, Gundobin, 1985). They grade southwards into continental and brackish and saline lake sediments of conglomerates and sandstones, and calcareous sandstones, sandy limestones with ostracods, characeae, bird's-eye structures, glauconite, bones and fish scales, superficial ooids, gypsum intercalations, and laminated dolomites.

According to Hey (1968) the late Quaternary non-marine deposits record the following history: 1) About 50,000 BP was a short period during which the climate was temperate and led to the deposition of tufa and calcareous marls in some wadis. 2) From 50,000–43,000 BP the climate was cold and frost-shattering and resulted in debris which formed talus. This talus was redeposited in wadis as the Younger Gravels which were later cemented. The sea level fall at that time resulted in the exposure of marine sands which were reworked and deposited as dunes by northwesterly prevailing winds. 3) Between 43,000–32,000 BP the climate was warm and dry and erosion of the Younger Gravels and talus took place. 4) From 32,000–12,000 BP the climate was cold and frost-shattering which produced a second generation of talus. 5) From 12,000-present the climate has been warm and dry with some downcutting but without deposition. At the end of the Pleistocene and beginning of the Holocene, the Tadrart-Acacus and the surrounding areas in Southwest Libya show wet conditions, interrupted by at least one arid oscillation, followed by progressive climatic decline probably between 8000 and 7300 BP (Cremaschi & Di Lernia, 1999). The majority of the North Africa's lakes were at a high level at 6000 BP, but show increasingly drier conditions after 6000 yr BP, with the minimum lake levels occurred just after 4000 BP Damnati, 2000). This wet phase resulted in the formation of the "West Nubian Lake" (Hoelzmann et al., 2001) North Africa that developed and became extinct in the period 9400–3800 BP, and was accompanied by human occupation from 6300 to 3500 BP.

# References

Aadland, A.J. & Hassan, A.A., 1972. Hydrocarbon potential of the Abu Gharadig basin in the Western Desert, 8th Arab Petrol. Congress, Algiers, No. 81 (B3).

Aadland, R.K. & Schamel, S., 1988 (Abs.). Mesozoic evolution of Northeast African shelf margin, Libya and Egypt, in Mediterranean Basin Conference & Exhibition, Nice, AAPG Bull., v. 72, No. 8, p. 982.

AAPG & AAPG European Region Energy Conference and Exhibition (November 18–21, 2007) (Abstracts): Regional Controls and Future Potential of the North African Petroleum Systems, Petroleum Geology, Exploration Successes and Future Potential in Algeria, Petroleum Geology, Exploration Successes and Future Potential of Egypt.

Aassoumi, H., Saber, H., Broutin, J. & El Wartiti, M., 2003 (Abs.). First spore, pollen and acritarch associations in the Ida Ou Ziki basin (southern slope of the Western High Atlas, Morocco). XVth International Congress on Carboniferous and Permian Stratigraphy, Utrecht, The Netherlands, 10–16 August 2003, Abstract, 195, pp. 1–2.

Abadi, A.M., van Wees, J.-D., van Dijk, P.M. & Cloetingh, S.A.P.L., 2008. Tectonics and subsidence evolution of the Sirt Basin, Libya. AAPG Bulletin, v. 92, no. 8, pp. 993–1027.

Abbass, H.L., 1961. A monograph of the Egyptian Cretaceous pelecypods, Egypt. Geol. Survey Paleontological Mon.1, p. 224.

Abd El-Naby, A., Abd El-Aal, M., Kuss, J., Boukhary, M. & Lashin, A. (subm.): Structural and basin evolution in Miocene time, southwestern Gulf of Suez, Egypt, N. Jb. Geol. Paläont.

Abd El-Naby, A., Frisch, W. & Hegner, E., 2000. Evolution of the Pan-African Wadi Haimur metamorphic sole, Eastern Desert, Egypt. J. Metamorphic Geology, v. 18, no. 6, pp. 639–652.

Abd el-Naby, H., Frisch, W. & Siebel, W., 2008. Tectono-metamorphic evolution of the Wadi Hafafit Culmination (central Eastern Desert, Egypt). Implication for Neoproterozoic core complex exhumation in NE Africa. Geologica Acta, v. 6, no. 4, pp. 293–312.

Abd El-Rahman, Y. Polat, A., Dilek, Y., Fryer, B., El-Sharkawy, M. & Sakran, S., 2009. Geochemistry and tectonic evolution of the Neoproterozoic Wadi Ghadir ophiolite, Eastern Desert, Egypt. Lithos, v 113, pp. 158–178.

Abd El-Rahman, Y. Polat, A., Fryer, B., Dilek, Y., El-Sharkawy, M. & Sakran, S., 2010. The provenance and tectonic setting of the Neoproterozoic Um Hassa Greywacke Member, Wadi Hammamat area, Egypt: Evidence from petrography and geochemistry. Journal of African Earth Sciences, v. 58, no. 2, pp. 185–196.

Abdallah, A.M. & Abdelhady, F.M., 1968. Geology of Sadat area, Gulf of Suez. Egypt. Journal of Geology, v. 10, no. 1, pp. 1–22.

Abdallah, A.M. & Eissa, R.A., 1971. The Campanian rocks of the southern Galala, Bull. Fac. Sci., Cairo University, v. 44, pp. 259–270.

Abdallah, A.M. & El Adindani, A., 1965. Stratigraphy of Upper Paleozoic rocks, western side of the Gulf of Suez, Geol. Surv. Min. Res. Dept., Cairo, Paper No. 25, p. 18.

Abdallah, A.M., El Adindani, A. & Fahmy, N., 1963. Stratigraphy of the Lower Mesozoic rocks, western side of the Gulf of Suez, Egypt, Geol. Surv. Min. Res. Dept., Cairo, Paper No. 27, p. 23.

Abdallah, A.M., El Sharkawi, M.A. & Marzouk, A., 1971. Geology of Mersa Thelmet area, Southern Galala, Eastern Desert, A.R.E. Bull. Fac. Sci. Cairo Univ., v. 47, p. 271–280.

Abdallah, H., Sassi, S., Meister, C. & Souissi, R., 2000. Stratigraphie séquentielle et paléogéographie á la limite Cénomanien-Turonien dans le région de Gafsa-Chotts (Tunisie centrale). Cretaceous Research, v. 21, pp. 35–106.

Abdallah, N., Liégeois, J.-P., De Waele, B., Fezaa, N., & Ouabadi, A., 2007. The Temaguessine Fe-cordierite orbicular granite (Central Hoggar, Algeria): U–Pb SHRIMP age, petrology, origin and geodynamical consequences for the late Pan-African magmatism of the Tuareg shield. Journal of African Earth Sciences, v. 49, pp. 153–178.

Abdallah, T. & Helal, H., 1990. Risk evaluation of rock mass sliding in El-Deir El-Bahary Valley, Luxor, Egypt. Bulletin of the International Association of Engineering Geology, Paris, no. 42, pp. 3–9.

Abdeen, M.M. & Greiling, R.O., 2005. A quantitative sructural study of Late Pan-African compressional deformation in the Central Eastern Desert (Egypt) during Gondwana assembly. Gondwana Research, v. 8, no. 4, pp. 457–471.

Abdeen, M.M., Sadek, M.F. & Greiling, R.O., 2008. Thrusting and multiple folding in the Neoproterozoic Pan-African basement of Wadi Hodein area, south Eastern Desert, Egypt. Journal of African Earth Sciences, v. 52, pp. 21–29.

Abdel Aal, A. & Moustafa, A.R., 1988. Structural framework of the Abu Gharadig Basin, Western Desert, Egypt. 9th Exploration and Production Conf., Egyptian General Petroleum Corp. (EGPC), Cairo, Egypt, November 1988, p. 38.

Abdel Aal, A., El Barkooky, A., Gerrits, M., Meyer, H., Schwander, M. & Kaki, H., 2000. Tectonic evolution of the Eastern Mediterranean Basin and its significance for hydrocarbon prospectivity in the ultradeepwater of the Nile Delta. The Leading Edge, Oct. 2000, pp. 1086–1102.

Abdel Gawad, G.I. & Gamiel, M., 1998 (Abs.). Middle Jurassic fauna from Minshera area, north central Sinai, Egypt, in Geology of the Arab World, Fourth International Conference, Cairo University, Cairo, pp. 17–18.

Abdelghany, O., 2002. Lower Miocene stratigraphy of the Gebel Shabrawat area, north Eastern Desert, Egypt. Journal of African Earth Sciences, v. 34, no. 3–4, pp. 203–212.

Abdelghany, O. & Piller, W.E., 1999. Biostratigraphy of Lower Miocene sections in the Eastern Desert (Cairo – Suez district, Egypt). Rev. Micropaléontol., v. 18, pp. 607–617.

Abdel-Karim, A.M., Azzaz, S.A., Moharem, A.F. & El-Alfy, H.M., 2008. Petrological and geochemical studies on the ophiolite and island arc association of Wadi Hammariya, Central Eastern Desert, Egypt. The Arabian Journal for Science and Engineering, v. 33, no. 1C, pp. 117–138.

Abdel-Kireem, M.R., Schrank, E., Samir, A.M. & Ibrahim, M.I.A., 1996. Cretaceous palaeoecology, palaeogeography and palaeoclimatology of the northern Western Desert, Egypt, Journal of African Earth Sciences, v. 22, no. 1, pp. 93–112.

Abdelmalik, W.M., Aboul Ela, N.A. & El-Shammaa, A.G., 1981. Upper Jurassic-Lower Cretaceous microflora from the North-Western Desert, Egypt, N. Jb. Geol. Paläont. Abh., 162, Stuttgart, pp. 244–263.

Abdel Rahman, A.M. & Doig, R., 1987. The Rb-Sr geochemistry evolution of the Ras Gharib segment of the northern Nubian Shield, J. Geol. Soc., London, v. 144, pp. 577–586.

Abdelsalam, M.G., Liégeois, J.-P. & Stern, R.J., 2002. The Saharan Metacraton. Journal of African Earth Sciences, v. 34, no. 3–4, pp. 119–136.

Abdel-Wahab, A. & McBride, E.F., 2001. Origin of calcite-cemented concretions, Temple Member, Qasr El Sagha Formation (Eocene), Faiyum Depression, Egypt. Journal of Sedimentary Research, v. 71, pp. 70–81.

Abdesselam-Rouighi, F. & Coquel, R., 1997. Palynology du Devonien terminal-Carbonifère Inferieur dans le sud-est du bassin D'Illizi (Sahara algérien). Position des premières lycospores dans la série stratigraphique. Annales de la Société Géologique du Nord, v. 5, pp. 47–57.

Abdine, A.S., 1981. Egypt's petroleum geology: good grounds for optimism, World Oil, December 1981, pp. 99–112.

Abdine, A.S., Homossani, A. & Lelek, J., 1992. October Field, the latest giant under development in the Gulf of Suez, Egypt. In Sadek, A. (Ed.), Geology of the Arab World, Cairo University, v. 1, pp. 61–86.

Abdine, A.S., Meshref, W., Shahin, A.N., Garossino, P. & Shazly, S., 1992. Ramadan Field -, Gulf of Suez Basin. In: Structural traps; VI. AAPG Treatise on Petroleum Geology, Atlas of Oil and Gas Fields, A24, pp. 113–139.

Abdine, A.S., Meshref, W., Shahin, A.N., Garissino, P. & Shazly, S., 1994. Ramadan Field – Egypt: Gulf of Suez Basin, in S.A. Landon (Ed.), Interior Rift Basins, AAPG Mem. No. 59, pp. 97–120.

Abdine, A.S., Meshref, W., Wasfi, S., Shahin, A.N., Aadland, A. & Abdel Aal, A., 1993. Razzak Field – Egypt; Razzak-Alamein Basin, Northern Western Desert. AAPG Treatise on Petroleum Geology, Structural Traps VIII, pp. 29–56.

Abdou, H.F. & Marzouk, A., 1969. Jurassic foraminifera and microfacies of Katib El-Makhazin western Sinai, U.A.R., Proc. 3rd African Micropaleontological Colloquium, Cairo, No. 3, pp. 13–23.

Abdulsamad, E.O., Bu-Argoub, F.M. & Tmalla, A.F.A., 2009. A stratigraphic review of the Eocene to Miocene rock units in the al Jabal al Akhdar, NE Libya. Marine and Petroleum Geology, v. 26, no. 7, pp. 1228–1239.

Abouessa, A. & Morad, S., 2009. An integrated study of diagenesis and depositional facies in tidal sandstones: the Hawaz Formation (Middle Ordovician), Murzuq Basin, Libya. J. Petroleum Geology, v. 32, no. 1, pp. 39–66.

Aboul Ela, N.M., Ibrahim, M.I. & Kholeif, S.E., 1998 (Abs.). Jurassic to Early Cretaceous Dinoflagellate cyst zonation from the north eastern Desert, Egypt, in Geology of the Arab World, Fourth International Conference, Cairo University, Cairo, pp. 21–22.

Aboussalam, S. & Becker, T.R., 2004. Givetian stratigraphy and faunas at Tiguisselt (Tata region, Dra Valley, Morocco). Devonian of the western Anti Atlas: correlations and events. Doc. Inst. Sci, Rabat, 19, pp. 60–63.

Aboussalam, S. & Becker, T.R., 2007. New upper Givetian to basal Frasnian conodont faunas from the Tafilalt (Anti-Atlas, Southern Morocco). Geological Quarterly, v. 51, no. 4, pp. 345–374.

Abramovich, Si. & Keller, G., 2002. High stress late Maastrichtian paleoenvironment: inference from planktonic foraminifera in Tunisia. Palaeogeography, Palaeoclimatology, Palaeoecology, v. 178, no. 3–4, pp. 145–164.

Abu El-Ella, R., 1990. The Neogene-Quaternary section in the Nile Delta, Egypt: Geology and hydrocarbon potential. Journal of Petroleum Geology, v. 13, no. 3, pp. 329–340.

Abul El-Ela, F.F., 1997. Geochemistry of an island-arc plutonic suite: Wadi Dabr intrusive complex, Eastern Desert, Egypt, J. African Earth Sciences, v. 24, pp. 473–496.

Abu-Zied, R.H., 2008. Lithostratigraphy and biostratigraphy of some Lower Cretaceous outcrops from Northern Sinai, Egypt. Cretaceous Research, v. 29, pp. 603–624.

Adaci, M., Tabuce, R., Mebrouk, F., Bensalah, M., Fabre, P.-H., Hautier, L., Jaeger, J.-J., Lazzari, V., Mhboubi, M., Marivaux, L., Otero, O., Peigné, S. & Tong, H., 2007. Nouveaux sites á vertébrés paléogènes dans la région des Gour Lazib (Sahara nord-occidental, Algérie). Comptes Rendus Palevol, v. 6, no. 8, pp. 535–544.

Adamson, D.A., Gasse, F., Street, F.A. & Williams, M.A., 1980. Late Quaternary history of the Nile, Nature, v. 288, pp. 50–55.

Adamson, K., Glover, T., Whittington, R. & Craig, J., 2000. The Lower Devonian succession of the Murzuq Basin – possible indicators of eustatic and tectonic controls on sedimentation. In: M.A. Sola & D. Worsley (Eds.), Geological Exploration in Murzuq Basin, pp. 431–447.

Adamson, K., Glover, T., Fitches, B., Whittington, R., Craig, J. & Rushworth, D., 1998 (Abs.). Controls in sequence development of Devonian units of the Murzuq intracratonic basin Southwest Libya, The Geological Conference on Exploration in Murzuq Basin, Sabha University, Sabha, Libya, p. 7.

Adatte, T., Keller, G. & Stinnesbeck, W., 2002. Late Cretaceous to early Paleocene climate and sea-level flucturations: the Tunisian record. Palaeogeography, Palaeoclimatology, Palaeoecology, v. 178, pp. 165–196.

Ade-Hall, J.M., Derstein, S., Gerstein, R.E., Reynolds, P.H., Dagley, P., Mussett, A.E. & Hubbard, T.P., 1975. Geophysical studies of North African Cenozoic volcanic areas III, Garian, Libya, Can. J. Earth Sci., v. 12, pp. 1264–1271.

Ade-Hall, J.M., Reynolds, P.H., Dagley, P., Mussett, A.E., Hubbard, T.P. & Klitzsch, E., 1974a. Geophysical studies of North African Cenozoic volcanic areas, part I, Haruj Assuad, Libya, Can. J. Earth Sci., v. 11, pp. 998–1006.

Ade-Hall, J.M., Reynolds, P.H., Dagley, P., Mussett, A.E., Hubbard, T.P. & Klitzsch, E., 1974b. Geophysical studies of North African Cenozoic volcanic areas, part II, Jebel Soda, Libya, Can. J. Earth Sci., v. 12, pp. 1257–1263.

Adloff, M.C., Doubinger, J., Massa, D. & Vachard, 1985. Trias de Tripolitain (Libye). Nouvelle donnees stratigraphiques et palynologiques. Rev Inst. Fr. Petrole, v. 40, no. 6, pp. 723–753.

Adloff, M.C., Doubinger, J., Massa, D. & Vachard, 1986. Trias de Tripolitain (Libye). Nouvelle donnees stratigraphiques et palynologiques (Deuxiè partie). Rév. Inst Fr. Pétrole, v. 41, no. 1, pp. 27–72.

Ahlbrandt, T.S., 2001. The Sirte Basin Province of Libya-Sirte-Zelten Total Petroleum System. U.S. Geological Survey Bulletin 2202–F, p. 33.

Ahlbrandt, T.S., Pollastro, R.M., Klett, T.R., Schenk, C.J., Lindquist, S.J. & Fox, J.E., 2000. Chapter R2: Region 2 assessment summary – Middle East and North Africa in U.S. Geological Survey Digital Data Series 60, p. 46.

Aigner, T., 1982. Event stratification in Nummulite accumulations and in shell bed from the Eocene of Egypt, in G. Einsele & A. Seilacher (Eds.), Cyclic and Event Stratification, Springer-Verlag, Berlin, pp. 248–262.

Aigner, T., 1983. Facies and origin of nummulitic buildups, an example from Giza Pyramids Plateau (Middle Eocene, Egypt), N. Jb. Geol. Paläont. Abh., 166, No. 3, pp. 347–368.

Aigner, T., 1984. Biofabrics as dynamic indicators in Nummulite accumulations, J. Sed. Pet., v. 55, No. 1, pp. 131–134.

Airaghi, C., 1934. Echinidi paleogenici della Sirtica e del Fezzan Orientale. Missione Scientifica della R. Accad. D'Italia a Cufra (1931-X), Roma, v. 3, pp. 63–81.

Aissaoui, D.M., 1989. Palégéographie du Jurassique supérieur au sud du choot El Hodna, Algérie. Journal of African Earth Sciences, v. 43, nos. 3/4, pp. 413–420.

Aït Chayeb, E.H., Youbi, N., El-Boukhari, A., Bouabdelli, M., & Amrhar, M. (1998). Le volcanisme Permien et Mésozoique inferieur du Bassin d'Argana (Haut-Atlas occidental, Maroc); un magmatisme intraplaque associé a l'ouverture de l'Atlantique central. Journal of African Earth Sciences, v. 26, no. 4, pp. 499–519.

Akaad, M.K. & Noweir, A., 1969. Lithostratigraphy of the Hammamat-Um Seleimat district, Eastern Desert, Egypt, Nature, v. 223, pp. 284–285.

Akarish, A.I.M. & El-Gohary, A.M., 2008. Petrography and geochemistry of lower Paleozoic sandstones, East Sinai, Egypt: Implications for provenance and tectonic setting. Journal of African Earth Sciences, v. 52, pp. 43–54.

Al Far, D.M., 1966. Geology and coal deposits of Gebel El Maghara, northern Sinai, Geol. Surv. Min. Res. Dept., Paper No. 37, p59.

Al Far, D.M., Hagemann, H.W. & Omara, S., 1965. Beitrage zur Geologie des kohle-fuhrend Gebietes von El Maghara, Nord-Sinai, Ägypten, Geol. Mitt., v. 4, No. 4, pp. 397–429.

Al-Ameri, T.K., 1983a. Acid resistant microfossils used in the determination of Paleozoic paleoenvironments in Libya, Palaeogeog., Palaeoclimat., Palaeoecol., v. 44, pp. 103–116.

Al-Ameri, T.K., 1983b. Palynological application for the palaeogeography of Tripolitania during the Silurian time, Proc. Indian Natn. Sci. Acad., v. 49, A, no. 6, pp. 637–646.

Albani, R., Massa, D. & Tongiorgi, M., 1991. Palynostratigraphy (Acritarchs) of some Cambrian beds from the Rhadames (Ghadamis) Basin (Western Libya – Southern Tunisia), Bolletino della Società Paleontologica Italiana, v. 30, No. 3, pp. 255–280.

Albritton, C.C. Jr., Brooks, J.E., Issawi, B. & Swedan, A., 1990. Origin of the Qattara Depression, Egypt, Geol. Soc. Am. Bull., v. 102, No. 7, pp. 952–960.

Albritton, C.C. Jr., Brooks, J.E., Issawi, B. & Swedan, A., 1991. Origin of the Qattara Depression, Egypt, Reply. Geol. Soc. Am. Bull., v. 103, No. 10, pp. 1375–1376.

Alegret, L., Ortiz, S., Orue-Etxebarria, X, Bernaola, G., Baceta, J.I., Monechi, S., Apellaniz, E. & Pujalte, V., 2009. The Paleocene-Eocene thermal maximum: New data on microfossil turnover at the zumaia section, spain. Palaios, v. 24, pp. 318–328.

Alem, N., Assassi, S., Benhebouche, S. & Kadi, B., 1998. Controls on hydrocarbon occurrence and productivity in the F6 reservoir, Tin Fouyé-Tabankort area, NW Illizi Basin. In: D.S. Macgregor, R.T.J. Moody, D.D. Clark-Lowes (Eds.), Petroleum Geology of North Africa, Geological Society of London, Special Publication, v. 132, pp. 75–186.

Alessio, M., Barich, B.E., Belluomini, G., Hassan, F.A., Mahmoud, A.A., Manfra, L., & Stoppiello, A., 1992. A Report on Farafra (Western Desert, Egypt): New Research and Radiation Dates. Nyame Akuma, v. 38, pp. 19–28.

Al-Far, D.M., 1966. Geology and coal deposits of Gebel Al Maghara North Sinai, Egypt. J. Geol. Surv. Egypt, 37, p. 59.

Al-Far, D.M., Hagemann, H.W. & Omara, S., 1965. Beitrage zur Geologie des Kohle-fuhrend Gebeites von El Maghara, Nord-Sinai, Ägypten. Geol. Mitt., 4(4), pp. 397–429.

Algouti, Ab., Algouti, Ah, Chbani, B. & Zaim, M., 2001. Sedimentation et volcanisme synsedimentaire de la serie de base de l'Adoudounien infra-cambrien a travers deux examples de l'Anti-Atlas du Maroc. Journal of African Earth Sciences, v. 32, no. 4, pp. 541–556.

Ali, O., 1975. El Agreb-El Gassi oil fields, central Algerian Sahara. AAPG Bulletin, v. 59, no. 9, pp. 1676–1684.

Allain, R., Aquesbi, N., Dejax, J., Meyer, C., Monbaron, M., Montenat, C., Richir, P., Rochdy, M., Russell, D. & Taquet, A., 2004. Basal sauropod dinosaur from the Early Jurassic of Morocco. C.R. Palevol, v. 3, pp. 199–208.

Allam, A., 1989. The Paleozoic sandstones in Wadi Feiran-El Tor area, Sinai, Egypt, J. Afr. Earth Sci., v. 9, No. 1, pp. 49–57.

Allix, P., Grosdidier, E., Jardine, S., Legoux, O. & Popoff, M., 1981. Découverte d'Aptien supérieur à Albien inférieur daté par microfossiles dans la série détritique Crétacé du fossé de la Bénoué (Nigéria), C.R. Acad. Sci. Paris, série 2, 292, pp. 1291–1294.

Almásy, L.E., 1936. Récent explorations dans le désert libyque. (See Geographical Journal, 1937, v. 89, pp. 265–267 for a review).

Almond, D.C., 1991. Anorogenic magmatism in Northeast Africa, in M.J. Salem, M.T. Busrewil & A.M. Ben Ashour (Eds.), The Geology of Libya, v. 1V, Amsterdam, Elsevier, pp. 2495–2509.

Almond, D.C., Busrewil, M.T. & Wadsworth, W.J., 1974. The Ghirian Tertiary volcanic province of Tripolitania, Libya, Geol. J., v. 9, pp. 17–28.

Almond, D.C., Darbyshire, D.P.F. & Ahmed, F., 1989. Age limit of major shearing episodes in the Nubian Shield of NE Sudan, J. Afr. Earth Sci., v. 9, pp. 489–496.

Al-Shanti, A.M. & Gass, I.G., 1983. The Upper Proterozoic ophiolite mélange zones of the easternmost Arabian shield. Journal of the Geological Society, v. 140, pp. 867–876.

Alsharhan, A.S. & Abd El-Gawad, E.A., 2008. Geochemical characterization of potential Jurassic/Cretaceous source rocks in the Shushan Basin, Northern Western Desert, Egypt. Journal of Petroleum Geology, v. 31, no. 2, pp. 191–212.

Alsharhan, A.S. & Salah, M.G., 1994. Geology and hydrocarbon habitat in rift setting: southern Gulf of Suez, Egypt, CSPG Bull., v. 42, pp. 312–331.

Alsharhan, A.S. & Salah, M.G., 1995. Geology and hydrocarbon potential in rift setting, northern and central Gulf of Suez, Egypt, CSPG. Bull., v. 43, pp. 156–176.

Alsharhan, A.S. & Salah, M.G., 1996. Geologic setting and hydrocarbon potential of north Sinai, Egypt, CSPG Bull., v. 44, pp. 615–631.

Alsharhan, A.S. & Salah, M.G., 1997a. A common source rock for Egyptian and Saudi hydrocarbons in the Red Sea. AAPG Bulletin, v. 81, no. 10, pp. 1640–1659.

Alsharhan, A.S. & Salah, M.G., 1997b. Lithostratigraphy, sedimentology and hydrocarbon habitat of the pre-Cenomanian Nubian Sandstone in the Gulf of Suez oil province, Egypt. GeoArabia, v. 2, no. 4, pp. 385–400.

Alsharhan, A.S., 2003. Petroleum geology and potential hydrocarbon plays in the Gulf of Suez rift Basin, Egypt. AAPG Bulletin, v. 87, no. 1, pp. 143–180.

Alvaro, J.J., Macouin, M., Ezzouhairi, H., Charif, A., Ait Ayad, N., Ribeiro, M.L., & Ader, M., 2008. Late Neoproterozoic carbonate productivity in a rifting context: the Adoudou Formation and its associated bimodal volcanism onlapping the western Saghro inlier, Morocco. Geological Society, London, Special Publications, v. 297. pp. 285–302.

Aly, S.A., Ali, H.A.M., Abou Ashour, N.M.H. & El-Gezeiry, M.M., 2003. Resistivity, radioactivity and porosity logs as tools to evaluate the organic content of Abu Roash "F" and "G" Members, North Western Desert, Egypt. Egyptian Geophysical Society EGS Journal, v. 1, no. 1, pp. 129–137.

Amard, B., 1996. Occurrence of Libycoceras ismaeli (Zittel) in the Upper Maastrichtian of Eastern Tademait, Algerian Sahara, Journal of African Earth Sciences, v. 22, no. 4, pp. 609–615.

Amard, B. & Blondeau, A., 1979. Le Paléocène supérieur à Ranikothalia bermudezi et l'Éocène inférieur Ilerdien basal à Nummulites fraasi et N. deserti du Tadémaït-E et Tinrhet-W. Geobios, v. 12, pp. 635–652.

Amari, A. & Bédir, M., 1989. Les bassins quaternaires du Sahel central de la Tunisie. Genèse et evolution des sebkhas en contexte décrochant compressif et distensif. Géodynamique, v. 4, no. 1, pp. 49–65.

Ambrose, G., 2000. Geology and hydrocarbon habitat of the Sarir Sandstone, SE Sirt Basin, Libya. J. Petroleum Geology, v. 23, no. 2, pp. 165–192.

Amenzou, M & Badra, L., 1996. Les granites d'Azegour et de Brikiine (Maroc): implication génétique d'après la typologie des zircons. Comptes Rendus de l'Academie des Sciences, Serie 2, Sciences de la Terre, v. 323, no. 3, pp. 213–220.

Amine, M.S., 1961. Subsurface features and oil prospects of the Western Desert, Egypt, U.A.R., Third Arab Petroleum Congress, Alexandria, Paper 16(B-3), p. 8

Ammar, A., Mauffret, A., Gorini, C. & Jabour, H., 2007. The tectonic structure of the Alboran margin of Morocco. Revista de la Sociedad Geológica de España, v. 20, pp. 3–4.

Anderson, R.V., 1947 (Abs.). Origin of the Libyan oasis basins. Bull. Geol. Soc. Am., v. 58, no. 12, pp. 1163.

Anderson, P.E., Benton, M.J., Trueman, C.N., Paterson, B.A. & Cuny, G., 2007. Palaeoenvironments of vertebrates on the southern shore of Tethys: The nonmarine Early Cretaceous of Tunisia. Palaeogeography, Palaeoclimatology, Palaeoecology, v. 243, pp. 118–131.

Andresen, A., Abu El-Rus, M.A., Myhre, P.I., Boghdady, G.Y. & Corfu, F., 2009. U-Pb TIMS age constraints on the evolution of the Neoproterozoic Meatiq Gneiss Dome, Eastern Desert, Egypt. International Journal of Earth Sciences, v. 98, pp. 481–497.

Andrawis, S.F. & Abdelmalik, W.M., 1981. Lower/Middle Miocene boundary in Gulf of Suez region, Egypt, Newsl. Stratigr., v. 10, pp. 156–163.

Andrews, C.W., 1906. Descriptive Catalogue of the Tertiary Vertebrata of the Fayum, Egypt. British Museum (Natural History), London, p. 324.

Anketell, J.M. & Ghellali, S.M., 1991a. The Jifarah Formation – aeolian and fluvial deposits of Quaternary age, Jifarah Plain, G.S.P.L.A.J., a redefinition in terms of composite strato-type, in M.J. Salem & M.N. Belaid (Eds.), The Geology of Libya, v. V, Elsevier, Amsterdam, pp. 1967–1986.

Anketell, J.M. & Ghellali, S.M., 1991b. Quaternary sediments of the Jifarah Plain, in M.J. Salem & M.N. Belaid (Eds.), The Geology of Libya, v. V, Elsevier, Amsterdam, pp. 1987–2013.

Anketell, J.M. & Ghellali, S.M., 1991c. Quaternary fluvio-aeolian sand/silt and alluvial gravel deposits of northern Libya – event stratigraphy and correlation, J. Afr. Earth Sci., v. 13, No. 3/4, pp. 457–469.

Anketell, J.M. & Mriheel, I.Y., 2000. Depositional environment and diagenesis of the Eocene Jdeir Formation, Gabes-Tripoli Basin, western offshore Libya. J. Petroleum Geology, v. 25, no. 3, pp. 425–447.

Anketell, J.M., 1996. Structure of the Sirt Basin and its relationships to the Sabrata and Cyrena-ican Platform, Northern Libya, in M.J. Salem, M.T. Busrewil, A.A. & Misallati, M.S. (Eds.), The Geology of the Sirt Basin, v. III, Elsevier, Amsterdam, pp. 57–88.

Ansary, D.E., Andrawis, S.F. & Fahmy, S.E., 1961. Biostratigraphic studies of the subsurface Eocene in the G.P.C. Bakr, Kareem, and Rahmi fields, Third Arab Petroleum Congress, Alexandria, p. 10

Antonovic, A., 1977. Geological map of Libya, 1:250,000, Explanatory Booklet, Sheet Mizdah NH 33-1, Industrial Research Centre, Tripoli, p. 68

Aouraghe, H., 2006. Histoire du peuplement paléolithique de l'Afrique du Nord et dynamique des interactions entre l'homme et son environnement. Paléontologie humaine et Préhistoire. C. R. Palevol, v. 5, pp. 237–242.

Arambourg, C., 1947. Les vertébrés fossiles des formations continentals des plateaux constan-tinois (note réliminaire). Bulletin de la Société d'Histoire Naturelle de l'Afrique du Nord, v. 38, pp. 45–48.

Arambourg, C., 1952. Les vertébrés fossiles des gisements de phosphates (Maroc-Algérie-Tunisie). Notes et Mémoires du Service géologique du Maroc, v. 92, pp. 1–372.

Arboleya, M.L., Babault, J., Owen, L., Teixell, A. & Finkel, B., 2008. Timing and nature of Quaternary fluvial incision in the Ouarzazate foreland basin, Morocco. Journal of the Geological Society, London, v. 165, pp. 1059–1073.

Arboleya, M.-L., Teixell, A., Charroud, M. & Julivert, M., 2004. A structural transect of the High and Middle Atlas of Morocco. Journal of African Earth Sciences, v. 39, nos. 3/4, pp. 319–327.

Ardagh, Colonel, 1886. The Red Sea petroleum deposits. Porceedings of the Royal Geographical Society and Monthly Record of Geography, New Monthly Series, v. 8, no. 8, pp. 502–507.

Aretz, M., Herbig, H.G. & Huck, S., 2007 (Abs.). Microbial reefs from the late Visean of the Jerada synclinorium (NE Morocco). MAPG First International Conference & Exbihition, Marrakech, Morocco, p. 149.

Argnani, A. & Torelli, L., 2001. Pelagian shelf and its graben system (Italy/Tunisia). In P.A. Ziegler, Cavazza, W., Robertson, A.H.F. & Crasquin-Soleau, S. (Eds.), Peri-Tethys Memoir 6, Peri-Tethyan Rift/Wrench Basins and Passive Margins, Memoire du Museum National d'Histoire Naturelle, Tome 186, pp. 529–544.

Arkell, W.J., 1956. Jurassic Geology of the World, London, Oliver & Boyd Ltd., p. 806.

Arni, P., 1963. L'évolution des Nummulites en tant que facteurs de modification des depots littoraux, Coll. Internat. Micropal., Dakar 1963, Mem. Bur. Rech. Géol. Min., v. 32, pp. 7–20.

Arni, P., 1967. A comprehensive graph of the essential diagnostics of the *Nummulites*, Micropaleontolgy, v. 13, pp. 41–54.

Arthur, T.J., MacGregor, S. & Cameron, S., 2003. Petroleum Geology of Africa: New Themes and Developing Technologies. Geological Society Special Publication No. 207, p. 289

Arthyushkov, E.V., Baer, M.A., Letnilov, F.A. & Ruzhich, v. V., 1991. On the mechanism of graben formation, in A.F. Gangi (Ed.), World Rift Systems, Tectonophysics, v. 197, pp. 99–115.

Askri, H., Belmecheri, A., Benrabah, B., Boudjema, A., Boumendjel, K., Daoudi, M., Drid, M., Ghalem, T., Docca, A.M., Ghandriche, H., Ghomari, A., Guellati, N., Khennous, M., Louici, R., Naili, H., Takherist, D. & Terkamani, M., 1995. Geology of Algeria. Schlumberger WEC Sonatrach, p. 93.

Assereto, R. & Benelli, F., 1971. Sedimentology of the pre-Cenomanian Formations of the Jebel Gharian, Libya, in C. Gray (Ed.), Symposium on the Geology of Libya, Tripoli, pp. 37–85.

Atef, A., 1988. Source rock evaluation of the Brown Limestone Formation (U.Cretaceous) of the Gulf of Suez, Egypt. 9th Exploration and Production Conf., Egyptian General Petroleum Corp. (EGPC), Cairo, Egypt, November 1988, p. 26.

Atherton, M.P., Lagha, S. & Flinn, D., 1991. The geochemistry and petrology of the Jabal Arknu and Jabal Al Awaynat alkaline ring complexes of SE Libya, in M.J. Salem, M.T. Busrewil & A.M. Ben Ashour (Eds.), v. VII, The Geology of Libya, Amsterdam, Elsevier, pp. 2559–2576.

Atrops, F. & Alméras, Y., 2005. Les Brachiopodes du Callovien de l Ouarsenis (Tell algérien): paléontologie, biostratigraphie et paléoenvironnements. Revue de Paléobiologie, Genève, v. 24, no. 2, pp. 563–595.

Attar, A., Fournier, J., Candilier, A.M. & Coquel, R., 1980. Etude palynologique du Dévonien terminal et du Carbonifère inférieur du Bassin d'Illizi (Fort Polignac) Algérie. Revue de l'Institut Français du Pétrole, v. 35, pp. 585–619.

Attar, M. & Hammat, M., 2001. Algeria's hydrocarbon potential (IV). Schlumberger WEC Sonatrach, p. 17.

Attia, M.I., 1954. Deposits in the Nile Valley and the Delta. Geol. Surv. Egypt, p. 356

Attia, M.I., 1955. Topography, geology and iron ore deposits of the district east of Aswan. Geol. Surv. Egypt, Cairo, p. 262.

Attia, M.I., 1956. Manganese deposits of Egypt, 20th Intern. Geol. Cong., Mexico, Sym. Sob. Yacim. Mang. v. 1 (Africa), pp. 143–171.

Attia, M.I. & Murray, G.M., 1951. Lower Cretaceous ammonites in marine intercalations in the "Nubian Sandstone" of the Eastern Desert of Egypt. Q.J. Geol. Soc., London, v. 107, pp. 442–443.

Aubry, M.P., 1995. Towards an upper Paleocene-lower Eocene high resolution stratigraphy based on calcareous nannofossil stratigraphy. Israel Journal of Earth Sciences, v. 44, pp. 239–253.

Aubry, M.P. & Ouda, K., 2003. Introduction to the upper Paleocene-Lower Eocene of the Upper Nile Valley: Part I Stratigraphy. Micropaleontology, v. 49, Suppl. 1, p. ii–iv.

Awad, G.H. & Ghobrial, M.G., 1966. Zonal stratigraphy of the Kharga oasis, Geol. Surv. Egypt, Cairo, Paper No. 34, p. 77.

Awad, G.H., 1946. On the occurrence of marine Triassic (Muschelkalk) deposits in Sinai. Bulletin Institut d'Egypte, v. 27, pp. 397–429.

Awad, G.M., 1984. Habitat of oil in Abu Gharadig and Fayium basins, Western Desert, Egypt, AAPG Bull., v. 68, No. 5, pp. 564–573.

Awad, G.M., 1985. A geophysical study on the Abu Gharadig basin Egypt: Geophysics. Soc. of Expl. Geophys., v. 50, pp. 5–15.

Awad, H. 1963. Some aspects of the geomorphology of Morocco related to the Quaternary climate. Geographical Journal, v. 129, pp. 129–139.

Awad, M., Gaber, N., El Sharkawi, E., Bedrous, M.A. & Attia, M.B., 2001. Strategic fuels of the future: Using oil shale as fuel for thermal power generation. World Energy Council, 18th Congress, Buenos Aires, October, 2001, p. 12.

Ayarza, P., Alvarez-Lobato, F., Teixell, A., Arboleya, M.L., Teson, E., Julivert, M. & Charroud, M., 2005. Crustal structure under the central High Atlas mountains (Morocco) from geological and gravity data. Tectonophysics, v. 400, pp. 67–84.

Ayyad, S.M., Abed, M.M. & Abu Zeid, R.H., 1996. Biostratigraphy and correlation of Cretaceous rocks in Gebel Arif El Naga, north-eastern Sinai, based on planktonic foraminifera. Cretaceous Research, v. 17, pp. 263–291.

Azer, M.K. & Stern, R., 2007. Neoproterozoic (835–720) serpentinites in the Eastern Desert, Egypt: Fragments of forearc mantle. Journal of Geology, v. 115, pp. 457–472.

Azer, M.K., 2006. The petrogenesis of late Precambrian felsic alkaline magmatism in south Sinai, Egypt. Acta Geologica Polonica, v. 56, no. 4, pp. 463–484.

Aziz, A. & Ghnia, S., 2009. Distribution of Infracambrian rocks and the hydrocarbon potential within the Murzuq and Al Kufrah basins, NW Africa. In J. Craig, J.W. Thurrow, B. & Y. Abutarruma. (Eds.): Global Neoproterozoic petroleum systems: the emerging potential in North Africa. Geol. Soc. Spec. Pub. No. 326, pp. 211–230.

Azizi-Samir, M.R., Ferrandini, J. & Tane, J.L., 1990. Late Pan-African (580–560 Ma.) tectonic activity and volcanism in the Anti-Atlas Mountains of Morocco: Geodynamic interpretation on the scale of north-west Africa. J. Afr. Earth Sci., v. 10, pp. 549–563.

Azzouni-Sekkal, A. & Bonin, B., 1998. Les granotoides de type A dans la chaine Pan Africaine, au Hoggar: L'exemple des complexes annulaires "Taourirts', in Geology of the Arab World, Fourth International Conference, Cairo University, Cairo, pp. 87–88.

Azzouni-Sekkal, A., Liégeois, J.P., Bechiri-Benmerzoug, F., Belaidi-Zinet, S. & Bonin, B., 2003. The "Taourirt" magmatic province, a marker of the closing stage of the Pan-African orogeny in the Tuareg Shield: review of available data and Sr-Nd isotope evidence. Journal of African Earth Sciences, v. 37, nos. 3–4, pp. 331–350.

B.R.P., C.E.P., C.F.P. (A). COPEFA. C.P.T. (L), et al., 1964. Essai de nomenclature lithostratigraphique du Cambro-ordovicien saharien, Colloque tenu par les compagnies pétrolieres, Paris (1962), Mem. Hors-ser. Soc. Géol. Fr., No. 2, p. 56, Paris.

Bachmann, M. & Kuss, J., 1998. The Middle Cretaceous carbonate ramp of the northern Sinai: sequence stratigraphy and facies distribution. In: V.P. Wright & T.P. Burchette (Ed.): Carbonate Ramps. Geological Society Special Publication 149, pp. 253–280.

Bachmann, M., Bassiouni, M.A.A. & Kuss, J., 2003. Stratigraphic modelling, graphic correlation and cyclicities of mid-Cretaceous platform carbonates – northern Sinai, Egypt. Palaeogeography, Palaeoclimatology, Palaeoecology, v. 200, 131–162.

Badalini, G., Redfern, J. & Carr, I., 2002. Introduction: A synthesis of current understanding of the structural evolution of North Africa. J. Petroleum Geology, v. 25, no. 3, pp. 249–258.

Badri, M., 2000. Shear anisotropy applications in production optimization, Western Desert, Egypt. 70th Ann. Internat. Mtg., Soc. Of Expl. Geophys., Calgary, pp. 1695–1698.

Bagge, A.M. & Keeley, M.L., 1994. The oil potential of Mid.-Jurassic coals in northern Egypt. In A.C. Scott & A.J. Fleet (Eds.), Coal and Coal-bearing Strata as Oil-prone Source Rocks?, Geological Society of London, Special Publication No. 77, pp. 183–200.

Bagnold, R.A., 1935. Libya Sands. travel in a dead world. Hodder and Stoughton, p. 351.

Bagnold, R.A., 1936. The Libyan Desert. Journal of the Royal African Society, v. 35, no. 140, pp. 294–305.

Bagnold, R.A., 1939. An Expedition to the Gilf Kebir and Uweinat, 1938. The Geographical Journal, v. 93, no. 4, pp. 281–313.

Bagnold, R.A., 1941. The Physics of Blown Sand and Desert Dunes, Chapman & Hall, p. 265. (reprinted 1954. new edition 1973).

Bagnold, R.A., 1945. Early days of the Long Range Desert Group. The Geographical Journal, v. 105, no. 1/2, pp. 30–42.

Bagnold, R.A. & Harding King, W.J., 1931. Journeys in the Libyan Desert. The Geographical Journal, v. 78, no. 6, pp. 524 and pp. 526–535.

Bagnold, R.A., Myers, O.H., Peel, R.F. & Winkler, H.S., 1939. An expedition to the Gilf Kebir and 'Uwenat, 1938. The Geographical Journal, v. 93, no. 4, pp. 281–312.

Baidder, L., Raddi, Y., Tahiri, T. & Michard, A., 2008. Devonian extension of the Pan-African crust north of the West African craton, and its bearing on the Variscan foreland deformation: evidence from eastern Anti-Atlas (Morocco). Geological Society, London, Special Publications, v. 297, pp. 453–465.

Baioumy, H.M., 2009. Mineralogical and geochemical characterization of the Jurassic coal from Egypt. Journal of African Earth Sciences, v. 54, pp. 75–84.

Baird, D.W., Aburawi, R.M. & Bailey, N.J.L., 1996. Geohistory and petroleum in the Central Sirt Basin, in M.J. Salem, M.T. Busrewil, A.A., Misallati, M.S. (Eds.), The Geology of Sirt Basin, v. III, Amsterdam, Elsevier, pp. 3–56.

Bakr, M.M.Y. & Wilkes, H., 2002. The influence of facies and depositional environment on the occurrence and distribution of carbazoles and benzocarbazoles in crude oils: a case study from the Gul of Suez, Egypt. Organic Geochemistry, v. 33, pp. 561–580.

Baldridge, W.S., Eyal, Y., Bartov, Y., Steinitz, G. & Eyal, M., 1991. Miocene magmatism of Sinai related to the opening of the Red Sea, in A.F. Gangi (Ed.), World Rift Systems, Tectonophysics, v. 197, pp. 181–201.

Balducchi, A. & Pommier, G., 1970. Cambrian oil field of Hassi Messaoud, Algeria. In: Geology of Giant petroleum Fields. AAPG Mem. No. 14, pp. 477–488.

Ball, J. & Beadnell, H.J.L., 1903. Baharia Oasis. its Topography and Geology. Egyptian Geological Survey Dept., Cairo, p. 84.

Ball, John, 1900. Kharga Oasis. its Topography and Geology, Part II. Egyptian Geological Survey Dept., Cairo, p. 116.

Ball, J., 1902. On the topographical and geological results of a reconnaissance Survey of Jebel Garra and the Oasis of Kurkur. Egypt Servey Department, p. 40.

Ball, J., 1907. A description of the First or Aswan cataract of the Nile. Egypt. Geol. Survey, p. 121.

Ball, J., 1910. On the origin of the Nile Valley and the Gulf of Suez. Geological Magazine, v. 7, no. 5, pp. 71–76.

Ball, J., 1911. The Gulf of Suez. Geological Magazine, v. 8, no. 5, pp. 1–10.

Ball, J., 1912a. The Geography and Geology of South-Eastern Egypt. Survey Department, Egypt, 12.

Ball, J., 1912b. The meteorite of El-Nakhla El-Baharia. Egypt. Survey Department, Cairo, Paper 25.

Ball, J., 1913. A brief note on the phosphate deposits of Egypt. Egypt. Survey Department, Cairo, 6.p.

Ball, J., 1916a. Military Notes on Western Egypt. Ministry of Defence, London.

Ball, J., 1916b. The geology and geography of west central Sinai, Egypt. Egypt. Geol. Survey, Cairo, p. 219.

Ball, J., 1920. The African Rift Valleys. The Geographical Journal, v. 56, no. 6, pp. 505–510.

Ball, J., 1927a. Problems of the Libyan Desert. The Geographical Journal, v. 70, no. 1, pp. 21–38.

Ball, J., 1927b. Problems of the Libyan Desert (Continued). The Geographical Journal, v. 70, no. 2, pp. 105–128.

Ball, J., 1927c. Problems of the Libyan Desert (Continued). The Geographical Journal, v. 70, no. 3, pp. 209–224.

Ball, J., 1928. Remarks on "Lost" Oases of the Libya Desert. The Geographical Journal, v. 72, no. 3, pp. 250–258.

Ball, J., 1939. Contributions to the geography of Egypt. Government Press, Bulaq, 308.

Balusseau, B. & Cariou, E., 1983. Sur l'âge des series du Jurassique moyen et supérieur du Djebel Zaress (Tunisie). Geobios, v. 16, no. 1, pp. 117–123.

Bamoumen, H., Aarab, El.M. & Soulaimani, A., 2008. Evolution tectono-sédimentaire et maagmatique des bassins viséen supérieur d'Azrou-Khénifra et des Jebilet orientales (Meseta marocaine). Estudios Geológicos, v. 64, no. 2, pp. 107–122.

Bandel, K., Kuss, J. & Malchus, N., 1987. The sediments of Wadi Qena (Eastern Desert, Egypt), J. Afr. Earth Sci., v. 6, pp. 427–455.

Banerjee, S., 1980. Stratigraphic Lexicon of Libya, Industrial Research Centre, Bull. 13, Tripoli, p. 300.

Baouche, R., Nedjari, A. & Eladj, S., 2003. Analysis and interpretation of environment sequence models in Hassi R'Mel Field in Algeria. Proceedings of the 3rd WSEAS Int. Conference on Energy Planning, Energy Saving, Environmental Education, pp. 38–48.

Barakat, A.O., El-Gayar, M.S. & Moustafa, A.R., 1997. An organic geochemical investigation of crude oils from Egypt. Fuel Processing Technology, v. 51, pp. 127–135.

Barakat, A.O., El-Gayar, M.S., Moustafa, A.R. & Rullkötter, J., 1997. Source-dependant biomarker properties of five crude oils from the Gulf of Suez, Egypt. Organic Geochemistry, v. 26, no. 7/8, pp. 441–450.

Barale, G. & Ouaja, M., 2001. Découverte de nouvelles flores avec des restes á affinities angiospermiennes dans le Crétacé inférieur du Sud Tunisien. Cretaceous Research, v. 22, no. 2, pp. 131–143.

Barathon, J.-J., El Abbasi, H., Joly-Saad, M.-C., Lechevalier, C., & Malek, F., 2000. Mise au point sur les formations holocènes dans le Rif oriental (Maroc)/A chronology of Holocene deposits in the eastern Rif mountains (Morocco). Géomorphologie: relief, processus, environnement, v. 6, no. 4, pp. 221–238.

Barber, P.M., 1981. Messinian subaerial erosion of the Proto-Nile Delta, Marine Geology, v. 44, pp. 253–272.

Barbero, L., Teixell, A., Arboleya, M.L., Del Rio, P., Reiners, P.W. & Bougadir, B., 2007. Jurassic-to-present thermal history of the central High Atlas (Morocco) assessed by low-temperature thermochronology. Terra Nova, v. 19, pp. 58–64.

Barbieri, R., 1996. Micropaleontology of the Rakb Group (Cenomanian to Early Maastrichtian) in the Hameimat Basin, Northern Libya, in The Geology of Sirt Basin, v. I, Elsevier, Amsterdam, pp. 185–194.

Barbieri, R. & Ori, G.G., 2000. Neogene palaeoenvironmental evolution in the Atlantic side of the Rifian Corridor (Morocco). Palaeogeography, Palaeoclimatology, Palaeoecology, v. 63, pp. 1–31.

Bard, K., 1994. The Egyptian Predynastic: A review of the evidence. Journal of Field Archaeology, Fall 1994, pp. 265–288.

Bardet, N., Pereda Suberbiola, X., Iarochène, M., Bouya, B. & Amaghzaz, M., 2005. A new species of Halisaurus from the Late Cretaceous phosphates of Morocco, and the phylogenetical relationships of the Halisaurinae (Squamata: Mosasauridae). Zoological Journal of the Linnean Society, v. 143, pp. 447–472.

Bardet, N., Houssaye, A., Rage, J.-C. & Suberbiola, X.P., 2008. The Cenomanian-Turonian (late Cretaceous) radiation of marine squamates (Reptelia): the role of the Mediterrean Tethys. Bulletin de la Societe Geologique de France, v. 179, no. 6, pp. 605–622.

Baric, G. & Cota, L., 2003. A geochemical investigation of hydrocarbons in the Belina-1 offshore well. In M.J. Salem, Khaled, M. and Seddiq, H.M. (Eds.), The Geology of Northwest Libya, v. III, pp. 79–90.

Baric, G., Spanic, D. & Maricic, M., 1996. Geochemical characterization of source rocks in NC157 block (Zaltan Platform), Sirt Basin, in M.J. Salem, A.J. Mouzughi & O.S. Hammuda (Eds.), The Geology of Sirt Basin, v. II, Elsevier, Amsterdam, pp. 541–553.

Barr, F.T. (Ed.), 1968a. Geology and archeology of northern Cyrenaica, Libya. Tenth Annual Field Conference, Petroleum Exploration Society, Tripoli.

Barr, F.T., 1968b. Late Cretaceous planktonic foraminifera from the coastal area east of Susa (Apollonia), north-east Libya, J. Paleontology, v. 42, No. 2, pp. 308–321.

Barr, F.T., 1972. Cretaceous biostratigraphy and planktonic foraminifera of Libya, Micropaleontology, v. 18, No. 1, pp. 1–46.

Barr, F.T. & Bergerren, W.A., 1980. Lower Tertiary biostratigraphy and tectonics of northeastern Libya, in M.J., Salem & M.T., Busrewil (Eds.), The Geology of Libya, v. I, London, Academic Press, pp. 163–192.

Barr, F.T. & Hammuda, O.S., 1971. Biostratigraphy and planktonic zonation of the Upper Cretaceous Atrun Limestone and Hilal Shale, Northeastern Libya, Proc. 2nd Intern. Conf. Planktonic Microfossils, Rome, pp. 27–38.

Barr, F.T. & Walker, B.R., 1973. Late Tertiary channel system in northern Libya and its implication on Mediterranean sea level changes. In W.B. Rayan & K.J., Hsü et al. (Eds.): Initial Report Deep Sea Project, Leg 13, pp. 1244–1251.

Barr, F.T. & Weegar, A.A., 1972. Stratigraphic nomenclature of the Sirte Basin, Libya. Petroleum Exploration Society, Tripoli, p. 179.

Barrat, J.A., Jahn, B.M., Amosse, J. & Diemer, E., 1997. Geochemistry and origin of Libyan Desert glasses. Geochimica et Cosmochimica Acta, v. 61, no. 9, pp. 1953–1959.

Barron, T. & Hume, W.F., 1900. Nôtes sur la géologie du Desert oriental de l'Egypte. C.R. Congr. Géol. Int., VIIIe session, Paris, pp. 867–912.

Barron, T. & Hume, W.F., 1901. The geology of the Eastern Desert of Egypt. Geol. Mag., London, v. 8, pp. 154–161.

Barron, T. & Hume, W.F., 1902. Topography and Geology of the Eastern Desert of Egypt, Central Portion. Geological Survey Report, Cairo, p. 331.

Barron, T., 1904. On the occurrence of lower Miocene beds between Cairo and Suez. Geol. Mag., v. 1, pp. 603–608.

Barron, T., 1905. On the age of the Gebel Ahmar sands and sandstones. Geol. Mag., v. 2, pp. 58–62.

Barron, T., 1907a. The Topography and Geology of the District between Cairo and Suez. Survey Department, Egypt, p. 133.

Barron, T., 1907b. The Topography and Geology of the Peninsula of Sinai (South-Eastern Portion). Survey Department, Egypt, p. 241.

Barsotti, G., 1963. Paleocenic ostracods of Libya (Sirte Basin) and their wide African distribution, Rev. Inst. Fr. Pétrole, v. 18, No. 11, pp. 1520–1535.

Barsoum, K., Aiolfi, C., Dalla, S. & Kamal, M., 1998. Evolution and hydrocarbon occurrence in the Plio-Pleistocene succession of the Egyptian Mediterranean margin: Examples from the Nile Delta. Proceedings of the 14th Petroleum Conference. The General Petroleum Corporation, Cairo, pp. 386–401.

Barsoum, T.K., 1995 (Abs.). Factors controlling the deposition of the pre-Upper Cretaceous clastics and hydrocarbon accumulation in the Sirte Basin, in First Symposium on the Hydrocarbon Geology of North Africa, Geol. Soc. of London Petroleum Group, p. 11.

Barthel, K.W. & Boettcher, R., 1978. Abu Ballas Formation (Tithonian/Berriasian. Southwestern Desert, Egypt) a significant stratigraphic unit of the former "Nubian Series", Mitt. Bayer. Saatsslg. Paläont. Hist. Geol., 18, pp. 153–166.

Barthel, K.W. & Herrmann-Degen, W., 1981. Late Cretaceous and Early Tertiary Stratigraphy in the Great Sand Sea and its SE margins (Farafra and Dakhla Oases), SW Desert, Egypt, Mitt. Bayer. Staatsslg. Paläont. Hist. Geol., v. 21, pp. 141–182.

Barthoux, J.C., 1922. Chronologie et déscription des roches ignées du Désert arabique. Inst. Égypte Mém., t.5, p. 262.

Barthoux, J.C. & Douvillé, H., 1913. Le Jurassic dans le desert a l'est l'Ismuthe de Suez. Compte Rendus de l'Academie des Sciences, Paris, 157, pp. 265–268.

Barthoux, J.C. & Frittel, P.H., 1925. Flore crétacée du grès de Nubie. Inst. Égypte Mém., t.7, pp. 65–119.

Barton, E.W., 1979. Samah Formation, Carbonate Accumulation on a Fault Block, Sirte Basin, Libya, Unpub. M.Sc. thesis, Juniata College, p. 64.

Bartov, Y. & Steinitz, G., 1977. The Judea and Mount Scorpus groups in the Negev and Sinai with trend surface analysis of the thickness data, Israel J. Earth Sci., v. 26, pp. 119–148.

Bartov, Y., Lewy, Z., Steinitz, G. & Zak, I., 1980. Mesozoic and Tertiary stratigraphy, paleogeography and structural history of the Gebel Arif en Naqa area, Eastern Sinai. Israel J. Earth Sci. 29, pp. 114–139.

Bassiouni, M.A.A., 1965. Ostracoden aus dem "Pliozän" von Kom el Shelul, Pyramiden-Plateau, Gizeh (Ägypten), Geol. Jb., v. 82, pp. 631–654.

Bassiouni, M.A.A., 1969a. Ostracoden aus dem Eozän von Ägypten 1: Trachyberidinae, Geol. Jb., v. 87, pp. 383–426.

Bassiouni, M.A.A., 1969b. Ostracoden aus dem Eozän von Ägypten 2: Die Unterfamilien Hemicythereinae Thaerocytherinae und Camplocytherinae, Geol. Jb., v. 88, pp. 203–234.

Bassiouni, M.A.A., 1971. Ostracoden aus dem Eozän von Ägypten 3: Die Unterfamilien Brachycytherinae und Buntoniinae, Geol. Jb., v. 89, pp. 169–192.

Bassiouni, M.A.A., Boukhary, M., Shama, K. & Blondeau, A., 1984. Middle Eocene Ostracodes from Fayoum, Egypt, Géol. Médit., v. 11, pp. 1–194.

Bassiouni, M.A.A. & Luger, P., 1990. Maastrichtian to Early Eocene ostracoda from southern Egypt, paleontology, paleoecology, paleobiogeography and biostratigraphy, Berliner geowiss. Abh.(A)120, pp. 755–928.

Basta, E.Z. & Amer, H.I., 1969. El-Gedida iron ores and their origin, Bahariya Oasis, Western Desert, U.A.R., Economic Geology, v. 64, pp. 424–444.

Basta, F.F., Maurice, A.E., Bakhit, B.R., Ali, K.A., Manton, W.I., 2011. Neoproterozoic contaminated MORB of Wadi Ghadir ophiolite, NE Africa: Geochemical and Nd and Sr isotopic constraints. Journal of African Earth Sciences, v. 59, pp. 227–242.

Basta, F. & Stern, R.J., 1998 (Abs.). Geochemistry and Rb/Sr geochronology of some gneisses from the Eastern Desert and Sinai, Egypt. In Geology of the Arab World, Fourth International Conference, Cairo University, Cairo, pp. 92–93.

Batten, D.J. & Uwins, P.J.R., 1985. Early Cretaceous (Aptian-Cenomanian) Palynomorphs, in B. Thusu & B. Owens (Eds.), Palynostratigraphy of North-East Libya. J. Micropaleontology, v. 4, part 1, pp. 151–167.

Baudet, D., 1988. Precambrian palynomorphs from northeast Libya, in A. El-Arnauti, B. Owens & B. Thusu (Eds.), Subsurface Palynostratigraphy of Northeast Libya, Benghazi, Libya, Garyounis University Publications, pp. 17–25.

Bauer, J., Kuss, J. & Steuber, T., 2002. Platform Environments, Microfacies and Systems Tracts of the Upper Cenomanian – Lower Santonian of Sinai, Egypt. Facies, 47, 1–26.

Bauer, J., Kuss, J. & Steuber, T., 2003. Sequence architecture and carbonate platform configuration (Late Cenomanian–Santonian), Sinai, Egypt. Sedimentology, v. 50, pp. 387–414.

Bauer, J., Marzouk, A.M., Steuber, T. & Kuss, J., 2001. Lithostratigraphy and biostratigraphy of the Cenomanian -Santonian strata of Sinai, Egypt. Cretaceous Research, v. 22, pp. 497–526.

Bauer, J., Steuber, T., Kuss, J. & Heimhofer, U., 2004. Distribution of shallow-water benthics: The Cenomanian–Turonian carbonate platform sequences of Sinai, Egypt. Proceedings of the 5th International Congress on Rudists. Courier Forschungsinstitut Senckenberg, v. 247, pp. 207–231.

Bauerman, H., 1869. Notes on a geological reconnaissance made in Arabia Petraea in the Spring of 1868. Q.J. Geological Society of London, v. 25, pp. 17–38.

Bauman, A., El Chair, M. & Thiedig, F., 1992. Pan-African granites from deep wells of Murzuk Basin, Fezzan, Western Libya, N. Jb. Geol. Paläont. Mh., v. 1, pp. 1–14.

Bayoumi, A.I. & Lotfy, H.I., 1989. Modes of structural evolution of Abu Gharadig Basin, Western Desert of Egypt as deduced from seismic data, J. Afr. Earth Sci., v. 9, No. 2, pp. 273–287.

Be'eri-Shlevin, Y., Katzir, Y. & Whitehouse, M., 2009. Post-collisional tectonomagmatic evolution in the northern Arabian-Nubian Shield: time constraints from ion-probe U-Pb dating of zircon. Journal of the Geological Society, v. 166, no. 1, pp. 71–85.

Beadnell, H.J.L., 1902. The Cretaceous Region of Abu Roash, near the Pyramids of Giza, Surv. Dept., Cairo, p. 48.

Beadnell, H.J.L., 1905a. The relations of the Eocene and Cretaceous systems in the Esna-Aswan reach of the Nile Valley, Quart. J. Geol. Soc. London, v. 61, pp. 667–678.

Beadnell, H.J.L., 1905b. The topography and geology of Fayium Province of Egypt, Surv. Dept., Cairo, p. 101.

Beadnell, H.J.Ll., 1901. On some recent discoveries in the Nile Valley and Libyan Desert. Geol. Magazine, v. 4, no. 8.

Beadnell, H.J.Ll., 1902. The Cretaceous region of Abu Roash, near the Pyramids of Giza. Geological Survey Report, 1900, Part II. Egypt. Geol. Survey Dept., Cairo, p. 48.

Beadnell, H.J.Ll., 1909a. An Egyptian Oasis: an account of the oasis of Kharga in the Libyan Desert, with special reference to its history, physical geography, and water supply. J. Murray, p. 248.

Beadnell, H.J.Ll., 1909b. The Peninsula of Sinai. Reviews. The Geographical Journal, v. 34, no. 6, pp. 667–668.

Beadnell, H.J.Ll., 1909c. The relation of the Nubian Sandstone and the crystalline rocks south of the Oasis of Kharga (Egypt). Q.J. Geol. Soc., London, v. 65, pp. 41–54.

Beadnell, H.J.Ll., 1910. The sand-dunes of the Libyan Desert. The Geographical Journal, v. 36, no. 3, pp. 367–368.

Beadnell, H.J.Ll., 1926. Central Sinai. The Geographical Journal, v. 67, no. 5, pp. 385–398.

Beadnell, H.J.Ll., 1927. The Wilderness of Sinai. Arnold, London, p. 180.

Beadnell, H.J.Ll., 1931. "Zerzura". The Geographical Journal, v. 77, pp. 245–250.

Beauchamp, J., Omer, M.K. & Perriaux, J., 1990. Provenance and dispersal of Cretaceous clastics in northeastern Africa: climatic and structural setting, J. Afr. Earth Sci., v. 10, No. 1/2, pp. 243–251.

Beauchamp, W., Barazangi, M., Demnati, A. & El Alji, M., 1996. Intracontinental Rifting and Inversion: Missour Basin and Atlas Mountains, Morocco. AAPG Bulletin, v. 80, no. 9, pp. 1459–1482.

Beauchamp, W., Barazangi, M., Demnati, A. & El Alji, M., 1997. Inversion of synrift normal faults in the High Atlas Mountains, Morocco. The Leading Edge, August, 1997, pp. 1171–1175.

Bebout, D.G. & Pendexter, C., 1975. Secondary carbonate porosity as related to Early Tertiary depositional facies, Zelten field, Libya, AAPG Bull., v. 59, pp. 665–693.

Beccaluva, L., Azzouni-Sekkal, A., Benhallou, A., Bianchini, G., Ellam, R.M., Marzola, Siena, M.F. & Stuart, F.M., 2007. Intracratonic asthenosphere upwelling and lithosphere rejuvenation beneath the Hoggar swell (Algeria): Evidence from HIMU metasomatised lherzolite mantle xenoliths. Earth and Planetary Science Letters, v. 260, nos. 3–4, pp. 482–494.

Beccaluva, L., Bianchini, G., Ellam, R.M., Marzola, M., Oun, K.M., Siena, F. & Stuart, F.M., 2008. The role of HIMU metasomatic components in the North African lithospheric mantle: petrological evidence from the Gharyan lherzolite xenoliths, NW Libya. In: M. Coltorti & M. Gregoire (Eds): Metasomatism in Oceanic and Continental Lithospheric Mantle. Geological Society, London, Special Publications, v. 293, pp. 253–277.

Becker, T.R., Aboussalam, S.Z., Bockwinkel, J., Ebbighausen, V., El Hassani, A. & Nübel, H., 2004a. Upper Emsian stratigraphy at Rich Tamelougou near Torkoz (SW Dra Valley, Morocco). Devonian of the western Anti Atlas: correlations and events. Doc. Inst. Sci, Rabat, v. 19, pp. 85–89.

Becker, T.R., Bockwinkel, J., Ebbighausen, V., Aboussalam, S.Z., El Hassani, A. & Nübel, H., 2004b. Lower and Middle Devonian stratigraphy and faunas at Bou Tserfine near Assa (Dra Valley, SW Morocco). Devonian of the western Anti Atlas: correlations and events. Doc. Inst. Sci, Rabat, v. 19, pp. 125–139.

Bédir, M. Tlig, S., Bobier, C. & Assaoui, N., 1996. Sequence stratigraphy, basin dynamics, and petroleum geology of the Miocene from Eastern Tunisia, AAPG Bull., v. 80, no. 1, pp. 63–81.

Beekman, F., Badsi, M. & van Wees, J.-D., 2000. Faulting, fracturing and in situ stress prediction in the Ahnet Basin, Algeria – a finite element approach. Tectonphysics, v. 320, pp. 311–329.

Belaid, A., Krooss, B.M. & Littke, R., 2010. Thermal history and source rock characterization of a Paleozoic section in the Awbari Trough, Murzuq Basin, SW Libya. Marine and Petroleum Geology, v. 27, no. 3, pp. 612–632.

Belkhodja, L. & Bignot, G., 2004. La transgression thanétienne (Paléocène supérieur) dans l'Aurès occidental (Algérie), daprès les associations de Foraminifères de la coupe d'El Kantara. Revue de Micropaléontologie, v. 47, no. 1, pp. 1–12.

Bellaiche, G. & Blanpied, C., 1979. Apercu néotectonique, in P.F. Burollet. P. Clairefond & E. Winnock (Eds.), Géologie Méditerranéenne, La Mer Pélagienne, Annales de l'Université de Provence, tome VI, No. 1, Marseille, pp. 50–59.

Bellardi, L., 1851. Liste de fossils Nummulitiques d'Egypte, de la collection de la Musée royale de Turin. Bull. Soc. Géol. Fr., t.VIII, pp. 261–262.

Bellardi, L., 1854. Catalogo regionato dei Fossili Nummulitici d'Egitto, della raccolta del regio Museo Mineralogico di Torino. Mem. R. Acad. Sci. Torino, ser. II, T. XV, pp. 171–204.

Bellini, E. & Massa, D., 1980. A stratigraphic contribution to the Paleozoic of southern basins of Libya, in M.J., Salem & M.T., Busrewil (Eds.), The Geology of Libya, v. I, London, Academic Press, pp. 3–56.

Bellini, E., Giori, I., Ashuri, O. & Benelli, F., 1991. Geology of Al Kufrah Basin, in M.J. Salem. A.M. Sbeta & M.R. Bakbak (Eds.), The Geology of Libya, v. VI, Amsterdam, Elsevier, pp. 2155–2184.

Bellion, Y., Saint-Marc, P. & Damotte, R., 1989. Contribution à la connaissance des dépots marins au passage Crétacé-Tertiaire dans la vallée du Tilemsi (Nord-Mali), J. Afr. Earth Sci., v. 9, No. 1, pp. 187–194.

Ben Avraham, Z. & Mart, Y., 1981. Late Tertiary structure and stratigraphy of north Sinai continental margin, AAPG Bull., v. 65, pp. 1135–1145.

Ben Ferjani, A., Burollet, P.F. & Mejir, F., 1990. Petroleum Geology of Tunisia, Entreprise Tunisienne d'Activités Petroliéres, Tunis, p. 194.

Ben Ismail, M.H. & M'Rabet, A., 1990. Evaporite, carbonate, and siliciclastic transitions in the Jurassic sequences of southeastern Tunisia. Sedimentary Geology, v. 66, pp. 65–82.

Bencherif, D., 2003. Giant Hassi R'Mel gas field. AAPG Hedberg Conference "Paleozoic and Triassic Petroleum Systems in North Africa" February 18–20, 2003, Algiers, Algeria.

Bender, A.A., Coelho, F.M. & Bedregal, R.P., 1996. A basin modelling study in the southeastern part of Sirt Basin, Libya, in The Geology of Sirt Basin, v. I, Elsevier, Amsterdam, pp. 139–155.

Benest, M., Atrops, F., Alméras, Y. & Benosman, B., 1991. Découverte et dynamique d'une transgression au Callovien sur le socle tardi-hercynien, dans le domaine sud-tellien (n, Algérie). Comptes Rendus de l'Académie des Sciences de Paris, v. 313, no. 2, pp. 1555–1562.

Benfield, A.C. & Wright, E.P., 1980. Post-Eocene sedimentation in the Eastern Sirt Basin, Libya Libya, in M.J. Salem & M.T. Busrewil (Eds.), The Geology of Libya, v. II, London, Academic Press, pp. 463–499.

Benfrika, E.M. & Raji, M., 2003. Analyse biostratigraphique des conodontd du Silurien supérieur de la zone de Rabat-Tiflet (Nord-ouest de la Meseta, Maroc). Bulletin de la Société Géde France, v. 174, no. 4, pp. 337–342.

Bennacef, A., Beuf, S.,Biju-Duval, B., De Charpal, O., Gariel, O. & Rognon, P., 1971. Example of cratonic sedimentation: Lower Paleozoic of Algerian Sahara, AAPG Bull., v. 55, No. 12, pp. 2225–2245.

Benniran, M.M., Taleb, M.T. & McCrossan, R.G., 1988 (Abs.). Geological history of the West Libya offshore and adjoining regions, in Mediterranean Basin Conf. & Exhib., Nice, AAPG Bull., v. 72, no. 8, p. 988.

Bensaid, M., Termier, H. & Termier, G., 1985. The Moroccan Varicides. In C.M. Diaz (Ed.), 1985: The Carboniferous of the World, II. Australia, Indian Subcontinent & North Africa: North Africa. IUGS Publication No. 20, Instituto Geológico y Minero de España and Empresa Nacional ADARO de Investigaciones Mineras, s.a., pp. 336–344.

Bensalah, M., Benest, M. & Truc, G., 1991. Continental detrital deposits and calcretes of Eocene age in Algeria (South of Oran and Constantine). Journal of African Earth Sciences, v. 12, pp. 247–252.

Bensalem, H., 2002. The Cretaceous-Paleogene transition in Tunisia: general overview. Palaeogeography, Palaeoclimatology, Palaeoecology, v. 178, nos. 3–4, pp. 139–143.

Benson, R.H., 1972. Ostracodes as indicators of threshold depth in the Mediterranean during the Pliocene, in D.J. Stanley (Ed.), The Mediterranean Sea, a Natural Sedimentation Laboratory, Dowden, Hutchinson & Ross, Inc., Stroudsburg, Pennsylvania, pp. 63–73.

Benssaou, M. & Hamoumi, N., 2001. L'Anti-Atlas occidental du Maroc: étude sedimentologique et reconstitutions paléogéographiques au Cambrien inferieur. Journal of African Earth Sciences, v. 32, no. 3, pp. 351–372.

Benssaou, M. & Hamoumi, N., 2003. The Lower-Cambrian western Anti-Atlasic graben: Tectonic control of paleogeography and sequential organisation, Comptes Rendus Géoscience, v. 335, pp. 297–305.

Benton, M.J., Bouaziz, S., Buffetaut, E., Martill, D., Ouaja, M., Soussi, M. & Trueman, C., 2000. Dinosaurs and other fossil vertebrates from fluvial deposits in the Lower Cretaceous of Southern Tunisia. Palaeogeogr. Palaeoclimatol. Palaeoecol. v. 157, pp. 227–246.

Bentor, Y.K. & Eyal, M., 1987. The Geology of Southern Sinai, its implication for the evolution of the Arabo-Nubian massif, The Israeli Academy of Sciences and Humanities, Jerusalem, p. 484.

Bentor, Y.K., 1985. The crustal evolution of the Arabo-Nubian massif with special reference to the Sinai Peninsula, Precambrian Research, v. 28, pp. 1–74.

Berggren, W.A., 1974. Paleocene benthonic foraminiferal biostratigraphy, biogeography and paleoecology of Libya and Mali, Micropaleontology, v. 20, no. 4, pp. 449–465.

Bergström, S.M. & Massa, D., 1991. Stratigraphic and biostratigraphic significance of Upper Ordovician conodonts from Northwestern Libya, in M.J. Salem. O.S. Hammuda & B.A. Eliagoubi (Eds.), The Geology of Libya, v. IV, Elsevier, Amsterdam, pp. 1323–1342.

Berkowski, B., 2006. Vent and mound rugose coral associations from the Middle Devonian of Hamar Laghdad (Anti-Atlas, Morocco). Geobios, v. 39, pp. 155–170.

Berkowski, B., 2008. Emsian deep-water Rugosa assemblages of Hamar Laghdad (Devonian), Anti-Atlas, Morocco. Palaeontographica, Abt. A, v. 284, 17–68.

Bermúdez-Lugo, O., 2005, The Mineral Industries of Morocco and Western Sahara, 2005 Minerals Yearbook. U.S. Department of the Interior, U.S. Geological Survey, pp. 1–5.

Bernasconi, A. Poliani, G. & Dakshe, A., 1991. Sedimentology, petrography and diagenesis of Metlaoui Group in the offshore northwest of Tripoli, in M.J. Salem & M.N. Belaid (Eds.), The Geology of Libya, v. V, Elsevier, Amsterdam, pp. 1907–1928.

Bernini, M., Boccaletti, M., Gelati, R., Moratti, G., Papani, G. & El Mokhtari, J., 1999. Tectonics and sedimentation in the Taza-Guercif Basin, Northern Morocco: Implications for the Neogene evolution of the Rif-Middle Atlas Orogenic System. Journal of Petroleum Geology, v. 22, no. 1, pp. 115–128.

Bernor, R.L. & Scott, R.S., 2003. New interpretations of the systematics, biogeography and paleoecology of the Sahabi hipparions (latest Miocene) (Libya). Geodiversitas, v. 25, no. 2, pp. 297–319.

Bertello, F., Visentin, C. & Ziza, W., 2003. An overview of the evolution of the petroleum systems of the Eastern Ghadames (Hamra) Basin – Libya. AAPG Hedberg Conference: Paleozoic and Triassic Petroleum Systems in North Africa, February 18–28, 2003, Algiers, Algeria, p. 4.

Bertrand, J.M., Michaud, A., Boullier, A.M. & Dautel, D., 1986. Structure and U/Pb geochronology of central Hoggar (Algeria): a reappraisal of its Pan-African evolution, Tectonophysics, v. 5, No. 7, pp. 955–972.

Bertrand-Sarfati, J. & Fabre, J., 1972. Les stromatolites des formations lacustres post-moscoviennes du Sahara septentrional (Algérie). 24 Congr. Géol. Int., Montreal, sect. 7, pp. 458–470.

Bettahar, Y., 2007. La géologie en Algérie (1880–1940); Enjeux coloniaux, démarche scientifique et dispositif académique. La Revue pour l'histoire du CNRS, no. 18, p. 8.

Bettar, I. & Meon, H., 2006. La palynoflore continentale de l'Albien du basin d'Agador-Essaouira (Maroc). Revue de Paléobiologie, Genève, v. 25, no. 2, pp. 593–631.

Beuf, S., Biju-Duval, B., De Charpal, O., Rognon, P., Gariel, O. & Bennacef, A., 1971. Les Grès du Paleozoique Inférieur au Sahara, éditions Technip, Paris, p. 464.

Beyrich, E., 1852. Bericht über die von Overweg auf der Reise von Tripoli nach Murzuk und von Murzuk nach Chat gefundenen Versteinerungen. Z. deutsch. Geol. Ges., 4, pp. 143–151.

Bezan, A.M., 1996. The Palaeocene sequence in Sirt Basin, in M.J. Salem. A.J. Mouzughi & O.S. Hammuda (Eds.), The Geology of Sirt Basin, v. I, Elsevier, Amsterdam, pp. 96–117.

Bezan, A.M. & Belhaj, F. & Hammuda, K., 1996. The Beda Formation in Sirt Basin, in M.J. Salem. A.S. El-Hawat & A.M. Sbeta (Eds.), The Geology of Sirt Basin, v. II, Elsevier, Amsterdam, pp. 135–152.

Bezan, A.M. & Malak, E.K., 1996. Oligocene sediments of Sirt Basin and their hydrocarbon potential, in M.J. Salem; A.J. Mouzughi & O.S. Hammuda (Eds.), The Geology of Sirt Basin, v. 1, Elsevier, Amsterdam, pp. 119–127.

Bezan, A.M., Eliagoubi, B.A., Hammuda, O.S. & Mirza, K., 1993. Maastrichtian subsurface stratigraphy of the Sirt basin, Libya, in Sedimentary Basins of Libya, first Symposium, Geology of Sirt Basin, Tripoli, Libya, p. 11.

Biely, A., Rakus, M., Robison, P. & Salaj, J., 1972. Essai de correlation des formations miocènes au Sud de la dorsale tunisienne. Notes Serv. Géol. Tunisie, v. 38, pp. 73–92.

Biely, A., Rakus, M., Robison, P. & Salaj, J., 1974. Le Négeone de la Tunisie. Ann. Geol. Surv. Egypt, p.?.

Bigazzi, G. & de Michele, V., 1996. New fission-track age determinations on impact glasses. Meteoritics Planetary Sci., v. 31, pp. 234–236.

Biju-Duval, B., Letouzey, J., Montadert, L., Courrier, P., Mugniot, J.F. & Sancho, J., 1974. Geology of the Mediterranean Sea Basins. In C.A. Burke & C.L. Drake (Eds.), The Geology of the Continental Margins, New York, Springer-Verlag, pp. 695–721.

Birchant, A.L., 1952. Sur la découverte du Triass au pied du Djebel Garian (Tripolitaine, Libye), Compte Rendus Acad. des Sciences, Paris, pp. 1456–1458.

Biron, P. & Courtinat, B., 1982. Contribution palynologique à la connaissance du Trias du Haut Atlas de Marrakech, Geobios (Villeurbanne), v. 15, no. 2, pp. 231–235.

Bishay, Y., 1961. Biostratigraphic study of the Eocene in the Eastern Desert between Samalut and Assiut by the large foraminifera, Third Arab Petroleum Congress, Alexandria, p. 7.

Bishay, Y., 1966. Studies on the large foraminifera of the Eocene (the Nile Valley between Assiut and Cairo and SW Sinai), unpub. Ph.D. thesis, Alexandria University, p. 244.

Bishop, W.F. & Debono, G., 1996. The hydrocarbon geology of southern offshore Malta, J. Peroleum Geology, v. 19, No. 2, pp. 129–160.

Bishop, W.F., 1975. Geology of Tunisia and adjacent parts of Algeria and Libya, AAPG Bull., v. 59, pp. 413–450.

Bishop, W.F., 1988. Petroleum geology of East-central Tunisia. AAPG Bull., v. 72, No. 9, pp. 1033–1058.

Bismuth, H. & Bonnefous, J., 1981. The biostratigraphy of carbonate deposits of the Middle and Upper Eocene in northeastern offshore Tunisia, Palaeogeography, Palaeoclimatology, and Palaeoecology, v. 36, pp. 191–211.

Bismuth, H., Cahuzac, B., Poignant, A., Hooyberghs, H.J.F., Saïd-Benzarti, R. & Trigui, A., 2009. Biostratigraphie d'après les foraminifères et paléoenvironnements des séries post-Éocène du sondage Ashtart 28, golfe de Gabès (Tunisie) Biostratigraphy based on Foraminifera and paleoenvironments of the post-Eocene series in the Ashtart 28 drilling, Gulf of Gabes (Tunisia). Revue de Micropaléontologie, v. 52, no. 4, pp. 289–315.

Bitner, M.A. & Boukhary, M., 2009. First record of brachiopods from the Eocene Of Egypt. Record of brachiopods from the Eocene of Egypt. Nat. Croat., Zagreb, v. 18, no. 2, pp. 393–400.

Black, R. & Liégeois, J.-P., 1993. Cratons, mobile belts, alkaline rocks and continental lithospheric mantle: the Pan-African testimony, J. Geol. Soc., London, v. 150, pp. 89–98.

Black, R., Caby, R., Moussine-Pouchkine, A., Bayer, R., Bertrand, J.M., Boullier, A.M., Fabre, J. & Lesquer, A., 1979. Evidence for late Precambrian plate tectonics in West Africa, Nature, 250, pp. 477–478.

Blanchard, D., El Melegi, M., Rabie, A. & Billman, B., 2002 (Abs.). Breakthrough horizontal well length in Gulf of Suez, Egypt. AAPG Cairo 2002: Ancient Oil-New Energy Technical Program.

Blanckenhorn, M., 1888. Die geognostischen Verhältnisse von Afrika. I. Teil: Der Atlas, das nord-afrikanische Faltengebirge.

Blanckenhorn, M., 1900. Neues zur Geologie und Palaontologie Aegyptens, II, Das Paleogen (Eocän und Oligocän), A. Deut. Geol. Ges., v. 52, pp. 403–479.

Blanckenhorn, M., 1901. Neues zur Geologie und Palaontologie Aegyptens, III Das Miozän, A. Deut. Geol. Ges., v. 53, pp. 52–132.

Blanckenhorn, M., 1902a. Neue geologisch-stratigraphische Beobachtungen in Aegypten, Separat-Abdruck aus den Sitzungberichten der mathem.-phys. Classe der Kgl. bayer. Akadamie der Wissenschaften Bd. XXXII, Heft III, pp. 353–433.

Blanckenhorn, M., 1902b. Geschichte des Nilstroms. Zeit. Ges. Erdk., Berlin, pp. 694–722 and pp. 753–762.

Blanpied, C. & Bellaiche, G., 1983. The Jarrafa Trough (Pelagian Sea), structural evolution and tectonic significance, Marine Geology, v. 52, pp. 1–10.

Blanpied, C. & Rubino, J., 1998 (Abs.). Sedimentology and sequence straigraphy of the Devonian – Base Carboniferous succession in the Gargaf High, The Geological Conference on Exploration in Murzuq Basin, Sabha University, Sabha, Libya, p. 9.

Blanpied, C., Deynoux, M., Ghienne, J.F. & Rubino, J.-L., 1998 (Abs.). Late Ordovician glaciation, depositional systems, a comparison from the Gargaf High (Libya) and the Taoudeni Basin (Mauritania). The Geological Conference on Exploration in Murzuq Basin, Sabha University, Sabha, Libya, p. 43.

Blasband, B., 1998 (Abs.). A Pan-African core complex in the Sinai, Egypt; implications for the tectonic history of the Arabian-Nubian Shield, in Geology of the Arab World, Fourth International Conference, Cairo University, Cairo, pp. 86–87.

Blasband, B., Brooijmans, P., Dirks, P., Visser, W. & White, 1997. A Pan-African core complex in the Sinai, Egypt, Geol. Mijnbouw, v. 76, no. 3, pp. 247–266.

Blayac, J. 1898. Sur l'existance probable du Trias gypso-salin dans le sud de la province de Constantine. Bull. Géol. Soc. Fr., v. 26, p. 305.

Blomeier, D.P.G. & Reijmer, J.G., 1999. Drowning of a Lower Jurassic carbonate platform: Jbel Bou Dahar, High Atlas, Morocco. Facies, v. 41, pp. 81–110.

Blondeau, A., 1972. Les Nummulites, Librairie Vuibert, Paris, p. 255.

Blondeau, A., Boukhary, M. & Shamah, K., 1984. L'évolution et la dispresion géographique de *Nummulites gizehensis* (Forskal), Géol. Méditerr., v. 11, pp. 173–179.

Blondeau, A., Boukhary, M. & Ambroise, D., 1987. Etude comparée de trois populations de *Nummulites gizehensis* (Forskal) en Egypte, Rev. Paléobiol., v. 6, pp. 131–138.

Boaz, N., Gaziry, A.W., Heizelin, J. de & El-Arnauti, A., 1982. Results from the International Sahabi Research Project (Geology and Paleontology), Garyounis Scientific Bull., Spec. Issue 4, Benghazi, p. 142.

Boaz, N.T., 1996. Vertebrate Palaeontology and terrestrial palaeoecology of As Sahabi and the Sirt Basin, in M.J. Salem. A.J. Mouzughi & O.S. Hammuda (Eds.), The Geology of Sirt Basin, v. 1, Elsevier, Amsterdam, pp. 531–539.

Bobier, C., Viguier, C., Chaari, A. & Chine, A., 1991. The post-Triassic sedimentary cover of Tunisia: Seismic sequences and structure, in R. Freeman. M. Huch & St. Mueller (Eds.), The European Geotraverse, Part 7, Tectonophysics, v. 195, pp. 371–410.

Bockwinckel, J. & Ebbighausen, V. 2006. A new ammonoid fauna from the Gattendorfia-Eocanites Genozone of the Anti-Atlas (Early Carboniferous, Morocco). Fossil Record, v. 9, pp. 87–129.

Bockwinkel, J., Korn, D. & Ebbighausen, V., 2010. The ammonoids from the Argiles de Timimoun of Timimoun (Early and Middle Viséan; Gourara, Algeria). Fossil Recod, v. 13, no. 1, pp. 215–278.

Bohannon, R.G. & Eittreim, S.L., 1991. Tectonic development of passive continental margins of the southern and central Red Sea with a comparison to Wilkes Land, Antarctica, in J. Markis; P. Mohr & R. Rihm (eds.), Red Sea: Birth and Early History of a New Oceanic Basin, Tectonophysics, v. 198, pp. 129–154.

Bojar, A.-V., Fritz, H., Kargl, S. & Unzog, W., 2002. Phanerozoic tectonothermal history of the Arabian-Nubian shield in the Eastern Desert of Egypt: evidence from fission track and paleostress data. Journal of African Earth Sciences, v. 34, no. 3–4, pp. 191–202.

Bonatti, E., Clocchiatti, R., Colantoni, P., Gelmini, R., Marinelli, G., Ottonello, G., Santacroce, R., Taviani, A.A., Abdel-Meguid, A.A., Assaf, H.S. & El Tahir, M.A., 1983. Zabargad (St. John's) Island: an uplifted fragment of sub-Red Sea lithosphere, J. Geol. Soc. London, v. 140, pp. 677–690.

Bond, D.P.G. & Wingall, P.B., 2008. The role of sea-level change and marine anoxia in the Frasnian-Famennian (Late Devonian) mass extinction. Palaeogeography, Palaeoclimatology, Palaeoecology, v. 263, pp. 107–118.

Bondon, J. & Clariond, E. 1934. Itinéraire géologique d'Aqqa à Tindouf (Sahara marocain). Compte Rendu hebdomadaires des séances, de l'Academie des Sciences, Paris, v. 199, sér. II, pp. 4–58.

Bonnefous, J., 1963. Synthese stratigraphique sur la Gothlandien des sondages du Sud Tunisien, Rev. Inst. Fr. Pétrole, v. 18, No. 10.

Bonnefous, J., 1972. Geology of the Quartzitic "Gargaf Formation" in Sirte Basin, Libya, Bulletin Centre Recherche. Pau-SNPA, v. 6–2, pp. 225–261.

Boote, D.R.D., Clark-Lowes, D.D. & Traut, M.W. 1998. Palaeozoic petroleum systems of North Africa. In: D.S. MacGregor, R.T.J. Moody, D.D. Clark-Lowes (eds): Petroleum Geology of North Africa. Geological Society of London, Special Publication No. 132, pp. 7–68.

Borghi, P. & Chiesa, C., 1940. Cenni geologici e paleontologici sul paleozoico dell'Egghidi Uan Caza nel desserto di Taita (Fazzan occidentale), Bolletino de la Societa Geologica Italiana, v. 2, Tripoli, pp. 123–127.

Bornemann, A., Schulte, P., Sprong, J., Steurbaut, E., Youssef, M. & Speijer, R.P., 2009. Latest Danian carbon isotope anomaly and associated environmental change in the southern Tethys (Nile Basin, Egypt). Journal of the Geological Society, London, v. 166, pp. 1135–1142.

Borocco, J. & Nyssen, R., 1959. Nouvelles observations sur les "grès inférieurs" cambro-ordoviciens du Tassili interne (Nord-Hoggar). Bull. Soc. géol. France, v. 7, no. 1, pp. 197–206.

Bosellini, A. & Hsü, K.J., 1973. Mediterranean plate tectonics and Triassic palaeogeography, Nature, v. 244, pp. 144–146.

Bossière, G., 1971. Les roches éruptives du champs d'Haoud Berkaoui et l'importance du volcanisme permo-triassique dans l'espace nord-saharien, Bull. Soc. His. Natur. Afr. Nord, v. 62, No. 1–2, Algiers, pp. 47–53.

Bossière, G. & Peucat, J.J., 1985. New geochronological information by Rb-Sr and U-Pb investigations from the pre-Alpine basement of Grande Kabylie (Algeria). Can. J. Earth Sci., v. 22, pp. 675–685.

Bossière, G. & Peucat, J.J., 1986. Structural evidence and Rb-Sr, 39–40 Ar mica age relationships for the existence of an Hercynian deep crustal shear zone in Grande Kabylie (Algeria) and its Alpine reworking. Tectonophysics, v. 121, pp. 277–294.

Bosworth, W., 1995. A high-strain rift model for the southern Gulf of Suez (Egypt). In J.J. Lambiase (Ed.), Hydrocarbon Habitat in Rift Basins, Geological Society Special Publication No. 80, pp. 75–102.

Bosworth, W. & McClay, K., 2001. Structural and stratigraphic evolution of the Gulf of Suez Rift, Egypt. In P.A. Ziegler, Cavazza, W., Robertson, A.H.F. & Crasquin-Soleau, S. (Eds.), Peri-Tethys Memoir 6, Peri-Tethyan Rift/Wrench Basins and Passive Margins, Memoire du Museum National d'Histoire Naturale, Tome 186, pp. 568–607.

Bosworth, W., Guiraud, R. & Kessler, L.G. II, 1999. Late Cretaceous (ca. 84 Ma) compressive deformation of the stable platform of northeast Africa (Egypt): Far-field stress effects of the "Santonian event" and origin of the Syrian arc deformation belt. Geology, v. 27, pp. 633–636.

Botros, N.S., 2002. Metallogeny of gold in relation to the evolution of the Nubian Shield in Egypt. Ore Geology Reviews, v. 19, nos. 3–4, pp. 137–164.

Botros, N.S., 2003. On the relationship between auriferous talc deposits hosted in volcanic rocks and massive sulphide deposits in Egypt. Ore Geology Reviews, v. 23, nos. 3–4, pp. 223–257.

Böttcher, R., 1982. Die Abu Ballas Formation (*Lingula* Shale) (Apt.?) der Nubischen Gruppe sudwest-Ägyptens. Eine Beschreibung der Formation unter besondern Beruckssichtigung der Paläontologie, Berliner geowiss. Abh. (A) 39, p. 145.

Bouabdelli, M., 1994. Tectonique de l'Est du Massif hercynien central (zone d'Azrou-Khénifra). *Bull. lnst. Sci.*, Rabat, no. 18, pp. 145–168.

Bouabdelli, M. & Piqué, A., 1996. Du bassin sur décrochement au bassin d'avant-pays: Dynamique du basin d'Azrou-Khénifra (Maroc Hercynien Central). Journal of African Earth Sciences, v. 23, pp. 213–223.

Bouaziz, S., Barrier, E., Soussi, M., Turki, M.M. & Zouari, H., 2002. Tectonic evolution of the northern African margin in Tunisia from paleostress data and sedimentary record. Tectonophysics, v. 357, pp. 227–253.

Bouaziz, S., Buffetaut, E., Ghanmi, M., Jaegar, J.-J., Martin, M., Mazin, J.-M. & Tong, H., 1988. Nouvelles découvertes de vertébrés fossils dans l'Albien du Sud Tunisien. Bull. Soc. Géol. Fr., v. 4, pp. 335–339.

Bouchareb Haouchine, F.-Z. & Boudoukha, A., 2009. Hydro-Geochemical and Litho-Structural Approach of Deep Circulations in the Mounts of Hodna, Algeria. European Journal of Scientific Research, v. 37, no. 1, pp. 12–20.

Bouchouata, A., Canerot, J., Souhel, A. & Almeras, Y., 1995. Stratigraphie séquentielle et évolution géodynamique du Jurassique dans la région de Talmest-Tazoult (Haut-Atlas central, Maroc). Comptes rendus de l'Académie des sciences de Paris. Série 2. Sciences de la terre et des planètes, v. 320, pp. 749–756.

Boudchiche, L., Nicollin, J.P. & Ruget, C., 1987. Evolution des assemblages de foraminifères pendant le Toarcien dans le massif des Beni Snassen (Maroc nord-oriental). Géologie Méditerranéenne, v. 14, pp. 161–166.

Boughdiri, M., Sallouhi, H., Cordey, F., Maalaoui, K. & Soussi, M., 2006a. Malm-Berriasian formations of northern Tunisia: updated biostratigraphy, regional correlations and western Tethys geodynamic context. 7th International Congress on the Jurassic System, Krakow, Poland, Abstract Volume, Session 4: Integrated Stratigraphy, pp. 148–149.

Boughdiri, M., Sallouhi, H., Haddad, S., Cordey, F. & Soussi, M., 2009. Integrated biostratigraphy and regional correlations of Upper Jurassic-lowermost Cretaceous series in northern Tunisia. GFF, v. 131, pt.1–2, pp. 71–81.

Boughdiri, M., Sallouhi, H., Maalaoui, K., Soussi, M. & Cordey, F., 2006b. Calpionellinid zonation of the Jurassic-Cretaceous transition in North-Atlasic, Tunisia. Updated Upper Jurassic stratigraphy of the "Tunisian Trough" and regional correlations. Comptes Rendus Geosciences, v. 338, no. 16, pp. 1250–1259.

Bouillin, J.P., 1986. Le «bassin Maghrébin»: une ancienne limite entre l'Europe et l'Afrique à l'ouest des Alpes. Bull. Soc. Geol. Fr., v. 8, no. 2, pp. 547–558.

Bouillin, J.P., Durand-Delga, M., Gélard, J.P. et al., 1970. Définition d'un flysch Massylien et d'un flysch Maurétanien au sein des flyschs allochtones de l'Algérie. Comptes Rendus de l'Académie des Sciences, v. 270, pp. 2249–2252.

Boujamaoui, M., Saadi, M., Inoubli, M.H., Zaghbib-Turki, D. & Turki, M.M., 2000. Geometrie des depots de la Formation Beglia (Miocene moyen) et ses equivalents lateraux en Tunisie et dans le Bloc Pelagien: sedimentologie et sequences de depots. Africa Geoscience Review, v. 7, no. 1, pp. 55–74.

Boukhary, M. & Abdelmalik, W.M., 1983. Revision of the stratigraphy of the Eocene deposits of Egypt, N. Jb. Geol. Paläont. Mh., v. 6, pp. 321–337.

Boukhary, M., Guernet, C.L. & Mansour, H., 1982. Ostracodes du Tertiare inférieur del l'Egypte, 8th Afr. Micropaleontol. Colloq., Paris, Cah. Micropaleontol. 1, pp. 13–20.

Boukhary, M., Kenawy, A. & Basta, R., 2009. Early Eocene Nummulitids from Gebel Umm Russeies, El Galala El Bahariya, Eastern Desert, Egypt. Geologia Croatica, v. 62, no. 1, pp. 1–18.

Boukhary, M., Kulbrok, F. & Kuss, J., 1998. New nummulitids from lower Eocene limestones of Egypt (Monastery of St. Paul, Eastern Desert). Micropaleontology, v. 44, pp. 99–108.

Boukhary, M., Morsi, A.M., Eissa, R. & Kerdany, M., 2009. Late Cenomanian ostracod faunas from the area south of Ain Sukhna, western side of the Gulf of Suez, Egypt. Geologia Croatica, v. 62, no. 1, pp. 19–30.

Boukhary, M., Blondeau, A. & Ambroise, D., 1982. Etude sur les nummulites de la region de Minia-Samalut, Vallée du Nil, Egypte. 1. Biometrie et biostratigraphie, 8th Afr. Micropaleontol. Colloq., Paris, Cah. Micropaleontol. 1, pp. 65–78.

Boukhary, M.A., Kuss, J.M. & Abdelraouf, M., 2008. Chattian Larger Foraminifera, northern Sinai, Egypt, and paleogeographic interpretations. Stratigraphy, v. 5. pp. 179–192.

Bourcart, J., 1927. Découverte du Cambrien à Archaeocyathus de l'Anti-Atlas marocain. Bull. Soc. Géol. Fr., v. 4, no. 27, C.R. som., pp. 10–11.

Bouougri, E.F. & Saquaque, A., 2004. Lithostratigraphic framework and correlation of the Neoproterozoic northern West African Craton passive margin sequence (Sirous-Zenaga-Bouazzer Elgraara Inliers, Central Anti-Atlas, Morocco): an integrated approach. Journal African Earth Sciences, v. 39, nos. 3–5, pp. 227–238.

Bourquin, S., Eschard, R. & Hamouche, B., 2010. High-resolution sequence stratigraphy of Upper Triassic succession (Carnian – Rhaetian) of the Zarzaitine outcrops (Algeria): a model of fluvio-lacustrine deposits, J. African Earth Sciences, v. 58, no. 2, pp. 365–386.

Bousquet, R. EL Mamoun, R., Sassiqi, O., Goffé, B., Möller, A. & Madi, A., 2008. Mélanges and ophiolites during the Pan-African orogeny: the case of the Bou-Azzer ophiolite suite

(Morocco). In: N. Ennih & J.-P. Liégeois (Eds.): The Boundaries of the West African Craton. Geological Society, London, Special Publications, v. 297, pp. 233–247.

Boussac, J., 1913. Observations nouvelles sur le Nummulitique de la Haute-Egypte, C.R.S., Soc. Géol. Fr., pp. 62–65 & pp. 109–110.

Bown, T., 1982. Ichnofossils and rhizoliths of the near shore fluvial Jebel Qatrani Formation (Oligocene), Fayum Province, Egypt, Palaeogeography, Palaeoclimatology, Palaeoecology, v. 40, pp. 255–307.

Bown, T.M. & Kraus, M.J., 1988. Geology and paleoenvironments of the Oligocene Jebel Qatrani Formation and adjacent rocks, Fayum Province, Egypt, US Geol. Surv., Prof. Paper 1452, pp. 1–60.

Bown, T.M., Kraus, M.J., Wing, S.L. Fleagle, J.G., Tiffaney, B.H., Simons, E.L. & Vondra, C.F., 1982. The Fayum Primate Forest revisited, Journal of Human Evolution, v. 11, pp. 503–560.

Brabers, P.M., 1988. A plate tectonic model for the Panafrican Orogeny in the Anti-Atlas, Morocco. In: V. Jacobschagen (Ed.): The Atlas System of Morocco. Springer-Verlag, Berlin, pp. 61–80.

Bracaccia, V., Carcano, C. & Drera, K., 1991. Sedimentology of the Silurian-Devonian Series in the southeastern part of the Ghadamis basin, in M.J. Salem & M.N. Belaid (Eds.), The Geology of Libya, v. V, Elsevier, Amsterdam, pp. 1727–1744.

Bracène, R., Bellahcene, A., Bekkouche, D., Mercier, E. & Frizon de Lamotte, D., 1998. The thin-skinned style of the South Atlas Front in central Algeria. In: D.S. Macgregor, R.T.J. Moody, D.D. Clark-Lowes (Eds.), Petroleum Geology of North Africa. Geol. Soc. Spec. Publ., v. 133, pp. 395–404.

Bracène, R. & Frizon de Lamotte, D., 2002. The origin of intraplate deformation in the Atlas system of western and central Algeria: from Jurassic rifting to Cenozoic-Quaternary inversion. Tectonophysics, v. 357, pp. 207–226.

Brachert, T.C., Buggisch, W., Fluegel, E., Huessner, H., Joachimski, M.M., Tourner, F. & Walliser, O.H., 1992. Controls of mud mound formation: the Early Devonian Kess-Kess carbonates of the Hamar laghdad, Antiatlas, Morocco. Geologishe Rundschau, v. 81, pp. 15–44.

Bradley, J., 1986. Dyke (Dike). In D.R, Bowes (Ed.), The Encyclopedia of Igneous and Metamorphic Petrology, pp. 134–136, New York, Van Nostrand Reinhold, p. 666.

Brady, T.J., Cambell, N.D.J. & Maher, C.E., 1980. Intisar 'D' Oil Field, Libya, in L.V. Illing & G.D. Hobson (Eds.), Giant Oil and Gas Fields of the Decade 1968–1978, AAPG Mem. No. 30, pp. 543–564.

Braithwaite, K., Racey, A. & Cowan, G., 1998 (Abs.). Palaeoenvironmental controls on the deposition of the Early Tertiary Eocene El Garia Formation of Tunisia, in Tertiary to Recent Larger Foraminifera, Their Depositional Environments and Importance as Petroleum Reservoirs, Kingston University, United Kingdom, p. 7.

Breccia, Buoncompani, Marchetti, Maugini, Petrocchi & Stefanini, 1935. Breve Guida alle escursioni geogische in Cirenaica in occasione della XLVIII riunione della Societá Geologica Italiana. Boll. Soc. Geol. Ital., v. 54, Roma, pp. 37–68.

Brede, R., 1992. Structural aspects of the Middle and the High Atlas (Morocco) – phenomena and causalities. Geologische Rundschau, v. 81, no. 1, pp. 171–184.

Brede, R., Hauptmann, M. & Herbig, H.-G. 1992. Plate tectonics and intracontinental mountain ranges in Morocco -The Mesozoic-Cenozoic development in the central High Atlas and the Middle Atlas, Geologische Rundschau, v. 81, no. 1, pp. 127–141.

Brenckle, P.L. & Marchant, T.R., 1987. Calcareous microfossils, depositional environments and correlation of the Lower Carboniferous Um Bogma Formation at Gebel Nukhul, Sinai, Egypt, J. Foraminiferal Research, v. 17, No. 1, pp. 74–91.

Brice, D., Legrand-Blain, M. & Nicollin, J.-P., 2007. Brachiopod faunal changes across the Devonian–Carboniferous boundary in NW Sahara (Morocco, Algeria). In: R.T. Becker & W.T. Kirchgasser, (Eds): Devonian Events and Correlations. Geological Society, London, Special Publications, No. 278, pp. 261–271.

Brichant, A.L., 1952. Sur la découverte du Trias au pie du Djebel Garian (Tripolitaine, Libye). C.R. Acad. Sci., Paris, pp. 1456–1458.

Brinkhuis, H. & Zachariasse, W.J., 1988. Dinofagellate cysts, sea level changes and planktonic foraminifer across the Cretaceous-Tertiary boundary at El Haria, northwest Tunisia. Marine Micropaleontol., v. 13, pp. 153–191.

Brives, A., 1905. Les terrains crétacés dans le Maroc occidental. Bull. Géol de Fr., sér. iv, pp. 81–96.

Bronner, G., Roussel, J. & Trompette, R., 1980. Genesis and geodynamic evolution of the Taoudeni cratonic basin (upper Precambrian and Palaeozoic), western Africa. In A.W. Bally, P.L. Bender, T.R. McGetchin & R.I. Walcott (Eds.). Dynamics of Plate Interiors. American Geophysical Union/Geological Society of America, Geodynamic series, v. 4, pp. 81–90.

Broughton, P.L., 1996. of permeability variations on water encroachment patterns at the Ar Raqubah Field, Sirt Basin, in M.J. Salem. A.S. El-Hawat & A.M. Sbeta (Eds.), The Geology of Sirt Basin, v. II, Amsterdam, Elsevier, pp. 391–418.

Broughton, P. & Trepanier, A., 1993. Hydrocarbon generation in the Essaouira Basin of Western Morocco. AAPG Bull., v. 77, pp. 999–1015.

Broutin, J., Ferrandini, J. & Saber, H., 1989. Implications stratigraphiques et paléogéographiques de la découverte d'une flore permienne euraméricaine dans le Haut-Atlas occidental (Maroc). Compte Rendu de l'Académie des Sciences, Paris, 308, II, pp. 1509–1515.

Brown, R.H., 1980. Triassic rocks of Argana Valley, southern Morocco, and their regional structural implications. AAPG Bull., v. 64, no. 7, pp. 988–1003.

Brown, R.N., 1980. History of exploration and discovery of Morgan, Ramadan and July oilfields, Gulf of Suez, Egypt, in A.D. Miall (Ed.), Facts and Principles of World Petroleum Occurrences, CSPG Mem. 6, Calgary, pp. 733–764.

Bruguier, O., Hammor, D., Bosch, D. & Caby, R., 2009. Miocene incorporation of peridotite into the Hercynian basement of the Maghrebides (Edough massif, NE Algeria): Implications for the geodynamic evolution of the Western Mediterranean. In: The Application and Interpretation of Micro-analytical Data in Geochronological Systems. Chemical Geology, v. 261, nos. 1–2, pp. 172–184.

Brugman, W.A., Eggink, J.W. & Visscher, H., 1985a. Middle Triassic (Anisian-Ladenian) palynomorphs, in B. Thusu & B. Owens (Eds.), Palynostratigraphy of North-East Libya, J. Micropaleontology, v. 4, pt.1, pp. 107–112.

Brugman, W.A., Eggink, J.W., Lobboziak, S. & Visscher, H., 1985b. Late Carboniferous-Early Permian (Ghezelian-Artinskian) palynomorphs, in B. Thusu & B. Owens (Eds.), Palynostratigraphy of North-East Libya, J. Micropaleontology, v. 4, pt.1, pp. 93–105.

Brugman, W.A. & Visscher, H., 1988. Permian and Triassic palynostratigraphy of northeast Libya, in A. El-Arnauti, B. Owens & B. Thusu (Eds.), Subsurface Palynostratigraphy of Northeast Libya, Benghazi, Libya, Garyounis University Publications, pp. 157–213.

Buforn, E., Martin-Davila, J. & Udias, A. (Eds.), 2004. Geodynamics of Azores-Tunisia, Series: Pageoph Topical Volumes, Birkhäuser Basel, p. 250.

Buist, G., 1846. Petrified forest near Cairo. American Journal of Science and Arts, v. 1, no. 3, pp. 433–434.

Bullen-Newton, R., 1911. Kainozoic mollusca from Cyrenaica. Q.J. Geol. Seol., v. 67, pp. 616–653.

Bumby, A.J. & Guiraud, R., 2005. The geodynamic setting of the Phanerozoic basins of Algeria. Journal of African Earth Sciences, v. 43, Issues 1–3, pp. 1–12.

Burchette, T.P., 1986. Mid.-Miocene tectonics and sedimentation, Esh Mellaha range, southwest Gulf of Suez, 8th EGPC Exploration Seminar, Egypt, v. 1, pp. 266–280.

Burckle, L.H., 1995 (Abs.). Review of Late Miocene/Early Pliocene paleoclimate, International Conference on the Biotic and Climatic Effects of the Messinian Event on the Circum Mediterranean, Technical Program and Abstracts, University of Garyounis, Benghazi, p. 26.

Burke, K. & Wells, G.L., 1989. Trans-African drainage system of the Sahara: Was it the Nile?, Geology, v. 17, pp. 743–747.

Burkhard, M., Caritg, S., Helg, U., Robert-Charrue, C. & Soulaimani, A., 2006. Tectonics of the Anti-Atlas of Morocco. Comptes Rendus Geosciences, v. 338, no. 1–2, pp. 11–24.

Burollet, P.G., 1956. Contribution à l'étude stratigraphique de la Tunisie centrale, Annales des Mines et de la Géologie No. 18, Ministère des Travaux Publiques, Services des Mines, de l'Industrie et de l'Energie, Tunis, p. 350.

Burollet, P.F., 1960. Léxique Stratigraphique International, v. IV, Afrique, part IV, Libye, Cong. Géol. Internat. Commission de Stratig., Rech. Sci., Paris, p. 62.

Burollet, P.F., 1963a. Field trip guidebook of the excursion to Jebel Nefusa 1st Saharan Symp. Petrol. Explor. Soc. Libya, p. 19.

Burollet, P.F., 1963b. Reconnaissance géologique dans le sud-est du bassin de Kufra. Premier Symposium Saharien, Tripoli, Revue Institut Francais du Pétrole, v. 18, no. 10, pp. 219–227.

Burollet, P., 1963c. Trias de Tunisie et de Libye, relation avec le Trias Européen et Saharien, in Colloque sur le Trias de la Frande et des régions limitrophes, Mémoire B.R.G.M., No. 15, éditions Téchnip, Paris, pp. 482–494.

Burollet, P.F., 1979. Résumé sur la géologie des régions voisines, in P.F. Burollet. P. Clairefond & E. Winnock (Eds.), Géologie Méditerranéenne, La Mer Pélagienne, Annales de l'Université de Provence, tome VI, No. 1, Marseille, pp. 23–27.

Burollet, P.F., 1991. Structure and tectonics of Tunisia, in R. Freeman. M. Huch & St. Mueller (Eds.), The European Geotraverse, Part 7, Tectonophysics, v. 195, pp. 359–369.

Burollet, P.F., 1993 (Abs.). Genesis and evolution of Sirt Basin in its African and Mediterranean setting, in Sedimentary Basins of Libya, First Symposium, Geology of Sirt Basin, Earth Science Society of Libya, Tripoli, p. 13.

Burollet, P.F., 1995. Travaux du Comité Français d'Histoire de la Geologie (COFRHIGEO) – Troisième série – T.IX, pp. 111–122.

Burollet, P.F. & Byramjee, R.S., 1964 (Abs.). Shape and structure of Saharah Cambro-Ordovician sand bodies – paleocurrents and depositional envrioment, AAPG Bull., v. 49, pp. 519–520.

Burollet, P.F. & Byramjee, R., 1969. Sedimentological remarks on Lower Paleozoic sandstones of South Libya, in W.H. Kanes (Ed.), Geology, Archaeology and Prehistory of Southwestern Fezzan, Libya, Petroleum Exploration Society, Eleventh Annual Field Conference, pp. 91–101.

Burollet, P.F. & Manderscheid, G., 1965. Le crétacé inférieur en Tunisie et en Libye, in Colloque sur le Crétacé Inférieur, Mémoire B.R.G.M., No. 34, Paris, pp. 769–794.

Burollet, P.F. & Mandersheid, G., 1967. Le Devonien en Libye et en Tunisie. In D.H. Oswald (Ed.), International Symposium on the Devonian System, Alberta Soc. Petrol. Geologists, Calgary, pp. 285–302.

Burollet, P.F., Magnier, Ph. & Manderscheid, G., 1971. La Libye, Tectonique de l'Afrique, Science de la Terre 6, UNESCO, pp. 409–416.

Burollet, P.F., Memmi, L. & M'Rabet, A., 1983. Le Crétacé inférieur de Tunisie: Aperçu stratigraphique et sédimentologique. Zitteliana, v. 10, pp. 255–264.

Burollet, P.F., Mugniot, J.M. & Sweeney, P., 1978. The geology of the Pelagian block. the margins and basins of southern Tunisia and Tripolitania, in A.E.M. Nairn & W.H. Kanes (Eds.), The Ocean Basins and Margins, v. 4B, New York, Plenum, pp. 331–359.

Burwood, R., Redfern, J. & Cope, M., 2003. Geochemical evaluation of East Sirte Basin (Libya) petroleum systems and oil provenance. In: Arthur, T.J., Macgregor, D.S. & Cameron, N. (Eds.), Petroleum Geology of Africa: New Themes and developing technologies, Geol. Soc. (London) Spec. Publ., v. 207, pp. 203–204.

Busnardo, R., Donze, P., Khessibi, Le Hegarat, G. & Memmi, L., 1980. Interpretation biostratigraphique nouvelle de la formation des "Argiles du Sidi Kralif" au Djebel Bou Hedma (Tunisie centrale). Geobios, v. 13, no. 3, pp. 459–463.

Busrewil, M.T. & Wadsworth, W.J., 1980. Preliminary chemical data on the volcanic rocks of Al Haruj area, central Libya, in M.J. Salem & M.T. Busrewil (Eds.), The Geology of Libya, v. III, London, Academic Press, pp. 1077–1080.

Busrewil, M.J. & Wadsworth, W.J., 1996. Tertiary-Quaternary alkaline-subalkaline magmatism in Gharyan area – Field aspects and petrography, Petroleum Research Journal, v. 8, Tripoli, pp. 13–23.

Busson, G., 1965. Sur les gisements fossilifères du Jurassique moyen et supérieur du Sahara tunisien, Ann. Paléontol. (Invertébrés), v. 1, no. 1, pp. 30–42.

Busson, G., 1967a. Mesozoic of southern Tunisia, in L. Martin (Ed.), Guide to the Geology and History of Tunisia, Petrol. Explor. Soc. Libya, 9th Annual Field Conf., pp. 131–151.

Busson, G., 1967b. La Mesozoic Saharien, Pt.1, L'Extreme-Sud Tunisien, Centre Recherches Zones Arides, Ser. Géol., No. 8, CNRS, Paris, p. 194.

Busson, G. & Burollet, P.F., 1973. La limite Permien-Trias sur la plate-form Saharienne (Algérie, Tunisie, Libye), in Permian and Triassic systems and their mutual boundary, CSPG, Mem. No. 2, pp. 74–88.

Busson, G. & Comee, A., 1989. Some data on climatic previous history of Sahara – signification of detrital red beds and evaporites from Trias to Lias-Dogger. Bull. Soc. Geol. France., v. 5, pp. 3–11.

Butt, A.A., 1984. Upper Cretaceous biostratigraphy of the Sirte basin, northern Libya, Géologie Méditerranéenne, v. 5, No. 2, pp. 237–242.

Buttler, C.J., Cherns, L. & Massa, D., 2007. Bryozoan mud-mounds from the Upper Ordovician Jifarah (Djeffara) Formation of Tripolitania, north-west Libya. Palaeontology, v. 50, no. 2, pp. 479–494.

Buttler, C.J. & Massa, D. 1996. Late Ordovician bryozoans from carbonate buildups, Tripolitania, Libya. In D.P. Gordon, A.M. Smith & J.A. Grant-Mackie (Eds). Bryozoans in space and time. Proceedings of the 10th International Bryozoology Conference. National Institute of Water and Atmospheric Research, Wellington, pp. 63–68.

Bütler, H., 1922. Contribution à la géologie de l'Ahaggar (Sahara central). Comptes Rendus XIIIème Congrés Géologique International, Bruxelles, no. 2, pp. 819–848.

Bütler, H., 1924. Dans les montagnes de l'Ahaggar (Sahara central). Écho des Alpes, Genève, no. 2, pp. 49–78.

Butzer, K.W. & Hansen, C.L., 1968. Desert and River in Nubia, University of Wisconsin Press, Madison, p. 562.

Butzer, K.W., 1960a. On the Pleistocene shore lines of Arab's Gulf, Egypt, J. Geol., v. 68, No. 6, pp. 626–637.

Butzer, K.W., 1960b. Archeology and geology in Ancient Egypt. Science, New Series, v. 132, no. 3440, pp. 1617–1624.

Butzer, K.W., 1976. Early hydraulic Civilization in Egypt, Chicago, University of Chicago Press, p. 134.

Caby, R., 2003. Terrane assembly and geodynamic evolution of central-western Hoggar: a synthesis. Journal of African Earth Sciences, v. 37, Issues 3–4, pp. 133–159.

Caby, R., Hammor, D. & Delor, C., 2001. Metamorphic evolution, partial melting and Miocene exhumation of lower crust in the Edough metamorphic core complex, west Mediterranean orogen, eastern Algeria. Tectonophysics, v. 342, no. 3–4, pp. 239–273.

Caby, R. & Andreopoulos-Renaud, U., 1987. Le Hoggar oriental, bloc cratonisé à 730 Ma dans la châine pan-africaine du nord du continent africain. Precambrian Research, v. 36, pp. 335–344.

Caillaud, F., 1826. Voyage a Me'roe, Paris.

Cameron, N., Zumberge, J. & Illich, H., 2007. Central Atlantic Petroleum Systems. The First MAPG International Convention, Conference & Exhibition, Marrakech, October 28–31, 2007, pp. 5–6.

Canudo, J.I., Keller, G., & Molina, E., 1991. Cretaceous/Tertiary boundary extinction pattern and faunal turnover at Agost and Caravaca, S.E. Spain, Marine Micropaleontology, v. 17, pp. 319–341.

Capot-Rey, R., 1927. Le relief de la Haut Zousfana à l'Est de Colomb-Bechar (Sud Oranais). In: Annales de Géographie, v. 36, no. 204. pp. 537–547.

Caputo, M.V. & Crowell, J.C., 1985. Migration of glacial centres across Gondwana during Paleozoic Era, Geol. Soc. Am. Bull., v. 96, pp. 1030–1036.

Carbonnel, G., Alzouma, K. & Dikouma, M., 1990. Les ostracodes Paléocenes du Niger, taxonomie – un témoignage de l'existence éventuelle de la Mer Transsaharienne?, Geobios, v. 6, pp. 671–697.

Carmignani, L., Giammarino, S., Giglia, G. & Pertusati, P.C., 1990. The Qasr As Sahabi succession and the Neogene evolution of the Sirte Basin, J. Afr. Earth Sci., v. 10, No. 4, pp. 753–769.

Caron, M., 1985. Cretaceous planktonic foraminifera. In: Bolli, H.M., Saunders, J.B., Perch-Nielsen, K. (Eds.), Plankton Stratigraphy. Cambridge Univ. Press, Cambridge, pp. 17–86.

Caron, M., Robaszynski, F., Baudin, F., Deconinck, J.-F., Hochuli, P.A., von-Salis-Perch Nielsen, K. & Tribovillard, N., 1999. Estimation de la duree de l'evenement anoxique global au passage Cenomenien/Turonien. approche cyclostratigraphique dans la formation Bahloul en Tunisie centrale. Bulletin de la Societe Geologique de France, v. 170, no. 2, pp. 145–160.

Carr, I.D., 2002. Second-order sequence stratigraphy of the Palaeozoic of North Africa. J. Petroleum Geology, v. 25, no. 3, pp. 259–280.

Cartwright, J., Haddock, R.C. & Pinherio, L.M., 1993. The lateral extent of sequence boundaries, in G.D. Williams & A. Dobb (Eds.), Tectonics and Seismic Sequence Stratigraphy, Geological Society Special Pub. No. 71, London, pp. 15–34.

Casati, L. & Craig, J., 2003 (Abs.). Petroleum Geology Overview of the North Africa Sahara Platform. AAPG Hedberg Conference "Paleozoic and Triassic Petroleum Systems in North Africa" February 18–20, 2003, Algiers, Algeria, p. 3.

Caton-Thompson, G. & Gardner, E.W., 1929. Recent work on the problem of Lake Moeris. The Geographical Journal, v. 73, no. 1, pp. 20–58.

Caton-Thompson, G. & Gardner, E.W., 1932. The prehistoric geography of Kharga Oasis. The Geographical Journal, v. 80, no. 5, pp. 369–406.

Caton-Thompson, G., 1952. Kharga Oasis in Prehistory, University of London, Athlone Press, London, p. 213.

Cavalazzi, B., 2006. Kess Kess carbonate mounds, Hamar Laghdad, Anti-Atlas, SE Morocco – A Field Guide, 01–05 December 2006 (prepared for the UNESCO Field Action), p. 20.

Cavaroc, V.V., Padgett, G., Stephens, D.G., Kanes, W.H., Boudda, A. & Woolen, I.D., 1976. Late Paleozoic of the Tindouf Basi, North Africa. Journal of Sedimentary Research, v. 46, no. 1, pp. 77–88.

Cayeux, L., 1935. Constitution des phosphates senoniens d'Egypte. Compte Rendus Hebdomadaires des Seances de l'Acadamie des Sciences, v. 200, no. 26, pp. 2134–2137.

Cecca, F., Martin-Garin, B., Marchand, D., Lathuiliere & Bartolini, A., 2005. Paleoclimatic control of biogeographic and sedimentary events in Tethyan and peri-Tethyan areas during the Oxfordian (Late Jurassic). Palaeogeography, Palaeoclimatology, Palaeoecology, v. 222, pp. 10–32.

Cepek, P., 1979. Geological Map of Libya, 1:250,000, Sheet-Al Qaryat Ash Sharqiyah, Industrial Research Centre, Tripoli, p. 95, 31 photos.

Cepek, P., 1980. The sedimentology and facies development of the Hasawnah Formation in Libya, in M.J. Salem & M.T. Busrewil (Eds.), The Geology of Libya, v. II, London, Academic Press, pp. 375–382.

Chaabani, F. & Razgallah, S., 2006. Aptian sedimentation: an example of interaction between tectonics and eustatics in Central Tunisia. In G. Moratti & A. Chalouan (Eds.), Tectonics of the Western Mediterranean and North Africa. Geol. Soc. London Spec. Pub. No. 262, pp. 55–74.

Chafiki, D., Canérot, J., Souhel, A. & Taj Eddine, K., 2004. Les monticules micritiques sinemuriens sur la transversale de Midelt-Errachidia (Haut Atlas Central, Maroc). Estudios Geol., v. 60, pp. 139–152.

Chalouan, A., El Mrihi, A., El Kadiri, Kh., Bahmad, A., Salhi, A. & Hlila, R., 2006. Mauretanian flysch nappe in the northwestern Rif Cordillera (Morocco): deformation chronology and evidence for a complex nappe emplacementIn G. Moratti G. & A. Chalouan (Eds), Tectonics of the Western Mediterranean and North Africa. Geological Society, London, Special Publications No. 262, pp. 161–175.

Chalouan, A., Michard, A., El Kadiri, Kh., Negro, F., Frizon de Lamotte, D., Soto, J.I. & Saddiqi, O., 2008. The Rif Belt. In A. Michard, O. Saddiqi, A. Chalouan & D. Frizon de Lamotte (Eds.), Continental Evolution: The Geology of Morocco. structure, stratigraphy, and Tectonism of the Africa-Atlantic-Mediterranean Triple Junction, Berlin, Springer-Verlag, pp. 203–302.

Chalouan, A., Saji, R., Michard, A. & Bally, A.W., 1997. Neogene tectonic evolution of the Southwestern Alboran Basin as inferred from seismic data off Morocco. AAPG Bull., v. 81, pp. 1161–1184.

Chanliau, M & Bruneton, A., 1988 (Abs.). Contribution to petroleum exploration of recent geophysical surverys offshore Egypt, in Mediterranean Basin Conf. & Exhib., Nice, AAPG Bull., v. 72, No. 8, p. 994.

Chanut, C. & Nyssen, R., 1958. Sur une "discordance" de ravinement dans les grès inférieurs de la région d'Amguid et de Takoumbaret (Mouydir). C.R. somm. Soc. géol. France, pp. 102–105.

Chaouchi, R., Mall, M.S. & Kechou, F., 1998. Sedimentological evolution of the Givetian-Eifelian (F3) sand bar of the West Alrar Field, Illizi Basin, Algeria, in D.J. Macgregor. R.T.J. Moody & D.D. Clark-Lowes (Eds.), Petroleum Geology of North Africa, Geological Society Spec. Pub. No. 132, London, pp. 187–200.

Chapman, F., 1911. Foraminifera, Ostracoda and Parasitic Fungi from the Kainozoic Limestone of Cyrenaica. Q.J. Geol. Soc. London, v. 67, no. 268, pp. 654–661.

Charlot, R., 1976. The Precambrian of the Anti-Atlas (Morocco): A geochronological synthesis. Precambrian Research, v. 3, no. 3, pp. 273–299.

Chatelier, J.Y. & Slevin, A., 1988. Review of African petroleum and gas deposits, Journal of African Earth Sciences, v. 7, No. 3, pp. 561–578.

Chaumeau, J., Legrand, P. & Renaud, A., 1961. Contribution à l'étude du Couvinien dans le bassin de Fort de Polignac (Sahara), Bull. Soc. géol. France, v. 7, no. 3, pp. 449–456.

Checchia-Rispoli, G., 1913. Sopra alcuni Echinidi oligocenici della Cirenaica. Giornale della Soc. Scienze Natur. ed Econom., v. 30, pp. 63–72.

Chen, Z. & Stanely, D.J., 1993. Alluvial stiff muds (late Pleistocene) underlying the Lower Nile delta plain, Egypt: petrology, stratigraphy and origin, Mediterranean Basin Program, Smithonian Institution, Washington, USA Journal of Coastal Research, v. 9, no. 2, pp. 539–576.

Chenet, P.Y., Colletta, B., Letouzey, J., Deforges, G., Ousset, E. & Zaghloul, E.A., 1987. Structures associated with extensional tectonism in the Suez rift. In M.P. Coward, J.F. Dewey & P.L. Hancock (Eds.), Continental Extensioanl Tectonics, Geological Society Special Publication No. 28, pp. 551–558.

Cherif, O.H. & Ismail, A.A., 1991. Late Senonian-Tertiary planktonic foraminiferal bios-tratigraphy and tectonism of the Esh-el-Mallaha and Gharamul areas, Egypt. M.E.R.C. Ain Shams Univ., Earth Sci. Ser., v. 5, pp. 146–159.

Cherif, O.H., Al-Rifaiy, I.A., Al Afifi, F.I. & Orabi, O.H., 1989. Foraminiferal biostratigraphy and paleoecology of some Cenomanian-Turonian exposures in west-central Sinai (Egypt), Revue de Micropaleontologie, v. 31, pp. 243–262.

Cherif, O.H., El-Sheikh, H.A. & Mohamed, S.I., 1994. Paleoecology of the Oligo-Miocene in the subsurface of the northern reaches of the Nile Delta, Egypt, N. Jb. Geol. Paläont. Mh. (12), pp. 703–725.

Chiarugi, A., 1929. Prime notizie sulle foreste pietrificate della Sirtica. N. Giorn. Bot. Ital., Firenze, v. 35, pp. 5558–566,

Chiarugi, A., 1934. Una tallofita arborea silicizzata del Deserto Libico: "*Nematophyton saharianum*" n.sp. Missione Scientifica della R. Accad. D'Italia a Cufra (1931-X), Roma, v. 3, pp. 293–319.

Chiesa, C., 1934. Su acuni nuovi fossili miocenici della Sirtica (Cirenaica). Missione Scientifica della R. Accad. D'Italia a Cufra (1931-X), Roma, v. 3, pp. 231–237.

Chiesa, C., 1940. La serie Eocenca del Gebel Uaddan nella Giofra (Libia), Ann. Mus. Libico Storia Nat., v. 2, pp. 189–202.

Chiesa, C., 1954. La serie stratigrafia dell'Uadi Tamet (Sirtica), Congrés Géologique International, Comptes Rendus de la 19ième Session, Alger 1952, Fascicule XX, pp. 145–153.

Chotin, P., Ait Brahim L. & Tabyaoui H., 2000. The southern Tethyan margin in north-eastern Morocco, sedimentary characteristics and tectonic control. Peri-Tethys Mémoire 5. Mémoires du Muséum National de l'histoire Naturelle, Paris, v. 182, pp. 107–128.

Choubert, G. 1963. Histoire géologique du Précambrien de l'Anti-Atlas. Tome I. Notes et Mémoires du Service Géologique du Maroc, v. 162, p. 352.

Choubert, G. & Faure-Muret, A. 1969. Sur la série stratigraphique précambrienne de la partie sud-ouest du massif du Bas Dra (Tarfaya, Sud marocain). Comptes Rendus de l'Acade´mie des Sciences, v. 269, pp. 759–762.

Chowdhary, L.R. & Taha, S., 1987. Geology and habitat of oil in Ras Budran Field, Gulf of Suez, Egypt. AAPG Bulletin, v. 71, no. 10, pp. 1274–1293.

Christie, A.M., 1955. Geology of the Garian Area, United Nations Technical Assistance Program, New York, Report TAA/LIB/2, 60p. Re-published 1966 by the Geological Survey, Ministry of Industry, Kingdom of Libya, Geological Section Bulletin No. 6, p. 64.

Church, W.R., 1976. Late Proterozoic ophiolites, Intel. Colloque CNRS 272, Associations mafiques-ultramafiques des orogenes, pp. 105–117.

Church, W.R., 1986. Comment on the paper "Pan African (late Precambrian) tectonic terrains and the reconstruction of the Arabian-Nubian shield" by J.R. Vail. Geology, v. 13, pp. 839–842.

Church, W.R., 1988. Ophiolites, sutures, and micro-plates of the Arabian-Nubian Shield: A critical comment, pp. 289–316, in S. El-Gaby & R.O. Greiling (Eds.), The Pan-African Belt of Northeast Africa and Adjacent Areas, Fiedr. Vieweg & Sohn, Baunschweig/Wiesbaden, p. 369.

Cita, M.B., 1982. The Messinian Salinity Crisis in the Mediterranean, A Review, in H. Berckhemer & K. Hsü (Eds.), Alpine-Mediterranean Geodynamics, Geodynamics Series, v. 7, Am. Geophys. Union/Geol. Soc. Am., pp. 113–140.

Clariond, L., 1932. Sur le Stéphanien des Ida ou Zal (Maroc occidental). Comptes Rendus de l'Académie des Sciences, Paris, v. 195, pp. 62–64.

Clarke, J.I., 1963. Oil in Libya: Some implications. Economic Geography, v. 39, no. 1, pp. 40–59.

Clauzon, G., Gautier, F., Berger, A. & Loutre, M.-F., 1995 (Abs.). An alternative model for a new interpretation of the Mediterranean Messinian salinity crisis. International

Conference on the Biotic and Climatic Effects of the Messinian Event on the Circum Mediterranean, Technical Program and Abstracts, University of Garyounis, Benghazi, p. 27.

Clauzon, G., Suc, J.-P., Gautier, F., Berger, A. & Loutre, M.-F., 1996. Alternative interpretation of the Messinian salinity crisis: controversy resolved?, Geology, v. 24, pp. 363–366.

Clayton, G., 1996. Mississippian miospores. In J. Jansonius & D.C. McGregor (Eds.), Palynology: principles and applications, American Association of Stratigraphic Palynologists Foundation, v. 2, pp. 589–596.

Clayton, G. & Lobboziak, S., 1985. Early Carboniferous (Early Visan-Serpukovian) palynomorphs setting, in B. Thusu & B. Owens (Eds.), Palynostratigraphy of North-East Libya, J. Micropalaeontology, v. 4, pt. 1, pp. 83–91.

Clayton, P.A. & Spencer, L.J., 1934. Silica-glass from the Libyan Desert. Mineralogical Magazine, v. 23, pp. 501–508.

Clayton, P.A., 1933. The western side of the Gilf Kebir. The Geographical Journal, v. 81, no. 3, pp. 254–259.

Clayton, P., 1998. Desert Explorer: A Biography of Colonel P.A. Clayton. Zerzura Press, Cargreen, p. 193.

Clerk, G., Kennedy Shaw, W.B., Shearer, J. & Sandford, K.S., 1945. Early days of the Long Range Desert Group: Discussion. The Geographical Journal, v. 5, no. 1/2, pp. 42–46.

Clifford, H., Grund, R. & Musrati, H., 1980. Geology of a stratigraphic giant, Messla oil field, Libya. In L.V. Illing & G.D. Hobson (Eds.), Giant Oil and Gas Fields of the Decade 1968–1978. AAPG Mem. No. 30, pp. 507–524.

Close, A., 1980. Current research and recent radiocarbon dates from Northern Africa. The Journal of African History, v. 21, no. 2, pp. 145–167.

Close, A., 1984. Current research and recent radiocarbon dates from Northern Africa II. The Journal of African History, v. 25, no. 1, pp. 1–24.

Clyde, W.C. & Gingerich, P.D., 1988. Mammalian community response to the latest Paleocene thermal maximum: An isotaphonomic study in the northern Bighorn Basin, Wyoming. Geology, v. 26, pp. 1011–1014.

Cochran, M.D. & Petersen, L.E., 2001, Hydrocarbon exploration in the Berkine Basin, Grand Erg Oriental, Algeria. In M.W. Downey, J.C. Threet & W.A. Morgan (Eds.), Petroleum Provinces of the Twenty-first Century. AAPG Memoir 74, pp. 531–557.

Coffield, D.Q. & Schamel, S., 1989. Surface expression of an accommodation zone within the Gulf of Suez rift, Egypt. Geology, v. 17, pp. 76–79.

Coggi, L., 1940. Fossile triassici della Gefara Tripolitania. Ann. Mus. Libico Stor. Nat., 2, pp. 139–195.

Cohen, A.J., 1959. Origin of Libyan Desert Silica-Glass. Nature, v. 183, No. 4674, pp. 1548–1549.

Cohen, C.R., Schamel, S. & Boyd-Lygi, P., 1980. Neogene deformation in northern Tunisia: Origin of the eastern Atlas by microplate – continental margin collision. Geological Society of America Bulletin, v. 91, no. 4, pp. 225–237.

Collignon, M. & Lefranc, J.P., 1974. Mise en evidence de la communication saharienne entre Tethys et Atlantique Sud d'apres les fossiles cenomaniens et turoniens du Tademait (Sahara Algerien), C.R. Acad. Sci., v. 278, Serie D, Paris, pp. 2257–2261.

Collomb, G.R., 1962. Etude géologique du Jebel Fezzan et de sa bordure Paléozoique, Notes et Mémoires No. 1, Compagnie Française des Pétroles, Paris, p. 35.

Comas, M.C., García-Dueñas, V. & Jurado, M.J., 1992. Neogene tectonic evolution of the Alboran Sea from MCS data. Geo-Marine Letters, v. 12, nos. 2–3, pp. 157–164.

Comas, M.C., Platt, J.P., Soto, J.I. & Watts A.B., 1999. The origin and tectonic history of the Alboran basin: insights from ODP leg 161 results. In Zahn R., Comas M.C., Klaus A. (Eds.), Proc. Ocean Drill. Program, Sci. Res., v. 161, pp. 555–580.

Compagnies Pétrolieres, 1964, Essai de nomenclature lithostratigraphique du Cambro-Orovicien Saharien, Soc. Géol. Fr. Mem. hors ser., 2, p. 54.

Compte, D. & Lehman, P., 1974. Sur les carbonates de l'Ypresien et du Lutétien basal de la Tunisie centrale, CFP, Notes et Mémoires, No. 11, Paris, pp. 275–202.

Conant, L.C. & Goudarzi, G.H., 1964. Geological map of Kingdom of Libya, U.S. Geol. Survey, Geological Investigation Map. 1-350A, scale 1:2,000,000, Washington.

Conant, L.C. & Goudarzi, G.H., 1967. Stratigraphic and tectonic framework of Libya, AAPG Bull., v. 51, No. 5, pp. 719–730.

Conely, C.D., 1971. Stratigraphy and lithofacies of Lower Paleocene rocks, Sirte Basin, L.A.R., in C. Gray (Ed.), Symposium on the Geology of Libya, Faculty of Science, U. of Libya, Tripoli, pp. 127–140.

Coney, P.J., 1987. The regional tectonic setting and possible causes of Cenozoic extension in the North American Cordillera, in M.P. Coward, J.F. Dewey & P.L. Hancock (Eds.), Continental Extensioanl Tectonics, Geological Society Special Publication No. 28, pp. 177–186.

Coniglio, M., James, N.P. & Aissaoui, D.M., 1988. Dolomitization of Miocene carbonates, Gulf of Suez, Egypt, J. Sed. Pet., v. 58, No. 1, pp. 100–119.

Conrad, J., 1985a. Timimoun Basin. In C.M. Diaz (Ed.), 1985: The Carboniferous of the World, II. Australia, Indian Subcontinent & North Africa: North Africa. IUGS Publication No. 20, Instituto Geológico y Minero de España and Empresa Nacional ADARO de Investigaciones Mineras, s.a., pp. 315–317.

Conrad, J., 1985b. Ahnet-Mouydir Area. In C.M. Diaz (Ed.), 1985: The Carboniferous of the World, II. Australia, Indian Subcontinent & North Africa: North Africa. IUGS Publication No. 20, Instituto Geológico y Minero de España and Empresa Nacional ADARO de Investigaciones Mineras, s.a., pp. 317–322.

Conrad, J., 1985c. Reggan Area. In C.M. Diaz (Ed.), 1985: The Carboniferous of the World, II. Australia, Indian Subcontinent & North Africa: North Africa. IUGS Publication No. 20, Instituto Geológico y Minero de España and Empresa Nacional ADARO de Investigaciones Mineras, s.a., pp. 322–323.

Conrad, J., 1985d. Tindouf Basin. In C.M. Diaz (Ed.), 1985: The Carboniferous of the World, II. Australia, Indian Subcontinent & North Africa: North Africa. IUGS Publication No. 20, Instituto Geológico y Minero de España and Empresa Nacional ADARO de Investigaciones Mineras, s.a., pp. 325–327.

Conway, A.M., Fahmy, K.H. & McGarva, R., 1988a. Sedimentation, diagenesis and reservoir properties of the Bahariya Formation (Cenomanian), Khalda Field, Western Desert, Egypt. 9th Exploration and Production Conf., Egyptian General Petroleum Corp. (EGPC), Cairo, Egypt, November 1988, p. 28.

Conway, A.M., Fahmy, K.A. & McGarva, R., 1988b (Abs.). Lower Bahariya Formation in the Khalda field, sedimentation, diagenesis, and reservoir properties. Mediterranean Basins Conference and Exhibition, Nice, France, AAPG Bull., v. 72, No. 8, pp. 995–996.

Cool, T., Katz, B., Dignes, T., Reimers, D. & Fleisher, R., 2008. Hydrocarbon source rock assessment and revised biostratigraphy of DSDP Site 369a, offshore Northwest African margin. J. Petroleum Geology, v. 31, no. 2, pp. 117–134.

Coquand, H., 1847. Description géologique de la partie septentrionale de empire du Maroc. Bull. Soc. Géol. France, sér 2, v. 4, pp. 1188–1249.

Coquel, R. & Abdesselam-Rouighi, F., 2000. Révision palynostratigraphique du Dévonien terminal-Carbonifère dans le Grand Erg Occidental (Bassin de Béchar) Sahara Algérien. Revue Micropaléontol., v. 43, no. 3, pp. 353–364.

Coquel, R. & Latreche, S. 1989. Etude palynologique de la Formation d'Illerene (dévono-carbonifère) du Bassin d'Illizi (Sahara Algerien Oriental). Palaeontographica, Abt. B 212, pp. 47–70.

Cordey, F., Boughridi, M. & Sallouhi, H., 2005. First direct age determination from the Jurassic radiolarian-bearing siliceous series (Jedidi Formation) of northwestern Tunisia. Comptes Rendus Geosciences, v. 337, no. 8, pp. 777–785.

Cornet, A. & Gouskov, N., 1959. Sur la présence d'une fauneYprésienne dans certains forages du bas Sahara. Compte rendu sommaire des séances de la Société géologique de France, v. 35, pp. 143–149.

Cornish, V., 1897. On the formation of sand dunes. The Geographical Jounal, v. 9, no. 3, pp. 278–309.

Cornish, V., 1900. On desert sand-dunes bordering the Nile Delta. The Geographical Journal, v. 15, no. 1, pp. 1–30.

Cornish, V., 1934. Ocean waves and kindered geophysical Phenomena, with additional notes by Dr. Harold Jeffreys.

Corriger, I.C. & Surcin, I., 1963. Les séries "cambro-ordovicienne" dans les sondages de la partie orientale du bassin de Fort Polignac, in First Saharan Symp., Tripoli, Revie de L'Institut du Pétrole, V. XVIII, No., éditions Technip, Paris, pp. 144–149.

Courel, L., Aït Salem, H., Benaouiss, N., Et-Touhami, M., Fekirine, B., Oujidi, M., Soussi, M. & Tourani, A., 2003. Mid-Triassic to Early Liassic clastic/evaporitic deposits over the Maghreb Platform. Palaeogeography, Palaeoclimatology, Palaeoecology, v. 196, pp. 157–176.

Courville, P., 2007. Échanges et colonisations fauniques (Ammonitina) entre Téthys et Atlantique sud au Crétacé supérieur: voies atlantiques ou sahariennes? In: G. Bulot, S. Ferry & D. Grosheny (Eds.), Relations entre les marges septentrionale et méridionale de la Téthys au Crétacé [Relations between the northern and southern margins of the Tethys ocean during the Cretaceous period]. Carnets de Géologie/Notebooks on Geology, Brest, Mémoire 2007/02, Résumé 02 (CG2007_M02/02), pp. 16–19.

Coutelle, A., Pautot, G. & Guennoc, P., 1991. The structural setting of the Red Sea axial valley and deeps, implications for crustal thinning processes, in J. Markis. P. Mohr & R. Rihm (Eds.), Red Sea: Birth and Early History of a New Oceanic Basin, Tectonophysics, v. 198, pp. 395–409.

Coutellier, V. & Stanley, D.J., 1987. Late Quaternary stratigraphy and paleogeography of the eastern Nile Delta, Egypt, Marine Geology, v. 77, pp. 257–275.

Coward, M.P., Purdy, E. G, Ries, A. C & Smith, D.G. 1999. The distribution of petroleum reserves in basins of the South Atlantic margins. In: N.R. Cameron, R.H. Bate, & V.S. Clure (Eds): The Oil and Gas Habitats of the South Atlantic. Geological Society, London, Special Publ. 153, pp. 101–132.

Coward, M.P. & Ries, A.C., 2003. Tectonic development of North African basins. In: Arthur, T.J., Macgregor, D.S., Cameron, N. (Eds.), Petroleum Geology of Africa: New Themes and developing technologies, Geol. Soc. (London) Sp. Publ. 207, pp. 61–83.

Craig, J., Rizzi, C., Said, F., Thusu, B., Lüning, S., Asbali, A.I., Keeley, M.L., Bell, J.F., Durham, M.J., Eales, M.H., Beswetherick, S. & Hamblett, C, 2004. Structural styles and prospectivity in the Precambrian and Palaeozoic hydrocarbon systems of North Africa. Conference proceedings. Geology of East Libya Symposium 2004, Benghazi (preprint).

Craig, J., Sutcliffe, O.E., Lüning, S., LeHeron, D. & Whittington, R., 2002. Ice Sheets and Hot Rocks: Unravelling the glacial signature in the Late Ordovician reservoirs of North Africa. Petroleum Exploration Society of Great Britain Newsletter December 2002, pp. 6–11.

Craig, J., Thurow, J., Thusu, B., Whitham, A. & Abutarruma, Y. (eds), 2009. Global Neoproterozoic Petroleum Systems: The Emerging Potential in North Africa. Geological Society, London, Special Publications, v. 326, pp. 1–25.

Crema, C., 1922. Il Maestrichtiano in Cirenaica, Rendiconto Reale Accad. Lincei, ser.5, v. 31, pp. 121–125.

Crema, C., 1926. Sulle manifestazioni di idrocarburi del pozzo artesiano di Sidi-Mesri presso Tripoli. La Miniera Italiano, v. 10, pp. 49–50.

Crema, C., Franchi, S. & Parona, C.F., 1913. Sulla Serie der terreni della Tripolitania Settentrionale, Communicazioni preliminare. Bull. Soc. Geol. Ital., v. 32, pp. 497–502.

Cremaschi, M. & Di Lernia, S., 1999. Holocene Climatic Changes and Cultural Dynamics in the Libyan Sahara. African Archaeological Review, v. 16, no. 4, pp. 211–238.

Crespo-Blanc, A. & Frizon de Lamotte, D., 2006. Structural evolution of the External Zones derived from the Flysch Trough and the South Iberian and Maghrebian paleomargins around the Gibraltar Arc: a comparative study. Bulletin de la Société géologique de France, v. 177, pp. 267–282.

Cronin, T.M. & Khalifa, H., 1979. Middle and Late Eocene Ostracoda from Gebel El Mereir, Nile Valley, Egypt, Micropaleontology, v. 25, pp. 397–411.

Crux, J., 1982. Upper Cretaceous (Cenomanian to Campanian) calcareous nannofossils, in A.R. Larson (Ed.), A Stratigraphical Index of Calcareous Nannofossils, British Micropaleontological Society, Ellis Horwood, Chichester, pp. 81–135.

Cunningham, K.J., Benson, R.H., Rakic-El Bied, K. & McKenna, L.W., 1997. Eustatic implications of late Miocene depositional sequences in the Melilla Basin, northeastern Morocco. Sedimentary Geology, v. 107, pp. 147–165.

Cunningham, K.J. & Collins, L.S., 2002. Controls on facies and sequence stratigraphy of an upper Miocene carbonate ramp and platform, Melilla basin, NE Morocco. Sedimentary Geology, v. 46, pp. 285–304.

Cunningham, S.M., 1988 (Abs.). Gothlandian source rock discovered north of the Talemzane Arch, central Tunisia. American Association of Petroleum Geologists Bulletin, v. 72, no. 8, pp. 996–997.

Cuny, G., Ouaja, M., Sparfi, D., Schmitz, L., Buffetaut, E. & Benton, M.J., 2004. Fossil sharks from the Early Cretaceous of Tunisia. Rev. Paléobiol., v. 9, pp. 127–142.

Cuvillier, J., 1930. Revision du nummulitique Egyptien, Mém. Inst. Egypte, v. 16, Cairo, p. 371.

Cuvillier, J., 1937. Présence du carbonifère marine dans l'Ouadi Abu Darag (Desert arabique), Compte Rendus, v. 204, pp. 1834–1835.

D'Archiac, E.J.A., 1868. Remarques à propos de la communication de Delanoüe sur les fossiles environs de Thebes. Compte Rendus 67, pp. 707–713.

D'Erasmo, G., 1934. Su alcuni avanzi di vertebrati terziari della Sirtica. Missione Scientifica della R. Accad. D'Italia a Cufra (1931-X), Roma, v. 3, pp. 63–81. Missione Scientifica della R. Accad. D'Italia a Cufra (1931-X), Roma, v. 3, pp. 259–279.

D'Hericourt, C.E.X.R., 1841. Observations géologiques recueillies en Egypte, sur la mer rouge, le golfe d'Aden, le pays d'Adel et le royaume de Choa. Bulletin de la Societé Géologique de France, v. 2, no. 3, pp. 541–546.

D'Hericourt, C.E.X.R., 1841. Observations géologiques recueillies en Egypte, sur la mer rouge, le golfe d'Aden, le pays d'Adel et le royaume de Choa. Compte Rendus Hebdomadaires des Seances de l'Academie de Sciences, v. 12.

D'Lemos, R.S., Inglis, J.D. & Samson, S.D., 2006. A newly discovered orogenic event in Morocco: Neoproterozoic ages for supposed Eburean basement of the Bou Azzer inlier, Anti-Atlas Mountains. Precambrian Research, v. 147, issues 1–2, pp. 65–78.

Dacqué, E., 1903. Mitteilungen über den Kreidecomplex von Abu Roash, bei Kairo. Palaeontographica 30, pp. 337–391.

Dacqué, E., 1912. Die fossilen Schildkröten Aegypten. Geol. Und Palaeont. Abh., Jena, Bd. 14, pp. 275–337.

Dakki, M., Hssain, M., El Alji, M. & El Abib, R., 2007. The Prerif domain and its forelands, Northern Morocco: Geology and petroleum play concepts. The First MAPG International Convention, Conference & Exhibition Marrakech Convention Center, October 28–31, 2007, Abstract Book, pp. 63–64.

Dalla, S., Harby, H. & Serazzi, M., 1997. Hydrocarbon exploration in a complex incised valley fill: An example from the late Messinian Abu Madi formation (Nile delta, Egypt). The Leading Edge, v. 16, no. 12, pp. 1819–1824.

Dalloni, M., 1934. Mission au Tibisti (1930–1931), Mémoire de l' Academie Sci. Inst. France, v. 61, Paris, p. 372.

Dalloni, M., 1945. La Mission Scientifique du Fezzan. Trav. Inst. Rech. Sahar., Alger, v. 3, pp. 176–178.

Dames, W., 1883. Über Zeuglodonten aus Aegypten und die Beziehungen der Achaeoceten zu den übrigen Cetacean. Palaeontologishe Abhandlungen 1(5), pp. 189–221.

Dames, W., 1894. Über eine teriäre wirbelthierfauna von der westlichen Insel des Birket-el-Qurun im Fajum (Aegypten). Sitzungsberichte der königlich-Preussichen Akademie der Wissenschaften, Berlin, 6, pp. 129–153.

Damnati, B., 2000. Holocene lake records in the Northern Hemisphere of Africa. Journal African Earth Sciences, v. 31, no. 2, pp. 253–262.

D'Archiac, E.J.A., 1868. Remarques à propos de la communication de Delanoüe sur les fossdes environs de Thèbes, Compte Rendus, v. 67, Paris, pp. 707–713.

Dardour, A.M., Boote, D.R.D. & Baird, A.W., 2004. Stratigraphic controls on palaeozoic petroleum systems, Ghadames Basin, Libya. J. Petroleum Geology, v. 27, no. 2, pp. 141–162.

Davies, W.C. & Bel Haiza, A., 1990. Sweeter E and P terms, Cretaceous Abiod chalk oil play lead to busier exploration in Tunisia. Oil & Gas Journal, v. 10, pp. 50–53.

Davison, I., 2005. Central Atlantic margin basins of North West Africa: Geology and hydrocarbon potential (Morocco to Guinea). Journal of African Earth Sciences, v. 43, nos. 1–3, pp. 254–274.

Davison, I. & Dailly, P., 2010. Salt tectonics in the Cap Boujdour Area, Aaiun Basin, NW Africa. Marine and Petroleum Geology, v. 27, pp. 435–441.

Dawson, J.W., 1884. Notes on the Geology of the Nile Valley. Geological Magazine, v. 1, pp. 289, p. 385.

Dawson, J.W., 1888. Modern Science in Bible Lands. Dawson & Brothers, Montreal, p. 606.

De Angelis, M., 1934. Osservazioni su alcune sabbie della Libia. Missione Scientifica della R. Accad. D'Italia a Cufra (1931-X), Roma, v. 3, pp. 431–450.

De Baets, K., Klug, C. & Plusqueelec, Y., 2010. Zlíchovian faunas with early ammonoids from Morocco and their use for the corrclation of the eastern Anti-Atlas and the western Dra Valley. Bulletin of Geosciences, Prague, v. 85, no. 2, pp. 317–352.

De Blieux, D.D., Baumrind, M.R., Simons, E.L., Chatrath, P.S., Meyer, G.E. & Attia, Y.S., 2006. Sexual dimorphism of the internal mandibular chamber in Fayum Pliohyracidae (Mammalia). Journal of Vertebrate Paleontology, v. 26, no. 1, pp. 160–169.

De Castro, J.C., Della Favera, J.C. & El-Jadi, M., 1991. Tempfacies, Murzuq basin, Great Socialist People's Libyan Arab Jamahiriya, their recognition and stratigraphic implications, in M.J. Salem & M.N. Belaid (Eds.), The Geology of Libya, v. V, Elsevier, Amsterdam, pp. 1757–1765.

De Castro, P. & Sirna, G., 1996. The *Durania arnaudi* biostrome of El-Hassana, Abu Roash area (Egypt).– Geo-logica Romana, v. 32, pp. 69–91.

De Foucauld, C.-E., 1883. Reconnaissance au Maroc, Paris.

De Gruyter, P. & Vogel, T.A., 1981. A model for the origin of the alkaline complexes of Egypt, Nature, v. 291, pp. 571–574.

De Heinzelin, J., 1968. Geological history of the Nile Valley in Nubia, in F. Wendorf (Ed.), The Prehistory of Nubia, 1, Southern Methodist University Press, Dallas, pp. 19–55.

De Heinzelin, J., El-Arnauti, A. & Gaziry, W., 1980. A preliminary revision of the Sahabi Formation, in M.J. Salem & M.T. Busrewil (Eds.), The Geology of Libya, v. I, London, Academic Press, pp. 127–133.

De la Harpe, P., 1883. Monographie der in Aegypten und der libyschen Wüste vorkommenden Nummuliten. Palaeontographica, 30, pp. 155–218.

De Lapparent, A.F., 1952. Stratigraphie du Trias de la Jeffara (Extreme Sud Tunisien et Triplitaine), Proc. 19th Int. Geol. Congress, Algeria, No. 21, pp. 129–134, Alger.

De Lapparent, A.F., 1954. Etat actuel de nos connaissance sur la stratigraphie, la paléontologie et al tectonique des "Grès de Nubie" du Sahara central, Proc. 19th Int. Geol. Congress, Algeria, No. 21, pp. 113–127, Alger.

De Lapparent, A.F., 1960. Les Dinosauriens du Continental Intercalaire du Sahara CentralS. Mémoires de la Société Géologique de France, Nouvelle série, no. 88A, p. 57.

De Lapparent, A.F. & Lelubre, M., 1948. Interpretation stratigraphique des series continentales entre Ohanet et Bourarhet (Sahara central), C.R. Acad. Sci., Séance du 22 novembre 1948, Paris.

De Larminat E., 1896. Étude sur les formes du terrain dans le Sud de la Tunisie. In: Annales de Géographie, v. 5, no. 22, p. 448.

De la Roche Jean, 1936. Concordance entre les techniques levalloisiennes de Berrouaghia (Algérie) et des environs du Caire (Egypte). Bulletin de la Société préhistorique française, v. 33, nos. 7–8. pp. 490–493.

De Lestang, J., 1965. Das Paleozoikum am Rande des Afro-Arabischen Gondwana-Kontinents, Mitteilung über das Erdi-Becken (Republic Tschad), Z. Dtsch. Geol. Ges., v. 117, pp. 479–488.

De Lestang, J., 1968. Das Paläozoikum am Rande des Afro-Arabischen Gondwana-Kontinents. Mitteilung über das Erdi-Becken (Republick Tschad). Z. dtsch. Geol. Ges. 117 (1965), 2–3, s. 479–488.

De Loriol, P., 1864. Monographie des Échinides contenus dans les couches nummulitique d'Égypte. Mém. Soc. Phys., Genève 28, pp. 59–148.

De Loriol, P., 1881. Description de deux Échinides nouveaux de l'Étage Nummulitique d'Égypte. Mém. Soc. Phys., Genève 17, pp. 103–107.

De Loriol, P., 1883. Eocaene Echinoiden aus Aegypten und der libyschen Wüste. Palaeontographica 30, pp. 5–59.

De Matos, J.E., 1994. Upper Jurassic-Lower Cretaceous stratigraphy: The Arab, Hith and Rayda formations in Abu Dhabi, in M.D. Simmons (Ed.), Micropalaeontology and Hydrocarbon Exploration in the Middle East, Chapman & Hall, London, pp. 81–113.

De Wall, H., Greiling, R.O. & Sadek, M.F., 2001. Post-collisional shortening in the late Pan-African Hamisana high strain zone, SE Egypt: field and magnetic fabric evidence. Precambrian Research, v. 107, pp. 179–194.

DeBari, S.M., 1997. Evolution of magmas in continental and oceanic arcs: The role of the lower crust, The Canadian Mineralogist, v. 35, part 2, pp. 501–519.

Debeis, S., Garcia, R. & Sundberg, K.R., 1988. Light hydrocarbon survey data and hydrocarbon accumulation in the Umbarka area, Western Desert, Egypt. 9th Exploration and Production Conf., Egyptian General Petroleum Corp. (EGPC), Cairo, Egypt, November 1988, p. 33.

Debrenne, F. & Debrenne, M., 1978. Archaeocyathid fauna of the lowest fossiliferous levels of Tiout (Lower Cambrian, Southern Morocco). Geol. Mag., v. 115, no. 2, pp. 101–119.

Debyser, J., de Charpal, O. & Merabet, O, 1965. Sur le caractère glaciaire de la sédimentation de l'Unité IV au Sahara Central. C.R. Acad. Sci. Paris, v. 261, pp. 5575–5576.

Decrée, S., Marignac, C., De Putter, T., Deloule, E., Liégeois, J.P. & Demaiffe, D., 2008. Pb-Zn mineralization in a Miocene regional extensional context: The case of the Sidi Driss and the Douahria ore deposits (Nefza mining district, northern Tunisia). Ore Geology Reviews, v. 34, pp. 285–303.

Decrouez, D. & Lanterno, E., 1979. Les "bancs a nummulites" de l'Eocene mésogéen et leurs implications, Arch. Sciences Genève, v. 32, pp. 67–94.

Dedok, T.A. & Hollard, H., 1980. Brachiopodes du Carbonifère inférieur du Maroc central. Notes Serv. Géol. Maroc, v. 41, no. 285, pp. 185–230.

Del Ben, A. & Finetti, I., 1991. Geophysical study of the Sirte Rise, in M.J. Salem, A.M. Sbeta & M.R. Bakbak (Eds.), The Geology of Libya, v. VI, Amsterdam, Elsevier, pp. 2417–2431.

Delanoüe, J., 1868. Note sur la constitution géologique des environs de Thèbes, Compte Rendus, v. 67, pp. 701–707, Paris.

Delesse, (A.E.O.J.) Prof., 1851. On the rose-coloured syenite of Egypt. Q.J. Geol. Soc. London, v. 7, pp. 9–13.

Demaison, G.J., 1965. The Triassic salt in the Algerian Sahara, Proc. Meet. Inst. Petrol. and Geol. Soc. on "Salt Basins around Africa", London, pp. 91–100.

Demathieu, G.R. & Wycisk, P., 1990. Tetrapod trackways from southern Egypt and northern Sudan, J. Afr. Earth Sci., v. 10, No. 3, pp. 435–443.

Demoulin, F., 1931. L'exploration du Sahara. Annales de Géographie, Année 1931, v. 40, no. 226, pp. 337–361.

Derbel-Damak, F. & Zaghbib-Turki, D., 2002. Identification des zones biostratigraphiques méditerranéennes dans le Pliocène du Cap-Bon (Tunisie) = Identification of the mediterranean biostratigraphical zones in the Cap-Bon (Tunisia) Pliocene deposits. Geobios, v. 35, no. 2, pp. 253–264.

Derin, B. & Garfunkel, Z., 1988 (Abs.). Late Permian to Mid-Cretaceous carbonate platform along the passive margin of the southeastern Mediterranean, in Mediterranean Basins Conf. and Exhib., Nice France, Mediterranean Conference, AAPG Bull., v. 72, p. 997.

Derradji, A., 2006. Le site acheulenn d'Errayah (Mostaganem, Algerie) dans son contexte géologique. Comptes Rendus, Palevol, v. 5, nos. 1–2, pp. 229–235.

Descossy, G. & Roch, E. 1934. Sur quelques fossiles de la basse vallée du Dra de la région de Tindouf. Comptes Rendus sommaires des séancsc, Société géologique de France, v. 4, série 5, pp. 104–105.

Desio, A., 1928. Resultati scientifica della Missione alla Oasi di Giarabub. Pt.I. La morfologia, Pub. Della R. Soc. Geogr. Ital., Roma.

Desio, A., 1931. Missione scientifica della Reale Accademia d'Italia a Kufra, v. 1, Roma.

Desio, A., 1933. Schizzo geologico della Libia alla scala di 1:4000.000. Note Illustrive, Firenze.

Desio, A., 1934. Missione scientifica della Reale Accademia d'Italia a Kufra, v. 3, Roma.

Desio, A., 1935. Studi geologici sulla Cirenaica, sul deserto libico, sulla Tripolitania e sul Fezzan orientale. Missione Scient. della R. Acc. d'Italia a Cufra (1931), v. 1, Roma, p. 480.

Desio, A., 1936. Prime notizie sulla presenza del Silurico fossilifero nel Fezzan, Boll. Soc. Geol. Ital., v. 55, pp. 116–120.

Desio, A., 1939a. Missione scientifica della Reale Accademia d'Italia a Kufra, v. 2, Roma.

Desio, A., 1939b. Carta geol. Della Libia scala 1:3,000,000. Ann. Museo Libico Storia Nat., v. 1, Tripoli.

Desio, A., 1942. Appunti geomorfologici sul Sahara Libico sud-occidentale. Ann. Museo Libico Storia Nat., v. 3, pp. 17–32.

Desio, A., 1943. L'esplorzione Minereraria della Libia, Collezione Scientifica e Documentaria, a cura del Minisero dell'Africa Italiana, v. 10, Milano, p. 333.

Desio, A., 1951. Cenno riassuntivo sulla costituzione geologica della Libia, Proc. 18th Int. Geol. Congress, London 1948, pp. 47–53.

Desio, A., 1967. Short history of the geological, mining and oil exploration in Libya. Atti della Accademia Nazionale dei Lincei, Mem. Ser. 8a, v. 8, no. 4, pp. 79–123.

Desio, A., 1968. Geology and Archaeology of Northern Cyrenaica, Libya. In F.T. Barr (Ed.), Petroleum Exploration Society of Libya Tenth Annual Field Conference, Tripoli, pp. 70–113.

Desio, A., 1971. Outlines and problems in the Geomorphological evolution of Libya from the Tertiary to the present day, in C. Gray (Ed.), Symposium on the Geology of Libya, Faculty of Science, U. of Libya, Tripoli, pp. 11–36.

Desio, A., Ronchetti, R.C., Pozzi, R., Clerici, F., Invernizzi, G., Pisoni, C. & Vigano, P.L., 1963. Stratigraphic studies in the Tripolitanian Jebel (Libya). Mem. Riv. Ital. Paleont., v. 9, p. 126.

Desio, A. & Rossi-Ronchetti, C., 1960. Sul Giurassic medio di Garet al-Bellaa (Tripolitania) e sulla posizione stratigrafica della formazione di Tachbal, Riv. Ital. Paleontol. Stratig., v. 66, pp. 173–196.

Desio, A., Rossi-Ronchetti, C. & Invernizzi, G., 1960. Il Giurassico dei dintorni di Jefren in Tripolitania, Riv. Ital. Paleontol. Stratig., v. 66, pp. 65–118.

Deunff, J. & Massa, D., 1975, Palynologie et stratigraphie Cambro-Ordovicien (Libya nord-occidentale), C.R. Acad. Sci. Paris, v. 281, Ser. D, pp. 21–24.

Deynoux, M., 1998 (Abs.). Earth's glacial record with special reference to Late Proterozoic and Late Ordovician glacial drifts in North Africa, The Geological Conference on Exploration in Murzuq Basin, Sabha University, Sabha, Libya, p. 48.

Di Caporiacco, l., 1934. Osservazioni sul Deserto Libico. L'Universo, v. 15, p. 9.

Di Cesare, F., Franchino, A. & Sommaruga, C., 1963. The Pliocene-Quaternary of Giarbub Erg region, Rev. Inst. Fr. Pétrole, 18, pp. 1344–1362.

Didon, J., Durand-Delga, M. & Kornprobst, J. 1973. Homologies géologiques entre les deux rives du détroit de Gibraltar. Bulletin de la Société Géologique de France, v. 15, pp. 77–105.

Dixon, T.H., 1981. Age and chemical characteristics of some pre-Pan-African rocks in the Egyptian Shield. Precambrian Res, v. 14, no. 2, pp. 119–133.

Dixon, R.J., Patton1, T.L., Hirst, J.P.P. & Diggens, J., 2008. Transition from Subglacial to Proglacial Depositional Systems: Implications for Reservoir Architecture, Illizi Basin, Algeria. AAPG Search and Discovery Article #50095. Adapted from oral presentation at AAPG Annual Convention, San Antonio, Texas, April 20–23, 2008.

Dolson et al, 2002. Regional Controls and Future Potential of the North African Petroleum Systems. Oil & Gas Journal, May 20, pp. 32–37.

Dolson, J., El Barkooky, A., Wehr, F., Gingerich, P.D., Prochazka, N. & Shann, M., 2002. The Eocene and Oligocene Paleo-Eology and Paleo-Geography of Whale Valley and the Fayoum Basins: Implications for Hydrocarbon Exploration in the Nile Delta and Eo-Tourism in the Greater Fayoum Basin. AAPG/EPEX/EGS/EAGE, Field Trip No. 7, Cairo, p. 79.

Dolson, J.C., Shann, M.V., Matbouly, S., Harwood, C., Rashed, R. & Hammouda, H., 2001. The petroleum potential of Egypt. In: Downey, M.W., J.C. Threet & W.A. Morgan (Eds.). Petroleum Provinces of the Twenty-first Century, AAPG Memoir 74, pp. 453–482.

Domaci, L., 1985. Geological map of Libya, 1:250,000, Sheet Bi'r Zaltan, NH34-14, Explanatory Booklet, Industrial Research Centre, Tripoli, p. 106.

Dominik, W. & Schaal, S., 1984. Notes on the stratigraphy of the Upper Cretaceous phosphates (Campanian) of the Western Desert (Egypt), Berliner geowiss. Abh.(A)50, pp. 153–175.

Dominik, W., 1985. Stratigraphie und sedimentologie (Geochemie, Schwermineral-analyse) der Oberkreide von Bahariya und ihre korrelation zum Dakhla-Becken (Western Desert, Ägypten), Berliner geowiss. Abh. (A) 62, p. 173.

Donze, P., Collin, J.P., Damotte, R., Oertli, H., Peypouquet, J.P. & Said, R., 1982. Les Ostracodes du Campanien Terminal à l'Eocene inférieur de la coupe de Kef, Tunisie nord-occidental. Bull. Centre Rech. Elf-Aquitaine, v. 6, pp. 273–355.

Donzeau, M., 1974. L'Arc Anti-Atlas-Ougarta (Sahara nord-occidental, Algérie-Maroc), C.R. Acad. Sci., sér. 2, v. 278, pp. 417–420.

Döring, S. & Kazmierczack, M., 2001. Stratigraphy, geometry, and facies of a Middle Devonian Ramp-to-Basin transect (Eastern Anti-Atlas, SE Morocco). Facies, v. 44, pp. 137–150.

Dostal, J., Caby, R., Dupuy, C., Mevel, C. & Owen, J.V., 1996. Inception and demise of a Neoproterozoic ocean basin: evidence from the Ougda complex, western Hoggar (Algeria), Geol. Rundsch., v. 85, pp. 619–631.

Dostal, J., Keppie, J.D., Hamilton, J., Aarab, E.M., Lefort, J.P. & Murphy, J.B., 2005. Crustal Xenoliths in Triassic lamprophyre dykes in western Morocco: tectonic implications for the Rheic Ocean suture. Geol. Mag., v. 142, no. 2, p. 159–172.

Doubinger, J. & Roy-Dias, C., 1985. La paléoflore du Stéphanien de l'Oued Zat (Haut Atlas de Marrakech-versant Nord – Maroc). Geobios, v. 18, pp. 573–586.

Douvillé, H., 1900. Découverte d'Orbitolina sur le versant Nord du Djebel Géneffé. Bull. Soc. Géol. Fr., sér.3, t.XXVIII, pp. 1001–1002.

Douvillé, H., 1901a. Découverte d'Orbitolina sur le versant Nord du Djebel Géneffé. Bull. Soc. Géol. Fr., sér.4, t.I, p. 156.

Douvillé, H., 1901b. Sur un foraminifère d'Egypte communiqué par Fourtau. C.R.S. Soc. Géol. Fr., février 1901, p. 156.

Douvillé, H., 1916. Les terrains secondaires dans le Massif de Moghara, á l'Est de l'Isthme de Suez, d'après les explorations de M. Couyat-Barthoux. Paleontologie, 1ère et 2ème parties. Mém. Ac. Sc., Paris, sér. 2, t.LIV, pp. 1–184.

Douvillé, H., 1920a. Le Lutetien inférieur à l'Est de l'Isthme de Suez. C.R.S. Soc. Géol. Fr, pp. 45–46.

Douvillé, H., 1920b. Les foraminfère de l'Eocène, dans la région de Suez, C.R.S. Soc. Géol. Fr, sér.4, t.XX, pp. 106–1076.

Douvillé, H., 1924. Le Crétacé et l'Eocène dans l'Est de l'Egypte. C.R.S. Soc. Géol. Fr, pp. 113–114.

Douvillé, H., 1934. Le Permien marin de l'Extrême-Sud Tunisien. II, Les fusulinidés de la Tunisie. Mém. Service Carte Géol. Tunisie, n. ser., no. 1, pp. 75–90.

Douvillé, H., Solignac, M., & Berkaloff, E., 1933. Découverte du Permien marin au Djebel Tebaga (Extrême-Sud Tunisien). Comptes Rendus Acad. Scéances, pp. 21–24.

Dragastan, O.N. & Herbig, H.-G., 2007. *Halimeda* (green siphonous algae) from the Paleogene of (Morocco) – Taxonomy, phylogeny and paleoenvironment. Micropaleontology, v. 53, nos. 1–2, pp. 1–72.

Dragastan, O.N. & Soliman, H.A., 2002. Paleogene calcareous algae from Egypt. Micropaleontology, v. 48, no. 1, pp. 1–30.

Du Dresnay, R., 1988. Répartition des dépots carbonatés du Lias inférieur et moyen le long de la cote atlantique du Maroc; conséquences sur la paléogéographie de l'Atlantique naissant. Journal of African Earth Sciences, v. 7, no. 2, pp. 395–396.

Druckman, Y., 1974a. The stratigraphy of the Triassic sequence in Southern Israel, Geol. Surv. Israel Bull. No. 64, p. 94.

Druckman, Y., 1974b. Triassic paleogeography of southern Israel and Sinai Peninsula, Proc. Intr. Symp. of the Alpine Mediterranean Triassic stratigraphy, Vienna, May 1973, Erdweiss. Komm. Osterr. Akad. Wiss., 2, pp. 79–86.

Druckman, Y., 1984. Evidence for Early-Middle Triassic faulting and possible rifting from the Helez Deep Borehole in the coastal plain of Israel, in J.E. Dixon & A.H.F. Robertson (Eds.), The Geological Evolution of the Eastern Mediterranean, Geol Soc. Spec. Pub. No. 17, Blackwell, pp. 203–212.

Druckman, Y., Buchbinder, B., Martinotti, G.M., Siman Tov, R. & Aharon, P. 1995. The buried Afiq Canyon (Eastern Mediterranean, Israel): a case study of a Tertiary submarine canyon exposed in Late Messinian times. Marine Geology, v. 123, pp. 167–185.

Dubar, G., 1960–62. Note sur la paléogéographie du Lias marocain (Domaine atlasique). Livre ala mémoire de P. Fallot, Mém. Soc. Géol. France, H. sér. 1, pp. 529–544.

Dubois, P., 1961. Stratigraphie du Cambro-Ordovicien du Tassili N'Ajjer (Sahara Central), Bull. Soc. Géol. Fr., pp. 206–209.

Dufaure, P., Fourcade, E. & Massa, D., 1984. Réalité des communication marines Transsaha-riennes entre la Téthys et l'Atlantique durant le crétacé superieur, C.R. Acad. Sci., Paris, v. 298, pp. 665–668.

Duffetaut, E., 1982. A ziphodont mesosuchian crocodile from the Eocene of Algeria and its implications for vertebrate dispersal. Nature, v. 300, pp. 176–178.

Duncan, P.M., 1869. Note on the Echinodermata, bivalve molluscaa, and some other fossil species from the Cretaceous rocks of Sinai. Q.J. Geol. Sco. London, v. 25, pp. 44–46.

Durand-Delga, M., 1969. Mise au point sur la structure du Nord-Est de la Berberie. Publ. Serv. Carte geol. Algerie, N.S., Bull. Soc. Géol. Fr., sér.7, no. xiii, pp. 328–337.

Durand-Delga, M., Esteras, M. & Olivier, P., 2007. Los "Taríquides" (Arco de Gibraltar): Problemas estructurales, paleogeográficos y consideración histórica. Revista de la Sociedad Geológica de España, v. 20, no. 3–4, pp. 119–134.

Durand-Delga, M., Rossi, P., Olivier, P. & Puglisi, D., 2000. Situation structurale et nature ophiolitique. de roches basiques jurassiques associées aux flyschs maghrébins du Rif (Maroc) et de Sicile (Italie). C.R. Acad. Sci. Paris, Sciences de la Terre et des planètes/Earth and Planetary Sciences, v. 331, pp. 29–38.

Durand-Delga, M. & Villiaumey, M. 1963. Sur la stratigraphie et la tectonique du Jebel Moussa. Bulletin de la Societé Géologique de France, séries 7, v. 5, no. 1, pp. 70–79.

Duronio, P., 1996. *Nummulites jamahiriae* n.sp. from the Sirt Basin, Libya, in M.J. Salem. A.J. Mouzughi & O.S. Hammuda (Eds.), The Geology of Sirt Basin, v. I, pp. 383–390.

Duronio, P., Dakshe, A. & Bellini, E., 1991. Stratigraphy of the offshore Cyrenaica (Libya), in M.J. Salem, O.S. Hammuda & B.A. Eliagoubi (Eds.), The Geology of Libya, v. IV, Amsterdam, Elsevier, pp. 1589–1620.

Dussert, D. 1924. Les gisements algeriens de phosphate de Chaux. Ann. Mines, v. 12, no. 6, pp. 135–221, pp. 229–325, pp. 333–398, and pp. 407–451.

Ebbighausen, V., Bockwinkel, J., Becker, R.T., Aboussalam, Z.S., Bultynck, P., El Hassani, A. & Nübel, H., 2004. Late Emsian and Eifelian stratigraphy at Oufrane (Tata region, eastern Dra Valley, Morocco). Documents de l'Institut Scientifique, v. 19, pp. 44–52.

Ebert, J., 1993. Globale Events im Grenz-Bereich Mittel-/Ober-Devon. Gött. Arb. Geol. Paläont., v. 59, pp. 1–106.

Echikh, K., 1998. Geology and hydrocarbon occurrences in the Ghadames Basin, Algeria, Tunisia, Libya, in D.J. Macgregor. R.T.J. Moody & D.D. Clark-Lowes (Eds.), Petroleum Geology of North Africa, Geological Society Spec. Pub. No. 132, London, pp. 109–129.

Echikh, K. & Sola, M.A., 2000. Geology and Hydrocarbon Occurrences in the Murzuq Basin, SW Libya. In: Sola, M.A., Worsley, D. (Eds.), Geological Exploration in the Murzuq Basin. Elsevier Science, v. 5., pp. 175–222.

Eck, O., 1914. Die Cephalopoden der Schweinfurthschen Sammlung aus der oberen Kreide Agyptens. Zeitschr. Dtsch. Geol. Ges., v. 66, pp. 179–216.

EGPC Stratigraphic Committee (1964), Oligocene and Miocene rock stratigraphy of the Gulf of Suez refion, EGPC, Cairo, p. 142.

EGPC Stratigraphic Committee (1974), Miocene rock stratigraphy of Egypt, Egypt. J. Geol., v. 18, pp. 1–69.

Egyptian General Petroleum Corporation, 1992. Western Desert: Oil and Gas Fields (A Comprehensive Review), Cairo, Egypt, p. 431.

Egyptian General Petroleum Corporation, 1994. Nile Delta and North Sinai: Fields, Discoveries and Hydrocarbon Potentials (A Comprehensive Review), Cairo, Egypt, p. 387.

Egyptian General Petroleum Corporation, 1996. Gulf of Suez Oilfields (A Comprehensive Review), Cairo, Egypt, p. 736.

Egyptian Geological Survey, 1981. Geological Map of Egypt, Scale 1:2,000,000, Cairo.

Eicher, D.B., 1946. Conodonts from the Triassic of Sinai, Egypt, Bull. AAPG, v. 30, pp. 613–616.

El Albani, A., Kuhnt, W., Luderer, F., Herbin, J.P. & Caron, M., 1999. Palaeoenvironmental evolution of the late Cretaceous sequences in the Tarfaya Basin (southwest of Morocco). In N.R. Cameron, R.H. Bate & V.S. Clure (Eds.): The Oil and Gas Habitats of the South Atlantic. Geological Society of London, Special Publication, v. 153, pp. 223–240.

El Ayouti, M.K., 1961. Geology of Belayim oilfield. 3rd Arab Petroleum Congress, Alexandria, p. 12.

El Ayouti, M.K., 1990. Petroleum Geology, in R. Said (Ed.), The Geology of Egypt, Balkema, Rotterdam/Brookfield, pp. 567–600.

El Bakai, M.T., 1991. Petrography and diagenesis of the Lidam Formation (carbonate units) – from selected wells in Sirt Basin, Libya, Petroleum Research Jorunal, v. 3, pp. 35–43.

El Bakai, M.T., 1996. Diagenesis and diagenetic history of the Lidam Formation in NW Sirt Basin, in M.J. Salem & M.N. Belaid (Eds.), The Geology of Libya, v. V, Elsevier, Amsterdam, pp. 83–97.

El Bayoumi, R.M., Hassan, M.A., Salman, A.B. & Abdallah, S.M., 1998 (Abs.). Subsurface geologic studies of Gebel El Missikat Uranium occurrence, Central Eastern Desert, Egypt, in Geology of the Arab World, Fourth International Conference, Cairo University, Cairo, pp. 115–116.

El Bchari, F. & Souhel, A., 2008. Stratigraphie séquentielle et évolution géodynamique du Jurassique (Sinémurien terminal – Aalénien) d'Ait Bou Guemmez (Haut Atlas central, Maroc). Estudios Geológicos, v. 64, no. 2, julio-diciembre 2008, pp. 151–160.

El Dakkak, M.W., 1988. Geological studies of subsurface Paleozoic strata of northern Western Desert, Egypt, Journal of African Earth Scieneces, v. 7, pp. 103–111.

El Deftar, T. & Issawi, B., 1977. Geological Map of Libya, 1:250,000: Sheet Al Bardia NH 35-1, Explanatory Booklet, Industrial Research Centre, Tripoli, p. 93.

El Hadi, H., Tahiri, A., Simancas Cabrera, F., Gonzalez Lodeiro, F., Azor Pérez, A. & Martinez Poyatos, D.J. 2006. Un exemple de volcanisme calco-alcalin de type orogénique mis en place en contexte de rifting (Cambrien de l'oued Rhebar, Meseta occidentale, Maroc). Comptes Rendus Géoscience, v. 338, pp. 229–236.

El Hammichi, F., Benshili, K. & Elmi, S., 2008. Les faunes d'Ammonites du Toarcien-Aalénien du Moyen Atlas sud-occidental (Maroc). Revue de Paléobiologie, Genève, v. 27, no. 2, pp. 429–447.

El Hammichi, F., S. Elmi, A. Faure-Muret & K. Benshili, 2002. Une plate-forme en distension, témoin de phases pré-accrétion téthysienne en Afrique du Nord pendant le Toarcien-Aalénien (synclinal Iguer Awragh-Afennourir, Moyen Atlas occidental, Maroc). Comptes Rendus Géosciences, v. 334, pp. 1003–1010.

El Hammichi, F., Tabyaoui, H., Chaouni, A., Ait Brahim, L. & Chotin, P., 2006. Mio-Pliocene tectonics in Moroccan Rifian foreland: Coexistence of compressive and extensional structures. Revista de la Sociedad Geológica de España, v. 19, nos. 1–2, pp. 143–152.

El Hatimi, N., Ben Yaich, A. & El Kadiri, K., 1988. Evolution meso-cenozoique a la limite Zones Internes-Zones Externes dans la Chaine rifaine. Bull. Inst. Sci., Rabat, no. 12, pp. 9–18.

El Hinnawy, M. & Cheshitev, G., 1975. Geological map of Libya, 1:250,000, Sheet Tarabulus NI33-13, Explanatory Booklet, Industrial Research Centre, p. 65.

El Kadiri, K., Hlila, R., Sanz de Galdeano, C., Lopez-Garrido, C., Chalouan, A., Serrano, F., Bahmad, A., Guerra-Merchan, A. & Liemlahi, H., 2006. Geological Society, London, Special Publications, v. 262, no. 1, pp. 193–215.

El Kamar, A., Boutakiout, M., Elmi, S., Sadki, D. & Ruget, C., 1998. Foraminifères et ostracodes du Lias supérieur et du Bajocien de la Ride de Talghemt (Haut-Atlas central, Maroc). Bulletin de l'Institut Scientifique, Rabat, n.21, pp. 31–41.

El Makhrouf, A.A., 1988. Tectonic interpretation of Jabal Eghei area and its regional applicationnn to Tibesti orogenic belt, south central Libya (S.P.L.A.J.), J. African Earth Sciences, v. 7, pp. 945–967.

El Nady, M.M., Harb, F.S. & Basta, J.S., 2003. Crude oil geochemistry and its relation to the potential source beds for some Meleiha oil fields in the northern Western Desert, Egypt. Petroleum Science and Technology, v. 21, nos. 1–2, pp. 11–36.

El Nagdy, S., 2004. Mapping the Neoproterozoic Um Nar banded iron formation, Egypt, usin Aster Data. Unpub. M.Sc. thesis, University of Texas, Dallas, p. 19.

El Ramly, M.F. & Akaad, M.K., 1960. The basement complex in the central Eastern Desert of Egypt between Lat. 24° 30' and 25°40'N., Geol. Surv. Egypt, Paper 8, p. 35.

El Safori, Y., Ziko, A. & Nabila, E., 1998 (Abs.). New and extant Miocene bryozoans of Matruh area, northern Western Desert, Egypt, in Geology of the Arab World, Fourth International Conference, Cairo University, Cairo, pp. 18–19.

El Shazly, E.M. & Saleeb, G.S., 1977. Metasomatism of Miocene sediments in St. John's Island and its bearing on the history of the Red Sea, Egypt. J. Geol., v. 21, pp. 140–153.

El Shazly, E.M., 1977. The Geology of the Egyptian region, in A.E.M. Nairn. W.H. Kanes & F. Stehli (Eds.), The Ocean Basins and Margins, v. 4A, New York, Plenum, pp. 379–444.

El Shazly, S.H., 1998 (Abs.). Coniacian-Santonian pelecypods and gastropods from Wadi Sudr and Wadi Matulla, western Sinai, Egypt, in Geology of the Arab World, Fourth International Conference, Cairo University, Cairo, p. 18.

El Wartiti, M., Broutin, J., Freytet, P., Larhrib, M. & Toutin-Morin, N., 1990. Continental deposits in Permian basins of the Mesetian Morocco, geodynamic history. Journal of African Earth Sciences, v. 10, no. 1/2, pp. 361–368.

El-Zouki, 1980. Stratigraphy and lithofacies of continental clastics (Upper Jurassic and Lower Cretaceous) of Jabal Nafusa, NW Libya, in M.J. Salem & M.T. Busrewil (Eds.), The Geology of Libya, v. II, London, Academic Press, pp. 393–418.

El-Akkad, S. & Issawi, B., 1963. Geology and iron ore deposits of Bahariya oasis, Geol. Surv. Egypt, Cairo, paper No. 18, p. 300.

El-Akkad, S.E. & Abdallah, A.M., 1971. Contribution to the geology of Gebel Ataqa area, Ann. Geol. Surv. Egypt 1, pp. 21–42.

El-Akkad, S.E. & Dardir, A., 1966. Geology of the Red Sea coast between Ras Shagra and Mersa Alam with short note on exploratory work at Gebel El Rusas lead-zinc deposits, Geol. Surv. Egypt, paper 35, p. 67.

Elakkari, T.S., 2005. Structural Configuration of the Sirt Basin. Unpub. M.Sc. Thesis, International Institute For Geo-information Science and Earth Observation, Enschede, the Netherlands, p. 58.

El-Alami, M.A., 1996. Habitat of oil in Abu Attifel area, Sirt Basin, Libya, in M.J. Salem. A.S. El-Hawat & A.M. Sbeta (Eds.), The Geology of Sirt Basin, v. II, Amsterdam, Elsevier, pp. 337–348.

El Arabi, H., Ouahhabi, B. & Charriere, A., 2001. Les séries du Toarcien-Aalénien du SW du Moyen Atlas (Maroc): précisions stratigraphiques et signification paléontologique. Bulletin de la Société géologique de France, Paris, v. 172, no. 6, pp. 723–736.

El-Araby, A. & Abdel-Motelib, A., 1999. Depositional facies of the Cambrian Araba Formation in the Taba region, east Sinai, Egypt. J. African Earth Sciences, v. 29, no. 3, pp. 429–447.

El-Arnauti, A. & Shelmani, M., 1985. Stratigraphic and structural setting, in B. Thusu & B. Owens (Eds.), Palynostratigraphy of North-East Libya, J. Micropalaeontology, v. 4, pt.1, pp. 1–10.

El-Arnauti, A. & Shelmani, M., 1988. A contribution to the northeast Libyan subsurface stratigraphy with emphasis on Pre-Mesozoic, in A. El-Arnauti, B.Owens & B. Thusu (Eds.), Subsurface Palynostratigraphy of Northeast Libya, Benghazi, Libya, Garyounis University Publications, pp. 1–16.

El-Arnauti, A., B. Owens, & B. Thusu (Eds.), 1988. Subsurface Palynostratigraphy of Northeast Libya, Garyounes University Publications, Benghazi, Libya.

El-Azabi, M.H. & El-Araby, A., 2007. Depositional framework and sequence stratigraphic aspects of the Coniacian–Santonian mixed siliciclastic/carbonate Matulla sediments in Nezzazat and Ekma blocks, Gulf of Suez, Egypt. Journal of African Earth Sciences, v. 47, pp. 179–202.

El-Baz, F., Boulos, L., Breed, C., Dardir, A., Dowidar, A., El-Etr, H., Embabi, N., Grolier, M., Haynes, V., Ibrahim, M., Issawi, B., Maxwll, T., McCauley, J., McHugh, W., Moustafa, A. & Yousif, M., 1980. Journey to the Gilf Kebir and Uweinat, Southwest Egypt, 1978. The Geographical Journal, v. 146, no. 1, pp. 51–59.

El-Beialy, S.Y. 1994. Palynostratigraphy and playnofacies of some subsurface Cretaceous formations in the Badr el Dein (Bed 1-1) borehole, North Western Desert, Egypt, N. Jb. Geol. Paläont. Abh., 192, No. 2, pp. 133–149.

El-Beialy, S.Y., Head, M.J. & El Atfy, H.S., 2010. Palynology of the Mid-Cretaceous Malha and Galala formations, Gebel El Minshera, North Sinai, Egypt. Palaios, v. 25, no. 8, pp. 517–526.

El-Beialy, S.Y., Zalat, A.A. & Ali, A.S., 2002. Palynology and palaeoenvironment of Bahrein Formation, Zetun-1 well, Western Desert, Egypt. Egypt. Journ. Paleontol., v. 2, pp. 371–384.

El-Dawy, M.H., 1994. The Coniacian-Santonian boundary in the Wadi El-Seig, western central Sinai, Egypt, Stratigraphy, foraminiferal faunas and sea level changes, N. Jb. Geol. Paläont. Abh., 192, No. 2, pp. 203–219.

Elewa, A.M., Ishizaki, K. & Nishi, H., 1995. Ostracoda from the El Sheikh Fadl – Ras Gharib stretch, the Eastern Desert, Egypt, with reference to distinguishing sedimentary environments. In: Riha (Ed.), Ostracoda and Biostratigraphy, Balkema, Rotterdam, pp. 203–213.

Elfigih, O.B., 1991; The sedimentology and reservoir characteristics of the lower Acacus Formation, NC2 Concession, Hamada Basin NW Libya. Unpub. M.Sc. thesis, Memorial University of Newfoundland, 150 p.

Elfigih, O.B., 2000. Regional diagensis and its relation to facies change in the Upper Silurian, lower Acacus Formation, Hamada (Ghadames) Basin, Northwestern Libya. Unpub. Ph.D. thesis, Memorial University of Newfoundland, p. 144.

El-Gaby, S., El-Nady, O. & Khudeir, A., 1984. Tectonic evolution of the basement complex in the Central Eastern Desert of Egypt. Geologische Rundschau, v. 73, no. 3, pp. 1019–1036.

El-Gaby, S., List, F.K. & Tehrani, R., 1990. The basement complex of the Eastern Desert and Sinai, in R. Said (Ed.), The Geology of Egypt, Balkema, Rotterdam/Brookfield, pp. 175–184.

El-Ghamri, M.A., Warburton, I.C. & Burley, S.D., 2002. Hydrocarbon generation and charging in the October Field, Gulf of Suez, Egypt. Journal of Petroleum Geology, v. 25, no. 4, pp. 433–464.

El-Ghoneimy, I. & El-Ghohary, Y., 1988. Factors controlling hydrocarbon accumulations in the Badr El Din Concession, Western Desert, Egypt. 9th Exploration and Production Conf., Egyptian General Petroleum Corp. (EGPC), Cairo, Egypt, November 1988, p. 27.

El-Ghoul, A., 1991. A modified Farwah Group type section and its application to understanding stratigraphy and sedimentation along an E-W section through NC35A, Sabratah basin, in M.J. Salem. O.S. Hammuda & B.A. Eliagoubi (Eds.), The Geology of Libya, v. IV, Elsevier, Amsterdam, pp. 1637–1655.

El-Haddad, A., Aissaoui, D.M. & Soliman, M.A.,1984. Mixed carbonate-siliciclastic sedimentation on a Miocene fault-block, Gulf of Suez, Sedimentary Geology, v. 37, pp. 185–202.

El Hammichi, F., Elmi, S., Faure-Muret, A. & Benshili, K., 2002. Une plate-forme en distension, témoin de phases pré-accrétion téthysienne en Afrique du Nord pendant le Toarcien-Aalénien (synclinal Iguer Awragh – Afennourir, Moyen Atlas occidental, Maroc). Comptes Rendus Géosciences, v. 334, pp. 1003–1010.

El-Hawat, A.S. & Bezan, A.H., 1998 (Abs.). Early Palaeozoic event stratigraphy of Western Liby: Glacio-eustatic and tectonic signatures and their impact on hydrocarbon exploration, The Geological Conference on Exploration in Murzuq Basin, Sabha University, Sabha, Libya, pp. 66–68.

El-Hawat, A.S. & Shelmani, M.A., 1993. Short Notes and Guidebook on the Geology of Al Jabal Al Akhdar, Cyrenaica, NE Libya, Earth Science Soc. Libya, p. 70.

El-Hawat, A.S., 1980a. Carbonate-terrigenous cyclic sedimentation and paleogeography of the Marada Formation (Middle Miocene), Sirte Basin, in M.J. Salem & M.T. Busrewil (Eds.), The Geology of Libya, v. II, London, Academic Press, pp. 427–448.

El-Hawat, A.S., 1980b. Intertidal and storm sedimentation from Wadi al Qattarah Member, Ar Rajmah Formation (Middle Miocene), Al Jabal al Akhdar, in M.J. Salem & M.T. Busrewil (Eds.), The Geology of Libya, v. II, London, Academic Press, pp. 449–461.

El-Hawat, A.S., 1992. The Nubian Sandstone sequence in the Sirte Basin, Libya: Sedimentary facies and events, Geology of the Arab World, Cairo University, pp. 317–327.

El-Hawat, A.S., Missalati, A.A., Bezan, A.M. & Taleb, T.M., 1996. The Nubian Sandstone in Sirt Basin and its correlatives, in M.J. Salem. A.S. El-Hawat & A.M. Sbeta (Eds.), The Geology of Sirt Basin, v. II, Elsevier, Amsterdam, pp. 3–30.

El-Hedeny, M.M., 2006. New systematic and biostratigraphic data on the Cenomanian-Turonian Radiolitidae (Bivalvi: hippuritoidea) of Abu Roash, Western Desert, Egypt. Poster. Interntional Congress of Bivalvia, July 22–27, Universidad Autónima de Barcelona Bellaterra, Catalunya, Spain.

El-Hedeny, M.M., El-Sabbagh, A.M., 2005. Eoradiolites liratus (Bivalvia, Radiolitidae) from the Upper Cenomanian Galala Formation at Saint Paul, Eastern Desert (Egypt). Cretaceous Research, Issue 4, p. 551–566.

El-Heiny, I. & and Enani, N., 1996. Regional stratigraphic interpretation pattern of Neogene sediments, Northern Nile Delta, Egypt. 13th Petrol. Conf., Cairo, October, 1996, v. 1, pp. 270–290.

El-Heiny, I. & Martini, E., 1981. Miocene foraminiferal and calcareous nanoplankton assemblages from the Gulf of Suez region and correlations, Géologie Méditerranéenne, v. VIII, No. 2, pp. 101–108.

Eliagoubi, B.A.H. & Powell, J.D., 1980. Biostratigraphy and palaeoenvironment of Upper Cretaceous (Maastrichtian) foraminifera of North-central and Northwestern Libya, in M.J. Salem & M.T. Busrewil (Eds.), The Geology of Libya, v. I, London, Academic Press, pp. 137–154.

Eliwa, H.A., Abu El-Enen, M.M, Khalaf, I.M., Itaya, T. & Murata, M., 2008. Metamorphic evolution of Neoproterozoic metapelites and gneisses in the Sinai, Egypt. Journal of African Earth Sciences, v. 51, pp. 107–122.

Eliwa, H.A., Kimura, J.-I. & Itaya, T., 2006. Late Neoproterozoic Dokhan Volcanics, North Eastern Desert, Egypt: Geochemistry and petrogenesis. Precambrian Research, v. 151, pp. 31–52.

El-Kazzaz, Y.A.H.A. & Taylor, W.E.G., 2001. Tectonic evolution of the Allaqi shear zone and implications for Pan-African terrane amalgamation in the south Eastern Desert, Egypt. J. African Earth Sciences, v. 33, no. 2, pp. 177–179.

El-Khoudary, R.H. & Helmdach, F.F., 1981. Biostratigraphy studies on the Upper Eocene Apollonia Formation of NW Al Jabal al Akhdar, NE Libya, Rev. Espan. Micropaleont., v. 13, pp. 5–23.

El Mahmoudi, A. & Gabr, A., 2009. Geophysical surveys to investigate the relation between the uaternary Nile channels and the Messinian Nile canyon at East Nile Delta, Egypt. Arab J. Geosci, v. 2, pp. 53–67.

Elmejdoub, N. & Jedoui, Y., 2009. Pleistocene raised marine deposits of the Cap Bon peninsula (NE Tunisia): Records of sea-level highstands, climatic changes and coastal uplift. Geomorphology, v. 112, nos. 3–4, pp. 179–189.

Elmejdoub, N., Mauz, B. & Jedoui, Y. 2011. Sea-level and climatic controls on Late Pleistocene coastal aeolianites in the Cap Bon peninsula, northeastern Tunisia. Boreas, Boreas, v. 40, pp. 198–207.

Elmi, S., Marok, A., Sebane, A., & Almeras, Y., 2010. Importance of the Mellala section (Traras Mountains, northwestern Algeria) for the correlation of the Pliensbachian/Toarcian boundary. Volumina Jurassica, V. VII, Polish Geological Institute – National Research Institute & Faculty of Geology, University of Warsaw, pp. 37–45.

El-Naggar, Z. & Haynes, J., 1967. *Globotruncana calciformis* in the Maastrichtian Sharawna Shale of Egypt, Contr. Cush. Found. Foram. Research, v. 18, pt. 1, pp. 1–13.

El-Naggar, Z.R., 1966. Stratigraphy and classification of type Esna Group of Egypt, AAPG Bull., v. 50, pp. 1455–1477.

El-Naggar, Z.R., 1970. On a proposed lithostratigraphic subdivision for the Late Cretaceous-Early Paleogene succession in the Nile Valley, Egypt, UAR, 7th Arab Petrol. Congr., Kuwait, Paper No. 64 (B-3), pp. 1–50.

EL-Sabbagh, A.M. & EL-Hedeny, M.M., 2003. Upper Turonian Radiolitidae (rudist bivalves) from the Actaeonella Series, El-Hassana dome, Abu Roash, Egypt. Egyptian J. Paleontology, v. 3, pp. 105–131.

El-Shinnawi, M.A. & Sultan, I.Z., 1972a. Biostratigraphy of some Upper Cretaceous sections in the Gulf of Suez area, Egypt, Proc. 5th Afr. Colloq. Paleont., Addis-Ababa, pp. 263–292.

El-Shinnawi, M.A. & Sultan, I.Z., 1972b. Biostratigraphy of some subsurface Lower Paleocene sections in the Gulf of Suez area, Egypt, Proc. 5th Afr. Colloq. Paleont., Addis-Ababa, pp. 225–262.

El Sisi, Z., Hassouba, M., Oldani, M.J. & Dolson, J.C., 2002. Geology of Bahariya Oasis in the Western Desert of Egypt and its archeological heritage. Field Trip No. 8, Cairo 2002 International Conference and Exhibition, p. 66.

El-Waer, A.A., 1991. Miocene Ostracoda from Al Khums Formation, Northwestern Libya, in M.J. Salem, O.S. Hammuda & B.A. Eliagoubi (Eds.), The Geology of Libya, v. IV, Amsterdam, Elsevier, pp. 1457–1481.

El-Waer, A.A., 1992. Tertiary and Upper Cretaceous Ostracoda from NW Offshore Libya, their Taxonomy, biostratigraphy and correlation with adjacent areas, Petroleum Res. Centre, Spec. Pub. 1, Tripoli, p. 445.

Elwarfalli, H.O. & Stow, D.A.V., 1998 (Abs.). Nummulitic bank-basin model for the paleogene of NE Libya, in Tertiary to Recent Larger Foraminifera, Their Depositional Environments and Importance as Petroleum Reservoirs, Kingston University, United Kingdom, p. 11.

Elzarka, M.H. & Mostafa, A.R., 1988. Oil prospects of the Gulf of Suez, Egypt – A case study. Organic Geochemistry, v. 12, no. 2, pp. 109–121.

Elzarka, M.H. & Wally, M., 1987. The physiographic features of the Miocene Ras Malaab Group Basin, South Bakr Oilfield, Gulf of Suez, Egypt. Journal of Petroleum Geology, v. 10, no. 3, pp. 295–318.

Enay, R., El Asmi, K., Soussi, M., Mangold, C. & Hantzpergue, P., 2002. Un *Pachyerymnoceras* arabique dans le Callovien supérieur du Dahar (Sud tunisien), nouvel élément de datation du membre Ghomrassène (formation Tataouine). corrélations avec l'Arabie Saoudite et le Moyen-Orient. Comptes Rendus Geoscience, v. 334, pp. 1157–1167.

Enay, R., Hantzpergue, P., Soussi, M. & Mangold, C., 2005. La limite Kimme´ridgien–Tithonien et l'âge des formations du Jurassique supérieur de la "Dorsale" tunisienne, comparaisons avec l'Algérie et la Sicile. Geobios, v. 38, pp. 437–450.

Energy Information Administration (EIA), 2001. Egypt, Official Energy Statistics from the USA Government, pp. 1–11, www.eia.doe.gov/emeu/cabs/egypt.html.

Energy Information Administration (EIA), 2002. Libya, Official Energy Statistics from the USA Government, pp. 1–11.

Energy Information Administration (EIA), 2008. Egypt: Country Analysis Briefs, 7p. www.eia. doe.gov.

Energy Information Administration (EIA), 2009. Algeria, Official Energy Statistics from the USA Government, pp. 1–3.

Engel, A.E.J., Dixon, T.H. & Stern, R.J., 1980. Precambrian evolution of Afro-Arabian crust from ocean-arc to craton, Geol. Soc. Am. Bull., v. 91, pp. 699–706.

Ennih, N. & Liégeois, P., 2001. The Moroccan Anti-Atlas: the West African craton passive margin with limited Pan-African activity. Implications for the northern limit of the craton. Precambrian Research, v. 112, issues 3–4, pp. 289–302.

Ensslin, R., 1992. Cretaceous synsedimentary tectonics in the Atlas system of Central Morocco. Geologische Rundschau, Berlin, Band 81, Heft 1, pp. 91–104.

Eppelbaum, L. & Katz, Y., 2011. Tectonic-Geophysical Mapping of Israel and the Eastern Mediterranean: Implications for Hydrocarbon Prospecting. Positioning, v. 2, pp. 36–54.

Eschard, R., Abdallah, H., Braik, F. & Guy Desaubliaux, G., 2005. The Lower Paleozoic succession in the Tassili outcrops, Algeria: sedimentology and sequence stratigraphy. first break, v. 23, pp. 27–36.

Eschard, A. & Hamel, A., 2002. Stratigraphic Architecture of the Triassic Basins in Algeria. AAPG Hedberg Conference "Paleozoic and Triassic Petroleum Systems in North Africa" February 18–20, Algiers, Algeria, p. 2.

Essaifi, A. & Hibti, M., 2008. The hydrothermal system of Central Jebilet (Variscan Belt, Morocco): A genetic association between bimodal plutonism and massive sulphide deposits? J. African Earth Sci., v. 50, nos. 2–4, pp. 188–203.

Essaifi, A., Potrel, A., Capdevila, R. & Lagarde, J.L., 2003. Datation U-Pb: âge de mise en place du magmatisme bimodal des Jebilet centrales (Chaîne varisque, Maroc). Implications géodynamiques. Comptes Rendus Géoscience, v. 35, pp. 193–203.

Ettachfini, El. M., Souhel, A., Andreu, B. & Caron, M., 2005. La limite Cenomanien-Turonien dans le Haut Atlas central, Maroc. The Cenomanian-Turonian boundary in the Central High Atlas, Morocco. Geobios, v. 38, pp. 57–68.

Ettalhi, J.A., Krokovic, d, & Banerjee, S., 1978. Bibliography of the Geology of Libya. Industrial Research Centre, Tarabulus, Bulletin No. 11, p. 135.

Ettaki, M. & Chellaï, E.H., 2005. Le Toarcien inférieur du Haut-Atlas de Todrha-Dadès (Maroc): sédimentologie et lithostratigraphie. C.R. Géosciences, v. 337, pp. 814–823.

Ettaki, M., Ibouh, H. & Chellai, E.H., 2007. Événements tectono-sédimentaires au Lias-Dogger de la frange méridionale du Haut-Atlas central, Maroc: Tectono-sedimentary events during Lias-Dogger at the southern margin of the Central High-Atlas, Morocco. Estudios Geológicos, v. 3, no. 2, pp. 103–125.

Evans, A.L., 1988. Neogene tectonic and stratigraphic events in the Gulf of Suez rift area, Egypt. Tectonophysics, v. 153, pp. 235–247.

Eyal, Y., Eyal, M. & Kröner, A., 1991. Geochronology of the Elat Terrain, metamorphic basement, and its implication for crustal evolution of the NE part of the Arabian-Nubian Shield, Israel J. Earth Sci., v. 40, pp. 5–16.

Fabre, J., 1976. Introduction à la géologie du Sahara algérien. S.N.E.D., Alger, p. 422.

Fabre, J., 1988. Les séries paléozoiques d'Afrique: une approche. J. African Earth Sci., v. 7, pp. 1–40.

Fabricius, F.H., 1984. Neogene to Quaternary geodynamics of the area of Ionian Sea and surrounding land masses, in J.E. Dixon & A.H.F. Robertson (Eds.), The Geological Evolution of the Eastern Mediterranean, Geological Society Special Pub. No. 17, Blackwell, pp. 819–824.

Fabricius, F.H., Berdau, D. & Münnich, K.-O., 1970. Early Holocene ooids in modern littoral sands reworked from a coastal terrace, Southern Tunisia, Science, v. 169, pp. 757–760.

Fabuel-Perez, I., Hodgetts, D., & Redfern, J., 2009. A new approach for outcrop characterization and geostatistical analysis of a low-sinuosity fluvial-dominated succession using digital outcrop models: Upper Triassic Oukaimeden Sandstone Formation, central High Atlas, Morocco. AAPG Bulletin, v. 93, no. 6, pp. 795–827.

Fallot, J., 1953. Ahnet et Mouydir. XIX Congr. Géol. Inter., Algeria, pp. 1–80.

Farag, I.A.M. & Ismail, M.M., 1959. Contribution to the stratigraphy of the Wadi Hof area (North East of Helwan), Bull. Fac. Sci., Cairo University, v. 34, pp. 147–168.

Farag, I.A.M. & Shata, A., 1954, Degeological survey of El Minshera area, Bull. Inst. Désert Egypte, v. 4, no. 2, pp. 5–82.

Farahat, E.S., El Mahalawi, M.M., Holinkes, G. & Abdel Aal, A.Y., 2004. Continental back-arc basin of some ophiolites from the Eastern Desert of Egypt. Mineralogy and Petrology, v. 82, nos. 1–2, pp. 81–104.

Farahat, E.S., El Mahalawi, M.M., Holinkes, G. & Abdel Aal, A.Y. & C.A. Hauzenberger, 2010. Reply to G.A. El Bahariya's Comments on: "Pillow form morphology of selected Neoproterozoic metavolcanics in the Egyptian Central Eastern Desert and their implications" [Journal of African Earth Sciences, 2010, v. 57, pp. 163–168].

Faris, M. & Abu Shama, A.M., 2007. Nannofossil biostratigraphy of the Paleocene-lower Eocene succession in the Thamad area, east central Sinai, Egypt. Micropaleontology, v. 53, nos. 1–2, pp. 12–144.

Faris, M., 1984. The Cretaceous-Tertiary boundary in central Egypt (Duwi region, Nile Valley, Kharga and Dakhla Oases), N. Jb. Geol. Paläont. Mh., v. 7, pp. 385–392.

Faris, M., Abd el Hameed, A.T., Marzouk, A.M. & Ghandour, I.M., 1998 (Abs.). Early Paleogene calcareous nannofossil and planktonic foraminiferal chronobiostratigraphy in central Egypt, in Geology of the Arab World, Fourth International Conference, Cairo University, Cairo, pp. 20–21.

Fatmi, A.N., Eliagoubi, B.A. & Hammuda, O.S., 1980. Stratigraphic of the pre-Upper Cretaceous Mesozoic rocks of Jabal Nafusah, NW Libya, in M.J. Salem & M.T. Busrewil (Eds.), The Geology of Libya, v. I, London, Academic Press, pp. 57–65.

Fauré, J.L. & Peybernès, B., 1986. Biozonation par ammonites et essai de corrélation des séries réduites liasiques de la "dorsale Tunisienne" Bulletin de la société d'histoire naturelle, Toulouse, v. 122, pp. 41–49.

Fekkak, A., Boualoul, M., Badra, L., Amenzou, M., Saquaque, A. & El-Amrani, I.E., 2000. Origin and geotectonic setting of lower Neoproterozoic Kelaat Mgouna detrital material [eastern Anti-Atlas, Morocco]. Journal of African Earth Sciences, v. 30, no. 2, pp. 295–311.

Felix, J.P., 1884. Korallen aus aegyptischen Tertiaerbildungen. Zeitschrift deutsch. Geol. Gesellsch., Berlin, pp. 415–453.

Felix, J.P., 1904. Studien über tertiäre und quatäre Korallen und Riffkalke aus Aegypten und der Sinaihalbinsel. Zeitschrift deutsch. Geol. Gesellsch., Bd. 56, Berlin, pp. 168–206.

Feller, G., 2009. Reward and risk of exploring in Egypt. World Oil, Demeber 2009, p. 13.

Fello, N., Lüning, S., Štorch, P., Redfern, J. (2006): Identification of early Llandovery (Silurian) anoxic palaeo-depressions at the western margin of the Murzuq Basin (southwest Libya), based on gamma-ray spectrometry in surface exposures. GeoArabia, v. 11, no. 3, pp. 101–118.

Ferjaoui, M., Meskini, A. & Acheche, M.H., 2001. Modeling hydrocarbon generation and expulsion from Tannezuft and Aouinet Ouinine Formations in southern Tunisia. CSPG Rock the Foundation Convention, Calgary, June 18–22, Extended Abstract No. 036, p. 10.

Fiduk, J.C., 2009. Evaporites, petroleum exploration, and the Cenozoic evolution of the Libyan shelf margin, central North Africa. Marine and Petroleum Geology, v. 26, no. 8, pp. 1513–1527.

Fiechtner, L., Friedrichsen, H., Hammerschmidt, K., 1992. Geochemistry and geochronology of Early Mesozoic tholeiites from Central Morocco. Geologische Rundschau, v. 81, pp. 45–62.

Figari, A., 1864. Studii scientifici sull'Egitto e sue adiacenze compresa la penisola dell'Arabia Petrea con accompagnamento di carta geografico-geologica.Locca, Giusti.

Fikirine, B. & Abdallah, H., 1998. Palaeozoic lithofacies correlatives and sequence stratigraphy of the Saharan Platform, Algeria, in D.J. Macgregor. R.T.J. Moody & D.D. Clark-Lowes (Eds.), Petroleum Geology of North Africa, Geological Society Spec. Pub. No. 132, London, pp. 97–108.

Fitches, R., 1998 (Abs.). The Saharan glaciation – an overview, The Geological Conference on Exploration in Murzuq Basin, Sabha University, Sabha, Libya, p. 45.

Flamand, G.B.M., 1911. Recherches Géologiques et géographiques sur le Haut-Pays de l'Oranie et sur le Sahara (Algérie et Territoires du Sud). Thèse Lyon, p. 1002.

Fleagle, J.G., Bown, T.M., Obradovich, J.D. & Simons, E.L., 1986. Age of the earliest African anthropoids, Science, v. 234, pp. 1247–1249.

Fleck, R.J., Coleman, R.G., Cornwall, H.R., Greenwood, W.R., Hadley, D.G., Schmidt, D.L., Prinz, W.C. & Ratte, J.C., 1976. Geochronology of the Arabian Shield, western Saudi Arabia – K-Ar results, Bull. Geol. Soc. Am., v. 87, pp. 9–21.

Flinn, D., Lagha, S., Atherton, M.P. & Cliff, R.A., 1991. The rock-forming minerals of the Jabal Arknu and Jabal Al Awaynat alkaline ring complexes of SE Libya, in M.J. Salem, M.T. Busrewil & A.M. Ben Ashour (Eds.), The Geology of Libya, v. VII, Elsevier, Amsterdam, pp. 2539–2557.

Floridia, G.B., 1933. Sull' esistena dell'Eocene superiore nel Gebel Cirenaico. Rend. R. Acc. Lincei, v. 17, ser. 6, pt. 9, pp. 735–736.

Floridia, G.B., 1934. Sopra alcune rocce nummulitiche del Gebel Cirenaico. Missione Scientifica della R. Accad. D'Italia a Cufra (1931-X), Roma, v. 3, pp. 33–41.

Floyer, E.A., 1893. Further routes in the Eastern Desert of Egypt. The Geographical Journal, v. 1, no. 5, pp. 408–431.

Flügel, E., 1982. Microfacies Analysis of Limestones, Springer-Verlag, Berlin, Heidelberg, p. 633.

Forbes, Rosita, 1922. Across the Libyan Desert to Kufara. The Geographical Journal, v. 58, no. 2, pp. 81–101.

Foucault, A. & Stanley, D.J., 1989. Late Quaternary palaeoclimatic oscillations in East Africa recorded by heavy minerals in the Nile delta, Nature, v. 339, pp. 1989.

Foureau, F., 1898. Documents scientifiques de la mission saharienne (Mission Foureau-Lamy), d'Alger au Congo par le Tchad, p. 1610.

Fournié, D., 1978. Nomenclature lithostratigraphique des séries du Crétacé superieur au Tertiare de Tunisie, Bulletin des Centres de Recherche Exploration-Production Elf-Aquitaine, v. 2, pp. 97–148.

Fourtau, R., 1899. Obsevations sur les terrains éocènes et oligocènes d'Egypte, Bull. Soc. Géol. France, Paris, séries 3, v. 27, pp. 382–383.

Fourtau, R., 1900. Sur la Crétacé du massif d'Abou Roach (Egypte), Compte Rendus, v. 131, pp. 629–631.

Fourtau, R., 1902. Sur les grès nubien. C.R. Acad. Sci., t.135, pp. 803–804.

Fowler, T.J. & Osman, A.F., 2001. Gneiss-cored interference dome associated with two phases of late Pan-African thrusting in the Central Eastern Desert, Egypt, Precambrian Research, v. 108, pp. 17–43.

Fraas, E., 1900. Geognostisches Profil vom Nil zum Rothen Meer. Zeitschr. D. Deutsch. Geol., Ges., pp. 2–54, pp. 569–618.

Fraas, E., 1904. Neue Zeuglodonten aus dem untern Mitteleocän vom Mokattam dei Cairo. Geologische und Paläontologische abhandlungen, Jena, Neue Folger, 6, p. 197–220.

Fraas, O., 1867. Aus dem Orient: geologische Beobachtungen am Nil, auf der Sinai Halbinsel und in Syrien. Ender u. Suebert, Stuttgart, p. 222.

Franchi, S., (13 papers between 1912–1913).

Francis, M. & Issawi, B., 1977. Geological map of Libya, 1:250,000, Sheet Soluq NH 34-2, Explanatory Booklet, Industrial Research Centre, Tripoli, p. 86.

Fraser, W.W., 1967. Geology of the Zelten field, Libya, north Africa. Proc. 7th Petroleum Congress, Mexico, Elsevier, Amsterdam, v. 2, pp. 259–264.

Freulon, J.M. & Lefranc, J.P., 1954a. Recente acquisitions de la géologie du Fezzan (1942–1952), 19th Int. Geol. Congress, Alger, No. 20, p. 39.

Freulon, J.M., 1953. Existence d'un niveau a stromatolithes (Collenia) dans le Carbonifère marin du Sahara Oriental. Soc. Géol. France, C.R. Somm. Seanc., nos. 11–12, pp. 233–234.

Freulon, J.M. & Lefranc, J.P., 1954b. Structure et stratigrapie du nord du Fezzan (Libye), 19th Int. Geol. Congress, Alger, No. 14, pp. 223–225.

Freulon, J.M., 1964. Etude géologique des séries primaires du Sahara central (Tassili n'Ajjar et Fezzan). Centre National Rech. Scient. Sér. Géologie no. 3, Paris, p. 198.

Freulon, J.M., Lefranc, J.P. & Lelubre, M., 1955. Carte géologique de reconaissance du Sahara Libyen, Feuille NG33-N-O, 'Sebha' au 1:500,000, Publ. Inst. Rech. Sahar., Alger.

Friedman, G.M., 1972. Significance of Red Sea in problems of evaporites and basinal limestones, AAPG Bull., v. 56, pp. 1072–1086.

Frisch, W., 1982 (Abs.). The Wadi Dib ring complex, Nubian Desert (Egypt) and its importance for the upper limit of the Pan-African orogeny, Precambrian Res., v. 16, A20.

Frisch, W. & Abdel-Rahman, A.M., 1999. Petrogenesis of the Wadi Dib alkaline ring complex, Eastern Desert of Egypt. Mineral. Petrology, v. 65, pp. 249–275.

Frischat, G.H., Heide, G., Müller, B. & Weeks, R.A., 2001. Mystery of the Libyan desert glasses. Physics and Chemistry of Glasses, v. 42, no. 3, pp. 179–183.

Fritz, H., Dallmeyer, D.R., Wallbrecher, E., Loizenbauer, J., Hoinkes, G., Neumayr, P. & Khudeir, A.A., 2002. Neoproterozoic tectonothermal evolution of the Central Eastern Desert, Egypt: a slow velocity tectonic process of core complex exhumation. Journal of African Earth Sciences, v. 34, nos. 3–4, pp. 137–155.

Fritz, H. & Messner, M., 1999. Intramontane basin formation during oblique convergence in the Eastern Desert of Egypt: magmatically versus tectonically induced subsidence. Tectonophysics, v. 315, pp. 145–162.

Fritz, H., Wallbrecher, E., Khudeir, A.A., Abu El Ela, F. & Dallmeyer, D.R., 1996. Formation of Neoproterozoic metamorphic core complexes during oblique convergence (Eastern Desert, Egypt), Journal of African Earth Sciences, v. 23, no. 3, pp. 311–329.

Frizon de Lamotte, D., Crespo-Blanc, A., Saint-Bezar, B., Comas, M., Fernandez, M., Zeyen, H., Ayarza, P., Robert-Charrue, C., Chalouan Zizi, M., Teixell, A., Arboleya, M.-L., Alvarez-Lobato, F., Julivert, M. & Michard, A. 2004. "TRANSMED Transect I". In W. Cavazza, Roure, F., Spakman, W., Stampfli, G.M. & Ziegler, P.A. (Eds.), The TRANSMED Atlas – The Mediterranean Region from Crust to Mantle, Springer, Berlin-Heildeberg.

Frizon de Lamotte, D., Zizi, M., Missenard, Y., Hafid, M., El Azzouzi, M., Maury, R.C., Charrière, A., Taki, Z., Benammi, M. & Michard, A., 2008. The Atlas System. In A. Michard, O. Saddiqi, A. Chalouan & D. Frizon de Lamotte (Eds.), Continental Evolution: The Geology of Morocco. structure, stratigraphy, and Tectonism of the Africa-Atlantic-Mediterranean Triple Junction, Berlin, Springer-Verlag, pp. 133–202.

Frizon de Lamotte, D., Leturmy, P., Missenard, Y., Khomsi, S., Ruiz, G., Saddiqi, O., Guillocheau, F. & Michard, A., 2009. Mesozoic and Cenozoic vertical movements in the Atlas system (Algeria, Morocco, Tunisia): An overview. Tectonophysics, v. 475, no. 1, pp. 9–28.

Fucek, V.P., Kucenjak, M.H. & Mesic, I.A., 1998 (Abs.). Eocene/Oligocene boundary (vascus Member) in the western part of the Tripolitania Basin, Libya, in Tertiary to Recent Larger Foraminifera, Their Depositional Environments and Importance as Petroleum Reservoirs, Kingston University, United Kingdom, p. 12.

Fürst, M. & Klitzsch, E., 1963. Late Caledonian paleogeography of the Murzuk basin, Rev. Inst. Fr. Petrol., v. 18, No. 10, pp. 1472–1484.

Fürst, M., 1964. Die Oberkreide-Paleozan-Transgresion im Ostlichen Fezzan, Geol. Rundschau, v. 54, pp. 1060–1088.

Fürst, M., 1993. The paleogeographic development of the Sirt Basin, Libya, since the Upper Cretaceous, a general review, in Sedimentary Basins of Libya, First Symposium, Geology of Sirt Basin, Earth Science Society of Libya, Tripoli, p. 20.

Fullagar, P.D., 1980. Pan-African age granites of northeastern Africa. New or reworked sialic materials? In M.J. Salem & M.T. Busrewil (Eds.), The Geology of Libya, v. III, pp. 1051–1058, London, Academic Press.

Furnes, H., Shimron, A.E. & Roberts, D., 1985. Geochemistry of Pan-African volcanic arc sequences in south-eastern Sinai Peninsula and plate tectonic implications, Precambrian Res., v. 29, pp. 359–382.

Furon, R., 1968. La Géologie de l'Afrique. Payot, Paris, p. 400.

Galal, G. & Kamel, S., 2007. Early Paleogene Planktic Foraminiferal Biostratigraphy at the Monastery of Saint Paul, Southern Galala, Eastern Desert, Egypt. Revue de Paléobiologie, Genève, v. 26, no. 2, pp. 391–402.

Galeazzi, S., Point, O., Haddadi, N., Mather, J. & Druesne, D., 2010. Regional geology and petroleum systems of the Illizi-Berkine area of the Algerian Saharan Platform: An overview. Marine and Petroleum Geology, v. 27, no. 1, pp. 143–178.

Galecic, M., 1984. Geological map of Libya, 1:25,000, Sheet Anay, Explanatory Booklet. Industrial Research Centre, Tripoli, p. 98.

Galil B.S & Zenetos, A., 2002. A sea change – Exotics in the Eastern Mediterranean. In: E. Leppäkoski, S. Gollasch & S. Olenin S (Eds.): Invasive Aquatic Species Of Europe: Distriution, impacts, and management. Kluwer Academic Publishers, Dordrecht, pp. 325–336.

Gallitelli, P., 1934. Contributo alla conoscenza delle rocce eruttive e metamorfiche della Libia. Missione Scientifica della R. Accad. D'Italia a Cufra (1931-X), Roma, v. 3, pp. 423–428.

Ganz, H.H., Luger, P., Schrank, E., Brooks, P.W. & Fowler, M.G., 1990. Facies evolution of Late Cretaceous black shales from Southeast Egypt. In Deposition of Organic Facies. AAPG Special Volumes, v. 30, pp. 217–229.

Gardin, S., 2002. Late Maastrichtian to early Danian calcareous nannofossils at Elles (Northwest Tunisia). A tale of one million years across the K-T boundary. Palaeogeography, Palaeoclimatology, Palaeoecology, v. 178, pp. 211–231.

Gardner, B., 1988. A new Cleithrolepis from the Triassic of central Cyrenaica, northeast Libya. In B. El-Arnauti. B. Owens & B. Thusu (Eds.), Subsurface Palynostratigraphy of Northeast Libya. Garyouunis University Publications, Benghazi, pp. 259–265.

Gardner, E.W. & Caton-Thompson, G., 1926. The recent geology and Neolithic industry of the northern Fayum Desert. The Journal of the Royal Anthropological Institute of Great Britain and Ireland, v. 56, pp. 301–323.

Garfunkel, Z., 1991. Darfur-Levant array of volcanics – A 140 Ma-long record of a hot spot beneath the African-Arabian continent, and its bearing on Africa's absolute motion, Israel J. Earth Sci., v. 40, pp. 135–150.

Garfunkel, Z. & Bartov, Y., 1977. The tectonics of the Suez rift, Geological Survey of Israel Bull., v. 71, pp. 1–44.

Garfunkel, Z. & Derin, B., 1984. Permian-early Mesozoic tectonism and continental margin formation in Israel and its implications for the history of the Eastern Mediterranean, in

J.E. Dixon & A.H.F. Robertson (Eds.), The Geological Evolution of the Eastern Mediterranean, Geol Soc. Spec. Pub. No. 17, Blackwell, pp. 187–201.

Garret, S.W. & Storey, B.C., 1987. Lithospheric extension on the Antarctic Peninsula during Cenozoic subduction, in M.P. Coward, J.F. Dewey & P.L. Hancock (Eds.), Continental Extensioanl Tectonics, Geological Society Special Publication No. 28, pp. 419–431.

Gasinski, M.A., 1988. Foraminiferal biostratigraphy of the Albian and Cenomanian sediments in the Polish part of the Pieniny Klippen Belt, Carpathian Mountains, Cretaceous Research, v. 9, pp. 217–247.

Gasquet, D., Ennih, N., Liégeois, J-P., Soulaimani, A. & Michard, A., 2008. The Panafrican belt. In Michard, A., Saddiqi, O., Chalouan, A., Frizon de Lamotte, D. (Eds.), Continental Evolution: The Geology of Morocco. Structure, Stratigraphy, and Tectonics of the Africa-Atlantic–Mediterranean Triple Junction. Springer Verlag, Berlin, Heidelberg, pp. 33–64.

Gass, I.G., 1977. The evolution of the Pan-African crystalline basement in NE Africa and Arabia, J. Geol. Soc. London, v. 134, pp. 129–138.

Gass, I.G., 1982. Upper Proterozoic (Pan-African) calc-alkaline magmatism in northeastern Africa and Arabia, in R.S. Thorpe (Ed.), Andesites, New York, John Wiley, pp. 591–609.

Gautier, E-F., 1922. Structure de l'Algérie. Société d'Éditions Géographiques et Scientifiques, Paris, p. 240.

Gebhardt, H., 1993. Neogene foraminifera from the Rabat area (Morocco: Stratigraphy, palaeobathymetry and palaeoecology). Journal of African Earth Sciences, v. 16, no. 4, 445–464.

Gebhardt, H., Kuhnt, W. & A.E. Holbourn, A.E., 2004. Foraminiferal response to sealevel change. organic carbon flux and oxygen deficiency in the Cenomanian of the Tarfaya Basin, southern Morocco. Marine Micropaleontology, v. 53, pp. 133–157.

Gelati, R., Moratti, G. & Papani, G., 2000. The Late Cenozoic sedimentary succession of the Taza-Guercif Basin, South Rifian Corridor, Morocco. Marine and Petroleum Geology, v. 17, no. 3, pp. 373–390.

Gendi, A., 1998 (Abs.). Facies relationships and mineralogical differences of Cenomanian sediments in northeastern and southeastern Sinai, in Geology of the Arab World, Fourth International Conference, Cairo University, Cairo, p. 3.

Gentil, L., 1906. Contribution à la géologie et à la géographie physique du Maroc. Annales de Géographie, v. 15, no. 80, pp. 133–151.

Gentil L., 1912. Le Maroc physique, F. Alcan (Ed.), France, 329 p.

Geological Map of Libya, 1985. Scale 1:2,000,000. Industrial Research Centre, Tripoli, Libya.

George, U., 2000. Der Stein des Tutanchamun., GEO, Heft10, pp. 18–46.

Geraads, D., 1989.Vertébrés fossiles du miocène supérieur du djebel krechem el artsouma (Tunisie centrale). Comparaisons biostratigraphiques. Geobios, Volume 22, Issue 6, 1989, Pages 777–801.

Geraads, D., 2008. Plio-Pleistocene Carnivora of northwestern Africa: A short review. Comptes Rendus Palevol, v. 7, no. 8, pp. 591–599.

Geraads, D., Raynal, J.P., & Eisenmann, V., 2004. The earliest human occupation of North Africa: A reply to Sahnouni et al. (2002). Journal of Human Evolution, v. 46, pp. 751–761.

Geyer, G. & Herbig, H.G., 1988. New Eocene oysters and the final regression at the southern rim of the central High Atlas (Morocco). Geobios, v. 21, pp. 663–691.

Geyer, G. & Landing, E., 1995. The Lower-Middle Cambrian Standard of Western Gondwana. Inst. für Paläontol., Univ. of Würzburg, Würzburg, Germany, 269 p.

Ghanoush, H. & Abubaker, H., 2007. Gravity and Magnetic Profile along Seismic Intersect Ku-89-04, Southern Kufra Basin – Libya. International Conference on Geo-resources in the Middle East and North Africa, Cairo University, 24–28 Feb 2007, p. 13.

Ghandour, I.M., Masuda, H. & Wataru Maejima, W., 2003. Mineralogical and chemical characteristics of Bajocian-Bathonian shales, G. Al-Maghara, North Sinai, Egypt: Climatic and environmental significance. Geochemical Journal, Vol. 37, pp. 87–108.

Ghienne, J.F. & Deynoux, M., 1998. Large-scale channel fill structures in Late Ordovician glacial deposits in Mauritania, Western Sahara. Sedimentary Geology, v. 119, pp. 141–159.

Ghienne, J.F., 2003. Late Ordovician sedimentary environments, glacial cycles and postglacial transgression in the Taoudeni Basin, West Africa. Palaeogeography, Palaeoclimatology, Palaeoecology, v. 1189, pp. 117–146.

Ghienne, J.F., Boumendjel, K., Paris, F., Videt, B., Racheboeuf, P. & Salem, H.A., 2007. Bulletin Geosciences, v. 83, no. 3, pp. 183–214.

Ghoneim, M.F., 1989. Mineral chemistry of some gabbroic rocks of the Central Eastern Desert, Egypt, J. Afr. Earth Sci., v. 9, No. 2, pp. 289–295.

Ghorab, M.A., 1961. Abnormal stratigraphic features in Ras Gharib oilfield, Third Arab Petroleum Congress, Alexandria, p. 10.

Ghori, K.A.R., 1991. Petroleum geochemical aspects of Cyrenaica, NE Libya. In M.J. Salem, M.T. Busrewil & A.M. Ben Ashour (Eds.): The Geology of Libya, V. VII, Elsevier, pp. 2743–2755.

Ghori, K.A.R. & Mohammed, R.A., 1996. The application of petroleum generation modeling to the eastern Sirt Basin, Libya. In M.J. Salem, A.S. El-Hawat & A.M. Sbeta (Eds)., The geology of Sirt Basin: Amsterdam, Elsevier, v. II, pp. 529–540.

Ghuma, M.A., 1975. The geology and geochemistry of the Ben Ghanima batholith, Tibisti massif, Southern Libya, L.A.R., Unpublished Ph.D. thesis, Rice University, p. 189.

Ghuma, M.A. & Rogers, J.J.W., 1980. Pan-African evolution in Jamahiriya and North Africa, in M.J. Salem & M.T. Busrewil (Eds.), The Geology of Libya, v. III, London, Academic Press, pp. 1059–1064.

Gill, W.D., 1965. The Mediterranean Basin, Proc. Meet. Inst. Petrol. and Geol. Soc. on "Salt Basin around Africa", London.

Gilles, C., Cobbett, A.M., Meunier, F.J. & Benton, M.J., 2010 (in print). Vertebrate microremains from the Early Cretaceous of southern Tunisia. Geobios.

Gillespie, J. & Sanford, R.M., 1967. The geology of the Sarir oilfield, Sirte Basin, Libya, Proc. Seventh World Petrol. Congress, v. 1A, pp. 182–193.

Gillhespy, L. & Exton, J., 2008 (Abs.). Petroleum Systems of the Zag Basin, Morocco. MAPG First International Conference & Exbihition, Marrakech, Morocco, p. 62.

Gindy, A.R. & El Askary, M.A., 1969. Stratigraphy, structure and origin of Siwa Depression, Western Desert of Egypt, AAPG Bull., v. 53, pp. 603–625.

Gindy, A.R., 1991. Origin of the Qattara Depression, Egypt, Discussion, Geol. Soc. Am. Bull., v. 103, No. 10, pp. 1374–1375.

Gingerich, P.D. 1992. Marine mammals (Cetacea and Sirenia) from the Eocene of Gebel Mokattam and Fayum, Egypt: Stratigraphy, age, and paleoenvironments. University of Michigan, Papers on Paleontology, No. 30, pp. 1–84.

Gingerich, P.D. 1993. Oligocene age of the Gebel Qatrani Formation, Fayum, Egypt. Journal of Human Evolution, v. 24, pp. 207–218.

Girdler, R.W., 1991a. The Afro-Arabian riftsystem – an overview, in A.F. Gangi (Ed.), World Systems, Tectonophysics, v. 197, pp. 139–153.

Girdler, R.W., 1991b. A case for ocean crust beneath the Red Sea (Extended Abstract), in J. Markis. P. Mohr & R. Rihm (Eds.), Red Sea: Birth and Early History of a New Oceanic Basin, Tectonophysics, v. 198, pp. 275–278.

Girgis, M.H., 1989. A morphometric analysis of the *Arkhangelskiella* group and its stratigraphical and paleoenvironmental importance, in J.A. Crux & S.E. Van Heck (Eds.), Nannofossils and Their Applications, Chichester, Ellis Horwood Ltd, pp. 327–339.

Glenn, C.R. & Arthur, M.A., 1990. Anatomy and origin of a Cretaceous phosphorite- green-sand giant, Egypt. Sedimentology, v. 37, pp. 123–154.

Glenn, C., 1990. Depositional sequences of the Duwi, Sibaiya and Phosphate Formations, Egypt, phosphogenesis and glauconitization in a Late Cretaceous epeiric sea, in A.J.G. Notholt & Jarvis, I. (Eds.), Phosphorite Research and Development, Geological Society Special Pub. No. 52, pp. 205–222.

Godinot, M. & Mahboubi, M., 1992. Earliest known simian primate found in Algeria. Nature, v. 357, pp. 324–326.

Gohrbandt, K.H.A., 1966a, Some Cenomanian foraminifera from Northwestern Libya, Micropaleontology, v. 12, No. 1, pp. 65–70.

Gohrbandt, K.H.A., 1966b. Upper Cretaceous and Lower Tertiary stratigraphy along the western and southwestern edge of the Sirte Basin, Libya, in South-central Libya and Northern Chad. In: A Guidebook to the Geology and Prehistory, J.J. Williams (Ed.). Petrol. Explor. Soc. Libya, pp. 331–341.

Goldberg, M. & Beyth, M., 1991. Tiran Island: an internal block at the junction of the Red Sea rift and Dead Sea transform, in J. Markis. P. Mohr & R. Rihm (Eds.), Red Sea: Birth and Early History of a New Oceanic Basin, Tectonophysics, v. 198, pp. 261–273.

Gomez, F., Barazangi, M. & Demnati, A., 2000. Structure and evolution of the Neogene Guercif Basin at the junction of the Middle Atlas Mountains and the Rif Thrust Belt, Morocco. AAPG Bulletin, v. 84, pp. 1340–1364.

Goodfriend, G.A. & Stanely, D.J., 1999. Rapid strand-plain accretion in the northeastern Nile Delta in the 9th Century A.D. & the demise of the port Pelusium, Geology, v. 27, No. 2, pp. 147–150.

Goudarzi, G.H., 1970. Geology and mineral resources of Libya, a reconnaissance, U.S. Geol. Surv., Professional Paper 660, p. 104.

Goudarzi, G.H., 1971. Geology of the Shati Valley area Iron deposits, in C. Gray (Ed.), Symposium on the Geology of Libya, U. of Libya, Tripoli, Fezzan, pp. 489–500.

Goudarzi, G.H., 1980. Structure – Libya, in M.J. Salem & M.T. Busrewil (Eds.), The Geology of Libya, v. III, London, Academic Press, pp. 879–892.

Gouwy, S., Haydukiewicz, J. & Bultynck, P., 2007. Conodont-based graphic correlation of upper Givetian-Frasnian sections of the Eastern Anti-At las (Morocco). Geological Quarterly, Warsaw, v. 51, no. 4, pp. 375–392.

Gras, R. & Thusu, B., 1998. Trap architecture of the Early Cretaceous Sarir Sandstone in the eastern Sirt Basin, Libya, in D.J. Macgregor; R.T.J. Moody & D.D. Clark-Lowes (Eds.), Petroleum Geology of North Africa. Geological Society Spec. Pub. No. 132, London, pp. 317–334.

Gras, R., 1996. Structural style of the southern margin of the Messlah High, in M.J. Salem; M.T. Busrewil; A.A. Misallati & M.A. Sola (Eds.). The Geology of Sirt Basin, v. III, Amsterdam, Elsevier, pp. 201–210.

Gray, C., 1971 (Ed.), Symposium on the Geology of Libya, U. of Libya, Tripoli.

Greco, B., 1915. Fauna cretacea dell'Egitto raccolta dal Figari Bey, Palaeontographica Italica, pp. 189–232, tav. XVII-XXII [I-VI].

Greco, B., 1916. Fauna cretacea dell'Egitto raccolta dal Figari Bey. Palaeontographica Italica, Parte seconda, pp. 103–170, tav. XV-XIX [VII-XI].

Greco, B., 1917. Fauna cretacea dell'Egitto raccolta dal Figari Bey. Palaeontographica Italica, Parte terza, fasc. 1, pp. 93–162, tav. XIII-XVII [XII-XVI].

Greco, B., 1918. Fauna cretacea dell'Egitto raccolta dal Figari Bey. Palaeontographica Italica, Parte terza, fasc. 2 (fine), pp. 1–58, tav. I-V [XVII-XXI].

Greenberg, J.K., 1981. Characteristics and origin of Egyptian Younger Granites, Summary. Geol. Soc. Am. Bull., v. 92, pp. 224–232.

Greenwood, W.R. Hadley, D.G., Anderson, R.E., Fleck, R.J. & Schmidt, D.L., 1976. Late Proterozoic cratonization in southwestern Saudi Arabia, Phil. Trans. Royal. Soc. Lond., v. 280, pp. 517–527.

Gregory, J.W., 1896. The Great Rift Valley, John Murry, London, p. 422.

Gregory, J.W., 1898. A collection of Egyptian fossil Echinoidea. Geol. Mag., London, v. 5, pp. 149–161.

Gregory, J.W., 1911a. Contributions to the Geology of Cyrenaica. Q.J. Geol. Soc., v. 67, pp. 572–615.

Gregory, J.W., 1911b. The fossil Echinoidea of Cyrenaica. Q.J. Geol. Soc., v. 67, pp. 661–680.

Gregory, J.W., 1916. Cyrenaica. The Geographical Journal, v. 67, no. 5, pp. 320–342.

Gregory, J.W., 1920a. The African Rift Valleys. The Geographical Journal, v. 56, no. 1, pp. 13–41.

Gregory, J.W., 1920b. The African Rift Valleys. The Geographical Journal, v. 56, no. 4, pp. 327–328.

Gregory, J.W., 1921. The African Rift Valleys. The Geographical Journal, v. 57, no. 3, pp. 238–239.

Greiling, R.O., Abdeen, M.M., Dardir, A.A., E1 Akhal, H., E1 Ramly, M.F., Kamal E1 Din, G.M., Osman, A.F., Rashwan, A.A., Rice, A.H.N. & Sadek, M.E., 1994. A structural synthesis of the Proterozoic Arabian-Nubian Shield in Egypt. Geol Rundsch, v. 83, pp. 484–501.

Greiling, R., Kröner, A. & El Ramly, M.F., 1984. Structural interference patterns and their origin in the Pan-African basement of the southeastern Desert of Egypt, in A. Kröner & R. Greiling (Eds.), Precambrian Tectonics Illustrated, pp. 401–412, E. Schweitzerart'sche Velangsbuchhandlung (Nagele u. Obermiller), Germany, Stuttgart, p. 419.

Griffin, D.L. 1999. The late Miocene climate of northeastern Africa: unraveling the signals in the sedimentary succession. Journal of the Geological Society, London, v. 156, pp. 817–826.

Griffin, D.L., 2002. Aridity and humidity: two aspects of the late Miocene climate of North Africa and the Mediterranean. Palaeogeography, Palaeoclimatology, Palaeoecology, v. 182, no. 102, pp. 65–91.

Griffin, D.L., 2006. The late Neogene Sahabi rivers of the Sahara and their climatic and environmental implications for the Chad Basin. J. Geological Society, v. 163, no. 6, pp. 905–921.

Grignani, D., Lanzoni, E. & Elatrash, H., 1991, Paleozoic and Mesozoic subsurface palynostratigraphy in the Al Kufra basin, Libya, in M.J. Salem, O.S. Hammuda & B.A. Eliagoubi (Eds.), The Geology of Libya, v. IV, Amsterdam, Elsevier, pp. 1159–1227.

Grosheny, D., Chikhi-Aouimeur, F., Ferry, S., Benkherouf-Kechid, F., Jati, M., Atrops, F. & Redjimi-Bourouiba, W., 2008. The Upper Cenomanian-Turonian (Upper Cretaceous) of the Saharan Atlas (Algeria). Bulletin de la Societé Géologique de France, v. 179, no. 6, pp. 593–603.

Grubic, A., 1984. Geological Map of Libya 1:250,000, Sheet South Anay, NF32-4, Industrial Research Centre, Tripoli, p. 61.

Grubic, A., Dimitrijevic, M., Galecic, M., Jakovljevic, Z., Komarnicki, S., Protic, D., Radulovic, P. & Roncevic, G., 1991. Stratigraphy of Western Fazzan (SW Libya). In M.J. Salem, M.T. Busrewil & A.M. Ben Ashour (Eds.), The Geology of Libya, v. 1V, Amsterdam, Elsevier, pp. 1529–1564.

Gruenwald, G., 2001. Hydrocarbon prospectivity of lower Oligocene deposits in the Maragh Trough, SE Sirt Basin, Libya. J. Petroleum Geology, v. 24, no. 2, p. 213–231.

Gruenwald, R.M., Buitrago, J., Dessay, J., Huffman, A., Moreno, C., Gonzalez Munoz, J.M., Diaz, C. & Tawengi, K.S., 2010. Pore Pressure Prediction Based on High Resolution Velocity Inversion in Carbonate Rocks, Offshore Sirte Basin -Libya. Search and Discovery Article #40551 (2010). Oral presentation at AAPG Annual Convention and Exhibition, New Orleans, Louisiana, April 11–14, 2010, p. 35.

Guasti, E. & Speijer, R.P., 2007. The Paleocene-Eocene Thermal Maximum in Egypt and Jordan: An overview of the planktic foraminiferal record. In: Monechi, S., Coccioni, R. & Rampino, M. (Eds.): Large Ecosystem Perturbations: Causes and Consequences, GSA Special Papers 424, pp. 53–67.

Gudjonsson, L., 1987. Local and global effects on the Early Pliocene Mediterranean stable isotope records, Marine Micropaleontology, v. 12, pp. 241–253.

Gueinn, K.J. & Rasul, S.M., 1986. A contribution to the biostratigraphy of the Palaeozoic of the Western Desert, using new palynological data from the sub-surface, EGPC VIII Exploration Conference, Cairo, pp. 1–22.

Guerrak, S., 1988. Geology of the Early Devonian oolitic iron ore of the Gara Djebilet Field, Saharan Platform, Algeria. Ore Geology Reviews, v. 3, no. 4, pp. 333–358.

Guerrak, S., 1989. Time and space distribution of Palaeozoic oolitic ironstones in the Tindouf Basin, Algerian Sahara, Geological Society of London Special Pub. No. 46, pp. 197–212.

Guerre, A., 1980. Hydrogeological study of the coastal karstic spring of Ayn-az Zayanah, eastern Libya, in M.J. Salem & M.T. Busrewil (Eds.), The Geology of Libya, v. 2, London, Academic Press, pp. 685–701.

Guiraud, R. & Bellion, Y., 1995. Late Carboniferous to Recent geodynamic evolution of the West Gondwanian cratonic Tethyan margins, in A. Nairn. J. Dercourt & B. Vrielynck (Eds.), The Ocean Basins and Margins, v. 8, The Tethys Ocean, Plenum, New York, pp. 101–124.

Guiraud, R., 1990. Tectono-sedimentary framework of the early Cretaceous Continental Bima Formation (Upper Benue Basin, NE Nigeria), J. African Earth Sciences, v. 10, pp. 341–353.

Guiraud, R., 1998. Mesozoic rifting and basin inversion along the Northern African Tethyan margin, an overview, in D.J. Macgregor. R.T.J. Moody & D.D. Clark-Lowes (Eds.), Petroleum Geology of North Africa, Geological Society Spec. Pub. No. 132, London, pp. 217–229.

Guiraud, R. & Bosworth, W., 1997; Senonian basin inversion and rejuvenation of rifting in Africa and Arabia: Synthesis and implications to plate-scale tectonics, Tectonophysics, v. 282, pp. 39–82.

Guiraud, R., Bosworth, W., Thierry, J. & Delplanque, 2005. Phanerozoic geological evolution of Northern and Central Africa: An overview. Journal of African Earth Sciences, v. 43, pp. 83–143.

Guiraud, R., Doumnang Mbaigane, J.-C., Carpentier, S. & Dominguez, S., 2000. Evidence of a 6000 km length NW-SE-striking lineament in northern Africa: the Tibesti Lineament. Journal of the Geological Society, v. 157, pp. 897–900.

Guiraud, R., Issawi, B. & Bosworth, W., 2001. Phanerozoic history of Egypt and Surrounding areas, In P.A. Ziegler, Cavazza, W., Robertson, A.H.F. & Crasquin-Soleau, S. (Eds.), Peri-Tethys Memoir 6, Peri-Tethyan Rift/Wrench Basins and Passive Margins, Memoire du Museum National d'Histoire Naturelle, Tome 186, pp. 469–509.

Guiraud, R. & Maurin, J.-C., 1991. Le rifting en Afrique au Crétacé inférieur synthése structurale, mise en évidence de deux étapes dans la genèse des bassins, relations avec les ouvertures océaniques péri-Africaine, Bulletin de la Société Géologique de France, 162, pp. 811–823.

Guiraud, R. & Maurin, J.-C., 1992. Early Cretaceous rifts of Western and Central Africa, an overview, Tectonophysics, v. 213, pp. 153–168.

Guiraud, R. & Maurin, J.-C., 1993. Early Cretaceous rifting in Africa, a regional framework for the first stages of the Sirt Basin, in Sedimentary Basins of Libya, First Symposium, Geology of Sirt Basin, Earth Science Society of Libya, Tripoli, p. 22.

Gumati, Y.D. & Kanes, W.H., 1985. Early Tertiary subsidence and sedimentary facies–Northern Sirte Basin, Libya. AAPG Bull. 69(1): 39–52.

Gumati, Y.D. & Nairn, A.E.M., 1991. Tectonic subsidence of the Sirte Basin, Libya, J. Petrol. Geology, v. 14, pp. 93–102.

Gundobin, V.M., 1985. Geological map of Libya, 1:250,000, Sheet Qararat al Marar NH 33–13, Explanatory Booklet, Industrial Research Centre, Tripoli, p. 166.

Gunnel, G.F., Simons, E.L. & Seifert, E.R., 2008. New bats (Mammalia: Chiroptera) from the Late Eocene and Early Oligocene, Fayum Depression, Egypt. Journal of Vertebrate Paleontology, v. 28, no. 1, pp. 1–11.

Gvirtzman, G. & Weissbrod, T., 1984. The Hercynian geanticline of Helez and the Late Palaeozoic history of the Levant, in J.E. Dixon & A.H.F. Robertson (Eds.), The Geological Evolution of the Eastern Mediterranean, Geol Soc. Spec. Pub. No. 17, Blackwell, pp. 177–186.

Habib, M.E., Ahmed, A.A. & El Nady, O.M., 1985. Two orogenies in the Meatiq area of the central Eastern Desert, Egypt, Precambrian Res., v. 30, pp. 83–111.

Haddoumi, H., Charriere, A., Andreu, B. & Mojon, P.-O., 2008. Les dépôts continentaux du Jurassique moyen au Crétacé inférieur dans le Haut Atlas oriental (Maroc): paléoenvironnements successifs et signification paléogéographique. Carnets de Géologie/Notebooks on Geology – Article 2008/06 (CG2008_A06).

Haddoumi, H., Charrière, A., Feist, M. & Andreu, B., 2002. Nouvelles datations (Hauterivien supérieur–Barrémien inférieur) dans les « Couches rouges » continentales du Haut Atlas central marocain. conséquences sur l'âge du magmatisme et des structurations mésozoïques de la chaîne Atlasique. C.R. Palevol 1, pp. 259–266.

Hadji, F., Marok, A., Sebane, A., Benyahia, M., Mehlaoui, R., Bounoua, B. & Soulimane, C., 2006. A preliminary results on the Toarcian Bio-Geo-Events in the Traras Mounts (Northwestern Algeria). Fifth International Symposium of IGCP 506 – Hammarnet (Tunisia). March, 28–31, 2006, pp. 35–36.

Hafid, M., Ait Salem, A. & Bally, A.W., 2000. The western termination of the Jebilet_High Atlas system (offshore Essaouira Basin, Morocco). Marine and Petroleum Geology, v. 17, pp. 431–443.

Hafid, M., Tari, G., Bouhadioui, D., El Moussaid, I., Echarfaoui, H., Aït Salem, A., Nahim, M. & Dakki, M., 2008. Atlantic Basins. In A. Michard, O. Saddiqi, A. Chalouan & D. Frizon de Lamotte (Eds.), Continental Evolution: The Geology of Morocco. structure, stratigraphy, and Tectonism of the Africa-Atlantic-Mediterranean Triple Junction, Berlin, Springer-Verlag, pp. 303–330.

Hagedorn, H., 1980. Geological and geomorphological observations on the northern slope of the Tibisti Mountains, Central Sahara, in M.J. Salem & M.T. Busrewil (Eds.), The Geology of Libya, v. II, London, Academic Press, pp. 823–835.

Hallett, D. & El Ghoul, A., 1996. Oil and Gas potential of the deep trough areas in the Sirt Basin, Libya, in The Geology of the Sirt Basin, v. II, pp. 455–484.

Hallett, D., 2002. Petroleum Geology of Libya, Elsevier, Amsterdam, p. 502.

Hallock, P. & Glenn, E.C., 1986. Larger Foraminifera: a tool for palaeoenviromental analysis of Cenozoic carbonate depositional facies, Palaios, 1, pp. 55–64.

Halpern, M. & Tristan, N., 1981. Geochronology of the Arabian-Nubian shield in southern Israel and eastern Sinai, J. Geology, v. 89, pp. 639–648.

Hamama, H., 2010. Barremian and Aptian Mollusca of Gabal Mistan and Gabal Um Mitmani, Al-Maghara Area, Northern Sinai, Egypt. Journal of American Science, v. 6, no. 12, pp. 1702–1714.

Hamblin, R.D., 1987. Stratigraphy and depositional environments of the Gebel el-Rus area, Eastern Fayum Egypt, Brigham Young University Geology studies, v. 34 (part 1), pp. 61–83.

Hamilton, W., 1987. Crustal extension in the Basin and Range Province, southwestern United States, in M.P. Coward, J.F. Dewey & P.L. Hancock (Eds.), Continental Extensional Tectonics, Geological Society Special Publication No. 28, pp. 155–176.

Hammama, H.H. & Kassab, A.S., 1990. Upper Cretaceous ammonites of Duwi Formation Gabal Abu Had and Wadi Hamama, Eastern Desert, Egypt, J. Afr. Earth Sci., v. 10, No. 3, pp. 453–464.

Hammill, M. & Robinson, N.D., 1992. Nearby oil discoveries add interest to licensing rounds in Algerian basins. Oil & Gas Journal, November 9, pp. 99–101.

Hammor, D. & Lancelot, J., 1998. Miocene metamorphism of Pan-African granites in the Edough Massif (NE Algeria). Comptes Rendus de l'Académie des Sciences – Series IIA – Earth and Planetary Science, v. 327, no. 6, pp. 391–396.

Hammuda, O.S., 1969. Jurassic and Lower Cretaceous rocks of central Jebel Nefusa, northwestern Libya, Petrol. Explor. Soc. Libya, p. 74.

Hammuda, O.S., 1971. Nature and significance of the Lower Cretaceous unconformity in Jebel Nefusa, northwest Libya, in C. Gray (Ed.), Symposium on the Geology of Libya, Faculty of Science, U. Libya, Tripoli, pp. 87–97.

Hammuda, O.S., 1980. Sediments and paleogeography of the Lower Campanian sand bodies along the southern tip of Ad Daffah-Al Waha Ridge, Sirt basin, in M.J. Salem & M.T. Busrewil (Eds.), The Geology of Libya, v. II, London, Academic Press, pp. 509–520.

Hammuda, O.S., Sbeta, A.M., Mouzughi, A.J. & Eliagoubi, B.A., 1985. Stratigraphic Nomenclature of the Northwestern Offshore of Libya. Earth Sci. Soc. Libya, p. 166.

Hammuda, O.S., van Hinte, J.E. & Nederbragt, S., 1991. Geohistory analysis in central and southern Tarablus basin, northern offshore of Libya, in M.J. Salem. O.S. Hammuda & B.A. Eliagoubi (Eds.), The Geology of Libya, v. IV, Elsevier, Amsterdam, pp. 1657–1680.

Hamy, E.-T., 1869. L'Egypte quaternaire et l'ancienneté de l'homme. Bulletins de la Société d'anthropologie de Paris, Année 1869, Volume 4, no. 1, pp. 711–719.

Hantar, G., 1990. North Western Desert, in R. Said (Ed.), The Geology of Egypt, Balkema, Rotterdam/Brookfield, pp. 293–319.

Haq, B.U., Hardenbol, J. & Vail, P.R., 1987. Chronology of fluctuating sea levels since the Triassic (250 million years ago to the present), Science, v. 235, pp. 1156–1167.

Hardie, L.A. & Lowenstein, T.K., 2004. Did the Mediterranean Sea dry out during the Miocene? A reassessment of the evaporite evidences from DSDP Legs and 42A cores. Journal of Sedimentary Research, v. 74, no. 4, pp. 453–461.

Harding King, W.J., 1916. The nature and formation of sand ripples and dunes. The Geographical Journal, v. 47, no. 3, pp. 189–209.

Harding King, W.J., 1918. Study of a dune belt. The Geographical Journal, v. 51, no. 1, pp. 16–33.

Harding King, W.J., 1928. Lost Oases of the Libyan Desert. The Geographical Journal, v. 72, no. 3, pp. 244–249.

Harding, T.P., 1984. Graben hydrocarbon occurrences and structural style, AAPG Bull., v. 68, No. 3, pp. 333–362.

Harland, W.B., Armstrong, R.L., Cox, A.V., Craig, L.E., Smith, A.G. & Smith, D.G., 1990. A Geologic Time Scale 1989, Cambridge, Cambridge University Press, p. 263.

Harms, J.C. & Wray, J.L., 1988 (Abs.). Pre-Pliocene history and depositional facies, Nile Delta, Egypt, in Mediterranean Basin Conference and Exhibition, Nice, France, AAPG Bull., v. 72, pp. 1005.

Harrell, J.A. & Lewan, M.D., 2002, Sources of mummy bitumen in ancient Egypt and Palestine: Archaeometry, v. 44, pt. 2, pp. 285–293.

Harrell, J.A., 2004. Archaeological geology of the world's first emerald mine. Geoscience Canada, v. 31, no. 2, pp. 69–76.

Harris, N.B.W. & Gass, I.G., 1981. Significance of contrasting magmatism in North East Africa and Saudi Arabia, Nature, v. 289, pp. 394–396.

Harris, N.B.W., Gass, I.G., Hawkesworth, C.J. & F.R.S., 1991. A geochemical approach to allochthonous terranes: A Pan-African case study, in J.D. Dewey. I.G. Gass. G.B. Curry. N.B.W. Harris & A.M.C. Sengör (Eds.), Allochthonous Terranes, Cambridge University Press, Cambridge, pp. 77–92.

Harris, N.B.W., Hawkesworth, C.J. & Ries, A.C., 1984. Crustal evolution in north-east and east Africa from model Nd ages, Nature, v. 309, pp. 773–776.

Hasi, I.A., 1995 (Abs.). Sequence stratigraphic analysis of the Tahara Formation, Hamada Basin, in First Symposium on the Hydrocarbon Geology of North Africa. Geol. Soc. of London Petroleum Group, pp. 31–32.

Hassan, A.A., 1967. A new carboniferous occurrence in Abu Durba, Sinai, Egypt, Sixth Arab Petroleum Congress, Baghdad, Paper No. 39 (B-3), p. 8.

Hassan, A.M., 1998 (Abs.). Geochemistry, Rb-Sr geochronology and petrogensis of the Late Pan-African Atalla Felsite, central Eastern Desert, Egypt, in Geology of the Arab World, Fourth International Conference, Cairo University, Cairo, pp. 89–90.

Hassan, M.A. & Hashad, A.H., 1990. Precambrian of Egypt, in R. Said (Ed.), The Geology of Egypt, Balkema, Rotterdam/Brookfield, pp. 201–245.

Hassanein Bey, A.M., 1924. Through Kufra to Darfur. The Geographical Journal, v. 64, no. 4, pp. 273–291 and pp. 353–393.

Hataba, H., 2005. Egypt hydrocarbon potential. The Hydrocarbon Reserves: Abundance or scarcity? OAPEC/IFP Joint Seminar.

Hatzfeld, D. & Frogneux, M., 1980. Structure and tectonics of the Alboran Sea area. Structure and Tectonics of the Alboran Sea Area. Instituto Geografico Nacional Spec. Pub. No. 201, Madrid, pp. 93–108.

Havlicek, V. & Massa, D., 1973. Brachiopodes de l'Ordovicien supérieur de Libye occidentale, implications stratigraphiques, Geobios, v. 6, fasc.4, pp. 267–290.

Hawkshaw, J.C., 1867. Geological description of the First Catarat, Upper Egypt. Q.J. Geol. Soc. London, v. 23, pp. 115–119.

Haynes, C.V., Jr., 1982. Great Sand Sea and Selima Sand Sheet, Eastern Sahara: Geomorphology of desertification, Science, v. 217, pp. 629–633.

Haynes, C.V., Jr., 2001. Geochronology and climate change of the Pleistocene-Holocene transition in the Darb el Arba'in Desert, Eastern Sahara. Geoarchaeology, v. 16, no. 1, pp. 119–141.

Haynes, C.V., Maxwell, T.A., El Hawary, A., Nicoll, K.A. & Stokes, S., 1997. An Acheulian site near Bir Kiseiba in the Darb el Arba'in Desert, Egypt. Geoarchaeology, v. 12, no. 8, pp. 819–832.

Haynes, C.V., Jr., 2001. Geochronology and climate change of the Pleistocene-Holocene transition in the Darb el Arba'in Desert, Eastern Sahara. Geoarchaeology: An International Journal, v. 16, no. 1, pp. 119–141.

Hea, J.P., 1971. Petrography of the Paleozoic-Mesozoic Sandstones of the southern Sirte Basin, Libya, in C. Gray (Ed.), Symposium on the Geology of Libya, U. of Libya, Tripoli, pp. 107–125.

Hecht, F., Fürst, M. & Klitzsch, E., 1963. Zur Geologie von Libyen. Geol. Rundschau, v. 53, pp. 413–470.

Hefferan, K.P., Admou, H., Hilal, R., Karson, J.A., Saquaque, A. Juteau, T., Marcel Bohn, M., Samson, S.D. & Kornprobst, J.M., 2002. Proterozoic blueschist-bearing mélange in the Anti-Atlas Mountains, Morocco. Precambrian Research, v. 118, nos. 3–4, pp. 179–194.

Hegab, A.A., 1989. New occurrence of Rhynchonellida (Brachiopoda) from the Middle Jurassic of Gebel El-Maghara, Northern Sinai, Afr. J. Earth Sci., v. 9, No. 3/4, pp. 445–453.

Hegazi, A.H., Andersson, J.T. & El-Gayar, M.S., 2003. Applicationn of gas chromatography with atomic emission detection to the geochemical investigation of polycyclic aromatic sulphur heterocycles in Egyptian crude oils. Fuel Processing Technology, v. 85, pp. 1–19.

Hegazi, A.H. & El-Gayar, M. Sh., 2009. Geochemical characterization of a biodegraded crude oil, Assran Field, Central Gulf of Suez. Journal of Petroleum Geology, v. 32, no. 4, pp. 343–355.

Heinl, M. & Brinkmann, P.J., 1989. A groundwater model of the Nubian aquifer system. Hydrological Sciences – Journal des Sciences Hydrologiques, v. 34, no. 4, pp. 425–447.

Heldt, M., Bahman, M. & Lehmann, J., 2008. Microfacies, biostratigraphy, and geochemistry of the hemipelagic Barremian-Aptian in north-central Tunisia: Influence of the OAE 1a on the southern Tethys margin. Palaeogeography, Palaeoclimatology, Palaeoecology, v. 261, pp. 246–260.

Helmy, H.M., 1990. Southern Gulf of Suez, Egypt: structural geology of the B-trend oil fields. Geological Society, London, Special Publication, v. 50, pp. 353–363.

Hemdan, M.A. & Aref, M.A., 1998 (Abs.). Modern and Quaternay evaporite deposits at the western side of the Qattara Depression and their environmental and climatic implications, in Geology of the Arab World, 4th International Conference, Cairo University, Cairo, p. 13.

Hendriks, F., 1985. Upper Cretaceous to Lower Tertiary sedimentary environments and clay mineral associations in the Kharga oasis area (Egypt), N. Jb. Geol. Paläont. Mh. (10), pp. 579–591.

Hendriks, F., 1986. The Maghrabi Formation of the El-Kharga area (SW Egypt), deposits from a mixed estuarine and tidal-flat environment of Cenomanian age, J. Afr. Earth Sci., v. 5, pp. 481–489.

Hendriks, F., Kallenbach, H. & Philobbos, E.R., 1990. Cretaceous to Early Tertiary continental and marginal marine sedimentary environments of southeastern Egypt. J. Afr. Earth Sci., v. 10, pp. 229–241.

Hendriks, F., Luger, P., Bowitz, J. & Kallenbach, H., 1987. Evolution of the depositional environments of SE-Egypt during the Cretaceous and Lower Tertiary. Berliner Geowiss. Abh. (A), v. 75, no. 1, pp. 49–82.

Hendriks, F., Luger, P., Kallenbach, H. & Schroeder, J.H., 1984. Stratigraphical and sedimento-logical framework of the Kharga-Sinn el-Kaddab Stretch (Wesern and Southern part of the Upper Nile Basin), Western Desert, Egypt. Berliner Geowiss. Abh.(A), v. 50, pp. 117–151.

Hendrickx, S. & Vermeersch, P., 2001. Prehistory: From the Palaeolithic to the Badarian Culture. In I. Shaw (Ed.), The Oxford History of Ancient Egypt. Oxford: Oxford University Press, pp. 17–43.

Hendrickx, S. & Vermeersch, P., 2001. Prehistory: From the Palaeolithic to the Badarian Culture. In I. Shaw (Ed.), The Oxford History of Ancient Egypt. Oxford: Oxford University Press, pp. 17–43.

Heppard, P.D. & Albertin, M.L., 1998. Abnormal Pressure Evaluation of the Recent Pliocene and Miocene Gas Discoveries from the Eastern Nile Delta, Egypt, Using 2D and 3D Seismic Data. Houston Geological Society, International Dinner Meeting, Nov. 16, 1998.

Herbig, H.-G., 1986. Lithostratigraphisch-fazielle Untersuchungen im marinen Alttertiär südlich des zentralen Hohen-Atlas (Marokko). Berliner Geowissenschaftliche Abhandlungen, v. 66, pp. 343–380.

Herbig, H.-G., 1991. Das Paläogen am Südrand des zentralen Hohen Atlas und im Mittleren Atlas Marokkos. Stratigraphie, Fazies, Paläogeographie und Paläotektonik. Berliner Geow-issenschaftliche Abhandlungen, (A), v. 135, p. 289.

Herbig, H.-G. & Aretz, M., 2007 (Abs.). Visean basin fill of the Jerada synclinorium (NE Morocco) – sedimentary dynamics and geodynamic implications. MAPG First International Conference & Exbihition, Marrakech, Morocco, pp. 46–47.

Hermina, M., 1990. The surroundings of Kharga, Dakhla and Farafra oases. In R. Said (Ed.), The Geology of Egypt, Balkema, Rotterdam/Brookfield, pp. 259–292.

Hermina, M. & Issawi, B., 1971. Rock stratigraphic classification of Upper Cretaceous-Lower Tertiary exposures in Southern Egypt. In: C. Gray (Ed.), Symposium on the Geology of Libya, U. of Libya, Tripoli, pp. 147–154.

Hernández, J. & Bellon, H., 1985. Chronologie K/Ar du volcanisme miocène du Rif orien-tal, Maroc. Implication géodynamiques tectoniques et magmatologiques. Revue Géologie Dynamique et Géographie Physique, v. 26, pp. 85–94.

Herodotus, translated by Aubrey de Selincourt, and edited by Marincola, J., 1996. The Histories. Penguin Books, London, p. 622.

Herngreen, G.F.W., Kedves, M., Rovnina, L.V. & Smirnova, S.B., 1996. Cretaceous palyno-floral provinces: a review. In: Jansonius, J., McGregor, D.C. (Eds.):, Palynology: Principles and Applications, vol. 3. American Association of Stratigraphic Palynologists Foundation, pp. 1157–1188.

Heselden, R., Madi, F.M. & Cubitt, J.M., 1996. Lithofacies study of the Lidam and Maragh formations (Late Cretaceous) of the Masrab Field and adjacent areas, Sirt Basin, Libya. In: M.J. Salem. A.S. El-Hawat & A.M. Sbeta (Eds.), The Geology of the Sirt Basin, v. II, Amsterdam, Elsevier, pp. 197–210.

Hesse, K.H., Hissene, A., Kheir, O., Schnäcker, E., Schneider, M. & Thorweihe, U., 1987. Hydrogeological investigations in the Nubian aquifer system, Eastern Sahara. Berliner Geowiss. Abhandl.5, v. 75, no. 2, pp. 397–464.

Hey, R.W., 1968. The Quaternary geology of the Jabal al Akhdar coast, in F.T. Barr (Ed.), Geol-ogy and Archeology of Northern Cyrenaica, Libya, Petrol. Explor. Soc. Libya, 10th Annual Field Conference, pp. 159–165.

Heybroek, F., 1965. The Red Sea Miocene evaporite basin, in Salt Basins around Africa. London, Inst. Petroleum, pp. 17–40.

Higazi, A.H. & El-Gayar, M.Sh., 2009. Geochemical characterization of bidegraded crude oil, Assaran Field, Central Gulf of Suez. Journal of Petroleum Geology, v. 32, no. 4, pp. 343–355.

Higazi, A.H., Andersson, J.T. & El-Gayar, M.Sh., 2003. Application of gas chromatography with atomic emission detection to the geochemical investigation of polycyclic aromatic sulfur heterocycles in Egyptian crude oils. Fuel Processing Technology, v. 85, pp. 1–19.

Hill, C.L., 2001. Geologic contexts of the Acheulian (Middle Pleistocene) in the Eastern Sahara. Geoarchaeology, v. 16, pp. 65–94.

Hill, C., 2009. Stratigraphy and sedimentology at Bir Sahara, Egypt: Environments, climate change and the Middle Paleolithic. Catena, v. 78, no. 3, pp. 250–259.

Hill, P.J. & Molyneux, S.G., 1988. Biostratigraphy, palynofacies and provincialism of Late Ordovician-Early Silurian acritarchs from northeast Libya, in A. El-Arnauti, B. Owens & B. Thusu (Eds.), Subsurface Palynostratigraphy of Northeast Libya, Benghazi, Libya, Garyounis University Publications, pp. 27–43.

Hill, P.J.; Paris, F. & Richardson, J.B., 1985. Silurian palynomorphs, J. Micropaleontol-ogy. In B. Thusu & B. Owens (Eds.), Palynostratigraphy of North-East Libya, v. 4, pt.1, pp. 27–48.

Hinz, K., Dortmann, H. & Fritsch, J., 1982. The continental margin of Morocco: Seismic sequences, structural elements of geological development. In U. von Rad, K. Hinz, M. Sarnthien & E. Seibold (Eds.): The Geology of Northwest African Continental Margin, Springer-Verlag, pp. 34–60.

Hirmer, M., 1925. Die fossilen Pflanzen aegyptens: D. Filicales-Ergebnisse der Forschungsreisen Prof. E. Strömers in den wüsten aegyptens, no. 3, Abhandl. Bayerisch. Akad. Wissensch., v. XXX, pt.3, pp. 1–18.

Hirsch, F., 1984. The Arabian sub-plate during the Mesozoic, in J.E. Dixon & A.H.F. Robertson (Eds.): The Geological Evolution of the Eastern Mediterranean. Geol Soc. Spec. Pub. No. 17, Blackwell, pp. 217–223.

Hirsch, F., Flexer, A., Rosenfeld, A. & Yellin-Dror, A., 1995. Palinspastic and crustal setting of the Eastern Mediterranean. J. Petroleum Geology, v. 18, pp. 149–170.

Hirst, J.P.P., Benbakir, A., Payne, D.F. & Westlake, I.R., 2002. Tunnel valley and density flow processes in the Upper Ordovician glacial succession, Illizi Basin, Algeria: Influence on res-ervoir quality. Journal of Petroleum Geology, v. 25, no. 3, pp. 297–324.

Hladil, J., Otava, J. & Galle, A., 1991. Oligocene carbonate buildups of the Sirt Basin, Libya. In M.J. Salem. O.S. Hammuda & B.A. Eliagoubi (Eds.), The Geology of Libya, v. IV, Amsterdam, Elsevier, pp. 1401–1420.

Hlustik, A., 1991. Late Paleozoic floras of the Wadi ash Shati area, Libya. In M.J. Salem, O.S. Hammuda & B.A. Eliagoubi (Eds.), The Geology of Libya, v. IV, Amsterdam, Elsevier. pp. 1275–1284.

Hmich, D., Schneider, J.W., Saber, H., Voigt, S. & El Wartiti, M., 2006. New continental Carboniferous and Permian faunas of Morocco: implications for biostratigraphy, palaeobiogeography and palaeoclimate. Geological Society, London, Special Publications, v. 265, no. 1, pp. 297–324.

Hoepffner, C., Houari, M.R. & Bouabdelli, M. 2006. Tectonics of the North African Variscides (Morocco, western Algeria): an outline. Comptes Rendus Géoscience, v. 338, pp. 25–40.

Hoepffner, C., Soulaimane, A. & Piqué, A., 2005. The Moroccan Hercynides. Journal of African Earth Sciences, v. 43, nos. 1–3, pp. 144–165.

Hoelzmann, P., Keding, B., Berke, H., Kröpelin, S. & Kruse, H.-J., 2001. Environmental change and archaeology: lake evolution and human occupation in the Eastern Sahara during the Holocene. Palaeogeography, Palaeoclimatology, Palaeoecology, v. 169, pp. 193–217.

Hoffman-Rothe, J., 1966. Stratigraphie und Tektonik des Palaozoikums der Algerische Ostsahara. Geol. Rundschau, v. 55, pp. 736–774.

Holik, J.S. & Rabinowitz, 1992. Structure of Oceanic crust within the Jurassic Quiet Zone, offshore Morocco. In Geology and Geophysics of Continental Margins, AAPG Special Volumes, v. 53, pp. 259–281.

Hollard, H. 1963. Un tableau stratigraphique du Dévonien du Sud de l'Anti-Atlas. Notes et Mémoires du Service Géologique du Maroc, v. 23, pp. 105–109.

Hollard, H. 1967. Le Dévonien du Maroc et du Sahara nord-occidental. In: Oswald, D.H. (ed.) International Symposium on the Devonian System, Calgary, v. 1, pp. 203–244.

Hollard, H. 1978. Corrélations entre niveaux à brachiopods et à goniatites au voisinage de la limite Dévonien inférieur – Dévonien moyen dans les plaines du Dra (Maroc pre´saharien). Newsletter on Stratigraphy, v. 7, no. 1, p. 8–25.

Hollard, H. 1981. Tableaux de corrélations du Silurien et du Dévonien de l'Anti-Atlas. Notes et Mémoires du Service Géologique du Maroc, v. 42, p. 23.

Hooghiemstra, H., Lezine, A.M., Leroy, S.A.G., Dupont, L. & Marret, F., 2006. Late Quaternary palynology in marine sediments: a synthesis of the understanding of pollen distribution patterns in the NW African setting. Quaternary International, v. 148, pp. 29–44.

Hooker J.D., Ball J., Maw G., 1878. Journal of a Tour in Marocco and the Great Atlas, Macmillan & Co., London, p. 499.

Hoque, M., 1975. An analysis of cross-stratification of Gargarish calcarenite (Tripoli, Libya) and Pleistocene Paleowinds, Geological Magazine, v. 112, No. 4, pp. 393–401.

Hornemann, F., 1802. Tagebuch seiner Reise von Cairo nach Murzuck der Hauptstadt des Königreiches Fezzan in Afrika in den Jahren 1797 und 1798. Aus der deutschen Handschrift herausgegeben von Karl König. Weimer.

Hosney, H.M., 2005. Geothermal investigations and hydrocarbon potentialities of the Northwestern Desert, Egypt. JKGU, v. 16, pp. 95–120.

Hsü, K.J., 1972a. When the Mediterranean dried up, Sci. Am., v. 277, pp. 27–36.

Hsü, K.J., 1972b. Origin of saline giants: A critical review after the discovery of the Mediterranean evaporite, Earth Science Reviews, v. 8, pp. 371–396.

Hsü, K.J., 1977. Tectonic evolution of the Mediterranean basins. In A.E.M. Nairn; W.H. Kanes & F.G. Stehli (Eds.), Ocean Basins and Margins, pp. 29–69.

Hsü, K.J., Cita, M.B. & Ryan, W.B.F., 1973. Origin of the Mediterranean evaporites, in: Initial Reports of the Deep Sea Drilling Project, v. 13, pp. 1203–1231.

Hughes, G.W., Abddine, S. &Girgis, M.H., 1992. Miocene biofacies development and geological history of the Gulf of Suez, Egypt. Marine and Petroleum Geology, v. 9, no. 1, pp. 2–27.

Hughes, P.D., Gibbard, P.L. & Woodward, J.C., 2004. Quaternary glaciation in the Atlas Mountains, North Africa. In Quaternary Glaciation: Extent and Chronology. Vol. 3: Asia, Latin

America, Africa, Australia, Antarctica, Ehlers J, Gibbard PL (Eds), Elsevier, Amsterdam, pp. 255–260.

Hull, E., 1886. Memoir on the geology and geography of Arabia Petraea. Survey of Western Palestine, 8. Palestine Exploration Fund, 9.

Hull, E., 1896. Observations on the geology of the Nile Valley, and on the evidence of the greater volume of the river at a former period. Q.J. Geol. Soc., London, v. 52, pp. 308–319.

Hume, W.F., 1901. The Rift Valleys of Eastern Sinai. Geological Magazine, v. 8, no. 4, pp. 198–200.

Hume, W.F., 1906b. The Topography and Geology of the Peninsula of Sinai (South-Eastern Portion). Survey Department, Egypt, p. 280.

Hume, W.F., 1911a. The first meteorite record in Egypt. Cairo Scientific Journal, v. 5, no. 59, p. 212.

Hume, W.F., 1911b. The effects of secular oscillation in Egypt during the Cretaceous and Eocene periods. Q.J. Geol. Soc. London, v. 67, pp. 118–148.

Hume, W.F., 1912. Explanatory notes to accompany the geological map of Egypt, Egypt. Survey Dept., Cairo, p. 50.

Hume, W.F., 1917. Movements in the oilfield region of Egypt. Geological Magazine.

Hume, W.F., 1921. The Egyptian Wilderness. The Geographical Journal, v. 58, no. 4, pp. 249–274.

Hume, W.F., 1924. Conclusions derived from the geological data collected by Hassanein Bey during his Kufra-Owenat expedition. The Geographical Journal, v. 64, no. 5, pp. 386–388.

Hume, W.F., 1925–1937. Geology of Egypt. Government Press, Cairo: V. I (1925), The Surface Features of Egypt, their determining causes and relation to geological structure, 408p, V. II, The Fundamental Pre-Cambrian Rocks of Egypt and the Sudan, their distribution, age and character. Part I (1934): The Metamorphic Rocks. Government Press, Cairo, 300p, Part II (1935): The later plutonic and minor intrusive rocks, pp. 301–688.

Hume, W.F., 1932. The Pre-Cambrian Rocks of Egypt: their nature, classification and correlation. Proc. Geol. Soc. London, v. 38, p.iii-iv.

Hume, W.F., 1934, 1935, 1937. Geology of Egypt, V. II, The fundamental Precambrian rocks of Egypt and the Sudan. Part I, The metamorphic rocks, pp. 1–300. Part II, The later plutonic and minor intrusive rocks, p. 301. Part III, The minerals of economic value, Egypt. Survey Dept., Cairo, pp. 689–990.

Hume, W.F., Moon, F.W. & Sadek, H., 1920. Preliminary geological report on Gebel Tanka area, Egypt. Survey Dept., Cairo, p. 16.

Hunter, A.W., Lefebvre, B., Nardin, E., Van Roy, P., Zamora, S. & Régnault, S., 2010. Preliminary report on new echinoderm Lagerstätten from the Upper Ordovician of the eastern Anti-Atlas, Morocco. In Harris et al. (eds): Echinoderms. Taylor & Francis Group, London, pp. 23–30.

Hunting Geology & Geophysics Ltd., 1974. Geology of Jabal Eghei area, Libyan Arab Republic, Unpubl. Rep., Industrial Res. Cent., Tripoli.

Hussein, A.A. & El Sharkawi, M.A., 1990. Minerals deposits. In R. Said (Ed.): The Geology of Egypt, Balkema, Rotterdam/Brookfield, pp. 511–566.

Hussein, Kamal el Din, 1928. L'exploration du désert de Libye. La Géographie, v. 50, pp. 171–183 and pp. 320–336.

Huvelin, P. & Mamet, B., 1997. Transgressions, faulting and redeposition phenomenon during the Visean in the Khenifra area, western Moroccan Meseta. Journal of African Earth Sciences, v. 25, pp. 383–388.

Ibouh, H., El Bchari, F., Bouabdelli, M., Souhel, A. & Youbi, N., 2001. L'accident Tizal-Azourki Haut Atlas Central du Maroc: Déformations synsédimentaires Liasiques en extension et conséquences du serrage Atlasique. Estudios Geol., v. 57, pp. 15–30.

Ibrahim, A.B., Baltesperger, P., Bohorich, M., Nessim, M. & Fleming, M., 1998. Optimization of 3-D acquisition design – From theory to practice, the Qarun discoveries in the Western Desert of Egypt, 68th Ann. Int. Mtg., Soc. Expl. Geophys., pp. 101–102.

Ibrahim, M.I.A. & Mansour, A.M.S., 2002. Biostratigraphy and palaeoecological interpretation of the Miocene-Pleistocene sequence at El-Dabaa, northwestern Egypt. J. PMicropalaeontolgy, v. 21, no. 1, pp. 51–65.

Ibrahim, M.I., Schrank, E. & Abdel-Karim, M.R., 1995. Cretaceous biostratigraphy and palaeogeography of Northern Egypt and NE Libya, Petroleum Research Journal, v. 7, Tripoli, pp. 75–93.

Ibrahim, M.M., 1943. The Petrified Forest. Institut d'Égypte Mém. 25, pp. 159–182.

Ince, D.M., Beibis, S., Mcsherry, A. & Seymour, W.P., 1988 (Abs.). Stratigraphic framework, sedimentology and structural setting of Miocene-Pliocene sediments of the Abu Qir area, offshore Nile delta, in Mediterranean Basins Conference and Exhibition, Nice, France, AAPG Bull., v. 72, pp. 1008.

Irwin, M., 1965. General theory of epeiric clear water sedimentation, AAPG Bull., v. 49, pp. 445–459.

Ismail, A.A. & Abdelghany, O., 1999. Lower Miocene foraminifera from some exposures in the Cairo-Suez district, Eastern Desert, Egypt. Journal of African Earth Sciences, v. 28, no. 3, pp. 507–526.

Ismail, A.A. & Soliman, S.I., 1997. Cenomanian-Santonian foraminifera and ostracodes from Horus Well-1, North Western Desert, Egypt. Micropaleontology, v. 43, no. 2, pp. 165–183.

Ismail, A.A., 1993. Lower Senonian foraminifera and ostracoda from Abu Zeneima area, west-central Sinai, Egypt. M.E.R.C. Ain Shams Univ., Earth Sci. Ser., Vol. 7, pp. 115–130.

Ismail, A.A., 2000. Upper Cretaceous stratigraphy and micropaleontology of the western part of the Gulf of Aqaba, East Sinai, Egypt. M.E.R.C. Ain Shams Univ., Earth Sci. Ser., Vol. 14, pp. 239–261.

Ismail, A.A., 2001. Correlation of Cenomanian-Turonian ostracods of Gebel Shabraweet with their counterparts in Egypt, North Africa and the Middle East. N. JB. Geol. Paläont. Mh., v. 9, pp. 513–533.

Ismail, A.A., Hussein-Kamel, Y., Boukhary, M. & Ghandour, A.A., 2007. Campanian-Early Eocene stratigraphy of the Southern Galala Plateau, Eastern Desert, Egypt. Geologia Croatica, v. 62, no. 2, pp. 115–137.

Issawi, B., 1968. The geology of Kurkur-Dungul area, Geological Survey of Egypt, Cairo, Paper 46, pp. 1–102.

Issawi, B., 1972. Review of Upper Cretaceous-Lower Tertiary stratigraphy in central and southern Egypt, AAPG Bull., v. 56, pp. 1448–1463.

Issawi, B., 1973. Nubia sandstone: type section. AAPG Bull., v. 57, no. 4, pp. 741–744.

Issawi, B. & Jux, U., 1982. Contribution to the stratigraphy of the Paleozoic rocks in Egypt, Geological Survey of Egypt, Cairo, Paper 64, pp. 1–28.

Issawi, B., El-Hinnawi, M., Francis, M. & Mazhar, A., 1999. The Phanerozoic geology of Egypt; a geodynamic approach. EGSMA Special publication No. 76, p. 462.

Jackson, C.A.-L, Gawthorpe, R.L., Carr, I.D. & Sharp, I.R., 2005. Normal faulting as a control on the stratigraphic development of shallow marine syn-rift sequences: the Nukhul and Lower Rudeis formations, Hammam Faraun fault block, Suez Rift, Egypt. Sedimentology, v. 52, no. 2, pp. 313–338.

Jackson, C.A.-L., 2008. Sedimentology and significance of an early syn-rift paleovalley, Wadi Tayiba, Suez Rift, Egypt. Journal of African Earth Sciences, v. 52, pp. 62–68.

Jackson, J.A., White, N.J. & Anderson, H., 1988. Relations between normal-fault geometry, tilting and vertical motions in extensional terrains: an example from the southern Gulf of Suez. Journal of Structural Geology, v. 10, pp. 155–170.

Jacobshagen V., Brede R., Hauptmann M., Heinitz W. & Zylka R., 1988a. Structure and post-Paleozoic evolution of the Central High Atlas, in V.H. Jacobshagen (Ed.), The Atlas system of Morocco, Lect. Notes Earth Sci., v. 15, Springer-Verlag, Berlin/Heidelberg, p. 245–271.

Jacobshagen, V.H. (Ed.), 1988. The Atlas System of Morocco: Studies of its Geodynamic Evolution. LecureNotes in Earth Sciences, v. 15, Springer-Verlag, Berlin/Heidelberg.

Jacobshagen, V.H., Görler, K. & Giese, P., 1988b. Geodynamic evolution of the Atlas System (Morocco) in post-Palaeozoic times. In V.H. Jacobshagen (Ed.), 1988. The Atlas System of Morocco: Studies of its Geodynamic Evolution. LecureNotes in Earth Sciences, v. 15, Springer-Verlag, Berlin/Heidelberg, pp. 481–499.

Jacqué, M., 1962. Reconnaissance géologique du Fezzan oriental, Compagnie Française des Pétroles, Notes et Mémoires 5, Paris, p. 44.

Jaeger, H., Bonnefous, J. & Massa, D., 1975. Le silurien en Tunisie, ses relations avec le silurien de Libye nord-occidentale, Bull. Soc. Géol. Fr., Sér.7, v. 17, No. 1, pp. 68–76.

Jaglin, J.C. & Paris, F., 2002. Biostratigraphy, biodiversity and palaeogeography of late Silurian chitinozoans from A1-61 borehole (north-western Libya). Review of Palaeobotany and Palynology, v. 118, no. 1–4, pp. 335–358.

Jakovljevic, A., Grubic, A. & Pantic, N., 1991. The age of Mesozoic Continental Formations on the Western Margin of the Murzuq Basin, in M.J. Salem. O.S. Hammuda & B.A.E. Eliagoubi (Eds.), The Geology of Libya, v. IV, pp. 1583–1587.

Jakovljevic, Z., 1984. Geological map of Libya, 1:250,000, Sheet Al Awaynat NG32-12, Explanatory Booklet, Industrial Research Centre, Tripoli, p. 140.

Jalil, N.E. & Dutuit, J.M., 1996. Permian Captorhinid reptiles from the Argana Formation, Morocco. Palaeontology, v. 39, no. 4, pp. 907–918.

Jallouli, C., Mickus, K., Turki, M.M & Rihane, C., 2003. Gravity and aeromagnetic constraints on the extent of Cenozoic rocks within the Nefza-Tabarka region, northwestern Tunisia. Journal of Volcanology and Geothermal Research, v. 122, pp. 51–68.

James, N.P., Coniglio, M., Aissaoui, D.M. & Purser, B.H., 1988. Facies and geologic history of an exposed Miocene Rift-margin carbonate platform: Gulf of Suez, Egypt, AAPG Bull., v. 72, pp. 555–572.

Jansen, U., Lazreq, N., Plodowski, G., Schemm-Gregory, M., Schindler, E. &Weddige, K., 2007. Neritic_pelagic correlation in the Lower and basal Middle Devonian of the Dra Valley (Southern Anti-Atlas, Moroccan Pre-Sahara). In: R.T. Becker & W.T. Kirchgasser (eds) Devonian Events and Correlations. Geological Society, London, Special Publications, v. 278, pp. 9–37.

Jardiné, S. & Yapaudjian, L., 1968. Lithostratigraphie et palynologie du Dévonien-Gothlandien gréseux du Bassin de Polignac (Sahara), Rev. Inst. Fr. Pét., v. 23, No. 4, pp. 439–469.

Jardiné, S., Combaz, A., Magloire, L., Peniguel, G. & Vachey, G., 1974. Distribution stratigraphique des Acritarches dans le Paléozoique de Sahara Algérien. Rev. Palaeobot. Palynol., 18, pp. 99–129.

Jarvis, I., Carson, G.A., Cooper, M.K.E., Hart, M.B., Leary, P.N., Tochter, B.A., Horne, D. & Rosenfeld, A., 1988. Microfossil assemblages and the Cenomanian-Turonian (late Cretaceous) Oceanic Anoxic Event, Cretaceous Research, v. 9, London, Academic Press, pp. 3–103.

Jarvis, I., Gale, A.S., Jenkyns, H.C. & Pearce, M.A., 2006, Secular variation in Late Cretaceous carbon isotopes: A new $\delta^{13}C$ carbonate reference curve for the Cenomanian–Campanian (99.6–70.6 Ma): Geological Magazine, v. 143, pp. 561–608.

Jasinski, S.M., 2008. Phosphate Rock. U.S. Geological Survey, Mineral Commodity Summaries, January 2008, pp. 124–125.

Jasinski, S.M. & Eldridge III, R.R., 2008. Mineral Industry Surveys: Marketable Phosphate Rock in September 2008. U.S. Department of the Interior, December 2008, U.S. Geological Survey, p. 3.

Jenkins, D.A., 1990. North and Central Sinai, in R. Said (Ed.), The Geology of Egypt, Balkema, Rotterdam/Brookfield, pp. 361–380.

Jenkyns, H.C., 1988. The early Toarcian anoxic events: stratigraphic, sedimentary and geochemical evidence. American Journal of Science, v. 288, pp. 101–151.

Jenkyns, H.C., 1991. Impact of Cretaceous sea level rise and anoxic events on the Mesozoic carbonate platform of Yugoslavia, AAPG Bull., v. 75, pp. 1007–1017.

Johnson, B.A. & Nicoud, D.A., 1996. Integrated exploration for Beda Formation reservoirs in the southern Zellah Trough (West Sirt Basin, Libya), in M.J. Salem. A.S. El-Hawat & A.M. Sbeta (Eds.), The Geology of Sirt Basin, v. II, Amsterdam, Elsevier, pp. 211–222.

Jones, R.W. & Racey, A., 1994. Cenozoic stratigraphy of the Arabian Peninsula and Gulf. In M.D. Simmons (Ed.), Micropalaeontology and Hydrocarbon Exploration in the Middle East, Chapman & Hall, London, pp. 273–306.

Jones, D.F., 1975. Stratigraphy, Environments of Deposition, Petrology, Age, and Provenance of the Red Beds of the Argana Valley, Western High Atlas Mountains, Morocco. Unpub. M.Sc. Thesis, New Mexico Institute of Technology, Socorro, New Mexico, p. 159.

Jones, L., 1996. Late Cretaceous-Early Tertiary stratigraphy of southern concession 6, Sirt Basin, Libya, in M.J. Salem. A.J. Mouzughi & O.S. Hammuda (Eds.), Geology of Sirt Basin, v. I, Amsterdam, Elsevier, pp. 169–183.

Jongsma, D., van Hinte, J.E. & Woodside, J.M., 1985. Geologic structure and neotectonic of the North African continental margin south of Sicily, Marine and Petroleum Geology, v. 2, pp. 156–179.

Jongsmans, W.J., 1940. Contributions to the flora of the Carboniferous of Egypt. MedEd. Geol. Bur. Heerlen, pp. 223–229.

Jongsmans, W.J. & van der Heide, S., 1953. Contribution à l'étude de la faune et de la flore du Carbonifère inférieur de l'Egypte, Compte rendue, 19th Congr. Geol. Intern., Algiers, Sec. 2, 2, pp. 65–70.

Jordi, H.A. & Lonfat, F., 1963. Stratigraphic subdivision and problems in Upper Cretaceous and Lower Tertiary deposits in Northwestern Libya, Rev. Inst. Fr. Pétrole, v. 8, No. 10–11, pp. 114–122.

Jorry, S., 2004. The Eocene Nummulite Carbonates (Central Tunisi and NE Libya): Sedimentology, Depositional Environments, and Application to Oil Reservoirs. Unpublished PPh.D. Thesis, Faculté des Sciences de l'Université de Genève, p. 93.

Jorry, S., Caline, B. & Davaud, E., 2001. Facies associations and geometry of nummulite bodies: Example of the Kesra Plateau in Central Tunisia (NW Kairouan). Géologie Méditerranéenne, v. 28, no. 1, pp. 103–106.

Jorry, S., Davaud, E. & Caline, B., 2003. Controls on the distribution of nummulite facies: A case study from the Late Ypresian El Garia Formation (Kesra Plateau, Central Tunisia). Journal of Petroleum Geology, v. 26, no. 3, pp. 283–306.

Jossen, J.-A., 1988. Carte géologique de Zawiat Ahançal. 1/100.000. Notes et Mémoires du Service Géologique du Maroc, Rabat, no. 355.

Jouve, S., Bardet, N., Jalil, N.-E., Suberbiola, X.P., Bouya, B. & Amaghzaz, M., 2008. The oldest African Crocodylian: phylogeny, paleobiogeography, and differential survivorship of marine reptiles through the Cretaceous-Tertiary boundary. Journal of Vertebrate Paleontology, v. 28, no. 2, pp. 409–421.

Jull, A.J.T., Beck, J.W. & Burr, G.S., 2000. Isotopic evidence for extraterrestrial organic material in the Martian meteorite, Nakhla. Geochimica et Cosmochimica Acta, v. 64, no. 21, pp. 3763–3772.

Jurak, L., 1978. Geological Map of Libya, 1:250,000, Sheet Jabal Al Hasawnah (NH33-14). Industrial Research Centre, Tripoli, p. 85.

Jurak, L., 1985. Geological Map of Libya, 1:250,000, Sheet Wadi Bu Ash Shaykh. Industrial Research Centre, Tripoli, p. 113.

Jux, U., 1954. Zur Geologie des Kreidegebietes von Abu Roasch bei Kairo, N. Jb. Geol. Paläont., Abh., 100, 159–207.

Jux, U., 1983. Diagenetic silica glass (formerly related to astroblems) from the Western Desert, Egypt. Ann. Geol. Surv. Egypt, v. 13, pp. 99–108.

Jux, U. & Issawi, B., 1983. Cratonic sedimentation in Egypt during the Paleozoic, Annals Geol. Survey of Egypt, Cairo, No. 13, pp. 223–245.

Kádár, L., 1934. A study of the Sand Sea in the Libya Desert. The Geographical Journal, v. 83, no. 6, pp. 470–478.

Kaiser, S.I., Becker, R.T. & El Hassani, A., 2007. Middle to Late Famennian successions at Ain Jemaa (Moroccan Meseta). implications for regional correlation, event stratigraphy and synsedimentary tectonics of NW Gondwana. Geological Society, London, Special Publications, v. 278, pp. 237–260.

Kallenbach, H. & Hendriks, F., 1986 (Abs.). Transgressive and regressive sedimentary environments of Cretaceous to Lower Tertiary age in upper Egypt: a case study from central Wadi Qena, International Assoc. Sediment. Congress, Canberra, p. 158.

Kalvacheva, R. & Kazandiev, H., 1987 (Abs.). Lower Palaeozoic acritarchs from Ghadames Basin, Northwest Libya, Third Symp. Geol. Libya, Tripoli, p. 78.

Kamal El-Din, M.M., 2003. Petrified wood from the Farafra Oasis, Egypt. IAWA Journal, v. 24, no. 2, pp. 163–172.

Kanes, W.H., 1969. Archeology an prehistory of southwestern Fezzan, Libya, 11th Ann. Field Conference. Petroleum Expl. Soc. Libya.

Karcz, I. & Zak, I., 1968. Paleocurrents in the Triassic Arayif En-Naqa, Sinai, Israel J. Earth Sci., v. 17, pp. 9–15.

Kardoes, M.A., 1993. A seismic-stratigraphic analysis of the Satal bank, Dahra Platform (western Sirt Basin), in Sedimentary Basins of Libya, First Symposium, Geology of Sirt Basin, Earth Science Society of Libya, Tripoli, p. 27.

Karoui-Yaakoub, N., Zaghbib-Turki & Keller, G., 2002. The Cretaceous-Tertiary (K-T) mass extinction in planktic foraminifera at Elles and El Melah, Tunisia. Palaeogeography, Palaeoclimatology, Palaeoecology, v. 178, pp. 233–255.

Kassab, A.S. & Ismael, M.I., 1994. Upper Cretaceous invertebrate fossils from the area northeast of Abu Zuneima, Sinai, Egypt, N. Jb. Geol. Paläont. Abh., 191, pp. 221–249.

Kassab, A.S. & Obaidalla, N.A., 2001. Integrated biostratigraphy and inter-regional correlation of the Cenomanian-Turonian deposits of Wadi Feiran, Sinai, Egypt. Cretaceous Research, v. 22, no. 1, pp. 105–114.

Kassab, A.S., 1994. Upper Cretaceous ammonites from the El Sheikh Fadl – Ras Gharib road, Northeastern Desert, Egypt, N. Jb. Geol. Paläont. Mh. (2), pp. 108–128.

Katzir, Y., Eyal, M., Litvinovsky, B.A., Jahn, B.M., Zanvilevich, A.N., Valley, J.W., Berri, Y., Pelly, I. & Shimshilashvili, E., 2007. Petrogenesis of A-type granites and origin of vertical zoning in the Katherina pluton, Gebel Mussa (Mt. Moses) area, Sinai, Egypt. Lithos, v. 95, Issues 3–4, pp. 208–228.

Keeley, M.L. & Massoud, M.S., 1998. Tectonic controls on the petroleum geology of NE Africa, in D.J. Macgregor. R.T.J. Moody & D.D. Clark-Lowes (Eds.), Petroleum Geology of North Africa, Geological Society Spec. Pub. No. 132, London, pp. 265–281.

Keeley, M.L. & Wallis, R.J., 1991. The Jurassic system in northern Egypt, II, Depositional and tectonic regimes, J. Petrol. Geology, v. 14, pp. 49–64.

Keeley, M.L., 1989. The Paleozoic history of the western Desert of Egypt, Basin Research, v. 2, pp. 35–48.

Keeley, M.L., 1994. Phanerozoic evolution of the basins of Northern Egypt and adjacent areas, Geol. Rundsch., v. 83, pp. 728–472.

Keeley, M.L., Dungworth, G., Floyd, C.S., Forbes, G.A., King, C., McGarva, R.M. & Shaw, D., 1990. The Jurassic system in northern Egypt, I, Regional strand implications for hydrocarbon prospectivity, J. Petrol. Geol., v. 13, pp. 397–420.

Keen, M.C., Al-Sheikly, S.S.J., Elsogher, A. & Ga, A.M., 1994. Tertiary ostracods of North Africa and the Middle East, pp. 371–401, in M.D. Simmons (Ed.), Micropaleontology and Hydrocarbon Exploration in the Middle East, Chapman & Hall, London, p. 418.

Keller, G., 1988. Extinction, survivorship and evolution of planktic foraminifera across the Cretaceous-Teritary boundary at El Kef, Tunisia. Mar. Micropaleontol., v. 13, pp. 239–263.

Keller, G., Adatte, T., Stinnesbeck, W., Stueben, D., Berner, Z. & Luciani, V., 2002. Paleoecology of the Cretaceous-Tertiary mass extinction in planktic foraminifera. Palaeogeogr. Palaeoclimatol. Palaeoecol., v. 178, pp. 257–297.

Keller, G., Adatte, T., Berner, Z. Chellai, E.H. & Stueben, D., 2008. Oceanic events and biotic effects of the Cenomanian-Turonian anoxic event, Tarfaya Basin, Morocco. Cretaceous Research, v. 29, pp. 976–994.

Keller, G., Li, L. & MacLeod, N., 1995. The Cretaceous/Tertiary boundary stratotype section at El Kef, Tunisia, how catastrophic was the mass extinction?, Paleogeo., Paleoclim., Paleoecology, v. 119, pp. 221–224.

Kelly, D.C., Bralower, T.J., Zachos, J.C., Premoli Silva, I. & Thomas, E., 1996. Rapid diversification of planktonic foraminifera in the tropical Pacific (ODP Site 865) during the late Paleocene thermal maximum. Geology, v. 4, no. 5, pp. 423–426.

Kennedy, W.O., 1964. The structural differentiation of Africa in the Pan-African tectonic episode, Annual Report, Res. Inst. Afr. Geol., Leeds University, pp. 48–49.

Kerdany, M.T. & Cherif, O.H., 1990. Mesozoic, in R. Said (Ed.), The Geology of Egypt, Balkema, Rotterdam/Brookfield, pp. 407–438.

Key, R.M., Loughlin, S.C., Gillespie, Del Rio, M., M., Horstwood, M.S.A., Crowley, Q.G., Darbyshire, D.P.F., Pitfield, P.E.J. & Henney, P.J., 2008. Two Mesoarchaean terranes in the Reguibat shield of NW Mauritania. Geological Society, London, Special Publication, v. 297, no. 1, pp. 33–52.

Khaled, K.A., 1999. Cretaceous source rocks at the Abu Gharadig oil- and gasfield, northern Western Desert, Egypt, J. Petroleum Geology, v. 22, No. 4, pp. 377–395.

Khalil, B. & Meshrif, W., 1988. Hydrocarbon occurrences and structural style of the Southern Suez rift basin. 9th Exploration and Production Conf., Egyptian General Petroleum Corp. (EGPC), Cairo, Egypt, November 1988, p. 31.

Khalil, S.M. & McClay, K.R.,2002. Extensional fault-reltaed folding, northwestern Red Sea. Journal of Structural Geology, v. 24, pp. 743–762.

Khalil, M.M. & Pigaht, J., 1991. Petrophysical Approach to Description of a Producing Fractured Basement Reservoir. In: Proc. SPE Middle East Oil Show, Bahrain, 16th–19th Nov., SPE 21444, pp. 903–914.

Kilani-Mazrzoui, F., Razgallah-Gargouri, S. & Mannai-Tayech, B., 1990. The Permo-Triassic of Southern Tunisia-biostratigraphy and palaeoenvironment. Review of Palaeobotany and Palynology, v. 66, nos. 3–4, pp. 273–291.

Kilenyi, T., Traynir, P., Doherty, M. & Jamieson, G., 1988 (Abs.). Seismic stratigraphy of the offshore Nile delta, in Mediterranean Basins Conf. and Exhib., Nice France, AAPG Bull., v. 72, pp. 1009–1010.

Kilian, C.-A. & Lelubre, M., 1946. De l'age des grès de Nubie a l'orient de l'Ajjer (Sahara central), C.R. Hebd. Seanc. Acad. Sci., v. 222, pp. 233–235.

Kilian, C.-A., 1928. Sur la presence du Silurien a l'Est et au Sud de l'Ahaggar, C.R. Acad. Sci., Paris, 20 fevrier 1928.

Kilian, C.-A., 1931. Des principaux complexes continentaux du Sahara, C.R. Sommaire Société Géologique de France,

Kim, J., Wagner, T. Bachmann, M. & Kuss, J., 1999. Organic facies and thermal maturity of Late Aptian to Early Cenomanian shelf deposits, Northern Sinai (Egypt). International Journal of Coal Geology, v. 39, no. 1–3, pp. 251–278.

Kirschbaum, M.A., Schenk, C.J., Charpentier, R.R., Klett, T.R., Brownfield, M.E., Pitman, J.K., Cook, T.A. & Tennyson, M.E., 2010. Assessment of Undiscovered Oil and Gas Resources of the Nile Delta Basin Province, Eastern Mediterranean. World Petroleum Resources Project, Fact Sheet 2010–3027, May 2010.

Kleinsmeide, W.F.J. & van den Berg., N.J., 1968. Surface geology of the Jabal al Akhdar, Cyrenaica, Libya, in Geology and Archeology of Northern Cyrenaica, Libya, F.T. Barr (Ed.), Petrol. Explor. Soc. Libya, pp. 115–123.

Klemm, D., Klemm, R. & Murr, A., 2001. Gold of the Pharaohs – 6000 years of gold minning in Egypt and Nubia. J. African Earth Sciences, v. 33, pp. 643–659.

Klemm, D.D. & Klemm, R., 2001. The building stones of ancient Egypt – a gift of its geology. J. African Earth Sciences, v. 33, nos. 3–4, pp. 631–642.

Klen, L., 1974. Geological map of Libya, 1:250,000, Sheet Benghazi NI34-14, Explanatory Booklet, Industrial Research Centre, Tripoli, p. 56.

Klerkx, J., 1980. Age and metamorphic evolution of the basement complex around Jabal al'Awaynat, in M.J. Salem & M.T. Busrewil (Eds.), The Geology of Libya, v. III, London, Academic Press, pp. 901–906.

Klett, T.R., 2000a. Total Petroleum Systems of the Grand Erg/Ahnet Province, Algeria and Morocco-The Tanezzuft-Timimoun, Tanezzuft-Ahnet, Tanezzuft-Sbaa, Tanezzuft-Mouydir, Tanezzuft-Benoud, and Tanezzuft-Béchar/Abadla. U.S. Geological Survey Bulletin 2202-B. http://greenwood.cr.usgs.gov/pub/bulletins/b2202-b/.

Klett, T.R., 2000b. Total Petroleum Systems of the Illizi Province, Algeria and Libya – Tannezzuft-Illizi. U.S. Geological Survey Bulletin 2202-A, p. 14.

Klett, T.R., 2001. Total Petroleum Systems of the Pelagian Province, Tunisia, Libya, Italy and Malta – Bou Dabbous – Tertiary and Jurassic-Cretaceous Composite. U.S. Geological Survey Bulletin 2202-D, p. 27.

Klitzsch, E., 1963. Geology of the North-East flank of the Murzuk basin (Djebel ben Ghnema-Dor el Gussa area), Rev. Inst. Fr. Petrol., 18, 10, pp. 1411–1427.

Klitzsch, E. 1965. Ein profil aus dem Typusgebiet gotlandischer und devonischer Schichten des zentral Sahara, Erdoel und Kohle, pp. 605–607.

Klitzsch, E., 1966. Comments on the geology of the central parts of southern Libya and northern Chad. In J.J. Williams (Ed.), South-Central Libya and Northern Chad, a Guidebook to the Geology and Prehistory. Petroleum Exploration Society of Libya, pp. 1–17.

Klitzsch, E., 1967. Der basalt vulkanismus des Djebel Haroudj Ostfezzan/Libyen, Geologische Rundschau, v. 57, pp. 585–601.

Klitzsch, E., 1969. A stratigraphic section from the type area of Silurian and Devonian strata at western Murzuk Basin (Libya), in W.H. Kanes (Ed.), Geology, Archaeology and Prehistory of Southwestern Fezzan, Libya, Petroleum Exploration Society, Eleventh Annual Field Conference, pp. 83–90.

Klitzsch, E., 1970. Die Strukturgeschichte der Zentral-Sahara, Neue Erkenntnisse zum Bau und zur Paleogeographie eines Tafellandes, Geologische Rundschau, 59(2), pp. 459–527.

Klitzsch, E., 1971. The structural development of parts of North Africa since Cambrian time, in C. Gray (Ed.), Symposium on the Geology of Libya, U. of Libya, Tripoli, pp. 253–262.

Klitzsch, E., 1972. Problems of continental Mesozoic strata of south-western Libya (with discussion), Proc. Conf. African Geology, Regional Geology, Univ. Ibadan, Dept. Geol., pp. 483–494.

Klitzsch, E., 1983. Paleozoic formations and a Carboniferous glaciation from the Gilf Kebir-Abu Ras Area in Southwestern Egypt, J. Afr. Earth Sci., v. 1, No. 1, pp. 17–19.

Klitzsch, E., 1990a. Paleozoic, in R. Said (Ed.), The Geology of Egypt, 2nd Ed., A.A. Balkema, Rotterdam/Brookfield, pp. 393–406.

Klitzsch, E., 1990b. Paleogeographical development and correlation of continental strata (former Nubian Sandstone) in northeast Africa, J. Afr. Earth Sci., v. 10, No. 1/2, pp. 199–213.

Klitzsch, E., 1994. Geological exploration history of the Eastern Sahara. Geologische Rundschau, v. 83, no. 3, pp. 475–483.

Klitzsch, E. & Baird, D.W., 1969. Stratigraphy and paleohydrology of the Germa (Jarma) area, Southwest Libya, in W.H. Kanes (Ed.), Geology, Archaeology and Prehistory of Southwestern Fezzan, Libya, Petroleum Exploration Society, Eleventh Annual Field Conference, pp. 67–80.

Klitzsch, E. & Lejal-Nicol, A. 1984. Flora and fauna from strata in Southern Egypt and Northern Sudan (Nubian and surrounding areas), Berliner Geowiss. Abh. (A) 50, pp. 47–80.

Klitzsch, E., Gröschke, M. & Herrmann-Degen, W., 1990. Wadi Qena: Paleozoic and pre-Campanian Cretaceous strata, in R. Said (Ed.), The Geology of Egypt, Balkema, Rotterdam/Brookfield, pp. 321–332.

Klitzsch, E., Harms, J.C., Lejal-Nicol, A. & List, F.K., 1979. Major subdivision and depositional environments of Nubia strata, southwestern Egypt, AAPG Bull., v. 63, pp. 967–974.

Klitzsch, E.H. & Smetner, A., 1993. Silurian paleogeography of NE Africa and Arabia, an updated interpretation, in U. Thornweihe & H. Schandelmeier (Eds.) Geoscientific Research in Northeast Africa (expanded abstracts), Balkema, Rotterdam, pp. 341–344.

Klitzsch, E.H. & Squyres, C.H., 1990. Paleozoic and Mesozoic geological history of northeastern Africa based upon new interpretation of Nubian strata, AAPG Bull., v. 74, No. 8, pp. 1203–1211.

Klitzsch, E. & Wycisk, P., 1987. Geology of the sedimentary basins of northern Sudan and bordering areas, Berliner Geowiss. Abh. (A) 75, pp. 97–136.

Klitzsch, E., Wycisk, P. & Léjal-Nicol, A. 1989. Carboniferous of Northern Sudan and Southern Egypt, XIe Congrès Internat. de Stratigraphie et de Géologie du Carbonifère, Beijing, 1987, Compte Rendu 2, t.2, pp. 74–79.

Knipper, A., Ricou, L.-E. & Dercourt, J., 1986. Ophiolites as indicators of the geodynamic evolution of the Tethyan Ocean, Tectonophysics, v. 123, pp. 213–240.

Koeberl, C., 2000 (Abs.). Confirmation of a meteoric component in Libyan Desert Glass from osmium isotope data. 63rd Meteoritical Society Meeting, Chicago, August 28–Sept. 1.

Koeberl, C., Rampino, M.R., Jalufka, D.A. & Winiarski, D.H., 2003. A 2003 Expedition into the Libyan Desert Glass strewn field, Great Sand Sea. The Meterite Impacts.

Koehler, R.P., 1982. Sedimentary Environment and Petrology of the Ain Tobi Formation, Tripolitania, Libya, Unpub. Ph.D. thesis, Rice University, p. 286.

Kogbe, C.A., 1980. The Trans-Saharan seaway during the Cretaceous, in M.J. Salem & M.T. Busrewil (Eds.), The Geology of Libya, v. I, London, Academic Press, pp. 91–96.

Kolonic, S., Sinninghe Damsté, J.S., Böttcher, M.E., Kuypers, M.M.M., Kuhnt, W., Beckmann, B., Scheeder, G. & Wagner, T., 2002. Geochemical characterization of Cenomanian/Turonian black shales from the Tarfaya Basin (SW Morocco): relationships between palaeoenvironmental conditions and early sulphurization of sedimentary organic matter. Journal of Petroleum Geology, v. 25, no. 3, pp. 325–350.

Kolonic, S., Wagner, T., Forster, A., Sinninghe Damsté, J.S., Walsworth-Bell, B., Erba, E., Turgeon, S., Brumsack, H.J., Chellai, E.H., Tsikos, H., Kuhnt, W. & Kuypers, M.M.M., 2005. Black shale deposition on the northwest African shelf during the Cenomanian-Turonian oceanic anoxic event: climate coupling and global organic carbon burial. Paleoceanography, v. 20, pp. 1006.

Konaté, M., Lang, J., Guiraud, M., Yahaya, M., Denis, M. & Alidou, S., 2006. Un bassin extensif formé pendant la fonte de la calotte glaciaire hirnantienne : le bassin ordovico-silurien de Kandi (Nord Bénin, Sud Niger). Africa Geoscience Review, v. 13, no. 2, pp. 157–183.

Konzalova, M., 1991. Palynological studies of the Mesak Formation (Central Libya), in M.J. Salem, O.S. Hammuda & B.A. Eliagoubi (Eds.), The Geology of Libya, v. IV, Amsterdam, Elsevier, pp. 1230–1241.

Kora, M. & Jux, U., 1986. On the Early Carboniferous macrofauna from the Um Bogma Formation, Sinai, N. Jb. Geol. Palaont. Mh. (2), pp. 85–98.

Kora, M., 1995. Carboniferous macrofauna from Sinai, Egypt: biostratigraphy and paleogeography, Journal of African Earth Sciences, v. 20, no. 1, pp. 37–51.

Kora, M., Shahin, A. & Semiet, A., 1994. Biostratigraphy and paleoecology of some Cenomanian successions in west central Sinai, Egypt, N. Jb. Geol. Paläont. Mh., v. 10, pp. 597–617.

Korab, T., 1984. Sheet: Tmassah, NG33-7, Geological Map of Libya, 1:250,000. Industrial Res. Centre, Tripoli, p. 87.

Korn, D., Ebbighausen, V. & Bockwinkel, J., 2010a. The ammonoids from the Grès du Kahla supérieur of Timimoun (Middle-early Late Tournaisian; Gourara, Algeria). Fossil Recod, v. 13, no. 1, pp. 13–34. 2010.

Korn, D., Bockwinkel, J. & Ebbighausen, V., 2010b. The ammonoids from the Argiles de Teguentour of Oued Temertasset (early Late Tournaisian; Mouydir, Algeria). Fossil Recod, v. 13, no. 1, pp. 35–152.

Korn, D., Ebbighausen, V. & Bockwinkel, J., 2010c. Ammonoids from the Dalle des Iridet of the Mouydir and Ahnet (Central Sahara) and the Formation d'Hassi Sguilma of the Saoura Valley (Late Tournaisian–Early Viséan; Algeria). Fossil Recod, v. 13, no. 1, pp. 203–214.

Korn, D. Klug, C. & Riesdorf, A., 2000. Middle Famennian ammonoid stratigraphy in the Amessoui Syncline (Late Devonian; eastern Anti-Atlas, Morocco). Trav, Inst. Sci., Rabat, Série Géol. & Géogr. Phys., v. 20, pp. 69–77.

Koscec, B.G. & Gherryo, Y.S., 1996. Geology and reservoir performance of the Messlah Oil Field, Libya, in M.J. Salem. A.S. El-Hawat & A.M. Sbeta (Eds.), The Geology of Sirt Basin, v. II, Amsterdam, Elsevier, pp. 365–389.

Kostandi, A.B., 1959. The Paleozoic and Mesozoic sedimentary basins of the Egyptian region, U.A.R., First Arab Petroleum Congress, Cairo, v. 2, pp. 54–62.

Kostandi, A.B., 1963. Eocene facies maps and tectonic interpretation in the Western Desert, U.A.R. First Saharan Symposium. Rev. Inst. Fran. Petrol., special volume 53, pp. 17–28.

Kotański, Z., Gierliński, G. & Ptaszyński, T., 2004. Reptile tracks (*Rotodactylus*) from the Middle Triassic of the Djurdjura Mountains in Algeria. Geological Quarterly, Warszaw, v. 48, no. 1, pp. 89–96.

Koufos, G.D., Kostopoulos, D.S. & Vlachou, T.D., 2005. Neogene/Quaternary mammalian migrations in Eastern Mediterranean. Belg. J. Zool., v. 135, no. 2, pp. 181–190.

Kriz, J. 2008. *Algerina* gen. nov. (Bivalvia, Nepiomorphia) from the Silurian of the North Gondwana margin (Algeria), peri-Gondwanan Europe (France, Italy), Perunica (Prague Basin, Bohemia) and the Siberian Plate (Tajmyr Basin, Russia). Czech Geological Survey, Prague. Bulletin of Geosciences, v. 83, no. 1, pp. 79–84.

Kröger, B., Klug, C. & Mapes, R., 2005. Soft-tissue attachments in orthocerid and bactritid cephalopods from the Early and Middle Devonian of Germany and Morocco. Acta Palaeontologica Polonica, v. 50, no. 2, pp. 329–342.

Kroh, A. & Nebelsick, J.H., 2003. Echinoid assemblages as a tool for palaeoenvironmental reconstruction – an example from the Early Miocene of Egypt. Palaeogeography, Palaeoclimatology, Palaeoecology, v. 201, pp. 157–177.

Kröner, A., 1980. Pan African crustal evolution, Episodes, v. 2, pp. 3–8.

Kröner, A., 1985. Ophiolites and the evolution of tectonic boundaries in the later Proterozoic Arabian-Nubian shield of NE Africa and Arabia, Precambrian Res., v. 27, pp. 277–300.

Kröner, A., 1991. Tectonic evolution in the Archean and Proterozoic, Tectonophysics, v. 187, No. 4, pp. 393–410.

Kröner, A., Eyal, E. & Eyal, Y., 1990. Early Pan-African evolution of the basement around Elat, Israel, and the Sinai Peninsula revealed by single-zircon evaporation dating, and implication for crustal arates, Geology, v. 18, pp. 545–548.

Kröner, A., Greiling, R., Reischmann, T.,, I.M., Stern, R.J., Dürr, S., Krüger, J. & Zimmer, M., 1987. Pan African crustal evolution of the Nubian segment of northeastern Africa, in A. Kröner (Ed.), Proterozoic lithospheric evolution, Am. Geophys. Union Geodynamics Series, v. 15, pp. 235–257.

Kroon, D., 1998. Oxygen isotope and sapropel stratigraphy in the Eastern Mediterranean during the last 3.2 million years. In A.H.F. Robertson, Emeis, K.C., Richter, C. & Camerlenghi, A. (Eds.), Proceedings of the Ocean Drilling Program, Scientific Results, v. 160, pp. 181–189.

Krumbeck, L., 1906. beitrage zur Geologie und Palaontologie von Tripolis, Palaeontographica, v. 53, Stuttgartt, pp. 51–136.

Kucenjak, M.H., Fucek, V.P. & Mesic, I.A., 1998 (Abs.). The Upper Oligocene and Lower Miocene of the western part of the Tripolitana Basin (Libya): Biostratigraphy and paleoecology. In Tertiary to Recent Larger Foraminifera, Their Depositional Environments and Importance as Petroleum Reservoirs, Kingston University, United Kingdom.

Kuehn, M., 1996. Palaeosols as an additool in defining reservoir characterisitics of pre-Upper Cretaceous sections in As Sarah and Tuama oil fields, in M.J. Salem. A.S. El-Hawat & A.M. Sbeta (Eds.), The Geology of Sirt Basin, v. II, Amsterdam, Elsevier, pp. 251–261.

Kuhnt, W., Herbin, J.P., Thurow, J. & Wiedmann, J., 1990. Distribution of Cenomanian-Turonian organic facies in the western Mediterranean and along the adjacent Atlantic Margin. In: A.Y. Huc (Ed.): Deposition of Organic Facies. AAPG Studies in Geology, v. 40, pp. 133–160.

Kuhnt, W., Holbourn, A., Gale, A., Chellai, E. & Kennedy, W.J., 2009. Cenomanian sequence stratigraphy and sea-level fluctuations in the Tarfaya Basin (SW Morocco). Geological Soc. Am. Bulletin, v. 121, nos. 11/12, pp. 1695–1710.

Kuhnt, W., Luderer, F., Nederbragt, S., Thurow, J. & Wagner, T., 2004. Orbital-scale record of the late Cenomanian-Turonian oceanic anoxic event (OAE-2) in the Tarfaya Basin (Morocco). International Journal of Earth Sciences, v. 94, pp. 147–159.

Kuhnt, W., Nederbragt, A. & Leine, L., 1997. Cyclicity of Cenomanian-Turonian organic-carbon-rich sediments in the Tarfaya Atlantic coastal basin (Morocco). Cretaceous Research, v. 18, pp. 587–601.

Kummel, B., 1960. Middle Triassic nautiloids from Sinai, Egypt and Israel. Bulletin Museum Comparative Zoology Harvard University, v. 123, no. 7, pp. 285–302.

Kuss, J., 1986. Facies development of upper Cretaceous-lower Tertiary sediments from the Monastery of St. Antony, Eastern Desert, Egypt, Facies 15, pp. 177–194.

Kuss, J. & Bachmann, M., 1996. Cretaceous paleogeography of the Sinai Peninsula and neighbouring areas. Comptes rendus de l'Academie des Sciences, Serie II, Sciences de la Terre et des Planetes, v. 322, pp. 915–933.

Kuss, J. & Boukhary, M.A., 2008. A new upper Oligocene marine record from northern Sinai (Egypt) and its paleogeographic context, Geoarabia v. 13, no. 1, pp. 59–84.

Kuss, J. & Leppig, U., 1989. The Early Tertiary (Middle-Late Paleocene) limestones from the western Gulf of Suez, Egypt. Neues Jahrbuch für Geologie und Paläontologie, Monatshefte, v. 177. no. 3, pp. 289–332.

Kuss, J., Scheibner C., & Gietl, R., 2000a. Carbonate platform to basin transition along an Upper Cretaceous to Lower Tertiary Syrian Arc Uplift, Galala Plateaus, Eastern Desert, Egypt. GeoArabia, v. 5, pp. 405–424.

Kuss, J. & Schlagintweit, F., 1988. Facies and stratigraphy of Early to Middle Cretaceous (Late Aptian-Early Cenomanian) strata from the northern rim of the African Craton (Gebel Maghara-Sinai), Egypt. Facies, v. 19, pp. 77–96.

Kuss, J., Westerhold, T., Groß, U., Bauer, J. & Lüning, S., 2000b. Mapping of Late Cretaceous stratigraphic sequences along a Syrian Arc uplift – Examples from the Areif el Naqa, Eastern Sinai. Middle East Research Center, Ain Shams University, Earth Science Series, v. 14, pp. 171–191.

Lahondere J.C. & Magne J., 1983. L'evolution du 'domaine neritique constantinois' dans la region de Guelma (Algerie), a la fin du Secondaire et au debut du Tertiaire: consequences Paleogeographiques. C.R. Acad. Sci. Paris., v. 229, pp. 775–778.

Lahr, M.M., 2010. Saharan Corridors and Their Role in the Evolutionary Geography of 'Out of Africa I'. In J.G. Fleagle, J.J. Shea, F.E. Grine, A.L. Baden, R.E. Leakey (Eds.),

Out of Africa I: The First Hominin Colonization of Eurasia, Vertebrate Paleobiology and Paleoanthropology, pp. 27–46.

Lang, J., Kogbe, C., Alidou, S., Alzouma, K.A., Bellion, G., Dubois, D., Durand, A., Guiraud, R., Houessou, A., De Klasz, I., Romann, E., Salard-Cheboldaeff, M. & Trichet, J., 1990. Continental Terminal in West Africa, J. Afr. Earth Sci., v. 10, No. 1/2, pp. 79–99.

Lanzoni, E. & Magloire, L., 1969. Associations palynologiques et leurs applications stratigraphiques dans le Dévonien supérieur et le Carbonifère inférieur du Grand Erg occidental (Sahara algérien). Rev. IFP, no. 24, pp. 441–448.

Larnaude, M., 1933. Géologie de l'Afrique du nord. Annales de Géographie, v. 42, no. 235, pp. 81–85.

Lashhab, M.I. & West, I.M., 1991. Sedimentology and geochemistry of the Jir Formation in Jabal al Jir and the Western Sirt Basin, in M.J. Salem & M.N. Belaid (Eds.), The Geology of Libya, v. V, Elsevier, Amsterdam, pp. 1855–1869.

Laubscher, H. & Bernoulli, D., 1977. Mediterranean and Tethys. In A.E.M. Nairn, W. Kanes & F.G., Stehli (Eds.), Ocean Basins and Margins, v. IV, pp. 1–28.

Laville, E. & Piqué, A., 1991. La distension crustale atlantique et atlasique au Maroc au début du Mésozoique: le rejeu des structures hercyniennes. Bulletin de la Société géologique de France, v. 162, pp. 1161–1171.

Laville, E. & Piqué, A., 1992. Jurassic penetrative deformation and cenozoic uplift in the central High Atlas (Morocco): a tectonic model; structural and orogenic inversions, Geologische Rundschau, v. 81, pp. 1170–1578.

Laville, E., Piqué, A., Amrhar, M. & Charroud, M., et al., 2004. A restatement of the Mesozoic Atlasic Rifting (Morocco). Journal of African Earth Sciences, v. 38, pp. 145–153.

Le Herisse, A., 2002. Paleoecology, biostratigraphy and biogeography of late Silurian to early Devonian acritarchs and prasinophycean phycomata in well A1-61, Western Libya, North Africa. Review of Palaeobotany and Palynology, v. 118, no. 1–4, pp. 359–395.

Le Heron, D.P., 2007. Late Ordovician south pole record of the Anti-Atlas, Morocco. Sedimentary Geology, v. 201, pp. 93–110.

Le Heron, D.P. & Dowdeswell, J.A., 2009. Calculating ice volumes and ice flux to constrain the dimensions of a 440 Ma North African ice Sheet. Journal of the Geological Society, London, v. 166, no. 2, pp. 277–281.

Le Heron, D.P. & Howard, J., 2010. Evidence for Late Ordovician glaciation of Al Kufrah Basin, Libya. Journal of African Earth Sciences, v. 58, no. 2, pp. 354–364.

Le Heron, D.P. & Thusu, B., 2007. Prospects in Libya's mature and frontier basins. First Break. v. 25, pp. 73–78.

Le Heron, D.P. & Craig, J., 2008. First-order reconstructions of a Late Ordovician Saharan ice sheet. Journal of the Geological Society, London, v. 165, no. 1, pp. 19–29.

Le Heron, D.P., Craig, J. & Etienne, J.L., 2009. Ancient glaciations and hydrocarbon accumulations in North Africa and the Middle East. Earth Science Reviews, v. 93, nos. 3–4, pp. 47–76.

Le Heron, D.P., Ghienne, J.-F., El Houicha, M., Khoukhi, Y. & Rubino, J.-L., 2007. Maximum extent of ice sheets in Morocco during the Late Ordovician glaciation. Palaeogeography, Palaeoclimatology, Palaeoecology, v. 245, pp. 200–226.

Le Pichon, X., 1988 (Abs.). Plate tectonics of the Mediterranean, in Mediterranean Basins Conf. and Exhib., Nice France, Mediterranean Conference, AAPG Bull., v. 72, pp. 1011.

Le Roy, P. & Piqué, A., 2001. Triassic-Liassic Western Moroccan synrift basins in relation to the Central Atlantic opening. Marine Geology, v. 172, pp. 359–381.

Le Roy, P., Guillocheau, F., Pique, A. & Morabet, A.M., 1998. Subsidence of the Atlantic Moroccan margin during the Mesozoic. Canadian Journal of Earth Sciences. v. 35, no. 4, pp. 476–493.

Leblanc, M., 1981. Ophiolites précambriennes et gétes arséniés de cobalt (Bou-Azzer, Maroc). Notes Mém. Serv. Géol. Maroc, v. 280, p. 306.

Leblanc, M. & Lancelot, J.R., 1980. Interprétation géodynamique du domaine pan-Africain (Précambrien terminal) de l'Anti-Atlas (Maroc) à partir des données géologiques et géodynamiques. Canadian Journal of Earth Sciences, v. 17, pp. 142–155.

Leblanc, M. & Moussine-Pouchkine, A., 1994. Sedimentary and volcanic evolution of a Neoproterozoic continental margin (Bleida, Anti-Atlas, Morocco). Precambrian Research, v. 70, pp. 25–44.

Lefebvre, B., Noailles, F., Franzin, B, Regnault, S., Nardin, E., Hunter, A.W., Zamora, S., Van Roy, P., El Hariri, K. & Lazreq, N., 2010. Les gisements à échinodermes de l'Ordovicien supérieur de l'Anti-Atlas oriental (Maroc): un patrimoine scientifique exceptionnel à preserver. Bulletin de l'Institut Scientifique, Rabat, section Sciences de la Terre, no. 32, pp. 1–17.

Lefeld, J. & Uberna, J., 1991. A new species of *Nostoceras* (Ammonittes: Nostoceratidae) in Northern Libya and its affinities with other global finds, in M.J. Salem, O.S. Hammuda & B.A. Eliagoubi (Eds.), The Geology of Libya, v. IV, Amsterdam, Elsevier, pp. 1383–1388.

Lefranc, J. Ph., 1958. Stratigraphie des séries continentales intercalcaires au Fezzan nord-occidental (Libye), C.R. Acad. Sci., Paris, 247, pp. 1360–1363.

Lefranc, J. Ph. & Guiraud, R., 1990. The Continental Intercalcaire of northwestern Sahara and its equivalents in the neighbouring regions, J. Afr. Earth Sci., v. 10, No. 1/2, pp. 27–77.

Legrand, P., 1967. Le Dévonien du Sahara Algérien, in D.H. Oswald (Ed.), International Symposium on the Devonian System, Calgary, v. I, pp. 245–272.

Legrand, P., 1988. The Ordovician–Silurian boundary in the Algerian Sahara. Bull. Br. Mus. Natl. Hist. Geol., v. 43, pp. 171–176.

Legrand, P., 2002. Bâtir une stratigraphie: les leçons de l'étude du Paléozoïque au Sahara algérien. Comptes Rendus, Palevol, v. 1, no. 6, pp. 383–397.

Legrand, P., 2003. Paléogéographie du Sahara algérien à l'Ordovicien terminal et au Silurien inferieur, Bull. Soc. géol. France, v. 174, no. 1, pp. 19–32.

Legrand-Blain, M., 1985a. Iullemedden Basin. In C.M. Diaz (Ed.), 1985: The Carboniferous of the World, II. Australia, Indian Subcontinent & North Africa: North Africa. IUGS Publication No. 20, Instituto Geológico y Minero de España and Empresa Nacional ADARO de Investigaciones Mineras, s.a., pp. 323–325.

Legrand-Blain, M., 1985b. Taoudenni Basin. In C.M. Diaz (Ed.), 1985: The Carboniferous of the World, II. Australia, Indian Subcontinent & North Africa: North Africa. IUGS Publication No. 20, Instituto Geológico y Minero de España and Empresa Nacional ADARO de Investigaciones Mineras, s.a., pp. 327–329.

Legrand-Blain, M., 1985c. Illizi Basin. In C.M. Diaz (Ed.), 1985: The Carboniferous of the World, II. Australia, Indian Subcontinent & North Africa: North Africa. IUGS Publication No. 20, Instituto Geológico y Minero de España and Empresa Nacional ADARO de Investigaciones Mineras, s.a., pp. 329–333.

Legrand-Blain, M., 1985d. Northern Algeria. In C.M. Diaz (Ed.), 1985: The Carboniferous of the World, II. Australia, Indian Subcontinent & North Africa: North Africa. IUGS Publication No. 20, Instituto Geológico y Minero de España and Empresa Nacional ADARO de Investigaciones Mineras, s.a., pp. 334–336.

Lehmann, J., Heldt, M., Bachman, M., Mohamed, E. & Negra, H., 2009. Aptian (Lower Cretaceous) biostratigraphy and cephalopods from north central Tunisia. Cretaceous Research, v. 30, no. 4, pp. 895–910.

Lejal-Nicole, A., 1990. Fossil flora. In R. Said (Ed.), The Geology of Egypt, Rotterdam, A.A. Balkema, pp. 615–625.

Lelek, J.J., Sheperd, D.B., Stone, D.M. & Abdine, A.S., 1992. October Field, the latest giant under development in Egypt's Gulf of Suez. In M.T. Halbouty (Ed.), Giant Oil and Gas Fields of the Decade 1978–1988. AAPG Mem.54, pp. 231–249.

Lelubre, M., 1946. Sur la Paleozoique du Fezzan, C.R. Hebd. Seanc. Acad. Sci., v. 222, pp. 1403–1404.

Lelubre, M., 1946. Sur les series antecambriennes du Tibesti septentrional. C.R. Acad. Sc., v. 223, pp. 429–431, Paris.

Lelubre, M., 1948. La Paléozoique du Fazzan sudoriental, Compte Rendus Société Géologique de France, v. 18, No. 4, Paris, pp. 79–81.

Lelubre, M., 1952a. Aperçu sur la géologie du Fezzan, Bull. Serv. Carte Géol. Algérie, v. 3, pp. 109–148.

Lelubre, M., 1952b. Conditions structurales et formes de relief dans le Sahara, Travaux Inst. Rech. Sahariens, t.8, pp. 189–238.

Lemosquet, Y. & Pareyn, C., 1985. Bechar Basin. In C.M. Diaz (Ed.), 1985: The Carboniferous of the World, II. Australia, Indian Subcontinent & North Africa: North Africa. IUGS Publication No. 20, Instituto Geológico y Minero de España and Empresa Nacional ADARO de Investigaciones Mineras, s.a., pp. 306–315.

Leroy, L.W., 1953. Biostratigraphy of the Maqfi section, Egypt, Geol. Soc. Am. Mem. 54, p. 54.

Lespès R., 1931. L'Atlas d'Algérie et de Tunisie. In: Annales de Géographie, v. 40, no. 227, pp. 519–526.

Lesquer, A., Bourmatte, S., Ly, S. & Dautria, J.M., 1989. First heat flow determination from the central Sahara, relationship with the Pan-African belt and Hoggar domal uplift. J. Afr. Earth Sci., v. 9, no. 1, pp. 41–48.

Levainville, J., 1926. Phosphates algériens. Annales de Géographie, v. 34, no. 190, pp. 369–370.

Levell, B.K., Braakman, J.H. & Rutten, K.W., 1988. Oil-bearing sediments of Gondwana glaciation in Oman, AAPG Bull., v. 72, no. 7, pp. 775–796.

Lewis, C.J., 1990. Sarir field. In A.E. Beaumont & N.H. Foster (Eds.) AAPG Treatise of petroleum geology, Atlas of oil and gas fields, Structural Traps, v. 2, pp. 253–267.

Lewy, Z., 1975. The geological history of southern Israel and Sinai during the Coniacian, Isr. J. Earth Sci., v. 24, pp. 19–43.

Li, I.Q., Keller, G., Adatte, T. & Stinnesbeck, W., 2000. Late Cretaceous sea-level changes in Tunisia: a multi-disciplanry approach. J. Geol. Soc. London, v. 157, pp. 447–458.

Li, I.Q., Keller, G., & Stinnesbeck, W., 1999. The Late Campanian and Maastrichtian in northwestern Tunisia: palaeoenvironmental inferences from lithology, macrofauna and benthic foraminifera. Cretaceous Research, v. 20, pp. 231–252.

Li, Z.X., Bogdanova, S.V., Collins, A.S., Davidson, A., De Waele, B., Ernst, R.E., Fitzsimons, I.C.W., Fuck, R.A., Gladkochub, D.P., Jacobs, J., Karlstrom, K.E., Lu, S., Natapov, L.M., Pease, V., Pisarevsky, S.A., Thrane, K. &Vernikovsky, V., 2008. Assembly, configuration, and break-up history of Rodinia: A synthesis. Precambrian Research, v. 160, pp. 179–210.

Liégeois, J.P., Diombana, D. & Black, R., 1996. The Tessalit ring complex (Adrar des Iforas, Malian Tuareg shield): a Pan-African, post-collisional, syn-shear, alkaline granite intrusion. In: D. Demaiffe (Ed.), Petrology and Geochemistry of Magmatic Suites of Rocks in the Continental and Oceanic Crust, A volume dedicated to J. Michot. ULB-MRAC, Brussels, pp. 227–244.

Liégeois, J.-P., Latouche, L., Boughrara, M., Navez, J. & Guiraud, M., 2003. The LATEA metacraton (Central Hoggar, Tuareg shield, Algeria): behaviour of an old passive margin during the Pan-African orogeny. Journal of African Earth Sciences, v. 37, Issues 3–4, pp. 161–190.

Liégeois, J.-P. & Stern, R.J., 2010. Sr–Nd isotopes and geochemistry of granite-gneiss complexes from the Meatiq and Hafafit domes, Eastern Desert, Egypt: No evidence for pre-Neoproterozoic crust. Journal of African Earth Sciences, v. 57, pp. 31–40.

Lindquist, S.J., 1998. The Red Sea basin province: Sudr–Nubia (!) and Maqna (!) petroleum systems. U.S. Geological Survey Open-File Report 99-50-A, p. 34.

Lipson-Benitah, S., Rosenfeld, A., Honigstein, A., Flexer, A. & Kashai, E., 1988. The Middle Turonian Daliyya type section in Israel: biostratigraphy, palaeoenvironment and sea-level changes, Cretaceous Research, v. 9, p. 336.

Llinás Agrasar, E., 2004. Crocodile remains from the Burdigalian (lower Miocene) of Gebel Zelten (Libya). Geodiversitas, v. 26, no. 2, pp. 309–321.

Loboziak, S. & Streel, M., 1989. Middle-Upper Devonian miospores from the Ghadamis Basin (Tunisia-Libya): systematics and stratigraphy, Rev. Palaeobot Palyn., v. 58, pp. 173–196.

Longacre, M., Bentham, P., Hanbal, I., Cotton, J. & Edwards, R., 2007. New Crustal Structure of the Eastern Mediterranean Basin: Detailed Integration and Modeling of Gravity, Magnetic, Seismic Refraction, and Seismic Reflection Data. EGM 2007 International Workshop: Innovation in EM, Grav and Mag Methods: A new Perspective for Exploration, Capri, Italy, April 15–18, 2007.

Loper, D.E., 1991. Mantle plumes, in J.L. Le Mouel (Ed.), Beyond Plate tectonics, Tectonophysics, v. 187, pp. 373–384.

Lorenz, G., 1980. Late Jurassic-Early Cretaceous sedimentation and tectonics of the Murzuq Basin, southwestern Libya, in M.J. Salem & M.T. Busrewil (Eds.), The Geology of Libya, v. I, London, Academic Press, pp. 383–392.

Lotfy, H.I., Van der Voo, R., Hall, C.M., Kamel, O.A. & Abdel Aal, A.Y., 1995. Palaeomagnetism of Early Miocene basaltic eruptions in the areas east and west of Cairo, Jorunal of African Earth Sciences, v. 21, no. 3, pp. 407–419.

Lottaroli, F., Craig, J.& Thusu, B., 2009. Neoproterozoic-Early Cambrian (Infracambrian) hydrocarbon prospectivity of North Africa: a synthesis. In: J. Craig, J. Thurow, B. Thusu, A. Whitham & Y. Abuturruma, (Eds.): Global Petroleum Systems: The Emerging Potential in North Africa. Geological Society, London, Special Publications, v. 326, pp. 137–156.

Loucks, R.G., Moody, R.T.J., Bellis, J.K. & Brown, A.A., 1998b. Regional depositional setting and pore network systems of the El Garia Formation (Metlaoui Group, Lower Eocene), offshore Tunisia. In D.J. Macgregor, R.T.J. Moody & D.D. Clark-Lowes (Eds.), Petroleum Geology of North Africa. Geological Society Spec. Pub. No. 132, London, pp. 355–374.

Loucks, R.G., Moody, R.T.J., Brown, A.A. & Bellis, J.K., 1998a (Abs.). Regional Depositional model for larger foraminiferal Nummulite deposits in the Lower Eocene Metlaoui Group, Tunisia, in Tertiary to Recent Larger Foraminifera, Their Depositional Environments and Importance as Pctroleum Reservoirs, Kingston University, United Kingdom, pp. 24–25.

Love, C.F., 1994. The palynostratigraphy of the Haushi Group (Westphalian-Artinskian) in Oman, in M.D. Simmons (Ed.), Micropalaeontology and Hydrocarbon Exploration in the Middle East, pp. 23–39, Chapman & Hall, London, p. 418.

Löwner, R., Souhel, A., Chafiki, D., Canérot, J. & Klitzsch, E., 2002. Structural and sedimentologic relations between the High and the Middle Atlas of Morocco during the Jurassic time. Journal of African Earth Sciences, v. 34, nos. 3–4, pp. 287–290.

Lubeseder, S., Carr, I.D. & Redfern, J., 2003. A Third-Order Sequence Stratigraphic Framework for the Devonian of Morocco: Its Implications for Enhanced Regional Correlation of the Devonian in North Africa. AAPG Hedberg Conference, "Paleozoic and Triassic Petroleum Systems in North Africa", February 18–20, 2003, Algiers, Algeria, pp. 1–4.

Lubeseder, S., Rath, J., Rücklin, M., & Messbacher, R., 2010. Controls on Devonian hemipelagic limestone deposition analyzed on cephalopod ridge to slope section, Eastern Anti-Atlas, Morocco. Facies, v. 56, pp. 295–315.

Lubeseder, S., Redfern, J. & Boutib, L., 2009. Mixed siliclastic-carbonate shelf sedimentation-Lower Devonian sequences of the SW Anti-Atlas, Morocco, Sedimentary Geology, v. 215, nos. 1–4, pp. 13–32.

Lüning, S. & Kolonic, S., 2003. Uranium spectral gamma-ray response as a proxy for organic richness in black shales: Applicability and limitations. J. Petroleum Geology, v. 26, pp. 153–174.

Lüning, S., Adamson, K. & Craig, J., 2003a. Frasnian 'Hot Shales' in North Africa: Regional Distribution and Depositional Model. In: Arthur, T.J., Macgregor, D.S., Cameron, N. (Eds.), Petroleum Geology of Africa: New Themes and developing technologies, Geol. Soc. (London) Special Publication No. 207, pp. 165–184.

Lüning, S., Archer, R., Craig, J. & Loydell, D.K., 2003b. The Lower Silurian 'Hot Shales' and 'Double Hot Shales' in North Africa and Arabia. In: Salem, M.J., Oun, K.M., Seddiq, H.M. (eds), The Geology of Northwest Libya (Ghadamis, Jifarah, Tarabulus and Sabratah Basins, Earth Science Society of Libya, Tripoli), v. 3, pp. 91–105.

Lüning, S., Bosence, D., Gräfe, K.-U., Luciani, V. & Craig, J., 2000a. Discovery of marine Late Cretaceous carbonates and evaporites in the Kufra Basin (Libya) redefines the southern limit of the Late Cretaceous transgression. Cretaceous Research, v. 21, pp. 721–731.

Lüning, S., Craig, J., Fitches, B., Mayouf, J., Busrewil, A., El Dieb, M., Gammudi, A. & Loydell, D.K., 2000b. Petroleum source and reservoir rock re-evaluation in the Kufra Basin (SE Libya, NE Chad, NW Sudan). In: M.A. Sola & D. Worsley (Eds.), Geological Exploration in Murzuq Basin. Elsevier, Amsterdam, pp. 151–173.

Lüning, S., Craig, J., Fitches, B., Mayouf, J., Busrewil, A., El Dieb, M., Gammudi, A., Loydell, D. & McIlroy, D., 1999. Reevaluation of the petroleum potential of the Kufra Basin (SE Libya, NE Chad): Does the source rock barrier fall? Marine and Petroleum Geology, v. 16, pp. 693–718.

Lüning, S., Craig, J., Loydell, D.K., Štorch, P. & Fitches, W.R., 2000c. Lowermost Silurian 'Hot Shales' in North Africa and Arabia: Regional Distribution and Depositional Model. Earth Science Reviews, v. 49, pp. 121–200.

Lüning, S., Kolonic, S., Belhadj, E.M., Belhadj, Z., Cota, L., Baric, G. & Wagner, T., 2004a. Integrated depositional model for the Cenomanian-Turonian organic-rich strata in North Africa. Earth-Science Reviews, v. 64, pp. 51–117.

Lüning, S., Kolonic, S., Geiger, M., Thusu, B., Bell, J.S. & Craig, J., 2009. Infracambrian hydrocarbon source rock potential and petroleum prospectivity of NW Africa. In: J. Craig, J. Thurow, B. Thusu, A. Whitham & Y. Abutarruma, (Eds.): Global Neoproterozoic petroleum systems: the emerging potential in North Africa. Geol. Soc. Spec. Pub. No. 326, pp. 157–180.

Lüning, S., Kolonic, S., Loydell, D. & Craig, J., 2003c. Reconstruction of the original organic richness in weathered Silurian shale outcrops (Murzuq and Kufra basins, southern Libya). GeoArabia, v. 8, pp. 299–308.

Lüning, S., Kuss, J., Bachmann, M., Marzouk, A.M. & Morsi, A.M., 1998a. Sedimentary response to basin inversion: Mid Cretaceous – Early Tertiary pre- to syndeformational deposition at the Areif El Naqa anticline (northern Sinai, Egypt ). Facies, v. 38, pp. 103–136.

Lüning, S., Loydell, D.K., Štorch, P., Shahin, Y., & Craig, J., 2006. Origin, sequence stratigraphy and depositional environment of an upper Ordovician (Hirnantian) deglacial black shale, Jordan – Discussion . Palaeogeogr., Palaeoclimatol., Palaeoecol., v. 230, pp. 352–355.

Lüning, S., Loydell, D.K., Sutcliffe, O., Ait Salem, A., Archer, R., Zanella, E., Craig, J. & Harper, D.A.T., 2000d. Silurian – lower Devonian black shales in Morocco: Where are the organically richest horizons? J. Petroleum Geology, v. 23, no. 3, pp. 293–311.

Lüning, S. & Maksoud, H., 2011 (Abs.). Syrian Arc Tectonics and Implications for Petroleum Prospectivity in the Eastern Mediterranean. Advances in Carbonate Exploration and Reservoir Analysis. Geological Society New and Emerging Plays in the Eastern Mediterranean, 23–25 February 2011, pp. 11–12.

Lüning, S., Marzouk, A.M. & Kuss, J., 1998b. The Paleocene of Central East Sinai, Egypt: 'Sequence stratigraphy' in monotonous hemipelagites. Journal of Foraminiferal Research, v. 28, no. 1, pp. 19–39.

Lüning, S., Marzouk, A.M. & Kuss, J., 1998c. Latest Maastrichtian high frequency litho- and ecocycles from the hemipelagic of Eastern Sinai, Egypt. J. African Earth Sciences, v. 27, nos. 3/4, pp. 373–395.

Lüning, S., Marzouk, A.M., Morsi, A.M., & Kuss, J., 1998d. Sequence stratigraphy of the Upper Cretaceous of central-east Sinai, Egypt. *Cretaceous Research*, v. 19, pp. 153–196.

Lüning, S., Schulze, F., Marzouk, A.M., Kuss, J., Gharaibeh, A. & Kolonic, S., 2004b. Uranium-enriched horizons refine stratigraphic framework in Cenomanian-Turonian organic-rich strata in central Jordan and northern Tuniasia. Zeitschr. Dt. Geol. Ges., 155/1, pp. 49–60.

Lüning, S., Wendt, J., Belka, Z. & Kaufmann, B., 2004c. Temporal-spatial reconstruction of the early Frasnian anoxia in NW Africa: New field data from the Ahnet Basin (Algeria). Sedimentary Geology, v. 163, pp. 237–256.

Lüning, S., Y.M. Shahin, D. Loydell, H.T. Al-Rabi, A. Masri, B. Tarawneh & S. Kolonic (2005): Anatomy of a world-class source rock: Distribution and depositional model of Silurian organic-rich shales in Jordan and implications for hydrocarbon potential. AAPG Bull., v. 89, pp. 1397–1427.

Luger, P. & Gröschke, M., 1989. Late Cretaceous Ammonites from the Wadi Qena area in Egyptian Eastern Desert, Paleontology, v. 32, pt.2, pp. 355–407.

Luger, P., 1985. Stratigraphie der marinen Oberkreide unde des Alttertiärs im sudwestlichen Obernil-Becken (SW Ägypten) unter besonderer Berücksichtigung der Micropaläontologie, Paläoekologie, Paläogeographie, Berl. Geowiss. Abh., 63(A), p. 151.

Lustrino, M. & Wilson, M., 2007. The circum-Mediterranean anorogenic Cenozoic igneous province. Earth-Science Reviews, v. 81, pp. 1–65.

Lyons, H.G., 1894. On the stratigraphy and physiography of the Libyan Desert of Egypt. Q.J. Geol. Soc. London, v. 50, pp. 531–546.

Lyons, H.G., 1906. The Physiography of the River Nile and its Basin. Survey Department, Egypt.

Lyons, H.G., 1908. Some geographical aspects of the Nile – Discussion. Geographical Journal, v. 32, no. 5, pp. 449–475.

Lyons, H.G., 1909. The longitudinal section of the Nile. Geographical Journal, v. 34, no. 1, pp. 36–51.

Lys, M., 1985. Biostratigraphy (Foraminifera) of the Tunisian Carboniferous marine succession. In C.M. Diaz (Ed.), 1985: The Carboniferous of the World, II. Australia, Indian Subcontinent & North Africa: North Africa. IUGS Publication No. 20, Instituto Geológico y Minero de España and Empresa Nacional ADARO de Investigaciones Mineras, s.a., pp. 351–364.

Maache, N., 1995 (Abs.). Applied sequence stratigraphy to the Ordovician in Bourarhet area (Illizi Basin, southern Algeria), environment and paleogeography, in First Symposium on the Hydrocarbon Geology of North Africa, Geol. Soc. of London Petroleum Group, p. 37.

Macaulay, C.I., Beckett, D., Braithwaite, K., Bliefnick, D. and Philps, B., 2001. Constraints on diagenesis and reservoir quality in the fractured Hasdrubal Field, Offshore Tunisia. Journal of Peroleum Geology, v. 24, no. 1, pp. 55–78.

MacFadyen, W.A., 1930. The undercutting of coral reef limestone on the coasts of some islands in the Red Sea. The Geographical Journal, v. 75, no. 1, pp. 27–34.

MacFadyen, W.A., 1931. Miocene foraminifera from the Clysmic area of Egypt and Sinai: with account of the stratigraphy and a correlation of the local Miocene succession. Geological Survey of Egypt, Cairo, p. 149.

Macgregor, D.J. & Moody, R.T.J., 1998. Mesozoic and Cenozoic petroleum systems of North Africa. In D.J. Macgregor. R.T.J. Moody & D.D. Clark-Lowes (Eds.), Petroleum Geology of North Africa, Geological Society Spec. Pub. No. 132, London, pp. 201–216.

Macgregor, D.J., 1998. Giant fields, petroleum systems and exploration maturity of Algeria. In D.J. Macgregor. R.T.J. Moody & D.D. Clark-Lowes (Eds.), Petroleum Geology of North Africa, Geological Society Spec. Pub. No. 132, London, pp. 79–96.

Macgregor, D.J., R.T.J. Moody & D.D. Clark-Lowes (Eds.), 1998. Petroleum Geology of North Africa, Geological Society Spec. Pub. No. 132, London, pp. 7–68.

MacLeod, N. & Keller, G., 1991. Hiatus distributions and mass extinctions at the Cretaceous/ Tertiary boundary, Geology, v. 19, pp. 497–501.

Madi, A., Savard, M.M., Bourque, P.-A. & Chi, G., 2000. Hydrocarbon potential of the Mississippian carbonate platform, Bechar Basin, Algerian Sahara. AAPG Bulletin, v. 84, no. 2, pp. 266–287.

Magloire, P.R., 1970. Triassic gas field of Hassi R'Mel, Algeria. In: M.T. Halbouty (Ed.): Geology of Giant Petroleum Fields. AAPG Mem. No. 14, pp. 489–501.

Magnier, Ph., 1963. Etude stratigraphique dans le Gebel Nefousa et le Gebel Garian (Tripolitaine, Libye), Bulletin de la Société Géologique de France, Ser.7, v. 5, no. 2, pp. 89–94.

Mahammed, F., Läng, É., Mami, L., Mekahli, L., Benhamou, M., Bouterfa, B., Kacemi, A., Chérief, S.-A., Chaouati, H. & Taquet, P., 2005. The 'Giant of Ksour', a Middle Jurassic sauropod dinosaur from Algeria. C.R. Palevol, v. 4, no. 8, pp. 707–714.

Mahmoud, M. & Soliman, H.A., 1994. Importance of Mid Cretaceous key miospores for the age-dating of the non-marine Six Hills Formation, west of the Abu Tartur Plateau, New Valley, Egypt, N. Jb. Geol. Paläont. Mh. (3), pp. 155–171.

Mahmoud, M.S. & Essa, M.A., 2007. Palynology of some Cretaceous mudstones from southeast Aswan, Egypt: significance to regional stratigraphy. Journal of African Earth Sciences, v. 47, pp. 1–8.

Makhous, M., 2001. The Formation of Hydrocarbon Deposits in the North African Basins. Lecture Notes in Earth Sciences, 89, Springer, Berlin, p. 329.

Makhous, M. & Galushkin, Yu.I., 2003. Burial history and thermal evolution of the northern and eastern Saharan basins. AAPG Bulletin, v. 87, no. 10, pp. 1623–1651.

Makris, J. & Rihm, R., 1991. Shear-controlled evolution of the Red Sea: pull apart model, in J. Markis. P. Mohr & R. Rihm (Eds.), Red Sea: Birth and Early History of a New Oceanic Basin, Tectonophysic, v. 198, pp. 441–466.

Makris, J., Hencke, C.H., Egolff, F. & Akamaluk, T., 1991. A gravity field of the Red Sea and East Africa. In J. Makris, P. Mohr, and R. Rihm (Eds.), Red Ses: Birth and Early History of a New Oceanic Basin. Tectonophysics, v. 198, pp. 369–281.

Maky, A.F. & Saad, S.A.M., 2009. Kerogen Composition and Hydrocarbon Potentiality of Some Cretaceous to Middle Eocene Rock Units and Their Depositional Environments at abu El-gharadig basin Western Desert, Egypt. Australian Journal of Basic and Applied Sciences, v. 3, no. 4, pp. 4675–4692.

Malpas, J.A., Gawthorpe, R.L., Pollard, J.E. & Sharp, I.R., 2005. Ichnofabric analysis of the shallow marine Nukhul Formation (Miocene), Suez Rift, Egypt: implications for depositional processes and sequence stratigraphic evolution. Palaeogeography, Palaeoclimatology, Palaeoecology, v. 215, pp. 239–264. Malti et al., 2008. The development of the Carboniferous Ben-Zireg-Zousfana Trough. Geological Journal, v. 43, no. 2–3, pp. 337–360.

Malville, J., Wendorf, F., Mazhar, A. & Schild, R., 1998. Megaliths and Neolithic astronomy in southern Egypt. Nature, v. 292, pp. 488–491.

Mamet, B. & Omara, S., 1969. Microfacies of the Lower Carboniferous dolomitic limestone formation of the Um Bogma terrane (Sinai, Egypt), Contribution form the Cushman Foundation for Foraminiferal Research, v. 20, pp. 106–109.

Mamet, B., Choubert, G. & Hottinger, L., 1966. Notes sur le Carbonifere du Jebel Ouarkziz. Etude du passage du Viseen au Namurien d'apres les Foraminiferes. Notes Mémoires Service Géologique Maroc, v. 27, pp. 4–21.

Mamgain, V.D., 1980. The Pre-Mesozoic (Precambrian to Paleozoic) Stratigraphy of Libya, A Reappraisal, Industrial Research Centre, Bull. No. 14, S.P.L.A.J., Tripoli, p. 104.

Mandic, O. & Piller, W.E., 2001. Pectinid coquinas and their palaeoenvironmental implications – examples from the early Miocene of northeastern Egypt. Palaeogeogr. Palaeoclimatol. Palaeoecol., v. 172, pp. 171–191.

Manger, W.L. & Pareyn, C., 1979. New Carboniferous Dimorphoceratid ammonoids from Algeria and Arkansa. J. Paleontology, v. 55, no. 3, pp. 657–665.

Mann, K., 1975a. Geological Map of Libya, 1:250,000, Sheet Al Khums NI33-14, Explanatory Booklet, Industrial Research Centre, Tripoli, p. 88.

Mann, K., 1975b. Geological Map of Libya, 1:250,000, Sheet Misrata NI33-15, Explanatory Booklet, Industrial Research Centre, Tripoli, p. 30.

Mann, P., Hempton, M.R., Bradley, D.C. & Burke, K., 1983. Development of pull apart basins, J. Geol., v. 91, pp. 529–554.

Mansour, A.S.M., 2004. Diagenesis of Upper Cretaceous Rudist Bivalves, Abu Roash Area, Egypt: A Petrographic Study. Geologia Croatica, v. 57, no. 1, pp. 55–66.

Mansour, A.T., Barakat, M.G. & Abdel Hady, Y. el S., 1969. Marine Pliocene planktonic foraminiferal zonation south-east of Salum, Egypt, Rev. Ital. Paleont., v. 75, No. 4, pp. 833–842.

Marchetti, M., 1934. Note illustrative per un abbozzo di carta geoloiga della Cirenaica. Boll. Soc. Geol. Ital., v. 53, pt.2, pp. 309–325, Roma.

Marchetti, M. 1935a. Sulla presenza del Cretaceo medio in Cirenaica. Rend. R. Acc. Lincei, v. 21, ser.6, pt.1, pp. 25–29.

Marchetti, M. 1935b. Sulla presenza di nuvi affioamenti oligocenici a sud del Gebel Cirenaico. Rend. R. Acc. Lincei, v. 21, ser.6, pt.3, pp. 187–191.

Marie, J., Trouvé, Ph., Deforges, G. & Duphaure, Ph., 1984. Nouveaux éléments de paléo-géographie du Crétacé de Tunisie, Notes et Mémoires No. 19, Compagnie Française des Pétroles, Paris, p. 37.

Marinelli, O., 1920. Sulla morfologia della Cirenaica. Riv. Geogr. Ital., v. 27, pp. 69–86.

Marinelli, O., 1921. I problemi morfologici della Cirenaica e la sua nuova carta al 50.000 del Instituto Geografico Militare. Riv. Geogr. Ital., v. 28, pp. 168–170.

Marinelli, O., 1923. La Cirenaica geografica economica, politica. Milano.

Marks, M.A.W., Schilling, J., Coulson, I.M., Wenzel, T. & Markl, G., 2008. The Alkaline-PeralkalineTamazeght Complex, High Atlas Mountains, Morocco: Mineral Chemistry and Petrological Constraints for Derivation from a Compositionally Heterogeneous Mantle Source. Journal of Petrology, v. 49, no. 6, pp. 1097–1131.

Mart, Y. & Ben Gai, Y., 1982. Some depositional patterns at continental margin of southeastern Mediterranean Sea, AAPG Bull., v. 66, pp. 460–470.

Mart, Y. & Sass, E., 1972. Geology and origin of the manganese ore of Um Bogma, Sinai, Econ. Geol., v. 67, pp. 145–155.

Mart, Y., 1991. The Dead Sea Rift, from continental rift to incipient ocean, in A.F. Gangi (Ed.), World Rift Systems, Tectonophysics, v. 197, pp. 155–179.

Marten. R., Shann, M., Mika, J., Rothe, S. & Quist, Y., 2004. Seismic challenges of developing the pre-Pliocene Akhen Field offshore Nile Delta. The Leading Edge. April 2004. v. 23. no. 4. pp. 314–320.

Marzouk, A.M. & Lüning, S., 1998. Comparative biostratigraphy of calcareous nannofossils and planktonic foraminifera in the Paleocene of Eastern Sinai, Egypt. Neues Jahrbuch für Geologie und Paläontologie, Abhandlungen, v. 207, no. 1, pp. 77–105.

Marzouk, A.M. & Lüning, S., 2005. Calcareous nannofossil Biostratigraphy and distribution patterns in the Cenomanian-Turonian of North Africa. Grmena (Egypt), v. 1, pp. 107–122.

Marzouk, I, 1970. Rock stratigraphy and oil potentialities of the Oligocene and Miocene in the Western Desert, UAR, 7th Arab Petroleum Congr., Kuwait, 54(B-3), p. 28.

Marzouk, L. & Ben Youssef, M., 2008. Relative Sea-level Changes of the Lower Cretaceous Deposits in the Chotts Area of Southern Tunisia. Turkish Journal of Earth Sciences, v. 17, pp. 835–845.

Massa, D., 1988. Paléozoique de Libye occidentale, Stratigraphie et Paléographie. Thesis, p. 514.

Massa, D. & Collomb, G.R., 1960. Observations nouvelles sur la région d'Aouinet Ouenine et du Djebel Fezzan (Libye), XXI Congrés Intern. VTFJ, Sec.-12, Copenhaguen, pp. 65–73.

Massa, D., Coquel, R., Lozobiak, S., & Taugourdeaulantz, J., 1980. Essai de synthèse stratigraphique et palynologique du Carbonifère en Libye occidentale. Ann. Soc. Géol. Nord., 99, pp. 429–442.

Massa, D. & Delort, T., 1984. Evolution du bassin de Syrte (Libye) du Cambrien au Cretacé basal, Bull. Soc. Géol. Fr., (9), 6, pp. 1087–1096.

Massa, D., Havelicek, V. & Bonnefous, J., 1977. Stratigraphic and faunal data on the Ordovician of the Radames basin (Libya and Tunisia), Bull. Cent. Rech. Explor. Prod. Elf-Aquitaine, v. 1, pp. 3–27.

Massa, D. & Jaeger, H., 1971. Données stratigraphiques sur le silurien de l'ouest de la Libye, in Colloque ordovicien-silurien, brest 1971, Fr. Bur. Rech. Géol. Minières, Mem. no. 73, Paris, pp. 313–321.

Massa, D. & Moreau-Benoit, A., 1976. Essai de synthèse stratigraphique et palynologique du systeme Dévonien en Libye occidentale, Rev. Fr. Pet., v. 31, pp. 287–333.

Massa, D., Termier, G. & Termier, H., 1974. Le Carbonifère de Libye occidentale, stratigraphie, paléontologie, Notes et Mémoires No. 11, Compagnie Française des Pétroles, pp. 139–206.

Masse, P. & Ben-Ahmed, M., 2003. Eocene biostratigraphy and its contribution to the stratigraphic nomenclature of Metlaoui (Farwah) Group in Tarabulus Basin, NW offshore Libya. In: M.J. Salem, Khaled, M. (Eds.): The Geology of Northwest Libya, v. 1, pp. 217–223.

Mastera, L., 1985. Geological map of Libya, 1:250,000, Explanatory booklet, Sheet Maradah NH34-9, Industrial Research Centre, Tripoli, p. 121.

Matsumoto, T., 1980. Inter-regional correlation of transgressions and regressions in the Cretaceous Period, Cretaceous Research, v. 1, pp. 359–373.

Mattauer, M, Proust, F & Tapponnier, P., 1972. Major strike-slip faults of Late Hercynian age in Morocco. Nature, v. 237, pp. 160–162.

Mattauer, M., Tapponnier, P. & Proust, F., 1977. Sur les mecanismes de formation des châines intracontinentales. L'exemple des châines atlasiques du Maroc. Bull. Soc. Géol. France, v. 19, no. 3, pp. 521–526.

Mattoussi Kort, H., Gasquet, D., Ikenne, M. & Laridhi Ouazaa, N., 2009. Cretaceous crustal thinning in North Africa: Implications for magmatic and thermal events in the Eastern Tunisian margin and the Pelagic Sea. Journal of African Earth Sciences, v. 55, pp. 257–264.

Maurin, J.C. & Guiraud, R., 1990. Relationships between tectonics and sedimentation in the Barremo-Aptian intracontinental basins of Northern Cameroon, J. African Earth Sciences, v. 10, pp. 331–340.

Mayer-Eymar, K., 1883. Die Versteinerungen der Tertiären Schichten von der westlichen Insel im Birket-el-Qurun-See (Mittel Aegypten). Palaeontographica 30, pp. 67–78.

Mayer-Eymar, K., 1886. Zur Geologie Aegyptens, Viert. Natur. Gesellsch. Zurich, Bd. 31, pp. 241–267.

Mayer-Eymar, K., 1898. Systematisches Verzeichniss der Fauna des unteren Saharianum (marines Quartär) der Umgegend von Kairo, nebst Beschreibung der neuen Arten. Palaeontographica 30, pp. 61–90.

Mazhar, A. & Issawi, B., 1977. Geological Map of Libya 1:250,000, Sheet Zt. Msus NH34-3, Explanatory Booklet, Indust. Res. Cent., Tripoli, p. 80.

McBride, E.F., Abdel-Wahab, A. & El-Younsy, A.R.M., 1999. Origin of spheroidal chert nodules, Drunka Formation (Lower Eocene), Egypt. Sedimentology, v. 46, no. 4, pp. 733–756.

McBurney, C.B.M. & Hey, R.W., 1955. Prehistory and Pleistocene Geology in Cyrenaican Libya, Museum Archeology Ethnology, v. 4, Cambridge University Press, p. 316.

McBurney, C.B.M., 1967. The Haua Fteah (Cyrenaica) and the Stone Age of the South-East Mediterranean. Cambridge University Press, p. 387.

McCauley, J.F., Breed, C.S., Issawi, B., Schaber, G.G., El-Hinnawi, M. & El-Kelani, A., 1996. Spaceborne imaging radar (SIR) geologic results in Egypt (a review: 1982–1997). The Geological Survey of Egypt Special Publication No. 75, pp. 489–527.

McCauley, J.F., Breed, C., Schaber, G.G., McHugh, W.P., Issawi, B., Haynes, C.V., Jr., Grolier, M.J. & El Kilani, A., 1986. Paleodrainages of the Eastern Sahara, The radar rivers revisited (SIR A/B implications for a mid-Tertiary trans-African drainage system), Institute of Electrical and Electronic Engineers (IEEE) Transactions on Geoscience and Remote sensing GE24, No. 24, pp. 624–648.

McCauley, J.F., Schaber, G.G., Breed, C.S., Grolier, M.J., Haynes, C.V., Jr., Issawi, B., Elachi, C. & Blom, R., 1982. Subsurface valleys and geoarchaeology of the eastern Sahara revealed by shuttle radar. Science, v. 218, pp. 1004–1020.

McDonald, M.M.A. 1998. Early African Pastoralism: View from Dakhleh Oasis (South Central Egypt). Journal of Anthropological Archaeology, v. 17, pp. 124–142.

McDougall, N. & Martin, M., 1998 (Abs.). Facies and sequence stratigraphy of Upper Ordovician outcrops, Murzuq Basin, Libya, The Geological Conference on Exploration in Murzuq Basin, Sabha University, Sabha, Libya, p. 17.

McDougall, N.D., Braik, F., Clarke, P. & Kaced, M., 2003. The Upper Ordovician of the Illizi Basin, Algeria: a core study of Unit IV palaeovalleys. AAPG Hedberg Conf., "Paleozoic and Triassic Petroleum Systems in North Africa", Feb.18–20, 2003, Algiers, Algeria, p. 2.

McHugh, W.P., McCauley, J.F., Haynes, C.V., Jr., Breed, C.S., Schaber, G.G., 1988. Paleorivers and geoarchaeology in the southern Egyptian Sahara. Geoarchaeology, v. 3, pp. 1–40.

McWilliams, A. & Harbury, N., 1995 (Abs.). Sedimentology of Late Jurassic marginal marine deposits in S.E. Tunisia, an outcrop study of the M'rabtine oil reservoir. First Symposium on the Hydrocarbon Geology of North Africa, Geol. Soc. of London Petroleum Group, p. 39.

Meckelein, W., 1975. Progress in the Exploration and in the Scientific Research of the Libya Desert since Gerhard Rohlfs. Annals of the Geological Survey of Egypt, v. 5, pp. 47–60.

Medina, F., 1995. Syn- and postrift evolution of the El Jadida-Agadir basin (Morocco): constraints for the rifting models of the central Atlantic. Canadian Journal of Earth Sciences, v. 32, pp. 1273–1291.

Meehan P., Majumder, S., Sauer, R. & Wiehart, H., 1993. Sedimentology and seismic models of the Upper Sabil Formation, Block NC163, southern Ajdabiya Trough, in Sedimentary Basins of Libya, First Symposium, Geology of Sirt Basin, Earth Science Society of Libya, Tripoli, pp. 30–31.

Megerisi, M. & Mamgain, V.D., 1980a. The Upper Cretaceous-Tertiary Formations of Northern Libya, A Synthesis, Dept. Geol. Res. & Mining, Ind. Res. Center Bull. No. 12, Tripoli, p. 85.

Megerisi, M. & Mamgain, V.D., 1980b. The Upper Cretaceous-Tertiary Formations of Northern Libya, in M.J. Salem & M.T. Busrewil (Eds.), The Geology of Libya, v. I, London, Academic Press, pp. 67–72.

Megerisi, M. & Mamgain, V.D., 1980c. Al Khowaymat Formation – An enigma in the stratigraphy of Northeastern Libya, in M.J. Salem & M.T. Busrewil (Eds.), The Geology of Libya, v. I, London, Academic Press, pp. 73–90.

Mehdi, D., Cirilli, S., Buratti, N., Kamoun, F. & Trigui, A., 2009. Palynological characterisation of the Lower Carnian of the KEA5 borehole (Koudiat El Halfa Dome. Central Atlas, Tunisia). Geobios, v. 42, no. 1, pp. 63–71.

Mehdi, M., Neuweiler, F. & Wilmsen, M., 2003. Lower Liassic formations of the Central High-Atlas near Rich (Morocco): lithostratigraphic specification and basin evolution. Bulletin de la Societe Geologique de France, v. 174, no. 3, pp. 227–242.

Meissner, R., 1986. The Continental Crust, A Geophysical Approach, International Geophysics Series, v. 34, San Diego, Academic Press, p. 426.

Mekahli, L., Elmi, E. & Benhamou, M., 2004. Biostratigraphy, Sedimentology and Tectono-eustatic Events of the Lower and Middle Jurassic of the Ksour Mountains (Western Saharian

Atlas, Southern Algeria). Field Trip Guide Book: 32nd International Geological Congress, Florence – Italy, August 20–28, 2004, pp. 37–54.

Memmi, L., 1999. L'Aptien et l'Albien de Tunisie biostratigraphie à partir des ammonites. Bulletin de la Societé Géologique de France, v. 170, no. 3, pp. 303–309.

Memmi, L., Donze, P., Combémorel, R. & Le Hégarat, G., 1989. Transition from Jurassic to Cretaceous in northeast Tunisia: biostratigraphic details and distribution of facies. Cretaceous Research, v. 10, no. 2, pp. 137–151.

Memoirs relative to Egypt, written in that country during the campaigns of General Bonaparte in the years 1798 and 1799, by The Learned and Scientific Men who accompanied the French Expedition, 1800. London.

Menard, H.W., 1967. Transitional types of crust under small ocean basins, J. Geophys. Res., v. 72, pp. 3061–3073.

Menchikoff, N., 1927a. Les roches cristallines et volcaniques du centre du désert de Libye. C.R. Acad. Sci., Paris, v. 184, pp. 215–217.

Menchikoff, N., 1927b. Etude pétrographique des roches cristallines et volcaniques dans la région d'Ouenat (Désert de Libye), Bulletin de la Société Géologique de France, v. 4, séries 27, Paris, pp. 337–354.

Menchikoff, N., 1944. Les grès de Serdelés (Fazzan), Compte Rendus Acad. Science, v. 9, Paris, pp. 292–293.

Meneisy, M.Y., 1990. Vulcanicity, in R. Said (Ed.), The Geology of Egypt, Balkema, Rotterdam/ Brookfield, pp. 1557–172.

Meneisy, M.Y. & Kreuzer, H., 1974a. Potassium-argon ages of Egyptian basaltic rocks, Geol. Jb., v. 8, pp. 21–31.

Meneisy, M.Y. & Kreuzer, H., 1974b. Potassium-argon ages of nepheline syenite ring complexes in Egypt, Geol. Jb., v. 9, pp. 33–39.

Mennig, J.J. & Vittimberga, P., 1962. Application des méthodes pétrographiques a l'étude du Paléozoique ancien du Fezzan, CFP, Notes et Mémoires No. 2, Paris, p. 63.

Mennig, J.J., Vittimberga, P. & Lehmann, E.F., 1963. Sédimentologie et pétrographie de la formation Ras Hamia, Trias moyen du Nord-Ouest de la Libye, Revue de l'Institut Français du Pétrole, v. 18, No. 10/11, pp. 186–201.

Mergl, M., Massa, D. & Plauchut, B., 2001. Devonian and Carboniferous Brachiopods and Bivalves of the Djado Sub-Basin (North Niger, SW Libya). Journal of the Czech Geological Society, v. 46, nos. 3–4, pp. 169–188.

Mesaed, A.A. & Surour, A.A., 1998 (Abs.). Mineralogy and geochemistry of the Bartonian stratabound diagenetic and lateritic glauconitic ironstones of El Gedida mine, El Bahariya Oases (sic.), Egypt, in Geology of the Arab World, Fourth International Conference, Cairo University, Cairo, pp. 114–115.

Meshref, W.M., 1990. Tectonic framework. In The Geology of Egypt, edited by R. Said. A.A. Balkema, Rotterdam, pp. 113–155.

Metwalli, F. & Pigott, J.D., 2005. Analysis of petroleum system critical of the Matruh-Shushan Basin, Western Desert, Egypt. Petroleum Geoscience, v. 11, no. 2, pp. 157–178.

Metwalli, M.H. & Abd El-Hady, Y.E., 1975. Petrographic characteristics of oil-bearing rocks in Alamein oil field, significance in source-reservoir relations in northern Western Desert, Egypt, AAPG Bull., v. 59, No. 3, pp. 510–523.

Meyer, C., 2004. II. Nakhla. Astronomical Research & Exploration Science (ARES), Lyndon Johnson Space Center, Houston, Texas.

Meyer-Berthaud, B., Rücklin, M., Soria, A., Belka, Z., & Lardeux, H., 2004. Frasnian plants from the Dra Valley, southern Anti-Atlas, Morocco. Geological magazine, v. 141, no. 6, pp. 675–686.

Meyers, P.A., 2006. Paleoceanographic and paleoclimatic similarities between Mediterranean sapropels and Cretaceous black shales. Palaeogeography, Palaeoclimatology, Palaeoecology, v. 235, no. 1–3, pp. 305–320.

Michard, A., Frizon de Lamotte, D., Liégeois, J.-P., Saddiqi, O. & Chalouan, A., 2008a. Conclusion: Continental Evolution inWestern Maghreb. In A. Michard, O. Saddiqi, A. Chalouan and D. Frizon de Lamotte (Eds.), Continental Evolution: The Geology of Morocco. structure, stratigraphy, and Tectonism of the Africa-Atlantic-Mediterranean Triple Junction, Berlin, Springer-Verlag, pp. 395–404.

Michard, A., Frizon de Lamotte, D., Saddiqi, O. & Chalouan, A., 2008b. An Outline of the Geology of Morocco. In A. Michard, O. Saddiqi, A. Chalouan and D. Frizon de Lamotte (Eds.) 2008. Continental Evolution: The Geology of Morocco. structure, stratigraphy, and Tectonism of the Africa-Atlantic-Mediterranean Triple Junction, Berlin, Springer-Verlag, pp. 1–32.

Michard, A., Hoepffner, C., Soulaimani, A. & Baidder, L., 2008c. The Variscan Belt. In A. Michard, O. Saddiqi, A. Chalouan and D. Frizon de Lamotte (Eds.), Continental Evolution: The Geology of Morocco. structure, stratigraphy, and Tectonism of the Africa-Atlantic-Mediterranean Triple Junction, Berlin, Springer-Verlag, pp. 65–132.

Michard, A., Saddiqi, O., Chalouan, A. and Frizon de Lamotte, D. (Eds.), 2008d. Continental Evolution: The Geology of Morocco. structure, stratigraphy, and Tectonism of the Africa-Atlantic-Mediterranean Triple Junction, Berlin, Springer-Verlag, p. 426.

Midnat-Reyes, B., 2000. The Prehistory of Egypt; From the First Egyptians to the First Pharaohs (translated by Ian Shaw). Blackwell Publishers, Oxford, p. 328.

Migliorini, C.I., 1914. Sulla geologia dei dintorni do Tobruk. Rend. R. Accad. Naz. Lincei, Ser. 5, vol. 23, pp. 833–839.

Migliorini, C.I., 1920. Geologia e Paleontologia dei dintorni do Tobruk. Paleont. Italica, v. 26, pp. 117–156.

Mijalkovic, N., 1977a. Geological Map of Libya, 1:250,000, Sheet Al Qaddahiyah, NH33-3, Explanatory Booklet, Industrial Research Centre, Tripoli, Libya, p. 70.

Mijalkovic, N., 1977b. Geological Map of Libya, 1:250,000, Sheet Qasr Sirt, NH33-4, Explanatory Booklet, Industrial Research Centre, Tripoli, Libya, p. 36.

Miller, P.M. & Barakat, H., 1988. Geology of Safaga Concession, northern Red Sea, Egypt. Tectonophysics, v. 153, pp. 123–136.

Miller, A.K. & Furnish, W.M., 1957. Permian Ammonoids from Tunisia. Journal of Paleontology, v. 31, no. 4, pp. 705–712.

Missenard, Y., Michard, A. & Durand-Delga, M., 2008. Major Steps in the Geological Discovery of Morocco. In A. Michard, O. Saddiqi, A. Chalouan & D. Frizon de Lamotte (Eds.), Continental Evolution: The Geology of Morocco. structure, stratigraphy, and Tectonism of the Africa-Atlantic-Mediterranean Triple Junction, Berlin, Springer-Verlag, pp. 377–394.

Mitchell, L.H., 1887. Ras Gemsah and Gabal Zeit, Report on their geology and petroleum. Egypt Min. Research Dept., Cairo, p. 58.

Mitchum, R.M., Jr., Vail, P.R. & Sangree, J.B., 1977. Seismic stratigraphy and global changes of sea level, Part 6: Stratigraphic interpretation of seismic reflection patterns in depositional sequences, in C.E. Payton (Ed.), Seismic Stratigraphy – application to hydrocarbon exploration, AAPG Mem.26, pp. 117–133.

Mittlefehldt, D.W. & Reymer, P.S., 1986. Sinai granites: a discussion of their origin based on petrological and Sr.-isotopic constraints, Israel J. Earth Sci., v. 35, pp. 40–50.

Mohamed, H., 2003 (Abs.). Geology and Petroleum Perspective Offshore Northwestern Morocco. AAPG International Conference, Barcelona, Spain, September 21–24, 2003, p. 1.

Mohamed, A.A., 2004. Reservoir characterization and depositional patterns of Pliocene and Miocene in the North Nile Delta, Mediterranean Sea, Egypt. AAPG International Conference: October 24–27, 2004, Cancun, Mexico, p. 4.

Mohr, B. & Fechner, G. 1986. Eine eozäne Mikroflora (Sporomorphae und Dinoflagellaten-Zysten) aus der Südatlas- Randzone westlich Boumalne du Dadès (Marokko). Berliner Geowissenschaftliche Abhandlungen, (A), v. 66, pp. 381–414.

Moissette, P., 1997. Bryozoaires récoltés dans les unités messiniennes de sondages offshore dans le golfe de Gabès (Tunisie). Revue de Micropaléontologie, v. 40, no. 2, pp. 181–203.

Moissette, P., Cornée, .-J., Mannaï-Tayech, B., Rabhi, M., André, J.-P., Koskeridou, E. & Méon, H., 2010. The western edge of the Mediterranean Pelagian Platform: A Messinian mixed siliciclastic–carbonate ramp in northern Tunisia. Palaeogeography, Palaeoclimatology, Palaeoecology, v. 285, nos. 1–2, pp. 85–103.

Mokaddem, O., 1995 (Abs.). The characteristics and origin of conglomerates in the Basal Triassic unit of the northwestern Ghadames Basin-Algeria, in First Symposium on the Hydrocarbon Geology of North Africa, Geol. Soc. of London Petroleum Group, p. 41.

Molyneux, S.G. & Paris, F., 1985. Late Ordovician palynomorphs, J. Micropaleontology, in B. Thusu & B. Owens (Eds.), Palynostratigraphy of North-East Libya, v. 4, pt.1, pp. 11–26.

Monbaron, M., Kubler, B., & Zweidler, D., 1990. Detrital rubified sedimentation in the High Atlas Trough in the Mesozoic: attempt at lateral correlations using correspondance factor analysis. Journal of African Earth Sciences, v. 10, nos. 1/2, pp. 369–384.

Monbaron, M., Russell, D.A. & Taquet, P., 1999. *Atlasaurus imelakei* n. g., n. sp., a brachio-saurid-like sauropod from the Middle Jurassic of Morocco, C.R. Acad. Sci. Paris, Ser. IIa, v. 329, pp. 519–526.

Monestier, J. (1930) – Observation nouvelles sur les Ammonites liasiques du Moyen Atlas marocain. Notes et Mémoires du Service géologique Maroc, v. 8.

Monod, T. & Bourcart, J., 1931. L'Adar Ahnet. Contribution à l'étude physique d'un district saharien, Rev. Géogr. phys. Géol. dynam, v. 4, pp. 107–150, 223–262.

Monod, T. & Bourcart, J., 1932. L'Adar Ahnet. Contribution à l'étude physique d'un district saharien, Rev. Géogr. phys. Géol. dynam, v. 5, pp. 245–297.

Monterin, U., 1935. Sulla transformazione delle dune trasversali in longitudinali nel Sahara Libico. Atti R. Acc. Sc.di Torino, v. 70, pp. 62–80.

Moody, R., 1998a. An introduction to the lithostratigraphy of northern and central Tunisia, in Tertiary to Recent Larger Foraminifera, Their Depostional Environments and Importance as Petroleum Reservoirs, Field Trip Guide, Northern and Central Tunisia, Part 3, pp. 1–23.

Moody, R., 1998b. An introduction to the lithostratigraphy of northern and central Tunisia, in Tertiary to Recent Larger Foraminifera, Their Depositional Environments and Importance as Petroleum Reservoirs, Field Trip Guide, Northern and Central Tunisia, Part 2, pp. 21–31.

Moody, R.T.J. & Sandman, R.I., 1998 (Abs.). Nummulite-rich accumulations: Associated biota and their depositional environments, in Tertiary to Recent Larger Foraminifera, Their Depositional Environments and Importance as Petroleum Reservoirs, Kingston University, United Kingdom, p. 31.

Moody, R.T.J., 1987. The Ypresian carbonates of Tunisia – a model of foraminiferal facies distribution. In: Hart, M.B. (Ed.), Micropalaeontology of Carbonate Environments. B.M.S. Series, Ellis Horwood, Chichester, pp. 82–92.

Moody, R.T.J., Grant, G.G., Loucks, R.G., Brown, A.A. & Belis, J.K., 1998 (Abs.). A revised model for Ypresian Nummulite accumulations within the Metlaoui carbonates (Eocene) of Tunisia, in Tertiary to Recent Larger Foraminifera, Their Depositional Environments and Importance as Petroleum Reservoirs, Kingston University, United Kingdom, pp. 28–29.

Moon, F.W. & Sadek, H., 1923. Preliminary geological report on Wadi Gharandal area (western Sinai), Petroleum Research Bulletin, v. 12, Government Press, Cairo, p. 42.

Moon, F.W., 1924. Notes on the geology of Hassanein Bey's Expedition, Sollum-Darfur, 1923. The Geographical Journal, v. 64, no. 5, pp. 388–393.

Morabet, M.A., Bouchta, A. & Jabour, A., 1998. An overview of the petroleum systems of Morocco, in: D.S. MacGregor, R.T.J. Moody, D.D. Clark Lowes (Eds.), Petroleum Geology of North Africa, Geological Society, London, pp. 283–296.

Moretti, I. & Chenet, O.Y., 1987. The evolution of the Suez rift: a combination of stretching and secondary convection. Tectonophysics, v. 133, nos. 3–4, pp. 229–234.

Moretti, I. & Colleta, B., 1987. Spatial and temporal evolution of the Suez Rift subsidence, Geodynamics, v. 7, pp. 151–168.

Moretti, I., & Colleta, B., 1988. Fault block tilting: The Gebel Zeit example, Gulf of Suez: Journal of Structural Geology, v. 10, pp. 9–20.

Moretti, I., Kerdraon, Y., Rodrigo, G., Huerta, F., Griso, J.J., Sami, M., Said, M. & Ali, H., 2010. South Alamein petroleum system (Western Desert, Egypt). Petroleum Geoscience, v. 16, no. 2, pp. 121–132.

Morgan, P., 1990. Egypt in the framework of global tectonics, in R. Said (Ed.), The Geology of Egypt, Balkema, Rotterdam/Brookfield, pp. 91–111.

Morgan, V.L. & Lucas, S.G., 2002. Walter Granger, 1872–1941, Palaeontologist. New Mexico Museum of Natural History and Science, Albuquerque, Bulletin 19, p. 14.

Morgan, W.J., 1983. Hotspot tracks and the early rifting of the Atlantic, Tectonophysics, v. 94, pp. 123–139, pp. 123–139.

Morlo, M., Miller, E.R., & El-Barkooky, A.N., 2007. Creodonta and Carnivora from Wadi Moghra, Egypt. Journal of Vertebrate Paleontology, v. 27, no. 1, pp. 145–159.

Morly, C.K., 1988. The tectonic evolution of the Zoumi Sandstone, western Rif. J. Geol. Society, v. 145, no. 1, pp. 55–63.

Morly, C.K., 1992. Tectonic and sedimentary evidence for synchronous and out-of-sequence thrusting, Larache-Acilah area, Western Moroccan Rif. J. Geol. Society, v. 149, no. 1, pp. 39–49.

Morsi, A.M., Faris, M., Zalat, A. & Salem, R.F.M., 2008. Maastrichtian-Early Eocene ostracodes from west-central Sinai, Egypt – taxonomy, biostratigraphy, paleoecology and paleobiogeography. Revue de Paléobiologie, Genève, v. 27, no. 1, pp. 159–189.

Morsi, A.M. & Scheibner, C., 2009. Paleocene–Early Eocene ostracodes from the Southern Galala Plateau (Eastern Desert, Egypt): Taxonomy, impact of paleobathymetric changes. Revue de micropaleontology, v. 52, pp. 149–192.

Moseley, F., 1965. Plateau calcrete, calcreted gravels, cemented dunes and related deposits of the Maalegh-Bomba region of Libya, Zeit. Geomorph., v. 9, pp. 166–185.

Mostafa, A.R., 1999. Organic geochemistry of the Cenomanian-Turonian sequence in the Bakr area, Gulf of Suez, Egypt. Petroleum Geoscience, v. 5, no. 1, pp. 43–50.

Mottequin, B. & Legrand-Blain, M., 2010. Late Tournaisian (Carboniferous) brachiopods from Mouydir (Central Sahara, Algeria). Geological Journal, v. 45, pp. 353–374.

Mounir, H.E., 1998 (Abs.). Facies analysis and palaeoenvironmental interpretation of Shabrawet Lower/Middle Cretaceous (Barremian-Cenomanian) succession and its sequence strati-graphic applications, north Eastern Desert, Egypt, in Geology of the Arab World, Fourth International Conference, Cairo University, Cairo, pp. 141–142.

Moustafa, A.G. 1976. Block faulting in the Gulf of Suez. 5th Petroleum Exploration and Pro-duction Conference, Cairo, pp. 14–38.

Moustafa, A.R. & Khalil, M.H., 1990. Structural characteristics and tectonic evolution of north Sinai fold belts. In R. Said (Ed.), The Geology of Egypt, Balkema, Rotterdam/Brookfield, pp. 381–389.

M'Rabet, A., 1987. Stratigraphie, sédimentation et diagenèse carbonatée des series du Crétacé Inférieur de Tunisie Centrale. Ann. Mines et Géol. Répub. Tunisienne, v. 30. p. 410.

Mridekh, A., Medina, F., Mhammdi, N., Samaka, F. & Bouatmani, R., 2009. Structure of the Kasbah fold zone (Agadir Bay, Morocco). Implications on the chronology of the recent tectonics of the western High Atlas and on the seismic hazard of the Agadir area. Estudios Geológicos, v. 65, no. 2, pp. 121–132.

Mrini, Z., Rafi, A., Duthou, J.L. & Vidal, P.H., 1992. Chronologie Rb-Sr des granitoïdes hercyniens du Maroc et conséquences. Bull. Soc. Géol. Fr., v. 163, pp. 281–291.

Müehle, G., 1998. Libyan Desert Glass. Meteorite, February 1998.

Murris, R.J., 1980. Middle East: stratigraphic evolution and oil habitat, AAPG Bull., v. 64, pp. 597–618.

Muttoni, G., Scardia, G., Dennis V. & Kent, D.V., 2010. Human migration into Europe during the late Early Pleistocene climate transition. Palaeogeography, Palaeoclimatology, Palaeoecology, v. 296, pp. 79–93.

Nagy, R.M., Ghuma, M.A. & Rogers, J.W., 1976. A crustal suture and lineament in North Africa, Tectonophysics, v. 31, pp. 67–72.

Naidoo, D.D., Bloomer, S.H., Saquaque, A. & Hefferan, K., 1991. Geochemistry and significance of metavolcanic rocks from the Bou Azzer-El Graara ophiolite (Morocco). Precambrian Research, v. 53, issue 1–2, pp. 79–97.

Nairn, A.E.M. & Salaj, J., 1991. Al Gharbiyah Formation, Upper Campanian-Upper Maastrichtian (Northwest Libya), in M.J. Salem. O.S. Hammuda & B.A.E. Eliagoubi (Eds.), The Geology of Libya, v. IV, Elsevier, Amsterdam, pp. 1621–1635.

Nakhla, F.M., 1961. The iron ore deposits of El-Bahariya oasis, Egypt, Economic Geology, v. 56, pp. 1103–1111.

Nakkady, S.E., 1950. A new foraminiferal fauna from the Esna Shales and Upper Cretaceous chalk of Egypt, J. Paleontology, v. 24, No. 6, pp. 675–692.

Nakkady, S.E., 1955. The stratigraphy and geology of the district between the northern and southern Galala plateaus (Gulf of Suez coast, Egypt), Bull. Inst. Egypte, 36, pp. 254–268.

Nakkady, S.E., 1957. Biostratigraphy and inter-regional correlation of the Upper Senonian and Lower Paleocene of Egypt, J. Paleontology, v. 31, pp. 428–447.

Nardine, E., 2007. New occurrence of the Ordovician Eocrinoid Cadiocystities. Acta Palaeontologica Pol., v. 52, no. 1, pp. 17–26.

Nashaat, M., 1998. Abnormally high fluid pressure and seal impacts on hydrocarbon accumulations in the Nile Delta and North Sinai Basins, Egypt. In: B.E. Law, Ulmishek, G.F. & Slavin, V. I. (Eds.): Abnormal Pressure in Hydrocarbon Environments. AAPG Memoir 70, pp. 161–180.

Nassim, G.L., 1950. The oolithic hematite deposits of Egypt. Economic Geology, v. 45, no. 6, pp. 578–581.

Navon, O. & Ryemer, A.P.S., 1984. Stratigraphy, struture and metamorphism of Pan-African age in central Wadi Kid, Southeastern Sinai, Israel J. Earth Sci., v. 33, pp. 135–149.

Nederbragt, A.J., Thurow, J, Vonhof, H. & Brumsack, H.-J., 2004. Modellling oceanic carbon and phosphorus fluxes: implications for the cause of the late Cenomanian Oceanic Anoxic Event (OAE2). Journal of the Geological Society, v. 161, no. 4, pp. 721–728.

Neev, D., 1977. The Pelusium Line – a major transconntinental shear, Tectonophysics, v. 38, pp. 1–8.

Neev, D. & Ben-Avram, Z., 1977. The Levantine countries: The Israeli coastal region. In: Nairn, E.M. & Kanes, W.H. (Eds.), The Ocean Basins and Margins. The Eastern Mediterranean, v. 4A. Plenum Press, New York, United States, 355–377.

Negri, A., Ferretti, A., Wagner, T. & Meyers, P.A., 2009. Organic-carbon-rich sediments through the Phanerozoic: Processes, progress, and perspectives. Palaeogeography, Palaeoclimatology, Palaeoecology, v. 273, pp. 2133–217.

Negri, G., 1934. Contribuzione alla conoscenza del Paleogene del Gebel Cirenaico. Atti Soc. Ital. Sc. Natur., Milano, v. 73, pp. 237–253.

Negro, F., Agard, P., Goffé, B. and Saddiqi, O., 2007. Tectonic and metamorphic evolution of the Temsamane units, External Rif (northern Morocco): implications for the evolution of the Rif and the Betic-Rif arc. Journal of the Geological Society, v. 164, no. 4, pp. 829–842.

Nelson, H.D. (Ed.), 1979. Libya: a country study. The American University, Washington, D.C., p. 350.

Nemec, M. & Colley, G., 1998. Qarun and Beni Suef oil discoveries, Western Desert, Egypt. Houston Geological Society Bull., International Explorationists Meeting, February 1998, p. 11.

Nemyrovska, T.I., Perret-Mirousse, M.-F. & Weynat, M., 2006. The early Visean (Carboniferous) conodonts from the Saoura Valley, Algeria. Acta Geologica Polonica, v. 56, no. 3, pp. 361–370.

Nesteroff, W.D., Ryan, W.B.F., Hsü, K., Pautot, G., Wezel, F.C., Lort, J.M., Cita, M.B., Maync, W., Stradner, H. & Dumitrica, P., 1972. Evolution de la sédimentation pendant le Néogène en Méditerranée d'après les forages JOIDES-DSDP, in D.J. Stanley (Ed.), The Mediterranean Sea, a Natural Sedimentation Laboratory, Dowden, Hutchinson & Ross, Inc., Stroudsburg, Pennsylvania, pp. 47–62.

Neumayr, P., Mogessie, A., Hoinkes, G. & Puhl, J., 1996. Geologic setting of the Meatiq metamorphic core complex in the Eastern Desert of Egypt based on amphibolite geochemistry, Journal of African Earth Sciences, v. 23, no. 3, pp. 331–345.

Newbold, Lieut., 1848a. On the geology of Egypt. Q.J. Geol. Soc. London, v. 4, pp. 324–349.

Newbold, Lieut., 1848b. On the geological position of the petrified wood of the Egyptian and Libyan Deserts, with a description of the "Petrified Forest" near Cairo. Q.J. Geol. Soc. London, v. 4, pp. 349–357.

Norman, S.E. & Chase, C.G., 1986. Uplift of the shores of the western Mediterranean due to Messinian desiccation and flexural isostasy, Nature, v. 322, pp. 450–451.

North American Commission on Stratigraphic Nomenclature, 1983. North American Stratigraphic Code, AAPG Bull., v. 67, pp. 841–875.

Norton, P., 1967. Rock stratigraphic nomenclature of the Western Desert, Egypt, Gupco, internal report, p. 18.

Nouidar, M. & El Chellaï, H., 2001. Facies and sequence stratigraphy of an estuarine incised-valley fill: Lower Aptian Bouzergoun Formation, Agadir Basin, Morocco. Cretaceous Research, v. 22, no. 1, pp. 93–104.

Novovic, T., 1977. Geological map of Libya, 1:250,000, Explanatory Booklet, Sheet Nalut NH 32-4, Industrial Research Centre, Tripoli, p. 74.

Obaidalla, N.A. & Kassab, A.S., 1998 (Abs.). Stratigraphy and depositional conditions of the Khoman Formation, Bahariya Oasis, Western Desert, Egypt, in Geology of the Arab World, Fourth International Conference, Cairo University, Cairo, p. 41.

Odin, G.S. & Matter, A., 1981. De Glauconiarum origine, Sedimentology, v. 28, pp. 611–641.

Oil & Gas Journal, 1996a. Egypt western desert activity hits high gear, Nov. 4, 1996, pp. 96–9853.

Oil & Gas Journal, 1996b. N. Tunisian Sahara hosts giant Triassic, L. Paleozoic prospects, Jan. 15, pp. 52–57.

Oil & Gas Journal, 1999, Industry briefs: Oil & Gas Journal, v. 97, no. 11, p. 30.

Omar, K.Z., El Ashmawy, T., Ibrahim, M.E. & Fouad, A.S., 1988. Petrophysical evaluation of Qantara Formation, Offshore Nile Delta, Egypt. 9th Exploration and Production Conf., Egyptian General Petroleum Corp. (EGPC), Cairo, Egypt, November 1988, p. 24.

Omara, S., 1972. An early Cambrian outcrop in southwestern Sinai, Egypt, N. Jb. Geol. Paläont., Mh., v. 5, pp. 306–314.

Omara, S. & Kenawy, A., 1966. Upper Carboniferous microfossils from Wadi Araba, Eastern Desert, Egypt, N. Jb. Geol. Paläont. Abh. 124, pp. 306–314.

Omara, S. & Ouda, K., 1972. Biostratigraphy of microfacies of Miocene rocks in the northwestern desert, Egypt. 8th Arab Petrol. Conf., Algiers, Paper No. 93 (B-3), p. 16.

Omara, S. & Schultz, G., 1965. A lower Carboniferous microflora from south-western Sinai, Egypt, Palaeontographica, B, v. 117, pp. 47–58.

Omara, S. & Vangerow, E.F., 1965. Carboniferous (Westphalian) foraminifera from Abu El Darag, Eastern Desert, Egypt, Geol Mijnbouw, v. 44, pp. 87–93.

Omara, S., Mansour, H.H., Youssef, M. & Khalifa, H., 1977. Stratigraphy, paleoenvironment and structural features of the area east of Beni Mazar, Upper Egypt, Bull. Fac. Sci. Assiut University, v. 6, pp. 171–197.

Omara, S., Vangerow, E.F. & Kenawy, A., 1966. Neue Funde von Foraminiferen im Oberkarbon von Abu Darag, Ägypten, Paläont. Z., v. 40, pp. 244–256.

Oppenheim, P., 1903. Zur Kenntnis alttertiärer Faunen in Ägypten. 1. Lieferung: Der Bivalven erster Teil (Monomyaria, Heteromyaria, Homomyaria und Siphonida integripalliata). Palaeontographica 30, pp. 1–164.

Oppenheim, P., 1906. Zur Kenntnis alttertiärer Faunen in Ägypten. 2. Lieferung: Der Bivalven zweiter Teil (Gastropoda und Cephalopoda). Palaeontographica 30, pp. 165–348.

Orlebar, A.B., 1845. Some observations on the geology of the Egyptian Desert. Journal of the Bombay Branch of the Royal Asiatic Society, v. 2, pp. 229–251.

Orszag-Sperber, F. & Plaziat, J.-C., 1990. La sédimentation continentale (Oligo-Miocène) des fossés du proto-rift du NW de la Mer Rouge (Egypte). Bull. Soc. Géol. France, 8ᵉ serie, v. 6, no. 3, pp. 385–396.

Orszag-Sperber, F., Purser, B.H., Rioual, M. & Plaziat, J.-C., 1998. Post-Miocene sedimentation and rift dynamics in the southern Gulf of Suez and northern Red Sea. In: B.H. Purser & D. Bosence (Eds.), Sedimentary and tectonic evolution of rift basins: The Red Sea-Gulf of Aden. Chapman & Hall, London, pp. 427–448.

Osman, A.F., El Kalioubi, B., Gaafar, A.Sh. & El Ramly, M.F., 2001. Provenance, geochemistry and tectonic setting of Neoproterozoic molasse-type sediments, Northern Eastern Desert, Egypt. Annals of the Geological Survey of Egypt, v. 24, pp. 93–114.

Osterlund, T., Conteras, J.A., Holm, C., Barrat, J.A., Jahn, B.M., Amosse, J., Rocchia, R., Keller, F., Poupeau, G.R., & Diemer, E., 1997. Geochemistry and origin of Libyan Desert glasses. Geochimica et Cosmochimica Acta, v. 61, no,9, pp. 1953–1959.

Ouanaimi, H. & Lazreq, N., 2008. The 'Rich' group of the Draa Basin (Lower Devonian, Anti-Atlas, Morocco): an integrated sedimentary and tectonic approach. In N. Ennih & J.-P. Liégeois (Eds.): The Boundaries of the West African Craton. Geological Society, London, Special Publications, 297, pp. 467–482.

Ouanaimi, H. & Petit, J.-P., 1992. La limite sud de la chaîne hercynienne dans le Haut Atlas marocain: reconstitution d'un saillant non déformé. Bulletin de la Société géologique de France, v. 163, pp. 63–72.

Ouazani-Touhami, A. & Chalouan, A., 1995. La distension de l'Oligocène-early Burdigalian extensional stages in the Ghomarides nappes (internal Rif), Maroc. Geogaceta, v. 17, pp. 113–116.

Oujidi, M. & Elmi, S., 2000. Evolution de l'architecture des monts d'Oujda (Maroc oriental) pendant le Trias et au debut du Jurassique. Bulletin de la Societe Géologique de France, v. 171, no. 2, pp. 169–179.

Oujidi, M., Azzouz, O. & Elmi, S., 2006. Synsedimentary tectonics of the Triassic carbonate formation of the Oujda Mountains (Eastern Meseta, Morocco): geodynamic implications. Geological Society, London, Special Publications, v. 262, no. 1, pp. 75–85.

Ould Bagga, M.A., Abdeljaouad, S. & Mercier, E., 2006. The Tunisian "zone des nappes": a slightly inverted mesocenozoic continental margin (Taberka/Jendouba. northwestern Tunisia). Bulletin de la Societe Geologique de France, v. 177, no. 3, pp. 145–154.

Oun, Kh.M., Hertogen, J., Pasteels, P. & Sanders, I.S., 1998b (Abs.). Geochemistry and K-Ar dating of the Tertiary undersaturated volcanic province of Jabal Al Hasawinah, Libya, in Geology of the Arab World, Fourth International Conference, Cairo University, Cairo, p. 91.

Oun, Kh.M., Liégeois, J.P. & Daly, S., 1998a (Abs.). Geology, geochemistry and geochronology of the Pan-African Jabal Al-Haswinah granites: evidence for reworking of Early Proterozoic crust, in Geology of the Arab World, Fourth International Conference, Cairo University, Cairo, p. 90.

Ouzegane, K., Liégeois, J.-P. & Kienast, J.R., 2003. The Precambrian of Hoggar, Tuareg shield: history and perspective. Journal of African Earth Sciences, v. 37, Issues 3-4, pp. 127–131.

Overweg, A., 1851. Geognostische Bemerkungen auf der Reise von Philippvillle uber Tunis nach Tripoli von hier nach Marzuch in Fezzan. Z. Dt. Geol. Ges., v. 3, pp. 93–102.

Owen, (Richard), Prof., 1875. On fossil evidence of a Sirenian mammal (*Eotherium aegyptia-cum*, Owen) from the nummulitic Eocene of the Mokattam cliffs, near Cairo. Q.J. Geol. Soc. London, v. 31, pp. 100–105.

Pachur, H.-J., 1980. Climatic history of late Qin southern Libya and the Western Libyan Desert, in M.J. Salem & M.T. Busrewil (Eds.), The Geology of Libya, v. 3, London, Academic Press, pp. 781–788.

Pachur, H.-J., 1996. Reconstruction of palaeodrainage systems in Sirt Basin and the area surrounding the Tibisti Mountains: Implications for the hydrological history of the region, in M.J. Salem, A.J. Mouzughi & O.S. Hammuda (Eds.), The Geology of Sirt Basin, v. I, Amsterdam, Elsevier, pp. 157–166.

Pachur, H.-J. & Hoelzmann, P., 2000. Late Quaternary palaeoecology and palaeoclimates of the eastern Sahara. Journal of African Earth Sciences, v. 30, no. 4, pp. 929–939.

Pachur, H.-J. & Rottinger, F., 1997. Evidence for a large extended paleolake in the Eastern Sahara as revealed by spaceborne Radar Lab Images. Remote Sensing Environ., v. 61, pp. 437–440.

Pachur, H.-J. & Wunnermann, B., 2002. Late Pleistocene lake deposits of the Great Sand Sea of Egypt. Annals of Geomorphology, v. 126, pp. 75–96.

Paillou, P., El Barkooky, A., Barakat, A., Malezieux, J.-M., Reynard, B., Dejax, J. & Heggy, E., 2004. Discovery of the largest crater impact field on Earth in the Gilf Kebir region, Egypt. C.R. Geoscience, 336, pp. 1491–1500.

Paillou, P., Schuster, M., Tooth, S., Farr, T., Rosenqvist, A., Lopez, S. & Malezieux, J.-M., 2009. Mapping of a major paleodrainage system in eastern Libya using orbital imaging radar: The Kufra River. Earth and Planetary Science Letters, v. 277, pp. 327–333.

Pallas, P., 1980. Water resources of the Socialist People's Libyan Arab Jamahiriya, in M.J. Salem & M.T. Busrewil (Eds.), The Geology of Libya, v. 3, London, Academic Press, pp. 539–594.

Pallister, J.S., Stacey, J.S., Fischer, L.B. & Wayne, R.P., 1987. Arabian Shield ophiolites and Late Proterozoic microplate accretion, Geology, v. 15, pp. 320–322.

Pancoast, R.D., Crawford, N., Magness, S., Turner, a., Jenkyns, H.C. & Maxwell, J.R., 2004. Further evidence for the development of photic-zone euxinic conditions during Mesozoic oceanic anoxic events. Journal of the Geological Society, v. 161, pp. 353–364.

Paris, F., 1988. Late Ordovician and Early Silurian chitinozoans from central and southern Cyrenaica, in A. El-Arnauti, B.Owens & B. Thusu (Eds.), Subsurface Palynostratigraphy of Northeast Libya, Benghazi, Libya, Garyounis University Publications, pp. 61–71.

Paris, F., 1990. The Ordovician chitinozoan biozones of the Northern Gondwana Domain, Review of Palaeobotany and Palynology, v. 66, pp. 181–209.

Paris, F., Bourahrouh, A. & Herisse, A.L., 2000. The effect of the final stages of the Late Ordovician glaciation on marine palynomorphs (Chitinozoans, acritarchs, leiosperes) in well NI-2 (NE Algerian Sahara). Review Palaeobotany and Palynology, v. 113, pp. 87–104.

Paris, F., Elaouad-Debbaj, Z., Jaglin, J.C., Massa, D. & Oulebsir, L., 1995. Chitinozoans and Late Ordovician glacial events on Gondwana, in C. Cooper. M.L. Droser & S. Finney (Eds.), Ordocivician Odyssey, short papers for the 7th, ISOS, Las Vagas, pp. 171–176.

Paris, F., Richardson, J.B., Riegel, W., Streel, M. & Vanguestaine, M., 1985. Devonian (Emsian-Famennian) palynomorphs, in B. Thusu & B. Owens (Eds.), Palynostratigraphy of North-East Libya, J. Paleontology, v. 4, part 1, pp. 49–81.

Parizek, A., Klen, L. & Röhlich, P., 1984. Geological map of Libya, 1:250,000, Sheet Idri NG 33-1, Explanatory Booklet, Industrial Research Centre, Tripoli, p. 108.

Parlak, O., Delaloye, M. & Bingöl, E., 1996. Mineral chemistry of ultramafic and mafic cumulates as an indicator of the arc-related origin of the Mersin ophiolite (southern Turkey), Geol. Rundsch., v. 85, pp. 647–661.

Parona, C.F., 1914. Per la geologia della Tripolitania. Atti Reale Accademia delle Scienze di Torino, 50, S. 16, Turin, p. 26.

Parona, C.F., 1928. Spigoloture paleontologiches. 2. Faunetta a rudiste dell Zavia di Assaba (Garian) in Tripolitania. 3. Serie Eocenica del pozzo di Sleiaia, alla quota 350 in Cirenaica. Boll. R. Ufficio Geol. Itl., v. 53, no. 8, pp. 3–7.

Parsons, M.G., Zagaar, A.M. & Curry, J.J., 1980. Hydrocarbon occurrences in the Sire Basin, Libya. In A.D. Miall (Ed.), Facts and Principles of World Petroleum Occurrence, CSPG Mem. No. 6, pp. 723–732.

Patton, T.L., Moustafa, A.R., Nelson, R.A. & Abdine, A.S., 1994. Tectonic evolution and structural setting of the Gulf of Suez Rift, in S.M. Landon (Ed.), Interior Rift Basins, AAPG Mem. No. 59, pp. 9–56.

Peck, J.M. & Horscroft, T.R., 2005. 'Bottom-up' analysis identifies eastern Mediterranean prospects,. Offshore Magazine, Pennwell, p. 3.

Pedley, H.M., House, M.R. & Waugh, B., 1976. The geology of Malta and Gozo, Proc. Geol. Ass., v. 87, No. 3, pp. 325–341.

Pegram, W.J., Register, J.K., Jr., Fullagar, P.D., Ghuma, M.A. & Rogers, J.J.W., 1976. Pan-African ages from a Tibesti massif batholith, southern Libya, Earth and Planetary Sciene Letters, v. 30, pp. 123–128.

Pérez-Belzuz, F., Alonso, B. & Ercilla, G., 1997. History of mud diapirism and trigger mechanisms in the Western Alboran Sea. Tectonophysics, v. 282, pp. 399–422.

Perrin, C., 2000. Changes of palaeozonation patterns within Miocene coral reefs, Gebel Abu Shaar, Gulf of Suez, Egypt. Lethaia, v. 33, no. 4, pp. 253–268.

Perry, S.K. & Schamel, S., 1990. The role of low-angle normal faulting and isostatic response in the evolution of the Suez rift, Egypt. Tectonophysics, .174, pp. 159–173.

Peryt, D. & Wyrwicka, K., 1991. The Cenomanian-Turonian Oceanic Anoxic Event in SE Poland, Cretaceous Research, v. 12, pp. 65–80.

Pesce, A., 1968. Gemini space photographs of Libya and Tibesti. A geological and geographical analysis. Petroleum Exploration Society, Special Publication, p. 81.

Peucat, J.-J. & Bossière, G., 1991. Age U-Pb fini-hercynien de la ceinture mylonitique de haute pression-haute temperature en Grande Kabylie (Algérie). CR Acad. Sci. Paris, v. 313, II, pp. 1261–1267.

Petrocchi, C., 1934. Iritrovamenti faunistici di Es-Sahabi. Riv. Delle colonie, v. 3, Bologne, pp. 773–842.

Petrocchi, C., 1941. Il giacimento fossilifero di Sahabi. Boll. Soc. Geol. Ital., v. 60, pt.1, pp. 107–114, Roma.

Petrocchi, C., 1943. Sahabi, eine neue Seite in der Geschichte der Erde. N.Jb. Min. Pal., Abt. B., Hefte 1, pp. 1–9.

Petters, S.W., 1979. West African cratonic stratigraphic Sequences. Geology, v. 7, no. 11, pp. 528–531.

Petters, S.W., 1991. Regional Geology of Africa. Lecture Notes in Earth Sciences 40, Berlin, Springer-Verlag, p. 722.

Peybernès, B., 1991. The Jurassic of Tunisia: an attempt at reconstruction of the south Neotethyan margin during and after the rifting phase. In M.J. Salem, M.T. Busrewil & A.M. Ben Ashour (Eds.), The Geology of Libya, v. 1V, Amsterdam, Elsevier, pp. 1681–1709.

Peybernès, B., Alméras, Y., Ben Youssef, M., Kamoun, F., Meloo, J., Rey, J. & Zargouni, F., 1985. Nouveaux elements de datation dans le Jurassique du Sud-Tunisien (Plate-forme saharienne). C.R. Acad. Sci., Paris, v. 3000, ser.II, no. 3, pp. 113–118.

Peypouquet, J.P., Grousset, F. & Mourguiart, P., 1986. Paleooceanography of the Meogen sea based on ostracods of northern Tunisian continental shelf between the late Cretaceous and early Paleogene. Geologische Rundschau, v. 75, no. 1, pp. 159–174.

Pfalz, R., 1938. Beiträge zur Geologie von Italienisch-Libyen. Z. Dtsch Geol. Ges., v. 90, pp. 473–500.

Pfannenstiel, M., 1953. die Entstehung der ägyptischen Oasendepressionen. Das Quartär der Levante II. Abh. Akad. Wiss. Liter. Math.-Naturw. Kl., Nr.7, pp. 344–406.

Philip, J., Floquet, M. et al., 2000. Late Cenomanian. In: Dercourt, J., et al. (Eds.), Atlas Peri-Tethys, Palaeogeographical maps. CCGM/CGMW, Paris, map. 14.

Phillip, G., Imam, M.M. & Abdel Gawad, G.I., 1997. Planktonic foraminiferal biostratigraphy of the Miocene sequence in the area between Wadi El-Tayiba and Wadi Sidri, west central Sinai, Egypt, Journal of African Earth Sciences, v. 25(3), pp. 435–451.

Pickford, M., 1991. Biostratigraphic correlation of the Middle Miocene Mammal locality of Jabal Zaltan, Libya, in M.J. Salem, O.S. Hammuda & B.A. Eliagoubi (Eds.), The Geology of Libya, v. IV, Amsterdam, Elsevier, pp. 1483–1490.

Pickford, M., Miller, E.R., & El–Barkooky, A.N. 2010. Suidae and Sanitheriidae from Wadi Moghra, early Miocene, Egypt. Acta Palaeontologica Polonica, v. 55, no. 1, pp. 1–11.

Pierobon, E.S.T., 1991. Contribution to the stratigraphy of the Murzuq basin, SW Libya, in M.J. Salem & M.N. Belaid (Eds.), The Geology of Libya, v. V, Elsevier, Amsterdam, pp. 1767–1783.

Pierre, C., Rouch, J.-M. and Blanc-Valleron, M.-M., 1998. Sedimentological and stable isotope changes at the Messinian/Pliocene boundary in the Eastern Mediterranean (Holes 968A, 969A, and 969B). In A.H.F. Robertson, Emeis, K.C., Richter, C. and Camerlenghi, A. (Eds.), Proceedings of the Ocean Drilling Program, Scientific Results, v. 160, pp. 3–7.

Pietersz, C.R., 1968. Proposed nomenclature for rock units in Northern Cyrenaica, in Geology and Archeology of Northern Cyrenaica, Libya, F.T. Barr (Ed.), Petrol. Explor. Soc. Libya, pp. 125–130.

Piqué, A., 2001. Geology of Northwest Africa, translated by M.S.N. Carpenter. Beiträge zur regionalen geologie der Erde, Band 9, Berlin, Gebrüder Borntraeger, p. 310.

Piqué, A., 2003. Evidence for an important extensional event during the Latest Proterozoic and Earliest Paleozoic in Morocco, C.R. Geoscience, v. 335, pp. 865–868.

Piqué, A., Bouabdelli, M., Soulaimani, A., Youbi, N. & Iliani, M., 1999. Les conglomérats du P III (Néoprotérozoïque supérieur) de l'Anti Atlas (Sud du Maroc): molasses panafricaines, ou marqueurs d'un rifting fini-protérozoïque? C.R. Acad. Sci., Paris, Sér. IIa, v. 328, pp. 409–414.

Piqué, A., Ait Brahim, L., Ait Ouali, R., Amrhar, M., Charroud, M., Gourmelen, C., Laville, Rekhiss, F. & Tricart, P., 1998. Evolution structurale des domains atlasiques du Maghreb au Meso-Cenozoique. le role des structures heritees dans la deformation du domaine atlasique de l'Afrique du Nord. Bulletin de la Societe Geologique de France, v. 169, no. 6, pp. 797–810.

Piqué, A. & Michard, A., 1989. Moroccan Hercynides. a synopsis. the Paleozoic sedimentary and tectonic evolution at the northern margin of West Africa. American Journal of Science, v. 289, pp. 286–330.

Pivnik, D.A., Ramzy, M., Steer, B.L., Thorseth, J., El Sisi, Z., Gaafar, I., Garing, J.D. & Tucker, R.S., 2003. Episodic growth of normal faults as recorded by syntectonic sediments, July oil field, Suez rift, Egypt. AAPG Bull., v. 87, no. 6, pp. 1015–1030.

Plaziat, J.-C., Montenat, C., Orszag-Sperber, F., Philobbos, E. & Purser, B.H., 1990. Geodynamic significance of continental sedimentation during initiation of the NW Red Sea rift (Egypt), J. African Earth Sciences, v. 10, pp. 355–360.

Plumb, K.A., 1991. New Precambrian time scale, Episodes, v. 14, No., pp. 139–140.

Pomel, A., 1877. Les grès dits nubiens sont de plusieurs âges . Bull. Soc. Géol. Fr., Paris, sér. 3, t.4, pp. 524–528.

Pomel, A., 1878. Sur la géologie de la Syrte. Bull. Soc. Géol. Fr., Paris, sér. 3, t.6, pp. 217–224.

Pomeyrol, R., 1968. "Nubian Sandstone". AAPG Bull., v. 52, no. 4, pp. 589–600.

Pomeyrol, R., 1969. A catharis on the term Nubian Sandstone, in W.H. Kanes (Ed.), Geology, Archaeology and Prehistory of Southwestern Fezzan, Libya, Petroleum Exploration Society, Eleventh Annual Field Conference, pp. 131–137.

Portolano, P., Schein, L. & Simonnot, A., 2000. 3-D geological modeling of Birsa field offshore Tunisia. World Oil, June 2000.

Pouclet, A., Ouazzani, H. & Fekkak, A., 2008. The Cambrian volcano-sedimentary formations of the westernmost High Atlas (Morocco): Their place in the geodynamic evolution of the west African Palaeo-Gondwana northern margin. Geological Society, London, Special Publications, v. 297, pp. 303–327.

Powers, R.W., 1962. Arabian Upper Jurassic carbonate reservoir rocks, in W.E. Ham (Ed.), Classification of Carbonate Rocks, a symposium, AAPG Mem.1, pp. 122–192.

Poyntz, I., 1995 (Abs.). Hydrocarbon potential of the Tadrart and Ouan Kasa formations (Lower Devonian), Ghadames Basin, NW Libya, in First Symposium on the Hydrocarbon Geology of North Africa, Geol. Soc. of London Petroleum Group, p. 45.

Pratsch, J.-C., 1996. Oil and gas potential of the Prerif Forland Basin onshore Northern Morocco. Journal of Petroleum Geology, v. 19, no. 2, pp. 199–214.

Pratz, E., 1883. Eocäne Korallen aus der libyschen Wüste und Aegypten. Palaontographica 30, pp. 219–238.

Preat, A., El Hassani, A. & Mamet, B., 2008. Iron bacteria in Devonian carbonates (Tafilalt, Anti-Atlas, Morocco). Facies, v. 54, pp. 107–120.

Prior, S.W., 1976. Matruh Basin, possible failed arm of Mesozoic crustal rift, Proc. Fifth EGPC Exploration Seminar, Cairo, p. 11.

Protic, D. 1985. Geological map of Libya. 1:250,000, Sheet Tikiumit NG 32-7, Explanatory Booklet, Industrial Research Centre, Tripoli, p. 120.

Purdy, E.G. & McGregor, D.S., 2003. Map compilations and syntheis of Africa's petroleum basins and systems. In: Arthur, T.J., Macgregor, D.S., Cameron, N. (Eds.), Petroleum Geology of Africa: New Themes and developing technologies, Geol. Soc. (London) Spec. Publ., v. 207, pp. 1–8.

Puri, S. & Aureli, A., 2005. Transboundary Aquifers: A Global Program to Assess, Evaluate, and Develop Policy. Ground Water, v. 43, no. 5, pp. 661–668.

Qarbous, A., Medina, F., & Hoepffner, C., 2003. Le bassin de Tizi n'Test (Haut Atlas, Maroc): example d'evolution d'un segment oblique au rift de l'Atlantique central au Trias. Can. J. Eart Sci., v. 40, pp. 949–964.

Quaas, A., 1902. Beitrag zur kenntiniss der Fauna der obersten Kreidbildungen in der libyschen Wüste (Overwegschichten und Blätterthone). Palaeontographica 30, pp. 153–336.

Quennell, A.M., 1958. The structural and geomorphic evolution of the Dead Sea Rift, Q.J. Geol. Soc. London, v. 114, pp. 1–24.

Quesada, S., Figari, E., Bolatti, N., Craik, D., Arnez, R., Jones, M., Jauregui, J.M., Serrano & Arregui, J., 2003 (Abs.). A comparative analysis of Paleozoic petroleum systems of Illizi and Murzuq basins (Algeria and Libya). In "Paleozoic and Triassic Petroleum Systems in North Africa", AAPG Hedberg Conference, February 18–20, Algiers, Algeria, p. 4.

Rabeh, T.T., 2003. Magneto-tectonic studies on the northern part of Sinai Peninsula, Egypt. Acta Geologica Universitatis Comenianae, no. 58, pp. 65–72.

Racey, A. Bailey, H.W., Beckett, D., Ghallagher, L.T., Hampton, M.J. & McQuilken, J., 2001. The petroleum geology of the Early Eocene El Garia Formation, Hasdrubal Field, Offshore Tunisia. J. Petroleum Geology, v. 24, no. 1, pp. 29–53.

Racey, A., 1994. Biostratigraphy and palaeobiogeographic significance of Tertiary nummulitids (foraminifera) from northern Oman, in M.D. Simmons (Ed.), Micropalaeontology and Hydrocarbon Exploration in the Middle East, pp. 343–367, Chapman & Hall, London, p. 418.

Racey, A., 2001. A review of Eocene nummulite accumulations, structures, formation and reservoir potential. J. Petroleum Geology, v. 24, no. 1, pp. 79–100.

Rachidi, M., Neuweiler, F. & Kirkwood, D., 2009. Diagenetic-geochemical patterns and fluid evolution history of a Lower Jurassic petroleum source rock, Middle Atlas, Morocco. J. Petroleum Geology, v. 32, no. 2, pp. 111–128.

Radulovic, P., 1984. Geological map of Libya, 1:250,000, Sheet Wadi Tanezzuft NG 32-11, Explanatory Booklet, Industrial Research Centre, Tripoli, p. 114.

Ragab, A.I., 1991. On the origin of the compositional variations of the post-collisional granitoids in arc-terranes and suture zones, Eastern Desert, Egypt, J. Afr. Earth Sci., v. 13, No. 3/4, pp. 333–341.

Ragab, A.I. & El-Gharabawi, R.I., 1989. Wadi El-Hudi migmatites, East of Aswan, Egypt, a geological study and some geotectonic implications for the Eastern Desert of Egypt, Precambrian Res., v. 44, pp. 67–79.

Ragab, A.I., Menesy, M.Y. & Diab, M.M., 1989. Petrology and petrogensis of the older and younger granitoids of Wadi Beizah Area, Central Eastern Desert, Egypt, J. Afr. Earth Sci., v. 9, no. 2, pp. 303–315.

Rajlich, P., 1983. Geology of Oued Mekta, a Mississippi valley-type deposit, Touissit-Bou Beker region, eastern Morocco. Economic Geology, v. 78, no. 6, pp. 1239–1254.

Rally Energy, 2007. Rally Energy Reports 104 Million BOE Proved and Probable Oil and Gas Reserves and Provides Operational Update. Press Release, February 15, 2007, p. 3.

Ramos, Emarzo, M., De Gibert, J.M., Tawengi, K.S., Khoja, A.A. & Bolatti, D.N. 2006. Stratigraphy and Sedimentology of the Middle Ordovician Hawaz Formation, Murzuq Basin, Libya. AAPG Bull., v. 90, pp. 1309–1336.

Ratschiller, L.K., 1967. Sahara, Correlazione geologico-litostratigrafiche fra Sahara centrale e occidentale, Memoire del Museo Tridentino di Scienze Naturali, v. 15, No. 1, Trieste, pp. 53–293.

Raynal, J.-P., Alaoui, S., Geraads, D., Magoga, L. & Mohi, A., 2001. The earliest occupation of North Africa: the Moroccan perspective, Quat. Int., v. 75, pp. 65–75.

Raynal, J.-P., Sbihi Alaoui, Magoga, L., Mohib, A. & Zouak, M., 2002. Casablanca and the earliest occupation of North Atlantic Morocco, Quaternaire, v. 13, pp. 65–77.

Reali, S. & Ronchi, P., 1998 (Abs.). Sedimentological model of Nummulitic limestone, El Garia Formation (central Tunisia and offshore Libya), in Tertiary to Recent Larger Foraminifera: Their Depositional Environments and Importance as Petroleum Reservoirs, Kingston University, United Kingdom, pp. 41–42.

Reali, S., Ronchi, P. & Borromeo, O., 2003. Sedimentological model of the El Garia Formation (NC41 Offshore Libya). In Salem, M.J., Oun, K.M. (Eds.), The Geology of North-West Libya. Sedimentary Basins of Libya, Earth Science Society of Libya, Tripoli, pp. 69–97.

Reclus, E., 1893. The Earth and Its Inhabitants, Africa, Volume II: North-West Africa. D. Appleton and Company, New York, p. 504.

Regagba, A., Mekahli, L., Benhamou, M., Hammadi, N. & Zekri, A., 2007. Discovery of dinosaurs footprints (theropods and sauropods) in the Ksour's Sandstone from Early Cretaceous in the El Bayadh Area (Central Saharian Atlas, Algeria). Bulletin du Service Géologique National, v. 18, no. 2, pp. 141–159.

Remack-Petitot, M.-L., 1960. Contribution à l'étude du Gothlandien du Sahara, bassin d'Adrar Reggane et de Fort-Polignac, Bull. Soc. géol. France, v. 7, no. 2, pp. 230–239.

Rey, J., Canérot, j., Peybernes, B., Taj-Edddine, K. & Thieuloy, J.P., 1988. Lithostratigraphy, biostratigraphy and sedimentary dynamics of Lower Cretaceous deposits of the northern side of the western High Atlas (Morocco). Cretaceous Research, v. 9, pp. 141–159.

Reyment, R.A., 1963. Studies on Nigerian Upper Cretaceous and Lower Tertiary Ostracods, Part 2, Danian, Paleocene and Eocene Ostracoda, Stockholm Contrib. Geol., 10, p. 286.

Reyment, R.A., 1966. Studies on Nigerian Upper Cretaceous and Lower Tertiary Ostracoda, III, Stratigraphical, paleoecological and biometrical conclusions, Stockholm Contrib. Geol., 14, Stockholm, p. 131.

Reyment, R.A., 1971. L'histoire de la mer transcontinentale saharienne pendant le Cenomanie, Bulletin de la Société Géologique de France, 7, 13, No. 5–6, pp. 528–531.

Reyment, R.A., 1981. The Ostracoda of the KalambFormation (Paleocene), northwestern Nigeria, Bulletin the Geological Institutions of the University of Uppsala, N.S., v. 9, Uppsala, pp. 51–65.

Reyment, R.A. & Reyment, E.R., 1980. The Paleocene Trans-Saharan transgression and its ostracod fauna. In M.J. Salem & M.T. Busrewil (Eds.), The Geology of Libya, v. I, London, Academic Press, pp. 245–254.

Reymer, A.P.S., 1983. Metamorphism and tectonics of a Pan-African terrain in southeastern Sinai, Precambrian Res., v. 19, pp. 225–238.

Reymer, A.P.S., 1984. Metamorphism and tectonics of a Pan-African terrain in southeastern Sinai – A Reply, Precambrian Res., v. 24, pp. 189–197.

Reymer, A.P.S. & Yogev, A., 1983. Stratigraphy and tectonic history of the Southern Wadi Kid metamorphic complex, southeastern Sinai, Israel J. Earth Sci., v. 32, pp. 105–116.

Richard, G.W., & Vita-Finzi, C., 1982. Marine deposits 35,000–25,000 years old in the Chott Djerid, southern Tunisia. Nature, v. 295, pp. 54–55.

Richardson, M. & Arthur, M.A., 1988. The Gulf of Suez – northern Red Sea Neogene rift: a quantitative basin analysis. Marine and Petroleum Geology, v. 5, pp. 247–270.

Richardson, M., Arthur, M.A., Quinn, J.S., Whelan, J.K. & Katz, B.J., 1988 (Abs.). Depositional setand hydrocarbon source potential of the Miocene Gulf of Suez syn-rift evaporites, Mediterranean Basin Conf. and Exhib., Nice, France, AAPG Bull., v. 72, No. 8, pp. 1020.

Richter, A & Schandelmeier, H., 1990. Precambrian basement inliers of Western Desert geology, petrology and structural evolution, in R. Said (Ed.), The Geology of Egypt, Balkema, Rotterdam/Brookfield, pp. 185–200.

Richter, A., 1986. Geologie der metamorphen und magmatischen Gesteine im Gebiet zwischen Gebel Uweinat und Gebel Kamel, SW Ägypten/NW-Sudan, Berliner Geowiss. Abh., 73(A), pp. 1–201.

Ricou, L.E., Marcoux, J. & Whitechurch, H., 1984. The Mesozoic organization of the Taurides: one or several ocean basins? In J.E. Dixon & A.H.F. Robertson (Eds.): The Geological Evolution of the Eastern Mediterranean. Geol Soc. Spec. Pub. No. 17, Blackwell, pp. 349–359.

Ries, A.C., Shackleton, R.M., Graham, R.H. & Fitches, W.R., 1983. Pan-African structures, ophiolites and mélange in the Eastern Desert of Egypt, a traverse at 26°N, J. Geol. Soc. London, v. 140, pp. 75–95.

Rigassi, D.A., 1969. "Nubian Sandstone", discussion, should the term "Nubian Sandstone" be dropped? AAPG Bull., v. 53, pp. 183–184.

Rigassi, D.A., 1970. Comments to discussions, AAPG Bulletin, v. 54, pp. 531–532.

Rigby, J.K., Newell, N.D. & Boyd, D.W., 1979 (Abs.). Marine Permian rocks of Tunisia. Bulletin of the American Association of Petroleum Geologists, v. 63, no. 3, p. 516.

Rihm, R., Makris, J. & Müller, L., 1991. Seismic surveys in the Northern Red Sea: asymmetric crustal structures, in J. Markis. P. Mohr & R. Rihm (Eds.), Red Sea: Birth and Early History of a New Oceanic Basin, Tectonophysics, v. 198, pp. 279–295.

Ritmann, A., 1958. Geosynclinal volcanism, ophiolites and Barramiya rocks, Egypt J. Geol., v. 2, pp. 61–65.

Rizzini, A., 1975. Sedimentary sequences of lower Devonian sediments (Uan Caza Formation), South Tunisia, in R.N. Ginsburg (Ed.), Tidal Deposits, Springer, Berlin, pp. 187–195.

Rizzini, A., Vezzani, F., Cococcetta, V. & Milad, G., 1978. Stratigraphy and sedimentation of a Neogene-Quaternary section of the Nile Delta area, A.R.E., Marine Geology, v. 27, No. 3/4, pp. 327–348.

Robaszynski, F., Caron, M., Dupuis, C., Amedro, F. & Gonalez Donoso, J.-M., 1990. A tentative integrated stratigraphy in the Turonian of central Tun isia – formations, zones and sequential stratigraphy in the Kalaat Senan area. Bull. du Centre de Recherche et d'Explorationnd'-Elf-Aquitaine, v. 14, pp. 213–284.

Robaszynski, F., Gonzales, J.M., Linares, D., Amédro, F., Caron, M., Dupuis, C., Dhondt, A. & Gartner, S., 2000. Le Crétacé Supérieur de la Région de Kalaat Senan, Tunisie Centrale. Lithobiostratigraphie intégrée: zones d'ammonites, de foraminifères planctoniques et de nannofossiles du Turonien supérieur au Maastrichtien. Bulletin des Centres de Recherche Exploration Production Elf-Aquitaine, Pau, v. 22, no. 2, pp. 359–490.

Robaszynski, F. & Mzoughi, M., 2010. The Abiod at Ellès (Tunisia): stratigraphies, Campanian-Maastrichtian boundary, correlation [L'Abiod d'Ellès (Tunisie): stratigraphies, limite Campanien-Maastrichtien et corrélation]. Carnets de Géologie/Notebooks on Geology, Brest, Article 2010/04. (CG2010_A04), p. 55.

Robert-Charrue, C. & Bukhard, M., 2005. The Anti-Atlas forld belt of Morocco: Variscan inversion tectonics and interference pattern of an "intracratonic" basin. 3rd Swiss Geoscience Meetiing, Zurich, pp. 59–60.

Roberts, R.M., 1970. Amal Field, Libya. In M.T. Halbbouty (Ed.), Geology of Giant Petroleum Fields. AAPG Memoir 14, pp. 439–448.

Robertson, A.H.F. & Dixon, J.E., 1984. Introduction: aspects of the geological evolution of the Eastern Mediterranean, in A.H.F. Robertson & J.E. Dixon (Eds.), The Geological Evolution of the Eastern Mediterranean, Geol. Society Spec. Pub. No. 17, London, pp. 1–74.

Robertson, A.H.F., 1994. Role of tectonic facies concept in orogenic analysis and its application to Tethys in the Eastern Mediterranean region, Earth Sci. Review, v. 37, pp. 139–213.

Robison, V.D., 1994. Source rock characterization of the Late Cretaceous Brown Limestone of Egypt. In B. Katz (Ed.), Petroleum Source Rocks, Heidelberg, Springer-Verlag, pp. 265–281.

Robinson, p. & Wiman, S.K., 1976. A revision of the stratigraphic subdivisionof the Miocene rocks of sub-Dorsale Tunisia. Notes Serv. Géol., Tunis, v. 42, pp. 71–86.

Robson, D.A., 1971. The structure of the Gulf of Suez (Clysmic) rift, with special reference to the eastern side. Journal of the Geological Society, v. 127, pp. 247–276.

Rocci, G., 1965. Essai d'interprétation de mesures géochronologique – la structure de l'Ouest Africain, Sci. Terre Nancy, v. 10, pp. 461–478.

Roddaz, M., Brusset, S., Soula, J.-C., Beziat, D., Ben-Abou, M., Debat, P., Driouch, Y., Christophoul, F., Ntarmouchant, A. & Deramound, J., 2002. Foreland basin magmatism in the western Moroccan Meseta and geodynamic interferences, Tectonics, v. 21, no. 5, pp. 1043.

Rodgers, M.R., Beahm, D.C. & Touati, M.A., 1990, Discovery of the Ezzaouia and Robbana accumulations, Gulf of Gabes, Tunisia. AAPG Bulletin, v. 74, p. 750.

Rodríguez-Tovar, F.J., Uchman, A., Martín-Algarra, A. & O'Dogherty, L., 2009. Nutrient spatial variation during intrabasinal upwelling at the Cenomanian-Turonian oceanic event in the westernmost Tethys: An ichnological and facies approach. Sedimentary Geology, v. 215, pp. 83–93.

Roe, D.A., Olsen, J.W., Underwood, J.R. & Giegengack, R.T., 1982. A handaxe of Libyan Desert Glass. Antiquity, v. 56, pp. 406–410.

Rogers, J.J.W., Ghuma, M.A., Nagy, R.M., Greenberg, J.K. & Fullagar, P.D., 1978. Plutonism in Pan-African belts and the geologic evolution of northeastern Africa, Earth Planet. Sci. Lett., v. 39, pp. 109–117.

Rognon, P., 1980. Comparison between the Late Quaternary Terraces around Atakor and Tibisti, in M.J. Salem & M.T. Busrewil (Eds.), The Geology of Libya, v. III, Amsterdam, Elsevier, pp. 815–821.

Rognon, P., 1987. Late Quaternary climatic reconstruction for the Maghreb (North Africa). Palaeogeography, Palaeoclimatology, Palaeoecology, v. 58, pp. 11–34.

Rohlfs, G., 1865–1866. Letter from M. Gerhard Rohlfs. Proceedings of the Royal Geographical Society of London, v. 10, no. 2, pp. 69–70.

Rohlfs, G., 1875. Dr. Rohlfs' exploration of the Libyan Desert. Journal of the American Geographical Society of New York, v. 7, pp. 171–173.

Röhlich, P., 1974. Geological map of Libya, 1:250,000, Explanatory Booklet, Sheet Al Bayda, NH 34-15, Industrial Research Centre, Tripoli, p. 70.

Röhlich, P., 1979. Geological map of Libya, 1:250,000, Explanatory Booklet, Sheet Ghadames NH 32-7, Industrial Research Centre, Tripoli, 63p, 21 photo.

Röhlich, P., 1980. Tectonic development of Al Jabal al Akhdar. In M.J. Salem & M.T. Busrewil (Eds.), The Geology of Libya, v. III, London, Academic Press, pp. 923–931.

Röhlich, P., Salaj, J. & Tröger, K.-A., 1996. Palaeontological dating of the pre-Campanian unconformity in the Ghawt Sas area. In M.J. Salem, A.J. Mouzughi & O.S. Hammuda (Eds.), Amsterdam, Elsevier, pp. 265–285.

Röhlich, P. & Youshash, B.M., 1991. The Ghadamis Fault – A disputed structure in NW Libya. In M.J. Salem. A.M. Sbeta & M.R. Bakbak (Eds.), The Geology of Libya, v. VI, Amsterdam, Elsevier, pp. 2371–2380.

Rohling, E.J. & Hilgen, F.J., 1991. The eastern Mediterranean climate at times of sapropel formation: a review, Geologie en Mijnbow, v. 70, pp. 253–264.

Rohrback, B.G., 1983. Crude oil geochemistry of the Gulf of Suez. In M. Bjoroy et al. (Eds.), Advances in Organic Geochemistry 1981, Chichster, John Wiley & Sons, pp. 39–48.

Rolland, M.G., 1880. Sur le terrain cretacé du Sahara septentrionale. Bull. Soc. Géol. Fr., v. 9, pp. 508–551.

Rolland, M.G., 1890. Chemin de Fer Transsaharien. Géologie du Sahara algérien et aperçu géologique sur le Sahara de l'Ocean Atlantique à la Mer Rouge., Paris, Imprimerie Nationale, 276 p.

Roman, F. & Solignac, M., 1934. Découverte d'un gisement de Mammifères pontiens à Douaria (Tunisie septentrionale). Comptes-Rendus de l'Académie des Sciences de Paris, v. 199, pp. 1649–1659.

Rose, E.P.F., 2004. Napoleon Bonaparte's Egyptian campaign of 1798: the first military operation assisted by geologists. Geology Today, v. 20, no. 1, pp. 24–29.

Rosenbaum, G., Lister, G.S. & Duboz, C., 2002. Relative motions of Africa, Iberia and Europe during Alpine orogeny. Tectonophysics, v. 359, pp. 117–129.

Rosenfeld, A., Gerry, E. & Honigstein, A., 1987. Jurassic ostracodes from Gebel Maghara, Sinai, Egypt, Revista Espan. Micropaleontologia, v. 19, pp. 251–280.

Rossi, C., Kälin, O., Arribas & Tortosa, A., 2002. Diagenesis, provenance and reservoir quality of Triassic TAGI sandstones from Ourhoud field, Berkine (Ghadames) Basin, Algeria. Marine and Petroleum Geology, v. 19, pp. 117–142.

Rossi, M.E., Tonna, M. & Labrash, M. 1991. Latest Jurassic-Early Cretaceous deposits in the subsurface of the eastern Sirt Basin (Libya): Facies an relationships with tectonics and sea-level changes. In M.J. Salem, A.M. Sbeta & M.R. Bakbak (Eds.), The Geology of Libya, v. VI, Amsterdam, Elsevier, pp. 2211–2225.

Rossignol-Strick, M., Nesteroff, W., Olive, P. & Vergnaud-Grazzini, C., 1982. After the deluge: Mediterranean stagnation and sapropel formation, Nature, v. 295, pp. 105–110.

Rouchy, J.-M., 1986. Les évaporites miocène de la Méditerranée et de la mer Rouge et leurs enseignements pour l'interprétation des grandes accumulations évaporatiques d'origine marine, Bulletin de la Société Géologique de France, 8, II, no. 3, pp. 511–520.

Rouchy, J.-M., Bernet-Rollande, M.C., Maurin, A.-F. & Monty, C., 1983. Signification sédimentologique et paléogéographique des divers types de carbonates bioconstruits associés aux évaporites du Miocène moyen près de Gebel Esh Mellaha (Egypte), C.R. Acad. Sci., II, v. 296, pp. 457–462.

Rouchy, J.-M., Iaccarino, S.M., Gennari, R., Vitale, F.P., Lucchi, F.R., 2006. Clastic vs. primary precipitated evaporites in the Messinian Sicilian basins. R.C.M.N.S. Interim Colloquium "The Messinian salinity crisis revisited-II", Parma (Italy), 7th-9th September 2006. Acta Naturalia de «L'Ateneo Parmense», v. 42, no. 4, 125–199.

Rouvier, H., 1977. Géologie de l'Extrême Nord Tunisien: tectoniques et paléogéographies superposées à l'extrémité orientale de la chaîne nord maghrébine. Thèse es Sci. Uni. P. & M. Curie, Paris VI, p. 703.

Rouvier, H., 1985. Géologie de l'extrême nord tunisien: tectoniques et paléogéographies superposées à l'extrémité orientale de la chaine nord-maghrébine. Thèse de Doctorat es Sciences Naturelles. Ann. Mines Géol., Tunis, v. 29, p. 427.

Roveri, M. & Manzi, V., 2006. The Messinian salinity crisis: Looking for a new paradigm? Palaeogeography, Palaeoclimatology, Palaeoecology, v. 238, pp. 386–398.

Roveri, M., Lugli, S., Manzi, V., Gennari, R., Iaccarino, S.M., Grossi, F. & Taviani, M., 2006. The record of Messinian events in the Northern Apenines Foredeep Basins. R.C.M.N.S. Interim Colloquium "The Messinian salinity crisis revisited-II" Parma (Italy), 7th–9th September 2006 Acta Naturalia de "L'Ateneo Parmense", v. 42, no. 3, pp. 47–123.

Ruban, D.A., 2007. Paleozoic palaeogeographic frameworks of the greater caucasus, a large gondwana-derived terrane: Consequences from the new tectonic model. Natura Nascosta, no. 34, pp. 16–27.

Ruban, D.A., Zorina, S.O. & Conrad, C.P., 2010. No global-scale transgressive–regressive cycles in the Thanetian (Paleocene): Evidence from interregional correlation. Palaeogeography, Palaeoclimatology, Palaeoecology, v. 295, pp. 226–235.

Rubino, J.-L., Deynoux, M., Ghienne, J.-F., Moreau, J., Blanpied, C., Lafont, F., Andres-Calatrava, R., Galeazzi, Mynth, T. & Sommer, F., 2003. Late Ordovician glaciations in Northern Gondwana, reappraisal and petroleum complications. AAPG Hedberg Conf., "Paleozoic and Triassic Petroleum Systems in North Africa", Feb.18–20, 2003, Algeries, Algeria, p. 4.

Russegger, J., 1837. Kreide und Sandstein, Einfluss von Granit auf letzeren Porphyry, Grunsteine, etc., in Ägypten und Nubien, bis nach Sennar. Neues Jahrb. Min., pp. 665–669.

Ryan, W.B.F. & Hsü, K.J. et al., 1973. Initial Reports of the Deep Sea Drilling Project, 13, Washington, U.S. Government Printing Office, p. 1447.

Ryan, W.B.F., 1978. Messinian badlands on the southeastern margin of the Mediterranean Sea, Marine Geology, v. 27, No. 3/4, pp. 349–363.

Sabaou, N., Lawton, D.E., Turner, P. & Pilling, D., 2005. Floodplain deposits and soil classification: The prediction of channel sand distribution within the Triassic Argilo-greseux Inferieur (TAG-I), Berkine Basin, Algeria. J. Petroleum Geology, v. 28, no. 3, pp. 223–239.

Saber, H., El Wartiti, M., Hmich, D. & Schneider, J.W., 2007. Tectonic evolution from the Hercynian shortening to the Triassic extension in the Paleozoic sediments of the Western High Atlas (Morocco). Journal of Iberian Geology, v. 33, no. 1, pp. 31–40.

Sadek, H., 1926. The geography and geology of the district between Gebel Ataqa and El Galala El Bahariya (Gulf of Suez). Geological Survey of Egypt, Cairo, Paper No. 40, p. 120.

Sadek, H., 1959. The Miocene in the Gulf of Suez region, Egypt. Geological Survey of Egypt Special Publication, p. 118.

Sahabi M., Aslanian D., Olivet J.-L., 2004. Un nouveau point de d'epart pour l'histoire de l'Atlantique central, C.R. Geosci., v. 336, pp. 1041–1052.

Sahnouni, M., 2006. Les plus vielles traces d'occupation humaine en Afrique du Nord: Perspective de l'Ain Hanecch, Algerie. Comptes Rendus, Palevol, v. 5, nos. 1–2, pp. 243–254.

Sahnouni, M. & de Heinzelin, J., 1998. The Site of Ain Hanech Revisited: New Investigations at this Lower Pleistocene Site in Northern Algeria. Journal of Archaeological Science, v. 25, pp. 1083–1101.

Said, I. & Rodríguez, S., 2007. A new genus of coral (Rugosa) from the Adarouch area (Brigantian, NE Central Morocco). Coloquios de Paleontología, v. 57, pp. 23–35.

Said, I., Rodríguez, S., Berkhli, M., Cózar, P. & Gómez-Herguedas, A., 2010. Environmental parameters of a coral assemblage from the Akerchi Formation (Carboniferous), Adarouch Area, central Morocco. Journal of Iberian Geology, v. 36, no. 1, pp. 7–19.

Said, M., 1979. Geological Map of Libya 1:250,000, Sheet Bunjim NH33-7, Ind. Res. Cent., Tripoli.

Said, R., 1960. Planktonic foraminifera from the Thebes Formation, Luxor, Egypt, Micropaleontology, v. 6, No. 3, pp. 277–286.

Said, R., 1961. Tectonic framework of Egypt and its influence on distribution of foraminifera, AAPG Bull., v. 45, pp. 198–218.

Said, R., 1962a. The Geology of Egypt, Amsterdam, Elsevier, p. 377.

Said, R., 1962b. Uber das Miozän in der westliche Wüste Ägyptens, Geol. Jb., v. 80, pp. 349–366.

Said, R., 1971. Explanatory Notes to Accompany Geological Map of Egypt, Geol. Surv. of Egypt, Cairo, Paper No. 56, p. 123.

Said, R., 1981. The Geological Evolution of the River Nile, New York, Springer, p. 151.

Said, R., 1983. Proposed classification of the Quaternary of Egypt, J. Afri. Earth Sci., v. 1, pp. 41–45.

Said, R. (Ed.), 1990a. The Geology of Egypt, in R. Said (Ed.), Balkema/Rotterdam/Brookfield, p. 734.

Said, R., 1990b. History of geological research, in R. Said (Ed.), The Geology of Egypt, Balkema, Rotterdam/Brookfiel, pp. 3–7.

Said, R., 1990c. Red Sea coastal plain, in R. Said (Ed.), The Geology of Egypt, Balkema, Rotterdam/Brookfield, pp. 345–359.

Said, R., 1990d. Cretaceous paleogeographic maps, in R. Said (Ed.), The Geology of Egypt, Balkema, Rotterdam/Brookfiel, pp. 439–449.

Said, R., 1990e. Cenozoic, in R. Said (Ed.), The Geology of Egypt, Balkema/Rotterdam/Brookfield, pp. 4–486.

Said, R., 1990f. Quaternary, in R. Said (Ed.), The Geology of Egypt, Balkema/Rotterdam/Brookfield, pp. 487–507.

Said, R., 1993. The River Nile: Geology, Hydrology, and Utilisation. Pergamon Press, Oxford, p. 332.

Said, R., 2004. Science and Politics in Egypt. A Life's Journey. The American University in Cairo Press, 288 p.

Said, R. & Andrawis, S.F., 1961. Lower Carboniferous microfossils from the subsurface rocks of Western Desert Egypt, Contr. Cush. Found. Foram. Research, v. 12, pp. 22–25.

Said, R. & Barakat, M.G., 1957. Lower Cretaceous foraminifera from Khashm el-Mistan, northern Sinai, Egypt, Micropaleontology, v. 3, pp. 39–47.

Said, R. & Barakat, M.G., 1958. Jurassic microfossils from Gebel Maghara, Sinai, Egypt. Micropaleontology, v. 4, no. 3, pp. 231–272.

Said, R. & Eissa, R.A., 1969. Some microfossils from upper Paleozoic rocks of western coastal plain of Gulf of Suez region, Egypt, Proc. 3ed African Micropaleont. Coll., Cairo, pp. 337–383.

Said, R. & El-Heiny, I., 1967. Planktonic foraminifera from the Miocene rocks of the Gulf of Suez region, Contr. Cush. Found. Foram. Research, v. 18, pt.1, pp. 14–26.

Said, R. & Issawi, B., 1963. Geology of northern plateau, Bahariya oasis, Egypt, Geol. Surv. Egypt., p. 41.

Said, R. & Issawi, B., 1964. Preliminary results of a geological expedition to lower Nubia and to Kurkur and Dungul, in F. Wendorf (Ed.), Contribution to the Prehistory of Nubia, Southern Methodist University Press, Dallas, pp. 1–20.

Said, R. & Kenawy, A., 1956. Upper Cretaceous and Lower Tertiary foraminifera from northern Sinai, Egypt, Micropaleontology, v. 2, No. 2, pp. 105–173.

Said, R. & Kerdany, M.T., 1961. The geology and micropaleontology of the Farafra Oasis, Egypt, Micropaleontology, v. 7, No. 3, pp. 317–336.

Said, R. & Martin, L., 1964. Cairo area geological excursion notes: Guidebook to the Geology and Archeology of Egypt, Petrol. Sixth Annual Field Conference, Explor. Soc. Libya, pp. 107–121.

Said, R. & Sabry, H., 1964. Planktonic foraminifera from the type locality of the Esna Shale in Egypt. Micropaleontology, v. 10, No. 3, pp. 375–395.

Said, R., Albritton, C., Wendorf, F., Schild, R. & Kobusiewicz, M., 1970. A preliminary report on the Holocene geology and archaeology of the northern Fayum Desert, in C.C. Reeves, Jr. (Ed.), Playa Lake Symposium, International Center for Arid and Semi-Arid Land Studies and Department of Geosciences, Texas Tech University, Article 6, pp. 41–62.

Said, R., Wendorf, F. & Schild, R., 1970. The geology and prehistory of the Nile Valley in Upper Egypt, Archeolgia Polona, 12, pp. 43–60.

Salah, M.G. & Alsharhan, A.S., 1996. Structural influence on hydrocarbon entrapment in the northwestern Red Sea, Egypt, AAPG Bull., v. 80, pp. 101–118.

Salah, M.G. & Alsharhan, A.S., 1997. The Miocene Kareem Formation in the Southern Gulf of Suez, Egypt: A review of stratigraphy and petroleum geology, J. Petroleum Geology, v. 20, pp. 327–346.

Salah, M.G. & Alsharhan, M.A., 1998. The Precambrian basement: A major reservoir in the rifted basin, Gulf of Suez. Journal of Petroleum Science and Engineering, v. 19, pp. 201–222.

Salahi, D., 1966. Ostracodes du Cretacé superieur et du tertiaire en provenance d'un sondage de la région de Zelten (Libye), Rev. Inst. Fr. Pétrole, Paris, v. 21, No. 1, pp. 3–43.

Salaj, J., 1979. Geological Map of Libya, 1:250,000, Sheet Al Qaryat Al Gharbiyah NH33-5, Explanatory Booklet, Industrial Research Centre, Tripoli, p. 61, 39 photo.

Salaj, J., 1978. The geology of the Pelagian Block: the eastern Tunisian platform. In Nairn, A.E.M., Kanes W.H.& Stehli, F.G. (Eds.), The Ocean Basins and Margins, v. 4B: The Western Mediterranean, pp. 361–416.

Salaj, J. & Nairn, A.E.M., 1987. Age and depositional environment of the Lower Tar "Member" of the Zimam Formation (upper Senonian) in the northern Hamadah al Hamra, Libya, Palaeogeography, Palaeoclimatology, Palaeoecology, v. 61, No. 3, pp. 121–143.

Salem, M.J. & Spreng, A.C., 1980. Middle Miocene stratigraphy, Al Khums area, northwestern Libya, in M.J. Salem & M.T. Busrewil (Eds.), The Geology of Libya, v. I, London, Academic Press, pp. 97–116.

Salem, R., 1976. Evolution of Eocene-Miocene sedimentation patterns in parts of Northern Egypt, AAPG Bull., v. 60, pp. 34–64.

Samir, A.M., 1994. Biostratigraphy and paleoecology of the Khoman Formation (Upper Cretaceous) between the Bahariya and the Farafra oases, Western Desert, Egypt, N. Jb. Geol. Paläont. Abh., v. 191, No. 2, pp. 271–297.

Samir, M., Hassan, W., Abugren, Y., Joshi, S. & Thabet, E., 2010. Reservoir characterization of the Issaran Heavy Oil Field. Petroleum Africa, May 2010, pp. 55–58. www.petroleumafrica.com.

Samson, S.D., Inglis, J.D., D'Lemos, R.S., Admou, H., Blichert-Toft, J. & Hefferan, K., 2004. Geochronological, geochemical, and Nd-Hf isotopic constraints on the origin of Neoproterozoic plagiogranites in the Tasriwine ophiolite, Anti-Atlas orogen, Morocco. Precambrian Research, v. 135, nos. 1–2, pp. 133–147.

Samuel, A., Kneller, B., Raslan, S., Sharp, A. & Parsons, C., 2003. Prolific deep-marine slope channels of the Nile Delta, Egypt. AAPG Bull., v. 87, no. 4, pp. 541–560.

Samuel, M.D. & Saleeb-Roufaiel, G.S., 1977. Lithostratigraphy and petrography of the Neogene sediments at Abu Ghusun, Um Mahara, Red Sea coast, Egypt, Beitrage zur Lithologie, Freiburg Forsch., 323(c), pp. 47–56.

Samuel, M.D., Ismail, A.A., Akarish, A.I.M. & Zaky, A.H., 2009. Upper Cretaceous stratigraphy of the Gebel Somar area, north-central Sinai, Egypt. Cretaceous Research, v. 30, pp. 22–34.

Sandford, K.S. & Arkell, W.J., 1929. The origin of the Fayium Depression: The Fayium and Uganda. The Geographical Journal, v. 74, pp. 578–584.

Sandford, K.S. & Arkell, W.J., 1939. Paleolithic Man and the Nile Valley in Lower Egypt. Chicago University, Oriental Institute Publication 35, pp. 1–105.

Sandford, K.S., 1929. The Oligocene and Pleistocene deposits of Wadi Qena and of the Nile Valley between Luxor and Assiut. Q.J. Geol. Soc. London, v. 75, pp. 493–548.

Sandford, K.S., 1935. Geological observations on the northwestern frontiers of the Anglo-Egyptian Sudan and the adjoining part of the southern Libyan Desert. Geological Society of London Quarterly Journal, v. 80, no. 3, pp. 323–381.

Sandford, K.S. & Arkell, W.J., 1939. Paleolithic man and the Nile Valley in Lower Egypt, Chicago Univ. Oriental Inst. Pub., v. 35, pp. 1–105.

Sanford, R.M., 1970. Sarir oil field, Libya Desert surprise. In M.T. Halbbouty (Ed.), Geology of Giant Petroleum Fields. AAPG Memoir 14, pp. 449–476.

Sanz de Galdeano, C. & Vera, J.A., 1992. Stratigraphic record and paleogeographical context of the Neogene basins in the Betic Cordillera, Spain. Basin Research, v. 4, pp. 21–36.

Saquaque, A, Admou, H., Karson, J., Hefferan, K. & Reuber, I., 1989. Precambrian accretionary tectonics in the Bou Azzer-El Graara region, Anti-Atlas, Morocco. Geology, v. 17, no. 12, pp. 1107–1110.

Saquaque, A., Benharref, M., Abia, H., Mrini, Z. & Reuber, I., 1992. Evidence for a Panafrican volcanic arc and wrench fault tectonics in the Jbel Saghro, Anti-Atlas, Morocco. Geologische Rundschau, 1992, v. 81, no. 1, pp. 1–13.

Sargeant, W.A.S., 1978. Hundredth year memoriam: Christian Gottfried Ehrenberg 1795–1877. Palynology, v. 2, pp. 209–211.

Sassi, P., 1942. Sui fossili di due giacimenti Wealdiani della Tripolitania, Ann. Museo Libico Storia Naturle, v. 3, Tripoli, pp. 41–51.

Sassi, S., Triat, J.M., Truc, G. & Millot, G., 1984. Découverte de l'Eocène continental en Tunisie centrale: la formation du Djebel Gharbi et ses encroûtements carbonatés. Comptes Rendu de l'Academie des Science Paris, v. 299, II, pp. 357–364.

Saunders, W.B., Manger, W.L. & Ramsbottom, W.H.C., 1979. Donetzoceras, a mid-Carboniferous (westphalian) index ammonoid. Journal of Paleontology, v. 53, no. 5, pp. 1136–1144.

Sautkin, A, Talukder, A.R., Comas, M.C., Soto, J.I. & Alekseev, A., 2003. Mud volcanoes in the Alboran Sea: evidence from micropaleontological and geophysical data. Marine Geology, v. 195, pp. 237–261.

Savage, R.J.G. & White, M.E., 1965. Two Mammal fauna from the early Tertiary of central Libya, Proc. Geol. Soc. London, No. 1623, London, pp. 89–91.

Savage, R.J.G., 1968. Contibution to the discussion of Selley, R.C., 1968: Near shore marine and continental deposits of the Sirte Basin, Libya, Q.J. Geol. Soc. London, v. 124, pp. 455–458.

Savage, R.J.G., 1971. Review of the fossil mammals of Libya, Symp. Geol. Libya, C. Gray (Ed.), Faculty of Science, University of Libya, Tripoli, pp. 217–225.

Savornin, J., 1924. Aperçu d'ensemble sur la géologie du Maroc. Annales de Géographie, v. 33, no. 183, pp. 234–243. http://www.persee.fr.

Sbeta, A.M., 1991. Petrography and facies of the Middle and Upper Eocene rocks (Tellil Group), offshore western Libya, in M.J. Salem & M.N. Belaid (Eds.), The Geology of Libya, v. V, Elsevier, Amsterdam, pp. 1929–1966.

Shackleton, R.M., 1994. Review of Late Proterozoic sutures, ophiolitic mélanges and tectonics of eastern Egypt and north-east Sudan. Geol. Rundsch., v. 83, pp. 537–456.

Shackleton, R.M., Ries, A.C., Graham, R.H. & Fitches, W.R., 1980. Late Precambrian ophiolitic mélange in the eastern desert of Egypt. Nature, v. 285, pp. 472–474.

Schandelmeier, H., 1988. Pre-Cretaceous intraplate basins of NE-Africa, Episodes, v. 11, pp. 270–274.

Schandelmeier, H., Klitzsch, E., Hendriks, F. & Wycisk, P., 1987. Structural development of North-East Africa since Precambrian times, Berliner Geowiss. Abh. (A), v. 75, pp. 5–24.

Scheibner, C., Kuss, J. & Marzouk A., 2000. Slope sediments of a Paleocene ramp-to-basin transition in NE-Egypt. International Journal of Earth Sciences, v. 88, pp. 708–724.

Scheibner, C., Kuss, J. & Speijer, R.P., 2003. Stratigraphic modelling of carbonate platform-to-basin sediments (Maastrichtian to Paleocene) in the Eastern Desert, Egypt. Palaeogeography, Palaeoclimatology, Palaeoecology, v. 200, nos. 1–4, pp. 163–185.

Scheibner, C., Marzouk, A.M. & Kuss, J., 2001a. Maastrichtian-Early Eocene lithostratigraphy and Palaeogeography of the N Gulf of Suez Region, Egypt, Journal of African Earth Sciences, v. 32, pp. 223–255.

Scheibner, C., Marzouk, A.M. & Kuss, J., 2001b. Shelf architectures of an isolated Late Cretaceous carbonate platform margin, Galala Mountains (Eastern Desert, Egypt). Sedimentary Geology, v. 145, pp. 23–43.

Scheibner, C., Reijmer, J.J.G., Marzouk, A.M., Speijer, R.P. & Kuss, J., 2003. From platform to basin: The evolution of a Paleocene carbonate margin (Eastern Desert, Egypt). International Journal of Earth Sciences/Geologische Rundschau, v. 92, no. 4, pp. 624–640.

Scheibner, C & Speijer, R.P., 2008. Decline of coral reefs during late Paleocene to early Eocene global warming. eEarth, v. 3, pp. 19–26.

Scheibner, C. & Speijer, R.P., 2009. Recalibration of the Tethyan shallow-benthic zonation across the Paleocene-Eocene boundary: the Egyptian record. Geologica Acta, v. 7, nos. 1–2, pp. 195–214.

Schellwien, E., 1894. Ueber eine angebliche Kohlenkalkfauna aus der ägyptisch-arabischen Wüste. Z. dtsch. Geol. Ges., v. 46, pp. 68–78.

Schenk, A., 1883. Fossile Hölzer. Palaeontographica 30, pp. 1–17.

Schettino, A. & Turco, E., 2009. Breakup of Pangaea and plate kinematics of the central Atlantic and Atlas regions. Geophysical Journal International, v. 178, no. 2, pp. 1078–1097.

Schild, R. & Wendorf, F., 2004. The Megaliths of Nabta Playa. Academia Focus on Archeology, v. 1, no. 1, pp. 10–15.

Schlager, W., 1981. The paradox of drowned reefs and carbonate platforms, Geol. Soc. Am. Bull., v. 92, pp. 197–211.

Schlanger, S.O. & Jenkyns, H.C., 1976. Cretaceous Oceanic anoxic events: causes and consequences, Géologie en Mijnbouw, v. 55, pp. 179–184.

Schlüter, T., 2008. Geological Atlas of Africa: With Notes on Stratigraphy, Tectonics, Economic Geology, Geohazards, Geosites and Geoscientific Education of Each Country. Springer, Berlin Heidelberg, p. 272.

Schlumberger, 1984. Well Evaluation Conference, Egypt, v. 1.

Schlumberger, 1995. Well Evaluation Conference Egypt, pp. 7–13l.

Schmieder, M., Buchner, E. & Le Heron, D.P., 2009. The Jebel Hadid structure (Al Kufrah Basin, SE Libya) – A possible impact structure and potential hydrocarbon trap? Marine and Petroleum Geology, v. 26, pp. 310–318.

Schnyder, J., Gorin, G., Soussi, M., Baudin, F. & Deconinck, J.-F., 2005. A record of the Jurassic/Cretaceous boundary climatic variation on the southern margin of the Tethys: clay minerals and palynofacies of the early Cretaceous Jebel Meloussi section (Central Tunisia, Sidi Kralif Formation). Bulletin de la Société Géologique de France, v. 176, no. 2, pp. 171–182.

Schofield, D.I. & Gillespie, M.R., 2007. A tectonic interpretation of "Eburnean terrane" outliers in the Reguibat Shield, Mauritania. Journal of African Earth Sciences, v. 49, pp. 179–186.

Schofield, D.I., Horstwood, M.S.A., Pitfield, P.E.J., Crowley, Q.G., Wilkinson & Sidaty, H. Ch.O., 2006. Timing aand kinematics of Eburnean tectonics in the central Reguibat Shield, Mauritania. Journal of the Geological Society, v. 163, no. 3, pp. 549–560.

Schrank, E. & Mahmoud, M.S., 1998a (Abs.). Palynology and Cretaceous stratigraphy of the Dakhla Oasis, Egypt, in Geology of the Arab World, Fourth International Conference, Cairo University, Cairo, pp. 23–24.

Schrank, E. & Mahmoud, M.S., 1998b. Palynology (pollen, spores and dinoflagellates) and Cretaceous stratigraphy of the Dakhla Oasis, central Egypt, Journal of African Earth Sciences, v. 26, no. 2, pp. 167–193.

Schrank, E., 1984. Paleozoic and Mesozoic palynomorphs from the Foram 1 well (Western Desert, Egypt), N. Jb. Geol. Paläont. Mh.v. 2, pp. 95–112.

Schrank, E., 1987. Paleozoic and Mesozoic palynomorphs from northeast Africa (Egypt and Sudan) with special reference to Late Cretaceous pollen and dinoflagellates, Berliner Geowissenschaftliche Abhandlungen, A75, pp. 249–310.

Schrank, E., 1991. Mesozoic palynology and continental sediments in NE Africa (Egypt and Sudan) – a review, J. Afr. Earth Sci., v. 12, No. 1/2, pp. 363–373.

Schrank, E., 1992. Non-marine Cretaceous correlations in Egypt and northern Sudan: palynological and palaeobotanical evidence, Cretaceous Research, v. 13, pp. 351–368.

Schreiber, B.C. & Ryan, W.B.F., 1995. Variations in the formative conditions of Messinian evaporites, water depths, and depositional cyclicity, in International Conference on the Biotic and Climatic Effects of the Messinian Event on the Circum Mediterranean, Technical Program and Abstracts, University of Garyounis, Benghazi, p. 55.

Schröter, T., 1996. Tectonic and sedimentary development of the central Zallah Trough (west Sirt Basin, Libya), in M.J. Salem, M.T. Busrewil, A.A. Misallati & M.A. Sola (Eds.), The Geology of Sirt Basin, v. III, Amsterdam, Elsevier, pp. 123–136.

Schulte, P., Scheibner, C. & Speijer, R.P., 2009. The Paleocene-Eocene Thermal Maximum (PETM) in the Dababiya Quarry Section, Egypt: New evidence for environmental changes from mineralogical and geochemical data. Geophysical Research Abstracts, v. 11, EGU2009-8751, p. 2.

Schürmann, H.M.E., 1974. The Precambrian in North Africa, E.J. Brill, Leiden, p. 352.

Schürmann, H.M.E., 1953. The Precambrian of the Gulf of Suez area, Compte rendue, 19th Congr. Geol. Intern., Algiers, Fasc. 1, pp. 115–135.

Schürmann, H.M.E., 1966. The Precambrian along the Gulf of Suez and the Northern Part of the Red Sea, E.J. Brill, Leiden, p. 404.

Schürmann, H.M.E., Burger, D. & Dijkstra, S.J., 1963. Permian near Wadi Araba Eastern Desert of Egypt. Geol. En Mijnbouw, v. 42, no. 10, pp. 329–336.

Schütz, K.I., 1994. Structure and stratigraphy of the Gulf of Suez Egypt. In S.M. Landon (Ed.), Interior Rift Basins, AAPG Mem.59, pp. 57–74.

Schull, T.J., 1988. Rift basins of interior Sudan: petroleum exploration and discovery, AAPG Bull., v. 72, no. 10, pp. 1128–1142.

Schuster, J.-M., 1977. Essai de reconstruction de l'histoire géologique et structurale de la Méditerranée centrale, Revue de l'Institut Francais du Pétrole, v. 32, No. 4, pp. 527–548.

Schwager, C., 1883. Die Foraminiferen aus den Eocänblagerungen der libyschen Wüste und Aegyptens. Palaeontographica, 30, pp. 79–154.

Schweinfurth, G., 1883. Ueber die Geologische Schichtengliederung des Mokattam bei Cairo. Zeitschr. Deutsch. Geol. Gesellsch., Berlin, Bd. XXXV, pp. 709–734.

Schweinfurth, G., 1885. Sur la découverte d'une faune paléozoique dans le grès d'Egypte. Bull. Inst. Egypte, sér.2, no. 6, pp. 239–255.

Schweinfurth, G., 1886. Reise in das Depressionsgebiet im Umkreise de Fajum im Januar 1886. Zeitschrift de Gesellshaft für Erkunde zu Berlin, 21, pp. 96–149.

Schweinfurth, G., 1887. Sur une récente exploration géologique de l'Ouady Arabah. Bull. Inst. Egypte, sér.2, no. 8, pp. 146–162.

Schweinfurth, G., 1889. Ueber die Kreideregion bei den Pyramieden von Gizeh. Petermanns Mitth., XXXV.

Schweinfurth, G., 1899–1910. Aufnahmen in der östlichen Wüste von Aegypten 1–4, Dietriech Reimer, Berlin.

Schweinfurth, G., 1921. Auf unbetreten Wegen in Aegypten. Hoffman und Campe, Hamburg and Berlin, p. 330.

Scotese, C.R., Bambach, R.K., Barton, C., van der Voo, R. & Ziegler, A.M., 1979. Palaeozoic base maps, J. Geology, v. 87, pp. 217–77.

Scotese, C.R., Boucot, A.J. & McKerrow, W.S., 1999. Gondwana palaeogeography and palaeoclimatology. Journal of African Earth Sciences, v. 28, pp. 99–114.

Sdzuy, K. & Geyer, G., 1988. The base of the Cambrian in Morocco. In V.H. Jacobshagen (Ed.), 1988. The Atlas System of Morocco: Studies of its Geodynamic Evolution. LecureNotes in Earth Sciences, v. 15, Springer-Verlag, Berlin/Heidelberg, pp. 91–106.

Searle, D.L., Carter, I.M., Shalaby, I.M. & Hussein, A.A., 1976 (Abs.). Ancient volcanism of island arc type in the Eastern Desert of Egypt, 25th Intl. Geol. Congr., Sydney, Sec.1, pp. 62–63.

Sebane, A., Morok, A. & Elmi, S., 2007. Évolution des peuplements de foraminifèrs pendant la crise toarcienne a l'exemple des donness des monts des Ksour (l'Atlas saharien occidental, Algérie). Comptes Rendus, Palevol, v. 6, no. 3, pp. 189–196.

Sebbar, A., 1997. Biostratigraphie (Foraminifère) du Carbonifère Moyen, Bassin de "Bechar-Mezarif". Annales de la Societé Géologique de Belgique, v. 120, no. 2, pp. 205–215.

Sebbar, A. & Mamet, B., 1996. Algues benthiques calcaires du Carbonifère inférieur et moyen, bassin de Béchar, Algérie. Rev. Micropaleontol., v. 39, pp. 153–167.

Seidl, K. & Röhlich, P., 1984. Geological Map of Libya, 1:250,000: Sheet Sabha NG 33-2, Explanatory Booklet, Industrial Research Centre, Tripoli, p. 138.

Seilacher, A., 1990. Paleozoic trace fossils, in R. Said (Ed.), Geology of Egypt, 2nd Edition, A.A. Balkema, Rotterdam/Brookfield, pp. 649–670.

Seilacher, A., Lüning, S., Martin, M.A., Klitzsch, F., Khoja, A. & Craig, J., 2002. Ichnostratigraphic correlation of Lower Palaeozoic clastics in the Kufra Basin (SE Libya). *Lethaia*, v. 35, no. 3, pp. 257–262.

Selim, A.A., 1974. Origin and lithification of the Pleistocene of the Salum area, Western coastal plain of Egypt, J. Sedimentary Petrology, v. 44, pp. 757–760.

Selley, R.C., 1968. Facies profile and other new methods of graphic data presentation: application in a quantitative study of Libyan Tertiary shoreline deposits, J. Sediment. Petrology, v. 38, pp. 363–372.

Selley, R.C., 1969. Near-shore marine and continental sediments of the Sirte Basin, Libya. Q.J. Geol. Soc. London, v. 124, pp. 419–460.

Selley, R.C., 1971. Structural control of Miocene sedimentation in the Sirte Basin, in C. Gray (Ed.), Symposium on the Geology of Libya, Faculty of Science, U. of Libya, Tripoli, pp. 99–106.

Sellwood, B. & Netherwood, R., 1984. Facies evolution in the Gulf of Suez area – sedimentation history as an indicator of rift initiation and development. Modern Geology, v. 9, pp. 43–69.

Sengör, A.M.C., Yilmaz, Y. & Sungurle, O., 1984. Tectonic of the Mediterranean Cimmerides: nature and evolution of the termination of the Paleotethys, in A.H.F. Robertson & J.E. Dixon (Eds.), The Geological Evolution of the Eastern Mediterranean, Geological. Soc. Spec. Pub. No. 17, London, pp. 77–112.

Senowbari-Daryan, B. & Rigby, J.K., 1991. Three additional thalamid sponges from the Upper Permian reefs of Djebel Tebaga (Tunisia). Journal of Paleontology, v. 65, no. 4, pp. 623–629.

Senturk, E., Majdoub, A., Elghmari, M. & Elghadban, A., 2007. Multi Lateral Horizontal Well Application to Enhance the Oil Recovery of a Mature Field, Intisar 103L Field. OAPEC Joint Seminar "Improved Oil Recovery (IOR) Techniques and Their Role In Boosting The Recovery Factor", Rueil-Malmaison, France, 26–28 June 2007, p. 3.

Septfontaine, M., 1986. Milieux de dépóts et Foraminiferes (lituolidés) de la plateforme carbonatée du Lias moyen au Maroc. Rev. Micropal., v. 28, pp. 268–289.

Sepúlveda, J., Wendler, J., Leider, A., Kuss, J., Summons, R. & Hinrichs, K.-U., 2009. Molecular-isotopic evidence of marine productivity changes across the C/T boundary in the Levant platform (Central Jordan). Organic Geochemistry, v. 40, pp. 553–586.

Sercombe, W., Golob, B.R., Kamel, M., Stewart, J.W., Smith, G.W. & Morse, J.D., 1997. Significant structural reinterpretation of the subsalt, giant October field, Gulf of Suez, Egypt, using SCAT, isogan-based sections and maps, and 3-D seismic. The Leading Edge, v. 16, no. 8, pp. 1143–1150.

Serencsits, C.M., Faul, H., Roland, K.A., Husssein, A.A. & Lutz, T.M., 1981. Alkaline ring complexes in Egypt, their age and relationships in time, J. Geophys. Res., v. 86(B4), pp. 3009–3013.

Serry Bey, H., 1929. The Qattara Power Scheme. Geographical Review, v. 19, no. 2, pp. 290–292.

Sestini, G., 1984. Tectonic and sedimentary history of the NE African margin (Egypt-Libya), in J.E. Dixon & A.H.F. Robertson (Eds.), The Geological Evolution of the Eastern Mediterranean, Geol Soc. Spec. Pub. No. 17, Blackwell, pp. 161–175.

Seward, A.C., 1907. Fossil plants from Egypt. Geol. Magazine, pp. 253–257.

Seward, A.C., 1935. Leaves of dictyoledons from Nubian sandstone of Egypt. Geol. Surv. Egypt, Cairo, p. 21.

Sghair, A.M.A. & El Alami, M.A., 1996. Depositional environment and diagenetic history of the Maragh Formation, NE Sirt Basin, Libya, in M.J. Salem. A.S. El-Hawat & A.M. Sbeta (Eds.), The Geology of Sirt Basin, v. II, Amsterdam, Elsevier, pp. 263–274.

Shackleton, R.M., Ries, A.C., Gr, R.H. & Fitches, W.R., 1980. Late Precambrian ophiolite melange in the Eastern Desert of Egypt, Nature, v. 285, pp. 472–474.

Shahin, A., El Halaby, O. & El Baz, S., 2008. Middle Eocene ostracodes of the Qattamiya area, northwest Eastern Desert, Egypt: Systematics, biostratigraphy and paleobiogeography. Revue de Paléobiologie, Genève, v. 27, no. 1, pp. 123–157.

Shahin, A. & Kora, M., 1991. Biostratigraphy of some Upper Cretaceous successions in the eastern central Sinai, Egypt. Neues Jahrb. Geol. Palaontol., Monatshefte., v. 11, pp. 671–692.

Shahin, A.N. & Shehab, M.M., 1988. Undiscovered hydrocarbon reserves and their preservation time limits in West Qarun area, Abu Gharadig Basin, Western Desert, Egypt. 9th Exploration and Production Conf., Egyptian General Petroleum Corp. (EGPC), Cairo, Egypt, November 1988, p. 23.

Shakoor, A., 1979. Geological Map of Libya, 1:250,000, Sheet-Hun NH33-6, Industrial Research Centre, Tripoli, p. 117.

Shalaby, A., 2010. The northern dome of Wadi Hafafit culmination, Eastern Desert, Egypt: Structural setting in tectonic framework of a scissor-like wrench corridor. Journal of African Earth Sciences, v. 57, pp. 227–241.

Shalaby, A., Stüwe, K., Makroum, F. & Fritz, H., 2006. The El Mayah molasse basin in the Eastern Desert of Egypt. Journal of African Earth Sciences, v. 45, pp. 1–15.

Shalaby, A., Stüwe, K., Makroum, F., Fritz, H., Kebede, T. & Klötzli, U., 2005. The Wadi Mubarak belt, Eastern Desert of Egypt: A Neoproterozoic conjugate shear system in the Arabian-Naubian Shield. Precambrian Research, v. 136, pp. 27–50.

Sharabi, S.A., 1998 (Abs.). Cenomanian-Turonian biostratigraphy in north Western Desert, Egypt, in Geology of the Arab World, Fourth International Conference, Cairo University, Cairo, pp. 40–41.

Sharaf, L.M., 2003. Source rock evaluation and geochemistry of condensates and natural gases, offshore Nile Delta, Egypt. J. Petroleum Geology, v. 26, no. 2, pp. 189–209.

Sharp, A., Maddox, S.J., Swallow, J., Wolfe, J.E. & Ramadan, I.R., 1998. Successful exploration of the Pliocene offshore Nile Delta, Egypt, using seismic anomalies: a case history. 60th Mtg. Eur. Assn. Geosci. Eng., Session 04–42.

Sharp, A. & Samuel, A., 2004. An example study using conventional 3D seismic data to delineate shallow gas drilling hazards from the West Delta Deep Marine Concession, offshore Nile Delta, Egypt. Petroleum Geoscience, v. 10, pp. 121–129.

Shaw, W.B.K., Sanford, K.S. & Mason, M., 1936. An expedition in the Southern Libyan Desert. The Geographical Journal, v. 87, no. 3, pp. 193–217.

Shelmani, M., Thusu, B. & El-Arnauti, A., 1992. Subsurface occurrences of Middle and Upper Triassic sediments in Eastern Libya. In A. Sadek (Ed.), Geology of the Arab World, Cairo University, v. 2, pp. 233–239.

Sherif, K.A.T., 1991. Biostratigraphy of the Miocene in Al Khums are, Northwestern Libya, in M.J. Salem, O.S. Hammuda & B.A. Eliagoubi (Eds.), The Geology of Libya, v. IV, Amsterdam, Elsevier, pp. 1421–1456.

Shimron, A.E., 1980. Proterozoic island arc volcanism and sedimentation in Sinai, Precambrian Res., v. 12, pp. 437–458.

Shimron, A.E., 1983. The Tarr Comloex revisited – Folding, thrusts, and melanges in the southern Wadi Kid region, Sinai Peninsula, Israel J. Earth Sci., v. 32, pp. 123–148.

Shimron, A.E., 1984. Evolution of the Kid Group, south-eastern Sinai Peninsula, thrusts, mélanges and implications for accretionary tectonics during the late Proterozoic of the Arabian-Nubian shield. Geology, v. 12, pp. 242–247.

Shimron, A.E., 1987. Pan-African metamorphism, Israel J. Earth Sci., v. 36, pp. 173–193.

Shukri, N.M., 1944. On the "living" petrified forest. Inst. Égypte Mém. 26, pp. 71–75.

Shukri, N.M., 1954. Remarks on the geologic structure of Egypt, Soc. Géogr. Egypte, Bull., v. 27, pp. 65–82.

Sigaev, N.A., 1959. The main teconic features of Egypt. An Explanatory Note to the Tectonic Map of Egypt, Scale 1:2,000,000. Geological Survey of Egypt, Paper No. 39, p. 26.

Sikander, A.H., Basu, S. & Rasul, S.M., 2003. Geochemical source-maturation and volumetric evaluation of Lower Palaeozoic source rocks in the West Libyan basins. In M.J. Salem, Khaled, M. & Seddiq, H.M. (Eds.), The Geology of Northwest Libya, v. III, pp. 3–53.

Silvestri, A., 1934. Su di alcuni foraminiferi terziari della Sirtica. Missione Scientifica della R. Accad. D'Italia a Cufra (1931-X), Roma, v. 3, pp. 7–30.

Simons, E.L. & Rasmussen, D.T., 1990. Vertebrate paleontology of Fayum: History of Research, faunal review and future prospects, in R. Said (Ed.), The Geology of Egypt, Balkema, Rotterdam/Brookfield, pp. 627–638.

Sims, P.K. & James, H.L., 1984. Banded iron-formations of late Proterozoic age in the central Eastern Desert. geology and tectonic setting. Economic Geology, v. 79, no. 8, pp. 1777–1784.

Sinha, R.N. & Eland, H.B., 1996. The pre-Cretaceous (Triassic) sequence in the subsurface of Marada Trough, Eastern Sirt Basin, Libya, Petroleum Research Journal, Tripoli, pp. 49–60.

Sinha, R.N. & Mriheel, I.Y., 1996. Evolution of subsurface Palaeocene sequence and shoal carbonates, south-central Sirt Basin, in M.J. Salem, A.M. Sbeta & M.R. Bakbak (Eds.), The Geology of Libya, v. VI, Amsterdam, Elsevier, pp. 153–195.

Sircar, A., 2004. Hydrocarbon production from fractured basement formations. Current Science, v. 87, no. 2, pp. 147–151.

Sluijs, A., 2008. Rapid Carbon Injection and Transient Global Warming During the Paleocene-Eocene Thermal Maximum. Gussow-Nuna Geoscience Conference, October 20–23, 2008, Banff, Alberta, p. 5.

Smale, J.L., Thunell, R.C. & Schamel, S., 1988. Sedimentologic evidence for early Miocene fault reactivation in the Gulf of Suez. Geology, v. 16, pp. 113–116.

Smetana, R., 1975. Geological map of Libya, 1:250,000, Sheet Ra's Jdeir NI 32-16, Explanatory Booklet, Industrial Research Centre, Tripoli, p. 60.

Smith, A.G. & Woodcock, N.H., 1982. Tectonic synthesis of the Alpine-Mediterranean region: a review, in H. Berckhemer & K. Hsü (Eds.), Alpine Mediterranean Geodynamics, Geodynamic Series Vol.7, Am. Geophys. Union/Geol. Soc. Am., pp. 15–38.

Smith, I.E.M., Worthington, J.J., Price, R.C. & Gamble, J.A., 1997. Primitive magmas in arc-type volcanic associations: Examples from the southwest Pacific, The Canadian Mineralogist, v. 35, part 2, pp. 257–273.

Smith, J.B., Lamanna, M.C., Lacovara, K.J., Dodson, P., Smith, J.R., Poole, K.J., Giegengack, R. & Attia, Y., 2001. A giant Sauropod dinosaur from an Upper Cretaceous mangrove deposit in Egypt. Science, v. 292, pp. 1704–1706.

Smith, K.T., Bhullar, B.-A.S. & Holroyd, P.A. 2008. Earliest African record of the *Varanus* stem-clade (Squamata: Varabude) from the Early Oligocene of Egypt. Journal of Vertebrate Paleontology, v. 28, no. 3 pp. 909–913.

Sneh, A. & Weissbrod, T., 1973. Nile Delta: defunct Pelusiuc Branch identified, Science, v. 180, pp. 59–61.

Snoke, A.W., Schamel, S. & Karasek, R.M., 1988. Structural evolution of Djebel Debadib Anticline: A clue to the regional tectonic style of the Tunisian Atlas, Tectonics, v. 7, no. 3, pp. 497–516.

Sola, M.A. & Worsley, D. (Eds.), 2000. Geological Exploration in Murzuq Basin. Elsevier, Amsterdam, p. 519.

Solignac., 1927: Etude géologique de la Tunisie septentrionale. Thèse Sci. Univ. Lyon, édit.: Dir. Trav. Publ., Tunis, p. 756.

Soliman, F., Ibrahim, M.E. & Trayner, P., 1988. Qanrara Formation, a promising oil potential in Nile Delta. 9th Exploration and Production Conf., Egyptian General Petroleum Corp. (EGPC), Cairo, Egypt, November 1988, p. 25.

Soliman, H.A. & Sultan, I., 1976. Spores et pollens des grès de Baharia, Desert Ouest, Egypte, Rév. Micropaléontologie, v. 19, pp. 108–111.

Soliman, H.A., 1975. Spores et pollens rencontrés dans le forage no. 8, El Kharga, Désert Ouest, Egypte, Rév. Micropaléontologie, v. 18, pp. 53–57.

Soliman, H.A., 1977. Foraminiféres et microfossiles végétaux du "Nubian Sandstone" de subsurface de l'Oasis El Kharga, Désert de l'Ouest, Egypte, Rév. Micropaléontologie, v. 20, pp. 114–124.

Soliman, M.A., Habib, M.E. & Ahmed, E.A., 1986. Sedimentologic and tectonic evolution of the Upper Cretaceous-Lower Tertiary succession at Wadi Qena, Egypt, Sedimentary Geology, v. 46, pp. 11–33.

Soliman, M.S. & El Fetouh, M.A., 1970. Carboniferous of Egypt, AAPG Bull., v. 54, No. 10, pp. 1918–1930.

Sonnenfeld, P., 1975. The significance of Upper Miocene (Messinian) evaporites in the Mediterranean Sea, J. Geology, v. 83, No. 3, pp. 287–307.

Soto, J.I., Comas, M.C. & de la Linde, J., 1996. Espesor de sedimentos en la cuenca de Alborán mediante una conversion sísmica corregida. Geogaceta, v. 20, pp. 382–385.

Souaya, F.J., 1963. On the foraminifera of Gebel Gharra (Cairo-Suez Road) and some other Miocene samples, J. Paleontology, v. 37, pp. 433–457.

Soussi, M. & Ben Ismail, M.H., 2000. Platform collapse and pelagic seamount facies: Jurassic development of central Tunisia. Sedimentary Geology, v. 133, pp. 93–113.

Soussi, M., 2003. Nouvelle nomenclature lithostratigraphique « événementielle » pour le Jurassique de la Tunisie atlasique. Gebios, v. 36, no. 6, pp. 761–773.

Spath, L.F., 1913. On Jurassic Ammonites from Jebel Zaghuan (Tunisia). J. Geol. Soc. London, v. 69, nos. 1–4, pp. 540–580.

Spath, L.F., 1946. The Middle Triassic cephalopod from Sinai. Bulletin Institute Egypt, v. 27, pp. 425–426.

Speijer, R.P., 2003. Danian-Selandian sea-level change and biotic excursion on the southern Tethyan margin (Egypt). Geological Society of America Special Paper 369, pp. 275–290.

Speijer, R.P., Schmitz, B., Aubry, M.P. & Charisi, S., 1995. The latest Paleocene benthonic extinction event: Punctuated turnover in outer neritic foraminiferal faunas from Gebel Aweina, Egypt, Israel J. Earth Sci., v. 44, pp. 207–222.

Speijer, R.P., van der Zwaan, G.J. & Schmitz, B., 1996.The impact of Paleocene/Eocene boundary events on middle neritic benthic foraminiferal assemblages from Egypt. Marine Micropaleontology, v. 28, no. 2, pp. 99–132.

Speijer, R.P, & Wagner, T., 2002, Sea-level changes and black shales associated with the late Paleocene thermal maximum: Organic-geochemical and micropaleontologic evidence from the southern Tethyan margin (Egypt-Israel). In C. Koeberl & K.G. MacLeod (Eds.): Catastrophic Events and Mass Extinctions: Impacts and Beyond: Boulder, Colorado. Geological Society of America Special Paper 356, pp. 533–549.

Speke, Captain & Captain Grant, 1863. The Nile and its sources. Proceedings of the Royal Geographical Society of London, v. 7, no. 5, pp. 217–224.

Spray, J.G., Bébien, J., Rex, D.C. & Roccick, J.C., 1984. Age constraints on the igneous and metamorphic evolution of the Hellenic-Dinaric ophiolites, in A.H.F. Robertson & J.E. Dixon (Eds.), The Geological Evolution of the Eastern Mediterranean, Geological Soc. Spec. Pub. No. 17, London, pp. 619–627.

Spring, D. & Hansen, O.P., 1998. The influence of platform morphology and sea level on the development of a carbonate sequence, the Harash Formation, Eastern Sirt Basin, Libya, in D.J. Macgregor. R.T.J. Moody & D.D. Clark-Lowes (Eds.), Petroleum Geology of North Africa, Geological Society Spec. Pub. No. 132, London, pp. 335–353.

Srivastava, P., 1980. Geological map of Libya, 1:250,000, Explanatory booklet, Sheet An Nuwfailyah NH 33-8, Industrial Research Centre, Tripoli, p. 87.

Stacey, G.B., 1867. On the geology of Benghazi, Barbary and an account of the subsidences in its vicinity. Q.J. Geol. Soc. London, v. 23, no. 92, pp. 384–386.

Stainforth, R.M., 1949. Foraminifera in the Upper Tertiary of Egypt, J. Paleontology, v. 23, pp. 419–422.

Stampfli, G.M. & Borel, G.D., 2002. A plate tectonic model for the Paleozoic and Mesozoic constrained by dynamic plate boundaries and restored synthetic oceanic isochrons". Earth and Planetary Science Letters 196 (1): 17–33.

Stampfli, G., Borel, G., Cavazza, W., Mosar, J. & Ziegler, P. (Eds.), 2001. Paleotectonic and palaeogeographic evolution of the western Tethys and PeriTethyan domain (IGCP Project 369). Episodes, v. 24, no. 4, pp. 222–228.

Stanley, D.J. & Goodfriend, G.A., 1997. Recent subsidence of the northern Suez canal, Nature, v. 388, pp. 335–336.

Stanley, D.J. & Maldonado, A., 1979. Levantine Sea – Nile Cone lithostratigraphic correlation with paleoclimatic and eustatic oscillations in the late Quaternary, Sedimentary Geology, v. 23, pp. 37–65.

Stanley, D.J., McRea, J.E., Jr. & Waldron, C., 1996. Nile Delta core and sample database for 1985–1994: Mediterranean Basin (MEDIBA) Program, Smithonian Institution Press, Washington, p. 428.

Stanley, D.J. & Warne, A.G., 1993. Nile Delta: Recent geological evolution and human impact, Science, v. 260, pp. 628–634.

Stanley, D.J. & Warne, A.G., 1994. Worldwide initiation of Holocene marine deltas by deceleration of sea-level rise, Science, v. 265, pp. 228–231.

Stanley, J.-D., Goddio, F., Jorstad, T.F. & Schnepp, G., 2004. Submergence of ancient Greek cities off Egypt's Nile Delta – A cautionary tale. GSA Today, v. 14, no. 1, pp. 4–10.

Stefanini, G., 1918. Echinidi Cretacci e Tertiari d'Egitto, raccolti da Antonio Figari Bey. Boll. Soc. Geol. Ital., v. 37, pp. 121–168.

Stefanini, G., 1921. Fossili terziari della Cirenaica. Paleont. Italica, v. 27, pp. 101–146.

Stefanini, G., 1923. Struttura geologica della Cirenaica e cenni descrittivi a corredo dello schizzo geologico dimostrativo della cirenaica. La Cirenaica geografica economica, politica. Milano, pp. 215–236.

Stefek, V. & Röhlich, P., 1984. Geological Map of Libya, 1:250,000. Sheet Awbary. Explanatory Booklet. Industrial Research Centre, Tripoli, p. 69.

Stein, R., 1986. Late Neogene evolution of paleoclimate and paleoceanic circulation in the Northern and Southern Hemispheres – A comparison. Geologische Rundschau, v. 75, no. 1, pp. 125–138.

Steinitz, G., Bartov, Y. & Hunziker, J.C., 1978. K/Ar age determinations of some Miocene-Pliocene basalts in Israel, their significance to the tectonics of the rift valley, Geol. Mag., v. 115, pp. 329–340.

Stern, R.J., 1985. The Najd Fault System, Saudi Arabia and Egypt: A late Precambrian rift-related transform system? Tectonics, v. 4, pp. 497–511.

Stern, R.J., 2002. Crustal evolution in the East African Orogen: a neodymium isotopic perspective. Journal of African Earth Sciences, v. 34, no. 3–4, pp. 109–117.

Stern, R.J. & Hedge, C.E., 1985. Geochronologic and isotopic constraints on late Precambrian crustal evolution in the Eastern Desert of Egypt, American Journal of Science, v. 285, pp. 97–127.

Stern, R.J. & Manton, W.I., 1987. Age of Feiran basement rocks, Sinai: Implications for late Precambrian crustal evolution in the northern Arabian-Nubian shield, Geological Society of London Journal, v. 144, pp. 569–575.

Stern, R.J., Gottfried, D., & Hedge, C.E., 1984. Late Precambrian rifting and crustal evolution in the north Eastern Desert of Egypt. Geology, v. 12, pp. 168–172.

Stern, R.J., Johnson, P.R., Kröner, A. & Yibas, B., 2004. Neoproterozoic ophiolites of the Arabian-Nubian Shield. Precambrian Ophiolites and Related Rocks Edited by Timothy M. Kusky. Developments in Precambrian Geology, Vol. 13 (K.C. Condie, Series Editor), pp. 95–128.

Stern, R.J., Sellers, G. & Gottfried, D.G., 1988. Bimodal dike swarms in the Northeastern Desert of Egypt: Significance for the origin of Late Precambrian "A-type" granites in northern Afro-Arabia. In S. El-Gaby & R.O. Greiling (Eds.), The Pan-African Belt of Northeast Africa and Adjacent Areas, Fiedr. Vieweg & Sohn, Baunschweig/Wiesbaden, pp. 147–179.

Stoeser, D.B. & Camp, V., 1985. Pan-African microplate accretion of the Arabian shield, Geological Society of America Bulletin, v. 96, pp. 817–826.

Streel, M., Paris, F., Riegel, W. & Vanguestaine, M., 1988. Acritarch, chitinozoan and spore stratigraphy from the Middle and Late Devonian of northeast Libya, in A. El-Arnauti, B.Owens & B. Thusu (Eds.), Subsurface Palynostratigraphy of Northeast Libya, Benghazi, Libya, Garyounis University Publications, pp. 11–128.

Strömer, E., 1914. Ergebnisse der Forschungsreisen Prof. E. Strömers in den Wüste Ägyptens, I. Die Topographie und geologie der Strecke Gharaq-Baharije nebst Ausfürungen über die geologische Gesschichte-Ägyptens. Abhandlungen der Königlich Bayerischen Akademie der Wissenschaften mathemaatisch-physikalische Klasse, v. 16, pp. 1–78.

Strömer, E., 1915. Ergebnisse der Forschungsreisen Prof. E. Strömers in den Wüste Ägyptens, II, Wirbeltier-Resten der Baharija-Stufe (unterstes Cenoman). 3, Das Original des *Theroden*

*Spinosaurus aegyptiacus* nov. gen., nov. spec. Abhandlungen der Königlich Bayerischen Akademie der Wissenschaften mathemaatisch-physikalische Klasse, 28, pp. 1–32.

Strömer, E., 1931. Wirbeltier-Resten der Baharija-Stufe (unterstes Cenoman). 10. Ein Skelett-Reste von *Carcharodontosaurus* nov. gen. Abhandlungen der Königlich Bayerischen Akademie der Wissenschaften mathemaatisch-naturwissenschaftlische Abteilung, 9, pp. 1–31.

Strömer, E., 1932. Ergebnisse der Forschungsreisen Prof. E. Strömers in den Wüste Ägyptens, II, Wirbeltier-Resten der Baharija-Stufe (unterstes Cenoman). II. Sauropoda. Abhandlungen der Königlich Bayerischen Akademie der Wissenschaften mathemaatisch-naturwissenschaftlische Abteilung, v. 10.

Strömer, E., 1934. Ergebnisse der Forschungsreisen Prof. E. Strömers in den Wüste Ägyptens, II, Wirbeltier-Resten der Baharija-Stufe (unterstes Cenoman). 13. Dinosauria. Abhandlungen der Königlich Bayerischen Akademie der Wissenschaften mathemaatisch-naturwissenschaftlische Abteilung, v. 22, pp. 1–79.

Strömer, E., 1936. Ergebnisse der Forschungsreisen Prof. E. Strömers in den Wüste Ägyptens, VII. Baharja-Kessel und Stufe mit deren Fauna und Flora. Eine ergänzende zusammenfassung. Abhandlungen der Königlich Bayerischen Akademie der Wissenschaften mathemaatisch-naturwissenschaftlische Abteilung, v. 33, pp. 1–102.

Strougo, A. & Azab, M.M., 1982. Middle Eocene mollusca from the basal beds of Gebel Qarara (upper Egypt) with remarks on the depositional environment of these beds, N. Jb. Geol. Paläont. Mh. (11), pp. 667–678.

Strougo, A. & Boukhary, M.A., 1987. The Middle Eocene-Upper Eocene boundary in Egypt, Present state of the problem, Revue de Micropaléontologie, v. 30, pp. 122–127.

Strougo, A. & Haggag, M.A.,Y., 1983. The occurrence of deposits of Paleocene age at Abu Roash, west of Cairo, Egypt, N. Jb. Geol. Paläont. Mh. (11), pp. 667–686.

Strougo, A. & Haggag, M.A.Y., 1984. Contribution to the age determination of the Gehannam Formation in the Fayium province, N. Jb. Geol. Paläont. Mh. (1), pp. 46–52.

Strougo, A., 1986. The *velascoensis* event, a significant episode of tectonic activity in the Egyptian Paleogene, N. Jb. Geol. Paläont. Abh., v. 173, No. 2, pp. 253–269.

Strougo, A., Bignot, G., Boukhary, M. & Blondeau, A., 1990. The Upper Libyan (possibly Ypresian) Carbonate platform in the Nile Valley, Egypt. Revue de Micropaléontologie, v. 33, no. 1, pp. 54–71.

Strougo, A., Haggag, M.A.Y. & Lutcrbacher, H., 1992. The basal Paleocene *"Globigerina" eugubina* Zone in the Eastern Desert (St. Paul's Monastery, South Galala), Egypt, N. Jb. Geol. Paläont. Mh. (2), pp. 97–101.

Sturchio, N.C., Sultan, M. & Batiza, R., 1983. geology and origin of Meatiq dome, Egypt, A Precambrian metamorphic core complex?, Geology, v. 11, pp. 72–76.

Suess, E., 1893. Are great ocean depths permanent? Natural Science 2, pp. 180–187.

Sultan, I.Z., 1985. Palynological studies in the Nubia Sandstone Formation, east of Aswan, southern Egypt, N. Jb. Geol. Paläont. Mh. (10), pp. 605–617.

Sultan, M., Arvidson, R.E., Duncan, I.J., Stern, R.J. & El Kaliouby, B., 1988. Extension of the Najd Shear System from Saudi Arabian to the central Eastern Desert of Egypt based on integrated field and Landsat observation. Tectonics, v. 7, pp. 1291–1306.

Sultan, M., Chamberlain, K.R., Bowring, S.A., Arvidson, R.E., Abu Zeid, H. & El Kaliouby, B., 1990. Geochronology and isotopic evidence for involvement of pre-Pan-African crust in the Nubian shield, Egypt, Geology, v. 18, pp. 761–764.

Sultan, N. & Abdel Halim, M., 1988. Tectonic framework of northern Western Desert, Egypt, and its effect on hydrocarbon accumulations. 9th Exploration and Production Conf., Egyptian General Petroleum Corp. (EGPC), Cairo, Egypt, November 1988, p. 24.

Sutcliffe, O.E., Dowdeswell, R.J., Whittington, R.J., Theron, J.N. & Craig, J., 2000. Calibrating the Late Ordovician glaciation and mass extinction by the eccentricity cycles of Earth's orbit. Geology, v. 28, pp. 967–970.

Sutcliffe, O.E., Harper, D.A.T., Ait Salem, A., Whittington, R.J. & Craig, J., 2001. The development of an atypical Hirnantian brachiopod Fauna and the onset of glaciation in the Late Ordovician of Gondwana. Transactions of the Royal Society of Edinburgh. Earth Sciences, v. 92, pp. 1–14.

Sutherland, E.L., 1994. Volcanism around K/T boundary time – its role in an impact senario for the K/T extinction events, Earth Science Reviews, v. 36, pp. 1–26.

Swedan, A. & Issawi, B., 1977. Geological Map of Libya 1:250,000, Sheet Bir Hacheim NH 34-4, Explanatory Booklet, Indust. Res. Cent., Tripoli, p. 80.

Swezey, C.S., 2009. Cenozoic stratigraphy of the Sahara, Northern Africa. Geological Society of Africa Presidential Review No. 13. Journal of African Earth Sciences, v. 53, pp. 89–121.

Synelnikov, A.S. & Kollerov, D.K., 1959. Palinologic (sic.) analysis and age of coal samples from El-Bedaa-Thora district, West Central Sinai. EGMS, Paper no. 4, p. 2.

Tahayt, A., Feigl, K.L., Mourabit, T., Rigo, A., Reilinger, R., McClusky, S., Fadil, A., Berthier, E., Dorbath, L., Serroukh, M., Gomez, F., Ben Sari, D., 2009. The Al Hoceima (Morocco) earthquake of 24 February 2004, analysis and interpretation of data from ENVISAT ASAR and SPOT5 validated by ground-based observations. Remote Sensing of Environment, v. 113, pp. 306–316.

Taher, M., Said, M. & El-Azhary, T., 1988. Organic geochemical study, Meleiha Area. 9th Exploration and Production Conf., Egyptian General Petroleum Corp. (EGPC), Cairo, Egypt, November 1988, p. 27.

Takla, M.A., Bakhit, F.S., El Mansi, M.M. & Arbab, A.A., 1998 (Abs.). The basement rocks of Gabal el Umrah area, Eastern Desert, Egypt, in Geology of the Arab World, Fourth International Conference, Cairo University, Cairo, pp. 88–89.

Talbi, F., Melki, F., Ben Ismail-Lattrache, K., Alouani, R. & Tlig, S., 2008. Le Numidien de la Tunisie septentrionale: données stratigraphiques et interprétation géodynamique. Estudios Geológicos, v. 64, no., pp. 31–44.

Tanfous Amri, D., Soussi, M., Bédir, M. & Azaiez, H., 2008. Seismic sequence stratigraphy of the Jurassic of the central Atlas, Tunisia. Journal of African Earth Sciences, v. 51, pp. 55–68.

Tantawy, A.A., Keller, G., Adatte, T., Stinnesbeck, W., Kassab, A. & Schulte, P., 2001. Maastrichtian to Paleocene depositional environment of the Dakhla Formation, Western Desert, Egypt: sedimentology, mineralogy, and integrated micro- and macrofossil biostratigraphies. Cretaceous Research, v. 22, pp. 795–827.

Taquet, P., 2007. On camelback: René Chudeau (1864–1921), Conrad Kilian (1898–1950), Albert Félix de Lapparent (1905–1975) and Théodore Monod (1902–2000), four French geological travellers cross the Sahara. Geological Society, London, Special Publications, v. 287, pp. 183–190.

Tari, G., Molnar, J., Ashton, P. & Hedley, R., 2000. Salt tectonics in t5the Atlantic margin of Morocco. The Leading Edge, v. 19, pp. 1074–1078.

Tate, C., 1871. On the age of the Nubian Sandstone. Geological Society of London Quarterly Journal, v. 27, pp. 404–406.

Tatsumi, Y. & Kimura, N., 1991. Backarc extension versus continental breakup, petrological aspects of active rifting, in A.F. Gangi (Ed.), World Rift Systems, Tectonophysics, v. 197, nos. 2–4, pp. 127–137.

Tavani, G., 1938. Fossili del Miocene della Cirenaica. Paleontol. Italica, v. 38, pp. 127–188.

Tavani, G., 1939. Fossili del Miocene della Cirenaica. Paleontol. Italica, v. 38, pp. 17–76.

Tavani, G., 1946. Fossili del eocenici della Cirenaica. Mem. Atti. Soc. Toscana Sci. Nat., v. 53, pp. 172–187.

Tawadros, E.E., 2001. Geology of Egypt and Libya, Balkema, Rotterdam, p. 468.

Tawadros, E., 2003. Geological history of Egypt and Libya and their hydrocarbon systems, presented to the Canadian Society of Petroleum Geologists (CSPG) February 11 Luncheon

Meeting, Calgary. Reservoir, v. 30, Issue 2, p. 8. (An updated version was presented at the Department of Geology, Ain Shams University in June 2009).

Tawadros, E., 2007. On Formalizing sequence stratigraphy in North Africa and the Middle East. International Scientific Journal: Stratigraphy and sedimentology of oil-gas basins, Azerbaijan National Academy of Sciences, pp. 28–31.

Tawadros, E., Elzaroug, R. & Rasul, S.M., 1999 (Abs.). Petrography, palynology and reservoir characteristics of the F3 Sand (Devonian), Al Wafa Field, West Libya, CSPG and Petroleum Society Joint Convention, Calgary.

Tawadros, E., Rasul, S.M., & Elzaroug, R., 2001. Petrography and palynology of qurartzites in the Sirte Basin, central Libya. Journal of African Earth Sciences, v. 32, no. 3, pp. 373–390.

Tawadros, E., Ruban, D. & Efendiyeva M., 2006. Evolution of NE Africa and the Greater Caucasus: Common Patterns and Petroleum Potential. The Canadian Society of Petroleum Geologists, the Canadian Society of Exploration Geophysicists, the Canadian Well Logging Society Joint Convention, May 15–18, 2006, Calgary, pp. 531–538.

Tawfik, M., Said, M. & Afify, H., 2005. Developing Powerful Treatment Program To Produce Very Difficult Paraffinic Crude With Lowest Cost: Case Study of East Beni Suef Field, Qarun Pet. Co., Egypt. Presentation at The SPE Work Shop Organized By SPE Egyptian Section and Khalda Petroleum Company, Egypt, 2005.

Teixell, A., Arboleya, M.L., Julivert, M. & Charroud, M., 2003. Tectonic shortening and topography in the central High Atlas (Morocco). Tectonics, v. 22, v. 5, pp. 1051–1064.

Teixell, A., Ayarza, P., Zeyen, H., Fernandez, M. & Arboleya, M.L., 2005. Effects of mantle upwelling in a compressional setting: the Atlas Mountains of Morocco. Terra Nova, v. 17, pp. 456–461.

Teixell, A., Ayarza, P., Tesón, E., Babault, J., Alvarez-Lobato, F., Charroud, M., Julivert, M., Barbero, L., Amrhar, M. & Arboleya, M.L., 2007. Geodinámica de las cordilleras del Alto y Medio Atlas: síntesis de los conocimientos actuales. Revista de la Sociedad Geológica de España, v. 20, nos. 3–4, pp. 333–350.

Tekbali, A.O., 1994. Palynological observations on the "Nubian Sandstone" southern Libya, Review of Palaeobotany and Palynology, 81, pp. 297–311.

Ten Brink, U.S. & Ben Avraham, Z., 1989. The anatomy of a pull apart basin: seismic reflection observations of the Dead Sea Basin, Tectonics, v. 8, pp. 333–350.

Termier, H., 1927. Sur le Toarcien du Maroc central. Comptes rendus sommaires de la Société géologique de France, v. 7, pp. 77–78.

Termier, H., 1936. Etudes géologiques sur le Maroc central et le Moyen Atlas septentrional. Notes et Mémoire du Service géologique Maroc, v. 33, no. 2, pp. 743–1082.

Termier, H., 1938. Nouveaux affleurements de Famennien dans le Maroc central. Compte Rendu Sociéte Géologique de France, 5e série, v. 8, pp. 40–42.

Termier, H. & Termier, G., 1958, Les Echinodermes Permiens du Djebel Tebaga (Extreme Sud Tunisien): Bulletin de la Société Géologique de France, ser.6, v. 8, pp. 51–64.

Terry, C.E. & Williams, J.J., 1969. The Idris "A" bioherm and oilfield, Sirte basin, Libya, its commercial development, regional Paleocene geologic setting and stratigraphy, in the Exploration for Petroleum in Europe and North Africa, London Inst. Petroleum, pp. 31–48.

Tewfik, N., 1988. An exploration outlook on the Northern Gulf of Suez. 9th Exploration and Production Conf., Egyptian General Petroleum Corp. (EGPC), Cairo, Egypt, November 1988, p. 26.

Thiele, J., Gramann, F. & Kleinsorge, H., 1970. Zur Geologie zwischen der Nordland der ostlichen Kattara-Senke und der Mittelmeer-Kuste, Aegypten, Westliche Wüste, Geol. Jb., v. 88, pp. 321–346.

Thomas, P., 1908. Essai d'une description géologique de la Tunisie. Stratigraphie des terrains paléozoiques et mésozzoiques, 2e partie, Paris.

Thomas, R.J., Chevallier, L.P., Gresse, P.G., Harmer, R.E., Eglington, B.M., Armstrong, R.A., de Beer, C.H., Martini, J.E.J., de Kock, G.S., Macey, P.H. & Ingram, B.A., 2002. Precambrian evolution of the Sirwa Window, Anti-Atlas Orogen, Morocco. Precambrian Research, v. 118, issues 1–2, pp. 1–57.

Thomas, R.J., Fekkak, A., Ennih, N., Errami, E., Loughlin, S.C., Gresse, P.G., Chevallier, L.P. & Liégeois, J.-P., 2004. A new lithostratigraphic framework for the Anti-Atlas Orogen, Morocco. Journal of African Earth Sciences, v. 39, pp. 217–226.

Thomson, J., 1899. The Geology of Southern Morocco and the Atlas Mountains. Quarterly Journal of the Geological Society, v. 55. pp. 190–213.

Thurmond, A.K., Stern, R.J., Abdelsalam, M.G., Nielsen, K.C., Abdeen, M.M. & Hinz, E., 2004. The Nubian Swell. Journal of African Earth Sciences, v. 39, pp. 401–407.

Thusu, B. & Mansouri, A., 1998 (Abs.). Reassignment of the upper Amal Formation to the Triassic and its implications for exploration in the south-east Sirt Basin, Libya, in First Symposium on the Hydrocarbon Geology of North Africa, Geol. Soc. of London Petroleum Group, p. 48.

Thusu, B. & Van der Eem, J.G.L.A., 1985. Early Cretaceous (Neocomian-Cenomanian) Palynomorphs, in B. Thusu & B. Owens (Eds.), Palynostratigraphy of North-East Libya, J. Micropalaeontology, v. 4, part 1, pp. 131–150.

Thusu, B. & Vigran, J.O., 1985. Middle-Late Jurassic (Late Bathonian-Tithonian) palynomorphs, in B. Thusu & B. Owens (Eds.), Palynostratigraphy of North-East Libya, J. Micropalaeontology, v. 4, part 1, pp. 113–129.

Thusu, B. & B. Owens (Eds.), 1985. Palynostratigraphy of North-East Libya. J. Micropalaeontology, v. 4, part 1, p. 182.

Thusu, B., 1996. Implication of the discovery of reworked and in-situ Late Paleozoic and Triassic palynomorphs on the evolution of Sirt Basin, Libya, in in M.J. Salem. A.J. Mouzughi & O.S. Hammuda (Eds.), The Geology of Sirt Basin, v. I, Amsterdam, Elsevier, pp. 455–474.

Thusu, B., Van der Eam, J.G.L.A., A., El-Mehdawi, A. & Bu-Argoub, F., 1988. Jurassic-Early Cretaceous palynostratigraphy in northeast Libya, in A. El-Arnauti, B.Owens & B. Thusu (Eds.), Subsurface Palynostratigraphy of Northeast Libya, Benghazi, Libya, Garyounis University Publications, pp. 171–257.

Tisserant, D., Thuizat, R. & Agard, J., 1976. Donées géochronologiques sur le complexe de roches alcalines du Tamazeght (Haut Atlas de Midelt, Maroc). Bull. BRGM, v. 3, pp. 279–283.

Tissot, B., Demaison, G., Masson, P., Deltaeil, J.R. & Combaz, A., 1980. Paleoenvironment and petroleum potential of Middle Cretaceous Black Shales in Atlantic Basins. AAPG Bulletin, v. 64, no. 12, pp. 2051–2063.

Tissot, B., Espitalie, J., Deroo, G., Tempere, C. & Jonathan, D., 1973. Origin and migration of hydrocarbons in the Eastern Sahara (Algeria). In G. Demaison & R.J. Murris (Eds.): Petrloeum Geochemistry and Basin Evolution. AAPG Mem.25, pp. 315–324.

Tjalsma, R.C. & Lohmann, G.P., 1983. Paleocene-Eocene bathyal and abyssal benthic foraminifera from the Atlantic Ocean. Micropaleontology Special Publication No. 4, pp. 1–90.

Tlatli, M., 1980. Étude des calcaires de l'Albo-Aptien des Djebel Serdj et Bellouta (Tunisie Centrale). Ph.D. thesis, Univ. Marseille, p. 187.

Tmalla, A.F.A., 1992. Stratigraphic position of the Cretaceous-Tetiary boundary in the northern Sirt Basin, Libya, Marine Petroleum Geology, v. 9, pp. 542–552.

Tmalla, A.F.A., 1996. Latest Maastrichtian and Palaeocene planktonic foraminiferal biostratigraphy of well A1a-NC29A, Northern Sirt Basin, Libya, in M.J. Salem. A.J. Mouzughi & O.S. Hammuda (Eds.), Geology of Sirt Basin, v. I, Elsevier, Amsterdam, pp. 195–232.

Tmalla, A.F.A., 2007. The stratigraphic positions of the Wadi Dukhan and Al Uwayliah forma-tions, northeast Libya – a review. Scripta Geologica, v. 134, pp. 119–130.

Tomkeief, S.I. & Peel, R.F., 1942. Igneous rocks from the Central Libyan Desert. Q.J. Geol. Soc., v. 98, pt.3, 4, pp. 223–234, London.

Torbi, A. & Gelard, J.P., 1994. Paléocontraintes enregistrés par la microfracturation, depuis l'Hercynien jusqu'à l'Actuel, dans les Monts du Sud-Est d.Oujda (Meseta oriental, Maroc). Comptes Rendus de l.Académie des Sciences Paris, v. 318, no. 2, pp. 131–135.

Torricelli, S. & Biffi, U., 2001. Palynostratigraphy of the Numidian Flysch of Northern Tunisia (Oligocene-early Miocene). Palynology, v. 25, pp. 29–55.

Tosson, S., 1953. The Rennebaum Volcano in Egypt. Bull. Volcanology, ser.2, v. 15, pp. 99–108.

Toto, E.A., Kaabouben, F., Zouhri, L., Belarbi, M., Benammi, Hafid, M. & Boutib, L., 2008. Geological evolution and structrual style of the Palaeozoic Tafilalt sub-basin, eastern Anti-Atlas (Morocco, North Africa). Geological Journal, v. 43, pp. 59–73.

Touil, A., Hafid, A., Moutte, J. & El Boukhari, A., 2008. Subduction-related to within-plate magmatism granitoids (central Anti-Atlas, Morocco): evolution from Petrology and geo-chemistry of the Neoproterozoic Siroua. Geological Society, London, Special Publications. v. 297. pp. 265–283.

Toumarkine, M. & Luterbacher, H., 1985. Paleocene and Eocene planktonic foraminifera, in H.M. Bolli. J.B. Saunders & K. Perch-Nielsen (Eds.), Planktonic Stratigraphy, Cambridge University Press, Cambridge, pp. 87–154.

Trabelsi, S., Bouri, S., Gasmi, M., Lahlou Mimi, A., Inoubli, N. & Ben Dhia, H., 2005. Geolog-ical and Hydrothermal Outlines of the Tunisian Fluorine Province (North-Eastern Tunisia). Proceedings World Geothermal Congress, Antalya, Turkey, 24–29 April 2005, pp. 1–3.

Trappe, J., 1991. Stratigraphy, facies distribution and paleogeography of the marine Paleogene from the western High Atlas, Morocco. Neues Jahrbuch für Geologie und Paläontologie, Abhandlungen, v. 180, pp. 279–321.

Trappe, J., 1992. Microfazies zonation and spatial evolution of a carbonate ramp: Marginal Moroccan phosphate sea during Paleogene. Geologische Rundschau, v. 81, no. 1, pp. 105–126.

Traut, M.W., Boote, D.R.D. & Clark-Lowes, D.D., 1998. Exploration history of the Palaeozoic Petroleum systems of North Africa, in D.J. Macgregor. R.T.J. Moody & D.D. Clark-Lowes (Eds.), Petroleum Geology of North Africa, Geological Society Spec. Pub. No. 132, London, pp. 69–78.

Trauth, M.H., Larrasoaña, J.C. & Mudelsee, M., 2009. Trends, rhythems and events in Plio-Pleistocene African climate. Quaternary Science Reviews, v. 28, pp. 399–411.

Tröger, K.-A. & Röhlich, P., 1991. Campanian-Maastrichtian Inoceramid (Bibalvia) assem-blages from NW Libya, in M.J. Salem. O.S. Hammuda & B.A. Eliagoubi (Eds.), Geology of Libya, v. IV, pp. 1357–1381.

Tromp, S.W., 1950. The age and origin of the Red Sea graben. Geological Magazine, v. 87, no. 6, pp. 385–392.

Tromp, S.W., 1951. The geological history of Egypt and of the Red Sea in particular. Türk. Jeol. Kur. Bült. 2, pp. 51–95.

Tucker, M.E., 2003. Mixed clastic-carbonate cycles and sequences: Quaternary of Egypt and Carboniferous of England. Geologia Croatica, v. 56, no. 1, pp. 19–37.

Turk, T.M., Doughri, A.K. & Banerjee, S., 1980. A review of the recent investigations on the Wadi Ash Shati iron ore deposits, Northern Fazzan, Libya, in M.J. Salem & M.T. Busrewil (Eds.), The Geology of Libya, v. III, London, Academic Press, pp. 1019–1043.

Turner, B.R., 1980. Paleozoic sedimentology of the southeastern part of Al Kufra basin, Libya, in M.J. Salem & M.T. Busrewil (Eds.), The Geology of Libya, v. II, London, Academic Press, pp. 351–374.

Turner, B.R., 1991. Paleozoic deltaic sedimentation in the southern part of Al Kufra basin, Libya, in M.J. Salem & M.N. Belaid (Eds.), The Geology of Libya, v. V, Amsterdam, Elsevier, pp. 1713–1726.

Turner, P. & Sherif, H., 2007. A giant Late Triassic-Early Jurassic evaporitic basin on the Saharan Platform, North Africa. In Tectonics, basin evolution and evaporites. Geological Society of London Special Publications, v. 285, pp. 87–105.

Turner, P., Pilling, D., Walker, D., Exton, J., Binnie, J. & Sabaou N., 2001. Sequence stratigraphy and sedimentology of the Late Triassic TAGI (blocs 401/402, Berkine basin, Algeria). Mar. Petrol. Geol., v. 18, pp. 959–981.

Underdown, R. & Redfern, J., 2008. Petroleum generation and migration in the Ghadames Basin, north Africa: A two-dimensional basin-modeling study. AAPG Bulletin, v. 92, no. 1, pp. 53–76.

Unger, Prof., 1859. On the fossil wood of Egypt. Q.J. Geol. Soc. London, v. 15, part II, p. 13.

Uwins, P.J.R. & Batten, D.J., 1988. Early to mid-Cretaceous palynology of northeast Libya, in A. El-Arnauti, B.Owens & B. Thusu (Eds.), Subsurface Palynostratigraphy of Northeast Libya, Benghazi, Libya, Garyounis University Publications, pp. 215–257.

Vaccari, D.A., 2009. Phosphorus: A Looming. Scientific American, June 2009, pp. 54–59.

Vachard, D., Orberger, B., Rividi, N., Pille, L. & Berkhli, M., 2006. New Late Asbian/Early Brigantian (Late Visean, Mississippian) dates in the Mouchenkour Formation (central Morocco): palaeogeographical consequences. Comptes Rendus Palevol, v. 5, no. 6, pp. 769–777.

Vadasz, E., 1933. Oolithische Roteisenerzlagerstaetten in Aegypten. Zentralblatt fuer Mineralogie, Geologie und Palaeontologie, Abteilung A, Mineralogie und Petrographie, v. 5, pp. 161–175.

Vail, J.R., 1971. Dike swarms and volcanic activity in northeastern Africa. In C. Gray (Ed.), Symposium on the Geology of Libya. U. of Libya, Tripoli, pp. 341–347.

Vail, J.R., 1976. Outline of the geochronology and tectonic units of the basement complex of Northeast Africa. *Proceedings of the Royal Society of London. Series A, Mathematical and Physical Sciences*, v. 350, no. 1660, pp. 127–141.

Vail, J.R., 1985. Pan-African (Late Precambrian) tectonic terranes and reconstruction of the Arabian-Nubian Shield, Geology, v. 13, pp. 839–842.

Vail, J.R., 1991. The Precambrian tectonic structure of North Africa, in M.J. Salem. A.M. Sbeta & M.R. Bakbak (Eds.), The Geology of Libya, v. VI, pp. 2259–2269.

Vail, P.R.,, Mitchum, R.M., Jr. & Thompson, III, S., 1977. Seismic stratigraphy and global changes of sea level, Part 4: Global cycles of relative changes of sea level, in C.E. Payton (Ed.), Seismic Stratigraphy – application to hydrocarbon exploration, AAPG Mem.26, pp. 83–97.

Van de Weerd, A.A. & Ware, P.L.G., 1994. A review of the East Algerian Sahara oil and gas province (Triassic, Ghadames and Illizi ), First Break, v. 12, No. 7, pp. 363–373.

Van de Weerd, A.A., 1995 (Abs.). Petroleum Geology of East Algeria, in First Symposium on the Hydrocarbon Geology of North Africa, Geol. Soc. of London Petroleum Group, p. 49.

Van den Berg, J. & Zijderveld, H., 1982. Palomagnetism in the Mediterranean Sea, in H. Berckhemer & K. Hsü (Eds.), Alpine-Mediterranean Geodynamics, Geodynamics Series, v. 7, Am. Geophys. Union/Geol. Soc. Am., pp. 83–112.

Van der Meer, F. & Cloetingh, S., 1996. Intraplate stresses and the subsidence history of the Sirt Basin, M.J. Salem, M.T. Busrewil, A.A. Misallati & M.A. Sola (Eds.), Geology of Sirt Basin, v. III, Amsterdam, Elsevier, pp. 211–230.

Van der Meer, F.D., Cloetingh, S.S.P.L., Abadi, A.M. & van Dijk, P.M., 1993 (Abs.). Intraplate stresses and subsidence history of the west Sirt Basin, Libya, in Sedimentary Basins of Libya, First Symposium, Geology of Sirt Basin, Earth Science Society of Libya, Tripoli, pp. 41–42.

Van der Ploeg, P., 1953. Egypt. In V.C. Illing (Ed.), The World's Oil Fields: The Eastern Hemisphere. The Science of Petroleum, Osford University Press, v. 6, no. 1, pp. 151–157.

Van Erve, A.W., 1993 (Abs.). Aspects of the Sirt Group stratigraphy (toward the solution of the Nubian problem in Libya), in 1st Symp. on the Geology of Sirt Basin, Tripoli, p. 42.

Van Houten, F.B. & Brown, R.H., 1977. Latest Paleozoic-Early Mesozoic paleography, Northwestern Africa, J. Geology, v. 85, pp. 143–156.

Van Houten, F.B. & Karasek, R.M., 1981. Sedimentologic framework of late Devonian oolitic formation, Shati Valley, West Central Libya, J. Sed. Pet., v. 51, pp. 415–427.

Van Houten, F.B., 1980. Latest Jurassic-Early Cretaceous regressive facies, northeast African craton, AAPG Bull., v. 64, pp. 857–867.

Van Houten, F.B., 1983. Sirte basin, north-central Libya, Cretaceous rifting above a fixed mantel hotspot?, Geology, v. 11, pp. 115–118.

Van Houten, F.B., Bhattacharyya, D.P. & Mansour, S.E.I., 1984. Cretaceous Nubia Formation and correlative deposits, Eastern Egypt: Major regressive-transgressive complex, Geol. Soc. Am. Bull., v. 95, pp. 397–405.

Vandre, C., Cramer, B., Gerling, P. & Winsemann, J., 2007, Natural gas formation in the western Nile delta (Eastern Mediterranean): Thermogenic versus microbial. Organic Geochemistry, v. 38, pp. 523–539.

Vaslet, D., Janou, D., Razin, Ph. & Halawani, M., 1998 (Abs.). Effects of the Late Ordovician glaciation in the deposits of Arabian Peninsula, The Geological Conference on Exploration in Murzuq Basin, Sabha University, Sabha, Libya, p. 47.

Vaughan, R.D., Atef, A. & El-Outefi, N., 2003 (Extended Abstract). Use of seismic sttributes and acoustic impedance in 3D reservoir modeling: an example from mature Gulf of Suez carbonate field. AAPG International Conference, Barcelona, Spain, September 21–24.

Vavrdova, M., 1991. Latest Devonian miospores and acritarchs from the surface samples of the Ashkidah Formation, in M.J. Salem, O.S. Hammuda & B.A. Eliagoubi (Eds.), The Geology of Libya, v. IV, Amsterdam, Elsevier, pp. 1285–1296.

Vecoli, M., 2000. Palaeoenvironmental interpretation of microphytoplankton diversity trends in the Cambrian-Ordovician of the northern Sahara Platform. Palaeogeography, Palaeoclimatology, Palaeoecology, v. 160, pp. 329–346.

Vecoli, M., Riboulleau, A. & Versteegh, G.J.M., 2009. Palynology, organic geochemistry and carbon isotope analysis of a latest Ordovician through Silurian clastic succession from borehole TT1, Ghadames Basin, southern Tunisia, North Afric: Palaeoenvironmental interpretation. Palaeogeography, Palaeoclimatology, Palaeoecology, v. 273, pp. 378–394.

Vermeersch, P., 1970. L'Elkabien, une novelle industrie épipaléolithique Elkab en Haute Egypte, sa stratigraphie, sa typologie, Chronique d'Egypte, v. 89, pp. 45–68.

Vesely, J., 1985. Geological Map of Libya 1:250,000, Sheet Zallah NH33-16, Explanatory Booklet, I.R.C., Tripoli, p. 125.

Vila, J.-M., Ben Yousef, M., Charrière, A., Chikhaoui, M., Ghanmi, M., Kamoun, F., Peybernès, B., Saadi, J., Souquet, P. & Zarbout, M., 1994. Découverte en Tunisie, au SW du Kef, de matériel triasique interstratifié dans l'Albien: extension du domaine à "glaciers de sel" sous-marins des confins algéro-tunisiens. C.R. Acad. Sci. Paris, v. 318 (II), pp. 1661–1667.

Villeneuve, M. & Cornee, J.J., 1994. Structure and évolution and palaeogeography of the West African craton and bordering belts during the Neoproterozoic. Precambrian Research, v. 69, pp. 307–326.

Vinassa de Regny, P., 1912. Cenni geologici sulla Libia italiana. Boll. Soc. Afr. Ital., v. 31, no. 1–4, pp. 3–35, carta geol. scala 1:5,000,000.

Vinassa de Regny, P., 1932. Breve Guida alle Escursioni Geologiche in Libia. Boll. Soc. Geol. Ital., v. 51, Roma.

Viotti, C. & Mansour, A., 1969. Tertiary planktonic foraminiferal zonation from the Nile delta, Egypt, U.A.R., pt.1, Miocene planktonic foraminiferal zonation, 3rd African Micropaleontology Colloq., Cairo, Proc., pp. 425–459.

Viotti, C. & El-Demerdash, G., 1969. Preliminary study in Eocene sediments of Wadi Nukhul area east coast – Gulf of Suez, 3rd African Micropaleontology Colloq., Cairo, Proc., pp. 403–423.

Visser, J.N.J. & Praekelt, H.E., 1996. Subduction, mega-shear systems and Late Palaeozoic basin development in the African segment of Gondwana, Geol. Rundschau, v. 85, pp. 632–646.

Vita-Finzi, C., 1971. Alluvial History of northern Libya since the last Interglacial, in C. Gray (Ed.), Symposium on the Geology of Libya, Beirut, Catholic Press, pp. 409–429.

Viterbo, I., 1969. Lower Cretaceous Charophyta from the subsurface "Nubian Complex" of the Sirte basin, Libya, Proc. 3rd African Micropaleontological Colloquium, Cairo, pp. 393–402.

Vittimberga, P. & Cardello, R., 1963. Sédimentologie et pétrographie du Paléozoique du bassin de Kufra, Rév. Inst. Fr. Pétrole, v. 18, No. 11, Paris, pp. 1546–1558.

Voigt, S., Hminna, A. Saber, H., Schneider, J.W. & Klein, H., 2009. Tetrapod Ikakern Formation (Argana Basin, Western High Atlas, Morocco). J. Afr. Earth Sc., v. 57, no. 5, pp. 470–478.

Von Bary, E., 1877. Reisebriefe aus Nord-Africa, Part 2. Geologische Beobachtungen. Zeitschrift des Gesellschaft für Erkunde, zu Berlin, v. 12, pp. 196–198.

Von Bary, E., 1898. Le dernier voyage d'un European sur Ghat et les Touareg de l'Air. Journal de voyage d'Erwin von Bary 1876–1877, traduit et annoté par M. Henri Schrimer, édit. Fischbacher, Paris, p. 221.

Von Nötling, F., 1885. Über Crustaceen aus dem Teriär Aegyptens. Sitz. K. Akad. Wissensch., Berlin, Bd.XXVI, pp. 487–500.

Vos, R.G., 1981a. Deltaic sedimentation in the Devonian of western Libya, Sedimentary Geology, v. 29, pp. 67–88.

Vos, R.G., 1981b. Sedimentology of an Ordovician fan delta complex, western Libya, Sedimentary Geology, v. 29, pp. 153–170.

Wacrenier, P., 1958a. Apercu sur l'Anticambrien du Tibesti, Afrique Equatoriale Francaise, 20th Int. Geol. Congr., Assoc. Serv. Geol. Afr., Mexico, pp. 281–288.

Wacronier, P., 1958b. Notice explicative de la carte geologique provisoire du Borkou-Ennedi-Tibesti aut 1/1,000,000, 24p. Direction des Mines et Geologie, Gouv. Gem., A.E.F., Brazzaville.

Waite, S.T. & Pooley, R.W., 1953. Report on the Nukhul Formation, Egyptian General Petroleum Corp., Cairo, Report GR 952.

Walley, C.D., 1985. Depositional history of southern Tunisia and northwestern Libya in Mid and Late Jurassic time, Geol. Magazine, v. 122, No. 3, pp. 233–247.

Walsh, G.J., Aleinikoff, J.N., Fouad Benziane, Abdelaziz Yazidi and Armstrong, T.R., 2002. U-Pb zircon geochronology of the Paleoproterzoic Tagragra de Tata inlier and its Neoproterozoic cover, western Anti-Atlas, Morocco. Precambrian Research, v. 117, issues 1–2, pp. 1–20.

Walther, J.K., 1887. L'apparition de la craie aux environs des Pyramides. Bull. Inst. Égypte, Le Caire, sér. 2, no. 8, pp. 3–13.

Walther, J.K., 1888. Die Korallenriffe de Sinaihalbinsel. Geologische und Biologische Beobachtungen. Akad. Wiss. Leipzig Abh., Math.-Naturw. Kl., Bd.14.

Walther, J.K., 1890. Über eine Kohlenkalk-Fauna aus de ägyptisch arabischen Wüste. Z. dtsch. Geol. Ges., 42, pp. 419–449.

Walther, J.K., 1900. Das Gesetz der Wüstenbildung in Gegenwart und Vorzeit, Dietrich Reimer, Berlin, p. 157.

Wanas, H.A., 2008. Cenomanian rocks in the Sinai Peninsula, Northeast Egypt: Facies analysis and sequence stratigraphy. Journal of African Earth Sciences, v. 52, pp. 125–138.

Wanner, J., 1902. Die Fauna der obersten weissen Kreide der lybische Wüste. Palaeontographica 30, pp. 91–152.

War Office, Bureau of the General Staff, Cairo, 1876. Summary of geographical and scientific results accomplished by expeditions made by the Government of the Khedive of Egypt during the three years 1874-5-6. Proceedings of the Royal Geographical Society of London, v. 21, no. 1, pp. 63–66.

Ward, J.V., Tekbali, A.O. & Doyle, J.A., 1987 (Abs.). Palynology of the Kiklah Formation (Cretaceous), Northwestern Libya, 3rd Symp. Geol. Libya, Tripoli, p. 123.

Ward, W.C. & McDonald, K.C., 1979, Nubian Formation of central Eastern Desert, Egypt – Major subdivisions and depositional setting, AAPG Bull., v. 63, pp. 975–983.

Watts, A.B., Platt, J.P. & Buhl, P., 1993. Tectonic evolution of the Alboran Sea basin. Basin Research, v. 5, pp. 153–177.

Webster, G.D. & Becker, R.T., 2009. Devonian (Emsian to Frasnian) crinoids of the Dra Valley, western Anti-Atlas Mountains, MoroccoIn Königshof, P. (Ed.), Devonian Change: Case Studies in Palaeogeography and Palaeoecology. The Geological Society, London, Special Publications No. 314, pp. 131–148.

Weill, R., 1908. La Presqu'île du Sinai: Etude de Géographie de d'Histoire. Bibliotèque de l'Ecole des Hautes Etudes. Paris, Honoré Champion, p. 380.

Weissbrod, T. & Horowitz, A., 1989. Carboniferous stratigraphy and distribution in the Middle East, a review, XIe Congrès Internat. de Stratigraphie et de Géologie du Carbonifère, Beijing, 1987, Compte Rendu 2, t.2, pp. 112–123.

Weissbrod, T. & Perath, I., 1990. Criteria for the recognition and correlation of sandstone units in the Precambrian and Paleozoic-Mesozoic clastic sequence in the Near East, J. Afr. Earth Sci., v. 10, No. 1/2, pp. 253–270.

Weissbrod, T., 1969. The Paleozoic of Israel and adjacent countries. Part II – The Paleozoic outcrops in southwestern Sinai and their correlation with those of southern Israel, Isr. Geol. Surv. Bull., No. 48, p. 32.

Weissbrod, T., 1970. "Nubian Sandstone", discussion. AAPG Bull., v. 54, no. 3, pp. 522–538.

Wells, N.A., 1986. Biofabrics as dynamic indications in nummulite accumulations – discussion, J. Sedimentary Petrology, v. 56, pp. 318–320.

Wendorf, F. & Schild, R., 1994. Are the Early Holocene Cattle in the Eastern Sahara Domestic or Wild? Evolutionary Anthropology, v. 3, no. 4, pp. 118–128.

Wendorf, F. & Schild, R., 1998. Nabta Playa and Its Role in Northeastern African Prehistory. Journal of Anthropological Archaeology, v. 17, no. 2, pp. 97–123.

Wenke, R.J., 1989. "Egypt"; Origins of complex societies. Annual Review of Anthropology, v. 18, pp. 129–155.

Wendorf, F. & the Members of the Combined Prehistoric Expedition, 1977. Late Pleistocene and Recent Climatic Changes in the Egyptian Sahara. The Geographical Journal, v. 143, no. 2, pp. 211–234.

Wendorf, F., Close, A.E., Schild, R., & Wasylikowa, K., 1991. The Combined Prehistoric Expedition: Results of the 1990 and 1991 seasons. Newsletter of the American Research Center in Egypt, 154, pp. 1–8.

Wendorf, F., Said, R. & Schild, R., 1970. New concepts in Egyptian Prehistory, Science, v. 169, pp. 1161–1171.

Wendorf, F. & Schild, R., 1976. Prehistory of the Nile Valley, Academic Press, New York, p. 404.

Wendorf, F. & Schild, R., 1980a. Prehistory of the Nile Valley, Academic Press, New York, p. 414.

Wendorf, F. & Schild, R., 1980b. Prehistory of the Eastern Sahara. Academic Press, New York.

Wendorf, F. & Schild, R., 1994. Are the Early Holocene Cattle in the Eastern Sahara Domestic or Wild? Evolutionary Anthropology, v. 3, no. 4, pp. 118–128.

Wendorf, F. & Schild, R., 1998. Nabta Playa and Its Role in Northeastern African Prehistory. Journal of Anthropological Archaeology, v. 17, no. 2, pp. 97–123.

Wendt, J., Aigner, T. & Neugebauer, J., 1984. Cephalopod limestone deposition on a shallow pelagic ridge: the Tafilalt Platform (upper Devonian, Eastern Anti-Atlas, Morocco). Sedimentology, v. 31, pp. 601–665.

Wendt, J. & Belka, Z., 1991. Age and depositional environments of Upper Devonian (Early Frasnian to Early Famennian) black shales and limestones (Kellwasser Facies) in the Eastern Anti-Atlas, Morocco. Facies, v. 25, pp. 51–90.

Wendt, J., Belka, Z. & Moussine-Pouchkine, A., 1993. New architectures of deep-water carbonate buildups: evolution of mud mounds into mud ridges (Middle Devonian, Algerian Sahara). Geology, v. 21, no. 8, pp. 723–726.

Wendt, J. & Kaufmann, B., 2006. Middle Devonian (Givetian) coral-stromatoporoid reefs in West Sahara (Morocco). Journal of African Earth Sciences, v. 44, pp. 339–350.

Wendt, J., Kaufmann, B. & Belka, Z., 2001. An exhumed Palaeozoic underwater scenery: the Visean mud mounds of the eastern Anti-Atlas (Morocco). Sedimentary Geology, v. 145, pp. 215–233.

Wendt, J., Kaufmann, B. & Belka, Z., 2009a. Carboniferous stratigraphy and depostional environments in the Ahnet Mouydir area (Algerian Sahara). Facies, v. 55, pp. 443–472.

Wendt, J., Kaufmann, B., Belka, Z. & Korn, D., 2010. Carboniferous stratigraphy and depostional environments in the Ahnet Mouydir area (Algerian Sahara): rely to the discussion Legrand-Blain et al. Facies, v. 56, pp. 477–481.

Wendt, J., Kaufmann, B. & Belka, Z., 2009b. Devonian stratigraphy and depositional environments in the southern Illizi Basin (Algerian Sahara). Journal of African Earth Sciences, v. 54, pp. 85–96.

Wendt, J., Kaufmann, B., Belka, Z., Klug C. & Lubeseder, S., 2006. SedimenPalaeozoic basin and ridge system: the Middle and Upper Devonian of the Ahnet and Mouydor (Algerian Sahara). Geological Magazine, v. 143, no. 3, pp. 269–299.

Wennekers, J.H.N., Wallace, F.K. & Abugares, Y., 1996. Geology and hydrocarbon occurrences of the Sirt Basin – A synopsis. In M.J. Salem, A.J. Mouzughi & O.S. Hammuda (Eds.). Geology of Sirt Basin 1, Amsterdam, Elsevier, pp. 3–56.

Wenzhe, G., 2009. Hydrocarbon generation conditions and exploration potential of the Taoudeni Basin, Mauritania. Petroleum Science, v. 6, pp. 29–37.

Wernli, R., 1988. Micropaléontologie du Néogène post-nappes du Maroc septentrional et description systématique des foraminifères planctoniques. Notes et Mémoires du Service Géologique du Maroc, v. 331, pp. 1–274.

Westphal, M., Montigny, R., Thuizat, R., Bardon, C., Bossert, A. & Hamzeh, R., 1979. Paléomagnetétisme et datation du volcanisme permien, triasique et crétacé du Maroc. Can. J. Earth Sci., v. 16, pp. 2150–2164.

Westermann, S., Föllmi, K.B., Matera, V. & Adatte, T., 2007. Phosphorus and trace-metal records during Cretaceous oceanic anoxic events: Example of the Early Aptian OAE in the western Tethys. In: L.G. Bulot, S. Ferry & D. Grosheny (Eds.), Relations entre les marges septentrionale et méridionale de la Téthys au Crétacé [Relations between the northern and southern margins of the Tethys ocean during the Cretaceous period] – Carnets de Géologie/Notebooks on Geology, Brest, Memoir 2007/02, Abstract 03 (CG2007_M02/03), pp. 20–22.

Wever, H.E., 2000. Petroleum and source rock characterization based on C7 Star Plot results: Examples from Egypt. AAPG Bulletin, v. 84, no. 7, pp. 1041–1054.

Weyant, M. & Massa, D., 1991. Contribution of conodonts to the Devonian Biostratigraphy of Western Libya, in M.J. Salem, O.S. Hammuda & B.A. Eliagoubi (Eds.), The Geology of Libya, v. IV, Amsterdam, Elsevier, pp. 1297–1322.

Whitebread, T. & Kelling, G., 1982. Mrar Formation of western Libya. evolution of an Early Carboniferous delta system, AAPG Bull., v. 66, No. 8, pp. 1091–1107.

Whitechurch, H., Juteau, T. & Montigny, R., 1984. Role of Eastern Mediterranean ophiolites (Turkey, Syria, Cyprus) in the history of the Neo-Tethys, in A.H.F. Robertson & J.E. Dixon (Eds.), The Geological Evolution of the Eastern Mediterranean, Geological Soc. Spec. Pub. No. 17, London, pp. 301–317.

Whiteman, A.J., 1970. Nubian Group: origin and status, discussion. AAPG Bull., v. 54, no. 3, pp. 522–538.

Wigger, S., Bailey, J., Larsen, M. and Wallace, M., 1997. Ha'py field: A Pliocene bright spot example from the Nile delta, Egypt. The Leading Edge, v. 16, no. 12, pp. 1827–1829.

Wight, A.W.R., 1980. Palaeogene vertebrate fauna and regressive sediments of Dur at Talhah, Southern Sirt Basin, Libya, in M.J. Salem & Busrewil (Eds.), The Geology of Libya, v. I, London, Academic Press, pp. 309–325.

Wilde, S.A. & Youssef, K. 2000. Significance of SHRIMP U-Pb dating of the Imperial Porphyry and associated Dokhan Volcanics, Gebel Dokhan, North Eastern Desert, Egypt. Journal of African Earth Sciences, v. 31, pp. 410–413.

Wilde, S.A. & Youssef, K., 2002. A re-evaluation of the origin and setting of the Late Precambrian Hammamat Group based on SHRIMP U–Pb dating of detrital zircons from Gebel Umm Tawat, North Eastern Desert, Egypt. Journal of the Geological Society, London, v. 159, pp. 595–604.

Williams, D.O., 1995 (Abs.). Reservoir heterogeneity of the Facha Member in Concession N27 4B, Sirte Basin, SPLAJ, in First Symposium on the Hydrocarbon Geology of North Africa, Geol. Soc. of London Petroleum Group, pp. 49–50.

Williams, G.D., 1993. Tectonics and seismic sequence stratigraphy: an introduction, in G.D. Williams & A. Dobb (Eds.), Tectonics and Seismic Sequence Stratigraphy, Geological Society Special Pub. No. 71, London, pp. 1–13.

Williams, J.J. (Fd.), 1966. South-Central Libya and Northern Chad, a Guidebook to the Geology and Prehistory. Petroleum Exploration Society of Libya.

Williams, J.J., 1968. The sedimentary and igneous reservoirs of the Augila oil field. In F.T. Barr (Ed.), Geology and archaeology of northern Cyrenaica, Libya. Petroleum Exploration Society of Libya, Tenth Annual Field Conference, pp. 197–206.

Williams, J.J., 1972. Augila Field, depositional environment and diagenesis of sedimentary reservoir and description of igneous reservoir, in Stratigraphic Oil and Gas Fields – Classification, Exploration Methods and Case Histories. AAPG Mem. 16, SEG Special Pub. No. 10, pp. 623–632.

Williams, M.A.J. & Hall, D.N., 1965. Recent expeditions to Libya from the Royal Military Academy, Sandhurst. The Geographical Journal, v. 131, no. 4, pp. 482–501.

Wilson, J.L., 1975. Carbonate Facies in Geologic History, Springer-Verlag, New York, Heidelberg, Berlin, p. 471.

Wiman, S.K., 1980. Stratigraphic and micropaleontologic expression of the Mediterranean Late Miocene (Messinian) regression and Early Pliocene transgression in Northern Tunisia, in M.J. Salem & Busrewil (Eds.), The Geology of Libya, v. I, London, Academic Press, pp. 117–126.

Wingate, O., 1934. In search of Zerzura. The Geographical Journal, v. 83, no. 4, pp. 281–303.

Winn, R.D., Jr., Crevello, P.D. & Bosworth, W., 2001. Lower Miocene Nukhul Formation, Gebel el Zeit, Egypt: Model for structural control on early synrift strata and reservoirs, Gulf of Suez. AAPG Bulletin, v. 85, no. 10, pp. 1871–1890.

Winnock, E., 1979. Les fosses du chenal de Sicile, in P.F. Burollet. P. Clairefond & E. Winnock (Eds.), Géologie Méditerranéenne, La Mer Pélagienne, Annales de l'Université de Provence, tome VI, No. 1, Marseille, pp. 41–49.

Winnock, E., 1981. Structure du bloc pélagien, in F.C. Wezel (Ed.), Sedimentary Basins of Mediterranean Margin, C.N.R. Italian Project of Oceanography, Tecnoprint, Bologne, pp. 445–464.

Winnock, E. & Bea, F., 1979. Structure de la mer Pélagienne, in P.F. Burollet; P. Clairefond & E. Winnock (Eds.), Géologie Méditerranéenne, La Mer Pélagienne. Annales de l'Université de Provence, tome VI, No. 1, Marseille, pp. 35–40.

Woller, F. & Fediuk, F., 1980. Volcanic rocks of Jabal as Sawda, in M.J. Salem & M.T. Busrewil (Eds.), The Geology of Libya, v. II, London, Academic Press, pp. 1081–1093.

Woller, F., 1978. Geological map of Libya 1:250,000, Sheet Al Washkah, NG 33-15 Explanatory Booklet, I.R.C., Tripoli, p. 104.

Wood, L.E., 1964. Pseudo-oolites of northern Libya, their occurrence and origin, J. Sediment. Petrol., v. 34, No. 3, pp. 661–663.

World Oil, 1995. Deep offshore prospecting begins, gas becomes more important, August, 1995, pp. 107–109.

World Oil, 1996. Africa, prosperous times, August, 1996, pp. 94–103.

World Energy Council, 2007. Survey of Energy Resources; Oil Shale, p. 16.

Wright, P.E. & Edmunds, W.M., 1971. Hydrogeological studies in central Cyrenaica, Libya, Symposium Geology of Libya, Fac. Sci. University of Libya, Tripoli, pp. 459–481.

Wycisk, P., 1990. Aspects of cratonal sedimentation: facies distribution of fluvial and shallow marine sequences in NW Sudan/SW Egypt since Silurian time, J.Afr. Earth Sci., v. 10, nos. 1/2, pp. 215–228.

Yahi, N., Schaefer, R.G. & Littke, R., 2001. Petroleum generation and accumulation in the Berkine Basin, eastern Algeria. AAPG Bull., v. 85, no. 8, pp. 1439–1467.

Yonan, A.A., 1998 (Abs.). petrography, geochemistry and petrogenesis of Urf Abu Hamam fluorite-bearing granites, south Eastern Desert, Egypt, a case of felsic A-type granite-fluorite-barite relationship, in Geology of the Arab World, Fourth International Conference, Cairo University, Cairo, pp. 127–128.

Younes, M.A., Hegazi, A.H., El-Gayar, M.SH. & Andersson, J.T., 2007. Petroleum biomarkers as environment and maturity indicators for crude oils from the central Gulf of Suez, Egypt. Oil Gas European Magazine, v. 33, pp. 15–21.

Younes, A.I. & McClay, K., 2002. Development of accommodation zones in the Gulf of Suez Red Sea rift, Egypt. AAPG Bull., v. 86, no. 6, pp. 1003–1026.

Younes, M.A., 2001. Application of biomarkers and stable carbon isotopes to assess depositional environments of source rocks and maturation of crude oil of East Zeit Field, Southern Gulf of Suez, Egypt. Petroleum Science & Technology, v. 19, nos. 9–10, pp. 1039–1061.

Younes, M.A., 2003a. Hydrocarbon seepage generation and migration in the southern Gulf of Suez, Egypt: insights from biomarker characteristics and source rock modeling. J. Petroleum Geology, v. 26, no. 2, pp. 211–224.

Younes, M.A., 2003b. Organic and carbon isotope geochemistry of crude oils from Ashrafi field, southern Gulf of Suez Province, Egypt: implication for the processes of hydrocarbon generation and maturation. Petroleum Science and Technology, v. 21, nos. 5 & 6, pp. 971–995.

Younes, M.A., 2003c (Extended Abstract). Alamein Basin Hydrocarbon Potentials, Northern Western Desert, Egypt. AAPG Annual Convention, May 11–14, 2003, Salt Lake City, Utah, p. 6.

Young, M.J., Gawthorpe, R.L. & Sharp, I.R., 2000. Sedimentology and sequence stratigraphy of a transfer zone coarse-grained delta, Miocene Suez Rift, Egypt. Sedimentology, v. 7, no. 6, pp. 1081–1104.

Young, M.J., Gawthorpe, R.L. & Sharp, I.R., 2002. Architecture and evolution of syn-rift clastic depositional systems towards the tip of a major fault segment, Suez rift, Egypt. Basin Research, v. 14, pp. 1–23.

Yousef, M., Moustafa, A.R. & Shann, M., 2010. Egypt Structural setting and tectonic evolution of offshore North Sinai. Geological Society, London, Special Publications, v. 341, pp. 65–84.

Youssef, M.I. & Abdel-Aziz, W., 1971. Biostratigraphy of the Upper Cretaceous-Lower Tertiary in Farafra Oasis, Libyan Desert, Egypt, in C. Gray (Ed.), Symposium on the Geology of Libya, Beirut, Catholic Press, pp. 227–249.

Youssef, M.I., 1954. Stratigraphy of Gebel Oweina section. Inst. Desert Egypte Bull., v. 4, no. 2, pp. 83–93.

Youssef, M.I., 1957. Upper Cretaceous rocks in Kosseir area. Inst. Desert Egypte Bull., v. 7, no. 2, pp. 35–53.

Youssef, M.I., 1968. Structural pattern of Egypt and its interpretation, AAPG Bull., v. 52, No. 4, pp. 601–614.

Youssef, M.I., Cherif, O.H., Boukhary, M. & Mohamed, A., 1984. Geological studies on the Sakkara area, Egypt. N. Jb. Geol. Paläont. Abh., v. 168, pp. 125–144.

Zaborski, P.M.P., 1983. Campano-Maastrichtian ammonites, correlation and palaeogeography in Nigeria, J. Afr. Earth Sci., v. 1, pp. 59–63.

Zaccagna, D., 1919a. Itinerari geologici nella Tripolitania Occidentale con appendice paleontological. Mem. Descritt. Carta Geol. d'Ital., v. XVIII, Roma, p. 126.

Zaccagna, D., 1919b. Itinerari geologici nella Tripolitania Occidentale. Mem. Descrittive Cart. Geol. D'Ital., v. 18, pp. 1–70.

Zachos, J.C., Pagani, M., Sloan, L., Thomas, E. & Billups, K., 2001. Trends, rythms, and aberrations in global climate 65 Ma to Present. Science, v. 292, pp. 686–693.

Zagarani, M.F., Negra, M.H. & Hanini, A., 2008. Cenomanian-Turonian facies and sequence stratigraphy, Bahloul Formation, Tunisia. Sedimentary Geology, v. 204, pp. 18–35.

Zahran, I. & Askary, S., 1988. Basement Reservoir in Zeit Bay Oilfield, Gulf of Suez. AAPG Bull., v. 72, no. 2, p. 261.

Zahran, M. & Meshrif, W., 1988. The Northern Gulf of Suez: Basin evolution, stratigraphy and facies relationship. 9th Exploration and Production Conf., Egyptian General Petroleum Corp. (EGPC), Cairo, Egypt, November 1988, p. 30.

Zakhera, M.S., 2010. Distribution and Abundance of Rudist Bivalves in the Cretaceous Platform Sequences in Egypt: Time and Space. Turkish Journal of Earth Sciences, v. 19, pp. 745–755.

Zalat, A.A., 1995a. Diatoms from the Quaternay sediments of the Nile Delta, Egypt, and their palaeoecological significance, Journal of African Earth Scieneces, v. 20, pp. 133–150.

Zalat, A.A., 1995b. Calcareous nannoplankton and diatoms from the Eocene/Pliocene sediments, Fayoum depression, Egypt, Journal of African Earth Scieneces, v. 20, pp. 227–244.

Zert, B., 1974. Geological Map of Libya, 1:250,000, Sheet Darnah NI34-16, Explanatory Booklet, Industrial Research Centre, Tripoli, p. 49.

Zeyen, H., Ayarza, P., Fernández, M. & Rimi, A., 2005. Lithospheric structure under the western African-European plate boundary: A transect across the Atlas Mountains and the Gulf of Cadiz. Tectonics, v. 24.

Zikmund, J., 1985. Geological Map of Libya, 1:250,000, Sheet Abu Na'im NH34-13, IRC, Tripoli, p. 114.

Ziko, A. & El Safori, Y., 1998 (Abs.). New and extant Miocene bryozoans of the Clysmic area, Egypt, and their paleobiogeographic significance, in Geology of the Arab World, Fourth International Conference, Cairo University, Cairo, p. 27.

Zittel, K.A., 1883a. Geologischer Theil (Sahara. libysche Wüste). Palaeontographica, v. 30, pp. 1–147.

Zittel, K.A., 1883b. Beitrage zur Geologie und Palaeontologie der Libyschen Wüste und der angrenzenden Gebiete von Aegypten, Palaeontographica, v. 30, no. 1, pp. 1–112.

Zivanovic, M., 1977. Geological map of Libya, 1:250,000, Explanatory Booklet, Sheet Bani Walid NH 33-2, Industrial Research Centre, Tripoli, p. 71.

Zobaa, M., Sanchez Botero, C., Browne, C., Oboh-Ikuenobe, F.E. & Ibrahim, M.I., 2008. Kerogen and palynomorph analyses of the mid-Cretaceous Bahariya Formation and Abu Roash "G" Member, North Western Desert, Egypt. Gulf Coast Association of Geological Societies Transactions, v. 58, pp. 933–943.

Zouhri, S., Kchikach, A., Saddiqi, O., El Haïmer, F.Z., Baidder, L. & Michard, M., 2008. The Cretaceous-Tertiary Plateaus. In A. Michard, O. Saddiqi, A. Chalouan and D. Frizon de Lamotte (Eds.), Continental Evolution: The Geology of Morocco. structure, stratigraphy, and Tectonism of the Africa-Atlantic-Mediterranean Triple Junction, Berlin, Springer-Verlag, pp. 331–358.

Zuffardi-Comerci, R., 1934. Su alcuni corallari terziari della Cirenaica e della Tripolitania Orientale. Missione Scientifica della R. Accad. D'Italia a Cufra (1931-X), Roma, v. 3, pp. 45–60.

Zuffardi-Comerci, R., 1940. Nuovo contributo allo studio di corallofaune cenozoiche della Sirtica (Libia), Annali del Museo Libico di Storia Naturale, v. II, pp. 203–210.

Zuffardi-Comerci, R., Silvestri, A., Desio, A., Cipolla, F., & Chiarugi, A., 1929. Resultati Scientifici della Missione alla oasi di Giarabub (1926–1927), Pt.3, La Paleontologia. Pub. Della R. Soc. Geogr. Ital., Roma.

# Subject index